010164   £270 (4 vols.)

635.094

96

THE EUROPEAN GARDEN FLORA

THE EUROPEAN GARDEN FLORA

*A manual for the identification of plants cultivated
in Europe, both out-of-doors and under glass*

VOLUME III

Dicotyledons (Part I)

edited by

S.M. Walters, J.C.M. Alexander, A. Brady,

C.D. Brickell, J. Cullen, P.S. Green, V.H. Heywood,

V.A. Matthews, N.K.B. Robson, P.F. Yeo and S.G. Knees

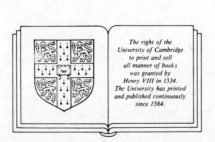

The right of the
University of Cambridge
to print and sell
all manner of books
was granted by
Henry VIII in 1534.
The University has printed
and published continuously
since 1584.

CAMBRIDGE UNIVERSITY PRESS

Cambridge

New York   New Rochelle

Melbourne   Sydney

Published by the Press Syndicate of the University of Cambridge
The Pitt Building, Trumpington Street, Cambridge CB2 1RP
32 East 57th Street, New York, NY 10002, USA
10 Stamford Road, Oakleigh, Melbourne 3166, Australia

First published 1989

Printed in Great Britain
at the University Press, Cambridge

British Library cataloguing in publication data

The European garden flora.
Vol. 3 : Dicotyledons (Part 1)
1. Europe. Gardens. plants – Field guides
I. Walters, S.M. (Stuart Max), *1920–*
635'.094

Library of Congress cataloguing in publication data
(Revised for volume 3)

The European garden flora.

Includes indexes.
Contents: v. 1. Pteridophyta, Gymnospermae,
Angiospermae–Monocotyledons (part I) — v. 2.
Monocotyledons (part II) — v. 3. Dicotyledons (part I)
1. Plants, Ornamental—Europe—Identification—
Collected works. 2. Fruit—Europe—Identification—
Collected works. 3. Nuts—Europe—Identification—
Collected works. I. Walters, S.M. (Stuart Max)
B406.93.E85E97   1984   635.9   83–7655

ISBN 0 512 36171 0

CONTENTS

MAPS AND FIGURES

ORGANISATION AND ADVISERS

Sponsor: The Royal Horticultural Society

Editorial Committee

J.C.M. Alexander, Royal Botanic Garden, Edinburgh
A. Brady, National Botanic Gardens, Glasnevin, Dublin
C.D. Brickell, Royal Horticultural Society, London
J. Cullen (Secretary), Royal Botanic Garden, Edinburgh
P.S. Green, Royal Botanic Gardens, Kew
V.H. Heywood, IUCN/CMC, London

V.A. Matthews, Royal Botanic Gardens, Kew
N.K.B. Robson, Natural History Museum, London
S.M. Walters (Chairman), Cambridge
P.F. Yeo, University Botanic Garden, Cambridge
S.G. Knees (Research Associate), Windsor

Advisers

Professor C.D.K. Cook, Zürich, Switzerland
Professor H. Ern, Berlin, Germany
Dr H. Heine, Paris, France

Professor P.-M. Jørgensen, Bergen, Norway
Dr D.O. Wijnands, Wageningen, Netherlands

EDITORS AND CONTRIBUTORS TO VOLUME III

The various sections of this volume were edited at the following institutions:

Royal Botanic Garden, Edinburgh: Fagaceae, Olacaceae. Santalaceae, Loranthaceae, Aizoaceae, Portulacaceae, Chenopodiaceae, Amaranthaceae, Cactaceae, Didiereaceae, Magnoliaceae, Annonaceae, Myristicaceae, Canellaceae, Lauraceae, Tetracentraceae, Trochodendraceae, Eupteleaceae, Cercidiphyllaceae, Lardizabalaceae, Menispermaceae, Nymphaeaceae, Ceratophyllaceae, Saururaceae, Piperaceae, Chloranthaceae.

Royal Botanic Gardens, Kew: Casuarinaceae, Myricaceae, Juglandaceae, Eucommiaceae.

National Botanic Gardens, Glasnevin, Dublin: Moraceae, Urticaceae, Proteaceae, Winteraceae, Aristolochiaceae.

University Botanic Garden, Cambridge: Salicaceae, Betulaceae, Ulmaceae, Polygonaceae, Phytolaccaceae, Nyctaginaceae, Basellaceae, Caryophyllaceae.

RHS Gardens, Wisley: Schisandraceae, Illiciaceae, Monimiaceae, Calycanthaceae, Ranunculaceae, Berberidaceae.

Contributors

J. Akeroyd (Department of Botany, University of Reading)

J.C.M. Alexander (Royal Botanic Garden, Edinburgh)

G. Argent (Royal Botanic Garden, Edinburgh)

K.B. Ashburner (Chagford)

P.G. Barnes (Royal Horticultural Society, Wisley)

A. Brady (National Botanic Gardens, Glasnevin, Dublin)

W. Brandenburg (Rijksinstuut voor het Rassenonderzoek van Cultuurgewassen, Wageningen, Netherlands)

C.D. Brickell (Royal Horticultural Society, London)

D.F. Chamberlain (Royal Botanic Garden, Edinburgh)

J. Cullen (Royal Botanic Garden, Edinburgh)

C. Gorman (National Botanic Gardens, Glasnevin, Dublin)

P.S. Green (Royal Botanic Gardens, Kew)

R.F.L. Hamilton (Hitcham, Ipswich)

E.H. Hamlet (Royal Botanic Garden, Edinburgh)

J. Howe (Royal Botanic Garden, Edinburgh)

D.R. Hunt (Royal Botanic Gardens, Kew)

J. Kendall (Edinburgh)

C.J. King (University Botanic Garden, Cambridge)

S.G. Knees (Windsor, Research Associate)

A.C. Leslie (Royal Horticultural Society, Wisley)

B.E. Leuenberger (Botanisches Museum und Botanischer Garten, Berlin-Dahlem, Germany)

F. McIntosh (Department of Botany, University of Edinburgh)

B. Mathew (Royal Botanic Gardens, Kew)

V.A. Matthews (Royal Botanic Gardens, Kew)

H.S. Maxwell (Royal Botanic Garden, Edinburgh)

D.M. Miller (Royal Horticultural Society, Wisley)

E.C. Nelson (National Botanic Gardens, Glasnevin, Dublin)

M.J.P. Scannell (National Botanic Gardens, Glasnevin, Dublin)

S.A. Spongberg (Arnold Arboretum, Harvard University, USA)

W.T. Stearn (Natural History Museum, London)

G.J. Swales (Sunderland)

N.P. Taylor (Royal Botanic Gardens, Kew)

S.M. Walters (Cambridge)

A.C. Whiteley (Royal Horticultural Society, Wisley)

D.O. Wijnands (Agricultural University, Wageningen, Netherlands)

P.F. Yeo (University Botanic Garden, Cambridge)

ACKNOWLEDGEMENTS

During the writing of this volume, The European Garden Flora has received substantial support from the following:

(a) The Council of the Royal Horticultural Society: financial support.

(b) The Stanley Smith Horticultural Trust (Director, Sir George Taylor, FRS): financial support.

(c) The institutions to which members of the Editorial Committee belong: staff time, support and services.

(d) The Cory Fund of Cambridge University Botanic Garden: financial support for travel by the Research Associate and other uses.

(e) The William Adlington Cadbury Charitable Trust: financial support.

(f) The John S. Cohen Foundation: financial support.

(g) The Humphrey Whitbread First Charitable Trust: financial support.

The Editorial Committee gratefully acknowledges all this generous support.

Particular thanks are due to Professor W. Barthlott (Botanisches Institut der Universität, Bonn), J.D. Donald (Worthing) and L.A. Lauener (Edinburgh) for specialist taxonomic advice; to Hazel Hamlet (Edinburgh), Julia Howe (Edinburgh) and Suzanne Maxwell (Edinburgh) for editorial support; and to M. Flanagan, Frances Hibbard, Julia Howe, Christabel King, Suzanne Maxwell and Sally Rae for the preparation of the illustrations.

INTRODUCTION

Amenity horticulture (gardening, landscaping, etc.) touches human life at many points. It is a major leisure activity for a very large number of people, and is a very important means of improving the environment. The industry that has grown up to support this activity (the nursery trade, landscape architecture and management, public parks, etc.) is a large one, employing a considerable number of people. It is clearly important that the basic material of all this activity, i.e. plants, should be readily identifiable, so that both suppliers and users can have confidence that the material they buy and sell is what it purports to be.

The problems of identifying plants in cultivation are many and various, and derive from several sources which may be summarised as follows:

(a) Plants in cultivation have originated in all parts of the world, many of them from areas whose wild flora is not well known. Many have been introduced, lost and re-introduced under different names.

(b) Plants in gardens are growing under conditions to which they are not necessarily well adapted, and may therefore show morphological and physiological differences from the original wild stocks.

(c) All plants that become established in cultivation have gone through a process of selection, some of it conscious (selection of the 'best' variants, etc.), some of it unconscious (by methods of cultivation and particularly propagation), so that, again, the populations of a species in cultivation may differ significantly from the wild populations.

(d) Many garden plants have been 'improved' by hybridisation (deliberate or accidental), and so, again, differ from the original stocks.

(e) Finally, and perhaps most importantly, the scientific study of plant classification (taxonomy) has concentrated mainly on wild plants, largely ignoring material in gardens.

Nevertheless, the classification of garden plants has a long and distinguished history. Many of the Herbals of pre-Linnaean times (i.e. before 1753) consist partly or largely of descriptions of plants in gardens, and this tradition continued, and perhaps reached its peak in the late eighteenth and early nineteenth centuries – the period following the publication of Linnaeus's major works, when exploration of the world was at its height. This is the period that saw the founding of *Curtis's Botanical Magazine* (1787) and the publication of J.C. Loudon's *Encyclopaedia of plants* (1829 and many subsequent editions).

The further development of plant taxonomy, from about the middle of the nineteenth century to the present, has seen an increasing divergence between garden and scientific taxonomy, leading on the one hand to such works as the *Royal Horticultural Society's dictionary of gardening* (1951 and reprinted, itself based on G. Nicholson's *Illustrated dictionary of gardening*, 1884–1888) and the very numerous popular, usually illustrated works on garden flowers available today, and, on the other hand, to the Floras, Revisions and Monographs of scientific taxonomy.

Despite this divergence, a number of plant taxonomists realised the importance of the classification and identification of cultivated plants, and produced works of considerable scientific value. Foremost among these stands L.H. Bailey, editor of *The standard cyclopedia of horticulture* (1900, with several subsequent reprints and editions), author of *Manual of cultivated plants* (1924, edn 2, 1949) and founder of the journals *Gentes Herbarum* and *Baileya*. Other important workers in this field are T. Rumpler (*Vilmorin's Blumengärtnerei*, 1879), L. Dippel (*Handbuch der Laubholzkunde*, 1889–93), A. Voss and A. Siebert (*Vilmorin's Blumengärtnerei*, edn 3, 1894–6), C.K. Schneider (*Illustriertes Handbuch der Laubholzkunde*, 1904–12), A. Rehder (*Manual of cultivated trees and shrubs*, 1927, edn 2, 1947), J.W.C. Kirk (*A British garden flora*, 1927), F. Enke (*Parey's Blumengärtnerei*, edn 2, 1958), B.K. Boom (*Flora Cultuurgewassen*, 1959 and proceeding), V.A. Avrorin & M.V. Baranova (*Decorativn'ie Travyanist'ie Rasteniya Dlya Otkritogo Grunta SSSR*, 1977) and R. Mansfeld (*Verzeichnis landwirtschaftlicher und gärtnerischer Kulturpflanzen*, edn 2, 1986).

The present Flora, which, of necessity, is based on original taxonomic studies by many workers, attempts to provide a scientifically accurate and up-to-date means for the identification of plants cultivated for amenity in Europe (i.e. it does not include crops, whether horticultural or agricultural, or garden weeds), and to provide what are currently thought to be their correct names, together with sufficient synonymy to make sense of catalogues and other horticultural works. The needs of the informed amateur gardener have been borne in mind at all stages of the work, and it is hoped that the Flora will meet his needs just as much as it meets the needs of the professional plant taxonomist. The details of the format and use of the Flora are explained in section 2 below (pp. xii–xiv).

In writing the work, the Editorial Committee has been fully aware of the difficulties involved. Some of these have been outlined above; others derive from the fact that herbarium material of cultivated plants is scanty and usually poorly annotated, so that material of many species is not available for checking the use of names, or for comparative purposes. Because of these facts, attention has been drawn to numerous problems which cannot be solved but can only be adverted to. The solution of such problems requires much more taxonomic work.

The form in which contributions appear is the responsibility of the Editorial Committee. The vocabulary and the technicalities of plant description are therefore not necessarily those endorsed by the contributors.

1. SELECTION OF SPECIES

The problem of determining which species are in cultivation is complex and difficult, and has no complete and final answer. Many species, for instance, are grown in botanic gardens but not elsewhere; others, particularly orchids, succulents and some alpines, are to be found in the collections of specialists but are not available generally. Yet others have been in cultivation in the past but are now lost, or perhaps linger in a few collections, unrecorded and unpropagated. Further problems arise from the fact that the identification of plants in collections is not always as good as it might be, and some less well-known species probably appear in published lists under the names of other, well-known species (and vice versa).

The Flora attempts to cover all those species that are likely to be found in general collections (i.e. excluding botanic gardens and specialist collections) in Europe, whether they are grown out-of-doors or under glass. In order to produce a working list of such species, a compilation of all European nursery catalogues available to us was made in 1978 by Margaret McDonald, a vacation student working at the Royal Botanic Garden, Edinburgh. Since then, numerous additions have been made. This list (known as the 'Commercial List'), which includes well over 12 000 specific names, forms the basis of the species included here. In addition to the 'Commercial List', several works on the flora of gardens have been consulted, and the species covered by them have been carefully considered for inclusion. These works are: Wehrhahn, H.R., *Die Gartenstauden* (1929–31); *The Royal Horticultural Society's dictionary of gardening*, edn 2 (1956, supplement revised 1969);

Encke, F. (ed.), *Parey's Blumengärtnerei* (1956); Boom, B.K., *Flora der Cultuurgewassen van Nederland* (1959 and proceeding); Bean, W.J., *Trees and shrubs hardy in the British Isles* (edn 8, 1970–81); Krüssmann, G., *Handbuch der Laubgeholze* (edn 2, 1976–78); Encke, F., Buchheim, G. & Seybold, S. (eds.) *Zander's Handwörterbuch der Pflanzennamen* (edn 13, 1984), and, since 1986, Mansfeld, R., *Verzeichnis landwirtschaftlicher und gärtnerischer Kulturpflanzen* (edn 2, 1986). Most of the names included in these works are covered by the present Flora, though some have been rejected as referring to plants no longer in general cultivation.

As well as the works cited above, several relating to plants in cultivation in North America have also been consulted: Rehder, A., *Manual of cultivated trees and shrubs* (edn 2, 1947) and *Bibliography of cultivated trees and shrubs* (1949); Bailey, L.H., *Manual of cultivated plants* (edn 2, 1949); *Hortus Third* (edited by the staff of the L.H. Bailey Hortorium, Cornell University, 1976).

The contributors have also drawn on their own experience, as well as that of the family editors, European advisers and other experts, in deciding which species should be included.

Most species have a full entry, being keyed, numbered and described as set out under section 2c below (p. xiii). A few, less commonly cultivated species are not keyed or numbered individually, but are described briefly under the species to which they are most likely to key out in the formal key. The system of asterisks used in volumes I and II has been misinterpreted and has now been abandoned.

2. USE OF THE FLORA

a. *The taxonomic system followed in the Flora*. Plants are described in this work in a taxonomic order, so that similar genera and species occur close to each other, rendering comparison of descriptions more easy than in a work where the entries are alphabetical. The families (and higher groups) follow the Engler & Prantl system as expressed in H. Melchior's edition (edn 12, 1964) of *Syllabus der Pflanzenfamilien*. The exceptions are minor, apart from the placing of the Monocotyledons (volumes I & II) before the Dicotyledons; this has been done purely for convenience, and has no other implications. The assignment of genera to families also usually follows the *Syllabus*.

The order of the species within each genus has been a matter for the individual author's discretion. In general, however, some established revision of the genus has been followed, or, if no such revision exists, the author's own views on similarity and relationships have governed the order used.

b. *Nomenclature*. The arguments for using Latin names for plants in popular as well as scientific works are often stated and widely accepted, particularly for Floras such as this, which cover an area in which several languages are spoken. Latin names have therefore been used at every taxonomic level. A concise outline of the taxonomic hierarchy and how it is used can be found in C. Jeffrey's *An introduction to plant taxonomy* (1968, edn 2, 1982). Because of the difficulties of providing vernacular names in all the necessary languages (not to say dialects), they have not been included. A supplementary volume including them may be possible after the systematic part of the Flora has been completed. Meanwhile, S. Priszter's *Trees and shrubs of Europe, a dictionary in eight languages* (1983) is a useful source for the vernacular names of woody plants.

Many horticultural reference works omit the authority which should follow every Latin plant name. Knowledge of this authority prevents confusion between specific names that may have been used more than once within the same genus, and makes it possible to find the original description of the species (using *Index Kewensis*, which lists the original references for all Latin names for higher plants published since 1753). Authorities are therefore given for all names at or below the genus level. These are unabbreviated to avoid the obscure contractions which mystify the lay reader, and, on occasions, the professional botanist. In most cases we have not thought it necessary to include the initials or qualifying words and letters which often accompany author names, e.g. A. Richard, Reichenbach filius, fil. or f. (the exceptions involve a few, very common surnames).

In scientific taxonomic literature, the authority for a plant name sometimes consists of two names joined together by *ex* or *in*. Such formulae have not been used here: the authority has been shortened in accordance with *The international code of botanical nomenclature*, e.g. *Capparis lasiantha* R. Brown ex de Candolle becomes *Capparis lasiantha* de Candolle; *Viburnum ternatum* Rehder in Sargent becomes *Viburnum ternatum* Rehder. The abbreviations *hort.* and *auct.*, which sometimes stand in place of the authority after Latin names, have not been used in this work as they are often obscure or misleading. The situations described by them can be clearly and unambiguously covered by the terms *invalid*, *misapplied* or *Anon*. *Invalid* implies that the name in question has not been validly published in accordance with the Code of Nomenclature, and therefore cannot be accepted. *Misapplied* refers to names which have been applied to the wrong species in gardens or in literature. *Anon*. is used with validly published names for which there is no apparent author.

Gardeners and horticulturists complain bitterly when long-used and well-loved names are replaced by unfamiliar ones. These changes are unavoidable if *The international code of botanical nomenclature* is adhered to. Taxonomic research will doubtless continue to unearth earlier names and will also continue to realign or split up existing groups, as relationships are further investigated. However, the previously accepted names are not lost; in this work they appear as synonyms, given in brackets after the currently accepted name; they are also included in the index. Dates of publication are not given either for accepted names or for synonyms.

c. *Descriptions and terminology*. Families, genera and species included in the Flora are generally provided with full-length descriptions. Shorter, diagnostic descriptions are, however, used for genera or species which differ from others already fully described in only a few characters, e.g.:

3. P. vulgaris Linnaeus. Like *P. officinalis* but leaves lanceolate and corolla red . . .
This implies that the description of *P. vulgaris* is generally similar to that of *P. officinalis* except in the characters mentioned; it should not be assumed that plants of the two species will necessarily look very like each other. Additional species (see p. xii), subspecies, varieties, formae and cultivars (see p. xiv) are described very briefly and diagnostically.

Unqualified measurements always refer to length (though 'long' is sometimes added in cases where confusion might arise); simi-

larly, two measurements separated by a multiplication sign indicate length and breadth respectively.

The terminology has been simplified as far as is consistent with accuracy. The technical terms which, inevitably, have had to be used are explained in the glossary (p. 419). Technical terms restricted to particular families or genera are defined in the observations following the family or genus description, and are also referred to in the glossary.

d. *Informal keys*. For most genera containing 5 to 20 species (and for most families containing 5 to 20 genera) an informal key is given; this will not necessarily enable the user to identify precisely every species included, but will provide a guide to the occurrence of the more easily recognised characters. A selection of these characters is given, each of which is followed by the entry-numbers of those species which show that character. In some cases, where only a few species of a genus show a particular character, the alternative states are not specified, e.g.:

Leaves. Terete: **18,19**.
This means that only species **18** and **19** in the particular genus have terete leaves; the other species may have leaves of various forms, but they are not terete.

e. *Formal keys*. For every family containing more than one genus, and for every genus containing more than one full-entry species, a dichotomous key is provided. This form of key, in which a series of decisions must be made between pairs of contrasting character-states, should lead the user step by step to an entry number followed by the name of a species. The reader should then check this identification with the description of that species: in some cases, other, less commonly cultivated species may be mentioned under the description of the full-entry species: the brief descriptions of these should also be scanned, so that a final identification can be made. A key to all the families of the Dicotyledons to be included in the Flora is provided (pp. 3–13); this is liable to modification as the rest of the Flora is written, and improved versions will be printed in volumes IV–VI.

f. *Horticultural information*. Notes on the cultural requirements and methods of propagation are generally included in the observation to each genus; more rarely, such information is given in the observations under the family description. These notes are generally brief and very generalised, and merely provide guidance. Reference to general works on gardening is necessary for more detailed information.

g. *Citation of literature*. References to taxonomic books, articles and registration lists are cited for each family and genus, as appropriate. No abbreviations are used in these citations (though very long titles have been shortened). The citation of a particular book or article does not necessarily imply that it has been used in the preparation of the account of the particular genus or family in this work. A list giving somewhat fuller bibliographical details of all books cited in volumes I–III is to be found on pp. 429–439.

h. *Citation of illustrations*. References to good illustrations are given for each species (or subspecies or variety); the names under which they were originally published (which may be different from those used here) are not normally given. The illustrations may be coloured or black and white, and may be drawings, paintings or photographs. Usually, up to four illustrations per species have been given, and an attempt has been made to choose from widely available, modern works. Where no illustrations are cited, they either do not exist, as far as we know, or those that do are considered to be of doubtful accuracy or of very restricted availability.

In searching for illustrations, use has been made of *Index Londinensis* (1929–31, supplement 1941) and R.T. Isaacson's *Flowering plant index of illustration and information* (1979). Readers are referred to these works if they wish to find further pictures.

Several pages of figures of diagnostic plant parts are included with various groups in the Flora, and should be particularly helpful when plants are being identified by means of the keys. Some of these are original, others have either been redrawn from various sources or are photocopies of leaves.

i. *Geographical distribution.* The wild distribution, as far as it can be ascertained, is given in italics at the end of the description of each species (or subspecies or variety). The choice and spelling of place names in general follows *The Times Atlas*, Comprehensive edition (1983 reprint), except:

(1) Well-established English forms of names have been used in preference to unfamiliar vernacular names, e.g. Crete instead of Kriti, Naples instead of Napoli, Borneo instead of Kalimantan;

(2) New names or spellings will be adopted as soon as they appear in readily available works of reference.

j. *Hardiness* (see map, p. xv). For every species a hardiness code is given. This gives a tentative indication of the lowest temperatures that the particular species can withstand:

G2 – needs a heated glasshouse even in south Europe.
G1 – needs a cool glasshouse even in south Europe.
H5 – hardy in favourable areas; withstands 0 to − 5 °C minimum.
H4 – hardy in mild areas; withstands − 5 to − 10 °C minimum.
H3 – hardy in cool areas; withstands − 10 to − 15 °C minimum.
H2 – hardy almost everywhere: withstands − 15 to − 20 °C minimum.
H1 – hardy everywhere; withstands − 20 °C and below.

The map of mean January minima (p. xv) shows the isotherms corresponding to these codes. It must be understood that H4 includes H5, etc.

k. *Flowering time.* The terms spring, summer, autumn and winter have been used as a guide to flowering times in cultivation in Europe. It is not possible to be more specific when dealing with an area extending from northern Scandinavia to the Mediterranean. In cases where plants do not flower in cultivation, or flower rarely, or whose time of flowering is not recorded, no flowering time is given.

l. *Subspecies, varieties and cultivars.* Subspecies and varieties are described, where appropriate. This is done in various ways, depending on the number of such groups; all the ways, are, however, self-explanatory.

No attempt has been made to describe the range of cultivars of a species, either partially or comprehensively. The former is scarcely worth doing, the latter virtually impossible. Reference to individual, commonly grown cultivars is, however, made in various ways:

(1) If a registration list of cultivars exists, it is cited in the 'Literature' paragraph (see section 2g) following the description of the genus.

(2) If a particular cultivar is very widely grown, it may be referred to, either in the description of the species to which it belongs (or most resembles), or in the observations to that species.

(3) If, in a particular species, cultivars are numerous and fall into reasonably distinct groups based on variation in some striking character, then these groups may be referred to, together with an example of each, in the observations to the species.

m. *Hybrids.* Many hybrids between species (interspecific hybrids) and some between genera (intergeneric hybrids) are in cultivation, and some of them are widely grown. Commonly cultivated interspecific hybrids are, where possible, included as though they were species. Their names, however, include the multiplication sign indicating their hybrid origin; the names of the parents (when known or presumed) are given at the beginning of the paragraph of observations following the description. Other hybrids which are less frequently grown are mentioned in the observations to the individual parent species they most resemble. In some genera where the number of hybrids is very large, only a small selection of those most commonly grown is mentioned.

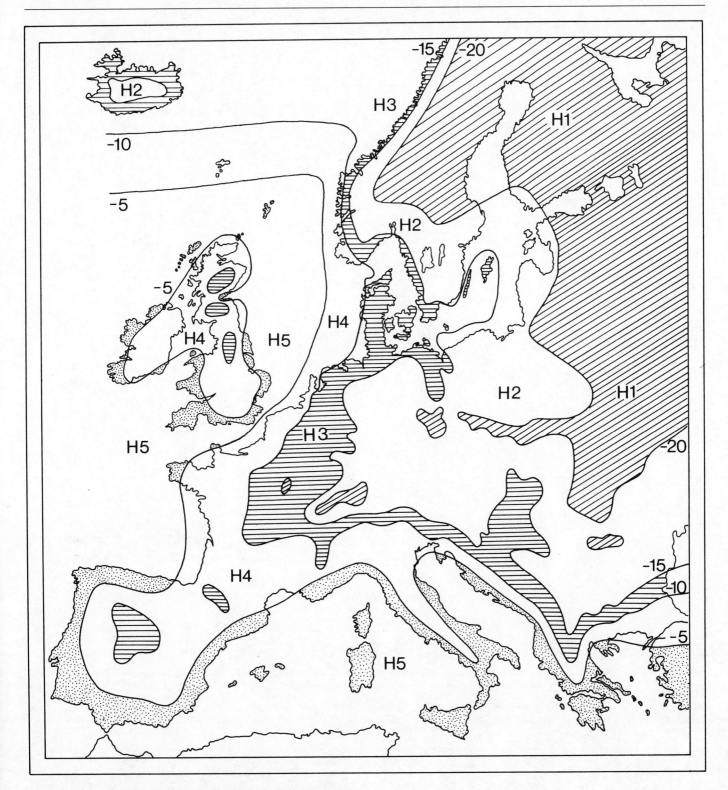

Map 1. Mean minimum January isotherms for Europe (hardiness
codes). (After Krüssmann, *Handbuch der Laubgehölze*, 1960 and
Mitteilungen der Deutsche Dendrologische Gesellschaft **75**:, 1983.)
Corrected from Europe Garden Flora volume II (1984).

ANGIOSPERMAE

Plants herbaceous or woody. Seedling leaves (cotyledons) 1 or 2. Stamens and ovary borne in unisexual or bisexual flowers which generally have protective and/or pollinator-attracting envelopes (perianth, often composed of differentiated sepals and petals). Ovules borne inside closed ovaries; seeds enclosed, until ripe, inside a fruit.

The flowering plants, as generally understood, to which most garden plants belong; there are about 300 000 species arranged in about 11 000 genera in about 400 families; some 15 000–20 000 species are in general cultivation. The group is arranged in 2 large classes, the Monocotyledons (53 families in all, covered by Volumes I & II) and Dicotyledons (about 350 families in all, part in this volume, the rest to appear in Volumes IV–VI).

KEY TO CLASSES

1a. Cotyledon 1, terminal; leaves usually with parallel veins, sometimes these connected by cross-veinlets; leaves without stipules, opposite only in some aquatic plants; flowers usually with parts in 3s; mature root system wholly adventitious   **Monocotyledons** (Volumes I & II)

b. Cotyledons usually 2, lateral; leaves usually net-veined, with or without stipules, alternate, opposite or whorled; flowers usually with parts in 2s, 4s or 5s or parts numerous; primary root (tap-root) usually persistent, branched   **Dicotyledons**

1

DICOTYLEDONS
(Part I)

Plants herbaceous or woody. Primary root (tap-root) often persisting, enlarging and branched. Seedling leaves (cotyledons) usually 2, lateral. Leaves alternate, opposite or whorled, rarely absent, veins usually forming a branched network. Parts of the flower usually in 2s, 4s or 5s, or numerous.

KEY TO FAMILIES

Families included in this volume are numbered and provided with page numbers; the other families will be covered in Volumes IV–VI.

KEY TO GROUPS

1a. Petals present, free from each other at their bases (rarely united above the base), usually falling individually, or petals absent   2

  b. Petals present, all united into a tube at their bases, sometimes shortly so, falling as a complete corolla   10

2a. Flowers unisexual and without petals, at least the males borne in catkins which are usually deciduous; plants always woody   **Group A**

  b. Flowers with or without petals, unisexual or bisexual, never in catkins; plants woody or not   3

3a. Ovary consisting of 2 or more carpels which are completely free from each other   **Group B** (p. 4)

  b. Ovary consisting of a single carpel or of 2 or more carpels which are united to each other wholly or in part (rarely the bodies of the carpels more or less free but the styles united)   4

4a. Perianth of 2 or more whorls, more or less clearly differentiated into calyx and corolla (calyx rarely very small and obscure; excluding aquatic plants with minute, quickly deciduous petals and branch-parasites with opposite, leathery leaves)   5

  b. Perianth of a single whorl (which may be petal-like) or perianth completely absent, more rarely the perianth of 2 or more whorls but the segments not differing from whorl to whorl   8

5a. Stamens more than twice as many as the petals   **Group C** (p. 4)

  b. Stamens twice as many as the petals or fewer   6

6a. Ovary partly or fully inferior   **Group D** (p. 6)

  b. Ovary completely superior   7

7a. Placentation axile, apical, basal or free-central   **Group E** (p. 6)

  b. Placentation parietal or marginal   **Group F** (p. 8)

8a. Stamens borne on the perianth or ovary inferior (perianth of female flowers sometimes very small)   **Group G** (p. 8)

  b. Stamens free from the perianth; ovary superior or naked (i.e. not surrounded by a perianth)   9

9a. Flowers unisexual   **Group H** (p. 9)

  b. Flowers bisexual   **Group I** (p. 10)

10a. Ovary partly or fully inferior   **Group J** (p. 10)

  b. Ovary completely superior   11

11a. Corolla radially symmetric   **Group K** (p. 11)

  b. Corolla bilaterally symmetric   **Group L** (p. 12)

Group A

1a. Stems jointed; leaves reduced to whorls of scales   **XXXIV. Casuarinaceae** (p. 14)

  b. Stems not jointed; leaves not as above   2

2a. Leaves pinnate   3

  b. Leaves simple and entire, toothed or lobed (sometimes deeply so)   4

3a. Leaves without stipules; fruit a nut   **XXXVI. Juglandaceae** (p. 17)

  b. Leaves with stipules; fruit a legume   **Leguminosae**

4a. Leaves opposite, evergreen, entire; fruit berry-like   **Garryaceae**

  b. Leaves alternate, deciduous or evergreen; fruit not berry-like   5

5a. Ovules many, parietal; seeds many, cottony-hairy; male catkins erect with the stamens projecting beyond the bracts or hanging and with fringed bracts   **XXXVII. Salicaceae** (p. 20)

  b. Ovules solitary or few, not parietal; seeds few, not cottony-hairy; male catkins not as above   6

6a. Leaves dotted with aromatic glands   **XXXV. Myricaceae** (p. 16)

  b. Leaves not gland-dotted   7

7a. Styles 3, each often branched; fruit splitting into 3 mericarps; seeds with appendages   **Euphorbiaceae**

  b. Styles 1–3, not branched; fruit and seeds not as above   8

8a. Plant with milky sap   **XLII. Moraceae** (p. 86)

  b. Plants with clear sap   9

9a. Male catkins compound, i.e. each bract with 2–3 flowers
    attached to it; styles 2   **XXXVIII. Betulaceae** (p. 45)

  b. Male catkins simple, i.e. each bract subtending a single
    flower; styles 1 or 3–6   **XXXIX. Fagaceae** (p. 59)

Group B

1a. Trees with bark peeling off in plates, palmately lobed leaves
    and unisexual flowers in hanging, spherical heads
                        Platanaceae

  b. Combination of characters not as above   2

2a. Perianth-segments and stamens borne independently below
    the ovary, or perianth absent   3

  b. Perianth-segments and stamens borne on a rim or cup
    which is borne below the ovary   23

3a. Aquatic plants with peltate leaves and 3 sepals
                 LXXVII. Nymphaeaceae (p. 400)

  b. Terrestrial plants, or, if aquatic then without peltate leaves
    and with more than 3 sepals   4

4a. Herbs, succulent shrubs or shrubs with yellow wood or
    climbers with bisexual flowers and opposite leaves   5

  b. Trees or shrubs which are neither succulent nor with
    yellow wood, if climbers then with unisexual flowers and
    alternate leaves   10

5a. Perianth absent   **LXXIX. Saururaceae** (p. 406)

  b. Perianth present   6

6a. Leaves succulent; stamens in 1 or 2 whorls   **Crassulaceae**

  b. Leaves not succulent; stamens spirally arranged, not
    obviously in whorls   7

7a. Petals fringed; fruits formed from each carpel borne on a
    common gynophore   **Resedaceae**

  b. Petals (when present) not fringed, but sometimes modified
    for nectar-secretion; fruits formed from each carpel not
    borne on a common gynophore   8

8a. Leaves opposite or whorled; flowers small, stalkless, in
    axillary clusters; ovule 1, placentation basal
                XLIX. Phytolaccaceae (p. 129)

  b. Combination of characters not as above   9

9a. Sepals differing among themselves, green; stamens ripening
    from the inside of the flower outwards, borne on a nectar-
    secreting disc   **Paeoniaceae**

  b. Sepals all similar, green or petal-like; stamens ripening from
    the outside of the flower inwards; nectar-secreting disc
    absent   **LXXIII. Ranunculaceae** (p. 325)

10a. Leaves simple   11

  b. Leaves compound   21

11a. Sepals and petals 5   12

  b. Sepals and petals not 5   14

12a. Leaves opposite; stamens 5–10   **Coriariaceae**

  b. Leaves alternate; stamens more numerous   13

13a. Anthers opening by pores   **Ochnaceae**

  b. Anthers opening by slits   **Dilleniaceae**

14a. Unisexual climbers   15

  b. Erect trees or shrubs, flowers usually bisexual   16

15a. Carpels many; seeds not U-shaped
                LXIV. Schisandraceae (p. 317)

  b. Carpels 3 or 6; seeds usually U-shaped
                LXXVI. Menispermaceae (p. 398)

16a. Stamens with a truncate connective which overtops the
    anthers; fruit usually a fleshy aggregate fruit; endosperm
    convoluted   **LXI. Annonaceae** (p. 315)

  b. Connective of stamens not as above; fruit not as above;
    endosperm not convoluted   17

17a. Carpels spirally arranged on a long receptacle; stipules
    large, united, early deciduous, leaving a ring-like scar
                LIX. Magnoliaceae (p. 302)

  b. Carpels in 1 whorl; stipules absent, minute or united to the
    leaf-stalk, not leaving a ring-like scar   18

18a. Petals present   19

  b. Petals absent   20

19a. Sepals free, overlapping, more than 6; ovules solitary in
    each carpel   **LXV. Illiciaceae** (p. 318)

  b. Sepals 2–6, united, or if free, then edge-to-edge in bud;
    ovules more than 1 in each carpel
                LX. Winteraceae (p. 314)

20a. Leaves in whorls; flowers bisexual; sepals minute or absent
                LXXI. Eupteleaceae (p. 324)

  b. Leaves opposite or alternate; flowers unisexual; sepals 4
                LXXII. Cercidiphyllaceae (p. 325)

21a. Unisexual climbers, or erect shrubs with blue fruits;
    perianth-parts in 3s   **LXXV. Lardizabalaceae** (p. 396)

  b. Erect shrubs, fruits not blue; perianth-parts not in 3s   22

22a. Flowers showy, bisexual; leaves not aromatic   **Paeoniaceae**

  b. Flowers inconspicuous, unisexual; leaves aromatic
                        Rutaceae

23a. Leaves modified into insectivorous pitchers   **Cephalotaceae**

  b. Leaves not modified into insectivorous pitchers   24

24a. Flowers unisexual; leaves evergreen
                LXVI. Monimiaceae (p. 319)

  b. Flowers bisexual; leaves usually deciduous   25

25a. Stamens all fertile; perianth-whorls with 4–9 segments;
    leaves usually alternate   **Rosaceae**

  b. Inner stamens sterile; perianth-whorls with more than 9
    segments; leaves opposite   **LXVII. Calycanthaceae** (p. 320)

Group C

1a. Perianth and stamens borne independently below the
    superior ovary   2

  b. Perianth and stamens either borne on the edge of a rim or
    cup which is itself borne below the superior ovary, or borne
    on the top or the sides of the (partly or fully) inferior ovary
                30

2a. Placentation axile or free-central   3

  b. Placentation marginal or parietal   20

3a. Placentation free-central; sepals usually 2
                LII. Portulacaceae (p. 171)

  b. Placentation axile; sepals more than 2   4

4a. Leaves all basal, tubular, forming insectivorous pitchers;
    style peltately dilated   **Sarraceniaceae**

b. Leaves not as above; style not dilated   5

5a. Leaves alternate   6

  b. Leaves opposite   17

6a. Anthers opening by terminal pores   7

  b. Anthers opening by longitudinal slits   9

7a. Shrubs with simple leaves without stipules, often covered with stellate hairs; stamens inflexed in bud; fruit a berry   **Actinidiaceae**

  b. Combination of characters not as above   8

8a. Ovary deeply lobed, borne on an enlarged receptacle or gynophore; petals not fringed   **Ochnaceae**

  b. Ovary not lobed, not borne as above; petals often fringed   **Elaeocarpaceae**

9a. Inner whorl of perianth-segments tubular or bifid, nectar-secreting; fruit a group of partly to fully united follicles   **LXXIII. Ranunculaceae** (p. 325)

  b. Combination of characters not as above   10

10a. Leaves with translucent, aromatic glands   **Rutaceae**

  b. Leaves without translucent, aromatic glands   11

11a. Large tropical trees; sepals 5, all, or 2–3 of them enlarged and wing-like in fruit; carpels 3   **Dipterocarpaceae**

  b. Combination of characters not as above   12

12a. Stipules absent; leaves evergreen   **Theaceae**

  b. Stipules present; leaves usually deciduous   13

13a. Filaments free; anthers 2-celled   14

  b. Filaments united into a tube, at least around the ovary, often also around the styles; anthers often 1-celled   15

14a. Disc absent; stamens more than 15; leaves simple   **Tiliaceae**

  b. Disc present, conspicuous; stamens 15; leaves dissected   **Zygophyllaceae**

15a. Styles divided, several; stipules often persistent; carpels 5 or more   **Malvaceae**

  b. Style 1, capitate or lobed, stigmas 1–several; stipules usually deciduous; carpels 2–5   16

16a. Stamens in 2 whorls, those of the outer whorl usually sterile   **Sterculiaceae**

  b. Stamens in several whorls, all fertile   **Bombacaceae**

17a. Sepals united, falling as a unit; fruit separating into boat-shaped units   **Eucryphiaceae**

  b. Sepals and fruit not as above   18

18a. Small trees; stamens with brightly coloured filaments which are at least twice as long as the petals, the anthers forming a ring   **Caryocaraceae**

  b. Combination of characters not as above   19

19a. Leaves simple, without stipules, often with translucent glands; stamens often united in bundles   **Guttiferae**

  b. Leaves pinnate, without translucent glands; stamens not united in bundles   **Zygophyllaceae**

20a. Aquatic plants with cordate leaves; style and stigmas forming a disc on top of the ovary   **LXXVII. Nymphaeaceae** (p. 400)

  b. Combination of characters not as above   21

21a. Carpel 1 with marginal placentation   22

  b. Carpels 2 or more, placentation parietal   23

22a. Leaves bipinnate or modified into phyllodes, with stipules   **Leguminosae**

b. Leaves various but not as above, without stipules   **LXXIII. Ranunculaceae** (p. 325)

23a. Leaves opposite   24

  b. Leaves alternate   26

24a. Styles numerous; floral parts in 3s   **Papaveraceae**

  b. Styles 1–5; floral parts in 4s or 5s   25

25a. Style 1; stamens not united in bundles; leaves without translucent glands   **Cistaceae**

  b. Styles 3–5, free or variously united; stamens united in bundles (sometimes apparently free); leaves with translucent glands   **Guttiferae**

26a. Small trees with aromatic bark; filaments united   **LXIII. Canellaceae** (p. 317)

  b. Herbs shrubs or trees, bark not aromatic; filaments free   27

27a. Trees; leaves with stipules; anthers opening by pore-like slits   **Bixaceae**

  b. Herbs or shrubs; leaves usually without stipules; anthers opening by longitudinal slits   28

28a. Sepals 2 or rarely 3, quickly deciduous   **Papaveraceae**

  b. Sepals 4–8, persistent in flower   29

29a. Ovary closed at its apex, borne on a gynophore; none of the petals fringed   **Capparidaceae**

  b. Ovary open at its apex, not borne on a gynophore; at least some of the petals fringed   **Resedaceae**

30a. Flowers unisexual; leaf-base oblique   **Begoniaceae**

  b. Flowers bisexual; leaf-base not oblique   31

31a. Placentation free-central; ovary partly inferior   **LII. Portulacaceae** (p. 171)

  b. Placentation not free-central; ovary either completely superior or completely inferior   32

32a. Aquatic plants with cordate leaves   **LXXVII. Nymphaeaceae** (p. 400)

  b. Terrestrial plants; leaves various   33

33a. Carpels 1 or 3, excentrically placed at the top of, the bottom of, or within, the tubular perigynous zone   **Chrysobalanaceae**

  b. Carpels and perigynous zone not as above   34

34a. Stamens united in bundles on the same radii as the petals; staminodes often present; plants usually rough with stinging hairs   **Loasaceae**

  b. Combination of characters not as above   35

35a. Sepals 2, united, falling as a unit; plant herbaceous   **Papaveraceae**

  b. Sepals 4–5, usually free, not falling as a unit; trees or shrubs   36

36a. Stamens united in several rings or sheets   **Lecythidaceae**

  b. Stamens not as above   37

37a. Carpels 8–12, one above the other   **Punicaceae**

  b. Carpels fewer, side by side   38

38a. Leaves with stipules   39

  b. Leaves without stipules   40

39a. Leaves opposite or in whorls; plant woody   **Cunoniaceae**

  b. Leaves alternate; plants woody or herbaceous   **Rosaceae**

40a. Leaves with translucent aromatic glands; style 1   **Myrtaceae**

  b. Leaves without translucent aromatic glands; styles usually more than 1   **Saxifragaceae**

Group D

1a. Petals and stamens numerous; plant succulent   2
 b. Petals and stamens each fewer than 10; plants usually not succulent   3
2a. Stems succulent, often very spiny, leaves absent, very reduced or falling early   **LVII. Cactaceae** (p. 202)
 b. Stems and leaves succulent, spines usually absent   **LI. Aizoaceae** (p. 133)
3a. Anthers opening by terminal pores   4
 b. Anthers opening by longitudinal slits   5
4a. Filaments with a knee-like joint below the anthers; leaves with 3 conspicuous main veins from the base   **Melastomataceae**
 b. Filaments straight; leaves with a single main vein   **Rhizophoraceae**
5a. Placentation parietal, placentas sometimes intrusive   6
 b. Placentation axile, basal, apical or free-central   7
6a. Climbing herbs with tendrils; flowers unisexual   **Cucurbitaceae**
 b. Erect herbs or shrubs, if climbing then without tendrils; flowers usually bisexual   **Saxifragaceae**
7a. Stamens on the same radii as the petals; trees or shrubs with simple leaves   **Rhamnaceae**
 b. Stamens not on the same radii as the petals; plants herbaceous or woody, leaves simple to compound   8
8a. Flowers borne in umbels, sometimes condensed into heads or layers of whorls; leaves usually compound   9
 b. Flowers not in umbels; leaves usually simple   10
9a. Fruit splitting into 2 mericarps; flowers usually bisexual; petals overlapping in bud; usually herbs without stellate hairs   **Umbelliferae**
 b. Fruit a berry; flowers often unisexual; petals edge-to-edge in bud; plants mostly woody, often with stellate hairs   **Araliaceae**
10a. Style 1   11
 b. Styles more than 1, often 2, divergent   19
11a. Floating aquatic herb; leaf-stalks inflated   **Trapaceae**
 b. Terrestrial herbs, trees or shrubs; leaf-stalks not inflated   12
12a. Small, low shrubs with scale-like, overlapping leaves; flowers in heads   **Bruniaceae**
 b. Trees, shrubs or herbs with expanded leaves; flowers not usually in heads   13
13a. Ovary 1-celled with 2–5 ovules; fruit leathery or drupe-like, 1-seeded   **Combretaceae**
 b. Ovary usually with 2–5 cells, ovules various; fruit not as above   14
14a. Ovule solitary in each cell of the ovary   15
 b. Ovules 2–numerous in each cell of the ovary   18
15a. Petals edge-to-edge in bud; flower usually bisexual   16
 b. Petals overlapping in bud; flowers often unisexual   17
16a. Stamens with swollen, hairy filaments; petals recurved   **Alangiaceae**
 b. Stamens without swollen, hairy filaments; petals not recurved   **Cornaceae**
17a. Flowers in heads subtended by 2 conspicuous, white bracts; ovary 6–10-celled   **Davidiaceae**

 b. Flowers various, but not as above; ovary 1-celled   **Nyssaceae**
18a. Sap milky; petals 5; ovary 3-celled   **Campanulaceae**
 b. Sap watery; petals 2 or 4; ovary usually 4-celled   **Onagraceae**
19a. Trees or shrubs; hairs often stellate; fruit a few-seeded woody capsule   **Hamamelidaceae**
 b. Herbs or shrubs; hairs simple or absent; fruit various, not a woody capsule   **Saxifragaceae**

Group E

1a. Perianth bilaterally symmetric   2
 b. Perianth radially symmetric (the stamens sometimes not radially symmetric due to deflexion)   14
2a. Anthers cohering, covering the ovary like a cap   **Balsaminaceae**
 b. Anthers free, not as above, filaments sometimes united   3
3a. Anthers opening by terminal pores   4
 b. Anthers opening by longitudinal slits   5
4a. Stamens 8, filaments united for at least half their length; fruit without barbed bristles   **Polygalaceae**
 b. Stamens 3 or 4, filaments free; fruit covered in barbed bristles   **Krameriaceae**
5a. Plants herbaceous   6
 b. Plants woody (trees, shrubs or climbers)   10
6a. Leaves with stipules   7
 b. Leaves without stipules   8
7a. Carpel 1; fruit a legume, sometimes 1-seeded **Leguminosae**
 b. Carpels 5; fruit a capsule or berry, or splitting into mericarps   **Geraniaceae**
8a. Sepals, petals and stamens borne on a rim, cup or tube which is borne below the ovary; leaves not peltate   9
 b. Sepals, petals and stamens borne independently below the ovary (rarely the petals and stamens somewhat united at the base); leaves usually peltate   **Tropaeolaceae**
9a. Leaves opposite   **Lythraceae**
 b. Leaves alternate or all basal   **Saxifragaceae**
10a. Stamens as many as or fewer than petals, borne on the same radii as the petals   **Sabiaceae**
 b. Stamens more than the petals, if as many or fewer then not on the same radii as the petals   11
11a. Carpel 1 with its style arising from near its base   **Chrysobalanaceae**
 b. Carpels 2 or more, styles not as above   12
12a. Leaves opposite, palmate; sepals united at the base   **Hippocastanaceae**
 b. Leaves alternate, usually pinnate; sepals free   13
13a. Stipules large, borne within the bases of the leaf-stalks   **Melianthaceae**
 b. Stipules absent, or, if present, minute, not borne as above   **Sapindaceae**
14a. Anthers opening by terminal pores   15
 b. Anthers opening by longitudinal slits, or by flaps   21

15a. Leaves with 3 more or less parallel veins from the base; each filament with a knee-like bend below the anther **Melastomataceae**

b. Leaves with 1 main vein; filaments straight or arched 16

16a. Low shrubs; leaves and stems covered in conspicuous, stalked glandular hairs on which insects are often caught 17

b. Shrubs or rarely low shrubs, not glandular hairy as above 18

17a. Carpels 2 **Byblidaceae**

b. Carpels 3 **Roridulaceae**

18a. Low shrub with unisexual flowers; stamens 4; petals 4, each usually 2–3-lobed, rarely a few unlobed **Elaeocarpaceae**

b. Combination of characters not as above 19

19a. Ovary lobed, consisting of several rounded humps, the style arising from the depression between them **Ochnaceae**

b. Ovary not lobed as above, style terminal 20

20a. Carpels 3; style divided above into 3 branches; nectar-secreting disc absent **Clethraceae**

b. Carpels 4–5 (rarely 3); style not divided; nectar-secreting disc usually present **Ericaceae**

21a. Placentation free-central (ovary sometimes with septa below), or basal 22

b. Placentation axile or apical 26

22a. Stamens as many as petals and on the same radii as them 23

b. Stamens more or fewer than the petals, if as many then not on the same radii as them 25

23a. Anthers opening by flaps; stigma 1 **LXXIV. Berberidaceae (p. 370)**

b. Anthers opening by longitudinal slits; stigmas more than 1 24

24a. Sepals 5; ovule 1, basal on a long, curved stalk; stipules absent **Plumbaginaceae**

b. Sepals 2 or rarely 3; ovules usually numerous, rarely 1, then not on a long, curved stalk; stipules usually present **LII. Portulacaceae (p. 171)**

25a. Ovary lobed, consisting of several humps, the style arising from the depression between them; leaves pinnatisect **Limnanthaceae**

b. Ovary not lobed, style terminal; leaves simple, entire **LIV. Caryophyllaceae (p. 177)**

26a. Petals and stamens both numerous; plants with succulent leaves and stems **LI. Aizoaceae (p. 133)**

b. Combination of characters not as above 27

27a. Small, hairless annual herbs growing in water or on wet mud; seeds pitted **Elatinaceae**

b. Combination of characters not as above 28

28a. Sepals, petals and stamens borne on a rim, cup or tube which is inserted below the ovary 29

b. Sepals, petals and stamens inserted individually below the ovary 34

29a. Stamens as many as the petals and borne on the same radii as them **Rhamnaceae**

b. Stamens more or fewer than the petals or if as many then not borne on the same radii as them 30

30a. Style 1 31

b. Styles more than 1, often 2, divergent 32

31a. Calyx tube not prominently ribbed; seeds with arils; mostly trees, shrubs or climbers **Celastraceae**

b. Calyx tube prominently ribbed; seeds without arils; mostly herbs **Lythraceae**

32a. Fruit an inflated, membranous capsule; leaves mostly opposite, compound **Staphyleaceae**

b. Combination of characters not as above 33

33a. Trees or shrubs; hairs often stellate; anthers often opening by flaps; fruit a few-seeded, woody capsule **Hamamelidaceae**

b. Herbs or shrubs; hairs simple or absent; anthers opening by longitudinal slits; fruit a non-woody capsule **Saxifragaceae**

34a. Leaves with translucent, aromatic glands **Rutaceae**

b. Leaves without translucent, aromatic glands 35

35a. Flowers with a well-developed disc (usually nectar-secreting) below and/or around the ovary 36

b. Flowers without a disc, nectar secreted in other ways 46

36a. Stamens as many as and on the same radii as the petals 37

b. Stamens more or fewer than the petals, if as many then not on the same radii as them 38

37a. Climbers with tendrils; stamens free **Vitaceae**

b. Erect shrubs without tendrils; stamens with the filaments united at least at the base **Leeaceae**

38a. Resinous trees or shrubs 39

b. Herbs, shrubs or trees, not resinous (sometimes aromatic) 40

39a. Ovules 2 per cell; fruit a drupe or capsule **Burseraceae**

b. Ovules 1 per cell; fruit a drupe **Anacardiaceae**

40a. Plant herbaceous 41

b. Plant woody (tree, shrub or climber) 42

41a. Petals long-clawed, united above the base; leaves not fleshy **Stackhousiaceae**

b. Petals entirely free, not long-clawed; leaves fleshy **Zygophyllaceae**

42a. Flowers, or at least some of them, functionally unisexual (i.e. apparent anthers not producing pollen or ovary containing no ovules) 43

b. Flowers bisexual 44

43a. Leaves alternate; ovary with 2–5 carpels, not flattened **Simaroubaceae**

b. Leaves opposite; ovary with 2 (rarely 3) carpels, flattened in the plane of the septum **Aceraceae**

44a. Leaves entire or toothed; stamens 4–5, emerging from the disc; seeds with arils **Celastraceae**

b. Combination of characters not as above 45

45a. Leaves without stipules, not fleshy; filaments of the stamens united into a tube **Meliaceae**

b. Leaves with stipules, fleshy; filaments of the stamens free **Zygophyllaceae**

46a. Plant herbaceous 47

b. Plant woody (tree, shrub or climber) 49

47a. Leaves always simple; ovary 6–10-celled by the development of 3–5 secondary septa between the original 3–5 septa during development of the flower **Linaceae**

b. Leaves lobed or compound; secondary septa absent 48

48a. Leaves with stipules   **Geraniaceae**
  b. Leaves without stipules   **Oxalidaceae**
49a. Filaments of the stamens united below   50
  b. Filaments of the stamens free   54
50a. Plant succulent, spiny; stamens 8 with woolly filaments;
     plants unisexual   **LVIII. Didiereaceae** (p. 301)
  b. Combination of characters not as above   51
51a. Stamens 2   **Oleaceae**
  b. Stamens 3 or more   52
52a. Leaves without stipules   **XLV. Olacaceae** (p. 118)
  b. Leaves with stipules though these are sometimes quickly
     deciduous   53
53a. Stipules persistent, borne between the bases of the leaf-
     stalks; petals with appendages   **Erythroxylaceae**
  b. Stipules quickly deciduous, not borne between the bases of
     the leaf-stalks; petals without appendages   **Sterculiaceae**
54a. Stamens 8–10   55
  b. Stamens 3–6   57
55a. Petals long-clawed, often fringed or toothed; stamens 10;
     some or all of the sepals with nectar-secreting glands
     outside   **Malpighiaceae**
  b. Petals neither clawed nor fringed nor toothed; stamens
     8–10; sepals without nectaries outside   56
56a. Ovules 1 per cell; sepals united at the base
           XLV. Olacaceae (p. 118)
  b. Ovules many per cell; sepals free   **Stachyuraceae**
57a. Staminodes present in flowers which also contain fertile
     stamens   58
  b. Staminodes absent from flowers with fertile stamens,
     present only in functionally female flowers   59
58a. Carpels 2, ovary containing a single apical ovule; stipules
     present, borne within the bases of the leaf-stalks
           Corynocarpaceae
  b. Carpels 2–4, each cell of the ovary containing 1–2 ovules;
     stipules absent   **Cyrillaceae**
59a. Trees with opposite, pinnate leaves; twigs tipped with large
     dark buds; fruit a samara   **Oleaceae**
  b. Combination of characters not as above   60
60a. Sepals united at the base   61
  b. Sepals free from each other at the base   62
61a. Carpels 2 or rarely 3, 1 or 2 of them sterile, the ovary
     containing 2 apical ovules   **Icacinaceae**
  b. Carpels 3–many, all fertile, the ovary containing 1–2 ovules
     per cell   **Aquifoliaceae**
62a. Ovules 1 per cell; petals 3–4   **Cneoraceae**
  b. Ovules many per cell; petals 5   **Pittosporaceae**

Group F

1a. Sepals, petals and stamens borne on a rim or cup which is
     inserted below the ovary   2
  b. Sepals, petals and stamens inserted individually below the
     ovary   4
2a. Trees; leaves bi- or tripinnate; flowers bilaterally symmetric;
     stamens 5, of different lengths   **Moringaceae**

  b. Combination of characters not as above   3
3a. Flower-stalks slightly united to the leaf-stalks so that the
     flowers appear to be borne on the latter; petals twisted (each
     overlapped by, and overlapping, 1 other) in bud; carpels 3
           Turneraceae
  b. Flower-stalks not united to leaf-stalks; petals not as above in
     bud; carpels 2 or 4   **Saxifragaceae**
4a. Perianth bilaterally symmetric   5
  b. Perianth radially symmetric   9
5a. Ovary of 1 carpel with marginal placentation   **Leguminosae**
  b. Ovary of 2 or more carpels with parietal placentation   6
6a. Ovary open at the apex; some or all of the petals fringed
           Resedaceae
  b. Ovary closed at the apex; no petals fringed   7
7a. Petals and stamens 5; carpels 3   **Violaceae**
  b. Petals and stamens 4 or 6; carpels 2   8
8a. Ovary borne on a stalk (gynophore); stamens projecting
     beyond the petals   **Capparidaceae**
  b. Ovary not borne on a stalk; stamens not projecting beyond
     the petals   **Papaveraceae**
9a. Petals and stamens numerous   **LI. Aizoaceae** (p. 133)
  b. Petals and stamens fewer than 7   10
10a. Stamens alternating with multifid staminodes
           Saxifragaceae
  b. Stamens not alternating with multifid staminodes   11
11a. Leaves insect-catching by means of stalked, glandular hairs
           Droseraceae
  b. Leaves not insect-catching   12
12a. Climbers with tendrils; ovary and stamens borne on a
     common stalk (androgynophore); corona present
           Passifloraceae
  b. Combination of characters not as above   13
13a. Petals 4, the inner pair trifid; sepals 2   **Papaveraceae**
  b. Petals not as above; sepals 4–5   14
14a. Stamens 6, 4 longer and 2 shorter; carpels 2; fruit with a
     secondary septum   **Cruciferae**
  b. Stamens 4–10, all more or less equal; carpels 2–5; fruit
     without a secondary septum   15
15a. Petals each with a scale-like appendage at the base of the
     blade; leaves opposite   **Frankeniaceae**
  b. Petals without appendages; leaves alternate or all basal   16
16a. Stipules present   17
  b. Stipules absent   18
17a. Stamens 10; flowers in dense, cylindric panicles
           Cunoniaceae
  b. Stamens 5; flowers not as above   **Violaceae**
18a. Leaves scale-like, alternate   **Tamaricaceae**
  b. Leaves normally developed, usually all basal   **Pyrolaceae**

Group G

1a. Aquatics or rhubarb (*Rheum*)-like marsh plants with cordate
     leaves   2
  b. Terrestrial plants, not as above   4
2a. Stamens 8, 4 or 2; leaves deeply divided or cordate
           Haloragaceae
  b. Stamen 1; leaves undivided, not cordate   3

3a. Leaves whorled; fruit small, indehiscent, dry, 1-seeded, not lobed   **Hippuridaceae**

b. Leaves opposite; fruit 4-lobed, up to 4-seeded

Callitrichaceae

4a. Trees or shrubs   5

b. Herbs, climbers or parasites   17

5a. Plant covered with scales; fruit enclosed in the berry-like, persistent, fleshy perianth   **Elaeagnaceae**

b. Plant not covered with scales; fruit not as above   6

6a. Stamen 1, or 1 whole stamen flanked by 2 half-stamens; leaves opposite   **LXXXI. Chloranthaceae** (p. 415)

b. Stamens not as above; leaves usually alternate   7

7a. Stamens on radii alternating with the sepals   **Rhamnaceae**

b. Stamens on the same radii as the sepals   8

8a. Ovary 2-celled, partly inferior; stellate hairs often present; fruit a woody capsule   **Hamamelidaceae**

b. Combination of characters not as above   9

9a. Stamens 4, situated at the top of the spoon-shaped, petal-like perianth-segments which split apart as the flower opens

XLIV. Proteaceae (p. 105)

b. Combination of characters not as above   10

10a. Ovary inferior   11

b. Ovary superior   14

11a. Placentation parietal   **Saxifragaceae**

b. Placentation axile or basal   12

12a. Styles 3–6; fruit a nut surrounded by a scaly cupule

XXXIX. Fagaceae (p. 59)

b. Style 1; fruit not as above   13

13a. Stamens 4–5; placentation basal

XLVI. Santalaceae (p. 119)

b. Stamens 5–10; placentation axile   **Cornaceae**

14a. Leaves aromatic, dotted with translucent glands; anthers opening by flaps   **LXVIII. Lauraceae** (p. 321)

b. Leaves neither aromatic nor gland-dotted; anthers not opening by flaps   15

15a. Stamens 2, or 8–10 borne at different levels in the perianth-tube; leaves simple, entire   **Thymelaeaceae**

b. Stamens not as above; leaves lobed or compound   16

16a. Inflorescence borne on the shoots of the current year; fruit a schizocarp of 2 (rarely 3) samaras   **Aceraceae**

b. Inflorescence borne on old wood; fruit a legume

Leguminosae

17a. Branch parasites with green, forked branches or with small, scale-like leaves joined in pairs

XLVII. Loranthaceae (p. 120)

b. Plants not parasitic, as above   18

18a. Perianth absent; flowers in spikes

LXXIX. Saururaceae (p. 406)

b. Perianth present; flowers usually not in spikes   19

19a. Leaf-base oblique; ovary inferior, 3-celled   **Begoniaceae**

b. Leaf-base not oblique; ovary not as above   20

20a. Ovary superior   21

b. Ovary inferior   26

21a. Carpel 1, ovule 1, apical; perianth tubular

Thymelaeaceae

b. Combination of characters not as above   22

22a. Carpels 3 (rarely 2), ovule 1, basal; perianth persistent in fruit; leaves usually alternate, entire   23

b. Combination of characters not as above   24

23a. Leaves without stipules; stamens 5

LIII. Basellaceae (p. 176)

b. Leaves with stipules united into a sheath (ochrea); stamens usually 6–9   **XLVIII. Polygonaceae** (p. 121)

24a. Leaves alternate, usually lobed or compound   **Rosaceae**

b. Leaves opposite, usually entire   25

25a. Ovule 1, fruit a nut; stipules scarious or rarely absent

LIV. Caryophyllaceae (p. 177)

b. Ovules numerous; fruit a capsule; stipules absent

Lythraceae

26a. Leaves pinnate; ovary open at the apex   **Datiscaceae**

b. Leaves not pinnate; ovary closed at apex   27

27a. Ovary 6-celled; perianth 3-lobed or tubular and bilaterally symmetric   **LXXXII. Aristolochiaceae** (p. 416)

b. Combination of characters not as above   28

28a. Ovules 1–5; seed 1   29

b. Ovules and seeds numerous   30

29a. Perianth-segments thickening in fruit; leaves alternate

LV. Chenopodiaceae (p. 195)

b. Perianth-segments not as above; leaves opposite or alternate

XLVI. Santalaceae (p. 119)

30a. Styles 2; placentation parietal   **Saxifragaceae**

b. Style 1; placentation axile   **Onagraceae**

Group H

1a. Aquatic herb; leaves divided into thread-like segments

LXXVIII. Ceratophyllaceae (p. 406)

b. Terrestrial plants; leaves not as above   2

2a. Trailing, heather-like shrublet; fruit a berry   **Empetraceae**

b. Combination of characters not as above   3

3a. Flowers in racemes or spikes; fruit a berry or drupe-like; leaves entire, alternate, without stipules; carpels more than 5   **XLIX. Phytolaccaceae** (p. 129)

b. Combination of characters not as above   4

4a. Ovary 3-celled; styles 3   5

b. Ovary 1-, 2- or 4-celled; styles 1–2   7

5a. Leaves with sheathing, membranous stipules; perianth-segments 6; fruit a nut   **XLVIII. Polygonaceae** (p. 121)

b. Combination of characters not as above   6

6a. Fruit schizocarpic; sap often milky; deciduous or evergreen herbs, trees or shrubs or stem-succulents; styles usually divided; seeds usually with appendages   **Euphorbiaceae**

b. Fruit a capsule splitting through the cells; sap not milky; evergreen shrubs; styles undivided; seeds black and shiny, without appendages   **Buxaceae**

7a. Resinous trees or shrubs; leaves simple or pinnate; flowers with a nectar-secreting disc; stamens 3–10; fruit 1-seeded, drupe-like   **Anacardiaceae**

b. Combination of characters not as above   8

8a. Stamens 2   **Oleaceae**

b. Stamens more than 2   9

9a. Leaves forming insectivorous pitchers   **Nepenthaceae**

b. Leaves not forming insectivorous pitchers   10

10a. Plants aromatic, dioecious; stamens 3–18, filaments united; ovary of 1 carpel containing a single, basal ovule
          LXII. Myristicaceae (p. 317)

  b. Combination of characters not as above   11

11a. Placentation parietal; stamens numerous; fruit a berry or capsule   **Flacourtiaceae**

  b. Combination of characters not as above   12

12a. Trees, shrubs or climbers, if herbaceous then flowers sunk in a fleshy receptacle; ovules apical   13

  b. Combination of characters not as above   16

13a. Ovules 4, of which only 1 develops; flowers in axillary racemes   **Daphniphyllaceae**

  b. Ovule 1; flowers not in axillary racemes   14

14a. Sap watery; fruit a drupe   **XL. Ulmaceae** (p. 79)

  b. Sap milky; fruit a syncarp or group of samaras   15

15a. Perianth present; fruit frequently a syncarp of drupes or achenes united with the flat to flask-shaped receptacle
          XLII. Moraceae (p. 86)

  b. Perianth absent; fruit a samara
          XLI. Eucommiaceae (p. 85)

16a. Stinging hairs present or plant rough to the touch; stamens touch-sensitive, inflexed in bud; leaves often with cystoliths; seed with a straight embryo   **XLIII. Urticaceae** (p. 102)

  b. Stinging hairs absent; stamens neither touch-sensitive nor inflexed in bud; cystoliths absent; seed often with a curved embryo   17

17a. Perianth scarious; stamens usually with the filaments united below   **LVI. Amaranthaceae** (p. 199)

  b. Perianth greenish or absent; stamens with free filaments 18

18a. Leaves all opposite; fruit splitting into 2 mericarps
          Euphorbiaceae

  b. Leaves alternate, at least above; fruit not as above 19

19a. Ovary with cross-walls, containing 4 ovules; leaves leathery
          Buxaceae

  b. Ovary 1-celled, without cross-walls, containing 1 ovule; leaves not leathery   20

20a. Leaves with stipules; ovule apical   **XLII. Moraceae** (p. 86)

  b. Leaves without stipules (sometimes stem succulent and continuous with the leaves); ovule basal
          LV. Chenopodiaceae (p. 195)

Group I

1a. Flowers in racemes or spikes; fruit a berry or drupe-like; leaves entire, alternate, without stipules
          XLIX. Phytolaccaceae (p. 129)

  b. Combination of characters not as above   2

2a. Trees or trailing, heather-like shrublets, rarely aromatic shrubs   3

  b. Herbs, climbers or non-aromatic shrubs   9

3a. Trailing, heather-like shrublet; fruit a drupe   **Empetraceae**

  b. Trees or aromatic shrubs; fruit a drupe, samara, nut or capsule   4

4a. Stamens numerous; ovary with 5 or more cells   5

  b. Stamens 12 or fewer; ovary with up to 4 cells   7

5a. Leaves in whorls   **LXX. Trochodendraceae** (p. 324)

  b. Leaves not in whorls   6

6a. Sepals 4; flowers in hanging spikes
          LXIX. Tetracentraceae (p. 324)

  b. Sepals not 4; flowers in cymes   **Tiliaceae**

7a. Leaves evergreen with translucent, aromatic glands; anthers opening by flaps   **LXVIII. Lauraceae** (p. 321)

  b. Leaves usually deciduous, without translucent, aromatic glands; anthers opening by slits   8

8a. Stamens 2; leaf-base not oblique   **Oleaceae**

  b. Stamens 4–8; leaf-base oblique   **XL. Ulmaceae** (p. 79)

9a. Perianth absent; flowers borne (and often sunk) in a fleshy spike; leaves well developed, often fleshy
          LXXX. Piperaceae (p. 407)

  b. Combination of characters not as above   10

10a. Leaves with stipules which are usually united into a sheath; fruit usually a 3-sided nut   **XLVIII. Polygonaceae** (p. 121)

  b. Leaves without stipules; fruit not a 3-sided nut   11

11a. Sepals falling as the flower opens; herbs with palmately-lobed leaves and orange sap   **Papaveraceae**

  b. Combination of characters not as above   12

12a. Ovary of 1 carpel; fruit 1-seeded; perianth usually petal-like, bracts sometimes calyx-like   **L. Nyctaginaceae** (p. 131)

  b. Ovary of 2 or more carpels, fruit 1–many-seeded; perianth not petal-like   13

13a. Ovary open at the apex; placentation parietal   **Resedaceae**

  b. Ovary closed at the apex; placentation basal, free-central or axile   14

14a. Ovule solitary, basal   15

  b. Ovules numerous, axile or free-central   16

15a. Perianth green, membranous or absent; stamens free
          LV. Chenopodiaceae (p. 195)

  b. Perianth scarious; stamens often united below
          LVI. Amaranthaceae (p. 199)

16a. Placentation axile; leaves alternate   **Saxifragaceae**

  b. Placentation basal or free-central; leaves usually opposite
          17

17a. Sepals free; stamens on the same radii as or more numerous than the sepals   **LIV. Caryophyllaceae** (p. 177)

  b. Sepals united; stamens as many as and alternating with the sepals   **Primulaceae**

Group J

1a. Leaves needle-like or scale-like; plants small, heather-like shrublets   **Bruniaceae**

  b. Combination of characters not as above   2

2a. Ovary divided into cells but placentation parietal; leaves very succulent   **LI. Aizoaceae** (p. 133)

  b. Combination of characters not as above   3

3a. Inflorescence a head subtended by an involucre of bracts; ovule always solitary   4

  b. Inflorescence and ovule not as above   5

4a. Each flower surrounded by a cup-like involucel; stamens 4, free; ovule apical   **Dipsacaceae**

  b. Involucels absent; stamens 5, their anthers united into a tube; ovule basal   **Compositae**

5a. Stamens 2; stamens and style united into a touch-sensitive column; leaves linear **Stylidiaceae**

b. Combination of characters not as above 6

6a. Leaves alternate or all basal 7

b. Leaves opposite or appearing whorled 15

7a. Anthers opening by pores; fruit a berry or drupe **Ericaceae**

b. Anthers opening by slits; fruit various 8

8a. Climbers with tendrils and unisexual flowers; stamens 1–5; placentation parietal; fruit berry-like **Cucurbitaceae**

b. Combination of characters not as above 9

9a. Stamens 10–many; plants woody 10

b. Stamens 4–5; plants mainly herbaceous 12

10a. Leaves gland-dotted, smelling of eucalyptus; corolla completely united, unlobed, falling as a whole **Myrtaceae**

b. Combination of characters not as above 11

11a. Hairs stellate or scale-like; stamens in 1 series; anthers linear **Styracaceae**

b. Hairs absent or not as above; stamens in several series; anthers broad **Symplocaceae**

12a. Stigma surrounded by a sheath **Goodeniaceae**

b. Stigma not surrounded by a sheath 13

13a. Stamens as many as, and on the same radii as the petals **Primulaceae**

b. Stamens not as above 14

14a. Stamens 2 or 4, borne on the corolla; sap not milky **Gesneriaceae**

b. Stamens 5 or more, free from the corolla; sap usually milky **Campanulaceae**

15a. Placentation parietal; stamens 2 or 4 and paired **Gesneriaceae**

b. Placentation axile or apical; stamens 1 or more, if 4 then not paired 16

16a. Stamens 1–3; ovary with 1 ovule **Valerianaceae**

b. Stamens 4 or more; ovary usually with 2 or more ovules 17

17a. Leaves divided into 3 leaflets; flowers in a few-flowered head **Adoxaceae**

b. Leaves simple or rarely pinnate; inflorescence various, usually not as above 18

18a. Stipules usually borne between the bases of the leaf-stalks (sometimes looking like the leaves); ovary usually 2-celled; flowers usually radially symmetric; fruit capsular, fleshy or schizocarpic **Rubiaceae**

b. Stipules usually absent, when present not as above; ovary usually with 3 cells (occasionally with 2–5 cells), sometimes only 1 cell fertile; flowers often bilaterally symmetric; fruit a berry or drupe **Caprifoliaceae**

Group K

1a. Stamens 2 **Oleaceae**

b. Stamens more than 2 2

2a. Carpels several, free; plants succulent **Crassulaceae**

b. Carpels united, or, if the bodies of the carpels are more or less free, the styles united, rarely the ovary of 1 carpel; plants usually not succulent 3

3a. Corolla scarious, 4-lobed; stamens 4, projecting from the corolla; leaves often all basal and with parallel veins **Plantaginaceae**

b. Combination of characters not as above 4

4a. Central flowers of the inflorescence abortive, their bracts forming nectar-secreting pitchers; petals completely united, falling as the flower opens **Marcgraviaceae**

b. Combination of characters not as above 5

5a. Stamens more than twice as many as the petals 6

b. Stamens up to twice as many as the petals 13

6a. Leaves evergreen, divided into 3 leaflets; filaments brightly coloured, at least twice as long as the petals **Caryocaraceae**

b. Leaves simple, entire or lobed, deciduous or evergreen; filaments not as above 7

7a. Leaves with stipules; filaments of stamens united into a tube around the ovary and style **Malvaceae**

b. Leaves without stipules; filaments free 8

8a. Anthers opening by pores **Actinidiaceae**

b. Anthers opening by longitudinal slits 9

9a. Leaves with translucent, aromatic glands; calyx cup-like, unlobed **Rutaceae**

b. Leaves without translucent, aromatic glands; calyx not as above 10

10a. Placentation parietal; leaves fleshy **Fouquieriaceae**

b. Placentation axile; leaves not fleshy 11

11a. Sap milky; ovules 1 per cell **Sapotaceae**

b. Sap not milky; ovules 2 or more per cell 12

12a. Ovules 2 per cell; flowers usually unisexual **Ebenaceae**

b. Ovules many per cell; flowers bisexual **Theaceae**

13a. Stamens as many as the petals and on the same radii as them 14

b. Stamens more or fewer than the petals, if of the same number then not on the same radii as them 20

14a. Tropical trees with milky sap and evergreen leaves **Sapotaceae**

b. Tropical or temperate trees, shrubs, herbs or climbers with watery sap and deciduous leaves 15

15a. Placentation axile 16

b. Placentation basal or free-central 17

16a. Climbers with tendrils; stamens free **Vitaceae**

b. Erect shrubs without tendrils; stamens with their filaments united below **Leeaceae**

17a. Trees or shrubs; fruit a berry or drupe 18

b. Herbs (occasionally woody at the extreme base); fruit a capsule 19

18a. Leaves with translucent glands; anthers opening towards the centre of the flower; staminodes absent **Myrsinaceae**

b. Leaves without translucent glands; anthers opening towards the outside of the flower; staminodes 5 **Theophrastaceae**

19a. Sepals 2, free **LII. Portulacaceae** (p. 171)

b. Sepals more than 4, united **Primulaceae**

20a. Flower compressed, with 2 planes of symmetry; stamens in 2 bundles **Papaveraceae**

b. Combination of characters not as above 21

21a. Leaves bipinnate or replaced by phyllodes; carpel 1, fruit a legume   **Leguminosae**

   b. Combination of characters not as above   22

22a. Anthers opening by pores   23

   b. Anthers opening by longitudinal slits, or pollen in masses (pollinia)   24

23a. Stamens free from the corolla-tube, often twice as many as petals   **Ericaceae**

   b. Stamens attached to the corolla-tube, as many as petals   **Solanaceae**

24a. Leaves alternate or all basal; carpels never 2 and almost free with a single, terminal style   25

   b. Leaves opposite or whorled, alternate only when carpels 2 and almost free with a single, terminal style   41

25a. Plant woody, leaves usually evergreen; stigma not stalked, borne directly on top of the ovary   **Aquifoliaceae**

   b. Combination of characters not as above   26

26a. Prostrate herbs with milky sap and stamens free from the corolla-tube   **Campanulaceae**

   b. Combination of characters not as above   27

27a. Ovary 5-celled   28

   b. Ovary 2-, 3- or 4-celled   30

28a. Placentation parietal; soft-wooded tree   **Caricaceae**

   b. Placentation axile; herbs   29

29a. Leaves fleshy; anthers 2-celled; fruit often deeply lobed, schizocarpic   **Nolanaceae**

   b. Leaves leathery; anthers 1-celled; fruit a capsule or berry   **Epacridaceae**

30a. Ovary 3-celled   31

   b. Ovary 2- or rarely 4-celled   32

31a. Dwarf, evergreen shrublets; 5 staminodes usually present; some petals overlapping on both sides in bud   **Diapensiaceae**

   b. Herbs or climbers with tendrils; staminodes absent; petals each overlapping 1 other and overlapped by 1 other in bud   **Polemoniaceae**

32a. Stamens with the filaments united into a tube; flowers in heads; stigma surrounded by a sheath   **Brunoniaceae**

   b. Combination of characters not as above   33

33a. Flowers in spirally coiled cymes or the calyx with appendages between the lobes; style terminal or arising from between the lobes of the ovary   34

   b. Flowers not in spirally coiled cymes, calyx without appendages; style terminal   35

34a. Style terminal; fruit a capsule, usually many-seeded   **Hydrophyllaceae**

   b. Style arising from the depression between the 4 lobes of the ovary; fruit of up to 4 nutlets, more rarely fruit a 1–4-seeded drupe   **Boraginaceae**

35a. Placentation parietal   36

   b. Placentation axile   37

36a. Corolla-lobes edge-to-edge in bud; leaves either of 3 leaflets or simple, cordate or peltate, hairless; aquatic or marsh plants   **Menyanthaceae**

   b. Corolla-lobes overlapping in bud; leaves never as above; plants not aquatic, not occurring in marshes   **Gesneriaceae**

37a. Ovules 1–2 in each cell of the ovary   38

   b. Ovules 3–many in each cell of the ovary   40

38a. Arching shrub with small purple flowers in clusters on last year's wood   **Buddleiaceae**

   b. Combination of characters not as above   39

39a. Sepals free; corolla-lobes each overlapping 1 other and overlapped by 1 other, and infolded in bud; twiners, herbs or dwarf shrubs   **Convolvulaceae**

   b. Sepals united; corolla-lobes not as above in bud; trees or shrubs   **Boraginaceae**

40a. Corolla-lobes folded, edge-to-edge or overlapping 1 other and overlapped by 1 other in bud; septum of ovary oblique   **Solanaceae**

   b. Corolla lobes variously overlapping, but not as above in bud; septum of ovary horizontal   **Scrophulariaceae**

41a. Trailing, heather-like shrublet   **Ericaceae**

   b. Plant not as above   42

42a. Milky sap usually present; fruit usually of 2 almost free 'follicles' and seeds with silky appendages   43

   b. Milky sap absent; fruit a capsule or fleshy; seeds without silky appendages   44

43a. Pollen granular; corona absent; corolla-lobes each overlapping 1 other and overlapped by 1 other in bud   **Apocynaceae**

   b. Pollen usually in pollinia; corona usually present; corolla-lobes as above or edge-to-edge in bud   **Asclepiadaceae**

44a. Herbs; flowers in spirally coiled cymes   **Hydrophyllaceae**

   b. Herbs or shrubs; flowers not in spirally coiled cymes   45

45a. Placentation parietal; carpels 2   46

   b. Placentation axile; carpels 2,3 or 5   47

46a. Leaves compound; epicalyx present   **Hydrophyllaceae**

   b. Leaves simple; epicalyx absent   **Gentianaceae**

47a. Stamens fewer than corolla-lobes   **Verbenaceae**

   b. Stamens as many as corolla-lobes   48

48a. Carpels 5; shrubs with leaves with spiny margins   **Desfontainiaceae**

   b. Carpels 2 or 3; herbs or shrubs, leaves without spiny margins   49

49a. Leaves without stipules; carpels 3; corolla-lobes each overlapping 1 other and overlapped by 1 other in bud; plants herbaceous   **Polemoniaceae**

   b. Leaves with stipules (often reduced to a ridge between the leaf-bases); carpels usually 2; corolla lobes variously overlapping or edge-to-edge in bud; plants usually woody   50

50a. Corolla usually 5-lobed; stellate and/or glandular hairs absent   **Loganiaceae**

   b. Corolla 4-lobed; stellate and/or glandular hairs present   **Buddleiaceae**

Group L

1a. Stamens more numerous than the corolla-lobes, or anthers opening by pores   2

   b. Stamens as many as corolla-lobes or fewer, anthers not opening by pores   6

2a. Anthers opening by pores; leaves undivided; ovary of 2 or more united carpels   3

   b. Anthers opening by slits; leaves dissected or compound; ovary of 1 carpel   5

3a. The 2 lateral sepals petal-like; filaments united   **Polygalaceae**

   b. No sepals petal-like; filaments free   4

4a. Shrubs with alternate or apparently whorled leaves; stamens 5–25   **Ericaceae**

   b. Herbs with opposite leaves; stamens 5   **Gentianaceae**

5a. Leaves pinnate or of 3 leaflets; perianth not spurred   **Leguminosae**

   b. Leaves laciniate; perianth spurred   **LXXIII. Ranunculaceae** (p. 325)

6a. Stamens as many as corolla-lobes; bilateral symmetry weak   7

   b. Stamens fewer than corolla-lobes; bilateral symmetry pronounced   12

7a. Stamens on the same radii as the petals; placentation free-central   **Primulaceae**

   b. Stamens on different radii from the petals; placentation axile   8

8a. Leaves of 3 leaflets, with translucent, aromatic glands; stamens 5, the 2 upper fertile, the 3 lower sterile   **Rutaceae**

   b. Combination of characters not as above   9

9a. Ovary of 3 carpels; ovules many   **Polemoniaceae**

   b. Ovary of 2 carpels; ovules 4 or many   10

10a. Flowers in spirally coiled cymes; fruit of up to 4 one-seeded nutlets   **Boraginaceae**

   b. Flowers not in spirally coiled cymes; fruit a many-seeded capsule   11

11a. Corolla-lobes each overlapping 1 other and overlapped by 1 other in bud; stamens 5, equal; leaves opposite; climber   **Loganiaceae**

   b. Corolla lobes overlapping in bud, but not as above; stamens 4, or 5 and unequal; leaves usually alternate   **Scrophulariaceae**

12a. Placentation axile; ovules 4 or many   13

   b. Placentation parietal, free-central, basal or apical; ovules many or 1–2   20

13a. Ovules numerous but not in vertical rows in each cell   14

   b. Ovules 4 or more numerous but then in vertical rows in each cell   16

14a. Seeds winged; mainly trees, shrubs or climbers with opposite, pinnate, palmate or rarely simple leaves   **Bignoniaceae**

   b. Seeds usually wingless; mainly herbs or shrubs with simple leaves   15

15a. Corolla-lobes variously overlapping in bud; septum of ovary horizontal; leaves opposite or alternate   **Scrophulariaceae**

   b. Corolla-lobes usually folded, edge-to-edge or overlapping 1 other and overlapped by 1 other in bud; septum oblique; leaves alternate   **Solanaceae**

16a. Leaves all alternate, usually with blackish, resinous glands; plants woody   **Myoporaceae**

   b. At least the lower leaves opposite or whorled, not glandular; plant herbaceous or woody   17

17a. Fruit a capsule; ovules 4–many, usually in vertical rows in each cell   18

   b. Fruit not a capsule; ovules 4, side by side   19

18a. Leaves all opposite, often prominently marked with cysto-liths; flower-stalks without swollen glands at the base; capsule opening elastically, seeds usually on hooked stalks   **Acanthaceae**

   b. Upper leaves alternate, cystoliths absent; flower-stalks with swollen glands at the base; capsule not elastic, seeds not on hooked stalks   **Pedaliaceae**

19a. Style arising from the depression between the 4 lobes of the ovary, or, if terminal then corolla with a reduced upper lip; fruit usually of 4 one-seeded nutlets; calyx and corolla often 2-lipped   **Labiatae**

   b. Style terminal; corolla with well-developed upper lip; fruit usually a berry or drupe; calyx often more or less radially symmetric   **Verbenaceae**

20a. Ovules 4–many; fruit a capsule, rarely a berry or drupe   21

   b. Ovules 1–2; fruit indehiscent, often dispersed in the persistent calyx   27

21a. Ovules 4, side by side   **Verbenaceae**

   b. Ovules many   22

22a. Placentation free-central; corolla spurred   **Lentibulariaceae**

   b. Placentation parietal or apical; corolla not spurred, rarely swollen at the base   23

23a. Leaves scale-like, never green; root parasites   24

   b. Leaves green, expanded; free-living plants   25

24a. Placentas 4; calyx laterally 2-lipped   **Orobanchaceae**

   b. Placentas 2; calyx 4-lobed   **Scrophulariaceae**

25a. Seeds winged; mainly climbers with opposite, pinnately divided leaves   **Bignoniaceae**

   b. Combination of characters not as above   26

26a. Capsule with a long beak separating into 2 curved horns; plant sticky-velvety   **Martyniaceae**

   b. Capsule without beak or horns; plants velvety or variously hairy or hairless   **Gesneriaceae**

27a. Flowers in heads surrounded by an involucre of bracts; ovule 1   **Globulariaceae**

   b. Flowers not in heads as above; ovary 2-celled, ovules 1 in each cell, often only 1 maturing   **Scrophulariaceae**

XXXIV. CASUARINACEAE

Evergreen trees or shrubs, dioecious or monoecious, branchlets jointed, deciduous. Leaves scale-like, whorled at the nodes, bases joined to form a short sheath. Inflorescence catkin-like, of alternating whorls of scale-like bracts each with 2 lateral bracteoles and a single unisexual flower; flowers wind-pollinated. Fruit a 'cone', with pairs of woody bracteoles which open to release a 1-seeded samara.

A tropical and subtropical family mainly from Australia but also SE Asia and the SW Pacific. The roots bear nodules containing nitrogen-fixing bacteria-like organisms. Dr L.A.S. Johnson (Royal Botanic Gardens, Sydney, Australia) is revising the family and is proposing 3 new genera separated from *Casuarina*: *Allocasuarina*, *Gymnostoma* and one other; but his complete revision has not yet been published, and *Casuarina* in the broad sense is treated here.

1. CASUARINA Adanson

P.S. Green

Evergreen trees or shrubs with slender, jointed, ridged branchlets, mostly shed after 1 or 2 years. Leaves scale-like in whorls of 4–*c.* 20. Male flowers with 1 or 2 scale-like perianth-segments and a single stamen. Female flowers without perianth but with a 2-celled ovary, of which only 1 cell is fertile, with 2 ovules; style with 2 long stigmatic lobes. Fruit with pairs of enlarged and woody bracteoles, each opening to release 1 terminally winged samara.

In the broad sense, a genus of about 90 species, many of them adapted to dryish or salty habitats. *C. equisetifolia* has been planted widely for stabilising sand dunes. Usually grown from seed, they can also be propagated under glass from semi-ripe cuttings. Being wind-pollinated, where closely related species have been grown near one another, hybrids have often appeared.

Literature: Franco, J.M.A.P. do A., Contribucão para o estudio sistematico das Casuarinas cultivadas en Portugal, *Anais do Instituto Superior de Agronomia* 14: 151–158 (1943); Johnson, L.A.S., Notes on Casuarinaceae II, *Journal of the Adelaide Botanic Garden* 6: 73–87 (1982).

Branchlets. More or less erect: **1,3,4**; more or less drooping: **2,5,6**.
Deciduous branchlets, length. 5–15 cm: **6**; 7–25 cm: **1,2,4**; 10–40 cm: **3,5**.
Branchlets, median node length. 3–10 mm: **1,2,4,6**; 8–20 mm: **3**; 1–4 cm: **5**.
Branchlet diameter. 0.4–0.8 mm: **1,4,6**; 0.5–1 mm: **2**; 0.7–1.5 mm: **3,5**.
Ridges and scale-like leaves. 4, rarely 5 per node: **6**; 6–8, rarely 9 per node: **1,2,4**; 9–13 per node: **5**; 12–17 per node: **3**.
Grooves between branchlet ridges. Hairy, sometimes minutely so: **2,4,5**; apparently hairless: **1,3,6**.
Male catkins, diameter. 0.7–1 mm: **6**; 1–1.5 mm: **1,4**; 2–4 mm: **2,3,5**.
Fruiting 'cones', length. 6–14 mm: **1**; 9–18 mm: **3**; 1–3 cm: **4**; 1–4 cm: **2**; 1.7–3 cm: **6**; 2–4 cm: **5**.

1a. Ridges and scale-like leaves at nodes of branchlets 6 or more   **2**
  b. Ridges and scale-like leaves at nodes of branchlets 4, rarely 5   **6. torulosa**
2a. Nodes with 6–8, rarely 9, scale-like leaves   **3**
  b. Nodes with 9–16 scale-like leaves   **5**
3a. Branchlets more or less erect, hairless between the ridges or minutely hairy; scale-like leaves shortly acute   **4**
  b. Branchlets more or less pendulous, downy between the ridges at least when young; scale-like leaves long acute   **2. equisetifolia**
4a. Hairless between the ridges of the branchlets; fruiting 'cones' 6–14 × 6–10 mm, cylindric to spherical, on stalks 2–9 mm   **1. cunninghamiana**
  b. Minutely hairy between the ridges of the branchlets; fruiting 'cones' 1–3 × 1–2 cm, cylindric to barrel-shaped, on stalks 5–20 mm   **4. littoralis**
5a. Branchlets erect, spreading or drooping; nodes with 12–17 scale-like leaves, 0.5–0.9 mm long; fruiting 'cones' 9–18 mm long   **3. glauca**
  b. Branchlets more or less drooping, nodes with 9–13 scale-like leaves, 0.7–1.2 mm long; fruiting 'cones' 2–4 cm long   **5. verticillata**

1. C. cunninghamiana Miquel (*C. equisetifolia* misapplied). Figure 1(1,7), p. 15. Illustration: Maiden, Forest flora of New South Wales 2: pl. 59 (1905); Costermans, Native trees and shrubs of south-eastern Australia, 148 (1981); Boland et al., Forest trees of Australia, edn 4, 97 (1984).

Tree to 35 m, dioecious. Deciduous branchlets 7–25 cm long, 0.4–0.7 mm in diameter, erect or drooping in vigorous specimens, hairless, with 7–8, rarely 6 or 9 ridges and scale-like leaves at the nodes; leaves 0.2–0.5 mm long; median nodes of branchlets 5–9 mm apart. Male catkins 1–4 cm × 1–1.5 mm. Fruiting 'cones' cylindric to almost spherical, 6–14 × 6–10 mm, stalks 2–9 mm long. *E Australia (Queensland, New South Wales)*. H5–G1. Autumn.

2. C. equisetifolia J.R. & G. Forster. Figure 1(2,8), p. 15. Illustration: Bailey, Standard cyclopedia of horticulture, 683 (1939); Whistler, Coastal flowers of the tropical Pacific, 14 (1980) – as C. litorea; Graf, Tropica, edn 3, 300 (1986).

Tree to 35 m, monoecious. Deciduous branchlets 7–20 cm long, 0.5–1 mm in diameter, pendulous, hairy between the ridges, with 6–8 ridges and scale-like leaves at nodes; leaves 0.3–1 mm long; median nodes of branches 5–10 mm apart. Male catkins 1–4 cm × 2–4 mm. Fruiting 'cones' cylindric to more or less spherical, 1–2.4 × 1–1.8 cm, stalks 3–12 mm long. *SE Asia to N & E Australia to Polynesia*. H5–G1. Autumn.

3. C. glauca Sprengel. Figure 1(3,9), p. 15. Illustration: Maiden, Forest flora of New South Wales 2: pl. 55 (1905); Costermans, Native trees and shrubs of south-eastern

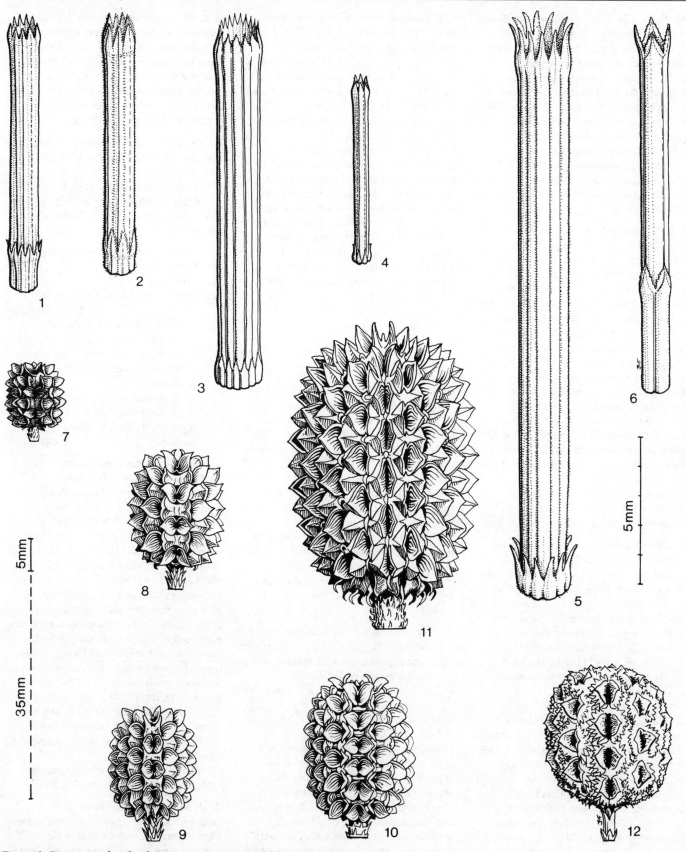

Figure 1. Diagnostic details of *Casuarina* species. Branchlet, node and internode: 1, *C. cunninghamiana*. 2, *C. equisetifolia*. 3, *C. glauca*. 4, *C. littoralis*. 5, *C. verticillata*. 6, *C. torulosa*. Fruiting 'cones': 7, *C. cunninghamiana*. 8, *C. equisetifolia*. 9, *C. glauca*. 10, *C. littoralis*. 11, *C. verticillata*. 12, *C. torulosa*.

Australia, 148 (1981); Boland et al., Forest trees of Australia, edn 4, 103 (1984). Tree 5–20 m, dioecious, frequently with root-suckers. Deciduous branchlets 1–3.5 cm long, 0.9–1.5 mm in diameter, hairless with 12–17 ridges and scale-like leaves at nodes; leaves 0.5–0.9 mm long; median nodes of branchlets 8–20 mm apart. Male catkins 1–4 cm long, about 4 mm in diameter. Fruiting 'cones' cylindric to almost spherical, 9–18 × 8–13 mm, stalks 3–10 mm long. *E Australia (Queensland, New South Wales).* H5–G1. Spring.

4. C. littoralis Salisbury (*Allocasuarina littoralis* (Salisbury) Johnson; *C. suberosa* Otto & Dietrich). Figure 1(4,10), p. 15. Illustration: Maiden, Forest flora of New South Wales 2: pl. 72 (1905); Heywood, Flowering plants of the world, 62 (1978); Costermans, Native trees and shrubs of south-eastern Australia, 146, 147 (1981); Stanley & Ross, Flora of south-eastern Queensland 1: 52, f. 5B (1983). Tree to 10 m, dioecious, less often monoecious, mature bark furrowed. Deciduous branchlets 7–20 cm long, 0.3–0.8 mm in diameter, mostly erect, with 6–8, rarely 5 or 9, ridges and scale-like leaves at nodes; leaves 0.4–0.9 mm long; median nodes of branchlets 4–10 mm apart. Male catkins 1–5 × *c.* 1 mm. Fruiting 'cones' more or less cylindric, 1–3 × 1–2 cm, stalks 5–20 mm long. *E Australia.* H5–G1. Late summer–autumn.

5. C. verticillata Lamarck (*Allocasuarina verticillata* (Lamarck) Johnson; *C. stricta* Aiton). Figure 1(5,11), p. 15. Illustration: Maiden, Forest flora of New South Wales 2: pl. 65 (1905); Costermans, Native trees and shrubs of south-eastern Australia, 146 (1981); Everett, New York Botanical Gardens illustrated encyclopedia of horticulture 2: 642 (1981); Morley & Toelken, Flowering plants in Australia, f. 33 (1983). Small tree, usually bushy, 4–10 m, dioecious, bark furrowed. Deciduous branchlets 10–40 cm long, 0.7–1.5 mm in diameter, usually drooping, minutely hairy between ridges, with 9–13 ridges and scale-like leaves at nodes; leaves 0.7–1.2 mm long; median nodes of branchlets 1–4 cm apart. Male catkins 3–12 cm × 2–2.5 mm. Fruiting 'cones' ovoid or ovoid-spherical, 2–4 × 2–3 cm, stalks 2–10 mm long. *SE Australia.* H5–G1. Winter.

6. C. torulosa Aiton (*Allocasuarina torulosa* (Aiton) Johnson). Figure 1(6,12), p. 15. Illustration: Maiden, Forest flora of New South Wales 2: pl. 63 (1905); Stanley & Ross, Flora of south-eastern Queensland 1: 52, f. 5A (1983); Boland et al., Forest trees of Australia edn 4, 107 (1984). Tree 5–20 m, dioecious, mature bark corky. Deciduous branchlets 5–15 cm long, 0.4–0.5 mm in diameter, more or less drooping, hairless, quadrangular with 4, rarely 5, ridges and scale-like leaves at nodes; leaves 0.3–0.8 mm long; median nodes of branchlets 3–6 mm apart. Male catkins 5–30 × 7–10 mm. Fruiting 'cones' more or less cylindric to almost spherical, pendent or spreading, warty, 1.7–3 × 1.7–2.5 cm, stalks 1–3.5 cm long. *E Australia (Queensland, New South Wales).* H5–G1. Autumn.

XXXV. MYRICACEAE

Dioecious or monoecious shrubs or trees. Leaves alternate, resinous gland-dotted, usually fragrant when crushed. Flowers wind-pollinated, usually in short catkins, usually unisexual, without sepals or petals. Male flowers with 2–16 stamens, though usually 4. Female flowers consisting of a 1-celled ovary containing 1 ovule; style short with 2 stigmatic branches. Fruit a small drupe, sometimes nut-like.

A small family of 3 genera mostly from temperate or subtropical regions of the Old and New Worlds. Their roots contain nitrogen-fixing bacteria in nodules. They are mainly cultivated for their fragrant foliage.

1a. Leaves entire or toothed, not incised, without stipules   **1. Myrica**
 b. Leaves pinnatifid, with pointed, more or less heart-shaped stipules
                                    2. Comptonia

1. MYRICA Linnaeus
P.S. Green

Deciduous or evergreen shrubs or small trees. Leaves entire or toothed, without stipules. Flowers unisexual. Male catkins ellipsoid-cylindric, rarely branched; filaments free or somewhat united at base. Female catkins usually ovoid, usually stalkless, ovary stalkless, with 2 or more rounded bractlets. Fruit spherical or ovoid, covered in wax in some species.

A genus of about 35 species, widespread throughout the world, except Australasia,

often on mountains in tropical latitudes. Although, in the wild, many species are characteristic of acid soils, nonetheless those in cultivation can usually thrive in ordinary neutral soil. Although Nos. 1 and 3 prefer damp soil, Nos. 2 and 4 will grow well in dry, acid soils. They may be propagated by seed or layering.

Leaves. Deciduous: **1,2**; evergreen: **3,4**.
Leaf-apex. Obtuse or broadly acute: **1,2**; acute: **3,4**.
Ripe fruit. Beaked, resin-dotted: **1**; spherical, strongly covered with whitish wax: **2,3**; spherical, dark purple, with thin wax covering: **4**.

1a. Deciduous or only semi-evergreen   2
 b. Evergreen   3
2a. Fruit resin-dotted   **1. gale**
 b. Fruit covered with whitish wax
                            2. pensylvanica
3a. Leaf oblanceolate, 3–9 cm, with scattered orange glandular dots beneath, entire or with a few acute teeth; plant dioecious; fruit 2–3 mm in diameter   **3. cerifera**
 b. Leaf narrowly elliptic to oblanceolate, 5–10 cm, black-dotted beneath; plant monoecious; fruit 4–6 mm in diameter   **4. californica**

1. M. gale Linnaeus (*Gale palustris* (Lamarck) Chevalier). Illustration: Schneider, Illustriertes Handbuch der Laubholzkunde 1: 73 (1904); Hegi, Illustrierte Flora von Mitteleuropa, edn 2, 3: 18 & 21, t. 77, 78 (1957); Ross-Craig, Drawings of British plants 27: t. 11 (1970); Krüssmann, Handbuch der Laubgehölze, edn 2, 2: t. 134 (1977). Deciduous shrub to 2 m, usually less, dioecious, branches ascending. Leaves oblanceolate, 2–6 cm, base tapering, entire, apex obtuse or broadly acute, toothed. Catkins on last year's naked wood; males somewhat crowded, 1–1.5 cm; females 3–5 mm, to 10 mm in fruit. Fruit brown, 3 mm wide, resin-dotted, with 2, thickened, fused scales. *Maritime Europe, N America (from Labrador west to Alaska and Oregon, south to Tennessee), Japan, Korea and eastern Siberia.* H1. Spring.

Material from the far east is sometimes recognised as var. *tomentosa* de Candolle.

2. M. pensylvanica Loiseleur-Deslongchamps (*M. carolinensis* misapplied). Illustration: Gleason, Illustrated Flora of the north-eastern United States and adjacent Canada 2: 25 (1952); Parey's Blumengärtnerei, edn 2, 2: t. 134 (1977); Everett, New York Botanical Gardens illustrated encyclopedia of horticulture **7**: 2261 (1981).

Deciduous or semi-evergreen shrub to 2 m, dioecious. Leaves oblanceolate-elliptic to narrowly obovate, 2–8 cm, base tapering, apex obtuse, minutely mucronate, entire or upper part with a few teeth. Catkins borne below the leaves, males 6–15 mm, females 5–10 mm. Fruit spherical, 3–4 mm in diameter, thickly covered with finely warty, whitish wax. *Eastern N America, coastal plain (S Newfoundland to N Carolina).* H2–3. Spring.

M. heterophylla Rafinesque. Illustration: Godfrey & Wooten, Aquatic and wetland plants of the southeastern United States **2**: 32, f. 8h–q (1981). Somewhat intermediate between **2** and **3** in characters and native distribution: a deciduous shrub or small tree with oblanceolate leaves 5–12 cm; fruit *c.* 3 mm in diameter, white-waxy. *USA (New Jersey to Louisiana).* H3. Spring.

3. M. cerifera Linnaeus (*M. caroliniensis* Miller). Illustration: Schneider, Illustriertes Handbuch der Laubholzkunde **1**: 71 (1904); Gleason, Illustrated Flora of the north-eastern United States and adjacent Canada **2**: 25 (1952); Krüssmann, Handbuch der Laubgehölze edn 2, **2**: t. 134 (1977).

Evergreen shrub or small tree to 12 m, dioecious. Leaves oblanceolate, 3–9 cm, base tapering, apex acute, entire or upper part with a few acute teeth. Catkins usually borne below the leaves, male 5–20 mm, female 5–10 mm. Fruit spherical, 2–3 mm in diameter, covered with whitish wax. *USA (New Jersey south to Florida and Texas).* H5. Spring.

M. rubra Siebold & Zuccarini. Illustration: Krüssmann, Handbuch der Laubgehölze edn 2, **2**: t. 123 (1977); Kitamura & Murata, Coloured illustrations of woody plants of Japan **2**: t. 130 (1977). Small evergreen tree; leaves oblanceolate, 6–12 cm, somewhat obtuse, entire; fruit spherical, 1–1.5 cm in diameter, dark red when ripe, edible. *Japan, Taiwan, E & S China.* G1. Spring.

4. M. californica Chamisso. Illustration: Sargent, Silva of N America **9**: t. 461 (1896); Munz, Californian Flora, 907 (1958); Hitchcock et al., Vascular plants of the Pacific Northwest **2**: 75 (1964); Krüssmann, Handbuch der Laubgehölze, edn 2, **2**: t. 134, 135 (1977).

Evergreen shrub to 4 m or tree to 10 m, monoecious. Leaves narrowly elliptic to oblanceolate, 5–10 cm, base tapering, apex acute, entire, margins mostly with teeth

pointing forwards. Male catkins 1–2 cm in axils of the lower leaves; female catkins 5–10 mm, axillary above the males. Fruit spherical, 4–6 mm in diameter, dark purple, thinly covered with wax. *W coastal USA (California to Washington).* H5. Spring.

A good hedge plant in temperate areas.

M. faya Aiton (illustration: Bramwell & Bramwell, Wild flowers of the Canary Islands, 47, f. 3, 1974; Kunkel, Arboles y arbustos de las Ilas Canarias, 89, 1981) is similar but is an evergreen, dioecious shrub or small tree; leaves oblanceolate, 4–12 cm; male catkins usually branched; fruit *c.* 5 mm in diameter, reddish to black, somewhat fleshy-waxy. *Canary Islands and other Atlantic islands.* G1. Spring.

2. COMPTONIA Aiton
P.S. Green

Deciduous shrubs to 1.5 m, dioecious or monoecious, stems hairy when young. Leaves linear to narrowly lanceolate, 5–10 cm, deeply pinnatifid into broad, obliquely rounded lobes, hairy, stipulate. Flowers unisexual; male catkins flexuous-cylindric, 2–3 cm, with 3–6 stamens per flower; female catkins spherical, ovary with 8 linear bractlets. Fruiting catkins spherical, bur-like, 1–2.5 cm, with long, linear, persistent bractlets. Nutlets ellipsoid, 4–5 mm. A genus of a single species grown for its sweetly fragrant foliage when brushed against or crushed. It grows well in dry, sandy, acid soils and is propagated by seed or by layering.

1. C. peregrina (Linnaeus) Coulter (*Myrica peregrina* (Linnaeus) Kuntze). Illustration: Schneider, Illustriertes Handbuch der Laubholzkunde **1**: 74 (1904); Gleason, Illustrated Flora of the north-eastern United States and adjacent Canada **2**: 25 (1952); Bean, Trees and shrubs hardy in the British Isles edn 8, **1**: 685 (1970); Krüssmann, Handbuch der Laubgehölze edn 2, **2**: t. 131d (1977).
E & SE USA. H2–3. Spring.

Var. **asplenifolia** (Linnaeus) Fernald (*Myrica asplenifolia* Linnaeus; *C. asplenifolia* (Linnaeus) Aiton) is a generally lower-growing shrub with stems and leaves hairless or sparingly hairy and slightly smaller leaves and catkins.

XXXVI. JUGLANDACEAE

Deciduous trees or shrubs with unisexual flowers, often aromatic. Leaves (in our

genera) pinnate with 5–21 opposite leaflets (reduced to 1 in some cultivars). Male flowers with or without an inconspicuous perianth, borne in catkins on the twigs of the previous year; female flowers in spikes, borne on the twigs of the current year; perianth present. Ovary inferior, 1-celled or incompletely 2-or 4-celled; ovule 1, erect from the base. Fruit a nut surrounded by a fleshy cover formed by the perianth and subtending bracts and bracteoles. Seed without endosperm; cotyledons often much contorted.

A family of 7–8 genera mainly from the northern hemisphere. All species included here flower in late spring and their fruits ripen in autumn.

Literature (& illustrations): Kindel, K.H., Nüsse in meiner Hand, *Mitteilungen der Deutsche Dendrologische Gesellschaft* **71**: 145–154 (1979), **75**: 141–158 (1984), **76**: 77–92 (1986); **77**: 111–117 (1987).

1a. Pith continuous; catkins 3 or more together   2
 b. Pith septate; catkins solitary   3
2a. Fruits more than 1 cm, few (1–20) on a pendent fruit-stalk   **2. Carya**
 b. Fruits less than 1 cm, many (*c.* 100) in an upright, cone-like structure bearing many persistent bracts   **1. Platycarya**
3a. Buds stalkless, with scales; fruit a large nut, without a wing   **3. Juglans**
 b. Buds stalked, often naked; fruit a winged nut   **4. Pterocarya**

1. PLATYCARYA Siebold & Zuccarini
D.O. Wijnands

Trees to 15 m in the wild, usually shrubs in cultivation. Pith continuous. Leaflets 7–15, stalkless, lanceolate to narrowly ovate, doubly toothed, 4–10 × 1–3 cm. Male catkins 5–8 cm, erect, shortly stalked, few together below the female catkin but overtopping it; flowers without perianths, stamens 8–10. Female catkins solitary, erect, ovoid-oblong, 3–4 cm, brown, with overlapping, narrowly lanceolate, persistent bracts; flowers without perianths, bracteoles 2, united to the ovary; styles 5. Fruit a winged nut, 5 mm, many together in an upright, cone-like structure.

A genus of a single species.

Literature: Scharrschmidt, H., Zur Verwandschaft von Carya Nutt. und Platycarya Sieb. & Zucc. und zur natürlichen Gliederung der Familie, *Feddes Repertorium* **96**: 345–361 (1985).

1. P. strobilacea Siebold & Zuccarini. Illustration: Krüssmann, Handbuch der

Laubgehölze edn 2, **2**: 439 (1977) & Manual of cultivated broad-leaved trees and shrubs 2: 418 (1986). *Japan, Korea, China, Formosa.* H2.

The leaves colour yellow in autumn. The bark is used for making a black dye.

2. CARYA Nuttall
D.O. Wijnands

Trees with fissured or stripping bark and solid pith. Male flowers in drooping catkins, female flowers in 2–10-flowered terminal spikes. Fruit a large nut surrounded by a thick, green husk that splits to a varying degree into 4 sections, borne 1–20 together on a pendent stalk.

Most of the 17 species are native to N America, a few in Mexico, China and Indo-China; only N American species are in general cultivation. *C. illinoinensis* and *C. ovata* are cultivated on a commercial scale for their edible nuts. All are good timber trees, especially *C. ovata* and *C. tomentosa*. They are rather slow-growing, but useful as park trees. *C. ovata* and *C. tomentosa* have particularly good yellow autumn colour; *C. laciniosa* and *C. ovata* have conspicuous stripping bark. Propagation is by seed; transplanting requires special care. The species hybridise easily, both in nature and in cultivation. The following hybrids are known to be in cultivation, especially the last: **C.** × **lecontei** Little (*C. aquatica* × *illinoinensis*), **C.** × **brownei** Sargent (*C. cordiformis* × *illinoinensis*) and **C.** × **laneyi** Sargent (*C. cordiformis* × *ovata*). It is not always possible to name a single, cultivated specimen in this genus.
Literature (& illustrations): Sargent, C.S., *Manual of the trees of North America* edn 2, 176–200 (1922); Brison, F.R., *Pecan culture* (1974); Krüssmann, *Handbuch der Laubgehölze*, edn 2, 1: 304–308 (1976) & *Manual of cultivated, broad-leaved trees and shrubs* 1: 283–287 (1984).

Bark. Fissured: **1,3,6–8,10**; fissured at first, with age developing small loose plates: **2,4,5,12**; with large more or less loose plates: **9,11**.
Length of terminal overwintering buds. 5–10 mm: **1,2,5,6,12**; 1–1.5 cm: **4,7,8,10,12**; 1.5–2.3 cm: **3,4,9,10,11**.
Overwintering bud-scales. Overlapping: **3,5–12**; not overlapping: **1,4**. Hairless: **7,12**; with scattered, peltate scales: **6,8**; scales soon shed: **2**; densely scaly: **3–5**; hairy: **9,11**; densely hairy: **1,10**.
Number of leaflets. 5–7: **3,5,7,8,11,12**; 5–9: **4,9,10**; 7–9: **6**; 7–11: **2**; 11–17: **1**.
Leaf-veins beneath. Hairless: **7**; hairy: **4,9,10**;

hairy, later hairless: **1,2,6,8,11,12**; with peltate scales: **2,3,5,6,8**.
Leaf-surface beneath. Hairless: **1,2,4,7,11**; hairy: **9,10**; with peltate scales: **3,5,6, 8,10**.
Nuts. Sweet: **1,3,5–12**; bitter: **2,4**; astringent: **7**.

1a. Leaflets asymmetric, somewhat sickle-shaped, 11–17, occasionally fewer   **2**
  b. Leaflets symmetric, rarely slightly asymmetric, usually 5–9, occasionally to 11   **3**
2a. Margin of leaflets distinctly toothed; nut cylindric, sweet   **1. illinoinensis**
  b. Margin of leaflets appearing entire, teeth shallow; nut ovoid or obovoid, bitter   **2. aquatica**
3a. Outer overwintering bud-scales with peltate scales (sometimes with hairs as well)   **4**
  b. Outer overwintering bud-scales without peltate scales, hairless or hairy   **9**
4a. Bud-scales not overlapping, completely covered with bright yellow peltate scales   **5**
  b. Bud-scales overlapping, peltate scales, if yellow, not completely covering the bud-scales   **6**
5a. Leaves with scattered, prominent peltate scales beneath, hairless   **3. cathayensis**
  b. Leaves without peltate scales beneath, hairless or the midrib downy   **4. cordiformis**
6a. Leaflets 5–7   **7**
  b. Leaflets 7–11   **8**
7a. Leaf-stalks and leaves beneath with peltate scales, not hairy; overwintering bud-scales not overlapping, densely brown-scaly   **5. myristiciformis**
  b. Leaf-stalks and at least the veins on the leaves beneath downy; overwintering bud-scales overlapping, silvery-scaly   **6. pallida**
8a. Young leaves, young branchlets and outer overwintering bud-scales more or less hairless; stalk of terminal leaflet 4–10 mm   **7. glabra**
  b. Young leaves, young branchlets and outer overwintering bud-scales rusty-hairy; stalk of terminal leaflet absent or to 5 mm   **8. texana**
9a. Leaves beneath persistently soft-hairy   **10**
  b. Leaves beneath becoming hairless   **11**
10a. Bark shaggy; branchlets and leaf-stalks hairless or lightly hairy   **9. laciniosa**

  b. Bark fissured; branchlets and leaf-stalks with densely felted hairs   **10. tomentosa**
11a. Leaf-margin persistently ciliate; overwintering terminal bud 1.5–3.5 cm; bark shaggy with thickish plates   **11. ovata**
  b. Leaf-margin not ciliate; overwintering terminal bud 5–15 mm; bark more or less fissured with small plates   **12. ovalis**

1. C. illinoinensis (Wangenheim) K. Koch (*C. oliviformis* (Michaux) Nuttall; *C. pecan* (Marshall) Engler & Graebner; *Hicoria pecan* (Marshall) Britton).
Large trees, bark deeply fissured; buds with densely felted, yellow hairs. Leaflets 11–17, oblong to lanceolate, to 17 cm, slightly sickle-shaped. Buds to 8 cm, ovoid to oblong, brown. *USA (N to Iowa), Mexico.* H3.

Grown for its nuts. Many named cultivars are grown, including hybrids with *C. ovata* (**C.** × **laneyi** Sargent).

2. C. aquatica (Michaux filius) Nuttall.
A smaller tree than *C. illinoinensis.* Leaflets 7–13, narrow, slightly sickle-shaped, marginal teeth obscure. Nuts bitter. *SE USA.* H2.
Uncommon in cultivation.

3. C. cathayensis Sargent.
Distinct on account of the dense orange-yellow peltate scales on young twigs and leaves beneath. *E China.* H2?
Rare in cultivation.

4. C. cordiformis (Wangenheim) K. Koch (*C. amara* (Michaux) Nuttall; *C. minima* (Marshall) Britton).
Tree to 30 m, bark fissured into thin scales. Leaflets 5–9, broadly lanceolate, to 15 cm, toothed, acuminate, downy beneath. Fruit approximately spherical, 4-winged above the middle, husk splitting, nut *c.* 2.5 cm, slightly flattened, abruptly contracted into a short point, smooth, grey, kernel bitter. *Eastern N America, from Quebec to Florida.* H1.

5. C. myristiciformis (Michaux filius) Nuttall.
Tree to *c.* 30 m. Young twigs, leaf-stalks and the leaves beneath with dense peltate scales, yellowish on the twigs, silvery on the leaves. *SE USA.* H4.
Uncommon in cultivation.

6. C. pallida (Ashe) Engler & Graebner.
Tree to 35 m. Leaflets silvery beneath with peltate scales, especially when young. *E & S USA.* H2.

7. C. glabra (Miller) Sweet (*C. porcina* (Michaux filius) Nuttall).
Large tree, bark dark-coloured, fissured. Leaflets mostly 5, the lower pair reduced, the others oblong, to 15 cm, acuminate, sharply toothed, hairless. Fruit obovoid, *c.* 2.5 cm, slightly winged at the top, nut obovoid, not angled, astringent, sometimes sweet. *Eastern N America and C Florida.* H1.

8. C. texana Buckley (*C. buckleyi* Durand; *C. arkansana* Sargent; *C. villosa* (Sargent) Schneider).
Small tree to 15 m. Leaves densely rusty-hairy when young. *C USA.* H2.
Uncommon in cultivation.

9. C. laciniosa (Michaux filius) Loudon (*C. sulcata* Nuttall; *C. pubescens* (Willdenow) Sweet).
Large tree, bark shaggy. Leaflets mostly 7, oblong to oblanceolate, to 20 cm, acuminate, toothed, downy beneath. Fruit 4–7 cm, approximately spherical with 4 ridges above, nut flattened, angled, pointed at both ends. *Eastern central N America.* H2.

10. C. tomentosa (Poiret) Nuttall (*C. alba* misapplied).
Tree to 30 m, bark ridged. Leaflets 7–9, oblong, to 18 cm, toothed, acuminate, downy and glandular beneath, fragrant. Fruit approximately spherical to pear-shaped, 3–5 cm, husk thick, nut slightly flattened, angled. *Eastern N America, from Massachusetts to Florida & Texas.* H1.
Grown for its excellent timber.

11. C. ovata (Miller) K. Koch (*C. alba* Nuttall; *Juglans alba* Linnaeus, in part).
Large tree, bark shaggy. Leaflets mostly 5, elliptic, to 15 cm, downy and glandular beneath when young, ciliate. Fruits approximately spherical, 3–7 cm, 4-ribbed at the top, husk thick, splitting to the base, nut ellipsoid, angled, white, kernel sweet. *Eastern & Central N America.* H1.
Grown for timber and, in a number of named cultivars, for nuts.

12. C. ovalis (Wangenheim) Sargent (*C. microcarpa* Nuttall, in part).
A tree much like *C. glabra* (No. 7), but the young shoots and leaves more downy, and bark tending to strip in older trees. The fruits are very variable in shape and the husk may split to the base. The nuts are sweet and edible. *Eastern USA, from New York to Florida.* H1.

3. JUGLANS Linnaeus
D.O. Wijnands
Trees (1 species occasionally shrubby) with furrowed bark and septate pith. Buds with scales. Leaves pinnate (may be reduced to 1 in some cultivars). Male flowers in drooping catkins, female in short, few-flowered spikes. Fruits solitary or in racemes, large nuts in a fleshy, indehiscent or eventually bilobed husk.

Mostly stately trees that allow little or no undergrowth, grown for timber and for their nuts. There are about 20 species in N & S America and from SE Europe to E Asia. They may suffer from leaf-spot disease, caused by *Gnomonia leptostyla* (in its sterile form, *Marssonella juglandis*).
Literature: Sargent, C.S., *Manual of the trees of North America*, edn 2, 169–176 (1922); Westeinde, J.C. van't, Juglans regia, *Dendroflora* 8: 36–41 (1971). Most of the species are illustrated in Krüssmann, G., *Handbuch der Laubgehölze* edn 2, **2**: 196–200 (1977) & *Manual of cultivated broad-leaved trees and shrubs* 2: 190–194 (1986).

Average leaflet number. 7–9: 1; 9–11: 3; 11–13: **2,3,5,8**; 13–15: **2,5–8**; 15–17: **2,4,6–8**; 17–19 or more: **2,4,6**.
Leaflet breadth. To 15 cm: **4**; 1.5–2.5 cm: **2,3**; 2–4 cm: **6**; 3–6 cm: **5,7,8**; 4–10 cm: **1**.
Underside of mature leaf. Without hairs, or hairs very few: **2**; hairs in small tufts in the axils of midrib and veins: **1,4,6**; hairs only on midrib and veins: **6**; hairs on leaf surface: **3,5,7,8**.
Leaflet margin. Entire: **1**; toothed: **2–8**.
Fruits. Solitary or paired: **1–4,6**; clustered: **5,7,8**. 1.5–2 cm across: **4**; 2.5–3.5 cm across: **2,3,7**; 3.5–4.5 cm across: **6,8**; 4–6 cm across: **1,5,6,8**.

1a. Margins of leaflets entire; leaflets up to 9, the terminal largest **1. regia**
b. Margins of leaflets toothed; leaflets usually more than 9, the terminal not the largest **2**
2a. Leaflets narrowly lanceolate, 1–2.5 cm broad; shrubs or small trees to 15 m **3**
b. Leaflets lanceolate or broadly lanceolate-elliptic, 3–6 cm broad; large trees to 50 m **5**
3a. Leaflets 17–23, very long, acute, 8–15 mm broad; fruit 1.5–2 cm across **4. microcarpa**
b. Leaflets 9–19, acute, 1.5–2.5 cm broad; fruit 2.5–3.5 cm across **4**
4a. Underside of leaflets becoming hairless except for small tufts of hairs in the vein-axils; leaflets 11–19, scarcely longitudinally asymmetric **2. hindsii**
b. Underside of leaflets with short fine hairs (soft to the touch), without tufts in the vein-axils; leaflets 9–13 (rarely more), slightly asymmetric to sickle-shaped **3. elaeopyren**
5a. Hairs on the leaf-stalk and underside of leaflets grouped; fruits clustered, 4–6 cm across **5. cinerea**
b. Hairs on the leaf-stalk and underside of leaflets solitary or in pairs; fruits solitary, paired or clustered, 2.5–4.5 cm across **6**
6a. Leaflets broadest about one-third from the base, narrowing from there to an acuminate apex, 2–4 cm broad; fruits 1 or 2 together, nut with numerous irregular ridges **6. nigra**
b. Leaflets broadest about half way up from the base and more or less parallel-sided for most of their length, apex acute or acuminate, 3–8 cm broad; fruits clustered, nut with 6–8 major ridges **7**
7a. Underside of leaflets densely hairy, apex acute or shortly acuminate; fruits clustered on a long stalk **7. ailantifolia**
b. Underside of leaflets becoming hairless, apex acute; fruits clustered on a stout stalk **8. mandshurica**

1. J. regia Linnaeus (*J. sinensis* (de Candolle) Dode; *J. orientis* Dode).
Wide-crowned tree to 30 m, bark silvery grey. Leaflets 1–9 or rarely more, elliptic, to 12 cm, the terminal the largest. Nut ovoid, pointed, wrinkled. *SE Europe, temperate Asia.* H2.
Grown for its nuts and its timber. This species has been so long and widely cultivated that the classification of its variation is immensely complex. Subspecies have been named but their limits are confused by old cultivated races which have re-adapted to the wild. There are many cultivars. Hybrids of *J. regia* which are sometimes seen in collections include those with *J. cinerea* (**J.** × **quadrangulata** (Carrière) Rehder), with *J. ailantifolia* (**J.** × **notha** Rehder) which resembles *J. regia* in its leaves but with fruits more like those of *J. ailantifolia* 'var. *ailantifolia*', and with *J. nigra* (**J.** × **intermedia** Carrière).

2. J. hindsii (Jepson) R.E. Smith.
Erect tree to 15 m, shoots downy. Leaves 25–30 cm, leaflets 13–19, ovate-lanceolate, to 10 cm. Fruits spherical, downy, 2.5–3 cm, nut shallowly grooved. *USA (California).* H4.
Sometimes grown under the name *J. californica*, of which it is sometimes treated as a variety.

3. J. eleaopyren Dode (*J. major* (Torrey)
Heller; *J. microcarpa* misapplied).
Tree to 15 m with rather narrow crown,
young shoots downy. Leaflets 9–13, soon
hairless, oblong-lanceolate, to 10 cm,
generally coarsely toothed, slightly
asymmetric, the lowest pair half as long as
the others. Fruits approximately spherical,
2.5–3.5 cm, rusty brown; nut thick-
shelled, slightly compressed, deeply
grooved. *Western Central N America,
Mexico.* H4.

4. J. microcarpa Berlandier (*J. rupestris*
Torrey; *J. nana* Engelmann).
Small tree or shrub, young shoots with
yellowish down. Leaves 15–30 cm, leaflets
15–23, lanceolate, to 8 cm, thin, finely
toothed, taper-pointed, slightly asymmetric.
Fruit solitary, spherical, 1.5–2 cm, nut
thick-shelled, spherical. *Western Central N
America, Mexico.* H2.

Rarely cultivated; most trees grown
under this name belong to *J. elaeopyren*.

5. J. cinerea Linnaeus.
Wide-crowned tree to 30 m, bark grey,
strongly fissured. Young shoots with rusty
brown felt. Leaves 25–50 cm, leaflets
11–19, oblong-lanceolate, to 12 cm, axis
and underside of leaflets sticky. Male
catkins to 10 cm. Fruits in drooping
clusters of 3–5, rounded at the base and
tapered to the top, sticky, not oblong,
shortly pointed, thick-shelled, with 8 ridges,
wrinkled. *Eastern Central N America.* H1.

This species has been crossed with *J.
ailantifolia* to give the hybrid **J. × bixbyi**
Rehder (*J. sargentii* Sudworth), an
intermediate tending towards *J. cinerea* but
with slightly larger fruits. Also, crossed
with *J. regia*, it has produced the hybrid **J. ×
quadrangulata** (Carrière) Rehder.

6. J. nigra Linnaeus.
Wide-crowned tree to 50 m, stem dark,
strongly fissured. Young shoots downy.
Leaves 30–60 cm, leaflets 13–23, the
terminal leaflet often small or lacking,
ovate-oblong, to 12 cm, irregularly toothed,
somewhat glossy above. Fruits solitary or
paired, approximately spherical and
pointed, 4–5 cm, downy, nut ovoid,
pointed, strongly and irregularly ridged.
Eastern and Central N America. H1.

The hybrid with *J. regia* has been called
J. × intermedia Carrière.

7. J. ailantifolia Carrière (*J. cordiformis*
Maximowicz not Wangenheim; *J.
sieboldiana* Maximowicz not Goeppert).
Wide-crowned tree to 20 m. Leaves to
50 cm or even more, leaflets 13–17,

oblong-elliptic, to 15 cm. Male catkins to
30 cm. Fruits clustered, nearly spherical,
sticky, downy; nut obovate, pointed and
wrinkled or smooth and flattened. *Japan,
USSR (Sakhalin).* H1.

Variants with the nut pointed and
wrinkled have been referred to as 'var.
ailantifolia', those with the nut smooth and
flattened as 'var. cordiformis'. The hybrid
with *J. cinerea* is called **J. × bixbyi** and that
with *J. regia*, **J. × notha** Rehder.

8. J. mandshurica Maximowicz (*J.
cathayensis* Dode; *J. stenocarpa*
Maximowicz).
Wide-crowned tree to 20 m. Young shoots
with brown glandular hairs. Leaves
60–90 cm, leaflets 11–17, oblong-elliptic,
to 18 cm, finely toothed, sticky beneath, as
is the axis. Fruits in a pendulous cluster of
6–10, ovoid, pointed, sticky, to 5 cm. Nut
3–4 cm with 8 strong ridges. *China, far
eastern USSR.* H1.

4. PTEROCARYA Kunth
D.O. Wijnands
Trees with furrowed bark and septate
pith. Leaves large, with 5–27 toothed
leaflets. Male flowers in short catkins,
female flowers in much longer catkins. The
fruits are winged nuts in long, pendent
spikes.

A genus with about 10 species restricted
to W & E Asia. They are moisture-loving,
handsome trees for parks. All may suffer
from late spring frosts. Propagation is
usually by seed.
Literature (& illustrations): Schneider, C.K.,
Illustriertes Handbuch der Laubholzkunde 1:
91–96 (1904), 2: 880–881 (1912);
Krüssmann, G., Handbuch der Laubgehölze
edn 2, 3: 59–62 (1978) & *Manual of
cultivated, broad-leaved trees and shrubs* 3:
61–63 (1986).

1a. Leaf-axis between the leaflets winged,
terminal leaflet usually absent; nuts
with lanceolate wings **2. stenoptera**
 b. Leaf-axis not winged, terminal leaflet
usually present; wings of nut about as
broad as long **2**
2a. Leaves with 11–21 leaflets **3**
 b. Leaves with 5–11 leaflets **4**
3a. Buds without scales; axis of leaf
hairless; leaflets hairless beneath,
except for tufts of hairs in the vein-
axils **1. fraxinifolia**
 b. Buds with 2–3 scales which fall early;
axis of leaf brown-downy, as are the
midrib and veins of the leaflet beneath
 3. rhoifolia
4a. Fruits including the 2 semicircular

wings 2.5–3 cm wide; leaflets in
strictly opposite pairs **4. hupehensis**
 b. Fruits including the single, disc-
shaped wing 3–7 cm wide; leaflets
not strictly opposite one another
 5. paliurus

1. P. fraxinifolia (Lamarck) Spach (*P.
caucasica* Meyer; *P. sorbifolia* Dippel).
Tree to 20 m or more, often many-
stemmed, bark deeply furrowed.
Leaves 20–50 cm, leaflets 9–21, ovate-
oblong to lanceolate, to 12 cm, axis terete
and hairless. Fruits in spikes of 20–45 cm,
each 1.5–2 cm broad and with
semicircular wings. *USSR (Caucasus), N
Iran.* H2.

'Dumosa' is a shrubby variant with small
leaflets.

P. × rehderiana Schneider, a hybrid of *P.
fraxinifolia* and *P. stenoptera*, differs in its
longer wings on the nuts and the (at least
in coppice-shoots), slightly winged leaf-axis.
The hybrid is more robust and suffers less
from late frosts than either parent.

2. P. stenoptera de Candolle (*P. japonica*
Dippel; *P. laevigata* Lavallée; *P. sinensis*
Rehder).
Tree to 25 m with a fissured trunk. Leaves
20–40 cm, leaflets 11–23, the terminal
leaflet often absent, the rest oblong, apex
usually rounded or blunt, to 10 cm, axis
winged. Fruits in racemes of 20–30 cm,
each 1.5–2 cm with oblong-lanceolate
wings. *C, S & W China.* H2.

3. P. rhoifolia Siebold & Zuccarini.
Tree to 30 m. Leaves 20–40 cm, leaflets
11–21, ovate-oblong, to 12 cm, sharply
and finely toothed, brown-downy on the
veins beneath. Fruits in racemes of
20–30 cm, each 2–2.5 cm across, with
broad, diamond-shaped wings. *Japan.* H2.

4. P. hupehensis Skan.
Like *P. rhoifolia* but tree to 20 m, leaflets
5–11. *China (Hubei).* H2.

Less common in cultivation than *P.
rhoifolia*.

5. P. paliurus Batalin (*Cyclocarpus paliurus*
(Batalin) Iljinskaya).
Tree to 20 m, distinct in its disc-like fruits.
C & W China. H2.

Not in general cultivation.

XXXVII. SALICACEAE
Dioecious, deciduous or rarely evergreen
trees and shrubs. Leaves undivided, usually
alternate and with stipules, commonly
developing after the flowers. Flowers in

spike- or raceme-like catkins, each flower subtended by a membranous catkin-scale; perianth reduced to a cup-shaped disk or to 1 or more nectary-scales, occasionally absent. Stamens 2–many; filaments sometimes united. Ovary 1-celled; stigmas 2–4; ovules numerous. Fruit a capsule opening by 2–4 segments. Seeds small, surrounded by a tuft of silky hairs; endosperm absent.

Two genera, mainly temperate, and rather sparsely represented in the southern hemisphere. Most members of the family are quick-growing and are valued by gardeners for providing screens and shelter-belts. The seeds are notably short-lived, and propagation is mainly from cuttings or suckers.

1a. Catkin-scales entire; buds generally with 1 cap-shaped scale, very rarely with 2 scales; catkins erect or spreading, rarely pendent **1. Salix**
  b. Catkin-scales toothed or fringed; buds with several overlapping scales; catkins pendent **2. Populus**

1. SALIX Linnaeus
R.D. Meikle

Trees, shrubs and dwarf shrubs with sympodial branching. Buds generally cap-shaped (calyptrate) with 1 apparent scale, rarely with 2 free scales. Leaves mostly lanceolate, oblong or obovate, entire or shortly toothed; leaf-stalk generally short, not compressed laterally; stipules often conspicuous and persistent. Catkins mostly erect or spreading, rarely pendent, stalked or almost stalkless, with a varying number of more or less leaf-like basal bracts; catkin-scales entire, persistent or sometimes soon falling; flowers primarily insect-pollinated (rarely wind-pollinated) with 1 or more small nectary-scales. Stamens usually 2 (sometimes up to 12), with free or occasionally united filaments. Ovary usually flask-shaped, style single, short or long, sometimes almost absent, stigmas 2, sometimes bifid. Fruit a capsule opening by 2 segments, the segments usually recurving from the apex at maturity.

A large and complex genus, probably including more than 300 species, chiefly in temperate regions of the northern hemisphere, and absent (as natives) from Australasia.

Most of the species are hardy and easy to grow, except in very dry or shady situations. The foliage is commonly disfigured by saw-flies during the summer. Propagation is either from seed (which is

very short-lived) or from cuttings; the latter generally root very readily and some species sometimes regenerate from robust branches used as stakes. Occasionally (as with *S. caprea*) cuttings are difficult to root satisfactorily.

In the following treatment the term 'shoot' is used for the unripe growths of the current season; 'twigs' for the ripened growths of the preceding year.

Many authors refer to what are here called 'catkin-scales' as 'bracts', a usage which tends to be confusing, since the term 'bract' is also often used for the modified leaf-like structures commonly found at the base of the catkin or along the catkin-stalk.

Specimens selected for the purpose of identification should be taken from normal growths, and not from abnormally vigorous coppice shoots, nor from plants growing in shaded situations. Catkins may be required, though the key is constructed to make the greatest possible use of foliage. Catkins should be in full flower, neither over-ripe nor immature.

Division of the genus poses many problems, nor is it felt that listing the very large number of sections within the genus, many based on slender differences, would be of much service here. The traditional subdivision of *Salix* into 3 subgenera: 1. *Salix*, 2. *Chamaetia* Dumortier and 3. *Vetrix* Dumortier is also open to objection; the subgenus *Salix* encompasses the true willows, trees and robust shrubs with catkin scales one-coloured, as in *S. alba* and *S. fragilis*, and includes species 1–14 in the following account; subgenus *Chamaetia* is more heterogeneous, but consists mainly of alpine and arctic dwarf shrubs and subshrubs, also usually with one-coloured catkin-scales (species 15–30 in this account); the remaining species fall into the large and ill-defined subgenus *Vetrix* (the sallows of popular English parlance) and all are shrubs or less often small trees, the catkin-scales usually with a dark brown or blackish tip. But the boundaries of the subgenera, especially those of *Chamaetia* and *Vetrix*, are not well defined.

Literature: There is no recent monograph of the entire genus. Up-to-date treatments on a regional basis are to be found in Fernald, M.L., *Gray's manual of botany*, ed. 8, 487–519 (1950); Gleason, H.A., *Illustrated Flora of the north-eastern United States and adjacent Canada* **2**: 2–23 (1952); Rechinger, K.H. in Tutin, T.G. et al., *Flora Europaea* **1**: 43–54 (1964); Cronquist, A. in Hitchcock, C.L. et al., *Vascular plants of the Pacific*

Northwest, **2**: 32–70 (1964); Skvortsov, A.K., *Ivi SSSR (Willows of the USSR)* (1968); Chmelar, J., Drobne druhy rodu Salix (The low-growing Salix species suitable for planting in rock gardens), *Skalky Skalnicky*, 3–149 (1977); but the species of parts of the Himalayan region and western China are still inadequately known.

1a. Leaves linear, lanceolate or oblanceolate, at least 3 times as long as wide 2
  b. Leaves ovate, oblong, elliptic or almost circular, less than 3 times as long as wide 54
2a. Leaves entire or almost so 3
  b. Leaves distinctly toothed 25
3a. Leaves hairy on 1 or both surfaces 4
  b. Leaves hairless or soon becoming so 17
4a. Small spreading shrubs, usually less than 1.5 m high 5
  b. Robust shrubs or trees more than 1.5 m high 7
5a. Stipules conspicuous, persistent, shortly stalked; leaves often opposite or almost opposite **72. subopposita**
  b. Stipules inconspicuous or absent; leaves neither opposite nor almost opposite 6
6a. Leaves oblanceolate, 1.5–2 cm wide; catkins cylindric **27. glaucosericea**
  b. Leaves narrowly lanceolate, less than 1 cm wide; catkins almost spherical **71. rosmarinifolia**
7a. Branches and twigs very slender, spreading; leaves narrowly linear-lanceolate, usually less than 1 cm wide 8
  b. Branches and twigs robust, erect or ascending; leaves usually more than 1 cm wide 10
8a. Leaves with a white- or greyish-felted undersurface **63. elaeagnos**
  b. Leaves thinly silky, the undersurface not felted 9
9a. Suckering shrubs; stamens with free filaments **17. exigua**
  b. Non-suckering shrubs; stamens with the filaments more or less united **79. wilhelmsiana**
10a. Leaves linear or linear-lanceolate, tapering from near the base to a slender acuminate apex 11
  b. Leaves lanceolate, oblanceolate or narrowly ovate-oblong 12
11a. Undersurface of leaves silvery-silky **57. viminalis**
  b. Undersurface of leaves greenish, thinly hairy **58. × rubra**

12a. Catkins developing with the leaves
4. × mollissima
b. Catkins developing before the leaves
13
13a. Undersurface of leaves white- or grey-felted
48. × seringeana
b. Undersurface of leaves adpressed hairy or silky, not felted
14
14a. Twigs soon hairless and reddish brown; anthers purple **61. rehderiana**
b. Twigs persistently hairy or downy
15
15a. Undersurface of leaves rather densely silky-hairy with prominent veins; wood of peeled twigs smooth
45. × sericans
b. Undersurface of leaves thinly hairy, veins not very prominent; wood of peeled twigs with short longitudinal ridges
16
16a. Catkins long-cylindric, 2–5 cm
46. × dasyclados group
b. Catkins shortly cylindric, usually less than 2 cm **52. × smithiana**
17a. Prostrate dwarf shrubs with rooting branches; leaves usually less than 15 × 5 mm **30. lindleyana**
b. Erect or spreading shrubs; leaves more than 1.5 cm
18
18a. Male catkins pendent; stamens 5
5. arbutifolia
b. Male (and female) catkins erect or spreading; stamens 2 or 3, sometimes with united filaments
19
19a. Catkins developing with the leaves 20
b. Catkins developing before the leaves
21
20a. Twigs and branches slender, spreading; leaves linear-lanceolate, less than 1 cm wide; stipules inconspicuous or absent
79. wilhelmsiana
b. Twigs and branches robust, erect or ascending; leaves more than 1 cm wide; stipules conspicuous, persistent
4. × mollissima
21a. Twigs with a whitish, waxy bloom
73. irrorata
b. Twigs without bloom
22
22a. Leaves bright shining green on the upper surface
23
b. Leaves dull, dark green or glaucous on the upper surface
24
23a. Catkins shortly ovoid-cylindric, *c.* 2.5 cm; anthers purplish or reddish
61. rehderiana
b. Catkins narrowly cylindric, *c.* 3–4 cm; anthers yellowish or at first tinged with red **60. udensis**
24a. Leaves 2–8 cm, often opposite or almost opposite, apex obtuse or

shortly acute; catkins very slender, usually less than 3 cm × 5 mm
75. purpurea
b. Leaves 3–12 cm, alternate, apex acute or shortly acuminate; catkins often more than 3 cm × 5 mm
59. × forbyana
25a. Mature leaves hairy on 1 or both surfaces
26
b. Mature leaves hairless or almost so
31
26a. Undersurface of leaves densely white- or grey-felted
48. × seringeana
b. Undersurface of leaves downy or adpressed-silky but not felted
27
27a. Catkins developing before the leaves
61. rehderiana
b. Catkins developing with the leaves 28
28a. Trees with a well-developed trunk
10. alba
b. Shrubs
29
29a. Stipules well-developed, persistent; branches and twigs robust
4. × mollissima
b. Stipules inconspicuous or absent; branches and twigs very slender 30
30a. Suckering shrubs; stamens 2 with free filaments
17. exigua
b. Non-suckering shrubs; stamens with united filaments **79. wilhelmsiana**
31a. Catkins developing before the leaves; catkin-scales dark-tipped
32
b. Catkins developing with the leaves; catkin-scales not dark-tipped
36
32a. Twigs with whitish, waxy bloom; catkins very large and showy
33
b. Twigs without bloom; catkins not very conspicuous
34
33a. Leaves long-acuminate; twigs slender, often pendent **69. acutifolia**
b. Leaves shortly acuminate or acute; twigs not slender and pendent
68. daphnoides
34a. Leaves narrowly lanceolate or linear-lanceolate; stamens with united filaments
35
b. Leaves more broadly lanceolate; stamens with free filaments
61. rehderiana
35a. Stipules small or absent
77. elbursensis
b. Stipules well developed, persistent
76. miyabeana
36a. Bud-scales free along their inner margins; buds not adpressed to the shoot
1. chilensis
b. Bud-scales united along their margins to form a cap (calyptrate); buds adpressed to the shoot
37
37a. Stipules ear-shaped, blunt or shortly acute, conspicuous, persistent 38

b. Stipules acute or acuminate, often inconspicuous and soon falling 42
38a. Apex of leaf-stalk conspicuously glandular
39
b. Apex of leaf-stalk not glandular or inconspicuously glandular 41
39a. Fully developed leaves with long, tail-like apices **9. lucida**
b. Fully developed leaves without tail-like apices
40
40a. Stamens 5–9; capsules long-stalked
8. lasiandra
b. Stamens 3; capules shortly stalked
3. triandra
41a. Leaves narrowly lanceolate or linear-lanceolate, often curved, deep green on both surfaces **2. nigra**
b. Leaves broadly lanceolate or ovate-lanceolate, pallid or glaucous on the undersurface **34. eriocephala**
42a. Branches pendent or much contorted
43
b. Branches spreading or erect 45
43a. Catkins almost stalkless, usually less than 2 cm **15. babylonica**
b. Catkins distinctly stalked, usually more than 2 cm
44
44a. Leaf-margins shortly and closely toothed; leaves at first downy or silky
12. × sepulcralis
b. Leaf-margins rather coarsely toothed; leaves at first hairless or very thinly hairy **14. × pendulina**
45a. Trees 46
b. Shrubs 50
46a. Leaves minutely toothed or almost entire; catkins without nectary-scales
5. arbutifolia
b. Leaves distinctly toothed; catkins with nectary-scales
47
47a. Twigs shining brown as if varnished; leaves with numerous, close, glandular teeth **7. × meyeriana**
b. Twigs not shining brown; leaves not as above
48
48a. Catkins stalkless or with very short stalks **15. babylonica** var. **pekinensis**
b. Catkins distinctly stalked 49
49a. Leaves at first silky-hairy but soon becoming hairless or almost so; leaf-margins with numerous, small, even teeth **11. × rubens**
b. Leaves hairless or almost so from the first; leaf-margins rather coarsely and unevenly toothed **13. fragilis**
50a. Stipules usually well developed and persistent **4. × mollissima**
b. Stipules inconspicuous or absent 51
51a. Suckering shrubs **17. exigua**
b. Non-suckering shrubs 52

52a. Leaves narrowly linear-lanceolate; stamens with united filaments **79. wilhelmsiana**

b. Leaves broadly lanceolate; stamens with free filaments 53

53a. Leaves and shoots at first silky-hairy, becoming hairless later; branches slender **16. japonica**

b. Leaves and shoots hairless or almost so from the first; branches robust **7. × meyeriana**

54a. Dwarf shrubs less than 1 m 55

b. Robust shrubs or trees more than 1 m high 75

55a. Creeping shrubs with underground, rooting stems, the aerial stems scarcely developed 56

b. Dwarf shrubs with well-developed aerial stems and branches 59

56a. Leaves short-stalked, green on both surfaces **21. herbacea**

b. Leaves long-stalked, whitish or ashy grey on the undersurface 57

57a. Ovaries and capsules densely hairy **18. reticulata**

b. Ovaries and capsules hairless or thinly hairy 58

57a. Leaves almost circular, 4–5 cm **20. nakamurana**

b. Leaves elliptic to obovate, 5–25 mm **26. uva-ursi**

59a. Leaf-margins sharply or bluntly toothed 60

b. Leaf-margins entire or almost so 65

60a. Undersurface of leaves shining green; stipules usually well developed and persistent; veins strongly impressed, forming a network **25. myrsinites**

b. Undersurface of leaves pale dull green or glaucous, not as above 61

61a. Stipules conspicuous, persistent 62

b. Stipules inconspicuous, absent or soon shed 63

62a. Branches prostrate or decumbent; catkin-scales dark-tipped; ovary hairless **36. apoda**

b. Branches ascending or erect; catkin-scales yellowish; ovary hairy **56. starkeana**

63a. Leaves closely glandular toothed **42. arbuscula**

b. Leaves bluntly or remotely and irregularly toothed 64

64a. Leaves bluntly toothed; catkin-scales yellowish or tinged with red towards apex **22. × grahamii**

b. Leaves sharply but remotely and irregularly toothed, often almost entire; catkin-scales dark-tipped **24. × cottetii**

65a. Leaves hairy on one or both surfaces 66

b. Leaves hairless or soon hairless 72

66a. Stipules conspicuous, persistent; catkin-scales with yellowish hairs **66. lanata**

b. Stipules inconspicuous or absent; catkin-scales with white or greyish hairs 67

67a. Leaves broadly obovate or almost circular; a small, rigid, erect, bonsai-like shrub **19. × boydii**

b. Leaves narrower, ovate, oblong or obovate 68

68a. Catkins appearing in the autumn; leaves densely and regularly arranged on the stems **78. bockii**

b. Catkins appearing in the spring 69

69a. Catkins 2.5–4 cm; ovaries densely whitish-felted 70

b. Catkins 1–2.5 cm; ovaries silky or almost hairless 71

70a. Leaves greyish on both surfaces **64. lapponum**

b. Leaves green on the upper surface **65. helvetica**

71a. Catkins developing before the leaves; branches usually sprawling or ascending **70. repens**

b. Catkins developing with the leaves; branches erect or almost so **29. × finmarchica**

72a. Leaves glaucous on one or both surfaces 73

b. Leaves green on both surfaces 74

73a. Branches spreading; leaves rigid, almost without stalks, glaucous on both surfaces; stamens with filaments wholly or partly united **74. caesia**

b. Branches erect; leaves not rigid, distinctly stalked, glaucous beneath; stamens with free filaments **28. myrtilloides**

74a. Branches prostrate; leaves indistinctly veined, often indented at apex **23. retusa**

b. Branches ascending; leaves distinctly veined, apex acute **38. myrsinifolia**

75a. Leaves hairy or downy on one or both surfaces 76

b. Leaves hairless or soon hairless 89

76a. Leaf-margins closely toothed 77

b. Leaf-margins entire or obscurely and irregularly toothed 80

77a. Stipules conspicuous, persistent **33. cordata**

b. Stipules inconspicuous or absent 78

78a. Catkins developing before the leaves **55. humilis**

b. Catkins developing with the leaves 79

79a. Leaves glossy green above; veins 15–25 pairs **31. fargesii**

b. Leaves dull green above; veins fewer than 15 pairs **38. myrsinifolia**

80a. Underside of leaves with numerous, prominent, parallel veins **62. gracilistyla**

b. Underside of leaves not as above 81

81a. Catkins developing with the leaves **38. myrsinifolia**

b. Catkins developing before the leaves 82

82a. Leaves glossy green above 83

b. Leaves not glossy green above 84

83a. Stipules small or absent **67. hookeriana**

b. Stipules conspicuous, persistent **47. 'Hagensis'**

84a. Stipules small or absent **55. humilis**

b. Stipules usually well developed, at least on robust shoots 85

85a. Leaves wrinkled; stipules conspicuous and persistent even on slender shoots **53. aurita**

b. Leaves scarcely wrinkled; stipules not as above 86

86a. Leaves broadly ovate, oblong or elliptic, densely and softly hairy beneath; wood of peeled twigs smooth, without longitudinal ridges 87

b. Leaves narrowly ovate, obovate or oblong, distinctly longer than wide, shortly and often sparsely hairy beneath; wood of peeled twigs with longitudinal ridges 88

87a. Stipules conspicuous, persistent; catkin-scales with yellow-tinged hairs **49. × balfourii**

b. Stipules often absent on slender shoots; catkin-scales with white or greyish hairs **44. caprea**

88a. Catkins 3–5 cm, conspicuous and showy **43. aegyptiaca**

b. Catkins 2–3 cm, not very showy **51. cinerea**

89a. Catkins developing before the leaves 90

b. Catkins developing with or after the leaves 93

90a. Twigs bloomed; leaves dark green above, glaucous beneath, distinctly and regularly toothed; stipules long-acuminate **68. daphnoides**

b. Twigs not bloomed; stipules abruptly acute 91

91a. Leaves bright shining green above, almost white beneath **54. discolor**

b. Leaves dull green above, paler but not white beneath 92

Figure 2. Leaves of *Salix* species. 1, *S. chilensis*. 2, *S. nigra*.
3, *S. triandra*. 4, *S.* × *mollissima*. 5, *S. pentandra*. 6, *S.* × *meyeriana*.
7, *S. lasiandra*. 8, *S. lucida*. Scale = 1 cm.

92a. Small shrub usually less than 2 m high; stipules often conspicuous and persistent; catkins to 8 cm; stamens with free filaments **35. hastata**

b. Large shrub or small tree, usually more than 2 m high; stipules inconspicuous or soon falling; catkins *c.* 3 cm; stamens with filaments wholly or partly united **50. × wimmeriana**

93a. Leaves very large, often 10–20 cm, margins entire or almost so; catkins 10–25 cm **32. magnifica**

b. Leaves much smaller; catkins less than 10 cm **94**

94a. Leaves closely and minutely toothed; catkin-scales uniformly pale yellow **95**

b. Leaves not closely toothed; catkin-scales dark at apex **96**

95a. Leaf-stalks *c.* 1.5 cm; stamens 2 **37. pyrifolia**

b. Leaf-stalks to 1 cm or less; stamens 5 or more **6. pentandra**

96a. Leaves blackening when bruised or dried **38. myrsinifolia**

b. Leaves not blackening when bruised or dried **97**

97a. Stipules usually well developed, persistent **35. hastata**

b. Stipules inconspicuous, soon falling **98**

98a. Leaves greenish beneath **39. mielichoferi**

b. Leaves glaucous beneath **99**

99a. Buds reddish brown; leaves almost hairless at all stages of development **40. phylicifolia**

b. Buds yellow or orange; leaves at first silky-hairy **41. schraderiana**

1. S. chilensis Molina (*S. humboldtiana* Willdenow). Figure 2(1), p. 24.
Tree to 20 m; branches and twigs slender, hairless or almost so, glossy brown; buds not adpressed to twigs, bud-scales 2, free along their inner margins. Leaves linear, 5–15 cm × 5–10 mm, at first thinly hairy, soon hairless and green on both surfaces, apex acuminate, margins closely and sharply toothed; leaf-stalk short, stipules small, acute, glandular-toothed, soon falling. Catkins stalked, with leafy bracts, narrowly cylindric, 2–5 cm, axis and scales densely hairy. Anthers yellow. Ovary ovoid, hairless, shortly stalked. Stigmas almost stalkless. *S America*. G1. Spring.

Tender and surviving outdoors only in the mildest areas. A variant with erect, fastigiate branching is of greater horticultural interest than the typical tree.

2. S. nigra Marshall. Figure 2(2), p. 24. Illustration: Systematic Botany Monographs 9: 42 (1986).
Large shrub or tree, 3–20 m with dark brown or blackish, scaly bark; twigs soon hairless, yellowish brown, brittle at junction with branch. Leaves lanceolate, sometimes sickle-shaped, narrowed to a slender apex, 5–15 cm × 5–20 mm, soon hairless or almost so, dark, shining green above, paler beneath, margins closely glandular-toothed; stipules ear-shaped, toothed, soon falling. Catkins developing with the leaves, slender, cylindric, 2–8 cm, stalked, scales pale yellow, downy, soon falling. Stamens 4–6. Ovary distinctly stalked, hairless. *East & central N America*. H1. Spring.

3. S. triandra Linnaeus (*S. amygdalina* Linnaeus). Figure 2(3), p. 24. Illustration: Meikle, Willows and poplars, 61 (1984).
Shrub or small bushy tree to 10 m; bark flaking off in large, irregular patches; shoots distinctly ridged or angled; twigs hairless, pliable, rather glossy. Leaves lanceolate or narrowly elliptic, acute or acuminate, hairless, 4–12 × 1–3 cm, margins toothed; stipules conspicuous, persistent, ear-shaped. Catkins appearing with or a little before the leaves, narrowly cylindric, usually stalked, 2.5–5 cm, scales uniformly pale yellow, thinly hairy. Stamens 3, anthers yellow. Ovary hairless, distinctly stalked. *Europe, Asia*. H1. Spring.

Much cultivated for basket-making, and with a wide range of cultivars, some with the leaves uniformly green, others with the lower surface distinctly glaucous.

4. S. × mollissima Elwert. Figure 2(4), p. 24. Illustration: Meikle, Willows and poplars, 67, 69 (1984).
Shrub to 5 m; bark sometimes flaking as in *S. triandra*; leaves lanceolate or linear-lanceolate, 8–13 × 1–1.5 cm, dark shining green above, paler or glaucous beneath, hairy beneath or soon hairless on both surfaces, margins entire or toothed, apex acuminate; stipules acute or acuminate, glandular, often persistent. Catkins developing with the leaves, cylindric or narrowly ovoid, to 4 cm, scales pallid, yellowish, hairy. Stamens 2 or 3. Ovary hairless or hairy. *Europe*. H1. Spring.

S. triandra × S. viminalis.

Var. **mollissima**, with persistently hairy entire or almost entire leaves and hairy ovaries is seldom seen; the most commonly cultivated varieties of this hybrid are var. **undulata** (Ehrhart) Wimmer (*S. undulata* Ehrhart), with toothed, long-acuminate

leaves and hairless ovaries and var. **hippophaifolia** (Thuillier) Wimmer (*S. hippophaifolia* Thuillier) with leaves entire or almost so and ovaries hairy.

5. S. arbutifolia Pallas (*S. eucalyptoides* Schneider; *Chosenia arbutifolia* (Pallas) Skvortsov). Illustration: Makino, Illustrated Flora of Japan, t. 2008 (1948).
Tree to 30 m; bark scaly in older trees, twigs hairless, often bloomed. Leaves lanceolate, 5–10 × 1–2.5 cm, hairless, at first bloomed, later bright green, margins entire or minutely toothed. Catkins developing with or before the leaves, slender, cylindric, the male pendent, 1–4 cm, scales pallid, thinly hairy, deciduous in female. Stamens 5, filaments joined to the catkin-scale. Ovary hairless, shortly stalked. Styles divided to base, deciduous. *NE Asia*. H1. Spring.

Remarkable for its pendent male catkins, filaments joined to the catkin-scale and deciduous styles; the flowers are said to be wind-pollinated, and lack the nectary-scales generally found in *Salix*. For these reasons the species is often placed in the separate genus, *Chosenia*.

6. S. pentandra Linnaeus (*S. laurifolia* Wesmael). Figure 2(5), p. 24. Illustration: Meikle, Willows and poplars, 25 (1984).
Large shrub or small tree, exceptionally to 17 m; twigs hairless, shining brown. Leaves ovate, obovate or broadly lanceolate, 5–12 × 2–5 cm, acute or shortly acuminate, hairless, dark glossy green above, paler beneath, margins minutely and closely glandular-toothed; stipules very small, often absent. Catkins shortly cylindric, 2–5 cm, developing with the leaves in late May or June, distinctly stalked, scales pale yellow. Stamens 5–8, anthers yellow. Ovary hairless. *N & C Europe*. H1. Late spring.

7. S. × meyeriana Willdenow. Figure 2(6), p. 24. Illustration: Meikle, Willows and poplars, 39 (1984).
A large shrub or small spreading tree; twigs hairless, shining brown. Leaves narrowly ovate-elliptic or broadly lanceolate, 5–12 × 1.5–4 cm, acuminate, dark shining green above, paler beneath, margins closely glandular-toothed; stipules absent or soon falling. Catkins developing with the leaves, cylindric, 4–5 cm, scales pale yellow, hairy. Stamens 2–5. Ovary hairless. *Europe*. H1. Late spring.

S. fragilis × S. pentandra.

S. × ehrhartiana Smith, which is *S. alba × S. pentandra*, is similar but with the leaves at first thinly adpressed hairy and with minutely toothed margins. *Europe*. H1. Spring. Illustration: Meikle, Willows and poplars, 39 (1984).

8. S. lasiandra Bentham. Figure 2(7), p. 24. Illustration: Hitchcock et al., Vascular plants of the Pacific Northwest **2**: 58 (1964).
Tree to 20 m, twigs at first hairy, becoming glossy brown with age. Leaves lanceolate, 10–13 × 1.3–4 cm, at first somewhat downy, soon dark shining green above, rather glaucous beneath, acuminate, margins closely glandular-toothed; stipules ear-shaped, glandular, often persistent. Catkins developing with the leaves, stalked, 3–7 cm, scales pale yellow, hairy, those of the male flowers commonly glandular-toothed at apex. Stamens 5–9. Ovary hairless. *Western N America*. H1. Spring.

9. S. lucida Muhlenberg. Figure 2(8), p. 24. Illustration: Gleason, Illustrated Flora of the north-eastern United States and adjacent Canada, **2**: 9 (1952); Systematic Botany Monographs **9**: 68 (1986).
Shrub or small tree to *c.* 7 m, twigs glossy yellow-brown, hairless. Leaves lanceolate or narrowly ovate, 10–13 × 1–4 cm, dark shining green above, paler beneath, hairless except on the midrib and veins, apex long-acuminate, margins closely glandular-toothed; leaf-stalk glandular; stipules large, bluntly ear-shaped, often persistent. Catkins developing with the leaves, 3–7.5 cm, shortly cylindric, scales pale, yellowish, hairy. Stamens usually 5. Ovary hairless. *N America*. H1. Spring.

10. S. alba Linnaeus.
Tree 10–30 m with deeply fissured bark, branches erect or ascending; twigs at first shortly hairy, soon almost hairless. Leaves lanceolate, acuminate, 5–10 cm × 5–15 mm, minutely toothed, silvery with adpressed silky hairs; leaf-stalks short; stipules small, soon falling. Catkins developing with the leaves, narrowly cylindric, 3–5 cm, scales pallid, thinly hairy. Stamens 2, anthers yellow. Ovary narrowly flask-shaped, hairless. *Europe, W Asia*. H1. Spring.
Var. **alba**. Figure 3(1), p. 27. Meikle, Willows and poplars, 41 (1984). Leaves thinly adpressed silky above, even at maturity; twigs brownish.
Var. **sericea** Gaudin (forma *argentea* Wimmer). Figure 3(2), p. 27. Less robust than var. *alba*, with densely silvery-silky leaves.
Var. **caerulea** (Smith) Smith. Illustration:

Meikle, Willows and poplars, 47 (1984). Leaves broader and more conspicuously toothed than in var. *alba*, becoming almost hairless with age. Often cultivated for the manufacture of cricket bats in the British Isles.
Var. **vitellina** (Linnaeus) Stokes. Figure 3(3), p. 27.
Illustration: Meikle, Willows and poplars, 45 (1984). Winter twigs bright orange-yellow, leaves rather narrow, thinly hairy at maturity. The cultivars 'Britzensis', 'Cardinal', 'Chermesina' and 'Chrysostela' resemble var. *vitellina* but have more richly coloured red or orange-red winter twigs.

11. S. × rubens Schrank. Figure 3(4), p. 27.
Tree to 30 m, branches spreading. Leaves lanceolate or linear-lanceolate, 7–15 cm × 8–20 mm, at first thinly hairy, soon green and shining above, paler beneath, margins closely toothed. Catkins narrowly cylindric, with pallid scales.
H1. Spring.
S. alba × S. fragilis.
Var. **rubens**, with olive-brown winter twigs, is seldom seen in cultivation, but two other variants are:
Forma **basfordiana** Meikle. Illustration: Meikle, Willows and poplars, 51 (1984). Winter twigs orange-yellow, leaves 9–15 × 1–2 cm, bright shining green, catkins narrowly cylindric and pendent, commonly to 10 cm, exceptionally to 15 cm.
Forma **sanguinea** Meikle. Illustration: Meikle, Willows and poplars, 53 (1984). Less vigorous than forma *basfordiana*; twigs dark red, leaves 7–8 × 1–1.5 cm, very shortly toothed, catkins 3–4 cm, erect or spreading.

12. S. × sepulcralis Simonkai (*S. × salamonii* Henry). Figure 3(5), p. 27.
Tree to 18 m or more, with fissured bark; branches strongly or rather weakly pendent, twigs olive-brown or yellowish, at first thinly silky, soon hairless. Leaves lanceolate, acuminate, 7–12 cm × 7–18 mm, at first silky, soon almost hairless, bright green above, glaucous beneath, margins minutely toothed; stipules small, ovate, glandular, soon falling. Catkins narrowly cylindric, 3–4 cm, commonly bisexual; scales pale yellow. Stamens 2, anthers yellow. Ovary almost stalkless, shortly flask-shaped, hairless. H1. Spring.
S. alba × S. babylonica. There are several clones under this parentage, depending on which varieties or cultivars of the parent

species have been crossed; the most important are:
Var. **sepulcralis**. Branches not strongly pendent, twigs olive-brown.
Var. **chrysocoma** (Dode) Meikle (*S. × chrysocoma* Dode) – *S. alba* var. *vitellina* × *S. babylonica*. Illustration: Meikle, Willows and poplars, 55 (1984). Branches strongly pendent, twigs golden-or greenish yellow, very slender. Now the most commonly planted Weeping Willow.
'Erythroflexuosa' (*S. × erythroflexuosa* Ragonese) – *S. babylonica* var. *pekinensis* 'Tortuosa' × *S. × sepulcralis* var. *chrysocoma*. With twisted orange-yellow twigs and twisted leaves. Originated in Argentina.

13. S. fragilis Linnaeus.
Tree 10–25 m with a stout, fissured trunk and spreading branches; twigs brittle at point of junction with branch, soon hairless. Leaves lanceolate, acuminate, hairless or at first thinly hairy, shining green above, rather glaucous beneath, margin distinctly glandular-toothed; leaf-stalk often glandular at apex; stipules usually small, soon falling. Catkins developing with the leaves, narrowly cylindric, distinctly stalked, scales uniformly pale yellow, thinly hairy. Stamens 2, occasionally 3. Ovary shortly stalked, flask-shaped, hairless.
Europe. H1. Spring.
This species includes the following 5 distinct varieties:
Var. **fragilis**. Figure 3(6), p. 27.
Illustration: Meikle, Willows and poplars, 29 (1984). Leaves and shoots at first thinly hairy, soon hairless; twigs olive-brown; leaves 9–15 × 1.5–3 cm; ovary very shortly stalked, 2.5–3 mm.
Var. **furcata** Gaudin. Illustration: Meikle, Willows and poplars, 31 (1984). Leaves at first thinly hairy, 3–4 or even to 5 cm wide, coarsely toothed; male catkins to 6 cm, often more than 1 cm in diameter, frequently 2-forked; stamens commonly 3; female catkins not recorded.
Var. **russelliana** (Smith) Koch (*S. russelliana* Smith). Figure 3(7), p. 27. Illustration: Meikle, Willows and poplars, 33 (1984). Leaves at first thinly hairy, *c.* 13–15 cm long, usually less than 3 cm wide, apex long-drawn out, margins coarsely and unevenly toothed; female catkins lengthening to 6 cm or more with age, often becoming somewhat pendent; ovary distinctly stalked, 6–7 m, tapered to apex; male catkins not recorded.
Var. **decipiens** (Hoffmann) Koch (*S.*

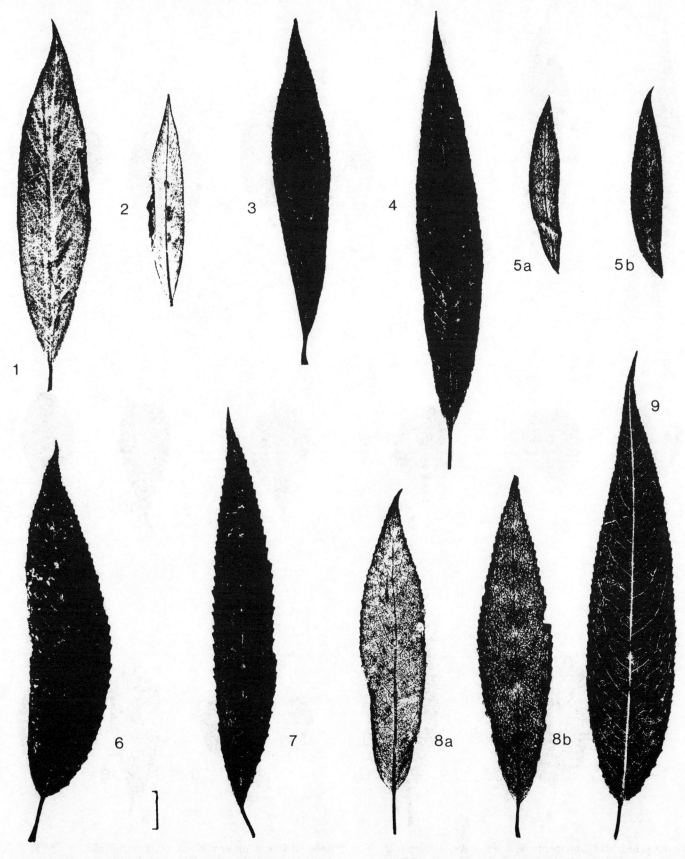

Figure 3. Leaves of *Salix* species. 1, *S. alba* var. *alba*. 2, *S. alba* var. *sericea*. 3, *S. alba* var. *vitellina*. 4, *S.* × *rubens*. 5, *S.* × *sepulcralis* (a, lower surface, b, upper surface). 6, *S. fragilis* var. *fragilis*. 7, *S. fragilis* var. *russelliana*. 8, *S. fragilis* var. *decipiens* (a, lower surface, b, upper surface). 9, *S. pendulina*. Scale = 1 cm.

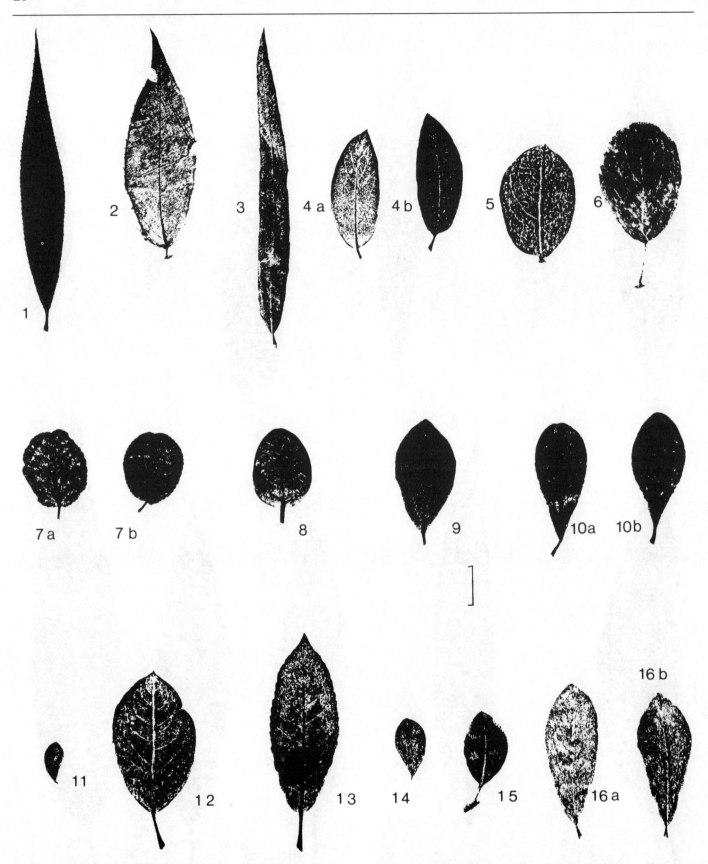

Figure 4. Leaves of *Salix* species. 1, *S. babylonica*. 2, *S. japonica*. 3, *S. exigua*. 4, *S. reticulata* (a, lower surface, b, upper surface). 5, *S.* × *boydii*. 6, *S. nakamurana*. 7, *S. herbacea* (a, lower surface, b, upper surface). 8, *S. polaris*. 9, *S.* × *grahamii*. 10, *S. retusa* (a, lower surface, b, upper surface). 11, *S. serpyllifolia*. 12, *S.* × *cottetii*. 13, *S. myrsinites*. 14, *S. alpina*. 15, *S. uva-ursi*. 16, *S. glaucosericea* (a, lower surface, b, upper surface). Scale = 1 cm.

decipiens Hoffmann). Figure 3(8), p. 27. Illustration: Meikle, Willows and poplars, 37 (1984). Twigs clay-coloured; leaves quite hairless, even when young, seldom more than 9 cm long, often 2–3 cm wide; male catkins rather puny, usually less than 3 cm, erect or almost so; ovary shortly stalked, 2–2.5 mm.

Var. **sphaerica** Hryniewiecki. Forming dense, many-branched, dome-like small trees looking as if they had been clipped; not uncommon as a planted tree in the USSR, but perhaps not cultivated in W Europe.

14. S. pendulina Wenderoth (*S. × blanda* Andersson; *S. × elegantissima* K. Koch). Figure 3(9), p. 27.
Tree 12–18 m with fissured bark; branches strongly or weakly pendent; twigs slender, olive-brown, soon hairless. Leaves lanceolate, acuminate, 10–12 × 1.5–2 cm, shining dark green above, paler or rather glaucous beneath, hairless or almost so, margins distinctly and rather unevenly toothed; leaf-stalk to 1.5 cm, stipules ovate, acute or acuminate, shortly toothed, glandular, soon falling. Catkins developing with the leaves, distinctly stalked, cylindric, 2–3 cm. Stamens 2. Ovary flask-shaped with a tapered apex, hairless or thinly hairy towards base. H1. Spring.

Var. **pendulina**. Branches strongly pendent; ovary shortly stalked, hairless.

Var. **elegantissima** (K. Koch) Meikle. Illustration: Meikle, Willows and poplars, 57 (1984). Branches strongly pendent; ovary shortly stalked, thinly hairy towards base.

Var. **blanda** Andersson. Branches not strongly pendent; stipules with a long-acuminate apex; ovary distinctly stalked, hairless.

15. S. babylonica Linnaeus. Figure 4(1), p. 28.
Tree 12–18 m, bark coarsely fissured; branches spreading or pendent; twigs slender, at first thinly downy, soon hairless and yellowish brown. Leaves lanceolate, acuminate, 7–10 × 1–2 cm, at first thinly hairy, soon hairless, bright green above, glaucous beneath, margins finely toothed; stalk *c.* 5 mm; stipules narrowly ovate, glandular, soon falling. Catkins developing with the leaves, almost stalkless, slender, cylindric, 2–3 cm; scales yellowish, thinly hairy. Stamens 2. Ovary narrowly ovoid, hairless or thinly hairy. Style short. *China.* Spring.

Var. **babylonica**. Illustration: Systematic Botany Monographs **9**: 82 (1986).

Branches conspicuously pendent; female flower with a single nectary-scale. *Widely cultivated, probably originated in China.* H4. Spring.

Now almost extinct in Europe and replaced by the hybrids *S. × pendulina* and *S. × sepulcralis*. The cultivar 'Crispa' (*S. annularis* Forbes), with curled and twisted leaves, is occasionally grown. 'Lavallei' (*S. japonica* misapplied) has less pendent branches than typical var. *babylonica*.

Var. **pekinensis** Henry (*S. matsudana* Koidzumi). Branches spreading (or pendent in 'Pendula'); female flower with 2 nectary-scales. *China.* H1. Spring.

Questionably distinct from var. *babylonica* even as a variety; the cultivar 'Tortuosa' with contorted twigs and branches is occasionally planted.

16. S. japonica Thunberg. Figure 4(2), p. 28.
Shrub *c.* 2 m, twigs slender, soon hairless. Leaves lanceolate, 5–8 × 1.5–2.5 cm, at first silky, soon hairless, bright green above, glaucous beneath, margins finely toothed; stipules small or absent. Catkins developing with the leaves, slender, loose-flowered, 3–12, usually 6–10 cm; catkin-scales blunt, pallid, shortly hairy. Stamens 2, free, filaments hairy. Ovary shortly stalked, hairless. Styles short. *Japan.* H1. Spring.

The name *S. japonica* is sometimes attached to a variant of *S. babylonica* ('Lavallei') with a less pendulous habit than typical *babylonica*. It is doubtful if genuine *S. japonica* is in cultivation in Europe.

17. S. exigua Nuttall. Figure 4(3), p. 28. Illustration: Hitchcock et al., Vascular plants of the Pacific Northwest 2: 54 (1964); Systematic Botany Monographs 9: 86, 87 (1986).
Suckering shrub or small tree to about 5 m, twigs slender, hairy or soon becoming hairless, greyish brown. Leaves linear-lanceolate, acuminate, 5–10 cm × 3–10 mm, almost stalkless, adpressed-silky on both surfaces, or becoming hairless, almost entire or distantly glandular-toothed; stipules small or sometimes absent. Catkins developing with the leaves, slender, cylindric, shortly stalked, 2.5–5 cm, scales yellowish, thinly hairy. Stamens 2. Ovary hairy or hairless, style almost absent. *North America, northern Mexico.* H2. Spring.

18. S. reticulata Linnaeus. Figure 4(4), p. 28. Illustration: Meikle, Willows and poplars, 163 (1984).
Prostrate shrub forming extensive mats with underground, rooting branches,

usually less than 10 cm high. Leaves broadly and bluntly obovate to almost circular, 1.2–4 × 1–3 cm, at first hairy, soon hairless and dull dark green above, ashy-grey beneath, margins entire, conspicuously net-veined; stalks long and slender, stipules absent. Catkins long-stalked, narrowly cylindric, 2–3.5 cm, developing after the leaves in summer, scales brownish or purplish. Stamens 2, anthers purplish. Ovary hairy. *Circumarctic and on the mountains of Europe.* H1. Summer.

19. S. × boydii E.F. Linton. Figure 4(5), p. 28.
Dwarf, erect, bonsai-like shrub, occasionally to 1 m high but usually much smaller, twigs at first downy, soon hairless. Leaves almost circular, indented at apex and base, 1–2 cm in diameter, at first downy all over, becoming dark green above, whitish and hairy beneath, margins recurved, entire, veins prominent beneath, stalk very short, stipules very small or absent. Catkins (female only known) developing with or a little before the leaves, 1–2 cm, scales dark-tipped. Ovary shortly flask-shaped, almost stalkless, hairy. *Scotland.* H1. Spring.

Parentage uncertain; some authorities regard it as *S. lapponum × S. reticulata*, others as *S. reticulata × S. lanata*, with the possibility that *S. herbacea* may also have entered into its composition. All 4 species are recorded from the area where it was first collected as a seedling (Scotland, Glen Fiagh).

20. S. nakamurana Koidzumi. Figure 4(6), p. 28. Illustration: Makino, Illustrated Flora of Japan, t. 2024 (1948).
Creeping, prostrate shrub, branches repeatedly forked, brown or purplish, rooting at the nodes. Leaves obovate, ovate or almost circular, 1–7 cm long and sometimes almost as wide, soon hairless, distinctly stalked, dull or shining green above, paler or rather glaucous beneath, with a prominent network of veins; margins almost entire, stipules absent. Catkins developing with the leaves, cylindric, 2–3 cm, the females distinctly stalked, scales dark red-brown, hairy. Stamens 2. Ovary hairy or hairless. Style long, stigmas slender. *Japan.* H1. Late spring.

Represented in cultivation by var. **yezoalpina** (Koidzumi) Kimura, with large, almost circular leaves 4–5 cm long and hairless ovaries.

21. S. herbacea Linnaeus. Figure 4(7), p. 28. Illustration: Raven & Walters, Mountain flowers, pl. 5 (1956); Meikle, Willows and poplars, 159 (1984); Polunin & Walters, Guide to the vegetation of Britain and Europe, pl. 64 (1985). Prostrate shrub with creeping underground branches, generally less than 6 cm high. Leaves almost circular, 3–20 mm in diameter, thinly hairy at first, soon glossy dark green, margins bluntly toothed. Catkins usually stalked, developing with the leaves, very small and few-flowered, scales yellowish or tinged with red. Stamens 2, anthers yellow or reddish. Ovary hairless, turning reddish as it matures. *Circumarctic.* H1. Late spring.

S. polaris Wahlenberg (Figure 4(8), p. 28) differs in having entire or almost entire leaves, blackish catkin-scales and hairy ovaries. *Arctic Europe and Asia.* H1. Late spring.

22. S. × grahamii Baker. Figure 4(9), p. 28. Low trailing or decumbent shrub, usually less than 30 cm high. Leaves broadly ovate-oblong, 1.5–4 × 1–3 cm, at first thinly hairy, soon glossy green above, paler beneath and prominently veined, margins bluntly toothed. Catkins developing with the leaves, cylindric, *c.* 1–1.5 cm, distinctly stalked, scales tinged with red towards the apex, thinly hairy. Ovary flask-shaped with a long, slender neck. Male catkins unknown. *Scotland & Ireland.* H1. Late spring.

S. aurita × S. herbacea × S. repens. Two varieties are found in cultivation:

Var. **grahamii**. Illustration: Journal of Botany **5**: 157, t. 66 (1867); Meikle, Willows and poplars, 127 (1984). Ovary quite hairless. *Scotland.*

Var. **moorei** (White) Meikle. Illustration: Meikle, Willows and poplars, 127 (1984). Ovary thinly or rather densely hairy. *Ireland.*

23. S. retusa Linnaeus. Figure 4(10), p. 28. Prostrate shrub, generally less than 10 cm high, with rooting, hairless or almost hairless, olive-brown branches. Leaves obovate, 5–30 × 3–8 mm, hairless or almost so, shining green above, paler beneath, apex obtuse, often indented, margins entire; stipules absent. Catkins developing with the leaves, stalked, shortly cylindric, to *c.* 2 cm, scales yellow or reddish, hairless or almost so. Stamens 2, anthers purplish. Ovary shortly flask-shaped, hairless. *Pyrenees, Alps, Apennines.* H1. Late spring.

Sometimes represented in gardens by the very closely allied **S. kitaibeliana** Willdenow, from the Carpathians, a taller shrub with larger leaves, or by the very small-leaved, compact **S. serpyllifolia** Scopoli (Figure 4(11), p. 28) from the east and central Alps.

24. S. × cottetii Kerner (*S. × gillotii* misapplied). Figure 4(12), p. 28. Illustration: Skalky Skalnicky, 54 (1977). Low, sprawling shrub; twigs at first thinly hairy, soon hairless or almost so and lustrous brown. Leaves oblong, obovate or elliptic, obtuse or shortly acute, *c.* 3–4 × 1.5–2 cm, hairless or almost so, green on both surfaces, margins entire or remotely toothed. Catkins shortly cylindric, to 2.5 cm, developing with the leaves, scales dark-tipped, hairy. Stamens 2. Ovary hairless. *Alps.* H2. Spring.

S. myrsinifolia × S. retusa. A male clone of this hybrid is occasionally cultivated under the misapplied name *S. gillotii* (correctly a hybrid between *S. lapponum* and *S. phylicifolia*).

25. S. myrsinites Linnaeus. Figure 4(13), p. 28. Illustration: Raven & Walters, Mountain flowers, pl. XXI (1956); Meikle, Willows and poplars, 157 (1984). Spreading or decumbent shrub, usually less than 40 cm high; twigs glossy red-brown. Leaves persistent when withered, ovate or obovate, 1.5–7 cm × 5–25 mm, dark shining green on both surfaces, hairless or almost so, margins shortly glandular-toothed, with a conspicuous network of veins; stipules often conspicuous and persistent. Catkins developing with the leaves, the female large, distinctly stalked, cylindric, 3–5 cm, the male rather smaller; scales dark reddish brown, hairy. Stamens 2, filaments and anthers purplish. Ovary thinly or densely hairy; style distinct. *N Europe.* H1. Spring.

S. alpina Scopoli (*S. jacquinii* Host). Figure 4(14), p. 28. Illustration: Skalky Skalnicky, 11 (1977). Similar but more prostrate, with entire or almost entire leaves which are deciduous and slender catkins. *E Alps and Carpathians.* H1. Spring.

26. S. uva-ursi Pursh. Figure 4(15), p. 28. Illustration: Gleason, Illustrated Flora of the north-eastern United States and adjacent Canada **2**: 15 (1952). Prostrate shrub forming extensive mats, with repeatedly divided, brown branches. Leaves elliptic or obovate, acute or obtuse, 5–25 × 3–10 mm, usually hairless, firm, shining green above, paler or slightly glaucous beneath, margins finely and remotely glandular-toothed; stipules very small or absent. Catkins developing with or after the leaves, cylindric, stalked, 1–2 cm, scales purplish or pink, hairy. Stamens 2, often apparently with united filaments. Ovary hairless, style short. *Greenland and eastern arctic America.* H1. Late spring.

27. S. glaucosericea Floderus. Figure 4(16), p. 28. Illustration: Skalky Skalnicky, 65 (1977). Shrub to *c.* 1 m, twigs densely hairy, becoming yellowish or brownish with age. Leaves oblanceolate, 5–8 × 1.5–2 cm, obtuse or shortly acute, at first adpressed-silky on both surfaces, becoming green and almost hairless above, glaucous and persistently hairy beneath, margins entire; stipules small or absent. Catkins developing with the leaves, erect, shortly stalked, cylindric, to *c.* 4.5 cm, scales dark-tipped, hairy. Stamens 2, anthers red or purple. Ovary narrowly flask-shaped, hairy. *Alps, Pyrenees.* H1. Late spring.

Closely related to the North Eurasian **S. glauca** Linnaeus, which has broader, less persistently hairy leaves.

28. S. myrtilloides Linnaeus. Figure 6(2), p. 33. Illustration: Skalky Skalnicky, 89 (1977). A small slender shrub with creeping underground stems, branches erect, less than 1 m high, hairless, brown. Leaves small, oblong-elliptic, obtuse or shortly acute, 1–4 cm × 5–20 mm, hairless, dark green above, often purplish or somewhat glaucous beneath, margins entire; stipules generally absent. Catkins developing with the leaves, stalked, cylindric, 1.5–2.5 cm, scales dark brown or reddish, hairless. Stamens 2, anthers at first reddish. Ovary long-stalked, hairless. *N Europe.* H1. Spring.

29. S. × finmarchica Willdenow. Figure 6(3), p. 33. Illustration: Skalky Skalnicky, 59 (1977). Like *S. myrtilloides* but a low spreading shrub with small, oblong or ovate leaves often persistently silky-hairy beneath. H1. Spring.

Sometimes planted as a ground-cover and much more common in cultivation than *S. myrtilloides.*

30. S. lindleyana Andersson (*S. hylematica* misapplied; *S. nepalensis* invalid). Figure 6(4), p. 33. Illustration: Skalky Skalnicky, 75 (1977). Prostrate dwarf shrub with rooting branches and short erect branchlets. Leaves very small, numerous, crowded, narrowly obovate or oblanceolate, usually less than

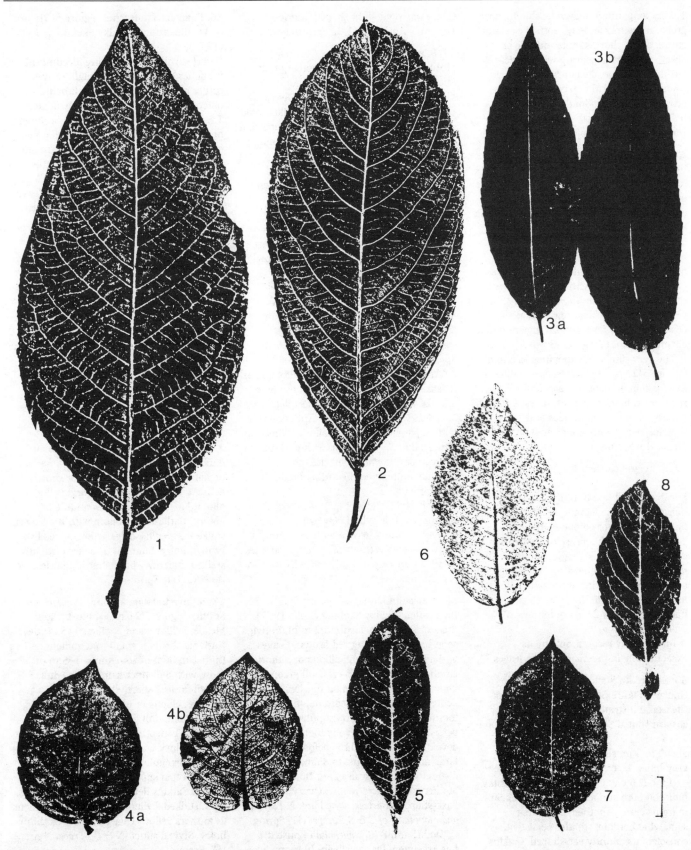

Figure 5. Leaves of *Salix* species. 1, *S. fargesii*. 2, *S. moupinensis*.
3, *S. cordata* (a, lower surface, b, upper surface). 4, *S. hastata* (a,
upper surface, b, lower surface). 5, *S. apoda*. 6, *S. pyrifolia*.
7, *S. myrsinifolia*. 8, *S. mielichoferi*. Scale = 1 cm.

1.5 cm long and 5 mm wide, shining dark green, margins remotely toothed or almost entire; stalks short, stipules minute or absent. Catkins appearing with the leaves, terminal on short, leafy branches, very small, ovoid, 1–1.5 cm, catkin-scales brownish, hairless. Stamens 2. Ovary hairless; style distinct, sometimes divided. *Himalaya*. H1. Late spring.

31. S. fargesii Burkill. Figure 5(1), p. 31. Robust shrub to *c.* 10 m high, twigs hairless, becoming shining brown with age; buds conspicuous, red. Leaves elliptic or oblong-elliptic, 8–18 × 3–8 cm, acute, dark shining green above with 15–25 deeply impressed pairs of lateral veins, paler beneath and often persistently silky, margins toothed; stipules generally absent. Catkins developing with the leaves, erect, shortly stalked, cylindric, to 16 cm, scales yellowish brown, silky-hairy. Stamens 2. Ovary narrowly flask-shaped, tapering to a distinct style. *C China*. H1. Spring.

The very similar **S. moupinensis** Franchet (Figure 5(2), p. 31) is often misidentified as *S. fargesii* in gardens. It has smaller, less silky (often hairless) leaves, hairless or almost hairless catkin-scales and a less tapering ovary with a shorter style. *W China*. H5. Spring.

32. S. magnifica Hemsley. Figure 6(1), p. 33. Shrub or small tree to 6 m high, twigs hairless, purple, ageing to red. Leaves hairless, oblong or elliptic-obovate, 10–20 × 8–14 cm, obtuse or shortly acute, dull bluish green above, glaucous beneath, margins entire or almost so; stipules absent or very small. Catkins developing with the leaves, stalked, cylindric, 10–25 cm, scales hairless. Stamens 2. Ovary hairless. *W China*. H2. Spring.

Immediately recognisable by its exceptionally large, *Magnolia*-like leaves.

33. S. cordata Michaux (*S. adenophylla* Hooker). Figure 5(3), p. 31. Illustration: Gleason, Illustrated Flora of the north-eastern United States and adjacent Canada 2: 13 (1952). Shrub to 3 m, shoots and twigs densely grey-hairy. Leaves crowded, oblong-ovate, 3–13 × 2–6 cm, acute or acuminate, silky-hairy on both surfaces, especially beneath, margins closely toothed, base often somewhat cordate; stipules persistent, conspicuous, bluntly ear-shaped. Catkins stalked, developing with or a little before the leaves, cylindric, 2–6 cm, scales blackish or brownish, hairy. Stamens 2.

Ovary narrowly flask-shaped, hairless. *Eastern N America*. H1. Late spring.

34. S. eriocephala Michaux (*S. rigida* Muhlenberg; *S. cordata* Muhlenberg not Michaux; *S. missouriensis* Bebb; *S. nicholsonii purpurascens* invalid). Figure 6(5), p. 33. Illustration: Gleason, Illustrated Flora of the north-eastern United States and adjacent Canada 2: 11 (1952); Systematic Botany Monographs 9: 130 (1986). Shrub or small tree to 5 m, shoots and young leaves hairless or almost so, often tinged with purple. Leaves distinctly stalked, rather crowded, green on both surfaces, oblong-lanceolate, 8–15 cm × 8–40 mm, acute or acuminate, margins closely toothed, base rounded or shallowly cordate; stipules conspicuous, often persistent, bluntly ear-shaped. Catkins developing before the leaves, cylindric, to 5 cm; scales blackish or brownish, hairy. Stamens 2. Ovary hairless, distinctly stalked. *Eastern and central N America*. H1. Spring.

'Americana' is a male clone cultivated as a basket-willow, and considered by some authors to be a hybrid of *S. eriocephala* with *S. petiolaris* Smith (*S. gracilis* Andersson) or with *S. purpurea*; it has smaller catkins than are usual in *S. eriocephala*, with dark-tipped hairy scales; the filaments are often partially united and the anthers reddish. H1. Spring.

S. × bebbii Gandoger (*S. myricoides* misapplied) is a hybrid between *S. eriocephala* and *S. sericea* with the young twigs and leaves thinly silky-hairy; stipules small; ovaries often thinly hairy. *N America*. H1. Spring.

35. S. hastata Linnaeus. Figure 5(4), p. 31. Illustration: Skalky Skalnicky, 69 (1977). Erect or spreading shrub to 2 m high; twigs soon hairless, shining red-brown. Leaves variable, ovate, oblong, elliptic or almost circular, 3–10 × 1.5–6 cm, dull green above, paler beneath, soon hairless, apex acute, base sometimes cordate, margins toothed or almost entire; stipules large, persistent, ovate, generally toothed. Catkins developing with or a little before the leaves, large and conspicuous, to 8 cm, females distinctly stalked, scales dark-tipped, hairy. Stamens 2, anthers yellow. Ovary narrowly flask-shaped, hairless, style long. *N Eurasia and mountains of C & S Europe*. H1. Spring.

'Wehrhahnii' (*S. wehrhahnii* Bonstedt) has exceptionally fine silvery hairy male catkins, and is now fairly frequent in gardens. It was found in the Engadine (Switzerland) around 1930.

36. S. apoda Trautvetter. Figure 5(5), p. 31. Illustration: Skalky Skalnicky, 13 (1977). Dwarf shrub with spreading, decumbent branches less than 30 cm high. Leaves rather dense, shortly stalked, broadly oblong-ovate, obtuse or abruptly acute, 1.5–5.5 × 1–3 cm, at first hairy, becoming hairless or almost so with age, dull green above, paler and rather glaucous beneath, margins shortly and irregularly toothed; stipules conspicuous, ear-shaped, persistent. Catkins developing a little before the leaves, cylindric, 2.5–3.5 cm, stalkless or shortly stalked, scales dark-tipped, silky-hairy. Stamens 2, anthers at first orange, becoming yellow. Ovary hairless, stigma linear. *USSR (Caucasus), Turkey*. H1. Spring.

37. S. pyrifolia Andersson (*S. balsamifera* Bebb). Figure 5(6), p. 31. Illustration: Gleason, Illustrated Flora of the north-eastern United States and adjacent Canada 2: 13 (1952). Balsam-scented shrub or occasionally small tree to 7 m high, twigs shining reddish brown, brittle at the junction with the branch; buds red. Leaves long-stalked, ovate or oblong-lanceolate, 3–10 × 2–4 cm, at first often tinged purple, later dark green above, pale and with a network of veins beneath, apex obtuse or shortly acute, base often cordate, margins finely glandular-toothed; stipules minute or absent. Catkins developing with the leaves, stalked, 2–5 cm, scales yellow or reddish brown, hairy. Stamens 2, ovary distinctly stalked, narrowly flask-shaped, hairless. *N America*. H1. Spring.

38. S. myrsinifolia Salisbury (*S. nigricans* Smith). Figure 5(7), p. 31. Illustration: Meikle, Willows and poplars, 135 (1984). Variable shrub or small tree, sometimes less than 1 m high or sometimes 3–5 m or more with a distinct trunk; twigs dull brown, at first shortly hairy. Leaves obovate, elliptic or oblong, 2–6.5 × 1.5–3.5 cm, dull green above, usually glaucous beneath, generally rather hairy, turning black when dried or bruised, margins irregularly toothed; stipules ear-shaped, often small, persistent or soon falling. Catkins developing with the leaves, often stalked, cylindric, 1.5–4 cm. Stamens 2, anthers yellow. Ovary hairless or thinly hairy. Style distinct. *N & C Europe, N Asia*. H1. Spring.

Figure 6. Leaves of *Salix* species. 1, *S. magnifica*. 2, *S. myrtilloides*.
3, *S. × finmarchica*. 4, *S. lindleyana*. 5, *S. eriocephala*.
Scale = 1 cm.

Figure 7. Leaves of *Salix* species. 1, *S. phylicifolia*.
2, *S. schraderiana* (a, upper surface, b, lower surface). 3, *S. arbuscula*
(a, upper surface, b, lower surface). 4, *S. foetida*. 5, *S. waldsteiniana*.
6, *S. aegyptiaca*. 7, *S. caprea* var. *caprea*. 8, *S. caprea* var. *sphacelata*.
9, *S. × sericans*. Scale = 1 cm.

39. S. mielichoferi Sauter. Figure 5(8), p. 31. Similar to *S. myrsinifolia* but with hairless or almost hairless leaves showing the same tendency to blacken. *C Alps*. H1. Spring.

40. S. phylicifolia Linnaeus. Figure 7(1), p. 34. Illustration: Meikle, Willows and poplars, 139 (1984).
Shrub or small tree, generally about 2–3 m high; twigs soon hairless, glossy reddish brown. Leaves ovate, obovate or elliptic, 2–6 × 1–5 cm, soon hairless, shining green above, glaucous beneath, margins toothed; stipules absent or very small. Catkins developing with the leaves in April or May, generally stalked, cylindric, 1–4 cm, scales dark brown, hairy. Stamens 2, anthers yellow. Ovary flask-shaped, ashy-hairy, style usually distinct. *N & C Europe*. H1. Spring.

41. S. schraderiana Willdenow (*S. bicolor* Willdenow not Ehrhart). Figure 7(2), p. 34. Very similar to *S. phylicifolia*, differing in its yellow or orange buds, leaves at first silky-hairy and anthers tinged with orange before opening. *Europe*. H1. Spring.

42. S. arbuscula Linnaeus (*S. prunifolia* Smith). Figure 7(3), p. 34. Illustration: Meikle, Willows and poplars, 153 (1984).
Creeping or sprawling shrub, usually less than 70 cm high, twigs soon hairless and shining reddish brown. Leaves small, ovate, oblong or elliptic, 1.5–3 × 1–1.5 cm, shining green above, rather glaucous beneath, closely glandular-toothed; stalks short, stipules small or absent. Catkins developing with the leaves, small, almost stalkless, narrowly cylindric, usually less than 2 cm, scales hairy, brownish or purplish. Stamens 2, anthers purplish. Ovary narrowly flask-shaped, densely grey-hairy. *N Europe to Siberia*. H1. Spring.

Sometimes replaced in cultivation by the closely related **S. foetida** Schleicher (Figure 7(4), p. 34) from the Pyrenees and W & C Alps; this has rather larger, conspicuously glandular-toothed leaves and proportionately broader catkins.

S. waldsteiniana Willdenow (Figure 7(5), p. 34) from the eastern Alps and N Balkan Peninsula has glandless, obscurely toothed or almost entire leaves to 5 cm, and distinctly stalked catkins; it is not infrequently cultivated under the name *S. arbuscula*.

43. S. aegyptiaca Linnaeus (*S. nitida* Gmelin; *S. medemii* Boissier). Figure 7(6), p. 34. Illustration: Botanical Magazine, n.s., 91 (1950); Townsend & Guest, Flora of Iraq 4(1): 39, pl. 7 (1980).
Large shrub or small tree; branches erect or spreading, twigs stout, at first downy, soon almost hairless; wood of peeled twigs longitudinally ridged. Leaves oblong-obovate, acute, 5–10 × 2–4 cm, wavy, irregularly toothed, becoming hairless and dull green above, persistently downy on midrib and veins beneath; leaf-stalks short, stipules ear-shaped, often large and persistent. Catkins developing before the leaves, showy, almost stalkless, cylindric or ovoid, 3–5 cm, scales silky, dark-tipped. Stamens 2, anthers yellow. Ovary narrowly flask-shaped, stalked, hairy. *Turkey & Israel to C Asia*. H1. Spring.

44. S. caprea Linnaeus. Figure 7(7), p. 34. Illustration: Meikle, Willows and poplars, 95 (1984).
Shrub or small tree, seldom exceeding 12 m; twigs at first downy, becoming hairless with age, wood not ridged. Leaves broadly oblong, obovate or elliptic, 5–12 × 2.5–8 cm, dull green and thinly downy above, densely and softly downy beneath with prominent veins, margins irregularly toothed, often wavy; stipules ear-shaped, conspicuous, sometimes persistent. Catkins developing before the leaves, ovoid or shortly cylindric, 1.5–2.5 cm, scales dark-tipped, hairy. Stamens 2, anthers yellow. Ovary narrowly flask-shaped, hairy. *Europe to C Asia*. H1. Spring.

'Kilmarnock' (male) and 'Weeping Sally' (female) are variants with somewhat pendent branches.

Var. **sphacelata** (Smith) Wahlenberg (Figure 7(8), p. 34) has almost entire, obscurely toothed, greyish-silky leaves.

45. S. × sericans Kerner. Figure 7(9), p. 34. Illustration: Meikle, Willows and poplars, 101 (1984).
Robust shrub to *c.* 9 m, twigs at first whitish-hairy, soon almost hairless, yellowish or reddish brown; wood not ridged. Leaves broadly lanceolate or ovate-lanceolate, acuminate, softly greyish-hairy and with prominent veins beneath; margins entire or almost so, stipules narrowly ear-shaped, acuminate. Catkins usually crowded towards the tips of the twigs, developing before the leaves, shortly cylindric, 2–5 cm, scales dark-tipped, hairy. Stamens 2, anthers yellow. Ovary narrowly flask-shaped, hairy. Stigmas narrowly oblong. *Europe*. H1. Spring.
S. caprea × S. viminalis.

46. S. × dasyclados Wimmer. Figure 8(1), p. 36.
A robust shrub to 10 m high, twigs at first densely velvety, becoming hairless with age. Leaves narrowly oblong-elliptic, 10–17 × 2.5–4 cm, dull green above, greyish and thinly hairy beneath, apex acute or shortly acuminate, margins obscurely toothed; stipules conspicuous, ear-shaped, persistent. Catkins developing before the leaves, almost stalkless, narrowly ovoid or shortly cylindric, 2–5 cm, scales dark-tipped, hairy. Stamens 2, anthers yellow. Ovary narrowly flask-shaped, hairy; style distinct, stigmas narrowly oblong. H1. Spring.

A group of closely related hybrids of obscure origin, but probably involving *S. caprea*, *S. cinerea* and *S. viminalis*. 'Grandis' (*S. aquatica grandis* invalid) is an exceptionally vigorous clone, stooled growths reaching 3 m or more in a single season. Other members of the group are:

S. × calodendron Wimmer (*S. acuminata* Smith). Figure 8(2), p. 36. Illustration: Meikle, Willows and poplars, 91 (1984). Twigs persistently velvety-hairy, wood ridged. Leaves oblong-elliptic, acute or shortly acuminate, usually *c.* 10 × 3 cm, ashy-grey beneath. Catkins narrowly cylindric, 2–5 cm, densely hairy. Stigmas narrowly oblong, equalling or shorter than style. H1. Spring. Mainly grown in Britain; only the female clone is known.

S. × stipularis Smith. Illustration: Smith & Sowerby, English Botany 17: 1214 (1803). Twigs at first velvety-hairy, becoming almost hairless; wood ridged, leaves persistently silky beneath. Stipules large, acuminate, persistent. Catkins ovoid or cylindric, 3–4 cm, hairy. Stigmas tapering, longer than style. H1. Spring. Also mainly grown in Britain.

47. S. 'Hagensis'.
A robust shrub with hairy twigs. Leaves oblong, acute or abruptly acuminate, 7–10 × 2–4 cm, glossy green above, whitish and hairy beneath; stipules conspicuous, acute. Catkins (female) 4–5 cm, with hairy scales and ovaries. H1. Spring.

Reportedly *S. gracilistyla × S. caprea*; raised at The Hague by S.G.A. Doorenbos.

S. × leucopithecia Kimura (*S. bakko* Kimura × *S. gracilistyla*) is similar to *S.* 'Hagensis', but male. It is said to be cultivated in Japan. *S. bakko* is a Japanese species very closely allied to the European *S. caprea*.

48. S. × seringeana Gaudin.
Tall shrub or small tree to 10 m; twigs at first thinly felted, becoming hairless and

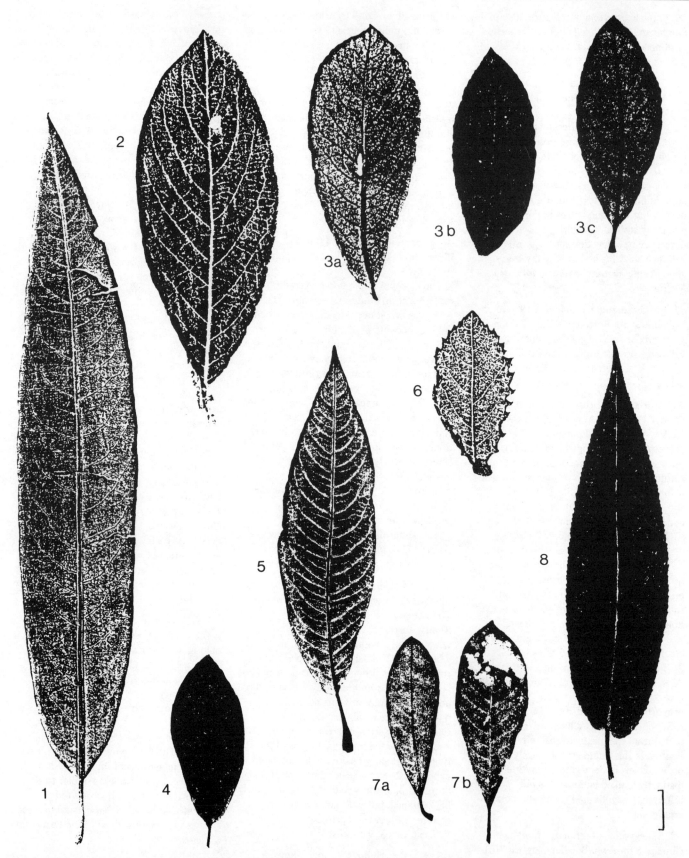

Figure 8. Leaves of *Salix* species. 1, *S.* × *dasyclados*.
2, *S.* × *calodendron*. 3, *S. cinerea* subsp. *cinerea* (a, b & c, three
different leaves). 4, *S. cinerea* subsp. *oleifolia*. 5, *S.* × *smithiana*.
6, *S. aurita*. 7, *S. discolor* (a, b, two different leaves). 8, *S. humilis*.
Scale = 1 cm.

yellowish brown. Leaves narrowly oblong, shortly acute, 8–12 × 2–2.5 cm, rather glossy dark green above, whitish- or greyish-felted beneath, margins minutely toothed; stipules small, ear-shaped. Catkins developing before the leaves, shortly stalked, cylindric, 3–4 cm, scales reddish, hairy. Stamens 2, filaments partly united. Ovary shortly stalked, thinly grey-hairy. H1. Spring.

S. elaeagnos × S. caprea.

49. S. × balfourii E.F. Linton.
Spreading bush to 3 m or more; twigs dark brown, buds yellowish or chestnut brown, downy. Leaves broadly elliptic, 3–8 × 2–6 cm, obtuse or shortly acute, entire or very obscurely and bluntly toothed, greyish above, whitish and shortly felted beneath, with prominent veins forming a network; stalks short, stipules conspicuous, persistent, rounded or ear-shaped. Catkins developing a little before the leaves, 3–5 × 2–3 cm, almost stalkless, showy, scales narrowly ovate, acute, densely silky with yellowish-tinged hairs. Stamens 2, anthers yellow or tinged with red. H1. Spring.

S. lanata × S. caprea. The male plant is not infrequent in gardens, and is valued for its exceptionally ornamental catkins. Cultivated stocks are derived from a plant artificially produced by E.F. Linton in his garden near Bournemouth (England).

50. S. × wimmeriana Grenier & Godron.
Robust shrub or small tree; twigs olive-brown, hairless. Leaves oblong or ovate, 5–10 × 2.5–4 cm, at first thinly felted, soon hairless, rather glossy green above, pale or glaucous beneath, apex shortly acute, margins remotely and obscurely toothed, stipules ovate, acute, sometimes absent. Catkins developing before the leaves, cylindric, c. 3 × 1–1.5 cm, scales blackish-tipped. Stamens 2, filaments partly united. Ovary hairy, shortly ovoid-conical, style very short. H1. Spring.

S. caprea × S. purpurea.

51. S. cinerea Linnaeus.
Shrub or small tree, occasionally to 15 m high; twigs reddish brown, persistently downy or soon hairless, wood ridged. Leaves oblong, obovate or oblanceolate, 2–9 × 1–3 cm, acute or obtuse, dull or somewhat lustrous green above, thinly or softly downy beneath; margins irregularly wavy-toothed; stipules ear-shaped, persistent or soon falling. Catkins developing before the leaves, stalkless or almost so, cylindric or ovoid, 2–3 cm, scales dark-tipped, hairy. Stamens 2,

anthers yellow. Ovary densely grey-hairy. *Europe, N Asia.* H1. Spring.

Subsp. **cinerea** (*S. aquatica* Smith). Figure 8(3), p. 36. Illustration: Meikle, Willows and poplars, 107 (1984). Twigs persistently downy; leaves greyish green, downy beneath without reddish hairs. *Europe, N Asia.*

Subsp. **oleifolia** Macreight. Figure 8(4), p. 36. Illustration: Meikle, Willows and poplars, 109 (1984). Twigs soon hairless; leaves rather lustrous green above, thinly downy with scattered reddish hairs beneath. *W Europe.* H1. Spring.

Subsp. *oleifolia* 'Tricolor' with the leaves blotched yellow and white, is occasionally grown.

52. S. × smithiana Willdenow. Figure 8(5), p. 36. Illustration: Meikle, Willows and poplars, 119 (1984).
Robust shrub to 9 m; twigs hairy, becoming hairless and reddish brown with age; wood ridged. Leaves narrowly oblong to lanceolate, shortly acuminate, dull green above, greyish and thinly hairy beneath, margins entire or almost so; stipules ear-shaped, acuminate, often persistent. Catkins rather crowded towards the tips of the twigs, developing before the leaves, shortly cylindric, 2–3 cm, scales dark-tipped, hairy. Stamens 2, anthers yellow. Ovary narrowly flask-shaped, hairy; stigmas narrowly oblong, about as long as style. H1. Spring.

S. cinerea × S. viminalis.

53. S. aurita Linnaeus. Figure 8(6), p. 36. Illustration: Meikle, Willows and poplars, 121 (1984).
Much-branched shrub, usually less than 2.5 m high; twigs dark reddish brown, at first downy, soon almost hairless, wood often prominently ridged. Leaves obovate, 2–6 × 1.5–5 cm, wrinkled, dull green above, paler and downy beneath, margins wavy-toothed; stalks rather short, stipules conspicuous, ear-shaped, persistent. Catkins developing before the leaves, almost stalkless, ovoid or shortly cylindric, 1–2 cm, scales dark-tipped, hairy. Stamens 2, anthers yellow. Ovary narrowly flask-shaped, hairy. *C & N Europe.* H1. Spring.

54. S. discolor Muhlenberg. Figure 8(7), p. 36. Illustration: Gleason, Illustrated Flora of the north-eastern United States and adjacent Canada **2**: 23 (1952); Systematic Botany Monographs **9**: 95 (1986).
Large shrub or small tree to 6 m high; twigs hairless or soon hairless, shining yellowish or purplish brown. Leaves

oblong, elliptic or obovate, 3–10 × 1–3 cm, at first downy, soon hairless, bright green above, whitish beneath, apex acute, margins toothed; stalk distinct, stipules ear-shaped, acute, toothed. Catkins developing before the leaves, stalkless, cylindric, 1.5–7 cm, scales blackish or dark reddish, hairy. Stamens 2. Ovary narrowly flask-shaped, downy, style distinct. *Eastern N America.* H1. Spring.

55. S. humilis Marshall (*S. tristis* Aiton). Figure 8(8), p. 36. Illustration: Gleason, Illustrated Flora of the north-eastern United States and adjacent Canada **2**: 23 (1952); Systematic Botany Monographs **9**: 103 (1984).
Variable shrub, 25 cm–3 m high; twigs slender, downy or becoming hairless and dull brown. Leaves oblanceolate or narrowly obovate, usually wider above the middle, 1.5–10 cm × 5–50 mm, stalkless or shortly stalked, downy or sometimes becoming hairless, dull green above, ashy-grey beneath, apex obtuse or shortly acute, margins toothed or sometimes almost entire; stipules small or absent. Catkins developing before the leaves, shortly ovoid or almost globular, 1–3 cm, scales brownish or blackish, hairy. Stamens 2, anthers reddish or purplish. Ovary stalked, flask-shaped, downy. Style very short. *Eastern & central N America.* H1. Spring.

56. S. starkeana Willdenow (*S. livida* Wahlenberg). Figure 9(1), p. 38. Illustration: Skalky Skalnicky, 127 (1977).
Slender shrub, usually less than 1 m high; twigs hairless or soon hairless, yellowish brown. Leaves broadly oblanceolate, obovate or almost circular, obtuse or acute, soon hairless, 2–6 cm × 8–40 mm, bright, rather glossy green above, dull green or glaucous beneath, apex often obliquely deflected, margins irregularly and remotely glandular-toothed; stipules usually present, broadly ear-shaped, toothed. Catkins developing before the leaves, stalked, shortly cylindric, 1.5–3 cm, scales greenish yellow, sparsely hairy. Stamens 2. Ovary long-stalked, hairy, style and stigmas short. *N Europe.* H1. Spring.

57. S. viminalis Linnaeus. Figure 9(2), p. 38. Illustration: Meikle, Willows and poplars, 87 (1984).
Tall shrub or small tree, to 10 m high, bark fissured; twigs pliable, at first ashy-hairy, becoming hairless and yellowish brown with age. Leaves linear, acuminate, 10–15 cm × 5–15 mm, dull green above, adpressed-silky beneath, margins recurved,

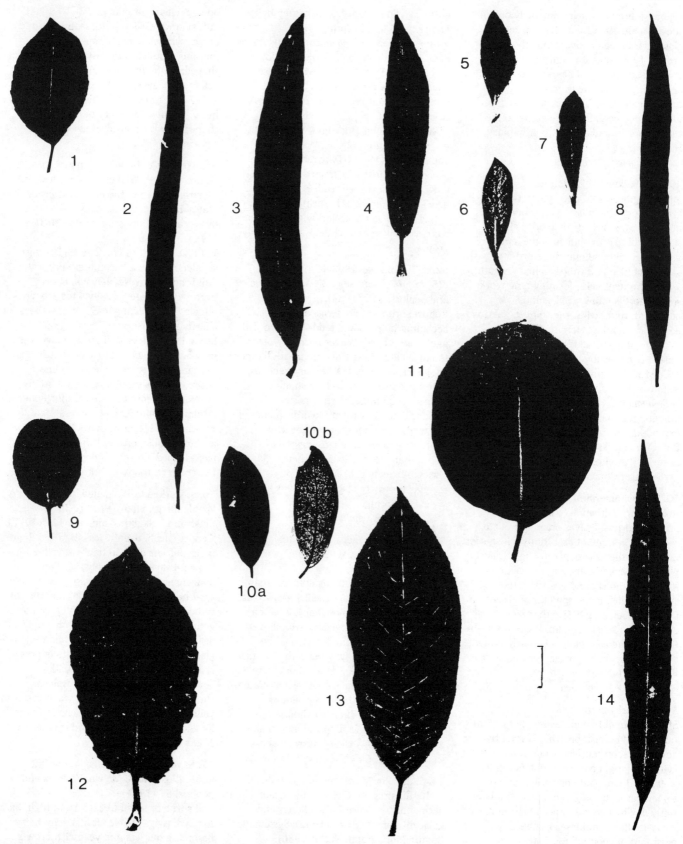

Figure 9. Leaves of *Salix* species. 1, *S. starkeana*. 2, *S. viminalis*.
3, *S.* × *rubra*. 4, *S.* × *forbyana*. 5, *S. udensis*. 6, *S. rehderiana*.
7, *S. gracilistyla*. 8, *S. elaeagnos*. 9, *S. lapponum*. 10, *S. helvetica* (a,
upper surface, b, lower surface). 11, *S. lanata*. 12, *S. hookeriana*.
13, *S. daphnoides*. 14, *S. acutifolia*. Scale = 1 cm.

entire or almost so; stipules narrow, long-acuminate. Catkins developing before the leaves, crowded towards the tips of the twigs, narrowly ovoid or cylindric, 1.5–3 cm, scales reddish or dark-tipped, hairy. Stamens 2, anthers yellow. Ovary narrowly flask-shaped, hairy, stigmas long, linear. *Europe, N Asia.* H1. Spring.

Formerly much cultivated for coarse basket-work.

58. S. × rubra Hudson. Figure 9(3), p. 38. Illustration: Meikle, Willows and poplars, 75 (1984).
Tall shrub or small tree, 3–6 m high; twigs tough, pliable, soon hairless, yellowish brown. Leaves lanceolate, acuminate, 4–15 cm × 8–15 mm, at first thinly felted, soon almost hairless above, sparsely hairy and greenish beneath, margins entire or almost so; stipules acuminate, often absent. Catkins developing before the leaves, stalkless or almost so, cylindric, 2–3.5 cm, scales blackish, hairy. Stamens 2, filaments free or often partly united, anthers often reddish. Ovary broadly flask-shaped, hairy. H1. Spring.

S. purpurea × S. viminalis. 'Eugenei' has slender, erect branches, glaucous leaves and pinkish male catkins with pale red anthers.

59. S. × forbyana Smith. Figure 9(4), p. 38. Illustration: Meikle, Willows and poplars, 81 (1984).
Similar to *S. × rubra* but exceptionally vigorous, with pale twigs and broader, glossier, almost hairless leaves. H1. Spring.

S. purpurea × S. viminalis, possibly also involving *S. cinerea* var. *oleifolia.*

60. S. udensis Trautvetter & Meyer (*S. sachalinensis* Schmidt). Figure 9(5), p. 38. Illustration: Makino, Illustrated Flora of Japan, t. 2019 (1948).
Large shrub or tree to 30 m high; twigs shining reddish or yellowish brown, soon hairless. Leaves lanceolate, acuminate, 10–15 × 1.5–2 cm, hairless or a little downy beneath, shining bright green above, glaucous beneath, margins obscurely toothed; stipules small or absent. Catkins developing before the leaves, narrowly cylindric, *c.* 4 cm, stalkless or almost so, scales dark-tipped, hairy. Stamens 2, anthers yellow or tinged with red. Ovary stalked, hairy; style and stigmas slender. *NE Asia.* H1. Spring.

Usually represented in gardens by 'Sekka' a vigorous male clone commonly producing flattened and contorted, fasciated growths.

61. S. rehderiana Schneider. Figure 9(6), p. 38.
Shrub or small tree to 9 m high, twigs at first hairy, soon hairless and reddish brown. Leaves lanceolate, shortly acuminate, 5–12 × 1–1.5 cm, green and hairless above, ashy and silky or almost hairless beneath, margins obscurely glandular-toothed; stipules very small, often absent. Catkins developing before the leaves, stalkless or almost so, shortly ovoid-cylindric, 2–3 cm, scales dark-tipped, hairy. Stamens 2, anthers at first purple. Ovary hairless or almost so, style distinct, stigmas short. *W China.* H1. Spring.

62. S. gracilistyla Miquel. Figure 9(7), p. 38. Illustration: Botanical Magazine, 9122 (1927); Bean, Trees and shrubs hardy in the British Isles, edn 8, **4**: 275 (1980).
Shrub to *c.* 3 m, twigs usually grey-felted. Leaves elliptic-oblong, ovate or obovate, 5–10 × 1–3 cm, at first silky hairy on both surfaces, becoming grey-green and hairless above, glaucous and persistently hairy beneath, apex acute, margins indistinctly and irregularly toothed, veins very conspicuous, acute. Catkins developing before the leaves, almost stalkless, cylindric, 3–5 cm, scales dark-tipped. Stamens 2, filaments wholly or partly united, anthers reddish. Ovary narrowly flask-shaped, hairy; style very long and slender. *China, Japan.* H1. Spring.

Commonly represented in gardens by var. **melanostachys** (Makino) Schneider (*S. melanostachys* Makino) with hairless or almost hairless twigs and leaves, and catkins with very dark, almost black scales. Var. **variegata** Kimura has white-bordered leaves.

63. S. elaeagnos Scopoli (*S. incana* Schrank; *S. lavandulifolia* misapplied). Figure 9(8), p. 38. Illustration: Meikle, Willows and poplars, 93 (1984).
Shrub or small tree to 6 m, twigs slender, at first greyish-felted, becoming hairless and yellow-brown or reddish with age. Leaves narrowly linear, 5–15 cm × 4–8 mm, dark shining green above, conspicuously white-felted beneath, margins narrowly recurved, entire or almost so; stalk very short, stipules usually absent. Catkins developing a little before the leaves, narrowly cylindric, 2–3.5 cm, almost stalkless, scales uniformly reddish or dark-tipped. Stamens 2, filaments united basally to beyond the middle, anthers yellow. Ovary narrowly flask-shaped,

hairless. *C & S Europe, western N Africa.* H1. Spring.

64. S. lapponum Linnaeus. Figure 9(9), p. 38. Illustration: Meikle, Willows and poplars, 149 (1984).
Low, much-branched shrub, 20–100 cm high, twigs at first thinly woolly, soon hairless and dark reddish brown. Leaves ovate, oblong or broadly lanceolate, 1.5–7 × 1–2.5 cm, dull greyish green and thinly or densely woolly, with entire margins; stipules small, often absent. Catkins developing with or a little before the leaves, almost stalkless, 2–4 cm, scales dark brown, hairy. Stamens 2, anthers yellow or reddish. Ovary densely white-woolly. *Europe & N Asia.* H1. Spring.

65. S. helvetica Villars. Figure 9(10), p. 38. Very similar to *S. lapponum* but with strongly contrasting leaf-surfaces, green and almost hairless above, white-hairy beneath. Catkins shortly stalked. *Alps of C Europe.* H1. Spring.

66. S. lanata Linnaeus. Figure 9(11), p. 38. Illustration: Raven & Walters, Mountain flowers, pl. XX (1956); Meikle, Willows and poplars, 151 (1984).
Low, gnarled bush, usually less than 1 m high, twigs at first thinly woolly, becoming hairless with age; catkin-buds conspicuously swollen. Leaves broadly ovate or almost circular, 3.5–7 × 3–6.5 cm, entire, at first thinly woolly or cobwebby, becoming thinly woolly or almost hairless, dark green above, distinctly glaucous beneath; stipules large, ovate, very persistent. Catkins developing before the leaves, erect, stalkless, shortly and broadly cylindric, 2.5–3.5 cm, scales dark brown, densely covered with long yellowish hairs. Stamens 2. Ovary hairless. *Subarctic Europe.* H1. Spring.

67. S. hookeriana Barratt. Figure 9(12), p. 38. Illustration: Hitchcock et al., Vascular plants of the Pacific Northwest **2**: 58 (1964).
Shrub to 2 m or occasionally a small tree; twigs stout, rigid, rather persistently greyish-felted. Leaves ovate or oblong-obovate, acute or obtuse, 5–16 × 2–4 cm, glossy green above, glaucous and often persistently whitish-felted beneath, margins obscurely toothed or almost entire; stipules small or absent. Catkins developing before or with the leaves, almost stalkless, shortly cylindric, 3–8 cm, scales dark brown, hairy. Stamens 2. Ovary stalked, hairless or sometimes hairy, style short. *Western N America.* H1. Spring.

S. piperi Bebb, also from western N America, is very similar but with the twigs soon becoming hairless, and with ovaries hairless or almost so.

68. S. daphnoides Villars. Figure 9(13), p. 38. Illustration: Meikle, Willows and poplars, 83 (1984).
Large shrub or small tree to 12 m; twigs at first covered with a glaucous bloom, becoming lustrous reddish brown with age. Leaves oblong or obovate, acute, 4–12 × 1–3 cm, margins glandular-toothed; stipules ovate-acuminate, glandular, often persistent. Catkins developing before the leaves, stalkless, shortly cylindric, 2–4 cm, scales densely silky-hairy. Stamens 2, anthers yellow. Ovary very narrowly flask-shaped, hairless. *N Europe*. H1. Spring.

One of the most popular of the cultivated willows, with ornamental, silky catkins. 'Aglaia' with broader leaves and large catkins is sometimes listed by nurserymen.

69. S. acutifolia Willdenow (*S. caspica* misapplied). Figure 9(14), p. 38. Illustration: Meikle, Willows and poplars, 85 (1984).
Large shrub or small tree; branches spreading or rather pendent, twigs hairless, bloomed. Leaves linear-lanceolate, acuminate, 8–15 × 1–2 cm, toothed, shining green above, greyish beneath; stalks rather long, slender, stipules lanceolate, toothed, persistent. Catkins developing before the leaves, stalkless, cylindric, 2–3 cm, scales with long silky hairs. Stamens 2, anthers yellow. Ovary narrowly flask-shaped, hairless. *USSR (to C Asia)*. H1. Spring.

70. S. repens Linnaeus. Figure 10(1), p. 41. Illustration: Meikle, Willows and poplars, 141 (1984).
Creeping, decumbent or rarely erect shrub, usually less than 1 m high; twigs slender, adpressed hairy or hairless and glossy red-brown or yellowish. Leaves lanceolate or oblong, 1–3.5 cm × 4–25 mm, almost hairless above or silky-hairy on both surfaces, blackening when dried, apex acute or obtuse, margins entire or minutely and irregularly toothed; stipules small or often absent. Catkins developing with or a little before the leaves, almost stalkless, ovoid or shortly cylindric, 1–2.5 cm, scales pale or reddish-stained. Stamens 2, anthers yellow. Ovary hairy or hairless. *Europe*. H1. Spring.

A variable species. Var. **argentea** (Smith) Wimmer & Grabowski (*S. arenaria* Linnaeus; *S. argentea* Smith). Figure 10(2),

p. 41. Illustration: Meikle, Willows and poplars, 143 (1984) is distinguished by its relatively large, silvery-silky leaves.

71. S. rosmarinifolia Linnaeus. Figure 10(3), p. 41.
Similar to and commonly confused with narrow-leaved variants of *S. repens*, but with linear-lanceolate, acute leaves and almost spherical catkins. *E Europe, N Asia*. H1. Spring.

72. S. subopposita Miquel (*S. repens* var. *subopposita* (Miquel) Seeman). Figure 10(4), p. 41. Illustration: Skalky Skalnicky, 131 (1977).
Small shrub less than 1 m high, branches erect or ascending, thinly hairy. Leaves rather crowded, often opposite or almost so, oblong-lanceolate, acute, 1–2.5 cm × 4–8 mm, at first densely silky-hairy, remaining thinly hairy beneath even when mature, margins entire; stipules ovate, distinctly stalked, often conspicuous and persistent. Catkins ovoid, 1–1.5 cm, developing before the leaves, scales yellowish or tinged with red. Ovary rather densely grey-hairy. *Japan, Korea*. H1. Spring.

73. S. irrorata Andersson. Figure 10(5), p. 41.
Shrub 3–4 m high, twigs hairless, purplish, coated with a persistent white bloom. Leaves narrowly oblong or lanceolate, acuminate, hairless, 6–10 cm × 5–10 mm, bright glossy green above, glaucous beneath, margins entire or almost so; stipules small or absent. Catkins developing before the leaves, almost stalkless, ovoid or shortly oblong, 1.5–2.5 cm, scales dark-tipped, hairy. Stamens 2, anthers reddish. Ovary hairless, style short. *Western N America*. H2. Spring.

S. lasiocarpa Bentham, from the same area, is a very close ally, but has less bloomy twigs, leaves at first downy, and rather longer catkins.

74. S. caesia Villars. Figure 10(6), p. 41. Illustration: Skalky Skalnicky, 39 (1977).
Low, straggling shrub, generally less than 1 m high, twigs dark brown, hairless; buds yellowish, hairless. Leaves oblong or obovate, rather rigid, 2–4 × 1–2 cm, hairless, dull bluish green above, glaucous beneath, margins entire; stalks very short, stipules minute or absent. Catkins appearing with the leaves, shortly stalked, 1–2 cm, scales pale, obtuse, thinly hairy. Stamens 2, filaments often wholly or partly united. Ovary almost stalkless, hairy, style

very short. *Europe to C Asia*. H1. Late spring.

75. S. purpurea Linnaeus. Figure 10(7), p. 41. Illustration: Meikle, Willows and poplars, 71, 73 (1984).
Much-branched bush, usually 2–3 m high, twigs slender, pliable, usually yellowish or greyish, hairless; bark yellow internally. Leaves often opposite or almost so, hairless, linear-oblong, oblanceolate or narrowly obovate, 2–8 cm × 5–30 mm, dark green or a little glaucous, entire or shortly toothed near apex; stipules usually absent. Catkins developing before the leaves, small, stalkless, narrowly cylindric, 1.5–3 cm, scales blackish, hairy. Stamens apparently 1 (strictly 2 with united filaments and anthers), anthers reddish or purplish. Ovary small, broadly flask-shaped, hairy. *Europe, Asia, N Africa*. H1. Spring.

Valued by basket-makers and represented by numerous cultivars. Subsp. **lambertiana** (Smith) Rechinger (*S. lambertiana* Smith) is a robust, broad-leaved variant (Figure 10(8), p. 41); var. **gracilis** Grenier & Godron (Figure 10(9), p. 41) is exceptionally slender, with narrow leaves. 'Pendula' has pendent branches, and is sometimes grafted to make a standard.

S. integra Thunberg is very similar but has almost stalkless, blunt leaves often clasping the stem basally. *Japan*.

76. S. miyabeana Seeman. Figure 10(10), p. 41.
Large shrub or small tree to 5 m, twigs hairless, shining brown, slender. Leaves narrowly lanceolate, 5–15 cm × 5–15 mm, acute, hairless, pale green above, glaucous beneath, margin bluntly toothed; stipules lanceolate, toothed, often persistent. Catkins developing before the leaves, stalkless, cylindric, 3–6 cm, scales dark brown, hairy. Stamens apparently 1, really 2 with united filaments; anthers yellow. Ovary ovoid or shortly flask-shaped, hairy. *NE Asia*. H1. Spring.

77. S. elbursensis Boissier. Figure 10(11), p. 41.
Shrub or small tree, branches spreading, flexible, twigs hairless, pale or dark brown. Leaves linear or narrowly oblanceolate, 3–10 cm × 5–15 mm, hairless or soon hairless, apex acute, margins distinctly toothed, at least in the upper half; stipules usually absent. Catkins developing a little before the leaves, cylindric, stalkless, 2–4 cm, scales blackish, hairy. Stamens 2, filaments united. Ovary ovoid, hairy. *Turkey & N Iran*.

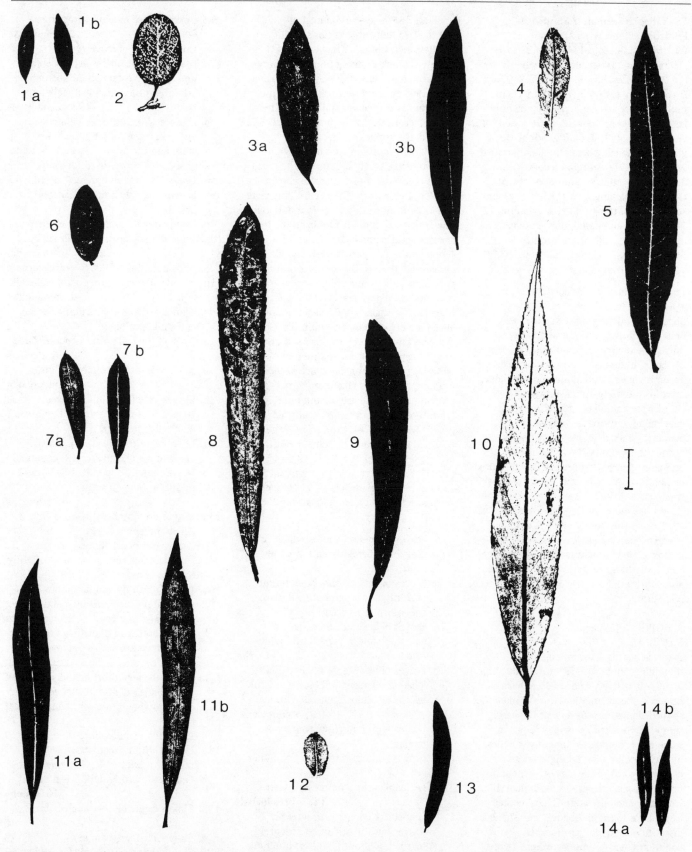

Figure 10. Leaves of *Salix* species. 1, *S. repens* var. *repens* (a, b, two different leaves). 2, *S. repens* var. *argentea*. 3, *S. rosmarinifolia* (a, b, two different leaves). 4, *S. subopposita*. 5, *S. irrorata*. 6, *S. caesia*. 7, *S. purpurea* subsp. *purpurea* (a, upper surface, b, lower surface). 8, *S. purpurea* subsp. *lambertiana*. 9, *S. purpurea* var. *gracilis*. 10, *S. miyabeana*. 11, *S. elbursensis* (a, upper surface, b, lower surface). 12, *S. bockii*. 13, *S. wilhelmsiana*. 14, *S. microstachya* (a, b, two different leaves). Scale = 1 cm.

78. S. bockii Seeman (*? S. variegata* Franchet). Figure 10(12), p. 41. Illustration: Botanical Magazine, 9079 (1925); Bean, Trees and shrubs hardy in the British Isles, edn 8, **4**: 263 (1980).
Dwarf shrub, generally less than 1.5 m high, twigs erect, slender, grey-downy. Leaves oblong or obovate, $5-20 \times 3-5$ mm, very shortly stalked, dark green above, glaucous and silky beneath, apex rounded or shortly acute, margins recurved, entire or almost so; stipules minute or absent. Catkins developing after the leaves in late summer or autumn, stalked, cylindric, $1-3.5$ cm, scales pallid, hairy. Stamens apparently 1, really 2 with united filaments. Ovary almost stalkless, downy. *China*. H3. Autumn.

79. S. wilhelmsiana Bieberstein. Figure 10(13), p. 41.
Shrub to 5 m, twigs very slender, silky at first, later brown and hairless. Leaves linear, often curved, acuminate, $2-5$ cm $\times$ $2-4$ mm, at first silky on both surfaces, becoming green and almost hairless later, margins minutely and distantly toothed; stipules very small or absent. Catkins developing with the leaves, slender, cylindric, stalked, $2-2.5$ cm, scales pale yellow, thinly hairy towards the base. Stamens apparently 1, really 2 with united filaments, anthers yellow. Ovary very small, silky. *SW & C Asia*. H2. Spring.

Perhaps more commonly represented in gardens by the closely allied **S. microstachya** Trautvetter (*S. wilhelmsiana* var. *microstachya* (Trautvetter) Herder), which has thinly hairy or almost hairless leaves and ovaries (Figure 10(14), p. 41). *NE Asia*. H1. Spring.

2. POPULUS Linnaeus
R.D. Meikle
Trees, generally with monopodial branching. Buds with several overlapping scales, often sticky and smelling of balsam. Leaves ovate, triangular-ovate or diamond-shaped, sometimes cordate at the base, rarely narrow; stalks usually long, sometimes compressed laterally; stipules inconspicuous, soon falling. Catkins developing before the leaves, without any obvious bracts, the males (and often the females) pendent; catkin-scales toothed or fringed. Perianth an oblique, cup-shaped disc. Stamens usually numerous, commonly with crimson anthers. Ovary broadly flask-shaped, style short or absent, stigmas 2–4. Fruit a capsule opening by 2–4 segments. Wind-pollinated.

About 35–40 species, almost all restricted to temperate regions of the northern hemisphere. They are valued chiefly as quick-growing timber, and are represented in plantations by a wide range of hybrids and cultivars, selected for their vigour and for their resistance to bacterial canker, a serious disease in *Populus*. While many of the species are decidedly ornamental, most are suitable only for large gardens, partly because of their size, but also because many produce copious suckers, and because their searching roots will block drains or damage foundations. The floss, shed in quantities from the fruits of some species and hybrids, can be a serious nuisance, particularly in built-up areas. Nonetheless, the genus is useful for providing screens and shelter-belts, and the buds and unfolding leaves of the Balsam Poplars (*P. trichocarpa*, *P. candicans* and allies) are delightfully aromatic. The species grow best in rich, well drained soils; they are unhappy and prone to disease in waterlogged situations, and are particularly intolerant of shade. Like many quick-growing trees, they are seldom long-lived. Literature: Gombocz, E., *Monographia generis Populi* (1908); Graf, J., Beiträge zur Kentniss der Gattung Populus, *Beihefte zum botanischen Centralblatt* **38**(1): 405–454 (1921); Houtzagers, G., *Het geschlacht Populus* (1937); Cansdale, G.S., *The black poplars and their hybrids cultivated in Britain* (1938).

1a. Leaves of leading shoots white- or grey-felted beneath; catkin-scales shortly toothed   2
 b. Leaves of leading shoots soon hairless or thinly downy on both surfaces; catkin-scales fringed   3
2a. Leaves of leading shoots deeply palmately lobed, persistently white-felted   **1. alba**
 b. Leaves of leading shoots irregularly lobed or coarsely and bluntly toothed, becoming almost hairless with age   **2. × canescens**
3a. Leaves with a narrow, translucent border   4
 b. Leaves without a translucent border   8
4a. Leaf-stalks not compressed laterally   **14. × berolinensis**
 b. Leaf-stalks compressed laterally   5
5a. Leaves bluntly and often obscurely toothed, without glands at junction with stalk   **22. nigra**
 b. Leaves conspicuously toothed, usually with glands at junction with stalk   6

6a. Undersurface of leaves greyish or whitish   **12. × generosa**
 b. Undersurface of leaves green   7
7a. Leaf-margins hairless or very shortly and sparsely ciliate; glands sometimes absent at junction of leaf-stalk and blade   **23. × canadensis**
 b. Leaf-margins distinctly ciliate; glands always present at junction of leaf-stalk and blade   **24. deltoides**
8a. Leaf-stalks compressed laterally   9
 b. Leaf-stalks not compressed laterally 13
9a. Leaves coarsely and bluntly toothed or lobed   10
 b. Leaves closely and shortly toothed   11
10a. Buds shining brown, rather sticky   **3. tremula**
 b. Buds dull, downy   **5. grandidentata**
11a. Glands absent at junction of leaf-stalk and blade   **4. tremuloides**
 b. Glands present at junction between leaf-stalk and blade   12
12a. Apex of the leaf abruptly acute; buds hairy   **6. sieboldii**
 b. Apex of leaf tapering; buds hairless   **7. adenopoda**
13a. Leaves green on both surfaces   14
 b. Leaves conspicuously pallid, whitish or greyish beneath   15
14a. Leaves hairless, $5-10$ cm; midrib, veins and stalk green   **18. acuminata**
 b. Leaves downy beneath, $15-30$ cm; midrib, veins and stalk red   **8. lasiocarpa**
15a. Leaves broadly heart-shaped with a deeply indented base   16
 b. Leaves not heart-shaped, with a rounded, tapering or, at most, slightly indented base   17
16a. Leaves abruptly acuminate (or cuspidate); stalks downy; catkin-scales and ovary hairless   **20. candicans**
 b. Leaves obtuse or shortly acute; stalks hairless; catkin-scales and ovary hairy   **9. wilsonii**
17a. Leaf-stalk very short, usually less than 2 cm; leaves diamond-shaped, elliptic or narrowly obovate or oblanceolate   18
 b. Leaf-stalk usually more than 2 cm 19
18a. Leaves bright green above with a reddish midrib   **15. yunnanensis**
 b. Leaves uniformly dark green above   **16. simonii**
19a. Shoots terete or very slightly angled   20
 b. Shoots distinctly angled   23
20a. Shoots downy; undersurface of leaves persistently downy along the veins   **19. maximowiczii**

b. Shoots and leaves hairless or soon
hairless   21
21a. Leaves very bluntly and obscurely
toothed; ovary hairy   **11. trichocarpa**
b. Leaves closely and distinctly toothed;
ovary hairless   22
22a. Leaves broadly ovate, upper surface
dark green   **21. balsamifera**
b. Leaves narrowly ovate or ovate-
lanceolate, bright shining green above
17. cathayana
23a. Leaves bluntly and obscurely toothed
11. trichocarpa
b. Leaves closely and distinctly
glandular-toothed   24
24a. Leaves 6–12 cm, stalks green
13. laurifolia
b. Leaves 11–30 cm, stalks reddish
10. szechuanica

Section **Populus**. Bark smooth, often with
conspicuous lenticels; buds frequently
hairy; leaves hairy or hairless, without
translucent margins, often coarsely toothed
or lobed, stalks compressed laterally or
terete; catkin-scales fringed with long hairs.

1. P. alba Linnaeus. Illustration: Meikle,
Willows and poplars, 167 (1984).
Suckering tree, usually 15–20 m, bark
blackish and fissured near base of trunk,
greyish and smooth above with
conspicuous horizontal bands of blackish
lenticels; twigs at first white-felted,
becoming hairless and shining brown.
Leaves of leading shoots conspicuously
palmatifid, 3–9 × 3–10 cm, at first white-
felted on both surfaces, becoming hairless
and dark green above, margins irregularly
toothed; stalk almost terete, felted, 5–6 cm.
Catkins 3–7 cm, scales pale brown, ovate
or obovate, minutely toothed, ciliate.
Anthers purple. Ovary hairless. *C & SE
Europe to C Asia*. H1. Spring.

Var. **pyramidalis** Bunge (*P. bolleana*
Carrière). Narrow, columnar tree; leaves of
short, lateral shoots roundish, shortly
lobed, thinly felted beneath. *C Asia*. H1.
Spring.

2. P. × canescens (Aiton) Smith.
Illustration: Meikle, Willows and poplars,
169 (1984); Polunin & Walters, Guide to
the vegetation of Britain and Europe, pl.
138 (1985).
Tree, usually 20–30 m, occasionally to
50 m, trunk with smooth greyish bark
banded with blackish lenticels as in
P. alba.
Leaves of 2 kinds: those of short lateral
spurs almost circular, usually 3.5–6 cm in
diameter, hairless or almost so with bluntly

toothed or sinuous margins; those of
leading shoots broadly ovate, often shortly
acute, 6–8 cm long and almost as wide,
green and hairless above, persistently grey-
felted beneath, coarsely and irregularly
toothed, ciliate. Anthers reddish. Ovary
hairless. H1. Spring.
P. alba × P. tremula.

3. P. tremula Linnaeus. Illustration: Meikle,
Willows and poplars, 171 (1984).
Suckering tree to 20 m, or a large shrub;
bark smooth, grey, buds glossy brown.
Leaves of normal growths almost circular,
1.5–8 cm in diameter, hairless or at first
thinly silky, dark green above, paler
beneath, margins irregularly and bluntly
toothed or lobed; stalks 4–7 cm, slender,
strongly laterally compressed. Catkins
5–8 cm, scales fan-shaped, dark brown,
hairy, deeply fringed. Anthers crimson.
Ovary hairless. *Temperate Europe and Asia to
China and Japan, reaching the tree-limit in the
Artic*. H1. Spring.

4. P. tremuloides Michaux.
Similar to *P. tremula* but with smaller, more
pointed, finely glandular-toothed leaves and
more slender catkins. *Western N America*.
H1. Spring.
Pendent cultivars of this species are
occasionally grown (e.g. 'Pendula').

5. P. grandidentata Michaux.
Tree 15–20 m with smooth grey bark.
Shoots at first thinly felted, buds downy.
Leaves broadly ovate, shortly acute,
4–12 × 4–10 cm, at first greyish-felted,
soon hairless, dark green, rather leathery,
margins coarsely and unequally toothed;
stalks laterally compressed towards
apex. Catkins 4–7 cm, scales with 5–15
short teeth. Anthers crimson. Ovary
thinly downy. *Eastern N America*. H1.
Spring.

6. P. sieboldii Miquel. Suckering tree to
20 m with spreading branches; shoots at
first whitish downy, buds thinly felted.
Leaves rather leathery, broadly ovate,
4–8 cm long and almost as wide, dark
green above, whitish beneath, downy at
first, becoming hairless, base truncate or
broadly wedge-shaped, apex shortly and
abruptly acute, margins minutely
glandular-toothed; stalks 1–4 cm, laterally
compressed, downy, glandular at junction
with blade. Catkins 6–10 cm, scales
brownish, fringed, hairy. Anthers crimson.
Ovary hairless. *Japan, USSR (Sakhalin)*. H1.
Spring.

7. P. adenopoda Maximowicz.
Similar to *P. sieboldii* but with hairless,
sticky buds and long-acuminate leaves.
China. H1. Spring.

Section **Leucoides** Spach. Bark rough,
flaking; leaves commonly hairy, at least on
the undersurface, often cordate at base,
without translucent margins; leaf-stalks not
compressed laterally except at apex;
perianth-disc lobed; ovary often downy.

8. P. lasiocarpa Oliver. Illustration:
Botanical Magazine, 8625 (1915).
Round-headed tree to 20 m, shoots angled,
felted when young; buds large, slightly
sticky, thinly downy. Leaves ovate,
acuminate, 15–30 × 10–20 cm, at first
downy, soon hairless and bright green
above, paler and persistently downy
beneath, with a red midrib; base cordate,
margins bluntly glandular-toothed; stalk
terete, 7–15 cm, red. Catkins 10–15 cm,
scales hairless, shortly fringed. Anthers
crimson. Ovary downy. *C China*. H1.
Spring.

9. P. wilsonii Schneider.
Tree to 25 m with a pyramidal crown;
shoots terete, hairless or soon hairless,
often purplish, buds shining brown, sticky.
Leaves broadly ovate, 8–20 × 7–15 cm,
obtuse, at first reddish and thinly downy,
becoming hairless, dull bluish green above,
pallid or greyish beneath, base cordate or
rounded, margins minutely and bluntly
toothed; stalk 6–15 cm. Catkins slender,
c. 7 cm, scales fringed, hairy. Ovary woolly.
C & W China. H1. Spring.

Section **Tacamahaca** Spach. Bark rough or
furrowed; buds large, sticky, balsam-
scented; leaves often with a conspicuously
pallid undersurface, without translucent
margins; leaf-stalk not compressed laterally;
catkin-scales fringed.

10. P. szechuanica Schneider.
Robust tree to 40 m, shoots angled,
becoming terete with age, hairless or at first
hairy; buds purplish, slender, rather sticky.
Leaves ovate-oblong or broadly lanceolate,
11–30 × 5–20 cm, acute or acuminate, at
first reddish, becoming dark green above,
whitish beneath, hairless or with the midrib
and veins hairy on the undersurface, base
cordate, rounded or broadly wedge-shaped,
margins bluntly glandular-toothed; stalks
reddish, 2–6 cm. Catkins 4–10 cm, scales
fringed, hairless. Ovary hairless. *W China*.
H2. Spring.
Hairy variants of the species have been
identified as var. **tibetica** Schneider, but the

validity of the distinction is questionable. *P. violascens* Dode, considered by Rehder to be related to *P. lasiocarpa*, appears to be indistinguishable from *P. lasiocarpa* var. *tibetica*.

11. P. trichocarpa Hooker. Illustration: Meikle, Willows and poplars, 183 (1984). Tree to 35 m or more with a rather narrow crown; bark fissured, grey-brown; shoots angled, at first thinly downy, twigs yellowish grey; buds shining brown, sticky, strongly balsam-scented. Leaves ovate, sometimes slightly heart-shaped, 5–20 × 4–15 cm, dark shining green above, conspicuously pallid beneath, bluntly toothed; stalk usually less than 4 cm, not compressed laterally. Catkins rather short and thick, usually less than 9 cm, scales fan-shaped, pale brown, deeply fringed, thinly hairy. Anthers crimson. Ovary densely grey-felted or sometimes almost hairless. *Western N America*. H1. Spring.

The most commonly cultivated of the Balsam Poplars, generally represented in gardens by the male.

12. P. × generosa Henry.
Tree to 40 m, bark fissured, shoots a little angled, hairless, buds large, glossy, sticky, balsam-scented. Leaves ovate-triangular, 10–13 × 8–10 cm on normal growths, light green above, paler or slightly greyish beneath, base truncate or a little cordate, margins narrowly translucent, distinctly glandular-toothed; stalk slightly compressed laterally, glandular at junction with blade. Catkins 10–13 cm, scales fringed. Anthers crimson. Ovary hairless. H1. Spring.
P. deltoides 'Cordata' × *P. trichocarpa*.

13. P. laurifolia Ledebour.
Tree to 20 m, with spreading branches which are pendulous at the tips; bark fissured, grey-brown; twigs yellowish grey, shoots angular, downy at least towards apex; buds sticky, balsam-scented. Leaves narrowly ovate, 6–12 × 2–6 cm, acuminate, dark green and hairless above, greyish, downy and prominently veined beneath, base rounded, margins minutely and regularly glandular-toothed; stalk 2–4 cm, not compressed laterally. Catkins 4–5 cm, scales large, fringed, downy. Anthers crimson. Ovary thinly downy. *C Asia*. H1. Spring.

Forma **lindleyana** (Carrière) Ascherson & Graebner, with small, lanceolate willow-like leaves, is occasionally cultivated.

14. P. × berolinensis Dippel.
Tree to 30 m, columnar with ascending or erect, yellowish grey branches; shoots somewhat angular, thinly downy; buds sticky, greenish. Leaves ovate or diamond-shaped, acuminate, 5–10 × 2–5 cm, bright green above, paler beneath, soon hairless, base rounded or wedge-shaped, margins narrowly translucent, minutely and closely toothed; stalk 2–4 cm, not compressed laterally, thinly downy. Catkins 4–7 cm, scales fringed, hairless. Anthers crimson. Ovary hairless. H1. Spring.

P. laurifolia × *P. nigra* var. *italica*, represented in cultivation by 'Frye', 'Rumford' and other named cultivars.

15. P. yunnanensis Dode.
Tree to 20 m or more, shoots strongly angled, hairless; buds hairless, sticky. Leaves ovate- or obovate-lanceolate, 6–15 × 4–8 cm, bright green above, whitish beneath, hairless, apex sharply acuminate, base somewhat cordate or wedge-shaped, margins bluntly glandular-toothed, midrib often reddish; stalk short, to 2 cm, often reddish. Catkins 4–8 cm, scales fringed, hairless. Anthers crimson. Ovary hairless. *SW China*. H2. Spring.

16. P. simonii Carrière.
Tree to 15 m, with slender, erect (var. **fastigiata** Schneider) or pendulous (var. **pendula** Schneider) branches; shoots angled, hairless, buds sticky, balsam-scented. Leaves diamond-shaped-ovate or diamond-shaped-elliptic, tapering to apex and base, 6–12 × 3–8 cm, dark green above, pallid beneath, margins minutely and bluntly toothed; stalk slender, 1–2 cm, reddish. Catkins 2–3 cm, scales fringed. Stamens 8, anthers reddish. Ovary hairless. *N & W China*. H1. Spring.

17. P. cathayana Rehder.
Tree to 30 cm, branches ascending, shoots terete, sparsely downy or almost hairless; buds sticky, balsam-scented. Leaves rather leathery, ovate or ovate-lanceolate, 6–10 × 3–6 cm, bright shining green above, whitish beneath, apex acuminate, base rounded or broadly wedge-shaped, margins ciliate, finely and bluntly toothed; stalk terete, hairless. Catkins 4–8 cm, scales fringed, hairless. Anthers crimson. Ovary hairless. *China*. H1. Spring.

Much confused with *P. suaveolens* Fischer (whose occurrence in cultivation is questionable) and *P. maximowiczii*.

18. P. acuminata Rydberg.
Tree to 15 m with a rounded crown, twigs rounded or occasionally somewhat angled, hairless or soon hairless, yellowish brown when mature; buds chestnut-brown, tapering, sticky. Leaves 5–10 × 2.5–6 cm, narrowly ovate or diamond-shaped, shining dark green above, paler beneath, base rounded or wedge-shaped, margins bluntly toothed; stalks rounded, 2–7 cm. Catkins slender, 2–7 cm, scales fringed, hairless. Ovary hairless, style very short. *Western N America*. H4.

Considered by some authorities to be of hybrid origin, resulting from a cross between *P. angustifolia* Torrey and *P. sargentii* Dode or *P. fremontii* Watson, three species which are not generally cultivated.

19. P. maximowiczii Henry.
Robust tree to 30 m or more, branches spreading; bark of young trees smooth and yellowish, becoming grey and fissured on mature trees; shoots terete, downy, reddish on the exposed side; buds sticky, balsam-scented. Leaves ovate-elliptic, 7–13 × 6–11 cm, rather leathery, thinly downy, bright green above, whitish beneath, base rounded or slightly cordate, apex abruptly narrowed, often obliquely deflexed, margins bluntly glandular-toothed, ciliate; stalks 1–4 cm, thinly downy. Catkins 5–15 cm, scales fringed, hairless. Anthers red. Ovary hairless. *E Asia*. H1. Summer.

The female parent of numerous, economically important, fast-growing hybrids, including 'Androscoggin' (*P. maximowiczii* × *P. trichocarpa*). 'Oxford' (*P. maximowiczii* × *P. × berolinensis*) and 'Rochester' (*P. maximowiczii* × *P. nigra* var. *plantierensis*).

20. P. candicans Aiton (*P. × gileadensis* Rouleau). Illustration: Meikle, Willows and poplars, 183 (1984).
Suckering tree, generally less than 25 m high with a broad crown, bark fissured on mature specimens; twigs at first thinly downy, becoming hairless and glossy brown with age; buds narrowly ovoid-acuminate, viscid, balsam-scented. Leaves broadly heart-shaped, 5–15 × 4.5–12 cm, dark shining green above, conspicuously white beneath with prominent veins, margins bluntly toothed, densely ciliate; stalks 3–7 cm, slightly flattened on the upper surface, downy, with 2 glands near apex. Catkins (female only known) at first 4–6 cm, lengthening to 16 cm, scales fan-shaped, pallid, hairless, fringed. Ovary hairless. *N America*. H1. Spring.

Variously regarded as a cultivated variant of *P. balsamifera* or as a hybrid between this and *P. deltoides*. Susceptible to bacterial canker and now rather rare. 'Aurora', with white-blotched leaves, is more common in gardens than typical *P. candicans*.

21. P. balsamifera Linnaeus (*P. tacamahacca* Miller). Illustration: Gleason, Illustrated Flora of the north-eastern United States and adjacent Canada 2: 4 (1952).
Suckering tree to 30 m, bark fissured at base of trunk, smooth above; shoots terete, buds large, sticky, balsam-scented. Leaves broadly ovate, acuminate, 5–12 × 4–8 cm on normal growths, hairless, dark green above, pallid with a metallic lustre beneath, bluntly toothed; stalk terete, 2–5 cm. Catkins 6–11 cm, scales fringed with numerous slender lobes. Anthers crimson. Ovary hairless. *N America*. H1. Spring.

Rare in cultivation and often confused with *P. candicans*. Var. **michauxii** Henry (*P. michauxii* Dode, in part; *P. balsamifera* var. *subcordata* Hylander) has leaves which are somewhat cordate and often asymmetric at base, with prominent veins beneath and the stalk downy. *Eastern N America*.

Section **Aegeiros** Duby. Bark furrowed; buds sticky; leaves often triangular-ovate or broadly diamond-shaped, green on both surfaces, with a narrow translucent cartilaginous margin; leaf-stalks long, slender, laterally compressed; catkin-scales fringed.

22. P. nigra Linnaeus. Illustration: Meikle, Willows and poplars, 175 (1984).
Robust tree 30–35 m, with a broad crown and massive, downcurved branches; trunk often with large swellings or burrs, bark coarsely fissured; twigs ochre-brown, buds dark glossy brown. Leaves ovate-triangular, 5–10 × 3–9 cm, acuminate, dark green above, paler beneath, hairless or almost so, margins bluntly toothed; stalk laterally compressed, 3–7 cm. Catkins 3–5 cm, scales greenish or brownish, fringed. Anthers crimson. Ovary almost spherical, hairless. *Europe & N Africa to C Asia*. H1. Spring.

Var. **afghanica** Aitchison & Hemsley (*P. thevestina* Dode; *P. uzbekistanica* Komarov 'Afghanica'). Female; habit narrowly columnar with erect branches; bark of trunk and main branches white; leaves ovate-diamond-shaped. *C Asia*. H1. Spring.

Var. **betulifolia** (Pursh) Torrey. Young leaves, shoots and axes of catkins thinly downy. *Europe*. H1. Spring.

Var. **italica** Muenchhausen. Male; habit narrowly columnar with erect branches; trunk buttressed, bark dark grey-brown; leaves ovate-triangular. *Europe*. H1. Spring.

'Gigantea' is similar to var. *italica* (the Lombardy Poplar) but is female and less narrowly columnar.

Var. **plantierensis** (Dode) Schneider (*P. plantierensis* Dode). Similar in habit to var. *italica* though less narrow in outline; shoots and young leaves thinly downy, stalks often reddish. Male or female. *Europe*. H1. Spring.

23. P. × canadensis Moench.
A robust tree, often to 30 m, with coarsely fissured grey-brown or whitish bark; shoots hairless or minutely downy, terete or sometimes obscurely angled, twigs olive-grey or yellowish; buds narrowly ovoid, acuminate, glossy, sticky. Mature leaves broadly ovate or triangular, acuminate, 4–12 × 4–10 cm, hairless, bright green, base broadly wedge-shaped, truncate or shallowly cordate, margins distinctly toothed; stalk 4–10 cm, laterally compressed, often glandular at junction with blade. Catkins 3–6 cm, scales fan-shaped, pale brown or tinged with purple, fringed, hairless. Anthers crimson; ovary hairless.

P. deltoides × *P. nigra*; a hybrid of considerable commercial importance, comprising a wide and ever-increasing range of cultivars, including:
'Serotina' (*P. × serotina* Hartig). Illustration: Meikle, Willows and poplars, 177 (1984). Male; crown fan-shaped with spreading branches; shoots hairless; young leaves reddish brown. H1. Spring.
'Gelrica'. Similar to 'Serotina' but with whitish bark and the leaves expanding about a fortnight earlier. H1.
'Regenerata' (*P. × regenerata* Henry). Female; branches arching outwards, tips pendent; shoots hairless, young leaves yellowish green, minutely ciliate. H1. Spring.
'Eugenei' (*P. × eugenei* Mathieu). Male; columnar with a very narrow crown; shoots hairless; young leaves coppery brown. H1. Spring.
'Robusta' (*P. × robusta* Schneider). Male; crown narrow, branches erect or ascending; shoots downy; young leaves dark reddish brown. H1. Spring.
'Marilandica' (*P. × marilandica* Poiret). Illustration: Meikle, Willows and poplars, 179 (1984). Female; crown rounded, branches spreading; shoots hairless; young leaves yellowish green, at first thinly hairy. H1. Spring.

24. P. deltoides Marshall (*P. monilifera* Aiton). Illustration: Gleason, Illustrated Flora of the north-eastern United States and adjacent Canada 2: 5 (1952).
Tree to 30 m or more with an open crown and spreading branches; bark grey, fissured, shoots angled or terete, hairless; buds shining brown, sticky, balsam-scented. Leaves broadly ovate-triangular, 7–12 cm long and almost as wide, bright green above, paler beneath, soon hairless, margins coarsely toothed, ciliate; stalks 5–10 cm, compressed laterally, glandular at point of junction with blade. Catkins 7–10 cm, the males dense, scales fringed with numerous slender lobes. Anthers crimson. Ovary hairless. *Eastern N America*. H1. Spring.

Var. **missouriensis** (Henry) Henry (*P. angulata* Aiton; *P. angulata* var. *missouriensis* Henry). Shoots strongly angled; leaves distinctly longer than wide. *South-eastern USA*. H2. Spring.

'Carolin' with reddish leaf-stalks, and 'Cordata' with yellow-green leaf-stalks are occasionally planted, especially in southern Europe.

XXXVIII. BETULACEAE

Monoecious, deciduous trees or shrubs. Leaves alternate, simple, toothed or shallowly lobed, with deciduous stipules. Flowers unisexual; male flowers in erect or pendent catkins, 3 to each bract, with perianth very small or absent and 2–15 stamens. Female flowers in usually erect catkins, clusters or short spikes, perianth present or absent, ovary inferior, 2-celled below, with 2 styles. Fruits (nuts, or if very small, sometimes called 'nutlets') in catkins, clusters or spikes, sometimes forming woody 'cones'; involucre made up of united bracts and bracteoles present or absent; fruits compressed and winged or not.

A small family of 6 genera with about 100 species mainly in north temperate zones but south to N India, SE Asia and the Andes.

1a. Nut small, compressed, winged, without an involucre; male flowers with 2–4 stamens, with a perianth 2
  b. Nut often large, not compressed or winged, with a leaf-like involucre made up of bracts and bracteoles; male flowers with 4–15 stamens, without a perianth 3
2a. Fruiting catkins ovoid, cone-like; fruiting catkin-scales 5-lobed, woody, persistent **1. Alnus**
  b. Fruiting catkins cylindric to narrowly ovoid, breaking up at maturity; fruiting catkin-scales 3-lobed, falling with the fruit **2. Betula**

3a. Female flowers in pendent catkins; anthers not hairy 4
b. Female flowers in erect clusters or short spikes; anthers hairy at apex 5
4a. Involucre of fruit flat, 3-lobed **3. Carpinus**
b. Involucre of fruit bladder-like **4. Ostrya**
5a. Flowers appearing before the leaves **6. Corylus**
b. Flowers appearing with the leaves **5. Ostryopsis**

1. ALNUS Miller
K.B. Ashburner & S.M. Walters
Trees or shrubs. Male flowers with 4 stamens, anther-lobes separated by a forked connective, and a 4-or 5-partite perianth. Female flowers 2 to each bract. Fruiting catkin an ovoid 'cone' with 5-lobed, persistent, woody scales from which the ripe, winged nuts are released (Figure 11(8, 9 & 18) p. 47).

A genus of about 35 species, mostly in N temperate regions, where they are often characteristic of wetlands. The persistent 'cones' on the leafless trees in winter make recognition of the genus very easy. Several species are commonly planted.
Literature: Clarke, D.L. (ed), Alnus in Bean, W.J., *Trees and shrubs hardy in the British Isles*, edn 8, 1: 271–284 (1970).

Shrub to 5 m. **1–6,10.**
Leaves. Obovate: **7**; cordate at base: **14,15**; lobed: **8–11**. Retained into December (at least in NW Europe): **4,5,12,14,15.**
Male catkins. Opening in early autumn: **13.**

1a. Buds almost stalkless, with several overlapping scales; shrubs or small trees not more than 10 m (Subgenus **Alnobetula**) 2
b. Buds stalked, with 2 opposite, more or less equal scales; mostly trees to at least 15 m (Subgenus **Alnus**) 7
2a. Leaves narrowly ovate to lanceolate, with 12–25 pairs of lateral veins 3
b. Leaves ovate to circular, with fewer than 12 pairs of lateral veins 5
3a. Leaves often lanceolate, with 15–25 pairs of lateral veins; 'cones' pendulous, in groups of 3–6 **6. pendula**
b. Leaves narrowly ovate, with 12–18 pairs of lateral veins; 'cones' solitary or paired, erect 4
4a. Shoots hairy; 'cones' paired **4. firma**
b. Shoots hairless; 'cones' solitary **5. sieboldiana**
5a. Leaves very shiny above; male catkins to 5 cm **2. sinuata**

b. Leaves dull or somewhat shiny above; male catkins to 2.5 cm 6
6a. Shrub to 2 m; leaf-teeth coarse **1. viridis**
b. Small tree or shrub to 10 m; leaf-teeth fine and sharp **3. maximowiczii**
7a. Leaves cordate at base 8
b. Leaves rounded or tapered at base 9
8a. Shoots and buds hairy; leaves dull **14. subcordata**
b. Shoots and buds usually hairless; leaves shiny **15. cordata**
9a. 'Cones' to 3 × 1.5 cm; buds hairy **14. subcordata**
b. 'Cones' not more than 2.5 × 1.2 cm; buds usually hairless 10
10a. 'Cones' with projecting scales; shrub or small tree to 10 m **10. rugosa**
b. 'Cones' smooth, without projecting scales; tree to 30 m 11
11a. Leaves ovate-elliptic or elliptic 12
b. Leaves ovate, obovate or almost circular 14
12a. Leaves thin in texture; male catkins to 8 cm, opening early in autumn **13. nitida**
b. Leaves thick in texture; male catkins to 4 cm, opening in winter or early spring 13
13a. Leaves shallowly lobed or double-toothed **11. rubra**
b. Leaves unlobed, single-toothed **12. japonica**
14a. Shoots hairless; leaves rounded at apex **7. glutinosa**
b. Shoots hairy; leaves more or less acute 15
15a. Leaves ovate; 'cones' to 1.2 cm × 9 mm **8. incana**
b. Leaves broadly ovate to almost circular; 'cones' to 2 × 1.2 cm **9. hirsuta**

Subgenus **Alnobetula** (Ehrhart) Persoon. Buds almost stalkless, with several overlapping scales.

1. A. viridis (Chaix) de Candolle. Figure 11(5), p. 47. Illustration: Mitchell, Field guide to the trees of Britain and northern Europe, 193, pl. 18 (1974); Polunin & Everard, Trees and bushes of Britain and Europe, 46 (1976).
Shrub to 2 m with erect stems and yellowish brown, shining bark. Shoots more or less hairless, reddish brown, with numerous translucent, inconspicuous white glands. Buds usually sticky. Leaves to 9 × 6.5 cm, ovate or circular-ovate, dull, dark green, hairless at least above, coarsely or doubly toothed, acute or rounded at apex and wedge-shaped to rounded at base.

Pairs of lateral veins 7–8, strongly impressed. Male catkins to 2.5 cm × 7 mm, opening with the leaves; scales often coated with white resin. 'Cones' to 1.2 cm × 9 mm, ovoid. Nut *c.* 2 × 1 mm, ovate-elliptic, with broad wing and persistent styles. *Mountains of C & E Europe; also Corsica.* H1. Spring.

Rarely grown in gardens. The N American subsp. **crispa** (Aiton) Turrill, a taller shrub with somewhat larger leaves, 'cones'and nuts, is also rarely cultivated (Figure 11(6), p. 47).

2. A. sinuata (Regel) Rydberg (*A. sitchensis* Sargent). Figure 11(7), p. 47. Illustration: Journal of the Royal Horticultural Society **94**: f. 33 (1969).
Like *A. viridis*, but often a large shrub or small tree to 10 m, with shining dark green leaves and very long male catkins to 5 cm. *Western N America, from Alaska to the Californian mountains.* H1. Spring.

Slow-growing in cultivation in Europe; the long catkins are very attractive in spring.

3. A. maximowiczii Callier. Figure 11(8), p. 47. Illustration: Takeda, Alpine flora of Japan, pl. 60 (1963) – as A. crispa subsp. maximowiczii; Bean, Trees and shrubs hardy in the British Isles edn 8, 1: pl. 11 (1976); Plantsman 8: pl. 3 (1986).
Like *A. viridis*, but often a small tree to 10 m with somewhat shiny leaves with fine, sharp teeth and long, sharp-pointed buds. *E Asia, from Kamtchatka to Korea and Japan.* H1. Spring.

Like the preceding species, rarely cultivated in Europe.

4. A. firma Siebold & Zuccarini. Figure 11(9), p. 47. Illustration: Kitamura & Okamoto, Coloured illustrations of trees and shrubs of Japan, pl. 20 (1958); Mitchell, The gardener's book of trees, 70 (1981).
Shrub or small tree to 10 m with grey bark and brown, hairy shoots with conspicuous lenticels. Leaves to 9 × 3.5 cm, narrowly ovate, somewhat shining, variably hairy, with simple, mucronate teeth pointing forwards. Pairs of lateral veins 12–18, impressed. Male catkins to 9 cm × 5 mm. 'Cones' to 2 × 1.5 cm, ovoid, erect, usually paired. Nut *c.* 2.5 × 1 mm, with broad, tapering wing. *Mountains of Japan.* H2. Spring.

5. A. sieboldiana Matsumura. Figure 11(2, 10), p. 47. Illustration: Kitamura & Okamoto, Coloured illustrations of trees and shrubs of Japan, pl. 20 (1958);

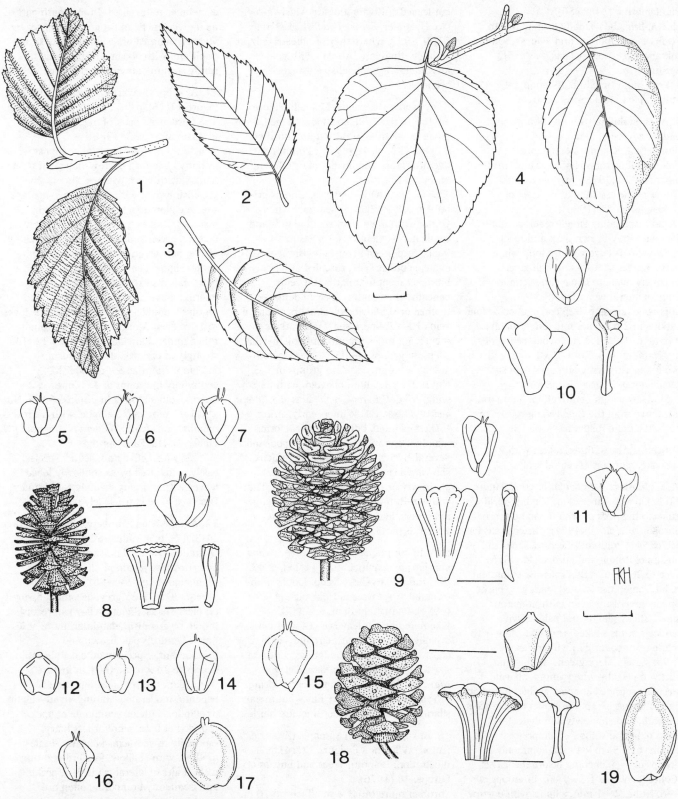

Figure 11. Diagnostic details of *Alnus* species. 1, Leaves of *A. incana*. 2, Leaf of *A. sieboldiana* (upper surface). 3, Leaf of *A. japonica* (lower surface). 4, Leaves of *A. cordata*. 5, Nut of *A. viridis*. 6, Nut of *A. viridis* subsp. *crispa*. 7, Nut of *A. sinuata*. 8, 'Cone', nut and cone-scales of *A. maximowiczii*. 9, 'Cone, nut and cone-scales of *A. firma*. 10, Nut and cone-scales of *A. sieboldiana*. 11, Nut of *A. pendula*. 12, Nut of *A. glutinosa*. 13, Nut of *A. incana*. 14, Nut of *A. rugosa*. 15, Nut of *A. hirsuta*. 16, Nut of *A. rubra*. 17, Nut of *A. japonica*. 18, 'Cone', nut and cone-scales of *A. subcordata*. 19, Nut of *A. cordata*. Scale for leaves = 1 cm, for nuts, etc., 5 mm.

Plantsman **8**: t. 3 (1986).
Like *A. firma* but usually a small tree to 10 m, shoots hairless and 'cones' usually solitary. *Japan, lowland and coastal.* H3. Spring.

Very handsome in cultivation but vulnerable to late spring frosts.

6. A. pendula Matsumara (*A. firma* var. *multinervis* Regel). Figure 11(11), p. 47. Illustration: Kitamura & Okamoto, Coloured illustrations of trees and shrubs of Japan, pl. 20 (1958); Miyabe & Kudo, Icones of the essential forest trees of Hokkaido, pl. 31 (1984).
Elegant, spreading small tree or shrub to 5 m with grey-brown bark and reddish brown shoots covered with long white hairs. Leaves to 9 × 4 cm, dull green, narrowly ovate to lanceolate, acuminate. Pairs of lateral veins 15–25, strongly impressed, causing deep corrugations. Male catkins to 2.4 cm × 4 mm, often solitary. 'Cones' to 1.5 × 1.2 cm, ovoid, pendulous, in groups of 3–6. Nut *c.* 3 × 1 mm, with broad tapering wing and persistent styles. *Mountains of N Japan.* H2. Spring.

Slow-growing in cultivation, but more attractive than the 2 preceding species in both habit and foliage.

Subgenus **Alnus**. Buds stalked with 2 opposite, more or less equal scales.

7. A. glutinosa (Linnaeus) Gaertner. Figure 11(12), p. 47. Illustration: Mitchell, Field guide to the trees of Britain and northern Europe, 193, pl. 18 (1974); Gartenpraxis for 1976: 541; Polunin & Everard, Trees and bushes of Britain and Europe, 44 (1976).
Tree to 20 m, of rather narrow, pyramidal habit, sometimes twisted, with grey bark. Shoots greenish brown with numerous small, sticky glands and pale brown lenticels. Buds stalked, resinous, often with white resin-coating in winter. Leaves to *c.* 9 × 8.5 cm, dark green, shining and sticky, especially when young, obovate-circular, rounded at apex, wedge-shaped at base, hairless except for brown tufts in axils beneath, and shallowly doubly toothed. Pairs of lateral veins 7–8, impressed. Male catkins to 2.5 cm × 4 mm, commonly in groups of 3–4, opening before the leaves. 'Cones' ovoid, to 1.5 × 1 cm, in groups of 3–6. Nut *c.* 2 × 1 mm, elliptic, with narrow wing and persistent styles. *Most of Europe, W Asia, mountains of N Africa.* H1. Early spring.

The typical tree characteristic of boggy ground and river-sides in the wild is rarely planted in gardens, but there are several

cut-leaved cultivars that are widely grown (see Hylander, *Svensk Botanisk Tidsskrift* **51**(2), 1957, where they are illustrated); in addition, there are 'Aurea', with golden leaves, and 'Pyramidalis', with erect branches.

8. A. incana (Linnaeus) Moench. Figure 11(1, 13), p. 47. Illustration: Mitchell, Field guide to the trees of Britain and northern Europe, 193, pl. 18 (1974); Polunin & Everard, Trees and bushes of Britain and Europe, 45 (1976).
Tree to 25 m with grey bark. Shoots grey-brown, velvety-hairy, especially when young, with a few very small glands and small, dot-like lenticels. Leaves to 13 × 9 cm, deep green above, greyish beneath, ovate, acute at apex, rounded at base, hairless except for midrib and veins beneath when mature, deeply double-toothed or shallowly lobed. Pairs of lateral veins 10–12, impressed. Male catkins to 4 cm × 5 mm, commonly on long, flexible stalks, in groups of 4–5. 'Cones' smooth 1.2 cm × 9 mm, ovoid, in groups of 3–5. Nut *c.* 2.5 × 1.5 mm, obovoid, with narrow wing. *NE & C Europe, mountains of S Europe and W Siberia.* H1. Winter–early spring.

Often planted, but in its typical form mainly used for quick growth in poor soils. Several cultivars are planted in gardens, showing a similar range to that in *A. glutinosa* (see Hylander, cited above). The cultivar 'Ramulis Coccineis', with orange-red shoots, leaf-buds and catkin-scales in winter, is particularly attractive.

9. A. hirsuta (Spach) Ruprecht (*A. incana* var. *hirsuta* Spach). Figure 11(15), p. 47. Illustration: Miyabe & Kudo, Icones of the essential forest trees of Hokkaido, pl. 29 (1984); Plantsman **8**: pl. 3 (1986).
Like *A. incana* but leaves to 14 × 14 cm, broadly ovate or almost circular, with short, slender point, and larger 'cones' to 2 × 1.2 cm. *E Asia, from Kamtchatka to Korea and Japan.* H2. Winter–early spring.

Typically the twigs are hairy but in var. **sibirica** (Spach) Schneider they are hairless.

10. A. rugosa (Duroi) Sprengel (*A. serrulata* (Aiton) Willdenow). Figure 11(14), p. 47. Illustration: Polunin, Trees and bushes of Europe, 46 (1976).
Shrub or more rarely a small tree to 10 m. Like *A. incana* but with darker bark, more or less unlobed, often finely single-toothed leaves, less strongly hairy twigs, and 'cones' with projecting scales. *Eastern N America.* H2. Winter–early spring.

Widely planted in C Europe and

sometimes naturalised. In the south part of its native American range, represented by var. **serrulata** Aiton, a shrub of suckering habit and with obovate-elliptic leaves, rarely seen in cultivation in Europe.

11. A. rubra Bongard (*A. oregona* Nuttall). Figure 11(16), p. 47. Illustration: Mitchell, The Gardener's book of trees, 71 (1981); Plantsman **8**: pl. 3 (1986).
Tree to 20 m, bark light grey (almost white in native habitat), narrow-ridged on older trunks, with black patches. Shoots shining, greenish brown with scant, grey wax layer, sparsely glandular. Buds red, curved, resinous. Leaves *c.* 15 × 10 cm, thick in texture, green and dull or somewhat shiny above, pale green to glaucous beneath, ovate-elliptic, acute at apex, rounded at base, hairless except for short hairs on midrib and veins beneath; strongly and regularly double-toothed or lobed, the lobes to 1 cm deep. Margins of leaves strongly rolled under. Pairs of lateral veins 12–14, strongly impressed. Male catkins to 3.5 cm × 5 mm, green or reddish, commonly in groups of 4. 'Cones' *c.* 2 × 1 cm, ovoid-ellipsoid, in groups of 3–5. Nut *c.* 1.5 × 1 mm, ovoid-elliptic, with narrow, membranous wing. *Western N America to N California.* H2. Early spring.

A very attractive and distinct species, easily recognised by its shallowly lobed leaves with margins strongly rolled under. Deserves to be more widely planted.

12. A. japonica (Thunberg) Steudel. Figure 11(3, 17), p. 47. Illustration: Kitamura & Okamoto, Coloured illustrations of trees and shrubs of Japan, pl. 20 (1958); Plantsman **8**: pl. 3 (1986).
Tree to 20 m. Bark grey, becoming fissured on older stems. Shoots yellowish, reddish brown or greenish, glandular, more or less hairy. Lenticels pale brown and protuberant, especially on older shoots. Leaves to 14 × 6 cm, thick in texture, wavy, dark green above, light green beneath, moderately shining, ovate-elliptic or elliptic, acute or acuminate at apex, rounded at base, more or less hairy, especially in vein-axils beneath; single-toothed, with shallow, forward-pointing teeth. Pairs of lateral veins 5–7, arching and frequently branching, often not reaching leaf-margin. Male catkins to 2.5 cm × 4 mm, in groups of 3–5. 'Cones' ovoid-globular, to 1.5 × 1.2 cm, commonly solitary. Nut *c.* 2.5 × 1.5 mm, elliptic, with narrow corky wing. *NE Asia, from NE China and N Japan to Taiwan and Philippines.* H2. Early spring.

A handsome tree, rarely planted. The hybrid with *A. subcordata* (*A.* × **spaethii** Callier), fast-growing with ovate-lanceolate leaves, purplish when unfolding, is increasingly popular.

13. A. nitida (Spach) Endlicher.
Tree to 20 m with rough, dark, furrowed bark. Shoots greenish or brownish, shiny, smooth and glandular, nearly hairless. Lenticels dot-like, inconspicuous. Leaves to 16 × 9 cm, thin, flat, shiny, ovate or ovate-elliptic, acute at apex and commonly wedge-shaped at base; entire or with shallow teeth. Pairs of lateral veins 3–9, usually much-branched and not reaching margin except in upper half. Male catkins 8 cm × 5 mm, thin and long. 'Cones' to 3 × 1.2 cm, ovoid or ovoid-cylindric, solitary and axillary below male catkin. Nut *c.* 4 × 1.5 mm, elliptic with wide, corky wing. *NW Himalaya.* H3. Early autumn.

Very attractive with long male catkins in early autumn, but not hardy in most of Europe. The only other autumn-flowering species, **A. maritima** (Marshall) Nuttall, from Eastern N America, and **A. nepalensis** D. Don from the E Himalayan foothills, are very rarely seen in cultivation.

14. A. subcordata Meyer. Figure 11(18), p. 47. Illustration: Mitchell, The gardener's book of trees, 71 (1981).
Tree to 30 m with broad crown and dark grey, furrowed bark. Shoots greenish brown with thin, grey waxy layer, moderately glandular and more or less hairy, with conspicuous lenticels. Buds stout, rounded, hairy. Leaves to 15 × 8 cm, dark green, somewhat wavy, ovate or oblong-ovate with abrupt, acute apex, cordate or rounded at base, singly or shallowly double-toothed, with broad, rounded, forward-pointing teeth. Pairs of lateral veins 8–10, well-defined and moderately arching. Male catkins to *c.* 3 cm × 4 mm, in groups of 3–5, often opening in midwinter. 'Cones' to 3 × 1.5 cm, ovoid, solitary, below male catkins. Nut *c.* 3 × 2 mm, ovoid, with narrow corky wing. *USSR (Caucasus), N Iran.* H2. Winter–early spring.

A handsome tree retaining its leaves very late. *A.* × **spaethii**, its hybrid with *A. japonica* is becoming popular.

15. A. cordata Desfontaines (*A. cordifolia* Tenore). Figure 11(4, 19), p. 47. Illustration: Mitchell, Field guide to the trees of Britain and northern Europe, 193, pl. 18 (1974); Bean, Trees and shrubs hardy in the British Isles edn 8, **1**: 272

(1976); Polunin & Everard, Trees and bushes of Europe, 46 (1976); Plantsman **8**: pl. 3 (1986).
Tree to 25 m with rounded crown. Bark greyish brown, smooth, becoming furrowed on older trees; shoots shiny, greenish brown, with grey waxy layer on upper side, sparsely glandular, with conspicuous lenticels. Leaves to 9 × 7 cm, deep green, wavy, shiny, broadly ovate or circular-ovate, cordate at base, short-pointed at apex, on long and flexible stalks and mobile in wind, single-toothed with flattened, forward-pointing teeth. Pairs of lateral veins 5–6, arching, often raised on upper surface. Male catkins to 3.5 cm × 6 mm, in groups of *c.* 5. 'Cones' to *c.* 2.2 × 1.7 cm, ovoid, patched with white resin, solitary and axillary below male catkins. Nut to *c.* 4 × 3 mm, obovoid, grey-brown, with thick, narrow wing. *S Italy & Corsica.* H2. Early spring.

A quick-growing tree, remarkably tolerant of a range of soils and climate, much planted in recent years. Its shiny leaves persist late in the season.

2. BETULA Linnaeus
K.B. Ashburner & S.M. Walters
Trees or shrubs, usually with slender twigs and thin, often papery bark. Male flowers with 2 stamens, each bifid below the anthers, and a minute perianth. Female flowers 3 to each bract. Fruiting catkins ('cones') cylindric to narrowly ovoid, with 3-lobed scales falling when the winged nuts are ripe (Figures 12 & 13, pp. 50, 51).

A genus of about 60 species confined to the northern hemisphere, where they are characteristic of pioneer or secondary woodland. Several tree species are widely grown; they are easily recognised collectively by their slender, graceful habit and the often decorative, thin, papery bark. Though the genus is unmistakable, the naming of *Betula* species in gardens is very difficult. Several of the widespread tree species of Eurasia and N America are unusually variable, and in this wind-pollinated genus hybridisation occurs freely in the wild and in gardens. Since self-sown seedlings are quickly established, many garden specimens are of unknown hybrid origin. Trees are relatively short-lived, rarely exceeding 100 years, so that new generations of hybrid origin are found in most long-established gardens & arboreta, whatever the original planting may have been. Selected clones are generally grafted on to the common European species *B. pendula*.

Literature: Clark, D.L. (ed), Betula in Bean, W.J., *Trees and shrubs hardy in the British Isles*, edn 8, **1**: 414–434 (1970); Ashburner, K.B., Betula – a survey, *The Plantsman* **2**: 31–53 (1980).

Habit. Shrub to 5 m: **6,7,10,16,21**.
Shoots. Very pendulous: **2**.
Leaves. More than 12 cm: **11,12**.
Leaf-tip. Long-acuminate: **3,4**.
'Cones'. More than 3 cm: **8,11,15–17**.

1a. Dwarf shrub to 1 m; leaves with 3 or 4 lateral veins only   **21. nana**
 b. Tree or shrub more than 1 m; leaves with more than 4 veins   **2**
2a. Leaves with 11–18 (rarely 10) pairs of strongly impressed lateral veins   **3**
 b. Leaves with 5–10 pairs of lateral veins, these often not impressed   **11**
3a. Lenticels on young shoots to 2 mm, conspicuous, white or pale brown   **4**
 b. Lenticels on young shoots not more than 1 mm, inconspicuous   **7**
4a. Shoots with brown glands; bark of older shoots tearing off easily   **5**
 b. Shoots without glands; bark of older shoots not tearing off easily   **6**
5a. Leaves thick in texture, shiny on upper surface, dark green; bark often very white   **17. utilis**
 b. Leaves thin in texture, dull on upper surface, light green; bark usually pinkish or orange-brown   **18. ermanii**
6a. Tree with ovate leaves; odour of oil of wintergreen in shoots and leaves   **12. alleghaniensis**
 b. Spreading shrub with broadly ovate leaves; shoots and leaves without odour of oil of wintergreen   **16. medwediewii**
7a. Bark peeling easily in sheets, often white, without odour of oil of wintergreen; cones pendulous   **17. utilis**
 b. Bark not peeling easily, pale or dark, with odour of oil of wintergreen; cones erect   **8**
8a. Leaves glaucous beneath; stipules persistent   **15. corylifolia**
 b. Leaves green beneath; stipules not persistent   **9**
9a. Young shoots densely hairy; bark flaky, pale beneath   **12. alleghaniensis**
 b. Young shoots somewhat hairy or hairless; bark not flaky, dark   **10**
10a. Leaves thin in texture; young shoots hairless   **13. lenta**
 b. Leaves thick in texture; young shoots somewhat hairy   **14. grossa**

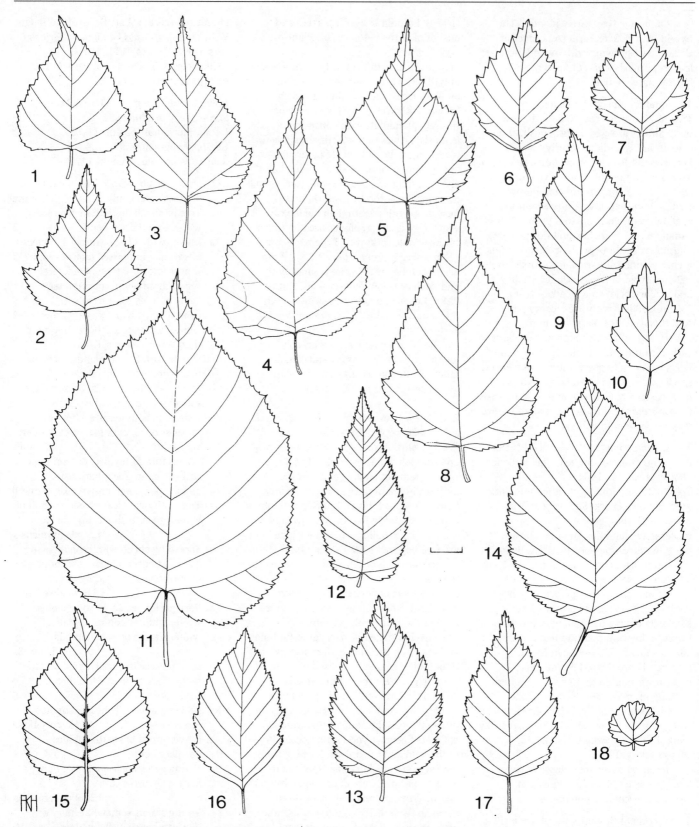

Figure 12. Leaves of *Betula* species. 1, *B. platyphylla*. 2, *B. pendula*. 3, *B. populifolia*. 4, *B. coerulea-grandis*. 5, *B. szechuanica*. 6, *B. celtiberica*. 7, *B. pubescens*. 8, *B. papyrifera*. 9, *B. davurica*. 10, *B. fontinalis*. 11, *B. maximowicziana*. 12, *B. alleghaniensis*. 13, *B. grossa*. 14, *B. medwediewii*. 15, *B. ermanii*. 16, *B. nigra*. 17, *B. albosinensis*. 18, *B. nana*. (All upper surface except 15). Scale = 1 cm.

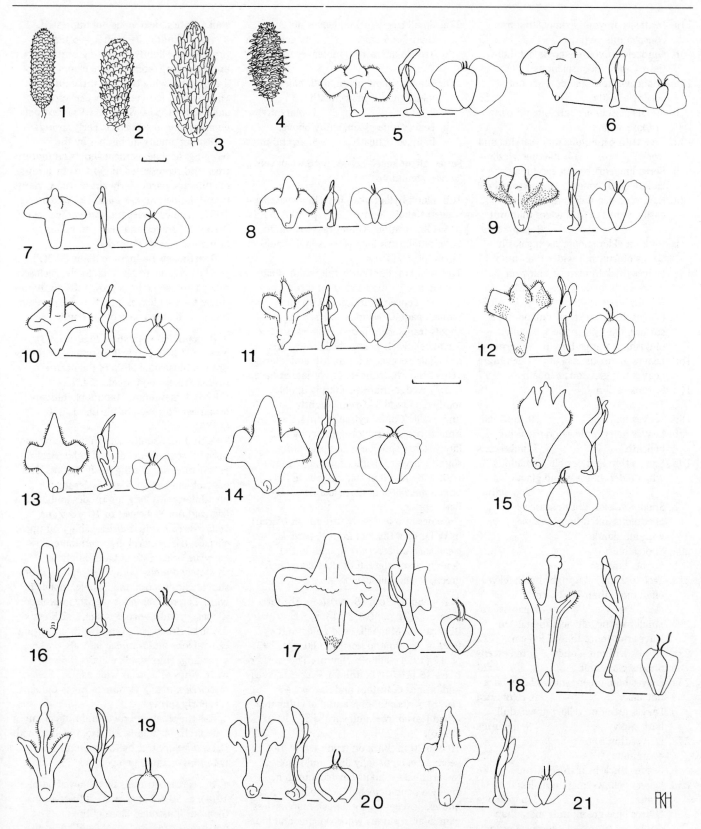

Figure 13. 'Cones' nuts and catkin-scales of *Betula* species. 1–4, 'cones': 1, *B. populifolia*. 2, *B. lenta*.

3, *B. ermanii*. 4, *B. albosinensis*. 5–21, Catkin-scales and nuts:

5, *B. platyphylla*. 6, *B. pendula*. 7, *B. aetnensis*. 8, *B. populifolia*.

9, *B. coeruleo-grandis*. 10, *B. szechuanica*. 11, *B. celtiberica*.

12, *B. pubescens*. 13, *B. papyrifera*. 14, *B. davurica*.

15, *B. fontinalis*. 16, *B. maximowicziana*. 17, *B. lenta*.

18, *B. medwediewii*. 19, *B. utilis*. 20, *B. ermanii*. 21, *B. albosinensis*. Scale for nos. 1–4 = 5 mm, for nos. 5–21 = 2 mm.

11a. Lenticels on young shoots to 2 mm,
     conspicuous, whitish   12
   b. Lenticels on young shoots not more
     than 1 mm, inconspicuous   16
12a. Shoots without glands; bark not
     peeling readily   13
   b. Shoots glandular; bark often peeling
     in loose sheets   14
13a. Tree with pendulous male catkins and
     'cones'   11. maximowicziana
   b. Spreading shrub with erect male
     catkins and 'cones'   16. medwediewii
14a. Bark on older shoots not peeling
     easily; wings of fruit wider than nut
                             8. papyrifera
   b. Bark on older shoots peeling easily;
     wings of fruit not wider than nut   15
15a. Leaves thick in texture, shiny on
     upper surface, dark green; bark
     usually very white   17. utilis
   b. Leaves thin in texture, dull on upper
     surface, light green; bark usually
     pinkish or orange-brown   18. ermanii
16a. Leaves glaucous beneath   19. nigra
   b. Leaves not glaucous beneath   17
17a. Shrubs to 7 m   18
   b. Trees to 30 m   19
18a. Leaves hairless   10. fontinalis
   b. Leaves hairy, at least on the veins
     beneath   7. pubescens
19a. Shoots thin and flexible, pendulous
     when older, hairless and glandular
                                2. pendula
   b. Shoots stouter, not or scarcely
     pendulous, often hairy and/or
     without glands   20
20a. Shoots hairy   21
   b. Shoots hairless   24
21a. Tree to 25 m, of upright habit; leaves
     often more than 10 × 6 cm
                             8. papyrifera
   b. Small tree, usually less than 20 m;
     leaves not more than 7 × 5 cm   22
22a. Shoots without glands   7. pubescens
   b. Shoots glandular   23
23a. Leaves blue-green, shiny above; bark
     not flaky   6. celtiberica
   b. Leaves green or yellow-green, dull;
     bark flaky   9. davurica
24a. Leaves thin in texture, long-
     acuminate   25
   b. Leaves thick in texture, acute   26
25a. Leaves yellow-green; nuts less than
     1 mm   3. populifolia
   b. Leaves blue-green; nuts more than
     1 mm   4. coerulea-grandis
26a. Shoots sparsely glandular or without
     glands, with few, pale brown warts
                             20. albosinensis
   b. Shoots densely glandular, with white
     resinous warts   27

27a. Small tree to 10 m; leaves not more
     than 7 × 5 cm   6. celtiberica
   b. Tree to at least 20 m; leaves to 10 ×
     7 cm   28
28a. Leaves yellow-green, dull, hairy on
     midrib and veins beneath
                              1. platyphylla
   b. Leaves blue-green, shiny above,
     hairless beneath   5. szechuanica

Series **Albae** Regel. Leaves with weak veins;
'cones' pendulous.

1. B. platyphylla Sukatchev (*B. mandshurica*
(Regel) Nakai). Figure 12(1), p. 50 & 13(5),
p. 51. Illustration: Kitamura & Okamoto,
Coloured illustrations of trees and shrubs of
Japan, pl. 18 (1958).

Tree to 20 m. Bark on trunk white, dusty,
peeling freely. Shoots stout and stiff,
hairless, densely glandular, with blue-
white, opaque, resinous warts; young
shoots translucent brown. Leaves to 10 ×
7 cm, yellowish green, thick in texture,
dull, hairless except for midrib and lateral
veins beneath, truncate or cordate at base
(rarely wedge-shaped), sharply double-
toothed. Lateral leaf-veins slightly
impressed. 'Cones' cylindric, to 4.5 × 1 cm.
Bracts with central lobe projecting and
lateral lobes spreading and projecting
slightly forward. Nuts elliptic, *c.* 2 mm,
with wing about as wide as nut. *E Asia
from Kamtchatka to Korea and Japan.* H1.
Early spring.

Comes into leaf early (usually in March)
in W Europe. Distinct from *B. pendula*,
populifolia, and *coerulea-grandis* in the
densely glandular, stout shoots and light
green, thick, dull leaves.

2. B. pendula Roth (*B. verrucosa* Ehrhart).
Figure 12(2), p. 50 & 13(6), p. 51.
Illustration: Mitchell, Field guide to the
trees of Britain and northern Europe, 192,
pl. 17 (1974); Edlin & Nimmo, The world of
trees, 19 (1974); Polunin & Everard, Trees
and bushes of Britain and Europe, 43
(1976); Krüssmann, Manual of cultivated
broad-leaved trees and shrubs 1: 230
(1986).

Tree to 20 m. Bark on trunk white, not
peeling freely, usually dark and fissured
towards base. Shoots flexible and thin,
giving a pendulous habit on mature trees;
hairless (except on seedlings), more or less
glandular; resinous warts opaque and blue-
white (except on seedlings and early in the
season). Leaves to 6 × 5 cm, thin, hairless,
shiny on upper surface (wavy on strong
shoots), triangular-acuminate; cordate,
truncate or wedge-shaped at base, sharply
double-toothed, with thin and flexible

stalks. Lateral leaf-veins not impressed.
'Cones' cylindric, 2–3 cm × 4–5 mm.
Bracts with slightly projecting central lobes
and spreading laterals. Nuts elliptic, *c.* 2 ×
1 mm, with wing twice or three times as
wide as nut. *Europe, from N Fennoscandia
and USSR to Spain, Sicily and N Greece, only
on mountains in south.* H1. Early spring.

Distinct among all birches by the
hanging, flexible, vertical shoots on mature
trees and represented in gardens by a range
of cultivars, particularly cut-leaved variants
(see Hylander, *Svensk Botanisk Tidsskrift* **51**,
1957). The very elegant 'Dalecarlica', with
deeply cut leaves and very pendulous
branches, is widely grown.

B. aetnensis Rafinesque (Figure 13(7),
p. 51), endemic to Mt Etna, Sicily, is closely
related and sometimes grown. It has almost
single-toothed leaves and does not develop
the dark, fissured base of the trunk.

3. B. populifolia Marshall. Figure 12(3),
p. 50, & 13(8), p. 51. Illustration: Britton &
Brown, Illustrated Flora of the northern
United States and Canada 1: f. 1121
(1896); Krüssmann, Manual of cultivated
broad-leaved trees and shrubs 1: 230
(1986).

Tree to 9 m, usually with many stems and
upright, scarcely pendulous habit. Bark
grey-white, not peeling freely. Shoots thin
but not flexible, hairless, moderately
glandular, with long internodes and angled
(zig-zag) nodes. Leaves to 10 × 8.5 cm,
yellow-green, thin, hairless, shiny on upper
surface, triangular, long-acuminate,
truncate or cordate at base; singly or
shallowly double-toothed (except on strong
shoots). Stalks thin and flexible, mobile in
wind. Lateral leaf-veins slightly impressed.
'Cones' cylindric-ovoid, small, *c.* 2.5 cm ×
3.5 mm. Bracts *c.* 2.5 mm, with projecting
central lobe and rounded laterals,
spreading. Nuts small, elliptic, 0.5 mm,
with wings at least as wide as nut. *Eastern
N America, from S Ontario to North Carolina.*
H1. Early spring.

Has the smallest bracts and nuts of any
tree-birch. Distinguished also, together with
B. coerulea-grandis, by the triangular leaf
with long-acuminate apex.

4. B. coerulea-grandis Blanchard. Figure
12(4), p. 50 & 13(9), p. 51. Illustration:
Gleason, Illustrated Flora of the
northeastern United States and adjacent
Canada 2: 35 (1952).

Tree to 25 m, usually single-stemmed. Bark
on trunk shining white, not peeling freely.
Shoots stout, hairless, moderately
glandular, with opaque, blue-white,

resinous warts. Internodes long; nodes angled (zig-zag). Leaves to *c.* 10 × 8 cm, blue-green, thin, shiny on upper surface, triangular, long-acuminate, cordate or truncate at base, singly or shallowly doubly toothed (except on strong shoots). Stalks flexible. Lateral leaf-veins slightly impressed. 'Cones' cylindric, *c.* 2.5 cm × 6 mm. Bracts with short central lobe and lateral lobes projecting slightly forward. Nuts elliptic, *c.* 2 × 1 mm, with wing nearly twice as wide as nut. *Eastern N America, from Quebec to New Hampshire.* H1. Early spring.

Very similar to *B. populifolia*, especially in the long-acuminate leaf-tip, but distinguished by taller stature, whiter bark and generally larger leaves, cones, bracts and nutlets.

5. B. szechuanica (Schneider) Jansson (*B. platyphylla* var. *szechuanica* (Schneider) Rehder; *B. mandshurica* var. *szechuanica* Schneider). Figure 12(5), p. 50 & 13(10), p. 51. Illustration: Plantsman 2: 37 (1980).
Tree to 20 m with spreading, stocky habit and sparse branching. Bark on main stems very white and dusty. Shoots stout, hairless, densely to sparsely glandular with opaque, blue-white, resinous warts. Internodes long; buds large, *c.* 8–10 × 3 mm. Lenticels conspicuous on second-year shoots. Leaves to 9 × 6 cm, deep blue-green, thick in texture, hairless, shiny on upper surface, ovate-acuminate, usually rounded or wedge-shaped at base, single-toothed with broad, fine-pointed teeth. Stalks not flexible. Lateral leaf-veins slightly raised on surface. 'Cones' cylindric, 3–5 cm × 3 mm. Bracts with central lobe projecting forward and spreading lateral lobes. Nuts rounded-elliptic, to 2.5 × 1.5 mm, with wing as wide as nut. *SW China.* H2. Early spring.

Distinct on account of its thick, blue-green leaves, sparse foliage, stocky habit and very white, dusty bark.

6. B. celtiberica Rothmaler & Vasconcellos. Figure 12(6), p. 50 & 13(11), p. 51.
Tree to 10 m, often shrubby. Bark on main trunk and branches white with conspicuous, wide lenticels. Shoots hairless or somewhat hairy, densely glandular, with opaque, blue-white, resinous warts. Internodes short; nodes not angled (zig-zag). Leaves to 7 × 5 cm, blue-green, rather thick in texture, shiny on upper surface, hairy on stalks and veins beneath, ovate, shortly acuminate, wedge-shaped or truncate at base, with coarse double toothing on strong shoots. Lateral leaf-

veins often raised on surface. 'Cones' cylindric, *c.* 3 cm × 4 mm. Bracts with projecting central lobe and spreading lateral lobes. Nuts elliptic, *c.* 2 × 1 mm, with wing at least as wide as nut. *Mountains of N & C Spain and N Portugal.* H2. Early spring.

Intergrades with *B. pendula* in the west Pyrenees. Resembles *B. pubescens* in leaf-shape, but shoots densely glandular.

7. B. pubescens Ehrhart. Figure 12(7), p. 50 & 13(12), p. 51. Illustration: Polunin & Everard, Trees and bushes of Britain and Europe, 44 (1976); Polunin & Walters, Guide to the vegetation of Britain and Europe, pl. 65, 69 (1985).
Tree to 20 m, sometimes shrubby; habit upright. Bark on main stems white or brownish, close or peeling freely, with wide and conspicuous lenticels. Shoots thin or stout, densely velvety-downy, usually without glands and straight, with nodes scarcely angled. Leaves to 7 × 5 cm, blue-green, rather thick in texture, shiny on upper surface, broadly ovate, acute; cordate, truncate or wedge-shaped at base, single-toothed except on strong shoots. Leaves softly hairy, at least on lower surface and on midrib and veins. Lateral leaf-veins slightly impressed to slightly raised on surface. 'Cones' cylindric, *c.* 2 cm × 4 mm. Bracts with rounded lateral lobes projecting slightly forward as far as central lobe, or spreading. Nuts elliptic, *c.* 1.5–2 × 1 mm, with wing as wide or wider than nut. *Europe & W Asia, from N Scandinavia and European Russia to the Pyrenees, Alps, Carpathians and Caucasus; eastwards at least to Lake Baikal.* H1. Early spring.

Not often planted, but often present as self-sown individuals or populations in woodland gardens. Very variable, intergrading with *B. pendula* throughout most of its range and with *B. ermanii* in Siberia. Populations in the west and north of the British Isles and in NW France have thin, reddish shoots, small, densely downy leaves and mostly brown bark; those of most of continental Europe have stout, blackish or brown shoots, larger and less downy leaves and white bark; some of these populations with large, shiny, glutinous, aromatic buds, twisted in habit and slower-growing, have been called **B. tortuosa** Ledebour.

8. B. papyrifera Marshall (*B. papyracea* Horton). Figure 12(8), p. 50 & 13(13), p. 51. Illustration: Britton & Brown, Illustrated Flora of the northern United

States and Canada 1: f. 1122 (1896); Krüssmann, Manual of cultivated broad-leaved trees and shrubs 1: 227, 230 (1986).
Tree to 25 m with upright habit. Bark on trunk white, pinkish or brown, peeling freely on mature trunks but not on young plants. Shoots usually both hairy and glandular, with opaque, blue-white, resinous warts; but sometimes hairless and rarely also lacking glands. Leaves to 11 × 8 cm, green or yellow-green, dull, ovate, short-acuminate, usually rounded at base, sometimes cordate, shallowly double-toothed or single-toothed, with sharp, projecting teeth at ends of lateral veins, hairy on midrib and veins beneath. Lateral leaf-veins weakly to moderately impressed. 'Cones' *c.* 4 cm × 7 mm, cylindric-ovoid, pendulous. Bracts with lateral lobes more or less projecting forwards nearly as far as central lobe. Nuts elliptic, 1.5–2 × 1 mm, with wing twice as wide as nut. *N America from Alaska and Labrador to N Colorado and North Carolina.* H1. Early spring.

Very variable: often easily distinguished in cultivation by its relatively large leaves, rounded but not usually cordate at base.

9. B. davurica Pallas (*B. dahurica* Regel). Figure 12(9), p. 50 & 13(14), p. 51. Illustration: Kitamura & Okamoto, Coloured illustrations of trees and shrubs of Japan, pl. 19 (1958).
Tree to 15 m with upright habit. Bark on main stem peeling but curling back and persisting, pinkish white beneath, with brownish red flakes. Shoots brownish, sparsely hairy, usually densely glandular, with translucent brown resinous warts, becoming white later in the season. Leaves to 8 × 6 cm, green or yellow-green, dull, ovate, shortly acuminate, rounded at base, with shallow double-toothing, hairless except for stalk and midrib above and beneath. Lateral leaf-veins scarcely impressed. 'Cones' 2–2.5 cm × 4 mm, ovoid-cylindric, pendulous. Bracts with central lobe and lateral lobes projecting slightly forwards. Nuts broadly elliptic, *c.* 2 × 2 mm, with wing as wide as nut. *E Asia, from Amur to Japan and Korea.* H1. Early spring.

Closely similar to *B. papyrifera*, but outer layers of bark persisting as curled-back flakes.

10. B. fontinalis Sargent (*B. occidentalis* Hooker). Figure 12(10), p. 50 & 13(15), p. 51. Illustration: Abrams & Ferris, Illustrated Flora of the Pacific States 1: f. 1244 (1923).

Large shrub to 7 m. Bark usually shiny brown (sometimes dull black), not peeling freely. Shoots reddish brown, glutinous at tips, hairless, with translucent, red-brown, glandular warts, becoming opaque and white later in the season. Leaves to 4.5 × 4 cm, pale green, thin and hairless, shiny on upper surface, glutinous, especially when young, ovate; wedge-shaped or truncate at base, short-pointed, doubly or coarsely singly toothed. 'Cones' 2–3 cm × 8–10 mm, cylindric, erect or pendulous. Bracts with lateral lobes projecting forwards, shorter than central lobe. Nuts ovate to obovate, 2 × 1.5 mm, with wing twice as wide as nut. *Western N America, from Alaska to Oregon, in Rocky mountains S to Colorado.* H1. Early spring.

Both this species and *B. dahurica* come into leaf early in western Europe, and are liable to be damaged by spring frosts. They are relatively little cultivated.

Series **Acuminatae** Regel. Leaves with strong, impressed veins ending in acuminate teeth. 'Cones' in groups, pendulous.

11. B. maximowicziana Regel (*B. candelae* Koidzumi). Figure 12(11), p. 50 & 13(16), p. 51. Illustration: Kurata, Illustrated important forest trees of Japan 1: pl. 36 (1968); Plantsman 2: 52 (1980); Krüssmann, Manual of cultivated broad-leaved trees and shrubs 1: 230 (1986).
Tree to 25 m with spreading crown. Bark on trunk white or pinkish, with long, conspicuous lenticels. Shoots stout, downy when young, without glands, greenish brown and shining, with conspicuous round or elliptic lenticels. Leaves to 16 × 13 cm, circular-ovate, acuminate, deeply cordate at base, bristly hairy, especially when young, and shallowly double-toothed with strong, impressed veins ending in acuminate teeth. Stalks flexible and therefore leaves mobile in wind. 'Cones' to 7 cm × 7 mm, long-cylindric, pendulous, in racemose groups. Nuts *c.* 2 × 1 mm, elliptic, with wing up to 4 times as wide as nut. *Japan & Kurile Islands.* H2. Early spring.

Very distinct, with its large, deeply cordate, mobile leaves. Fast-growing in NW Europe.

Series **Costatae** Regel. Leaves with close, parallel, deeply impressed veins. 'Cones' mostly erect.

12. B. alleghaniensis Britton (*B. lutea* Michaux). Figure 12(12), p. 50. Illustration: Britton & Brown, Illustrated Flora of the northern United States and Canada 1: f. 1216 (1896); Krüssmann, Manual of cultivated broad-leaved trees and shrubs 1: 222, 230 (1986).
Tree to 25 m. Outer bark peeling in thin sheets, translucent yellow or brown; flakes persistent. Shoots without glands, greenish brown, densely hairy, especially when young, with moderately conspicuous lenticels. Leaves to 14 × 7 cm, dull, yellow-green, ovate or elliptic, pointed; rounded or cordate at base; sharply toothed and densely downy along midrib and veins beneath, especially when young. Veins strongly impressed. 'Cones' to 2 cm × 8 mm, ovoid-globular, erect. Bracts with lateral lobes projecting forwards, as long as central lobe. Nuts 4 × 2 mm, ovoid-elliptic, with wing *c.* 1 mm. *Eastern N America, from Newfoundland to Georgia.* H2. Early spring.

13. B. lenta Linnaeus (*B. carpinifolia* Ehrhart). Figure 13(17), p. 51. Illustration: Bean, Trees and shrubs hardy in the British Isles edn 8, 1: 421 (1976); Krüssmann, Manual of cultivated broad-leaved trees and shrubs 1: 222 (1986).
Tree to 25 m with blackish bark on young trees, becoming reddish black with thick, persistent flakes on older trunks. Shoots shining yellow-brown, hairless and without glands, with small, dot-like lenticels. Leaves to 10 × 6 cm, yellow-green, thin in texture, shiny, hairless except for vein-axils beneath, ovate, pointed; rounded or shallowly cordate at base, single-toothed with fine, forward-pointing teeth. 'Cones' to 2 × 1.4 cm, ovoid, erect. Bracts with short lobes, lateral lobes rounded, spreading. Nuts *c.* 2 × 1.5 mm, ovoid, with narrow wing. *Eastern USA, from Maine to Georgia.* H1. Early spring.

14. B. grossa Siebold & Zuccarini (*B. carpinifolia* Siebold & Zuccarini; *B. ulmifolia* Siebold & Zuccarini). Figure 12(13), p. 50. Illustration: Krüssmann, Manual of cultivated broad-leaved trees and shrubs 1: 224, 230 (1986).
Tree to 25 m with grey bark, smooth even on mature trees. Shoots greyish brown, smooth, without glands and moderately hairy, with small, dot-like lenticels. Leaves to 9 × 5 cm, dark green, thick in texture, dull, hairy, ovate, pointed; cordate or truncate at base, coarsely and sharply single- or double-toothed; veins strongly impressed. 'Cones' to 2 × 1 cm, cylindric-ovoid, erect. Bracts with short, rounded lobes; lateral lobes half as long as the broad central lobe. Nuts *c.* 2 mm, ovoid, with very narrow wing. *Japan.* H2. Early spring.

This species is closely related to *B. lenta*. Neither is commonly planted.

15. B. corylifolia Regel & Maximowicz. Illustration: Krüssmann, Manual of cultivated broad-leaved trees and shrubs 1: 224, 230 (1986).
Small tree to 20 m with upright, spreading habit. Bark greyish white with conspicuous lenticels. Shoots reddish brown, nearly hairless, smooth and without glands. Leaves to 8 × 5 cm, pale green, dull above and glaucous beneath, hairy on midrib and veins beneath, ovate or elliptic, rounded or shallowly cordate at base, with characteristic coarse, forward-pointing teeth. Main veins impressed. 'Cones' to 4 cm × 4 mm, cylindric-ovoid, erect. Bracts with deep, narrow lobes, the central twice as long as the diverging laterals. Nuts *c.* 3 × 2 mm, ovoid-circular, with narrow wing. *Japan.* H2. Early spring.

Distinguished by the characteristic, coarse, forward-pointing teeth of the leaves. Nos. **12–15** contain oil of wintergreen in the bark; it can usually be detected by the characteristic smell.

16. B. medwediewii Regel. Figure 12(14), p. 50 & 13(18), p. 51. Illustration: Botanical Magazine, 9569 (1939); Plantsman 2: 52 (1980); Gartenpraxis for 1987(12): 10, 11; Krüssmann, Manual of cultivated broad-leaved trees and shrubs 1: 230 (1986).
Large spreading shrub to 5 m, with silvery brown, scarcely peeling bark. Shoots stout, brownish green, without glands, hairy, especially when young, with conspicuous rounded lenticels. Leaves to 11 × 8 cm, deep green, shining, thick in texture, hairy on veins beneath, ovate or elliptic, rounded or cordate at base, short-pointed, with sharp, forward-pointed teeth. Veins strongly impressed. 'Cones' to 3.5 × 1.2 cm, cylindric, stout and persistent, erect, with strongly reflexed bract-lobes. Nuts *c.* 4 × 1.5 mm, narrowly elliptic, with narrow wing. Bracts narrow, with thin lobes; laterals diverging, half as long as the central. *USSR (Caucasus), NE Turkey.* H2. Spring.

Very distinct, with its stout, shrubby habit, resembling an *Alnus*.

17. B. utilis D. Don (*B. bhojpattra* Wallich). Figure 13(19), p. 51. Illustration: Mitchell, Field guide to the trees of Britain and northern Europe, 192, pl. 17 (1974) – var. jacquemontii; Bean, Trees and shrubs hardy in the British Isles edn 8, 1: pl. 20 (1976) – var. jacquemontii; Krüssmann,

Manual of cultivated broad-leaved trees and shrubs 1: 230, 232 (1986).

Tree to 20 m with upright habit. Bark brown or red-brown, with or without bloom, sometimes very white, loose and peeling in sheets. Shoots hairless or hairy, with brown, translucent, glandular warts and conspicuous round or elliptic, cream-coloured lenticels. Leaves to 12 × 6 cm, dark green, shining, thick in texture, ovate, acuminate or acute, more or less hairy, especially on midrib and veins beneath; singly or shallowly doubly toothed. Veins more or less impressed. 'Cones' to 3.5 × 1 cm, cylindric-ovoid, pendulous. Bracts with diverging lateral lobes half as long as central lobe. Nuts 2 × 1 mm, elliptic, with relatively broad wing. *Mountains of SW China to the Himalaya and Afghanistan.* H2. Early spring.

Variable; NW Himalayan plants, var. **jacquemontii** (Spach) Winkler (*B. jacquemontii* Spach), characteristically with very white bark and less strongly impressed leaf-veins (illustration: Bean, Trees and shrubs hardy in the British Isles edn 8, 1: 433–434 (1970); Gartenpraxis for 1976(10): 479; Plantsman 2: 45–47 (1980), have given rise to several clones in wide cultivation (such as 'Jermyns').

18. B. ermanii Chamisso (*B. costata* misapplied). Figure 12(15), p. 50 & 13(20), p. 51. Illustration: Takeda, Alpine Flora of Japan, pl. 60 (1963); Miyabe & Kudo, Icones of the essential forest trees of Hokkaido, pl. 26 (1984); Krüssmann, Manual of cultivated broad-leaved trees and shrubs 1: 230, 232 (1986).

Tree to 20 m with spreading crown. Bark loose, peeling in sheets on young trees and flaking on mature trees, pinkish white, with conspicuous lenticels. Shoots hairless, densely glandular, with translucent brown warts and conspicuous round or elliptic lenticels. Leaves to 10 × 7 cm, thin in texture, dull, broadly ovate, acuminate; rounded or cordate at base, sharply doubly toothed, hairless except for sparse hairs in axils of the veins beneath. Veins strongly impressed. 'Cones' to 3 × 1 cm, ovoid-cylindric, erect or nearly so. Bracts with long, thin central lobe and diverging, shorter lateral lobes. Nuts *c.* 2 mm, rounded, with narrow wing. *E Asia, from Kamtchatka to Korea and Japan.* H2. Early spring.

Very variable. Plants of Japanese origin are hardy in W Europe. Plants called '*B. costata*' in gardens are usually this species; the true *B. costata* Trautvetter (from NE Asia) is hardy, but very rare in cultivation.

19. B. nigra Linnaeus. Figure 12(16), p. 50. Illustration: Gartenpraxis for 1976(10): 481; Mitchell, The gardener's book of trees, 75 (1981); Krüssmann, Manual of cultivated broad-leaved trees and shrubs 1: 230 (1986).

Tree to 20 m with upright or slightly pendulous habit. Bark pinkish white on young trees, with persistent flakes; ultimately dark and coarsely scaly on older trees. Shoots thin, brown, sometimes hairy, sparsely glandular, with small, white warts. Leaves to 11 × 7 cm, thin in texture, shining above, glaucous beneath, diamond-shaped-ovate, with tapering base, regularly double-toothed on strong shoots, hairy on midrib and veins beneath. Veins somewhat impressed. 'Cones' to 2.5 × 1.2 cm, cylindric, erect. Bracts with diverging lateral lobes nearly as long as the central lobe. Nuts *c.* 3 mm, broadly ovate with relatively broad wing. *Eastern USA from Massachusetts to Florida.* H1. Early spring.

Distinct on account of pink or orange flaky bark and leaves with tapering base, glaucous beneath. The only species ripening its fruit in early summer.

20. B. albosinensis Burkill. Figure 12(17), p. 50 & 13(21), p. 51. Illustration: Mitchell, Field guide to the trees of Britain and northern Europe, 192, pl. 17 (1974) – var. septentrionalis; Bean, Trees and shrubs hardy in the British Isles edn 8, 1: pl. 19 (1976); Krüssmann, Manual of cultivated broad-leaved trees and shrubs 1: 232 (1986).

Tree to 25 m with broad crown and somewhat pendulous branches. Bark orange-brown or brown with bloom when young, loose and peeling in sheets. Shoots thin, hairless, with angled internodes, without glands or sparsely glandular, with few pale brown warts and small, dot-like lenticels. Leaves to 11 × 6.5 cm, thick in texture, rather shiny, deep green, ovate or ovate-elliptic, acuminate, commonly double-toothed, hairless except for lower part of midrib beneath. Veins somewhat impressed. 'Cones' *c.* 2.5 cm × 7 mm, cylindric-ovoid, pendulous. Bracts with spreading lateral lobes much shorter than the narrow central lobe. Nuts *c.* 1 × 0.5 mm, elliptic, with relatively broad wing. *Mountains of China.* H1. Spring.

Var. **septentrionalis** Schneider, with glandular shoots, paler green, dull leaves and beautiful orange-brown bark with white bloom, is represented by a grafted clone in cultivation.

Series **Humiles** Koch. Low shrubs to 3 m; leaves 2–5-veined.

21. B. nana Linnaeus. Figure 12(18), p. 50. Illustration: Ross-Craig, Drawings of British plants 27: t. 14 (1970); Polunin & Walters, Guide to the vegetation of Britain and Europe, pl. 62 & 24, f. 4 (1985).

Low shrub to 1 m with brown, dull bark; shoots velvety-hairy, without glands or glandular. Leaves to 2 × 1.5 cm, somewhat shining above, circular or kidney-shaped, hairless except for a slight down beneath, with rounded teeth. Pairs of lateral veins 3–4, somewhat branched. 'Cones' *c.* 8 × 4 mm, ovoid-cylindric, erect. Bracts with diverging lateral lobes nearly as long as central lobe. Nuts *c.* 1.5 × 1.5 mm, broadly ovate to circular, with very narrow wing. *Circumpolar throughout subarctic Eurasia and N America, on mountains in north temperate areas.* H1. Spring.

An attractive, neat, small shrub, easy to cultivate in streamsides and damp places. The allied **B. humilis** Schrank, native to peaty ground in W Siberia and E Europe, is a shrub to 3 m; it is very rarely grown; illustration: Gartenpraxis for 1987(12): 11.

3. CARPINUS Linnaeus

S.G. Knees

Deciduous, much-branched trees generally without central stems, rarely shrubs. Bark grey, smooth or scaly. Branches slender, buds acute with many overlapping scales. Leaves alternate, arranged in 2 rows, margins toothed, veins 6–24 pairs. Flowers unisexual, appearing with the leaves. Male catkins pendulous, enclosed in buds in winter, borne on short shoots; flowers lacking perianth or bracteoles, stamens 4–12 borne in the bract axils, filaments bifid almost to the base. Female flowers in terminal catkins, 2 per bract, each with 6 bracteoles; perianth with 6–10 teeth, joined to 2-celled ovary; styles short with 2 stigmas. Fruit a 1-seeded, ribbed nut with a large 3-lobed solitary bract, which continues to grow after fertilisation.

A genus of about 35 species occurring throughout the north temperate world, but chiefly in eastern Asia and North America. All species are best raised from seed though grafting onto *C. betulus* has proved successful especially for the rarer species. Hornbeams thrive in any soil type and grow well on chalk.

Literature: Rushforth, K., Key to Carpinus and Ostrya, *Plantsman* 8: 249–256 (1987); Furlow, J.J., The Carpinus caroliniana complex in North America. II. Systematics, *Systematic Botany* **12**: 416–434 (1987).

Fruit bracts. Symmetric: **6**; asymmetric: **1–5,7**.

Vein-pairs. 15 or fewer: **1–5**; 15 or more: **6,7**.

1a. Leaves with 15–24 pairs of veins;
     bark scaly; fruit bracts symmetric   5
 b. Leaves with 6–15 pairs of veins; bark
     smooth; fruit bracts asymmetric   2
2a. Leaves acute, 2.5–5 cm   **3. orientalis**
 b. Leaves acuminate, 4–12 cm   3
3a. Branches densely covered with soft
     hairs   **5. tschonoskii**
 b. Branches hairless or with few hairs
     when young   4
4a. Fruiting bracts 1.6–1.9 cm
                              4. laxiflora
 b. Fruiting bracts 2–5 cm   6
5a. Leaves cordate at base, with 15–20
     pairs of veins   **6. cordata**
 b. Leaves rounded at base, with 20–24
     pairs of veins   **7. japonica**
6a. Buds *c.* 3 mm, softly hairy; fruiting
     bracts 2–3 cm   **1. caroliniana**
 b. Buds *c.* 5 mm, hairless; fruiting bracts
     3–5 cm   **2. betulus**

1. C. caroliniana Walter (*C. americana* Michaux, in part). Illustration: Sargent, Silva of North America **9**: t. 447 (1896); Krüssmann, Manual of cultivated broad-leaved trees & shrubs **1**: 279, 281 & pl. 94 (1986); Systematic Botany **12**: 420 (1987). Small tree, rarely exceeding 10 m. Trunk fluted, bark smooth, young branches hairy, becoming hairless with age. Buds *c.* 3 mm, softly hairy. Leaf-stalks 1.2–2 cm, downy, blades 5–11 × 2.5–3.5 cm, ovate-oblong, cordate at base, covered with silky hairs when young, becoming hairless except on veins beneath, margins with 2 rows of sharp, forward-pointing teeth, veins 6–12 pairs. Male catkins 2.5–3.5 cm. Fruiting clusters 5–10 cm, on slender stalks, bracts 2–3.5 cm, ovate to ovate-lanceolate, with 2 short lateral lobes, central lobe often with irregularly shaped marginal teeth. *E & Central S USA, NE Mexico.* H2–3. Spring.

2. C. betulus Linnaeus. Illustration: Das Pflanzenreich **61**: 29, 39 (1904); Elwes & Henry, Trees of Great Britain and Ireland **3**: t. 148–152 (1908); Humphries et al., Trees of Britain and Europe, 119 (1981); Krüssmann, Manual of cultivated broad-leaved trees & shrubs **1**: 279 (1986). Tree commonly 20–30 m, similar to *C. caroliniana*, but branches ultimately pendulous. Buds *c.* 5 mm, hairless. Leaves also similar but often with one side longer

than the other and 10–13 pairs of veins. Fruiting clusters 7–14 cm, bracts 3–5 cm, central lobe ovate-lanceolate, often toothed, lateral lobes ovate, teeth absent or distant. *Europe, SW Asia, eastwards to Iran.* H1. Spring.

Widely cultivated with several cultivars regularly offered. Among the commonest are 'Pendula' (selected from plants known as forma **pendula** Kirchner) and 'Fastigiata' (selected from var. **fastigiata** Jaeger and including forma **pyramidalis** Moehl).

3. C. orientalis Miller. Illustration: Reichenbach, Icones florae Germanicae et Helveticae **12**: t. 634 (1850); Hora, Oxford encyclopaedia of trees of the world, 142 (1981); Krüssmann, Manual of cultivated broad-leaved trees and shrubs **1**: 281 & pl. 94, 95 (1986).
Small tree or shrub to 5 m. Young branches with silky hairs. Leaves 2.5–5 × 1.2–2.5 cm, ovate to ovate-elliptic, acute, rounded at the base, margins with 2 rows of forward-pointing teeth, veins 12–15 pairs. Male catkins 1.2–2 cm. Fruiting clusters 2.7–6 cm, bracts 1–2.3 cm, ovate, margins toothed on both sides. Nuts 5 mm, hairy at apex. *SE Europe, SW Asia.* H2. Spring.

4. C. laxiflora (Siebold & Zuccarini) Blume (*Distegocarpus laxiflora* Siebold & Zuccarini). Illustration: Hooker's Icones Plantarum **20** t. 1989 (1891); Elwes & Henry, Trees of Great Britain & Ireland **3**: t. 201 (1908); Nakai, Flora sylvatica Koreana **2**: t. 12 (1915); Krüssmann, Manual of cultivated broad-leaved trees & shrubs **1**: 279, 280 (1986).
Tree 13–17 m. Leaves 3.5–8 × 2.5–4 cm, ovate-elliptic, cordate or slightly rounded at base, abruptly narrowed at apex into a long slender point, margins with 2 rows of forward-pointing teeth, veins 8–10 pairs. Fruiting clusters 5–7 cm, bracts 1.3–2 cm, central lobe toothed on 1 side only. *W & C China, Japan.* H1–2. Spring.

More commonly grown is var. **macrostachya** Oliver (*C. fargesii* Franchet) which has leaves to 10 cm and fruiting clusters to 13 cm.

5. C. tschonoskii Maximowicz (*C. yedoensis* Maximowicz; *C. fauriei* Nakai). Illustration: Nakai, Flora sylvatica Koreana **2**: t. 10 (1915); Krüssmann, Manual of cultivated broad-leaved trees & shrubs **1**: 279 (1986). Tree to 15 m. Young branches with soft hairs, sometimes persisting through the winter. Leaf-stalks 8 mm, blades 4–9 × 1–4 cm, ovate, acuminate, rounded at

base, marginal teeth mucronate, irregular or in 2 rows, veins 9–15 pairs. Fruiting clusters 5–7 cm, with slender hairy stalks 8 mm long; bracts 1.5–2 cm, ovate to lanceolate, acute, toothed on 1 side. Nuts 4–6 mm, hairless. *NE Asia, N China and Japan.* H1. Spring.

C. henryana Winkler is closely related and is sometimes treated as a variety. However the leaves are slightly shorter and have only 1 row of marginal teeth. The nuts are softly hairy. Another species in the same group is **C. fargesii** Winkler. The only reliable distinctions are its bluntly toothed leaf-margins and shorter blades (4–6.5 cm).

6. C. cordata Blume. Illustration: Das Pflanzenreich **61**: 25, 27 (1904); Nakai, Flora sylvatica Koreana **2**: t. 7 (1915); Krüssmann, Manual of cultivated broad-leaved trees & shrubs **1**: 279–281 & pl. 94, 95 (1986).
Tree 12–17 m, with scaly, furrowed bark. Young branches hairy at first, soon becoming hairless, terminal buds to 2 cm. Leaf-stalks 1.2–2 cm, blades 7–12 × 3.5–8 cm, ovate to oblong-ovate, acuminate, cordate, margins irregularly toothed, teeth sometimes in 2 distinct rows, veins 15–20 pairs. Male catkins 2.5–5 cm. Fruiting clusters 7–12 cm, bracts closely overlapping, margins with few distant, sharply pointed teeth. Nuts partially covered by overlapping bract lobes. *N & W China, NE Asia, Japan & Korea.* H1. Spring.

7. C. japonica Blume (*C. carpinoides* Makino; *Distegocarpus carpinus* Siebold & Zuccarini). Illustration: Das Pflanzenreich **61**: 25, 27 (1904); Botanical Magazine, 8534 (1914); Bean, Trees & shrubs hardy in the British Isles **1**: 508 (1970); Krüssmann, Manual of cultivated broad-leaved trees & shrubs **1**: 279 & pl. 94 (1986).
Tree 12–17 m, with scaly furrowed bark. Young shoots downy at first. Leaf-stalks 6–12 mm, downy, blades 5–10 × 1.8–4.5 cm, oblong-ovate to oblong lanceolate, acuminate, rounded or cordate at base, margins irregularly toothed, veins 20–24 pairs. Male catkins 2.5–5 cm. Fruiting clusters 5–6.5 cm, bracts overlapping. Nuts covered by a bractlet and part of the bract. *Japan.* H1. Spring.

4. OSTRYA Scopoli
S.G. Knees
Deciduous trees with rough scaly bark. Buds pointed, covered in overlapping scales. Leaves alternate, stalked, ovate to ovate-

oblong, acuminate, margins sometimes with 2 rows of forward-pointing teeth, blades hairless to softly hairy with 9–15 pairs of veins. Flowers unisexual, appearing in spring with the leaves. Male catkins pendulous, developing in autumn, flowers lacking perianth but with bracts surrounding 5–15 stamens. Female catkins erect with 2 flowers in each bract-axil, calyx closely adpressed to ovary. Bract and 2 bracteoles joined to form tubular involucre. Stigmas 2, linear. Fruit a ribbed nut, enclosed by the involucre, which expands considerably after fertilisation.

A genus of 10 species from temperate North America, Europe and Asia. All are quite hardy and can be grown in most soil types. Propagation is normally from seed. Literature: Rushforth, K., Key to Carpinus and Ostrya, *Plantsman* **8**: 249–256 (1987).

1a. Leaves softly hairy with 9–12 pairs of veins **3. japonica**
  b. Leaves hairless or with few hairs, vein-pairs 11–15 **2**
2a. Fruiting bracts 1.3–1.7 cm; leaves rounded at the base, stalks 5–10 mm, glandless; bark grey **1. carpinifolia**
  b. Fruiting bracts 2–2.5 cm; leaves cordate at base, stalks 2–5 mm, often with glands; bark brown **2. virginiana**

1. O. carpinifolia Scopoli (*O. vulgaris* Willdenow; *O. italica* Winkler; *Carpinus ostrya* Linnaeus, in part). Illustration: Revue Horticole **5**: 188–189 (1905); Nicholson & Clapham, Oxford book of trees, 152 (1975); Humphries et al., Trees of Britain and Europe, 119 (1981); Krüssmann, Manual of cultivated broad-leaved trees & shrubs **2**: 340 (1986). Tree to 20 m with grey bark. Young branches hairy. Leaf-stalks 5–10 mm, without glands, blades 5–10 cm, ovate-acuminate and rounded at the base with few hairs above, hairs confined to veins beneath, veins 11–15 pairs, margins with 2 distinct rows of forward-pointing teeth. Male catkins 3–7.5 cm × 2.5–3 mm, bracts finely pointed. Fruiting clusters 3.5–5 × 2 cm, nuts 4–5 mm, ovoid, with a tuft of hairs at the apex, enclosed in a bladder-like involucre 1–1.5 cm long. *SE Europe, W Asia.* H2. Spring.

2. O. virginiana (Miller) Koch (*Carpinus virginiana* Miller; *Carpinus ostrya* Linnaeus, in part; *Ostrya virginica* Willdenow). Illustration: Schneider, Illustriertes Handbuch der Laubholzkunde **1**: 138 (1906); Elwes & Henry, Trees of Great

Britain & Ireland **3**: t. 201 (1908); Krüssmann, Manual of cultivated broad-leaved trees & shrubs **2**: 340 (1986). Tree to 20 m with brown bark. Young branches with glandular hairs. Leaf-stalks 2–5 mm, with glandular hairs, blades 5–12 × 2–5 cm, ovate-lanceolate, cordate at base, acuminate, softly hairy beneath, veins 12–15 pairs, margins with sharp forward-pointing teeth, not in 2 distinct rows. Male catkins 5 cm. Fruiting clusters 3–6 × 1.5–3 cm, nuts 6–9 mm, spindle-shaped, hairless at apex; involucre 2–2.5 cm. *Eastern N America.* H2. Spring.

3. O. japonica Sargent. Illustration: Sargent, Forest flora of Japan, t. 22 (1894); Schneider, Illustriertes Handbuch der Laubholzkunde **1**: 137 (1906); Elwes & Henry, Trees of Great Britain and Ireland **3** t. 201 (1908); Nakai, Flora sylvatica Koreana **2**: t. 6 (1915). Tree to 25 m with greyish bark. Young branches densely covered with soft hairs. Leaf-stalks 6–9 mm, blades 7.5–12.5 × 3–5 cm, ovate to ovate-oblong, rounded or cordate at base, long-acuminate, coarsely and irregularly toothed, softly hairy on both surfaces, veins 9–12 pairs. Male catkins 5–6 cm. Fruiting clusters 3–4.5 × 2 cm, nuts 5–6.5 mm, oblong-ovoid, hairless at the apex. Involucre 1.4–1.8 cm. *China, Japan, Korea.* H2. Spring.

5. OSTRYOPSIS Decaisne
S.G. Knees
Deciduous shrubs with hairy branches and pointed buds covered by overlapping scales. Leaves alternate, simple, with toothed margins. Flowers unisexual, appearing with the leaves. Male flowers in pendulous catkins, perianth absent, filaments divided into 2 at apex, anthers hairy at apex. Female flowers in very short spikes, each bract subtending 2 flowers, involucre with 3 teeth, calyx united to ovary, ovule 1 per cell, style bifid. Fruit a nut enclosed in a tubular involucre.

A genus of just 2 species, only one of which is regularly cultivated. Both are Chinese and are closely related to *Corylus* and *Ostrya*, being somewhat intermediate between the 2 genera. Propagation is normally by seed.

1. O. davidiana (Baillon) Decaisne (*Corylus davidiana* Baillon). Illustration: Das Pflanzenreich **61**: 19 (1906); Schneider, Illustriertes Handbuch der Laubholzkunde **1**: 137 (1906); Krüssmann, Manual of cultivated broad-leaved trees & shrubs **2**: 340 (1986).

Deciduous suckering shrub to 3 m. Young branches hairy. Leaf-stalks 5 mm, blades 3–7 cm, broadly ovate, cordate at base, margins with 2 rows of forward-pointing teeth, densely hairy and with stalkless red glands on underside. Fruits 6–12 together in long stalked clusters. Involucre 1.2–1.8 cm, splitting on 1 side to release nuts. Nuts 8 mm, ovoid, beaked. *N & W China.* H2. Spring.

6. CORYLUS Linnaeus
S.G. Knees
Deciduous shrubs, more rarely trees. Winter buds broadly ovoid, obtuse, with many overlapping scales. Leaves broadly ovate, softly hairy, margins with 2 rows of forward-pointing teeth. Flowers unisexual, usually appearing before the leaves. Male catkins pendulous, flowers lacking perianth, each bract with 4–8 stamens, filaments bifid, anthers with long hairs at apex. Female flowers in capitate clusters, enclosed in small scaly buds, with only the styles projecting, ovaries with 1–2 ovules per cell, styles bifid to base, usually red. Fruit a spherical or ovoid nut, with woody pericarp, enclosed or surrounded by a leafy involucre; involucral bracts toothed or dissected, joining to form a tube.

A small genus of shrubby plants from temperate Asia, Europe and North America grown for both their ornamental value and their edible nuts. About 15 species are usually recognised, several with many cultivars and selected forms. Propagation is normally from suckers or else by seed. Literature: Schneider, B., Clavis analyticus sectionum specierumque Asiae orientalis Himalayaeque, in Sargent, C.S. (ed.), *Plantae Wilsonianae* **2**: 447–455 (1916); Kasapligil, B., Corylus colurna and its varieties, *Journal of the Californian Horticultural Society* **24**(4): 94–105 (1963); Kasapligil, B., Chromosome studies in the genus Corylus, *Scientific Reports of the Faculty of Science, Ege University No. 59*: 3–14 (1968); Kasapligil, B., A bibliography of Corylus (Betulaceae) with annotations, *63rd Annual report of the northern Nut Growers Association*, 107–162 (1972).

1a. Winter buds acute; involucre with spiny teeth **2**
  b. Winter buds obtuse; involucre without spiny teeth **3**
2a. Involucre hairless; leaves broadly ovate **1. tibetica**
  b. Involucre with soft hairs; leaves narrowly oblong **2. ferox**

3a. Leaf-stalks 1.7–2 cm; involucre
divided into linear lobes   4
  b. Leaf-stalks 8–11 mm; involucre
irregularly divided into lanceolate or
ovate lobes   6
4a. Involucre cylindric, constricted
sharply above the nut into a short
tube, lobes recurved   **5. chinensis**
  b. Involucre bell-shaped with deeply
divided lobes   5
5a. Involucre covered with fine down and
gland-tipped bristly hairs   **3. colurna**
  b. Involucre lacking glandular hairs
                                4. jacquemontii
6a. Involucre tubular, the join between
bracts indistinct   7
  b. Involucre openly bell-shaped,
consisting of 2 overlapping bracts   9
7a. Involucre slightly constricted above
nut   **9. maxima**
  b. Involucre abruptly constricted above
nut   8
8a. Leaf-stalks 6–12 mm; anthers
yellowish   **10. cornuta**
  b. Leaf-stalks 1.5–2.5 cm; anthers red
                                11. sieboldiana
9a. Leaves irregularly and sometimes
deeply lobed; involucral teeth broadly
triangular   **8. heterophylla**
  b. Leaves shallowly and regularly lobed;
involucre with forward-pointing,
narrowly triangular teeth   10
10a. Involucre 1.5–2 cm, about as long as
nut   **6. avellana**
  b. Involucre 2.5–3 cm, about twice as
long as nut   **7. americana**

1. C. tibetica Batalin (*C. ferox* var. *tibetica*
Franchet). Illustration: Revue Horticole,
203 (1910); Hora, Trees of the world, 144
(1981).
Tree or shrub to 8 m. Branches hairless,
brown, with conspicuous lenticels. Leaf-
stalks 1.5–2.5 cm, blades 5–12 × 4–7 cm,
broadly obovate or ovate, cordate at base,
apex long-acuminate, margins with
irregular, sharp, forward-pointing teeth.
Male catkins 5–7 cm, bracts acuminate.
Nuts 1–1.5 cm, compressed, in groups of
3–6, with spiny hairless involucres, to
4–5 cm across. *C & W China*. H1. Spring.

2. C. ferox Wallich. Illustration: Das
Pflanzenreich **61**: 45 (1904).
Very similar to *C. tibetica* but with leaves
6–14 × 3–6 cm, narrowly oblong.
Involucre with soft hairs between the
spines. *E Himalaya*. H3. Spring.

3. C. colurna Linnaeus. Illustration: Das
Pflanzenreich **61**: 49 (1904); Botanical
Magazine, 9469 (1937); Bean, Trees &

shrubs hardy in the British Isles 1: 724
(1970); Krüssmann, Manual of cultivated
broad-leaved trees & shrubs 1: 378–379
(1986).
Tree to 25 m, trunk with peeling yellowish
grey bark. Branches glandular hairy when
young, becoming corky and furrowed with
age. Leaf-stalks 1.2–2.5 cm with glandular
hairs when young, blades 8–12 × 5–9 cm,
broadly ovate to obovate, cordate at base,
margins with 2 rows of forward-pointing
teeth, sometimes lobed. Male catkins
5–7 cm. Fruits clustered; involucre deeply
divided into linear, recurved lobes,
glandular hairy; nuts 1.5–2 cm across. *SE
Europe, W Asia*. H1. Spring.

4. C. jacquemontii Decaisne (*C. lacera*
Wallich). Illustration: Botanical Magazine,
n.s., 391 (1960).
Very similar to *C. colurna* but with
distinctly obovate leaves which are more
deeply lobed and sharply toothed. Involucre
without glandular hairs. *Himalaya*. H1.
Spring.

5. C. chinensis Franchet (*C. colurna* var.
chinensis (Franchet) Burkhill). Illustration:
Das Pflanzenreich **61**: 49–50 (1904).
Tree to 40 m, closely related to *C. colurna*,
with branches persistently glandular-hairy.
Leaves 10–18 cm, very oblique, long-
acuminate, margins finely and evenly
toothed but not lobed. Fruits in clusters of
4–6; involucre 4 × 2.5 cm, hairless, bottle-
shaped, constricted sharply above the nut
into a short tube, lobes recurved,
irregularly toothed. Nuts 1.5 cm across,
almost spherical. *C & W China*. H1. Spring.

6. C. avellana Linnaeus. Illustration: Hora,
Trees of the world, 144 (1981); Humphries
et al., Trees of Britain and Europe, 121
(1981); Krüssmann, Manual of cultivated
broad-leaved trees & shrubs 1: 378–379
(1986).
Shrub 4–6 cm. Branches glandular hairy.
Leaf-stalks 6–15 mm, blades 5–10 ×
4–7 cm, rounded, obovate, cordate at base,
margins shallowly lobed at apex,
irregularly toothed, downy on both sides.
Male catkins 3–6 cm. Fruits in clusters of
1–4, involucre shorter or about the same
length as nut, deeply and irregularly
divided into narrow toothed lobes. Nuts
1.5–2 cm, spherical to ovoid. *SE Europe,
SW Asia, N Africa*. H1. Spring.

  Many cultivars of ornamental and
culinary value have been selected over the
centuries; among the most widely grown
are: 'Aurea' (forma *aurea* Kirchner) with
yellowish branches and leaves; 'Contorta'

(var. *contorta* Bean) with curiously twisted
and curled branches (illustration:
Krüssmann, Manual of cultivated broad-
leaved trees & shrubs 1: pl. 141, 1986);
'Pendula' (forma *pendula* Dippel) with
weeping branches (illustration: Krüssmann,
Manual of cultivated broad-leaved trees and
shrubs 1: pl. 143 (1986)).

7. C. americana Marshall. Illustration: Das
Pflanzenreich **61**: 49 (1904); Krüssmann,
Manual of cultivated broad-leaved trees &
shrubs 1: 379 (1986).
Very similar to *C. avellana*, possibly its
American representative, but generally
smaller, forming a much-branched shrub to
3m. Young shoots with glandular hairs.
Leaf-stalks 8–15 mm, blades 6–12 ×
4–8 cm, broadly ovate to elliptic, rounded
at base, shortly acuminate. Male catkins
3.5–7 cm. Fruits in clusters of 2–6,
occasionally solitary; involucre 2.5–3 cm,
about twice as long as nut, deeply and
irregularly lobed. *N America, from
Saskatchewan to Florida*. H1. Spring.

8. C. heterophylla Fischer & Trautvetter.
Illustration: Das Pflanzenreich **61**: 49
(1904).
Tree or shrub to 7 m, similar to *C. avellana*
and probably replacing it in the far east.
Branches with glandular hairs. Leaf-stalks
1–2 cm, blades 5–10 × 3–8 cm, broadly
obovate, cordate at base, margins
irregularly toothed. Male catkins 2–4 cm.
Fruits solitary or in pairs, rarely 3 per
cluster. Involucre bell-shaped, divided into
large triangular teeth, longer than nut,
with glandular hairs near the base. Nuts
1.5 cm across, almost spherical. *NE Asia,
Japan*. H1. Spring.

9. C. maxima Miller (*C. tuberosa*
Willdenow). Illustration: Das Pflanzenreich
61: 49 (1904); Humphries et al., Trees of
Britain and Europe, 121 (1981).
Small tree or shrub, closely related to *C.
avellana* but larger, reaching 7–8 m. Leaf-
stalks 6–12 cm, blades 5–12 × 4–10 cm,
rounded or obovate, cordate at base,
margins with 2 rows of teeth. Male catkins
5–8 cm. Fruits solitary or in clusters of
2–3; involucre 4–5 cm, tubular, completely
enclosing nut and about twice as long.
Nuts 1.5–2.5 cm, ovoid to oblong. *SE
Europe, SW Asia*. H2. Spring.

  The most ornamental variety is var.
purpurea (Loudon) Rehder, usually sold as
'Purpurea': illustration: Botanical
Magazine, n.s., 268 (1956); Hora, Trees of
the world, 144 (1981). Leaves and catkins
deep reddish purple.

10. C. cornuta Marshall (*C. rostrata* Aiton). Illustration: Hora, Trees of the world, 144 (1981); Krüssmann, Manual of cultivated broad-leaved trees and shrubs **1**: 379 (1986).

Shrub 2–3 m, with erect, much-branched stems. Branches slightly hairy. Leaf-stalks 6–12 mm, blades 4–11 × 2.5–7.5 cm, ovate or obovate, cordate at base, margins with fine teeth, sometimes slightly lobed, underside with soft hairs. Male catkins 2.5–3 cm, anthers yellowish. Fruits solitary or in pairs; involucre 4 cm, abruptly constricted into a narrow tube above the nut, with fine down and bristly hairs. Nuts 1.2–1.8 cm. *Eastern N America*. H1. Spring.

11. C. sieboldiana Blume (*C. rostrata* Aiton var. *sieboldiana* (Blume) Maximowicz). Illustration: Nakai, Flora sylvatica Koreana **2**: t. 4 (1915); Krüssmann, Manual of cultivated broad-leaved trees & shrubs **1**: 379 & pl. 142 (1986).

Very similar to *C. cornuta* and probably its Asian representative. Shrub to 5 m with hairy branches. Leaf-stalks 1.5–2.5 cm, blades 10–15 × 8–10 cm, rounded to obovate, cordate at base, margins lobed with 2 rows of forward-pointing teeth. Male catkins with red anthers. Fruits solitary or in clusters of 2–3; involucre 1.5–4 cm, tubular, constricted above the nut, divided at apex into short, entire lobes, densely covered with short bristles. Nuts 1.2–1.4 cm, conical. *N China, Japan & Korea*. H1. Spring.

This species is most commonly represented in cultivation by var. **mandshurica** (Maximowicz) Schneider from northern China.

XXXIX. FAGACEAE

Monoecious trees or shrubs. Buds with overlapping scales, small to large and conspicuous. Leaves alternate, sometimes 2-ranked, stalked, entire to toothed, scolloped or lobed, evergreen or deciduous. Stipules usually deciduous. Flowers unisexual: male flowers in erect or pendent catkins with a single flower to each bract, or in stalked heads or solitary or in 3-flowered cymes, perianth 4–7-lobed, stamens 8–40; female flowers solitary or in 3s enclosed by a cupule made up of overlapping and partly fused scales, these solitary or in spikes, rarely at the bases of the male catkins, ovary inferior, 3–6-celled,

styles 3, ovules 2 per cell. Fruit of 1–3 nuts enclosed in the enlarged cupule.

A family of 8 genera and about 600 species found all over the world except in Africa south of the Sahara.

1a. Male flowers solitary or in 3s or in long-stalked heads   2
  b. Male flowers in erect or pendent catkins   3
2a. Male and female flowers solitary or in 3s; cupule with 1–3 nuts   **2. Nothofagus**
  b. Male flowers in heads; female flowers in pairs; cupule with 2 nuts   **1. Fagus**
3a. Male flowers in pendent catkins   **7. Quercus**
  b. Male flowers in stiffly erect catkins   4
4a. Leaves deciduous, toothed; ovary 6-celled   **3. Castanea**
  b. Leaves evergreen, usually not toothed; ovary 3-celled   5
5a. Nut solitary, clearly projecting from the cupule   **6. Lithocarpus**
  b. Nuts 1–3, more or less completely enclosed in the cupule   6
6a. Young shoots and undersides of the leaves covered in bright yellow hairs (which age to tawny)   **5. Chrysolepis**
  b. Young shoots and undersides of leaves hairy or not, hairs, if present, not bright yellow   **4. Castanopsis**

1. FAGUS Linnaeus
J. Cullen & H.S. Maxwell

Usually large, monoecious trees with smooth, grey bark. Buds large, spindle-shaped. Leaves in 2 ranks, entire, toothed or scolloped (lobed in some cultivars), usually silky-hairy, at least when young. Male flowers in long-stalked, more or less spherical heads; perianth 4–6-lobed, stamens 8–12. Female flowers usually 2 together surrounded by a stalked, 4-lobed, scaly cupule made up of needle-like or somewhat flattened appendages; perianth 4–6-lobed, ovary 3-celled. Fruit a nut, released on the splitting of the cupule into 4 segments.

A genus of about 10 species from temperate parts of the northern hemisphere. The common beech, *F. sylvatica* is widespread as a native tree in Europe and is also widely cultivated both within and beyond its natural distribution. All grow in most soils, though a light soil is preferred. They are best propagated from fresh nuts, but cultivars have to be grafted on *F. sylvatica* stocks.

Literature: Wyman, D., Registration list of cultivar names in Fagus, *Arnoldia* **24**: 1–8

(1964), **27**: 62 (1967); Grootendorst, H.J., Fagus, *Dendroflora* **11/12**: 3–17 (1975).

1a. Cupule-stalk to 2.5 cm, usually downy; leaves green beneath   2
  b. Cupule-stalk longer than 2.5 cm, hairless; leaves somewhat glaucous or greyish green beneath   6
2a. Veins running to the sinuses between the scollops, sometimes projecting there as small teeth   3
  b. Veins running to the points of the teeth or the middles of the scollops, not to the sinuses between them   4
3a. Veins running out as small teeth in the sinuses between the scollops   **5. lucida**
  b. Veins running to the sinuses or the scollops but not projecting as small teeth   **4. crenata**
4a. Leaves broadest at or below the middle; vein-pairs up to 8; leaves finely toothed (deeply so or lobed in some cultivars)   **1. sylvatica**
  b. Leaves broadest above the middle, distinctly toothed; vein-pairs more than 8   5
5a. Some appendages at the base of the cupule flattened, linear or spathulate   **2. orientalis**
  b. All appendages on the cupule needle-like   **3. grandifolia**
6a. Nuts considerably longer than the small cupule and projecting from it   **8. japonica**
  b. Nuts about as long as the cupule, not projecting   7
7a. Some appendages at the base of the cupule spathulate, leaf-like; leaf-stalk 4–13 mm; vein-pairs 10–12   **6. engleriana**
  b. All appendages of the cupule needle-like; leaf-stalk 8–30 mm; vein-pairs 12–15   **7. longipetiolata**

1. F. sylvatica Linnaeus. Illustration: Garcke, Illustrierte Flora, edn 23, 428 (1972); Hart & Raymond, British trees in colour, 13, 14 (1973); Polunin & Everard, Trees and bushes of Europe, 50 (1976); Il Giardino Fiorito **48**: 612, 613 (1982).

Tree of 30–50 m, with very smooth grey bark. Young shoots silky-hairy at first. Leaves ovate to elliptic, acute at the apex, rounded to the base, remotely and finely toothed (entire or variably toothed or lobed in cultivars, see below), veins running out to points of the teeth, dark green above, paler beneath and silky-hairy at least when young, usually persistently so along the veins; 3.5–10 × 2.5–4 cm, stalk 3–10 mm, vein-pairs 5–8. Stalk of cupule to 2.5 cm, densely hairy; appendages of cupule all needle-like. Nuts about as long as cupule. *Most of Europe, Turkey.* H1.

A very widely grown species which is extremely variable in gardens; numerous cultivars for use as specimen trees or for street plantings or plantations are available, varying in form and leaf coloration, toothing, shape and hairiness. Only a small number can be included here; for further information see the papers by Wyman and Grootendorst cited above.

Variants in the form of the tree.

'Aurea Pendula': like 'Pendula' but leaves yellow when young.

'Borneyensis': a small conical tree with pendulous branches.

'Dawyck': a tall, narrowly columnar tree, often grown under the name 'Fastigiata'.

'Miltonensis': branches gracefully drooping.

'Pendula': main branches more or less horizontal, smaller branches more or less drooping.

'Purpurea Pendula': slow growing and bushy with a rounded crown, drooping branches and brown-purple leaves.

Variants with coloured leaves.

'Aurea Variegata': leaves edged with yellow.

'Purpurea': leaves pale red in spring, purple when mature; the most widely grown coloured-leaved beech; others with purple or brown-purple foliage are available.

'Purpurea Tricolor': like 'Purpurea' but leaves narrower and edged with pink or white.

'Rohanii': leaves brownish purple with the margins deeply and irregularly toothed.

Variants in leaf-shape and toothing.

'Asplenifolia': leaves variously lobed and/or toothed, sometimes long and narrow without any lobing.

'Cristata': leaves shortly stalked and clustered, crumpled, the apex curved downwards.

'Grandidentata': leaves coarsely but regularly toothed.

'Laciniata': leaves deeply and regularly toothed or lobed.

'Rotundifolia': leaves entire and almost circular, closely set.

2. F. orientalis Lipsky (*F. sylvatica* var. *macrophylla* Hohenacker; *F. macrophylla* (Hohenacker) Koidzumi; *F. asiatica* (de

Candolle) Winkler). Illustration: Botanischer Jahrbücher **29**: 286 (1900); Schneider, Illustriertes Handbuch der Laubholzkunde **1**: 152 (1904); Polunin & Everard, Trees and bushes of Europe, 51 (1976); Phillips, Trees in Britain, Europe and North America, 116 (1978).

Like *F. sylvatica* but leaves mostly obovate, distinctly toothed, 6.5–15 × 3.5–8 cm, with 8–13 pairs of veins; some scales at the base of the cupule flattened, linear or spathulate. *SE Europe, SW Asia.* H3.

Intermediates between this and *F. sylvatica* are found in the Balkan peninsula and have been named **F. moesiaca** (Maly) Czeczott.

3. F. grandifolia Ehrhart (*F. ferruginea* Aiton; *F. americana* Sweet). Illustration: Gleason, Illustrated flora of the eastern United States and adjacent Canada **2**: 38 (1952).

Similar to *F. sylvatica* but usually less tall, leaves mostly obovate, distinctly toothed, 5–14 × 1.8–6.5 cm, stalk 5–12 mm, vein-pairs 9–15. *Eastern N America.*

4. F. crenata Blume (*F. sieboldii* de Candolle; *F. sylvatica* var. *sieboldii* (de Candolle) Maximowicz). Illustration: Botanischer Jahrbücher **29**: 286 (1900); Makino's new illustrated Flora of Japan, 87, f. 347 (1963); Kitamura & Okamoto, Coloured illustrations of trees and shrubs of Japan, f. 112, 113 (1977).

Tree to 30 m. Young shoots sparsely hairy. Leaves ovate or elliptic or diamond-shaped, shortly acuminate at the apex, rounded at the base, shallowly scolloped, the veins running to the sinuses between the scollops but projecting as teeth; 2.5–10 × 1.8–5.6 cm, stalks 2–10 cm, vein-pairs 7–10. Stalk of cupule to 2.5 cm, hairy. Some appendages at the base of the cupule flattened, linear or spathulate. *Japan.* H3.

5. F. lucida Rehder & Wilson.

Like *F. crenata* but leaves ovate to narrowly ovate, the veins projecting as small teeth in the sinuses between the scollops, 5–9 × 2–4.5 cm, stalk 3–13 mm, vein-pairs 8–10; appendages of the cupule broadly triangular with recurved, free tips. *W China.* H5.

6. F. engleriana Seemen (*F. sylvatica* var. *chinensis* Franchet). Illustration: Botanischer Jahrbücher **29**: 286 (1900).

Tree to 25 m, usually much smaller, with several trunks from the base. Young shoots hairless. Leaves ovate to elliptic or slightly obovate, apex acuminate, base tapered or rounded, margins wavy or slightly toothed

or scolloped, dark green above, greyish green beneath, with silky hairs along the veins, 4–10 × 1.5–5 cm, stalk 4–13 mm, vein-pairs 10–12. Stalk of cupule more than 2.5 cm, slender and hairless. Some appendages at the base of the cupule flattened, leaf-like, spathulate. *C China.* H3.

7. F. longipetiolata Seemen (*F. sylvatica* var. *longipes* Oliver).

Like *F. engleriana* but generally a taller tree with a single trunk, leaves finely and distantly toothed, usually rather densely hairy beneath, 7–13 × 3–7.5 cm, stalk 8–30 mm, vein-pairs 12 or more; all appendages of the cupule needle-like. *C & W China.* H3.

8. F. japonica Maximowicz. Illustration: Botanischer Jahrbücher **29**: 286 (1900); Schneider, Ilustriertes Handbuch der Laubholzkunde **1**: 155 (1904); Makino's new illustrated Flora of Japan, 87, t. 348 (1963).

Tree to 25 m with usually several trunks. Young shoots hairy when young, soon hairless. Leaves ovate or ovate-elliptic, acuminate at the apex, tapered or somewhat rounded at the base, margin wavy or scolloped, upper surface dark green, lower surface greyish green, persistently hairy. Stalk of cupule slender, hairless, more than 2.5 cm. Cupule small, the nuts much longer than it and projecting. *Japan.* H4.

2. NOTHOFAGUS Blume

J. Cullen & H.S. Maxwell

Deciduous or evergreen, monoecious trees or sometimes shrubs. Leaves alternate, often asymmetric, pleated, flat or the margins rolled under in bud; buds small, ovoid. Male flowers solitary or in 3-flowered cymes with 4–6-lobed perianth and 8–40 stamens. Female flowers 1–3 in a stalkless or shortly stalked cupule which is 2–4-lobed and made up of entire, toothed or laciniate scales, often with glandular tips or appendages. Nuts 1–3 per cupule.

A genus of 20 or more species from the temperate and tropical areas of the southern hemisphere. Several species are grown in Europe as ornamentals and a few species (especially *N. obliqua* and *N. alpina*) are grown on an increasing scale for forestry. They are relatively easy to grow in suitable climates (all are somewhat tender in most of Europe) but will not tolerate calcareous soils. They are also very prone to damage by strong winds. Propagation is by fresh seed.

Literature: Van Steenis, G.C.C.J., Results of the Archbold expeditions: Papuan Nothofagus, *Journal of the Arnold Arboretum* **24**: 301–374 (1953); Clarke, D., Notes on South American species of Nothofagus with a key to the species of Nothofagus cultivated in the British Isles, *International Dendrology Society Yearbook for 1970*, 24–29 (1971); Phillipson W.R. & Phillipson M.N., Leaf vernation in Nothofagus, *New Zealand Journal of Botany* **17**: 417–421 (1971); Wigston, D.L., Nothofagus Blume in Britain, *Watsonia* **12**: 344–345 (1978); Romero, E.J., Arquitectura foliar de las species sudamericanas de Nothofagus Bl., *Boletin de la Sociedad Argentina de Botanica* **19**: 289–308 (1980); Romero, E.J. & Carrasco Aguirre, A., Arquitectura foliar de las species australianas y neocelandesas de Nothofagus, *Boletin de la Sociedad Argentina de Botanica* **20**: 227–240 (1982).

Leaves. Evergreen: **1–7**; deciduous: **8–11**. Entire: **7**: toothed: **3–6,10,11**; scolloped (sometimes the scollops themselves toothed): **1,2,8,9**. Closely and densely hairy beneath: **7**. Finely and persistently hairy on the surface above: **6**.

Vein-pairs. Up to 8: **1,2,4–7,10,11**; 8–10: **8**; 10 or more: **3,9**.

Young leaves in bud. Flat: **1–5**; margins rolled downwards: **6,7**; folded upwards along the midrib and between the veins: **8–11**.

Cupule. 3-lobed: **7**; 4-(rarely 2-)lobed: **1–6,8–11**.

1a. Leaves evergreen   2
  b. Leaves deciduous   8
2a. Leaves completely entire   **7. solandri**
  b. Leaves variously toothed or scolloped   3
3a. Leaves 4.5–7 cm, acute, vein-pairs usually more than 10   **3. moorei**
  b. Leaves smaller, usually rounded or obtuse, vein-pairs less than 10   4
4a. Leaves scolloped, not gland-dotted   5
  b. Leaves toothed, gland-dotted   6
5a. Leaves doubly scolloped, with hair-fringed pits towards the base beneath   **1. menziesii**
  b. Leaves simply scolloped, without hair-fringed pits towards the base beneath   **2. cunninghamii**
6a. Leaves with 4–7 deep, forwardly-pointing, acute teeth on either side, finely hairy above   **6. fusca**
  b. Leaves with more numerous, small, triangular teeth on either side, hairless except on the midrib above   7
7a. Male flowers solitary; most leaves doubly toothed, broad, ovate or broadly ovate, length/breadth ratio up to 1.4 (narrower leaves may occur near the tips of the branches)   **4. betuloides**
  b. Male flowers in 3s; all leaves mostly simply toothed, lanceolate or narrowly ovate, length/breadth ratio more than 2   **5. dombeyi**
8a. Leaves with at most 6 vein-pairs   9
  b. Leaves with at least 8 vein-pairs   10
9a. Margin of leaf between adjacent veins with 2 teeth   **11. pumilio**
  b. Margin of leaf between adjacent vein-pairs with several teeth   **10. antarctica**
10a. Vein-pairs 8–9, each vein running to the middle of a scollop on the leaf margin   **8. obliqua**
  b. Vein-pairs more than 10, each vein running to a sinus between 2 scollops on the leaf margin   **9. alpina**

1. N. menziesii (J.D. Hooker) Oersted. Figure 14(1), p. 62. Illustration: Everett, Living trees of the world, 114 (1969); Leathart, Trees of the world, 112 (1977); Salmon, The native trees of New Zealand, 209–211 (1980); Boletin de la Sociedad Argentina de Botanica 20: pl. I, 4 (1982).
Evergreen tree to 20 m or more in the wild. Young shoots hairy. Leaves flat in bud, broadly ovate to almost circular, rounded at the apex, truncate or very broadly wedge-shaped at base, margins coarsely and doubly scolloped, hairless above except at the base of the midrib, hairless beneath except for the base of the midrib and 2 or 3 hair-fringed pits near the base; 6–18 × 5–15 mm, stalk hairy, 1–3 mm, vein-pairs 2–3. Male flowers solitary. Cupules with 4 lobes, each covered with glandular appendages and ending in a gland-tipped point. *New Zealand.* H5.

2. N. cunninghamii (J.D. Hooker) Oersted. Figure 14(2), p. 62. Illustration: Cochrane et al., Flowers and plants of Victoria, f. 432 (1973); Boland et al., Forest trees of Australia, 132 (1984).
Similar to *N. menziesii* but leaves triangular-ovate, the base usually truncate, simply scolloped and without hair-fringed pits beneath, 4–15 × 5–11 mm, stalk hairy, 1–2 mm, vein-pairs 2–4. *Australia (Tasmania).* H5.
N. nitida (Philippi) Krasser is very similar to *N. cunninghamii* but has almost circular leaves, rounded at the base, 2–3.5 × 1.5–2.5 cm. *Chile.* Not in general cultivation in Europe, though it may be found in some specialist collections.

3. N. moorei (Mueller) Krasser. Figure 14(3), p. 62. Illustration: Boland et al., Forest trees of Australia, 132 (1984); Boletin de la Sociedad Argentina de Botanica 20: pl. I, 5 (1982).
Evergreen tree to 30 m or more in the wild. Young shoots hairy. Leaves flat in bud, glossy and dark green, ovate or ovate-lanceolate, acute at apex, tapered or rounded at the base, margins sharply and regularly toothed, hairless above and beneath except for a few hairs along the midrib; 4–7.5 × 2–4.5 cm, stalk hairy, 2–5 mm, vein-pairs 10 or more. Cupule with 4 lobes, scaly and with glandular appendages. *Australia (New South Wales and Queensland).* H5.

4. N. betuloides (Mirbel) Blume. Figure 14(4), p. 62. Illustration: Boletin de la Sociedad Argentina de Botanica 19: pl. II, 7, III, 3 (1980); Rodriguez, Matthei & Quezada, Flora arbórea de Chile, 237, 238 (1983).
Dense, evergreen tree to 30 m in the wild, usually much less in cultivation. Young shoots hairy. Leaves flat in bud, ovate to broadly ovate to almost circular (those towards the tips of the twigs often relatively narrower) rounded to obtusely pointed at the apex, broadly rounded to the base, margins irregularly doubly toothed, hairless above and beneath except towards the base of the midrib, gland-dotted; 1.1–2.5 cm × 8–22 mm, stalk hairy, 1–3 mm, vein-pairs 3–5. Male flowers solitary. Cupule with 4 lobes covered with short, tooth-like appendages. *Chile, Argentina.* H5.

5. N. dombeyi (Mirbel) Blume. Figure 14(5), p. 62. Illustration: Phillips, Trees in Britain, Europe and North America, 141 (1978); Boletin de la Sociedad Argentina de Botanica 19: pl. II, 5, 6 (1980); Rodriguez, Matthei & Quezada, Flora arbórea de Chile, 240, 241 (1983).
Very similar to *N. betuloides* but leaves all relatively narrower, lanceolate or lanceolate-ovate, somewhat irregularly but mostly simply toothed, 2.2–2.7 cm × 9–11 mm, stalk hairy, 1–3 mm, vein-pairs 4–6. Male flowers in 3s. *Chile, Argentina.* H5.
Very difficult to distinguish from *N. betuloides* in the absence of male flowers. The leaves are generally narrower than those of *N. betuloides* (though this may have equally narrow leaves towards the ends of the twigs) and the toothing is usually simple.

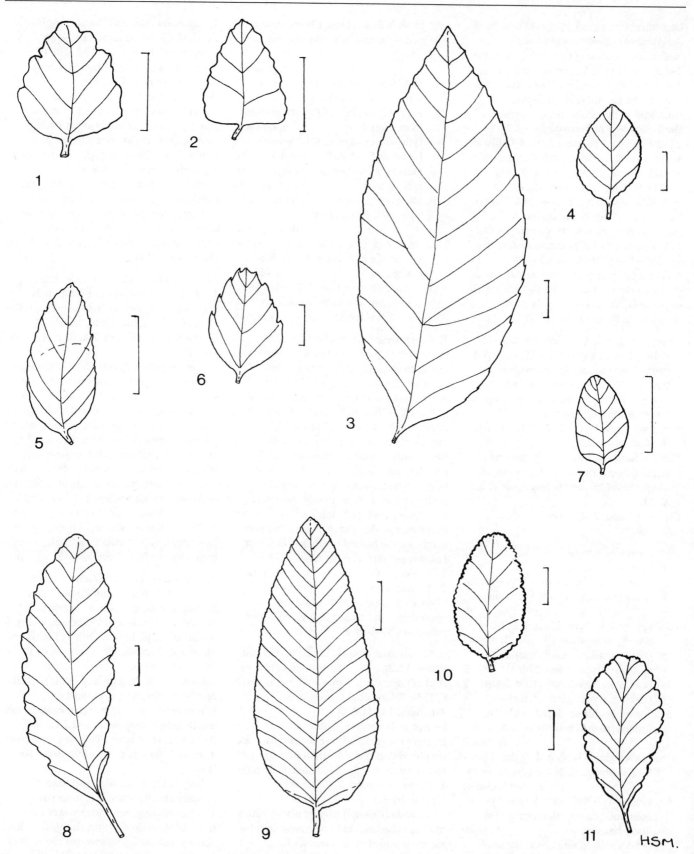

Figure 14. Leaves of *Nothofagus* species. 1, *N. menziesii*.
2, *N. cunninghamii*. 3, *N. moorei*. 4, *N. betuloides*. 5, *N. dombeyi*.
6, *N. fusca*. 7, *N. solandri*. 8, *N. obliqua*. 9, *N. alpina*. 10, *N. antartica*.
11, *N. pumilio*. Scales = 1 cm.

6. N. fusca (J.D. Hooker) Oersted. Figure
14(6), p. 62. Illustration: Bean, Trees and
shrubs hardy in the British Isles edn 8, **3**:
12 (1976); Salmon, The native trees of
New Zealand, 214, 215 (1980).
Large evergreen tree to 30 m or more in
the wild, usually much less in cultivation.
Young shoots hairy, often conspicuously
zig-zag. Leaves with margins rolled
downwards in bud, ovate to broadly ovate
or almost circular, apex rounded or
somewhat acute, base rounded or tapered,
margins deeply toothed with 4–7 acute,
forwardly-directed teeth on each side, finely
hairy above, hairless beneath except for the
margins and the base of the midrib, gland-
dotted; 1.6–4.5 × 1.2–1.9 cm, stalk hairy,
2–4 mm, vein-pairs 3–5. Cupule with 4
lobes covered with transverse, hairy plates.
New Zealand. H5.

7. N. solandri (J.D. Hooker) Oersted. Figure
14(7), p. 62. Illustration: Barber & Philips,
The trees around us, 101 (1975); Bean,
Trees and shrubs hardy in the British Isles
edn 8, **3**: 19 (1976); Salmon, The native
trees of New Zealand, 216, 217 (1980);
Boletin de la Sociedad Argentina de
Botanica **20**: pl. I, 2 (1982).
Evergreen tree to 25 m in the wild, usually
much smaller in cultivation. Young shoots
hairy. Leaves with the margins rolled
downwards in bud, ovate-elliptic to elliptic-
oblong, very rounded at the apex, rounded
at the base, margins entire, thinly hairy
above when young, later hairless, densely
and closely hairy beneath; 4–12 ×
3–6 mm, stalks hairy, 1–2 mm, vein-pairs
3–6. Cupule with 3 lobes which are
covered in transverse plates. *New Zealand.*
H4.

Var. **cliffortioides** (J.D. Hooker) Poole (*N.
cliffortioides* (J.D. Hooker) Oersted) has more
ovate, acute leaves.

8. N. obliqua (Mirbel) Blume. Figure 14(8),
p. 62. Illustration: Barber & Philips, The
trees around us, 101 (1975); Phillips, Trees
in Britain, Europe and North America, 142
(1978); Boletin de la Sociedad Argentina de
Botanica **19**: pl. I, 3,4 (1980); Rodriguez,
Matthei & Quezada, Flora arbórea de Chile,
252, 253 (1983).
Deciduous tree to 30 m or more in the wild.
Young shoots hairless. Leaves folded
upwards and between the veins in bud,
oblong or oblong-lanceolate, rounded to
obtuse at the apex, tapered or rounded and
often very unequal at the base, dark green
and hairless above (except along the
midrib), paler and sparsely hairy, especially
along the veins beneath, margin scolloped,

the scollops finely and irregularly toothed,
the veins running to the middles of the
scollops; 2–9 × 1–3.6 cm; stalk hairy,
3–12 mm; vein-pairs 8–10. Cupule with 4
lobes covered with broad, hairy, gland-
tipped scales. *Chile, Argentina.* H5.

N. glauca (Philippi) Mirbel is very similar
but has a 2-lobed cupule. It is from Chile
and is probably not in general cultivation
in Europe.

9. N. alpina (Poeppig & Endlicher) Oersted
(*N. procera* (Poeppig & Endlicher) Oersted).
Figure 14(9), p. 62. Illustration: Barber &
Philips, The trees around us, 101 (1975);
Leathart, Trees of the world, 112 (1977);
Boletin de la Sociedad Argentina de
Botanica **19**: pl. I, 7, II, 4 (1980);
Rodriguez, Matthei & Quezada, Flora
arbórea de Chile, 231, 232 (1983).
Deciduous tree to 25 m in the wild. Young
shoots hairy. Leaves folded upwards and
between the veins in bud, lanceolate-elliptic
to lanceolate-oblong, obtuse at the apex,
truncate or rounded at the base, margins
shallowly scolloped, the scollops finely
toothed, veins running to the sinuses
between the scollops, hairless except along
the midrib above, finely and sparsely hairy,
mainly along the veins and gland-dotted
beneath; 2–12 × 1–3.6 cm, stalk hairy,
2–10 mm, vein-pairs 14–22. Cupule with
4 lobes covered by pinnately laciniate hairy
scales which are glandular at the tips of all
the lobes. *Chile, Argentina.* H4–5.

N. alessandri Espinosa, also from Chile, is
similar but has a 2-lobed cupule covered
with simple appendages. It is not in general
cultivation in Europe.

10. N. antarctica (Forster) Oersted. Figure
14(10), p. 62. Illustration: Barber & Philips,
The trees around us, 101 (1975); Phillips,
Trees in Britain, Europe and North
America, 141 (1978); Boletin de la
Sociedad Argentina de Botanica **19**: pl. II,
1,2 (1980); Rodriguez, Matthei & Quezada,
Flora arbórea de Chile, 234, 235 (1983).
Deciduous tree to 16 m in the wild. Young
shoots hairy. Leaves folded upwards and
between the veins in bud, ovate to broadly
ovate, apex rounded, base truncate or
cordate, margin finely toothed with several
teeth between the endings of adjacent
veins, hairless above (except along the
midrib), sparsely hairy along the main
veins beneath; 1–4.5 cm × 8–25 mm, stalk
hairy, 3–6 mm, vein-pairs 4–6. Cupule
4-lobed, containing 3 nuts, the lobes with a
few, entire, transverse scales. *Chile.* H4.

11. N. pumilio (Poeppig & Endlicher)
Krasser. Figure 14(11), p. 62. Illustration:
Boletin de la Sociedad Argentina de
Botanica **19**: pl. II, 3, III, 2 (1980);
Rodriguez, Matthei & Quezada, Flora
arbórea de Chile, 256, 257 (1983).
Similar to *N. antarctica* but leaves oblong or
somewhat obovate, toothed with 2 teeth
between the endings of adjacent veins,
2–4.2 × 1.3–2.4 cm, stalk hairy 3–5 mm,
vein-pairs 5–7. Cupule containing a single
nut. *Chile, Argentina.* H4.

3. CASTANEA Miller
J. Cullen & H.S. Maxwell
Deciduous, monoecious trees or shrubs.
Buds small, ovoid. Leaves in 2 ranks,
toothed, often hairy and often with short
glands which are translucent when young,
yellowish or orange when mature and
whitish when old, on the surface. Male
catkins erect, often showy, with a 6-lobed
perianth and 10–12 (rarely to 20) stamens.
Female flowers borne towards the bases of
the male catkins or on separate catkins,
usually grouped in 3s, each group
surrounded by a cupule (involucre) made
up of spine-like, spreading or reflexed
scales. Ovary 6-celled. Nuts large, 1–3
(rarely more) borne within the enlarged
cupule which splits at maturity into 2–4
flaps or segments.

A genus of about 12 rather similar
species from temperate regions of the
northern hemisphere, several cultivated for
ornament as well as for their edible nuts.
Many hybrids have been raised or have
arisen spontaneously, rendering the
identification of plants of unknown origin
very difficult; for further details of these
hybrids see the work by Camus cited below.

The plants are reasonably tolerant as to
soil, but most thrive best in areas with hot,
dry summers. Propagation is most
successful with fresh nuts, but some
cultivars must be grafted on *C. sativa*
stocks.
Literature: Camus, A., *Les Chataignier*,
1–241 (1929).

1a. Plant a shrub reaching at most 60 cm
    6. alnifolia
  b. Plant a tree or shrub of more than
    1 m   2
2a. Leaves completely hairless beneath 3
  b. Leaves hairy beneath, sometimes on
    the veins only   4
3a. Nut 1 in each cupule; vein-pairs in
    each leaf 14–22; teeth always bristle-
    tipped; male catkins to 9 cm
    3. henryi

b. Nuts 2–3 in each cupule; vein-pairs in each leaf 22 or more; teeth mucronate, not bristle-tipped; male catkins more than 9 cm   **4. dentata**

4a. Leaves hairy on the veins only beneath   **5. seguinii**

b. Leaves hairy all over the surface beneath, at least when young   5

5a. Nut 1 in each cupule; leaves 4.5–10 cm, stalk 3–7 mm   **2. pumila**

b. Nuts 2–3 in each cupule; leaves 9 cm or more, stalk 8–30 mm   **1. sativa**

1. C. sativa Miller (*C. vesca* Gaertner; *C. vulgaris* Lamarck). Illustration: Edlin & Nimmo, The world of trees, 16, 17 (1974); Leathart, Trees of the world, 118, 119 (1977); Barber & Philips. The trees around us, 73 (1975).

Large tree to 40 m or more; bark ridged longitudinally, the ridges often spiralling, dark. Young shoots downy at first or hairless, ultimately hairless. Leaves oblong to oblong-lanceolate, acuminate at apex, tapered, rounded, truncate or somewhat cordate at base, toothed or scolloped with bristle-tipped teeth, densely hairy at least when young and with yellow glands beneath, midrib more or less hairless, 9–29 × 3–9 cm, stalk 1.5–3 cm, vein-pairs 20–24. Nuts 2–3 in each cupule. *S Europe, SW Asia, N Africa*. H2.

The Sweet or Spanish Chestnut, widely cultivated for its edible nuts as well as for ornament. Numerous cultivars are available. Related to it are 2 species which are difficult to distinguish from it or each other unless their wild origins are known:

C. mollissima Blume (*C. bungeana* Blume). Illustration: Everett, Living trees of the world, 111 (1969). Young shoots densely hairy; leaves 9–21 × 3.5–9.5 cm, stalk 1.5–2 cm, vein-pairs 13–20, base usually somewhat rounded, densely white-hairy beneath with no visible glands, midrib covered with long and short hairs. *China, Korea*. H3.

C. crenata Siebold & Zuccarini (*C. japonica* Blume). Illustration: Kurata, Illustrated important forest trees of Japan 1: pl. 41 (1971). Young shoots hairy; leaves 7.5–15 × 2.5–4 cm, stalk 8–15 mm, vein-pairs 17–26, base usually truncate or cordate, whitish or greyish hairy beneath with visible glands, midrib with long hairs only. *China, Japan*. H3.

2. C. pumila (Linnaeus) Miller. Illustration: Schneider, Illustriertes Handbuch der Laubholzkunde **1**: 155 (1904), **2**: f. 91 (1912); Justice & Bell, Wild flowers of North Carolina, 52 (1968); Elias, The complete trees of North America, 305 (1980).

A shrub to 1 m or more with suckers or a small tree. Young shoots persistently hairy. Leaves elliptic-oblong or oblong-obovate, bluntly pointed at the apex, tapered or rounded towards the base, toothed, the teeth mucronate but not bristle-tipped, 4.5–10 × 2–3.3 cm, stalk 3–7 mm, vein-pairs 14–17, greyish hairy with glands scarce or not visible beneath. Nut 1 (rarely 2) in each cupule. *Eastern N America*. ?H3.

Var. **ozarkensis** (Ashe) Tucker (*C. ozarkensis* Ashe) is taller and has somewhat larger leaves; it is found in the southern part of the species' range.

3. C. henryi (Skan) Rehder & Wilson. Tree to 30 m; young shoots hairless, almost black. Leaves oblong-ovate to elliptic, apex acuminate, base rounded to truncate, toothed with wide, shallow, forwardly-pointing, bristle-tipped teeth, completely hairless (or rarely with a few sparse hairs on the midrib) but sparsely glandular beneath, 8–17.5 × 2–5 cm, stalk 1.3–1.6 cm, vein-pairs 14–22. Male catkins to 9 cm. Nut 1 per cupule. *C & W China*. H3.

4. C. dentata (Marsh) Borkhausen (*C. americana* Rafinesque). Illustration: Justice & Bell, Wild flowers of North Carolina, 51 (1968); Journal of the Royal Horticultural Society 99: pl. 73 (1974); Elias, The complete trees of North America, 304 (1980).

Tree to 30 m; young shoots completely hairless or soon becoming so. Leaves oblong-elliptic to oblong-lanceolate, apex acuminate, base rounded, toothed with mucronate, sometimes hooked teeth, hairless but glandular beneath, 11–24 × 3.5–7.5 cm, stalk 6–25 mm, vein-pairs 22–27. Male catkins more than 9 cm. Nuts 2–3 per cupule. *Eastern N America*. H3.

Almost extinct in the wild due to the ravages of Chestnut Blight.

5. C. seguinii Dode. Tree to 10 m. Young shoots hairy. Leaves elliptic-oblong to ovate-oblong, rarely somewhat obovate, apex acuminate, base rounded or truncate, widely and shallowly toothed with mucronate teeth, hairy only on the veins beneath, the surface conspicuously glandular, 7–17 × 2.5–7 cm, stalk 4–12 mm, vein-pairs 12–22. Nuts 2–3 per cupule. *E, C & S China*. H5.

6. C. alnifolia Nuttall (*C. nana* Muehlenberg). Illustration: Schneider, Illustriertes Handbuch der Laubholzkunde 2: f. 564 (1912).

Small suckering shrub to 60 cm. Leaves oblong or oblong-obovate, mostly obtuse or rounded at the apex, rounded at the base, toothed with distant, wide, mucronate teeth, brownish hairy and with few glands beneath, 5–15 × 3–6 cm, vein-pairs 12–16. *SE USA*. H5.

4. CASTANOPSIS (D.Don) Spach

J. Cullen & H.S. Maxwell

Large, evergreen, monoecious trees with scaly bark. Buds small, ovoid. Leaves usually very leathery, usually toothed, more rarely entire. Male catkins borne at the ends of the branches: female catkins distinct. Male flowers with a 5–6-lobed perianth and 10–12 stamens. Female flowers usually 3 together within a common, often spiny cupule which ultimately splits into segments. Nuts ripening in their second year.

A genus of about 100 species from SE Asia, most not hardy in Europe; only a small number is cultivated, and these only in the most climatically favourable areas. Cultivation as for *Castanea*.

The plant often grown as *Castanopsis chrysophylla* (or *C. sempervirens*) is now placed in the genus *Chrysolepis*.

1a. Leaves hairy beneath, the hairs often plastered down tightly on to the surface   2

b. Leaves hairless beneath   3

2a. Leaves greyish white beneath, rounded or shortly acute at the apex; cupule covered with branched spines   **4. delavayi**

b. Leaves greenish or brownish beneath, conspicuously and abruptly acuminate; cupule smooth with transverse ridges or bands of tubercles   **3. cuspidata**

3a. Leaf-stalk 1–3 cm   **1. chinensis**

b. Leaf-stalk 2–10 mm   **2. concolor**

1. C. chinensis Hance. Tree to 12 m in the wild. Young shoots hairless. Leaves ovate, lanceolate or ovate-oblong to lanceolate-oblong, very leathery, apex acuminate, base tapered to rounded, toothed in the upper half, 5–13 × 1.5–3.5 cm, stalk 1–3 cm, vein-pairs 9–15. Male catkins long and interrupted, the axis visible. Cupule covered with branched spines. *S China*. H5.

2. C. concolor Rehder & Wilson. Very similar to *C. chinensis* but leaves ovate-elliptic to oblong, 6.5–11 × 1.5–5 cm, stalk 2–10 mm, vein-pairs 9–12. *W China*. H5.

3. C. cuspidata (Thunberg) Schottky (*Quercus cuspidata* Thunberg). Illustration: Kitamura & Okamoto, Coloured illustrations of trees and shrubs of Japan, f. 128 (1977).

Tree to 15 m with widely spreading then pendent branches. Young shoots hairy at first. Leaves very leathery, ovate to lanceolate or ovate-oblong, abruptly long-acuminate at the apex, tapered at the base, margins more or less entire or wavy or toothed in the upper half, hairy beneath, the hairs plastered tightly down to the surface giving it a brownish green, metallic tinge, 4–13 cm × 8–75 mm, stalk 5–15 mm, vein-pairs 8–12. Cupules ovoid, rather smooth, with transverse ridges or bands of tubercles. *Japan, China*. H4.

'Variegata' has smaller leaves with at least the margins creamy yellow.

4. C. delavayi Franchet.
Tree to 25 m in the wild. Young shoots hairy at first. Leaves leathery, elliptic to elliptic-obovate, apex rounded or shortly pointed, base tapering or somewhat rounded, margins coarsely toothed, greyish white beneath with dense, tightly plastered hairs, 4–11 × 1.5–6.5 cm, stalk 6–15 mm, vein-pairs 6–8. Cupules covered with short, stiff spines. *W China*. H5.

5. CHRYSOLEPIS Hjelmqvist
J. Cullen & H.S. Maxwell
Similar to *Castanopsis* but leaves with bright yellow hairs beneath (fading to tawny as the leaves age), the cupule made up from its inception of 7 free, spiny segments, 5 surrounding the 3 nuts, 2 others within and between the nuts; female flowers borne at the bases of the male catkins.

A genus of a single species (sometimes treated as 2) from western N America (Oregon & California); it is grown as for *Castanopsis*.

1. C. chrysophylla (Hooker) Hjelmqvist (*Castanopsis chrysophylla* (Hooker) de Candolle; *Castanea chrysophylla* Hooker). Illustration: Botanical Magazine, 4953 (1856); Bean, Trees and shrubs hardy in the British Isles edn 8, 1: 611 (1970); Elias, The complete trees of North America, 308 (1980); Krüssmann, Manual of cultivated broad-leaved trees and shrubs 1: 326 (1984).

Small tree to 10 m in cultivation. Young shoots densely hairy with yellow hairs. Leaves ovate, lanceolate or more rarely obovate, acute at apex, tapered to the base, margins entire, bright yellow-hairy beneath, 1.5–7 cm × 8–25 mm, stalk

4–15 mm, vein-pairs 4–10. Cupule covered with slender, branched spines. *Western USA (California, Oregon)*. H4.

A variable species. 'Obovata' is a dwarf shrub. Var. **minor** (Bentham) de Candolle is also dwarf, with small leaves. Var. **sempervirens** (Kellogg) Henry (*C. sempervirens* (Kellogg) Hjelmqvist; *Castanopsis sempervirens* (Kellogg) Dudley) is a dwarf shrub with rather oblong, obtuse leaves.

6. LITHOCARPUS Blume
J. Cullen & H.S. Maxwell
Trees, like *Quercus* but leaves always evergreen, usually very leathery, entire or toothed. Male catkins upright, stiff. Stamens 10–12.

A genus of about 100 species, mostly from the tropics and subtropics. Only a small number is grown, and these only in the most favourable climates. They will grow in most soils; propagation is by fresh acorns.
Literature: Camus, A., *Les chênes* 3: 511–1196 (1952–54).

1a. Leaves toothed, at least above the
    middle   **1. densiflorus**
 b. Leaves entire   2
2a. Leaves hairless or with very small,
    scale-like hairs and shining beneath
         6. edulis
 b. Leaves hairy beneath   3
3a. Young shoots hairy   4
 b. Young shoots hairless   5
4a. Acorns in groups of 3, fused together
    by their cups into barnacle-like
    masses   **3. pachyphyllus**
 b. Acorns in a raceme but not grouped
    in 3s, the cups free from each other
         2. glabra
5a. Leaves greyish green to glaucous
    beneath; acorns enclosed for most of
    their length in the cup
         4. cleistocarpa
 b. Leaves green beneath; most of the
    acorn projecting from the cup
         5. henryi

1. L. densiflorus (Hooker & Arnott) Rehder (*Quercus densiflora* Hooker & Arnott; *Pasania densiflora* (Hooker & Arnott) Oersted). Illustration: Sargent, Silva of North America 8: t. 438 (1895); Elias, The complete trees of North America, 310 (1980); University of Washington Arboretum Bulletin 44(4): 3,4 (1981). Tree to 30 m in the wild, usually much less in cultivation. Young shoots densely hairy. Leaves lanceolate-oblong, acute, toothed above the middle, hairy beneath, 6–12 ×

2–4 cm, vein pairs 9–19, stalk 1–1.5 cm. Acorns solitary or in pairs, the base enclosed in a cup made up of spreading or reflexed scales. *Western USA (Oregon, California)*. H4.

Var. **echinoides** (R. Brown) Abrams is shrubby with smaller, scarcely toothed, obtuse leaves.

2. L. glabra (Thunberg) Nakai (*Quercus glabra* Thunberg; *Pasania glabra* (Thunberg) Oersted). Illustration: Elwes & Henry, Trees of Great Britain and Ireland 5: t. 338 (1910); Makino's new illustrated Flora of Japan, f. 368 (1963); Kitamura & Okamoto, Coloured illustrations of trees and shrubs of Japan, f. 129 (1977).

Tree to 15 m, usually smaller in cultivation. Young shoots hairy. Leaves elliptic-oblong to lanceolate, entire, apex abruptly acuminate, hairy beneath, 6–12 × 2–3.5 cm, stalk 1–2 cm, vein-pairs 6–9. Acorns in a raceme, cup enclosing only the base; the individual cups free from each other, made up of adpressed scales. *Japan, China*. ?H3.

3. L. pachyphyllus (Kurz) Rehder (*Quercus pachyphyllus* Kurz). Illustration: Brandis, Indian trees, 631 (1906).

Very like *L. glabra* but leaves 14–21 × 4–6.5 cm, vein pairs 8–11, stalk 1–2 cm, acorns large, borne in groups of 3, each largely covered by the cup, the cups of each group of 3 united to form a barnacle-like mass. *E Himalaya, W China*. H5.

4. L. cleistocarpa (Seemen) Rehder & Wilson (*Quercus cleistocarpa* Seemen; *Q. wilsonii* Seemen).

Tree to 15 m in the wild. Young shoots hairless. Leaves elliptic-oblong to elliptic-lanceolate, entire, acuminate, 6–15 × 2–5.5 cm, stalk 1–2.5 cm, vein-pairs 8–14, greyish green to glaucous and hairy beneath. Acorn almost completely enclosed in the cup which is made up of adpressed scales. *W China*. H5.

5. L. henryi (Seemen) Rehder & Wilson (*Quercus henryi* Seemen). Illustration: Journal of the Royal Horticultural Society 58: f. 38 (1933).

Tree to 18 m in the wild. Young shoots hairless. Leaves oblong-lanceolate to oblong-elliptic, 8–20 × 3–8 cm, entire, abruptly acuminate, green and hairy beneath, stalk 1–4 cm, vein-pairs 9–16. Acorn almost spherical, enclosed only at the base by the cup which is made up of adpressed scales. *S & W China*. H5.

6. L. edulis (Makino) Nakai (*Pasania edulis* Makino; *Quercus edulis* Makino).
Illustration: Makino's new illustrated Flora of Japan, f. 367 (1963).
Tree to 10 m in the wild. Young shoots hairless. Leaves oblong-obovate or oblanceolate, 8–15 × 2.5–7 cm, abruptly acute, green, hairless and shining beneath (though with small, scale-like hairs when young); stalk 1–2 cm, vein-pairs 8–12. Acorns in 3s, ovoid, the lower quarter enclosed in the cup which is made up of adpressed scales. *Japan.* ?H3.

7. QUERCUS Linnaeus
J. Cullen & H.S. Maxwell

Monoecious trees or shrubs, often with stellate or tree-like hairs. Leaves evergreen, semi-evergreen (persisting from one spring to the next) or deciduous, alternate, entire, toothed or variously lobed. Male flowers in long, slender, hanging catkins, each flower with a 4–7-lobed perianth, 4–6, rarely to 12 stamens and, often, a rudimentary ovary in the centre. Female flowers solitary or a few together on a short or long common stalk; perianth inconspicuous, 6-lobed, ovary inferior, usually 3-celled, with 3 long or short styles; ovules 2 per cell. Fruit a nut (acorn) containing a single seed, surrounded at least at the base by a cup (cupule) made up of scales which are closely adpressed at least in their lower parts.

A genus of about 600 species, mostly from the northern hemisphere, but reaching the southern in S America and SE Asia. Numerous species are cultivated as specimen trees or in plantations; many colour well in autumn. They are usually long-lived trees, sometimes slow-growing, but generally are easy to grow in a suitable climate on most soils. Propagation is by seed, which must be sown as fresh as possible, or by grafting. The relative hardiness in Europe of the species is very difficult to establish, and the codings given here are speculative.

Oaks are extremely variable, particularly in terms of their hairiness, leaf-shape and lobing. This, together with the frequent occurrence of hybrids or intermediate forms (especially in the wild but also in gardens) makes their identification difficult. The account below, especially the key, is based, as far as possible, on easily seen vegetative characters and must be used with caution; mature foliage (preferably with several leaves) and branches from the crown of the tree are required; sucker shoots and branches arising directly from the trunk are often atypical and their use will lead to incorrect identifications.

In the key and descriptions, leaves are decribed as toothed if the marginal indentations are acute, and reach only a short distance from margin to midrib; they are described as lobed if the indentations reach further; lobes are usually oblong or triangular, usually have rounded sinuses between them, and are often rounded at the apex (though sometimes tipped with a bristle or mucro). In identifying a plant it may sometimes be necessary to key it twice, once treating the leaves as toothed, a second time treating them as lobed. The illustrations (Figures 15–19, pp. 69–78) should be of some help in this regard.

The classification of the genus into subgenera and sections is difficult, being based mainly on characters of the female flower (especially the nature of the styles) and the acorn (size, shape, hairiness, position of the abortive ovules and type of cup). There is no widely accepted system; in this account the subgenera and sections recognised by Camus (reference below) are interpolated for guidance. Camus's scheme, though not entirely satisfactory, is the only one to cover species from the whole range of the genus.
Literature: Camus, A., *Les chênes: monographie du genre Quercus* (1934–54); Schwarz, O., Monographie der Eichen Europas und des Mittelmeergebietes, *Feddes Repertorium*, Sonderbeiheft D (1936–39); Miller, H. & Lamb, S., *The oaks of North America* (1985).

1a. Leaves evergreen, persisting for more than one summer 2
　b. Leaves deciduous or semi-evergreen (i.e. falling in the spring following that in which they developed) 15
2a. Leaves hairy over most of the surface beneath 3
　b. Leaves completely hairless beneath or with tufts of hairs in the axils of the main veins 11
3a. Leaves almost stalkless, broadly elliptic to almost circular, brown-hairy beneath **4. semecarpifolia**
　b. Leaves stalked, variously shaped but not as above, not brown-hairy beneath 4
4a. Leaves with yellowish hairs beneath 5
　b. Leaves with white or grey hairs beneath 7
5a. Leaves mostly entire, lanceolate to narrowly elliptic, greyish green or silvery **42. hypoleucoides**

　b. Leaves not as above 6
6a. Leaves circular to obovate, margins entire **5. alnifolia**
　b. Leaves narrowly ovate or elliptic, margins toothed, at least in the upper half **39. chrysolepis**
7a. Leaves entire or toothed, if toothed the teeth not spine-tipped 8
　b. Leaves toothed, the teeth spine-tipped 10
8a. Leaves glaucous, obovate, acuminate, toothed in the upper part; vein-pairs 10 or more **2. glauca**
　b. Combination of characters not as above 9
9a. Leaves conspicuously tapered to the base, apex rounded **29. virginiana**
　b. Leaves rounded or slightly tapered to the base, apex usually acute **20. ilex**
10a. Leaves distinctly lobed (see under 14.) × **hispanica**
　b. Leaves toothed **7. suber**
11a. Leaves conspicuously acuminate, 7–20 cm 12
　b. Leaves acute or obtuse, 2.5–7 cm 13
12a. Leaves 4–6 cm broad; margins entire, wavy **3. acuta**
　b. Leaves 2–3 cm broad; margins toothed **1. myrsinifolia**
13a. Large tree; leaves with tufts of hairs in the axils of the major veins beneath **40. agrifolia**
　b. Shrub or small tree (in gardens); leaves completely hairless beneath 14
14a. Buds ovoid, pointed **41. wislizenii**
　b. Buds almost spherical, rounded **6. coccifera**
15a. Leaves absolutely entire, neither toothed nor lobed; occasionally slightly 3-lobed towards the apex 16
　b. Leaves distinctly lobed or toothed 20
16a. Leaves persistently hairy above and beneath **33. douglasii**
　b. Leaves hairless at least above 17
17a. Leaves hairy beneath all over the surface **45. imbricaria**
　b. Leaves completely hairless beneath or with tufts of hairs in the axils of the major veins 18
18a. Leaves broadly obovate **48. arkansana**
　b. Leaves linear to oblong or narrowly elliptic 19
19a. Leaves linear to oblong, never lobed **44. phellos**
　b. Leaves narrowly elliptic, rarely oblong, often obscurely 3-lobed towards the apex **43. laurifolia**
20a. Lobes or teeth with bristle-like points 21

b. Lobes or teeth entire or mucronate but not with bristle-like points   36
21a. Leaves distinctly lobed   22
b. Leaves toothed but not lobed   32
22a. Leaves obovate to spoon-shaped with 3 or rarely 5 distinct lobes towards the apex   23
b. Leaves of other shapes, usually with 5 or more evenly distributed lobes   24
23a. Leaves spoon-shaped, hairless   **46. nigra**
b. Leaves very broadly obovate, hairy beneath at least when young   **47. marilandica**
24a. Leaves persistently whitish or greyish downy beneath   25
b. Leaves green beneath, hairy or hairless, not persistently downy   26
25a. Lobes of leaves broadly triangular   **50. ilicifolia**
b. Lobes of leaves narrowly triangular, long, often the terminal lobe sickle-shaped   **49. falcata**
26a. Leaves hairy all over the surface beneath, at least when young   27
b. Leaves hairless beneath except for tufts of hairs in the axils of the main veins when young   28
27a. Mature leaves with persistent tufts of hairs in the axils of the main veins beneath   **56. velutina**
b. Mature leaves completely hairless beneath   **57. kelloggii**
28a. Sinuses between the leaf-lobes deep, the longest lobe longer than the width of the narrowest part of the leaf; leaves shiny above   29
b. Sinuses not so deep, the longest lobe not longer than the width of the narrowest part of the leaf; leaves dull above   **55. rubra**
29a. Tufts of hairs in the axils of the main veins small and inconspicuous   **52. coccinea**
b. Tufts of hairs in the axils of the main veins large and conspicuous   30
30a. Bases of most leaves tapering   **51. palustris**
b. Bases of most leaves truncate   31
31a. Winter buds chestnut brown   **53. ellipsoidalis**
b. Winter buds greyish or yellowish brown   **54. shumardii**
32a. Leaves white- or yellow-hairy above and beneath; vein-pairs 4–8; base auriculate; young shoots densely white- or yellow-downy; winter buds densely white-hairy   **9. macrolepis**
b. Combination of characters not as above   33
33a. Upper surface of leaves dark green,

lower surface white- or yellowish-downy   **13. variabilis**
b. Leaves green on both surfaces, hairless or sparsely hairy   34
34a. Leaves up to 3 times longer than broad, lanceolate   **12. acutissima**
b. Leaves 3–6 times longer than broad, linear-oblong   35
35a. Leaf-stalk 8–20 mm   **8. libani**
b. Leaf-stalk to 8 mm   **11. trojana**
36a. Thread-like, persistent stipules present below the main terminal and axillary buds   37
b. All stipules deciduous, not persisting below main buds   40
37a. Leaves with triangular teeth   **10. castaneifolia**
b. Leaves with distinct lobes   38
38a. Leaves elliptic to olong, mostly less than 10 cm   **14. cerris**
b. Leaves obovate, mostly 10 cm or more   39
39a. Leaf-base narrowly or broadly tapered   **17. macranthera**
b. Leaf-base rounded or auriculate   **18. frainetto**
40a. Leaves toothed, or lobed to less than one-third of the distance from margin to midrib   41
b. Leaves deeply lobed, the sinuses extending for more than one-third of the distance from margin to midrib   51
41a. Leaves with small, mucronate teeth   42
b. Leaves with shallow, rounded, obtuse or rarely somewhat acute lobes   46
42a. Most leaves more than 10 cm; vein-pairs 15–20   **16. pontica**
b. Most leaves less than 10 cm; vein-pairs up to 12   43
43a. Leaves glaucous beneath; stalks more than 1 cm   **21. glandulifera**
b. Leaves not glaucous beneath; stalks usually less than 1 cm   44
44a. Leaves persistently hairy beneath   **24. fruticosa**
b. Leaves completely hairless beneath   45
45a. Leaves 4–7 cm, elliptic to obovate   **25. infectoria**
b. Leaves 8–10 cm, obovate   **32. warburgii**
46a. Shoots hairy into their second year   **15. dentata**
b. Shoots hairless or sometimes with hairs when young, the hairs soon falling   47
47a. Leaves more or less hairless beneath   48
b. Leaves persistently hairy beneath   49

48a. Leaf-stalks very short   **22. mongolica**
b. Leaf-stalks 8–20 mm   **23. canariensis**
49a. Leaves persistently hairy above   **33. douglasii**
b. Mature leaves hairless above   50
50a. Leaves greyish white beneath   **30. bicolor**
b. Leaves green or brownish beneath   **31. prinus**
51a. Branches hairless, or, if hairy when young, then hairs falling rapidly   52
b. Branches persistently hairy into their second year   54
52a. Leaf-stalks less than 1 cm   **27. robur**
b. Leaf-stalks more than 1 cm   53
53a. Leaves glaucous beneath, with 3–4 lobes on each side   **37. alba**
b. Leaves not glaucous beneath, with 4–6 lobes on each side   **26. petraea**
54a. Leaves persistently hairy above   55
b. Mature leaves hairless above   57
55a. Leaves usually 5-, rarely 7-lobed   **36. stellata**
b. Leaves usually with more than 7 lobes   56
56a. Lobes parallel-sided, then abruptly tapered towards the apex   **19. pyrenaica**
b. Lobes parallel-sided, either truncate or broadening, often toothed or lobed towards the apex   **35. lobata**
57a. Leaf-stalks more than 1 cm   **34. garryana**
b. Leaf-stalks less than 1 cm   58
58a. Leaves 5–10 cm   **28. pubescens**
b. Leaves more than 10 cm   59
59a. The median sinuses of the leaf deeper than the others, and reaching almost to the midrib; base wedge-shaped   **38. macrocarpa**
b. All sinuses more or less equally deep, not as above; base rounded or auricled   **18. frainetto**

Subgenus **Cyclobalanopsis** (Oersted) Schneider. Scales of the cup joined to form concentric circles; leaves leathery, evergreen, entire or slightly scolloped; style short, widened towards the apex; acorns ripening in their first year; aborted ovules towards the apex of the fruit.

1. Q. myrsinifolia Blume (*Q. vibreyana* Franchet & Savatier). Figure 15(1), p. 69. Illustration: Schneider, Illustriertes Handbuch der Laubholzkunde 1: 189 (1904); Makino, Illustrated Flora of Japan, t. 1131 (1924); Krüssmann, Manual of broad-leaved trees and shrubs, pl. 29 (1986).

Tree to 25 m in the wild, usually much smaller in cultivation; bark blackish. Leaves evergreen, lanceolate, acuminate, tapered to the base, 7–12 × 2–3 cm, margins very weakly toothed or scolloped, shining above, pale glaucous green beneath, entirely hairless; stalk 1–1.8 cm. Acorn ovoid-oblong, 1.5–2 cm; cup hairless, enclosing about the lower third of the acorn. *S China, Japan, Laos.* H5.

2. Q. glauca Thunberg. Figure 15(2), p. 69. Illustration: Makino, Illustrated Flora of Japan, t. 1132, 1133 (1924); Kitamura & Okamoto, Coloured illustrations of trees and shrubs of Japan, f. 116 (1977).
Tree to 17 m; young shoots hairy at first. Leaves evergreen, elliptic-oblong to ovate-oblong, acuminate, tapered to base, 5–13 × 3–4.5 cm, with mucronate teeth, green above, glaucous and white-hairy beneath; stalks 1–2 cm. Acorn somewhat flattened, enclosed for half to two-thirds of its length in a downy cup. *Himalaya, SW China, Korea, Japan, Taiwan.* H5.

3. Q. acuta Thunberg. Figure 15(3), p. 69. Illustration: Schneider, Illustriertes Handbuch der Laubholzkunde 1: 170, 179 (1904).
Tree to 25 m in the wild, usually much smaller in cultivation; young shoots hairy at first, soon hairless. Leaves hairless, very leathery, oblong-ovate or lanceolate-oblong, 10–17 × 4–6 cm, acuminate, tapered to the base, margins entire, finely wavy; stalks 2–3.5 cm. Acorn ellipsoid, to 2 cm; cup downy. *Japan.* H5.

Subgenus **Quercus** (Subgenus *Euquercus* Hickel & Camus). Scales of cup not fused to form concentric circles; leaves deciduous or evergreen, entire, toothed or lobed; style generally elongate; acorns ripening in the 1st or 2nd year; aborted ovules towards the apex or base of the fruit.

Section **Cerris** Spach. Style long, not or scarcely expanded at the apex; inner wall of acorn usually not hairy; abortive ovules towards the base of the fruit.

4. Q. semecarpifolia Smith. Figure 15(4), p. 69. Illustration: Journal of the Royal Horticultural Society 40: f. 25, 26 (1914).
A large tree to 30 m in the wild, usually much smaller in cultivation; bark greyish, rough; young shoots hairy at first. Leaves variable, evergreen, very leathery, at least some broadly elliptic to almost circular, others elliptic-oblong, 5–9 × 3.5–6 cm, margins with spiny teeth when the plant is young, later entire; apex rounded, base rounded or auriculate; hairless above, brownish green beneath with dense,

plastered, brownish hairs; stalk very short, at most 5 mm. Acorn spherical or ovoid, borne in a shallow cup made up of erect, triangular, ciliate scales. *Afghanistan, Himalayas, W China.* H4.

5. Q. alnifolia Poech. Figure 15(5), p. 69. Illustration: Megaw, Wild flowers of Cyprus (1973).
Shrub or tree, 2–8 m; trunk rough, grey; young shoots covered with persistent yellowish grey down. Leaves evergreen, broadly obovate to almost circular, 2.5–5 × 1.5–3.5 cm, dark brownish green and hairless above, densely yellowish grey felted beneath; margins rolled downwards, entire or with small teeth; stalk 4–10 mm. Acorn 2.5–3.8 cm, obovoid, broadening above, pointed; cup rather small with spreading, downy scales. *Cyprus.* H5.

6. Q. coccifera Linnaeus. Figure 15(6), p. 69. Illustration: Bonnier, Flore Complète 10: t. 559 (1929); Polunin, Flowers of Europe, pl. 5 (1969); Gartenpraxis for 1977, 548, 549; Polunin & Walters, Guide to the vegetation of Britain and Europe, pl. 105 (1985).
Shrub or small tree to 10m; young shoots densely hairy, brownish; bark grey, smooth; buds more or less spherical, rounded. Leaves evergreen, ovate to oblong-lanceolate, flat or wavy, 1.5–5 × 1.5–2.5 cm, completely hairless, apex rounded to a fine point, base cordate or auriculate, margins with teeth which project as small spines; stalk 2–6 mm. Acorn spherical or ovoid, 1.5–3 cm; cup enclosing one-third to one-half of the acorn, made up of downy, spreading or recurved scales. *Mediterranean area.* H4.

7. Q. suber Linnaeus. Figure 15(7), p. 69. Illustration: Schneider, Illustriertes Handbuch der Laubholzkunde 1: 187 (1904); Bonnier, Flore Complète 10: t. 559 (1929); Phillips, Trees in Britain, Europe and North America, 190 (1978); Krüssmann, Manual of cultivated broad-leaved trees and shrubs, pl. 31 (1986).
Tree to 20 m with thick, corky bark; young shoots downy. Leaves evergreen, ovate-oblong, dark green above, grey-hairy beneath, 2–7 × 1.5–3.5 cm, apex acute, base tapered, margins with several bristle-tipped teeth; stalk 3–17 mm. Acorn ovoid-oblong, 1.5–3 cm, cup with the uppermost scales long and spreading, the lower adpressed. *S Europe, N Africa.* H4.
Widely grown in the Mediterranean area for the production of cork, as well as for ornament.

8. Q. libani Olivier. Figure 15(8), p. 69. Illustration: Schneider, Illustriertes Handbuch der Laubholzkunde 1: 179, 180 (1904); Bean, Trees and shrubs hardy in the British Isles, edn 8, 3: 491 (1976); Phillips, Trees in Britain, Europe and North America, 184 (1978); Krüssmann, Manual of cultivated broad-leaved trees and shrubs, pl. 31 (1986).
Shrub or tree to 6 m or more; bark grey, at first smooth, later becoming fissured; young shoots reddish brown downy at first, soon hairless. Leaves semi-evergreen or deciduous, hairless or almost so above and beneath, oblong to oblong-lanceolate, 4.5–11 × 1–3.5 cm, apex acuminate, base tapering, rounded or almost auriculate, vein-pairs more than 8; margin with numerous bristle-tipped teeth; stalk 8–20 mm. Acorn cylindric or slightly tapering; cup enclosing about the lower half of the acorn, made up of scales which are entirely adpressed or the upper spreading and the lower adpressed. *Turkey, Syria, N Iran, N Iraq.* H4.

9. Q. macrolepis Kotschy (*Q. aegilops* Willdenow not Linnaeus; *Q. graeca* Kotschy). Figure 15(9), p. 69. Illustration: Schneider, Illustriertes Handbuch der Laubholzkunde 1: 179 (1904); Polunin, Flowers of Europe, pl. 5 (1969); Krüssmann, Manual of cultivated broad-leaved trees and shrubs, pl. 31 (1986).
Tree to 15 m, often with a broad crown; trunk brown, young shoots persistently white or yellowish white hairy; buds large, white-hairy. Leaves semi-evergreen or deciduous, variable in shape, usually ovate to oblong, 4–8.5 × 2–4 cm, densely white- or yellow-hairy above and beneath, vein pairs 4–8, margins with large, triangular, bristle-tipped teeth, apex acute, base truncate to auriculate; stalk 1–2.5 cm. Acorn to 4.5 cm, hairy, enclosed for most of its length in the large cup which is made up of many long scales, free in their upper parts, hairy all over. *Italy, Balkans, Turkey.* H4.

A very variable species, cultivated for a long time in SE Europe for its tannin content.

10. Q. castaneifolia Meyer. Figure 15(10), p. 69. Illustration: Schneider, Illustriertes Handbuch der Laubholzkunde 1: 179, 180 (1904); Bean, Trees and shrubs hardy in the British Isles, edn 8, 3: 466 & pl. 57 (1976); Phillips, Trees in Britain, Europe and North America, 181 (1978); Krüssmann, Manual of cultivated broad-leaved trees and shrubs, pl. 31 (1986).

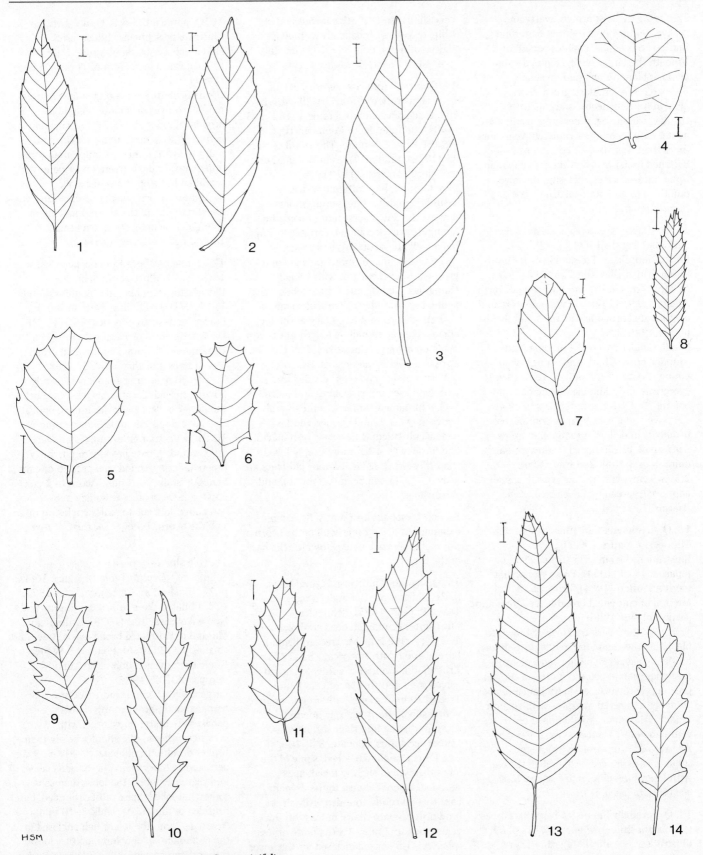

Figure 15. Leaves of *Quercus* species. 1, *Q. myrsinifolia*.
2, *Q. glauca*. 3. *Q. acuta*. 4, *Q. semecarpifolia*. 5, *Q. alnifolia*.
6, *Q. coccifera*. 7, *Q. suber*. 8, *Q. libani*. 9, *Q. macrolepis*.
10. *Q. castaneifolia*. 11, *Q. trojana*. 12, *Q. acutissima*. 13, *Q. variabilis*.
14, *Q. cerris*. Scales = 1 cm.

Tree to 25 m, bark brown, warty and corky; young shoots hairy at first, soon hairless; thread-like stipules persistent beneath the larger buds. Leaves deciduous, elliptic-oblong to oblong-lanceolate, 6–16 × 2.5–4 cm, dark green above, greyish hairy beneath, margins with regular, triangular, mucronate teeth; apex acute, base tapering or rounded; vein-pairs more than 8; stalk 1–2 cm. Acorn 2–3 cm, flattened or depressed at the top, cup made up of narrow scales enclosing the lower one-third to half of the acorn. *USSR (Caucasus), Iran.* H3.

11. Q. trojana Webb (*Q. macedonica* de Candolle). Figure 15(11), p. 69. Illustration: Bean, Trees and shrubs hardy in the British Isles, edn 8, 3: 517 (1976); Phillips, Trees in Britain, Europe and North America, 191 (1978); Krüssmann, Manual of cultivated broad-leaved trees and shrubs, pl. 31 (1986).
Small, slender tree to 18 m, often with a rounded crown; bark greyish or brown; young shoots downy at first. Leaves semi-evergreen or deciduous, oblong or ovate-oblong, 3–8 × 1–3 cm, hairless or almost so above and beneath, apex pointed, base usually rounded or auriculate, margins with small, bristle-tipped teeth; vein-pairs more than 8; stalk 2–8 mm. Acorn 2.5–3.5 cm, cylindric, cup variable, scales adpressed, spreading or reflexed. *Italy, Balkans, Turkey.* H4.

12. Q. acutissima Carruthers (*Q. serrata* Thunberg). Figure 15(12), p. 69. Illustration: Makino, Illustrated Flora of Japan, t. 1129 (1924); Poor, Plants that merit attention (1984); Krüssmann, Manual of cultivated broad-leaved trees and shrubs, pl. 32 (1986).
Tree 15–20 m; bark corky, grey-brown to black, furrowed and fissured; young shoots hairless. Leaves deciduous, lanceolate or rarely lanceolate-oblong, 6–16 × 3–6.5 cm, green on both surfaces, hairy beneath; apex acute, base tapered, margins with shallow, broadly triangular, bristle-tipped teeth, vein-pairs more than 8; stalk 2.5–5 cm. Acorn to 2.5 cm, ovoid with the top sometimes depressed or flattened, the cup with long, spreading, hairy scales. *Himalaya and China to Japan.* H3.

13. Q. variabilis Blume (*Q. bungeana* Forbes; *Q. chinensis* Bunge). Figure 15(13), p. 69. Illustration: Botanical Magazine Tokyo 9: pl. 7 (1895); Krüssmann, Manual of cultivated broad-leaved trees and shrubs, pl. 32 (1986).

Very similar to *Q. acutissima* but leaves white or yellowish beneath with dense, persistent hairs, 6.5–15 × 3–5 cm; stalk 2–4 cm. *N China, Korea, Japan.* H3–4.

14. Q. cerris Linnaeus. Figure 15(14), p. 69. Illustration: Schneider, Illustriertes Handbuch der Laubholzkunde 1: 182, 183 (1904); Bonnier, Flore Complète 10: t. 558 (1929); Edlin & Nimmo, The world of trees, 11 (1974); Phillips, Trees in Britain, Europe and North America, 181 (1978).
Tree to 25 m; bark ultimately deeply fissured, greyish white; young shoots downy, the down persistent or sometimes falling. Leaves deciduous, variable in shape, oblong-elliptic to oblong-lanceolate, 5–11 × 1.5–3 cm or more, pale green to greenish white beneath, with broadly triangular, slightly mucronate lobes, apex acute, base rounded or truncate; stalk 1–2 cm. Acorn to 4 cm, hairy at the top, enclosed for up to half its length in the cup made up of long, adpressed scales. *C & S Europe, E Mediterranean area.* H3.
Very variable and often divided into varieties based on leaf-lobing and hairiness.

Q. × hispanica Lamarck is the hybrid between *Q. cerris* and *Q. suber*, and is widely cultivated. It has evergreen leaves like those of *Q. suber*, more deeply lobed than those of *Q. cerris*. Various cultivars are grown, e.g. 'Lucombeana', 'Dentata' and 'Diversifolia'.

Section **Mesobalanus** Camus. Style long, expanded at apex; inner wall of the acorn not hairy; abortive ovules towards the base of the fruit.

15. Q. dentata Thunberg. Figure 16(1), p. 71. Illustration: Botanical Magazine Tokyo 9: pl. 7 (1895); Schneider, Illustriertes Handbuch der Laubholzkunde 1: 210 (1904); Phillips, Trees in Britain, Europe and North America, 182 (1978); Krüssmann, Manual of cultivated broad-leaved trees and shrubs, pl. 32 (1986).
Tree to 20 m or more; bark brown, fissured; young shoots stout, densely downy, the down persisting into their second year. Leaves deciduous, obovate, 8.5–20 × 5–12.5 cm, shallowly lobed, some of the lobes themselves slightly lobed, apex rounded, tapered almost to the extreme base then abruptly rounded, greyish or brownish beneath, hairy at least on the veins; stalk to 7 mm. Acorns more or less spherical, 1.5–2 cm, enclosed for the lower half or more in the cup which is made up of narrow, spreading scales. *Japan, Korea, China.* H2.

16. Q. pontica C. Koch. Figure 16(2), p. 71. Illustration: Schneider, Illustriertes Handbuch der Laubholzkunde 1: 171 (1904); Davis (ed), Flora of Turkey 7: 665 (1982).
Open shrub to 5 m; young shoots hairless, reddish. Leaves deciduous, elliptic to broadly elliptic, 15–25 × 5–13 cm, vein-pairs 15–20, apex acute, base tapered, margins with numerous mucronate teeth, upper surface dark green, lower paler with scattered hairs on the veins; stalk 3–20 mm. Acorn ovoid; cup enclosing one-quarter to half of the acorn, made up of triangular, acuminate, downy scales. *Turkey, USSR (Caucasus).* H4.

17. Q. macranthera Hohenacker. Figure 16(3), p. 71. Illustration: Schneider, Illustriertes Handbuch der Laubholzkunde 1: 179 (1904); Phillips, Trees in Britain, Europe and North America, 185 (1978); Davis (ed), Flora of Turkey 7: 669 (1982); Krüssmann, Manual of cultivated broad-leaved trees and shrubs, pl. 32 (1986).
Tree to 20 m; young shoots persistently downy; thread-like stipules persisting below the larger buds. Leaves deciduous, obovate, 6–19 × 2.5–11 cm, mostly more than 10 cm, with triangular, mucronate lobes, apex rounded, base broadly to narrowly tapered, dark green above, greyish downy beneath; stalk 5–17 mm. Acorn to 2 cm; cup made up of erect or somewhat spreading, lanceolate scales enclosing up to half the acorn. *USSR (Caucasus), N Iran.* H2.

18. Q. frainetto Tenore (*Q. conferta* Schultes; *Q. farnetto* Tenore). Figure 16(4), p. 71. Illustration: Gartenpraxis for 1977, 473; Phillips, Trees in Britain, Europe and North America, 182 (1978); Krüssmann, Manual of cultivated broad-leaved trees and shrubs, pl. 32 (1986); Polunin & Walters, Guide to the vegetation of Britain and Europe, pl. 107 (1985).
Large tree to 30 m or more in the wild; trunk smooth, grey; young shoots persistently hairy; buds large, with persistent, thread-like stipules below them. Leaves deciduous, obovate, 9–20 × 5–9 cm, persistently hairy beneath; margins deeply and regularly lobed, the lobes themselves somewhat lobed, apex rather rounded, base rounded or auriculate, stalk 3–10 mm. Acorn 1–2 cm, the lower half enclosed in the cup made up of oblong, obtuse, loose, rather downy scales. *S & SE Europe, Turkey.* H2.

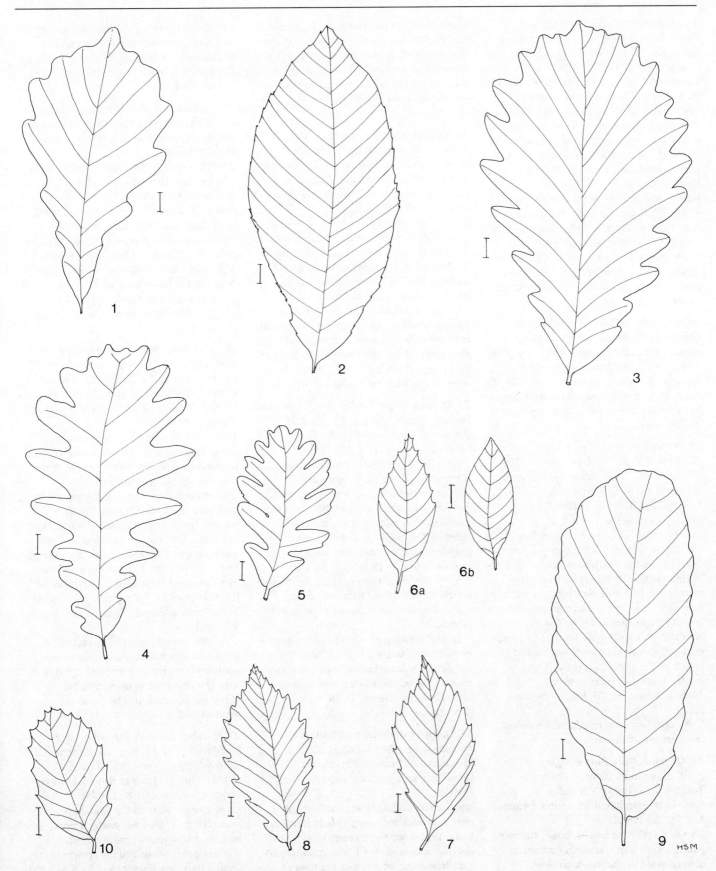

Figure 16. Leaves of *Quercus* species. 1, *Q. dentata*. 2, *Q. pontica*.
3, *Q. macranthera*. 4, *Q. frainetto*. 5, *Q. pyrenaica*. 6, *Q. ilex* (a, b, two
different leaves). 7, *Q. glandulifera*. 8, *Q. mongolica*.
9, *Q. canariensis*. 10, *Q. fruticosa*. Scales = 1 cm.

19. Q. pyrenaica Willdenow (*Q. toza* Bastard). Figure 16(5), p. 71. Illustration: Schneider, Illustriertes Handbuch der Laubholzkunde 1: 179, 180 (1904); Polunin & Smythies, Flowers of SW Europe, pl. 1 (1973); Phillips, Trees in Britain, Europe and North America, 187 (1978).
Tree to 20 m, suckering freely; bark brown or blackish, with long, longitudinal fissures; young shoots persistently downy, pendent. Leaves deciduous, obovate, elliptic or broadly oblong, 7–16 × 4–8 cm, persistently downy above, more so beneath; deeply lobed, lobes parallel-sided at base then tapering, apex rounded, base broadly rounded or auriculate, stalk 5–15 mm. Acorn enclosed in its lower half in the cup which is made up of narrow, lanceolate, loose, hairy scales. *France, Spain, Portugal, Morocco.* H3.

Section **Quercus** (Section *Lepidobalanus* Endlicher). Styles very short, abruptly expanded into the stigmas; inner wall of the acorn usually hairless; abortive ovules towards the base of the fruit. Species **36–38** are known informally as white oaks; they have acorns with a sweet taste, ripening in the first year.

20. Q. ilex Linnaeus. Figure 16(6), p. 71. Illustration: Bonnier, Flore Complète 10: t. 558 (1929); Edlin & Nimmo, The world of trees, 12 (1974); Gartenpraxis for 1977, 549; Phillips, Trees in Britain, Europe and North America, 182 (1978).
Tree to 25 m; bark greenish brown or black; young shoots densely grey-downy. Leaves thick but flexible, evergreen, oblong-ovate to lanceolate, 3–7.5 × 1–2.5 cm, apex acute, base rounded, margins entire or variably toothed, dark green and soon hairless above, densely grey-downy beneath; stalk 4–8 mm. Acorn 2.5–3 cm, enclosed for one-quarter to one-half its length in the cup which is made up of tightly adpressed, hairy scales. *Mediterranean area, SW France, Portugal.* H4.

A variable species in which numerous cultivars are available.

21. Q. glandulifera Blume. Figure 16(7), p. 71. Illustration: Botanical Magazine Tokyo 9: t. 7 (1895); Schneider, Illustriertes Handbuch der Laubholzkunde 1: 179, 205 (1904).
Tree to 15 m; bark longitudinally furrowed; young shoots hairy. Leaves deciduous, oblong-ovate to obovate-lanceolate, 5–15 × 2–4.5 cm, margins with forwardly-pointing, mucronate, triangular teeth, apex acute, base tapered, green above, glaucous and slightly hairy beneath; vein-pairs less than 12; stalk 1.2–2 cm. Acorn 1–1.5 cm, enclosed for about one-third of its length by the shallow cup which is made up of closely adpressed scales. *Japan, Korea, China.* H2.

22. Q. mongolica Turczaninow. Figure 16(8), p. 71. Illustration: Schneider, Illustriertes Handbuch der Laubholzkunde 1: 199 (1904); Kitamura & Okamoto, Coloured illustrations of trees and shrubs of Japan, f. 119 (1977).
Tree to 30 m or more in the wild; trunk grey, smooth or furrowed; young shoots hairless, shining, grey. Leaves obovate, 7–20 × 2.5–9 cm, margin lobed, the lobes usually rounded but sometimes rather acute, triangular and tooth-like (var. **grosseserrata** (Blume) Rehder & Wilson), apex acute, base auriculate, all more or less hairless; stalk to 6 mm. Acorns ovoid or ellipsoid, to 2 cm, the lower third enclosed in the cup which is made up of triangular, more or less adpressed scales. *E Asia.* H2.

23. Q. canariensis Willdenow (*Q. mirbeckii* Durieu). Figure 16(9), p. 71. Illustration: Schneider, Illustriertes Handbuch der Laubholzkunde 1: 179, 180 (1904).
Tree to 30 m or more in the wild; bark black, deeply fissured; young shoots hairless. Leaves ovate-oblong to obovate, 5–18 × 4–12 cm, margins with shallow, rounded, obtuse lobes, hairy at first beneath, the hair soon falling, apex rounded, base tapering then abruptly rounded; stalk 8–18 mm. Acorns to 2.5 cm, the lower third enclosed in the cup which is formed from flattened, downy scales. *Portugal, Spain, Morocco, Algeria, Tunisia.* H4.

Q. hartwissiana Steven may occasionally be cultivated and will key out here. It has distinctive, reddish brown young shoots, obovate leaves and fruits borne on long stalks. *Bulgaria, Turkey, USSR (Transcaucasia).* H4.

24. Q. fruticosa Brotero (*Q. lusitanica* misapplied). Figure 16(10), p. 71. Illustration: Schneider, Illustriertes Handbuch der Laubholzkunde 1: 179, 180 (1904).
Small shrub to 2 m (often smaller) with stolons or suckers; young shoots shining, hairy. Leaves semi-evergreen, oblong or obovate-oblong, 3–6 × 1.5–3 cm, margins with triangular, mucronate teeth, apex rather rounded, base rounded, persistently hairy but not glaucous beneath; vein-pairs less than 12; stalk 5–10 cm. Acorns ovoid, to 2.5 cm, the cup made up of small, triangular, overlapping scales, enclosing up to two-thirds of the acorn. *Portugal, Spain, Morocco.* H4.

25. Q. infectoria Olivier. Figure 17(1), p. 73. Illustration: Krüssmann, Manual of cultivated broad-leaved trees and shrubs, pl. 31 (1986).
Small tree to 4 m (or more); young shoots hairy at first, the hair soon falling. Leaves semi-evergreen, elliptic to ovate or narrowly oblong, 3–7 × 1–4.5 cm, margins with shallow, rather acute, triangular, mucronate lobes, rounded at apex, tapered at the base, hairless when mature; stalk 3–10 mm. Acorn ellipsoid, 4–5 cm, cup covering the lower third of the acorn, made up of strongly adpressed, greyish downy scales. *E Mediterranean area, N Iran, N Iraq.* H5.

26. Q. petraea (Mattuschka) Lieblein (*Q. sessilis* Ehrhart; *Q. sessiliflora* Salisbury). Figure 17(2), p. 73. Illustration: Edlin & Nimmo, The world of trees, 11 (1974); Phillips, Trees in Britain, Europe and North America, 186 (1978).
Tree to 30 m; bark greyish or brown to black, fissured; young shoots hairless; trunk penetrating into the crown. Leaves ovate, obovate or oblong, deciduous, 6–17 × 3–9 cm, margins with 4–6 deep lobes on each side, apex rounded, base wedge-shaped, truncate or somewhat cordate but not auriculate, hairy or hairless and pale or glaucous green beneath, hairless and bright green above; stalk 1 cm or more. Acorns ovoid to ovoid-oblong, to 3 cm, the lower third enclosed by the cup which is formed from closely adpressed, hairy scales. *Europe, W Asia.* H1.

A very variable and widely cultivated species in which numerous varieties and cultivars have been recognised. Hybrids with *Q. robur* are common in parts of Europe and may also be found in gardens.

27. Q. robur Linnaeus (*Q. pedunculata* Ehrhart). Figure 17(3), p. 73. Illustration: Edlin & Nimmo, The world of trees, 5 (1974); Phillips, Trees in Britain, Europe and North America, 187 (1978).
Large tree, to 40 m in the wild; bark brownish grey, fissured; young shoots hairless; trunk usually rather short, scarcely penetrating into the crown. Leaves deciduous, ovate-oblong, 5–14 × 3.5–6 cm, margins with deep, rounded lobes, apex rounded, base auriculate; stalk less than

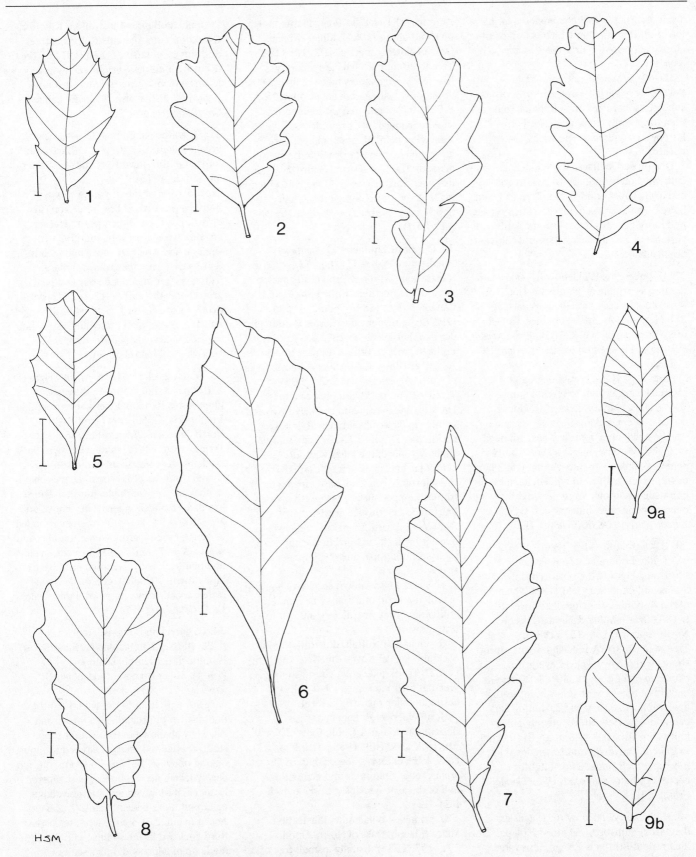

Figure 17. Leaves of *Quercus* species. 1, *Q. infectoria*. 2, *Q. petraea*.
3, *Q. robur*. 4, *Q. pubescens*. 5, *Q. virginiana*. 6, *Q. bicolor*.
7, *Q. prinus*. 8, *Q. warburgii*. 9, *Q. douglasii* (a, b, two different leaves).
Scales = 1 cm.

1 cm. Acorn 1.5–2.5 cm, enclosed for its lower third in the cup which is formed from closely adpressed, hairy scales. *Europe, SW Asia.* H1.

Like *Q. petraea*, a common oak in N Europe, very variable and with numerous subspecies, varieties and cultivars, some of them (especially those from southern Europe) recognised as species by some authors.

Q. × turneri Willdenow. A widely cultivated plant thought to be a hybrid between *Q. robur* and *Q. ilex*. Tree to 15 m; leaves semi-evergreen, lobed, leathery, pale and hairy beneath; most flowers sterile. It is variable in leaf-shape, and several cultivars are available.

28. Q. pubescens Willdenow (*Q. lanuginosa* (Lamarck) Thuillier). Figure 17(4), p. 73. Illustration: Coste, Flore de la France 3: 258 (1906); Polunin & Smythies, Flowers of SW Europe, pl. 1 (1973); Phillips, Trees in Britain, Europe and North America, 187 (1978).
Small tree to 10 m, crown open and rounded; bark brown to black; young shoots densely hairy. Leaves deciduous, 4–9 × 2–5 cm, oblong-obovate, margins deeply lobed, apex rounded, base rounded or slightly cordate, hairless above, hairy beneath; stalk 5–12 mm. Acorn 1.5–2 cm, ovoid, its lower third to half enclosed in the cup which is formed from very tightly adpressed, narrow, hairy scales. *C & S Europe, Turkey, USSR (Crimea).* H2.

29. Q. virginiana Miller. Figure 17(5), p. 73. Illustration: Britton & Brown, Illustrated Flora of the northern United States and Canada 1: 523 (1896); Memoirs of the National Academy of Sciences 20: t. 189 (1924); Miller & Lamb, Oaks of North America, 136, 137 (1985).
Tree to 20 m; trunk reddish brown; young shoots white-hairy. Leaves evergreen, elliptic to oblong, 3–12 cm × 8–20 mm, margins entire or with a few blunt teeth, apex rounded, base tapered, shining dark green and hairless above, densely white-hairy beneath; stalk 4–6 mm. Acorn ovoid, to 2.5 cm, the lower quarter enclosed in the cup which is made up of tightly adpressed, hairy scales. *SE USA (Florida, Georgia).* H5.

30. Q. bicolor Willdenow (*Q. platanoides* (Lamarck) Sudworth). Figure 17(6), p. 73. Illustration: Britton & Brown, Illustrated Flora of the northern United States and Canada, edn 2, 1: 623 (1913); Phillips, Trees in Britain, Europe and North

America, 181 (1978); Poor, Plants that merit attention (1984); Miller & Lamb, Oaks of North America, 177, 178 (1985).
Tree to 20 m; bark pale grey, deeply fissured; young shoots hairy at first. Leaves deciduous, oblong-obovate or obovate, 6.5–16 × 3–7 cm, margin toothed or shallowly lobed, apex rounded or somewhat pointed, base tapered, dark green above, greyish- or white-hairy beneath; stalk 8–20 mm. Acorn 2–3 cm, its lower third enclosed by the cup which is made up of numerous, narrow, hairy, tightly adpressed scales. *Northeastern N America.* H1.

31. Q. prinus Linnaeus (*Q. montana* Willdenow). Figure 17(7), p. 73. Illustration: Britton & Brown, Illustrated Flora of the northern United States and Canada 1: 522 (1896), edn 2, 1: 624 (1913); Schneider, Illustriertes Handbuch der Laubholzkunde 1: 202, 208 (1904); Phillips, Trees in Britain, Europe and North America, 186 (1978); Miller & Lamb, Oaks of North America, 181, 182 (1985).
Tree to 25 m; bark dark grey to almost black, in age deeply and coarsely furrowed; young shoots reddish, hairy at first. Leaves deciduous, 5–23 × 2.8–11.5 cm, margins shallowly lobed, the lobes without glandular tips; apex acute or tapered and then abruptly rounded, base tapered or rounded, upper surface yellowish green, hairless, lower surface greyish-hairy; stalk 2–25 mm. Acorn 2.5–4 cm, the lower quarter or third enclosed in the cup made up from numerous, narrow, adpressed scales. *E USA.* H1.

This is the best known species of a group of which several others may be found in cultivation; they are all very difficult to distinguish:

Q. michauxii Nuttall. Illustration: Miller & Lamb, Oaks of North America, 185, 186 (1985). Trunk pale grey; young shoots green at first, later grey, lobes of leaves with glandular tips. *SE USA.* H4.

Q. muehlenbergii Engelmann. Illustration: Miller & Lamb, Oaks of North America, 188, 189 (1985). Trunk grey, flaky; lobes of leaves deeper than in *Q. prinus*, more triangular and acute; lower half of the acorn enclosed in the cup. *E USA.* H2.

Q. prinoides Willdenow. Illustration: Miller & Lamb, Oaks of North America, 191–193 (1985). Usually a small tree or shrub to 4 m; bark grey. *E USA.* H1.

Also keying out here, but not closely related is:

Q. aliena Blume. Illustration: Kitamura & Okamoto, Coloured illustrations of trees and shrubs of Japan, f. 120 (1977). Tree to 20 m with hairless, bright yellow-green leaves which are coarsely toothed and hairy beneath; acorns distinctly stalked. *China, Japan, Korea.* H2.

32. Q. warburgii Camus (*Q. genuensis* invalid). Figure 17(8), p. 73. Illustration: Journal of the Royal Horticultural Society **41**: 8, f. 5 (1915).
Tree to 20 m or more; young shoots hairless, greenish at first, later greyish brown. Leaves semi-evergreen, leathery, obovate to oblanceolate, margins with shallow and irregular mucronate teeth, 8–12 × 3–8 cm, apex rounded, base truncate or rounded, hairless and pale green beneath; stalk 8–12 mm. Acorn ovoid, its lower third enclosed in the cup which is made up of many adpressed, hairy scales. *Known only in cultivation.* H4?
Possibly of hybrid origin.

33. Q. douglasii Hooker & Arnott. Figure 17(9), p. 73. Illustration: Schneider, Illustriertes Handbuch der Laubholzkunde 1: 189 (1904); Miller & Lamb, Oaks of North America, 268, 269 (1985).
Tree to 18 m with a compact crown; bark thick, greyish or brown, scaly; young shoots stout, reddish brown. Leaves oblong to obovate, very variable, margins entire or lobed, dark bluish green, hairy above and beneath, 2–7 × 1.2–3.3 cm, apex rounded or acute, base tapering (sometimes broadly so); stalk 3–7 mm. Acorn 2–3 cm, ovoid and pointed, sweet, enclosed at the extreme base only in a shallow cup made up of rather broad, very adpressed, hairy scales. *USA (California).* H5.

34. Q. garryana Douglas. Figure 18(1), p. 76. Illustration: Schneider, Illustriertes Handbuch der Laubholzkunde 1: 199 (1904); Miller & Lamb, Oaks of North America, 275, 276 (1985).
Large tree to 30 m in the wild; bark thick, light brown or grey, divided into broad ridges by shallow fissures; young shoots stout, orange-red, hairy. Leaves deciduous, oblong-obovate, 7–15 × 5.5–8 cm, margins deeply lobed, apex rounded, base tapered or rounded; dark green and hairless above, paler and hairy beneath; stalk 1–2.5 cm. Acorn ovoid, 2.5–3 cm, sweet, its lower third enclosed in the shallow cup which is made up of adpressed, hairy scales with thin, free, pointed tips. *Western N America (British Columbia to California).* H4.

35. Q. lobata Née. Figure 18(2), p. 76. Illustration: Schneider, Illustriertes Handbuch der Laubholzkunde 1: 171, 205 (1904); Miller & Lamb, Oaks of North America, 272, 273 (1985).

Very large tree to 30 m or more; bark thick, grey, broken into plates on older trees; young shoots reddish brown to grey, hairy. Leaves deciduous, obovate, 3.5–8 × 2–5 cm, margins deeply lobed, apex blunt, base tapering, dark green and hairy above, paler and hairy beneath; stalk 1–1.5 cm. Acorn 2.5–4 cm, ovoid, pointed, sweet, its basal third enclosed in the cup which is made up of numerous adpressed scales, the lower ones large and thick, the upper with free tips, all hairy. *USA (California).* H5.

Very slow-growing in cultivation.

36. Q. stellata Wangenheim. Figure 18(3), p. 76. Illustration: Britton & Brown, Illustrated Flora of the northern United States and Canada, edn 2, 1: 622 (1913); Miller & Lamb, Oaks of North America, 162, 163 (1985).

Tree to 20 m, but very variable in size; bark grey or reddish, ridged; young shoots grey, hairy. Leaves deciduous, obovate, 8–20 × 4–8 cm, with 5 (rarely 7) very deep lobes, the lobes themselves occasionally very shallowly lobed, apex blunt, base broadly tapered, dark green and hairy above, densely brownish-downy beneath; stalk 8–15 mm. Acorn 1.5–2.5 cm, obtuse, its lower third enclosed by the cup which is made up of tightly adpressed, hairy scales. *Southern part of E USA.* H3.

37. Q. alba Linnaeus. Figure 18(4), p. 76. Illustration: Schneider, Illustriertes Handbuch der Laubholzkunde 1: 203, 204 (1904); American Horticulturist 54: 18 (1975); Phillips, Trees in Britain, Europe and North America, 180 (1978); Miller & Lamb, Oaks of North America, 170–172 (1985).

Tree to 30 m or more in the wild; bark pale grey with plates or ridges; young shoots more or less reddish brown, hairless. Leaves obovate, oblong or elliptic, 10–22 × 7–14 cm, margins with 3–4 deep lobes on each side, apex bluntly pointed, base tapering, hairy when young, the hairs soon shed, bright green above, glaucous beneath; stalk 1–2.5 cm. Acorn 1–3 cm, ovoid-oblong, its basal third or less enclosed in the cup which is made up of adpressed, thickened, hairy scales. *E USA.* H2.

Slow-growing in cultivation.

Q. lyrata Walter. Illustration: Miller & Lamb, Oaks of North America, 158–160 (1985). Similar to *Q. alba* but leaves obovate with 4 or more deep, rounded lobes on either side, silvery white-hairy beneath; acorn more or less spherical, enclosed for two-thirds or more of its length in a deep cup. *SE USA.* H3.

38. Q. macrocarpa Michaux. Figure 18(5), p. 76. Illustration: Schneider, Illustriertes Handbuch der Laubholzkunde 1: 201, 202 (1904); Phillips, Trees in Britain, Europe and North America, 185 (1978); Poor, Plants that merit attention (1984); Miller & Lamb, Oaks of North America, 153, 154 (1985).

Tree to 24 m or occasionally more in the wild; bark thick, dark brown, deeply furrowed into vertical ridges; young shoots persistently hairy. Leaves obovate to oblong-obovate, 10–24 × 5–12 cm, deeply lobed, the median sinuses deeper than the others and reaching almost to the midrib, apex rounded, base conspicuously tapered, dark green above, whitish-hairy beneath. Acorn broadly ovoid, 1.5–5 cm, with usually its lower half (or more) enclosed in the deep cup which is made up of hairy scales which are adpressed in the lower part of the cup, the tips in the upper part free and somewhat spreading. *C & E USA.* H3.

Section **Protobalanus** Trelease. Style short, very abruptly expanded into the stigmas; inner wall of the acorn hairy; abortive ovules towards the base of the fruit.

39. Q. chrysolepis Liebmann. Figure 18(6), p. 76. Illustration: Schneider, Illustriertes Handbuch der Laubholzkunde 1: 179, 205 (1904); Phillips, Trees of Britain, Europe and North America, 182 (1978); Miller & Lamb, Oaks of North America, 282, 283 (1985); Krüssmann, Manual of cultivated broad-leaved trees and shrubs, pl. 31 (1986).

Tree to 20 m or more in the wild; bark thick, grey-brown, forming scales; young twigs at first hairy, later hairless. Leaves evergreen, oblong-ovate to elliptic, margins with a few sharp teeth mostly in the upper half, apex acute, broadly tapering to cordate at the base, bright or yellowish green above, with persistent yellow-brown hairs beneath; stalk 3–10 mm. Acorn ovoid or oblong-ovoid, 1–5 cm, enclosed only at the base by the shallow, thick-walled cup which is made up of small adpressed, hairy scales, the hairs sometimes very dense and yellowish. *Western USA (California, Oregon).* H4.

Section **Erythrobalanus** Spach. Styles long, capitate or expanded above; inner wall of the acorn downy or velvety; abortive ovules usually towards the apex of the fruit.

Species **48–57** are known informally as red oaks; they have leaves with bristle-tipped teeth or lobes and the acorn is generally bitter, with its inner wall hairy. The leaves often colour well in autumn.

40. Q. agrifolia Née. Figure 18(7), p. 76. Illustration: Schneider, Illustriertes Handbuch der Laubholzkunde 1: 186 (1904); Phillips, Trees in Britain, Europe and North America, 180 (1978); Miller & Lamb, Oaks of North America, 298, 299 (1985).

Tree to 25 m or more in the wild; bark thick, variable, whitish, reddish, brown or black, smooth or ridged; young shoots downy. Leaves evergreen, ovate-elliptic, elliptic or broadly elliptic, 2.5–7 × 2.5–4 cm, with spine-tipped teeth, apex acute, base rounded, dark green above, paler and hairless beneath except for tufts of hairs in the axils of the major veins; stalk 5–8 mm. Acorn ovoid, narrowed above, 2–3.5 cm, the lower quarter enclosed in the cup which is bowl-shaped and downy inside and out. *USA (California).* H4.

41. Q. wislizenii A. de Candolle. Figure 18(8), p. 76. Illustration: Schneider, Illustriertes Handbuch der Laubholzkunde 1: 185f (1904); Miller & Lamb, Oaks of North America, 287–289 (1985).

Shrub or tree to 20 m; bark thick, blackish or reddish, deeply furrowed; young shoots slender, rigid, often hairy at first. Buds ovoid, pointed. Leaves evergreen, leathery, broadly lanceolate to broadly elliptic, 2.5–3.8 × 1.3–3 cm, margins with few to many shallow teeth, apex somewhat rounded, base rounded or cordate, dark green above, paler and completely hairless beneath; stalk 2–3 cm. Acorn slender, oblong-cylindric, pointed, 2–4 cm, the cup shallow or deep (enclosing up to two-thirds of the acorn), made up of thin, flattened, downy scales. *USA (California).* H5.

Q. × kewensis Osborn is a hybrid raised from acorns of *Q. wislizenii* which had probably been fertilised by pollen from *Q. cerris*. The leaves are evergreen, but lobed like those of *Q. cerris*; the persistent, thread-like stipules found in that species do not occur in the hybrid.

42. Q. hypoleucoides Camus (*Q. hypoleuca* Engelmann not Miquel). Figure 18(9), p. 76. Illustration: Miller & Lamb, Oaks of North America, 251, 252 (1985).

Shrub or small tree, 2–5 m, rarely a tree to 10 m or more; bark blackish, deeply

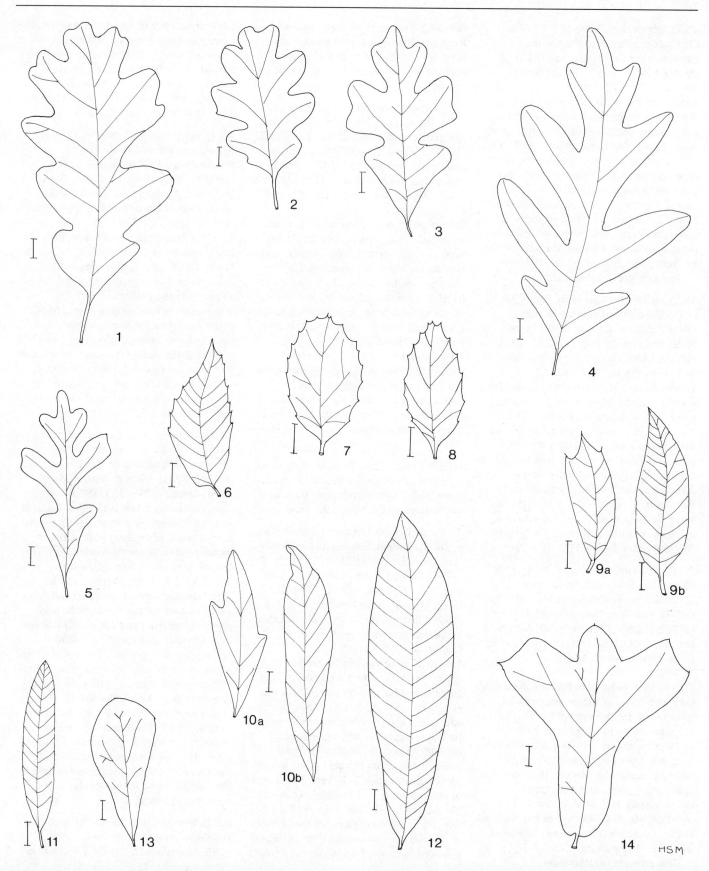

Figure 18. Leaves of *Quercus* species. 1, *Q. garryana*. 2, *Q. lobata*. 3, *Q. stellata*. 4, *Q. alba*. 5, *Q. macrocarpa*. 6, *Q. chrysolepis*. 7, *Q. agrifolia*. 8, *Q. wislizenii*. 9, *Q. hypoleucoides* (a, b, two different leaves). 10, *Q. laurifolia* (a, b, two different leaves). 11, *Q. phellos*. 12, *Q. imbricaria*. 13, *Q. nigra*. 14, *Q. marilandica*. Scales = 1 cm.

furrowed; young shoots densely grey-downy, the hairs ultimately falling, the shoots then reddish. Leaves evergreen, oblong-lanceolate to narrowly obovate, 4–10 × 1.2–2.5 cm, margins entire or with a few sharp teeth, apex acute, base abruptly tapered or rounded, bright green and ultimately hairless above, densely grey or silvery downy beneath; stalk 4–12 mm. Acorns to 1.5 cm, pointed, the lower third enclosed in the cup which is hairy inside and made up of adpressed, downy scales. *S USA (New Mexico, Arizona, Texas), N Mexico.* ?H5.

43. Q. laurifolia Michaux. Figure 18(10), p. 76. Illustration: Schneider, Illustriertes Handbuch der Laubholzkunde **1**: 166 (1904); Memoirs of the National Academy of Sciences **20**: t. 304–306 (1924); Miller & Lamb, Oaks of North America, 116, 117 (1985).

Tree to 25 m; bark dark brown, more or less smooth, in age becoming darker, almost black, and furrowed; young shoots dark red, hairless. Leaves deciduous, narrowly oblong to obovate, 6–13 × 1.2–4 cm, acute at apex, tapered at the base, margins entire or with 3 rather obscure lobes towards the apex, shiny green above, paler beneath with a conspicuous, yellow midrib, completely hairless; stalk to 4 mm. Acorn ovoid, dark brown to almost black, 1–1.5 cm, enclosed only at the base in the shallow cup which is made up of adpressed, blunt scales. *SE USA.* ?H5.

44. Q. phellos Linnaeus. Figure 18(11), p. 76. Illustration: Phillips, Trees in Britain, Europe and North America, 186 (1978); Miller & Lamb, Oaks of North America, 112, 113 (1985); Krüssmann, Manual of cultivated broad-leaved trees and shrubs, pl. 29 (1986).

Tree to 20 or rarely to 30 m; bark grey to reddish brown, smooth, in age breaking up into rough ridges; young shoots reddish brown, slender. Leaves deciduous, linear to narrowly oblong, 7–12 cm × 8–25 mm, margins entire, sometimes wavy, apex acute, base tapered, dark green above, paler beneath, hairless; stalk 5–10 mm. Acorns more or less spherical, pale, enclosed at the base in the thin, shallow cup which is made up of grey-hairy, adpressed scales. *E USA.* H4.

45. Q. imbricaria Michaux. Figure 18(12), p. 76. Illustration: Phillips, Trees in Britain, Europe and North America, 183 (1978); Poor, Plants that merit attention (1984);

Miller & Lamb, Oaks of North America, 123, 124 (1985); Krüssmann, Manual of cultivated broad-leaved trees and shrubs, pl. 29 (1986).

Tree to 15 or 20 m, rarely more; bark grey-brown, divided by irregular, shallow fissures; young shoots pale brown, hairy at first, soon hairless. Leaves deciduous, oblong to lanceolate or obovate, 7–18.5 × 1.5–5 cm, apex acute, base tapered, margins entire but slightly wavy, green and shining above, paler and persistently hairy beneath; stalk to 1.5 cm. Acorn more or less spherical, 1–1.5 cm, the lower third to half enclosed in the cup which is made up of hairy, adpressed, scales. *C & E USA.* H3.

Q. × leana Nuttall. A hybrid between *Q. imbricaria* and *Q. velutina*, which occurs wild in the USA. It is like *Q. imbricaria* but the leaves are irregularly lobed and less hairy.

46. Q. nigra Linnaeus (*Q. aquatica* Walter). Figure 18(13), p. 76. Illustration: Schneider, Illustriertes Handbuch der Laubholzkunde **1**: 167 (1904); Miller & Lamb, Oaks of North America, 180, 109 (1985); Krüssmann, Manual of cultivated broad-leaved trees and shrubs, pl. 29 (1986).

Tree to 25 m; bark at first brown, later blackish and furrowed; young shoots hairless, reddish. Leaves deciduous, 3–15 × 1.5–6 cm, spoon-shaped (rarely some elliptic or narrowly elliptic), most 3-lobed towards the apex, the lobes with bristle-like points; apex acute or blunt, base tapered, hairless; stalk to 1 cm. Acorn more or less spherical, black, the lower half or less enclosed in the thin cup which is made up of adpressed scales. *SE USA.* H5.

Very variable in leaf-shape, size and lobing.

47. Q. marilandica Muenchhausen. Figure 18(14), p. 76. Illustration: Schneider, Illustriertes Handbuch der Laubholzkunde **1**: 168 (1904); Gartenpraxis for 1977, 474; Miller & Lamb, Oaks of North America, 99–101 (1985); Krüssmann, Manual of cultivated broad-leaved trees and shrubs, pl. 29 (1986).

Small tree to 15 m with an irregular crown; bark black, often very rough; young shoots stout, hairy at least at first. Leaves deciduous, broadly obovate, 6–17 × 5.5–11 cm, with 3 large lobes towards the apex (sometimes these slightly lobed on the outer margin), the lobes with bristle-like points, apex obtuse, mucronate, base usually rounded or cordate, yellowish green above, brownish and hairy beneath;

stalk 5–20 mm. Acorn ovoid, 1–2 cm, its lower third to two-thirds enclosed in the cup which is made up of broad, adpressed, hairy scales. *SE USA.* ?H5.

48. Q. arkansana Sargent. Figure 19(1), p. 78. Illustration: Miller & Lamb, Oaks of North America, 104–106 (1985).

Tree to 18 m, crown narrow; bark thick, black, deeply furrowed. Leaves deciduous, broadly obovate, 3–14 × 2.5–6 cm, entire or obscurely 3-lobed towards the apex, apex rounded, base tapered, yellowish green above, paler and with tufts of hairs in the axils of the main veins beneath. Acorn almost spherical or ovoid, to 1.2 cm, its lower quarter enclosed in the shallow cup. *Scattered in the southeastern part of the USA.*

49. Q. falcata Michaux. Figure 19(2), p. 78. Illustration: Miller & Lamb, Oaks of North America, 86, 87 (1985); Krüssmann, Manual of cultivated broad-leaved trees and shrubs, pl. 30 (1986).

Tree to 30 m; bark thick, dark brown to black, with deep, narrow fissures; young shoots dark red, downy. Leaves deciduous, elliptic, 9.5–22 × 5–17 cm, margins with several shallow to deep, narrowly triangular, often curved (especially the terminal) lobes which are bristle-pointed; upper surface dark green and shining, lower surface persistently white- or grey-downy; stalk 2.5–4.5 cm. Acorn to 2 cm, with up to the lower third enclosed in a thin, shallow cup made up of reddish brown, adpressed scales. *Southern and eastern USA.* ?H5.

Var. **pagodifolia** Elliott has more regular lobing of the leaves.

50. Q. ilicifolia Wangenheim. Figure 19(3), p. 78. Illustration: Britton & Brown, Illustrated Flora of the northern United States and Canada, edn 2, **1**: 620 (1913); Miller & Lamb, Oaks of North America, 97, 98 (1985); Krüssmann, Manual of cultivated broad-leaved trees and shrubs, pl. 30 (1986).

In general similar to *Q. falcata* but a small tree or shrub to 6 m, leaves with fewer, broadly triangular, straight lobes; acorns with up to the lower half enclosed in the shallow cup. *E USA.*

Produces acorns in great abundance in some years.

51. Q. palustris Muenchhausen. Figure 19(4), p. 78. Illustration: Gartenpraxis for 1977, 474; Phillips, Trees in Britain, Europe and North America, 185 (1978); Miller & Lamb, Oaks of North America, 69, 70 (1985); Krüssmann, Manual of

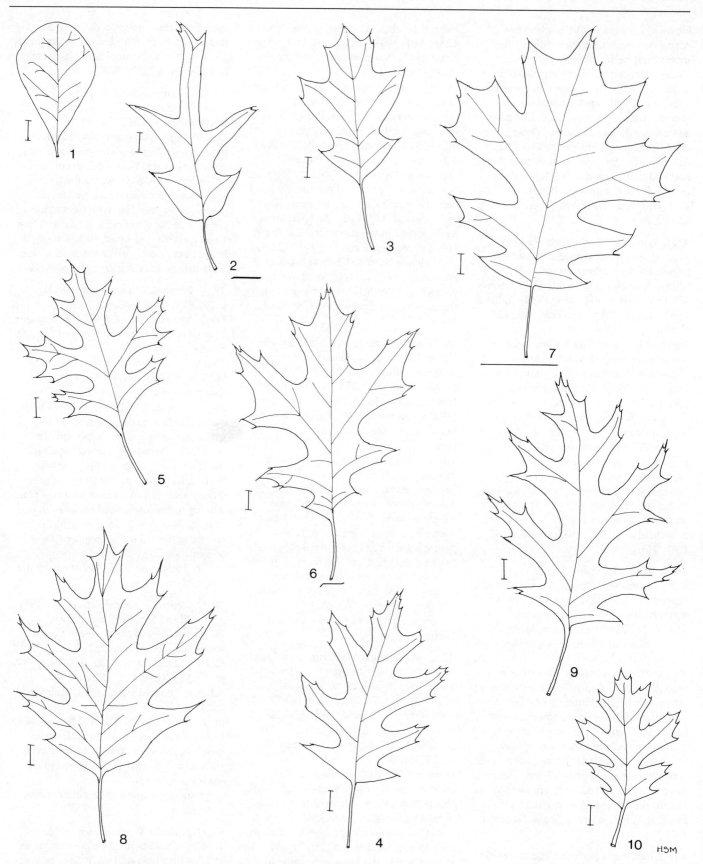

Figure 19. Leaves of *Quercus* species. 1, *Q. arkansana*. 2, *Q. falcata*.
3, *Q. ilicifolia*. 4, *Q. palustris*. 5, *Q. coccinea*. 6, *Q. ellipsoidalis*.
7, *Q. shumardii*. 8, *Q. rubra*. 9, *Q. velutina*. 10, *Q. kelloggii*.
Scales = 1 cm.

cultivated broad-leaved trees and shrubs, pl. 30 (1986).

Tree to 25 m or rarely more; bark greyish brown, smooth but in age becoming fissured and scaly; young shoots shining, hairless, reddish brown. Leaves deciduous, elliptic, very deeply lobed, the lobes bristle-tipped, the longest lobe longer than the width of the narrowest part of the leaf, apex acute, base broadly tapered; green and shining above, paler and hairless except for the large, conspicuous tufts of hairs in the axils of the main veins beneath, often turning brilliant scarlet in autumn; stalk 1.5–4 cm. Acorns almost hemispherical, 1.2–1.7 cm, somewhat ribbed, base enclosed in a thin, saucer-like cup. *E USA.* H2.

52. Q. coccinea Muenchhausen. Figure 19(5), p. 78. Illustration: Phillips, Trees in Britain, Europe and North America, 182 (1978); Miller & Lamb, Oaks of North America, 60, 61 (1985); Krüssmann, Manual of cultivated broad-leaved trees and shrubs, pl. 30 (1986).

Very similar to *Q. palustris* but axillary tufts of hairs rather small and inconspicuous, acorns 1.2–1.5 cm, usually with concentric rings at the apex. *E USA.* H2.

Colours more reliably in autumn than does *Q. palustris.*

53. Q. ellipsoidalis Hill. Figure 19(6), p. 78. Illustration: Britton & Brown, Illustrated Flora of the northern United States and Canada, edn 2, 1: 618 (1913); Miller & Lamb, Oaks of North America, 73, 74 (1985).

Also similar to *Q. palustris* but the bases of most leaves truncate; winter buds chestnut brown; acorns with the lower third to half enclosed in the cup which is made up of closely adpressed, brownish, downy scales. *Central northern USA.* ?H2.

54. Q. shumardii Buckley. Figure 19(7), p. 78. Illustration: Phillips, Trees in Britain, Europe and North America, 190 (1978); Poor, Plants that merit attention (1984); Miller & Lamb, Oaks of North America, 64, 65 (1985); Krüssmann, Manual of cultivated broad-leaved trees and shrubs, pl. 30 (1986).

Also similar to *Q. palustris* but the bases of most leaves truncate; winter buds greyish or yellowish brown; acorns 2–3 cm, enclosed for the lower third in the thick, saucer-like cup formed from adpressed, blunt scales. *SE USA.* H4.

55. Q. rubra Linnaeus (*Q. borealis* Michaux). Figure 19(8), p. 78. Illustration:

Gartenpraxis for 1977, 472; Phillips, Trees in Britain, Europe and North America, 190 (1978); Miller & Lamb, Oaks of North America, 55, 56 (1985).

Like *Q. palustris* but leaves dull, less deeply lobed, the longest lobe not as long as the width of the narrowest part of the leaf. Acorns 1.2–2.5 cm, the cup shallow or deep, hairy inside. *E USA.* H2.

The leaves frequently turn dull reddish or yellowish in autumn.

56. Q. velutina Lamarck. Figure 19(9), p. 78. Illustration: Phillips, Trees in Britain, Europe and North America, 191 (1978); Miller & Lamb, Oaks in North America, 81, 82 (1985); Krüssmann, Manual of cultivated broad-leaved trees and shrubs, pl. 30 (1986).

Tree to 25 m; bark thick, black or almost so, deeply furrowed, inner bark bright orange or yellow; young shoots hairy at first. Leaves deciduous, elliptic, 6–22.5 × 4.5–15.5 cm, rather deeply lobed with bristle-tipped lobes, dark shining green above, yellowish and densely hairy when young beneath, the mature leaves with rather sparse but persistent hairs on the surface and tufts of hair in the axils of the major veins; apex acute, base truncate; stalks 2.5–7 cm. Acorn almost spherical, 1.5–2.5 cm, pale brown, the lower half enclosed in the cup which is made up of loosely adpressed, hairy scales. *E USA.* H2.

57. Q. kelloggii Newberry. Figure 19(10), p. 78. Illustration: Phillips, Trees in Britain, Europe and North America, 184 (1978); Miller & Lamb, Oaks of North America, 279, 280 (1985); Krüssmann, Manual of cultivated broad-leaved trees and shrubs, pl. 30 (1986).

Like *Q. velutina* but mature leaves hairless beneath, inner bark not yellow. Acorns ovoid-oblong, rounded or pointed, 2.5–3.5 cm, variably enclosed (up to two-thirds) in the deep cup made up of loosely adpressed, hairless scales. *W USA (California, Oregon).* H4.

XL. ULMACEAE

Trees or rarely shrubs, without latex. Leaves alternate, 2-ranked, simple, usually toothed, usually with the 2 halves unequal, with deciduous stipules. Hairs simple. Flowers bisexual or unisexual on monoecious plants. Perianth of 1 whorl of 3–8 segments. Stamens the same number as the perianth-segments or twice as many.

Ovary superior, 1-celled, with 1 ovule attached at the top; style-branches 2. Fruit a nutlet or winged nutlet, or a drupe.

A family of about 15 genera in both northern and southern hemispheres.

1a. Veins of leaves not running into teeth, looped and joining other veins
    1. Celtis
  b. Veins of leaves running direct to teeth, sometimes forking before entering them   **2**
2a. Teeth of leaves uniform, mucronate; fruit a small drupe produced in autumn   **2. Zelkova**
  b. Teeth of leaves usually of 2 sizes, the larger alternating with 1 or more smaller ones, not mucronate; fruit winged, produced in late spring (in our species)   **3. Ulmus**

1. CELTIS Linnaeus
P.F. Yeo

Trees. Leaves simple, toothed and deciduous in our species, asymmetric, with veins looped near the margin, not running directly into teeth. Flowers on the current year's growth, appearing with the leaves, in clusters with male flowers below and bisexual flowers above. Sepals and stamens 4–5. Fruit a drupe with thin flesh, ripening in autumn.

About 70 species in North Temperate regions and the tropics. They are propagated by seed or by grafting on to *C. occidentalis. C. australis* is an attractive shade tree grown mainly within its native range in S Europe.

1a. Leaf-blades softly hairy beneath; lateral veins curving away from the margin, the first pair traversing half the length of the blade or more
    1. australis
  b. Leaf-blades hairless beneath or hairy on veins only; lateral veins straight nearly to the margin, the first pair not traversing half the length of the blade except in the smaller leaves
    2. occidentalis

1. C. australis Linnaeus. Figure 20(2), p. 81. Illustration: Guinea López & Vidal Box, Parques y jardines de España: arboles y arbustos, 260 (1969); Hegi, Illustrierte Flora von Mitteleuropa, edn 2, 3(2): 265 (1957–58); Kunkel, Flowering trees in subtropical gardens, 48 (1978); Mitchell, Field guide to the trees of Britain and northern Europe, 258 (1974).

Tree to 25 m. Bark smooth and grey. Twigs and leaves often drooping. Leaf-stalks

5–15 mm. Leaf-blades 5–12 cm, rather narrowly ovate, often acuminate at apex, rounded at base on both sides or tapered on one, with 3–5 pairs of main veins, the lower ones curving away from the margin as they approach it and traversing half the length of the blade or more; margin saw-toothed, entire at base; upper surface rough, lower softly hairy and grey-green. Fruit-stalks 1.5–2.5 cm; drupe 1–1.2 cm, yellowish white, becoming purplish or black, sweet-tasting. *S Europe, N Africa, W Asia.* H4. Spring.

2. C. occidentalis Linnaeus. Figure 20(1), p. 81. Illustration: Hough, Handbook of the trees of the northern states and Canada, 192, 193 (1947); Boom, Nederlandse dendrologie, edn 5, 154 (1965); Krüssmann, Handbuch der Laubgehölze, edn 2, **1**: 304 (1976); Hosie, Native trees of Canada, 206, 207 (1979).

Tree to 40 m. Bark rough. Leaf-stalks 1–1.5 cm. Leaf-blades 5–10 cm, ovate or ovate-oblong, gradually acuminate, slightly lobed or rounded at the base on both sides, or lobed on 1 side and rounded on the other, with 5–8 pairs of main veins running straight almost as far as the margin, the lowest not traversing as much as half the length of the blade except in the smaller leaves; margin saw-toothed, entire at base; upper surface smooth or rough, slightly glossy, lower hairless or slightly hairy on the veins. Fruit-stalks 1.2–1.8 cm, drupe 7–10 mm, yellowish or reddish, becoming dark purple with yellowish flesh. *USA except the W & SW, SE Canada.* H1.

2. ZELKOVA Spach
P.F. Yeo

Trees or shrubs. Leaves deciduous, simple, toothed, with veins running direct to teeth. Flowers on current year's growth, solitary or few together, male in lower axils of the shoot, bisexual higher up. Sepals and stamens 4–5. Fruit a green drupe, wider than long, almost stalkless, ripening in autumn.

Four or 5 species in W and E Asia. Propagation is by seed or by grafting on to *Ulmus*. Of the 2 species described here, *Z. serrata* is susceptible to elm-disease (see *Ulmus*); the leaves colour yellow in autumn. *Z. carpinifolia* is a picturesque, slow-growing tree.
Literature: Czerepanov, S., Revisio specierum generis Zelkova Spach et Hemiptelea Planchon, *Botanicheskii Materialy* **18**: 58–72 (1957).

1a.   Leaf-blade not acuminate; leaf-stalk

not more than 2 mm; leaf-teeth obtuse and mucronate   **1. carpinifolia**
 b.   Leaf-blade acuminate; leaf-stalk usually more than 2 mm; leaf-teeth acute and mucronate   **2. serrata**

1. Z. carpinifolia (Pallas) K. Koch (*Z. crenata* (Michaux) Spach. Figure 20(3), p. 81. Illustration: Mitchell, Field guide to the trees of Britain and northern Europe, 254, 255 (1974); Nicholson & Clapham, The Oxford book of trees, 160, 161 (1975); Krüssmann, Handbuch der Laubgehölze, edn 2, **3**: 433, 494, 495 (1978); Bean, Trees and shrubs hardy in the British Isles, edn 8, **4**: 778 (right hand piece) & t. 111 (1980).
Tree to 25 m. Trunk furrowed, split into numerous erect branches at base of crown. Bark smooth, flaking in age to leave crumbling orange patches. Leaf-stalks not more than 2 mm. Leaf-blades to 5 or rarely to 9 cm, elliptic or oblong, shortly tapered at apex, asymmetric at base, with 6–12 pairs of veins running out into coarse, obtuse and mucronate teeth; upper surface sometimes rough, lower with hairs on veins. Fruit *c.* 5 mm wide, irregular or ridged. *USSR (Transcaucasia), NE Turkey, N Iran.* H1.

2. Z. serrata (Thunberg) Makino. Figure 20(4), p. 81. Illustration: Boom, Nederlandse dendrologie, edn 5, 154 (1965); Fitschen, Gehölzflora, edn 7, 51, fig. 132 (1977); Krüssmann, Handbuch der Laubgehölze, edn 2, **3**: 433, 494, 495 (1978); Bean, Trees and shrubs hardy in the British Isles, edn 8, **4**: 778 (left hand piece) (1980).
Tree to 30 m. Branches rather widely spreading. Bark smooth, striped horizontally, when old flaking into strips that leave pale brown patches. Leaf-stalks usually to 6 mm. Leaf-blades 2–6 cm (to 12 cm on long shoots), ovate, acuminate, rounded or weakly lobed at base, asymmetric or nearly symmetric, with 8–12 pairs of veins (only 6 in the smallest leaves) that run out into coarse, acute, mucronate teeth; upper surface with scattered hairs, lower hairless or with hairs on veins. Fruit *c.* 4 mm wide. *Japan.* H1.

3. ULMUS Linnaeus
P.F. Yeo

Deciduous, rarely half-evergreen trees. Leaves shortly stalked; blades toothed, usually doubly (with small teeth on the sides of the major teeth), the 2 halves unequal in size and shape (at least in a proportion of the leaves). Flowers in

stalkless clusters appearing in spring before the leaves or rarely in autumn. Perianth *c.* 3 mm, of 4–9 segments joined at the base. Stamens the same number as and longer than the perianth-lobes. Fruits flattened, circular to elliptic, with an apical notch, green during development but dry, papery and brownish when ripe, with a single seed occupying a swelling on the mid-line, ripening within a few weeks of flowering.

A genus variously estimated to contain between 20 and 45 species. In much of Europe numerous vegetatively propagated clones occur, some of which produce little viable seed. Single such clones or groups of similar ones have been made the basis for a number of specific names; disagreement as to their acceptability accounts for most of the doubt about the number of species in the genus. Many of these plants belong to a group centred on *U. minor*, which is a suckering species, while others are clearly hybrids between *U. minor* and *U. glabra* (non-suckering). These hybrids, most of which are also suckering, are assigned to *U. × hollandica*, though inevitably the position of some clones is debatable.

Elms have been important trees in roadside planting, and in avenues, parks and cemeteries but are susceptible to attack by the fungus *Ceratocystis ulmi* which is spread by bark beetles and periodically breaks out in lethal epidemics. Successive epidemics are caused by different strains of the fungus, and a particularly devastating one appeared in Europe in the 1960s and has since ravaged wild and cultivated elms over large areas. It is too soon to know which elms will survive the present outbreak and this account covers the elms that have been important up to recent times.

Shrubby variants arise occasionally, such as *U × hollandica* 'Jacqueline Hillier'. Other elms suitable for the smaller garden are *U. glabra* 'Camperdownii' and *U. parvifolia*. There are a few coloured-leaved cultivars but these are rarely seen.
Propagation is by suckers or by grafting.

In our descriptions the length of the leaf-stalk is taken from the base to the beginning of the shorter side of the blade, while the veins are counted on the longer side; no attempt is made to count veins in the narrow acuminate tip. Sucker shoots and new twigs breaking out directly from the trunk or limbs bear uncharacteristic leaves; our descriptions refer to the leaves of the crown; the leaf sizes given refer to the majority of leaves and extreme sizes are

Figure 20. Leaves of species of Ulmaceae. 1, *Celtis occidentalis*.
2. *C. australis*. 3, *Zelkova carpinifolia*. 4, *Zelkova serrata*. 5, *Ulmus
americana*. 6, *U. rubra* (a, b, two different leaves). 7, *U. glabra*
'Camperdownii'. 8, *U. glabra* 'Exoniensis' (a–e, five different
leaves). Scale = 2 cm.

thus omitted. Identification is difficult and is more likely to succeed if a few different trees are tackled simultaneously.
Literature: Green, P.S., Registration of cultivar names in Ulmus, *Arnoldia* **24**: 41–80 (1964); Richens, R.H., *Elm* (1983).

Leaf-blades. Glossy: **1,5,6,9**; not glossy: **2,4,5,8**. Rough: **2–5,7**. With prominent point(s) next to apex: **4**.
Leaf-veins. Mostly 12 or more: **1–6,8**; fewer than 12: **5–9**.
Leaf-teeth. Simple: **8,9**. Blunt: **3,6,9**.
Fruit. Ciliate: **1,2**. Developing in autumn: **9**.

1a. Many of the leaves with 12 or more veins   2
  b. Most leaves with fewer than 12 veins   8
2a. Long side of leaf-blade with a lobe at base that overlies the stalk and frequently the adjacent twig; upper surface of blade very rough, dark green, not shiny   **4. glabra**
  b. Not with this combination of characters   3
3a. Upper surface of leaf-blades smooth, shiny and hairless, lower surface hairless; leaf-teeth often rather blunt   **6. minor**
  b. Upper surface of leaf-blades rough or dull or hairy or some combination of these, lower surface hairy or rough; leaf-teeth often sharp   4
4a. Buds and centre of fruit with brown hairs; leaf-teeth very shallow and rather blunt; leaves and bark fragrant; inner bark mucilaginous   **3. rubra**
  b. Buds and centre of fruit without brown hairs; leaf-teeth acute; leaves and bark not fragrant; inner bark not mucilaginous   5
5a. Leaves mostly 3.5–10 cm   **8. pumila**
  b. Leaves mostly 6–12 or 8–16 cm   6
6a. Flower-stalks not longer than the flowers; fruits not ciliate
                                          5. × hollandica
  b. Flower-stalks 1–2 cm, many times longer than the flowers; fruit ciliate
                                                                                                  7
7a. Leaf-blades broadest near the middle, often more than twice as long as broad, hairy beneath mainly on the veins   **2. americana**
  b. Leaf-blades broadest above the middle, mostly less than twice as long as broad; usually densely soft-hairy all over the lower surface   **1. laevis**
8a. Serration of leaf-blade mostly simple, only a few subordinate teeth present

(but see remarks on **8. pumila**)
                                          9. parvifolia
  b. Serration of leaf-blades constantly double, almost all teeth with subordinate teeth   9
9a. Leaf-blades thick, with upper surface smooth and shiny   10
  b. Leaf-blades not especially thick, with upper surface dull, sometimes glossy when old, often rough   11
10a. Leaf-blades 12–16 cm
                                5. × hollandica 'Vegeta'
  b. Leaf-blades not more than 7 cm
                                                  6. minor
11a. Leaf-blades 3–7 cm   12
  b. Leaf-blades commonly exceeding 7 cm
                                                                 13
12a. Leaves densely arranged
                   5. × hollandica 'Jacqueline Hillier'
  b. Leaves normally spaced   **8. pumila**
13a. Main branches divergent, forming a round crown; trunk not continued far into crown   **5. × hollandica**
  b. Main branches mostly spreading and arching, forming a columnar crown; trunk continued well up into the crown   **7. procera**

1. U. laevis Pallas (*U. effusa* Willdenow; *U. racemosa* Borkhausen). Illustration: Šavulescu, Flora republicii populare Romane 1: 339 (1952); Krüssmann, Handbuch der Laubgehölze, edn 2, **3**: 421 (1978).
Tree to 35 m. Crown wide, open. Buds conical, acute. Leaf-stalks 4–8 mm. Leaf-blades mostly 5–15 × 2.5–9 cm, obovate or broadly elliptic, mostly less than twice as long as broad, weakly acuminate, the larger side rounded at the base and forming a shallow lobe joined to the stalk near the base, the smaller side tapered; margin rather shallowly but sharply doubly saw-edged; veins usually 12–17; upper surface bright green, glossy, hairless to sparsely hairy or slightly rough; lower surface more or less hairy. Flower-stalks to 2 cm. Stamens 5–8. Stigmas white. Fruit 1–1.5 cm, ovate, ciliate, the notch not reaching the seed. *C & SE Europe, Turkey, USSR (Caucasus)*. H1.

2. U. americana Linnaeus. Figure 20(5), p. 81. Illustration: Hough, Handbook of the trees of the northern states and Canada, 182, 183 (1947); Boom, Nederlandse dendrologie, edn 5, 154 (1965).
Similar to *U. laevis* but with branches drooping at tips, buds usually more or less obtuse, leaf-blades 6–15 × 2–7.5 cm, ovate or elliptic, often more than twice as long as broad, the larger side joining the stalk half

way up or higher, subsidiary teeth very small, fruit *c.* 10 mm, ciliate, the apical notch reaching the seed. *N America E of the Rocky mountains*. H1.
Cultivated occasionally.

3. U. rubra Muhlenberg (*U. fulva* Michaux). Figure 20(6), p. 81. Illustration: Hough, Handbook of the trees of the northern states and Canada, 188, 189 (1947); Gleason, New Britton & Brown illustrated flora of the northeastern United States and adjacent Canada 2: 49 (1952); Krüssmann, Handbuch der Laubgehölze, edn 2, **3**: 427, t. 143 (1978).
Tree to 25 m. Branches spreading, forming a broad, open crown. Buds obtuse, with orange-brown hairs. Inner bark mucilaginous and fragrant. Leaf-stalks 4–8 mm. Leaf-blades 7–20 × 4–10 cm, thick and firm, fragrant in drying, ovate, sometimes narrowly so, gradually to abruptly narrow-acuminate, the larger side rounded at the base, usually forming a shallow lobe which may be wide or have its sharpest curvature near the stalk, and joining the stalk at the top or some other point above the middle, the smaller side straight-tapered to rounded; margin very shallowly and rather bluntly doubly saw-edged; veins 10–19; upper surface very rough; lower surface softly and densely hairy, the hairs sometimes lost in age. Flowers crowded. Stamens 5–9. Stigmas pink. Fruit 1–2 cm, nearly circular to broadly elliptic, lightly notched, with red-brown hairs over the central seed. *N America*. H1.
Grown more widely on the European continent than in the British Isles.

4. U. glabra Hudson (*U. campestris* Linnaeus, in part; *U. montana* Withering). Illustration: Clapham et al., Flora of the British Isles, edn 1, 718 (1952), edn 2, 564 (1962); Hegi, Illustrierte Flora von Mitteleuropa, edn 2, **3**(2): 260 (1957–58); Nicholson & Clapham, The Oxford book of trees, 160 (1975); Edlin, The tree key, 115 (1978).
Tree to 40 m. Suckers absent. Crown wide, lobed. Tips of branches retaining smooth grey bark for several years. Twigs often bearing leaves in 2 distinct rows. Leaf-stalks usually 5–10 mm. Leaf-blades 6–16 × 3.5–10 cm, elliptic to obovate or oblong-obovate, strongly acuminate, sometimes with shoulders produced into 1 or 2 points, the larger side at base forming a rounded lobe lying over the stalk and often over the twig bearing it and free or attached to the stalk for half its length, the

smaller side with base weakly lobed to narrowly wedge-shaped; margins deeply and sharply doubly saw-edged, each main tooth often with more than 1 subsidiary tooth on the lower edge, sometimes also with a tooth on the upper edge (more than 1 in 'Exoniensis'); veins 12–18; upper surface dark green, dull, very rough; lower surface hairy, especially on the veins. Flowers in dense clusters. Stamens 5–6. Stigmas red. Fruit 2–2.5 cm, obovate or broadly elliptic, with a small notch at apex, sometimes hairy at the notch; seed at or below the middle. *Europe*. H1.

'Camperdownii' (*U.* × *vegeta* 'Camperdownii'). Figure 20(7), p. 81. Illustration: Mitchell, Field guide to the trees of Britain and northern Europe, 249 (1974); Edlin, The tree key, 114 (1978). Branches drooping (grafted on a standard; height rarely more than 2 or 3 m above the graft). Twigs flexuous, bearing leaves in 2 rows. Leaf-blades 10–14 × 5.5–8.5 cm, or occasionally to 20 × 18 cm, a few of them with shoulders produced into points; main teeth sometimes with subsidiary teeth on both edges.

'Exoniensis' (*U. glabra fastigiata* invalid). Figure 20(8), p. 81. Illustration: Krüssmann, Handbuch der Laubgehölze, edn 2, 3: t. 143, 146 (1978). Medium-sized tree, narrow when young, with erect branches; twigs turned up at tips. Leaf-blades to *c.* 10 cm, erect, broadly obovate, twisted or folded; main teeth deep, mostly with 3 or 4 teeth on the lower edge and 1 or 2 on the upper.

'Pendula' ('Horizontalis'). Illustration: Mitchell, Field guide to the trees of Britain and northern Europe, 249 (1974). Branches (from a graft on a standard, making a tree to 15 or 20 m) spreading widely, slightly and stiffly descending in a herring-bone pattern and exposed above the leaves. Leaf-blades mostly 8–13 cm, many of them with shoulders produced into points.

5. U. × hollandica Miller.
Tree to 40 m. Suckers present. Buds acute. Leaf-stalks to 6–20 mm. Leaf-blades to 15 cm, broadly ovate or broadly obovate to narrowly elliptic-obovate, acuminate, the larger side at base rounded and usually forming a lobe partly attached to the stalk, the smaller side rounded; margins either sharply or obtusely doubly saw-edged; veins 10–18; upper surface more or less rough, hairless or nearly so, sometimes glossy, lower hairy. Flowers crowded on short stalks. Fruit 1.5–2 cm, ovate or

obovate; apical notch not or scarcely reaching the seed. *Europe*. H1.

Possibly hybrids between *U. glabra* and *U. minor*, often growing apparently naturally in hedgerows in Europe, but many such occurrences represent descendants of planted specimens.

The following are or have been important in urban planting:

'Belgica' (*U.* × *belgica* Burgsdorf). Illustration: Elwes & Henry, Trees of Great Britain and Ireland 7: t. 142 (1913); Boom, Nederlandse dendrologie, edn 5, 154 (1965). Large, broad-crowned tree. Leaf-stalks to 6 mm. Leaf-blades 8–12 cm, elliptic-obovate or narrowly obovate, with 14–18 veins, slightly rough above, similar to the narrower type of leaf of *U. glabra*. Fruit with the seed above the middle.

Very susceptible to elm disease.

'Hollandica' (*U.* × *hollandica* var. *major* (Smith) Rehder; *U.* × *hollandica* 'Major'). Illustration: Moss, The Cambridge British Flora 2: t. 96, 97 (1914); Mitchell, Field guide to the trees of Britain and northern Europe, 252 (1974); Nicholson & Clapham, The Oxford book of trees, 160, 161 (1975). Large tree with upper branches divergent, lower spreading. Sucker shoots numerous, becoming corky. Leaf-stalks to *c.* 10 mm. Leaf-blades 6–12 cm, broadly ovate to broadly obovate with 8–12, rarely 14 veins; teeth not very sharp but rather coarse; margins often puckered; upper surface rough, deep green, becoming glossy. Apical notch of fruit reaching the seed.

'Vegeta' (*U.* × *hollandica* var. *vegeta* (Loudon) Rehder). Figure 21(1), p. 84. Illustration: Moss, The Cambridge British Flora 2: t. 94, 95 (1914); Mitchell, Field guide to the trees of Britain and northern Europe, 241, 253 (1974); Nicholson & Clapham, The Oxford book of trees, 160, 161 (1975); Richens, Elm, 95 (1983). Large tree with rather narrowly divergent branches. Crown rounded. Sucker shoots few, not corky. Leaf-stalks 1–2 cm. Leaf-blades 8–16 cm, elliptic, ovate or obovate, with 12–16 veins; teeth moderately sharp, rather fine; upper surface slightly rough, glossy. Fruit with the seed above the middle.

Found in gardens is 'Jacqueline Hillier' (*U.* × *elegantissima* 'Jacqueline Hillier'), a Shrub to 2 m, with small elliptic-lanceolate leaves, rough above, very densely arranged in two rows.

6. U. minor Miller (*U. angustifolia* (Weston) Weston; *U. campestris* misapplied; *U. carpinifolia* Suckow; *U. nitens* Moench;

U. stricta (Aiton) Lindley. Illustration: Clapham et al., Flora of the British Isles, edn 1, 720 (1952), edn 2, 565 (1962); Boom, Nederlandse dendrologie, edn 5, 154 (1965); Nicholson & Clapham, The Oxford book of trees, 160 (1975); Fitschen, Gehölzflora, edn 7, 270 (1977).
Tree to 30 m. Suckers present. Branches mostly erect with drooping tips. Twigs hairless except on suckers. Buds ovoid, acute. Leaf-stalks 1–2 cm, occasionally to 3 cm. Leaf-blades mostly 5–13 × 2.5–6 cm, rather thick, elliptic or obovate to oblanceolate, gradually acuminate, the larger side at the base forming a more or less narrow and shallow lobe, joined to the stalk about halfway up, the smaller side tapered from a point one-third to one-half the way along the blade; margins shallowly and almost obtusely doubly saw-edged; veins 12–16 (some cultivars have smaller leaves and fewer veins); upper surface hairless, bright green, glossy; lower surface hairless except for the vein axils. Flowers crowded. Stamens 4–5. Stigmas usually white. Fruit 1.2–1.7 cm, obovate with the seed near the notch. *Europe, N Africa, SW Asia*. H1.

Inclusion of *U. coritana* Melville with 8–10 veins and *U. plotii* Druce with nearly equal halves of the leaf and 7–10 veins is probably justified, but would disrupt the description.

'Pendula' (*U. carpinifolia* forma *pendula* (Henry) Rehder). Figure 21(2), p. 84. Large tree with tips of branches and twigs drooping. Leaf-stalks 1.5–3 cm. Leaf-blades to *c.* 8 cm long and 5.5 cm wide, only slightly acuminate.

The following have dense crowns and small leaves:

'Cornubiensis' (*U. angustifolia* var. *cornubiensis* (Weston) Melville); *U. carpinifolia* var. *cornubiensis* (Weston) Rehder). Figure 21(3), p. 84. Illustration: Mitchell, Field guide to the trees of Britain and northern Europe, 251 (1974); Nicholson & Clapham, The Oxford book of trees, 160 (1975); Edlin, The tree key, 117 (1978). Medium-sized tree with crown conical at first, columnar and round-topped when old. Leaves in clusters separated by gaps from clusters on other branches. Leaf-blades mostly less than 6.5 cm, thick, glossy, sometimes curved up towards the edges, with up to 10 veins and obtuse teeth. *Possibly native in SW England*.

Sometimes confused with 'Sarniensis'.
'Dampieri' (*U. carpinifolia* var. *dampieri* (Wesmael) Rehder; *U.* × *hollandica* 'Dampieri'). Illustration: Fitschen,

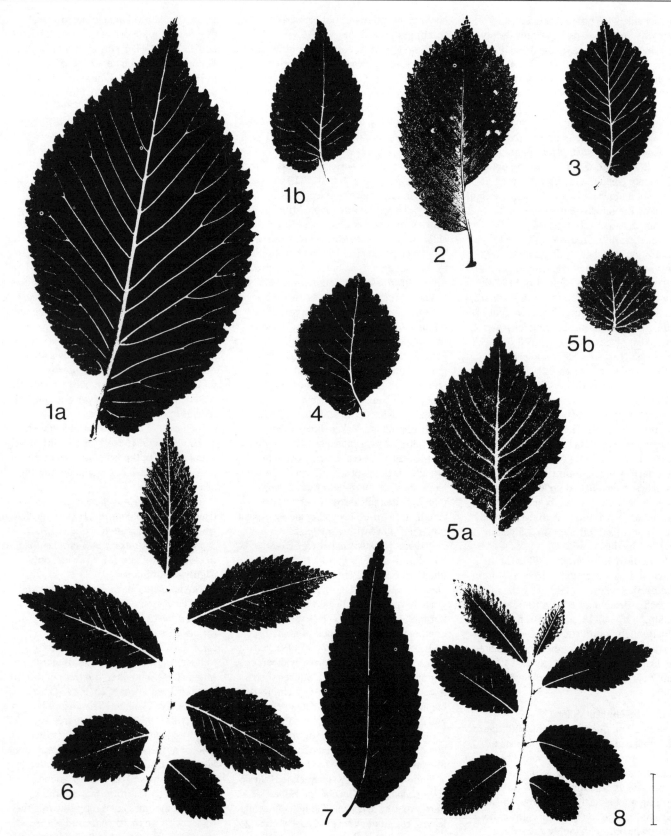

Figure 21. Leaves of species of Ulmaceae. 1, *Ulmus* × *hollandica* 'Vegeta' (a, b, two different leaves). 2, *U. minor* 'Pendula'. 3, *U. minor* 'Cornubiensis'. 4, *U. minor* 'Sarniensis'. 5, *U. procera* (a, b, two different leaves). 6, *U. pumila* var. *pumila*. 7. *U. pumila* var. *arborea*. 8, *U. parvifolia*. Scale = 2 cm.

Gehölzflora, edn 7, 272 (1977); Lancaster, Trees for your garden, 139 & 141 (1974). Medium-sized tree with conical crown, narrow when young. Leaves of short twigs crowded. Leaf-blades to 6 cm, firm, broadly ovate or broadly elliptic, less asymmetric at base than in other cultivars, with *c.* 10 veins; margins wavy; main teeth with subsidiary teeth on both edges; colour bright green or (in a branch-sport, 'Wredei' (*U. carpinifolia* forma *wredei* (Juehlke) Rehder), golden. Grown more on the European continent than in the British Isles.

'Sarniensis' (*U. carpinifolia* var. *wheatleyi* (Simon-Louis) Bean; *U. carpinifolia* var. *sarniensis* (Loudon) Rehder; *U. sarniensis* invalid; *U. stricta* var. *sarniensis* (Loudon) Moss; *U. wheatleyi* invalid). Figure 21(4), p. 84. Illustration: Mitchell, Field guide to the trees of Britain and northern Europe, 251 (1974); Edlin, The tree key, 117 (1978); Krüssmann, Handbuch der Laubgehölze, edn 2, **3**: t. 143 (1978); Bean, Trees and shrubs hardy in the British Isles, edn 8, **4**: t. 90 (1980). Similar to 'Cornubiensis' but remaining conical until very old, evenly and densely leafy; lower branches spreading; leaf-blades scarcely curled, with the teeth still more obtuse, coarser and rather irregular, the subsidiary teeth sometimes unusually large. Widely planted in England, Wales and S Scotland, especially on roadsides and in avenues.

'Umbraculifera' (*U. carpinifolia* var. *umbraculifera* (Trautvetter) Rehder). Illustration: Boom, Nederlandse dendrologie, edn 5, 153 (1965); Krüssmann, Handbuch der Laubgehölze, edn 2, **3**: t. 144, 145 (1978). Crown spherical or hemispherical, dense and uniform. Leaf-blades not more than 7 cm, elliptic, slightly rough above. Grown in continental Europe.

7. U. procera Salisbury (*U. campestris* Linnaeus in part). Figure 21(5), p. 84. Illustration: Clapham et al., Flora of the British Isles, edn 1, 718 (1952), edn 2, 564 (1962); Mitchell, Field guide to the trees of Britain and northern Europe, 241 (1974); Nicholson & Clapham, The Oxford book of trees, 16 (1975); Edlin, The tree key, 117 (1978).
Tree to 40 m. Crown broadly columnar with the trunk continued more than halfway through it; branches few, massive. Leaf-stalks to *c.* 1.1 cm. Leaf-blades to 9.5 cm, broadly ovate to broadly obovate, width two-thirds to equalling the length,

the larger ones abruptly acuminate, the larger side usually broadly rounded, forming a lobe that joins the stalk at or below the middle, the smaller rounded or tapered, sometimes forming a shallow lobe; margins rather shallowly doubly saw-edged with acute teeth; veins 10–13; upper surface dull, dark green, very rough; lower surface thinly hairy. Flowers crowded. Stamens 3–5. Stigmas white. Fruit 1–1.7 cm, circular with the seed in the upper half, usually sterile. *England, possibly introduced.* H1.

Apparently a single clone or small number of closely similar clones.

The leaves come within the range of shape of *U.* × *hollandica* but the habit is distinct. It is notorious for dropping large limbs in calm weather, and is very susceptible to elm-disease.

8. U. pumila Linnaeus. Illustration: Mitchell, Field guide to the trees of Britain and northern Europe, 254 (1974); Krüssmann, Handbuch der Laubgehölze, edn 2, **3**: 430, 432 (1978).
Tree to 30 m with a domed crown. Twigs drooping, growing continuously throughout the summer. Leaves persisting late in autumn. Leaf-stalks 5–7 mm, very thin. Stipules half as long to as long as leaf-stalks, with ovate base and tongue-like tip. Leaf-blades 3–10 × 2–4 cm, with matted hairs in vein-axils beneath, elliptic to elliptic-lanceolate or ovate-lanceolate; margin doubly toothed; veins 7–15; upper surface dark green, matt. Flowers crowded. Fruit 1–1.5 cm, circular, seed near the middle. *C & E Asia.* H1.

Resistant to elm-disease. Fast-growing despite the small leaves. Although the illustrations cited above show doubly-toothed leaves, most descriptions refer also to simply toothed margins. The name may be misapplied to very small-leaved plants related to *U. minor*.

Var. **pumila**. Figure 21(6), p. 84. Stipules as long as leaf-stalks. Leaf-blades to 7 × 3 cm, broadest near the middle, thin, with the veins and teeth directed strongly forward, the larger side at base narrowly rounded, joining the stalk at or near the apex, the smaller side rounded or tapered; veins 7–10. *E Asia.*

Var. **arborea** Litvinow. Figure 21(7), p. 84. Stipules half as long as leaf-stalks. Leaf-arrangement conspicuously 2-rowed. Leaf-blades to 10 × 4 cm, the larger ones broadest near the base, thick, the larger side at base usually rounded, joining the stalk above the middle, the smaller side

tapered or slightly rounded; veins 11–15. *C Asia.*

Used as a street-tree.

9. U. parvifolia Jacquin. Figure 21(8), p. 84. Illustration: Mitchell, Field guide to the trees of Britain and northern Europe, 254 (1974); Krüssmann, Handbuch der Laubgehölze, edn 2, **3**: 432 (1978).
Tree to 15 or 25 m with broad crown. Leaves persisting late in autumn. Leaf-stalks 3–6 mm, very thin. Stipules much longer than leaf-stalks, linear. Leaf-blades 1.5–4 × 1–2 cm, thick, with matted hairs at base beneath when young, elliptic-ovate to elliptic-obovate, the 2 sides nearly equal in length, one usually much broader than the other; margins simply toothed; teeth rounded; veins 7–11; upper surface bright yellow-green, waxily glossy; lower surface moderately glossy. *N & C China, Korea, Japan.* H1. Late summer-autumn.

XLI. EUCOMMIACEAE

Deciduous, dioecious trees. Stems hollow, pith septate. Leaves alternate, simple, without stipules. Flowers solitary, without petals or sepals, in the axils of bracts below the upper leafy portions of the shoots, the males crowded, the females separated. Fruit a samara.

A family of a single genus, said to be related on the one hand to the Hamamelidaceae and on the other to the Ulmaceae.

1. EUCOMMIA Oliver
P.S. Green
Deciduous trees to 20 m, dioecious. Leaves stalked, simple, narrowly ovate to elliptic, 7–15 cm, acuminate, pinnately veined, hairy along the veins when young, margins with fine teeth pointing forwards. Male flowers stalked, consisting of 6–12 linear stamens, *c.* 1 cm. Female flowers stalked, ovary long, *c.* 1 cm, 1-celled with 2 hanging ovules, style terminal, 2-lobed, reflexed. Fruit ellipsoid, 3–4 cm with narrow, longitudinal wings, terminally notched.

A genus of a single species. It may be propagated by seed or by half-ripe cuttings placed in gentle heat.

1. E. ulmoides Oliver. Illustration: Boom, Nederlandse dendrologie, edn 2, **1**: 152, f. 48d (1949); Bean, Trees and shrubs hardy in the British Isles, edn 8, **2**: t. 23 (1973); Krüssmann, Handbuch der

Laubgehölze, edn 2, **2**: 49 & t. 10i (1977);
Everett, New York Botanical Gardens
illustrated encyclopedia of horticulture,
1283 (1981). *C China*. H1. Spring.

Grown as a specimen tree, this species
resembles an elm. Even in China it is said to
be known only from cultivation where it is
grown for its medicinally prized bark. It
also produces rubber but in insufficient
quantity for commerce.

XLII. MORACEAE

Trees, shrubs, herbs or vines, evergreen or
deciduous, dioecious or monoecious, most
with milky sap containing latex. Leaves
alternate or rarely opposite, simple, entire
or lobed or toothed, with 2 (rarely 1)
deciduous stipules. Flowers unisexual,
reduced and small, radially symmetric, in
globular heads or catkins, sometimes
densely packed inside hollow receptacles.
Calyx-lobes 4–5, sometimes absent. Petals
absent. Stamens 1–5, opposite the calyx-
lobes. Ovary usually 1-celled, ovules mostly
pendulous. Styles simple or branched. Fruit
an achene, nut or drupe, sometimes
enclosed in the fleshy, hollow receptacle.
Seed with usually curved embryo.

There are 50 genera and 1200–1500
species in this mainly tropical family. Two
genera (*Cannabis* and *Humulus*) that do not
have milky sap and possess 5 (not 4) sepals
and stamens are sometimes separated into
a distinct family, Cannabaceae (Cannabin-
aceae, Cannabidaceae).

1a. Annual or perennial herbs or vines   2
 b. Trees or shrubs, occasionally climbing
                                                4
2a. Vines, perennial or annual
                                  8. Humulus
 b. Erect herbs   3
3a. Perennial herbs; leaves entire or
    rarely lobed; stamens 1–3; sap milky
                                 6. Dorstenia
 b. Annual herbs; leaves palmately
    divided; stamens 5; sap not milky
                                 9. Cannabis
4a. Tropical trees with large, lobed or
    pinnate leaves; inflorescence and
    fruiting head spherical or oblong
                                3. Artocarpus
 b. Combination of characters not as
    above   5
5a. Trees with spines on the branches   6
 b. Trees without spines, or shrubs   7
6a. Male flowers in a loose raceme; leaves
    always entire   **4. Maclura**

 b. Male flowers in spherical heads;
    leaves sometimes lobed   **5. Cudrania**
7a. Receptacle (fig) fleshy and urn-shaped
    with the flowers borne inside
                                     7. Ficus
 b. Fruit not as above   8
8a. Fruiting head succulent, juicy; female
    flowers with 4 distinct sepals
                                     1. Morus
 b. Fruiting head hairy, not as above;
    female flowers with a tubular calyx
                               2. Broussonetia

1. MORUS Linnaeus
E.C. Nelson

Deciduous trees, monoecious or dioecious.
Leaves alternate, simple, entire or lobed,
toothed. Flowers in axillary spikes. Male
flowers with 4 calyx-segments and 4
stamens. Female flowers with 4 calyx-
segments, style with 2 branches. Fruits
juicy, seeds enclosed by the enlarged,
succulent calyx.

About 12 species are now recognised
within this genus, although many more
names have been published. The commonly
cultivated trees have edible fruits.

1a. Leaves distinctly hairy beneath   2
 b. Leaves not hairy beneath (sometimes
    with hairs in vein-axils)   3
2a. Leaves densely hairy beneath; tall tree
    (15–20 m)   **5. rubra**
 b. Leaves downy beneath; tree rarely
    over 10 m   **4. nigra**
3a. Young shoots downy   4
 b. Young shoots hairless   **2. australis**
4a. Leaves rough above   **1. alba**
 b. Leaves not rough above
                                 3. cathayana

1. M. alba Linnaeus (*M. multicaulis*
Perrotet). Illustration: Mitchell, The
complete guide to trees of Britain and
northern Europe, 57 (1985).
Tree to 15 m; young shoots downy,
becoming hairless. Leaves ovate, usually
entire but occasionally with 3 lobes,
7–30 × 5–15 cm, base cordate, apex acute
or obtuse, hairless above and beneath but
with few hairs in axils of veins beneath;
stalk 1.5–2.5 cm. Male spike *c*. 2.5 cm.
Fruit white or pink (almost black in var.
multicaulis (Perrotet) Loudon), sweet but
insipid. *C Asia eastwards to China*. H3. Late
spring.

A very variable species; many of its
variants have been described as distinct
species, e.g. plants labelled *M. bombycis*
Koidzumi (*M. kagayamae* Koidzumi) have
hairless shoots and leaves occasionally with
3 lobes, and probably represent a variant of

M. alba. A distinctive variant is var.
multicaulis, a multi-stemmed shrub with
leaves to 30 cm, and almost black fruits.
There are also numerous cultivars
including 'Pendula' with pendulous
branches and 'Laciniata' with deeply lobed
leaves. This is the species used as a fodder
for silkworms.

M. mongolica (Bureau) Schneider (*M.
alba* var. *mongolica* Bureau). Illustration:
Iconographia cormophytorum sinicorum **1**:
479 (1972). Resembles *M. alba* but the
leaves have coarse marginal teeth, each
one with a long bristle. *China*. H3. Late
spring.

2. M. australis Poiret (*M. acidosa* Griffith).
Illustration: Li, Woody Flora of Taiwan, 14
(1963); Iconographia cormophytorum
sinicorum **1**: 479 (1972).
Bushy shrub or small tree to 8 m. Young
shoots hairless. Leaves ovate, 2–15 ×
2–15 cm, entire or deeply lobed, base
cordate or truncate, upper surface rough
with minute warts, downy beneath,
becoming hairless; lobes (if present)
scolloped; stalk 2–4 cm. Male spikes to
2 cm. Fruit dark red, sweet. *China, Japan,
Taiwan*. H3. Late spring.

3. M. cathayana Hemsley. Illustration:
Iconographia cormophytorum sinicorum **1**:
480 (1972); Krüssmann, Handbuch der
Laubgehölze **2**: t. 131 (1976).
Tree to 8 m. Young shoots downy,
becoming hairless. Leaves on adult plant
round to ovate, 7–15 cm long and broad,
base cordate, apex abruptly acuminate,
acute, with rounded teeth, hairy above and
beneath; stalk to 2.5 cm, hairy. Male spikes
2 cm, stalk to 2 cm. Fruit white, red or
black. *China*. H3. Late spring.

4. M. nigra Linnaeus. Illustration: Mitchell,
The complete guide to trees of Britain and
northern Europe, 57 (1985).
Tree to 10 m. Young shoots downy. Leaves
ovate or with 2–3 lobes, 7–12 × 5–10 cm
(larger on vigorous shoots), base cordate,
apex tapering to a short point, rough and
hairy above, downy beneath. Male spikes to
2.5 cm. Fruit juicy, slightly acidic, red-
black. *Asia (long cultivated)*. H3. Late spring.

This is the common mulberry with
edible, juicy fruits; they may be eaten raw
but unripe fruit causes illness. It forms a
rugged spreading tree. No named cultivars
of importance are recorded.

5. M. rubra Linnaeus. Illustration: Everett,
The New York Botanical Garden illustrated
encyclopedia of horticulture, **7**: 2237
(1981).

Tree to 20 m. Leaves ovate, entire or occasionally with 3 lobes, 7–25 × 5–17 cm, base cordate, margins with acute teeth, rough with few scattered coarse hairs above, very downy beneath. Male spikes to 5 cm. Fruit red becoming dark purple. *E USA, Canada* H3. Late spring.

2. BROUSSONETIA Ventenat
E.C. Nelson
Trees or shrubs, monoecious or dioecious. Leaves alternate, toothed, sometimes 3-lobed. Inflorescence axillary. Male flowers in cylindrical catkins, with 4 calyx-lobes, stamens 4, curved inwards in bud. Female flowers in spherical heads; calyx tubular or ovoid, with 3–4 teeth, persistent in fruit; style simple, Fruiting heads small, spherical, hairy. Fruit fleshy, projecting from calyx.

A genus of about 8 species distributed through eastern Asia and Polynesia; only 2 are in general cultivation in Europe. Propagation is by cuttings taken in July and August of young wood with a heel, or from seed.

1a. Tree with stout hairy branches; leaf-stalk 2–10 cm; leaves hairy
2. papyrifera
  b. Shrub with slender hairless branches; leaf-stalk 3–20 cm; leaves hairless or slightly downy   **1. kazinoki**

1. B. kazinoki Sieber (*B. sieboldii* Blume; *B. kaempferi* misapplied). Illustration: Iconographia cormophytorum sinicorum 1: 481 (1972); Krüssmann, Handbuch der Laubgehölze 1: 255 (1976).
Monoecious shrub. Branches slender, tinted with purple. Leaves downy when young, becoming hairless, ovate, 5–15 × 3–8 cm, not lobed, apex acuminate, base cordate; stalk less than 2 cm. Male catkins 1–1.5 cm, stalk to 1 cm. Fruit red. *China, Japan, Korea.* H3. Spring.

B. kaempferi Sieber. Illustration: Li, Woody Flora of Taiwan, 108 (1963). Similar but dioecious and with climbing branches, leaves with rounded bases. *China, Japan, Korea, Taiwan.* H3. Spring.

2. B. papyrifera (Linnaeus) Ventenat (*Morus papyrifera* Linnaeus). Illustration: Botanical Magazine, 2358 (1822); Krüssmann, Handbuch der Laubgehölze 1: 256 (1976); Mitchell, The complete guide to trees of Britain and northern Europe, 56 (1985).
Tree to 15 m. Branches stout, with thick grey-brown hairs when young. Leaves coarsely hairy, obliquely ovate, 7–18 × 5–15 cm, often with 3–5 deep lobes, apex acuminate, base cordate; stalk more than 2 cm. Male catkins woolly, often curled, yellow-green, 4–8 cm. Fruiting heads orange with red fruits. *China, Japan, Korea, Polynesia.* H3. Spring.

In Japan the tree is cultivated for its bark which is used to make paper, hence the specific name; it is used to manufacture cloth in Polynesia.

Several named cultivars are grown, including 'Laciniata', a dwarf tree with leaves reduced to a stalk and veins.

3. ARTOCARPUS Forster & Forster
E.C. Nelson
Monoecious trees. Leaves alternate, entire or pinnately lobed or pinnate. Flowers in spherical heads or catkins; male flowers with 2–4-lobed calyx, 1 stamen, ovary absent; female flowers with tubular calyx, apical pore 3–4-lobed, ovary enclosed in juicy receptacle, style simple or with 2–3 lobes. Aggregate fruit often large and edible.

A tropical genus of perhaps 30 species including several important food-plants (e.g. breadfruit and jackfruit). In Europe these generally must be cultivated in heated glasshouses. Propagation is best from seed, although some may be propagated vegetatively.
Literature: Jarrett, F.M., Studies in Artocarpus and allied genera, *Journal of the Arnold Arboretum* 40: 1–29, 113–155, 298–368 (1959).

1a. Leaves pinnatifid; stipules 10–25 cm, lanceolate   **1. altilis**
  b. Leaves entire; stipules 1.5–8 cm, ovate   **2. heterophyllus**

1. A. altilis (Parkinson) Fosberg (*A. communis* Forster & Forster; *A. incisus* Linnaeus filius; *Sitodium altile* Parkinson). Illustration: Botanical Magazine, 2869, 2870 (1828); Masefield, The Oxford book of food plants, 115 (1969).
Evergreen or deciduous tree to 35 m. Shoots hairy or hairless, 5–15 mm in diameter. Stipules lanceolate, 10–25 cm, with grey or brown hairs. Leaves downy, diamond-shaped to ovate in outline, pinnatifid (rarely entire), 5–25 × 3.5–12 cm with 1–5 (rarely to 9) pairs of lateral lobes, lanceolate, attenuate. Fruits 15–30 cm in diameter, spherical or cylindric, green ripening to yellow, and black when dried, surface covered with acute, conical projections, 1.5 cm × 5 mm, downy to hairless. *Probably originated in Indonesia (Moluccas) and New Guinea.* G2. Summer.

The breadfruit is one of the most famous examples of plants deliberately transported from one hemisphere to another as an economic crop; Captain Bligh and HMS Bounty figure prominently in its story. Now widely cultivated throughout the tropics, where numerous cultivars exist, including variants without seeds. The fruit (either with or without the seeds) is eaten after roasting or boiling; it may also be dried for storage. The latex is also used as birdlime, and cloth can be woven using the fibres from this tree.

The name of this plant has been the subject of much discussion: the earliest available one is *Sitodium altile*, although Jarrett disputed its validity.

2. A. heterophyllus Lamarck (*A. jaca* Lamarck; *A. integrifolius* misapplied). Illustration: Botanical Magazine, 2833, 2834 (1828); Masefield, The Oxford book of food plants, 115 (1969); Everett, New York Botanical Gardens illustrated encyclopedia of horticulture, 260 (1981).
Evergreen tree, 10–15 m. Shoots hairless, 2–6 mm in diameter. Stipules ovate, 1.5–8 cm, apex acute, downy or hairless. Adult leaves hairless, ovate to elliptic, 5–25 × 3.5–12 cm, base wedge-shaped, apex obtuse or acuminate, margins entire. Juvenile leaves elongated or with 1–2 pairs of lateral lobes. Fruits 30–100 × 25–60 cm, cylindric, yellow (brown when dried), smelling sweetly of ripe bananas; surface hairy, covered with closely set, acute, conical projections, 4–10 × 4–5 mm. *Probably originated in India (western Ghats).* G2. Summer.

The jackfruit is not as widely cultivated as the breadfruit; the fleshy perianth is eaten and the seeds roasted or boiled. The wood is a valuable timber, and when made into wood chips yields a yellow dye.

A. integer (Thunberg) Merrill (*A. integrifolius* Linnaeus filius). Like *A. heterophyllus* but with hairy shoots and smaller fruits. *Malaysia, Indonesia, New Guinea.* G2. Summer.

4. MACLURA Nuttall
E.C. Nelson
Dioecious trees to 15 m. Branches with spines (to 4 cm long). Leaves deciduous, alternate, simple, entire, 5–12 × 2–6 cm, ovate to oblong, acuminate, dark green and hairless above, paler and downy beneath. Flowers green, axillary, inconspicuous; males in a loose raceme. Fruit a spherical aggregate fruit to 14 cm in diameter, resembling an orange.

A genus of a single species from E USA. It is hardy but is rarely seen in fruit because trees are not planted in sexed pairs; the fruit, despite its appearance, is inedible, being full of latex. *M. pomifera* can be used as a hedge-plant. It is propagated by cuttings, layers or seeds.

1. M. pomifera (Rafinesque) Schneider (*M. aurantiaca* Nuttall; *Ioxylon pomiferum* Rafinesque). Illustration: Sargent, Silva of North America **7**: pl. 322, 323 (1895); Everett, The New York Botanical gardens illustrated encyclopedia of horticulture **6**: 2092 (1981); Mitchell, The complete guide to trees of Britain and northern Europe, 56 (1985).
USA. H2. Early summer.

A variant without branch-thorns is cultivated ('Inermis') and was a parent, with *Cudrania tricuspidata* of the bigeneric hybrid X *Macludrania hybrida* André.

5. CUDRANIA Trécul
E.C. Nelson
Small trees or climbing shrubs, dioecious. Branchlets often reduced to spines. Leaves alternate, entire or 3-lobed. Inflorescences usually spherical, small. Male flowers with 4 calyx-lobes, stamens 4, anthers erect; ovary absent or rudimentary. Female flowers with 4 calyx-lobes closely surrounding the ovary; style undivided or 2-branched. Fruit with enlarged bracts, fleshy.

A genus of 4–8 species distributed in tropical and subtropical regions, from Korea and Japan south to Australasia. It is not common in cultivation, only 1 species being generally present in European gardens. Propagation is by seed or cuttings.

1. C. tricuspidata (Carrière) Bureau (*Maclura tricuspidata* Carrière). Illustration: Bean, Trees and shrubs hardy in the British Isles, edn 8, **1**: 795 (1970).
Deciduous tree to 7 m. Leaves ovate, 6–10 × 3–6 cm, sometimes 3-lobed, lobes abruptly narrowing to the acute apex, hairless or very sparsely downy on veins; stalk 1.5–2.5 cm. Flowers green. *Korea, China.* H3. Summer.

It is possible that only male plants are in cultivation in Britain. In China the leaves are used to feed silkworms.

6. DORSTENIA Linnaeus
E.H. Hamlet
Perennial herbs with rhizomes, occasionally woody below. Stems sometimes fleshy or thickened and lower leaves reduced to scales or plants stemless with long-stalked leaves from a thick tuber. Stipules generally falling early, sometimes persistent. Leaves stalked or almost stalkless, alternate, with margin entire, irregularly toothed, or lobed, acute or acuminate. Inflorescences solitary or sometimes 2 or 3, in the upper leaf axils. Flowering-stem generally shorter than the leaves, except in stemless species, when elongated, and sometimes exceeding the leaves. Flowers numerous, unisexual, sunk in receptacles of variable shape and outline, with few or many marginal appendages differing in length, appendages sometimes absent. Male flowers with 2-or 3-lobed perianth and 1–3 free stamens. Female flowers with ovary concealed in the receptacle; style simple or bifid. Fruit rounded, seed ejected when ripe.

A genus of 170 species from tropical Africa, America and Asia, which are popular because of their unusual inflorescences. They can be grown in a warm greenhouse in ordinary moist soil, preferably with plenty of humus. Shading from strong sun is needed and a minimum temperature of 13° C is desirable. Propagation is by division in late winter or spring, before growth begins or from seeds sown in spring with a temperature of 15-21° C. Self-sown seedlings often appear.
Literature: Carauta, J.P.P., Valente, M.da C.D., Sucre B., Dorstenia L.(Moraceae) dos Estados da Guanabara e do Rio de Janeiro, *Rodriguesia* **39**: 225–296 (1974); Carauta, J.P.P., Dorstenia L.(Moraceae) do Brasil e paises limitrofes, *Rodriguesia* **44**: 53–223 (1978).

Leaves. Almost stalkless: **6**.
Stems and receptacle appendages. Purple: **8**.
Receptacle appendages. Two, opposite: **3**.
　Inflexed and overlapping: **10**.
Style simple. **7**.
Mature receptacle. Vertical: **1,3**.

1a. Leaves more than 10 cm wide　　　　　　2
　b. Leaves less than 10 cm wide　　　　　　3
2a. Receptacle margin regular with triangular appendages　　**1. arifolia**
　b. Receptacle margin wavy, irregularly divided　　　　**2. contrajerva**
3a. Receptacle with 1 or more appendages of 2 cm or longer　　4
　b. Receptacle with appendages to 1.5 cm, or appendages lacking　　6
4a. Receptacle appendages 2, opposite
　　　　　　　　　　　　3. psilurus
　b. Receptacle with more than 2 appendages　　　　　　5

5a. Longest receptacle appendage at least 3 times the receptacle diameter
　　　　　　　　　4. yambuyaensis
　b. Longest receptacle appendage not more than twice the receptacle diameter　　　　**5. multiradiata**
6a. Leaves almost stalkless
　　　　　　　　　6. hildebrandtii
　b. Leaves distinctly stalked　　　7
7a. Style simple　　　**7. foetida**
　b. Style divided　　　　　8
8a. Receptacle with hairy purple appendages　　　**8. argentata**
　b. Receptacle appendages not as above
　　　　　　　　　　　　　　9
9a. Inflorescence stalk to 1 cm
　　　　　　　　　9. convexa
　b. Inflorescence stalk 2 cm or longer　10
10a. Receptacle appendages inflexed and overlapping　　**10. urceolata**
　b. Receptacle appendages not as above
　　　　　　　　　　　　　11
11a. Receptacle appendages of unequal length (at the base of the longer and between them are shorter appendages)　　**5. multiradiata**
　b. Receptacle appendages not as above
　　　　　　　　　11. hirta

1. D. arifolia Lamarck. Illustration: Botanical Magazine, 2476 (1824); Rodriguesia **39**: f. 14 (1974), **44**: 195 (1978) same in part.
Rhizomes scaly, aromatic. Stem short and woody, 1–2 cm across, simple or branched. Stipules triangular, leathery, covering the stem. Leaf-stalks 10–20 cm with some glandular hairs; blades 5–20 × 12–25 cm, variable, oblong to ovate, base sagittate or hastate, apex acute or acuminate. Margin entire, lobed or divided. Flowering-stem 5–18 cm with glandular hairs. Receptacle circular in outline when young, later oval to elliptic or sometimes rectangular and almost lyrate. Margin divided into an inner reddish brown area and an outer dark green area; appendages small, triangular. Mature receptacle vertical. Style bifid. *Brazil.* G2. Flowering sporadically.

2. D. contrajerva Linnaeus. Illustration: Rodriguesia **39**: f.15 (1974), **44**: 199 (1978) same in part; Everett, New York Botanical Gardens illustrated encyclopedia of horticulture **4**: 1117 (1981); Graf, Exotica edn 11, 1561 (1982).
Stems absent or very short, 3–4 × 1–2 cm. Stipules triangular-lanceolate, 3–4 mm. Leaf-stalks 10–20 cm; blades 10–23 × 15–27 cm, triangular, deeply pinnately or palmately lobed; base sagittate or hastate; apex acuminate, margin entire, toothed,

wavy or with 2–9 lobes. Receptacle angular, 1–2.5 cm across, margin wavy and irregularly divided. Style bifid. *C America, West Indies to Peru*. G2. Flowering sporadically.

3. D. psilurus Welwitsch. Illustration: De Wildeman, Études Flore du Bas- et du Moyen-Congo **2**: pl. 60 (1907); Planta **13**: 61 (1931), **52**: 281 (1958); Morat (ed), Flore du Gabon **26**: 79 (1984).
Stem erect, 30–80 cm, somewhat succulent, hairless below, downy towards the apex. Stipules narrowly lanceolate or linear-acuminate, 2.5–4 mm, downy, more or less persistent. Leaf-stalks sparsely downy, 5 cm in lower leaves, shorter above; blades 6.5–17 × 3–7 cm, membranous, variable, ovate to oblong-elliptic, acuminate, weakly toothed, sometimes obovate and more or less deeply cut towards the apex or with 3–several lobes. Inflorescences solitary or in pairs, vertical, stalks 2.5–6 cm. Receptacle linear-lanceolate, green, to 9.5 cm with 2 opposite appendages, apical 5–9 cm, tapering, basal to 1.5 cm tending to curve upwards. Male flowers many, with 1–3 stamens, females fewer, in a row on either side of the middle line. Style bifid. Fruit ovoid to ellipsoid, 2 mm across. *Tropical Africa*. G2. Spring–autumn.

4. D. yambuyaensis De Wildeman. Illustration: Botanical Magazine, 8616 (1915); Graf, Tropica edn 1, 650 (1978); Everett, New York Botanical Gardens illustrated encyclopedia of horticulture **4**: 1118 (1981); Morat (ed), Flore du Gabon **26**: 75 (1984).
Stems 30–60 cm, downy when young. Stipules 2–10 mm, subulate, persistent. Stalks 3–15 mm, downy when young; blades 5–16 × 2–7 cm, oblong to elliptic, shortly acuminate, paler green beneath, tapered to the obtuse or nearly cordate base, margin entire, wavy or irregularly toothed. Inflorescences solitary, stalk 2.5–6 cm, shortly downy. Receptacle angular to rounded, 1.5–2 cm across, margin 2–8 mm wide, some appendages as long as 5–12 cm, hairless. Male flowers with 2–3 stamens, females with bifid style. Fruit 3 mm across. *W & C Tropical Africa*. G2. Flowering time unknown.

5. D. multiradiata Engler (*D. barteri* var. *multiradiata* (Engler) Hijman & Berg). Illustration: Planta **13**: 56 (1931); Everett, New York Botanical Gardens illustrated encyclopedia of horticulture **4**: 1117 (1981); Graf, Exotica edn 11, 1560 (1982).

Stems to 60 cm or more, erect, from a creeping or rooting base, woody and hairless below, often branched above, shortly downy. Stipules 5–15 mm, linear-subulate. Leaf-stalks 1.5 cm, with long scattered hairs; blades 15–20 × 5–8 cm, narrowly elliptic, shortly acuminate, hairless, paler green beneath; margin slightly wavy. Inflorescence solitary, axillary, stalk 2–5 cm, hairless. Receptacle grey-green, angular, 2–3 cm across; margin green, leaf-like, 4–10 mm across; appendages linear, of unequal length, the longest 4 cm. Flowers fragrant, male 2 or 3 lobed, with 2 or 3 stamens, female with bifid style. Fruit smooth, 4 cm across. *W Tropical Africa*. G2. Spring to Autumn.

6. D. hildebrandtii Engler. Illustration: Jacobsen, Lexicon of succulent plants **4**: pl. 52 (1974); Lamb & Lamb, The illustrated reference on cacti and other succulents **5**: 1427 (1978); Graf, Exotica, edn 11, 1561 (1982).
Stems 15–20 cm, erect, succulent, 5 mm across, hairless. Stipules 2 mm, subulate. Leaf-blades 4 × 1 cm, almost stalkless, fleshy, lanceolate, acute, narrowed to the base; margin shortly toothed. Flowering-stem 5–10 mm, minutely and densely downy. Receptacle dark purple, 5 mm across, almost round; appendages linear-tapering, to 5 mm, minutely downy. Female flowers numerous with bifid style. Fruit ovoid, warty. *E Tropical Africa*. G2. Spring to late Autumn.

7. D. foetida (Forsskål) Schweinfurth. Illustration: Cactus and Succulent Journal **38**(5): 175 (1966); Lamb & Lamb, The illustrated reference on cacti and other succulents **5**: 1425 (1978); Rowley, The illustrated encyclopedia of succulents 241 (1978); Riha & Subik, The illustrated encyclopedia of cacti and other succulents 342 (1981).
Stems 1–4, swollen at base, succulent, to 15 cm and to 1.5 cm across with prominent leaf and stipule scars. Leaves crowded near stem apex, stalks 1–3.6 cm, finely downy, blades variable, 1.8–18 × 1–2.5 cm, obovate, lanceolate to elliptic or almost circular, with tapered, truncate or rounded bases; margin entire, scolloped, crisped or toothed, apex rounded, obtuse or acuminate; leaf surface rough with short stiff hairs above, sparsely downy beneath. Inflorescences 1–3, stalks 1–6 cm; receptacle circular in outline, 6–15 mm across with 7–10 linear appendages to 15 × 1.5 mm. Flowers scattered on receptacle surface, male with 2–3 stamens,

female with simple style and unbranched stigma. Fruit 1–1.4 mm. *Tropical Arabia & E Tropical Africa*. G2. Spring–autumn.

8. D. argentata J.D. Hooker. Illustration: Botanical Magazine, 5795 (1869); Rodriguesia **44**: 163 (1978); Everett, New York Botanical Gardens illustrated encyclopedia of horticulture **4**: 1117 (1981); Graf, Exotica edn 11, 1560 (1982).
Main stem horizontal, rooting for *c.* 30 cm, bearing a few ascending leafy stems, 15–30 cm, dull purple, downy. Leaves numerous, stalks 1.3–2.5 cm, blades 7.5–12 × 2.5–3.5 cm, oblong or narrowly lanceolate, apex acuminate, sinuous-toothed, deep green at margins with a broad central silvery zone, marbled with green where the colours join, sometimes unmarked, midrib purple beneath. Flowering-stem axillary, dull purple, finely downy. Receptacle 2–2.5 cm across, dark green, spherical, slightly concave with a series of short, obtuse, conical, purple appendages, each tipped with a few short hairs. Male flowers at margin and female in the centre. Style bifid, recurved. *Brazil*. G2. Flowering time unknown.

9. D. convexa De Wildeman. Illustration: De Wildeman, Études Flore du Bas- et du Moyen-Congo **3**: pl. 4 (1909); Planta **13**: 55 (1931), **52**: 282 (1958).
Stems to 30 cm or more, shortly hairy. Stipules 10 mm, linear-tapering, persistent. Leaf-stalks 7–17 mm, hairy; blades 4–13 × 2.5–6 cm, obovate to elliptic, narrowing to the base and sometimes obscurely cordate, shortly acuminate, margin slightly wavy or with irregular short broad teeth. Inflorescences solitary, stalk to 6 mm, hairless. Receptacle 4 × 1.5–2 cm, elliptic to oblong, convex, with numerous, small, fleshy, linear marginal appendages to 3 mm, the terminal slightly longer and often irregularly 2-lobed. Male flowers with 2–3 stamens, female with bifid style. *Congo*. G2. Flowering time unknown.

10. D. urceolata Schott. Illustration: Rodriguesia **39**: f. 4 (1974), **44**: 149 (1978), same in part.
Rhizomes 5–10 cm; aerial stem erect or suckering, 10–20 cm. Leaf-stalks 2.5–3.5 cm, blades 2–19 × 3–5 cm, papery, lanceolate, ovate or elliptic to almost oblong, base tapered or almost rounded, apex acute or acuminate, margin entire, distantly scolloped or toothed; both surfaces hairless. Flowering-stem 6–10 cm. Receptacle spherical to urn-shaped, 1–2 cm across with marginal inflexed, overlapping

appendages. Outer surface greenish, finely downy, inner dark purple. Flowers uniformly scattered across receptacle surface, male with 2 stamens, female with bifid style, violet-blue. *S Brazil*. G2. Flowering time unknown.

'Variegata' has leaves with raised veins, surface mottled light and yellowish green with silvery band along centre, and sometimes margin. Illustration: Parey's Blumengärtnerei, 521 (1958); Graf, Exotica edn 11, 1560 (1982).

11. D. hirta Desvaux (*D. erecta* Vellozo). Illustration: Planta **13**: 43 (1931), **52**: 282 (1958); Rodriguesia **39**: f. 6 (1974), **44**: 159 (1978), same in part.
Rhizomes 5–10 cm; aerial stems to 1 m, erect, ascending or decumbent, stiff, densely hairy, sparsely downy or hairless. Stipules subulate, 1-veined. Leaf-stalks 1–1.5 cm, blades 6–30 × 3.5–9 cm, various shades of green, sometimes with silvery patches, lanceolate-elliptic to obovate or spathulate, membranous; base wedge-shaped to cordate; apex acute, acuminate or notched. Margin wavy or doubly toothed, upper surface stiffly hairy or as lower surface, downy or hairless. Flowering-stem 2–8 cm. Receptacle circular in outline, 1–3 cm across, marginal appendages close together when young, spaced in mature state. Flowers uniformly distributed over receptacle surface, male with 2 stamens, female having a markedly white bifid style. *Brazil*. G2. Flowering time unknown.

7. FICUS Linnaeus
J. C. M. Alexander
Evergreen (except Nos. **43** & **44**) trees, shrubs, trailers or climbers with milky sap. Many trees initially epiphytic (in the wild), producing vigorous aerial roots which may eventually crush the host and form a hollow trunk (strangling figs); further aerial roots may then proliferate forming multiple trunks. Leaves alternate (sometimes opposite in No. **48**), usually stalked, often large and showy with prominent veins, membranous or papery to thinly or thickly leathery, sometimes abrasive, simple or lobed, with an entire, toothed or irregular margin. Stipules paired (solitary in No. **21**), enclosing the developing bud. Climbers usually seen in cultivation as the juvenile, non-reproductive phase with creeping stems bearing numerous, clinging, adventitious roots and two ranks of small, stalkless, often asymmetric leaves; mature, reproductive phase with erect or trailing stems and foliage generally similar to trees

and shrubs. Flowers minute, densely packed inside hollow receptacles (figs) inside which pollination and fruit development take place. Flowers of three types: male, functional female, and non-functional female (gall flowers), modified for the nourishment and reproduction of minute wasps intimately involved in pollination. In most species, plants are either male (bearing figs with male and gall flowers) or female (producing fertile figs), though some species are bisexual having all three types of flower within each fig. (Details of flower structure and arrangement are not required for the key, and are not given in the descriptions.)

A cosmopolitan tropical and subtropical genus of about 700 species, richly represented in Africa and SE Asia, several of which are grown, usually as pot-plants, for their large, decorative foliage, and in some cases for their red or orange figs. Though many species are vast trees in the wild, they grow slowly as pot-plants and many do not produce figs. They are easily grown from seed but are more commonly propagated from cuttings or by layering or air-layering. Some species, (e.g. *F. carica* are grown for their edible figs, and a few (e.g. *F. elastica*) were previously grown for rubber. Several are of religious significance in India and SE Asia (e.g. *F. religiosa* and *F. benghalensis*).
Literature: Corner, E.J.H., Check-list of Ficus in Asia and Australia, *The Gardens' Bulletin Singapore* **21**(1): 1–186 (1965); Hill, D.S., *Figs of Hong Kong* (1967); Condit, I.J., *Ficus: The exotic species* (1969).

1a. Plant trailing or climbing, often clinging to support with short adventitious roots in the manner of ivy (*Hedera*)   **2**
  b. Plant an erect shrub or tree, often epiphytic, sometimes with long aerial roots   **10**
2a. Leaves simple; margins entire   **3**
  b. Leaves lobed or indented   **9**
3a. Leaves densely hairy below or with long hairs on the veins only **1. villosa**
  b. Leaves hairless or with short sparse hairs   **4**
4a. Leaf-tips acute or acuminate   **5**
  b. Leaf-tips obtuse or rounded   **8**
5a. Lower leaf surface almost obscured by net-like greyish or brownish veins   **4. sarmentosa**
  b. Lower leaf surface not obscured   **6**
6a. Leaves variegated   **2. sagittata**
  b. Leaves not variegated   **7**
7a. Leaf-bases narrowly cordate
                      2. sagittata

  b. Leaf-bases tapered   **7. subulata**
8a. Leaves elliptic to broadly ovate, midrib central or nearly so   **3. pumila**
  b. Leaves oblong, midrib not central   **5. punctata**
9a. Leaf-blades 5–15 cm, irregularly lobed or indented   **6. montana**
  b. Leaf-blades very small, star-shaped   **4. sarmentosa**
10a. Leaves simple, margins entire, sometimes wavy   **11**
  b. Leaves lobed or margins toothed or scalloped   **64**
11a. Buds, leaves and stems without hair or rusty covering even when young (hand-lens); fruits and fruit-stalks sometimes hairy or rusty   **12**
  b. Buds, leaves or stems hairy or rusty, sometimes only when young (hand-lens)   **36**
12a. Terminal buds enclosed in a single large stipule   **21. elastica**
  b. Terminal buds enclosed in two stipules or stipules soon falling and buds appearing naked   **13**
13a. Leaf-tips acute or acuminate   **14**
  b. Leaf-tips obtuse or rounded, sometimes prolonged   **26**
14a. Leaf-blades (excluding tips) about as long as broad   **15**
  b. Leaf-blades longer than broad   **16**
15a. Leaf-tip abruptly narrowed, long and slender, at least a quarter of blade length; lowest lateral vein at more than 45° to midrib   **12. religiosa**
  b. Leaf-tip gradually narrowed, triangular, at most a fifth of blade length, lowest lateral vein at 45° to midrib or less   **13. rumphii**
16a. Lateral veins distinctly raised on lower leaf surface   **17**
  b. Lateral veins almost flat on lower leaf surface   **21**
17a. Terminal buds to 1.8 cm   **18**
  b. Terminal buds 2.5 cm or more   **20**
18a. Stipules persistent; leaves broadest above middle   **15. cyathistipula**
  b. Stipules soon falling; leaves elliptic or ovate, sometimes narrowly so   **19**
19a. Leaves leathery, at least 3 × longer than wide, base narrowly tapered   **16. barteri**
  b. Leaves papery to slightly leathery, usually about 2 × longer than wide, base broadly rounded to truncate   **14. virens**
20a. Leaves leathery   **17. macrophylla**
  b. Leaves papery   **13. rumphii**
21a. Lateral veins 15 or more   **22**
  b. Lateral veins less than 15   **25**

Figure 22. Leaves of *Ficus*. 1, *F. racemosa*. 2, *F. elastica*.
3, *F. sycomorus*. 4, *F. hispida*. 5, *F. macrophylla*. 6, *F. callosa*.
7, *F. sagittata*. 8, *F. carica*. 9, *F. palmata* (a, unlobed, b, lobed leaf).
10, *F. villosa* (a, adult leaf, b, juvenile leaf). Scale = 12 cm.

22a. Main lateral veins at least 30, very
    fine   **21. elastica**
  b. Main lateral veins fewer   23
23a. Leaf-blades less than 2 times longer
    than wide   **20. sundaica**
  b. Leaf-blades at least 2 times longer
    than wide   24
24a. Leaf-tips acuminate, point narrow
    and rounded   **19. stricta**
  b. Leaf-tips not or slightly acuminate,
    point broad and rounded
            18. watkinsiana
25a. Leaf-tips strongly acuminate, tip more
    than 1 cm   **22. benjamina**
  b. Leaf-tips weakly acuminate
            24. obliqua
26a. Lateral veins distinctly raised on
    lower leaf surface   27
  b. Lateral veins almost flat on lower leaf
    surface   30
27a. Leaf-blades 10 cm or less, bases
    tapered   **33. leprieurii**
  b. Leaf-blades more than 20 cm, bases
    cordate   28
28a. Leaf-stalk 15 cm or more
            28. eetveldiana
  b. Leaf-stalk 12 cm or less   29
29a. Leaves ovate, obtuse
            27. nymphaeifolia
  b. Leaves obovate, broader in upper half,
    rounded or bluntly acuminate
            11. lyrata
30a. Terminal buds usually less than
    2.5 cm   31
  b. Terminal buds more than 2.5 cm   33
31a. Leaves papery to thinly leathery
            25. aurea
  b. Leaves leathery or rubbery   32
32a. Leaf-blades usually less than 5 cm
            26. deltoidea
  b. Leaf-blades usually more than 5 cm
            23. microcarpa
33a. Leaf-blades less than 15 cm   34
  b. Leaf-blades more than 15 cm   35
34a. Leaf-base narrowly tapered or
    rounded, tip acute to obtuse (figs
    spherical, to 8 mm across, weakly
    flecked)   **24. obliqua**
  b. Leaf-base broadly tapered or rounded,
    tip obtuse (figs spherical to oblong, to
    1.5 cm across, strongly flecked)
           30. rubiginosa
35a. Leaves broadly ovate
            17. macrophylla
  b. Leaves elliptic   **18. watkinsiana**
36a. Hairs or rusty covering present on
    mature leaves, veins or leaf-stalks   37
  b. Hairs or rusty covering only present
    on young leaves, veins, leaf-stalks or
    stems   50
37a. Lateral veins almost flat on lower leaf
    surface   38

  b. Lateral veins distinctly raised on
    lower leaf surface   40
38a. Lateral veins 16–20   **17. macrophylla**
  b. Lateral veins 8–12   39
39a. Leaves wrinkled, 12–30 × 6–10 cm
           9. dryepondtiana
  b. Leaves flat, 7.5–17 × up to 6.2 cm
           30. rubiginosa
40a. Leaf-base asymmetric   41
  b. Leaf-base symmetric   42
41a. Leaves usually less than 7.5 cm wide,
    base strongly asymmetric
           10. ulmifolia
  b. Leaves usually more than 8 cm wide,
    base slightly asymmetric   **45. nota**
42a. Leaf-blades (excluding tips) *c.* as long
    as broad   43
  b. Leaf-blades longer than broad   44
43a. Leaf-tips finely tapered
           32. benghalensis
  b. Leaf-tips rounded   **31. sycomorus**
44a. Leaf-tips obtuse or rounded   45
  b. Leaf-tips acute or acuminate   48
45a. Terminal buds to *c.* 1 cm   46
  b. Terminal buds to 2.5 cm or more   47
46a. Leaf-stalks to 4 cm   **37. habrophylla**
  b. Leaf-stalks to 15 cm   **35. vogelii**
47a. Leaves glossy above   **42. altissima**
  b. Leaves dull above   **32. benghalensis**
48a. Terminal buds 1.8 cm or more
           39. drupacea
  b. Terminal buds 1.5 cm or less   49
49a. Leaf-stalks 3.5 cm or more
           35. vogelii
  b. Leaf-stalks 2 cm or less   **38. parietalis**
50a. Leaf-blades broadest at or near tip   51
  b. Leaf-blades broadest near or below
    middle   54
51a. Leaves 20–45 cm, bases cordate   52
  b. Leaves to 8 cm, bases tapered   53
52a. Lateral veins 4–6; leaf-stalks to
    7.5 cm, hairless   **11. lyrata**
  b. Lateral veins 10–17; leaf-stalks
    2–4 cm, brown-hairy **37. habrophylla**
53a. Leaf-tips rounded; stipules persistent
           34. lingua
  b. Leaf-tips truncate; stipules soon falling
           33. leprieurii
54a. Lateral veins almost flat on lower leaf
    surface   55
  b. Lateral veins distinctly raised on
    lower leaf surface   57
55a. Lateral veins 15 or more; leaf-stalks
    5 cm or more   **18. watkinsiana**
  b. Lateral veins less than 15; leaf-stalks
    less than 5 cm   56
56a. Lateral veins paler than lower leaf-
    surface   **25. aurea**
  b. Lateral veins very indistinct
           24. obliqua
57a. Leaves papery   58
  b. Leaves more or less leathery   60

58a. Leaf-stalks 3 cm or less   **29. erecta**
  b. Leaf-stalks 5 cm or more   59
59a. Leaf-tip abruptly acuminate
           14. virens
  b. Leaf-tip acute or acuminate
           40. racemosa
60a. Terminal buds *c.* 2 cm or less   61
  b. Terminal buds 2.5 cm or more   63
61a. Leaf-stalk brown; midrib green   62
  b. Leaf-stalk and midrib similarly
    coloured   **41. callosa**
62a. Lateral veins usually 3.5 cm or more
    apart   **35. vogelii**
  b. Most lateral veins less than 3.5 cm
    apart   **36. nekbudu**
63a. Leaves dull above   **32. benghalensis**
  b. Leaves glossy above   **42. altissima**
64a. Leaves with 3–5 palmate lobes   65
  b. Leaves not palmately lobed   66
65a. Lobes with obtuse or rounded tips
           43. carica
  b. Lobes with narrowly tapered tips
           10. ulmifolia
66a. Leaves rough (at least when young)
           67
  b. Leaves not rough to the touch   71
67a. Some leaves opposite   **48. hispida**
  b. All leaves alternate   68
68a. Leaves symmetric at base, margin
    with regular wide teeth   **44. palmata**
  b. Leaves usually asymmetric at base,
    margin irregularly toothed or
    scalloped   69
69a. Leaves usually more tham 8 cm wide,
    broadly elliptic, occasionally slightly
    asymmetric at base   **45. nota**
  b. Leaves usually less than 7.5 cm wide,
    ovate to oblong or narrowly elliptic,
    usually strongly asymmetric at base
           70
70a. Leaves variegated (in cultivation),
    young growth hairy   **8. aspera**
  b. Leaves not variegated, young growth
    bristly   **10. ulmifolia**
71a. Leaf-blades longer than broad
           47. sus
  b. Leaf-blades *c.* as long as broad
           46. auriculata

1. F. villosa Blume (*F. barbata* Miquel).
Figure 22(10), pp. 91. Illustration: Boom,
Flora cultuurgewassen 3: 74 (1975); Graf,
Exotica, edn 11, 1575 (1982).
Climbing or trailing shrub. Branchlets
stout, purplish brown, ridged, densely
hairy. Stipules to 4 cm, persistent for a few
nodes. Juvenile leaf-blades to 17 × 9.5 cm,
broadly ovate, acute to slightly acuminate,
cordate at base, papery to thinly leathery,
hairless above or with a few long thin hairs
near the margin, hairy beneath, especially
on the veins; margin entire, with spreading

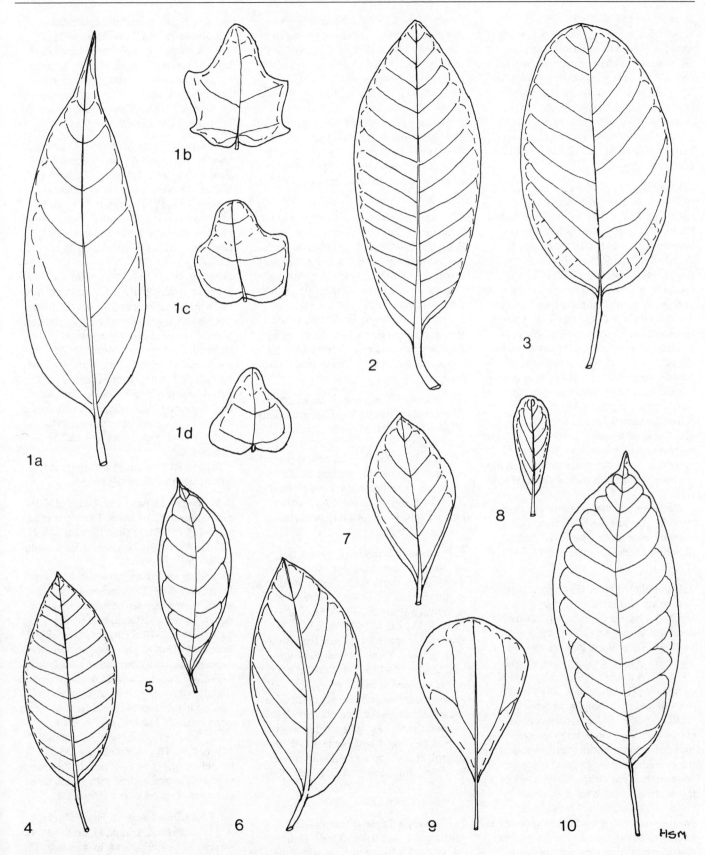

Figure 23. Leaves of *Ficus*. 1, *F. sarmentosa* (a, adult leaf; b, c, d, juvenile leaves × 2.5). 2, *F. rubiginosa*. 3, *F. nekbudu* (× 0.4). 4, *F. obliqua*. 5, *F. cyathistipula*. 6, *F. pumila*. 7, *F. microcarpa*. 8, *F. lingua*. 9. *F. deltoidea*. 10, *F. dryepondtiana* (× 0.3).

hairs. Adult leaf-blades 12–25 ×
7.5–12.5 cm, ovate, acute to acuminate,
rounded to cordate at base, sometimes
asymmetric, hairless or very sparsely hairy
above, densely hairy beneath, or with long
hairs on the veins only; margin entire.
Lateral veins 6–9, raised below. Figs
clustered in leaf-axils, spherical, to 1 cm,
green to brown, moderately to densely
hairy; stalks 5–7 mm. *NE India, Andaman
Islands, SE Asia.* G2.

2. F. sagittata Vahl (*F. radicans*
Desfontaines; *F. rostrata* Anon. not
Lamarck). Figure 22(7), p. 91. Illustration:
Rochford & Gorer, The Rochford book of
house plants, 44 (1963); Seabroook &
Rochford, Plants for your home, 52 (1971);
Graf, Exotica, edn 11, 1575 (1982).
Climbing or trailing shrub, rooting on walls
or trees. Branchlets slightly hairy, green.
Terminal buds *c.* 1 cm, stipules *c.* 8 mm,
persistent for a few nodes. Leaf-blades
5–10.6 × 2.5–3.6 cm, narrowly ovate to
elliptic, acute or slightly acuminate,
rounded to slightly cordate at base, papery,
hairless above and below, margin entire
(rarely with irregular lobes in some
cultivated forms). Lateral veins 6–8,
slightly raised beneath. Leaf-stalk to *c.*1 cm,
softly hairy. Figs 1 or 2 per leaf-axil,
spherical, 1–1.4 cm, red; stalks 2–3 mm.
NE India, Andaman Islands, S China, SE Asia.
G2.

'Variegata', the form in cultivation, has
irregular patches of white along the leaf-
margins, often extending to the midrib.
Illustration: Graf, Exotica, edn 11, 1574
(1982).

3. F. pumila Linnaeus (*F. repens* misapplied;
F. stipulata Thunberg; *F. scandens*
Lamarck). Figure 23(6), p. 93. Illustration:
Botanical Magazine, 6657 (1882);
Seabrook & Rochford, Plants for your
home, 53 (1971); Graf, Exotica, edn 11,
1574, 1575 & 1586 (1982); Hay et al., The
dictionary of indoor plants, 239 (1983).
Juvenile stage climbing; stem brown,
rooting; stipules to 1 cm, persistent. Leaf-
blades to *c.* 5 × 3 cm, broadly ovate,
rounded, asymmetrically cordate at base,
thinly leathery, dark green above, paler
beneath, hairless, entire. Midrib more or
less central; lateral veins 3–5, raised
beneath. Leaf-stalks to 4 mm. Adult
branches erect or spreading, ridged; stipules
silky. Leaf-blades to *c.* 10 × 5 cm, ovate to
elliptic, rounded or very bluntly obtuse,
rounded to very slightly cordate at base,
leathery, dull and hairless above, slightly
hairy on veins beneath; margin entire.

Lateral veins 5–8, raised beneath; lower
leaf-surface almost concealed by lattice of
fine veins. Leaf-stalks to 2.5 cm, hairy. Figs
usually solitary in leaf-axils, pear-shaped,
to 6 × 4 cm, hairy; stalks to 1 cm. *N
Vietnam to China, Taiwan, Japan & Ryukyu
Islands.* H5–G1.

'Minima' (*F. minima* Anon.) is very small
and slow growing and has leaves to 1 cm.
Illustration: Graf, Exotica, edn 11, 1574
(1982). 'Variegata' has leaves marbled with
white or cream. Illustration: Graf, Exotica,
edn 11, 1575 (1982).

4. F. sarmentosa J.E.Smith var. **nipponica**
(Franchet & Savatier) Corner (*F. nipponica*
Franchet & Savatier; *F. foveolata* Wallich
var. *nipponica* (Franchet & Savatier) King).
Figure 23(1), p. 93. Illustration: Makino,
New illustrated Flora of Japan, 97 (1963);
Graf, Exotica, edn 11, 1574 (1982).
Differs from *F. pumila* in its lobed or star-
shaped juvenile leaves to 2 × 2 cm and its
narrowly lanceolate adult leaves to 15 ×
6 cm. Figs to 7 mm. *E Himalaya to China &
Japan.* G1.

5. F. punctata Thunberg (*F. falcata* Miquel;
F. macrocarpa Blume). Illustration: Annals
of the Royal Botanic Garden, Calcutta, **1**:
t. 88 (1888).
Differs from *F. pumila* in its oblong juvenile
leaves to 2.5 cm with strongly asymmetric
bases and excentric midribs. Adult leaves to
5 × 2.5 cm, elliptic to obovate, rounded,
tapered at base. Figs solitary or clustered, to
1.5 cm. *SE Asia.* G2.

6. F. montana Burman (*F. quercifolia*
Blume; *F. heterophylla* Linnaeus). Figure
24(2), p. 95. Illustration: Condit, Ficus: the
exotic species, 307 (1969); Boom, Flora
cultuurgewassen 3: 74 (1975); Graf,
Exotica, 1574 & 1575 (1982).
Prostrate or sprawling shrub; bark brown,
slightly hairy. Terminal buds to 8 mm.
Leaf-blades 7.5–15 × 5–8.5 cm, oblong,
acute, rounded at base, papery, hairless
above, rough beneath, deeply lobed as in
some *Quercus* species. Lateral veins 7–11,
raised beneath. Leaf-stalks to 1.8 cm, hairy,
channelled. Figs 1 or 2 per leaf-axil, urn-
shaped, to 8 × 5 mm, green with white
tubercles; stalks 5–10 mm. *SE Asia, ?India.*
G2.

Often unlobed in the wild.

7. F. subulata Blume (*F. acuminata*
Roxburgh; *F. salicifolia* Miquel). Figure
24(3), p. 95. Illustration: Annals of the
Royal Botanic Garden, Calcutta, **1**: t. 6
(1887); Hill, Figs of Hong Kong, pl. 56
(1967).

Trailing shrub, sometimes epiphytic.
Stipules to 1.5 cm. Leaf-blades to 25 ×
8.5 cm, elliptic, acuminate or long-tailed,
broadly tapered at base, hairless. Lateral
veins 7–13, very regular, slightly raised
beneath. Leaf-stalks to 1.5 cm. Figs paired
or clustered in leaf-axils or on older wood.
NE India to China & SE Asia. G2.

8. F. aspera Forster (*F. parcellii* Cogniaux &
Marchal; *F. cannonii* (Van Houtte) N. E.
Brown). Figure 25(6), p. 97 Illustration:
Kunkel, Flowering trees in subtropical
gardens, 55 (1978); Everett, New York
botanical gardens illustrated encyclopedia
of horticulture, 4: 1366 & 1368 (1981);
Graf, Exotica, edn 11, 1563, 1573 &
2112A (1982).
Shrub or small tree to 4 m. Branchlets
green or pink, slightly hairy; terminal buds
to 1 cm. Leaf-blades to 35 × 18 cm, ovate,
acuminate, asymmetrically cordate at base,
papery, rough above, slightly hairy
beneath, light green, variegated and
speckled with white, margin coarsely and
irregularly toothed. Lateral veins 7–9,
raised beneath. Leaf-stalks 1–2 cm, hairy.
Figs 1–many in leaf axils or on older wood,
spherical, *c.* 1.8 cm, green, often striped
with white or pink; stalks to 1 cm. *New
Hebrides.* G2.

Some wild forms have unvariegated
leaves, often with entire margins.

9. F. dryepondtiana Gentil. Figure 23(10),
p. 93. Illustration: Condit, Ficus: the exotic
species, 309 (1969); Graf, Exotica, edn 11,
1570 & 1576 (1982); Morat (ed), Flore du
Gabon **26**: 249 (1984).
Strangling epiphyte or free-standing tree.
Branchlets ridged, hairless or sparsely and
shortly hairy; stipules persistent for a few
nodes, linear, 3–20 mm. Leaf-blades
12.5–30 × 6.2–10.5 cm, oblong or elliptic
to ovate or obovate, acute or acuminate,
rounded to slightly cordate at base, papery
to slightly leathery, wrinkled, entire or
slightly wavy, shiny dark green and
hairless above, reddish purple and hairy on
veins beneath. Lateral veins 7–9, looped
some distance from margin. Leaf-stalks
4.3–6.2 cm. Figs clustered on older
branches, spherical, 3–5 cm, slightly hairy,
yellowish brown flecked with white; stalks
c. 1.8 cm. *Tropical W & C Africa.* G2.

10. F. ulmifolia Lamarck. Figure 24(9),
p. 95. Illustration: Condit, Ficus: the exotic
species, 311 (1969); Graf, Exotica, edn 11,
1570 (1982).
Small tree to 4.5 m. Branchlets slender,
brownish green, rough; terminal buds to

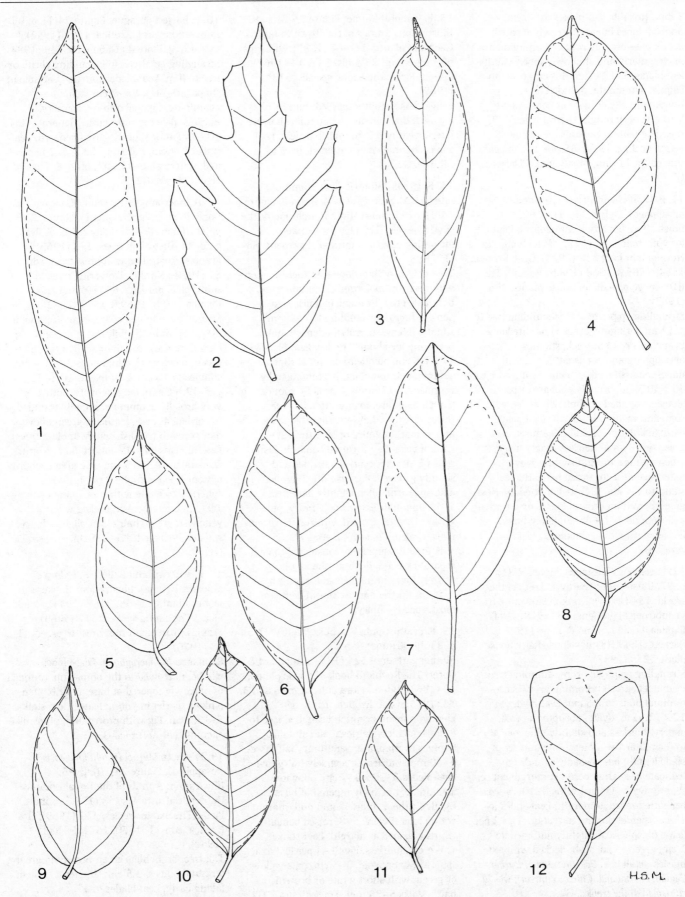

Figure 24. Leaves of *Ficus*. 1, *F. barteri*. 2, *F. montana*.
3, *F. subulata*. 4, *F. virens*. 5, *F. sundaica*. 6, *F. aurea*.

7, *F. capensis*. 8, *F. benjamina*. 9, *F. ulmifolia*. 10, *F. stricta*.
11, *F. erecta*. 12, *F. leprieurii*. Scale = 2 cm.

5 mm. Juvenile leaf-blades to 16 cm, 3-lobed; lobes linear, narrowly tapered, acute; base broadly rounded; margin entire or irregularly and shallowly lobed. Adult leaf-blades 5–15 × 2.5–4.3 cm, ovate to elliptic, acuminate, rounded or asymmetrically cordate at base, slightly leathery, very rough. Lateral veins 4–9, raised beneath. Leaf-stalks 5–11 mm. Figs 1 or 2 per leaf-axil, spherical to pear-shaped, to 11 mm, bristly, red. *Philippines*. G2.

11. F. lyrata Warburg (*F. pandurata* misapplied). Figure 25(11), p. 97.
Illustration: Seabrook & Rochford, Plants for your home, 52 (1971); Wickham, The House-plant book, 38 (1977); Graf, Exotica, 1564, 1565 & 1568 (1982); Beckett, The RHS encyclopaedia of house plants, 255 (1987).
Strangling epiphyte or free-standing tree to *c.* 12 m; bark grey, cracked or shredding. Branchlets *c.* 1 cm thick; stipules overlapping and persistent for several nodes, boat-shaped, *c.* 5 cm. Leaf-blades to 45 × 30.5 cm, broadly obovate or pear-shaped, rounded or shallowly indented, sometimes with a small rounded point, cordate at base, hard and leathery, dark green, glossy and hairless (finely hairy when young), margin slightly wavy. Lateral veins 4–6, raised beneath. Leaf-stalks to 7.5 cm, slightly flattened, hairless, green becoming brown. Figs 1 or 2 per leaf-axil, spherical, 2.5–4 cm, finely hairy, flecked with white, stalkless. *Tropical W Africa*. G2.

12. F. religiosa Linnaeus. Figure 25(1), p. 97. Illustration: Leathart, Trees of the world, 134 (1977); Kunkel, Flowering trees in subtropical gardens, 75 (1978); Graf, Exotica, edn 11, 1564 & 1568 (1982); Beckett, The RHS encyclopaedia of house plants, 255 (1987).
Strangling epiphyte or free-standing tree. Branchlets green becoming grey, hairless; terminal buds to 2.5 cm. Leaf-blades to 17 × 12.5 cm, ovate, abruptly tapered, papery, hairless and slightly glaucous; tip to 6 cm, narrow, at least one-quarter of blade length; base broadly rounded to truncate or slightly cordate; margin entire, slightly wavy. Lateral veins 8–10, raised beneath, looped at margin. Leaf-stalks to 15 cm, slender, round in section. Figs 2 per leaf-axil, spherical, slightly flattened, to 1 cm, green to purplish, flecked with red, hairless, stalkless. *Pakistan, India, Burma, Ceylon, N Thailand, China (Yunnan); widely cultivated in the tropics*. G2.

13. F. rumphii Blume. Figure 25(2), p. 97. Illustration: Annals of the Royal Botanic Garden, Calcutta **1**: t. 67 (1887); Hill, Figs of Hong Kong, 26 & pls. 13 & 14 (1967); Condit, Ficus: the exotic species, 313 (1969).
Leaves less abruptly tapered than in *F. religiosa*; tips narrowly triangular. Lowest lateral vein at 45° to midrib or less. Leaf-stalks channelled or flattened. *India to SE Asia*. G2.

14. F. virens Aiton (*F. lacor* Hamilton; *F. infectoria* Miquel; *F. glabella* Blume). Figure 24(4), p. 95. Illustration: Condit, Ficus: the exotic species, 315 (1969); Kunkel, Flowering trees in subtropical gardens, 83 (1978).
Tree to 9 m with drooping branches, often with aerial roots forming multiple trunks; bark pale grey. Branchlets hairless to densely hairy; terminal buds 7–8 mm; stipules linear, sparsely to densely hairy, persisting for about 7 nodes. Leaf-blades to 17 × 7.5 cm, elliptic to ovate, abruptly acuminate into a long, narrow, usually rounded, 7–11 mm tip, broadly rounded to truncate at base, membranous to very thinly leathery, hairless, slightly glossy above; margin entire or slightly wavy. Lateral veins 8–11, raised beneath. Leaf-stalks 5–10 cm, hairless, slender and slightly grooved. Figs usually 2 per leaf-axil, spherical, often slightly flattened, 7–11 mm, hairless or finely hairy, pale green to white, spotted with red. *Pakistan, India, Ceylon, China & SE Asia*. G2.

F. lacor is sometimes considered to be a separate species, distinguished by its densely hairy branchlets and stipules; *F. virens* (in the narrow sense) being, at most, sparsely hairy.

15. F. cyathistipula Warburg. Figure 23(5), p. 93. Illustration: Parey's Blumengärtnerei, 524 (1958); Rochford & Gorer, The Rochford book of house plants, 43 (1963); Boom, Flora cultuurgewassen **3**: 72 (1975); Graf, Exotica, 1568 (1982).
Epiphytic shrub or much branched tree to 8 m. Branchlets hairless, pale and flaky. Stipules overlapping, persistent, half fused, 5–20 mm, hairless or sparsely hairy, pale. Leaf-blades 6–20 × 3–7 cm, obovate, acuminate, narrowly tapered at base, rigidly leathery, hairless and dull, margin entire; lateral veins 5–8, raised beneath, whitish, looped at margin. Leaf-stalks 1.5–4 cm, hairless. Figs 1–3 per leaf axil, spherical to obovoid, 3–5 cm, pale yellow or green, with short white or brownish hairs; stalks 5–25 mm. *Tropical Africa*. G2

16. F. barteri Sprague. Figure 24(1), p. 95. Illustration: Graf, Exotica, 1572 (1982); Morat (ed), Flore du Gabon **26**: 249 (1984).
Strangling epiphyte, free-standing shrub or tree to 8 m. Leaf-blades narrow, variable in shape, 10–30 × 1.5–7 cm, acute or acuminate, tapered at base, leathery, hairless, pale beneath, entire. Lateral veins 10–20, flat or slightly raised beneath. Leaf-stalks 1–4 cm, hairless. Figs 1 or 2 per leaf-axil, spherical, 6–11 mm, orange. *Tropical W & C Africa*. G2.

17. F. macrophylla Persoon (*F. magnolioides* Borzi; *F. macrocarpa* Bouché). Figure 22(5), p. 91. Illustration: Holiday & Hill, A field guide to Australian trees, 143 (1969); Francis, Australian rainforest trees, edn 3, 81 (1970); Kunkel, Flowering trees in subtropical gardens, 69 (1978); Graf, Exotica, 1568, 1569 & 1573 (1982).
Strangling epiphyte or free-standing tree to 20 m (60 m in the wild) with aerial roots; bark dark grey, rough or scaly; branchlets green, hairless; terminal buds 3.6–7.6 cm; stipules to 15 cm. Leaf-blades 10.5–22.7 × 7.5–12.6 cm, oblong to ovate, obtuse or very broadly acuminate, broadly rounded to cordate at base, leathery, green above, pale beneath (reddish-scurfy in older trees). Lateral veins 17–20, almost flat beneath. Leaf-stalks 10.5–15 cm, pale green, slightly flattened. Figs usually 2 per leaf-axil, spherical to ovate with protruding opening, 1.1–1.8 cm, greenish, flecked with yellowish green, hairless or slightly hairy; stalk club-shaped, 1.1–1.5 cm. *E Australia*. G1.

18. F. watkinsiana Bailey (*F. bellingeri* Moore & Betcha). Illustration: Hooker's Icones Plantarum **35**: 3187 (1933); Condit, Ficus: The exotic species, 317 (1969); Francis, Australian rainforest trees, edn 3, 83 (1970).
Similar to *F. macrophylla*. Leaf-blades elliptic, narrowed to the obtuse tip, rounded to narrowly tapered at base; midrib often pink or scarlet in young leaves; leaf-stalk 5–12.6 cm. Fig with protruding, nipple-like opening. *NE Australia*. G1.

19. F. stricta Miquel (*F. philippinensis* misapplied). Figure 24(10), p. 95. Illustration: Annals of the Royal Botanic Garden, Calcutta **1**: t. 53 (1887); Condit, Ficus: The exotic species, 317 (1969); Graf, Exotica, edn 11, 1570, 1571 & 1583 (1982).
Tall tree; branchlets green becoming grey; terminal buds *c.* 2.5 cm; stipules *c.* 3.6 cm, falling early. Leaf-blades to 17 × 5 cm,

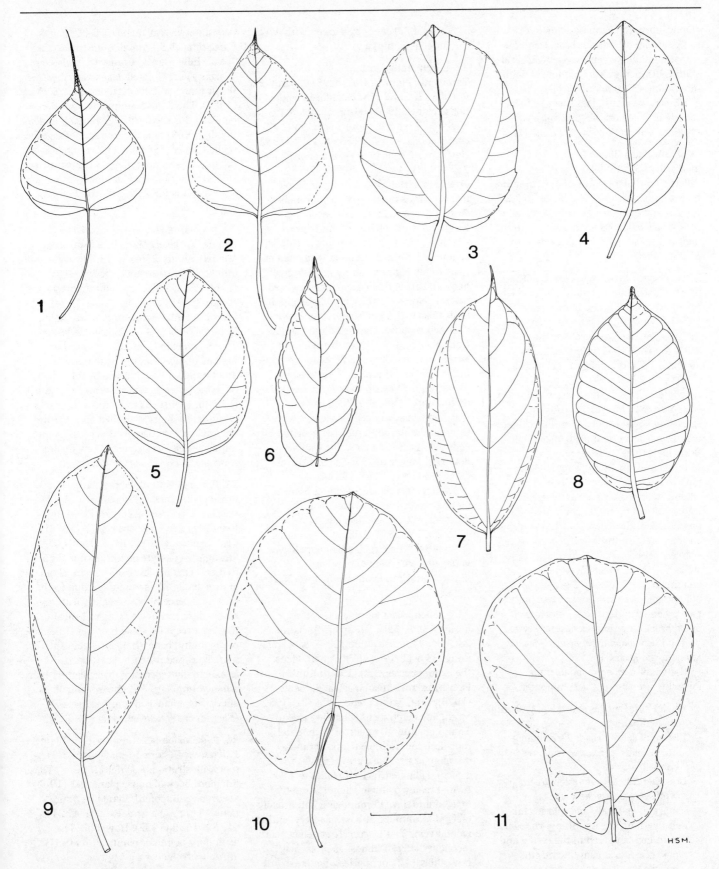

Figure 25. Leaves of *Ficus*. 1, *F. religiosa*. 2, *F. rumphii*.
3, *F. auriculata*. 4, *F. altissima*. 5, *F. benghalensis*. 6, *F. aspera*.
7, *F. parietalis*. 8, *F. drupacea*. 9, *F. vogelii*. 10, *F. nymphaeifolia*.
11, *F. lyrata*. Scale = 5 cm

elliptic, abruptly acuminate into a tip
c. 1 cm long, rounded at base, leathery,
smooth and glossy, entire, wavy. Lateral
veins indistinct, at least 15; lowest pair at
more than 45° to midrib. Leaf-stalks
c. 1.8 cm, brownish. Figs 2 per leaf-axil,
spherical, c. 1.8 cm, yellow, hairless,
stalkless. *China (Yunnan) & SE Asia*. G2.

20. F. sundaica Blume (*F. indica*
misapplied). Figure 24(5), p. 95.
Differs from *F. stricta* in its larger terminal
buds to 3.6 cm, larger leaf-blades to 22.7 ×
6.2 cm and more prominent lateral veins;
lowest pair of lateral veins at c. 45° to the
midrib. Leaf-stalks 2.5–5 cm, often curved.
SE Asia, ?India. G2.

21. F. elastica Roxburgh. Figure 22(2),
p. 91. Illustration: Rochford & Gorer, The
Rochford book of house plants, pl. 38, 39
(1963); Wickham, The house-plant book,
127 (1977); Graf, Exotica, edn 11, 1563,
1564 & 1569 (1982); Hay et al., The
dictionary of indoor plants, 237 & 238
(1983).
Epiphyte or free-standing tree to 60 m (in
the wild); with buttressed trunk and
spreading surface roots; in humid climates
aerial roots form multiple trunks.
Branchlets thick, hairless and green;
terminal buds to 15 cm; stipules 1 per bud,
pink to red. Leaf-blades to 30 × 15 cm,
elliptic, acuminate or abruptly narrowed
into a pointed tip to 2.5 cm, rounded at
base, thick and rubbery, hairless and shiny,
entire. Lateral veins 25–30, indistinct,
scarcely raised, at c. 90° to midrib. Leaf-
stalks 5–10 cm, greenish white,
channelled. Figs usually 2 per leaf-axil,
oblong and often angled, c. 1 cm, hairless,
pale green with dark flecks, stalkless (a spur
of old bract bases may persist for several
years). *E Himalaya to peninsular Malaya,
Sumatra and Java*. G1.

Many cultivars are available; the
following are among the commonest:

'Belgica': Internodes short; leaves slightly
corrugated.
'Decora': Leaves broad, dark shining
green; midrib creamy white, reddish
beneath.
'Doescheri': Leaves marbled green, grey,
white and pale yellow near the
margin; leaf-stalk and midrib pink.
'Schrijveriana': Leaves shiny, variegated
with greyish green in the centre and
pale green or cream towards the
margin.
'Tricolor': Leaves greyish green,
variegated with pink and creamy
white; midrib red.

'Variegata': Leaves pale green with white
or yellow margin.

22. F. benjamina Linnaeus (*F. nitida*
Thunberg). Figure 24(8), p. 95. Illustration:
Rochford & Gorer, The Rochford book of
house plants, 39 (1963); Seabrook &
Rochford, Plants for your home, 53 (1971);
Graf, Exotica, edn 11, 1565 & 1571 (1982);
Hay et al., The dictionary of indoor plants,
237 (1983).
Tree with aerial roots forming multiple
trunks; bark smooth and grey; branches
drooping; branchlets green becoming
greyish; terminal buds c. 8 mm, green.
Leaf-blades to 11 × 5 cm, ovate to elliptic,
acuminate, rounded to broadly tapered at
base, thinly leathery, shiny greyish green,
hairless, entire; tip to 2 cm, often twisted.
Lateral veins 8–12, obscure, flat beneath.
Leaf-stalks to 2.5 cm, slender, hairless,
channelled above. Figs usually 2 per leaf-
axil, spherical to slightly ellipsoid, c. 11 ×
8 mm, hairless, deep red flecked with white.
India, S China, SE Asia & N Australia. G1.

Var. **nuda** (Miquel) Barrett (*F. comosa*
Roxburgh; *F.* 'Nuda'; *F.* 'Comosa';
F. philippinensis Anon.). Has leaves with
transparent margins clustered near the
ends of slender branches. Illustration:
Botanical Magazine, 3305 (1834); Graf,
Exotica, edn 11, 1570 (1982).

'Exotica' has slender, arching branches
and narrower leaves twisted near the tip.
Illustration: Graf, Exotica, edn 11,
1567,1571 & 1574 (1982).

'Variegata' has small pale green leaves
variegated with white or cream.
Illustration: Graf, Exotica, edn 11, 1563
(1982).

23. F. microcarpa Linnaeus (*F. retusa*
misapplied; *F. nitida* misapplied). Figure
23(7), p. 93. Illustration: Hill, Figs of Hong
Kong, 33 & pl. 19 (1967); Condit, Ficus:
the exotic species, 320 (1969); Kunkel,
Flowering trees in subtropical gardens, 71
(1978); Graf, Exotica, edn 11, 1571 (1982).
Strangling epiphyte or spreading tree to
15 m high and 30 m across with aerial
roots forming multiple trunks; branches
drooping at tips; bark grey; stipules to
1.5 cm, falling early. Leaf-blades to 11 ×
6 cm, broadly ovate to elliptic, abruptly
tapered into a small rounded point, broadly
tapered to rounded at base, leathery, entire.
Lateral veins 7–12, scarcely raised beneath,
secondary laterals almost as prominent.
Leaf-stalks 1–2 cm, hairless, flattened
above. Figs 1 or 2 per leaf-axil, spherical,
red to black, flecked with greenish brown,
stalkless (a spur of old bract bases may

persist for several years). *India, Ceylon &
China through SE Asia to E Australia*. G2.

Var. **hillii** (Bailey) Corner (*F. schlechteri*
Warburg) has 'jointed' leaf-stalks.
Illustration: Graf, Exotica, edn 11, 1570
(1982). There has been much confusion
between this species and *F. retusa* Linnaeus.
Condit (1969) puts the two species together
while Corner (1965) separates *F. retusa* as
having the secondary lateral veins much
less prominent. If they are regarded as
separate, it is unlikely that true *F. retusa* is
in cultivation.

24. F. obliqua Forster (*F. eugenioides*
Miquel; *F. obliqua* Miquel; *?F. platypoda*
Miquel). Figure 23(4), p. 93. Illustration:
Condit, Ficus: the exotic species, 319
(1969); Francis, Australian rainforest trees,
edn 3, 79 (1973).
Tree; branchlets with short sparse hairs;
terminal buds to 3.6 cm. Leaf-blades to
15 × 6 cm, elliptic, obtuse, rounded or
tapered at base, slightly leathery, glossy
and hairless above, hairless or softly hairy
on veins beneath. Lateral veins 10–12, flat
beneath. Leaf-stalks to 5 cm, hairless or
very slightly hairy, flattened and slightly
channelled. Figs usually 2 per leaf-axil, to
8 mm. *Indonesia (Sulawesi), New Guinea,
Australia & Pacific Islands*. G2.

25. F. aurea Nuttall. Figure 24(6), p. 95.
Illustration: Condit, Ficus: the exotic
species, 320 (1969); Kunkel, Flowering
trees in subtropical gardens, 57 (1978);
Graf, Exotica, edn 11, 1573 (1982).
Strangling epiphyte or tree; terminal buds
to 3 cm. Leaf-blades to 17 × 4 cm, elliptic,
obtuse or slightly acuminate, rounded or
slightly tapered at base, slightly leathery,
glossy dark green above, paler beneath,
hairless except for some hairs on either side
of the midrib beneath. Lateral veins 9–15,
scarcely raised beneath. Leaf-stalks
c. 2.5 cm, flattened and thick, slightly hairy
when young. Figs 1 or 2 per leaf-axil,
almost spherical, c. 8 mm, yellow. *USA
(Florida) & West Indies*. G1.

26. F. deltoidea Jack (*F. diversifolia* Blume;
F. lutescens Desfontaines). Figure 23(9),
p. 93. Illustration: Rochford & Gorer, The
Rochford book of house plants, 41 (1963);
Seabrook & Rochford, Plants for your
home, 53 (1971); Graf, Exotica, edn 11,
1576 & 1580 (1982); Hay et al., The
dictionary of indoor plants, 236 (1983).
Shrub with slender zig-zag twigs,
sometimes epiphytic; bark green becoming
grey and flaky; terminal buds c. 1 cm.
Leaves thick and rubbery, hairless above
and beneath, entire, of two forms: either

obovate to spathulate, or elliptic (both often on the same twig). Obovate or spathulate blades to 6.3 × 4.7 cm, broadly rounded, narrowly tapered at base; midrib divided *c*. 1 cm from the leaf-base and the branches then further subdivided. Elliptic blades to 10.5 cm, obtuse, narrowed at base; midrib not divided; lateral veins *c*.5. Leaf-stalks to 1.5 cm, sparsely hairy, flaky. Figs 1 per leaf-axil, produced freely in young specimens, pear-shaped, to 8 × 11 mm, yellow, hairless; stalks to 1 cm. *Malaysia*. G1.

27. F. nymphaeifolia Miller. Figure 25(10), p. 97. Illustration: Annales des Sciences Naturelles, ser. 10, **4**: 178 (1922); Condit, Ficus: the exotic species, 309 (1969). Leaf-blades to 22.5 × 18 cm, broadly elliptic, rounded, with a short, broad, obtuse point, deeply cordate at base with a sinus to 5 cm or more, thinly leathery, hairless, shiny dark green above, much paler beneath, entire. Lateral veins 5–7, raised beneath, greenish white to pink above. Leaf-stalks to 11.5 cm, hairless. Figs 2 per leaf-axil, spherical; stalk to 7 mm or absent. *Panama to Brazil*. G2.

28. F. eetveldiana André. Illustration: Gardeners' Chronicle **18**: 303 (1900); Revue Horticole **75**: 421 (1903). Tree with greyish bark. Leaf-blades *c*. 30 × 20 cm, broadly ovate to elliptic, rounded or obtuse, very broadly cordate at base, bright green, paler beneath, hairless, entire; lateral veins 7 or 8, widely spaced, raised beneath; leaf-stalks 15–20 cm. *C Africa*. G2.

This name is sometimes found in commerce.

29. F. erecta Thunberg. Figure 24(11), p. 95. Illustration: Botanical Magazine, 7550 (1897); Makino, New illustrated Flora of Japan, 97 (1963); Condit, Ficus: the exotic species, 319 (1969). Low straggling shrub (in cultivation). Leaf-blades *c*. 17 × 8 cm, elliptic to obovate, acute to slightly acuminate, tapered, rounded, truncate or slightly cordate at base, membranous to papery, hairless and slightly rough above, slightly hairy below when young. Lateral veins 7–9, raised beneath. Leaf-stalks slender, to 2.5 cm. Figs 1 per leaf-axil, spherical to urn-shaped, to 8 mm; stalks *c*. 1 cm. China, Korea, Japan, *Taiwan & Hong Kong*. G1.

30. F. rubiginosa Ventenat (*F. australis* Willdenow). Figure 23(2), p. 93. Illustration: Botanical Magazine, 2939 (1829); Kunkel, Flowering trees in subtropical gardens, 77 (1978); Everett, New York Botanical Gardens illustrated encyclopedia of horticulture 4: 1365 (1981); Graf, Exotica, edn 11, 1569–1571 & 1582 (1982). Tree to 12 m (in cultivation); bark dark grey, aerial roots present; branchlets flaky, flattened; terminal buds 2.5–5 cm, rusty-hairy; stipules to 12.5 cm, unequal. Leaf-blades 7.5–17.2 × *c*. 6 cm, elliptic, obtuse, broadly tapered to rounded at base, leathery, rusty-hairy above and below when young, becoming hairless above with age, entire. Lateral veins 9–14, scarcely raised. Leaf-stalks to 4.5 cm, slightly hairy when young, later rusty, flattened. Figs 2 per leaf-axil, variable, spherical to oblong, 1.1–1.5 cm, hairy or hairless, rusty, yellow flecked with white. *E Australia*. G1.

A variegated form ('Variegata') is also in cultivation. Illustration: Graf, Exotica, edn 11, 1563 (1982). Condit (1969) considers that the cultivated *F. australis* Anon., usually separated from *F. australis* Willdenow, is a form of *F. rubiginosa* lacking rusty hair.

31. F. sycomorus Linnaeus. Figure 22(3), p. 91. Illustration: Condit, Ficus: The exotic species, 338 (1969); Palmer & Pitman, Trees of Southern Africa, 444 & 446 (1972); Palmer, A field guide to the trees of Southern Africa, pl. 4 (1977); Graf, Exotica, edn 11, 1569 & 1574 (1982). Spreading tree to 2.5 m; bark yellow, flaky; branchlets brown, silky at nodes; terminal buds *c*. 1 cm, green, silky; stipules falling early. Leaf-blades to 15 × 13.5 cm, broadly ovate to almost circular, obtuse or rounded at apex, rounded to cordate at base, hard and slightly leathery, dark green and hairless above, paler and very slightly hairy below, especially on the veins, margin entire or slightly wavy. Lateral veins 3–6, raised beneath. Leaf-stalks to 5.6 cm, sparsely silky, brown, contrasting with green midrib. Figs solitary, in pairs or clusters, on trunk and main branches, spherical to slightly flattened, to 2.5 cm, velvety-white; stalks 2–3 cm with white hairs. *E Mediterranean area, Arabia & NE Africa to SE Africa*. G1.

32. F. benghalensis Linnaeus (*F. indica* Linnaeus; *F. chauvienii* Gillaumin). Figure 25(5), p. 97. Illustration: Kunkel, Flowering trees in subtropical gardens, 59 (1978); Everett, New York Botanical Gardens illustrated encyclopedia of horticulture 4: 1363 (1981); Graf, Exotica, edn 11, 1564 (1982); Beckett, The RHS encyclopaedia of house plants, 255 (1987). Strangling epiphyte or tree with prolific aerial roots forming multiple trunks, covering large areas of ground in warm climates; bark pale grey. Branchlets hairy; terminal buds *c*. 3 cm. Leaf-blades 15–25 × 12–17 cm, broadly ovate, obtuse or rounded, broadly rounded to truncate or very slightly cordate at base, stiff and leathery, dull and hairless above, very sparsely hairless beneath, more so on veins, entire. Lateral veins 5–8, pale above, raised beneath. Leaf-stalks to 7.5 cm, slightly flattened. Figs 2 per leaf-axil, *c*. 1.8 cm, spherical to slightly flattened, hairy, red flecked with white. *Pakistan, India & Bangladesh*. G2.

Var. **krishnae** (de Candolle) Corner has cup-shaped leaves with distorted lateral veins. *India*. Illustration: Botanical Magazine, 8092 (1906); Graf, Exotica, edn 11, 1564 (1982).

33. F. leprieurii Miquel (*F. triangularis* Warburg; ?*F. buxifolia* De Wildeman: *F. natalensis* Hochstetter subsp. *leprieurii* (Miquel) Berg). Figure 24(12), p. 95. Illustration: Seabrook & Rochford, Plants for your home, 53 (1971); Boom, Flora cultuurgewassen 3: 74 (1975); Graf, Exotica, edn 11, 1571, 1582 & 1584. Tree to 25 m. Branchlets short, hairless; stipules rusty-hairy, falling early. Leaf-blades 4–8 × 3–7 cm, triangular, narrowly tapered at base, truncate and often broadly indented at tip, rarely rounded, papery to slightly leathery, hairless and dull on both surfaces; margin entire, slightly recurved. Midrib divided well below tip; lateral veins 4–7, raised beneath, looped at margin. Leaf-stalks 5–20 mm, hairless. Figs 2 per leaf-axil, spherical, hairless; stalks to 1 cm. *Tropical Africa*. G2.

Some authors include this species in the more widespread *F. natalensis* which has rounded, ovate to elliptic leaves. *F. buxifolia*, orginally described as having abruptly truncate, triangular leaves may be a synonym of *F. leprieurii*.

34. F. lingua De Wildeman & Durand. Figure 23(8), p. 93. Illustration: Morat (ed), Flore du Gabon 26: 169, lower (1984). Epiphytic shrub; branchlets brownish-hairy. Leaf-blades 18–25 × 5–8 mm, obovate, rounded, tapered at base, papery, dull and hairless on both surfaces; lateral veins 5 or 6. Leaf-stalks *c*. 5 mm. Figs 1 or 2 per leaf-axil, *c*. 5 mm. *W & C Tropical Africa*. G2.

F. buxifolia is sometimes quoted as a synonym for this species, but the original description shows it to be more like *F. leprieurii* in leaf-shape.

35. F. vogelii Miquel (*F. nekbudu* misapplied). Figure 25(9), p. 97. Illustration: Condit, Ficus: The exotic species, 333 (1969).
Epiphyte or free-standing tree to 15 m with rough bark. Branchlets stout, grey, with prominent white hairs at base of terminal bud and first node. Leaf-blades 10–30 × 4–13 cm, ovate, rounded to bluntly acuminate, rounded to almost cordate at base, leathery, hairless above and beneath except for white hairs along veins; margin entire; lateral veins 5–7, raised beneath, usually 3.5 cm or more apart. Leaf-stalks 5–15 cm, brown, contrasting with green midrib. Figs 3–7 in leaf-axils or on old leaf-scars on branches of all ages, oblong to pear-shaped or almost spherical, c. 1 cm, yellowish orange flecked with brown, densely hairy, stalkless. *Tropical W Africa.* G2.

Recent work suggests that *F. lutea* Vahl may be the correct name for this species.

36. F. nekbudu Warburg (*F. utilis* Sim; *F. quibeba* Fical). Figure 23(3), p. 93. Illustration: Rochford & Gorer, The Rochford book of house plants, 42 (1963); Palmer & Pitman, Trees of Southern Africa, 460 & 461 (1972); Graf, Exotica, edn 11, 1564 (1982).
Very similar to *F. vogelii.* Branchlets densely silky; leaf-blades to 20 cm wide. Lateral veins 2.5–3.5 cm apart. Figs densely clustered and angled, c. 2 cm. *Tropical & S Africa.* G2.

37. F. habrophylla Seemann (*F. edulis* Bureau). Illustration: Graf, Exotica, edn 11, 1572 (1982).
Small shrub or tree; branchlets velvety. Leaf-blades to 35 × 17 cm, oblong to obovate, obtuse or bluntly acuminate, cordate at base, slightly leathery, hairless when mature, slightly shiny above, dull beneath; lateral veins 10–17, raised beneath. Leaf-stalks 2–4 cm, brown-hairy, flaky. Figs in dense clusters on lower stem, to 3 cm (larger in cultivated forms). *New Caledonia.* G2.

Grown for its fruit in parts of the tropics.

38. F. parietalis Blume (*F. cerasiformis* Desfontaines; *F. acuminata* misapplied). Figure 25(7), p. 97. Illustration: Botanical Magazine, 3282 (1833); Annals of the Royal Botanic Garden, Calcutta 1: t. 8 (1887); Graf, Exotica, edn 11, 1568 (1982).
Epiphytic shrub to large free-standing tree; bark brownish red; branchlets brown, ridged, densely yellow-hairy; terminal buds c. 1 cm. Leaf-blades to c. 23 × 10 cm, elliptic, narrowly acuminate with tail to 2 cm, broadly tapered to rounded at base, leathery, hairless and shiny above, hairy beneath, margin entire; lateral veins c. 4, raised beneath, strongly curved. Leaf-stalks to 2 cm, ridged, densely hairy. Figs 1 or 2 per leaf-axil, spherical to slightly flattened, c. 1.5 cm, hairy, orange and warty when ripe; stalks c. 1 cm. *SE Asia.* G2.

39. F. drupacea Thunberg var. **drupacea.** Figure 25(8), p. 97. Illustration: Graf, Exotica, edn 11, 1583 (1982).
Epiphytic strangler or free-standing tree to 20 m or more. Branchlets stout, hairy, yellowish brown, becoming grey and smooth; terminal buds c. 2.5 cm, silky. Leaf-blades to 25 × 14 cm, elliptic to slightly ovate or obovate, abruptly pointed with tip to 8 mm, shallowly cordate at base, slightly leathery, hairy on veins and near base when young, hairless later, margin entire. Lateral veins 7–12, pale and raised beneath, looped at margin. Leaf-stalks to 2.5 cm, hairy, becoming less so. Figs 2 per leaf-axil, 1.8–3 cm, orange-red, hairless, stalkless. *India & Sri Lanka through SE Asia to N Australia.* G2.

Var. **pubescens** (Roth) Corner (*F. mysorensis* Roth) has leaves with 9–15 lateral veins and figs which are hairy when young. Illustration: Graf, Exotica, edn 11, 1576 (1982).

40. F. racemosa Linnaeus (*F. glomerata* Roxburgh not Blanco; *F. vesca* Miquel). Figure 22(1), p. 91. Illustration: Condit, Ficus: the exotic species, 334 (1969); Boom, Flora cultuurgewassen 3: 74 (1975), Graf, Exotica, edn 11, 1567 (1982).
Tree to 25 m with large buttresses; bark rough and scaly when old. Branchlets brown, slightly hairy when young, soon becoming hairless. Terminal buds to c. 1 cm; stipules persistent or falling early. Leaf-blades to 20 × 11 cm, elliptic to ovate, obtuse to acuminate, sometimes with a narrowly rounded tip, rounded to tapered at base, papery to leathery, hairless and glossy with a metallic lustre above, sparsely hairy on veins beneath, margins entire to irregularly wavy; lateral veins 8–12, raised beneath. Leaf-stalks 6–11 cm, brown. Figs clustered on short spurs borne on trunk and large branches, spherical to pear-shaped, to 3.5 cm, red flecked with white; stalks to 1.8 cm. *India & Sri Lanka to China & SE Asia.* G2.

41. F. callosa Willdenow (*F. porteana* Regel). Figure 22(6), p. 91. Illustration: Annals of the Royal Botanic Garden, Calcutta 1: t. 85 (1887); Condit, Ficus: the exotic species, 323 (1969); Graf, Exotica, edn 11, 1574 (1982).
Large tree with smooth, grey bark; branchlets dark brown, slightly hairy; terminal buds c. 7 mm. Leaf-blades to 27.5 × 15 cm, broadly ovate to elliptic, oblong or obovate, obtuse to rounded at tip, broadly tapered, rounded or cordate at base, leathery, hairless, shiny and dark green above, paler beneath, margin entire, slightly rolled. Lateral veins 10–12, raised beneath, curved near margin. Leaf-stalks to 7 cm. Figs usually 1 per leaf-axil, spherical to oblong, c. 2 cm, green; stalks c. 8 mm. *India, Sri Lanka & SE Asia.* G2.

42. F. altissima Blume. Figure 25(4), p. 97. Illustration: Hill, Figs of Hong Kong, pl. 17 (1967); Kunkel, Flowering trees in subtropical gardens, 55 (1978); Graf, Exotica, edn 11, 1564 (1982).
Strangling epiphyte or free-standing tree with aerial roots forming multiple trunks; branchlets hairy; bark green to pale grey; terminal buds to 3.6 cm, silky. Leaf-blades to 28 × 15 cm, oblong to elliptic or almost circular, obtuse or bluntly acuminate, broadly tapered to rounded at base, leathery, hairless and shiny, paler beneath, margin entire. Lateral veins 6–10, raised beneath. Leaf-stalks to 11.5 cm, flattened, hairless. Figs usually 2 per leaf-axil, spherical, to 1.5 cm, red and hairless, stalkless. *India, China & SE Asia.* G2.

43. F. carica Linnaeus. Figure 22(8), p. 91. Illustration: Leathart, Trees of the world, 134 (1977); Kunkel, Flowering trees in subtropical gardens, 63 (1978); Everett, New York Botanical Gardens illustrated encyclopedia of horticulture 4: 1369 (1981); Graf, Exotica, edn 11, 1566 & 1576 (1982).
Deciduous shrub or tree to 10 m, bark grey and smooth; branchlets green to brown, slightly hairy. Leaf-blades to c. 35 × 35 cm, with 3 or 5 (rarely 7), scalloped or toothed lobes with obtuse or rounded tips, cordate at base, papery to slightly leathery, rough above, shortly hairy beneath; veins palmate, 1 per lobe. Leaf-stalks to 10 cm. Figs spherical to pear-shaped, to 8 cm across, yellow to purplish, hairless or slightly hairy. *E Mediterranean area & W Asia; widely grown for its fruit in warm temperate and subtropical areas.* H5–G1.

F. palmata may also key out here.

44. F. palmata Forsskål (*F. pseudocarica* Miquel). Figure 22(9), p. 91.
Very similar to *F. carica* and sometimes

united with it. Secondary leaves often unlobed, broadly ovate and regularly toothed, base broadly tapered, rounded or cordate. Mature leaves usually with 3, 5 or 7 lobes, sometimes unlobed and regularly toothed. Figs smaller than in *F. carica*. *NE Africa and Arabia to N India*. G1.

45. F. nota (Blanco) Merrill. Illustration: Condit, Ficus; the exotic species, 309 (1969).
Shrub or tree to *c.* 10 m; branchlets densely silky becoming brown; terminal buds *c.* 2.5 cm. Leaf-blades to 30 × 16 cm, ovate or obovate, acute or acuminate with a narrow, rounded tip, tapered or cordate at base, papery and rough, hairy on veins above, more uniformly so beneath; margin entire to coarsely toothed. Lateral veins 8–10, raised beneath. Leaf-stalks to 5 cm, silky, rusty-scurfy, contrasting with green midrib. Figs in clusters on trunk and branches, spherical or slightly flattened, *c.* 3.5 cm, red; stalks *c.* 1 cm. *Philippines & N Borneo*. G2.

46. F. auriculata Loureiro (*F. roxburghii* Miquel; *F. macrophylla* misapplied). Figure 25(3), p. 97. Illustration: Condit, Ficus: the exotic species, 11 & 332 (1969); Graf, Exotica, edn 11, 1576 (1982).
Bushy shrub or tree; branchlets hollow, green becoming grey; terminal buds thick, *c.* 2.5 cm, brown or green. Leaf-blades to 40 × 35 cm, very broadly ovate to almost circular, rounded, obtuse, acute or acuminate, shallowly cordate at base, papery to slightly leathery, hairless above, softly hairy on veins beneath, margin entire to widely scalloped. Lateral veins 5 or 6, raised beneath, lowest pair reaching well over half way up leaf. Leaf-stalks to 22.5 cm, hairless, green or brown. Figs clustered on spurs on main branches, also on short leafless branches at base of trunk, pear-shaped, flattened, to 6 cm, green to reddish brown, silky; stalks *c.* 2.5 cm. *Himalaya to S China, Thailand & Vietnam*. G2.

47. F. sus Forsskål (*F. capensis* Thunberg *F. brassii* R. Brown). Figure 24(7), p. 95. Illustration: Palmer & Pitman, Trees of Southern Africa, 447–449 (1972); Palmer, A field guide to the trees of Southern Africa, pl. 4 (1977).
Shrub or small tree; branchlets slightly hairy, becoming hairless; terminal buds *c.* 2 cm, flattened, hairy at base. Leaf-blades to 23 × 13 cm, ovate to elliptic, obtuse or acuminate with a narrow rounded tip, rounded to slightly cordate at base, slightly

leathery, hairless (sometimes slightly hairy on midrib beneath), margin broadly scalloped or toothed, sometimes almost entire. Lateral veins 4–6, raised beneath. Leaf-stalks to 10.2 cm, sparsely hairy when young, becoming flaky. Figs in bunches on branched stems hanging from trunk and main branches, spherical to pear-shaped, to 2.5 cm, green flecked with white. *Tropical & S Africa*. G2.

The name *F. capensis* is widely misapplied.

48. F. hispida Linnaeus. Figure 22(4), p. 91. Illustration: Hill, Figs of Hong Kong, 83 & pl. 63 (1967); Condit, Ficus: the exotic species, 310 (1969).
Spreading bush or tree; branchlets green becoming brown, softly hairy; terminal buds 1–2.5 cm, densely hairy. Leaves opposite, blades to 32 × 15 cm, ovate to obovate, obtuse to acuminate, rounded to shallowly cordate at base, membranous to papery, rough and bristly above, margin with regular or irregular distant teeth. Lateral veins 6–10, raised beneath. Leaf-stalks thin, to 7.5 cm, hairy to bristly. Figs clustered on main branches, or on branched stems hanging from trunk and main branches, spherical or flattened, 1–2.5 cm. *India & S China to SE Asia & N Australia*. G2.

8. HUMULUS Linnaeus
E.C. Nelson
Monoecious vine-like herbs. Leaves opposite, with stalks, cordate at base. Flowers unisexual, inflorescence pendulous, male branched. Male flowers with 5-lobed calyx and 5 stamens. Female flowers in cone-like inflorescences; calyx membranous, ovary stalkless and style with 2 branches.

A genus of 3 species from temperate Eurasia; two of these are in cultivation white the third (*H. yunnanensis* Hu) is a recently discovered plant from C China of apparently neither ornamental nor economic interest. The hop used in brewing is the fruiting head of *H. lupulus*. Some authorities place *Humulus*, with *Cannabis*, in the Cannabaceae because of the 5-partite flowers.
Literature: Small, E.A., A numerical and nomenclatural analysis of morpho-geographic taxa of Humulus, *Systematic Botany* 3: 37–76 (1978).

1a. Rhizomatous climbing perennial; leaves 1–3-lobed   **1. lupulus**
  b. Annual climbing herb; leaves 5–7-lobed   **2. japonicus**

1. H. lupulus Linnaeus (*H. americanus* Nuttall). Illustration: Bonnier, Flore complète 10: pl. 555 (1929); Keble Martin, Concise flora of the British Isles, 76 (1965); Everett, The New York Botanical Gardens illustrated encyclopedia of horticulture 5: 1732 (1981).
Perennial vine with rhizome. Shoots deciduous, prickly. Leaves entire or with 3 lobes, ovate to circular, 5–12 × 5–12 cm, base cordate, irregularly toothed. Female cones 1.5–2 cm enlarging to 5 cm in diameter when the fruits mature. *Europe, Asia*. H3. Late summer.

The fruiting heads (hops) are picked for flavouring beer. In ornamental gardens the variant with golden leaves ('Aureus') forms an attractive climbing plant. It is propagated vegetatively.

2. H. japonicus Siebold & Zuccarini (*H. scandens* (Loureiro) Merrill). Illustration: Gleason, Illustrated Flora of the north-eastern United States and adjacent Canada 2: 53 (1952); Systematic Botany 3: 58 (1978).
Climbing annual herb, stems spiny. Leaves with 5–7 lobes, 5–12 × 5–12 cm, cordate at base, lobes acuminate, regularly toothed. Female cone not enlarged at maturity, with prominent bristles. *China, Japan*. H3. Late summer.

The name *H. scandens* is of doubtful significance; Merrill argued that Loureiro's *Antidesma scandens* could only be a species of *Humulus*, but Small and others reject this, saying that Loureiro's description is too vague. If Merrill's views are accepted, the valid name for the plant is *H. scandens*.

A cultivar 'Variegatus' with leaves spotted and splashed with white is also grown.

9. CANNABIS Linnaeus
E.C. Nelson
Erect annual herb to 2.5 m, dioecious. Leaves alternate (lower leaves may be opposite), stalked, palmately divided; segments 3–9, linear-lanceolate, margins deeply toothed. Inflorescences erect, glandular; males branched. Male flowers with 5-lobed calyx, stamens 5; female inflorescence a raceme, female flowers with membranous calyx closely adhering to ovary; style with 2 branches. Embryo curved.

A genus of a single species, sometimes placed with *Humulus* in a separate family, Cannabaceae; arguments that more than 1 species is recognisable within *Cannabis* are spurious according to Small (cited below).

The plant is probably native in C Asia but is now cultivated throughout temperate and subtropical regions for the fibres (hemp) which can be extracted from the stems, and for the narcotic drug. Cultivation of this plant is illegal in most European countries.

Literature: Small, E., Jui, P.Y. & Lefkovitch, L.P., A numerical taxonomic analysis of Cannabis with special reference to species delimitation, *Systematic Botany* 1: 67–84 (1976); Small, E., The forensic taxonomic debate on Cannabis: semantic hokum, *Journal of Forensic Science* 21: 239–251 (1976).

1. C. sativa Linnaeus (*C. gigantea* Vilmorin; *C. ruderalis* Janischewsky). Illustration: Bonnier, Flore complète 10: pl. 555 (1929); Stodola, Volak & Bunney, The illustrated book of herbs, 94 (1984).
C Asia. H2. Summer.

XLIII. URTICACEAE

Dioecious or monoecious herbs, undershrubs or rarely trees with very soft wood. Leaves alternate or opposite, sometimes with stinging hairs; stipules present but often deciduous; cystoliths abundant in stems and leaves, usually spindle-shaped. Flowers mostly unisexual, in cymes, sometimes crowded on a common, enlarged receptacle. Male flowers with 4–5 perianth-lobes and 4–5 stamens inflexed in bud, exploding when ripe. Female flowers with 2–4-lobed perianth, often with staminodes, ovary superior, free from or fused to perianth, 1-celled. Fruit an achene or a fleshy berry. Seed with oily endosperm.

A family of about 50 genera and 2000 species from temperate and tropical regions throughout the world, but poorly represented in Australia.

In European gardens only 8 genera are commonly seen; most are cultivated for their attractive foliage, but the exploding anthers also provide novelty. Outside Europe many other species are grown to produce fibres although these are now largely superseded by artificial fibres. The commonly cultivated plants are propagated vegetatively by cuttings or division, and by seed. Some species can be noxious weeds.

1a. Leaves alternate, base with distinctly unequal sides   **4. Pellionia**
 b. Leaves opposite or if alternate then base with equal sides   2

2a. Flowers inconspicuous, solitary; stems creeping; leaves small, more or less circular; stipules absent; mat-forming herb   **8. Soleirolia**
 b. Flowers in conspicuous inflorescences; leaves and habit not as above   3
3a. Herbs (annual or perennial); fruits not fleshy   4
 b. Shrubs or trees; fruits sometimes fleshy   6
4a. Stinging hairs present; leaves opposite   **1. Urtica**
 b. Stinging hairs absent; leaves opposite or alternate   5
5a. Stigma brush-like; perianth 3-lobed, not tubular   **3. Pilea**
 b. Stigma not brush-like; perianth tubular, with usually 4 lobes   **5. Boehmeria**
6a. Stinging hairs present; fruit white and waxy   **2. Urera**
 b. Stinging hairs absent; fruits red or yellow or dry   7
7a. Fruits red or yellow; leaves with 3 main veins   **7. Debregeasia**
 b. Fruit dry; leaves pinnately veined   **6. Myriocarpa**

1. URTICA Linnaeus
M.J.P. Scannell

Annual or perennial herbs, usually with stinging hairs; stems ridged or 4-angled. Leaves opposite, margins toothed, stipules free. Flowers unisexual, green, in axillary cymes. Perianth with 4 lobes, which are unequal in the female flowers. Stigma brush-like. Fruit a shining, flattened nut.

There are some 100 species in temperate regions. Three are found in European gardens, usually by invasion rather than deliberate cultivation. Propagation may be by division or by seed.

1a. Plant perennial, dioecious   **1. dioica**
 b. Plant annual, monoecious   2
2a. Female flowers in separate spherical heads; male in branched cymes   **2. pilulifera**
 b. Female and male flowers intermixed in cymes   **2. urens**

1. U. dioica Linnaeus. Illustration: Ross-Craig, Drawings of British plants 2: t. 876 (1960); Garrard & Streeter, The wild flowers of the British isles, 118 (1983).
Dioecious perennial, stems to 1.5 m, square in cross-section, rhizome yellow. Leaves ovate, acuminate, base cordate, with copious stinging hairs. *Northern hemisphere.* H1. Summer.

Not usually cultivated, except perhaps in its variegated variant; a stingless variant is also known. The young shoots are rich in iron and the plant is sometimes eaten as a vegetable. Otherwise a painful and unwelcome weed, like the two succeeding annual species.

2. U. pilulifera Linnaeus. Illustration: Coste, Flore de la France 1: 248 (1906). Monoecious annual herb, stems to 60 cm. Leaves ovate, deeply and regularly toothed or entire, with copious stinging hairs. Male flowers in distinct clusters (branched cymes) along the inflorescence-stalks; female flowers in dense, spherical heads less than 1 cm in diameter. *S Europe.* H5. Summer–autumn.

3. U. urens Linnaeus. Illustration: Coste, Flore de la France 1: 248 (1906); Garrard & Streeter, The wild flowers of the British Isles, 118 (1983).
Like *U. pilulifera* but leaves blue-green and male and female flowers intermixed in axillary clusters. *Northern hemisphere.* H2. Summer–autumn.

2. URERA Gaudichaud-Beaupré
M.J.P. Scannell & E.C. Nelson

Shrubs or small trees, usually dioecious, often with very powerful stinging hairs. Leaves alternate, toothed or lobed, with 3–5 veins or pinnately veined; stipules free or united. Inflorescence a loose, branched, axillary panicle. Male flowers with 4–5 slightly overlapping perianth-segments and 4–5 stamens. Female flowers with 4 perianth-lobes; stigma stalkless on top of the ovary. Fruit berry-like, an achene in the fleshy, enlarged, persistent perianth.

About 35 species are recognised, inhabiting tropical and subtropical regions.

1. U. baccifera (Linnaeus) Weddell. Illustration: Everett, The New York Botanical Gardens illustrated encyclopedia of horticulture 10: 3449 (1981).
Tree or shrub to 6 m; trunk and branches with recurved prickles. Leaves ovate, to 30 × 25 cm, base cordate or rounded, with scattered stinging hairs on the upper surface, margins with shallow teeth 1–2 cm broad, veins with prickles. Flowers white or pink. Fruits white, waxy. *C & S America, West Indies.* G2. Spring.

The sting of this plant is viciously painful, and it should not be grown in public places. In S America this tree is used for hedges on cattle ranches.

3. PILEA Lindley
E.C. Nelson & M.J.P. Scannell

Annual or perennial herbs, monoecious or

dioecious, stems sometimes woody or creeping. Leaves opposite, those of each pair sometimes very different in size (spirally arranged in No.1) entire, occasionally fleshy, sometimes with cystoliths; stipules fused. Male flowers with 4 concave perianth-segments each with a protruberance on the outer side; stamens 4. Female flowers with 3 hooded, unequal perianth-lobes; staminodes scale-like; stigma brush-like. Fruit enclosed in the persistent perianth; seeds ovoid to spherical.

Over 600 species have been described in *Pilea*, making it the largest genus in the family. Most inhabit tropical regions and are not of horticultural value. The cultivated species are grown for their foliage and are easily propagated by cuttings. Most are probably represented in gardens by clones selected for their foliage characters which may not come true if the plants are raised from seed.
Literature: Killip, E.P., The Andean species of Pilea, *Contributions from the U.S. National Herbarium* **26**(10) (1939).

1a. Leaves peltate, hairless, uniformly light green; plant resembling species of *Peperomia* (p. 407)
                                1. peperomioides
  b. Leaves not peltate, not uniformly light green or not hairless   2
2a. Leaves uniformly green or red-green, less than 1.5 cm   3
  b. Leaves not uniformly green or red-green, more than 2 cm   5
3a. Plant tinted with red on stems and leaves, succulent; leaves almost globular   **2. serpyllacea**
  b. Plant not red-tinted; leaves not globular   4
4a. Leaves of each pair very unequal in size   **3. microphylla**
  b. Leaves of each pair equal
                          4. nummulariifolia
5a. Leaves ovate, scolloped, *c.* 3 × 2 cm, stems downy, becoming hairless, uniform in colour   **5. repens**
  b. Leaves larger or coloured   6
6a. Leaves puckered or hairy   7
  b. Leaves not puckered, hairless   8
7a. Leaves variously marked with red, bronze and green; stems hairy
                        6. involucrata
  b. Leaves green (only veins red); stems hairless   **7. crassifolia**
8a. Leaves silver and green, not more than 8 × 5 cm   **8. cardierei**
  b. Leaves green, to 22 × 16 cm
                        9. grandifolia

1. P. peperomioides Diels. Illustration: Kew Magazine 1: t. 5 (1984).
Superficially resembling a species of *Peperomia*. Erect herb, somewhat succulent, monoecious, hairless. Leaves pale green, peltate with long, erect stalks, blade elliptic to circular, *c.* 4–9 × 4–9 cm, margins entire, conspicuous veins 8–10, radiating from top of stalk. Stipules persistent, brown. Male and female flowers on separate branched inflorescences; flowers cream-white to pale green. *China (Yunnan)*. H5–G1. Summer.

A deceptive plant which can be mistaken when not in flower for a species of *Peperomia*. This was first cultivated in Europe in 1946, having been brought to Norway from China by a Norwegian missionary (see Kew Magazine 2(3): 334–336, 1985). In Norway the species is called 'Misjonaerplanten' ('missionary plant'). Its succulent peltate leaves are most attractive. It is easily cultivated in well-lit, unheated rooms away from direct sunlight, but does not tolerate temperatures below 0° C. Propagation by cuttings (side-shoots root easily).

2. P. serpyllacea (Knuth) Liebman (*P. serpyllifolia* misapplied). Illustration: Everett, The New York Botanical Gardens illustrated encyclopedia of horticulture **8**: 2637 (1981).
Succulent herb, hairless, tinted red throughout. Leaves pale red-green, almost globular, 1–5 mm in diameter, margins entire or very shallowly scolloped. *Tropical S America*. G1. Summer.

This species resembles *P. microphylla* but may be distinguished by the well-developed inflorescence-stalks and the succulent, almost spherical leaves. Red-tinged plants grown as *P. microphylla* probably belong here.

3. P. microphylla (Linnaeus) Liebman (*P. muscosa* Lindley; *P. callitrichoides* (Knuth) Knuth). Illustration: Everett, The New York Botanical Gardens illustrated encyclopedia of horticulture **8**: 2637 (1981); Krempin, 1000 decorative plants, 192, 193 (1983).
Succulent, usually monoecious herb to 30 cm, hairless. Leaves pale green, minute, those of a pair very unequal in size, obovate to circular, 3–10 × not more than 3 mm. Flowers inconspicuous. *Tropical America (Mexico to Brazil)*. G1. Summer.

The 'artillery plant' is cultivated for its foliage and because of its diverting characteristic, the explosive discharge of the anthers which release clouds of pollen. It is sometimes mistaken for a fern (and it is commonly known as the artillery fern).

Variegated and red-foliaged cultivars are available, but the red ones are usually *P. serpyllacea*. It prefers light shade and a moist, peaty compost.

4. P. nummulariifolia (Swartz) Weddell. Illustration: Everett, The New York Botanical Gardens illustrated encyclopedia of horticulture **8**: 2737 (1981); Krempin, 1000 decorative plants, 193 (1983).
Creeping herb, hairy. Leaves pale green, minute, less than 1.5 cm in diameter; margins toothed and ciliate. Flowers minute, green. *Tropical S America, West Indies*. G1. Summer.

5. P. repens (Swartz) Weddell. Illustration: Everett, New York Botanical Gardens illustrated encyclopedia of horticulture **8**: 2638 (1981).
Creeping herb, stems downy at first, later almost hairless, rooting at the nodes. Leaves broadly ovate, to 3 × 2 cm, margins scolloped, base truncate. *West Indies*. G1. Summer.

6. P. involucrata (Sims) Urban (*P. pubescens* misapplied; *P. spruceana* misapplied). Illustration: Botanical Magazine, 2481 (1824); Everett, The New York Botanical Gardens illustrated encyclopedia of horticulture **8**: 2637 (1981); Krempin, 1000 decorative plants, 192, 193 (1983).
Branched herb with hairy stems. Leaves ovate to obovate, to 6 × 3 cm, margins toothed, often puckered, with 3 prominent veins, variously coloured and marked with red, bronze and silver. *C & S America*. H5–G1. Summer.

A plethora of cultivars of *Pilea* with attractive foliage are in cultivation and most have obscure origins. Moreover, their precise botanical status is uncertain. Plants grown under the name *P. spruceana* can be accommodated within *P. involucrata*, as can cultivars named 'Moon Valley', 'Curly Top' and 'Norfolk'. 'Moon Valley' superficially resembles a cultivar of *Coleus*. Originating in the collections of Mr L.M. Mason, it was introduced by Rochford in 1970 under the name *P. repens*, and has also been placed in *P. crassifolia* but this latter species has hairless leaves, whereas 'Moon Valley' is hairy.

7. P. crassifolia (Willdenow) Blume. Bushy herb *c.* 50 cm, hairless. Leaves glossy green above, veins often pink beneath, elliptic to lanceolate, 3–13 × 1–4 cm, often puckered between veins. Flowers pink or crimson. *Jamaica*. G1. Summer.

Resembles *P. grandifolia* but with smaller leaves. Some authorities place the cultivar

'Moon Valley' here, but see *P. involucrata* above.

8. P. cardierei Gagnepain & Guillaumin. Illustration: Everett, The New York Botanical Gardens illustrated encyclopedia of horticulture 8: 2638 (1981); Krempin, 1000 decorative plants, 176 (1983). Bushy herb to 25 cm, hairless. Leaves dark green along margins and veins, with silver-white patches between, obovate, *c.* 8 × 5 cm, with 3 conspicuous veins, base wedge-shaped, tip acuminate, margins toothed towards apex. *Vietnam.* H5–G1. Summer.

Introduced into cultivation in France during 1938, and now widely cultivated; several cultivars of doubtful distinction have been named, especially one that is reputed to be dwarf, although all plants in cultivation at present may derive from the original introduction. The silver and green foliage is attractive. *P. cardierei* prefers semi-shade and requires a minimum temperature of 12° C. Plants should not be kept for more than 1 year; cuttings taken in late summer root easily and rapidly form bushy young plants.

9. P. grandifolia (Linnaeus) Blume. Illustration: Everett, The New York Botanical Gardens illustrated encyclopedia of horticulture 8: 2638 (1981). Shrubby herb to 2 m, hairless. Leaves glossy dark green with red veins, ovate-acuminate to 22 × 16 cm (smaller in cultivation), margins toothed towards the tip, sometimes puckered between veins. *Jamaica.* G1. Summer.

This is the largest-leaved species in cultivation; *P. crassifolia* is similar but has smaller leaves.

4. PELLIONIA Gaudichaud-Beaupré
E.C. Nelson & M.J.P. Scannell
Herbs or rarely undershrubs, monoecious or dioecious, stems sometimes succulent. Leaves alternate, distinctly asymmetric at base, stipules small or absent. Flowers in dense cymes. Male flowers with 4–5 overlapping perianth-segments; stamens 4–5. Female flowers with 3–5 perianth-segments, staminodes scale-like. Achene compressed, with tubercles.

There may be as many as 80 species all inhabiting tropical or subtropical regions of Asia, from India to Japan. The cultivated species require minimum temperatures of 12° C. and moist, peaty loam. They can be propagated very easily by cuttings and make excellent house-plants.

1a. Leaves dark green and purple
                            1. pulchra
  b. Leaves pale green with bronze-green margins   **2. repens**

1. P. pulchra N.E. Brown (*Elatostema pulchra* (N.E. Brown) Hall). Illustration: Everett, The New York Botanical Gardens illustrated encyclopedia of horticulture 8: 2536 (1981); Krempin, 1000 decorative plants, 176 (1983). Creeping herb, stems with purple tint, becoming hairless. Leaves green with dull black-green marks along the veins above, purple beneath, oblong to elliptic, 2–5 × 1–3 cm, apex obtuse. *Vietnam.* G1. Summer.

2. P. repens (Loureiro) Merrill (*P. argentea* invalid; *P. daveauana* (Carrière) N.E. Brown; *Elatostema repens* (Loureiro) Hall). Illustration: Krempin, 1000 decorative plants, 176 (1983); Huxley, World guide to house plants, 134 (1983). Creeping herb, stems succulent, pink, hairless. Leaves dark bronze-green with purple margins and pale green or grey-green central stripe above, pink beneath, almost stalkless, oblong to elliptic or circular, 1–6 cm, margins scolloped. *SE Asia (Vietnam to Malaysia and Burma).* G1. Summer.

A clone with pale green leaves marked with white veins is in cultivation under the invalid name *P. argentea*.

5. BOEHMERIA Jacquin
M.J.P. Scannell & E.C. Nelson
Small trees, shrubs, undershrubs or perennial herbs, monoecious or dioecious. Leaves alternate or opposite, toothed, with 3 veins, stipules mostly free. Flowers usually unisexual, in axillary clusters. Male flowers with 4-lobed perianth, stamens 4. Female flowers with tubular perianth with 2–4 lobes at the apex, opening often constricted; stigma hairy, persistent. Ovary stalked. Fruits sometimes enlarged, with 2 acute wings, enclosed in the persistent perianth.

Some 100 species are named and many are grown commercially in China and neighbouring regions for fibre; those species cultivated in European gardens are grown for their foliage as the flowers are not ornamental.

1a. Leaves alternate with white woolly hairs beneath   **2. nivea**
  b. Leaves opposite, not white beneath **2**
2a. Leaves entire or 2-lobed at apex, margins with minute teeth   **1. biloba**

  b. Leaves coarsely toothed, with 3–5 pointed lobes at apex   **3. platanifolia**

1. B. biloba Weddell. Illustration: Everett, The New York Botanical Gardens illustrated encyclopedia of horticulture 2: 438 (1981). Perennial herb, 30–70 cm, stems stout, rough, tinged red. Leaves broadly ovate, 6–15 × 4–10 cm, puckered, often with 2 (or rarely 3) lobes at the apex, with sparse, short hairs beneath on veins only, margins with minute teeth. Inflorescence a short, thick spike. Perianth with short, erect hairs. Achene narrow, 1.5 cm. *Japan.* H5. Summer.

2. B. nivea (Linnaeus) Gaudichaud-Beaupré. Illustration: Iconographia cormophytorum sinicorum 1: 518 (1972); Everett, The New York Botanical Gardens illustrated encyclopedia of horticulture 2: 439 (1981). Perennial herb, monoecious, 1–2 m; stems densely covered with grey-white hairs, occasionally branched, sometimes becoming woody at base. Leaves alternate, oval, abruptly acuminate, 7–15 × 6–12 cm, with white, woolly hairs beneath, margins regularly toothed. Inflorescence branched, axillary. Female flowers with ovoid, 3–4-lobed perianth. Achene 1 mm. *E Asia (Japan to Malaysia).* H5–G1. Autumn.

3. B. platanifolia Franchet & Savatier (*B. japonica* Miquel var. *platanifolia* (Franchet & Savatier) Maximowicz). Illustration: Iconographia cormophytorum sinicorum 1: 519 (1972). Perennial herb, monoecious, to 1 m. Leaves circular to elliptic in outline, membranous, hairy, base rounded, margins with coarse, irregular teeth, apex with 3–5 prominent, toothed lobes. Inflorescence axillary, slender, 7–8 cm. Female flowers stalkless, with hairy, tubular, 2–4-lobed perianth. Male flowers with 4-lobed perianth. Achene covered with persistent perianth. *China, Japan, Korea.* H3. Autumn.

6. MYRIOCARPA Bentham
M.J.P. Scannell & E.C. Nelson
Small trees or shrubs, dioecious (rarely monoecious). Leaves alternate, often large, margins toothed, with pinnate veins, stipules often with 2 lobes; cystoliths present on the leaf-surface. Flowers in slender, mostly axillary spikes or racemes. Male flowers with 4–5-(rarely 3)-lobed perianth, lobes broad and overlapping; stamens 4–5. Female flowers without

perianth, or with 2 opposite bracts, stigma velvety, ovary compressed. Achene ovoid, margins often thickened.

About 15 species are known, from C & S America. The cultivated species requires a heated glasshouse.

1. M. stipitata Bentham (*Boehmeria argentea* misapplied). Illustration: Everett, The New York Botanical Gardens illustrated encyclopedia of horticulture 7: 2262 (1981).

Shrub or small tree to 12 m. Leaves blue-green, ovate to obovate, 10–30 × 10–20 cm, margins usually toothed and silvery, apex obtuse or pointed, thickly covered with hairs when young, cystoliths very dense and arranged radially. Inflorescence a loose, long raceme, green-white. *S America*. G2. Summer.

7. DEBREGEASIA Gaudichaud-Beaupré
M.J.P. Scannell & E.C. Nelson
Shrub or tree, monoecious or dioecious. Leaves alternate, with 3 veins, rough, margins finely toothed, stipules fused and 2-lobed. Flowers in spherical, stalkless, axillary cymes. Perianth ovoid, constricted at mouth, persistent and becoming succulent in fruit. Stigma resembling a tuft of hairs. Male flowers with 4-lobed perianth. Achene enclosed in fleshy, persistent perianth.

Three species are known from subtropical and temperate regions of Asia and Africa.

1. D. longifolia (Burman) Weddell (*D. velutina* Gaudichaud-Beaupré). Illustration: Polunin & Stainton, Flowers of the Himalaya, pl. 1248 (1985).
Shrub to 3 m. Leaves linear to oblong-lanceolate, apex slender, 10–20 cm, with white woolly hairs on the lower surface. Fruits red or yellow, to 1 cm in diameter in erect, shortly stalked heads. *SE Asia, Himalaya*. H4. Summer.

8. SOLEIROLIA Gaudichaud-Beaupré
M.J.P. Scannell
Slender, creeping, hairy, perennial herb, forming dense, evergreen mats. Stem to 20 cm × 0.5 mm, rooting at nodes. Leaves pale green, alternate, circular, entire, shortly stalked, 2–6 mm, without stipules. Flowers unisexual, axillary, solitary, inconspicuous. Female flowers enclosed in an involucre of 1 bract and 2 bracteoles, perianth tubular with 4 narrow lobes. Male flowers with 4 perianth-lobes. Fruit enclosed by persistent perianth and involucre; achene ovoid, shining.

A genus of a single species from the islands of the western Mediterranean. Cultivated originally in rock gardens and cool glasshouses, it is now naturalised in western Europe. This plant prefers damp, shady places and can become a serious weed.

1. S. soleirolii (Requien) Dandy (*Helxine soleirolii* Requien). Illustration: Roles, Flora of the British Isles, illustrations 2: t. 874 (1960); Garrard & Streeter, The wild flowers of the British Isles, 118 (1983). *Spain (Balearic Islands, France (Corsica) & Italy (Sardinia)*. H5–G1. Summer.

A variant with yellow-green foliage is sometimes cultivated.

XLIV. PROTEACEAE

Evergreen trees and shrubs, occasionally monoecious or dioecious, sometimes with woody tubers (lignotubers). Leaves alternate (rarely opposite or in whorls), simple or pinnate, often toughened (sclerified), toothed or spiny. Flowers unisexual or bisexual, radially symmetric or bilaterally symmetric, in complex inflorescences. Perianth-segments 4, petal-like, coloured. Stamens 4, anthers usually stalkless, rarely 1–3 infertile, 1–4 hypognous scales present or absent. Ovary superior; stigma 1, often flattened and modified as a pollen-presenter; carpel 1, ovules 1–many, marginal. Fruit a woody 'cone' or many-seeded follicle or 1-seeded achene, often fire-resistant.

About 75 genera and perhaps 1300 species, almost entirely confined to the southern hemisphere where they are becoming increasingly popular as ornamental garden plants. Many species from semi-arid climates are suitable for cultivation in southern Europe, but few are grown in gardens; in northwestern Europe most species are too tender.

In general, the Australian and southern African species are not tolerant of lime and are also sensitive to fertilisers, which should not be applied. They are also susceptible to root damage caused either by over-watering or soil disturbance.

This treatment is based on a survey of Proteaceae growing in European gardens during 1985 and 1986; 16 genera are treated here, and 12 other genera are recorded in cultivation in European Botanic Gardens but are not considered to be in general cultivation at this time.

Literature: Lamont, B., The ecology of the Australian Proteaceae with implications for their cultivation, *Australian Plants* 7: 47–68 (1973); Johnson, L.A.S. & Briggs, B.G., On the Proteaceae - the evolution and classification of a southern family, *Botanical Journal of the Linnaean Society* 70: 83–182 (1975); Wrigley, J.W. & Fagg, M., *Australian native plants* (1979); Vogts, M., *South Africa's Proteaceae* (1982); Nelson, E.C., *Proteaceae cultivated in European gardens* (in preparation).

For identification, flowering plants are essential.

1a. Trees or shrubs with hairless, paired, red flowers within unbranched inflorescences 2
  b. Trees or shrubs with hairy flowers, or flowers not paired, or inflorescences conspicuously branched 3
2a. Flowers in loose axillary panicles **11. Embothrium**
  b. Flowers in erect, terminal heads often with coloured bracts **12. Telopea**
3a. Flowers very numerous (up to 3000) in dense cylindric spikes with a woody core **16. Banksia**
  b. Flowers not as above 4
4a. Flowers unisexual, plants dioecious 5
  b. Flowers bisexual 6
5a. Hypogynous scales absent; flowers yellow, not surrounded by brightly coloured leaves **5. Aulax**
  b. Hypogynous scales present; flowers variously coloured, inflorescence often surrounded by brightly coloured leaves **4. Leucadendron**
6a. Inflorescences terminal, surrounded by brightly coloured bracts **2. Protea**
  b. Inflorescences axillary or not surrounded by brightly coloured bracts 7
7a. Flowers not in pairs within inflorescences 8
  b. Flowers in pairs within inflorescences 9
8a. Inflorescences axillary; 3 upper perianth-segments remaining fused after flowering; anthers stalkless **3. Leucospermum**
  b. Inflorescences terminal; all perianth-segments separating after flowering; stamens with free filaments **1. Bellendena**
9a. Adult foliage pinnate, leaflets broad and toothed; fruit an indehiscent follicle; hypogynous glands 2 **10. Gevuina**
  b. Adult foliage entire or fruit a dehiscent follicle or hypogynous scales not 2 10

10a. Hypogynous scales 4, free
                          14. Knightia
  b. Hypogynous scales fewer than 4 or
     absent   11
11a. Flowers in clusters arranged like the
     spokes of a wheel in branched
     inflorescence   **15. Stenocarpus**
  b. Flowers in long spikes of spherical
     heads, not branched   12
12a. Foliage of 2 forms, juvenile pinnate,
     adult simple   **8. Roupala**
  b. Foliage not as above   13
13a. Leaves entire, whorled, margins
     toothed; inflorescence a long panicle;
     flowers radially symmetric; seeds
     spherical   **9. Macadamia**
  b. Leaves entire or divided, alternate;
     flowers bilaterally symmetric; seeds
     compressed, winged   14
14a. Hypogynous scales 3 or 4 (the fourth
     minute); seeds numerous in each
     follicle   **13. Lomatia**
  b. Hypogynous scale 1; seeds 1 or 2   15
15a. Leaf surfaces dissimilar; flowers
     always in pairs; follicle thin-walled
                               6. Grevillea
  b. Leaf surfaces similar or leaves terete
     (not grooved); flowers not regularly
     paired; follicle very woody   **7. Hakea**

1. BELLENDENA R. Brown
E.C. Nelson

Shrubs to 70 cm. Leaves toughened,
alternate, entire or lobed at apex, obovate
or broadly tapering, *c.* 3 × 1.5 cm, bright
green and sometimes glaucous. Racemes
terminal, on stout, erect stems; flowers
usually bisexual (some do not produce
viable pollen and are functionally female);
perianth pale pink or white, radially
symmetric, to 4 mm; segments separating
at maturity. Stamens not attached to
perianth-segments; filaments free, *c.* 4 mm;
hypogynous scales absent. Styles straight.
Fruit flattened, winged, red and fleshy
when young, membranous and red-orange
when mature, with 1 seed.

A genus of a single species, restricted to
the Tasmanian mountains, where it is
locally abundant above 1000 m; in its
native habitat it withstands frost and light
snow. It does not tolerate lime, requiring a
moist, acid, peaty soil. Propagation is by
seed.

1. B. montana R. Brown. Illustration:
Curtis & Stones, Endemic flora of Tasmania
3: 106 (1971); Elliott & Jones, Encyclopedia
of Australian plants **2**: 316 (1982).
Australia (Tasmania). H5. Summer.

2. PROTEA Linnaeus
E.C. Nelson

Shrubs or rarely small trees, erect or
prostrate, some with woody underground
lignotubers. Leaves toughened, entire.
Inflorescences terminal, solitary, rarely
clustered and axillary, surrounded by an
involucre of coloured bracts; bracts hairless
or hairy, often with bearded apices;
receptacle flat. Flowers bisexual, bilaterally
symmetric; perianth of 4 segments, tubular
in bud, 3 upper segments remaining fused
throughout, forming a sheath, the other
separating at flowering and soon
deciduous. Anthers stalkless, attached to
perianth-limb, all 4 fertile, or the
uppermost sterile and reduced to a linear
staminode. Style straight or curved,
tapering to a terminal pollen presenter with
stigmatic groove at apex. Fruit an achene.

A genus of about 115 species, of which
at least 82 occur in South Africa; the
others occur north of the Limpopo river in
tropical Africa. Many of the species from
the Cape Province are especially attractive
as garden plants, but they require acid,
well-drained, frost-free conditions.
Propagation is easiest by seed.
Literature: Rourke, J.P., *The Proteas of
Southern Africa* (1982).

1a. Leaves giving off sulphurous odour
    when crushed   **15. susannae**
  b. Leaves not as above   2
2a. One anther reduced to a linear
    staminode, the other 3 fertile   3
  b. Anthers all fertile   9
3a. Leaves narrow-linear, 9–20 × 2 cm;
    flowers densely covered with long
    silky white and purple-black hairs
                          8. longifolia
  b. Leaves broader and not linear; flowers
    not as above   4
4a. Leaves with long, distinct stalks
                          4. cynaroides
  b. Leaves without distinct stalks   5
5a. Mature inflorescences cup- or bowl-
    shaped (diameter greater than height)
                                         6
  b. Mature inflorescences goblet-shaped
    (diameter less than height)   7
6a. Bracts of inflorescence hairless
                             2. caffra
  b. Bracts of inflorescence hairy
                          11. punctata
7a. Stigmas longer than inflorescence
    bracts   **1. aurea**
  b. Stigmas shorter than inflorescence
    bracts   8
8a. Bracts prominently bearded at tips
                          6. grandiceps
  b. Bracts not bearded, but margins with
    cilia   **3. compacta**
9a. Mature inflorescences not more than
    6 cm in diameter   10
  b. Mature inflorescences more than
    6 cm in diameter   11
10a. Leaves less than 1 cm broad
                        14. scolymocephala
  b. Leaves broader than 1 cm
                           7. lacticolor
11a. Perianth with black hairs
                           8. longifolia
  b. Perianth without black hairs   12
12a. Inner bracts distinctly spathulate 13
  b. Inner bracts not spathulate   14
13a. Perianth with purple awns; leaves
     broad, base cordate   **5. eximia**
  b. Perianth without awns; leaves
     narrow, base not cordate
                          13. roupelliae
14a. Bracts recurved at tips, bearded
                           9. neriifolia
  b. Bracts with erect tips   15
15a. Leaves 5–15 cm × 5–20 mm; bracts
     with silky hairs   **10. obtusifolia**
  b. Leaves 10–15 × 2–4 cm; bracts
     hairless   **12. repens**

1. P. aurea (Burman) Rourke (*P. longiflora*
Lamarck). Illustration: Botanical Magazine,
2720 (1827); Flowering Plants of Africa
43: t. 1704 (1974); Rourke, The Proteas of
southern Africa, 173, 175 (1982); Vogts,
South Africa's Proteaceae, 87 (1982).
Shrub to 5 m. Leaves oblong to ovate,
1.5–4 cm × 4–9 mm, apex acute or obtuse,
hairless or becoming so. Inflorescence
terminal, cylindric in bud, to 12 cm,
obconical when open; inner involucral
bracts to 9 × 1.5 cm, shorter than styles,
fringed with silky hairs, reflexing at
maturity, cream-green to crimson. Fertile
stamens 4. Styles to 10 cm, straight, cream
to crimson. *South Africa (Cape Province).*
H5. Spring–summer.

One of the most attractive species,
growing rapidly and flowering within a few
years of germination; however, rather
short-lived. Cream-, pink- and crimson-
flowered variants are available.

2. P. caffra Meisner. Illustration: Rourke,
The Proteas of southern Africa, 51 (1982);
Vogts, South Africa's Proteaceae, 88
(1982).
Shrub or small tree to 3 m. Leaves elliptic,
7–25 cm × 5–45 mm, apex acute or
obtuse, hairless, pale to dark green,
sometimes glaucous, clustered.
Inflorescences solitary or 3–4 clustered at
the tips of the shoots, spherical to ovoid in
bud, to 8 cm in diameter, shallowly cup-

shaped when open; involucral bracts to 5 cm, shorter than styles, pink or red towards the tips, green below. Fertile stamens 4. Styles to 6 cm, curved inwards, pink. *South Africa (Cape Province, Natal, Transvaal), Lesotho.* H5. Spring–summer.

A variable species with a range of flower-colours, but not spectacular. Some variants may tolerate low temperatures and dry, cold winters.

3. P. compacta R. Brown. Illustration: Flowering Plants of South Africa **3**: t. 84 (1923); Rourke, The Proteas of southern Africa, 109 (1982); Vogts, South Africa's Proteaceae, 89 (1982).
Erect shrub to 3.5 m. Leaves stalkless, oblong to elliptic, 5–13 × 2–5.5 cm, becoming hairless, leathery, margins prominently horny, apex obtuse. Inflorescences oblong in bud, 9–12 cm, obovoid when open, 7–10 cm in diameter; involucral bracts to 10 × 1.5 cm, exceeding the styles, bright pink (rarely white), margins ciliate. Fertile stamens 4. Styles slightly curved inwards, to 7 cm, pink. *South Africa (Cape Province).* H5. Spring–summer.

This beautiful species is easily raised from seeds, quick to flower and it is exploited commercially in South Africa.

4. P. cynaroides (Linnaeus) Linnaeus. Illustration: Flowering Plants of South Africa **6**: t. 231 (1926); Rourke, Proteas of southern Africa, 83, 85 (1982); Vogts, South Africa's Proteaceae, 91, 93 (1982); Batten, Flowers of southern Africa, 2 (1986).
Shrub 30–200 cm, with few, massive stems, with lignotuber. Leaves leathery, hairless, blade round or elliptic, 8–14 × 2–13 cm, stalk 4–18 cm. Inflorescence goblet- or bowl-shaped when open, 12–30 × 12–20 cm, involucral bracts lanceolate, to 12 cm, rich crimson to pink or cream-green. Fertile stamens 4. Style 8–9.5 cm, shorter than involucral bracts, straight. *South Africa (Cape Province).* H5. Summer.

The most widespread of the Cape species, its flowering head resembles the Globe Artichoke (species of *Cynara*). The colour varies from deep red to almost white.

5. P. eximia (Knight) Fourcade. Illustration: Rourke, The Proteas of southern Africa, 107 (1982); Vogts, South Africa's Proteaceae, 94 (1982).
Shrub or small tree to 5 m, sparsely branched, stems upright. Leaves ovate, 6–10 × 3–6.5 cm, base cordate, leathery, hairless, usually glaucous grey-green to

purple-green. Inflorescence oblong to obconical, to 14 × 12 cm at flowering; involucral bracts 4–10 cm × 8–15 mm, inner and outer series clearly differentiated, margins ciliate; inner bracts spathulate, splayed at flowering, red-pink. Fertile stamens 3. Style 6–7.5 cm, pink, shorter than bracts. *South Africa (Cape Province).* H5. Spring–summer.

A fast-growing species that benefits greatly from the removal of old flower-heads, and the nipping out of shoot-tips to encourage branching.

6. P. grandiceps Trattinick. Illustration: Rourke, The Proteas of southern Africa, 139 (1982); Vogts, South Africa's Proteaceae, 95, 96 (1982); Batten, Flowers of southern Africa, 322 (1986).
Shrub to 2 m. Leaves erect, stalkless, ovate to obovate, 8–13 × 3–8.5 cm, leathery, hairless, often with red margins. Inflorescence oblong, goblet-shaped, to 14 × 8 cm; involucral bracts coral-pink, to 8 × 2 cm, spathulate, tips incurved and prominently bearded with white and/or purple hairs. Fertile stamens 4. Style curved, to 7.5 cm, shorter than bracts. *South Africa (Cape Province).* H5. Summer.

A spectacular plant though the flower colour seems to depend on light intensity – plants raised inside tend to have green bracts, so they should be kept outdoors during summer. In the wild, reaching altitudes of 1700 m, and thus tolerant of sharp, brief frosts.

7. P. lacticolor Salisbury. Illustration: Rourke, The Proteas of southern Africa, 167 (1982); Vogts, South Africa's Proteaceae, 98 (1982).
Shrub or small tree to 6 m. Leaves blue-green, ascending, lanceolate, with truncate base, 7–11 × 2.5–5 cm, apices more or less acute. Inflorescence oblong, to 8 × 6 cm; involucral bracts ivory, cream or rich pink, outer sides downy, margins with long silky hairs, especially at apices, to 5 × 1.5 cm, incurved at tips. Fertile stamens 4. Style straight, to 7 cm, equalling bracts. *South Africa (Cape Province).* H5. Spring–summer.

A species hitherto much confused and plants grown under this name may not represent the true species.

8. P. longifolia Andrews (*P. minor* (Phillips) Compton). Illustration: Flowering Plants of Africa **37**: t. 1426 (1966); Rourke, The Proteas of southern Africa, 117 (1982); Vogts, South Africa's Proteaceae, 101 (1982).
Shrub to 1.5 m, sprawling. Leaves ascending, linear, 9–20 cm × 5–17 mm.

Inflorescence variable in size, to 16 × 9 cm, oblong in bud, conical when open; involucral bracts green, white or pink, hairless except for ciliate margins, to 12 × 1.5 cm, acute, papery when dry. Flowers with a dense covering of white silky hairs (those on perianth-limb purple-black towards the tips). Fertile stamens 3. Styles to 6 cm, straight. *South Africa (Cape Province).* H5. Summer.

Hybridises freely with other species. Short-lived in cultivation although easy to grow.

9. P. neriifolia R. Brown. Illustration: Rourke, The Proteas of southern Africa, 124 (1982); Vogts, South Africa's Proteaceae, 105 (1982).
Shrub to 3 m, erect. Leaves dark to bright green, ascending, stalkless, narrow, oblong with parallel sides, 10–18 × 1.4–3 cm, apex obtuse to acute, hairless when mature. Inflorescence goblet-shaped when open, to 13 × 8 cm; involucral bracts white to deep pink, oblong to spathulate, to 14 × 1.5 cm; tips incurved and very densely bearded with black (and occasionally white) hairs. Fertile stamens 3. Style straight, to 7 cm, shorter than bracts. *South Africa (Cape Province).* H5. Spring–summer.

Has a range of flower colours and a wide tolerance of growing conditions.

10. P. obtusifolia Meisner. Illustration: Rourke, The Proteas of southern Africa, 111 (1982); Vogts, South Africa's Proteaceae, 107 (1982).
Shrub or small tree to 4 m. Leaves ascending, lanceolate to elliptic, broadest towards the tip, becoming hairless, 10–15 × 2–4 cm, leathery. Inflorescence goblet-shaped when open; involucral bracts to 10 × 1.2 cm, with fringed margins, cream-green to deep pink. Fertile stamens 3. Style to 7 cm, straight, shorter than bracts. *South Africa (Cape Province).* H5. Spring–summer.

Displays a wide range of flower-colour. It grows in coastal regions in the wild and will tolerate alkaline soils (pH 8.4 is reported); Rourke records that it is one of the few lime-tolerant species, but cautions that it is not frost-hardy. It does grow well in coastal gardens.

11. P. punctata Meisner. Illustration: Rourke, The Proteas of southern Africa, 169 (1982); Vogts, South Africa's Proteaceae, 167 (1982).
Shrub or small tree, to 4 m, erect. Leaves stalkless, hairy when young but becoming hairless, grey-green, oval to obovate, 3.5–8.5 × 2–4.5 cm. Inflorescence-bud

cylindric, to 6 × 2–2.5 cm, opening into a bowl-shaped head to 9 cm in diameter, with bracts spreading horizontally; bracts pale pink or whitish, oblong, to 4 × 1.5 cm, with ciliate margins. Fertile stamens 4. Style straight, to 5 cm, remaining erect after flowering. *South Africa (Cape Province)*. H5. Spring–summer.

Not tolerant of humid conditions; it is most suited to cultivation in semi-arid gardens on well-drained sites in full sun.

12. P. repens (Linnaeus) Linnaeus. Illustration: Rourke, The Proteas of southern Africa, 97 (1982); Vogts, South Africa's Proteaceae, 112 (1982).
Shrub or small tree to 4 m; stems upright. Leaves erect, linear to lanceolate, 5–15 cm × 5–20 mm, hairless, with acute or rounded apices. Inflorescence bud obovoid, to 6 cm, opening into a goblet-shaped head to 9 cm in diameter; involucral bracts hairless but coated with sticky resin, to 11 × 2 cm, acute, uniformly cream-white or tipped with dark red to pink. Fertile stamens 3. Style slightly curved, to 9 cm, shorter than bracts. *South Africa (Cape Province)*. H5. Spring–summer.

Despite its name, this is not a creeping plant. It was probably the first *Protea* species to flower in cultivation in Europe, growing at Kew about 1780. It tolerates a wide range of soil conditions.

13. P. roupelliae Meisner. Illustration: Rourke, The Proteas of southern Africa, 105 (1982); Vogts, South Africa's Proteaceae, 114 (1982).
Tree to 8 m, or (subsp. **hamiltonii** Rourke) a low sprawling shrub to 30 cm. Leaves ascending, lanceolate to obovate, 8–17 × 1.5–5 cm, stalkless, varying from hairy and silvery to hairless and green. Inflorescences goblet-shaped when open, to 12 × 10 cm; involucral bracts with silky hairs, outer bracts short, to 1.5 cm, apex becoming brown, dry and ragged; inner bracts 4–10 × 1–1.5 cm, innermost spathulate with acute apices, creamy pink to red-pink. Fertile stamens 3. Styles curved inwards, to 6.5 cm, shorter than bracts. *South Africa (Cape Province, Natal, Transvaal), Lesotho, Swaziland*. H5. Spring–summer.

14. P. scolymocephala (Linnaeus) Reichard. Illustration: Rourke, The Proteas of southern Africa, 221 (1982); Vogts, South Africa's Proteaceae, 178 (1982).
Shrub, erect and well branched, to 1.5 m. Leaves linear to spathulate, 3.5–9 cm × 3–6 mm, hairless, with acuminate apices. Inflorescence bud spherical, to 3 cm in diameter, opening into a bowl-shaped flower-head to 4.5 cm in diameter, bracts eventually becoming slightly reflexed; involucral bracts ciliate, to 2.5 cm × 7 mm, apices rounded and concave, creamy green with pink tips. Fertile stamens 4. Style strongly curved inwards, to 2.5 cm. *South Africa, (Cape Province)*. H5. Spring–summer.

Full sun and good drainage are essential.

15. P. susannae Phillips. Illustration: Rourke, The Proteas of southern Africa, 113 (1982); Vogts, South Africa's Proteaceae, 122 (1982).
Shrub to 4 m, loose. Leaves oblong, 8–16 × 1.5–3 cm, apex obtuse, becoming hairless, leathery with horny margins. Inflorescence goblet-shaped at flowering, to 10 × 11 cm; involucral bracts brownish pink with sticky brown resin on outer surfaces, to 8 × 2 cm, innermost spathulate with rounded, slightly concave apices. Fertile stamens 3. Style straight, to 7 cm, shorter than bracts. *South Africa (Cape Province)*. H5. Spring–summer.

The flower-heads are succinctly described as appearing to 'have been lightly brushed with brown shoe polish'. Tolerant of alkaline sandy soil and coastal conditions, though the leaves exude an offensive sulphurous odour.

3. LEUCOSPERMUM R. Brown
E.C. Nelson
Small trees or shrubs, erect or sometimes prostrate. Leaves alternate, toughened. Inflorescences axillary heads. Perianth bilaterally symmetric, tubular in bud, claws of the 3 upper segments remaining fused after flowering; anthers stalkless or almost so, all fertile; hypogynous scales 4. Style straight or curved with terminal pollen presenter and stigmatic groove. Fruit a 1-seeded nut.

This Southern African genus contains 47 species, native to South Africa and Zimbabwe. They require well-drained, lime-free soil. Propagation by seed.
Literature: Rourke, J.P., Taxonomic studies on Leucospermum R. Br., *Journal of South African Botany* supplement volume **8** (1972).

1a. Leaves 2–4.5 cm broad, ovate; inflorescences at right angles to stem **1. cordifolium**
b. Leaves 5–30 mm broad, linear to lanceolate; inflorescences not at right angles to stem **2. cuneiforme**

1. L. cordifolium (Knight) Fourcade. Illustration: Vogts, South Africa's Proteaceae, 132 (1982); Batten, Flowers of southern Africa, 258 (1986).
Shrub to 2 m; flowering stems spreading. Leaves ovate, entire or with 3–6 apical teeth, 2–8 × 2–4.5 cm, base cordate, downy becoming hairless. Inflorescence spherical, c. 12 cm in diameter, borne at right angles to the stem. Perianth yellow, orange or crimson, tube hairless, claws slightly hairy; pollen-presenter obliquely top-shaped. *South Africa (Cape Province)*. H5. Summer.

2. L. cuneiforme (Burman) Rourke (*L. attenuatum* R. Brown; *L. ellipticum* (Thunberg) R. Brown). Illustration: Vogts, South Africa's Proteaceae, 133 (1982).
Shrub to 3 m, with lignotuber; flowering stems erect. Leaves linear to lanceolate, 4.5–11 cm × 5–30 mm, base tapering, apex with 3–11 teeth, hairless. Inflorescence ovoid, c. 9 cm in diameter; perianth yellow becoming orange, tube hairless, claws slightly hairy; style slightly curved, yellow becoming orange; pollen-presenter conical. *South Africa (Cape Province)*. H5. Summer.

4. LEUCADENDRON R. Brown
E.C. Nelson
Dioecious shrubs or trees to 10 m, occasionally with lignotubers. Leaves toughened, entire, downy when young. Inflorescences terminal, solitary, surrounded by involucral leaves which are larger than the stem-leaves and coloured at flowering. Male flowers with 4 stalkless anthers; hypogynous scales 4 (rarely absent); style slender, downy or hairless, with terminal pollen-presenter, stigma abortive. Female flowers with 4 rudimentary staminodes which are stalkless at the base of the perianth; hypogynous scales 4 (rarely absent); style slender, hairless. Fruiting head cone-like, with thick woody bracts which conceal the fruits. Fruit a nut or samara, released at maturity or after a few years.

There are over 90 species in this genus, which is endemic to South Africa. Many of the species are cultivated in Africa.

Like other genera in the family, *Leucadendron* requires acid, well-drained soil and is propagated principally by seed.
Literature: Williams, I.J.M., A revision of the genus Leucadendron (Proteaceae), *Contributions from the Bolus Herbarium* **3** (1972).

1a. Leaves large, c. 15 × 2 cm, silvery, hairy; inflorescence also silvery **1. argenteum**
b. Leaves and inflorescence not as above

2a. Inflorescence with conspicuous brown, oily bracts **5. microcephalum**
 b. Inflorescence without brown, oily bracts 3
3a. Leaves terete, linear, mucronate **8. teretifolium**
 b. Leaves not terete 4
4a. Leaves over 10 cm, not more than 8 mm broad **3. eucalyptifolium**
 b. Leave shorter and broader 5
5a. Male plants with leaves less than 3.5 cm × 5 mm, hairless; involucral leaves absent **6. rubrum**
 b. Male plants with leaves larger; involucral leaves present 6
6a. Leaves more than 1 cm broad **4. laureolum**
 b. Leaves less than 1 cm broad 7
7a. Leaves soft, silvery, sharply pointed **9. xanthoconus**
 b. Leaves leathery, not silvery 8
8a. Leaves rarely more than 6 cm; shoots hairless **7. salignum**
 b. Leaves to 8.5 cm; young shoots downy **2. coniferum**

1. L. argenteum (Linnaeus) R. Brown. Illustration: Rousseau, The Proteaceae of South Africa, 91 (1970); Vogts, South Africa's Proteaceae, 160 (1982).
Tree to 10 m. Leaves lanceolate with acute, mucronate apex, with adpressed silver hairs, to 15 × 2 cm. Male inflorescence spherical, 4 × 5 cm; female inflorescence spherical, to 5 × 4 cm. Fruiting cones silvery, to 9 × 6 cm, bracts in 3 ascending spirals; fruits retained for several years. *South Africa (Cape Province, Cape peninsula).* H5. Summer.
Cultivated principally for its silver foliage.

2. L. coniferum (Linnaeus) Meisner. Illustration: Vogts, South Africa's Proteaceae, 166 (1982).
Shrub to 2 m, rarely a small tree. Leaves crowded at tips of branches, yellow-green (involucral leaves yellow), linear-lanceolate, acute, hairless, twisted through 90° at base, in male plants to 7.5 cm × 7 mm, in female to 8.5 × c. 1 cm. Male inflorescence 2.8 × 1.8 cm; female 2.7 × 1.4 cm, silvery green with yellow-green flowers. Fruiting cone red, 4–5 × 3 cm, ellipsoid, not dehiscent. *South Africa (Cape Province).* H5. Summer.
This species thrives in alkaline, sandy soil; it was first cultivated in European gardens early in the 18th century.

3. L. eucalyptifolium Meisner. Illustration: Rousseau, The Proteaceae of South Africa, 82 (1970); Vogts, South Africa's Proteaceae, 173 (1982).

Tall shrub to 5 m. Leaves linear-lanceolate, to 11 cm × 8 mm, acute with mucronate tip, downy when young, twisted through 90° at base; involucral leaves few, broadened at base, yellow. Male inflorescence conical, to 2.5 × 1.6 cm; female ovoid, silver-green with yellow flowers, to 2.3 × 1.2 cm. Fruiting cone to 4.5 × 2 cm, not dehiscent; fruits retained for many years. *South Africa (Cape Province).* H5. Summer.
Differs from *L. coniferum* in having narrower, downy leaves.

4. L. laureolum (Lamarck) Fourcade. Illustration: Rousseau, The Proteaceae of South Africa, 73 (1970); Vogts, South Africa's Proteaceae, 180 (1982).
Shrub to 2 m, male plants forming larger, more conspicuous bushes; female plants sparse. Leaves in male plants to 7.5 × 1.5 cm, in female to 9.5, oblong, obtuse, with red, recurved tip, becoming hairless. Male inflorescence spherical, to 2.3 × 2 cm, surrounded by a yellow involucral leaves (innermost broader than the stem-leaves); female inflorescence long, to 2.7 × 1.4 cm, concealed by the green-yellow involucral leaves. Fruiting cone to 4.5 × 3.5 cm, distinctively 8-sided; fruits retained for several years. *South Africa (Cape Province).* H5. Summer.

5. L. microcephalum (Gandoger) Gandoger & Schinz (*L. stokoei* Phillips). Illustration: The Flowering Plants of South Africa 1: t. 7, 8 (1921); Vogts, South Africa's Proteaceae, 184 (1982).
Shrub to 1.5 m; foliage of male plants bright yellow in flowering season, female plants only with yellow involucral leaves. Leaves oblong, obtuse, hairless, to 9 × 2 cm; involucral leaves longer and broader, crowded. Inflorescence spherical, with conspicuous, oily brown bracts around base in both sexes, buds very sticky; male to 1.5 × 1.8 cm, female to 1.6 × 1 cm. Fruiting cone 4 × 3 cm, bracts hairless, brown, shining; fruit heart-shaped, retained in cone. *South Africa (Cape Province).* H5. Summer.

6. L. rubrum Burman. Illustration: Vogts, South Africa's Proteaceae, 196 (1982).
Shrub to 2 m with erect branches. Male plant more densely branched and with smaller foliage than female. Leaves obovate, in male to 3.4 cm × 5 mm, in female to 7 cm × 8 mm, twisted towards base, becoming hairless, involucral leaves absent. Male inflorescences numerous, c. 1 cm × 5 mm, clustered; female inflorescences ovoid, to 4 × 2 cm, tinted purple, with

acute apex, stigmas protruding from apex in a tuft. Fruiting cone conical, bracts acute with red margins, fruits plumed. *South Africa (Cape Province).* H5. Summer.

7. L. salignum Bergius (*L. adscendens* R. Brown). Illustration: Rousseau, The Proteaceae of South Africa, 74 (1970); Vogts, South Africa's Proteaceae, 197–199 (1982).
Shrub to c. 1 m, with lignotuber. Leaves linear-lanceolate, acute, hairless, twisted below, in male plants 2–5 cm × 5 mm, in female plants to 6 cm. Male inflorescence spherical, to 1.5 cm in diameter, surrounded by yellow involucral leaves longer than stem-leaves; female inflorescence ovoid, to 1.5 × 1.2 cm, surrounded by ivory-coloured involucral leaves, often broader than stem-leaves and then concealing the inflorescence. Fruiting cone spherical, to 2 cm in diameter. *South Africa (Cape Province).* H5. Summer.
Plants with red flower buds are also reported; the flower buds are yellow in the predominant variant.

8. L. teretifolium (Andrews) Williams. Illustration: Vogts, South Africa's Proteaceae, 204 (1982).
Shrub to 1 m. Leaves terete, linear, mucronate, in male plants to 8 × 1 mm, in female to 2.2 cm × 1.3 mm. Inflorescences without involucral leaves; basal bracts brown, lanceolate; male inflorescences spike-like, to 1.4 cm × 7 mm; female inflorescence ovoid, to 2 × 1.2 cm, green. Fruiting cone to 3.5 × 2.5 cm, hairless, green; fruit retained. *South Africa (Cape Province).* H5. Summer.

9. L. xanthoconus (Kuntze) Schumann. Illustration: Rousseau, The Proteaceae of South Africa, 75 (1970); Vogts, South Africa's Proteaceae, 209 (1982).
Shrub to 2 m. Leaves soft, linear-oblong, to 6.5 cm × 6 mm, when young with silvery hairs, becoming hairless; involucral leaves yellow, crowded, longer than inflorescences. Male inflorescence ellipsoid, to 1.3 × 1 cm; female ovoid, to 1.8 × 1.1 cm. Fruiting cone to 3 × 2 cm, bracts bilobed. *South Africa (Cape Province).* H5. Summer.

5. AULAX Bergius
E.C. Nelson
Dioecious shrubs. Leaves toughened, alternate, entire, hairless. Male inflorescence a loose, terminal raceme; perianth yellow, tubular, segments separating at maturity; anthers more or less stalkless; hypogynous scales absent;

style terete with an abortive, terminal stigma. Female inflorescence a dense, cone-like raceme surrounded by a persistent whorl of flattened bract-like branchlets; perianth yellow, tubular, lower half of segments remaining fused at maturity; staminodes 4; ovary densely downy, style slightly curved, stigma enlarged, oblique. After fertilisation the involucre becomes woody, enclosing the compressed fruits; each achene fringed with longitudinal lines of silky hairs.

Three species found only in South Africa (Cape Province), requiring acid, well-drained soil and protection from frost. Propagation by seed.

1a. Leaves broad, spathulate, to 1.5 cm across **1. umbellata**
 b. Leaves narrow, linear **2. cancellata**

1. A. umbellata (Thunberg) R. Brown (*A. cneorifolia* Knight). Illustration: Vogts, South Africa's Proteaceae, 212 (1982). Leaves obovate to spathulate, to 1.5 cm broad, apex obtuse. *South Africa (Cape Province)*. H5. Summer.

2. A. cancellata (Linnaeus) Druce (*A. pinifolia* Bergius). Illustration: Flowering Plants of South Africa 7: pl. 244 (1927); Vogts, South Africa's Proteaceae, 211 (1982). Leaves linear, broadening slightly towards tip, curved, apex acute. *South Africa (Cape Province)*. H5. Summer.

A. pallasia Stapf. Illustration: Vogts, South Africa's Proteaceae, 211 (1982). Leaf-shape intermediate between **1 & 2**. Female flowers produced on involucral branches and on central axis of the inflorescence. *South Africa (Cape Province)*. H5. Summer.

6. GREVILLEA R. Brown

E.C. Nelson

Trees and shrubs, some with persistent lignotubers. Leaves alternate, toughened, simple or divided. Flowers bisexual, paired in terminal or axillary inflorescences. Perianth radially symmetric, segments remaining fused at maturity; hypogynous scale 1, more or less ring-like. Style with conical or disc-shaped pollen-presenter. Ovary shortly stalked, with 2 ovules. Fruit a follicle, not woody, seeds with or without wings.

Grevillea contains perhaps 250 species mostly natives of Australia; there are a few species in New Guinea, Indonesia and New Caledonia. Some are easy to cultivate; the silky oak (*G. robusta*) is treated as an annual in northern Europe and widely used as a plant for interior landscaping. Other species are more difficult, requiring well-drained, lime-free soil; they are sensitive to root damage and to fertilisers. Most are easily propagated from seed, but named cultivars should only be raised vegetatively.

Some excellent cultivars have been produced in Australia during the last decade, but few of these are currently available in Europe.

Literature: this treatment is based on works available in September 1987; a review of *Grevillea* will be included in a forthcoming volume of the *Flora of Australia* and may result in changes of names.

1a. Leaves entire 2
 b. Leaves pinnate 6
2a. Leaves rigid with sharply pointed tips 3
 b. Leaves leathery, not sharply pointed 5
3a. Leaves *c.* 5 cm, margins rolled under; perianth hairless outside, ovary hairless **8. rosmarinifolia**
 b. Leaves and perianth not as above 4
4a. Leaves *c.* 1.2 cm; perianth *c.* 5 mm, cream-white **2. australis**
 b. Leaves 1–2 cm; perianth more than 1 cm, red **6. juniperina**
5a. Leaves 1–2 cm; perianth orange-red, downy **1. alpina**
 b. Leaves 3–10 cm; perianth red with brown hairs **11. victoriae**
6a. Leaves more than 10 cm 7
 b. Leaves less than 10 cm 9
7a. Leaf-segments linear, with parallel sides 8
 b. Leaf-segments broad, sides not parallel **7. robusta**
8a. Flowers red in broad, cylindric spikes **3. banskii**
 b. Flowers white in dense spikes **5. hilliana**
9a. Flowers white; leaves divided at least twice into numerous terete segments **4. biternata**
 b. Flowers red; leaves not as above 10
10a. Style and ovary hairless **9. thelmanniana**
 b. Style and ovary downy **10. thyrsoides**

1. G. alpina Lindley (*G. alpestris* Meisner). Illustration: Botanical Magazine, 5007 (1857); Cochrane et al., Flowers and plants of Victoria, 311 (1967). Shrub. Leaves entire, linear to round, *c.* 1–2 cm × 3 mm, densely hairy above, not sharply pointed. Flowers fewer than 20, in loose clusters; stalks slender. Perianth sparsely downy outside, bearded inside, usually orange-red (occasionally pink), ovary hairy. *Australia (Victoria, New South Wales)*. H5. Most of the year.

A very variable species in both foliage and flower-colour, and one of the hardiest.

2. G. australis R. Brown. Illustration: Costin, Kosciusko alpine flora, 149, 150 (1979). Shrub, erect, sprawling or prostrate; shoots with brown hairs. Leaves leathery, rigid, oblong to linear, 1–2 cm × 1–5 mm, margins recurved, hairless above, downy beneath, with sharply pointed apex. Flowers fragrant, in umbel-like racemes. Perianth cream-white, 4–6 mm, with brown hairs outside. Style hairless, curved, less than 1 cm. Fruit hairless, ovoid, with persistent style. *Australia (alpine regions of New South Wales, Victoria, Tasmania)*. H5. Summer.

3. G. banksii R. Brown. Illustration: Botanical Magazine, 5870 (1870); Wrigley & Fagg, Australian native plants, 217 (1979). Shrub to 3 m (prostrate variants are also known). Leaves leathery, pinnate, to 15 cm, segments linear, entire (rarely lobed), silky beneath. Flowers in erect, comb-like racemes to 18 × 4 cm. Perianth hairless outside, downy inside, bright red (rarely cream). Style hairless, ovary downy. *Australia (Queensland, northern New South Wales)*. H5. Summer.

As well as prostrate variants, a cultivar 'Kingaroy Slipper' has been named in Australia; it retains the perianth attached to the style after flowering; a beautiful cultivar with dark foliage and large, conspicuous flowers.

4. G. biternata Meisner. Illustration: Wrigley & Fagg, Australian native plants, 64 (1979). Prostrate or erect shrub to 1 m. Shoots with red-brown hairs. Leaves rigid, 4–6 cm, hairless, divided into 2–3 terete segments; segments also divided 2–3 times, ultimate divisions terete, 2–4 cm long, each with sharply pointed tip. Fragrant flowers in stalkless, axillary racemes. Perianth white, straight, hairless, to 5 mm. Ovary hairless. *Australia (South Australia)*. H5. Spring.

The application of this name seems to be confused.

5. G. hilliana Mueller. Illustration: Botanical Magazine, 7524 (1897); Maiden, Forest Flora of New South Wales 5: 159 (1913).

Tree to 20 m. Leaves leathery, entire or

pinnately lobed, to 30 cm, glossy and dark green above, with silky silvery hairs beneath; if lobed, lobes 3 or more, linear, 7–12 × 1.5 cm, obtuse. Inflorescence dense, to 25 cm, erect. Perianth white, hairy outside, hairless inside, *c.* 5 mm. Style and ovary hairless. *Australia (Queensland, New South Wales).* H5. Summer.

A rainforest species similar to *G. robusta* but with white flowers.

6. G. juniperina R. Brown (*G. sulphurea* Cunningham). Illustration: Edwards's Botanical Register **13**: t. 1089 (1829); Botanical Magazine, n.s., 761 (1978); Plantsman **6**(3): frontispiece (1984).
Shrub 1–2.5 m. Shoots becoming hairless. Leaves clustered on lateral shoots, leathery, narrow, linear-lanceolate, 1–1.7 cm, margins strongly rolled under, covering midrib beneath, hairless, apex sharply pointed. Flowers in terminal or lateral umbel-like racemes. Perianth pale to bright red or yellow, *c.* 1.2 cm, downy outside, densely hairy inside below middle. Style hairless. Fruits slender, spindle-shaped. *Australia (New South Wales).* H5. Most of the year.

This is represented in European gardens principally by the elegant yellow-flowered forma **sulphurea** (Cunningham) Ferguson (*G. juniperina* var. *sulphurea* (Cunningham) Bentham). The red-flowered form is not so common. The species apparently tolerates a wide range of conditions but not lime-rich soil.

The yellow form may be raised from seed or propagated by cuttings (bottom heat is required).

G. × semperflorens Mulligan. Illustration: Botanical Magazine, n.s., 353 (1960). An artifical hybrid between *G. juniperina* forma *sulphurea* and *G. thelmanniana*, raised in Devon. Shrub to 2 m. Leaves entire or divided; perianth orange-yellow and green; style rose pink. H5. Winter–spring.

7. G. robusta Cunningham. Illustration: Botanical Magazine, 3184 (1832); Maiden, Forest Flora of New South Wales **1**: 1 (1904); Krüssmann, Handbuch der Laubgehölze **2**: t. 3 (1976).
Tree to 35 m in the wild. Leaves leathery, obovate in outline, pinnate, to 30 cm, hairless above, with silky hairs beneath, segments 11–12 pairs, obovate to linear, deeply and irregularly toothed or lobed. Inflorescences axillary racemes, often 2–3 together, to 15 cm. Perianth orange-red, hairless outside. Style and ovary hairless. *Australia (Queensland).* H5. Spring.

This is raised in quantity every year from seed for use as a foliage plant indoors; it is rarely seen growing in gardens but forms a magnificent tree that blossoms profusely.

8. G. rosmarinifolia Cunningham. Illustration: Botanical Magazine, 5971 (1872); Cochrane et al., Flowers and plants of Victoria, 310 (1967).
Shrub to 1 m. Leaves leathery, blade flat, linear and needle-like to lanceolate, to 5 cm, margin rolled under, apex tapering to a sharp point. Flowers in axillary clusters. Perianth hairless outside, bearded inside, green, cream, pink or crimson, *c.* 7 mm. Style red, hairless, 1.5 cm. Ovary hairless. *Australia (New South Wales).* H5. Most of the year.

9. G. thelmanniana Heugel (*G. preissii* Meisner). Illustration: Botanical Magazine, 5837 (1870).
Prostrate or erect shrub, 20–100 cm. Leaves leathery, pinnate, segments linear, terete, erect or spreading, *c.* 3 cm, neither pointed nor rigid, green or blue-green. Flowers in long racemes to 8 cm. Perianth green in bud, becoming red, hairless or slightly hairy outside, bearded inside. Style red with green pollen-presenter, hairless. Ovary hairless. *Australia (Western Australia).* H5. Summer.

10. G. thyrsoides Meisner. Sprawling shrub. Leaves pinnate. Flowers in terminal recemes, perianth pink, hairy outside, style and ovary downy; pollen-presenter disc-shaped. *Australia (Western Australia).* H5. Summer.

11. G. victoriae Mueller. Illustration: Cochrane et al., Flowers and plants of Victoria, 506 (1967); Wrigley & Fagg, Australian native plants, 228 (1979); Costin, Kosciusko alpine flora, 151, 152 (1979).
Shrub to 2 m; shoots with grey-silver hairs. Leaves elliptic to lanceolate, 3–10 cm × 7–25 mm, with distinct stalk, margins slightly recurved, apex acute or obtuse, becoming hairless above, grey-silver hairs persisting beneath. Flowers in racemes 1–7 cm long. Perianth red with dense brown hairs outside and white beard inside below middle, 1.5–2 cm. Style hairless *c.* 1.5 cm. Ovary hairless. Fruit hairless. ellipsoid, style persistent. *Australia (alpine regions of New South Wales, Victoria).* H5. Summer.

Very variable in the wild.

7. HAKEA Schrader
E.C. Nelson
Trees or shrubs without lignotubers. Leaves toughened, alternate, entire or divided,

often sharply pointed. Inflorescence axillary, umbel- or raceme-like. Perianth-segments spreading at maturity. Anthers stalkless. Hypogynous gland 1, a semi-annular ring. Style curved, with terminal, disc-shaped or conical pollen-presenter and stigma. Fruit a follicle, often enlarged and woody with beak.

There are perhaps 150 species of *Hakea* all native to Australia, with a concentration in the southwest of Western Australia. Literature: the names used here follow works published before September 1987 but the publication of a review of *Hakea* in the forthcoming *Flora of Australia* could result in changes.

1a. Leaves entire, not divided   2
  b. Leaves divided into terete segments
                              14. suaveolens
2a. Leaves terete   3
  b. Leaves with flat blade   8
3a. Perianth hairless   4
  b. Perianth hairy   **3. epiglottis**
4a. Inflorescences with *c.* 6 flowers, stalks hairy   5
  b. Inflorescences with more than 6 flowers, or stalks not hairy   6
5a. Leaves 2–5 cm   **13. sericea**
  b. Leaves 3–15 cm   **7. lissosperma**
6a. Fruit with 2 beaks   **14. suaveolens**
  b. Fruit with a single beak   7
7a. Leaves with hairy apex **6. leucoptera**
  b. Leaves hairless   **8. microcarpa**
8a. Leaves very thick, ovate to oblong, without stalks   9
  b. Leaves leathery yet not thick, lanceolate or with stalks   10
9a. Leaves oblong, apex mucronate; flowers brown and cream
                              1. crassifolia
  b. Leaves oval, apex obtuse; flowers white   **2. elliptica**
10a. Leaves ovate with distinct stalk
                              11. petiolaris
  b. Leaves not as above   11
11a. Leaves lanceolate with single vein
                              4. eriantha
  b. Leaves with more than 1 vein or not lanceolate   12
12a. Leaves elliptic; flowers red, in spherical axillary clusters **5. laurina**
  b. Leaves and flowers not as above   13
13a. Leaves lanceolate, without visible veins   **12. salicifolia**
  b. Leaves with visible veins, or oblong
                                        14
14a. Leaves oblong; flowers white, in axillary clusters   **10. oleifolia**
  b. Leaves lanceolate; flowers white or red, in long racemes **9. minyma**

1. H. crassifolia Meisner. Illustration: George, An introdcution to the Proteaceae of Western Australia, 76 (1984).
Shrub to 5 m. Leaves leathery, very thick, oblong, 5–10 cm, smooth, mucronate. Flowers in small axillary clusters. Perianth brown and cream. Fruit spherical, 4–5 cm in diameter, often with smooth, speckled bark. *Australia (Western Australia)*. H5. Summer.

2. H. elliptica (J.E. Smith) R. Brown.
Shrub to 3 m. Leaves very leathery and thick, ovate, to 7 cm, stalkless, sometimes wavy, young foliage bronze. Flowers in axillary clusters. Perianth white. Style less than 1 cm. Pollen-presenter conical. *Australia (Western Australia)*. H5. Summer.

3. H. epiglottis Labillardière. Illustration: Curtis, Student's Flora of Tasmania **3**: 608 (1967); Curtis & Stones, Endemic Flora of Tasmania **6**: 466 (1978).
Dioecious shrub or small tree; young shoots downy, becoming hairless. Leaves rigid, terete, curved upwards, 2–10 cm, mucronate and sharply pointed. Sweetly scented flowers in stalkless axillary clusters. Perianth cream-yellow, silky outside, 4–5 mm. Style and ovary hairless. Pollen-presenter disc-shaped. Fruit S-shaped, recurved at base with short, erect beak, 2 × 1 cm, smooth or rough. *Australia (Tasmania)*. H5. Spring.

The flowers are functionally unisexual; the females have a short, broadly cylindric style in the centre of the disc. This is one of the hardiest species, growing out of doors in western Britain and Ireland.

4. H. eriantha R. Brown. Illustration: Maiden, Forest Flora of New South Wales **5**: 175 (1912).
Tree to 7 m. Leaves leathery, blade flat, narrow, lanceolate to elliptic, 7–12 × *c.* 1 cm, with single conspicuous mid-vein, shining. Perianth silky. Style with disc-shaped pollen presenter. Fruit *c.* 2.5 × 1.5 cm. *Australia (Victoria, New South Wales, Queensland)*. H5. Summer.

5. H. laurina R. Brown. Illustration: Botanical Magazine, 7127 (1890); Growing Native Plants **2**: 39 (1972); Gardner, Wild flowers of Western Australia, edn 2, 26 (1973); Erickson, Flowers and plants of Western Australia, 83 (1973).
Shrub to 6 m. Leaves leathery, narrow, elliptic, *c.* 10–15 cm, with several prominent veins. Flowers in globular heads 4–6 cm in diameter. Perianth cream at first, turning pink or red. Style white. Fruit 2–5 cm, with short, stumpy beak. *Australia (Western Australia)*. H5. Summer.

6. H. leucoptera R. Brown. Illustration: Maiden, Forest Flora of New South Wales, **6**: 198 (1914).
Shrub or small tree. Leaves rigid, terete, *c.* 8 cm × 2 mm, often hairy, with sharp apex. Inflorescence a short raceme. Perianth hairless, to 4 mm. Fruit to 2.5 cm, smooth or warty, with conical beak. *Australia (most of eastern part)*. H5. Summer.

7. H. lissosperma R. Brown (*H. acicularis* Knight var. *lissosperma* (R. Brown) Bentham; *H. sericea* var. *lissosperma* (R. Brown) Maiden & Betche).
Shrub ranging from dwarf, alpine plant to 6 m. Young shoots hairy. Leaves rigid, terete, 3–5 cm, upright, curved or spreading, with sharp points. Flowers in clusters of *c.* 6–8. Perianth white, 5–6 mm, hairless or nearly so. Flower-stalks 4–6 mm, downy. Fruit with warty surface and small beak. *Australia (Victoria, New South Wales, Tasmania)*. H5. Summer.

Similar to *H. sericea*, but in this species the leaves are generally thicker.

8. H. microcarpa R. Brown (*H. microphylla* invalid). Illustration: Loddiges' Botanical Cabinet, 219 (1818); Edwards's Botanical Register **6**: t. 475 (1820); Burbidge & Gray, Flora of the Australian Capital Territories, 145 (1970).
Shrub to 2 m. Leaves terete or with flat blade, 3–6 cm, ascending or erect, hairless, with sharp points. Inflorescence-axis slightly downy. Flower-stalks hairless. Perianth 4–5 mm, hairless. Style and ovary hairless. Pollen-presenter disc-shaped. Fruit 2 cm × 5 mm. *Australia (Victoria, Tasmania)*. H5. Summer.

H. ulicina R. Brown. Illustration: Cochrane et al., Flowers and plants of Victoria, 12 (1968). Similar but leaves sometimes silky, 2.5–20 cm and pollen-presenter conical. *Australia (Victoria, Tasmania, New South Wales, South Australia)*. H5. Summer.

9. H. minyma Maconchie. Illustration: Transactions of the Royal Society of South Australia **97**: 131 (1973); Jessop, Flora of central Australia, 21 (1981).
Shrub 1–2 m. Leaves leathery, lanceolate to narrowly elliptic, flat, 8–15 cm × 5–8 mm, hairless, with 14–17 veins. Inflorescence a raceme, axis hairless, 3–5 cm. Perianth hairless, creamy white. Style and ovary hairless. Fruit ovoid to spherical with very prominent beak. *Australia (central)*. H5. Summer.

The specific name is an Australian Aboriginal word for woman, and alludes to

the fruit's resemblance to a woman's breast.

H. bucculenta Gardner. Illustration: Gardner, Wild flowers of Western Australia, edn 11, 26 (1973); Erickson, Flowers and plants of Western Australia, 108 (1973); Growing Native Plants **12**: 286 (1983).
Shrub to 4 m; leaves to 15 cm with 1 distinct central vein; inflorescence to 15 cm; perianth red. *Australia (Western Australia N of Perth)*. H5. Summer.

H. multilineata Meisner. Illustration: Gardener's Chronicle **19**: 85 (1896); Maconchie, Australian plants, 7: 304–306 (1974). Like *H. bucculenta* but leaves with numerous veins. *Australia (SW Western Australia)*. H5. Summer.

10. H. oleifolia (J.E. Smith) R. Brown.
Shrub to 6 m. Leaves leathery, oblong, to 7 cm, green. Flowers in axillary clusters. Perianth white. *Australia (Western Australia)*. H5. Summer.

11. H. petiolaris Meisner. Illustration: Erickson, Flowers and plants of Western Australia, 133 (1973); George, An introduction to the Proteaceae of Western Australia, 75 (1984).
Shrub to 6 m. Leaves leathery, elliptic to broadly ovate, 6–12 cm, grey-green. Flowers in dense spherical heads 3.5–5 cm in diameter, on old wood. Perianth dull red. Style white. Fruit 2.5–3 cm, beaked. *Australia (Western Australia)*. H5. Summer.

12. H. salicifolia (Ventenat) Burtt (*H. saligna* (Knight) Salisbury). Illustration: Maiden, Forest Flora of New South Wales **5**: 171 (1912).
Bushy tree to 8 m. Young shoots silky. Leaves leathery, lanceolate, 10–15 cm, obtuse, pale green without visible veins. Flowers in axillary clusters, stalks hairless. Perianth white, hairless. Fruit spherical with short beak, covered with warty protruberances. *Australia (Queensland, New South Wales)*. H5. Summer.

13. H. sericea Schrader (*H. tenuifolia* (Salisbury) Britten; *H. acicularis* Knight). Illustration: Botanical Magazine, n.s., 229 (1954); Growing Native Plants **3**: 57 (1973).
Shrub to 4 m. Leaves toughened, rigid, terete, 2.5–6 cm, with sharp points. Inflorescence *c.* 1 cm, with *c.* 6 flowers, axis and flower-stalks with silky hairs. Perianth white, hairless. Style and ovary hairless. Pollen-presenter disc-shaped. *Australia (Victoria, New South Wales, Tasmania)*. H5. Winter–summer.

Now naturalised in New Zealand, South

Africa, Spain and Portugal, where it is often a serious weed.

14. H. suaveolens R. Brown.

Shrub to 3 m. Leaves rigid, terete, entire or divided into linear segments with sharp points. Flowers in spherical, axillary clusters. Perianth-tube straight, erect, white. Fruit with 2 beaks. *Australia (Western Australia)*. H5. Summer.

This species is a serious weed in South Africa.

8. ROUPALA Aublet
E.C. Nelson

Trees. Foliage of 2 forms: juvenile leaves pinnate, adult leaves alternate, entire or toothed, simple. Flowers bisexual, in pairs in axillary or terminal racemes, bracts absent. Perianth radially symmetric. Hypogynous scales 4. Ovary hairless or downy. Fruit a woody follicle with 2 compressed, winged seeds.

This genus of tropical trees restricted to S America comprises perhaps 90 species; the spelling *Rhopala* is invalid. Species from other regions (e.g. Australia) previously placed within *Roupala* have all been transferred to other genera. These plants have to be cultivated in heated greenhouses; they prefer acid soil. Propagation by seed.

1a. Juvenile leaves hairless
                                    2. longipetiolata
  b. Juvenile leaves hairy   2
2a. Perianth to 6 mm; juvenile leaves
    very rapidly becoming hairless
                                    3. macrophylla
  b. Perianth *c.* 1 cm; juvenile leaves
    retaining their hairs   **1. brasiliensis**

1. R. brasiliensis Klotzsch (*R. pohlii* Meisner in part).

Leaves leathery, ovate to oblong, apex acute, 8–14 × 5–7 cm, downy when young, becoming almost completely hairless, margins shallowly toothed; stalk 3–6 cm. Inflorescence cylindric, erect, to 15 cm. Perianth densely downy, *c.* 1 cm. Ovary downy. *Brazil (São Paulo, Paraná)*. G2. Summer.

2. R. longepetiolata Pohl (*R. elegans* Pohl).

Like *R. brasiliensis* but leaves hairless when young, 6–10 × 1.5–3 cm; stalk 1.5–2.5 cm; inflorescences 8–12 cm; perianth not more than 7 mm long. *Brazil (Rio de Janeiro)*. G2. Summer.

3. R. macrophylla Pohl (*R. pohlii* Meisner in part). Illustration: Botanical Magazine, 6095 (1874).

Like *R. brasiliensis* but young leaves very

rapidly becoming hairless; perianth *c.* 6 mm, with orange-red hairs outside. *Brazil (Rio de Janeiro)*. G2. Summer.

The illustration cited shows a shoot with juvenile, pinnate leaves.

9. MACADAMIA Mueller
E.C. Nelson

Trees or shrubs. Leaves in whorls, simple, entire or toothed. Inflorescence terminal or axillary, flowers in pairs. Perianth radially symmetric, white, segments spreading and recurving at flowering. Anthers 4, each with apical gland or appendage. Hypogynous scales 4 or fused into a ring. Ovules 2. Fruit a drupe, indehiscent, with fleshy exocarp, with 1 globular or 2 hemispherical seeds.

There are about 6 species native to eastern Australia and Sulawesi; some species previously placed within the genus have been transferred recently to other genera. *M. integrifolia* Maiden & Betche, *M. tetraphylla* Johnson and their hybrid are cultivated, especially in Hawaii and California, producing the edible Macadamia nut.

1. M. ternifolia Mueller. Illustration: Maiden, Forest Flora of New South Wales 1: 40 (1912).

Small tree. Leaves leathery, stalkless or with very short stalks, oblong to lanceolate, 10–30 cm, hairless and shiny, margins toothed, teeth sharply pointed. Inflorescence downy, to 30 cm, with numerous flowers more or less in whorls. Perianth downy or hairless. Hypogynous scales forming a ring. *Australia (New South Wales, Queensland)*. H5. Summer.

10. GEVUINA Molina
E.C. Nelson

Tree or suckering shrub; young shoots densely covered with brown hairs. Leaves leathery, alternate, pinnate (occasionally bipinnate), to 50 cm; leaflets 5–30, to 20 × 10 cm, ovate, toothed, dark green, hairless. Flowers in pairs in axillary racemes. Perianth radially symmetric, white to yellow or green, sometimes red, segments separating at maturity. Stamens attached to perianth-segments, filaments to 1 cm; hypogynous scales 2. Styles curved. Fruit woody, hard, to 2 cm in diameter, becoming red then black, with 1 edible seed.

A genus of a single species from southern S America. The spelling 'Guevina' is invalid. It is cultivated in western Europe for its elegant, dark green foliage; the flowers are relatively inconspicuous. It requires plenty of moisture and is damaged by frosts and cold, dry winds; it does not

tolerate lime, and may be propagated from seed or from suckers.

1. G. avellana Molina. Illustration: Gardeners' Chronicle 90: 176 (1906); Botanical Magazine 9161 (1928).

Chile, Argentina. H5. Summer.

11. EMBOTHRIUM Forster & Forster
E.C. Nelson

Trees with a single trunk and without suckers, or tall shrubs suckering to form thickets. Leaves leathery, alternate, lanceolate to ovate, to 15 × 4 cm, hairless, green to dark green above, pale beneath; stalk to 1.5 cm. Flowers in pairs arranged in loose, terminal racemes. Perianth to 3 cm, bilaterally symmetric, scarlet to orange-red, very rarely pale yellow or white; segments separating at maturity. Anthers 4, stalkless. Style slightly curved with conical pollen-presenter. Fruit a woody pod, greenish yellow becoming red-brown when mature, with persistent style, containing numerous winged seeds.

A genus of a single species confined to temperate South America from Tierra del Fuego (*c.* 56° S) north to *c.* 35° S. Numerous minor variants have been given specific rank, but there is continuous variation within wild populations. Several cultivar names are applied to plants in cultivation in European gardens but the precise application of those names is uncertain, and seedlings abound to confuse further the nomenclature.

Embothrium apparently tolerates a wide range of conditions, succeeding best in acid or neutral soil and in areas where severe frosts are rare, but well-established plants can survive relatively severe winters out of doors in northwestern Europe (Ireland, Britain), where protected from cold, east winds. Propagation is by seed, which germinates readily, or by removal of suckers, or by cuttings.

1. E. coccineum Forster & Forster (*E. lanceolatum* Ruiz & Pavon; *E. coccineum* var. *lanceolatum* (Ruiz & Pavon) Kuntze; *E. longifolium* invalid). Illustration: Botanical Magazine, 4856 (1855); The Garden 102: 472 (1977); Moore, Flora of Tierra del Fuego, 156 (1983).

Chile, W Argentina. H4. Late spring–early summer.

12. TELOPEA R. Brown
E.C. Nelson

Shrubs or small trees. Leaves alternate, entire or toothed. Flowers bisexual, in pairs, in dense, umbel-like racemes, terminal, surrounded by an involucre of bracts (sometimes coloured). Perianth radially symmetric, hairless. Stamens 4, all fertile, anthers stalkless. Hypogynous glands united into a ring. Style curved with disc-shaped pollen-presenter. Ovary shortly stalked. Fruit a follicle containing winged seeds.

An Australian genus composed of 3 species, including the spectacular Waratah (*T. speciosissimum*) which is the floral emblem of New South Wales. They require some protection in northern countries, being sensitive to frost, but most can be cultivated outside in well-drained, lime-free soil. Propagation by seed is most successful.

1a. Shoots downy or leaves entire   2
 b. Shoots hairless; leaves prominently
    veined, toothed   **2. speciosissimum**
2a. Leaves hairy, with silver hairs
    beneath   **3. truncata**
 b. Leaves hairless   **1. oreades**

1. T. oreades Mueller (*T. mongaensis* Cheel). Illustration: Botanical Magazine, 8684 (1916), n.s., 851 (1984); Maiden, Forest Flora of New South Wales **5**: 159 (1913).
Shrub to 4 m; branches downy with red-brown hairs. Leaves obovate to oblong or lanceolate, 10–20 × 1–5 cm, acute or obtuse, tapering into a long stalk, margins entire or rarely toothed, glaucous. Flowers in a spherical raceme; involucral bracts 2–5 cm, not coloured. Perianth red. *Australia (Victoria, New South Wales)*. H5. Spring–summer.

T. mongaensis, from Monga in New South Wales, is not clearly distinct from *T. oreades* and is best regarded as a minor variant of it.

2. T. speciosissimum (J.E. Smith) R. Brown (*Embothrium speciosissimum* J.E. Smith). Illustration: Botanical Magazine, 1128 (1808); Maiden, Forest Flora of New South Wales **6**: 218 (1917); The Garden **108**: 455 (1983).
Shrub to 3 m; shoots hairless. Leaves obovate, 12–25 × 1–4 cm, margins toothed above the middle, tapering into a long stalk. Flowers in a dense, domed umbel to 15 cm in diameter, outer flowers maturing first; involucral bracts coloured, ovate-lanceolate, *c.* 8 cm. Perianth red, 2.5 cm. *Australia (New South Wales)*. H5. Summer.

White-flowered plants are occasionally reported. To get good blooms, plants should be carefully pruned. This species is not generally suitable for cultivation out-of-doors in NW Europe where it requires glasshouse protection.

A hybrid between this species and *T. oreades* (as *T. mongaensis*) is reported; it is grown in Australia under the cultivar name 'Braidwood Brilliant'.

3. T. truncata (Labillardière) R. Brown. Illustration: Botanical Magazine, 9660 (1944); Curtis & Stones, Endemic Flora of Tasmania **4**: 141 (1973).
Shrub 1–3 m, in cultivation occasionally a tree to 8 m. Leaves lanceolate to obovate, 5–18 × 1–4 cm, acute or notched, tapering into a stalk, margins recurved, green and hairless above, with silver hairs beneath. Inflorescence short, flat-topped, *c.* 8 cm in diameter, with 3–20 flowers all maturing simultaneously; involucral bracts petal-like, innermost membranous and hairy in the upper parts. Perianth red or rarely yellow. *Australia (Tasmania)*. H5. Early summer.

White-flowered plants are referred to as forma *lutea* Gray (illustration: Curtis & Stones, Endemic Flora of Tasmania **4**: 141, 1973).

13. LOMATIA R. Brown
E.C. Nelson
Trees or shrubs. Leaves toughened, leathery, alternate (rarely opposite), entire or divided. Flowers white to cream, in pairs in axillary or terminal raceme-like inflorescences. Perianth bilaterally symmetric; anthers all fertile; hypogynous scales 3 (if 4 then the 4th scale minute). Fruit a woody follicle with numerous winged seeds.

A genus of perhaps 12 species from eastern Australia and Tasmania (8 species) and S America (4 species); none of the species occurs in both continents. Those from temperate habitats are easy to cultivate throughout southern and western Europe; generally they prefer moist, acid (usually peaty) soils. Propagation is by seeds, but cuttings may be tried.
Literature: Sleumer, H., Proteaceae Americanae, *Botanischer Jahrbücher* **76**: 196–198 (1954).

1a. Leaves simple, entire or toothed   2
 b. Leaves pinnate   5
2a. Leaves linear, less than 1 cm broad
                              6. myricoides
 b. Leaves more than 1 cm broad   3
3a. Perianth hairless outside **5. ilicifolia**
 b. Perianth hairy outside   4
4a. Leaves obovate, tapering gradually
    into stalk, broadest above middle;
    style and ovary hairy   **1. dentata**
 b. Leaves ovate, base abrupt, broadest
    below middle; style and ovary hairless
                                 4. hirsuta

5a. Shoots with red-brown hairs
                               2. ferruginea
 b. Shoots hairless   6
6a. Leaves more than 10 cm   **7. silaifolia**
 b. Leaves all less than 10 cm   7
7a. Leaf-segments narrow, linear,
    1–4 cm, entire   **8. tinctoria**
 b. Leaf-segments broad, ovate-
    lanceolate, coarsely toothed
                              3. fraxinifolia

1. L. dentata (Ruiz & Pavon) R. Brown. Illustration: Muñoz Pizarro, Sinopsis de la Flora Chilena, pl. 173 (1966).
Shrub to at least 4 m, young shoots with pale fawn hairs. Leaves entire, obovate, 3–9 × 2–5 cm, with irregular teeth above the middle, base tapered, very dark green above, glaucous green beneath. Perianth hairy, creamy white; style and ovary hairy. *Chile, Argentina*. H5. Summer.

2. L. ferruginea (Cavanilles) R. Brown (*L. pinnatifolia* invalid; *L. pinnatifida* invalid). Illustration: Annesley, Beautiful rare trees and plants, 9 (1903); Botanical Magazine, 8192 (1907).
Shrub to 3 m, young shoots with red-brown hairs. Leaves bipinnate (rarely tripinnate), 7–50 × 5–25 cm, dark green on upper surface, with white-fawn felt beneath; leaflets ovate-lanceolate, apex acute. Inflorescences to 12 cm; perianth with orange hairs outside, yellow or scarlet outside, bright scarlet inside. Style and ovary hairless. *Chile, Argentina*. H5. Summer.

The fern-like leaves and dense rust-coloured hairs make this a distinctive species. It is tolerant of slight frost.

3. L. fraxinifolia Bentham.
Tree or shrub; branches hairless. Leaves pinnate, 5–10 cm, segments 3–7, ovate to lanceolate, coarsely toothed and contracted at base into a distinct stalk which is 5–10 cm. Inflorescences solitary or grouped in a broad terminal panicle, axis with sparse red-brown hairs; perianth hairless. *Australia (Queensland)*. G1. Summer.

4. L. hirsuta (Lamarck) Diels (*L. obliqua* R. Brown). Illustration: Botanical Magazine, n.s., 335 (1959); Muñoz Pizarro, Sinopsis de la Flora Chilena, pl. 173 (1966); Krüssmann, Handbuch der Laubgehölze **2**: t. 93 (1977).
Shrub or tree to 20 m. Leaves entire, ovate, 5–13 × 2–8 cm, base abruptly rounded or truncate, margins with obtuse, coarse, unequal teeth, apex obtuse, downy when young but becoming hairless, dark glossy green above, sometimes coloured beneath.

Inflorescences to 8 cm, terminal and axillary, axis with red-brown hairs; perianth pale yellow-green, hairy. Style and ovary hairless. *Ecuador, Peru, Argentina & Chile.* H5. Summer.

Not common in cultivation.

5. L. ilicifolia R. Brown. Illustration: Botanical Magazine, 4023 (1843); Cochrane et al., Flowers and plants of Victoria, 33 (1968).
Shrub to 2 m. Leaves entire (rarely with lobes near base), ovate to lanceolate, to 10 cm, upper surface with conspicuous net-like veins, hairless and shining when mature, margins with coarse, sharply pointed teeth. Inflorescence branched, axis hairless or slightly downy; perianth hairless, cream-white. *Australia (Queensland, New South Wales, Victoria).* H5. Summer.

6. L. myricoides (Gaertner) Domin (*L. longifolia* R. Brown). Illustration: Botanical Magazine, 7698 (1900); Burbidge and Gray, Flora of the Australian Capital Territories, 149 (1970).
Shrub or small tree to 5 m. Leaves entire, linear-lanceolate, 5–20 × c. 1 cm, hairless, margins sometimes with few, widely spaced teeth towards apex, or entire, dull green when mature. Inflorescences terminal and axillary, axis hairless; perianth white, hairless. *Australia (New South Wales, Victoria).* H5. Summer.

7. L. silaifolia (J.E. Smith) R. Brown. Illustration: Botanical Magazine, 1272 (1810).
Erect shrub to 1.5 m. Leaves bi- or tripinnate, to 30 cm, almost as broad, hairless when mature, ultimate segments lanceolate, broadening towards base, to 6 mm. Inflorescences terminal and axillary, to 30 cm, erect; perianth cream, hairless. *Australia (New South Wales, Queensland).* H5. Summer.

A tender species; plants named *L. silaifolia* but grown outside in northern and western Europe may be *L. tinctoria*.

8. L. tinctoria (Labillardière) R. Brown. Illustration: Botanical Magazine 4110 (1844); Curtis & Stones, Endemic Flora of Tasmania 3: 107 (1971); Plantsman 6: 144 (1984).
Shrub to 1.5 m, suckering, shoots hairless. Leaves pinnate (rarely simple or bipinnate), ovate to triangular in outline, 4–8 cm, with 3–7 pairs of lobes, usually hairless but sometimes downy when young; lobes linear, entire or toothed, 1–4 cm, acute. Inflorescence loose, erect, axis hairless or

rarely downy. Perianth white or cream in bud with red or green apex. *Australia (Tasmania).* H5. Summer.

The hardiest species, thriving in moist, sheltered gardens.

14. KNIGHTIA R. Brown
E.C. Nelson
Trees with upright branches; shoots covered with red-brown hairs. Leaves toughened, alternate, oblong, to 15 × 4 cm, margins toothed. Flowers in pairs in dense, axillary or terminal racemes to 10 cm. Perianth red, with red-brown hairs on outside; segments separating at maturity and coiling spirally. Stamens attached to perianth-segments; anthers almost stalkless. Hypogynous scales 4. Style terete, straight, with terminal, club-shaped stigma. Fruit woody, to 4 cm, hairy with persistent style. Seeds numerous, winged.

The single species is native to New Zealand and is now considered to be the only one in the genus; 2 other species recorded from New Caledonia have been transferred to *Eucarpha* (R. Brown) Spach.

This tree is uncommon in European gardens and is usually treated as a plant for cool glasshouses; it will survive out-of-doors in moist, mild, western regions. *K. excelsa* requires acid, humus-rich soil and protection from cold, dry winds. Propagation is from seed.

1. K. excelsa R. Brown. Illustration: Salmon, The native trees of New Zealand, 126, 127 (1980); Goulding, Fanny Osborne's flower paintings, 22 (1983). *New Zealand (both islands).* H5. Summer.

15. STENOCARPUS R. Brown
E.C. Nelson
Trees. Leaves alternate or scattered, entire or pinnatifid, lobes irregular. Flowers bisexual, in compound inflorescences composed of several umbel-like clusters. Perianth bilaterally symmetric, segments separating at maturity. Anthers all fertile, stalkless. Hypogynous glands united into a single disc, or almost absent. Style with disc-shaped pollen-presenter. Fruit a leathery follicle containing several winged seeds.

A genus of about 4 species from northern and eastern Australia, New Guinea and New Caledonia. In northern and northwest Europe *Stenocarpus* requires greenhouse protection, but may be cultivated outside in southern parts; propagated by seed.

1a. Leaves entire, lanceolate to elliptic; flowers white **1. salignus**

b. Leaves pinnatifid; flowers red **2. sinuatus**

1. S. salignus R. Brown. Illustration: Edwards's Botanical Register 13: t. 441 (1820); Maiden, Forest Flora of New South Wales 1: 23 (1904).
Leaves leathery, elliptic or lanceolate, 5–20 × 2–4 cm, base tapering to short stalk, hairless, a few leaves on young trees larger and pinnatifid. Umbels solitary or 2–3 together, each with 12–25 flowers. Perianth c. 1 cm, hairless, white or green-white, slightly fragrant. Style hairless, ovary and its stalk downy. *Australia (Queensland, New South Wales).* H5–G1. Summer.

2. S. sinuatus Endlicher (*S. cunninghamii* invalid). Illustration: Botanical Magazine, 4263 (1846); Maiden, Forest Flora of New South Wales 6: 210 (1917); Morley & Toelken, Flowering plants in Australia, 241 (1983).
Tree to 35 m. Leaves entire or lobed or pinnatifid, to 60 cm, base tapering into stalk, hairless, sometimes coloured red beneath; with up to 4 pairs of lobes, entire, lanceolate, to 10 × 4 cm. Compound inflorescence with c. 5 rays arranged like the spokes of a wheel on a common stalk; individual umbels with 12–18 flowers spreading horizontally. Perianth c. 2.5 cm, hairless, in bud yellow-green, opening red to orange-red. Style red, hairless, pollen presenter gold-yellow. Ovary and its stalk hairy. *Australia (Queensland, New South Wales).* H5–G1. Summer.

16. BANKSIA Linnaeus filius
E.C. Nelson
Shrubs or trees, some with fire-resistant lignotubers. Leaves arranged in spirals or whorls or scattered, toughened. Inflorescence terminal, cylindric or spherical with woody central axis and up to 3000 flowers; flowers stalkless, in pairs, arranged in vertical spiralling rows; perianth radially or bilaterally symmetric, segments separating at flowering except at base. Anthers 4, all fertile, stalkless. Hypogynous scales 4. Ovary stalkless, style terete, straight or curved; stigma a small groove on the style-end. Fruit few or many follicles in a woody cone, each containing 2 winged seeds; follicles of some species not opening until subjected to fire.

There are 71 species recognised in a recent revision, and all but 1 are confined to Australia; the single exception is recorded from N Australia through New Guinea to the Aru Islands.

The species in cultivation in Europe are from southern parts of Australia. In general they are intolerant of lime, requiring well-drained acid soil; fertilisers should not be applied and over-watering must be avoided. In northern Europe all require protection in a greenhouse, but in southern Europe many species will thrive out of doors. Propagation from seed is best, and plants will be short-lived unless they are trees. Literature: George, A.S., The genus Banksia L. f. (Proteaceae), *Nuytsia* 3(3): 239–473 (1981); George, A.S., *The Banksia book* (1984).

Unless otherwise stated, colours of perianth and style noted below are the colours after flowering (see George, 1981, 247–8, cited above).

1a. Leaves needle-like or linear, less than 2.5 mm broad   2
 b. Leaves with broad blade more than 1 cm across or not linear   5
2a. Inflorescences pendulous   **8. nutans**
 b. Inflorescences erect   3
3a. Leaves 4–13 cm, in whorls
                                   9. occidentalis
 b. Leaves less than 3 cm, not in whorls
                                                        4
4a. Leaves usually toothed above middle
                                     17. spinulosa
 b. Leaves not toothed above middle
                                        3. ericifolia
5a. Leaves usually lobed or toothed   6
 b. Leaves usually entire, without teeth or lobes   15
6a. Leaves lobed and hairless; inflor-escence and fruiting head shaggy
                                   13. quercifolia
 b. Combination of characters not as above   7
7a. Style red; perianth grey with white and brown hairs; leaves oblong with toothed margins, apex truncate
                                       2. coccinea
 b. Style and perianth usually yellow; leaves not as above   8
8a. Leaves wedge-shaped, margins flat   9
 b. Leaves with recurved margins or not wedge-shaped   10
9a. Leaves remaining downy beneath; inflorescence small   **11. pilostylis**
 b. Leaves becoming hairless beneath; inflorescence large   **7. media**
10a. Leaf-apex truncate   **14. sceptrum**
 b. Leaf-apex not truncate   11
11a. Leaves 25 cm or more   12
 b. Leaves less than 25 cm   14
12a. Leaves lobed; lobes rounded, 2–5 × 1–5 cm   **4. grandis**
 b. Leaves toothed, teeth acute   13

13a. Teeth not indented to midrib
                                     12. prionotes
 b. Teeth indented to midrib **16. speciosa**
14a. Leaves 3–11 cm × 5–25 mm, obovate
                                        10. ornata
 b. Leaves 7–20 × 2–4 cm, oblong
                                       15. serrata
15a. Leaves hairless or becoming so   16
 b. Leaves hairy when mature, at least beneath   17
16a. Leaves 2–5 cm, scattered on shoots
                                         1. canei
 b. Leaves 3–9 cm, in whorls
                                   18. verticillata
17a. Leaf-margins recurved; hairs brown
                                     6. marginata
 b. Leaf-margins not recurved; hairs white-brown   **5. integrifolia**

1. B. canei Willis. Illustration: Wrigley & Fagg, Australian native plants, 172 (1979); George, The Banksia book, 61 (1984).
Shrub to 3 m. Leaves scattered, elliptic to tapering, 2–5 cm × 5–20 mm, margins recurved and entire (rarely toothed), truncate with mucronate apex, hairy beneath, becoming hairless; stalk 4 mm, hairy. Inflorescence cylindric, 4–6 cm in diameter, 5–10 cm tall; perianth and style yellow, deciduous. Fruiting cone cylindric to ovoid; follicles numerous (to 150), hairy, some opening spontaneously but most remaining closed for several years until burned. *Australia (Victoria, New South Wales)*. H5. Spring–summer.

Similar to *B. marginata* but with mucronate leaves, deciduous perianth and style and fruiting cones with numerous hairy follicles.

2. B. coccinea R. Brown. Illustration: Botanical Magazine, n.s., 630 (1973); Erickson, Flowers and plants of Western Australia, 61 (1973); George, The Banksia book, 176, 177 (1984).
Shrub to 3 m, without lignotuber, branches erect. Leaves broadly oblong, 3–9 × 2–7 cm, margins toothed, with truncate or mucronate apex, downy when young, hairless above. Inflorescence squat, cylindric, *c.* 8 cm tall, to 10 cm in diameter; perianth grey with white or red-brown hairs. deciduous; styles to 5 cm, bright red, rarely orange, deciduous. Fruiting cone small, 2–4 × 2–4 cm, upper part without follicles; follicles small, less than 8 mm, remaining closed until burned. *SW Australia*. H5. Spring–summer.

A beautiful species, the predominant colour of the inflorescences at flowering is red, coming from the prominent styles.

3. B. ericifolia Linnaeus filius. Illustration: Botanical Magazine, 738 (1804); Elliott & Jones, Encyclopaedia of Australian plants 2: 292 (1982); George, The Banksia book, 183, 184 (1984).
Shrub without lignotuber, to 6 m. Leaves crowded, linear, 1–2 cm × 1 mm, margins recurved, concealing the densely downy undersurface, hairless above, apex mucronate. Inflorescence cylindric, to 30 × 6 cm; perianth gold-brown with mauve base, persistent on fruiting cone; style gold-orange or orange-red, deciduous. Follicles numerous, hairy, remaining closed until burned. *Australia (E New South Wales)*. H5. Spring–summer.

A cultivar with dark purple leaves is reported in Australia, and there is a range of flower colours from orange to dark red.

4. B. grandis Willdenow. Illustration: Elliott & Jones, Encyclopedia of Australian plants 2: 293 (1982); George, The Banksia book, 77–79 (1984).
Tree to 10 m, or sometimes a spreading shrub, without a lignotuber. Leaves narrowly oblong in outline, to 45 × 3–11 cm, divided to midrib into triangular, opposite to alternate lobes (2–5 × 1–5 cm). Inflorescences to 40 cm, cylindric, narrow; perianth pale yellow, style cream, hairless, both deciduous. Fruiting cone massive with numerous follicles which open spontaneously. *SW & W Australia*. H5. Spring–summer.

Coastal plants form tall spreading shrubs, not tall trees.

5. B. integrifolia Linnaeus filius. Illustration: Botanical Magazine, 2770 (1827); George, The Banksia book, 49 (1984).
Tree. Leaves in whorls or scattered, elliptic to lanceolate, 4–20 cm × 5–35 mm, entire (rarely with a few apical teeth), margins not recurved, with persistent pale, white-brown hairs beneath. Inflorescences cylindric, to 12 × 7 cm; perianth and style pale yellow, deciduous. Fruiting cone with numerous small follicles which open spontaneously. *Australia (eastern seaboard from N Queensland to Victoria)*. H5. Spring–summer.

Three varieties are recognised in the wild. Generally cultivated in glasshouses in northern Europe, but seed from southern parts of its range (var. **integrifolia**) may yield plants suitable for outdoor cultivation in milder parts of Britain and Ireland.

6. B. marginata Cavanilles. Illustration: Cochrane et al., Flowers and plants of Victoria, 83 (1968); Burbidge & Gray, Flora of the Australian Capital Territory, f. 128 (1970); Elliott & Jones, Encyclopedia of Australian plants **2**: 296 (1982); George, The Banksia book, 59 (1984).

Tree or shrub, sometimes with a lignotuber. Leaves scattered, linear-oblong, 1.5–6 cm × 5–15 mm, margins often recurved and usually entire, hairless above, with brown hairs beneath. Inflorescence cylindric, to 10 × 6 cm; perianth and style yellow, persistent. Fruiting cone small, cylindric, follicles few to many, often opening spontaneously. *Australia (South Australia, Victoria, Tasmania, New South Wales)*. H5. Spring–summer.

Perhaps the species most widely cultivated in Europe; it is grown out of doors in NE Ireland where it flowers regularly but it does require a place sheltered from cold winds.

7. B. media R. Brown. Illustration: Botanical Magazine, 3120 (1831); Erickson, Flowers and plants of Western Australia, 81 1973); Elliott & Jones, Encyclopedia of Australian plants **2**: 297 (1982); George, The Banksia book, 127 (1984).

Shrub without lignotuber, rarely a small tree. Leaves tapering, 4–10 × 1–2 cm, margins flat or slightly recurved, with obtuse teeth, sinuses V-shaped, hairy above becoming hairless, hairless below except for woolly hairs in the sinuses. Inflorescence cylindric, to 15 × 8 cm; perianth gold-yellow becoming brown, style cream, both persistent. Fruiting cone robust, follicles concealed by persistent perianths and styles at first, opening after fire. *Australia (SE Western Australia)*. H5. Spring–summer.

8. B. nutans R. Brown. Illustration: Erickson, Flowers and plants of Western Australia, 81 (1973); George, An introduction to the Proteaceae of Western Australia, 13 (1984); George, The Banksia book, 227–229 (1984).

Shrub without lignotuber, to 1.5 m. Leaves scattered, linear with sharp tip, 1–3 cm × 1 mm, margins recurved concealing hairy lower surface, often blue-green. Inflorescences pendent, almost spherical, to 7 cm in diameter, onion-scented; perianth pink-purple in bud becoming brown-purple, style cream. Fruiting cone robust, woody, spherical to ovoid, to 10 cm in diameter; follicles enlarged, with flattened tops. *Australia (SW Western Australia)*. H5. Spring–summer.

Among those presently in general cultivation in Europe this is the only species with pendulous inflorescences; 4 other Western Australian species (e.g. *B. caleyi* R.

Brown, *B. lehmànniana* Meisner, *B. aculeata* George and *B. elderiana* Mueller & Tate) have a similar type of inflorescence but they are not grown except in botanic gardens.

9. B. occidentalis R. Brown. Illustration: Botanical Magazine, 3535 (1836); Erickson, Flowers and plants of Western Australia, 64 (1973); George, An introduction to the Proteaceae of Western Australia, 14 (1984); George, The Banksia book, 189 (1984).

Small tree or shrub without lignotuber. Leaves in whorls, linear, 4–13 cm × 2 mm, margins recurved, sparsely toothed, lower surface hairy. Inflorescence cylindric, to 14 cm; perianth yellow, style yellow at base, with shining red upper part and cream stigma, both persistent. Fruiting cone with numerous follicles, opening after fire. *Australia (southern coastal Western Australia)*. H5. Spring–summer.

B. occidentalis inhabits sandy swamps in its native habitat and should therefore be quite amenable to cultivation, but even in Australia it is reluctant to flower in gardens.

10. B. ornata Meisner. Illustration: Elliott & Jones, Encyclopedia of Australian plants **2**: 298 (1982); George, The Banksia book, 99 (1984).

Shrub without a lignotuber. Leaves obovate with obtuse apex, 3–11 cm × 5–25 mm, margins usually not recurved, with numerous triangular teeth (1–3 mm), hairy becoming hairless above. Inflorescence cylindric, broad, 5–8 × 5–8 cm; perianth and style cream, stigma pink, persistent in fruit. Fruiting cone with up to 50 follicles, hairy, opening only after fire. *Australia (South Australia, Victoria)*. H5. Spring–summer.

11. B. pilostylis Gardner. Illustration: George, The Banksia book, 131 (1984).

Shrub without a lignotuber. Leaves scattered, narrow, tapering, 5–16 cm × 5–20 mm, mucronate, margins recurved, with numerous sharp teeth, hairy above, becoming almost hairless, remaining hairy beneath. Inflorescence small, cylindric, 5–10 × 5 cm, flowers densely packed. Perianth pale yellow, style cream, persistent. Fruiting cone robust with *c.* 25 large follicles opening after fire. *Australia (SE Western Australia)*. H5. Spring–summer.

12. B. prionotes Lindley. Illustration: George, An introduction to the Proteaceae of Western Australia, 7 (1984); George, The Banksia book, 116 (1984).

Tree or shrub without lignotuber. Leaves linear, to 25 × 2 cm, margins toothed, not recurved, hairy above becoming hairless, hairy beneath. Inflorescence large, cylindric and tapering towards apex, to 10 cm in diameter; flowers white, hairy in bud; perianth bright orange, style yellow-orange, deciduous. Fruiting cones usually few, opening spontaneously or after fire. *Australia (W Western Australia)*. H5. Spring–summer.

13. B. quercifolia R. Brown. Illustration: Erickson, Flowers and plants of Western Australia, 92 (1973); George, The Banksia book, 83 (1984).

Shrub without lignotuber, branchlets hairless. Leaves scattered, thin, tapering, 3–15 × 1–4 cm, margins flat with prominent lobes (2–9 mm long), hairless except for some woolly hairs in the sinuses. Inflorescence cylindric; flowers orange or brown, densely hairy outside; perianth and style persistent, giving the fruit a shaggy appearance. Fruiting cone to 6 cm in diameter, with *c.* 30 follicles, opening after fire. *Australia (SW Western Australia)*. H5. Spring–summer.

In the wild, plants flower within 3 years of germination. The species inhabits damp, peaty sand.

14. B. sceptrum Meisner. Illustration: Nuytsia 3: 336 (1981); George, The Banksia book, 113–115 (1984).

Bushy shrub without lignotuber. Leaves scattered, stiff, oblong, 4–9 × 1–3 cm, margins flat, with small triangular teeth, apex truncate, hairy becoming hairless except in the sinuses beneath. Inflorescence cylindric, to 20 × 10 cm; perianth and style yellow, persistent. Fruiting cone massive with *c.* 50 follicles, opening after fire. *Australia (W Western Australia)*. H5. Spring–summer.

15. B. serrata Linnaeus filius. Illlustration: Botanical Magazine, 9642 (1942); Cochrane et al., Flowers and plants of Victoria, 23 (1968); George, The Banksia book, 92, 93 (1984).

Tree or shrub. Leaves crowded at tips of branches, oblong, 7–20 × 2–4 cm, margins with regular, triangular teeth, hairy or downy when young, becoming almost hairless. Inflorescence cylindric; perianth cream-grey, style cream, both persistent. Fruiting cone massive with *c.* 30 follicles, opening after fire or spontaneously. *Australia (Victoria, Tasmania, New South Wales, Queensland)*. H5. Spring–summer.

This species is frequently stated to have

been the first Australian plant cultivated in England; that claim is erroneous (see Nelson, E.C., *Australian plants cultivated in England before 1788*, *Telopea* 2(4): 347–353, 1983).

16. B. speciosa R. Brown. Illustration: Botanical Magazine, 3052 (1831); Edwards's Botanical Register **7**: t. 1728 (1834); Erickson, *Flowers and plants of Western Australia*, 81 (1973); George, *An introduction to the Proteaceae of Western Australia*, 9 (1984); George, *The Banksia book*, 106, 107 (1984).

Shrub without a lignotuber. Leaves scattered, linear in outline, 20–45 × 2–4 cm, divided to midrib by triangular, alternate or opposite lobes each 1–2 cm long. Inflorescence cylindric, to 20 × 10 cm; perianth and style cream, persistent. Fruiting cone spherical with *c.* 20 massive follicles, opening after fire. *Australia (S Western Australia)*. H5. Spring–summer.

17. B. spinulosa J.E. Smith. Illustration: Botanical Magazine, n.s., 498 (1967); Cochrane et al., *Flowers and plants of Victoria*, 128 (1968); Elliott & Jones, *Encyclopedia of Australian plants* **2**: 304 (1982); George, *The Banksia book*, 179–181 (1984). Shrub with or without a lignotuber. Leaves linear, to 10 × 2 mm, margins flat or recurved, toothed at least in the upper portion. Inflorescence cylindric, to 15 × 7 cm; perianth yellow, persistent; style all yellow or with dark red or purple-black upper portion, deciduous. Fruiting cone with numerous follicles opening spontaneously or after fire. *Australia (Queensland, New South Wales, Victoria)*. H5. Spring–summer.

A complex species in the wild with 3 varieties recognised: var. **collina** (R. Brown) George is recorded as cultivated in Europe; it has a lignotuber and leaves with prominent veins and toothing for the entire length; var. **spinulosa** has toothing only towards the apex.

18. B. verticillata R. Brown. Illustration: George, *The Banksia book*, 195 (1984). Shrub without a lignotuber. Leaves in whorls, narrow, elliptic, 3–9 × 1 cm, margins entire, recurved. Inflorescence cylindric; perianth gold-yellow, style cream, both deciduous. Fruiting cone cylindric; follicles opening spontaneously after several years. *Australia (SW Western Australia)*. H5. Spring–summer.

XLV. OLACACEAE

Shrubs, climbers or trees, sometimes half-parasitic on the roots of other vegetation. Leaves alternate, simple and entire, sometimes more or less in 2 ranks, sometimes with latex ducts and/or resin canals or translucent spots, without stipules. Flowers usually small, in axillary clusters, cymes or racemes or solitary, usually bisexual, radially symmetric, often fragrant. Sepals 4–6, joined into a cup below, the lobes sometimes obscure. Petals 3–6, usually free, sometimes united into a tube below, usually linear, oblong or spoon-shaped, edge-to-edge in bud, sometimes 2 or 3 of them deeply bilobed. Stamens usually twice as many as petals, sometimes 5 or 6 of them reduced to infertile staminodes. Ovary superior, 1-celled or partially 3-celled below; ovules 2 or 3, pendent from the apex of a central placenta. Fruit a drupe, often surrounded by the enlarged and persistent calyx.

A family of 27 genera and about 230 species from the tropics. Some 6 genera and 10 species have been reported from cultivation, though this seems unlikely in view of their half-parasitic habit and lack of outstanding ornamental features. However, they are included below in case any remain in gardens in Europe.

Little seems to be recorded of the cultivation conditions required for these plants. Most need glasshouse protection and relatively high humidity, and, presumably, access to the roots of other plants.

1a. Fertile stamens 3, staminodes 5 or 6
                                        2
  b. Fertile stamens 6 or more, staminodes lacking   3
2a. Petals 3, free, sometimes 2 or 3 of them deeply bilobed so that they may appear superficially as 5 or 6; fruit free from the persistent calyx   **4. Olax**
  b. Petals 6, united into a tube at the base, none bilobed; fruit united to the persistent calyx for most of its length   **5. Dulacia**
3a. Stamens united into a tube; large tree   **6. Ongokea**
  b. Stamens free; shrubs or small trees 4
4a. Petals each with a large patch of bristles on the inner surface
                              2. Ximenia
  b. Petals hairless or finely downy on the inner surface   5
5a. Calyx not enlarging in fruit; style columnar, long   **3. Ptychopetalum**

b. Calyx enlarging and becoming brightly coloured in fruit; style short, conical   **1. Heisteria**

1. HEISTERIA Jacquin
J. Cullen

Shrubs or small trees. Leaves with small translucent dots. Flowers in axillary clusters. Calyx with 5 small lobes, enlarging and becoming brightly coloured in fruit. Petals 5, hairless inside (ours). Stamens 10, all fertile. Ovary 3-celled below, 1-celled above; style short and conical. Fruit surrounded by the persistent, enlarged and brightly coloured calyx.

A genus of about 35 species, mostly from S America, a few from Africa.
Literature: Sleumer, H., *Flora Neotropica* **38**: 42–82 (1984).

1. H. coccinea Jacquin.
Tree with hard, reddish wood. Petals white, *c.* 1.5 cm. Fruiting calyx to 2 cm in diameter, membranous, bright purplish red. *West Indies, Venezuela*. G2.

2. XIMENIA Linnaeus
J. Cullen

Spiny shrubs or small trees. Flowers solitary or in clusters in the axils of the leaves. Calyx small with 4–5 small lobes, persistent but not or scarcely enlarging in fruit. Petals 4–5, each with a patch of bristles on the inner surface. Stamens 8–10 (in ours), all fertile. Ovary 3-celled below, 1-celled above.

A genus of 8 species from the tropics and southern Africa; their fruits are edible but tart.

1a. Twigs and leaves (except for the midribs) hairless; flowers usually 2–10 in axillary cymes, rarely solitary when the stalk bears a bract about halfway up   **1. americana**
  b. Twigs and leaves hairy; flowers solitary, bract at the base of the stalk   **2. caffra**

1. X. americana Linnaeus. Illustration: Die Natürlichen Pflanzenfamilien, edn 2, **16b**: 23 (1935); Flora Malesiana **10**: 12 (1984); Flora Neotropica **38**: 92 (1984).
Shrub or small tree to 12 m. Twigs and leaves (except for the midrib) hairless. Flowers 2–10 in axillary cymes (rarely 1-flowered when the stalk bears a bract about halfway up). Petals whitish to yellowish green. Fruit 1.7–3.5 × 1.5–3 cm, yellow to orange. *Pantropical*. G2.

2. X. caffra Sonder. Illustration: Flowering Plants of Africa **28**: pl. 1081 (1950–51).

Shrub to 7 cm. Twigs and leaves (at least the lower surface) hairy. Flowers solitary, each with a bract at the base of the stalk (the flowers sometimes appear clustered due to the shortness of the shoot). Petals bright green. Fruit to 3 × 2.4 cm, orange to scarlet. *South Africa (Transvaal)*. G1?

3. PTYCHOPETALUM Bentham
J. Cullen

Hairless shrubs or trees. Flowers in short, few-flowered axillary racemes. Calyx small, 5–6-lobed, not enlarged in fruit. Petals usually 5, not hairy within. Stamens usually 10. Ovary 1-celled with 2 or 3 ovules. Style long, stigma slightly 3-lobed.

A genus of 4 species from tropical America and Africa: 2 have reputedly been in cultivation.

1a. Leaves lanceolate-oblong, rather gradually tapering at the apex; petals without appendages   **1. olacoides**
  b. Leaves ovate, ovate-elliptic or ovate-oblong, rather abruptly tapered at the apex; petals each with a bifid appendage which is recurved in a hook-shape when the flower is fully open   **2. uncinatum**

1. P. olacoides Bentham. Illustration: Die Natürlichen Pflanzenfamilien, edn 2, **16b**: 24 (1935); Flora Neotropica **38**: 134 (1984).
Tree to 15 m. Leaves lanceolate-oblong, gradually tapering at the apex. Racemes with 5–8 flowers. Petals 1–1.3 cm, white, without appendages. Drupe oblong-ellipsoid. *N Brazil, Surinam, French Guiana*. G2.

2. P. uncinatum Anselmino.
Differs from *P. olacoides* in the characters given in the key and in the ovoid or pear-shaped drupe. *N Brazil*. G2.

4. OLAX Linnaeus
J. Cullen

Trees or shrubs, sometimes climbing, often with thorns. Flowers in racemes, panicles or spikes, rarely solitary. Calyx cup-shaped, not distinctly lobed, conspicuously enlarged in fruit. Petals 3, entire or 2 or 3 of them deeply bilobed. Fertile stamens 3, staminodes 5. Ovary 3-celled below, 1-celled above. Drupe included in the persistent, enlarged calyx, but free from it.

A genus of about 25 species from the old world tropics.
Literature: Sleumer, H., A taxonomic account of the Olacaceae of Asia, Malesia and the adjacent areas, *Blumea* **26**:

145–168 (1980); Sleumer, H., *Flora Malesiana* **10**: 4–10 (1984).

1a. Flowers solitary in the leaf-axils   **3. stricta**
  b. Flowers in axillary racemes   **2**
2a. Thorns absent; branchlets more or less hairless; petals 1–1.2 cm   **1. imbricata**
  b. Thorns usually present; branchlets downy; petals 7–9 mm   **2. scandens**

1. O. imbricata Roxburgh. Illustration: Matthew, Illustrations on the flora of the Tamil Nadu Carnatic, 135 (1983).
Shrub, usually climbing, branchlets hairless and without thorns. Flowers in racemes or panicles. Petals not bilobed, white or pinkish, 1–1.2 cm. Drupe almost covered by the enlarged, persistent, orange calyx. *India and Sri Lanka to Micronesia, northwards to SW China*. G2.

2. O. scandens Roxburgh. Illustration: Flora Malesiana **10**: 8 (1984).
Hanging or climbing shrub, branches downy, with curved thorns. Racemes 1–3 in each leaf-axil. Two (rarely all) of the petals deeply bilobed, so that superficially there appear to be 5 or 6, white, 7–9 mm. Drupe orange to yellow, 8–15 × 6–10 mm. *From the W Himalaya through India to Vietnam, south to Java and Bali*. G2.

3. O stricta R. Brown. Illustration: Rotherham et al., Flowers and plants of New South Wales and southern Queensland, 50 (1975).
Erect, slender shrub to 2 m. Leaves yellowish green. Flowers solitary in the leaf axils. Petals yellowish white, 2 or 3 of them bilobed. *Australia (Queensland, New South Wales, Victoria)*. G2.

5. DULACIA Vellozo
J. Cullen

Shrubs or trees, Flowers few in raceme- or cyme-like axillary clusters. Calyx cup-like, very obscurely 5-lobed, enlarging considerably in fruit. Petals 6, entire, united into a tube in the lower half. Fertile stamens 3, staminodes 6. Ovary 3-celled below, 1-celled above. Fruit drupe-like, united to the enlarged calyx for most of its length.

A genus of about 13 species from tropical America; one is reputedly cultivated, usually under the name *Liriosma ovata*.

1. D. inopiflora (Miers) Kuntze (*Liriosma ovata* Miers).
Small tree (which may reach 18 m in the wild), branchlets hairy. Flowers 2–5 in short racemes, of which there may be 1 or

2 in each leaf-axil. Petals white or pale green, 3–4 mm. Filaments of fertile stamens and staminodes hairy. Drupe 1.7–2 cm, orange or yellow. *Colombia, Venezuela, Brazil, Peru*. G2.

6. ONGOKEA Pierre
J. Cullen

A large tree with slightly winged branchlets. Leaves oblong-elliptic, somewhat acuminate. Flowers small, in clusters. Calyx saucer-shaped, not or obscurely lobed, enlarging in fruit. Petals 5, free. Stamens 5, the filaments united into a tube around the style. Ovary 2-celled below, 1-celled above. Fruit a drupe surrounded by the enlarged calyx which eventually splits into 3 segments.

A genus of 2 species from West tropical Africa: 1 is said to have been in cultivation, though, as it is a large tree, this seems unlikely.

1. O. gore (Hua) Pierre. Illustration: Die Natürlichen Pflanzenfamilien, edn 2, **16b**: 29 (1935).
Tree to 40 m with a straight trunk without buttresses. Leaves 4–7 × 2.5–3.5 cm. Petals to 3 mm, white. Fruit yellow. *West tropical Africa*. G2.

XLVI. SANTALACEAE

Herbs, shrubs or trees, parasitic or partially parasitic on the roots of other plants. Leaves opposite or alternate, simple, entire (rarely scale-like and falling early), without stipules. Flowers in spikes, racemes, cymes, clusters or solitary, unisexual or bisexual, radially symmetric. Sepals 3–6, sometimes petal-like. Petals absent. Stamens 3–6. Ovary inferior or at least partly so (in all cultivated genera), 1-celled with 2 or 3 (rarely 1) ovules borne at the apex of a basal, central placenta; style absent or present, stigmas 2–6. Fruit a nut or drupe.

A family of 35 genera and about 400 species, mostly found in the tropics, though some extend into temperate regions. All are partially or completely parasitic, and few are found in gardens, though 6 genera and 7 species have been recorded. Little is known of their cultivation requirements; some are host-specific (e.g. *Buckleya*, see below), while others are apparently able to grow on a range of host shrubs or trees.

1a. Leaves small, scale-like, soon falling, most of the plant leafless   **1. Leptomeria**
  b. Leaves well-developed, persistent   **2**

2a. Leaves opposite or almost so   3
 b. Leaves all alternate   4
3a. Flowers unisexual, males and females
    on separate plants   **2. Buckleya**
 b. Flowers bisexual   **6. Santalum**
4a. Plant with herbaceous stems, woody
    only at or near ground level
                              5. Comandra
 b. Plant with persistent, aerial, woody
    branches   5
5a. Male and female flowers in spikes
                              3. Pyrularia
 b. Male flowers in axillary racemes;
    female flowers solitary or in clusters of
    up to 3   **4. Osyris**

1. LEPTOMERIA R. Brown
J. Cullen

Small shrubs. Leaves scale-like, small, soon
falling, most of the plant leafless; ends of
the twigs somewhat spine-like. Flowers in
racemes, small, bisexual. Sepals 5. Stamens
5. Fruit a succulent drupe.

A genus of about 15 species from
Australia.

1. L. billardieri R. Brown. Illustration:
Hodgson & Paine, A field guide to
Australian wild flowers **2**: pl. 83 (1977).
Erect shrub. Perianth white. Drupe red.
Australia (New South Wales, Queensland).
G1.

2. BUCKLEYA Torrey
J. Cullen

Shrubs. Leaves opposite, in 2 ranks, shortly
stalked, lanceolate. Flowers unisexual,
males and females on separate plants. Male
flowers in umbel-like racemes, with 4 ovate
sepals and 4 stamens, without patches of
hairs on the sepals behind the stamens.
Female flowers solitary, terminal or
axillary, with 4 sepal-like bracts and 4
sepals; style 1, stigmas 4. Fruit a nut.

A genus of about 5 species from eastern
N America and China.

1. B. distichophylla (Nuttall) Torrey.
Illustration: Gleason, Illustrated Flora of the
north-eastern United States and adjacent
Canada **2**: 61 (1952).
Shrub to 4 m. Leaves 2–6 cm. Flowers
greenish. Fruit ellipsoid, 1–1.5 cm. *Eastern
USA (Virginia, N Carolina, Tennessee).* ?H5.
Parasitic on species of *Tsuga.*

3. PYRULARIA Michaux
J. Cullen

Large shrubs, downy when young. Leaves
alternate, shortly stalked. Flowers
unisexual, males and females on separate
plants. Male flowers with 5 sepals and 5
stamens, each sepal with a patch of hairs
just behind the attachment of the stamen.
Female flowers with 5 sepals and an ovary
that tapers strikingly towards the base,
making the flowers appear to be
superficially stalked. Fruit a pear-shaped
drupe.

A genus of 2 or 3 species, 1 from N
America, the others from the Himalaya.

1. P. pubera Michaux. Illustration: Gleason,
Illustrated Flora of the north-eastern United
States and adjacent Canada **2**: 59 (1952).
Shrub to 4 m. Leaves obovate-oblong,
acute. Spikes few-flowered. Drupe
2–2.5 cm or more. *Eastern USA
(Pennsylvania to Georgia and Alabama).* ?H5.
Parasitic on other shrubs. The whole
plant produces an acrid, poisonous oil.

4. OSYRIS Linnaeus
J. Cullen

Shrubs. Leaves alternate, shortly stalked.
Flowers unisexual, males and females on
separate plants. Male flowers in axillary
racemes, with 3–4 sepals and 3–4 stamens,
each sepal with a patch of hairs just behind
the attachment of the stamen. Female
flowers solitary or in clusters of 2 or 3 in
the leaf-axils, ovary tapering towards the
base. Fruit a drupe.

A genus of about 10 species extending
from the Mediterranean area across Asia to
China; parasitic on other shrubs.

1. O. alba Linnaeus. Illustration: Bonnier,
Flore complète **9**: pl. 537 (1957);
Guittoneau & Huon, Connaitre et
reconnaitre la flore et la vegetation
méditerranéennes, 75 (1983).
Shrub to 1.2 m with many slender
branches. Leaves linear-lanceolate,
leathery. Sepals 3, whitish. Drupe red,
5–7 mm. *Mediterranean area.* H5.

5. COMANDRA Nuttall
J. Cullen

Plants with woody, creeping stocks at
ground level and erect, simple, herbaceous
flowering stems. Leaves alternate, almost
stalkless. Flowers bisexual, in terminal
panicles. Sepals 5, petal-like, white.
Stamens 5, each with a tuft of hairs behind
its attachment. Style long. Fruit a nut.

A genus of 6 species, 1 in Europe, the
rest in N America.

1. C. pallida de Candolle. Illustration:
Gleason, Illustrated Flora of the north-
eastern United States and adjacent Canada
2: 59 (1952); Rickett, Wild flowers of the
United States **3**: pl. 17 (1969).

Leaves acute, glaucous, firm. Sepals
3–5 mm, white. Fruit 6–10 mm.
N America. H3.

6. SANTALUM Linnaeus
J. Cullen

Trees or shrubs. Leaves opposite or more or
less so, somewhat fleshy or leathery.
Flowers bisexual in axillary or terminal
panicles, cymes or racemes. Sepals 4–5,
each with a tuft of hairs at the insertion of
the stamens. Stamens 4–5. Ovary half-
inferior, style short, 2–4-lobed. Fruit a
drupe.

A genus of about 20 species from tropical
Asia to Australia and extending to Hawaii;
perhaps 2 are in cultivation.

1a. Trees; leaves oblong, margins wavy
                                  1. album
 b. Shrub to 2 m; leaves oblong-linear to
    lanceolate, margins not wavy
                              2. obtusifolium

1. S. album Linnaeus. Illustration:
Botanical Magazine, 3235 (1833).
Tree with very fragrant wood. Leaves
oblong and with wavy margins, 4–8 ×
2–4 cm. Sepals yellowish when young,
later brownish red. Drupe ellipsoid, dark
red, *c.* 1 mm. *Origin uncertain, probably
Malaysia or Indonesia, widely cultivated in the
Asiatic tropics.* G2.
Cultivated for its scented wood and oil,
from which a perfume (Sandalwood) is
made.

2. S. obtusifolium R. Brown. Shrub to 2 m.
Leaves oblong-linear to lanceolate, margins
not wavy, 5–8 cm, glaucous. Sepals
whitish. Drupe 6–8 mm, black. *Australia
(New South Wales).* G1.

XLVII. LORANTHACEAE

Green shrubs parasitic on various trees.
Leaves opposite or more rarely whorled,
sometimes scale-like, entire, without
stipules. Flowers small, unisexual (in all
cultivated species), radially symmetric.
Perianth of 1 or 2 whorls. Calyx 2–6-lobed.
Petals 4–6 or absent. Stamens 2–6,
attached to the petals, at least at their
bases. Ovary inferior, 1-celled, the ovules
not differentiated from the placenta. Fruit a
dry or sticky berry; seed 1, embryos 1–3.

A mainly tropical, parasitic family of
about 40 genera and 1400 species. Only a
few are grown for ornament, though 1,
Viscum album, has been an important cult-
object in the past and is still grown for
house-decoration at Christmas.

1a. Leaves scale-like, joined in pairs, each pair forming a sheath around the stem **1. Arceuthobium**
  b. Leaves with broad blades, not joined in pairs 2
2a. Leaves evergreen with parallel veins **2. Viscum**
  b. Leaves deciduous with pinnate veins **3. Loranthus**

1. ARCEUTHOBIUM Bieberstein
J. Cullen

Evergreen shrub. Leaves scale-like, joined at the base in pairs, each pair forming a sheath around the stem. Flowers solitary or in pairs. Perianth of 1 whorl, 2–5-lobed. Male flowers stalkless with 2–5 stamens. Female flowers shortly stalked. Fruit a dry, green berry which dehisces explosively.

A genus of 15 species from the northern hemisphere; only 1 is cultivated as a curiosity.

1. A. oxycedri (de Candolle) Bieberstein. Illustration: Fiori & Paoletti, Iconographia Florae Italicae 1: 110 (1895–99); Coste, Flore de la France 3: 220 (1906).
Found on species of *Juniperus*. Stems tufted, jointed. Leaves 0.5–1 mm. Fruit 2–3 mm, ovoid, acute. *Widely distributed from the Azores through the Mediterranean area east to India and south to Kenya.* H5–G1. Late summer–autumn.

2. VISCUM Linnaeus
J. Cullen

Evergreen shrub with hard wood. Leaves well developed, with parallel veins. Flowers crowded in cymes. Calyx 4-lobed or 4-toothed in the female flowers, absent in the male. Petals usually 4. Stamens 4, joined to the petals for most of their length, anthers opening by pores. Fruit a sticky, white or yellow berry.

A widespread genus of about 65 species, parasitic on various trees.

1. V. album Linnaeus. Illustration: Fiori & Paoletti, Iconographia Florae Italicae 1: 110 (1895–99); Coste, Flore de la France 3: 219 (1906); Ceballos et al., Plantas silvestres de la peninsula Iberica, 66 (1980).
Stem to 1 m, yellowish green. Leaves 2–8 cm, occasionally whorled, obtuse, with 3–5 veins. Cymes stalkless. Berry 6–10 mm, spherical or pear-shaped, white or occasionally yellow. *Temperate Eurasia, N Africa.* Spring. H2.

A variable species, divided usually into 3 units which are mostly treated as subspecies: subsp. **album** occurs on

dicotyledonous trees, and is found throughout the range of the species; subsp. **abietis** (Wiesbaur) Abromeit is usually found on species of *Abies* in C & S Europe and N Turkey; and subsp. **austriacum** (Wiesbaur) Vollman is found on species of *Pinus* and *Larix* in C & S Europe, Turkey and the Caucasus.

3. LORANTHUS Jacquin
J. Cullen

Deciduous shrub. Leaves well-developed, with pinnate venation. Flowers in racemes or spikes. Calyx small with 4 fringed teeth. Petals 4–6. Stamens attached to the petals only at their bases. Fruit a sticky berry.

A mainly tropical genus of 500 or more species.

1. L. europaeus Jacquin. Illustration: Fiori & Paoletti, Iconographia Florae Italicae 1: 109 (1895–99).
Parasitic on members of the *Fagaceae* (see p. 59). Stem to 50 cm, brown. Leaves 1–6 cm × 5–15 mm, obtuse, dull green. Berry to 1 cm, spherical to pear-shaped, yellow. *C & SE Europe, Turkey.* H5. Summer.

XLVIII. POLYGONACEAE

Annual to perennial herbs, shrubs or climbers, rarely trees. Leaves mostly alternate, simple, variable in outline; stipules often united into a membranous sheath (*ochrea*). Flowers bisexual or unisexual, radially symmetric, often in racemes or raceme-like inflorescences. Perianth-segments 3–6, sepal- or petal-like, free or united at the base, often enlarging in fruit and becoming hardened, papery or fleshy. Fruit a 3-angled or flattened and 2-angled nut, usually enclosed by the persistent perianth.

A family of about 30 genera and 800 species, mostly in north temperate regions. Relatively few species are cultivated (most of these in the genus *Polygonum*), but the family includes several important weeds. Literature: Roberty, G. & Vautier, S., Les genres de Polygonacés, *Boissiera* 10: 7–128 (1964).

1a. Erect, branched shrubs or trees 2
  b. Herbs, dwarf shrubs or woody climbers 5
2a. Branches flattened, 1–2 cm wide; flowers in stalkless clusters **9. Homalocladium**

  b. Branches more or less cylindric; flowers in stalked racemes or panicles 3
3a. Often spiny shrubs; leaves less than 3 cm **5. Atraphaxis**
  b. Spineless trees or shrubs; leaves usually at least 3 cm 4
4a. Flowers in broad panicles; perianth papery in fruit **2. Rumex**
  b. Flowers in narrow racemes; perianth fleshy in fruit **10. Coccoloba**
5a. Perianth-segments 4; nut flattened, 2-angled **3. Oxyria**
  b. Perianth-segments 5 or 6; nut usually not flattened, usually 3-angled 6
6a. Woody climbers 7
  b. Herbs or dwarf shrubs 9
7a. Plant climbing by tendrils; perianth-segments usually bright pink **7. Antigonon**
  b. Plant not climbing by tendrils; perianth-segments greenish or white, sometimes flushed with pink 8
8a. Perianth not fleshy in fruit **6. Polygonum**
  b. Perianth fleshy in fruit **8. Muehlenbeckia**
9a. Stem-leaves few or absent; stamens always 9 10
  b. Stem-leaves usually numerous; stamens not more than 8 11
10a. Leaves palmately lobed; flowers in long panicles **4. Rheum**
  b. Leaves not palmately lobed; flowers in compact heads surrounded by whorls of bracts **1. Eriogonum**
11a. Perianth white and fleshy in fruit **8. Muehlenbeckia**
  b. Perianth usually not fleshy in fruit, if fleshy then reddish 12
12a. Perianth-segments 5, white or pink, equal in fruit or the outer larger **6. Polygonum**
  b. Perianth-segments 6, greenish, the inner much larger than the outer in fruit **2. Rumex**

1. ERIOGONUM Michaux
J.R. Akeroyd

Annual to perennial herbs or small shrubs. Leaves mostly basal, small, alternate or in whorls, entire. Ochreae absent. Flowers bisexual, in compact heads or umbels partly enclosed by a tubular or bell-shaped involucre of 4–8-lobed bracts. Perianth-segments 6, petal-like, united in lower part, enlarged in fruit. Stamens 9. Stigmas 3. Fruit a 3-angled nut.

A genus of some 150 species mostly from Western N America. A number of species has been cultivated, especially in rock

gardens, but few are at all widely grown in Europe. The majority are rather uninteresting, although some are valued for their grey or silvery leaves. They require well-drained soils and a sunny position as the leaves rot easily under damp winter conditions. Propagation is by seed.
Literature: Stokes, S.G., *The genus Eriogonum* (1936).

1a. Small shrub; leaves more or less linear **6. fasciculatum**
 b. Herbaceous perennial, often woody at base; leaves obovate, elliptic or ovate to circular 2
2a. Flowers in solitary terminal heads or clusters; leaves less than 2 cm 3
 b. Flowers in simple or compound umbels; leaves usually more than 2 cm 4
3a. Flowering stem rarely more than 8 cm; flowers often orange or reddish **1. caespitosum**
 b. Flowering stems usually more than 8 cm; flowers cream or yellow **3. ovalifolium**
4a. Leaves at least 5 cm **4. compositum**
 b. Leaves rarely as much as 5 cm 5
5a. Leaves green above; flowers in simple umbels **2. umbellatum**
 b. Leaves white-hairy on both surfaces; flowers in compound umbels **5. niveum**

1. E. caespitosum Nuttall. Illustration: Bulletin of the Alpine Garden Society **53**: 253 (1985).
Compact, mat-forming perennial, woody at base. Leaves crowded, 5–10 mm, elliptic to spathulate, with short stalks, densely covered with silvery grey hairs. Flowering stems 3–8 cm, bearing solitary terminal heads of yellow to orange or reddish flowers. *W USA*. H1. Summer.
Subsp. **douglasii** (Bentham) Stokes (*E. douglasii* Bentham). Leaves 1–2 cm; flowering stems to 12 cm, with a whorl of bracts below the flowers and larger perianth. Infrequent in cultivation.

2. E. umbellatum Torrey (*E. speciosum* Drew; *E. subalpinum* Greene). Illustration: Journal of the Royal Horticultural Society **93**: t. 153 (1968).
Low, spreading, rather woody perennial. Leaves in rosettes, 2.5–5 cm, ovate to spathulate, green above, white-hairy beneath, the stalk about as long as the blade. Flowering stems 8–30 cm bearing simple, 5–10-rayed umbels of 20–30 sulphur-yellow or (var. **subalpinum** (Greene) Jones) cream flowers. *Western N America*. H1. Summer.

A very variable species with several subspecies and varieties. Var. **torreyanum** (Gray) Jones (*E. torreyanum* Gray) has hairless, dark green leaves. *USA (N California, S Oregon)*. Var. **polyanthum** (Bentham) Jones (*E. polyanthum* Bentham) has deep yellow to reddish flowers. *USA (California)*.

3. E. ovalifolium Nuttall (*E. proliferum* Torrey & Gray). Illustration: Gardeners' Chronicle 7: 260 (1890); Journal of the Royal Horticultural Society **70**: 292 (1945).
Compact, mat-forming perennial. Leaves 5–18 mm, spathulate, blade elliptic to almost circular, covered with silvery grey hairs. Flowering stems 8–20 cm, bearing globular heads *c.* 2 cm across, with up to 25 greenish cream to yellow flowers. *Mountains of western N America*. H1. Summer.
Subsp. **vineum** (Small) Stokes (*E. vineum* Small; *E. ovalifolium* var. *purpureum* (Nuttall) Nelson). Leaves more or less circular; flowers pink to purplish. *USA (California, Oregon)*.

4. E. compositum Bentham. Illustration: Die Natürlichen Pflanzenfamilien 3(1A): 13 (1893); Bulletin of the Alpine Garden Society **47**: 46 (1979).
Rather loose perennial. Leaves 5–20 cm with long stalks, blade ovate to ovate-lanceolate, wedge-shaped to somewhat cordate at base, green above, white-hairy beneath. Flowering stems 20–40 cm, rather stout, weakly ascending to erect, bearing compound umbels of pale yellow flowers. *Western N America*. H1. Summer.

5. E. niveum Bentham.
Compact, white-hairy perennial. Leaves in whorls, 1.2–2.5 cm, ovate to ovate-oblong, stalk about as long as blade. Flowering stems 10–25 cm, bearing compound umbels of cream or pink flowers. *Western N America*. H1. Summer.
Infrequent in cultivation.

6. E. fasciculatum Bentham.
Small shrub to 1 m with leafy, branched stems. Leaves in clusters, 5–15 mm, linear-lanceolate to linear-oblong, inrolled, green above, white-woolly beneath. Flowers in loose, terminal heads at the ends of small, leafless branches, white or pinkish. *Western USA (California to Utah)*. H3. Summer.
Infrequent in cultivation.

2. RUMEX Linnaeus
J.R. Akeroyd
Annual or perennial herbs, rarely shrubs, hairless or inconspicuously hairy, usually

with stout roots or rhizomes. Flowers bisexual or unisexual, in whorls arranged in panicles, green or reddish, wind-pollinated. Perianth-segments 6 in 2 whorls of 3, the outer whorl usually inconspicuous, the inner whorl enlarged in fruit and becoming hardened or papery, usually 1, 2 or all 3 segments with corky tubercles developing (the ripe inner perianth-segments are often known as 'valves'). Stamens 6. Stigmas 3, feathery. Fruit a 3-angled nut, enclosed by the enlarged inner perianth-segments.

A genus of some 150 species, mainly from North Temperate regions. Few are cultivated; several are notorious weeds of cultivation, which is probably why the genus is unpopular with gardeners. The flowers are small and uninteresting, but the plants are decorative in fruit. Propagation is by seed or by division of the rootstock in perennial species.
Literature: Lousley, J.E., Notes on British Rumices, I, II, *Botanical Society and Exchange Club Reports* **12**: 118–157 (1939), 547–585 (1944); Lousley, J.E. & Kent, D.H., *Docks and Knotweeds of the British Isles* (1981).

1a. Flowers unisexual, male and female usually on separate plants; leaves with an acid taste 2
 b. Flowers bisexual; leaves without an acid taste 4
2a. Shrub; leaves frequently as wide as or wider than long **1. lunaria**
 b. Perennial herb, sometimes slightly woody at the base; leaves usually longer than wide 3
3a. Stems many, branched near base; leaves shield-, kidney- or fiddle-shaped **2. scutatus**
 b. Stems few, branched in upper part; leaves ovate-oblong **3. rugosus**
4a. Basal leaves almost circular, less than twice as long as broad; perianth-segments without corky tubercles in fruit **4. alpinus**
 b. Basal leaves lanceolate, at least 3 times as long as broad; perianth usually with 1–3 corky tubercles in fruit 5
5a. Stem usually less than 1 m; perianth less than 5 mm long in fruit, with 0–1 corky tubercles 6
 b. Stems 1–2 m; perianth at least 5 mm long in fruit, with 1–3 corky tubercles 7
6a. Leaves green with purple or reddish veins; perianth with 1 almost spherical tubercle in fruit **7. sanguineus**

b. Leaves reddish brown; perianth without tubercles in fruit **8. flexuosus**

7a. Basal leaves usually at least 50 cm; perianth with 3 equal corky tubercles in fruit **5. hydrolapathum**

b. Basal leaves not more than 50 cm; perianth with 1 or 3 unequal corky tubercles in fruit **6. patientia**

1. R. lunaria Linnaeus. Illustration: Bramwell & Bramwell, Wild flowers of the Canary Islands, t. 127 (1974); Pignatti, Flora d'Italia 1: 150 (1982).
Dioecious shrub to 1.5 m with flexuous branches. Leaf-blades 3–6 × 3–7 cm, broadly ovate, truncate to slightly cordate at base, more or less acute or obtuse, thick, entire; stalk shorter than blade. Ochreae to 8 mm, papery. Flowers in broad, compound panicles to 15 cm. Inner perianth-segments 5 × 7 mm, rounded to kidney-shaped, cordate, rounded or truncate at base, entire, papery when ripe, each with a small tubercle near the base. Nut 3 mm. *Canary Islands.* H5. Winter–early spring.

Often planted as a hedge in mild regions of S Europe.

2. R. scutatus Linnaeus. Illustration: Bonnier, Flore complète 9: 523 (1927); Hess et al., Flora der Schweiz 1: 721 (1967); Pignatti, Flora d'Italia 1: 150 (1982).
Rather untidy, often glaucous perennial herb, with tough, branched rootstock. Stems 20–60 cm, ascending, branched mostly near base, with long internodes. Leaves mostly basal, 1–6 cm, shield-, fiddle- or kidney-shaped, obtuse, fleshy, the uppermost sometimes hastate; stalk about as long as blade. Ochreae silvery. Flowers unisexual and bisexual on the same plant, in loose, little-branched racemes. Flower-stalks 2–4 mm, thread-like. Perianth-segments 4–7 mm in fruit, more or less circular, purplish pink or red, ripening to pale brown and papery; tubercles absent. *Europe, mainly in the mountains, W Asia, N Africa.* H1. Summer.

Grown both as a culinary herb and as an ornamental for its glaucous foliage.

3. R. rugosus Campdera (*R. acetosa* misapplied). Illustration: Botaniska Notiser for 1944, 251 – as R. acetosa subsp. ambiguus.
Dioecious perennial herb. Stems 50–160 cm, erect, stout. Basal leaves to 15 × 12 cm, usually 2–4 times as long as wide, ovate-oblong, with a pair of prominent, deflexed basal lobes, rounded at

apex, somewhat fleshy. Leaves smaller up the stem, the upper leaves stalkless, triangular, more or less acute. Flowers in rather loose, repeatedly branched panicles. Inner perianth-segments 3–4 mm in fruit, reddish at first, brown and papery when ripe and soon falling. Nut *c.* 2 mm, brown, glossy. *Known only from cultivation.* H1.

Grown as a vegetable and not usually encouraged to flower. It is similar to *R. acetosa* Linnaeus, the wild sorrel of grasslands, which occurs occasionally as a garden weed in W Europe, and from which it may have been derived.

4. R. alpinus Linnaeus. Illustration: Bonnier, Flore complète 9: 525 (1927); Hess et al., Flora der Schweiz 1: 724 (1967); Polunin, Flowers of Europe, pl. 9 (1969); Pignatti, Flora d'Italia 1: 153 (1982).
Robust perennial herb with stout, creeping rhizome to 3 cm thick. Stems 50–150 cm, erect. Basal leaves with long stalks; blade 15–35 × 12–25 cm, broadly ovate to almost circular, cordate at base, rounded, entire, shortly and roughly hairy on veins of lower surface. Stem-leaves few, smaller. Flowers in dense, broad, much-branched, almost leafless pyramidal panicles to 35 cm. Inner perianth-segments 3–5 mm in fruit, ovate-triangular, obtuse, hardened and brown when ripe; tubercles absent. *Mountains of Europe and SW Asia.* H1. Summer.

5. R. hydrolapathum Hudson. Illustration: Polunin, Flowers of Europe, pl. 9 (1969); Ross-Craig, Drawings of British plants 26: pl. 16 (1969); Pignatti, Flora d'Italia 1: 153 (1982).
Robust perennial herb with stout rootstock. Stems 1–2 m, erect. Basal leaves 50–100 cm (rarely as little as 30 cm), more or less erect, broadly lanceolate, equally tapered at both ends, about 5 times as long as wide, acute, entire. Flowers in rather dense panicles with spreading branches. Inner perianth-segments 5–7 mm in fruit, triangular, more or less entire, each with an elongate, ovoid tubercle, hardened and dark brown when ripe. Nut 3–5 mm, brown. *Europe.* H1. Summer.

This plant requires moisture throughout the year and grows best next to water.

6. R. patientia Linnaeus.
Robust perennial herb. Stems 1–2 m, erect. Basal leaves with blade 15–45 × 5–12 cm, ovate- to oblong-lanceolate, truncate or wedge-shaped at base, acute; stalk about as long as blade. Flowers in dense, branched

panicles, leafless above. Inner perianth-segments 6–10 mm in fruit, broadly ovate, cordate at base, entire or minutely toothed, hardened and brown when ripe; 1 ovoid tubercle developed (occasionally 2 smaller tubercles also). Nut 3–3.5 mm, pale brown, glossy. *C & E Europe, Asia, NW Africa.* H2. Summer.

Formerly widely cultivated as a vegetable.

7. R. sanguineus Linnaeus. Illustration: Bonnier, Flore complète 9: 528 (1927) – var. sanguineus; Ross-Craig, Drawings of British plants 26: pl. 21 (1969).
Perennial herb. Stems 30–100 cm, erect, slender, often purple or reddish. Leaves 5–15 × 2–5 cm, oblong, acute, green with purple or blood-red veins. Flowers in a more or less leafless panicle with few rather upright branches; whorls of flowers widely spaced. Inner perianth-segments 2.5–3 mm in fruit, oblong, 1 with a more or less spherical tubercle which is shorter than the segment, red while developing; perianth hardening and becoming brown when ripe, soon falling. *Europe, SW Asia, N Africa.* H1. Summer.

The cultivated plant is var. **sanguineus**, long grown both as a pot-herb and as an ornamental. The wild plant, var. **viridis** Sibthorp, from which var. *sanguineus* was probably selected, lacks the purple coloration; it is sometimes a weed in shady gardens.

8. R. flexuosus J.D. Hooker.
Perennial herb. Stems 20–50 cm, slender, flexuous, branched. Leaves narrowly lanceolate to linear, acute, reddish brown, with wavy margins. Flowers in very loose panicles with spreading branches; whorls of flowers widely spaced. Inner perianth-segments *c.* 4 mm, ovate to diamond-shaped, acuminate, with 2–4 hooked, spine-like teeth on each margin, hardened and brown when ripe; tubercles absent. Nut *c.* 2 mm, pale brown. *New Zealand.* H2. Summer.

3. OXYRIA Hill
J.R. Akeroyd
Hairless perennial herbs with stout, scaly rootstocks. Leaves almost all basal, with long stalks. Flowers in loose, leafless panicles. Perianth-segments sepal-like, in 2 pairs, those of the inner pair enlarged in fruit. Flowers bisexual. Stamens 6. Stigmas 2. Fruit lens-shaped, broadly winged.

A genus of 2 species, widespread in the arctic and mountainous regions of the northern hemisphere. One species is cultivated in alpine and rock gardens, in shady but well-drained sites. Propagation is by seed or division of the rootstock.

1. O. digyna (Linnaeus) Hill. Illustration: Bonnier, Flore complète 9: 521 (1927); Keble Martin, The concise British Flora in colour, 74 (1965); Hess et al., Flora der Schweiz 1: 732 (1967); Ross-Craig, Drawings of British plants 26: pl. 15 (1969).
Stems 10–40 cm, erect. Leaf-blades 1–6 cm, cordate at base, kidney-shaped, rounded, entire, somewhat fleshy. Ochreae 2.5–5 mm, papery, pale brown. Inner perianth-segments 3–4 mm in fruit, adpressed to one another, circular, notched at apex, ripening from green through red or purplish to pale brown, papery when ripe. Flower-stalks to 5 mm, thread-like. *Europe southwards to C Spain and Greece, Asia, N America.* H1. Summer.

4. RHEUM Linnaeus
J.R. Akeroyd
Robust perennial herbs with stout, woody rhizomes. Leaves mostly basal, large, palmately lobed. Ochreae loose, persistent. Flowers bisexual, in panicles, wind-pollinated. Perianth-segments 6, free, not enlarging in fruit. Stamens 9. Stigmas 3, almost stalkless. Fruit a 3-winged nut.

A genus of about 50 species from temperate and subtropical Asia, just extending to E Europe. The genus has long been in cultivation in European gardens as a source of drugs, notably purgatives, and for the edible leaf-stalks; several species are grown as ornamentals. Their large size makes them unsuitable for smaller gardens, but they are handsome in a spectacular way. Propagation is by division of the rhizome in winter or spring, or from seed. Hybridisation between species can be a problem for purity of seed-stock. They do best in fairly moist soil (Nos. **5** and **7** perhaps require drier conditions).

1a. Leaf-blade less than 30 cm long or wide; branches of panicle shorter than or only slightly longer than bracts **2**
b. Leaf-blade usually more than 30 cm long and wide; branches of panicle longer than bracts **3**
2a. Bracts not overlapping, partly obscuring flowers **5. alexandrae**
b. Bracts overlapping, completely obscuring the flowers **6. nobile**
3a. Leaves shallowly lobed, entire or minutely toothed **4**
b. Leaves deeply lobed and coarsely toothed **6**

4a. Leaves circular to kidney-shaped, minutely toothed **7. ribes**
b. Leaves ovate to circular, entire **5**
5a. Leaves hairy on both surfaces **1. australe**
b. Leaves hairless above **4. × cultorum**
6a. Leaf-stalks cylindric; leaves multi-lobed **2. palmatum**
b. Leaf-stalks channelled; leaves 5-lobed, the lobes deeply toothed **3. officinale**

1. R. australe D. Don (*R. emodi* Wallich). Illustration: Botanical Magazine, 3508 (1836).
Plant roughly hairy. Leaf-blades 40–75 × 30–60 cm, broadly ovate to almost circular, 5–7-veined, shallowly lobed, cordate at base, obtuse, entire with wavy margins; stalk about as long as blade, stout, channelled above, ridged, reddish. Flowering stems 2–3 m, stout, ridged, reddish. Flowers white in dense panicles with upright branches to 1 m. Fruit 7–10 mm, ovate-triangular, obtuse. *Himalaya.* H1. Summer.

2. R. palmatum Linnaeus. Illustration: Die Natürlichen Pflanzenfamilien 3(1A): 20 (1893); Botanical Magazine, 9200 (1927).
Differs from *R. australe* by the larger leaves: blade 50–90 × 50–70 cm, 3–5-veined, multi-lobed, lobes ovate to lanceolate, acute, often coarsely toothed, glossy and usually hairless above, roughly hairy beneath; stalk almost cylindric. Fruits almost circular, rounded. *NW China (including Xizang).* H1. Summer.
Var. **tanguticum** Regel has very deeply lobed leaves. 'Atrosanguineum' is flushed throughout with reddish crimson. 'Bowles Crimson' has leaves crimson beneath and crimson fruits. The species was formerly widely grown as a purgative.

3. R. officinale Baillon. Illustration: Botanical Magazine, 6135 (1874); Die Natürlichen Pflanzenfamilien 3(1A): 21 (1893).
Differs from *R. australe* by the more or less ovate, 5-veined, 5-lobed leaves which are coarsely toothed with irregular, acute teeth and paler green. Flowers greenish cream or pinkish. *W China (including Xizang).* H1. Summer.
Formerly widely grown as a purgative.

4. R. × cultorum Thorsrud & Reisaeter (*R. rhabarbarum* misapplied; *R. rhaponticum* misapplied; *R. undulatum* misapplied). Illustration: Wehrhahn, Die Gartenstauden 1: 20 (1931); Hess, Landoldt & Hirzel, Flora der Schweiz, 1: 733 (1967).

Differs from *R. australe* by the 5-veined, less hairy leaves (hairy on the veins beneath), leaf-stalks not ridged, shallowly channelled; flowering stems 1–1.5 m, little ridged, branches of panicle to 30 cm, less upright. Fruit ovate, cordate. *Garden origin.* H1. Early summer.

The vegetable rhubarbs have been cultivated for nearly 2 centuries for their edible leaf-stalks; they have a complex hybrid origin involving *R. rhabarbarum* Linnaeus (from E Asia) and other species, and many cultivars are known. The panicle is not usually allowed to develop.

5. R. alexandrae Batalin. Illustration: Journal of the Royal Horticultural Society 76: pl. 151 (1951), 97: pl. 201 (1972).
Leaves dark green, in compact rosettes. Leaf-blade to 20 × 12 cm, ovate-oblong, cordate at base, rounded, hairless; stalk about as long as blade. Flowering stems 1–1.2 m. Ochreae to 10 cm. Flowers greenish, in rather distant axillary panicles, partly hidden by stalkless, ovate, slightly cordate, greenish yellow bracts. *W China (Xizang, Sichuan).* H1. Summer.

6. R. nobile Hooker & Thomson. Illustration: Gardeners' Chronicle 13: 793 (1880); Die Natürlichen Pflanzenfamilien 3(1A): 22 (1893); Morley & Everard, Wild flowers of the world, pl. 96 (1977); Quarterly Bulletin of the Alpine Garden Society 52: 203 (1984).
Like *R. alexandrae* but leaves glossy, with red stalks; ochreae pink; stems 1–2 m; panicles less distant, completely hidden by overlapping, broadly ovate, cream or yellowish bracts. *India (Sikkim).* H1. Summer.

7. R. ribes Linnaeus. Illustration: Botanical Magazine, 7591 (1898).
Leaf-blades 15–40 × 20–70 cm, circular to kidney-shaped, cordate at base, minutely toothed, roughly and shortly hairy beneath, rather greyish green. Flowering stems 1–1.5 m, leafless in upper part, bearing broad, dense panicles. Flowers white. Fruits 8–15 mm, reddish brown. *SW Asia.* H1. Summer.

The leaf-stalks have a flavour reminiscent of currants (*Ribes*).

5. ATRAPHAXIS Linnaeus
J.R. Akeroyd
Dwarf, often spiny shrubs with spreading or twisted branches. Leaves small, deciduous. Ochreae cartilaginous, bifid. Flowers bisexual in short racemes. Perianth-segments 4–5, in 2 whorls, those of the

inner whorl enlarged in fruit. Stamens 6 or 8. Stigmas 2–3. Nut 2-or 3-angled.

A genus of about 20 species in the Near East and Asia. Propagation is by seed or by layering of adult plants. They require well-drained soil and a sunny position.

1a. Perianth-segments 4; leaves 6–12 mm **1. spinosa**
  b. Perianth-segments 5; leaves usually more than 12 mm **2**
2a. Branches not spiny; inner perianth-segments scarcely longer than the nut **2. frutescens**
  b. Branches spiny; inner perianth-segments about twice as long as the nut **3. billardieri**

1. A. spinosa Linnaeus. Illustration: Die Natürlichen Pflanzenfamilien 3(1A): 23 (1893).
Erect shrub 30–80 cm. Branches slender, often spiny, whitish. Leaves 6–12 mm, ovate or elliptic, mucronate; stalks short. Racemes axillary. Perianth-segments 4, pink with white margins, the inner 2 almost circular, 4–5 × 5–6 mm in fruit, the 2 outer much smaller, reflexed. Nut ovate, 2-angled, flattened, pale greenish brown. *W & C Asia, extending to S Russia.* H5. Late summer.

2. A. frutescens (Linnaeus) K. Koch. Illustration: Edwards's Botanical Register, pl. 254 (1818) – as Polygonum frutescens.
Erect shrub 20–70 cm. Branches slender, whitish; older branches with grey bark, not spiny. Leaves 1–2.5 cm × 2–8 mm, lanceolate to oblong or obovate, mucronate, greyish green; stalks short. Racemes terminal, 2–6 cm. Perianth-segments 5, greenish or pinkish white, the inner 3 almost circular, 4–5 × 5–6 mm in fruit, the outer much smaller, reflexed. Nut 3-angled, not flattened, dark brown, glossy, slightly shorter than the enclosing perianth. *W & C Asia, extending to S Russia.* H5. Late summer.

3. A. billardieri Jaubert & Spach. Illustration: Botanical Magazine, 8820 (1919); Strid, Mountain flora of Greece 1: 80 (1986).
Differs from *A. frutescens* by the spiny branches, leaves ovate to oblong and usually acute but not mucronate; inner perianth-segments pink, 5–9 × 4–8 mm in fruit, about twice as long as the nut. *SW Asia, extending to S Greece.* H5. Summer.

6. POLYGONUM Linnaeus
J.R. Akeroyd
Annual to perennial herbs, dwarf shrubs or climbers. Leaves usually distinctly longer than wide. Flowers bisexual, in spikes or panicles. Perianth-segments usually 5 (rarely 4), usually equal, free or united at the base, petal-like. Stamens 5–8. Stigmas 2–3. Nut 3-angled or 2-angled and flattened, enclosed by the persistent, sometimes winged, perianth-segments, or protruding by up to half its length.

A genus of about 300 species, mostly in north temperate regions; some species are serious weeds of cultivation, but a number are widely cultivated in gardens. Most require fairly moist conditions. Propagation is by seed or division of the rootstocks.

The genus is often divided up into a number of smaller genera. There are correlated differences between these in habit, ochreae, inflorescence, floral parts, chromosome numbers and especially pollen, but these are not consistent and some further research is needed. Despite good evidence for acceptance of these segregate genera, *Polygonum* forms a more or less natural and traditional unit. The present account takes a broad view of the genus, not least in order to simplify the names used.

1a. Plant usually at least 2 m; outer perianth-segments winged or keeled in fruit **2**
  b. Plant rarely more than 2 m; outer perianth-segments neither winged nor keeled in fruit **5**
2a. Twining perennials **3**
  b. Erect, rhizomatous perennials **4**
3a. Perianth in fruit 4–6 mm wide, white or rarely pinkish; axes of panicles roughly hairy **22. aubertii**
  b. Perianth in fruit 6–8 mm wide, frequently pink; axes of panicles hairless **23. baldschuanicum**
4a. Leaves broadly ovate, truncate at base; flowers white or cream **24. japonica**
  b. Leaves ovate-oblong, somewhat cordate at base; flowers greenish **25. sachalinense**
5a. Leaves triangular, about as wide as long; fruit about twice as long as perianth **6**
  b. Leaves not triangular, usually distinctly longer than wide; fruit shorter than perianth or about half of it protruding **7**
6a. Annual; panicles dense, somewhat corymbose **26. fagopyrum**
  b. Perennial; panicles rather loose, spike-like **27. dibotrys**
7a. Flowers in wide, diffuse panicles; perianth usually white or cream **8**
  b. Flowers in spikes, sometimes branched; perianth usually pink, red or purple **12**
8a. Leaves not more than 3 cm wide; ochreae whitish, soon disappearing **17. alpinum**
  b. Leaves 3–20 cm wide; ochreae brown, papery, persistent **9**
9a. Leaves with dense, pinkish brown or white hairs beneath; plant with stolons **20. campanulatum**
  b. Leaves hairy but not densely so beneath; plant with rhizomes **10**
10a. Flowers greenish; stalks of lower stem-leaves more than 3 cm **21. weyrichii**
  b. Flowers white or cream; stalks of lower stem-leaves 1–3 cm **11**
11a. Flowers in loose, leafy panicles; perianth papery in fruit **18. polystachyum**
  b. Flowers in dense, more or less leafless panicles; perianth fleshy in fruit **19. molle**
12a. Plant *Juncus-* or *Equisetum*-like, with very loose spikes of white flowers; leaves usually absent during flowering **1. scoparium**
  b. Plant not as above, with loose to dense spikes or usually pink, red or purple flowers; leaves present during flowering **13**
13a. Plant aquatic with floating leaves **2. amphibium**
  b. Plant terrestrial **14**
14a. Stems prostrate, creeping, often forming mats **15**
  b. Stems erect, never forming mats **18**
15a. Plant hairy **16**
  b. Plant hairless **17**
16a. Inflorescence-stalk 1–3 cm; leaves rarely more then 2 cm wide **5. capitatum**
  b. Inflorescence-stalk more then 3 cm; leaves usually more than 2 cm wide **6. alatum**
17a. Leaves at least 3 cm; flowers in dense spikes **7. affine**
  b. Leaves not more than 3 cm; flowers in loose spikes **8. vacciniifolium**
18a. Plant annual, densely hairy **3. orientale**
  b. Plant perennial **19**
19a. Flowers in very loose spikes 10–30 cm; stigmas conspicuously persistent in fruit **16. virginianum**
  b. Flowers in more or less dense spikes usually less than 10 cm; stigmas not conspicuously persistent in fruit **20**
20a. Plant densely white-hairy; rootstock creeping **4. lanigerum**

b. Plant hairless or hairy only on the lower surfaces of the leaves; rootstock short and thick **21**

21a. Flowers in lower part of inflorescence replaced by purplish brown bulbils **13. viviparum**

b. Bulbils not present **22**

22a. Stems less than 20 cm; flowers white **15. tenuicaule**

b. Stem at least 20 cm; flowers pink, red or purple (rarely white) **23**

23a. Spikes of flowers often in pairs; stem-leaves many; ovate **14. amplexicaule**

b. Spikes of flowers usually single; stem-leaves few **24**

24a. Flowers in loose, somewhat drooping spikes; perianth-segments 5–8 mm **12. griffithii**

b. Flowers in dense spikes; perianth-segments 3–5 mm **25**

25a. Blades of basal leaves wedge-shaped; flowers crimson **11. milletii**

b. Blades of basal leaves truncate; flowers pink or rarely white **26**

26a. Leaf-stalks winged in upper part; flowers almost stalkless **9. bistorta**

b. Leaf-stalks not winged; flowers with distinct stalks *c.* 5 mm **10. macrophyllum**

1. P. scoparium Loiseleur (*P. equisetiforme* misapplied). Illustration: Coste, Flore de la France **3**: 208 (1906); Bolletino della Società Sardi de Scienze Naturali **17**: 290 (1977); Plantsman **8**: 230 (1987).
Hairless perennial with woody rootstock. Stems numerous, *Juncus*- or *Equisetum*-like, 30–80 cm, erect, with few branches. Ochreae 3–5 mm, truncate, brown. Leaves to 1.5 cm, narrowly elliptic, more or less acute, fallen before flowering. Flowers axillary, solitary or in pairs in very loose terminal spikes. Perianth-segments 2–3 mm, broadly elliptic, white. Nut rarely produced. *Corsica, Sardinia.* H3. Late summer–autumn.

Plants in cultivation (in Britain, at least) appear to belong to one or a very few clones.

2. P. amphibium Linnaeus (*Persicaria amphibia* (Linnaeus) Gray). Illustration: Polunin, Flowers of Europe, pl. 8 (1969); Ross-Craig, Drawings of British plants **26**: t. 5 (1969).
Hairless, aquatic perennial with creeping rootstock. Stems 30–100 cm, rooting at the nodes. Leaves floating, with long stalks; blade 5–15 cm, ovate-oblong, truncate to slightly cordate at base, more or less acute at apex. Flowers in dense, ovoid, terminal spikes 2–4 cm. Perianth-segments pink or

red. Stamens projecting. Nut *c.* 2 mm, brown, rather shiny. *Europe: widely distributed in temperate regions.* H1. Summer.

Can be grown on land, where it has an erect habit and stalkless, oblong-lanceolate leaves, but this growth-form is untidy, produces fewer flowers, and behaves as a weed.

3. P. orientale Linnaeus (*Persicaria orientalis* (Linnaeus) Vilmorin).
Densely hairy annual herb. Stems 50–100 cm, erect, robust. Leaves to 25 × 12 cm, ovate, truncate to slightly cordate at base, acuminate at apex; stalk much shorter than blade. Ochreae brown, often with green, leaf-like lobes on upper margin. Flowers red or purple, in rather dense, branched spikes 2–8 cm. Nut 3 mm, more or less lens-shaped, brown, glossy. *E & SE Asia.* H2. Summer–early autumn.

4. P. lanigerum R. Brown (*P. senegalense* misapplied). Illustration: Botanical Journal of the Linnaean Society **95** (1987).
Differs from *P. orientale* by the perennial habit, narrower, white-hairy leaves and pink or white flowers. *Tropical Africa & Asia.* H3. Autumn.

Infrequently cultivated; sometimes grown for foliage in summer bedding.

5. P. capitatum Buchanan-Hamilton (*Persicaria capitata* (Buchanan-Hamilton) Gross). Illustration: The Garden **106**: 509 (1981); Polunin & Stainton, Flowers of the Himalaya, pl. 107 (1984); Sjogren, Açores Flores, pl. 86 (1984).
Glandular-hairy, creeping perennial. Stems to 20 cm, sometimes weakly ascending, slender, rooting at lower nodes. Leaves 2–5 × 1–2.5 cm, ovate to elliptic, often reddish, with dark V-shaped stripe. Flowers pink, in dense, almost spherical heads 5–10 mm in diameter; inflorescence-stalk 1–3 cm. Nut *c.* 2 mm. *Himalaya.* H5. Summer.

6. P. alatum D. Don (*Persicaria alata* (D. Don) Gross). Illustration: The Garden **108**: 110 (1983).
Prostrate perennial; stems 1–2 m, rooting at the nodes, reddish. Leaves all along the stems; blades 3–7 × 2–4 cm, ovate-oblong, cordate at base, acuminate at apex, shortly hairy beneath, pale green with paler V-shaped markings, the margins minutely crisped, ciliate; stalk 2–3 cm, narrowly winged, reddish. Ochreae 1.5–3 cm, hairless or shortly downy, pale brown, striate. Flowers pale pink in loose heads 1–1.5 cm in diameter. Inflorescence-stalk

4–8 cm, glandular-hairy above. *Himalaya.* H1. Summer.

A vigorous plant increasingly grown for ground-cover.

7. P. affine D. Don (*P. brunonis* Meissner; *Bistorta affinis* (D. Don) Greene). Illustration: Botanical Magazine, 6472 (1880); RHS Dictionary of gardening, 1628 (1956); Everard & Morley, Wild flowers of the world, 96 (1970); Polunin & Stainton, Flowers of the Himalaya, pl. 106 (1984).
Hairless perennial, forming mats of prostrate stems. Leaves mostly basal, reddish bronze in autumn; blade 5–15 × 1–2.5 cm, elliptic- to oblong-lanceolate, wedge-shaped at base, acute at apex; stalk shorter than blade. Ochreae to 3 cm, shredded, brown. Flowering stems 10–30 cm, erect. Flowers pink or red, slightly fragrant, in dense terminal spikes 4–8 cm. Perianth-segments 4–5 mm, spreading, broadly elliptic, obtuse to somewhat acute; flower-stalks to 7 mm. Stamens projecting. *Himalaya.* H1. Late summer–early autumn.

A variable species with several cultivars that differ in shape and size of the leaves, density and length of the flower-spike and size and colour of flowers. 'Donald Lowndes' from Nepal, with robust spikes of deep pink flowers, is widely grown.

8. P. vacciniifolium Meissner (*Bistorta vacciniifolia* (Meissner) Greene). Illustration: Botanical Magazine, 4622 (1851); RHS Dictionary of gardening, 1629 (1956); Hay & Synge, Dictionary of garden plants in colour, pl. 145 (1969); Grierson & Long, Flora of Bhutan **1**: t. 16 (1983).
Hairless perennial forming mats of prostrate, slender stems. Leaves all along the stems, 1–2.5 cm × 5–10 mm, ovate-elliptic, wedge-shaped at base, acute at apex, crimson in autumn; stalks to 5 mm, slender. Ochreae *c.* 1 cm, very shredded, brown. Flowering stems 10–20 cm, ascending to erect, branched. Flowers pink, in somewhat loose terminal spikes 3–8 cm. Perianth-segments 4–6 mm, erect to spreading, elliptic; flower-stalks very short. Stamens projecting. *Himalaya.* H1. Late summer–autumn.

9. P. bistorta Linnaeus (*Bistorta major* Gray). Illustration: Bonnier, Flore complète **9**: pl. 530 (1927); Ross-Craig, Drawings of British plants **26**: t. 4 (1969); The Garden **102**: 388 (1977); Annales Musei Goulandris **4**: pl. 1A (1978).
Perennial with stout, twisted rootstock, forming clumps. Stems 30–100 cm, erect.

Leaves mostly basal; blade 10–30 cm, ovate to oblong, truncate at base, obtuse at apex, paler beneath; stalk to one and a half times the length of the blade, winged in the upper part. Upper stem-leaves triangular. Ochreae to 6 cm, truncate, brown. Flowers almost stalkless, pink or white, in dense, cylindric, terminal spikes 2–8 × 1–1.5 cm. Perianth-segments 4–5 mm. Nut *c.* 5 mm, pale brown, glossy. *Mountains of Eurasia.* H1. Summer.

'Superbum' (Illustration: Hay & Synge, Dictionary of garden plants in colour, 1325, 1969), to 100 cm tall with rose-pink flowers in spikes 5–8 cm, is widely grown.

Subsp. **carneum** (Koch) Coode & Cullen (*P. carneum* Koch) has flowers with distinct stalks in conical to almost spherical spikes 2–3 cm. *USSR (Caucasus), NE Turkey.*

10. P. macrophyllum D. Don (*P. sphaerostachyum* Meissner; *Bistorta macrophylla* (D. Don) Sojak).
Perennial with stout rhizome. Leaves mainly basal: blade 8–18 × 2–6 cm, lanceolate to oblong, more or less truncate at base, acute at apex, paler beneath, marginal veins prominent; stalk about as long as the blade. Stem-leaves few, slightly clasping the stem, cordate at base, acute at apex. Ochreae 1–5 cm, brown, papery. Flowering stems 20–40 cm, erect. Flowers rose-pink in dense terminal spikes 2.5–6 × 1.5 cm. Perianth-segments 3–4 mm; flower-stalks *c.* 5 mm. *Himalaya to W China.* H1. Summer–early autumn.

11. P. milletii (Léveillé) Léveillé (*P. sphaerostachyum* misapplied; *Bistorta milletii* Léveillé). Illustration: Botanical Magazine, 6847 (1885).
Differs from *P. macrophyllum* by the linear-lanceolate leaves, wedge-shaped at the base with less prominent marginal veins, stalkless; flowers crimson, perianth-segments 4–5 mm. *Himalaya to W China.* H1. Summer–early autumn.

12. P. griffithii J.D. Hooker (*Bistorta griffithii* (J.D. Hooker) Grierson; *P. chlorostachyum* Diels).
Differs from *P. macrophyllum* by its usually broader, elliptic, basal leaves, hairier beneath. Flowers dark crimson in looser, somewhat drooping spikes to 10 cm. Perianth-segments 5–8 mm; flower-stalks *c.* 10 mm. *Himalaya, W China (Yunnan).* H1. Summer.

13. P. viviparum Linnaeus (*Bistorta vivipara* (Linnaeus) Gray). Illustration: Bonnier, Flore complète 9: pl. 350 (1927); Hess, Landoldt & Hirzel, Flora der Schweiz 1: 737 (1967); Polunin, Wild flowers of Europe, pl. 9 (1969); Ross-Craig, Drawings of British plants 26: t. 6 (1969).
Rhizomatous perennial. Stems 10–30 cm, erect, unbranched. Leaves 3–10 cm × 5–20 mm, ovate-oblong, acute, margins recurved. Upper stem-leaves stalkless, linear. Ochreae entire. Flowers white or pink in loose terminal spikes 2–6 cm; flowers of lower part of spike replaced by bulbils; bulbils 2–4 mm, ovoid, beaked, purplish brown. Stamens projecting. *Europe, widespread in arctic and mountainous regions of the northern hemisphere.* H1. Summer.

Propagated readily by means of the bulbils.

14. P. amplexicaule D. Don (*Bistorta amplexicaulis* (D. Don) Greene). Illustration: Botanical Magazine, 6500 (1880); Hay & Synge, Dictionary of garden plants in colour, 1324 (1969).
More or less hairless perennial; rootstock woody. Stems 50–120 (rarely to 200) cm, erect. Leaves 8–25 × 4–10 cm, ovate to broadly lanceolate, strongly cordate at base, acuminate at apex, downy beneath; stalks of basal leaves to 3 times as long as blade. Upper stem-leaves clasping the stem. Ochreae 4–6 mm, brown. Flowers in rather loose terminal spikes 3–10 cm, often in pairs; spike-stalks long, slightly hairy. Perianth-segments 5–6 mm, red, purplish, pink or white; flower-stalks *c.* 4 mm. Stamens slightly projecting. Nut 4–5 mm, pale brown. *Himalaya.* H1. Summer–early autumn.

There is a number of cultivars differing in flower-colour and size of the flower-spike.

15. P. tenuicaule Bisset & Moore (*Bistorta tenuicaule* (Bisset & Moore) Nakai). Illustration: RHS Dictionary of gardening, 1629 (1956).
Hairless, rhizomatous perennial. Leaves mostly basal; blades 3–10 × 2–6 cm, ovate-oblong, shallowly cordate or wedge-shaped at base, acute at apex; stalk to one and a half times the length of the blade, stout, narrowly winged. Ochreae to 5 mm, brown, acute. Flowering stem 5–15 cm, simple or little-branched, with 1–2 small stem-leaves. Flowers fragrant, in a rather loose terminal spike 2–4 cm. Perianth-segments *c.* 3 mm, narrowly elliptic, white. Stamens projecting. Nut slightly longer than perianth. *Japan, Korea* H1. Spring, early autumn.

16. P. virginianum Linnaeus (*Tovara virginiana* (Linnaeus) Rafinesque; *P. virginianum* var. *filiforme* (Thunberg) Merrill; *Tovara filiforme* (Thunberg) Nakai).

Illustration: Journal of the Royal Horticultural Society 102: 388 (1977); The Garden 111: 20 (1986).
Hairless or somewhat hairy perennial. Stems 40–120 cm, erect. Leaves 8–25 × 3–9 cm, ovate to elliptic, slightly acuminate, with dark blotches; stalks 1–3 cm. Ochreae with a fringe of hairs. Flowers in slender, very loose, terminal and axillary spikes 10–30 cm. Perianth-segments 4, 2–3 mm, green becoming red. Nut 2–3 mm, 2-angled and flattened, pale brown, glossy; stigmas conspicuously persistent. *Eastern N America; Himalaya; Japan.* H1. Late summer–early autumn.

The Asian plants have been regarded as a distinct species by some authors. Usually grown as one of its variegated cultivars. 'Variegata' has leaves spotted and splashed with cream. 'Painter's Palette' has leaves marked with cream, pale green and red. These variegated plants are less hardy than the unvariegated.

17. P. alpinum Allioni (*P. undulatum* Murray; *P. sericeum* misapplied; *Pleuropteropyrum alpinum* (Allioni) Nakai). Illustration: Bonnier, Flore complète 9: pl. 525 (1927); Polunin, Flowers of Europe, pl. 9 (1969); Hess, Landolt & Hirzel, Flora der Schweiz 1: 740 (1967).
Rhizomatous perennial, forming clumps. Stems to 1 m, erect, usually branched, hairless or with addressed or spreading hairs. Ochreae papery, pale brown, with long spreading hairs. Leaves 3–8 × 1–3 cm, almost stalkless, ovate to lanceolate, acute, ciliate. Flowers cream or white, in erect terminal panicles. Perianth-segments 2–3 mm, ovate-oblong, obtuse. Nut 2–3.5 mm, pale brown, longer than perianth. *N & C Asia, extending to the mountains of SW Asia and Europe.* H1. Summer.

18. P. polystachyum Meissner (*Aconogonon polystachyum* (Meissner) Haraldson). Illustration: Hess, Landolt & Hirzel, Flora der Schweiz 1: 741 (1967).
Hairless or softly hairy perennial with creeping rhizomes, forming clumps. Stems 80–220 cm, erect, stout, branched. Leaves 10–30 × 3–10 cm, oblong- to ovate-lanceolate, truncate or slightly wedge-shaped at base, acuminate at apex, usually with red midrib; stalk 1–3 cm, usually red. Ochreae *c.* 5 cm, thick, entire, persistent. Flowers white, fragrant, in loose, leafy panicles. Perianth-segments *c.* 4 mm, broadly obovate to circular. Styles long, slender. Fruit pale brown, much longer than perianth. *Himalaya.* H1. Late summer.

19. P. molle D. Don (*P. rude* Meissner; *P. paniculatum* Blume; *Aconogonon molle* (D. Don) Hara). Illustration: The Garden **108**: 175 (1983).
Rhizomatous, softly hairy perennial. Stems 1–2.5 m, branched. Leaves with blade 10–20 × 4–12 cm, ovate to ovate-lanceolate, more or less cordate at base, acuminate at apex; stalk 1–3 cm, green or pink. Ochreae short, truncate. Flowers slightly fragrant, white or cream, in branched, dense, more or less leafless panicles to 30 cm. Perianth-segments 1.5–2.2 mm, oblong, more or less acute, enlarged in fruit to form a fleshy, almost spherical, berry-like structure 2.5–4 mm, purplish black, enclosing the nut. *Himalaya*. H1. Late summer–early autumn.

20. P. campanulatum J.D. Hooker (*Aconogonon campanulatum* (J.D. Hooker) Hara). Illustration: Botanical Magazine, 9098 (1925); Hay & Synge, Dictionary of garden plants in colour, 1326 (1969); The Garden 102: 386 (1977); Polunin & Stainton, Flowers of the Himalaya, pl. 106 (1984).
Hairy perennial, with stolons, forming extensive patches. Stems 50–100 cm, erect, much-branched. Leaves 5–15 × 2–4 cm, lanceolate to ovate or ovate-elliptic, wedge-shaped at base, slightly acuminate at apex, with conspicuous buff or pinkish brown hairs beneath. Ochreae *c.* 1 mm, papery. Flowers in a compact terminal panicle of 3–5-flowered cymes. Perianth 4–5 mm, pink or white. Nut *c.* 2 mm, brown, glossy, enclosed within perianth. *Himalaya, W China*. H1. Summer–early autumn.

Var. **lichiangense** (W.W. Smith) Steward (*P. lichiangensis* W.W. Smith) has white hairs on the undersurfaces of the leaves, and white flowers. *W China*.

21. P. weyrichii Schmidt (*Aconogonon weyrichii* (Schmidt) Hara).
Robust, rhizomatous perennial. Stems 1–1.8 m, shortly hairy, green. Leaves 15–30 × 12–20 cm, ovate to ovate-lanceolate, slightly truncate at base, acuminate at apex, softly hairy beneath; stalks to 12 cm, those of upper leaves usually only 1–2 cm. Flowers in large, dense, terminal, branched panicles. Perianth-segments *c.* 1.5 cm, triangular-obtuse, greenish white, enlarged considerably in fruit. Fruit 5–7 × *c.* 4 mm, pale brown. Inflorescence- and flower-stalks with short, rather rough hairs. *USSR (Sakhalin)*. H1. Summer–early autumn.

22. P. aubertii L. Henry (*Bilderdykia aubertii* (L. Henry) Dumortier; *Fallopia aubertii* (L. Henry) Holub). Illustration: Lousley & Kent, Docks and knotweeds of the British Isles, 87 (1981).
Vigorous, mostly hairless woody climber to 15 m, forming a mass of entwined stems. Leaves 3–10 cm, ovate to ovate-oblong, cordate at base, acute at apex, with wavy margins. Flowers in large, erect, branched axillary and terminal panicles, with roughly and minutely hairy axes. Perianth in fruit 4–6 mm across, white or greenish, rarely becoming pinkish, the outer segments winged. Nut rarely produced. *China (Sichuan)*. H1. Late summer–autumn.

23. P. baldschuanicum Regel (*Bilderdykia baldschuanica* (Regel) Webb; *Fallopia baldschuanica* (Regel) Holub). Illustration: Botanical Magazine, 7544 (1897).
Differs from *P. aubertii* by the more woody habit, broader and less shiny leaves, broader, drooping panicles with hairless axes and perianth in fruit 6–8 mm across, white to pink. *USSR (Tadzhikistan)*. H1. Late summer–autumn.

P. aubertii and *P. baldschuanicum* are very similar, and the differences between them are not apparently always consistent: they should probably be regarded as subspecies of the same species. Their vigour makes them problematic garden subjects, but they are frequently used to cover unsightly trees or buildings such as sheds.

24. P. japonicum Meissner (*P. cuspidatum* Siebold & Zuccarini; *P. sieboldii* de Vriese; *Reynoutria japonica* (Meissner) Houttuyn). Illustration: Botanical Magazine, 6476 (1880); Wehrhahn, Die Gartenstauden 1: 307 (1931); Polunin, Flowers of Europe, pl. 8 (1969).
Dioecious, rhizomatous perennial forming dense and extensive clumps. Stems 1–3 m, erect, very robust, glaucous or reddish. Leaves 6–15 × 5–12 cm, broadly ovate, truncate at base, acuminate at apex, hairless. Flowers in axillary branched panicles 6–10 cm, white to cream. Nut 3–4 mm, dark brown. *Japan*. H1. Late summer–early autumn.

P. japonicum is an unsuitable subject for most gardens on account of its vigour and its invasive potential. Two smaller variants may be appropriate for limited cultivation: 'Spectabile', a variegated plant with leaves marked with yellow and pale green is somewhat smaller and less vigorous. Var. **compactum** (J.D. Hooker) Bailey (*P. compactum* J.D. Hooker; *P. reynoutria* misapplied; *Reynoutria japonica* var.

compacta (J.D. Hooker) Buchheim) is a compact plant to 80 cm with almost circular leaves 4–7 cm and reddish flowers in denser inflorescences to 6 cm. *E Asia*.

25. P. sachalinense Schmidt (*Reynoutria sachalinensis* (Schmidt) Nakai). Illustration: Botanical Magazine, 6540 (1881).
Like *P. japonicum* but stems taller and more robust, 2–4 m. Leaves 15–40 × 8–25 cm, ovate-oblong, slightly cordate at base, acute at apex. Flowers greenish, in panicles 3–8 cm, denser than those of *P. japonicum*. Nut 4–5 mm. *USSR (Sakhalin)*. H1. Late summer–early autumn.

As with *P. japonicum*, an unsuitable subject for most gardens and now rarely planted.

26. P. fagopyrum Linnaeus (*Fagopyrum esculentum* Moench; *F. sagittatum* Gilibert). Illustration: Polunin, Flowers of Europe, pl. 9 (1969); Lousley & Kent, Docks and knotweeds of the British Isles, 95 (1981).
Stems 25–100 cm, erect, hollow, hairless or slightly hairy, often reddish. Leaves to 12 cm, often only a little longer than wide, arrowhead-shaped, triangular, cordate at base, entire to sinuate, dark green. Leaf-stalk shorter than the blade. Upper stem-leaves stalkless. Ochreae short, entire. Flowers in compact, terminal and axillary, almost corymb-like panicles with long stalks. Perianth-segments 3–5 mm, greenish white, white or pink. Nut 5–8 mm, dark brown, dull, smooth, much longer than the perianth. *Asia, long cultivated as a food crop, and naturalised in Europe*. H1. Late summer.

27. P. dibotrys D. Don (*Fagopyrum dibotrys* (D. Don) Hara; *F. cymosum* (Treviranus) Meissner; *P. chinensis* misapplied). Illustration: Lousley & Kent, Docks and knotweeds of the British Isles, 97 (1981).
Differs from *P. fagopyrum* by the more robust perennial habit with stems to 1.2 m, leaves broadly triangular, the stalk often longer than the blade; flowers white, in looser, spike-like clusters. *Himalaya to China*. H2. Late summer.

7. ANTIGONON Endlicher
J.R. Akeroyd
Perennial climbers with angular stems. Leaves cordate at the base. Flowers bisexual, in branched racemes that end in tendrils. Perianth-segments 5, heart-shaped, brightly coloured. Stamens 7–9. Stigmas 3. Nut 3-angled, enclosed by the enlarged perianth.

A genus of 3 species from C America. The species cultivated in Europe can be grown outside in the south, but otherwise requires a heated glasshouse. Propagation is by cuttings and seed.

1. A. leptopus Hooker & Arnott.
Illustration: Botanical Magazine, 5816 (1870); Die Natürlichen Pflanzenfamilien **3(1A)**: 31 (1893); Menninger, Flowering vines of the world, pl. 148, 150 (1970); Lötschert & Beese, Collins' Guide to tropical plants, t. 98 (1983).
Root tuberous. Stems slender, to 12 m, climbing by tendrils. Leaves 5–12 × 3–8 cm, broadly ovate, heart- or spear-shaped, slightly acuminate. Flowers pink or white ('Album'), 6–20 in axillary racemes. Nut strongly angled at apex. *Mexico, widely cultivated elsewhere*. G1. Summer–early autumn.

8. MUEHLENBECKIA Meissner
J.R. Akeroyd
Shrubs or climbers, usually hairless, mostly deciduous. Ochreae flimsy, soon disappearing. Flowers unisexual, with males and females on separate plants, or with unisexual and bisexual flowers on the same plant. Perianth-segments 5, united at base, enlarged and fleshy in fruit, usually white. Stamens 8. Stigmas 3, more or less stalkless. Nut 3-angled, partly fused with the persistent perianth.

A genus of about 20 species from Australasia and temperate S America. The climbing species are best grown supported by other trees or shrubs. Propagation is by seeds or cuttings.

1a. Prostrate shrubs with stems not more than 1 m 2
  b. Climbing shrubs with stems more than 1 m 3
2a. Leaves linear to linear-lanceolate; ochreae *c*. 1 mm **4. ephedroides**
  b. Leaves ovate-oblong to almost circular; ochreae 2–3 mm **5. axillaris**
3a. Leaves usually less than 2.5 cm; flowers in racemes not more than 2.5 cm long **3. complexa**
  b. Leaves 2.5–8 cm; flowers in racemes to 8 cm long 4
4a. Leaves ovate, 3-lobed or fiddle-shaped 5
  b. Leaves lanceolate, often spear-shaped 6
5a. Leaves circular to ovate, not lobed **1. adpressa**
  b. Leaves fiddle-shaped or 3-lobed **6. australis**

6a. Flowers in racemes 2.5–8 cm long **2. gunnii**
  b. Flowers in racemes less than 2.5 cm long **7. sagittifolia**

1. M. adpressa (Labillardière) Meissner.
Illustration: Botanical Magazine, 3145 (1832) – as Polygonum adpressum.
Spreading or climbing shrub to 2 m. Leaves 2–7 cm, ovate to circular, cordate at base, leathery, margins wavy. Flowers in axillary racemes 2.5–8 cm long. Perianth-segments 2.5–3 mm. Nut more or less globular, bluntly 3-angled. *Temperate Australia*. G1. Winter.

2. M. gunnii (J.D. Hooker) Walpers (*M. adpressa* var. *hastata* Meissner). Illustration: RHS Dictionary of gardening, 1325 (1956) – as M. adpressa.
Differs from *M. adpressa* by the longer stems to 10 m, leaves broadly lanceolate, perianth-segments 4–5 mm and nut ovoid-oblong. *Australia (S Australia, Tasmania)*. G1. Winter.

3. M. complexa (Cunningham) Meissner.
Illustration: Botanical Magazine, 8449 (1912).
Straggling or climbing shrub to 5 m, with a mass of slender, wiry, twining branches. Leaves 5–25 (rarely to 40) mm, oblong to almost circular, slightly apiculate. Flowers yellowish green in short racemes. Perianth-segments oblong, berry-like in fruit, white and somewhat translucent. Nut 2–2.5 mm, black, glossy. *New Zealand*. H4. Summer–early autumn.

A variable species that has been divided into a number of varieties by several authors. Var. **triloba** (Colenso) Cheeseman (*M. triloba* Colenso; *M. varians* Meissner) has leaves 1.2–4 × 1.2–2.5 cm, fiddle-shaped, slightly cordate at base. Very small, compact plants have been called var. **microphylla** (Colenso) Cockayne.

4. M. ephedroides J.D. Hooker.
Prostrate, greyish shrub. Stems *Juncus*-like, to 1 m, much branched, with long internodes. Leaves 5–25 mm, stalkless, linear to narrowly linear-lanceolate. Ochreae *c*. 1 mm. Flowers in small axillary clusters. Perianth-segments 2–3 mm, narrowly triangular, greenish. Nut *c*. 3 mm, black, glossy. *New Zealand*. H3. Summer–early autumn.

Var. **muriculata** (Colenso) Cheeseman, with very slender stems and perianth membranous in fruit, may be a hybrid between *M. ephedroides* and *M. complexa*. It is rare in cultivation, but has curiosity value.

5. M. axillaris (J.D. Hooker) Walpers (*M. nana* misapplied).
Small prostrate or straggling shrub. Stems to 50 cm, very slender, much-branched, often rooting at the nodes, minutely hairy when young. Leaves 5–10 × 3–6 mm, broadly ovate-oblong to almost circular; stalk to 3 mm, slender. Ochreae 2–3 mm, truncate. Flowers solitary or paired, axillary, on tiny stalks. Perianth-segments *c*. 2 mm, narrowly triangular, yellowish green. Nut *c*. 3 mm, black, glossy. *New Zealand, SE Australia*. H4. Summer–early autumn.

6. M. australis (Forster) Meissner.
Rather robust climbing shrub to 10 m, similar in general appearance to *M. complexa*. Leaves to 8 × 3 cm, somewhat fiddle-shaped to 3-lobed. Flowers in racemes to 5 cm long. *New Zealand*. H3. Summer–early autumn.
  Infrequently cultivated.

7. M. sagittifolia (Ortega) Meissner.
Spreading or climbing shrub. Leaves *c*. 5 cm, lanceolate, spear-shaped or truncate at base, acuminate or mucronate. Flowers and fruits similar to those of *M. complexa*. *Temperate S America*. H3. Summer.
  Infrequently cultivated.

9. HOMALOCLADIUM (Mueller) Bailey
J.R. Akeroyd
Erect, evergreen shrub to 1.2 m (taller in the tropics). Branches flattened, 1–2 cm wide, green. Leaves sparsely scattered on angles of branches, 1.5–6 cm, lanceolate to diamond-shaped, tapered at both ends; falling before flowering. Flowers in stalkless clusters on angles of branches, greenish white. Stamens 8. Stigmas 3. Perianth-segments 5, united at base, enlarged and fleshy in fruit, purplish red, enclosing a 3-angled nut.

The single species requires glasshouse cultivation. Propagation is by cuttings.

1. H. platycladum (Mueller) Bailey (*Muehlenbeckia platyclada* (Mueller) Meissner). Illustration: Botanical Magazine, 5382 (1863) – as Coccoloba platyclada; Die Natürlichen Pflanzenfamilien 3(1A): 32 (1893); Heywood, Flowering plants of the world, 78 (1978) – as Coccoloba platyclada; Krussman, Manual of cultivated broad-leaved trees and shrubs **2**: pl. 53 (1986). *Solomon Islands*. G2. Spring.

10. COCCOLOBA Browne
J.R. Akeroyd
Dioecious trees, shrubs or climbers. Leaves large. Flowers in terminal or axillary

panicles. Perianth-segments 5, greenish white. Stamens 8, longer than perianth-segments. Nut surrounded by fleshy, enlarged perianth.

A genus of some 150 species from tropical and subtropical America. Only 1 species is cultivated in Europe; it can be grown outside in the south but elsewhere requires glasshouse cultivation. Propagation is by seed and cuttings.

1. C. uvifera (Linnaeus) Linnaeus. Illustration: Botanical Magazine, 3130 (1832); Hay & Synge, Dictionary of garden plants in colour, 475 (1969); Kunkel, Flowering trees in subtropical gardens, 107 (1978); Lötschert & Beese, Collins' guide to tropical plants, t. 23 (1983).
Evergreen, branched tree to 10 m with thick trunk and greyish bark; smaller and shrubby in indoor cultivation. Leaves to 20 cm, broadly ovate to circular, cordate at base, rounded, hairless, leathery and glossy, with red veins; stalks short. Flowers fragrant, in erect racemes 10–20 cm long that become pendulous in fruit. Fruit 1.5–2 cm, almost spherical to pear-shaped, purplish, edible. *West Indies, tropical C & S America.* G1. Spring–summer.

XLIX. PHYTOLACCACEAE

Trees, shrubs (some climbing) and herbs. Leaves alternate, simple, entire, stalked or not, usually without stipules, or if stipules present then very small. Flowers usually bisexual and radially symmetric, mostly in terminal or axillary racemes, often subtended by bracts and bracteoles. Perianth-segments 4 or 5, usually free and persistent. Stamens as many as the perianth-segments or more numerous. Ovary usually superior; carpels 1–many, free or united, each containing a single ovule; styles short, as many as carpels, or absent. Fruit a fleshy berry, dry nut or rarely a capsule.

A family of about 20 genera and 100 species, native to the warmer parts of the world, mainly America and Africa. Literature: Walter, H., Phytolaccaceae, *Das Pflanzenreich* **39**: 1–154 (1909); Nowicke, J.W., Palynological study of the Phytolaccaceae, *Annals of the Missouri Botanical Garden* **55**(3): 294–364 (1968), **56**(2): 288–90 (1969).

1a. Leaves reddish purple beneath
4. Trichostigma
b. Leaves green beneath 2

2a. Perianth-segments 4; stamens 4
3. Rivina
b. Perianth-segments 5; stamens 5 or more 3
3a. Racemes axillary; leaf-blades not more than 7 cm **1. Ercilla**
b. Racemes terminal or opposite the leaves; leaf-blades usually more than 7 cm **2. Phytolacca**

1. ERCILLA Jussieu
C.J. King
Climbing evergreen shrubs to 20 m with slender, sparsely branched, very leafy stems bearing aerial roots. Leaves 2.5–7 × 2–5.5 cm, ovate to oblong, blunt at apex, tapered to cordate at base, margin somewhat curled, hairless, stalk 3–6 mm. Flowers in dense axillary racemes, 1–8 cm, each flower 6–7 mm in diameter, shortly stalked and subtended by a bract and 2 bracteoles. Perianth-segments 5, elliptic, green with purplish margins. Stamens 6–10, white, protruding. Fruit a red or purple berry, rarely produced in cultivation.

A genus of 1 or 2 species from Chile and Peru, tolerant as to soil, but requiring partial shade and protection from cold winds. They will grow up naturally through trees but need some attachment if grown against a wall. Propagation is by stem-cuttings or layering. The taxonomic and nomenclatural history of the genus is somewhat confused.

1. E. volubilis Jussieu (*E. spicata* (Hooker & Arnott) Moquin). Illustration: Gardeners' Chronicle **87**: 405 (1930); Botanical Magazine, n.s., 780 (1979); Everett, New York Botanical Gardens illustrated encyclopedia of gardening **4**: 1229 (1981). *Chile.* H4. Spring.

2. PHYTOLACCA Linnaeus
C.J. King
Perennial herbs, shrubs (rarely climbing) or trees, hairless or nearly so, sometimes dioecious. Leaves ovate, elliptic or lanceolate, usually stalked. Racemes dense or open, erect or drooping, terminal or opposite the leaves. Flowers usually bisexual, sometimes unisexual; 1 bract and 2 bracteoles usually present. Perianth-segments 5, greenish to white or pink. Stamens 5–30, in 1 or 2 whorls. Carpels 5–16, free or united. Fruit a somewhat flattened spherical berry, usually glossy black with purple juice.

Identification in this mainly tropical and subtropical genus is difficult, due in part to

the numerous species described, which often have overlapping or intergrading characters. In addition there may be considerable variation within a single species, and also hybridisation may occur when different species are grown together. The genus is now thought to contain about 25 species, several of which are found in cultivation and occasionally escape and become naturalised. The herbaceous species are usually grown singly or in groups in sunny or partially shaded borders. Propagation is by seed or division, and the stems of established plants should be cut down in November. The more tender, woody species, *P. dioica*, may be grown outdoors in favourable areas and will form a wide-spreading tree in Mediterranean gardens. Elsewhere it will grow less vigorously and will need to be overwintered in a greenhouse.

1a. Tree with leaves not larger than 12 × 6 cm; racemes not longer than 15 cm; flowers unisexual (plants dioecious) **2. dioica**
b. Herb with leaves often larger than 12 × 6 cm; racemes often longer than 15 cm; flowers bisexual 2
2a. Stamens c. 10, in 1 whorl; fruiting racemes usually pendent
3. americana
b. Stamens 7–16, usually in 2 whorls; fruiting racemes erect 3
3a. Carpels free or united only at base
1. acinosa
b. Carpels more or less completely united
4. polyandra

1. P. acinosa Roxburgh (*P. esculenta* Van Houtte). Illustration: Coventry, Wild flowers of Kashmir 1: t. 41 (1923); Polunin & Stainton, Wild flowers of the Himalaya, t. 106 (1984); Mansfeld, Verzeichnis landwirtschaftlicher und gärtnerischer Kulturpflanzen, 124–5 (1986).
Herb to 3 m with branched stems, somewhat woody at base. Leaves to 35 × 16 cm, ovate, ovate-elliptic or rarely ovate-lanceolate, acute or sometimes mucronate at apex, tapered to slightly rounded at base; stalks to 6 cm. Raceme to 30 cm, more or less erect. Bract 2–4 mm, bracteoles c. 1.5 mm. Perianth-segments 3–4 mm. Stamens 7–15, sometimes in 2 whorls. Carpels 6–9, somewhat united at the base in flower, mostly free in fruit. Fruit c. 7 mm in diameter, the fruiting racemes remaining erect. *Himalaya to China and Japan.* H4. Summer–early autumn.

2. P. dioica Linnaeus (*P. arborea* Moquin). Illustration: Das Pflanzenreich **39**: 48 (1909); Moeller, What's blooming where on Tenerife?, 82, 96 (1968); Guinea López and Vidal Box, Parques y jardines de España, 104,105 (1969); Kunkel, Flowering trees in subtropical gardens, 105 (1978).
Dioecious evergreen tree to 25 m. Leaves to 12 × 6 cm, ovate, acute at the apex, rounded or sometimes decurrent at the base; stalks to 7 cm. Racemes to 15 cm, somewhat erect or drooping. Male flowers with perianth-segments *c.* 3 mm, white; stamens 20–30 in 2 whorls; ovary of 2–4 abortive carpels occasionally present; bract and bracteoles *c.* 1 mm; flower-stalks to 4 mm. Female flowers with perianth-segments 2–3 mm, white, persistent in fruit; rudimentary stamens *c.* 10; ovary of 8–12 carpels, united at their bases; bract and bracteoles *c.* 0.5 mm; flower-stalks *c.* 3 mm. Fruit 6–8 mm in diameter, the fruiting racemes pendent. *S America.* H5. Spring–early summer.

Often planted as a quick-growing shade tree in parks and large gardens in southern Europe. Female trees bear hanging racemes of fruits from November to January.

3. P. americana Linnaeus (*P. decandra* Linnaeus). Illustration: Botanical Magazine, 931 (1786); Das Pflanzenreich **39**: 52 (1909); Bonnier, Flore complète 9: pl. 509 (1927); Rickett, Wild flowers of the United States **2**: 145 (1967).
Herb to *c.* 3 m, somewhat woody at the base, with more or less succulent, branched stems. Leaves to 30 × 12 cm, lanceolate, elliptic or rarely ovate, acute or sometimes mucronate; stalks to *c.* 6 cm. Raceme somewhat arching, to 30 cm. Bracts 2–3 mm, bracteoles 1–1.5 mm. Perianth-segments *c.* 2.5 mm. Stamens *c.* 10, in 1 whorl. Carpels *c.* 10, united or sometimes the apices free. Fruit 6–8 mm in diameter, the fruiting racemes more or less pendent. *Eastern N America.* H3. Summer–early autumn.

A variant from the southeastern USA with shorter, permanently erect racemes and fruits longer than their stalks has sometimes been separated as **P. rigida** Small. This may represent an ecological variant or may be the result of hybridisation with another species.

4. P. polyandra Batalin (*P. clavigera* W.W. Smith). Illustration: Gardners' Chronicle **71**: 39 (1922); Botanical Magazine, 8978 (1923); Everett, New York Botanical Gardens illustrated encyclopedia of horticulture **8**: 2626 (1981).

Herb to 2 m with robust, branched stems woody at the base. Leaves to 32 × 13 cm, ovate to elliptic, acute to mucronate at the apex, tapered at the base; stalks to 3 cm. Raceme to 30 cm, densely flowered, erect. Bract *c.* 2 mm, bracteoles *c.* 1 mm. Perianth-segments 3–3.5 mm. Stamens 12–16, in 2 whorls. Carpels 7–9, more or less completely united. Fruit 7–10 mm in diameter, the fruiting racemes erect. *China.* H4. Summer–early autumn.

3. RIVINA Linnaeus
C.J. King
Erect, perennial herbs with branched stems, sometimes woody at base. Leaves long-stalked, blades ovate to lanceolate, acute to acuminate at apex, rounded or truncate at base, stalks slender. Flowers bisexual, in slender terminal or axillary racemes, with minute bracts and bracteoles. Perianth-segments 4, concave, white or pink. Stamens 4, shorter than perianth. Ovary 1-celled. Fruit a spherical berry, red or yellow, 1-seeded.

A genus of 1 (or possibly 3) species from the new world tropics; often cultivated in warm greenhouses for its ornamental fruits. The plants should be grown in a moderately moist soil, with a night temperature of 16–18 °C and a daytime temperature 3–9 °C higher. They should be shaded from strong sun. Propagation is by seeds or cuttings.

1. R. humilis Linnaeus. Illustration: Botanical Magazine, 1781 (1815); Addisonia **12**: pl. 410 (1927); Graf, Tropica, 812 (1978); Graf, Exotica edn 11, 1922–23 (1982).
A bushy-stemmed plant to *c.* 1 m, the vegetative parts with short hairs. Leaves ovate, to 12 × 6 cm; leaf-stalks to 6 cm. Flowers in more or less erect racemes to 20 cm which become pendent in fruit. Perianth-segments *c.* 2.5 mm, pinkish white, sometimes reflexed, becoming green as fruit develops. Berry *c.* 4 mm in diameter, red, orange or yellow. *S USA to tropical S America, West Indies.* G2. Flowering throughout most of the year, berries at their best in winter.

Var. **glabra** Linnaeus (*R. laevis* Linnaeus). Illustration: Botanical Magazine, 2333 (1822). Differs in having the stems and leaves hairless.

4. TRICHOSTIGMA Richard
C.J. King
Erect or climbing shrubs with stalked, ovate or elliptic leaves. Flowers in loose, many-

flowered racemes. Perianth-segments 4, free, concave, ovate or elliptic, reflexed in fruit. Stamens 8–25. Ovary 1-celled; style very short. Fruit globular, berry-like, black to reddish purple, 1-seeded.

A genus of 3 or 4 species from tropical S America. *T. peruvianum* is cultivated in warm greenhouses primarily for its attractive foliage. This plant is tolerant as to soil, but needs to be shaded slightly in summer. It should be watered moderately from spring to autumn, but more sparingly in winter, Propagation is normally by cuttings, though seeds may be used when available.

1. T. peruvianum (Moquin) Walter (*Ledenbergia roseo-aenea* Lemaire). Illustration: Illustration Horticole **16**: pl. 591 (1869); Parey's Blumengärtnerei **1**: 551 (1958); Graf, Exotica edn 11, 1923 (1982); Everett, New York Botanical Gardens illustrated encyclopedia of horticulture **10**: 3397 (1982).
Semi-climbing shrub to about 2 m with slender, more or less downy branches. Leaves to 30 × 12.5 cm, elliptic, acuminate at apex, oblique and slightly auricled at base, glossy metallic green above, reddish purple beneath, conspicuously veined, stalk to 3 cm, more or less downy. Racemes to 30 cm, slender, terminal or axillary, erect at first, later arching. Perianth-segments 5 mm, concave, the outer pair reddish purple, the inner pair whitish. Stamens 12. Fruit 4–6 mm in diameter, black. *Peru, Ecuador.* G2. Flowering throughout the year.

L. NYCTAGINACEAE
Annual or perennial herbs, woody vines, shrubs or trees. Leaves usually entire, alternate, opposite or whorled, without stipules. Inflorescence usually cymose. Flowers bisexual or unisexual, sometimes subtended by coloured bracts. Perianth usually 5-lobed, united below into a tube. Stamens 1–many, usually 5 and then alternate with the perianth-lobes, free or united at base, and sometimes branched above. Ovary superior, 1-celled, containing a single basal ovule, and bearing a long style. Fruit an achene, enclosed in the persistent perianth.

A family of 30 genera and about 290 species from the tropics and subtropics, especially America. Although the 4 genera usually cultivated comprise about 150 species, only 8 of these are generally represented in European gardens.

Literature: Heimerl, A., Nyctaginaceae in Engler, A. & Prantl, K., *Die Natürlichen Pflanzenfamilien*, edn 2, **16c**: 86–134 (1934).

1a. Plant woody   2
  b. Plant herbaceous   3
2a. Plant usually climbing; flowers subtended by large, coloured bracts
                                                            1. Bougainvillea
  b. Plant usually erect; flowers subtended by small bracts or bracts absent
                                                                    2. Pisonia
3a. Leaves at least slightly fleshy, the blades of each pair a different size; involucral bracts free   **3. Abronia**
  b. Leaves not fleshy, the blades of each pair equal in size; involucral bracts united   **4. Mirabilis**

1. BOUGAINVILLEA Jussieu
C.J. King

Shrubs, vines or small trees, often spiny. Leaves alternate, ovate-oblong to elliptic-lanceolate, stalked. Inflorescences solitary or in groups in the leaf-axils or at the ends of the branches, each inflorescence composed of 1–3 flowers subtended by 3 large, coloured bracts. Perianth tubular with shallowly 5-lobed limb. Stamens 5–10, usually 7 or 8, very unequal, scarcely protruding. Fruit spindle- to pear-shaped, 5-ribbed.

A genus of 18 species from S America. Only 3 species are of any importance, but a hybrid and numerous cultivars are grown. In southern Europe the plants can be used out-of-doors as an ornamental covering for walls, pergolas, fences and other supports, where they will flourish in any well-drained soil in full sun. Elsewhere they must be grown under glass. Propagation is by cuttings, which when rooted can be placed in pots and trained up canes, or planted in a border and trained along wire or strings. Frequent watering is desirable throughout the summer, but during the winter the soil should be kept much drier with a minimum temperature of about 10° C.
Literature: Pal, B.P. & Swarup, V., *Bougainvilleas* (1974).

1a. Leaf-blades broadly ovate to almost circular; perianth hairless
                                                            1. peruviana
  b. Leaf-blades elliptic to broadly ovate; perianth hairy   2
2a. Leaf-blades almost elliptic; perianth with very short, upward-curved hairs
                                                            2. glabra
  b. Leaf-blades broadly ovate to elliptic; perianth with longer, spreading hairs
                                                            3. spectabilis

1. B. peruviana Humboldt & Bonpland. Shrub 3–7 m, erect or climbing, spines numerous, 1–2.5 cm; branches sparsely downy. Leaves hairless or nearly so, blades to 10 × 7.5 cm, broadly ovate to almost circular, stalks slender. Bracts 1.5–3.5 cm, rose to pale magenta-pink, obtuse or rounded at the apex, hairless except along the downy midrib. Perianth 1.6–2 cm, white or whitish, hairless. Fruit 10 mm, hairless. *Colombia to Peru.* H5–G1. Summer.

This species is probably best known as a parent of *B. × buttiana*, though several cultivars are grown.

2. B. glabra Choisy. Illustration: Revue Horticole **61**: 276 (1889); Hay et al., Dictionary of indoor plants in colour, t. 72 (1974); Graf, Tropica, 689, 691, 692, 694 (1978); Graf, Exotica, edn 11, 1616, 1620 (1982).
A woody, somewhat spiny climber to 4 m, the branches downy or hairless. Leaves downy when young, quickly becoming hairless, blades to 13 × 6.5 cm, almost elliptic, acute or acuminate at the apex and tapered at the base. Bracts 2.5–4.5 cm, purple or magenta, sparsely downy or hairless. Perianth 1.5–2 cm, white or yellowish, finely downy. Fruit 7–13 mm, sparsely downy or hairless. *Brazil.* H5–G1. Summer–early autumn.

The species and its many cultivars are widely grown as pot plants, since they will flower over a long period when the plants are quite small. The colour of the bracts ranges from bright purple through coral red and pink to white. There are also cultivars with multiple bracts or with leaves variegated with creamy white.

B. × buttiana Holttum & Standley. Illustration: Hay et al., Dictionary of indoor plants in colour, t. 71 (1974); Graf, Tropica, 689, 690, 692, 694 (1978); Graf, Exotica, edn 11, 1952A (1982); Hessayon, The indoor plant spotter, 76 (1985). This hybrid between *B. glabra* and *B. peruviana* arose in a garden in Colombia, but numerous cultivars are now grown. It is a sturdy climber with broadly ovate leaves to 13 × 9 cm, which are hairless except for the downy midrib. The colour of the bracts varies from crimson to scarlet, orange, yellow and pinkish white, and there are downy hairs on the veins. The perianth is also sparsely downy.

3. B. spectabilis Willdenow. Illustration: Illustration Horticole **42**: pl. 30 (1895); Hay et al., Dictionary of indoor plants in colour, t. 73 (1974); Graf, Tropica,

689–91, 693 (1978); Lötschert & Beese, Collins' guide to tropical plants, t. 96 (1983).
A woody climber to 10 m with stout spines, the branches densely covered with soft hairs. Leaves softly hairy, densely so beneath, blades 5–10 cm, broadly ovate to elliptic, acute or acuminate at the apex and rounded or tapered at the base. Bracts 2–4.5 cm, purplish red, sparsely downy. Perianth 1.5–3 cm, yellowish, with soft spreading hairs. Fruit 1.1–1.4 cm, densely soft-hairy. *Brazil.* H5–G1. Summer–early autumn.

A more vigorous species with a number of cultivars having bracts ranging through magenta, pink and various shades of red to white. Several of these have multiple bracts. 'Variegata' has the leaves bordered with creamy white.

2. PISONIA Linnaeus
C.J. King

Trees and shrubs, rarely woody climbers, sometimes spiny, usually dioecious. Leaves opposite, whorled or alternate. Flowers in corymb-like cymes, usually subtended by 2–4 bracteoles. Perianth with 4 or 5 lobes, pink, greenish or yellow. Stamens 3–40. Fruit normally oblong, sometimes 5-ribbed, often glandular and sticky, sometimes with short spines on the ribs.

A genus of 50 species, chiefly maritime, 2 of which are grown for their foliage in ground beds or pots in greenhouses, or as house-plants. They prefer a moderately humid atmosphere, with protection from strong sun. The soil should be well-drained, and kept fairly moist from spring to autumn, but drier in winter. Flowering may take place at any time of the year. Propagation is by cuttings.

1a. Leaves rounded to cordate at base; flowers usually unisexual, subtended by 2–4 bracteoles; fruit spiny
                                                            1. grandis
  b. Leaves usually tapering at base; flowers usually bisexual, bracteoles absent; fruit not spiny   **2. umbellifera**

1. P. grandis R. Brown (*P. alba* Spanoghe). Tree to 13 m. Leaves opposite, mainly hairless but with spreading downy hairs on the midrib beneath. Leaf-blades 9–24 × 3.5–16 cm, elliptic to ovate-oblong, rounded to cordate at the base, acute to acuminate at the apex, stalks 1.5–6 cm. Flowers usually unisexual, 4–6 mm, fragrant, subtended by 2–4 bracteoles. Stamens 8–10. Stigma entire. Fruit 1–1.5 cm, shortly spiny on the ribs.

Tanzania, coastal regions and islands of S & SE Asia to Polynesia. G2.

In the tropics a pale-leaved variant is cultivated for food as well as for ornament.

2. P. umbellifera (Forster & Forster) Seemann (*P. excelsa* Blume). Illustration: Moore & Irwin, The Oxford book of New Zealand plants, 60, 61 (1978); Salmon, The native trees of New Zealand, 130 (1980); Graf, Exotica, edn 11, 1615 (1982). Tree to 20 m. Leaves opposite or whorled, hairless, the blades 10–35 × 4–12 cm, elliptic to lanceolate, usually tapered at the base; stalks 2–4 cm. Flowers usually bisexual, 4–7 mm, fragrant; bracteoles absent. Stamens 5–13. Stigma fringed. Fruit 3–3.75 cm, without spines but glandular and very sticky. *Islands of the Indian Ocean, Malaysia to New Zealand.* G1.

The New Zealand variant is sometimes treated as a separate species, *P. brunoniana* Endlicher (*Heimerliodendron brunonianum* (Endlicher) Skottsberg) which is distinguished by a constricted perianth-tube and other floral characters.

'Variegata' has leaf-blades irregularly variegated and margined with creamy white and the stalks tinged pinkish to red. Illustration: Graf, Tropica, 692, 694, 698 (1978); Graf, Exotica, edn 11, 320A, 1620, 1621 (1982); Hessayon, The indoor plant spotter, 61 (1985).

3. ABRONIA Jussieu
C.J. King
Annual or perennial herbs, often with sticky stems. Leaves opposite, stalked, slightly fleshy, the blades of each pair being of a different size. Flowers bisexual, fragrant, red, pink, yellow or white, few to many in a head, subtended by an involucre of 5–8 thin, dry bracts and borne on a terminal or axillary inflorescence-stalk. Perianth with 4 or 5 notched lobes, united below into a slender tube. Stamens 4 or 5, of unequal length. Fruit usually top-shaped, ribbed or winged.

A genus of about 30 species from western N America and Mexico, only 1 of which is in general cultivation in Europe, as a perennial in warmer regions or in greenhouses, and as a half-hardy annual in the colder areas. It is suitable for flower beds, rock gardens and as a basket-plant, and succeeds best in a light, sandy, well-drained soil fully exposed to the sun. Propagation is usually by seed, although cuttings may be used where the plant is perennial.

1. A. umbellata Lamarck. Illustration: Hooker, Exotic Flora, 194 (1826); Flore des Serres **11**: t. 1095 (1856); Rickett, Wild flowers of the United States **4**: 109 (1970). A prostrate or weakly climbing, hairless to glandular-hairy herb, with simple or branched stems to 1 m. Leaves stalked, the blades 2.5–7 cm, ovate to lanceolate-oblong, the stalks usually as long as or longer than the blades. Inflorescence-stalks 2–12 cm. Bracts lanceolate, 4–6 mm. Flowers 8–15 in a head. Perianth pink, rarely white, the tube 1–1.5 cm, the limb 8–10 mm in diameter with deeply notched lobes. Fruit usually winged. *Coasts of western N America (British Columbia to Baja California).* H5–G1. Summer–early autumn.

A variable plant whose different forms have sometimes been treated as varieties, subspecies or even species. A variant with larger flowers and leaves is known as 'Grandiflora'.

4. MIRABILIS Linnaeus
C.J. King
Annual or perennial herbs, often with tuberous roots. Stems usually branched, hairless or glandular-hairy. Leaves opposite, the lower stalked, the upper almost stalkless. Flowers fragrant or scentless in axillary cymes, surrounded by a tubular or narrowly bell-shaped, calyx-like involucre of 5 bracts. Perianth funnel-shaped, with a long tube, contracted above the ovary, and a spreading, slightly 5-lobed limb. Stamens 3–5, filaments hairless. Fruit ellipsoid to spherical or ovoid, often angled or ribbed.

A genus of about 60 species native to western N America, C and S America. They are usually treated as annuals, the seeds being sown in a warm frame in early spring, and the seedlings planted out in sunny, open ground. The tuberous roots may be lifted in autumn and planted again the following spring. Plants grown from overwintered roots will flower earlier and more abundantly.

1a. Perianth-tube not more than 5 cm, hairless **1. jalapa**
 b. Perianth-tube 6–15 cm, glandular-hairy **2. longiflora**

1. M. jalapa Linnaeus. Illustration: Botanical Magazine, 371 (1797); Hay & Synge, Dictionary of garden plants, 44 (1969); Heywood, Flowering plants of the world, 69 (1978); Hay & Beckett, Reader's Digest encyclopaedia of garden plants and flowers, 446 (1978).

Stems to 1 m, branched, hairless or nearly so. Lower leaves with stalks about half as long as the blades, upper leaves almost stalkless; blades 5–15 cm, ovate, entire, hairless or slightly hairy on the margins, acute to acuminate at the apex, shallowly cordate to rounded at the base. Flowers in dense clusters at the ends of the branches, opening in the afternoon, fragrant, subtended by an involucre 7–15 mm. Perianth red, yellow, white or variegated, the tube 2.5–5 cm, hairless, the limb 2–3.5 cm in diameter. Stamens 5, not or slightly protruding. Fruit dark brown or black, 5-angled, 7–10 mm. *Tropical America.* H5. Summer–autumn.

This species was formerly thought to be Peruvian in origin but its native country may have been Mexico. It is now not known to occur wild but it is widely cultivated in the tropics and often escapes. A variant of normal size but with variegated leaves is known as 'Variegata' and another, also with variegated leaves but growing only to about 30 cm has been named 'Pumila Variegata'.

2. M. longiflora Linnaeus. Illustration: Rickett, Wild flowers of the United States **4**: 111 (1970); Graf, Tropica, 693 (1978). Plant glandular-hairy throughout. Stems to 1 m, much-branched, very sticky. Leaves shortly stalked to stalkless, blades 6–11.5 cm, ovate to ovate-lanceolate, acute to acuminate at apex, cordate at base. Flowers clustered at the ends of the branches, opening in the evening, strongly fragrant. Perianth white, sometimes tinged or coloured with pink or violet, the tube 6–15 cm, glandular-hairy, the limb 2–3 cm in diameter. Stamens 5, usually protruding. Fruit *c.* 8 mm, 5-angled. *SW USA, Mexico.* Summer–autumn.

Var. **wrightiana** (Gray) Kearney & Peebles differs in having only slightly sticky stems and all leaves stalked. 'Violacea' has the perianth pale violet and is usually late in flowering.

LI. AIZOACEAE
Succulent annual or perennial herbs or small shrubs. Leaves usually opposite, fleshy, without stipules, often the leaves of each pair slightly to almost completely joined at the base (in some species the plant-body consists only of a single pair of closely adpressed or divergent leaves); leaves very variable in shape, often papillose, sometimes toothed, sometimes

with translucent 'windows' at the broad, flat apex. Flowers bisexual, radially symmetric, axillary or terminal and solitary or in terminal cymes, often with fleshy bracts and bracteoles. Sepals 4–15, mostly free, sometimes united into a tube at the base. Petals very numerous in 1–several rows, mostly free, rarely joined into a short tube at the base. Stamens numerous; staminodes often present between the petals and stamens and transitional between the two. Nectary usually present, either as 5 or 6 distinct glands, or as a continuous, sometimes scolloped ring. Most or all of the ovary inferior, 4–20-celled; ovules usually numerous, usually parietal (but see below), rarely axile. Stigmas usually free, more rarely united into a columnar or disc-shaped structure. Fruit a capsule, usually opening when wetted and closing when dry, very complex in structure (see below), more rarely breaking up irregularly or fleshy and indehiscent. Seeds numerous, usually with long funicles.

A large family, sometimes divided into 3 (Aizoaceae in the strict sense, Molluginaceae and Mesembryanthemaceae), with about 120 genera (most of them split off from the old genus Mesembryanthemum during the present century) and about 2500 species, mostly from South Africa, though with a few native in other arid parts of the world.

The classification of the family as currently understood is based largely on the structure of the fruits, which are in many ways remarkable and complex structures. In a few genera the fruits are fleshy and indehiscent or break up irregularly, but in most (those genera which form the Mesembryanthemaceae) the capsules open when wetted, releasing the seeds gradually, and then close again when dry; the organs responsible for this opening and closing are unique to the family and are particularly important in classification at the genus level.

The ovary is for the most part inferior, and the placentation is usually parietal, that is, the ovules are borne in the inner surface of the outer walls of the cells into which the ovary is divided (Figure 26(1, 2)). During development of the flower and fruit, however, the cells may change shape due to differential growth, so that the ovules appear to be borne on the basal parts of the ovary (Figure 26(3, 4)); they are still, however, parietal. In the few genera that have axile placentation, the ovules are clearly attached to the central axis (Figure 26(5, 6)).

During development of the fruit various layers and structures develop in its walls, particularly in its top. The top wall of each cell splits (i.e. the dehiscence is loculicidal) forming a generally triangular or linear segment (the 'valve') between adjacent splits. When wetted, these segments open outwards like a flower; the sides of the valves are often thickened and keeled ('expanding keels'), the thickenings or keels functioning in the absorption of water and expansion, causing the opening of the segments (Figure 26(7, 8)); the inner surfaces of the valves often also bear 2 expanded wings ('valve-wings') which may be of various sizes; they may also be toothed or end in bristle-tips.

Variation in the presence or absence of these structures and in their form, if present, is the basis of the classification of the family; a detailed account can be found in Herre, The genera of the Mesembry-anthemaceae (1971). Because fruits are not often found in cultivated specimens, these characters are not stressed in the present account, and are not used at all in the key to the genera (they are mentioned, where significant, in the genus descriptions).

The classification based on these fruit characters (and others) is complex, and the identification of genera and species is difficult. This difficulty has many causes: the plants are often sparsely distributed in the wild, their succulence makes them poor subjects for the herbarium, the fruit characters are often difficult to interpret, and the earlier workers on the classification (N.E. Brown, H.M.L. Bolus, G. Schwantes) took a very narrow view of both genera and species, producing a plethora of names of uncertain or dubious application and confusing and sometimes contradictory descriptions. In recent years, some of these plants have been re-studied in great detail in Hamburg by H. Ihlenfeldt, H. Hartmann, H.H. Poppendieck, G. Leide and others. Their painstaking work has resulted in substantial reductions in size of the genera they have worked on (e.g. Pleiospilos reduced from 35 to 7 species (in 2 genera), and Cheiridopsis from about 150 to 23 species). Unfortunately, such detailed work is slow and many genera have not yet been revised. Some of the larger genera, e.g Lithops, Conophytum and Gibbaeum are particularly difficult and will probably undergo similar reduction in the number of species. In the meantime, we have merely attempted to summarise present knowledge and have not made significant changes or reductions. As mentioned above, the key to

the genera presented here does not make use of fruit characters; instead, it relies on more easily observed characters of the leaves and flowers. It must be borne in mind that such a key, in avoiding the difficult but important fruit characters, may not be totally effective; careful checking of generic identifications with the genus descriptions and with good illustrations will often be necessary.

All members of the Aizoaceae require as much light as possible, a dry atmosphere and very good drainage which can be achieved by mixing a high proportion of sharp sand, perlite or crushed brick into the potting compost. High nitrogen levels which cause abnormally soft growth should be avoided and the medium should be low in organic matter. The timing and amount of water given is critical. In general, plants should be well watered when actively growing, though the pot should be allowed to dry out between waterings. When growth ceases, water can be almost completely withheld and just enough given to avoid damage from dessication. Most genera manage to adapt their growing (and flowering) season to coincide with the European summer; others, including Conophytum, Dactylopsis, Frithia, Gibbaeum, Oophytum and Ophthalmophyllum persist in growing between October and February. These are the most difficult to grow as they have to be watered when there is relatively little light so their growth tends to be sluggish and they are prone to rot. Difficult species can be plunged (in their pots) in ash or gravel and water applied around the pots rather than directly to them. Care should always be taken to avoid wetting the foliage. The shrubby perennials and annuals will perform well out-of-doors, even in N Europe, once the danger of frost is over, though the perennials will have to be moved back under cover in the autumn. The dwarf succulents such as Lithops and Pleiospilos always need greenhouse cultivation in N Europe, but will also do well on a draught-free windowsill. Resting plants can withstand temperatures as low as 5° C and should not be kept in centrally heated rooms. Most genera will grow adequately from seed and the perennials also from stem cuttings. If a plant is affected by rot the branches can be pulled off, allowed to dry for a few days, and then treated as cuttings. Repotting should be carried out just as growth is beginning.

As cultivation details are relatively uniform throughout the family, they

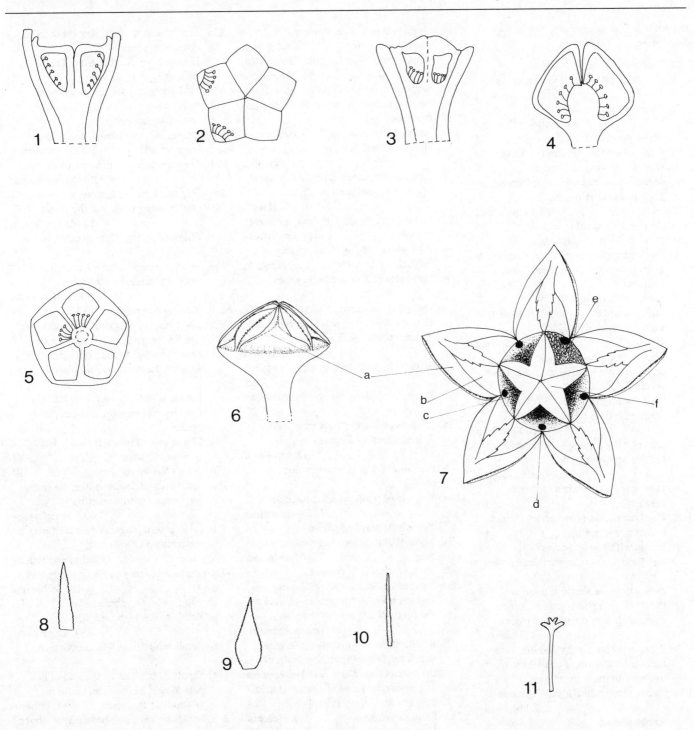

Figure 26. Diagrams to illustrate characters used in the identification of the Aizoaceae. 1, longitudinal section of an ovary with parietal placentation; 2, transverse section of the same ovary; 3, longitudinal section of an ovary with parietal placentation in which differential growth has caused the ovules to appear basal; 4, longitudinal section of an ovary with axile placentation; 5, transverse section of the same ovary; 6, unopened (dry) fruit; 7, opened (wetted) fruit seen from above; 8, a subulate stigma; 9, an ovate or triangular stigma; 10, a linear or thread-like stigma; 11, style as in *Conophytum*, with stigmatic lobes at the top (a, valve of fruit; b, valve-wing; c, cell of ovary; d, tubercle; e, seeds; f, cell-roofs)

are not repeated under each genus description.

1a. Placentation axile   2
  b. Placentation parietal or appearing basal   4
2a. Plant annual or biennial
    2. Mesembryanthemum
  b. Plant perennial   3
3a. Leaves flat, though fleshy **1. Aptenia**
  b. Leaves rounded, circular to semicircular in section **3. Dactylopsis**
4a. Fruit juicy, indehiscent
    66. Carpobrotus
  b. Fruit a dry capsule   5
5a. Plant annual or rarely biennial   6
  b. Plant perennial, sometimes shrubby   9
6a. Leaves 3-angled in section, thick, rounded **64. Conicosia**
  b. Leaves flat, though fleshy   7
7a. Stigmas and ovary-cells 10–20; leaves hairy; stigmas thread-like
    61. Carpanthea
  b. Stigmas and ovary-cells 5; leaves papillose; stigmas broadened at base or apex   8
8a. Anthers bright red **57. Pherelobus**
  b. Anthers brown or black
    56. Dorotheanthus
9a. Leaves with a conspicuous cluster of bristles around the tip
    29. Trichodiadema
  b. Leaves without such a cluster of bristles   10
10a. Plants with stem-internodes visible between the leaf-pairs   11
  b. Plants with the leaf-pairs close together, so that the stem-internodes are not visible between them, or consisting of a single leaf-pair   32
11a. Petals divided sharply into a very narrow claw and a distinct blade
    31. Kensitia
  b. Petals not sharply divided into claw and blade, tapering evenly from near the apex to the narrow base   12
12a. Spines present in the inflorescence
    5. Eberlanzia
  b. Spines absent   13
13a. Stigmas and ovary-cells 8 or more   14
  b. Stigmas and ovary-cells 4–7   20
14a. Stigmas shorter than the ovary depth   15
  b. Stigmas equal to or longer than the ovary depth   17
15a. Margins and keels of leaves cartilaginous and finely toothed
    30. Semnanthe
  b. Margins and keels of leaves not as above   16

16a. Stigmas thread-like; flowers 7 cm or more in diameter **64. Conicosia**
  b. Stigmas broadly triangular; flowers to 6 cm in diameter **54. Malephora**
17a. Stigmas thread-like, parallel-sided 18
  b. Stigmas subulate, broadened towards the base   19
18a. Stigmas and ovary-cells 9–10; leaves all dotted; leaf-pairs of 2 forms
    19. Vanzijlia
  b. Stigmas and ovary-cells 12–24; leaves not dotted; leaf-pairs all similar
    65. Herrea
19a. Main stems short, with long, prostrate branches **14. Cephalophyllum**
  b. Main stem long, with ascending or decumbent branches **13. Leipoldtia**
20a. Leaf-pairs of 2 or 3 different forms   21
  b. Leaf-pairs all more or less similar   22
21a. Flowers red **60. Meyerophytum**
  b. Flowers white, pink or yellow
    58. Mitrophyllum
22a. Stigmas broadly triangular or ovate   23
  b. Stigmas subulate, linear or thread-like   27
23a. Stigmas pinnatifid; flowers in many-flowered terminal cymes
    62. Stoeberia
  b. Combination of characters not as above   24
24a. The whole plant without papillae
    21. Lampranthus
  b. The whole plant papillose   25
25a. Nectary-glands forming a continuous ring **26. Jacobsenia**
  b. Nectary-glands 5, distinct   26
26a. Papillae conspicuously glistening; stigmas broadly triangular, as long as the depth of the cells of the ovary
    28. Drosanthemum
  b. Papillae fine, not glistening; stigmas subulate, shorter than the depth of the cells of the ovary **27. Delosperma**
27a. Stigmas and ovary-cells 6 **7. Astridia**
  b. Stigmas and ovary cells 4–5   28
28a. Flowers yellow **9. Hereroa**
  b. Flowers white, pink or purple   29
29a. Leaves with conspicuous, white, cartilaginous margins **24. Braunsia**
  b. Leaves not as above   30
30a. Nectary-glands 5, distinct
    22. Oscularia
  b. Nectary-glands forming a continuous, scolloped ring   31
31a. Leaves usually dotted; ovary markedly conical **4. Ruschia**
  b. Leaves not dotted; ovary cup-shaped
    63. Amphibolia

32a. Leaves 2 in a single pair on each flowering shoot   33
  b. Leaves in more than a single pair on each flowering shoot   48
33a. Leaf-margins toothed, sometimes finely so   34
  b. Leaf-margins entire   35
34a. Leaves with windows at the apex; teeth small **17. Vanheerdea**
  b. Leaves without windows; teeth large
    20. Odontophorus
35a. Stigmas forming a more or less stalkless disc on top of the ovary
    32. Argyroderma
  b. Stigmas separate and distinct, not as above   36
36a. Flowers lateral to the 2-leaved plant-body, subtended by 2 bracts
    48. Didymaotus
  b. Flowers terminal to the 2-leaved plant-body, appearing from between the leaves, with or without bracts 37
37a. Leaves united for half or more of their lengths, forming distinct plant-bodies   38
  b. Leaves united for less than half their lengths, not forming distinct plant-bodies   43
38a. Sepals united into a parallel-sided tube at the base   39
  b. Sepals free to the base   40
39a. Style present, column-like; nectaries forming a continuous ring
    49. Conophytum
  b. Style absent, stigmas free; nectaries 6, separate and distinct
    52. Ophthalmophyllum
40a. Stigmas broadly ovate or triangular
    46. Gibbaeum
  b. Stigmas linear, thread-like   41
41a. Petals white with pink tips
    51. Oophytum
  b. Petals white or yellow, never pink   42
42a. Sepals 5–6; plant-body looking like a pair of pebbles, each leaf with a window at the apex **41. Lithops**
  b. Sepals 6–15; plant-body not as above, leaves without windows at the apex
    44. Dinteranthus
43a. Stigmas ovate or triangular, shorter than the depth of the ovary   44
  b. Stigmas subulate, linear or thread-like, equalling or longer than the depth of the ovary   46
44a. Nectary-glands 5–6, separate; leaves rarely finger-like   45
  b. Nectary-glands forming a continuous ring; leaves finger-like, rounded
    18. Cylindrophyllum

45a. Stigmas 6–11; nectary-glands 6 or more   **46. Gibbaeum**
  b. Stigmas and nectary-glands 5
      25. Cerochlamys
46a. Petals white   **50. Herreanthus**
  b. Petals yellow, pink or purple   47
47a. Leaves with conspicuous dark green dots   **35. Pleiospilos**
  b. Leaves not dotted, or with few, inconspicuous dark green dots
      47. Antegibbaeum
48a. Leaves spathulate, flattened   49
  b. Leaves not as above   50
49a. Sepals all similar, all with membranous margins; petals yellow, orange, brown or pink
      33. Aloinopsis
  b. Sepals dissimilar, only some with membranous margins; petals always yellow   **34. Titanopsis**
50a. Leaves erect, finger-like, round in section, their tips covered with conspicuous, coralline warts
      39. Neohenricia
  b. Leaves not as above   51
51a. Leaves finger-like, round to almost square in section, with windows at their apices   52
  b. Leaves not finger-like, generally 3-angled in section, without windows at their apices   54
52a. Leaves conspicuously dotted with dark green dots   **35. Pleiospilos**
  b. Leaves not dotted   53
53a. Stigmas 5   **45. Frithia**
  b. Stigmas 10–16   **15. Fenestraria**
54a. Stigmas broadly ovate, triangular or columnar and blunt, shorter than or equalling the depth of the ovary-cells   55
  b. Stigmas linear or subulate, acute, longer than the depth of the ovary-cells   60
55a. Leaves toothed   56
  b. Leaves not toothed   58
56a. Stigmas 5 or 6   57
  b. Stigmas 8–11   **20. Odontophorus**
57a. Stigmas columnar, blunt; petals yellow or white   **36. Stomatium**
  b. Stigmas broadly ovate; petals white with red margins   **6. Acrodon**
58a. Leaves all similar   **25. Cerochlamys**
  b. Leaves differing in form on the same plant   59
59a. Some leaf-pairs with reduced blades, looking like stem-segments, others with fleshy blades   **59. Monilaria**
  b. Leaves within each pair differing, all with fleshy blades   **55. Glottiphyllum**
60a. Nectary-glands separate and distinct   61

  b. Nectary-glands forming a continuous ring   66
61a. Leaves with backwardly-directed, bristle-tipped teeth   **53. Faucaria**
  b. Leaves entire, or if toothed then teeth not as above   62
62a. Leaves with dark green or white dots or white warts   63
  b. Leaves without dots or warts   65
63a. Top of ovary not centrally domed
      37. Chasmatophyllum
  b. Top of ovary centrally domed   64
64a. Leaves each with a toothed, lobed or axe-like chin   **10. Rhombophyllum**
  b. Leaves not as above   **8. Bergeranthus**
65a. Leaves irregularly club-shaped, tips rounded   **11. Bijlia**
  b. Leaves long-cylindric, 3-angled, tips pointed   **12. Machairophyllum**
66a. Leaves without dots or warts   67
  b. Leaves with pale or dark dots or warts   69
67a. Flowers white or pink
      40. Juttadinteria
  b. Flowers yellow   68
68a. Leaves to 1.5 cm, about as broad as long   **43. Lapidaria**
  b. Leaves 4 cm or longer, much longer than broad   **12. Machairophyllum**
69a. Stigmas and ovary-cells 8 or more   70
  b. Stigmas and ovary-cells 5 or 6   71
70a. Flowers stalked   **16. Cheiridopsis**
  b. Flowers stalkless or stalks completely hidden   **35. Pleiospilos**
71a. Flowers yellow   72
  b. Flowers white, pink or purple   73
72a. Flowers to 2 cm in diameter
      38. Rhinephyllum
  b. Flowers 3.5 cm in diameter or more
      42. Schwantesia
73a. Flowers white; sepals and stigmas 6
      50. Herreanthus
  b. Flowers pink or purple; sepals and stigmas 5   **23. Ebracteola**

1. APTENIA N.E. Brown

J.C.M. Alexander & J. Kendall

Short-lived perennial, succulent, prostrate, herbs covered in minute shiny papillae. Leaves opposite, stalked, flat, ovate. Flowers 1–3, terminal or lateral, short-stalked, without bracts. Sepals 4, free, 2 often larger and leaf-like. Petals numerous, shortly fused at base, shorter than sepals. Stamens many, borne on corolla-tube; filaments hairy, surrounded by staminodes. Ovary inferior with axile placentas; style absent, stigmas 4, minute. Cells of capsule 4, broader than long, much thickened above. Cells open; seeds compressed.

A genus of 2 species from Transvaal and

the eastern coastal districts of South Africa. Poorly grown specimens often behave as annuals.

1. A. cordifolia (Linnaeus filius) N.E. Brown (*Mesembryanthemum cordifolium* Linnaeus filius). Illustration: Bichard & McClintock, Wild flowers of the Channel Islands, 87 (1975); Huxley & Taylor, Wild flowers of Greece and the Aegean, pl. 23 (1977); Graf, Tropica, 42, 49 (1978); Everett, New York Botanical Garden illustrated encyclopedia of horticulture, 211 (1980).

Stems to 60 cm, round in section, much-branched, green becoming grey. Leaves to 2.5 cm and almost as wide, broadly ovate, obtuse to acute, heart-shaped at base, fleshy, bright green; midrib prominent beneath. Flowers solitary, terminal or lateral, shortly stalked; petals red to purple. *South Africa (Cape Province); naturalised in S Europe & the Channel Islands.* H5–G1. Late Summer–autumn.

'Variegata' has a cream-coloured border to the leaves. Illustration: Graf, Tropica, 49 (1978).

2. MESEMBRYANTHEMUM Linnaeus

J.C.M. Alexander

Annual or biennial, usually prostrate, succulent herbs, covered with glistening papillae; stems and branches sometimes partially erect. Leaves opposite below, alternate above, stalkless or shortly stalked, round in section or flat. Flowers stalkless or shortly stalked, borne singly opposite the leaves or in terminal branched cymes lacking bracts. Sepals 4–5, shortly joined, 2 often larger and leaf-like. Petals numerous, narrowly linear to thread-like, shortly joined at base. Stamens many, borne on the corolla-tube, filaments long; staminodes sometimes present. Ovary partly inferior with axile placentas. Nectary-glands 5, stigmas 5 (rarely 4), erect, thread-like. Cells of capsule 5 (rarely 4), open. Seeds numerous, small, spherical or compressed.

A genus of about 70 species of annual and biennial succulents from S Europe, SW Asia, S Africa and the Atlantic Islands, with some species naturalised in warmer areas. *M. crystallinum* is sometimes grown for its edible leaves. The genus was formerly very much larger, containing species now accommodated in other genera.

1a. Plant prostrate, glistening green or greyish green   **1. crystallinum**
  b. Plant at least partially erect, glaucous or purplish   **2. barklyi**

1. M. crystallinum Linnaeus (*Cryophytum crystallinum* (Linnaeus) N.E. Brown). Illustration: Herre, The genera of the Mesembryanthemaceae, 205 (1971); Jacobsen, Lexicon of succulent plants, pl. 183 (1974); Heywood, Flowering plants of the world, 66 (1978); Buishand et al., The complete book of vegetables, 12 (1986). Annual or biennial with much-branched, prostrate stems to 60 cm. Leaves to 15 × 8 cm, ovate to spathulate, stalked. Flowers solitary or 3–5 together, almost stalkless, 2–3 cm across, white or reddish. *Southwest Africa; naturalised from Arabia through the Mediterranean area to the Canary Islands and Portugal, also in Australia and USA (California).* H5–G1. Late summer.

2. M. barklyi N.E. Brown (*Cryophytum barklyi* (N.E. Brown) N.E. Brown; *Mesembryanthemum grandifolium* misapplied).
Erect or partially so, to 60 cm, glaucous or purplish. Stems 4-angled. Leaves broadly spathulate, to 28 × 18 cm. Flowers *c.* 3.5 cm across, whitish. *Southwest Africa, South Africa (Cape Province).* H5–G1.

3. DACTYLOPSIS N.E. Brown
J.C.M. Alexander & J. Kendall
Dwarf, succulent, clump-forming perennials. Leaves 2–4 per shoot, alternate, thick, round in section, with basal sheaths concealing the soft, pulpy stem which withers and disappears in the resting stage. Flowers solitary, stalkless, terminal. Sepals 5, shortly joined at the base. Petals numerous in several ranks. Staminodes petal-like but stiff and dry, the inner concealing the stamens and joined at the base. Stamens numerous in 4 ranks, borne on the corolla-tube, not projecting. Ovary half-superior with axile placentas. Cells of capsule 5, open; central axis ending in a 5-rayed star above. Seeds numerous, small, compressed, pointed.
A genus of 2 species from the Cape Province of South Africa. Plants should be kept quite dry when resting in summer and watered very moderately in winter, with the addition of 2% kitchen salt in the soil.

1. D. digitata (Aiton) N.E. Brown (*Mesembryanthemum digitatum* Aiton; *M. digitiforme* Thunberg). Illustration: Herre, The genera of the Mesembryanthemaceae, 125 (1971); Innes, The complete handbook of cacti and succulents, 177 (1977); Rauh, The wonderful world of succulents, pl. 87 (1984).
To 8 cm, densely clumped; stems hidden. Leaves finger-like, unequal, 8–12 ×

2–2.5 cm, obtuse, soft, hairless, greyish green, sometimes purple-tipped. Flowers 1.5–2 cm across; petals white. *South Africa (Cape Province).* H5–G1. Winter.

4. RUSCHIA Schwantes
J.C.M. Alexander
Dwarf mat-forming or clump-forming to ascending or erect shrubs; internodes visible; stems with old leaf-remains. Leaf-pairs fused at base, sometimes shortly so; free parts boat-shaped, keeled beneath, spreading or pressed together, often mucronate at tip and broadly toothed on the keel, sometimes finely hairy, with papillae, often with dark, translucent dots. Flowers stalked or stalkless, terminal or axillary, solitary or in few-flowered to many-flowered inflorescences. Sepals 4–5, free; petals numerous in 1–several series, thread-like to spathulate. Stamens many, often pressed together, usually hairy at base; staminodes few–many. Ovary inferior, with parietal placentas. Stigmas 4–5, linear to narrowly subulate; cells of capsule 4–5; valve-wings fused to cell-wall, inconspicuous; keels diverging; cell-roofs present; tubercle large. Seeds spherical, flattened, yellow to brown, patterned.
A genus of about 350 species from South and Southwest Africa, occasionally grown for their brightly coloured flowers. The dwarf species are sensitive to excess moisture when resting and must be kept under glass; the taller shrubs are less demanding and flower more easily.

Habit. Erect or ascending: **1–5,7–10,19**; clump-, mat- or cushion-forming: **6,11–13,15–18**.
Leaves. Less than 10 cm: **5–7, 15–17**; 10 cm or more: **1–4,7–14,16,18,19**.
Flower-colour. White: **1–3**; red, pink or purple: **4–19**.

1a. Erect or ascending shrubs more than 25 cm tall 2
 b. Clump-, mat- or cushion-forming or dwarf shrubs less than 25 cm tall 11
2a. Flowers white 3
 b. Flowers red, pink or purple 5
3a. Plant smelling of fish **1. odontocalyx**
 b. Plant not smelling of fish 4
4a. Flower-stalks 1.5–2 cm; young branches 6-angled; leaves greyish green **2. multiflora**
 b. Flower-stalks *c.* 4 cm; young branches flattened; leaves bright green **3. umbellata**
5a. Flowers solitary 6
 b. Flowers many, in inflorescences 10
6a. Flowers 3.5 cm or more across; leaves 2–3 cm **4. semidentata**

 b. Flowers 2.5 cm or less across; leaves 2 cm or less 7
7a. Leaves 9 mm long or less 8
 b. Leaves 1.1 cm long or more 9
8a. Robust erect shrubs **5. robusta**
 b. Small shrub with spreading or prostrate stems **6. uncinata**
9a. Leaves pale reddish grey, fused for 2–2.5 cm **7. perfoliata**
 b. Leaves glaucous, fused for 1–1.3 cm **8. vulvaria**
10a. Leaves 4–6 cm; flowers *c.* 3.2 cm across **9. strubeniae**
 b. Leaves 2–3 cm; flowers *c.* 2 cm across **10. tumidula**
11a. Longest leaves 2 cm or more 12
 b. Longest leaves 1.5 cm or less 16
12a. Flowers 3.5 cm across or more **4. semidentata**
 b. Flowers 2.5 cm across or less 13
13a. Flowers stalkless 14
 b. Flower-stalks 1 cm or more 15
14a. Leaves *c.* 2 cm × 5 mm, greyish white **11. dualis**
 b. Leaves *c.* 3 × 1 cm, greyish green **12. nobilis**
15a. Dense cushion-forming shrub; shoots with 1–2 pairs of leaves **13. amoena**
 b. Loosely branched shrub; stems densely leafy **14. macowanii**
16a. Leaf-pairs 2 per shoot, dissimilar; lower pair 4–5 mm, almost completely joined, upper pair with free, spreading blades **15. pygmaea**
 b. Leaf-pairs 2–many per shoot, all similar 17
17a. Flowers 2 cm or more across 18
 b. Flowers 1.7 cm or less across 19
18a. Prostrate; leaves 8 mm or less, greyish green **6. uncinata**
 b. Erect; leaves 1.1 cm or more, reddish grey **7. perfoliata**
19a. Leaf-pairs about as thick as long 20
 b. Leaf-pairs half as thick as long 21
20a. Leaf-pairs many per shoot; flower-stalks *c.* 1.5 cm **16. dolomitica**
 b. Leaf-pairs 2 or 3 per shoot; flowers stalkless **17. evoluta**
21a. Prostrate; flowers *c.* 1.5 cm across **18. mollis**
 b. Erect; flowers *c.* 7 mm across **19. quartzitica**

1. R. odontocalyx (Schlechter & Diels) Schwantes (*Mesembryanthemum odontocalyx* Schlechter & Diels; *R. rupicola* (Engler) Schwantes). Illustration: Jacobsen, Handbook of succulent plants, 1400 (1960).
Erect, sparingly branched shrub. Leaves spreading; free part 1.1–1.5 cm, 3-angled,

narrowed towards the hard, red tip, distantly toothed on keel, pale reddish grey, joined basal part 2–2.5 cm. Flowers 2–2.5 cm across, solitary, pinkish red. *South Africa (Cape Province).* H5–G1.

2. R. multiflora (Haworth) Schwantes (*Mesembryanthemum multiflorum* Haworth). Shrub to 1 m, and to 2 m across. Branches erect, much divided, brown becoming grey, 6-angled at first. Leaves spreading; free part *c.* 3 cm × 3–4 mm, 3-angled, curved, shortly mucronate, greyish green with translucent dots; joined basal part about equalling internode. Flowers many, 2.5–3 cm across in branched inflorescences, white; stalks 1.5–2 cm. *South Africa (Cape Province).* H5–G1.

3. R. umbellata (Linnaeus) Schwantes (*Mesembryanthemum umbellatum* Linnaeus; *M. tumidulum* according to N.E. Brown & Bolus).
Shrub to 80 cm with ascending branches; young shoots flattened, greenish becoming brown to grey. Leaves distant, 5–7 cm × 4–6 mm, 3-angled to almost round in section, curved, blunt at tip with a short point, bright green with transparent dots; sheath short and thick, clasping the stem. Flowers numerous, *c.* 3 cm across, in umbel-like inflorescences, white; stalks *c.* 4 cm. *South Africa (Cape Province).* H5–G1.

4. R. semidentata (Salm-Dyck) Schwantes (*Mesembryanthemum semidentatum* Salm-Dyck). Illustration: Herre, The genera of the Mesembryanthemaceae, 269 (1971). Erect shrub with stiff, forked branches. Leaves distant, 2–3 cm, 3-angled, straight to slightly curved, pale grey; keel broader towards the tapered tip and with 2–4 broad teeth. Flowers 3.5–4 cm across, red. *South Africa (Cape Province).* H5–G1.

5. R. robusta Bolus. Illustration: Jacobsen, Lexicon of succulent plants, pl. 195 (1974). Robust, erect shrub to 50 cm; branches slender, 2–4 cm, covered in withered leaves. Leaves spreading, 6–9 × 3.5 mm, 3-angled, obtuse, indistinctly keeled, dark bluish green, dotted; upper surface flat, lower surface convex; fused basal part *c.* 25 mm. Flowers solitary on closely packed shoots, *c.* 1.8 cm across, purple; stalks absent to 2 mm. *South Africa (Cape Province).* H5–G1.

6. R. uncinata (Linnaeus) Schwantes (*Mesembryanthemum uncinatum* Linnaeus). Small shrub with long curved or prostrate stems, bearing branches and shoots on 1 side only. Leaves spreading, 4–8 mm, 3-angled to round in section, tapered to a short point and with 1 or 2 teeth on the keel, greyish green with darker dots; sheath long, enclosing the branch. Flowers solitary, terminal on short shoots, *c.* 2 cm across, pink. *South Africa (Cape Province).* H4–5.

7. R. perfoliata (Miller) Schwantes (*Mesembryanthemum perfoliatum* Miller). Illustration: Rauh, The wonderful world of succulents, pl. 95 (1984). Erect, sparingly branched shrub. Leaves spreading, 1.1–1.5 cm, 3-angled, narrowed towards the hard red tip, distantly toothed on keel, pale reddish grey; sheath 2–2.5 cm. Flowers solitary, *c.* 2.5 cm across, pinkish red. *South Africa (Cape Province).* H5–G1.

8. R. vulvaria (Dinter) Schwantes (*Mesembryanthemum vulvarium* Dinter). Illustration: Jacobsen, Handbook of succulent plants, 1409 (1960); Jacobsen, Lexicon of succulent plants, pl. 195 (1974). Dense hemispherical shrub to *c.* 40 cm. Free part of leaf 1.5–2 cm, bluntly 3-angled with curved sides, mucronate and with a broad tooth on the keel near the tip, glaucous with very small green papillae and a few translucent dots; sheath 1–1.3 cm. Flowers solitary, terminal, *c.* 2 cm across, pink to purple; stalks *c.* 7 mm. *Southwest Africa.* H5–G1.

9. R. strubeniae (Bolus) Schwantes (*Mesembryanthemum strubeniae* Bolus). Erect, hairless shrub to *c.* 75 cm with red, flattened branches. Leaves 4–6 cm × 5–6 mm × 1.2 cm, 3-angled, flattened sideways, green, upper surface flat or slightly grooved; lateral surfaces flat or rounded; margin toothed; sheath short. Flowers many, *c.* 3.2 cm across, in branched inflorescences; petals pink with a purple stripe; stalks 1.8–2.5 cm. *South Africa (Cape Province).* H5–G1.

10. R. tumidula (Haworth) Schwantes (*Mesembryanthemum tumidulum* Haworth according to Schwantes). Much-branched shrub to 60 cm; branches slightly compressed, reddish becoming grey. Leaves 2–3 cm × *c.* 4 mm, linear to narrowly 3-angled, slightly curved, blunt and shortly mucronate at tip, greyish green with translucent dots; upper surface flat to slightly concave, lower surface convex, angles smooth; sheath slightly inflated. Flowers many, *c.* 2 cm across, in branched inflorescences, pink; stalks 1.5–2 cm. *South Africa (Cape Province).* H5–G1.

11. R. dualis (N.E. Brown) Bolus (*Mesembryanthemum duale* N.E. Brown). Illustration: Barkhuizen, Succulents of southern Africa, 146 (1978); Rauh, The wonderful world of succulents, pl. 95 (1984). Clump-forming, to *c.* 5 cm; later with branchlets covered in dry leaves. Leaves to *c.* 2 cm × 5 mm, upper surface flat, lower surface rounded, keeled towards the tip; margins cartilaginous. Flower *c.* 1.5 cm across, pink, stalkless. *South Africa (Cape Province).* H5–G1.

12. R. nobilis Schwantes. Clump-forming; stems short with very small branches. Leaves *c.* 3 × 1 cm, 3-angled, greyish green, finely and densely hairy, joined for 5–7 mm at base; upper surface flat to convex, triangular, lower surface keeled, rounded. Flowers terminal, *c.* 1.7 cm, purplish pink, stalkless. *South Africa (Cape Province).* H5–G1.

13. R. amoena Schwantes. Illustration: Jacobsen, Lexicon of succulent plants, pl. 194 (1974). Forming hemispherical cushions to 15 cm across; shoots with 1 or 2 leaf-pairs. Leaves spreading, to 2.5 cm × 8 mm, finger-like, obscurely 3-angled, slightly curved, mucronate, bright bluish green with translucent dots, very finely and shortly hairy; upper surface narrowly triangular, flat or slightly concave, lower surface rounded and keeled; sheath to 8 mm. Flowers solitary, terminal, *c.* 1.5 cm across, reddish purple; stalks to 2 cm. *South Africa (Cape Province).* H5–G1. Summer.

14. R. macowanii (Bolus) Schwantes (*Mesembryanthemum macowanii* Bolus). Loosely branched shrub to 20 cm with long prostrate stems; branches ascending to erect, very leafy. Leaves 2–2.5 cm × 4 mm, boat-shaped, asymmetric, acute; upper surface flat, lower surface keeled; sheath *c.* 5 mm. Flowers *c.* 2.2 cm across, petals pink with a darker stripe; stalks 1–2 cm. *South Africa (Cape Province).* H5–G1.

15. R. pygmaea (Haworth) Schwantes (*Mesembryanthemum pygmaeum* Haworth). Illustration: Jacobsen, Handbook of succulent plants, 1398 (1960); Jacobsen, Lexicon of succulent plants, pl. 194 (1974). Mat-forming; stems very short, each with 2 dissimilar leaf-pairs; lower pair 4–5 × 2–3 mm, fused almost to tip, green, drying and enclosing the bases of the upper pair with spreading ovate to lanceolate blades, rounded or keeled beneath. Flower *c.* 1.8 cm across. *South Africa (Cape Province).* H5–G1.

16. R. dolomitica (Dinter) Dinter & Schwantes (*Mesembryanthemum dolomiticum* Dinter).
Forming cushions to *c.* 20 cm across and to 10 cm high, later with several erect stems to 25 cm. Leaf-pairs many on terminal branchlets, obovoid, 6–15 × 8–15 mm. Leaves broadly boat-shaped, mucronate, with small white papillae which are longer on keel and margins; upper surface curved, lower surface rounded and keeled; fused portion 5–8 mm. Flowers solitary, terminal, *c.* 1 cm across, pink; stalk *c.* 1.5 cm. *Southwest Africa.* H5–G1.

17. R. evoluta (N.E. Brown) Bolus (*Mesembryanthemum evolutum* N.E. Brown). Illustration: Barkhuizen, Succulents of southern Africa, 147 (1978).
Forming a densely tufted mat; shoots numerous, each with 2 or 3 very small hemispherical leaf-pairs and white chaffy remains of dried leaves. Leaves glaucous, ciliate. Flowers solitary, to 1.7 cm across, reddish purple. *South Africa (Cape Province).* G1.

18. R. mollis (Berger) Schwantes (*Mesembryanthemum molle* Berger).
Small shrub with prostrate branches. Leaves 1.2–1.5 cm × 4–6 mm, 3-angled, obtuse, with rounded surfaces, greyish green sometimes tinged with red, minutely hairy. Flowers solitary, *c.* 1.5 cm across, red. *South Africa (Cape Province).* G1.

19. R. quartzitica (Dinter) Dinter & Schwantes (*Mesembryanthemum quartziticum* Dinter).
Much-branched shrubs to *c.* 15 cm with dense cushions of terminal branchlets, each with 2–4 leaf-pairs. Leaves 1–1.2 cm × 4–5 mm, indistinctly 3-angled, with slightly convex surfaces, bluish green, finely papillose except on basal part of upper surface; tip slightly recurved, red, mucronate. Flowers 6–8 mm across, purplish red; stalks 1–2 mm. *Southwest Africa.* G1.

5. EBERLANZIA Schwantes
J.C.M. Alexander & J. Kendall
Erect hairless shrubs, with visible internodes. Leaves obscurely 3-angled to cylindric, long, or short and thick, obtuse and minutely mucronate at the apex, fused at the base, greyish to bluish green, with papillae and small, dark dots. Flowers stalked, in branched, spiny inflorescences. Sepals 5, 3-angled; petals several in 1 row or absent and replaced by staminodes, hairy at base or higher up, as are the

stamens. Ovary inferior, with parietal placentas. Stigmas and cells of capsule 5; valve-wings fused to cell-wall and apparently absent; keels much diverging; cell-roofs and tubercles present. Seeds spherical to pear-shaped, rough or smooth, yellow to yellowish brown.
A genus of 28 species from South Africa and Southwest Africa.

1. E. spinosa (Linnaeus) Schwantes (*Mesembryanthemum spinosum* Linnaeus). Illustration: Jacobsen, Handbook of succulent plants, 1133 (1960); Herre, The genera of the Mesembryanthemaceae, 145 (1971); Jacobsen, Lexicon of succulent plants, pl. 162 (1974).
To *c.* 60 cm, much branched. Leaves erect or curved inwards, 12–24 × 3–4 mm, greyish green. Inflorescences with thorns 10–20 cm long; flowers paired, 1.5–2 cm across, pink. *South Africa (Cape Province), Southwest Africa.* G1.

6. ACRODON N.E. Brown
J.C.M. Alexander
Perennial, branched at base into short leafy shoots; internodes not visible. Leaves crowded, spreading or curved, 3-angled, acuminate and toothed on margin at top, greyish green, hairless, sheathed at base. Flowers solitary on long stalks; sepals 5–6, ovate; petals many in 2 rows, strap-shaped to narrowly lanceolate, notched at tip; stamens leaning inwards, anthers reddish; staminodes absent. Ovary half-inferior with parietal placentas; nectary-glands dark green, toothed. Stigmas 5, broadly ovate, feathery, shorter than depth of ovary-cells. Cells of capsule 5. Valve-wings bristle-tipped; cell-roofs present; tubercles large. Seeds pear-shaped, rough, dark brown.
A genus of 3 species from South Africa.

1. A. bellidiflorus (Linnaeus) N.E. Brown (*Mesembryanthemum bellidiflorum* Linnaeus). Illustration: Jacobsen, A handbook of succulent plants, 978 (1960); Herre, The genera of the Mesembryanthemaceae, 63 (1971); Everett, New York Botanical Garden illustrated encyclopedia of horticulture, 44 (1980).
Leaves 3–5 cm, flattened sideways and toothed near the tip. Flowers *c.* 4 cm across; petals white with red margins; flower-stalks to *c.* 5 cm. *South Africa (Cape Province).* G1. Summer.

7. ASTRIDIA Dinter & Schwantes
J. Kendall & F. McIntosh
Erect shrub with visible internodes; green parts hairless or softly and shortly hairy.

Leaves erect, ascending or spreading, 3-angled, long-persistent, shortly fused at base. Flowers solitary or 3 together and terminal, shortly stalked. Sepals 6, flattened; petals in 2–5 rows, the inner much narrower. Stamens in 6–8 rows, leaning inwards; anthers yellow; staminodes papillose beneath. Ovary inferior with parietal placentas, lobed; nectary-glands inconspicuous. Stigmas 6, subulate; cells of capsule 6. Valve-wings bristle-tipped; cell-roofs and tubercles well-developed. Seeds hairy, spiny or warty, pear-shaped to horseshoe-shaped, brown.
A genus of 10 species from South Africa; requires a long dry rest in summer.

1. A. dinteri Bolus (*A. velutina* (Dinter) Dinter). Illustration: Jacobsen, A handbook of succulent plants, 997 (1960); Jacobsen, Lexicon of succulent plants, pl. 146 (1974).
Compact, to *c.* 30 cm; branches with 2 or 3 leaf-pairs. Leaves arched, rounded, 3–3.5 × 1.5 × 1.1–2.1 cm, narrower near the top, greyish white. Flowers *c.* 4 cm, white or pink. *Southwest Africa.* G2. Winter.

8. BERGERANTHUS Schwantes
J.C.M. Alexander & J. Kendall
Clumped stemless perennial succulents with dense leafy shoots, internodes not visible. Leaves crowded, 3-angled and obscurely keeled at base, semi-cylindric towards the tip, sometimes sharply bent upwards and chin-like, pale to dark green with dark dots, shortly fused at base; upper surface linear, slightly broader in the middle. Flowers solitary, or several in cymes, long-stalked. Sepals 5, 3-angled or lanceolate, acuminate and spine-tipped; margins membranous. Petals in several rows, linear to narrowly lanceolate. Stamens numerous, erect. Ovary domed, inferior with parietal placentas; nectary-glands large, separate; stigmas 5, linear, longer than the depth of the ovary-cells. Capsule 5-celled; valve-wings very broad with membranous points; cell roofs stiff; tubercles large, 3-sided. Seeds pear-shaped, 3-sided or rounded.
A genus of 11 species from South Africa.

1a. One of each pair of leaves sharply bent upwards, and chin-like near the tip **2. scapiger**
 b. Leaves straight or gently curved **2**
2a. Leaves 2.5–3 cm × 8–10 mm, lacking translucent dots on the upper surface **1. multiceps**
 b. Leaves *c.* 6 cm × 5–6 mm, finely tapered; upper surface with dark translucent dots **3. vespertinus**

1. B. multiceps (Salm-Dyck) Schwantes. Illustration: Jacobsen, Handbook of succulent plants, 998 (1960); Jacobsen, Lexicon of succulent plants, pl. 147 (1974); Rauh, The wonderful world of succulents, pl. 85 (1984).
Tufted, suckering and mat-forming. Leaves 6–8 mm, 3-angled, acuminate and bristle-tipped, green, lacking translucent dots; upper surface flat; lower surface obscurely keeled, flattened towards the tip. Flowers *c.* 3 cm across, yellow, reddish outside; stalk 3–4 cm. *South Africa (Cape Province)*. G1. Summer.

2. B. scapiger (Haworth) N.E. Brown. Illustration: Schwantes, The cultivation of Mesembryanthemaceae, f. 64 (1954); Herre, The genera of the Mesembry-anthemaceae, 91 (1971).
Similar in habit to No. **1**. Leaves 7–12 × 1–1.7 cm, dark green; one of each pair longer, sharply bent upwards and chin-like near the tip. Flowers 3 or 4 together, deep yellow, reddish outside; stalks *c.* 4 cm. *South Africa (Cape Province)*. G1. Summer.

3. B. vespertinus (Berger) Schwantes. Illustration: Jacobsen, A handbook of succulent plants, 999 (1960).
Similar in habit to Nos. **1 & 2**. Leaves erect at first, later prostrate, *c.* 6 cm × 5–6 mm, 3-angled, finely tapered; upper surface flat, greyish green with darker translucent dots, slightly wrinkled; lower surface semi-cylindric at base, keeled towards tip. Flowers 3–5 together, yellow; stalks 2–3 cm. *South Africa (Cape Province)*. G1. Summer.

9. HEREROA (Schwantes) Dinter & Schwantes
J.C.M. Alexander
Small mat-forming or taller shrubby succulents with short prostrate stems and ascending branches, with visible internodes. Leaves spreading to ascending, rounded or 3-angled with a keel, often broader and club-shaped or flattened towards the top, shortly fused at base, dark green with large dark spots. Flowers yellow, several together (rarely solitary); stalks often long. Sepals 5, unequal, acuminate. Petals in 2–4 rows, acute or acuminate, narrowed at base. Stamens erect, rarely leaning inwards. Ovary concave above, inferior, with parietal placentas; nectary-glands often distinct. Stigmas 5, thread-like. Cells of capsule 5; cell-roofs present; tubercle small. Seeds pear-shaped, slightly concave, small, pale brown.

A genus of about 35 species from South Africa.

1a. Leaves bilobed at tip   **3. dyerii**
  b. Leaves sometimes broadened but not bilobed at tip   2
2a. Leaves rough with prominent transparent dots or warts   3
  b. Leaves with less prominent dark dots   4
3a. Leaves 2.5–3 cm × 5 mm   **1. nelii**
  b. Leaves 4–6 cm × 6 mm   **2. granulata**
4a. Leaves 6–7 cm   **4. puttkameriana**
  b. Leaves less than *c.* 3.5 cm   5
5a. Erect shrub; leaves broadest at tip; flower-stalks *c.* 1.5 cm
     5. hesperantha
  b. Spreading shrub; leaves broadest at base; flower-stalks *c.* 5 cm   **6. incurva**

1. H. nelii Schwantes.
Shoots each with 1–3 leaf-pairs; leaves curved, 2.5–3 cm × 5 mm, to 10 mm thick, green, with transparent warts; upper surface flat; lower surface rounded near the base, distinctly keeled towards the tip. Flowers 1–3 together, yellow. *South Africa (Cape Province)*. G1–2. Summer.

2. H. granulata (N.E. Brown) Dinter & Schwantes. Illustration: Schwantes, The cultivation of the Mesembryanthemaceae, f. 13 (1954); Jacobsen, Handbook of succulent plants, 1172 (1960).
Leaves spreading, often prostrate, 4–6 cm × 6 mm, *c.* 2 mm thick, slightly curved, mucronate, dark green, rough with transparent warts; lower surface rounded near the base, indistinctly keeled towards the tip. Flowers solitary, *c.* 4 cm across, yellow. *South Africa (Cape Province)*. G1–2. Summer.

3. H. dyerii Bolus.
Compact shrub to *c.* 10 cm. Branches short, covered in persistent leaf-remains, densely leafy. Leaf-pairs 2–6 per shoot, unequal; leaves ascending to erect, *c.* 5 cm × 3–10 mm, semi-cylindric near the base, broader and unequally bilobed at the tip, green with prominent dots. Flowers 3 together, *c.* 2.5 cm, golden yellow; stalks 5–6 cm. *South Africa (Cape Province)*. G1–2. Summer.

4. H. puttkameriana (Berger & Dinter) Dinter & Schwantes.
Compact shrub, densely leafy. Leaf-pairs 4–8 per shoot; leaves incurved in lower part, curved outwards near the tip, 6–7 cm × 6–7 mm, obscurely 3-angled, rounded near the tip, pale shiny green, redder near the tip, with dark dots,

especially on the keel. Flowers 1–3 together, to 3 cm, deep yellow to orange. *Southwest Africa*. G1–2. Summer.

5. H. hesperantha (Dinter) Dinter & Schwantes. Illustration: Jacobsen, Lexicon of succulent plants, pl. 168 (1974).
To 30 cm, with erect, stiff, leafy branches to 9 cm. Leaves spreading, later curved upwards, *c.* 3.5 cm × 4–5 mm, 3-angled, rounded, flattened and axe-shaped at tip, greyish green with large dark dots. Flowers 1–several together, 1.5–3 cm, yellow; stalks *c.* 1.5 cm. *Southwest Africa*. G1–2. Summer.

6. H. incurva Bolus.
Spreading shrub; leaves incurved, *c.* 2.5 cm × 7 mm, to 6 mm thick, acute to acuminate, broadest at base, bluish green or reddish, covered in small, dark dots. Flowers solitary, 3–6 cm, yellow; stalk *c.* 5 cm. *South Africa (Cape Province)*. G1–2. Summer.

10. RHOMBOPHYLLUM Schwantes
J.C.M. Alexander & F. McIntosh
Small tufted or mat-forming shrubby succulents; shoots short, densely leafy, internodes not visible. Leaves semi-cylindric, keeled towards the tip, with a prominent, toothed, lobed or axe-like 'chin', shortly joined at the base, smooth, dark shining green with pale translucent dots; upper surface narrow, broader towards the tip. Flowers 3–7 together (rarely solitary), stalked. Sepals 5, about equal. Petals many in *c.* 3 rows, linear to narrowly lanceolate. Stamens many, erect. Ovary inferior, flat above, with parietal placentas; nectary-glands distinct. Stigmas 5, linear, longer than depth of ovary-cells. Capsule 5-celled; cell-roofs stiff; tubercle large or small, flat, with 2 warts. Seeds pear-shaped, pointed, light brown.

A genus of 3 species from South Africa.

1a. Leaf-keel chin-like near tip, but lacking lobes or teeth
     1. rhomboideum
  b. Leaf-keel axe-like near tip, toothed or 2-lobed   2
2a. Leaves 2.5–3 cm; tip axe-like and toothed   **2. dolabriforme**
  b. Leaves *c.* 1.5 cm; tip 2-lobed   **3. nelii**

1. R. rhomboideum (Salm-Dyck) Schwantes (*Bergeranthus rhomboideus* (Salm-Dyck) Schwantes). Illustration: Jacobsen, Handbook of succulent plants, 1365 (1960); Jacobsen, Lexicon of succulent plants, pl. 193 (1974).
Stemless and tufted. Rosettes prostrate with

4–5 unequal leaf-pairs. Leaves 2.5–3 × 1–2 cm, dark greyish green with many white dots; upper surface narrowly elliptic, lower surface rounded near the base, deeply keeled and chin-like at tip, unlobed, usually lacking teeth. Flowers *c.* 3 cm across, deep yellow, reddish outside. *South Africa (Cape Province)*. G1–2. Summer.

2. R. dolabriforme (Linnaeus) Schwantes (*Hereroa dolabriformis* (Linnaeus) Bolus). Illustration: Jacobsen, Handbook of succulent plants, 1365 (1960); Herre, The genera of the Mesembryanthemaceae, 267 (1971); Jacobsen, Lexicon of succulent plants, pl. 193 (1974); Rauh, The wonderful world of succulents, pl. 95 (1984).
Tufted when young, later erect, to 30 cm. Leaves 2.5–3 cm, mid-green with translucent dots; upper surface narrowly tapered, lower surface with a narrow and axe-like 'chin' and tooth-like tip. Flowers to 4 cm, yellow. *South Africa (Cape Province)*. G1–2. Summer.

3. R. nelii Schwantes. Illustration: Barkhuizen, Succulents of southern Africa, 146 (1978).
Similar to *R. dolabriforme*. Leaves 1.5 cm × 4–8 mm, 2-lobed at tip, pale bluish green. *South Africa (Cape Province)*. G1–2. Summer.

11. BIJLIA N.E. Brown
J.C.M. Alexander
Mat-forming or tufted perennial succulent. Rosettes prostrate, with 2–4 unequal leaf-pairs, internodes not visible. Leaves very irregularly shaped, 3–5 × 1.5 cm, to 2.5 cm wide near the tip, smooth, pale greyish green, tips rounded; upper surface triangular to diamond-shaped, acuminate; lower surface strongly keeled to chin-like. Flowers 1–3 together, *c.* 3 cm across, yellow, stalks 2–10 mm. Sepals 5, about equal, keeled. Petals in 2 rows, linear to narrowly obovate. Stamens numerous, leaning inwards. Ovary inferior with parietal placentas; nectary-glands 5, separate; stigmas 5, thread-like, longer than depth of ovary-cells. Capsule 5-celled, valves small, lacking wings; cell-roofs pointed, stiff; tubercle large. Seeds ovoid, pointed.
A genus of one species from South Africa.

1. B. cana N.E. Brown. Illustration: Jacobsen, A handbook of succulent plants, 1000 (1960); Herre, The genera of the Mesembryanthemaceae, 95 (1971);

Barkhuizen, Succulents of southern Africa, 113 (1978). *South Africa (Cape Province)*. G2. Winter.

12. MACHAIROPHYLLUM Schwantes
J.C.M. Alexander
Densely tufted perennial, to 1.2 m across. Leaves crowded, concealing internodes, narrowly boat-shaped, pointed, keeled towards tip, shortly joined at base, smooth and shiny, pale or bluish grey. Flowers 1–3 together, stalked, night- or evening-opening. Sepals 5–8, unequal, with membranous margins. Petals in 3–5 dense rows. Stamens erect, filaments without papillae, staminodes absent. Ovary convex above, inferior, with parietal placentas; nectary-glands small, separate or merged; stigmas 5–15, linear, longer than depth of ovary-cells. Cells of capsule 5–15, valve-wings parallel below, wider and diverging above, fringed or finely and irregularly toothed, bristle-tipped; cell-roofs present; tubercles small. Seeds obovoid.
A genus of 9 species from South Africa, closely related to *Bergeranthus*.

1a. Leaves 4–4.5 cm; flowers solitary
                                    1. acuminatum
  b. Leaves 7–10 cm; flowers 3 together
                                    2. albidum

1. M. acuminatum Bolus. Illustration: Jacobsen, Handbook of succulent plants, 1252 (1960); Jacobsen, Lexicon of succulent plants, pl. 182 (1974).
Leaves pale green, finely tapered, 4–4.5 cm × 8–11 mm, 6–10 mm thick, joined for 1 mm; upper surface flat, lower surface rounded near base, keeled towards tip. Flowers solitary, *c.* 5 cm, golden yellow; stalks 4–6 cm. *South Africa (Cape Province)*. G2. Summer.

2. M. albidum (Linnaeus) Schwantes. Illustration: Botanical Magazine, 1824 (1816).
Leaves incurved, 7–10 × 2 cm, whitish, unspotted; upper surface flat, lower surface rounded near base, keeled towards tip. Flowers 3 together, *c.* 6 cm, yellow inside, reddish outside. *South Africa (Cape Province)*. G2. Summer.

13. LEIPOLDTIA Bolus
J.C.M. Alexander & F. McIntosh
Erect or straggling hairless shrubs; internodes visible. Leaves 3-angled, with slightly rounded sides, broadest and slightly joined at base. Flowers solitary or 2–5 in cymes, usually stalked, opening in the day; stalks narrowed below calyx, sometimes

persistent and spine-like. Sepals 5, rarely 6, almost equal, margins membranous. Petals in 1 or 2 series, linear to obovate. Stamens erect or leaning inwards, inner filaments bearded; staminodes few or absent. Ovary slightly convex above, inferior with parietal placentas; glands scalloped. Stigmas 10–12, subulate, acuminate, longer than ovary-depth. Cells of capsule 10–12; valve-wings touching below, diverging above; cell-roofs touching, reaching beyond middle of cell; tubercle obovoid, curved. Seeds rough, pointed.
A genus of 20 species from South Africa and Southwest Africa.

1a. 1 leaf of each pair spreading, with an axillary shoot, the other ascending; flower-stalks *c.* 5 mm **1. weigangiana**
  b. Leaves symmetric; flower-stalks to 4 cm   **2. jacobseniana**

1. L. weigangiana (Dinter) Dinter & Schwantes.
To 50 cm; leaves *c.* 1.5 cm × 5 mm, boat-shaped, densely dotted; 1 of each pair ascending, the other spreading and with an axillary shoot. Flowers solitary, terminal, *c.* 2 cm across, pink to violet; stalks *c.* 5 mm. *Southwest Africa*. G1. Summer.

2. L. jacobseniana Schwantes.
Leaves to 1.5 cm × 4 mm, boat-shaped, curved at tip; upper side flat, lower side sharply keeled. Flowers 1.2–1.5 cm across; stalks to 4 cm. *South Africa (Cape Province)*. G1. Summer.

14. CEPHALOPHYLLUM N.E. Brown
J.C.M. Alexander & J. Kendall
Perennial, main stem short; branches prostrate (erect in Nos. **1 & 2**) rooting, bearing clumps of crowded leaf-pairs, often separated by long internodes. Leaves 3-angled to rounded with convex sides, curved, finely dotted, free or shortly fused at base. Flowers terminal, 1–3 together in a branched cyme, long-stalked. Sepals 5, unequal; petals linear; stamens numerous, erect. Ovary convex above, inferior; placentas parietal (apparently basal); glandular wall undivided. Stigmas 8–20, subulate, longer than ovary-depth. Cells of capsule 8–20; wings well-developed; cell-roofs present; tubercles often large, sometimes concealed by cell-roofs. Seeds ovoid, brown.
A genus of about 60 species from South Africa.

1a. Main stems erect, to 30 cm or more

2

b. Main stems prostrate or apparently absent; plant less than 15 cm **3**

2a. Leaves unequal, 6–9 mm wide, glaucous; flowers yellow **1. frutescens**

b. Leaves equal, 1–1.9 cm wide, scaly-white; flowers scarlet **2. spongiosum**

3a. Petals yellow at least in part **4**

b. Petals red, pink or purplish **8**

4a. Petals yellow throughout **5**

b. Petals red or purple in part **6**

5a. Leaves usually less than 3.5 cm × 4 mm, yellowish green **3. gracile**

b. Leaves usually more than 5 cm × 6 mm, bright green **4. decipiens**

6a. Flower-stalks more than 2.8 cm **5. pillansii**

b. Flower-stalks less than 2.6 cm **7**

7a. Leaves tapered at tip, pale or greyish green; petal tips yellow **6. tricolorum**

b. Leaves bluntly rounded at tip, glaucous; petal tips red **7. aureorubrum**

8a. Leaves 5 cm or more **9**

b. Leaves 4 cm or less **10**

9a. Leaf-rosettes *c.* 5 cm apart on prostrate branches **8. alstonii**

b. Leaf-rosettes crowded; branches very short **9. subulatoides**

10a. Leaves bluish green; sheath *c.* 7 mm; flower-stalks *c.* 1.2 cm **10. pulchrum**

b. Leaves dull green; sheath *c.* 3 mm; flower-stalks *c.* 2.5 cm **11. framesii**

1. C. frutescens Bolus.
Erect shrub to *c.* 60 cm and 90 cm across; internodes 2–5 cm; branches 3–6 mm thick; short shoots with 2 leaf-pairs. Leaves ascending, 5–8.5 cm × 6–9 mm, rounded and bluntly keeled, narrowed to the mucronate tip, glaucous; upper surface flat to convex; sheath 2–8 mm. Flowers *c.* 6 cm across, yellow; stalks 2.5–9.5 cm. *South Africa (Cape Province).* G1. Summer.

2. C. spongiosum (Bolus) Bolus.
Illustration: Everett, New York Botanical Gardens illustrated encyclopedia of horticulture, 681 (1981); Court, Succulent flora of southern Africa, 91 (1981).
Erect shrub to *c.* 30 cm. Stems ascending, *c.* 35 × 1 cm, thicker at nodes, shiny yellowish brown, white and spongy within; internodes 2–5 cm. Branches with 1–3 unequal leaf-pairs. Leaves to 11 × 1.2 cm and 1–1.9 cm thick, curved, blunt or shortly acuminate at tips, scaly-white; upper surface flat; lower surface convex, indistinctly keeled. Flowers *c.* 6 cm across, scarlet; anthers yellow; flower-stalks 5–5.5 cm. *South Africa (Cape Province).* G1. Summer.

3. C. gracile Bolus. Illustration: Bolus, Notes on Mesembryanthemum 2: 119 (1928); Everett, New York Botanical Gardens illustrated encyclopedia of horticulture, 682 (1981).
Flowering-stems erect, to *c.* 11 cm. Leaves almost cylindric, slender, 2–3.5 (rarely 6) cm × 2–4 mm, yellowish green. Flowers *c.* 3 cm across, golden yellow. *South Africa (Cape Province).* G1. Summer.

4. C. decipiens (Haworth) Bolus.
Branches prostrate, 30–40 cm, greenish red, later grey. Leaves crowded, to 5 (or more) cm × 6–8 mm, 3-angled, rounded and shortly tapered near the mucronate tip, bright green with small rough dots; sheath short, reddish. Flowers solitary, yellow; stalks 3–4 cm. *South Africa (Cape Province).* G1. Summer.

5. C. pillansii Bolus. Illustration: Court, Succulent flora of southern Africa, 91 (1981).
Branches prostrate, ascending at tips. Leaves 5–6 (rarely 16) cm, acutely tapered, dark green, dotted. Flowers many, 4–6 cm across, yellow with red centres. *South Africa (Cape Province).* G1. Summer.

6. C. tricolorum (Haworth) N.E. Brown.
Stems prostrate; branches curved with prominent nodes, red becoming grey. Leaves crowded, 4–5 cm × 4–6 mm, almost cylindric, tapered at the tip, pale green to greyish green, with minute dots; sheath short. Flowers solitary (or 6 on same branch over a period of time), 4–5 cm across; petals yellow, purple at base, red at tip outside, filaments red, anthers brown; stalks very short. *South Africa (Cape Province).* G1. Summer.

7. C. aureorubrum Bolus. Illustration: Everett, New York Botanical Gardens illustrated encyclopedia of horticulture, 682 (1981).
Stems prostrate, 14–17 cm. Leaves 4.5–6 cm × 6–8 mm, 5–8 mm thick, almost cylindric, bluntly rounded at tip (from above), tapered in side view, narrowed at base, glaucous; lower surface rounded; short, *c.* 6 mm. Flowers *c.* 3.5 cm; petals yellow at base, red above, red outside; stalks 1.3–2.5 cm. *South Africa (Cape Province).* G1. Summer.

8. C. alstonii Marloth. Illustration: Jacobsen, A handbook of succulent plants, 1008 (1960); Jacobsen, Lexicon of succulent plants, pl. 150 (1974).
Stems prostrate, to 50 cm or more, becoming grey; internodes *c.* 5 cm. Leaves in clusters, to 7 cm × 8 mm, *c.* 7 mm thick, almost cylindric, finger-like, slightly curved, greyish green with dark transparent dots; sheath short. Flowers 5–8 cm across, dark red with violet stamens. *South Africa (Cape Province).* G1. Summer.

9. C. subulatoides (Haworth) N.E. Brown. Illustration: Jacobsen, A handbook of succulent plants, 1015 (1960); Jacobsen, Lexicon of succulent plants, pl. 150 (1974).
Stemless; branches short; leaf-rosettes dense. Leaves crowded, spreading, 5–7 cm × 9–10 mm, semi-cylindric, keeled and acuminate at tip, greyish green with many translucent dots; keel cartilaginous; sheathing at base. Flowers *c.* 4 cm, purplish red; stalks 5–7 cm. *South Africa (Cape Province).* G1. Summer.

10. C. pulchrum Bolus.
Compact; flowering-branches with 2 leaf-pairs. Lower leaves 3.5–4 cm × 7 mm, *c.* 7 mm thick, blunt at tip, indistinctly keeled, bluish green; sheath *c.* 7 mm, upper leaves longer and more slender. Flowers *c.* 4.5 cm across, pink; stalk *c.* 1.2 cm. *South Africa (Cape Province).* G1. Summer.

11. C. framesii Bolus.
Stems prostrate, to *c.* 20 cm; internodes 5–10 mm; branches crowded, with 1 or 2 unequal leaf-pairs and old leaf remains below. Leaves erect, 2–3 cm × 3–5 mm, slightly swollen at tip, dull green with small loose scales; keel sharp; side margins indistinct; sheath *c.* 3 mm. Flowers solitary, *c.* 4 cm, pink; stalk *c.* 2.5 cm. *South Africa (Cape Province).* G1. Summer.

15. FENESTRARIA N.E. Brown
J.C.M. Alexander & J. Kendall
Dwarf, stemless, perennial succulents; internodes not visible. Leaves in rosettes, thick, club-shaped, with broad, flat, window-like tips. Flowers solitary, terminal, long-stalked. Sepals 5, free. Petals in 1 row, linear, narrowed at base, often notched at tip. Stamens erect or slightly incurved; staminodes absent. Ovary half-inferior, with parietal placentas. Stigmas 10–16, spreading, feathery. Cells of capsule 10–16; wings parallel, diverging at tips, toothed; cell-roofs well developed, with tubercles. Seeds pale, with brown points.

A genus now considered to contain only 1 species from South and Southwest Africa. In the wild, the plants are subjected to extreme temperatures and the leaves are normally buried so that the 'windows' are level with the soil surface. In cultivation there is a high risk of rotting so the leaves should mostly be clear of the soil. Re-potting should be avoided; the plant still in its pot should be put into a larger pot. Propagation is by seed, leaf-cuttings or division.

Literature: Hartmann, H., Monographie der Gattung Fenestraria, *Botanische Jahrbücher* **103**: 145–183 (1982).

1. F. rhopalophylla (Schlechter & Diels) N.E. Brown subsp. **aurantiaca** (N.E. Brown) Hartmann (*F. aurantiaca* N.E. Brown). Illustration: Herre, The genera of the Mesembryanthemaceae, 159 (1971); Rowley, The illustrated encyclopaedia of succulents, 144 (1978); Court, Succulent flora of Southern Africa, 91 (1981). Tufted cushion-forming succulent to 10 cm across. Leaves inclined, 2–3 cm, club-shaped, indistinctly 3-angled, smooth, whitish, reddish below; upper surface convex, slightly flattened, lower surface rounded; tip flat to convex, circular to almost triangular, to 6–8 mm across, translucent and lacking chlorophyll. Flowers to 6 cm across; petals 43–70, yellow to orange. Filaments lacking papillae. *South Africa, Southwest Africa (restricted to a small area around Alexander Bay)*. G2. Summer.

Subsp. **rhopalophylla** (*F. aurantiaca* forma *rhopalophylla* (Schlechter & Diels) Rowley). Illustration: Parey's Blumengärtnerei, 564 (1958); Martin & Chapman, Succulents and their cultivation, centre pages (1977). Differs from subsp. *aurantiaca* in its often smaller, white flowers; petals 35–55; filaments always papillose at base. *Southwest Africa* G2. Summer.

16. CHEIRIDOPSIS N.E. Brown
J.C.M. Alexander
Dwarf, tufted, perennial succulents, forming clumps or mats; internodes not visible. Shoots short, with 1 or 2 leaf-pairs (often longer and with up to 4 leaf-pairs in cultivation). Leaf-pairs all similar or of 2 alternating types: the shorter much-fused pairs each enclosing a developing pair of longer, less-fused leaves. Leaves compact and club-shaped or longer, semi-cylindric or 3-angled, green to greyish or white, usually dotted; keel often developed only towards the tip, sometimes with a few small teeth. Flowers terminal, solitary, usually stalked; bracts present. Sepals 4–5, petals numerous, in several rows; stamens numerous, erect or leaning inwards. Nectary-glands forming a continuous ring. Ovary partly to half-inferior with parietal placentas. Stigmas 8–19, subulate, longer

than depth of ovary-cells, acute. Cells of capsule 8–19; wings bristle-tipped, cell-roofs present; tubercle large. Seeds ovoid, rough or smooth, angled.

A genus previously considered to contain as many as 150 species; recent research has however reduced this to 23. Cuttings are hard to establish and propagation is best effected from seed.

1a. Leaf-pairs of 2 alternating types differing in length and degree of fusion   2
  b. Leaf-pairs all similar   4
2a. Stems not or scarcely branched
                                              2. peculiaris
  b. Stems branched more than 5 times   3
3a. Valve-wings small, flag-shaped; maximum cell number more than 10
                                              1. cigarettifera
  b. Valve-wings bristle-shaped; maximum cell number less than 10   **3. rostrata**
4a. Mat-forming (slightly shrubby in cultivation); leaves fused for less than one-sixteenth; valve-wings broad
                                              8. acuminata
  b. Clump-forming; leaves fused for more than one-quarter; valve-wings bristle-like   5
5a. Leaves hood-like, without pointed tips or toothed keels   **5. pillansii**
  b. Leaves with pointed tips and sometimes toothed keels   6
6a. Leaves without toothed keels   7
  b. Some leaves with toothed keels   8
7a. Keels *c.* half the leaf length; bracts reaching base of capsule   **6. purpurea**
  b. Keels half the leaf length or less; bracts not reaching base of capsule
                                              7. robusta
8a. Capsule almost spherical below; some leaves with toothed keels
                                              6. purpurea
  b. Capsule cone-shaped to hemispherical below; most leaves with toothed keels
                                              4. denticulata

1. C. cigarettifera (Berger) N.E. Brown (*C. brevis* Bolus; *C. duplessii* Bolus; *C. framesii* Bolus; *C. longipes* Bolus; *C. luckhoffii* Bolus; *C. marlothii* Bolus; *C. scabra* Bolus). Much-branched. Leaf-pairs 1–3 per shoot, of 2 slightly different types, shorter, *c.* one-third fused pairs alternating with longer, less fused pairs. During the resting period the underdeveloped longer pairs are concealed within the fused part of the shorter pair immediately below them. Leaves to *c.* 7.5 cm, rounded or indistinctly 3-angled at base, 3-angled and keeled towards the tip, glaucous, dotted; tip pointed; upper surface flat, lower surface of longer leaves (and

sometimes shorter) drawn forward into a chin; keel minutely toothed near the top. Flowers to *c.* 6 cm across, yellow. Capsule cup-shaped below, flat to conical above; stalk long; bracts attached in lower quarter of stalk. Valve-wings small, flag-shaped. Cells usually *c.* 10. *South Africa (Cape Province)*. G1. Winter.

2. C. peculiaris N.E. Brown. Illustration: Jacobsen, Handbook of succulent plants, 1028 (1960).
With a single shoot, rarely branched. Leaf-pairs 1–2 per shoot, of 2 very different types (see No. 1); shorter pairs totally fused; longer pairs shortly fused, prostrate, 4–5 cm long and wide, abruptly pointed, smooth with darker dots; upper surface flat to convex, lower surface similar but keeled towards tip. Flowers *c.* 3.5 cm across, yellow. Bracts usually reaching capsule base; capsule cup-shaped below, flat above. Cells usually fewer than 10. *South Africa (Cape Province)*. G1.

3. C. rostrata (Linnaeus) N.E. Brown (*C. rostratoides* (Haworth) N.E. Brown; *C. tuberculata* Miller; *C. velutina* Bolus). Illustration: Jacobsen, Handbook of succulent plants, 1032 (1960); Herre, The genera of the Mesembryanthemaceae, 115 (1971).
Similar to *C. cigarettifera*, with leaf-types more differing. Longer leaves to *c.* 12 × 1.8 cm; shorter leaves fused to *c.* 3.5 cm; free parts spreading, often curved; edges and keel rough to slightly toothed; lower surfaces rarely chin-like; sheaths often flushed with violet. Flowers to *c.* 6.5 cm across, yellow. Valve-wings reduced to small bristles, sometimes with broader bases. Cells usually 8. *South Africa (Cape Province)*. G1. Autumn.

4. C. denticulata (Haworth) N.E. Brown (*C. candidissima* (Haworth) N.E. Brown; *C. inconspicua* N.E. Brown). Clump-forming, with short stems. Leaves to *c.* 6.5 × 1 cm, fused for up to 1 cm, 3-angled towards tip, grey to bluish green with papillae of 2 sizes, dotted; tip pointed, keel less than one-quarter of the leaf length; at least some leaves on each plant with toothed keels. Flowers to 8 cm, pale yellow. Bracts attached in lower one-third of flower-stalk. Capsule shortly funnel-shaped below, flat, rarely cone-shaped above. Cells more than 14 (usually 17–19). *South Africa (Cape Province)*. G1.

5. C. pillansii Bolus (*C. crassa* Bolus; *C. gibbosa* Schick & Tischer; *C. pillansii* var. *crassa* (Bolus) Rowley). Illustration:

Jacobsen, Handbook of succulent plants, 1028 (1960).

Very short-stemmed, forming dense clumps; shoots with 1 or 2 leaf-pairs. Leaves very compact, to 4.5 × 4.5 cm and to 2.5 cm thick, to two-thirds fused, little separated above, bluntly keeled at tip, pale grey to bluish green with slender, curved papillae, upper surface flat, lower surface drawn forward and hood-like. Flowers *c.* 6 cm across, pale to bright yellow; bracts attached to middle or upper part of flower-stalk, usually longer than capsule. Capsule nearly spherical below, almost flat above. Valve-wings reduced to small bristles with narrow bases. *South Africa (Cape Province).* G1.

6. C. purpurea Bolus (*C. splendens* Bolus; *C. 'purpurata'* in error). Illustration: Jacobsen, Handbook of succulent plants, 1030 (1960); Jacobsen, Lexicon of succulent plants, pl. 151 (1974).

Much-branched. Leaf-pairs 1 or 2 per shoot, leaves to 3.5 × 1 cm, pointed at tip, fused for *c.* 1 cm, bluish grey to pinkish, upper surface flat; lower surface strongly keeled, rounded near base, sometimes chin-like near tip; at least some leaves on each plant with toothed keels. Flowers to *c.* 6 cm across, purplish red; bracts attached to upper half of stalk, reaching or enclosing the capsule. Capsule almost spherical below, flat above. Cells usually *c.* 10. *South Africa (Cape Province).* G1.

7. C. robusta (Haworth) N.E. Brown (*C. aurea* Bolus; *C. brevis* Schwantes). Leaves 5–8 cm, pointed at tip; keel less than one-quarter of the leaf length. Flowers to *c.* 6 cm across, whitish to yellow, often reddish-flushed. Bracts attached to lower one-third of flower-stalk. Capsule funnel-shaped to hemispherical below, flat with a central depression above. Valve-wings bristle-shaped with slightly broader bases. Cells 10–13. *South Africa (Cape Province).* G1.

8. C. acuminata Bolus (*C. carinata* Bolus). Mat-forming (slightly shrubby in cultivation); leaves keeled for less than one-sixteenth, 4.5–8.5 × 1.2–1.8 cm. Flowers to *c.* 6 cm across, pale yellow. Bracts overtopping the capsule. Valve-wings broad. *South Africa (Cape Province).* G1.

17. VANHEERDEA Bolus
J.C.M. Alexander

Dwarf, clump-forming, perennial succulents. Leaf-bodies crowded, rarely solitary, spherical to ovoid, composed of 2 leaf-pairs. Leaves hemispherical to 3-angled, joined below for half to two-thirds of their length, pale or greyish green, undotted, sometimes with an apical window; edges often finely toothed. Flowers 1–3, stalked. Sepals 5–9, free, rounded, unequal, broad at base. Petals in *c.* 3 rows, linear, rounded or notched, narrowed at base. Stamens erect or leaning inwards; staminodes absent. Ovary inferior with parietal (apparently basal) placentas. Stigmas 7–15, erect, thread-like, papillose. Cells of capsule 7–15, shallow, flat above; valves spreading or reflexed; wings more or less diverging, fringed; cell-roofs flexible, membranous, sometimes narrow; tubercle absent. Seeds many, very small, almost spherical, pointed, deep brown.

A genus of 4 lime-loving species from South Africa, flowering in early spring and requiring a rest in summer and autumn.

1a. Leaf-bodies *c.* 3.5 cm, pale violet; tips with windows **2. primosii**
 b. Leaf-bodies more than 4 cm, reddish or greyish green 2
2a. Leaves flattened above; keel and angles with several fine teeth **3. divergens**
 b. Leaves not flattened above; keel and angles with 1–3 teeth **1. angusta**

1. V. angusta (Bolus) Bolus.
Leaf-bodies narrow, to 7 cm, 1.5–2 cm wide at junction, 5–10 mm across at base. Free part of leaves 8–14 mm, obtuse to acute, truncate or rounded in profile, reddish or brownish green; keel and edges with 1–3 small teeth. Flowers *c.* 3.5 cm across, yellow, 3 together; common stalk *c.* 4.7 cm. *South Africa (Cape Province).* G1–2. Spring.

2. V. primosii Bolus.
Leaf-bodies *c.* 3.5 × 2.3 cm, pale violet. Free part of leaf *c.* 3 cm; tip rounded to truncate, with pale green window. Flowers *c.* 2.5 cm yellow. *South Africa (Cape Province).* G1–2. Spring.

3. V. divergens (Bolus) Bolus. Illustration: Barkhuizen, Succulents of southern Africa, f. 152a (1978).
Similar to *V. angusta* but leaf-bodies shorter and broader, 4–6 × 2.5–3 cm; free parts compressed at apex; keel and edges finely toothed. Flowers 2–4 cm across, yellow. *South Africa (Cape Province).* G1–2. Spring.

18. CYLINDROPHYLLUM Schwantes
J.C.M. Alexander

Dwarf tufted perennial succulents with short branchlets. Leaf-pairs 1 per shoot; leaves thick, indistinctly 3-angled to cylindric; acute, obtuse or truncate at apex, greyish green with fine dots; fused at base. Flowers solitary, terminal, shortly stalked. Sepals 5, unequal, 3 larger and keeled. Petals in several rows. Stamens and staminodes many, inward-leaning, orange. Ovary half-inferior with parietal placentas; nectary-glands forming a continuous ring. Stigmas 5–8, ovate, pointed, shorter than depth of ovary. Cells of capsule 5–8, wings narrow, bristle-tipped. Seeds pear-shaped, warted, dark brown.

A genus of 5 species from South Africa.

1a. Leaf-clusters forming cushions to *c.* 13 cm high **1. comptonii**
 b. Leaf-clusters solitary or in mats to *c.* 5 cm high 2
2a. Shoots with 3 or 4 leaf-pairs **2. calamiforme**
 b. Shoots with 1 leaf-pair **3. tugwelliae**

1. C. comptonii Bolus.
Forming dense cushions to 13 cm high and 25 cm across. Leaves erect to spreading, *c.* 9 × 1 cm, cylindric to slightly flattened, indistinctly keeled, acute to acuminate. Flowers *c.* 7.5 cm across, silvery white; stalks *c.* 1.5 cm. *South Africa (Cape Province).* G2. Summer.

2. C. calamiforme (Linnaeus) Schwantes. Illustration: Jacobsen, Handbook of succulent plants, 1091 (1960); Succulenta 60: 280 (1981); Riha & Subik, The illustrated encyclopaedia of cacti and other succulents, pl. 389 (1981).
Stemless or very short-stemmed, to 5 cm, eventually forming small mats; shoots with 3 or 4 leaf-pairs. Leaves clustered, ascending to arched, 5–7 cm × *c.* 8 mm, almost cylindric, obtuse, narrow and curved near the tip, greyish green with fine dots, fused at base. Flowers 5–7 cm across, very pale pink with a pale yellow centre; stalks *c.* 2 cm. *South Africa (Cape Province).* G2. Summer.

3. C. tugwelliae Bolus. Illustration: Herre, The genera of the Mesembryanthemaceae, 123 (1971).
Forming dense mats; shoots with 1 leaf-pair. Leaves erect, *c.* 8 × 1 cm, almost cylindric, tapered, bluish green. Flowers *c.* 5 cm across, pale pink to yellow, stalk 1–2 cm. *South Africa (Cape Province).* G2. Summer.

19. VANZIJLIA Bolus
J.C.M. Alexander & F. McIntosh

Dwarf, hairless, clump-forming, perennial succulents; branches 5–10, ascending, rarely prostrate and rooting, with visible internodes. Leaf-pairs of 2 different types: shorter, much fused pairs, to *c*. 1.4 cm, alternating with longer, less than half-fused pairs to *c*. 4 cm. All leaves 3-angled, acuminate, mucronate at first, pale green, dotted. Flowers terminal, solitary, stalked or stalkless, to 6 cm across. Sepals 5 with thick, membranous margins. Petals in 2 rows, white to pale pinkish purple, narrowly linear. Stamens leaning inwards, hiding stigmas; staminodes many in 2–3 rows. Ovary flat above, conical below, inferior, with parietal placentas. Stigmas 9–10, thread-like, acute, longer than ovary-depth. Cells of capsule 9–10. Cell-roofs present, with tubercles. Seeds smooth, pointed.

A genus now considered to contain only one species from South Africa.
Literature: Hartmann, H., Monographie der Gattung Vanzijlia, *Botanische Jahrbücher* **103**: 499–538 (1983).

1. V. annulata (Berger) Bolus (*V. angustipetala* (Bolus) N.E. Brown).
Illustration: Jacobsen, Handbook of succulent plants, 1437 (1960); Herre, The genera of the Mesembryanthemaceae, 307 (1971); Jacobsen, Lexicon of succulent plants, pl. 200 (1974).
South Africa (W Cape Province). G1.

20. ODONTOPHORUS N.E. Brown
J.C.M. Alexander
Dwarf, clump-forming shrubs with prostrate or ascending branches; shoots with 1–2 leaf-pairs. Leaves very thick and soft, 3-angled, obtuse or acuminate, with large teeth on margin, greyish green with warts and fine dots, softly hairy. Flowers solitary, stalked. Sepals 5, the inner with membranous margins. Petals in 3–6 rows, narrow. Stamens erect, in 6 rows, papillose; staminodes absent. Ovary inferior with parietal placentas; nectary-glands toothed or scolloped. Stigmas 8–11, narrowly ovate. Cells of capsule 8–11; cell-roofs stiff; tubercle large, almost closing the opening. Seeds disc-shaped, ovate, pale brown.

A genus now considered to contain only 3 species from South Africa.
Literature: Hartmann, H., Monographie der Gattung Odontophorus, *Botanische Jahrbücher* **97**: 161–225 (1976).

1a. Leaves with short hairs; leaf-tip narrowly triangular (viewed from above) with 6 or 7 bristle-tipped teeth on each side   **1. marlothii**
  b. Leaves with long hairs; leaf-tip rounded (viewed from above), with 4–5 teeth on each side   **2. nanus**

1. O. marlothii N.E. Brown. Illustration: Jacobsen, Handbook of succulent plants, 1328 (1960); Succulenta **50**: 66 (1971); Court, Succulent flora of South Africa, 91 (1981).
Leaf-pairs 2 or 3 per short shoot, bearing longer shoots in their axils. Leaves 2.5–3.5 × 7–8 mm, swollen and fused at base, dark greyish green, rough with white-hairy warts, hairs short; upper surface slightly convex, broadly triangular at tip, lower surface rounded, keeled and flattened towards the top, slightly chin-like; edges with 6 or 7 bristle-tipped teeth. Flowers *c*. 3 cm across, yellow; stalks *c*. 1.2 cm. *South Africa (W Cape Province)*. G2. Autumn–winter.

2. O. nanus Bolus (*O. primulinus* Bolus). Illustration: Jacobsen, Handbook of succulent plants, 1328 (1960).
Leaves to 4 × 1.8 cm, appearing white from dense, hairy warts, hairs long; upper surface rounded at tip; lower surface almost semi-cylindric; edges with 4–5 broad teeth. Flowers *c*. 4.3 cm across, pale yellowish brown; stalks *c*. 3 cm. *South Africa (W Cape Province)*. G2.

21. LAMPRANTHUS N.E. Brown
J.C.M. Alexander
Low shrubs with erect, spreading or prostrate branches, with visible internodes. Leaves many, cylindric, semi-cylindric or 3-angled, blunt or tapered at tip, usually curved, rarely with sparse dots, shortly fused at base. Flowers terminal or axillary, solitary or in cymes, usually stalked. Sepals 5. Petals in 2 or 3 rows. Stamens erect, to inward-leaning; staminodes usually absent. Stigmas 4–7, ovate, small. Ovary flat, convex or slightly concave, inferior, with parietal placentas. Nectary-glands forming a continuous ring. Cells of capsule 4–7. Cell-roofs present, tubercle lacking. Seeds pear-shaped, pointed, rough, very dark.

A genus of about 180 species from South Africa; 1 in Australia. They perform well out-of-doors in the summer and in milder years can remain out as they will withstand a certain amount of winter moisture. Most of the plants available are of uncertain hybrid origin.

1a. Flowers yellow or orange   2
  b. Flowers red, pink or purple   5
2a. Flowers pale yellow throughout   **2. glaucus**
  b. Flowers orange, sometimes yellow outside   3

3a. Flowers *c*. 2 cm across, orange and yellow turning pale red   **1. brownii**
  b. Flowers 4 cm across or more, orange throughout   4
4a. Stems rust-coloured; leaves 2–3 cm   **3. aurantiacus**
  b. Stems dark brown; leaves 5 cm or more   **4. aureus**
5a. Stems prostrate   6
  b. Stems erect, arching or tangled   7
6a. Leaves 5–8 cm × 6 mm; flowers 5–7 cm across   **5. spectabilis**
  b. Leaves 1–2.5 cm × 3 mm; flowers 3–3.5 cm across   **6. sociorum**
7a. Leaves 1.8 cm or less   8
  b. Leaves 2.5 cm or more   12
8a. Stems thread-like, tangled   9
  b. Stems thicker, erect, arching or spreading   10
9a. Leaves 4–6 mm; flowers 1.2–1.6 cm across   **7. falcatus**
  b. Leaves 1–1.5 cm; flowers *c*. 4 cm across   **8. falciformis**
10a. Flowers reddish orange inside, yellow outside, becoming pale red throughout   **1. brownii**
  b. Flowers pinkish purple to violet   11
11a. Leaves green to reddish green, not fused at base   **9. glomeratus**
  b. Leaves greyish green, shortly fused at base   **10. emarginatus**
12a. Stems to 2 cm thick; leaves 6–7 cm   **11. conspicuus**
  b. Stems thinner; leaves 4 cm or less   13
13a. Leaves without transparent dots; flowers *c*. 7 cm across; flower-stalks *c*. 4 cm   **12. haworthii**
  b. Leaves with transparent dots; flowers 6 cm across or less; flower-stalks 5 cm or more   14
14a. Stems arched; leaves almost cylindric, bright green   **13. zeyheri**
  b. Stems erect or ascending; leaves distinctly 3-angled   15
15a. Flowers *c*. 6 cm across, pinkish red   **14. blandus**
  b. Flowers *c*. 4 cm across, pale pink   **15. roseus**

1. L. brownii (J.D. Hooker) N.E. Brown.
Erect to *c*. 30 cm, much-branched; branches slender, brown, angled, with many short shoots. Leaves 8–10 mm, semi-cylindric, broadest towards the tip, compressed, greyish green with prominent transparent dots and a red point, shortly fused at base. Flowers usually 3 together, terminal, *c*. 2 cm across, at first reddish orange inside, yellow outside, later pale red throughout; stalks 2–4 cm. *South Africa (Cape Province)*. H5–G1. Summer.

2. L. glaucus (Linnaeus) N.E. Brown.
Illustration: Rice & Compton, Wild flowers of the Cape of Good Hope, pl. 64 (1950); Eliovson, South African wild flowers for the garden, 291 (1965).
Erect to *c*. 30 cm, branches stiff and grey. Leaves 1.5–3 cm, 3-angled, compressed, minutely pointed, dark green with a greyish bloom, rough with prominent, transparent dots, especially on the keel; shortly fused at base. Flowers solitary, 4–5 cm across, pale yellow, stalk 1.5–3 cm. *South Africa (Cape Province).* H5–G1. Summer.

3. L. aurantiacus (de Candolle) Schwantes.
Illustration: Jacobsen, Handbook of succulent plants, 1192 (1960).
To 45 cm, sparsely branched, branches ascending to erect, rust-coloured. Leaves 2–3 cm × 4 mm, bluntly 3-angled, broadest in upper half, slightly tapered and finely pointed at tip, green with a greyish bloom, rough with transparent dots, shortly fused at base. Flowers solitary, 4–5 cm across, orange; stalks 4–5 cm. *South Africa (Cape Province).* H5–G1. Summer.

4. L. aureus (Linnaeus) N.E. Brown.
Illustration: Jacobsen, Handbook of succulent plants, 1193 (1960); Mason, Western Cape sandveld flowers, pl. 42 (1972); Barkhuizen, Succulents of southern Africa, f. 124 (1978); Il Giardino Fiorito **13**: 24 (1987).
To 40 cm, branches erect, dark brown. Leaves to 5 cm or more, with rounded sides, gradually tapered, finely pointed at tip, bright green with a greyish bloom, smooth, with fine transparent dots, shortly fused at base. Flowers *c*. 6 cm across, bright orange; stalks *c*. 6 cm. *South Africa (Cape Province).* H5–G1. Summer.

5. L. spectabilis (Haworth) N.E. Brown.
Illustration: Batten & Bokelmann, Wild flowers of the eastern Cape Province, pl. 53 (1966); Martin & Chapman, Succulents and their cultivation, pl. 16 (1977); Noailles & Lancaster, Mediterranean plants and gardens, 100 (1977).
Stems prostrate with long internodes and clusters of leaves. Leaves ascending, 5–6 cm × 6 mm, 3-angled, curved, keeled, flattened sideways, green with a pale green bloom; tip with a short red spine. Flowers 5–7 cm across, purple; stalks 8–15 cm. *South Africa (Cape Province).* H5–G1. Summer.

6. L. sociorum (Bolus) N.E. Brown.
Similar to *L. spectabilis.* Leaves 1–2.5 cm × 3 mm, 3-angled to semi-cylindric, bluish green. Flowers 3–3.5 cm across, purple; stalks 6–20 mm. *South Africa (SW Cape Province).* H5–G1. Summer.

7. L. falcatus (Linnaeus) N.E. Brown.
Small, much-branched, tangled shrub with brown thread-like branches. Leaves 4–6 mm, 3-angled, curved, slightly compressed, tapered and with a short point at the tip, greyish green with transparent dots. Flowers 3 together, 1.2–1.6 cm across, pink, scented; stalks 4–5 cm. *South Africa (Cape Province).* H5–G1. Summer.

8. L. falciformis (Haworth) N.E. Brown.
Similar to *L. falcatus* but generally larger. Leaves 1–1.5 cm × 3–5 mm, with acute edges. Flowers *c*. 4 cm across, pale pink, numerous. *South Africa (Cape Province).* H5–G1. Summer.

9. L. glomeratus (Linnaeus) N.E. Brown.
To 30 cm, branches slender, brown turning grey, with numerous short shoots. Leaves 1.2–1.8 cm, 3-angled, flattened, broader towards tip, green or reddish green, with prominent transparent dots, not fused at base. Flowers very numerous, *c*. 2.5 cm across, pinkish purple to violet; stalks *c*. 2.5 cm. *South Africa (Cape Province).* H5–G1. Summer.

10. L. emarginatus (Linnaeus) N.E. Brown.
Illustration: Jackson, Wild flowers of the fairest Cape, 50 (1980).
To 40 cm, branches erect, brown, with many short shoots. Leaves 1.2–1.6 cm × 1–2 mm, semi-cylindric, curved and flattened, broader towards the tip, obtuse, with a short point, greyish green, rough with numerous transparent dots. Flowers many in groups of 1–3, 3 cm across, pinkish purple. *South Africa (Cape Province).* H5–G1. Summer.

11. L. conspicuus (Haworth) N.E. Brown.
Illustration: Jacobsen, Handbook of succulent plants, 1195 (1960).
To 45 cm, stems to 2 cm thick, branches with conspicuous leaf-scars, curved. Leaves clustered at tips of branches, 6–7 cm × 4–5 mm, semi-cylindric, indistinctly keeled, tapered and with a red point at the tip, green, smooth, often dotted, shortly fused at base. Flowers 1–3 together, *c*. 5 cm across, purplish red. *South Africa (Cape Province).* H5–G1. Summer.

12. L. haworthii (G. Don) N.E. Brown.
To 60 cm, freely branched; stems almost erect, brown. Leaves spreading, 2.5–4 cm × 4–6 mm, semi-cylindric, curved, tapered to base and tip, smooth, pale green with dense grey bloom, shortly fused at base. Flowers to *c*. 7 cm across, pale purple, stalks *c*. 4 cm. *South Africa (Cape Province).* H5–G1. Summer.

13. L. zeyheri (Salm-Dyck) N.E. Brown.
Illustration: Eliovson, South African wild flowers for the garden, 291 (1965); Eliovson, Wild flowers of South Africa, 291 (1980).
Branches arched, with many short shoots. Leaves arched and spreading, 4 cm × 3–4 mm, cylindric, bluntly tapered, soft, smooth, shiny green, with transparent dots. Flowers solitary, 5–6 cm across, purple; stalks 6–8 cm. *South Africa (Cape Province).* H5–G1. Summer.

14. L. blandus (Haworth) Schwantes.
Illustration: Eliovson, South African wild flowers for the garden, 291 (1965); Noailles & Lancaster, Mediterranean plants and gardens, 100 (1977); Barkhuizen, Succulents of southern Africa, f. 125 (1978).
To 50 cm, branches ascending to erect, deep red. Leaves 3–4 cm × 3 mm, 3-angled, shortly tapered, pale greyish green with fine transparent dots. Flowers 3 together, *c*. 3 cm across, pale pinkish red; stalks 6–7 cm. *South Africa (Cape Province).* H5–G1. Summer.

15. L. roseus (Willdenow) Schwantes.
Illustration: Polunin & Huxley, Flowers of the Mediterranean, pl. 14 (1965); Bichard & McClintock, Wild flowers of the Channel Islands, pl. 69 (1975); Rowley, Illustrated encyclopaedia of succulents, f. 11.2 (1978).
Erect or slightly spreading, to 60 cm. Leaves 2.5–3 cm × 4 mm, 3-angled, flattened, tapered and shortly pointed at the tip, with prominent transparent dots. Flowers *c*. 4 cm across, pale pink; stalks *c*. 5 cm. *South Africa (Cape Province).* H5–G1. Summer.

22. OSCULARIA Schwantes
J.C.M. Alexander & F. McIntosh
Small, shrubby succulents; branches erect to spreading, with visible internodes. Leaves more or less equal, 3-angled, compact, obovoid, shortly pointed, greyish or bluish green, waxy, rarely dotted, shortly fused at base; margins toothed or not; keel toothed or entire. Flowers 1–3 together, usually stalked, remaining open. Sepals 5. Petals short, very narrowly elliptic. Stamens erect or inward-leaning, hiding stigmas; staminodes sometimes present. Ovary inferior, with parietal placentas; nectary-glands 5, separate, dark green. Stigmas usually 5, subulate. Cells of

capsule usually 5; cell-roofs present; tubercle absent. Seeds pear-shaped, pointed, rough, very dark.

A genus of 3 species from South Africa. They do well out-of-doors in summer.

1a. Branches erect; leaves 6–10 × 5–8 mm, broadest near tip, always toothed   **1. deltoides**
  b. Branches spreading; leaves *c.* 2 × 1 cm, broadest at base, sometimes toothed   **2. caulescens**

1. O. deltoides (Linnaeus) Schwantes. Illustration: Jacobsen, Handbook of succulent plants, 1338 (1960); Herre, The genera of the Mesembryanthemaceae, 245 (1971); Kakteen und andere Sukkulenten **36**: 39 (1985).
Small, erect, much-branched shrub; branches reddish. Leaves incurved, 6–10 × 5–8 mm, narrow and shortly fused at base, with 2–4 reddish, acute teeth on edges and keel, greyish blue with a bloom. Flowers 3 together, *c.* 1.2 cm across, rose-pink; stalks 2.5–4 cm. *South Africa (Cape Province).* H5–G1. Spring–summer.

2. O. caulescens (Miller) Schwantes. Illustration: Rice & Compton, Wild flowers of the Cape of Good Hope, pl. 65 (1950); Jacobsen, Handbook of succulent plants, 1337 (1960); Eliovson, South African flowers for the garden, 178 (1965).
Branches spreading. Leaves *c.* 2 × 1 cm, slightly tapered, toothless or with 2 or 3 reddish teeth on edges and keel, pale green with a grey bloom. Flowers similar to those of *O. deltoides*. *South Africa (Cape Province).* H5–G1. Spring–summer.

23. EBRACTEOLA Dinter & Schwantes
J.C.M. Alexander & F. McIntosh
Dwarf, tufted or mat-forming, perennial succulents; branchlets short and thick, densely leafy, internodes hidden. Leaves 3-angled to cylindric, fused at base, pale bluish green, usually with dots or warts. Flowers terminal, solitary, stalkless or shortly stalked. Sepals 5, free. Petals linear, rounded. Filaments broad and curved. Ovary inferior, with parietal placentas. Nectary-glands forming a continuous ring. Stigmas 5, subulate, curved at tip. Cells of capsule 5; cell-roofs stiff, tubercle small, often scolloped. Seeds almost spherical, smooth, pale.

A genus of 2 species from Southwest Africa.

1. E. montis-moltkei (Dinter) Dinter & Schwantes. Illustration: Jacobsen, Handbook of succulent plants, 1340

(1960); Herre, The genera of the Mesembryanthemaceae, 147 (1971); Succulenta **61**: 175 (1981).
Branches 1–2 cm. Leaves curved, 2–3 cm, *c.* half-fused, broadest towards tip, 3-angled, greyish green, finely dotted; free part 3–5 mm thick at base, 6–7 mm thick towards tip. Flowers *c.* 1.5 cm across, pale pinkish purple, almost stalkless. *Southwest Africa.* G1. Summer.

24. BRAUNSIA Schwantes
J.C.M. Alexander
Small prostrate to erect woody succulents with visible internodes; nodes often rooting. Leaves 3-angled, curved, up to half-fused, velvety or hairless, often dotted; margins white and cartilaginous. Flowers solitary or in cymes, terminal on short shoots, usually stalked. Sepals 5, unequal, outer with wide membranous margins. Petals in 4–5 rows, narrowly obovate, rounded. Staminodes curved outwards away from stamens; stamens erect, papillose below. Nectary-glands small, finely scalloped. Ovary inferior, with parietal placentas. Stigmas 5, subulate, acute. Cells of capsule 5; cell-roofs sometimes vestigial; tubercles absent. Seeds almost spherical, with long spine-like appendages.

A genus of 5 species from South Africa, originally illegitimately named *Echinus* Bolus.

1a. Leaves *c.* one-quarter fused; keel hairy; flowers pink, *c.* 2 cm across   **1. apiculata**
  b. Leaves *c.* half-fused; keel smooth; flowers white, *c.* 4 cm across   **2. geminata**

1. B. apiculata (Kensit) Bolus. Illustration: Herre, The genera of the Mesembryanthemaceae, 197 (1971); Barkhuizen, Succulents of southern Africa, 113 (1978); Court, Succulent flora of southern Africa, 93 (1981).
To 20 cm, green parts velvety; stems covered in withered leaves; branches short, with 1 or 2 leaf-pairs. Leaves crowded, ascending to spreading, 1.5–3 cm × 6–9 mm, *c.* one-quarter fused, with a sharp or blunt brown point; keel acute, with small brown hairs; upper surface flat to convex, sides slightly convex. Flowers *c.* 2 cm across, pink; stalks *c.* 5 mm. *South Africa (Cape Province).* G1.

2. B. geminata (Haworth) Bolus. Illustration: Jacobsen, Lexicon of succulent plants, pl. 147 (1974).
Branches ascending to 15 cm, regularly

forked. Leaves erect, more than half-fused, scarcely diverging, *c.* 2.5 cm, *c.* 1.5 cm wide in fused part, 6–7 mm wide above, smooth; keel and margins white. Flowers *c.* 4 cm across, white. *South Africa (Cape Province).* G1.

25. CEROCHLAMYS N.E. Brown
J.C.M. Alexander
Dwarf perennial succulents, stemless at first, later tufted and developing very short branched stems; internodes not visible; shoots with 1–3 leaf-pairs. Leaves very firm, 5–6 cm, 8–10 mm wide at base, to 1.6 cm wide above, club-shaped, bluntly and very asymmetrically 3-angled, greenish brown with a waxy bloom, lacking dots or warts; upper surface flat to convex; 1 side broad and convex, the other narrow and flat. Flowers 1–3, terminal, *c.* 3 cm across, reddish purple, stalked. Sepals 5, almost equal, some with narrow membranous margins. Petals numerous in 2 or 3 rows, linear. Stamens numerous, leaning inwards, staminodes thread-like. Nectary-glands 5, separate, dark green. Ovary inferior, with parietal placentas. Stigmas 5, narrowly triangular, curved and toothed, shorter than depth of ovary. Cells of capsule 5, with raised seams; wings with free, curved, bristle-like tips; cell-roofs with stiff wings; tubercle absent or indistinct. Seed ovoid, smooth.

A genus of a single species from South Africa.

1. C. pachyphylla (Bolus) Bolus. Illustration: Jacobsen, Handbook of succulent plants, 1017 (1960); Herre, The genera of the Mesembryanthemaceae, 111 (1971); Riha & Subik, The illustrated encyclopaedia of cacti and other succulents, pl. 391 (1981); Everett, The New York Botanical Gardens illustrated encyclopedia of horticulture, 696 (1981). *South Africa (Cape Province).* G1.

26. JACOBSENIA Bolus & Schwantes
J.C.M. Alexander
Papillose perennial, non-flowering shoots short, clustered; flowering shoots erect, taller, developing long internodes; green parts hairless or slightly velvety. Leaf-pairs similar on flowering and non-flowering shoots. Leaves cylindric. Flowers solitary, terminal, stalked. Sepals 5, 3 with membranous margins. Petals in 3 or 4 rows, narrowly linear, acute. Stamens erect; staminodes absent. Ovary inferior with parietal placentas; nectary-glands forming a continuous ring; stigmas 5,

triangular to ovate, curved at tip. Cells of capsule 5; wings broad, curved, diverging, toothed, bristle-tipped; cell-roofs present; tubercles absent. Seeds brown, spherical.

A genus of 2 species from South Africa.

1. J. kolbei (Bolus) Bolus & Schwantes (*Drosanthemum kolbei* invalid). Illustration: Jacobsen, Handbook of succulent plants, 1184 (1960); Herre, The genera of the Mesembryanthemaceae, 181 (1971).
To 30 cm; internodes to *c.* 5 cm. Leaves ascending to erect, 2–3.5 cm × 8–9 mm, fused for *c.* 5 mm, cylindric, slightly flattened on inner (upper) surface, obtuse to acute at tip, green, with many fine papillae. Flowers to 1.5 cm across, white; stalks 2.5–5 cm. *South Africa (Cape Province).* G1.

27. DELOSPERMA N.E. Brown
J. Cullen

Low shrubs or perennial herbs with fine papillae, some with tuberous, swollen roots. Stem internodes usually visible. Leaves opposite, stalkless, slightly joined at their bases, flat or almost cylindric. Flowers solitary or in few- or many-flowered cymes. Sepals 4–6 (in cultivated species), often unequal. Petals in few series. Stamens usually erect. Ovary inferior, 5-celled; stigmas 5, subulate, shorter than depth of ovary-cells; placentation parietal. Capsule 5-celled. Cell-roofs absent or poorly developed.

A genus in which over 140 species have been named from southern and eastern Africa to Arabia. Its classification is disorganised and the identification of individual plants is very difficult. Only about 11 species are in general cultivation, but these are all poorly known in the wild, and poorly described in the literature. The account given here must be used with caution.

1a. Leaves thickest just below the apex, tapering from there (when seen from the side) towards the base; stem very short, internodes not visible, the plant appearing as a tuft of erect or erect-spreading leaves **1. pergamentaceum**
  b. Leaves and plant not as above 2
2a. Plant herbaceous, stems dying off annually 3
  b. Plant a low shrub, stems persistent through the winter 5
3a. Leaves clearly 3-dimensional, thicker than wide, 3-sided, the upper angles white-margined **4. litorale**
  b. Leaves more or less flat, wider than thick, without white-margined angles 4

4a. Flowers 2–2.5 cm in diameter; leaves linear, to 5 mm wide **3. lydenburgense**
  b. Flowers 5–7 cm in diameter; leaves oblanceolate, to 2 cm wide **2. sutherlandii**
5a. Flowers white to pale yellow 6
  b. Flowers pink, magenta or purplish 8
6a. Flowers *c.* 4 cm in diameter; sepals 6; leaves with compressed, 2-leaved short shoots in their axils **5. lehmannii**
  b. Flowers to 2 cm in diameter; sepals 5; leaves without compressed short shoots in their axils 7
7a. Leaves lanceolate, tapering to a short point (both from above and from the side), not hairy when mature **6. ecklonis**
  b. Leaves not 3-angled, ovoid-hemispherical, blunt, hairy **7. pruinosum**
8a. Flowers to 2.5 cm in diameter 9
  b. Flowers more than 3 cm in diameter **10. cooperi**
9a. Flowers purplish; leaves papillose **8. obtusum**
  b. Flowers pale pink; leaves not obviously papillose **9. subpetiolatum**

1. D. pergamentaceum Bolus. Illustration: Flowering plants of South Africa **7**: t. 278 (1927).
Plant hairless, to 30 cm, stems very short, the plant consisting of mainly crowded branches and erect or erect-spreading leaves. Leaves 3-dimensional, thickest just below the apex, tapering (when viewed from the side) to the base, glaucous, suffused with purple towards the base. Flowers solitary, to 4.4 cm in diameter, borne on very short stalks. Petals white or very pale magenta. *South Africa (Cape Province).* G1. Flowering most of the year.

2. D. sutherlandii (J.D. Hooker) N.E. Brown. Illustration: Botanical Magazine, 6299 (1877).
Perennial herb, branches dying back annually. Leaves green, thick but flat, oblanceolate, to 2 cm wide, papillose and hairy. Flowers solitary, borne on obvious stalks, 5–7 cm in diameter. Petals violet or magenta. *South Africa (Natal, Transvaal).* G1.

3. D. lydenburgense Bolus.
Hairless perennial herb, branches dying back annually. Leaves thick but flat, linear, to 5 mm wide, papillose. Flowers in 2–3-flowered cymes, shortly stalked, 2–2.5 cm in diameter. Petals purplish. *South Africa (Transvaal).* G1.

4. D. litorale (Kensit) Bolus.
Diffuse hairless herb, branches dying back annually, with distinct, visible internodes. Leaves 3-dimensional, 3-angled in section, the angles white-margined, thicker than wide, glaucous. Flowers in 3-flowered cymes, shortly stalked, to 2 cm in diameter. Petals white. *South Africa (Cape Province).* G1.

5. D. lehmannii (Ecklon & Zeyher) Schwantes. Illustration: Jacobsen, Handbook of succulent plants, 1100 (1960).
Hairless, sprawling shrub, stems with visible internodes. Leaves 3-angled, to 4 cm long × 7 mm wide × 7 mm thick, tapering evenly towards the acute apex from above, more abruptly so from the side, each leaf with a developed, 2-leaved, short shoot in its axil, glaucous, not papillose. Flowers *c.* 4 cm in diameter in a stalked, several-flowered cyme. Sepals 6. Petals yellow. *South Africa (Cape Province).* G1.

6. D. ecklonis (Salm-Dyck) Schwantes. Sprawling shrub, finely hairy at least when young; internodes visible. Leaves 3-angled, to 1.8 cm long × 3 mm wide and thick, tapering evenly to the acute apex, green, hairless when mature, not papillose. Flowers to 1.6 cm in diameter, in a cyme. Petals white. *South Africa (Cape Province).* G1.

7. D. pruinosum (Thunberg) Ingram (*D. echinatum* (Aiton) Schwantes). Illustration: Jacobsen, Handbook of succulent plants, 1097 (1960).
Bushy shrub; stems short, internodes not very distinct. Leaves thick, ovoid-hemispherical, blunt, bright green, densely papillose, many papillae prolonged into long bristles. Flowers solitary, shortly stalked, 1.2–1.5 cm in diameter. Petals white or yellowish. *South Africa (Cape Province).* G1. Flowering throughout the year or intermittently.

8. D. obtusum Bolus.
Sprawling shrub; branches with visible internodes. Leaves 3-sided, narrowly lanceolate, to 1.7 cm long × 3 mm wide and thick, tapering to a downwardly-curved point, glaucous, papillose but not hairy. Flowers in few- to many-flowered cymes, shortly stalked, 1.5–1.8 cm in diameter. Petals purple. *South Africa (Orange Free State).* G1.

9. D. subpetiolatum Bolus.
Hairless, compact shrub, branches with visible internodes. Leaves more or less

terete, dull glaucous tinged with red, not papillose, 1.5–2 cm long × 5–6 mm wide and thick, blunt or shortly pointed. Flowers solitary, stalked, to 1.8 cm in diameter. Petals pink. *South Africa (Natal)*. G1.

10. D. cooperi (J.D. Hooker) Bolus. Illustration: Botanical Magazine, 6312 (1877); Letty, Wild flowers of the Transvaal, pl. 65 (1962); Fabian & Germishuizen, Transvaal wild flowers, pl. 48 (1982).
Hairless, branched low shrub; internodes visible but usually shorter than the leaves. Leaves more or less cylindric, somewhat tapering, glaucous, papillose, with short shoots in their axils, to 3.5 cm long × 6 mm wide and thick. Flowers in a few-flowered cyme, 4.5–5 cm in diameter. Petals purple. *South Africa (Orange Free State, Transvaal)*. G1.

D. ashtonii Bolus. Illustration: Flowering plants of South Africa 26: t. 1023 (1947). Similar but plant often hairy, roots tuberous, internodes longer than the green leaves. *South Africa (Orange Free State, Transvaal), Botswana*. G1.

28. DROSANTHEMUM Schwantes
J. Cullen
Erect or prostrate shrubs, with conspicuously glistening papillae. Stems with distinct internodes. Leaves opposite, 3-angled or cylindric in section, covered with glistening papillae. Flowers solitary or up to 3 together, in cymes. Sepals 5. Petals numerous, spreading. Staminodes sometimes black. Stamens numerous. Ovary 5-celled, with parietal placentas. Stigmas broadly triangular, as long as depth of ovary-cells; cell-roofs present.

A genus of almost 100 species from South Africa and Southwest Africa. Only a few are widely grown. They are easily cultivated and, in suitable areas, may be moved out-of-doors in summer.
Literature: Bolus, L., Notes on Mesembryanthemum and allied genera, *Journal of South African Botany* **30**: 36–38 (1964).

1a. Papillae on at least the stems drawn out into fine, acute hairs   2
b. Papillae not drawn out as hairs   4
2a. Leaves cylindric in section; flower-stalks 1–5 cm   3
b. Leaves semi-cylindric in section, flattened above; flower-stalks to 8 mm   **1. archeri**
3a. Petals pale pink; flowers numerous, each terminating a short shoot   **2. floribundum**

b. Petals purple; flowers 1–3, terminating the main shoots   **3. hispidum**
4a. Petals purple; staminodes not black   **4. ambiguum**
b. Petals yellow, red, orange or pink; staminodes black   5
5a. Petals yellow, sometimes with red tips   **5. hallii**
b. Petals yellow pink, red or orange   6
6a. Petals pale yellow-pink   **7. bellum**
b. Petals red or orange, sometimes suffused with purplish brown   **6. micans**

1. D. archeri Bolus.
Branches prostrate, papillae drawn out into long hairs. Leaves semi-cylindric, broadening towards the blunt apex, green, shiny papillose, to 1.5 cm × 2–3.5 mm. Flower-stalks to 8 mm. Flowers 2–4 cm in diameter, petals purple. *South Africa (Cape Province)*. G1. Summer.

2. D. floribundum Schwantes.
Cushion-forming shrub. Branches prostrate, green when young, turning brownish, covered with papillae which are drawn out into long hairs. Leaves cylindric, broadening towards the blunt apex, light green, 1.2–1.4 cm × 2–5 mm. Flower-stalks to 3 mm. Flowers very numerous, terminating short, lateral shoots, to 1.8 cm in diameter, petals pale pink. *South Africa (Cape Province)*. G1. Summer.

3. D. hispidum (Linnaeus) Schwantes. Illustration: Eliovson, Namaqualand in flower, pl. 48 (1972); Le Roux & Schelpe, Namaqualand and Clanwilliam, 81 (1981). Like *D. floribundum* but leaves 1.5–2.5 cm × 3–4 mm, flowers to 3 cm in diameter, terminating the main branches. Petals purple, white at the base. *South Africa (Cape Province), Southwest Africa*. G1. Summer.

4. D. ambiguum Bolus.
Branches prostrate, papillose. Leaves semi-cylindric in section, narrowing to a blunt tip, shining papillose, 1.5–2 mm thick. Flower-stalks 3–7 mm. Flowers to 2.4 cm in diameter. Petals purple. *South Africa (Cape Province)*. G1. Summer.

5. D. hallii Bolus. Illustration: Court, Succulent flora of South Africa, 48 (1981). Branches more or less erect. Leaves semi-cylindric in section, narrowed to the apex, green, covered with papillae which are scarcely shining, 1–2.5 cm × 1.5–2.5 mm. Flowers 3.5–4 cm in diameter. Petals yellow, sometimes with red tips.

Staminodes black. *South Africa (Cape Province)*. G1. Summer.

6. D. micans (Linnaeus) Schwantes. Illustration: Botanical Magazine, 448 (1799).
Branches erect, shining papillose when young, later yellow-brown with white crystalline papillae. Leaves more or less cylindric, green, covered in crystalline papillae, 1.2–2.5 cm × 2–4 mm. Flowers terminating the branches, to 4.5 cm in diameter. Petals orange to red, sometimes suffused with brown-purple. Staminodes black. *South Africa (Cape Province)*. G1. Summer.

D. splendens Bolus. Illustration: Botanical Magazine, n.s., 44 (1948). Leaves yellowish green, 1.7–3 cm × 4–5 mm, flowers 3–5 cm in diameter, golden yellow. *South Africa (Cape Province)*. G1. Summer.

D. speciosum (Haworth) Schwantes. Illustration: Court, Succulent flora of South Africa, 91 (1981); Herre, The genera of the Mesembryanthemaceae, 143 (1971). Branches grey at maturity, leaves 1.2–1.6 cm × 4–6 mm, petals orange-red at the base. *South Africa (Cape Province)*. G1. Summer.

7. D. bellum Bolus. Illustration: Flowering Plants of Africa 28: pl. 1106 (1950–51). Like *D. micans* but petals pale yellowish pink. *South Africa (Cape Province)*. G1. Summer

29. TRICHODIADEMA Schwantes
J. Cullen
Shrubs, tufted or with long, slender branches. Leaves scarcely joined at the base, cylindric or half-cylindric in section, papillose, each with a cluster of spreading bristles at the tip. Flowers solitary, shortly stalked, red or white. Stigmas 5–8. Ovary with 5–8 cells, and parietal placentas. Capsules with or without cell roofs.

A genus of about 30 species from South Africa, Southwest Africa and Ethiopia.

1a. Leaf-bristles 12 or more; plant tufted   2
b. Leaf-bristles up to 11; plants not tufted   3
2a. Bristles 20–25; papillae acute, hair-like   **5. densum**
b. Bristles 12–15; papillae rounded   **6. stellatum**
3a. Leaf-bristles white   **1. bulbosum**
b. Leaf-bristles pale to dark brown or black   4
4a. Flowers completely stalkless   **2. peersii**
b. Flowers shortly stalked   5

5a. Flowers to 3 cm in diameter; petals
    intense red   **3. barbatum**
 b. Flowers to 2 cm in diameter; petals
    magenta, pink, white or yellow
                             4. intonsum

1. T. bulbosum (Linnaeus) Schwantes.
Plant to *c.* 20 cm. Leaves inclined, half-
cylindric, grey-green due to the presence of
many papillae, 5–8 × 2.5–3 mm, each with
8–11 white bristles at the tip. Flower to
2 cm in diameter, petals deep red. *South
Africa (Cape Province).* G1. Spring–autumn.

2. T. peersii Bolus.
Plant more or less erect, to 10 cm,
branches spreading to ascending. Leaves
erect, almost spherical to half-cylindric,
covered with long papillae, 5–8 × 4 mm,
each with 8–9 pale brown bristles at the
tip. Flowers completely stalkless, to 3.8 cm
in diameter, petals white. *South Africa (Cape
Province).* G1. Spring–summer.

3. T. barbatum (Linnaeus) Schwantes.
Illustration: Botanical Magazine, 70
(1788).
Plant to 20 cm, branches prostrate. Leaves
distant, half-cylindric in section, stiffly
inclined and slightly recurved, grey-green
due to the presence of many papillae,
8–12 × 3–4 mm, each with 8–10 black
bristles at the top. Flowers shortly stalked,
to 3 cm in diameter, petals intense red.
South Africa (Cape Province). G1.
Spring–autumn.

4. T. intonsum (Haworth) Schwantes.
Illustration: Botanical Magazine, 6057
(1873).
Plant erect. Leaves distant, slightly
recurved, half-cylindric, narrowed above,
covered with acute papillae which render
the margins almost ciliate, 1.2–1.3 cm ×
4 mm, each with 8–10 dark brown bristles
at the tip. Flowers with long stalks, to 2 cm
in diameter, petals magenta, pink, white or
yellow. *South Africa (Cape Province).* G1.
Spring–autumn.

5. T. densum (Haworth) Schwantes.
Illustration: Botanical Magazine, 1220
(1809).
Stems short, tufted. Leaves crowded, green,
covered with acute papillae, 1.5–2 cm ×
4–5 mm, each with 20–25 long, white
bristles at the tip. Flowers shortly stalked,
4–5 cm in diameter, petals deep red. *South
Africa (Cape Province).* G1. Spring–autumn.

6. T. stellatum (Miller) Schwantes.
Illustration: Everard & Morley, Wild flowers
of the world, pl. 73 (1970).
Plant to 10 cm, tufted. Leaves crowded,

inclined, half-cylindric, grey-green due to a
covering of rough papillae, *c.* 1 cm ×
3–4 mm, each with 12–15 white bristles
which are 3–4 mm long at the tip. Flowers
shortly stalked, to 3 cm in diameter, petals
pale violet-red. *South Africa (Cape Province).*
G1. Winter–early spring.

30. SEMNANTHE N.E. Brown
F. McIntosh
Shrubs, 60–80 cm with spreading,
compressed and angled branches and
visible internodes. Leaves boat-shaped,
3-angled in section, but compressed, the
margins and keel cartilaginous and finely
and irregularly toothed, incurved,
mucronate, light grey-green, with
transparent dots, 3–5 cm × 8–11 mm.
Flowers 1 or 2 together, shortly stalked,
4–7 cm in diameter. Sepals 5–6, unequal,
3-angled in section, with margins like those
of the leaves. Petals numerous, violet or
rose-red, hiding the numerous stamens.
Ovary 10-celled, with parietal placentas,
stigmas 10, much shorter than the ovary
depth. Capsule with stiff cell roofs,
remaining open after wetting. Seeds
numerous.
  A genus of a single species from South
Africa.

1. S. lacera (Haworth) N.E. Brown.
Illustration: Flowering plants of Africa 33:
pl. 1286 (1959); Jacobsen, Handbook of
succulent plants, f. 1590 (1960); Herre,
The genera of the Mesembryanthemaceae,
285 (1971); Japan Succulent Society,
Colour encyclopedia of succulents, pl. 809
(1981).
South Africa (Cape Province). G1.

31. KENSITIA Fedde
F. McIntosh
Erect shrub 30–60 cm, much-branched;
stems reddish, internodes visible. Leaves
glaucous, laterally compressed, 3-angled,
incurved and acute with a fine point,
2.8–3.3 cm × 5–7 mm wide, 7–10 mm
thick. Flowers solitary, terminal, to 5 cm in
diameter, on a stalk 8–10 mm. Sepals
5. Petals numerous, the outer spreading in
different planes, to 1.5 cm, clearly divided
into blade and claw, the claw whitish, the
blade purplish pink; inner petals shorter
and united at the base with the stamens
and bent inwards. Stigmas 8–10. Ovary
with 8–10 cells, placentation parietal. Cells
roofed with flexible, membranous wings.
  A genus of a single species from South
Africa.

1. K. pillansii (Kensit) N.E. Brown.
Illustration: Flowering Plants of South
Africa 3: pl. 110 (1923); Jacobsen, Das
Sukkulentenlexicon, t. 171 (1970); Lamb &
Lamb, The illustrated reference on cacti
and other succulents, pl. 517 (1973); Japan
Succulent Society, Colour encyclopedia of
succulents, pl. 139 (1981).
South Africa (Cape Province). G1.

32. ARGYRODERMA N.E. Brown
J. Cullen
Stemless, very tufted, very succulent
perennials. Leaves 2 per flowering stem,
united at the base, short and thick or long,
the upper surface flat or slightly convex,
the lower convex and often forming a
somewhat hooded apex. Flower solitary,
terminal, stalked or stalkless. Ovary with
10–24 cells. Capsule with 10–24 cells;
placentation parietal. Stigmas forming a
stalkless disc on top of the ovary. Cell-roofs
well-developed, each cell with the opening
almost completely closed by a well-
developed tubercle.
  A genus of 48 species from a small area
of the Cape Province (South Africa).
Literature: Hartmann, H., Monographie der
Gattung Argyroderma, *Mitteilungen der
Institut fur Allgemeine Botanik Hamburg* **15**:
121–235 (1977).

1a. Leaves finger-like, rounded at the tip,
    about 2–4 times longer than broad,
    green, spreading   **1. fissum**
 b. Leaves ovoid or almost spherical, tip
    hood-like, silver-white or dark green,
    erect   2
2a. Older leaves yellow   3
 b. Older leaves rust-brown   4
3a. Older leaves dying off, passing from
    yellow to grey and crumbling away
                               2. delaetii
 b. Older leaves persisting, remaining
    yellow for some time, later becoming
    black   **3. testiculare**
4a. Leaves hood-like, with a distinct
    division between those of a pair   5
 b. Leaves ovoid or spherical, their upper
    surfaces pressed together, without a
    division between the pair   6
5a. Leaf-pair 2–4 × 1.5–2 cm, division
    between the leaves a little wider than
    the leaf-breadth   **4. patens**
 b. Leaf-pair 1–2 cm × 5–12 mm,
    division between the leaves narrower
    than the leaf-breadth   **5. framesii**
6a. Leaf-pair 2–4 × 1.5–3 cm; plant
    usually unbranched   **6. pearsonii**
 b. Leaf-pair 1–2 × *c.* 1 cm; plant usually
    highly branched   **7. subalbum**

1. A. fissum (Haworth) Bolus (*A. brevipes* (Schlechter) Bolus; *A. braunsii* (Schwantes) Schwantes; *A. orientale* Bolus). Illustration: Kakteen und andere Sukkulenten **32**: 13 (1981); Flowering plants of southern Africa **2**: t. 78 (1922).
Plant forming clumps, sometimes with a short, prostrate stem when old. Leaves 1.5–10 cm × 8–12 mm, half-cylindric (sometimes cylindric at the base), upper surface flat or convex, tip rounded and often hooded, bluish green to whitish grey, surface smooth. Flower-stalk to 4 cm. Flowers *c*. 4 cm in diameter, petals pink to magenta-red. *South Africa (Cape Province)*. G1. Summer.

2. A delaetii Maass (*A. aureum* Bolus; *A. roseum* Schwantes; *A. schildtii* Schwantes). Illustration: Kakteen und andere Sukkulenten **32**: 13, 300 (1981).
Plant variably branched. Leaves variable, 2–5 × 1.3–3 cm, very variable in shape, hemispherical, outer surface rounded or keeled, 1.5–3.5 mm apart at the apex, silver-white, older leaves becoming yellow, then grey and crumbling. Flower-stalks to 8 mm. Flowers 3–4 cm in diameter, petals white, yellow, pink or violet. *South Africa (Cape Province)*. G1. Summer.

3. A. testiculare (Aiton) N.E. Brown. Illustration: Gardeners' Chronicle **159**: 103 (1966).
Plant little-branched. Leaves 2.5–3.5 × 1.5–2.5 cm, half-ovoid, distinctly keeled, separated by 4–10 mm at the apex, dark green, the old leaves turning yellowish and then eventually black, persisting. Flowers stalkless, 2.5–3.5 cm in diameter, petals usually violet (occasionally yellow). Fruit long-stalked. *South Africa (Cape Province)*. G1. Summer.

4. A. patens Bolus. Illustration: Kakteen und andere Sukkulenten **33**: 174 (1982).
Plant branched, forming mats. Leaves 2–4 × 1.5–2 cm, half-ovoid, compressed and keeled, greyish blue, older leaves rusty brown, 1.5–4.5 cm apart at the apex. Flowers stalkless, *c*. 3 cm in diameter, petals yellow or violet. *South Africa (Cape Province)*. G1. Summer.

5. A. framesii Bolus. Illustration: Kakteen und andere Sukkulenten **32**: 13 (1981), **33**: 172–3 (1982).
Plant tufted, very branched. Leaves 1–2 cm × 6–12 mm, half-ovoid, hooded, usually keeled, 4.5–10 mm apart at the apex, older leaves rusty-brown. Flower-stalks 3–4 mm. Flowers 2–2.5 cm in diameter, petals usually violet, rarely yellow. *South Africa (Cape Province)*. G1. Summer.

Two subspecies are cultivated: subsp. **framesii** has leaves less than 1 × 1 cm, the gap between them up to 5 mm, and subsp. **hallii** (Bolus) Hartmann has larger leaves with the gap between them more than 5 mm.

6. A. pearsonii (N. E. Brown) Schwantes (*A. luckhoffii* Bolus). Illustration: Botanical Magazine, 8463 (1912); Kakteen und andere Sukkulenten **32**: 13 (1981).
Plant unbranched. Leaves 1.5–4 × 1.5–3 cm, half-ovoid, obscurely keeled to rounded on the back, pressed together before flowering, 2–8 mm apart at the apex, older leaves rusty brown. Flower stalkless, 2–3.5 cm in diameter, petals yellow flushed with violet or orange, sometimes pure violet or violet-white. *South Africa (Cape Province)*. G1. Summer.

7. A. subalbum (N.E. Brown) N.E. Brown. Plant clump-forming, branched. Leaves 1.6–1.8 cm × 5–12 mm, hemispherical or elongate-ovoid, pressed together before flowering, smooth and whitish, older leaves rusty-brown. Petals violet or white. *South Africa (Cape Province)*. G1. Summer.

33. ALOINOPSIS Schwantes
J. Cullen
Small, tufted perennials; internodes not visible. Leaves 4–6 to a flowering branch. Leaves and stems hairless, when variously tuberculate, or downy or velvety. Leaves fleshy, spathulate, the lower surface flat, convex or slightly keeled. Flowers opening in the afternoon, closing at night, solitary or rarely 2–4 together, shortly stalked or stalkless. Sepals 5–6, all similar. Petals yellow, orange, pale brown or pink. Ovary with 6–14 cells; placentation parietal. Cells completely roofed.

A genus of about 15 species from the Cape Province of South Africa. They are easily grown, though require deep pots.

1a. Leaves downy   **1. peersii**
  b. Leaves hairless   2
2a. Leaves with tubercles of 2 sizes, the larger white or pink; sepals 6   3
  b. Leaves with tubercles all the same size; sepals 5   6
3a. Leaves flat on both surfaces, fan-shaped; petals pale brown to pink   **2. malherbei**
  b. Leaves keeled or convex on the lower surface   4
4a. Some of the tubercles long and bristle-like   **3. setifera**
  b. No tubercles long and bristle-like   5

5a. Margin of upper part of leaf with regular, pink tubercles   **4. luckhoffii**
  b. Margin of upper part of leaf entire   **5. villetii**
6a. Leaves flattened; petals rose-purple   **6. spathulata**
  b. Leaves not flattened; petals yellow or brownish red   7
7a. Leaves conspicuously broadened above the base; petals yellow   **7. jamesii**
  b. Leaves tapering from the base; petals yellow with brownish tips   **8. rubrolineata**

1. A. peersii (Bolus) Bolus. Illustration: Jacobsen, Handbook of succulent plants, f. 1197 (1960).
Leaves 2–4 to each flowering shoot, at first ascending, later recurved, 2–2.2 cm × *c*. 8 mm at the base, widening to 1.5 cm at the centre, to 5 mm thick, tapering regularly to a blunt apex, upper surface flat, lower surface rounded towards the base, bluntly keeled towards the apex, bluish grey-green, downy, with distinct tubercles. Flowers solitary, to 2.5 cm in diameter. Petals yellow. *South Africa (Cape Province)*. G1.

2. A. malherbei (Bolus) Bolus (*Nananthus malherbei* Bolus). Illustration: Flowering plants of southern Africa **26**: pl. 1035 (1947); Jacobsen, Handbook of succulent plants, f. 1196 (1960). Lamb & Lamb, Illustrated reference on cacti and other succulents, pl. 522 (1966); Riha & Subik, The illustrated encyclopaedia on cacti and other succulents, pl. 411 (1981).
Leaves 1.8–2.5 cm × *c*. 4 mm at the base, widening to 1.2–2.2 cm at the apex, erect, fan-shaped, glaucous, the apex more or less truncate, both surfaces with a slight covering of tubercles, the margins and the apex with thick, white, distant tubercles. Flower-stalks 2–2.5 cm. Flowers to 2.5 cm in diameter. Petals pale brown to pink. *South Africa (Cape Province)*. G1.

3. A. setifera (Bolus) Bolus.
Plant clump-forming, with 2–3 pairs of leaves to a rosette. Leaves to 2 cm × 5–6 mm, 4 mm thick, upper surface flat, lower surface convex, roundly keeled towards the apex, all glaucous to dark reddish, uniformly rough with tubercles some of which are drawn out as bristles. Flowers to 2.5 cm in diameter. Petals golden yellow to orange. *South Africa (Cape Province)*. G1.

4. A. luckhofii (Bolus) Bolus. Illustration: Jacobsen, Handbook of succulent plants, f. 1196 (1960).

Plant rosette-forming. Leaves to 1.8 cm × 4–5 mm at the base, widening to 1.2 cm towards the triangular, recurved apex, the lower surface rounded towards the base, keeled towards the apex, the keel drawn forward to a blunt point, the whole bluish or grass green, both surfaces covered with coarse grey-green tubercles, the margins with 5–6 large, pink, regularly spaced tubercles. Flowers to 2.5 cm in diameter. Petals pale yellow. *South Africa (Cape Province)*. G1.

5. A. villetii (Bolus) Bolus. Illustration: Jacobsen, Handbook of succulent plants, f. 1202 (1960).
Leaves erect, to 2.2 × 1.6 cm, 7 mm thick, the blade more or less circular, all glaucous and densely covered with large, white tubercles. Flower-stalk to 8 mm. Flowers to 2 cm in diameter. Petals pale yellow with brownish tips. *South Africa (Cape Province)*. G1.

6. A. spathulata (Thunberg) Bolus. Illustration: Jacobsen, Handbook of succulent plants, f. 1201 (1960); Herre, The genera of the Mesembryanthemaceae, 67 (1971).
Plant with several short branches. Leaves crowded, erect, narrowed at the base, rounded at the apex, the area near the apex and the margins covered in minute tubercles, the whole grey-green with reddened margins. Flowers solitary, stalkless, 2.8–3 cm in diameter. Petals deep pink, paler on the underside. *South Africa (Cape Province)*. G1.

7. A. jamesii Bolus. Flowering branches with 4–6 leaves together. Leaves 1.5–1.8 cm × 5 mm at the base, widening to 1 cm at the middle, 3 mm thick, tapering to the apex; upper surface concave, lower surface rounded towards the base, keeled towards the tip, the whole grey-green, covered with rough tubercles. Flowers to 2.3 cm in diameter. Petals golden yellow, each with a red midrib. *South Africa (Cape Province)*. G1. Winter.

8. A. rubrolineatus (N.E. Brown) Schwantes. Illustration: Jacobsen, Handbook of succulent plants, f. 1199 (1960); Ashingtonia 1(1): 10 (1973).
Plant clump-forming, each branch with 4–6 leaves. Leaves to 2.5 cm, 1 cm wide at the base, widening to 2 cm in the middle, to 5 mm thick, grey-green, slightly recurved towards the apex, upper surface flat, lower surface rounded towards the base, bluntly keeled towards the apex, smooth at the base, rough above with

whitish, flat tubercles. Petals yellow, each with a red midrib. *South Africa (Cape Province)*. G1.

34. TITANOPSIS Schwantes
J. Cullen & F. McIntosh
Short-stemmed perennials forming dense mats or tufts; internodes not visible. Stems with 6–8 or more opposite, densely crowded leaves. Leaves narrowly or broadly spathulate, covered with raised, whitish or coloured tubercles. Flowers solitary, stalked or stalkless, *c.* 2 cm in diameter. Sepals 5 or 6, with tubercles like those on the leaves, some with membranous margins. Petals numerous, in 1 or 2 whorls, reflexed, yellow or orange. Stigmas 5 or 6. Ovary with 5 or 6 cells, placentation parietal; cells with membranous roofs. Seeds many.

A genus of 6 species found in the southern part of Southwest Africa and in South Africa.

1a. Leaves whitish to greenish blue with reddish to greyish white tubercles **1. calcarea**
  b. Leaves reddish, grey or bluish green, tubercles greyish brown or yellow **2**
2a. Tubercles greyish brown on a purple ground colour **2. fulleri**
  b. Tubercles yellowish, or yellowish or greyish brown **3. schwantesii**

1. T. calcarea (Marloth) Schwantes. Illustration: Herre, The genera of the Mesembryanthemaceae, 301 (1971); Japan Succulent Society, Colour encyclopedia of succulents, pl. 160 (1981).
Rosettes of leaves 6–8 cm wide. Leaves spreading, to 2.5 cm × 8 mm, broad towards the base, to 1.2 cm broad above, whitish to greenish blue with reddish to greyish white tubercles. Flowers to 2 cm in diameter, petals golden yellow to orange. *South Africa (Cape Province)*. G1.

2. T. fulleri Tischer.
Rosettes of leaves 4–5 cm wide, each containing 5 or 6 pairs of leaves. Leaves 2–2.2 cm × 4 mm broad towards the base, to 1 cm broad above, bluish green to reddish, with faint, dark dots and greyish brown tubercles on a purple ground colour along the margins. Flowers to 1.6 cm in diameter, petals dark yellow. *South Africa (Cape Province)*. G1.

3. T. schwantesii (Dinter) Schwantes (*T. luederitzii* Tischer).
Rosettes of leaves to 7 cm across. Leaves to 3 cm × 3–7 mm broad towards the base, to 1.2 cm broad above, light grey, upper and lower surfaces covered with yellow,

yellowish brown or greyish yellow tubercles. Flowers 1.5–1.8 cm in diameter, petals pale yellow. *Southwest Africa*. G1.

In cultivar 'Primosii' (*Titanopsis primosii* Bolus) the petals are bright yellow with pink tips.

35. PLEIOSPILOS N.E. Brown
J. Cullen
Stemless or somewhat branched succulent perennial herbs, the plant body formed from clumps of up to 4 leaf-pairs. Leaves very thick and fleshy, upper surface usually flat, lower surface very convex, grey-green to brownish green, usually covered with conspicuous dark green dots. Flowers solitary or in cymes of up to 4 together, stalkless or shortly stalked. Sepals 5–6. Petals numerous, yellow to orange or pink. Ovary with 9–15 cells, placentation parietal. Capsule with 9–15 cells, cell-roofs usually present, rigid; tubercles present, large or small.

A genus of 7 species from the Cape Province of South Africa. In their recent revision (reference below), Hartmann & Leide reduce the number of species recognised from about 35 to 7, but distribute them over 2 genera, *Pleiospilos* in the strict sense (*P. bolusii, nelii, compactus*) and *Tanquana* Hartmann & Leide (*P. hilmarii,* and *prismatica* and others). This distinction is not maintained in the present account.

The plants are easily grown in even the poorest soils; as the roots tend to be large, pot-grown specimens should be planted in large pots.
Literature: Hartmann, H. E. & Leide, S., Die Gattung Pleiospilos s. lat. (Mesembryanthemaceae), *Botanische Jahrbücher* **106**: 433–485 (1986).

1a. Leaves finger-like; petals up to 65; stamens papillose at base **2**
  b. Leaves swollen or half-spherical, not finger-like; petals 80 or more; stamens not papillose **3**
2a. Plant unbranched, completely sunk in the ground, only the tops of the leaves visible **4. hilmarii**
  b. Plant branched above ground **5. prismaticus**
3a. Flowers mostly solitary; capsules with very large tubercles **3. compactus**
  b. Flowers usually in 2–4-flowered cymes; capsule with small tubercles **4**
4a. Flowers stalked **2. nelii**
  b. Flowers stalkless **1. bolusii**

1. P. bolusii (J.D. Hooker) N.E. Brown.
Illustration: Botanical Magazine, 6664
(1882).
Plant branched in age. Growths usually
with 1 pair of leaves. Leaves 4–7 cm, half-
cylindric, less than half-fused, broader than
long, upper surface flat, lower surface
drawn up over the upper, often by 2–3 cm
to form a hood, reddish brown-green to
pale grey-green, covered with dark green
dots. Flowers stalkless in 2–3-flowered
cymes, 6–8 cm in diameter, petals 80 or
more, yellow. Stamens not papillose.
Capsules with small tubercles. *South Africa
(Cape Province)*. G1. Summer–autumn.

2. P. nelii Schwantes. Illustration: Lamb &
Lamb, The illustrated reference on cacti
and other succulents **2**: 526 (1973); Court,
Succulent flora of southern Africa, pl. 92
(1981); Flowering plants of southern Africa
47: pl. 1865A (1981).
Growths with up to 3 pairs of leaves.
Leaves widely gaping, 4–7 cm, half-
cylindric, usually broad, lower surface
drawn forward over the upper to form a
hood, smooth, olive green. Flowers in
2–3-flowered cymes, stalks 1.5–2 cm.
Flowers *c.* 7 cm in diameter, petals 80 or
more, orange-pink. Capsules with small
tubercles. *South Africa (Cape Province)*. G1.
Summer–autumn.

3. P. compactus (Aiton) Schwantes.
Growths with 1–3 leaf-pairs. Leaves
swollen or half-spherical, 2.2–8 cm ×
5–30 mm, very variable in shape, usually
greyish green but covered with dark green
dots. Flowers mostly solitary, usually
stalkless though sometimes stalks to 8 mm.
Petals more than 80, yellow, pale yellow or
pink. Stamens not papillose below. Capsules
with large tubercles. *South Africa (Cape
Province)*. G1. Summer–autumn.

A very variable species; Hartmann &
Leide recognise 5 subspecies, of which the
following are cultivated: Subsp. **compactus**
(*P. optatus* (N.E. Brown) Schwantes).
Illustration: Flowering plants of southern
Africa **20**: pl. 769 (1940); Botanische
Jahrbücher **106**: 439 (1986). Ovary-cells
10 in most flowers; capsule shortly stalked.
Subsp. **canus** (Haworth) Hartmann &
Leide (*P. magnipunctatus* (Haworth)
Schwantes; *P. rouxii* Bolus; *P. leipoldtii*
Bolus). Illustration: Jacobsen, Handbook of
succulent plants, f. 1556 (1960). Ovary-cells
12 in most flowers; capsule shortly stalked.
Subsp. **sororius** (N.E. Brown) Hartmann
& Leide (*P. dimidiatus* Bolus). Illustration:
Jacobsen, Handbook of succulent plants,
f. 1554 (1960). Stalk of capsule 1–2 cm.

P. × purpusii Schwantes is known only in
cultivation; it is a hybrid between *P.
compactus* and *P. bolusii* and is intermediate
between its parents.

4. P. hilmarii Bolus (*Tanquana hilmarii*
(Bolus) Hartmann & Leide). Illustration:
Jacobsen, Handbook of succulent plants,
f. 1555 (1960); Lamb & Lamb, The
illustrated reference on cacti and other
succulents **3**: 827 (1971).
Plants borne mostly below the ground, only
the tops of the leaves visible. Growths with
1–2 pairs of leaves. Leaves *c.* 2.5 × 1.6 cm
at the base, narrowing towards the apex,
finger-like, half-cylindric, lower surface
drawn forward over the upper at the apex,
dark green dots coalescing to form a
window at the apex. Flowers stalkless, to
2.4 cm in diameter, petals to 65, yellow.
Stamens papillose at the base. *South Africa
(Cape Province)*. G1. Summer–autumn.

5. P. prismaticus (Marloth) Schwantes
(*Tanquana prismatica* (Schwantes)
Hartmann & Leide). Illustration: Flowering
plants of southern Africa **7**: pl. 269 (1927).
Plant branched above the ground. Growths
with usually 1 pair of leaves. Leaves
3.5–4 × 3 cm at the base, widening slightly
towards the middle, thickening in profile
towards the apex, finger-like, green with
dark green dots. Flowers stalkless, to 4 cm
in diameter. Petals to 65, yellow. Stamens
papillose at the base. *South Africa (Cape
Province)*. G1. Summer–autumn.

36. STOMATIUM Schwantes
J. Cullen
Tufted, branched perennials with very
short stems; internodes not visible. Leaves
4–6 or more to a stem, those of each pair
often unequal, swollen at the base,
3-angled in section, keeled, the keel and
margins toothed, the surface often rough.
Flowers stalkless or shortly stalked. Petals
usually yellow. Ovary with 5–6 cells;
placentation parietal. Stigmas 5 or 6,
columnar and blunt. Capsule with or
without cell roofs.

A genus of about 40 species from the
Cape Province of South Africa.

1a. Flowers to 2 cm in diameter   2
  b. Flowers more than 2 cm in diameter
                                           3

2a. Leaf-margins with 4–6 teeth which
     are tipped with white or red bristles;
     petals yellow with reddish tips
                               1. geoffreyi
  b. Teeth on leaf-margins not tipped with
     bristles; petals white   **2. niveum**

3a. Leaf-margins and keel with distinct
     teeth; petals yellow   **3. trifarium**
  b. Leaf-margins with indistinct teeth,
     keel without teeth; petals white, pink
     on the lower surface   **4. meyeri**

1. S. geoffreyi Bolus.
Leaves to 2 cm × 9–12 mm, 4–5 mm thick,
spreading, keeled, keel with teeth at the
apex, margins with 4–6 small teeth which
are tipped with red or white bristles.
Flowers shortly stalked or stalkless, to
1.5 cm in diameter. Petals pale yellow,
reddish at the tips. *South Africa (Cape
Province)*. G1.

2. S. niveum Bolus.
Leaves to 2 × 1 cm, to 6 mm thick, those of
each pair unequal, expanded towards the
apex, glaucous, covered with large, white
tubercles, and with 3 indistinct marginal
teeth. Flower-stalks to 3 mm. Flowers to
1.8 cm in diameter, scented. Petals white.
South Africa (Cape Province). G1.

3. S. trifarium Bolus. Illustration: Jacobsen,
Handbook of succulent plants, f. 1601
(1960).
Leaves 1.5–2.5 cm × 6–9 mm, to 8 mm
thick, ascending to erect, keeled, keel with
2–4 very small teeth near the apex,
margins with 3–4 teeth; the whole
glaucous, covered with crowded tubercles.
Flowers almost stalkless, to 2.4 cm in
diameter, opening in the evening. Petals
yellow. *South Africa (Cape Province)*. G1.

4. S. meyeri Bolus. Illustration: Jacobsen,
Handbook of succulent plants, f. 1597
(1960).
Leaves 1.8–2 cm × *c.* 6 mm, widening
slightly towards the apex, upper surface
flat, lower surface convex, indistinctly
keeled towards the apex, tip drawn
forwards, the whole grey-green, rough with
coarse tubercles. Flowers to 2.4 cm in
diameter, opening in the evening. Petals
white above, pale pink beneath. *South
Africa (Cape Province)*. G1.

37. CHASMATOPHYLLUM Dinter &
Schwantes
F. McIntosh
Small shrub, branches erect at first, later
prostrate or creeping, each branch with
6–8 leaves; internodes not visible. Leaves
spathulate, half-circular or keeled in
section, with 1 or 2 blunt teeth on the
margins and lower surface towards the
apex, both surfaces covered with whitish
tubercles. Flower solitary, terminal, shortly
stalked. Sepals 5, unequal. Petals in 2 or 3
whorls. Stigmas 5. Ovary 5-celled, not

centrally domed, placentation parietal. Stigmas linear, longer than depth of ovary-cell. Nectary-glands distinct. Cell-roofs present, but seeds only partly covered.

A genus of 6 species from Southwest Africa and South Africa (Cape Province, Orange Free State).

1. C. musculinum (Haworth) Dinter & Schwantes. Illustration: Herre, The genera of the Mesembryanthemaceae, 113 (1971); Japan Succulent Society, Colour encyclopedia of succulents, pl. 118 (1981). Low shrub with prostrate branches forming dense mats. Leaves 1.5–2 cm × 4–6 mm, upper surface grey-green with rough, transparent dots. Flowers to 1.5 cm in diameter, petals yellow, red at the tips on the lower surface. *Southwest Africa.* G1.

38. RHINEPHYLLUM N.E. Brown
F. McIntosh
Small, short-stemmed shrub, internodes hidden by the leaves, branches with 2–4 pairs of leaves. Leaves thick, thicker towards the apex, upper surface flat, lower surface rounded or keeled, rough due to small, white tubercles. Flowers solitary, terminal. Sepals 5. Petals in 1–5 series. Stigmas 5, linear, longer than depth of ovary-cells. Ovary 5-celled, placentation parietal. Nectary-glands forming a continuous ring. Cell-roofs absent or reduced to a limb. Seeds smooth or with small tubercles along the margins.

A genus of 12 species from South Africa.

1. R. broomii Bolus. Illustration: Jacobsen, Handbook of succulent plants, f. 1573 (1960); Riha & Subik, The illustrated encyclopaedia of cacti and other succulents, pl. 412 (1981).
Plant very compact, to 3 cm high (including flowers). Internodes of the stems hidden by the leaves. Leaves spathulate, flat, the top finely tuberculate, with 4 or 5 marginal teeth, the upper leaves densely tuberculate, indistinctly keeled towards the tip, olive green, to 1 cm × 7 mm. Flower-stalks to 5 mm. Flowers 1–2 cm in diameter, petals yellow. *South Africa (Cape Province).* G1.

39. NEOHENRICIA Bolus
F. McIntosh
Freely branching dwarf perennials, forming mats; internodes not visible. Leaves 4 on each branch, erect, finger-like, 10 mm long, to 2 mm wide at base, to 5 mm wide above, upper surface flat, the lower convex; apex covered with pale warts which may form rows. Flowers solitary, to 1.2 cm in

diameter, on slender stalks to 2 cm long. Sepals 5, more or less equal, obtuse, red-brown. Petals in 2 series, loose, white. Stamens few. Stigmas 5. Ovary 5-celled, placentation parietal. Cell-roofs reduced. Seeds rough.

A genus of a single species from South Africa.

1. N. sibbetii (Bolus) Bolus. Illustration: Herre, The genera of the Mesembryanthemaceae, 230 (1971); Lamb & Lamb, The illustrated reference on cacti and other succulents, pl. 519 (1973).
South Africa (Cape Province). G1.
The flowers open at night.

40. JUTTADINTERIA Schwantes
F. McIntosh
Small, extremely succulent shrubs forming tufts or clumps; internodes not visible. Leaves broadly boat-shaped, crowded, 3-angled in section, sometimes toothed. Flowers solitary, terminal. Sepals 4 or 5, unequal, sometimes toothed. Petals in 1–3 series. Stamens erect, forming a tuft. Stigmas 5–13 (or rarely more), linear, longer than depth of ovary-cells. Ovary with 5–13 cells; placentation parietal. Nectary-glands forming a distinct ring. Cell-roofs reduced to a limb. Seeds covered with warts.

A genus of 12 species, mostly from Southwest Africa, but just extending into adjacent South Africa.

1a. Petals white   **1. albata**
  b. Petals pink   **2. longipetala**

1. J. albata (Bolus) Bolus.
Plant to 30 cm. Leaves 2–7 × 1–2.5 cm broad and thick, upper surface flat or slightly convex, lower surface rounded towards the base, keeled towards the apex, all smooth, whitish grey to bright grey-green, with transparent dots. Flowers 3.5–4 cm in diameter. Petals white. *Southwest Africa.* G1.

2. J. longipetala Bolus (*Namibia ponderosa* (Dinter) Dinter & Schwantes).
Plant cushion-forming, cushions to 20 cm wide and 10 cm high. Leaves boat-shaped, to 3 cm × 1.4–2 cm broad and thick, whitish or greyish green with distinct dark dots, surfaces covered with small, hair-like papillae. Flowers 3–5 cm in diameter. Petals bright pink. *Southwest Africa.* G1.

41. LITHOPS N.E. Brown
J. Cullen
Perennial plants forming a body consisting of 2 tightly adpressed or somewhat gaping

leaves, united for more than half their length, more or less flat to the soil surface or raised above it. Apex of the leaf flat or convex, truncate, often marked with ridges, wrinkles or coloured spots or lines, with or without a more or less translucent window. Flowers usually solitary (rarely 2 or 3 together) arising from the fissure between the leaves. Sepals free, 5–6. Petals many in 1–4 series, white, yellow orange or golden yellow. Stamens many, free, close together and forming a column. Ovary inferior with 4–7 cells and 4–7 linear stigmas; placentation parietal. Cell-roofs absent.

A genus of perhaps 40 rather ill-defined species from South Africa and Southwest Africa, widely grown because of their resemblance to the stones among which they grow. The genus is divided into 2 subgenera, **Lithops**, in which the petals are almost invariably yellow, golden yellow or orange and **Leucolithops** Schwantes in which the petals are white. A seedling character reinforces this division: in the white-flowered species the fissure between the leaves in the seedling extends right across the body from one side to the other, whereas in the yellow-flowered species the fissure is smaller and does not extend from side to side. This character is not used for identification purposes in this account.

The species are mainly distinguished by petal colour and by the size, form and colour of the tops of the leaves. Such characters are often difficult to describe and many species appear to be variable in them, leading to the recognition of numerous varieties; these are indicated in the account below but are not individually distinguished, as their status is uncertain (many varieties of the one species appear to grow together in wild populations).

Seed size may be taxonomically useful in the genus (see the paper by Jump cited below) but it is not used here because seeds are infrequently available when identifications are being made.
Literature: Nel, G.C., *Lithops* (undated, appeared 1946); De Boer, H.W. & Boom, B.K., An analytical key for the genus Lithops, *National Cactus and Succulent Journal* 19: 34–37, 51–55 (1964) – reprinted with amendments in Jacobsen, H., *Lexicon of succulent plants*, 504–511 (1974); Fearn, B., New combinations and an analytical key for the genus Lithops, *Cactus & Succulent Journal of the Cactus and Succulent Society of America* 42: 89–93 (1970) – reprinted in Jacobsen, H., *Lexicon of succulent plants*, 511–513 (1974); Cole, D.T., Lithops: a checklist and index, *Excelsa*

3: 37–71 (1973); Jump, J.A., The seed as a criterion in Lithops classification, *Cactus & Succulent Journal (U.S.)* **53**: 197–200 (1981).

1a. Tops of leaves green, greyish green, whitish green, olive green or purplish green (i.e. basically green) 2
 b. Tops of leaves some other colour 17
2a. Pellucid dots present 3
 b. Pellucid dots absent 6
3a. Tops of leaves strongly convex; leaves not tightly pressed together, the fissure slightly gaping 4
 b. Tops of leaves not strongly convex; leaves tightly pressed together, the fissure not gaping 5
4a. Pellucid dots regularly scattered over the top surface of the leaves **9. localis**
 b. Pellucid dots on the top surface of the leaves aggregated into indistinct lines **14. franciscii**
5a. Red or brown lines present, often joined up to form tree-like markings **12. pseudotruncatella**
 b. Red or brown lines absent **13. lesliei**
6a. Tops of leaves strongly convex 7
 b. Tops of leaves not strongly convex 11
7a. Leaves tightly pressed together, fissure not gaping **26. olivacea**
 b. Leaves not tightly pressed together, fissure gaping 8
8a. Fissure less than 8 mm deep **2. optica**
 b. Fissure more than 8 mm deep 9
9a. Flowers white, parts in 6s **1. marmorata**
 b. Flowers yellow, parts in 5s 10
10a. Plants always greyish green; window sometimes reduced by islands and marginal ingrowths **24. helmutii**
 b. Plants sometimes brownish or purplish green; window with a few small islands **25. comptonii**
11a. Red dots and/or red or brown lines present 12
 b. Red dots and/or red or brown lines absent 14
12a. Tops of leaves with small raised red dots **16. verruculosa**
 b. Not as above 13
13a. Window present; petals white **8. salicola**
 b. Window absent; petals yellow **12. pseudotruncatella**
14a. Petals white; flowers with parts in 6s **7. villetii**
 b. Petals yellow; flowers with parts in 5s 15
15a. Window large and distinct, with a few large islands **23. otzeniana**
 b. Window obscured by many small islands and marginal ingrowths 16

16a. Window not distinct, with vague islands **27. meyeri**
 b. Window with distinct islands **28. herrei**
17a. Tops of leaves white or pinkish white 18
 b. Tops of leaves not white or pinkish white 24
18a. Tops of leaves strongly convex 19
 b. Tops of leaves not strongly convex 20
19a. Window absent; petals yellow **20. ruschiorum**
 b. Window obscured by islands and marginal ingrowths; petals white **1. marmorata**
20a. Pellucid dots present **12. pseudotruncatella**
 b. Pellucid dots absent 21
21a. Tops of leaves wrinkled with very fine chalky ridges **21. vallis-mariae**
 b. Tops of leaves not as above 22
22a. Window present **3. julii**
 b. Window absent 23
23a. Tops of leaves flat **5. karasmontana**
 b. Tops of leaves slightly convex **6. erniana**
24a. Leaf entirely purplish red **2. optica**
 b. Tops of leaves not purplish red 25
25a. Tops of leaves brownish 26
 b. Tops of leaves not brownish 45
26a. Pellucid dots present 27
 b. Pellucid dots absent 38
27a. Window present 28
 b. Window absent 31
28a. Tops of leaves flat 29
 b. Tops of leaves not flat 30
29a. Red or brown lines present on top of leaf **18. bromfieldii**
 b. Red or brown lines absent from top of leaf **13. lesliei**
30a. Pellucid dots numerous, regularly scattered over the top of the leaf; window small **9. localis**
 b. Pellucid dots few; window often large **14. franciscii**
31a. Red or brown lines absent from the top of the leaf 32
 b. Red or brown lines present on top of the leaf 33
32a. Tops of leaves flat **15. aucampiae**
 b. Tops of leaves convex **9. localis**
33a. Lines joined up to form tree-like markings 34
 b. Lines not joined up to form tree-like markings 36
34a. Tops of leaves convex **11. werneri**
 b. Tops of leaves flat 35
35a. Flowering in July **12. pseudotruncatella**
 b. Flowering in September–October **22. schwantesii**

36a. Lines and markings impressed so that the top of the leaf is bullate and grooved 37
 b. Tops of leaves not as above **18. bromfieldii**
37a. Tops of the leaves with numerous large bluish green pellucid dots up to 0.5 mm in diameter **10. fulviceps**
 b. Pellucid dots inconspicuous **19. turbiniformis**
38a. Window present 39
 b. Window absent 43
39a. Red dots and lines absent **7. villetii**
 b. Red dots and lines present 40
40a. Lines and markings impressed so that the top of the leaf is bullate and grooved 41
 b. Lines and markings not impressed, top of leaf smooth **8. salicola**
41a. Tops of leaves with small raised red dots **16. verruculosa**
 b. Not as above 42
42a. Tops of leaves flat **3. julii**
 b. Tops of leaves slightly convex **4. bella**
43a. Petals yellow; flowers with parts in 6s **19. turbiniformis**
 b. Petals white; flowers with parts in 5s 44
44a. Tops of leaves flat **5. karasmontana**
 b. Tops of leaves slightly convex **6. erniana**
45a. Pellucid dots absent; petals white **7. villetii**
 b. Pellucid dots present; petals yellow 46
46a. Red dots absent from leaves **11. werneri**
 b. Red dots present on leaves 47
47a. Window present **17. dinteri**
 b. Window absent **22. schwantesii**

1. L. marmorata (N.E. Brown) N.E. Brown (*L. framesii* Bolus; *L. diutina* Bolus). Illustration: Nel, Lithops, pl. 25 & f. 72 (1946).
Bodies solitary or in clumps, to 3 cm high and broad. Tops of leaves strongly convex, grey-green, with a large window which has irregular margins and islands, and is crossed by fine lines; fissure to 1 cm deep. Flowers usually with parts in 6s, to 3 cm in diameter; petals white. *South Africa (Cape Province).* G1.

 Var. **elisae** (De Boer) Cole (*L. elisae* De Boer) is smaller and has narrower windows.

2. L. optica (Marloth) N.E. Brown. Illustration: Nel, Lithops, pl. 31 & f. 82, 83 (1946); Rauh, Die grossartige Welt der Sukkulenten, t. 92 (1967).

Bodies forming clumps of up to 30, to 3 cm high, the leaves spreading away from the fissure which is less than 8 mm deep. Tops of the leaves convex, greenish white with a large window which may contain some islands (leaves entirely purplish in forma **rubra** Tischer). Flowers to 2 cm in diameter, usually with parts in 6s, petals white or white tinged with red. *South Africa (Cape Province)*. G1. Autumn.

3. L. julii (Dinter & Schwantes) N.E. Brown. Illustration: Nel, Lithops, pl. 19 & f. 56 (1946); Excelsa **3**: 51 (1973); Rowley, The illustrated encyclopedia of succulents, 148 (1978); Riha & Subik, The illustrated encyclopedia of cacti and other succulents, f. 401 (1981).
Bodies forming small groups, each 2–3 cm across. Tops of the leaves flat, conspicuously pale, whitish, pink or reddish with distinct, somewhat impressed furrows which form a network, and a brown line along the lip of the fissure. Flowers usually with parts in 6s, to 3 cm in diameter, petals white. *Southwest Africa*. G1.

A very variable species, in which several varieties have been recognised: var. **julii**, var. **reticulata** De Boer, var. **rouxii** De Boer and var. **littlewoodii** De Boer. These are distinguished by the coloration of the tops of the leaves, and are of uncertain status.

L. **fulleri** N.E. Brown (*L. julii* subsp. *fulleri* (N.E. Brown) Fearn) is sometimes regarded as a separate species: it has large windows with islands and transverse strips (Illustration: Rauh, Die grossartige Welt der Sukkulenten, t. 92, 1967). L. **hallii** De Boer is also sometimes treated as a separate species: it has darker tops to the leaves and never has a brown line along the lip of the fissure (Illustration: Lamb & Lamb, The illustrated reference on cacti and other succulents, 1137, 1966).

4. L. bella N.E. Brown. Illustration: Nel, Lithops, pl. 3, 4 & f. 17A (1946); Lamb & Lamb, The illustrated reference on cacti and other succulents, 1133 (1966); Rauh, Die grossartige Welt der Sukkulenten, t. 92 (1967); Riha & Subik, The illustrated encyclopedia of cacti and other succulents, f. 400 (1981).
Bodies forming clumps, each 2.5–3 cm. Tops of the leaves somewhat convex, pale brown or yellowish (var. **bella**) or grey, bluish grey, red brown or purple (var. **eberlanzii** (Dinter & Schwantes) De Boer & Boom), with a large, irregular, dark green window which is below the level of the edges, and has lobed margins and islands. Flowers white, usually with parts in 6s. *Southwest Africa*. G1.

5. L. karasmontana (Dinter & Schwantes) N. E. Brown (*L. jacobseniana* Schwantes). Illustration: Nel, Lithops, pl. 20 & f. 59 (1946); Excelsa **3**: 51 (1973).
Bodies solitary or in groups, to 4 cm high. Tops of leaves flat, grey to bluish yellow, with branched, yellow-brown pits or wrinkles which form a network, window absent or not obvious. Flowers with parts in 5s, petals white. *Southwest Africa*. G1.

Variable. Several varieties – var. **karasmontana**, var. **mickebergensis** (Dinter) De Boer & Boom, var. **lericheana** (Dinter & Schwantes) Cole, var. **opalina** (Dinter) De Boer & Boom and var. **tischeri** Cole – have been described, based on the colour of the tops of the leaves. These, however, apparently always occur in mixed populations, and their status is uncertain.

6. L. erniana Jacobsen (*L. erniana* Loesch & Tischer, invalid name). Illustration: Nel, Lithops, pl. 10 & f. 36 (1946).
Very similar to *L. karasmontana* but tops of the leaves convex, the brown wrinkles on the surface slightly branched but the major ones not linking up with each other, all with grey edges. *Southwest Africa*. G1.

Like many other species, variable in coloration: several varieties – var. **erniana**, var. **aiaisensis** De Boer and var. **witzputensis** De Boer – have been described on the basis of this variation, but their status is uncertain.

7. L. villetii Bolus. Illustration: Excelsa **2**: 25 (1972).
Bodies to 5 cm high. Tops of leaves variable in colour, mostly greenish yellow to brownish, slightly convex, with a completely clear window, or the window with a few small islands in it. Flowers to 3 cm in diameter, usually with parts in 6s. Petals white. *South Africa (Cape Province)*. G1.

Variable in colour like many other species: the following varieties have been described: var. **deboeri** (Schwantes) Cole (*L. deboeri* Schwantes; var. **kennedyi** (De Boer) Cole (*L. fulleri* var. *kennedyi* De Boer).

8. L. salicola Bolus. Illustration: Nel, Lithops, pl. 35 & f. 94–97 (1946); Rauh, Die grossartige Welt der Sukkulenten, t. 92 (1967).
Bodies solitary or in clumps, to 2.5 cm high. Tops of leaves slightly convex, olive or greyish green or reddish, each with a large window divided into a large number of confluent areas by small, slightly raised, white islands. Flowers to 2.5 cm in diameter, petals white. *South Africa (Orange Free State)*. G1.

9. L. localis (N.E. Brown) Schwantes. Illustration: Rauh, Die grossartige Welt der Sukkulenten, t. 91 (1967); Excelsa **6**: 61 (1976).
Bodies solitary or forming groups, to 3.5 cm high. Tops of leaves green or variously coloured, strongly convex, with pellucid dots scattered regularly and with a small window; leaves spreading, fissure widely gaping. Flowers 2.5–3 cm in diameter, petals yellow. *South Africa (Cape Province)*. G1.

10. L. fulviceps (N.E. Brown) N.E. Brown. Illustration: Botanical Magazine, 8776A (1918); Nel, Lithops, pl. 14, 14A & f. 39, 40 (1946); Excelsa **3**: 51 (1973); Riha & Subik, The illustrated encyclopedia of cacti and other succulents, f. 400 (1981).
Bodies usually solitary, more rarely in groups of up to 4, to 3 cm high. Tops of the leaves pale brown to orange-brown marked with numerous large, bluish green, pellucid dots up to 0.5 mm in diameter, flat or slightly convex. Flowers to 3 cm in diameter, petals yellow, whitish on the underside. *Southwest Africa*. G1.

Var. **lactinea** Cole has the tops of the leaves bluish white.

11. L. werneri Schwantes & Jacobsen. Bodies solitary or in groups, to 1 cm high. Tops of the leaves greyish brown with several dark greenish brown lines and small pellucid dots, convex. Petals yellow. *Southwest Africa*. G1.

12. L. pseudotruncatella (Berger) N.E. Brown. Illustration: Nel, Lithops, pl. 33 & f. 89–91 (1946); Herre, The genera of the Mesembryanthemaceae, 196 (1971); Excelsa **5**: 43 (1975); Court, Succulent flora of southern Africa, 92 (1981).
Bodies solitary or in groups, to 4 cm high. Tops of leaves flat or convex, pressed tightly together, green, brownish grey or pale grey, with red or brown lines forming tree-like patterns, without a distinct window. Flowers to 3.5 cm in diameter, usually with parts in 6s, petals yellow or occasionally white. *Southwest Africa*. G1. Summer (July).

A very variable species in which a number of varieties has been described, on the basis of the coloration of the tops of the leaves. The varieties are: var. **archerae** (De Boer) Cole (*L. archeri* De Boer), var. **dendritica** (Nel) De Boer & Boom (*L. dendritica* Nel, *L. farinosa* Dinter, *L. pseudotruncatella* var. *pulmonuncula* (Jacobsen) Jacobsen), var. **edithae** (N.E. Brown) De Boer, var. **elisabethae** (Dinter) De Boer & Boom and var. **volkii** De Boer & Boom. The status of these is uncertain.

L. gracilidelineata Dinter (*L. pseudotruncatella* var. *gracilidelineata* (Dinter) Fearn) is sometimes recognised as a distinct species: it has the tops of the leaves wrinkled with red-brown lines in the furrows.

13. **L. lesliei** (N.E. Brown) N.E. Brown. Illustration: Nel, Lithops, pl. 23 & f. 66 (1946); Excelsa **5**: 43 (1975); Riha & Subik, The illustrated encyclopedia of cacti and other succulents, f. 406 (1981).
Bodies solitary or in small groups, to 4.5 cm high. Tops of the leaves greenish, flat or slightly convex, pressed tightly together, without obvious red-brown lines but with pellucid dots. Flowers to 3 cm in diameter, petals golden yellow or rarely white. *South Africa (Cape Province), Southwest Africa*. G1. Autumn.

14. **L. franciscii** (Dinter & Schwantes) N.E. Brown. Illustration: Nel, Lithops, pl. 11 & f. 37 (1946).
Bodies usually forming groups, to 3 cm high. Tops of leaves convex, pale green or whitish green, with pellucid dots aggregated into indistinct lines; window present but indistinct. Leaves rather divergent, the fissure between them wide. Flowers 1–1.8 cm in diameter, petals yellow. *Southwest Africa*. G1.

L. gesinae De Boer (*L. franciscii* var. *gesinae* (De Boer) Fearn) has a more distinct window, and the plants are generally brownish in colour.

15. **L. aucampiae** Bolus. Illustration: Nel, Lithops, pl. 1 & f. 14–16 (1946); Rauh, Die grossartige Welt der Sukkulenten, t. 8 (1967); Excelsa **1**: 44 (1971) & **5**: 43 (1975); Riha & Subik, The illustrated encyclopedia of cacti and other succulents, f. 406 (1981).
Bodies usually in groups, to 3.5 cm high. Tops of leaves reddish, reddish brown or pale brown, without reddish or brown lines, window obscure or absent, flat. Flowers to 2.5 cm in diameter, petals yellow. *South Africa (Cape Province, Transvaal), Southwest Africa*. G1.

Variable in leaf coloration. Several varieties – var. **aucampiae**, var. **euniciae** De Boer, var. **fluminalis** Cole and var. **koelemanii** (De Boer) Cole (*L. koelemanii* De Boer) – have been described but their status is doubtful.

16. **L. verruculosa** Nel. Illustration: Nel, Lithops, pl. 42 & f. 114–116 (1946); Lamb & Lamb, The illustrated reference on cacti and other succulents, 1133 (1966); Excelsa **1**: 63 (1971) & **5**: 43 (1975).

Bodies usually in groups, to 3 cm high, bluish. Tops of leaves flat, with bluish margins, the fissure bordered by a line, the centre without an obvious window, marked by small, raised, red dots, wrinkled, the dots mainly in the wrinkles. Flowers orange-yellow. *South Africa (Cape Province)*. G1.

Var. **glabra** De Boer has very few red dots on the tops of the leaves.

17. **L. dinteri** Schwantes. Illustration: Nel, Lithops, pl. 6 & f. 28, 29 (1946); Excelsa **3**: 52 (1973).
Bodies solitary or in small groups, to 3 cm high. Tops of the leaves convex, reddish, greyish yellow or yellow, with a conspicuous window which contains 5–15 red spots and some pellucid dots. Fissure 5–7 mm deep. Petals yellow. *South Africa (Cape Province), Southwest Africa*. G1.

Variable in coloration and the numbers of red dots on the tops of the leaves; varieties have been described but these, again, are of doubtful status: var. **dinteri**, var. **frederici** Cole, var. **brevis** (Bolus) Fearn (*L. brevis* Bolus) and var. **multipunctata** De Boer.

18. **L. bromfieldii** Bolus. Illustration: Nel, Lithops, pl. 5 & f. 19 (1946); Rauh, Die grossartige Welt der Sukkulenten, t. 91 (1967); Rowley, Illustrated encyclopedia of succulents, 148 (1978).
Bodies solitary or in groups, to 1.5 cm high. Tops of the leaves flat, irregularly ridged or wrinkled, window present and irregular in shape or more or less absent, dark greyish with red or brown lines which branch towards the edges. Flowers to 4 cm in diameter, petals yellow. *South Africa (Cape Province)*. G1.

As with many other species, varieties have been described on the basis of variation in the colour of the tops of the leaves; the status of these is uncertain: var. **glaudinae** (De Boer) Cole (*L. glaudinae* De Boer), var. **insularis** (Bolus) Fearn (*L. insularis* Bolus), and var. **mennellii** (Bolus) Fearn (*L. mennellii* Bolus).

19. **L. turbiniformis** (Haworth) N.E. Brown. Illustration: Nel, Lithops, pl. 38, 38A & f. 104 (1946); Excelsa **1**: 63 (1971) & **6**: 61 (1976).
Bodies solitary or in small groups, to 2.5 cm high. Tops of leaves somewhat convex, furrowed and with a network of dark brown lines, the whole reddish brown, window absent. Flowers mostly with parts in 6s (the calyx may have up to 8 sepals), to 4 cm in diameter, petals yellow. *South Africa (Cape Province)*. G1.

Again, a number of varieties has been described on the basis of variation in the colour of the leaves. These are of uncertain status: var. **dabneri** (Bolus) Cole (*L. dabneri* Bolus), var. **elephina** Cole, var. **lutea** De Boer, var. **marginata** (Nel) Cole (*L. marginata* Nel), and var. **subfenestrata** De Boer (var. *brunneoviolacea* De Boer).

20. **L. ruschiorum** (Dinter & Schwantes) N.E. Brown (*L. nelii* Schwantes). Illustration: Nel, Lithops, pl. 34, 35 & f. 93, 93A (1946).
Bodies solitary or in small groups, to 4.5 cm high. Tops of leaves convex, greyish, yellowish or brownish white, without windows, either unmarked or with small furrows which may be lined with brown; fissure to 2 cm deep. Flowers to 2.5 cm in diameter, petals yellow. *Southwest Africa*. G1.

21. **L. vallis-mariae** (Dinter & Schwantes) N.E. Brown (*L. vallis-mariae* var. *margarethae* De Boer). Illustration: Nel, Lithops, pl. 40 & f. 107, 108 (1946).
Bodies usually forming groups, to 4 cm high. Tops of leaves flat, yellowish to bluish white, without a window but with papillae which may be aligned as irregular, chalky ridges, without coloured markings or these very indistinct. Flowers to 3.5 cm in diameter, petals yellow. *Southwest Africa*. G1.

Var. **groendraaiensis** (Jacobsen) De Boer has the tops of the leaves bluish white and the ridges rather indistinct.

22. **L. schwantesii** Dinter. Illustration: Nel, Lithops, pl. 35 A & f. 99 (1946); Lamb & Lamb, The illustrated reference on cacti and other succulents, 1141 (1966).
Bodies forming groups, to 4 cm high. Tops of leaves more or less flat, without windows, grey, yellowish brown or reddish with red dots and lines which are slightly sunk below the surface in depressions. Flowers to 3 cm in diameter, petals yellow. *Southwest Africa*. G1. Autumn.

As with many other species, several varieties have been described based on the variation of the colour of the tops of the leaves. Their status is uncertain: var. **schwantesii** (*L. kiubisensis* Jacobsen, *L. schwantesii* var. *triebneri* (Bolus) De Boer & Boom (*L. triebneri* Bolus), var. **christinae** (De Boer) Fearn (*L. christinae* De Boer; *L. schwantesii* var. *nutupsdriftensis* De Boer), var. **gebseri** De Boer, var. **kunjasensis** (Dinter) De Boer & Boom, var. **marthae** (Loesch & Tischer) Cole, var. **rugosa** (Dinter) De Boer & Boom (*L. rugosa* Dinter), and var. **urikosensis** (Dinter) De Boer & Boom (*L. urikosensis* Dinter).

23. L. otzeniana Nel. Illustration: Nel, Lithops, pl. 32 & f. 84–87 (1946); Lamb & Lamb, The illustrated reference on cacti and other succulents, 1142 (1966); Rauh, Die grossartige Welt der Sukkulenten, t. 8 (1967).
Bodies mostly forming groups, to 3 cm high. Tops of leaves more or less flat, with a large, distinct, greenish or purplish window surrounded by a pale grey, scalloped margin and with a few large islands. Flowers to 2 cm in diameter, usually with parts in 5s, petals golden yellow. *South Africa (Cape Province)*. G1.

24. L. helmutii Bolus. Illustration: Nel, Lithops, pl. 16 & f. 48 (1946).
Bodies solitary or forming groups, to 3 cm high. Tops of leaves strongly convex, greyish green, window conspicuous, pale green, sometimes reduced by islands and marginal indentations. Flowers with parts in 5s, to 3 cm in diameter, petals golden yellow. *South Africa (Cape Province)*. G1.

25. L. comptonii Bolus (*L. viridis* Luckhoff). Illustration: Nel, Lithops, pl. 7 & f. 23–26 (1946).
Very similar to *L. helmutii* but tops of leaves green or brownish or purplish green, the windows with a few small islands. Flowers to 2.5 cm in diameter. *South Africa (Cape Province)*. G1.

26. L. olivacea Bolus. Illustration: Nel, Lithops, pl. 30 & f. 80, 81 (1946); Court, The succulent flora of southern Africa, 92 (1981).
Bodies usually in groups, to 2 cm high. Leaves tightly pressed together, fissure not gaping. Tops of leaves strongly convex with large, olive green windows which contain a few whitish islands. Flowers yellow. *South Africa (Cape Province)*. G1.

27. L. meyeri Bolus. Illustration: Nel, Lithops, pl. 27, 28A & f. 76, 77 (1946); Rauh, Die grossartige Welt der Sukkulenten, t. 91 (1967).
Bodies usually forming groups, to 3 cm high, the leaves widely spreading, the fissure gaping. Tops of the leaves more or less flat, green, with indistinct windows which contain inconspicuous islands. Flowers to 3.5 cm in diameter, petals yellow. *South Africa (Cape Province)*. G1.

28. L. herrei Bolus. Illustration: Nel, Lithops, pl. 17, 17A & f. 49, 50 (1946); Lamb & Lamb, The illustrated reference on cacti and other succulents, 1138 (1966).

Very similar to *L. meyeri* but window more distinct and islands more conspicuous. Flower to 1.8 cm in diameter, petals yellow or rarely whitish. *South Africa (Cape Province)*. G1.

A number of doubtful varieties has been described within this species.

42. SCHWANTESIA Dinter
J. Cullen
Clump- or cushion-forming perennials, internodes not visible. Leaves several, opposite, those of each pair somewhat unequal, boat-shaped, sometimes with teeth at the apex, whitish-dotted, hairy or not, usually greyish. Flowers solitary, with parts in 5s. Petals numerous in a single series. Ovary with 5 cells, placentation parietal. Stigmas 5, linear to narrowly triangular, longer than the depth of ovary-cells. Nectary-glands forming a continuous ring. Cell-roofs very reduced.
A genus of 11 species mostly from Southwest Africa, but also extending into the northern part of Cape Province of South Africa.

1a. Plant velvety (especially when young) due to short hairs which terminate the papillae, leaves without teeth at the apex **1. triebneri**
 b. Plant hairless, leaves with 3–7 teeth at the apex **2. ruedebuschii**

1. S. triebneri Bolus.
Leaves whitish to yellowish glaucous, 4–6 cm × 6–10 mm, to 5 mm thick, half-circular beneath near the base, keeled towards the apex, the margins bordered with red at the tip, velvety, especially when young due to minute hairs which terminate the papillae, without teeth at the apex. Flowers 4–5 cm in diameter, stalked. Petals yellow. *South Africa (Cape province)*. G1.

2. S. ruedebuschii Dinter. Illustration: Jacobsen, Handbook of succulent plants, f. 1588 (1960); Herre, The genera of the Mesembryanthemaceae, 281 (1971).
Leaves glaucous, mottled with white, 3–5 × 1–1.2 cm, 1 cm thick at the base but tapering upwards, half circular in section towards the base, the apex expanded into 3–7 thick, blue teeth which are up to 4 mm long, all hairless. Flowers 3.5–4 cm in diameter, stalked. Petals pale yellow. *Southwest Africa*. G1.

43. LAPIDARIA Schwantes
F. McIntosh
Small, stemless perennial herb; internodes not visible. Leaves 6–8 together, opposite,

shortly united at the base, flat or slightly concave on the upper surface, the lower surface sharply keeled, bluntly 3-angled towards the apex, 1–1.5 × up to 1 cm wide and thick, surfaces smooth, whitish or reddish white, margins reddish. Flowers solitary, stalk concealed in the leaves, 3–5 cm in diameter. Sepals 7, dotted. Petals in 3 whorls, golden yellow above, whitish yellow beneath, reddish when fading. Stamens numerous, the inner shorter and inflexed. Stigmas 6 or 7, free, linear to subulate, acute. Ovary 6–7-celled, borne on an elongate, winged stalk; placentation parietal. Nectary-glands forming a continuous ring. Cell-roofs present. Seeds small, pear-shaped, light brown.
A genus of a single species from Southwest Africa.

1. L. margaretae (Schwantes) Dinter & Schwantes. Illustration: Jacobsen, Handbook of succulent plants, f. 1424, 1425 (1960); Lamb & Lamb, The illustrated reference on cacti and other succulents, **5**: pl. 1131, 1134 (1978); Herre, The genera of the Mesembryanthemaceae, 193 (1971); Japan Succulent Society, Colour encyclopedia of succulents, pl. 139 (1981). *Southwest Africa*. G1.

44. DINTERANTHUS Schwantes
J. Cullen
Stemless perennials, bodies composed of 2 leaves united for more than half their length, solitary or forming groups. Leaves usually greyish, hairless or very finely hairy, tops convex or flat. Flowers solitary between the leaves, the compressed flower-stalk often hidden. Sepals free, 6–15. Petals numerous in several series, usually yellow. Ovary 6–15-celled, placentation parietal. Stigmas 6–15, linear. Capsule 6–15-celled, cell-roofs reduced to small rims. Seeds very numerous, extremely small.
A genus of 5 species from Southwest Africa and the adjacent part of Cape Province of South Africa. The flowers open in the afternoon.
Literature: Sauer, N., *Flowering Plants of Africa* **45**: t. 1778–1780 (1978).

1a. Bodies obconical, upper surface flat or slightly convex, with purple or red-brown dots or lines, bodies often more or less totally submerged in the ground **3. vanzylii**
 b. Bodies obovate to spherical or cordate when viewed from the side, mostly above ground level **2**

2a. Fissure between the leaves gaping,
    V-shaped; sepals 6
        1. microspermus subsp. **puberulus**
  b. Fissure between the leaves not
    gaping, leaves more or less adpressed;
    sepals more than 6   3
3a. Outer petals white, inner yellow;
    surface of the leaves rough; stigmas
    8–13   **4. pole-evansii**
  b. All petals yellow; surface of the leaves
    smooth; stigmas 6–8   **2. wilmotianus**

1. D. microspermus (Dinter & Derenberg)
Derenberg subsp. **puberulus** (N.E. Brown)
Sauer (*D. puberulus* N.E. Brown; *D.
punctatus* Bolus). Illustration: Lamb &
Lamb, The illustrated reference on cacti
and other succulents, 670 (1966);
Flowering plants of Africa **45**: t. 1779B
(1978); Japan Succulent Society, Colour
encyclopedia of succulents, f. 661 (1981).
Bodies obovate to cordate (when viewed
from the side), the fissure between the
leaves widely gaping, 2.5–4 cm high,
2.5–3.5 cm thick, 3.5–5 cm wide. Leaves
keeled near the apex, very finely velvety,
greyish with coloured dots. Flowers 3.5–5
cm in diameter. Sepals 6. Petals numerous,
yellow. Stigmas 6–7. *Southwest Africa*. G1.
  Subsp. *microspermus* is not in cultivation.

2. D. wilmotianus Bolus. Illustration:
Jacobsen, Handbook of succulent plants,
f. 1328 (1960); Lamb & Lamb, The
illustrated reference on cacti and other
succulents, 671 (1966); Flowering plants of
Africa **45**: t. 1780A & B (1978); Japan
Succulent Society, Colour encyclopedia of
succulents, f. 663 (1981).
Bodies solitary or forming groups. Leaves
united for more than half their length,
smooth, hairless, rounded, with 1 or 2
keels on the back towards the apex,
2.5–4.5 cm high, grey sometimes tinged
with pink, with dark violet dots or dots
absent. Flowers 3.5–4.5 cm in diameter.
Sepals 8. Petals all golden yellow. Stigmas
6–8. *South Africa (Cape Province)*. G1.
  Subsp. **wilmotianus** has leaves with
conspicuous dots, while subsp. **impunctatus**
Sauer (*D. inexpectatus* Jacobsen) has leaves
without dots.

3. D. vanzylii (Bolus) Schwantes.
Illustration: Jacobsen, Handbook of
succulent plants, f. 1327 (1960); Flowering
plants of Africa **45**: t. 1778B (1978);
Barkhuizen, Succulents of southern Africa,
122 (1978); Japan Succulent Society,
Colour encyclopedia of succulents, f. 662
(1981).
Bodies forming groups, to 4 cm high,

rather like those of a species of *Lithops*, the
2 leaves joined for more than half their
length and with flattish tops which are
greyish or yellowish green marked with red
dots which coalesce into indistinct lines.
Flowers to 1.5 cm in diameter. Sepals
7–8. Petals all orange-yellow. Stigmas
7–8. *South Africa (Cape Province)*. G1.

4. D. pole-evansii (N.E. Brown) Schwantes.
Illustration: Jacobsen, Handbook of
succulent plants, f. 1326 (1960); Flowering
plants of Africa **45**: t. 1778A (1978); Japan
Succulent Society, Colour encyclopedia of
succulents, f. 660 (1981).
Bodies usually solitary. Leaves united for
more than half their length, to 4.5 cm
high, 2.2–2.4 cm thick with rounded
backs, rather rough, grey often tinged with
red and yellow, without dots. Flowers to
4 cm in diameter. Sepals 8 or 9. Petals of
outer series white, of inner series yellow.
Stigmas 8–13. *South Africa (Cape Province)*.
G1. Spring.

45. FRITHIA N.E. Brown
F. McIntosh
Small, stemless perennial; internodes not
visible. Leaves 5–9 together, forming a
rosette, erect, finger-like, truncate at the
apex and with a transparent window, to
2 cm long × 3–6 mm thick, grey-green.
Flowers solitary, terminal, stalkless or
shortly stalked, 9–25 mm in diameter.
Sepals 5, joined into a short tube above the
ovary. Petals linear, in several series,
purplish red, violet or white. Stamens more
or less concealed by the petals. Nectary of 5
glands. Ovary 5-celled, placentation
parietal, stigmas 5. Tubercles and cell-roofs
absent. Seeds small with fine tubercles.
  A genus of a single species from South
Africa (Transvaal), which should be kept
quite dry when resting in summer.

1. F. pulchra N.E. Brown. Illustration:
Flowering Plants of Africa **7**: pl. 275
(1927); Rowley, The illustrated
encyclopedia of succulents, f. 1122 (1978);
Riha & Subik, The illustrated encyclopedia
of cacti and other succulents, pl. 396
(1981); Japan Succulent Society, Colour
encyclopedia of succulents, pl. 675 (1981).
South Africa (Transvaal). G1.

46. GIBBAEUM Haworth
V.A.Matthews
Dwarf, usually stemless succulent
perennials with prostrate branches forming
clumps or mats composed of a number of
growths. Each growth composed of 2 equal
or unequal leaves which are variably joined

and form an almost spherical or ovoid body
with a fissure at or below the tip, or with
the leaves very unequal, adpressed and
somewhat incurved so that in profile the
growth resembles a shark's head, or with
the leaves joined at the base and spreading
widely. Leaves hairless or, more usually,
minutely hairy (use of a lens is necessary to
see the hairs). Flowers solitary, terminal,
borne on a short stalk; bracts absent. Sepals
usually 6, free, unequal, the 2 lateral
longer and often keeled. Petals numerous in
1–3 rows, free, linear, white to red or
purplish. Stamens numerous, erect.
Staminodes present. Nectary composed of 6
separate glands. Stigmas usually 6,
sometimes to 11, broadly ovate, shorter
than depth of ovary. Ovary usually
6-celled; placentation parietal.
  A genus of about 21 species from the
Cape Province of South Africa. Most species
grow actively from October to May, during
which time they should be watered
carefully, and from June to September they
should be rested and water almost
withheld. During the resting period the
bodies shrink and shrivel; when watering is
resumed, the new bodies break through the
old skins and the plants will flower.
G. album and *G. pachypodium* tend to grow
and flower later than the other species, but
in all, the growing and flowering time
varies with environmental conditions and
the plants should be carefully observed to
see when water should be applied or
withheld. Propagation is by seed or
cuttings.
Literature: Nel, G.C. (ed Jordaan, P.G. &
Shurly, E.W.), *The Gibbaeum handbook*
(1953).

1a. Bodies spherical to ovoid; leaves equal
    or unequal, rounded on the back, or,
    if keeled, then hairless, joined for two-
    thirds or more of their length   2
  b. Bodies not spherical to ovoid; leaves
    unequal, keeled on the back at least
    towards the tip, joined for less than
    two-thirds of their length, or if joined
    for more than two-thirds then larger
    leaf with 2 keels   7
2a. Leaves hairless   3
  b. Leaves hairy (use lens)   4
3a. Leaves not keeled; flowers white,
    cream or pinkish; stigmas 6–11,
    usually 7–9   **13. heathii**
  b. Leaves keeled; flowers magenta to
    reddish; stigmas 6   **14. petrense**
4a. Leaves whitish to greyish green,
    sometimes tinged with red or brown   5

b. Leaves yellowish green, sometimes
tinged with red or purple   6
5a. Flowers usually white, sometimes
pink; leaves whitish   **7. album**
b. Flowers bright purplish pink; leaves
greyish green, often tinged with red or
brown   **8. dispar**
6a. Leaves with short white hairs, or
hairless at the top; flowers *c.* 2.5 cm
in diameter; stigmas 6
2. cryptopodium
b. Leaves with long fine white hairs;
flowers 6–16 mm in diameter;
stigmas 7   **3. pilosulum**
7a. Larger leaf of a pair with 2 keels
beneath   8
b. Larger leaf of a pair with 1 keel
beneath, or the keel running over the
tip   **1. gibbosum**
8a. Leaves of each pair adpressed for most
of their length   9
b. Leaves of each pair adpressed only at
the base, spreading above   11
9a. Leaves club-shaped, broader towards
the tip; plant with prostrate branches
which spread in all directions to form
an irregular mat   **4. geminum**
b. Leaves broader at the base that the
tip; plant forming compact cushions
10
10a. Leaves covered with simple, dense,
adpressed hairs which point
downwards   **5. pubescens**
b. Leaves with sparser, stellate hairs
6. shandii
11a. Larger leaf of a pair 1.5–4 cm; flower-
stalk *c.* 1.5 cm   **12. haagei**
b. Larger leaf of a pair 5–10 cm; flower-
stalk 2–7.5 cm   12
12a. Flowers to 1.5 cm in diameter with
the petals in 3 rows
11. pachypodium
b. Flowers 3–5 cm in diameter with the
petals in 2 rows   13
13a. Leaves grey-green to grey; flower-
stalk 2–2.5 cm   **9. velutinum**
b. Leaves green, yellow-green or
brownish; flower-stalk 3–5 cm
10. schwantesii

1. G. gibbosum (Haworth) N.E. Brown
(*G. perviride* (Haworth) N.E. Brown).
Illustration: Flowering Plants of Africa 29:
pl. 1153 (1952–3); Nel, The Gibbaeum
handbook, f. 26–29 (1953); Everard &
Morley, Wild flowers of the world, pl. 73
(1970); Jacobsen, Lexicon of succulent
plants, pl. 166, f. 2 (1974).
Plant with many branches, forming
clumps. Bodies green to yellowish green,
2–6 cm, 1–2 cm thick at the base. Leaves

very unequal, finely hairy, the larger
slightly incurved, somewhat flattened
above, with 2 keels beneath, the smaller
c. one-third as long. Flower-stalk
8–12 mm. Flowers 1–3 cm in diameter,
petals pale pink to purplish or magenta.
Stigmas 6. *South Africa (Cape Province).* G1.
Winter–spring.

2. G. cryptopodium (Kensit) Bolus.
Illustration: Nel, The Gibbaeum handbook,
f. 15–17 (1953).
Bodies almost spherical to ovoid, 1.5–2.5 ×
1–2.5 cm, pale or yellowish green,
sometimes red-tinged. Leaves almost equal
to distinctly unequal, covered with very
short white hairs, sometimes hairless at the
top. Flower-stalk 5–12 mm. Flowers
c. 2.5 cm in diameter; petals *c.* 35, in 2
rows. Stigmas 6. *South Africa (Cape
Province).* G1. Winter.

3. G. pilosulum (N.E. Brown) N.E. Brown.
Illustration: Brown et al., Mesembryanthema, 220 (1931); Nel, The Gibbaeum
handbook, f. 12–14 (1953).
Differs from *G. cryptopodium* in the leaves
having long, fine, white hairs, the flowers
6–16 mm in diameter with petals bright
pink to mauvish red, and stigmas 7. *South
Africa (Cape Province).* G1. Winter.

4. G. geminum N.E.Brown. Illustration:
Bolus, Notes on Mesembryanthemum, **1**:
58 (1928); Nel, The Gibbaeum handbook,
f. 37–39 (1953); Jacobsen, Handbook of
succulent plants, pl. 166, f. 1 (1974);
Barkhuizen, Succulents of southern Africa,
f. 115 (1978).
A very dwarf shrub with prostrate
branches forming an irregular mat;
branches themselves branched. Bodies pale
grey-green, often reddish in the upper half.
Leaves very unequal, club-shaped, whitish-
velvety, the larger 3–4 times longer than
the smaller; larger leaf 1–3 cm × 4–6 mm,
with a keel running over the tip; smaller
leaf 4–8 mm. Flower-stalk 1.2–1.5 cm.
Flowers 1.2–1.5 cm in diameter, petals in 1
row, purple or reddish. Stigmas 6. *South
Africa (Cape Province).* G1. Winter–spring.

5. G. pubescens (Haworth) N.E. Brown.
Illustration: Nel, The Gibbaeum handbook,
f. 33, 34 (1953); Herre, The genera of the
Mesembryanthemaceae, 163 (1973);
Jacobsen, Lexicon of succulent plants, pl.
166, f. 5 (1974); Rauh, Schöne Kakteen
und andere Sukkulenten, f. 222 (1978).
Plant with many branches forming
compact cushions. Bodies greyish white,
with the persistent remains of old leaves at
the base. Leaves unequal, cylindric but

broader at the base, with simple, white,
dense, adpressed hairs which point
downwards, the larger 2–3 times longer
than the smaller; larger leaf *c.* 3 × 1.5 cm,
somewhat compressed laterally and with a
keel beneath towards the tip. Flower-stalk
1–1.5 cm. Flowers *c.* 1.5 cm in diameter,
petals lilac to purple or violet-red. Stigmas
6. *South Africa (Cape Province).* G1.
Winter–spring.

6. G. shandii (N.E. Brown) N.E. Brown.
Illustration: Gardeners' Chronicle 71: 129
(1922); Brown et al., Mesembryanthema,
223 (1931); Nel, The Gibbaeum handbook,
pl. 7, f. 35, 36 (1953).
Differs from *G. pubescens* in the leaves
having sparser, stellate hairs. *South Africa
(Cape Province).* G1. Winter–spring.

7. G. album N.E. Brown. Illustration:
Brown et al., Mesembryanthema, 218
(1931); Nel, The Gibbaeum handbook, pl.
5, 6, f. 7, 8 (1953); Aloe **12**: 67 (1974);
Jacobsen, Lexicon of succulent plants, pl.
164, f. 3 (1974).
Plant forming a compact clump. Bodies
almost spherical to obliquely ovoid, 2–3 ×
1.2–1.5 × *c.* 1 cm, whitish to pale grey.
Leaves almost equal or unequal, elliptic-
ovoid, velvety. Flower-stalks 5–6 mm.
Flowers *c.* 2.5 cm in diameter, petals in
several rows, white. Stigmas 6. *South Africa
(Cape Province).* G1–2. Spring–summer.
Forma **rosea** (N.E. Brown) Rowley has
pink flowers.

8. G. dispar N.E.Brown. Illustration: Nel,
The Gibbaeum handbook, f. 23–25 (1953);
Jacobsen, Lexicon of succulent plants, pl.
164, f. 6 (1974); Barkhuizen, Succulents of
southern Africa, f. 114 (1978); Succulenta
62(4): front cover (1983).
Plant forming a clump. Bodies almost
spherical to obliquely ovoid, 1–1.5 ×
1–1.4 cm, grey-green, often flushed with
red or brown. Leaves almost equal or
unequal, minutely velvety. Flower-stalks
2–4.5 mm. Flowers 8–25 mm in diameter,
petals in 1 row, bright purplish red.
Stigmas 6. *South Africa (Cape Province).* G1.
Spring.

9. G. velutinum (Bolus) Schwantes.
Illustration: Nel, The Gibbaeum handbook,
f. 45–47 (1953); Cactus and succulent
Journal of Great Britain 29: 24 (1967);
Jacobsen, Lexicon of succulent plants, pl.
166, f. 6 (1974); Rauh, The wonderful
world of succulents, pl. 89 (1984).
Plants forming a mat, the branches
prostrate, with the remnants of old leaves,
each branch ending in 2 pairs of leaves.

Leaves grey-green to pale grey, minutely velvety, unequal, joined in the basal third, strongly keeled beneath; the older leaf-pair with the larger leaf 5–6 cm with a hooked tip, the smaller 2–4 cm. Flower-stalk 2–2.5 cm. Flowers 4–5 cm in diameter, petals in 2 rows, pink or white. *South Africa (Cape Province)*. G1. Spring.

10. G. schwantesii Tischer. Illustration: Nel, The Gibbaeum handbook, f. 48–50 (1953).
Differs from *G. velutinum* in having green, yellow-green or brownish leaves and the flower-stalks 3–5 cm. *South Africa (Cape Province)*. G1. Spring.

11. G. pachypodium (Kensit) Bolus. Illustration: Nel, The Gibbaeum handbook, f. 40, 41 (1953).
Differs from *G. velutinum* in having the larger leaf 6–10 cm, the flower-stalks 4–7.5 cm and the flowers to 1.5 cm in diameter with pink to reddish petals in 3 rows. *South Africa (Cape Province)*. G1. Spring.

12. G. haagei Schwantes.
Differs from *G. velutinum* in having longer leaves 1.5–4 cm, flower-stalks *c.* 1.5 cm and flowers 1.5–3 cm in diameter, red or lilac-red. *South Africa (Cape Province)*. G1. Spring.

13. G. heathii (N.E. Brown) Bolus (*G. comptonii* (Bolus) Bolus; *G. luckhoffii* (Bolus) Bolus; *Rimaria heathii* (N.E. Brown) N.E. Brown). Illustration: Nel, The Gibbaeum handbook, pl. 1–4, f. 3–6 (1953); Aloe **12**: 67 (1974); Jacobsen, Lexicon of succulent plants, pl. 166, f. 4 (1974); Barkhuizen, Succulents of southern Africa, f. 116 (1978).
Plant forming a compact clump or mat. Bodies almost spherical to obovoid, 1–5 cm, green to whitish or greyish, sometimes tinged with yellow, red or purple. Leaves equal to slightly unequal, joined for up to three-quarters of their length, rounded beneath, hairless. Flower-stalks 1–1.2 cm. Flowers 1–4 cm in diameter, petals white, cream or pinkish. Stigmas 6–11, usually 7–9. *South Africa (Cape Province)*. G1. Spring.

14. G. petrense (N.E. Brown) Tischer. Illustration: Nel, The Gibbaeum handbook, f. 7, 9–11 (1953); Jacobsen, Lexicon of succulent plants, pl. 166, f. 3 (1974); Barkhuizen, Succulents of southern Africa, f. 117 (1978).
Differs from *G. heathii* in the leaves being 5–12 mm, joined only in the basal third,

with the free part keeled beneath; flowers 1–1.5 cm in diameter, petals magenta to reddish. Stigmas 6. *South Africa (Cape Province)*. G1. Spring.

47. ANTEGIBBAEUM Schwantes
F. McIntosh
Thick, succulent, stemless perennials to 8 cm high, branching from the base and forming clumps; internodes not visible, each shoot with 2 leaves. Leaves of each pair somewhat unequal, curved, upper surface flat or slightly convex, lower surface rounded or angled and compressed towards the apex, all grey-green or reddish, smooth or somewhat rough, to 3 cm × 6–8 mm. Flowers 5–6 cm in diameter, solitary, terminal, shortly stalked, the stalk bearing 2 bracts. Sepals 6, keeled, often with membranous margins. Petals in 1–3 whorls, pale red to violet-red. Stamens numerous. Ovary 6–7-celled, placentation parietal; stigmas 6 or 7, linear, longer than ovary-depth. Capsule 6–7-celled, keels ending in large, broad wings; cell-roofs present, tubercles absent. Seeds large, dark brown, minutely spiny.
A genus of a single species from South Africa.

1. A. fissoides (Haworth) Schwantes. Illustration: Jacobsen, Handbook of succulent plants, f. 1203 (1960); Herre, The genera of the Mesembryanthemaceae, 75 (1971).
South Africa (Cape Province). G1.

48. DIDYMAOTUS N.E.Brown
F. McIntosh
Stemless, very succulent perennials forming clumps, with 2 pairs of leaves. Leaves very thick, ovate, broader than long to twice as long as broad, roughly dotted, upper surface flat or a little concave, 1.5–2 × up to 3 cm, the lower surface keeled, drawn forward over the tip like a chin, whitish grey-green. Flowers solitary, axillary from a shoot which consists of 2 bracts and a flower-stalk, borne at each side of the body, to 4 cm in diameter. Sepals 6, rounded, keeled, 2 of them larger than the others. Petals in several whorls, pink to dark violet-red, the base always darker than the rest. Stamens with violet filaments and yellow anthers. Nectary forming a continuous ring. Ovary 6-celled, placentation parietal; stigmas 6. Capsule 6-celled, cell-roofs present. Seeds bulb-shaped, light brown.
A genus of a single species from South Africa, remarkable for the fact that the flowers are borne on lateral shoots and not between the leaves.

1. D. lapidiformis (Marloth) N.E. Brown. Illustration: Lamb & Lamb, Illustrated reference on cacti and other succulents, pl. 253 (1966); Rowley, The illustrated encyclopedia of succulents, pl. 1118 (1978); Barkhuizen, Succulents of southern Africa, pl. 103 (1978); Court, Succulent flora of southern Africa, pl. 92 (1981). *South Africa (Cape Province)*. G1.

49. CONOPHYTUM N.E. Brown
J. Cullen & F. McIntosh
Very small stemless perennials, usually growing in groups. Bodies composed of 2 almost completely united leaves, spherical, ovoid, oblong or cylindric, the top flat, depressed, notched or 2-lobed, with a small fissure at the centre, very rarely with a window at the apex. Flowers solitary. Calyx with a distinct tube, 4–7-lobed. Petals numerous, in several series, united into a tube at the base. Stamens collected into a column in the centre of the flower. Nectaries forming a continuous ring. Ovary 4–7-celled, placentation parietal; stigmas 4–7, thread-like, united into a columnar style below. Cell-roofs absent. Seeds very small, smooth.
A genus of about 290 named species from South Africa (Cape Province) and Southwest Africa. Many of the species are poorly known, and the total number is likely to be considerably reduced as research is undertaken. A complex hierarchy of sections and subsections has been devised by Tischer (see Jacobsen, Lexicon of succulent plants, 430–433, 1974), but this is of little help in practical identification, so is not used here. The species are distinguished mainly by the form of the body (see Gardeners' Chronicle **78**: 450, f. 187, 1925), and its coloration and markings. About 30 species are reputed to be in general cultivation.
Literature: Boom, B.K., Het geschlacht Conophytum N.E. Br., *Succulenta* **52**: 22–26, 42–45, 62–65, 82–86, 102–105, 122–126, 142–146 (1973); Rawe, R., The genus Conophytum Section 3: Wettsteiniana Subsection 2: Minuta (Littlewood) Tischer, *Cactus & Succulent Journal (U.S.)* **47**: 209–212 (1975); The genus Conophytum Section Costata Schwantes amended Tischer Subsection Verrucosa Schwantes amended Tischer, *Cactus & Succulent Journal (U.S.)* **47**: 126–132 (1975); The genus Conophytum Section Cataphracta Schwantes, *Cactus & Succulent Journal (U.S.)* **47**: 180–186 (1975); The genus Conophytum Section Saxetana, *Cactus & Succulent Journal (U.S.)*

53; 183–186 (1981); The genus Conophytum Section Ovigera, *Cactus & Succulent Journal (U.S.)* 54: 218–222 (1982); The genus Conophytum Section Truncatella, *Cactus & Succulent Journal (U.S.)* 54: 165–169 (1982).

1a. Top surface with a window which is often band-shaped   **1. pillansii**
 b. Top surface without a window, but often with translucent dots   2
2a. Flowers opening at night (rarely in the evening) and sometimes remaining open during the following day   3
 b. Flowers opening in the morning, sometimes remaining open during the subsequent night   15
3a. Body shortly 2-lobed, compressed from side to side; flowers usually 5–8 mm in diameter   4
 b. Body not 2-lobed, not or only slightly compressed, flowers usually 1 cm or more in diameter   5
4a. Main colour of body green, yellow-green or blue-green   **10. saxetanum**
 b. Main colour of body grey, grey-green or reddish   **9. hians**
5a. Body without dark dots or stripes (rarely with very scanty spots), ribbed   **19. calculus**
 b. Body with spots and/or stripes, ribs absent   6
6a. Flowers reddish or lilac, sometimes white when opening or fading   7
 b. Flowers white, yellowish, straw-coloured, brownish or pale coppery   8
7a. Body without lobes; flowers 5–8 mm in diameter   **20. advenum**
 b. Body shortly lobed; flowers 1.5–2 cm in diameter   **11. placitum**
8a. Top surface with green or red-brown spots in stripes or lines   9
 b. Top surface without stripes or lines, dots evenly distributed, occasionally in irregular rows   10
9a. Top surface with brown, branched stripes which often slant on the margins of the fissure   **23. scitulum**
 b. Top surface with brown lines which are not or little branched, not arranged as above   **21. minimum**
10a. Top surface arched or vaulted, dots (at least for the most part) arranged in lines   11
 b. Top surface flat or sometimes concave, dots mostly not arranged in lines   14
11a. Top surface of body 4–6 mm in diameter   **22. piluliforme**
 b. Top surface of body more than 6 mm in diameter   12

12a. Petals white often suffused with red, at least 1 mm broad   **11. placitum**
 b. Petals whitish yellow, less than 1 mm broad   13
13a. Top surface 1.5–2 cm in diameter, oval, dull green with numerous dots more or less in lines   **13. uviforme**
 b. Top surface to 1.5 cm in diameter, round or slightly oval, grey- or blue-green with numerous scattered dots   **12. truncatum**
14a. Top surface apple green, without dots or with few, very small dots   **14. viridicatum**
 b. Top surface green, with large dots (*c.* 1 mm in diameter)   **24. obcordellum**
15a. Top surface sometimes shortly lobed, laterally compressed   16
 b. Top surface not or very shortly lobed, when mature not or little compressed   22
16a. Petals red or lilac   17
 b. Petals yellow or orange   18
17a. Body finely hairy   **8. velutinum**
 b. Body not hairy, at most somewhat papillose   **7. polyandrum**
18a. Body obscurely lobed, the lobes less than 2 mm   19
 b. Body clearly 2-lobed, the lobes more than 2 mm   20
19a. Body hairless   **5. luisae**
 b. Body finely hairy   **6. meyeri**
20a. Lobes erect, when viewed from the top forming half-circles   **2. bilobum**
 b. Lobes not as above   21
21a. Body to 3 cm high, usually with many dark spots; keel of the lobes more or less horizontal   **3. elishae**
 b. Body to 5 cm high, with a few spots or without, the keels of the lobes arched outwards   **4. meyerae**
22a. Staminodes present, orange   **18. minutum**
 b. Staminodes absent   23
23a. Top surface of body with a border of raised spots; old bodies persistent as withered sheaths   24
 b. Bodies not as above   25
24a. Top surface of body with a shining green border   **25. ectypum**
 b. Top surface of body without a shining green border   **26. fulleri**
25a. Petals red or violet   **16. kubusanum**
 b. Petals yellow   26
26a. Top surface of body 5–8 mm in diameter; corolla-tube projecting 1–1.5 cm above the body   **17. novicium**
 b. Top surface more than 8 mm in diameter; corolla-tube projecting no

more than 1 cm above the body   **15. flavum**

1. C. pillansii Lavis. Illustration: Jacobsen, Lexicon of succulent plants, pl. 156 (1974); Barkhuizen, Succulents of southern Africa, f. 101 (1978); Japan Succulent Society, Colour encyclopedia of succulents, pl. 591 (1981).

Bodies 1–3 together, to 2.2 × 2.2 cm wide and 2 cm thick, light yellowish green, reddened at the sides, the upper surface with large, green, translucent dots which usually merge to form a band-shaped window. Flowers to 2.5 cm in diameter, petals violet-red. *South Africa (Cape Province)*. G1.

2. C. bilobum (Marloth) N.E. Brown. Illustration: Lamb & Lamb, The illustrated reference on cacti and other succulents, 249 (1955); Herre, The genera of the Mesembryanthemaceae, 121 (1971); Jacobsen, Lexicon of succulent plants, pl. 153 (1974); Rowley, The illustrated encyclopedia of succulents, f. 11.1 (1978); Japan Succulent Society, Colour encyclopedia of succulents, pl. 602 (1981). Plants forming mats with age. Bodies slightly flattened, compressed, 3–5 × 2–2.5 cm wide, to 2 cm across, clearly 2-lobed, the lobes blunt or rounded, keeled, erect, forming half-circles when viewed from the top, all greyish green or whitish green, the tips and keels of the lobes reddish, all slightly papillose. Flowers to 3 cm in diameter, opening in the morning, petals yellow-orange. *South Africa (Cape Province)*. G1.

3. C. elishae (N.E. Brown) N.E. Brown. Illustration: Rowley, The illustrated encyclopedia of succulents, 134 (1978). Plants forming mats with age. Bodies to 2.5 × 1.4 wide × 1–1.2 cm thick, compressed, distinctly 2-lobed, keeled, the keels more or less horizontal, green to bluish, rough, covered with minute dots, the sides of the fissure with dark, more or less translucent dots, reddish on the ridges. Flowers to 2 cm in diameter, opening in the morning, petals golden yellow. *South Africa (Cape Province)*. G1.

   C. andausanum N.E. Brown is very similar but has less distinct lobes.

4. C. meyerae Schwantes. Illustration: Jacobsen, Lexicon of succulent plants, pl. 155 (1974); Court, Succulent flora of southern Africa, 92 (1981). Bodies forming small clumps, to 5 × 4 cm wide, laterally compressed, 2-lobed, the lobes keeled and arched backwards, dull

green to bluish green, often rough, the tips
of the lobes often reddish, with a line of
dark dots along the margins and undersides
of the lobes, and very indistinct spots
elsewhere. Flowers opening in the morning,
petals yellow. *South Africa (Cape Province).*
G1.

5. C. luisae Schwantes. Illustration: Lamb &
Lamb, The illustrated reference on cacti
and other succulents, 252 (1955).
Plants forming loose mats. Bodies to
1.5 cm × 1 cm wide, to 8 mm thick,
shortly 2-lobed, the lobes 0.5–2 mm, the
surface with a few distinct dots, the fissure
with a dark spot on either side. Flowers to
1.8 cm in diameter, opening in the
morning, petals yellow to orange. *South
Africa (Cape Province).* G1.

6. C. meyeri N.E. Brown. Illustration:
Jacobsen, Lexicon of succulent plants, pl.
155 (1974); Cactus & Succulent Journal
(U.S.) **54**: 218, 221 (1982).
Plants forming mats with age. Bodies
2–2.5 × 1.3–1.5 cm thick, compressed,
2-lobed, the lobes to 2 mm, the surface
greyish green, rough with short hairs.
Flowers 1.4–1.6 cm in diameter, opening
in the morning, petals yellow to orange.
South Africa (Cape Province). G1.

7. C. polyandrum Lavis. Illustration: Cactus
& Succulent Journal (U.S.) **54**: 220 (1982).
Bodies to 2.2 × 1.2 cm wide, compressed,
olive green, hairless, without spots or with
very few indistinct spots. Flowers to 2 cm
in diameter, opening in the morning, petals
whitish with light pink margins. *South
Africa (Cape Province).* G1.

8. C. velutinum (Schwantes) Schwantes
(*C. tischeri* Schick). Illustration: Jacobsen,
Lexicon of succulent plants, pl. 157 (1974);
Japan Succulent Society, Colour
encyclopedia of succulents, pl. 614 (1981);
Cactus & Succulent Journal (U.S.) **54**: 218
(1982).
Bodies to 1.3 cm × 8-10 mm wide ×
7–9 mm thick, compressed, 2-lobed with
roundish lobes, all olive green or greyish
green with dark green dots, minutely hairy.
Flowers to 2 cm in diameter, opening in the
morning, petals pinkish purple. *South Africa
(Cape Province).* G1.

9. C. hians N.E. Brown. Illustration: Lamb
& Lamb, The illustrated reference on cacti
and other succulents, 251 (1955).
Plant forming clumps with age. Bodies to
1.2 cm × 8 mm in diameter, laterally
compressed, 2-lobed, green to greyish green
or reddish, with scattered dots merging into

lines along the margins, hairy. Flowers to
8 mm in diameter, opening at night, petals
reddish yellow to white. *South Africa (Cape
Province).* G1.

10. C. saxetanum (N.E. Brown) N.E.
Brown. Illustration: Rowley, The illustrated
encyclopedia of succulents, f. 11.26 (1978);
Cactus & Succulent Journal (U.S.) **53**: 185
(1981).
Plant forming very large clumps with age.
Bodies 6–10 × 2–4 mm in diameter,
compressed above, very shortly 2-lobed
(lobes to 1.5 mm), green with a few
indistinct spots, fissure with white hairs.
Flowers to 8 mm in diameter, opening at
night, petals white. *Southwest Africa.* G1.

11. C. placitum (N.E. Brown) N.E. Brown.
Plant forming mats with age. Bodies to
1.5 × 1 cm in diameter, slightly bilobed
above (the lobes less than 2 mm), vaulted,
ridged, green to reddish with dark dots.
Flowers 1.5–2 cm in diameter, opening at
night, petals white often suffused with red,
1 mm or more broad. *South Africa (Cape
Province).* G1.

12. C. truncatum (Thunberg) N.E. Brown.
Illustration: Jacobsen, Lexicon of succulent
plants, pl. 157 (1974); Cactus & Succulent
Journal (U.S.) **54**: 165 (1982).
Bodies to 1.5 × 1.3 cm in diameter, round
or slightly oval in section, top surface
slightly rounded or vaulted, greyish green
to bluish green with large and small dark
green dots (dark blue-green with large,
regular dots in var. **brevitubum** (Lavis)
Tischer), the lips of the fissure protruding
and usually surrounded by a ring of dots.
Flowers 1.4–1.6 cm in diameter, opening
at night, petals straw-coloured to almost
white, to 1 mm broad. *South Africa (Cape
Province).* G1.

   C. notatum N.E. Brown is very similar
but has scattered, reddish brown dots on
the surface.

13. C. uviforme (Haworth) N.E. Brown.
Illustration: Flowering plants of South
Africa **13**: t. 509A (1933).
Plants forming small clumps. Bodies
1–1.2 cm tall and wide, 6–20 mm thick,
pale yellowish green with dark or reddish
spots, the sides sometimes reddish, top
surface flat with numerous dots arranged
more or less in lines. Flowers to 1 cm in
diameter, opening at night, petals whitish
yellow. *South Africa (Cape Province).* G1.

14. C. viridicatum (N.E. Brown) N.E.
Brown. Illustration: Cactus & Succulent
Journal (U.S.) **54**: 166 (1982).

Bodies 1.2–2.3 cm tall, 1–1.5 cm wide,
9–13 mm thick, top surface very slightly
vaulted, all green with scarcely visible dots
or dots absent. Flowers opening at night,
more than 1 cm in diameter, petals white.
South Africa (Cape Province). G1.

15. C. flavum N.E. Brown. Illustration:
Jacobsen, Lexicon of succulent plants, pl.
154 (1974).
Plants forming mats with age. Bodies
1.8–2 cm × 6–12 mm in diameter, flat or
very slightly convex on the top, fissure
bordered by a dark green line, the whole
green with a few dark green spots. Flowers
1.2–1.6 cm in diameter, opening during
the morning, corolla tube projecting at
most to 1 cm above the body, petals yellow.
South Africa (Cape Province). G1.

16. C. kubusanum N.E. Brown.
Bodies to 1 cm high, 4–6 mm in diameter,
the upper surface flat or slightly depressed
in the centre, dark greyish green with
numerous dark green dots and with a
darker line surrounding the fissure. Flowers
to 2 cm in diameter, opening in the
morning, petals light pinkish mauve. *South
Africa (Cape Province).* G1.

17. C. novicium N.E. Brown.
Plants forming loose cushions with age.
Bodies to 1.2 cm tall, to 9 mm in diameter,
slightly constricted below the top which is
flat or slightly convex with a depressed
fissure, all light grey to greyish green with
darker green dots, those around the fissure
sometimes merging to form a ring. Flowers
to 2 cm in diameter, corolla tube projecting
for up to 1.5 cm above the body, opening
during the morning, petals yellow to golden
yellow. *South Africa (Cape Province).* G1.

18. C. minutum (Haworth) N.E. Brown.
Illustration: Gardeners' Chronicle **119**: 231
(1922); Herre, The genera of the
Mesembryanthemaceae, 121 (1971);
Jacobsen, Lexicon of succulent plants, pl.
155 (1970); Barkhuizen, Succulents of
southern Africa, f. 99 (1978).
Plant forming circular mats with age. Bodies
pear-shaped, 1–1.2 cm tall, 6–10 mm in
diameter, top flat or slightly depressed,
bluish to greyish green, sometimes faintly
spotted. Flowers 1.2–1.5 cm in diameter,
opening in the morning, with orange
staminodes. Petals pinkish mauve. *South
Africa (Cape Province).* G1.
   Var. **sellatum** (Tischer) Boom (*C. sellatum*
Tischer) has hairs along the fissure, and
var. **pearsonii** (N.E. Brown) Boom (*C.
pearsonii* N.E. Brown) has larger flowers (to
2.2 cm in diameter).

19. C. calculus (Berger) N.E. Brown.
Plants forming mats with age. Bodies compressed, 1.6–2 cm high, to 2.4 cm in diameter, chalky greyish green, without dots, sometimes ribbed. Flowers to 1.2 cm in diameter, opening at night, petals deep yellow with brownish tips. *South Africa (Cape Province).* G1.

20. C. advenum N.E. Brown (*C. leightoniae* Bolus).
Bodies 5–7 mm tall, 4–6 mm wide, 4–5 cm thick, top surface flat or slightly vaulted and with scalloped margins, smooth, greyish green with a transverse line of dark green or brown spots and some solitary spots above which sometimes merge into lines. Flowers 4–5 mm in diameter, opening at night, scented, petals yellowish pink, sometimes coppery red outside. *South Africa (Cape Province).* G1.

21. C. minimum (Haworth) N.E. Brown.
Bodies to 1 cm high, to 6 mm in diameter, the upper surface flat to slightly vaulted, depressed in the centre and with a dark zone surrounding the fissure, which is ciliate; surface dark grey-green with darker, brownish red spots above which sometimes merge into short lines, the sides often reddish. Flowers opening at night, petals white to cream. *South Africa (Cape Province).* G1.

There are several very similar species which may or may not be distinct: **C. occultum** Bolus has no hairs around the fissure; **C. labyrintheum** (N.E. Brown) N.E. Brown has the bodies 6–8 mm in diameter with the upper surface more or less flat, and the flowers to 8 mm in diameter; **C. praecinctum** N.E. Brown has the bodies 6–8 mm in diameter with the upper surface vaulted and the flowers more than 8 mm in diameter; **C. vagum** N.E. Brown has the bodies 9–15 mm in diameter.

22. C. piluliforme (N.E. Brown) N.E. Brown.
Bodies 8–10 mm tall, 5–7 mm in diameter, fissure compressed and surrounded by a dark line, the top flat, the surface grey-green with rather darker dots and minutely hairy. Flowers opening at night, more than 1 cm in diameter, petals copper-coloured. *South Africa (Cape Province).* G1.

C. brevipetalum Lavis is very similar, but has the top surface vaulted, the dots more conspicuous and the petals reddish.

23. C. scitulum (N.E. Brown) N.E. Brown. Illustration: Riha & Subik, The illustrated encyclopedia of cacti and other succulents, f. 387 (1981).

Bodies crowded, 1.2–1.4 cm tall, 8–14 mm wide, 7–13 mm thick, the top surface slightly vaulted, the surface greyish green with several reddish brown, branched, vein-like lines above which slant at the fissure. Flowers 6–16 mm in diameter, opening at night, petals white. *South Africa (Cape Province).* G1.

24. C. obcordellum (Haworth) N.E. Brown. Illustration: Jacobsen, Lexicon of succulent plants, pl. 156 (1974).
Plants forming mats with age. Bodies 6–9 mm tall, 4–9 mm in diameter, the top surface flat or depressed with large (to 1 mm in diameter), slightly raised dots often arranged in lines, the sides mostly pink to dark red, the surface otherwise light green to greyish or brownish green. Flowers 6–15 mm in diameter, opening at night, petals white or straw-coloured. *South Africa (Cape Province).* G1.

Two very similar species are reputedly cultivated: **C. giftbergense** Tischer. Illustration: Barkhuizen, Succulents of southern Africa, f. 98 (1978). Bodies taller and compressed, with dark green or crimson spots. **C. mundum** N.E. Brown. Illustration: Jacobsen, Lexicon of succulent plants, pl. 156 (1974); Barkhuizen, Succulents of southern Africa, f. 100 (1978); Japan Succulent Society, Colour encyclopedia of succulents, pl. 645 (1981). Surface greyish green with large raised dots which sometimes merge into lines; petals cream.

25. C. ectypum N.E. Brown. Illustration: Jacobsen, Lexicon of succulent plants, pl. 154 (1974); Cactus & Succulent Journal (U.S.) **47**: 129 (1975); Rowley, The illustrated encyclopedia of succulents, f. 11.25 (1978).
Plants forming small clumps. Bodies enclosed in papery sheaths, 5–10 mm high, 4–6 mm in diameter, the upper surface smooth, green with a prominent line around the fissure, marked with distinct ridges or raised dark green dots. Flowers 8–10 mm in diameter, opening in the morning, petals pale pink. *South Africa (Cape Province).* G1.

26. C. fulleri Bolus. Illustration: Cactus & Succulent Journal (U.S.) **47**: 130 (1975); Japan Succulent Society, Colour encyclopedia of succulents, pl. 627 (1981).
Very similar to *C. ectypum*, but the top surface without a prominent green line around the fissure. *South Africa (Cape Province).* G1.

50. HERREANTHUS Schwantes
F. McIntosh
Small, stemless, clump-forming perennials to *c.* 10 cm, internodes not visible. Leaves 2 per shoot, united for less than half their length, thick, triangular in section, upper surface tapering, flat, sides keeled towards the apex, half-circular at the base, all light blue-green or yellow-green, sometimes slightly dotted, spiny at the tip, to 4 cm long × 2 cm wide and 1.5 cm thick. Flowers solitary, terminal, with a very short stalk bearing 2 bracts, to 3 cm in diameter, remaining open day and night for about 2 weeks. Sepals 6. Petals in several series, reflexed, white. Ovary 6-celled, placentation parietal; stigmas 6, linear, longer than ovary-depth. Capsule 6-celled, the cells deep, without cell-roofs or tubercles. Seeds almost smooth, brown.

A genus of a single species from South Africa.

1. H. meyeri Schwantes. Illustration: The flowering plants of Africa **45**: pl. 1777 (1978); Barkhuizen, Succulents of southern Africa, pl. 123 (1978); Court, Succulent flora of southern Africa, 93 (1981); Riha & Subik, The illustrated encyclopedia of cacti and other succulents, pl. 395 (1981). *South Africa (Cape Province).* G1.

51. OOPHYTUM N.E. Brown
F. McIntosh
Small perennial herbs forming tufts or clumps. Shoots with 2 leaves forming small, ovoid, fleshy bodies, the leaves united for more than half their length, the fissure between them gaping or not; bodies green to reddish, often with dots, occasionally hairy. Flowers solitary, terminal, without bracts. Sepals free, 6. Petals in 2 or 3 series, free at the base, generally white below, reddish above. Nectary-glands forming a continuous ring. Ovary 5–6-celled, placentation parietal; stigmas 5–6, linear. Capsule 5–6-celled with broad wings, without cell-roofs or tubercles. Seeds compressed.

A genus of 2 species from South Africa, which should be kept quite dry when resting in summer.
Literature: Ihlenfeldt, H.D., Morphologie und Taxonomie der Gattung Oophytum N.E. Br., *Botanische Jahrbücher* **99**: 303–328 (1978).

1a. Bodies 5–7 mm in diameter; flowers to 1 cm in diameter   **1. nanum**
 b. Bodies 1–1.2 cm in diameter; flowers to 2.2 cm in diameter   **2. oviforme**

1. O. nanum (Schlechter) Bolus.
Illustration: Herre, The genera of the
Mesembryanthemaceae, 237 (1971);
Barkhuizen, Succulents of southern Africa,
140, f. 137 (1978).
Plant to 2 cm high, bodies almost spherical,
5–7 mm in diameter, green, covered with
minute hairs. Flowers to 1 cm in diameter,
petals white with reddened tips. *South
Africa (Cape Province)*. G1. Winter.

2. O. oviforme (N.E. Brown) N.E. Brown.
Illustration: Barkhuizen, Succulents of
southern Africa, 143, f. 138 (1978).
Bodies to 2 cm tall, 1–1.2 cm in diameter,
olive green to bright reddish. Flowers to
2.2 cm in diameter, petals white, purplish
pink towards the tips. *South Africa (Cape
Province)*. G1. Winter.

52. OPHTHALMOPHYLLUM Dinter &
Schwantes
F. McIntosh
Small, stemless perennials, forming clumps.
Shoots 2-leaved, forming bodies, the 2
leaves of a pair united for most of their
length, often enclosed in the withered
remains of old leaves. Bodies almost
spherical, usually dotted, hairy or papillose,
with a gaping fissure at the apex. Each leaf
with 1 or more translucent windows at the
apex, the rest brownish green, reddish or
purplish red. Flowers solitary, terminal,
shortly stalked, the stalks bearing fleshy,
whitish bracts. Sepals 4–7, equal, reddish
brown, united at the base to form a short
tube. Petals in several series, white to
violet. Ovary 4–7-celled, placentation
parietal; stigmas free, 4–7. Nectary-glands
6, separate. Capsule 4–7-celled, keels
present, cell-roofs reduced to a limb or
absent. Seeds numerous, brownish yellow.

A genus of 19 species from South Africa
and Southwest Africa, which should be
kept dry when resting in summer.

1a. Petals pink to violet   **1. praesectum**
 b. Petals white   2
2a. Windows translucent; fissure between
    the leaves extending completely
    across the body   **2. friedrichii**
 b. Windows pale but not translucent;
    fissure not extending completely
    across the body   **3. schlechteri**

1. O. praesectum (N.E. Brown) Schwantes
(*O. jacobsenianum* Schwantes). Illustration:
Jacobsen, Handbook of succulent plants,
f. 1541 (1960); Japan Succulent Society,
Colour Encyclopedia of succulents, pl. 800
(1981).
Bodies solitary or in small clumps,

1.7–3 cm high, 8–18 mm wide, 4–12 mm
thick, compressed, cylindric when viewed
from the side; apex obscurely windowed,
slightly translucent, the rest dotted, green
or brownish above. Flowers to 3 cm in
diameter, petals pink to violet. *South Africa
(Cape Province)*. G1.

2. O. friedrichii (Dinter) Dinter &
Schwantes (*Derenbergia friedrichii* (Dinter)
Schwantes). Illustration: Jacobsen,
Handbook of succulent plants, f. 1534,
1535 (1960); Japan Succulent Society,
Colour encyclopedia of succulents, pl. 795
(1981).
Bodies solitary, cylindric, 2–3 cm high,
1.4–1.6 cm wide, 1–1.2 cm thick, with the
fissure extending completely across,
smooth, hairless, green with a reddish
tinge, coppery in the resting season, lobes
distinctly but roundly keeled, windows very
translucent. Flower 1.2–2 cm in diameter,
petals white. *South Africa (Cape Province)*.
G1.

3. O. schlechteri Schwantes. Illustration:
Jacobsen, Handbook of succulent plants,
f. 1543, 1544 (1960).
Bodies solitary or 2 together, oblong to
oval, 3–4 cm long, to 1.5 cm wide,
somewhat less thick, smooth, hairless, dull
green to pink to reddish, covered with pale
dots, the fissure between the lobes not
extending completely across the body.
Flowers to 2.5 cm in diameter, petals
white. *South Africa (Cape Province)*. G1.

53. FAUCARIA Schwantes
F. McIntosh
Small perennial rosette- or tuft-forming
plants, without a stem when young,
internodes not visible. Leaves 4–6 or more
together on a shoot, crowded, very thick
and fleshy, half-cylindric in section towards
the base, triangular towards the apex,
margins with cartilaginous or thick,
backwardly-directed bristle-tipped, teeth,
often with white scattered spots or warts.
Flowers solitary, terminal, stalkless or
shortly stalked, without bracts. Sepals 5,
free, unequal, often with membranous
margins and keels. Petals in 2 or 3 series,
yellow to orange above, reddish to copper-
coloured beneath. Nectary of 5 separate,
distinct glands. Ovary 5–6-celled,
placentation parietal; stigmas 5–6, linear,
longer than depth of ovary-cells. Capsule
5–6-celled, with expanding keels whose
wings almost completely cover the cells,
though without genuine cell-roofs;
tubercles absent. Seeds reddish brown.

A genus of about 35 species from the
Cape Province of South Africa.

1a. Leaves without dots or warts   2
 b. Leaves with white or grey dots or
    warts   3
2a. Tubercles present on the upper leaf-
    surface   **5. tuberculosa**
 b. Tubercles absent from upper leaf-
    surface   **6. bosscheana**
3a. Leaves bright, fresh or pale green   4
 b. Leaves greyish or glaucous   5
4a. Leaf-margins without teeth, or rarely
    teeth 1–3 per side   **4. subintegra**
 b. Leaf-margins with 3–5 teeth per side
                             5. albidens
5a. Leaf-margins with 3–4 teeth midway
    along each side; leaves grey-green
    with grey dots   **2. britteniae**
 b. Leaf-margins with 5-10 evenly
    distributed teeth on each side; leaves
    glaucous or grey-green with white
    dots   6
6a. Teeth ending in reddish bristles;
    leaves glaucous with scattered white
    dots   **1. kingiae**
 b. Teeth ending in whitish bristles;
    leaves grey-green with white dots
    arranged in rows   **7. tigrina**

1. F. kingiae Bolus.
Leaves lanceolate, 4.5–5 cm long, to
1.5 cm wide at the middle, to 1 cm wide
towards the tip, to 1.2 cm thick, lower
surface keeled, margins with 5–10 teeth on
each side, each tooth tipped with a red
bristle to 9 mm long, glaucous, densely
covered with small white dots. Flowers to
6 cm in diameter, petals yellow. *South
Africa (Cape Province)*. G1.

2. F. britteniae Bolus.
Leaves crowded, diamond-shaped to ovate,
to 3.5 cm long, to 2 cm wide at the middle,
tapering to 1.5 cm wide at the base,
sharply and obliquely keeled beneath, the
tip drawn forward, grey-green covered with
grey dots, the margins and the keel (partly)
bordered with a whitish or reddish horny
band to 1 mm wide, the margins with 3–4
teeth midway along each side. *South Africa
(Cape Province)*. G1.

3. F. albidens N.E. Brown. Illustration:
Jacobsen, Handbook of succulent plants,
f. 1344 (1960).
Leaves 5–6 pairs, to 2.5 cm long, to 1 cm
or more wide at the middle, 6–7 mm thick,
triangular, long-tapering, keeled towards
the apex beneath, surfaces smooth and
shiny, fresh green with small, scattered,
white dots, margins with 3–5 teeth on
each side, each tooth arising from a

whitish, cartilaginous projection and ending in a whitish bristle, the keel with a horny, whitish edge. Flowers 3–4 cm in diameter, petals golden yellow. *South Africa (Cape Province)*. G1.

4. F. subintegra Bolus.
Leaves 2–2.5 cm long, to 1.3 cm wide and 1 cm thick, ovoid above, blunt, almost square in section towards the base, the lower surface with a rounded keel above, margins usually without teeth, occasionally with 1–3 teeth per side, teeth very rarely tipped with bristles. Flowers to 4.5 cm in diameter, petals golden yellow. *South Africa (Cape Province)*. G1.

5. F. tuberculosa (Rolfe) Schwantes.
Illustration: Jacobsen, Handbook of succulent plants, f. 1353 (1960); Rowley, The illustrated encyclopedia of succulents, 140 (1978).
Leaves very thick, to 2 cm long and 1.6 cm wide, dark green, the upper surface with several tooth-like tubercles, the margins with up to 3 stout teeth and a few rudimentary teeth on each side. Flowers to 4 cm in diameter, petals yellow. *South Africa (Cape Province)*. G1.

6. F. bosscheana (Berger) Schwantes.
Illustration: Jacobsen, Handbook of succulent plants, f. 1345 (1960).
Leaves narrowly lanceolate to narrowly diamond-shaped, to 3 cm long and 1 cm wide, glossy green, the margins somewhat white-cartilaginous and with 2–3 rather irregularly distributed teeth on each side, each tooth 2–3 mm long. Flowers 3–3.5 cm in diameter, petals shiny golden yellow. *South Africa (Cape Province)*. G1.

Var. **haagei** (Tischer) Jacobsen (*F. haagei* Tischer) has larger leaves with margins and keel white-cartilaginous, somewhat wavy, with fewer teeth. Illustration: Riha & Subik, The illustrated encyclopedia of cacti and other succulents, pl. 392 (1981).

7. F. tigrina (Haworth) Schwantes.
Illustration: Flowering plants of Africa 7: pl. 267 (1927); Jacobsen, Handbook of succulent plants, f. 1352 (1960); Herre, The genera of the Mesembryanthemaceae, 157 (1971); Barkhuizen, Succulents of southern Africa, 126 (1978).
Leaves crowded, ovate to diamond-shaped, 3–5 cm long, 1.6–2.5 cm wide, keeled beneath towards the apex, the tip drawn forwards, grey-green with numerous white dots aligned in rows, margins with 9 or 10 teeth on each side. Flowers to 5 cm in diameter, petals golden yellow. *South Africa (Cape Province)*. G1.

54. MALEPHORA N.E. Brown
F. McIntosh
Erect or creeping branched shrubs with distinct, visible internodes. Leaves triangular or half-cylindric in section, smooth, often with a bluish waxy coating. Flowers solitary, terminal or axillary, stalked, without bracts. Sepals 4–6. Petals yellow to red. Nectary glands forming a continuous ring. Ovary 8–11-celled, placentation parietal; stigmas 8–11, broadly triangular, shorter than ovary-depth. Valves winged, cell-roofs present, tubercles absent or very small. Seeds flat, very rough with tubercles in rows.

A genus of 15 species from South Africa and South West Africa.

1a. Flowers to 6 cm in diameter; leaves to 1.4 cm wide **1. crassa**
  b. Flowers to 3 cm in diameter; leaves to 6 mm wide **2**
2a. Petals bright red **2. crocea** var. **purpureocrocea**
  b. Petals yellow to orange **3. lutea**

1. M. crassa (Bolus) Jacobsen & Schwantes. Stem prostrate, internodes 2.5–4 cm. Leaves erect, keeled, thick, green or pale green, reddish towards the apex, to 4 cm long, 1.3 cm wide, 1.4 mm thick. Flowers shortly stalked, to 6 cm in diameter, petals golden yellow. *South Africa (Cape Province)*. G1.

2. M. crocea (Jacquin) Schwantes var. **purpureocrocea** (Haworth) Jacobsen & Schwantes (*Mesembryanthemum purpureocroceum* Haworth; *Hymenocyclus purpureocroceus* (Haworth) Schwantes). Stems very branched, leaves crowded on short shoots. Leaves 2.5–4.5 cm × 6 mm wide, slightly 3-angled, compressed, pale green, bloomed. Flower-stalk 2.5–4.5 cm. Flowers solitary, terminal, to 3 cm in diameter, petals intense red. *South Africa (Cape Province)*. G1.

Var. **crocea** has golden yellow petals which are slightly reddish beneath, but is not in general cultivation.

3. M. lutea (Haworth) Schwantes (*Hymenocyclus luteus* (Haworth) Schwantes). Illustration: Jacobsen, Handbook of succulent plants, f. 1502 (1960); Herre, The genera of the Mesembryanthemaceae, 201 (1971); Japan Succulent Society, Colour encyclopedia of succulents, pl. 782 (1981).
Branches bearing numerous short shoots. Leaves spreading, 2.5–4.5 cm long, to 4 mm wide, narrowed towards the apex, 3-angled in section, yellowish green,

whitish-bloomed. Flower-stalks 2.5–3.5 cm. Flowers terminal, to 2.5 cm in diameter, petals orange to yellow. *South Africa (Cape Province)*. G1.

55. GLOTTIPHYLLUM (Haworth) N.E. Brown
J. Cullen & F. McIntosh
Perennial, tuft-forming, very succulent plants, branching at ground level, internodes not visible. Each shoot with 4 or more leaves which are often in unequal pairs, frequently asymmetric and lying obliquely, in 2 ranks or with those of alternate pairs crossing one another; tips of the leaves of a pair frequently differing from each other, the upper surface often bulging towards the base; all green or pale green or brownish green, sometimes with clear dots visible. Flowers solitary, stalked or stalkless, bracts absent. Sepals 4, often angular, united into a short tube below. Petals more or less free, yellow. Ovary 8–20-celled, placentation parietal; stigmas 8–20, broadly ovate, shorter than depth of ovary-cells. Cell-roofs and a large tubercle present. Seeds small, ovoid.

A genus of about 55 species from South Africa.

1a. Leaves distinctly keeled 2
  b. Leaves not keeled 7
2a. Flowers not stalked 3
  b. FLowers distinctly stalked 6
3a. Leaves less than 1.5 cm wide; flowers not fragrant **1. parvifolium**
  b. Leaves more than 1.5 cm wide; flowers fragrant 4
4a. Leaves 3–5 cm long **2. suave**
  b. Leaves 6–8 cm long 5
5a. Keels ciliate towards the base **3. herrei**
  b. Keels not ciliate towards the base **4. fragrans**
6a. Leaves to 1.4 cm wide, without tooth-like projections on the margins **5. davisii**
  b. Leaves to 6 mm wide, with tooth-like projections on the margins **6. semicylindricum**
7a. Leaves with hooked tips 8
  b. Leaves without hooked tips 9
8a. Leaves pale green, 4–5 cm; flowers to 4 cm in diameter **7. nelii**
  b. Leaves grey-green, 6.5–9 cm; flowers 6.5–7 cm in diameter **8. muirii**
9a. Young leaves ciliate **9. praepingue**
  b. Young leaves not ciliate 10
10a. Leaves not clearly in 2 ranks **10. longum**
  b. Leaves clearly in 2 ranks 11

11a. Leaves of each pair distinctly differing, both covered with dots
11. oligocarpum

b. Leaves of each pair more or less similar, not covered with dots
12. linguiforme

1. G. parvifolium Bolus. Illustration: Jacobsen, Handbook of succulent plants, f. 1394 (1960).
Leaves not clearly in 2 ranks, 3–4 cm long × 1–1.2 cm wide and thick, with an acute apex terminated by a short spine, rather roundly keeled towards the apex on the lower surface, green, without visible dots. Flowers stalkless, to 8 cm in diameter. Petals glossy golden yellow. *South Africa (Cape Province)*. G1.

2. G. suave N.E. Brown.
Leaves 3–5 cm long × 1.5–2 cm wide and 6–8 mm thick, one leaf of each pair obliquely keeled on the lower surface, fleshy, grass green, not clearly in 2 ranks, without visible dots. Flowers stalkless, to 6.5 cm in diameter, fragrant. Petals glossy, pale yellow. *South Africa (Cape Province)*. G1.

3. G. herrei Bolus.
Leaves to 6 cm long, to 2.2 cm wide and 1 cm thick, more or less in 2 distinct ranks, keeled towards the apex on the lower surface, the keel finely ciliate towards the base, all bluish green to reddish. Flower stalkless, to 7 cm in diameter, fragrant. Petals yellow. *South Africa (Cape Province)*. G1.

4. G. fragrans (Salm-Dyck) Schwantes. Illustration: Jacobsen, Handbook of succulent plants, f. 1384 (1960).
Leaves 6–8 cm long, to 2.5 cm wide and 1.2 cm thick, not in 2 distinct ranks, keeled towards the apex. Flowers stalkless, 8–10 cm in diameter, fragrant. Petals yellow. *South Africa (Cape Province)*. G1.

5. G. davisii Bolus.
Leaves in 2 distinct ranks, to 3.5 cm long, 1.4 cm wide, 8 mm thick, keeled, yellowish green or green. Flower-stalk to 1.7 cm. Flowers 5–6 cm in diameter. Petals yellow. *South Africa (Cape Province)*. G1.

6. G. semicylindricum (Haworth) N.E. Brown. Illustration: Jacobsen, Handbook of succulent plants, f. 1398 (1960).
Leaves 4–5 cm long, 5–6 mm wide and thick, slightly bent inwards, keeled towards the apex, with small, tooth-like projections on the margins towards the middle, fresh, glossy green, with numerous slightly translucent dots. Flower-stalks to 2.5 cm.

Flowers to 4 cm in diameter, petals golden yellow. *South Africa (Cape Province)*. G1.

7. G. nelii Schwantes. Illustration: Jacobsen, Handbook of succulent plants, f. 1391 (1960); Barkhuizen, Succulents of southern Africa, 132 (1978); Court, Succulent flora of southern Africa, pl. 93 (1981); Japan Succulent Society, Colour encyclopedia of succulents, pl. 686 (1981).
Leaves very clearly arranged in 2 ranks, in unequal pairs, 4–5 cm long, to 2 cm wide and 1.2 cm thick, the longer leaf flat on the upper surface and with a hook-like tip, the shorter leaf somewhat bent upwards and with rounded margins; all pale green. Flowers stalkless, to 4 cm in diameter. Petals golden yellow. *South Africa (Cape Province)*. G1.

8. G. muirii N.E. Brown. Illustration: Jacobsen, Handbook of succulent plants, f. 1389 (1960).
Leaves clearly in 2 ranks, 6.5–9 cm long, 1.5–2 cm wide, 1–1.1 cm thick, strap-shaped, curving upwards at the tip to a blunt hook, rounded on the back, grass green. Flowers stalkless, 6.5–7 cm in diameter. Petals pale yellow. *South Africa (Cape Province)*. G1.

9. G. praepingue (Haworth) N.E. Brown. Illustration: Rowley, The illustrated encyclopaedia of succulents, 141 (1978).
Leaves more or less in 2 ranks, incurved, 4–7 cm long, 1.2–1.5 cm wide, half-cylindric in section, the tip curved upwards, light green, with fine hairs at the margins when young. Flowers shortly stalked, to 5 cm in diameter. Petals yellow. *South Africa (Cape Province)*. G1.

10. G. longum (Haworth) N.E. Brown.
Leaves not in 2 ranks, erect, tongue-shaped, 7–10 cm long, to 2 cm wide, upper surface flat, apex blunt, all dark green and without visible dots. Flowers 6–8 cm in diameter on stalks 7–8 cm long. Petals golden yellow. *South Africa (Cape Province)*. G1.

11. G. oligocarpum Bolus. Illustration: Jacobsen, Handbook of succulent plants, f. 1393 (1960).
Leaves clearly in 2 ranks, those of each pair clearly unequal, the longer flat with a blunt apex, the smaller rounded, 4–4.5 cm long to 2.2 cm wide and 1 cm thick, all whitish grey-green with conspicuous dots which are crowded towards the apex. Flowers 5–6 cm in diameter, on stalks to 1.2 cm long. Petals yellow. *South Africa (Cape Province)*. G1.

12. G. linguiforme (Linnaeus) N.E. Brown. Illustration: Jacobsen, Handbook of succulent plants, f. 1387 (1960); Lamb & Lamb, The illustrated reference on cacti and other succulents 3: 820 (1963); Barkhuizen, Succulents of southern Africa, 132 (1978); Japan Succulent Society, Colour encyclopedia of succulents, pl. 685 (1981).
Leaves distinctly in 2 ranks, 5–6 cm long, 3–4 cm wide, slightly curved upwards, fresh green, without dots. Flowers shortly stalked, 5–7 cm in diameter. Petals golden yellow. *South Africa (Cape Province)*. G1.

The most widely cultivated of the species, though there has been some suggestion that plants in cultivation under this name may be hybrids.

56. DOROTHEANTHUS Schwantes
F. McIntosh
Much-branched annual plants, the branches often reddish and markedly papillose. Leaves at first in rosettes, later opposite or alternate, flat, narrowed to the base, entire, soft, usually spathulate. Flowers terminal or axillary, stalked, without bracts. Sepals 5, unequal, covered with clear papillae. Petals in 2 whorls. Anthers brown or black. Ovary 5-celled, placentation parietal; stigmas 5, hardened and persistent in fruit. Cell-roofs present; tubercles absent. Seeds pear-shaped, smooth.

A genus of about 10 species from South Africa (Cape Province).

1a. Leaves linear, 3–5 mm wide
1. gramineus
b. Leaves spathulate or oblanceolate, more than 5 mm wide   2
2a. Petals white below, purplish pink above; stamens and stigmas purplish black, giving the flower a 3-coloured effect   **2. tricolor**
b. Flowers not as above   3
3a. Flowers 3–4 cm in diameter, petals obtuse   **3. bellidiformis**
b. Flowers to 6 cm in diameter, petals acute   **4. hallii**

1. D. gramineus (Haworth) Schwantes. Illustration: Jacobsen, Lexicon of succulent plants, pl. 161 (1974).
Plant to 10 cm, branching from the base, branches reddened. Leaves united at the base, linear, 3–5 cm × 3–5 mm, upper surface flat, lower surface rounded and very papillose, all fresh green. Flowers terminal, 2–2.5 cm in diameter, on stalks 3–6 cm long. Petals brilliant crimson with darker bases. *South Africa (Cape Province)*. G1.

2. D. tricolor (Willdenow) Bolus.
Leaves to 1.2 cm wide, spathulate, blunt to slightly pointed. Petals white in the lower half, purplish pink in the upper half, the stamens and stigmas purplish black giving the flower a 3-coloured effect. *South Africa (Cape Province).* G1.

3. D. bellidiformis (Burman filius) N.E. Brown. Illustration: The Flowering Plants of Africa **35**: pl. 1365 (1962); Herre, The genera of the Mesembryanthemaceae, 139 (1971).
Plant branching from the base. Leaves usually basal, alternate, spathulate, 2.5–7 cm × 6–10 mm, fleshy, roughly papillose, narrowed towards the base. Flowers 3–4 cm in diameter, petals obtuse, white, pale pink, red, orange or white with red or pink tips. *South Africa (Cape Province).* H5–G1.

A widely cultivated annual, often used for bedding schemes.

4. D. hallii Bolus.
Leaves narrowly spathulate to oblanceolate, 3.5–6.5 cm × 5–10 mm. Flowers to 6 cm in diameter, petals acute, white to pale straw-coloured, each edged with red at the base. *South Africa (Cape Province).* G1.

57. PHERELOBUS N.E. Brown
F. McIntosh
Annual plants, 5–10 cm, with decumbent branches and long internodes. Leaves flat, spathulate, entire, upper surface convex, papillose. Flowers solitary, terminal, stalked, to 5.5 cm in diameter. Petals pale violet or pale yellow fading to whitish, darker at the base. Stamens numerous, anthers red or purple. Ovary 5-celled, placentation parietal; stigmas 5. Capsule 5-celled; cell-roofs present, tubercles absent. Seeds rough with prominent papillae, brown.

A genus of a single species from South Africa.

1. P. maughanii N.E. Brown. Illustration: Herre, The genera of the Mesembryanthemaceae, 249 (1971); Court, Succulent flora of southern Africa, 93 (1981). *South Africa (Cape Province).* G1.

Two varieties occur: var. **maughanii** has petals pale violet, reddish purple at the base, and var. **stayneri** Bolus with petals pale yellow fading to whitish, the base red with a small orange line just above the red patch.

58. MITROPHYLLUM Schwantes
F. McIntosh
Sparingly branched perennial herbs with visible internodes, the branches covered by dry leaf-remains. Leaves of two forms: some larger, free for most of their length, the upper surface more or less flat, the others smaller, united into bodies for most of their length with only the tips free; these two types of leaf-pairs alternate, though the alternation may sometimes be irregular. The flowers are subtended by leaf-like bracts which are different in form from the leaves, thus giving a plant in flower the appearance of having leaves of three forms. Flowers solitary, terminal, stalked or not. Sepals 5, almost equal. Petals in 3–4 whorls, white, pink or yellow. Nectaries forming a continuous ring. Ovary 5–7-celled; placentation parietal. Stigmas 5–7. Cell-roofs present but tubercles absent. Seeds ovoid or pear-shaped.

A genus of 6 species from South Africa (Cape Province).
Literature: Poppendieck, H.-H., Untersuchungen zur Morphologie und Taxonomie der Gattung Mitrophyllum Schwantes s. lat., *Botanische Jahrbücher* **97**: 339–413 (1976).

1a. Internode below the bracts absent
                                  1. mitratum
  b. Internode below the bracts at least 3 mm, often much more   2
2a. Anthers golden yellow; bodies more than 4.5 cm long × 1.5 cm thick
                                     2. grande
  b. Anthers whitish; bodies less than 4.5 cm long × 1.5 cm thick
                                  3. clivorum

1. M. mitratum (Marloth) Schwantes. Illustration: Jacobsen, Handbook of succulent plants, f. 1163, 1183, 1514, 1515, 1516 (1960); Botanische Jahrbücher **97**: t. 2, f. 5 (1976).
Tuft- or mat-forming plants with thick, soft stems that become woody with age, the nodes thickened, forming expanded rings. Bodies formed from shorter leaves 7–8 cm or more long, to 2 cm thick, the bluntly triangular tips free for 1–1.2 cm, pale green; large leaves 8–10 cm long, to 1 cm thick at the base, undersurface with a rounded keel, all shining glossy green papillose when young. Flower-stalks to 1 cm, internode below the bracts absent. Flower 2.5–3 cm in diameter, petals white with reddish tips. *South Africa (Cape Province).* G1.

2. M. grande N.E. Brown. Illustration: Jacobsen, Lexicon of succulent plants, pl. 185.3 (1974).
Similar to *M. mitratum*: flowering branches 2-angled, internode beneath the bract at least 3 mm long and often more; bodies more than 4.5 cm long and 1.5 cm thick; large leaves to 20 × 4–5 cm wide at the base. Flowers 4–4.5 cm in diameter, petals glossy white, anthers golden yellow. *South Africa (Cape Province).* G1.

3. M. clivorum (N.E. Brown) Schwantes (*M. framesii* Bolus). Illustration: Botanische Jahrbücher **97**: t. 2, f. 4 (1976).
Plant 25–30 cm tall, stems 6–10 mm in diameter, nodes mostly thickened, internodes 1–3 cm long, reddish brown, later pale grey. Bodies less than 4.5 cm long and 1.5 cm thick. Longer leaves cylindric, 2–3 cm long × 3.5–6 mm thick, pale green, glossy, papillose. Internode below the bracts at least 3 mm, often more. Petals white, anthers whitish. *South Africa (Cape Province).* G1.

59. MONILARIA (Schwantes) Schwantes
F. McIntosh
Small clump-forming perennials with crowded, erect branches which are constricted at the nodes, the internodes not visible, covered by leaf-sheaths. Leaves opposite, papillose, of 2 forms, one pair of each kind produced annually. The leaves of the first pair are rudimentary, consisting of scarcely more than a fleshy sheath; those of the second pair are borne between the first pair and are longer and more or less cylindric but somewhat flattened on the upper surface; sheaths persistent. Flowers solitary, terminal, stalked. Sepals 5, unequal, papillose. Petals in 3 or 4 whorls, white, yellow or violet-red. Nectaries forming a continuous ring. Ovary 5–7-celled, placentation parietal; stigmas 5–7, broadly ovate, shorter than depth of ovary-cells. Cell-roofs present, tubercles absent. Seeds each pointed at one end.

A genus of 6 species from South Africa.
Literature: Ihlenfeldt, H.-D. & Jorgensen, S., Morphologie und Taxonomie der Gattung Monilaria (Schwantes) Schwantes, s. str., *Mitteilungen der Staatsinstitut für allgemeine Botanik, Hamburg,* **14**: 49–94 (1973).

1a. Petals yellowish or reddish, at least at the base   **3. pisiformis**
  b. Petals white throughout   2
2a. Joints of the stem 2 times broader than long   **1. moniliformis**
  b. Joints of the stem longer than broad
                                  2. chrysoleuca

1. M. moniliformis (Thunberg) Ihlenfeldt & Jorgensen (*Conophyllum moniliforme* (Haworth) Schwantes). Illustration: Gardeners' Chronicle **86**: 32 (1929). Plant 7.5–10 cm, stem and branches 8–11 mm thick, internodes twice as wide as long, constricted. Expanded leaves 10–15 cm × 4–5 mm, united at the base for *c.* 1.2 cm, this united part more or less concealed in the sheath of the smaller leaves, half-cylindric in section, blunt, papillose when young. Flower-stalks to 5 cm. Flowers 3–5 cm in diameter, petals white, stamens deep yellow. *South Africa (Cape Province)*. G1.

2. M. chrysoleuca (Schlechter) Schwantes. Plant 6–10 cm tall, internodes longer than broad. Expanded leaves erect, cylindric, blunt, finely papillose, 3–7.5 cm × 4–5 mm. Flower-stalks 3.5–7 cm. Flowers to 3 cm in diameter, petals white, stamens yellow. *South Africa (Cape Province)*. G1.

The above description refers to var. **chrysoleuca**; var. *polita* (Bolus) Ihlenfeldt & Jorgensen is not cultivated.

3. M. pisiformis (Haworth) Schwantes. Illustration: Gardeners' Chronicle **86**: 33 (1929).

Stem very branched with short branches, joints longer than broad. Expanded leaves glossy papillose when young, cylindric, 5–6 cm long × up to 4 mm in diameter. Flower *c.* 3 cm in diameter. Petals yellow, the base reddish with a white margin. *South Africa (Cape Province)*. G1.

60. MEYEROPHYTUM Schwantes
F. McIntosh

Small, much-branched perennials, branches slender, with visible internodes. Leaves dark green, yellowish green or reddish, in differing pairs, the lower leaves thicker than the upper and united for half or more of their length, the upper leaves united for not more than a third of their length. Flowers solitary, terminal, stalked. Sepals 5, 3 of them with membranous margins. Petals in 3–4 whorls, linear, bright red. Nectaries forming a continuous ring. Ovary 5–7-celled, placentation parietal; stigmas 5–7. Cell-roofs present, tubercles absent. Seeds ovoid or pear-shaped.

A genus of 4 species from South Africa.

1. M. meyeri (Schwantes) Schwantes (*Mitrophyllum meyeri* Schwantes). Dwarf plant forming clumps. Leaf-pairs forming bodies of two types, the shorter round with the leaves scarcely separated,

the longer to 1 cm × 4 mm, the leaves separated for 2–3 mm, all deep yellowish green, becoming reddish with age. Flowers shortly stalked, petals fiery red. *South Africa (Cape Province)*. G1.

61. CARPANTHEA N.E. Brown
F. McIntosh

Annual plants with stems with distinct internodes. Leaves opposite on the lower parts of the flowering branches, united at the base, flat, spathulate to lanceolate, margins hairy. Flowers 1–3, terminal, stalked. Sepals 5, unequal, 2 leaf-like. Petals in several series, yellow. Ovary 12–20-celled, placentation parietal; stigmas 12–20, thread-like. Seeds small, round.

A genus of 2 species from South Africa.

1. C. pomeridianum (Linnaeus) N.E. Brown. Illustration: Herre, The genera of the Mesembryanthemaceae, 101 (1971); Succulenta **60**: 66–67 (1981). Plant erect, branched, to 30 cm high, stem, inflorescences and calyx covered with shaggy, white hairs. Leaves 4–10 × 1.2–2.5 cm wide, with minute hairs at the margins. Flowers 4–7 cm in diameter, on stalks 3–10 cm long. Petals golden yellow. *South Africa (Cape Province)*. G1.

62. STOEBERIA Dinter & Schwantes
F. McIntosh

Perennial, somewhat shrubby plants with visible internodes. Leaves opposite, free or slightly joined at the base, boat-shaped, almost flat above, keeled towards the apex beneath. Flowers in a much-branched, terminal inflorescence, each flower-stalk subtended by 2 bracts. Sepals 5 or 6, triangular, pale pink. Petals in a single series (or absent in some species which are not cultivated), white or red. Nectaries forming a continuous ring. Ovary 5–6-celled, placentation parietal; stigmas 5 or 6, pinnatifid. Large tubercles present. The capsule, once opened, does not close again on drying. Seeds pear-shaped, rough.

A genus of 2 species from South Africa and Southwest Africa.

1a.   Flowers 7–14 mm in diameter; leaves 5–7 mm broad   **1. beetzii**
  b.   Flowers 1.5–2.2 cm in diameter; leaves 8–15 mm broad   **2. carpii**

1. S. beetzii (Dinter) Dinter & Schwantes. Plant much-branched, spreading, to 50 cm high, internodes smooth, 2.2–5 cm long. Leaves greyish green with small, darker dots, club-shaped, 1–3.5 cm × 5–7 mm broad × 5–8 mm thick. Flower 7–14 mm in

diameter, petals white. *Southwest Africa*. G1.

2. S. carpii Friedrich. Plant to 1 m, stem spreading, angular, internodes 1–9 cm long. Leaves more or less club-shaped, 2–6 cm × 8–15 mm wide × 8–20 mm thick, soft, whitish grey. Flower 1.5–2.2 cm in diameter, petals white. *Southwest Africa*. G1.

63. AMPHIBOLIA Bolus
F. McIntosh

Perennial plants; branches with distinct, visible internodes. Leaves erect or spreading, acute or blunt, cylindric or triangular in section, bluish green, the younger ones reddish. Flowers solitary or 1–3 together, shortly stalked, with bracts. Sepals 5, almost equal. Petals in 2 series, pink. Nectaries forming a continuous ring. Ovary 5-celled, cup-shaped, placentation parietal; stigmas 5, subulate. Cell-roofs present, tubercles small. Seeds round, brown.

A genus of about 5 species from South Africa.

1. A. littlewoodii (Bolus) Bolus (*Stoeberia littlewoodii* Bolus). Plant erect, to 30 cm. Branches rigid, internodes 1.5–2.5 cm. Leaves tapering towards the tip, 1–2.2 cm × 3–4 mm wide and thick. Flowers to 4.5 cm in diameter. Petals purplish pink. *South Africa (Cape Province)*. G1.

64. CONICOSIA N.E. Brown
F. McIntosh

Perennial or rarely annual or biennial plants with erect stems; internodes visible. Leaves forming a dense rosette, opposite, slightly stem-clasping at the base, erect except when old, long and narrow, either sharply 3-angled or almost circular in section, slightly channelled on the upper surface, persisting on the stem after withering, dotted with pale or dark dots. Flowers solitary, terminal or axillary, stalked, without bracts, often unpleasantly scented. Sepals 5, unequal, broadly based. Petals in several whorls, often ciliate towards their bases, yellow. Ovary 10–20-celled, placentation parietal; stigmas 10–20, thread-like, shorter than ovary-depth. Cell-roofs and tubercles absent. Seeds large, usually smooth.

A genus of 10 species from South Africa.

1a.   Leaves to 40 cm; petals pale yellow   **1. capensis**
  b.   Leaves to 20 cm; petals bright yellow   **2. pugioniformis**

1. C. capensis (Haworth) N.E. Brown.
Plant lasting for not much more than 2–3
years. Stem 15 cm or more. Leaves in a tuft
at the apex of the stem, to 40 cm long,
compressed, 3-angled, slightly grooved
above and along the sides, bluish. Flower-
stalks 12–15 cm, angled and rough. Flower
to 7.5 cm in diameter, petals pale yellow.
South Africa (Cape Province). G1.

2. C. pugioniformis (Linnaeus) N.E. Brown.
Illustration: Jacobsen, A handbook of
succulent plants, f. 1248, 1249 (1960);
Herre, The genera of the Mesembryanthemaceae, 117 (1971).
Stem erect, simple or sparingly branched,
15–30 cm high, 1–2 cm thick. Leaves
3-angled, upper side deeply grooved, all
greyish green, reddish at the base, 15–20 ×
1.2 cm. Flower-stalks 10–14 cm. Flowers
1–3 together from lateral branches, to
7 cm in diameter. Petals bright, glossy
sulphur yellow. *South Africa (Cape
Province)*. G1.

65. HERREA Schwantes
F. McIntosh
Perennial plants with long, prostrate
branches, internodes distinct and visible;
annual growth dying off after fruiting.
Leaves opposite or alternate on the same
branch, circular or semicircular in cross-
section, acute or blunt, rarely somewhat
keeled. Flowers stalked, opening in the
afternoon. Sepals 5 (rarely 4), flattened
below. Petals in 4 or 5 series, spreading in
different planes, pale to golden yellow or
white above, dull pink beneath. Nectaries
inconspicuous, forming a continuous ring.
Ovary 12–24-celled, placentation parietal;
stigmas 12–24, thread-like, as long as
ovary-depth. Cell-roofs and tubercles
absent. Fruit breaking up into separate
segments when mature. Seeds lens-shaped,
brown.
A genus of 24 species from South Africa.

1. H. grandis Bolus.
Leaves blunt, rounded beneath, to
1.6 cm × 5–7 mm wide × 4–5 mm thick.
Flower-stalks stiff, to 14 cm. Flower to
10 cm in diameter, petals yellow with pink
tips above, pale pink beneath.
South Africa (Cape Province). G1.

66. CARPOBROTUS N.E. Brown
F. McIntosh
Small perennial herbs with long,
2–3-angled, stout, trailing branches.
Leaves opposite, thick, 3-angled in section,
margins entire or toothed, smooth, grey-

green with small translucent spots and
often swollen below at the base. Flowers
solitary, terminal, stalked. Sepals 5. Petals
numerous, purple, red, pink, yellow or
white. Ovary 10–16-celled, placentation
parietal; stigmas 10–16. Fruit fleshy and
indehiscent, without valves. Seeds slightly
compressed.
A genus of 23 species from South Africa,
of which a few have become naturalised in
various parts of Europe. The fruits of some
species are edible.

1a. Flowers to 12 cm in diameter; leaves
    broadest at or above the middle
                        1. acinaciformis
  b. Flowers to 8.5 cm in diameter; leaves
    broadest at the base   2
2a. Petals pinkish purple; leaves greyish
    green   **2. deliciosus**
  b. Petals yellow at first, later pink or
    purple; leaves dull green   **3. edulis**

1. C. acinaciformis (Linnaeus) Bolus.
Illustration: Fenaroli, Flora Mediterranea 1:
72 (1962); Taylor, Wild flowers of Spain
and Portugal, 39 (1974); Jacobsen, Lexicon
of succulent plants, pl. 149 (1974); Huxley
& Taylor, Flowers of Greece and the
Aegean, pl. 20 (1977).
Stems to 1.5 m long, with short lateral
branches. Leaves curved, greyish green, to
9 × 1 cm broad × 1.5–2 cm thick, broadest
at or above the middle, angles
cartilaginous, entire or slightly wavy and
rough. Flowers to 12 cm in diameter, petals
bright crimson purple. Stamens purple.
South Africa (Cape Province). H5–G1.

2. C. deliciosus (Bolus) Bolus. Illustration:
Fenaroli, Flora Mediterranea 1: 72 (1962);
Herre, The genera of the Mesembryanthemaceae, 103 (1971); Huxley & Taylor,
Flowers of Greece and the Aegean, pl. 21
(1977).
Stems long, creeping, branches bearing
8–10 leaves. Leaves curved, widest at the
base, to 11 × 1.6 cm thick, dark greyish
green, keel slightly horny and toothed.
Flower-stalks 7–9 cm. Flowers 7–8 cm in
diameter, petals pinkish purple. *South Africa
(Cape Province)*. H5–G1.

3. C. edulis (Linnaeus) Bolus.
Branches to 1 m, angled, 8–13 mm thick.
Leaves curved inwards, widest at the base,
4–12 cm × 8–17 mm wide and thick, dull
grass green, keel minutely toothed. Flowers
to 8.5 cm in diameter, stalked, petals
yellow at first, becoming pale pink to purple
later. Stamens yellow. *South Africa (Cape
Province)*. H5–G1.

LII. PORTULACACEAE
Herbs or rarely shrubs or trees, often fleshy.
Leaves alternate, opposite or all basal,
entire, with or without stipules, usually
somewhat fleshy (leaves sometimes very
reduced and hidden between the enlarged,
parchment-like stipules). Flowers bisexual,
usually radially symmetric, in racemes or
cymes (often condensed and cluster-like),
rarely solitary; bracts often conspicuous.
Sepals usually 2, free or united at base,
persistent or deciduous. Petals 3–18,
usually 5, free or rarely slightly united at
the base. Stamens 3–many, anthers
opening by slits. Ovary superior or at least
partly inferior, of 2, 3 or rarely more
carpels, 1-celled, ovules 1–many,
placentation basal when ovule 1, free-
central when more. Fruit usually a capsule
opening by lobes at the apex or by a split
developing along the circumference, the top
falling like a lid. Seeds 1–many, with or
without appendages.
A cosmopolitan family of 19 genera (see
McNeill, J., Synopsis of a revised
classification of the Portulacaceae, *Taxon*
23: 725–728, 1974) and about 500
species, centred in South Africa and
America.

1a. Plant a soft-wooded shrub or tree;
     fruit indehiscent, 3-winged
                     2. Portulacaria
  b. Plant a herb; fruit dehiscent, not
     3-winged   2
2a. Ovary partly or wholly inferior
                       1. Portulaca
  b. Ovary fully superior   3
3a. Capsule splitting around its
     circumference, the top falling like a
     lid; petals usually more than 5
                     3. Lewisia
  b. Capsule splitting into lobes at the top;
     petals usually 5   4
4a. Stigmas 2; capsule opening by 2 lobes
                     7. Spraguea
  b. Stigmas 3; capsule opening by 3 lobes
     or by loss of the outer skin   5
5a. Ovules 3–6   6
  b. Ovules more than 6   7
6a. Stem leaves several, alternate or
     opposite; basal rosette absent
                     9. Montia
  b. Stem leaves 2, opposite; basal rosette
     usually present   **8. Claytonia**
7a. Sepals deciduous   **5. Talinum**
  b. Sepals persistent   8
8a. Stamens 3–14   **4. Calandrinia**
  b. Stamens 15–60   **6. Anacampseros**

1. PORTULACA Linnaeus

J. Cullen

Usually sprawling annual herbs. Leaves fleshy, flat or terete, alternate or some opposite. Flowers solitary, terminal, stalkless. Sepals 2. Petals 6 or 7. Stamens 7–many. Ovary at least partly inferior, 1-celled, with many ovules. Style 1, stigmas 3–9. Fruit a capsule opening by splitting around its circumference above the point of attachment of the sepals. Seeds many.

A genus of over 100 species mostly from the tropics. Propagation is by seed.

1a. Leaves flat (though fleshy); plant hairless   **1. oleracea**
  b. Leaves terete; plant hairy at the nodes and around the flowers
         2. grandiflora

1. P. oleracea Linnaeus. Illustration: Polunin, Flowers of Europe, pl. 11 (1969); Bichard & McClintock, Wild flowers of the Channel Islands, 76 (1975).
Erect or spreading hairless herb. Leaves flat though fleshy, obovate or spoon-shaped. to 3.5 cm, rounded at the apex, tapered to the base. Flowers 1.2–1.6 cm in diameter, yellow. Petals erect-spreading, notched. Stamens 11 or more. Stigmas 5–6. *Original distribution unknown, widely cultivated.* H2. Summer–autumn.

Most often seen as subsp. **sativa** (Haworth) Čelakovsky, described above, which is used as a salad. Double-flowered variants also occur.

2. P. grandiflora Hooker. Illustration: Botanical Magazine, 2885 (1829), 3084 (1831); Perry, Flowers of the world, 238 (1972); American Horticulturist **64**(6): 5 (1985).
Sprawling herb, hairy around the nodes and flowers. Leaves terete, to 2.5 cm. Flowers to 2.5 cm in diameter, yellow, pink, red or white, sometimes striped. Petals spreading, notched. *Brazil, Uruguay, naturalised in parts of C Europe.* H1. Summer–autumn.

2. PORTULACARIA Jacquin

J. Cullen

Succulent shrubs or small trees. Leaves deciduous, fleshy, opposite. Flowers in small clusters. Sepals 2, persistent. Petals 5, pink, persistent. Stamens 4–7. Ovary superior, 3-angled, ovule 1; stigma stalkless. Fruit indehiscent, 3-winged. 1-seeded.

A genus of 2 species from South Africa.

1. P. afra (Linnaeus) Jacquin. Illustration: Lamb & Lamb, The illustrated reference on cacti and other succulents **3**: 830 (1963); Journal of the Cactus & Succulent Society (US) **45**(2): 71 (1973).
Soft-wooded tree to 4 m, stems jointed. Leaves to 1.8 cm. Flowers to 2 mm in diameter, pink. *South Africa.* H5–G1.

3. LEWISIA Pursh

J. Cullen & H.S. Maxwell

Hairless perennial herbs with corms or fleshy roots and a crown which may be small or large. Leaves all or mostly basal, evergreen or deciduous. Scapes long or short, bearing 1–3 flowers or many-flowered panicles; bracts occasionally borne close to the sepals and similar to them. Sepals usually 2, occasionally to 8. Petals 4–18, often unequal in width, sometimes so in length. Stamens 5–many. Ovary 1-celled with 3–8 stigmas. Fruit a thin-walled capsule, splitting around its circumference near the base and also splitting into 3–8 flaps at the apex. Seeds numerous, sometimes with arils.

A genus of 16 species from western N America (extending into Mexico). They are prized as ornamentals for growing in walls or in pots in an alpine house, but hybridise freely and are difficult to maintain as separate species. Many hybrids are in cultivation and their parentage is often uncertain. Lewisias require a sunny position (though with partial shade), rich soil and extremely good drainage; the crowns tend to rot if they become too wet. Propagation is by division of the crowns or by seed (garden seed is often hybrid).
Literature: Elliott, R., *The genus Lewisia* (1977) – reprinted from *Bulletin of the Alpine Garden Society* 34 (1966).

1a. Petals 2.5 cm or more; seeds with conspicuous arils   **1. tweedyi**
  b. Petals at most 2 cm; seeds without arils   2
2a. Flowers in panicles on scapes longer than the basal leaves   3
  b. Flowers 1–3 on scapes not or little longer than the basal leaves   7
3a. Leaves distinctly toothed   4
  b. Leaves entire, margins sometimes wavy towards the base   5
4a. Leaves 1.7–5 cm, broadest at apex, gradually tapering below; petals *c.* 6 mm   **3. cantelovii**
  b. Leaves 6–12 cm, broadest below the somewhat acute apex, tapering rather abruptly below; petals 1–1.7 cm
         2. cotyledon var. **heckneri**
5a. Leaves terete   **4. leana**
  b. Leaves flat   6

6a. Petals 5–9 mm; leaves oblong-oblanceolate   **5. columbiana**
  b. Petals 1–1.7 cm; leaves broadly spoon-shaped   **2. cotyledon**
7a. Bracts borne immediately below the 2 sepals and similar to them, so there appear superficially to be 4 sepals
         8. brachycalyx
  b. Bracts remote from the sepals and not similar to them; sepals 2 or 4 or more
         8
8a. Sepals 4–8   **7. rediviva**
  b. Sepals 2   **6. pygmaea**

1. L. tweedyi (Gray) Robinson. Illustration: Botanical Magazine, 7633 (1899); Elliott, The genus Lewisia, 61, 70 (1977); Gartenpraxis for 1983: 17, for 1984: 60; Pacific Horticulture **46** (2): 48, 49 (1985).
Perennial with branching root and short, thick crown. Leaves broadly obovate, rounded or notched at the apex, tapered to a long, narrow stalk. Scapes several, equalling or exceeding the leaves, 1–3-flowered. Sepals 2. Petals 10–12, white, yellow or orange-yellow, 2.5–3 cm. Seeds with a conspicuous, scale-like aril. *Western N America (British Columbia, Washington).* H4–5. Late spring–summer.

2. L. cotyledon (Watson) Robinson. Illustration: Abrams & Ferris, Illustrated Flora of the Pacific states **2**: 136 (1944); Hay & Synge, The dictionary of garden plants in colour, pl. 101, 102 (1971); Elliott, The genus Lewisia, 20, 21 (1977); Pacific Horticulture **46**(2): 44, 45 (1985).
Perennial herb with thickened root and crown. Leaves flat, 6–12 cm, broadly spoon-shaped, broadest just below the somewhat acute apex (sometimes the apex with a fine point), margins entire, wavy or toothed. Scapes several, to 30 cm, flowers numerous in a panicle. Sepals 2. Petals 8–10, 1–1.7 cm, white tinged pink, becoming pink in age, veins often deep pink. *Western N America (British Columbia, Oregon, California).* H4. Spring–summer.

Var. **cotyledon** has entire leaves and is little cultivated. More frequently seen are var. **heckneri** (Morton) Munz with toothed leaves and var. **howellii** (Watson) Jepson with the leaf-margin wavy.

3. L. cantelovii Howell. Illustration: Bulletin of the Alpine Garden Society **40**: 17 (1972); Elliott, The genus Lewisia, 10 (1977).
Like *L. cotyledon* var. *heckneri* but leaves shorter (to 5 cm), broadest at the rounded or shallowly notched apex, gradually tapering from there, petals pink with deeper pink veins, *c.* 6 mm. *NW USA (California).* H5. Spring–summer.

4. L. leana (Porter) Robinson. Illustration: Abrams & Ferris, Illustrated Flora of the Pacific states **2**: 136 (1944); Elliott, The genus Lewisia, 28, 34 (1977).
Perennial with branched roots and thick, fleshy crown. Leaves many, fleshy, terete, acute, 1.5–3 cm. Scapes several, to 18 cm; flowers in panicles, numerous. Sepals 2. Petals magenta or white with pink veins, 5–6.5 mm. *NW USA (Oregon, California).* H5. Summer.

5. L. columbiana (Howell) Robinson. Illustration: Abrams & Ferris, Illustrated Flora of the Pacific states **2**: 136 (1944); Elliott, The genus Lewisia, 12, 15 (1977); Gartenpraxis for 1984: 60; Bulletin of the Alpine Garden Society **53**: 260 (1985).
Perennial with a thick root and short crown. Leaves 3–8 cm, flat, oblong-oblanceolate. Scapes several, 20–30 cm; flowers numerous in loose panicles. Sepals 2. Petals up to 10, bright pink or white with pink veins. *Western N America (British Columbia south to California).* H4–5. Spring–summer.

A variable species in which a number of varieties has been described. The plant in cultivation as *L. columbiana* 'Rosea' is of uncertain status.

6. L. pygmaea (Gray) Robinson.
Perennial with an unbranched, spindle-shaped root. Leaves linear, 3–6 × cm × 1.5–3 mm. Scapes not or scarcely longer than the leaves, bearing 1–3 flowers. Sepals 2. Petals 5–8, 6–10 mm, white or pink. *Western and southern USA.* H4. Summer.

Two varieties are grown:

Var. **pygmaea.** Illustration: Abrams & Ferris, Illustrated Flora of the Pacific states **2**: 136 (1944); Elliott, The genus Lewisia, 46, 49 (1977). Sepals with glandular teeth.

Var. **nevadensis** (Gray) Fosberg (*L. nevadensis* Gray). Illustration: Gartenpraxis for 1977: 345; Bulletin of the Alpine Garden Society **54**: 312 (1986). Sepals either without teeth, or teeth, if present, without glands.

7. L. rediviva Pursh. Illustration: Abrams & Ferris, Illustrated Flora of the Pacific states **2**: 136 (1944); Elliott, The genus Lewisia, 50, 52 (1977); Pacific Horticulture **46**(2): 40, 41 (1985).
Perennial with branching root and thick crown. Leaves broadly linear to narrowly club-shaped, thick, 2–5 cm. Scapes numerous, not exceeding the leaves, each 1-flowered. Sepals 4–8. Petals 12–18, 2–2.5 cm, pink or sometimes white. *Western N America.* H5. Spring–summer.

8. L. brachycalyx Gray. Illustration: Abrams & Ferris, Illustrated Flora of the Pacific states **2**: 140 (1944); Elliott, The genus Lewisia, 8, 9 (1977); Gartenpraxis for 1977: 345, for 1984: 60.
Perennial with large branching root and thickened crown. Leaves broadly oblanceolate, 3–6 cm. Scapes several, shorter than the leaves, 1-flowered. Bracts 2, borne just below the flower and closely resembling the 2 sepals so that superficially there appear to be 4 of these. Petals 5–9, white or white flushed with pink, 1.2–1.8 cm. *Western USA, Mexico.* H5. Spring–summer.

4. CALANDRINIA Humboldt, Bonpland & Kunth
J. Cullen & H.S. Maxwell
Annual or perennial herbs, more or less succulent. Leaves alternate or all basal. Flowers in racemes, panicles or umbel-like clusters, remaining open for 1 day or less. Sepals 2, persistent. Petals usually 5, red or purple, rarely white. Stamens 3–14. Ovary superior, style short, stigmas 3. Fruit a capsule opening by 3 flaps at the apex. Seeds numerous.

A genus of about 150 species mostly from S America and Australia, a few extending to western N America; only 3 are commonly grown. They are easy to grow in a sunny position and are often treated as annuals.

1a. Flowers in umbel-like clusters
　　　　　　　　　　　　3. umbellata
　b. Flowers in panicles or racemes　　2
2a. Leaves flat, thick, 10–20 cm, acute
　　　　　　　　　　　　1. grandiflora
　b. Leaves narrowly spoon-shaped to linear, less than 10 cm, obtuse
　　　　　　　　　2. ciliata var. **menziesii**

1. C. grandiflora Lindley. Illustration: Edwards's Botanical Register **14**: t. 1194 (1828); Botanical Magazine, 3369 (1834).
Perennial herb, often grown as an annual. Stems to 1 m. Leaves 10–20 cm, thick, flat, ovate, acute, abruptly narrowed into the stalk, mostly concentrated at the base of the stem. Flowers in a panicle, 2.5–5 cm in diameter, rose-purple. *Chile.* H3. Summer–autumn.

2. C. ciliata de Candolle var. **menziesii** (Hooker) Macbride (*C. speciosa* misapplied). Illustration: Edwards's Botanical Register **19**: t. 1598 (1833); Abrams & Ferris, Illustrated Flora of the Pacific states **2**: 120 (1944).
Annual herb. Stems many from the base, to 40 cm. Leaves narrowly spoon-shaped to

linear, tapering gradually into the stalk, obtuse, less than 10 cm, numerous and distributed evenly along the flowering stem. Flowers in a raceme, 1.2–1.8 cm in diameter, purple-red, rarely white. *Western N America.* H3. Summer.

3. C. umbellata de Candolle. Illustration: Flore des Serres, ser. 1, **2**: t. 5 (1846); Step, Favourite flowers of garden and greenhouse **1**: t. 44 (1896).
Perennial herb, often grown as an annual. Stem erect to 15 cm, reddish. Leaves mostly basal, linear, acute, hairy, 1.5–2 cm. Flowers in umbel-like clusters, numerous, bright red-violet, to 2 cm in diameter. *Chile, Peru.* H3. Summer–autumn.

5. TALINUM Adanson
J. Cullen & H.S. Maxwell
Rhizomatous perennial herbs sometimes woody at the base. Leaves mostly alternate, without stipules, flat or terete. Flowers solitary, axillary, or in cymes or panicles. Sepals 2, deciduous. Petals 5 or rarely 8–10, soon withering and falling. Stamens 5–many. Capsule opening by 3 apical flaps.

A genus of about 50 species from the tropics and subtropics. Only a few are grown and these generally require cool glasshouse or alpine house conditions. Propagation is by seed.
Literature: von Poellnitz, K., Monographie der Gattung Talinum Adans., *Feddes Repertorium* **35**: 1–34 (1934).

1a. Leaves flat, usually at least 8 mm broad, often much broader　　2
　b. Leaves semicircular to circular in section, rarely more than 8 mm broad
　　　　　　　　　　　　　　3
2a. Inflorescence a many-flowered panicle
　　　　　　　　　　　1. paniculatum
　b. Inflorescence 1–3-flowered
　　　　　　　　　　　　2. caffrum
3a. Main veins of the older leaves persisting as spines　**3. spinescens**
　b. Main veins of the older leaves not persisting as spines, sometimes persisting as curved bristles　　4
4a. Stamens 30 or more　**5. calycinum**
　b. Stamens 12–20　　　　　5
5a. Petals purple or bluish　**6. teretifolium**
　b. Petals white tinged with yellow or pink　　**4. okanoganense**

1. T. paniculatum (Jacquin) Gaertner.
Rhizome tuberous. Stems to 1 m, woody below. Leaves alternate or almost opposite, shortly stalked or stalkless, flat, oblanceolate, ovate or spoon-shaped, acute, to 10 × 5 cm. Panicle terminal, large, many-flowered. Flowers 1.2–2.5 cm in

diameter, red or yellow. Stamens
15–20. *Southern USA, C America, widely
introduced elsewhere.* H5. Summer.

2. T. caffrum (Thunberg) Ecklon & Zeyher.
Rhizome thick. Stems to 50 cm. Leaves
variable, fleshy, more or less flat,
2.5–13 cm × 8–15 mm, narrowed at the
base to a short stalk. Flowers usually
solitary (rarely 1–3 together), axillary,
1–2 cm in diameter. Petals pale yellow or
orange. Stamens numerous. *East and
southern Africa.* G1.

3. T. spinescens Torrey. Illustration:
Abrams & Ferris, Illustrated Flora of the
Pacific states **2**: 120 (1944).
Rhizome long. Aerial branches short,
numerous, the older parts covered with
spines formed from the persistent midribs of
the older leaves. Leaves almost terete,
1–3 cm × 1–2 mm, obtuse, narrowed to
the base. Inflorescence cymose, few-
flowered. Flowers deep red, 1.2–1.6 cm in
diameter. Stamens 20–30. *Western N
America (British Columbia, Washington).*
H5–G1. Summer.

4. T. okanoganense English. Illustration:
Abrams & Ferris, Illustrated Flora of the
Pacific states **2**: 120 (1944).
Like *T. spinescens* but leaf-midribs scarcely
persistent (if so as curved bristles), flowers
white tinged with pink or yellow. *Western
N America (British Columbia, Washington).*
H5–G1. Summer.

5. T. calycinum Engelmann.
Rhizome spherical, flattened. Stems to
10 cm. Leaves almost terete, 1–8 cm ×
1–3 mm, acute, rather sparse.
Inflorescence a raceme. Flowers 2–3 cm in
diameter with 8–10 pink or red petals.
Stamens 30–45. *C & S USA.* H5–G1.
Summer.

6. T. teretifolium Pursh.
Rhizome tuberous. Stems to 50 cm, much
branched. Leaves terete, 1–6 cm ×
2–3 mm, clustered at the ends of the
branches. Inflorescence a many-flowered
cyme. Flowers 1–1.2 cm in diameter, petals
purple or blue. Stamens 12–20. *Eastern
USA.* H5–G1. Summer.

6. ANACAMPSEROS Linnaeus
J. Cullen & H.S. Maxwell
Small succulent herbs, sometimes with a
well-developed crown above ground.
Leaves either large, fleshy and conspicuous
or small and hidden by large, parchment-
like stipules; when the leaves are large the
stipules are reduced to conspicuous hairs.

Flowers either solitary or in few-flowered
racemes, stalked or not, sometimes
cleistogamous. Sepals 2, persistent. Petals
5. Stamens 15–60. Ovary superior. Stigmas
3. Fruit a thinly membranous capsule
whose outer skin falls away leaving a
number of fibres between which the seeds
escape; the capsule may open by flaps at
the apex as well. Seeds numerous.

A genus of about 50 species, most from
southern Africa, a few from Australia. The
genus is divided into 4 subgenera by von
Poellnitz (reference below), of which 2 are
represented here, differing markedly in
their form. Subgenus **Avonia** Meyer (Nos.
1–5) has very small leaves hidden by
enlarged, parchment-like stipules and
usually solitary, terminal flowers which are
often cleistogamous. Subgenus
Anacampseros (subgenus *Telephiastrum*
Dillenius) – Nos. **6–10** – has well-
developed, fleshy leaves and stipules
represented by a number of hairs; the
flowers are borne in a stalked raceme and,
in many species, are open for only a few
hours. The dehiscence of the fruit in both
subgenera is remarkable and the seeds are
very variable, providing important
characters in the classification of the genus;
these, however, are not dealt with here, as
seed is not usually available.

The species are grown as for most South
African succulents, requiring a sunny
position and light, well-drained soil;
watering should be greater in summer than
in winter. Propagation is by seed, when
available, or by cuttings.
Literature: Berger, A., *Mesembrianthemen
und Portulacaceen*, 295–306 (1908); von
Poellnitz, K., Anacampseros L.: Versuch
einer Monographie, *Botanische Jahrbücher*
65: 382–448 (1933); Mieras, J.,
Aantekeningen over Anacampseros,
Succulenta **58**: 163–4, 193–5, 248–50
(1979), **59**: 131–5 (1980).

1a. Leaves small, hidden by the large,
parchment-like, overlapping stipules
    2
 b. Leaves well-developed, large, stipules
reduced to hairs   6
2a. Plants with a conspicuous fleshy
crown above ground   3
 b. Plant without a conspicuous fleshy
crown above ground   4
3a. Flowers white, *c.* 3 cm in diameter;
leaves in 5 rows on the stem
    4. alstonii
 b. Flowers purple, 1.2–1.5 cm in
diameter; leaves in a spiral
    5. quinaria

4a. Branches 6–10 mm thick
    1. papyracea
 b. Branches at most 5 mm thick   5
5a. Flowers red; inner bases of the stipules
hairless   **3. dinteri**
 b. Flowers white; inner bases of the
stipules hairy   **2. albissima**
6a. Flowers to 2 cm in diameter
    9. densifolia
 b. Flowers more than 2 cm in diameter
    7
7a. Leaves hairless when mature   8
 b. Leaves with cobweb-like hairs when
mature   9
8a. Leaves rounded at the apex, ovoid to
almost spherical   **7. telephiastrum**
 b. Leaves lanceolate-obovoid, acute
    6. rufescens
9a. Leaves *c.* 2 cm; stipule-hairs shorter
than the leaves   **10. arachnoides**
 b. Leaves much smaller; stipule-hairs
longer than the leaves   **8. filamentosa**

1. A. papyracea Sonder. Illustration:
Berger, Mesembrianthemen und
Portulacaceen, 299 (1908); Lamb & Lamb,
The illustrated reference on cacti and other
succulents **1**: 110 (1955); Jacobsen,
Handbook of succulent plants **1**: 221
(1960); Court, Succulent flora of South
Africa, 129 (1981).
Crown small, below ground. Branches
numerous, 6–10 mm thick. Stipules
spirally arranged, broadly ovate, adpressed,
blunt, whitish, entire, with small hairs at
the base within. Leaves hidden. Flowers
solitary, stalkless, greenish white, not
opening. Stamens 16. Capsule *c.* 4 mm.
South Africa (Cape Province). H5–G1.

2. A. albissima Marloth. Illustration: Court,
Succulent flora of South Africa, 129 (1981).
Crown small, below ground. Branches
numerous, spreading or erect, 3–4 mm in
diameter. Stipules entire or irregularly
toothed, ovate, obtuse, whitish, each often
with a brown spot, somewhat reflexed at
the tip, hairy within at the base. Leaves
hidden. Flowers 1–3 together, petals white,
opening widely. Capsule spherical, *c.* 3 mm.
South Africa, Southwest Africa. H5–G1.

A. buderiana von Poellnitz is very similar
with the hairs at the base of the stipules
very few or absent. *South Africa (Cape
Province).* H5–G1.

3. A. dinteri Schinz.
Crown small, below ground. Branches
numerous, to 4 mm thick. Stipules acute,
silvery, without hairs at the base within.
Leaves hidden. Flowers red. Capsule
stalked. *Southwest Africa.* H5–G1.

4. A. alstonii Schonland. Illustration: Lamb & Lamb, The illustrated reference on cacti and other succulents **2**: 444 (1959); Jacobsen, Handbook of succulent plants **1**: 217 (1960); Barkhuizen, Succulents of southern Africa, 157 (1978); Court, Succulent flora of South Africa, 110 (1981).

Crown spherical, above ground, flattened above, to 6 cm in diameter. Branches very numerous, to 2 mm thick. Stipules silver-white, triangular, acute, not hairy within at the base; leaves hidden but in 5 rows on the stem. Flowers solitary, *c.* 3 cm in diameter, white. Stamens to 60. Capsule to 7 mm. *South Africa (Cape Province).* H5–G1.

5. A. quinaria Sonder. Illustration: Jacobsen, Handbook of succulent plants **1**: 222 (1960).

Crown above ground, to 2.5 cm in diameter. Branches numerous, to 2 mm thick. Stipules spirally arranged, silvery with brownish spots, broadly ovate to triangular, not hairy within at the base. Leaves hidden. Flowers 1.2–1.5 cm in diameter, purple. *South Africa (Cape Province).* H5–G1.

6. A. rufescens de Candolle. Illustration: Berger, Mesembrianthemen und Portulacaceen, 304 (1908); Lamb & Lamb, The illustrated reference on cacti and other succulents **1**: 172 (1955); Jacobsen, Handbook of succulent plants **1**: 224 (1960).

Clump- or mat-forming plant. Stems 5–8 cm, erect or creeping. Leaves unequal in size, the larger to 2 × 1 cm, all obovate-lanceolate, acute, somewhat recurved at the apex, shining green or reddish on the back, hairless when mature. Stipule-hairs white, numerous, bristle-like, often wavy, white or yellow. Flowers 3–4 together, 2–3 cm in diameter, pink or purple-red. *South Africa (Cape Province).* H5–G1.

The flowers open only for a short period in the afternoon.

7. A. telephiastrum de Candolle. Illustration: Jacobsen, Handbook of succulent plants **1**: 224 (1960); Batten & Bokelman, Wild flowers of the eastern Cape Province, 61 (1966); Gledhill, Eastern Cape veld flowers, pl. 25 (1977); Rauh, The wonderful world of succulents, pl. 81 (1984).

Plant eventually forming mats. Leaves ovoid or almost spherical, shortly tapering, to 2 cm long and broad, green or brownish, often reddish on the back, hairless when mature. Stipule-hairs few, shorter than the leaves. Inflorescence to 10 cm, with 1–4 flowers. Flowers opening in the afternoon, 3–3.5 cm in diameter, pink. Stamens numerous. *South Africa (Cape Province).* H5–G1.

8. A. filamentosa Sims. Illustration: Botanical Magazine, 1367 (1811); Rauh, The wonderful world of succulents, pl. 5 (1984).

Stems to 5 cm, covered with dense, overlapping leaves. Leaves ovoid to spherical, thick, 6–10 cm long, rough at the tip, covered with fine, cobweb-like hairs when mature. Stipule-hairs longer than the leaves, numerous, whitish. Inflorescence to 8 cm, with 3–5 flowers. Flowers 3 cm in diameter, pink. *South Africa (Cape Province).* H5–G1.

9. A. densifolia Dinter.

Plant low and branched. Leaves obovoid, shining reddish green, persistently felted and covered with small warts, somewhat pointed at the apex, *c.* 8 × 5 mm. Stipule-hairs numerous, longer than the leaves, whitish with brown tips. Flowers *c.* 2 cm in diameter, intense pink. *Southwest Africa.* H5–G1.

10. A. arachnoides Sims. Illustration: Botanical Magazine, 1368 (1811); Batten & Bokelman, Wild flowers of the eastern Cape Province, 61 (1966).

Stem to 5 cm or more, covered with overlapping leaves. Leaves unequal in size, ovoid, acute or acuminate, shining green with sparse, cobweb-like hairs. Stipule-hairs white, shorter than the leaves. Inflorescence 5–8 cm, with 3–4 flowers. Flowers *c.* 3 cm in diameter, pale pink or almost white. *South Africa (Transvaal, Cape Province).* H5–G1.

7. SPRAGUEA Torrey
J. Cullen
Plants mostly perennial, from a deep, woody taproot. Leaves mostly basal, in rosettes. Flowering stems very short. Inflorescence compact, head-like, made up of umbels of cymes. Sepals 2, persistent, conspicuous, scarious. Petals 4. Stamens 3. Ovary with 1–8 ovules; style 1, stigmas 2. Fruit a membranous capsule opening by 2 lobes at the apex.

A genus of a few species from western N America; only 1 is found in gardens. It is difficult to grow.

1. S. umbellata Torrey (*S. multiceps* Howell; *Calyptridium umbellatum* (Torrey) Greene). Illustration: Botanical Magazine, 5143 (1859); Abrams & Ferris, Illustrated Flora of the Pacific states **2**: 128 (1944).

Perennial (ours). Stems to 10 cm. Leaves spoon-shaped, rounded to the apex, tapered to the base, to 2.5 cm. Inflorescence 1–3.5 cm in diameter, the scarious sepals the most conspicuous parts. *Western N America.* H5.

The variant normally grown is var. **caudicifera** Gray, described above.

8. CLAYTONIA Linnaeus
J. Cullen
Annual or perennial herbs, usually with a basal rosette of leaves and with 2 opposite leaves on the stem, these sometimes united, stem-leaves otherwise absent. Flowers in long or contracted racemes or in panicles. Sepals 2. Petals 5, white or pinkish. Stamens 5. Ovary containing 3–6 ovules. Style 3-lobed. Fruit a capsule opening by 3 lobes at the apex.

A genus of 24 species mainly from N America but just extending into E Asia; a few species have been introduced into parts of Europe and have naturalised there. The genus is similar to *Montia* and several species have been placed in both genera; the generic distinctions used here follow those proposed by McNeill (reference below). The species are easily grown; propagation is by seed.
Literature: McNeill, J., A generic revision of Portulacaceae tribe Montieae using techniques of numerical taxonomy, *Canadian Journal of Botany* **53**: 789–809 (1975).

1a. Stem-leaves united at the base, forming a sometimes lobed, disc-shaped structure through which the stem apparently passes   **4. perfoliata**
  b. Stem-leaves free, not as above   **2**
2a. Stems arising from a more or less spherical corm; flowers in a long raceme   **1. virginica**
  b. Stems arising from a massive, erect, underground stock; flowers in a loose raceme or panicle or in a corymbose raceme   **3**
3a. Stem-leaves oblanceolate or linear, to 1 cm; flowers in a corymbose raceme; ovules 6   **2. megarhiza**
  b. Stem-leaves broadly ovate to ovate, more than 1 cm; flowers in loose panicles or racemes; ovules 3   **3. sibirica**

1. C. virginica Linnaeus. Illustration: Botanical Magazine, 941 (1806); Rickett, Wild flowers of the United States **1**: pl. 68 (1966); Courtenay & Zimmerman, Wildflowers and weeds, 28 (1972); American Horticulturist **64**(6): 4 (1985).

Perennial herb, stems arising from a more or less spherical corm. Stems to 30 cm. Basal leaves (sometimes absent at flowering) linear-lanceolate, acute, 2–7 cm × 4–10 mm, with long stalks. Stem leaves similar or sometimes linear. Raceme long with up to 15 flowers. Petals 7–12 mm, white, sometimes tinged with pink. Ovules 6. *Eastern N America from Quebec to Texas, Mexico.* H3. Spring.

2. C. megarhiza (Gray) Watson. Perennial from a massive, erect, underground root. Basal leaves oblanceolate to obovate, 4.5–10 cm, long-stalked. Stem-leaves linear to linear-lanceolate, bract-like, to 1 cm. Flowers 3–10 in a corymbose, contracted raceme. Petals white to pink, 1.2–1.4 cm. Ovules 6. *Western USA.* H3. Spring.

Usually seen in gardens as the deep pink-flowered var. **nivalis** (English) Hitchcock (*C. nivalis* English); illustration: Abrams & Ferris, Illustrated Flora of the Pacific states 2: 125 (1944).

3. C. sibirica Linnaeus (*C. alsinoides* Sims; *Montia sibirica* (Linnaeus) Howell). Illustration: Botanical Magazine, 2243 (1821); Ross-Craig, Drawings of British plants **6**: pl. 1 (1952); Roles, Illustrations to Clapham, Tutin & Warburg, Flora of the British Isles **1**: 95 (1957).
Plant usually perennial, occasionally annual. Basal leaves broadly to narrowly ovate, blades 1.5–6 cm, stalks very long. Stem-leaves stalkless, broadly ovate to ovate, 2–5 cm. Flowers in a loose raceme or panicle. Petals 7–10 mm, white with pink veins or pink. Ovules 3. *Western N America, E Asia; introduced and naturalised in parts of Europe.* H3. Spring.

4. C. perfoliata Willdenow (*Montia perfoliata* (Willdenow) Howell). Illustration: Illustrated Flora of the Pacific states 2: 128 (1944); Ross-Craig, Drawings of British plants **6**: pl. 2 (1952); Reader's Digest field guide to the wild flowers of Britain, 88 (1981); American Horticulturist **64**(6): 6 (1985).
Annual herb. Basal leaves broadly ovate, acute or somewhat rounded, long-stalked. Stem-leaves joined at their bases to form a disc-shaped, sometimes lobed structure through which the stem apparently passes. Flowers in a congested raceme. Petals white, notched at the apex, 3–4.5 mm. Ovules 3. *Western N America, Mexico; introduced and naturalised in parts of Europe and elsewhere.* H3. Spring.

9. MONTIA Linnaeus
J. Cullen
Annual or perennial herbs similar to *Claytonia*, but stems with numerous opposite or alternate leaves; ovules always 3. (There are also important differences in the structure of the pollen but these are not recorded here; see McNeill, cited above under *Claytonia*, for details.)

A genus of 37 species from most parts of the world though with main centres in western N America, S America and Australasia. They are generally plants of damp places and are best grown on streamsides. Propagation is by seed. Literature: as for *Claytonia*.

1. M. australasica (J.D. Hooker) Pax & Hoffmann (*Claytonia australasica* J.D. Hooker). Illustration: Hooker's Icones Plantarum 3: 293 (1840); Aston, Aquatic plants of Australia, 155 (1973); Moore & Irwin, The Oxford book of New Zealand plants, 51 (1978); Elliot & Jones, Encyclopaedia of Australian plants 3: 43 (1984).
Succulent, hairless perennial with creeping stems rooting at the nodes, forming mats. Leaves alternate with sheathing bases, shortly stalked, 1–5 cm, thread-like or linear or narrowly spoon-shaped. Flowers solitary, paired or in few-flowered racemes, stalked, to 2 cm in diameter. Petals white or pink. Capsule spherical. *Australia, New Zealand.* H5.

LIII. BASELLACEAE

Perennial hairless herbs with slender, twining, climbing or decumbent stems growing from tuberous rootstocks. Leaves alternate, entire, fleshy. Flowers radially symmetric, bisexual or unisexual, in axillary spikes, racemes or panicles. Bracteoles 2, often united with the perianth. Perianth-segments 5, free or basally united. Stamens 5, opposite the perianth-segments. Ovary superior, 1-celled, ovule 1. Style branched or simple. Stigmas 3 or 1. Fruit a drupe, usually enclosed by the persistent, fleshy perianth.

A family of 4 genera and 12–17 species, distributed throughout tropical regions, but mainly in C & S America.
Literature: Ulbrich, E., Basellaceae in Engler, A. & Prantl, K., *Die natürlichen Pflanzenfamilien* edn 2, **16c**: 263–71 (1934).

1a. Filaments curved outwards in bud
1. Anredera
b. Filaments erect and straight in bud 2
2a. Flowers in spikes; stigmas 3
2. Basella
b. Flowers in racemes; stigma 1
3. Ullucus

1. ANREDERA Jussieu
C.J. King
Twiners with much-branched stems arising from tuberous roots. Leaves ovate, heart-shaped or elliptic, more or less fleshy. Flowers bisexual or unisexual, in spikes or racemes. Perianth-segments united at base. Filaments curved outwards in bud. Style simple or 3-branched. Fruit spherical, enclosed in the perianth.

Between 5 and 10 species are recognised in this genus from tropical America. Only 1 species is commonly cultivated, usually as a climber and this is sensitive to frost. But even if the stems are killed, the plant will survive the winter provided that its roots are not frozen. It requires a humus-rich, well-drained soil, and may be grown outdoors as a screening plant in warmer regions but needs the protection of a greenhouse in cooler climates. Propagation is by division of the roots and by the small aerial tubers which often form in the non-flowering leaf-axils.

1. A. cordifolia (Tenore) van Steenis (*Boussingaultia cordifolia* Tenore; *B. baselloides* misapplied; *B. gracilis* Miers; *B. gracilis* var. *pseudobaselloides* (Hauman) Bailey; *B. gracilis* forma *pseudobaselloides* Hauman). Illustration: Botanical Magazine, 3620 (1837); Heywood, Flowering plants of the world, 76 (1978).
Root oblong, fleshy, somewhat scaly when young. Stems to 6 m. Leaves ovate to lanceolate, cordate at base, blades 2.5–10 cm; stalks to 2.5 cm. Racemes to 30 cm. Flowers fragrant. Perianth white, subtended by 2 ovate bracteoles. Style 3-branched. Fruits not known to be produced. *Paraguay to S Brazil & N Argentina.* G1. Autumn.

2. BASELLA Linnaeus
C.J. King
Twining herbs with much-branched stems to 9 m. Leaves oblong, ovate to ovate-lanceolate, or circular, fleshy, stalked. Flowers bisexual in usually simple spikes. Perianth urn-shaped, white, red or purple. Filaments erect and straight in bud. Style branched, stigmas 3.

A genus of 5 variable species of vines,

native in tropical Africa and Asia, one of which, *B. alba*, is cultivated for ornament in Europe and as a pot-herb in the tropics. In temperate regions the plant is usually treated as an annual. Seed should be sown indoors. The seedlings may either be planted out in a sunny position in ordinary fertile soil, or grown in a greenhouse as climbers or basket plants.

1. B. alba Linnaeus. Illustration: Addisonia **13**: pl. 420 (1928); Heywood, Flowering plants of the world, 76 (1978); Mansfeld, Verzeichnis landwirtschaftlicher und gärtnerischer Kulturpflanzen, 135 (1986). Stems to 1 m or more. Leaves usually ovate to ovate-lanceolate, blades 5–15 × 4–12 cm, shallowly cordate to truncate at base; stalks to 3 cm. Flower-spikes 7.5–15 cm. Flowers white, or pinkish to red or pale purple. Fruit black, sometimes red or white, glossy, spherical, *c.* 6 mm. *Probably native in Africa and SE Asia, but distribution now pantropical.* H5. Summer.

The reddish-flowered variant is often distinguished as a separate species, *B. rubra* Linnaeus, or as a variety or cultivar of *B. alba*. The stems of this variant may be purplish, and the leaves marked with white and pink.

3. ULLUCUS Caldas
C.J. King
Perennial herbs with slender underground rhizomes bearing edible, fleshy tubers. Stems twining, angular, 30–60 cm. Leaves entire, heart-shaped or kidney-shaped, somewhat fleshy, glossy, blades to 20 cm broad; stalks thick, grooved above, longer than the blades. Flowers bisexual, in axillary racemes, each flower subtended by 2 circular red bracteoles. Perianth-segments yellow, glossy, tapering into a needle-like, flexuous apex. Stamens 5. Filaments erect and straight in bud. Style simple, stigma 1.

A single species, native in the northern Andes but cultivated for food elsewhere in S America. In the wild, the underground tubers are 1–3 cm and usually rose-violet. The tubers of cultivated varieties may exceed 3 cm and may be white, yellow or various shades of red. Small aerial tubers may also develop in the leaf-axils. The plant grows best in a light, rich soil with plenty of leaf-mould, but is sensitive to frost. The tubers should be planted in spring and lifted in October or November.

1. U. tuberosus Lozano (*Basella tuberosa* (Caldas) Humboldt, Bonpland & Knuth). Illustration: Botanical Magazine, 4617

(1851); Nicholson et al., The Oxford book of food plants, 179 (1969); Heywood, Flowering plants of the world, 76 (1978); Mansfeld, Verzeichnis landwirtschaftlicher und gärtnerischer Kulturpflanzen, 136, 137 (1986). *S America (Andes).* G1. Summer.

LIV. CARYOPHYLLACEAE
Herbs, more rarely small shrubs, with simple, entire, usually opposite leaves, sometimes with small stipules. Stems often weak and brittle with prominent nodes. Flowers usually bisexual, radially symmetric, often in terminal dichasia. Sepals 4 or 5, free or joined, sometimes in a long tube ending in calyx-teeth (Figure 27(2, 3, 5), p. 178). Petals 4 or 5, free, sometimes absent. Stamens usually 8 or 10. Ovary superior, 1-celled (at least above), of 2–5 fused carpels, with 1–many ovules on a basal or free-central placenta (Figure 27(6, 7), p. 178). Styles 2–5, free. Fruit usually a capsule opening by teeth or flaps, more rarely a berry or 1-seeded nutlet. Seeds usually kidney-shaped with a curved embryo.

A cosmopolitan family of about 90 genera and more than 2000 species, well represented in north temperate regions. Several familiar European genera are important in gardens, among them *Dianthus*, *Gypsophila* and *Silene*. Some of these familiar garden plants are commonly seen as double variants; since it is not possible to decide the genus to which such plants belong by using the ordinary key, a special key is provided.
Literature: Pax, F. & Hoffmann, K., Caryophyllaceae, *Die Natürlichen Pflanzenfamilien* edn 2, **16c**: 275–364 (1934). This work contains a very complete bibliography.

Habit. Dense cushion plants with more or less stalkless flowers: species of **1,6,9,12**.
Leaves. With 3 prominent more or less parallel veins (Figure 27 (1), p. 178): **7** (*Saponaria officinalis*), **9** (*Dianthus barbatus*).
Petals. Scarlet or crimson: species of **1,2**. Deeply cut into narrow segments: species of **1,2,9**.
Flowers. Unisexual: **1** (*Silene dioica, S. nutans*), **14** (*Minuartia sedoides*); 'double': **see separate key.**
Fruit. A berry: **4**; a 1-seeded nutlet: **16–18**.

1a. Sepals free (Subfamily *Alsinoideae*, see
Figure 27(2), p. 178)   2

1b. Sepals joined, sometimes into a long calyx-tube (Subfamily *Silenoideae*, see Figure 27(3), p. 178)   7
2a. Stipules present; petals very small or absent   3
b. Stipules absent; petals usually well-developed   4
3a. Bracts surrounding flowers conspicuous, silvery   **18. Paronychia**
b. Bracts inconspicuous, greenish
                                    17. Herniaria
4a. Flowers without petals; fruit a nutlet
                                    16. Scleranthus
b. Flowers (with rare exceptions) with petals; fruit a capsule   5
5a. Petals 2-toothed or 2-lobed
                                    11. Cerastium
b. Petals more or less entire   6
6a. Leaves linear-subulate, joined at base round stem; flower-buds spherical
                                    15. Sagina
b. Leaves sometimes linear but not joined round stem; flower buds more or less elongated
        12. –14. Arenaria and related genera
                                    (joint key)
7a. Fruit a berry   **4. Cucubalus**
b. Fruit a capsule   8
8a. Styles 2   9
b. Styles 3–5   13
9a. Epicalyx present at base of calyx (Figure 27(5), p. 178)   10
b. Epicalyx absent   11
10a. Calyx with membranous bands of tissue between the veins
                                    10. Petrorhagia
b. Calyx without membranous bands
                                    9. Dianthus
11a. Petals with coronal scales (Figure 27(4), p. 178)   **7. Saponaria**
b. Petals without coronal scales   12
12a. Calyx unwinged, with membranous bands of tissue between the veins
                                    6. Gypsophila
b. Calyx winged, without membranous bands   **8. Vaccaria**
13a. Calyx-teeth linear, more than 3 cm; coronal scales absent **5. Agrostemma**
b. Calyx-teeth much less than 3 cm; coronal scales usually present   14
14a. Seeds with a tuft of hairs; styles 5
                                    3. Petrocoptis
b. Seeds without a tuft of hairs; styles 3 or 5
        1. Silene and **2. Lychnis** (joint key)
        (see also *Saponaria pumilio*, p. 184)

Key to double-flowered variants

1a. Sepals free (Subfamily *Alsinoideae*)
                        Minuartia verna (p. 193)
b. Sepals joined (Subfamily *Silenoideae*) 2

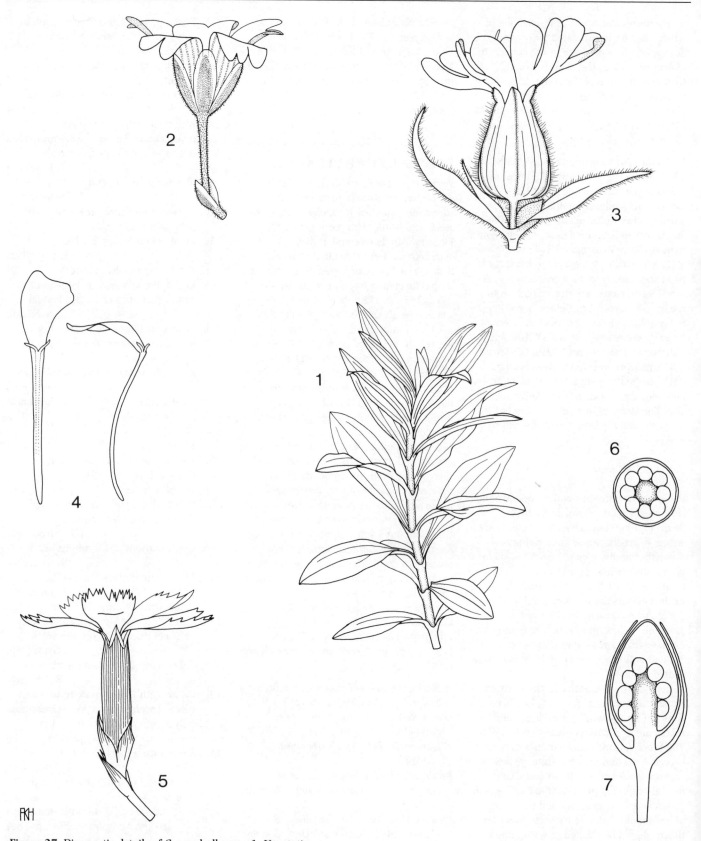

Figure 27. Diagnostic details of Caryophyllaceae. 1, Vegetative shoot of *Saponaria officinalis*. 2, Flower of *Cerastium tomentosum*. 3, Female flower of *Silene dioica*. 4, Petal of *Saponaria officinalis*. 5, Flower of *Dianthus deltoides*. 6, Transverse section of capsule of *Silene dioica* showing free-central placentation. 7, Longitudinal section of calyx, carpophore and capsule of *Silene dioica*.

2a. Epicalyx present at base of calyx (see Figure 27(6), p. 178)
   Dianthus species and cultivars (p. 185)
 b. Epicalyx absent   3
3a. Perennial with long white stolons and ovate or ovate-lanceolate leaves with 3 more or less parallel veins
   Saponaria officinalis (p. 185)
 b. Annual or perennial without stolons; leaves without 3 more or less parallel veins   4
4a. Cushion plant with more or less stalkless flowers   **Silene acaulis** (p. 181)
 b. Plant not cushion-forming; flowers more or less stalked or inflorescence capitate   5
5a. Annual   **Silene pendula** (p. 181)
 b. Perennial   6
6a. Inflorescence very large and intricately branched with numerous small flowers   **Gypsophila paniculata** (p. 183)
 b. Inflorescence capitate or with relatively few flowers or flowers solitary   7
7a. Calyx not more than 7 mm; petals 4-toothed   **Silene pusilla** (p. 181)
 b. Calyx more than 7 mm; petals not 4-toothed   8
8a. Calyx strongly inflated
   Silene vulgaris subsp. **maritima** (p. 180)
 b. Calyx not inflated   9
9a. Inflorescence capitate
   Lychnis chalcedonica (p. 182)
 b. Inflorescence not capitate   10
10a. Leaves hairy   **Silene dioica** (p. 181)
 b. Leaves hairless or nearly so   11
11a. Inflorescence narrow, spike-like; petals entire or shallowly 2-lobed
   Lychnis viscaria (p. 182)
 b. Inflorescence a loose, few-flowered dichasium; petals deeply cut
   Lychnis flos-cuculi (p. 182)

1. SILENE Linnaeus
S.M. Walters

Herbs, or sometimes small shrubs with woody stocks. Leaves opposite, linear to ovate or obovate, entire. Flowers usually conspicuous, solitary or in few- to many-flowered inflorescences. Calyx tubular, sometimes very persistent and strongly inflated in fruit, with usually 10–30 (rarely 5) veins and 5 short teeth. Petals free, usually with a narrow claw and wide, spreading blade. Coronal scales (Figure 27(4), p. 178) often present at junction of claw and blade. Stamens 10. Styles 3 (rarely 5). Fruit a capsule with variably-

developed basal septa, dehiscing apically with teeth twice the number of styles. A stalk of variable length (carpophore) is often present between calyx and capsule (Figure 27(7), p. 178). Seeds numerous, usually 1–2 mm.

A wide definition of the genus is adopted here (as in *Flora Europaea*), including *Melandrium* Röhling and *Heliosperma* (Reichenbach) Reichenbach, covering some 500 species in the northern hemisphere, with many in the Mediterranean area and the European mountains. Relatively few are grown in gardens, and their habits and horticultural requirements are diverse (see individual species accounts).

Because of the difficulties in discriminating the genera without dehisced fruit, species of *Lychnis* (p. 182) are also included in the key below.

1a. Annual or biennial   2
 b. Perennial   7
2a. Inflorescence corymbose or capitate; annual or biennial   3
 b. Inflorescence neither corymbose nor capitate; strictly annual   5
3a. Plant densely hairy
   Lychnis coronaria (p. 182)
 b. Plant not densely hairy   4
4a. Uppermost stem-leaves closely enclosing the base of the capitate inflorescence   **17. compacta**
 b. Uppermost stem-leaves not closely enclosing the base of the inflorescence
   16. armeria
5a. Flowers less than 10 mm in diameter
   21. gallica var. **quinquevulnera**
 b. Flowers more than 1.5 cm in diameter   6
6a. Flowering stems erect, hairless
   19. coeli-rosa
 b. Flowering stems spreading, hairy
   20. pendula
7a. Plant with thick woody stock   8
 b. Plant with little or no woody stock   9
8a. Plant hairless except for the glandular-hairy calyx   **1. fruticosa**
 b. Plant hairy   **2. mollisima**
9a. Plant more than 30 cm, with well-developed stem-leaves   10
 b. Plant less than 30 cm, often with few or no stem-leaves   20
10a. Plant covered with a dense felt of hairs   11
 b. Plant not densely hairy   12
11a. Inflorescence capitate
   Lychnis flos-jovis (p. 182)
 b. Inflorescence a loose panicle
   Lychnis coronaria (p. 182)

12a. Petals deeply fringed, cut or lobed into 4 or more lobes   13
 b. Petals entire or bifid (sometimes with short side-teeth)   15
13a. Petals fringed; calyx strongly inflated
   8. fimbriata
 b. Petals cut or lobed into 4 or more; calyx not inflated   14
14a. Plant glandular-hairy; petals crimson
   4. californica
 b. Plant hairy but not glandular; petals pale purple or white
   Lychnis flos-cuculi (p. 182)
15a. Inflorescence capitate
   Lychnis chalcedonica (p. 182)
 b. Inflorescence not capitate   16
16a. Flowers 4 cm or more in diameter   17
 b. Flowers less than 4 cm in diameter   18
17a. Petals shallowly lobed or toothed
   Lychnis × haageana (p. 182)
 b. Petals deeply bifid   **5. virginica**
18a. Inflorescence a spike-like panicle; petals nearly entire
   Lychnis viscaria (p. 182)
 b. Inflorescence not spike-like; petals deeply bifid   19
19a. Flowers inclined and directed to 1 side of the inflorescence; plants not dioecious (though flowers sometimes unisexual)   **3. nutans**
 b. Flowers more or less erect and not directed to 1 side of the inflorescence; plant dioecious   **18. dioica**
20a. Calyx strongly inflated   21
 b. Calyx not or only slightly inflated   23
21a. Petals purplish red   **6. elisabethae**
 b. Petals white   22
22a. Calyx 20-veined   **9. vulgaris**
 b. Calyx 10-veined   **7. zawadzkii**
23a. Dense cushion-plant of moss-like appearance   **13. acaulis**
 b. Dwarf, sometimes mat-forming, but not moss-like   24
24a. Petals 4-toothed; seeds winged
   15. pusilla group
 b. Petals entire to deeply lobed, but not 4-toothed; seeds not winged   25
25a. Flowers more than 4 cm in diameter, on very long stalks   **11. hookeri**
 b. Flowers less than 4 cm in diameter, on short stalks   26
26a. Calyx more than 2 cm   27
 b. Calyx less than 2 cm   28
27a. Petals shallowly bifid, with conspicuous, broad coronal scales
   14. schafta
 b. Petals deeply bifid, with inconspicuous coronal scales
   10. vallesia

28a. Plant softly hairy on stems
　　　　　　　　　　　　　　　12. keiskii
　　b. Plant hairless or nearly so　　　　　29
29a. Inflorescence narrow, rather spike-
　　　like; petals entire or nearly so
　　　　　　　　　Lychnis viscaria (p. 182)
　　b. Inflorescence corymbose, sometimes
　　　almost capitate; petals deeply bifid
　　　　　　　　　　Lychnis alpina (p. 182)

1. S. fruticosa Linnaeus. Illustration:
Sibthorp & Smith, Flora Graeca 5: t. 428
(1825); Fiori & Paoletti, Iconographia
Florae Italicae, edn 3, 147 (1933).
Robust, almost completely hairless plant
with thick woody stock, shiny, lanceolate
to spathulate basal leaves, and flowering-
stems to 50 cm. Inflorescence dense, with
several large flowers held erect on short
stalks. Calyx *c.* 2.5 cm, glandular; petals
pink or red, shallowly bifid, with
conspicuous coronal scales. *Mediterranean
sea-cliffs from Sicily to Crete and Carpathos; N
African coast.* H5–G1. Late spring.

2. S. mollissima group (including
S. mollissima (Linnaeus) Persoon; *S. hifacensis*
Willdenow; *S. sicula* misapplied). Illustration:
Pignatti, Flora d'Italia 1: 242 (1982).
Like *S. fruticosa* but hairy and with a looser
inflorescence with several 3-flowered
opposite branches. *Mediterranean area from
Spain to S Italy.* H5–G1. Late spring.

The taxonomy of this group of species is
uncertain; Balearic plants are in
cultivation. For further information see
Jeanmonod, D., Candollea 39: 195–259
(1984).

3. S. nutans Linnaeus. Illustration: Hegi,
Illustrierte Flora von Mitteleuropa 3: 279,
293, (1910); Garrard & Streeter, Wild
flowers of the British Isles, 49 (1983).
Perennial, slightly woody at the base, with
slender flowering stems to 50 cm, hairy
below, sticky and hairless above. Basal
leaves long-stalked, spathulate, hairy.
Inflorescence a loose, 1-sided panicle with
opposite 3–7-flowered branches. Flowers
sometimes unisexual, drooping on short,
sticky stalks, opening and fragrant in the
evening. Calyx 9–12 mm, glandular; petals
whitish, often tinged with pink or greenish
yellow, deeply bifid with narrow inrolled
lobes and small coronal scales. *Most of
Europe, except the Mediterranean islands and
the arctic.* H2. Summer.

4. S. californica Durand. Illustration:
Abrams & Ferris, Illustrated Flora of the
Pacific States 2: f. 1727 (1944); Rickett,
Wild flowers of the United States 4: pl. 80
(1970).

Perennial herb with weak, branched,
glandular-hairy, leafy stems to 40 cm.
Leaves ovate to narrowly obovate, hairy.
Inflorescence few- to many-flowered; calyx
1.5–2 cm; petals large, crimson, with long
claw and blade to 1.5 cm, very deeply 4-fid
with conspicuous coronal scales. *Western N
America.* ?H4. Summer.

5. S. virginica Linnaeus. Illustration:
Rickett, Wild flowers of the United States 1:
pl. 53 (1965); Justice & Bell, Wild flowers of
North Carolina, 58 (1968).
Perennial herb with weak, glandular-hairy,
sticky stems to 50 cm. Basal leaves
narrowly obovate to narrowly spathulate,
almost hairless; stem-leaves 2–4 pairs,
long, narrow, almost stalkless.
Inflorescence loose, with leafy bracts and
few large, long-stalked flowers. Calyx
c. 2 cm; petals large, deep red to crimson,
with blade to 2 cm, deeply bifid with
variably-developed lateral teeth. *Eastern N
America.* ?H4. Late summer–autumn.

Nos. 4 & 5 are somewhat difficult to
overwinter in cultivation in Europe, but are
prized by rock-gardeners for their striking
flower colour.

6. S. elisabethae Jan (*Melandrium elisabethae*
(Jan) Rohrbach). Illustration: Hegi,
Illustrierte Flora von Mitteleuropa 3: 303
(1910).
Tufted perennial with a rather woody stock
and hairy flowering stems to 25 cm. Basal
leaves lanceolate, acute, thick in texture,
almost hairless. Flowers large, often
solitary; calyx 1.5–2 cm, densely
glandular-hairy; petals with dark purplish
red blade to 1.5 cm, bifid and toothed on
the margin, and deeply-cut coronal scales.
Italian Alps. H2. Summer.

A plant of limestone screes, grown in
rock-gardens and very tolerant of lime.

7. S. zawadzkii Herbich (*Melandrium
zawadzkii* (Herbich) Braun). Illustration:
Everett, New York Botanical Gardens
illustrated encyclopedia of horticulture 9:
3153 (1982).
Like *S. elisabethae*, but flowers smaller,
white, in few-flowered inflorescences, and
calyx not or only slightly glandular. *E
Carpathian mountains.* H2. Summer.

8. S. fimbriata Sims (*S. multifida* (Adams)
Rohrbach, not Edgeworth). Illustration:
Botanical Magazine, 908 (1806).
Perennial herb with erect, leafy stems to
60 cm, hairy throughout. Leaves ovate-
cordate, upper stalkless, lower stalked.
Inflorescence of few, large flowers in a loose
panicle. Calyx 1.5–2 cm, inflated and

persistent round the ripe capsule. Petals
white with narrow, projecting claw and
deeply cut (fringed) blade; coronal scales
present. *USSR (Caucasus).* H2. Summer.

Formerly grown in gardens; now rarely
seen.

9. S. vulgaris Linnaeus subsp. **maritima**
(Withering) Löve & Löve (*S. maritima*
Withering; *S. uniflora* Roth). Illustration:
Beckett, Concise encyclopaedia of garden
plants, 382 (1983); Garrard & Streeter,
Wild flowers of the British Isles, 48 (1983).
Perennial, with branching, prostrate,
hairless vegetative stems arising from a
woody stock, often forming a loose cushion
or mat. Leaves lanceolate, fleshy, grey-
green, hairless except for marginal cilia.
Flowering stems to 20 cm, ascending, with
1–4 large flowers. Calyx 20-veined, inflated
and persistent around the ripe capsule.
Petals with deeply bifid white blade and
small coronal scales. *Coasts of Atlantic and
arctic Europe from Murmansk to Spain.* H3.
Summer.

Long cultivated in NW Europe and
available in several cultivars, including
double-flowered variants. The typical, erect,
subsp. **vulgaris** is a very familiar,
widespread and variable European plant,
not in cultivation.

10. S. vallesia Linnaeus subsp. **vallesia**.
Illustration: Hegi, Illustrierte Flora von
Mitteleuropa 3: 285 (1912).
Mat-forming perennial with oblong-
lanceolate leaves, ascending, glandular-
hairy stems to 15 cm, and large flowers,
solitary or in 2-or 3-flowered inflorescences.
Calyx 2–2.5 cm, glandular-hairy. Petal-
blade *c.* 10 mm, bifid, usually pink above,
reddish beneath; coronal scales present.
W Alps and Apennines. H3. Summer.

11. S. hookeri Torrey & Gray. Illustration:
Botanical Magazine, 6051 (1873); RHS
Dictionary of gardening 4: 1955 (1956).
Tufted, grey-hairy perennial with ovate-
lanceolate, acute leaves on loose, prostrate
stems. Flowers very large, axillary on long,
slender stalks. Calyx *c.* 2.5 cm; petal-blade
1.5–2 cm, very deeply bifid, each lobe with
further deep cuts, pink with white rays and
white coronal scales. *Western N America
(Oregon, California).* H4. Summer.

Needs a sunny, well-drained position to
overwinter outside: better grown in an
alpine house with winter protection.

12. S. keiskii Miquel (*Melandrium keiskii*
(Miquel) Ohwi). Illustration: Kitamura &
Murata, Coloured illustrations of
herbaceous plants of Japan, pl. 58 (1961).

Tufted perennial with lanceolate, almost hairless leaves and short, hairy flowering stems to 15 cm. Inflorescence a few-flowered cyme. Calyx 1–1.2 cm; petals 1–1.5 cm, pink with obovate, shallowly bifid blade and prominent, oblong coronal scales. *N Japan (mountains)*. H3. Late summer–autumn.

The plant usually grown in Europe is a dwarf variant, forma **minor** Takeda.

13. S. acaulis (Linnaeus) Jacquin (*S. exscapa* Allioni; *S. longiscapa* Kerner). Illustration: Hegi, Illustrierte Flora von Mitteleuropa **3**: 279, 295 (1912); Garrard & Streeter, Wild flowers of the British Isles, 49 (1983). Very dwarf, hairless, mat-forming or tufted, moss-like perennial with densely packed, narrowly subulate leaves and short, erect flowering stems to 5 cm. Flowers solitary, erect, with very short stalks elongating in fruit. Calyx 8–10 mm, often reddish; petals deep pink (sometimes white), with short (*c.* 5 mm), obovate, notched or shallowly bifid blade and small coronal scales. *Widespread and variable arctic-alpine, in both the Old and New World*. H1. Summer.

Easy to cultivate in rock-gardens but not often flowering freely. **S. elongata** Bellardi is the name applied to a variant with somewhat larger flowers and stems rising 1–2 cm above the cushion of leaves. 'Correvoniana' is a double-flowered cultivar.

14. S. schafta Hohenacker. Illustration: Edwards's Botanical Register **32**: t. 20 (1846); Beckett, Concise encyclopaedia of garden plants, 382 (1983). Loosely mat-forming, hairy perennial with short, non-flowering shoots and flowering shoots to 25 cm, ascending or prostrate but turning upwards at their ends. Basal leaves to 4 cm, linear-lanceolate, stalked; stem-leaves shorter, lanceolate, stalkless. Flowers few in a loose dichasium. Calyx 2.2–2.5 cm, narrow, with short triangular teeth. Petals reddish purple; blade *c.* 10 mm, obovate, shallowly bifid; coronal scales broad, forming an obvious coronal 'eye'. *USSR (Caucasus)*. H3. Late summer–autumn.

Easy to cultivate, and valuable in rock-gardens because of its late flowering.

15. S. pusilla group (including *S. pusilla* Waldstein & Kitaibel & *S. alpestris* Jacquin; synonyms include *S. quadrifida* Linnaeus; *S. quadridentata* in the sense of Hayek; *Heliosperma quadrifidum* in the sense of Hegi). Illustration: Coste, Flore de la France **1**: 175 (1901). Perennial with slightly woody stock and slender, weak, variably hairy, branched stems to 30 cm. Inflorescence a spreading dichasium of small flowers on slender stalks; branches rather sticky, sometimes with stalked glandular hairs. Calyx 4–7 mm, hairless or glandular-hairy; petal-blade *c.* 5 mm, usually 4-toothed, white or pinkish. Seeds winged. *European mountains, from the Pyrenees to the Carpathians*. H4. Summer.

A very variable group, difficult to classify. The plants in cultivation are generally considered to be *S. alpestris* Jacquin from the E Alps and Balkans, distinguished from *S. pusilla* (in the strict sense) by the more robust habit, larger flowers and ciliate (not hairless) petal-claw. This plant is grown also in the double-flowered variant.

16. S. armeria Linnaeus. Illustration: Beckett, Concise encyclopaedia of garden plants, 382 (1983). Annual or biennial; stems to 40 cm, hairless, glaucous, usually unbranched, sticky above. Basal leaves spathulate, withering early; stem-leaves ovate to lanceolate, clasping the stem. Inflorescence corymbose. Calyx 1.2–1.5 cm; petal-blade *c.* 6 mm, pink, obovate with shallow indentation; coronal scales obvious, acute. *C & S Europe, widely naturalised further north*. H3. Summer.

17. S. compacta Fischer. Like *S. armeria* but inflorescence densely capitate, with upper stem-leaves closely enclosing its base; calyx 1.4–2 cm; petal-blade entire. *SE Europe*. H3. Summer.

18. S. dioica (Linnaeus) Clairville (*Lychnis dioica* Linnaeus; *Melandrium rubrum* (Weigel) Garcke; *Melandrium dioicum* (Linnaeus) Cosson & Germaine). Figure 27(3, 6, 7), p. 178. Illustration: Ross-Craig, Drawings of British plants **5**: pl. 16 (1951); Keble Martin, The concise British Flora in colour, pl. 14 (1965) – male and female plants; Garrard & Streeter, Wild flowers of the British Isles, 48 (1983). Dioecious perennial herb with spreading vegetative shoots and erect flowering stems to 80 cm. Basal leaves obovate, long-stalked; stem-leaves oblong-obovate, acute, decreasing gradually into leafy bracts in the dichasial inflorescence. Calyx 1–1.5 cm, in male flowers cylindric and faintly 10-veined, in female flowers ovoid and 20-veined. Petal-blade 8–10 mm, bright pink, broadly obovate, deeply bifid; coronal scales narrow, acute. Styles 5 (not 3). *NW Europe*. H2. Summer.

Often present and tolerated in wild woodland gardens, though rarely deliberately cultivated. A double-flowered variant has long been in cultivation, and dwarf variants from exposed rock habitats make attractive rock plants. Where *S. dioica* and the widespread related, often annual weed *S. alba* grow together, pink-flowered hybrid populations frequently occur, and may be encouraged in woodland gardens.

19. S. coeli-rosa (Linnaeus) Godron (*Agrostemma coeli-rosa* Linnaeus; *Eudianthe coeli-rosa* (Linnaeus) Reichenbach; *Lychnis coeli-rosa* (Linnaeus) Desrousseaux). Illustration: Fiori & Paoletti, Iconographia Florae Italicae, edn 3, 142 (1933). Hairless annual to 50 cm, with linear-lanceolate leaves and a loose, irregular dichasial inflorescence with several large flowers on long stalks. Calyx 1.5–2.5 cm, deeply grooved between the 10 veins and with long, acute, spreading teeth; petal-blade 1–2 cm, shallowly bifid; coronal scales acute. Styles 5 (not 3). *W Mediterranean area, from S Spain & N Africa to Italy; casual elsewhere*. H4. Summer.

Typically pink-flowered, but several colour-variants exist, from purple to white; 'Oculata' has a dark 'eye' to the flower.

20. S. pendula Linnaeus. Illustration: Fiori & Paoletti, Iconographia Florae Italicae, edn 3, 146 (1933). Glandular-hairy annual with weak, branched, ascending stems to 20 cm. Leaves ovate to ovate-lanceolate, acute. Inflorescence a loose, raceme-like cyme; flower-stalks at first erect, spreading and eventually deflexed in fruit. Calyx 1.2–1.8 cm, inflated, with wide colourless bands between the prominent veins; petal-blade 7–10 mm, usually pink, shallowly bifid. *Mediterranean area; native to N Africa, Italy and Sicily, elsewhere often naturalised*. H4. Summer–autumn.

Commonly used in S & W Europe as a summer bedding plant and available in colour and double-flowered variants.

21. S. gallica Linnaeus var. **quinquevulnera** (Linnaeus) Mertens & Koch. Hairy annual, with simple or branched stems to 40 cm, and long, narrow, leafy, raceme-like cymose inflorescences, usually sticky in the upper part. Basal leaves spathulate, stalked; stem-leaves lanceolate, stalkless. Calyx 7–10 mm; petal-blade 3–5 mm, entire or notched, pinkish or white with a deep crimson blotch; coronal scales present. *Widespread in S & C Europe, casual further north*. H2. Summer.

2. LYCHNIS Linnaeus
S.M. Walters

Like *Silene*, but styles always 5 and ripe capsule opening apically with 5 teeth only.

As here defined, a small genus of some 15 species of north temperate regions. Several species are old, familiar, garden plants; all are biennial or perennial herbs. Cultivation as for *Silene*.

Since it is impossible to distinguish a *Lychnis* from a *Silene* without ripe, dehiscing fruit, a joint key has been provided under the genus *Silene* (p. 179).

1. L. chalcedonica Linnaeus. Illustration: Jelitto et al., Die Freiland Schmuckstauden, edn 2, 383 (1985).
Stems simple, to 60 cm, roughly hairy, with ovate basal leaves and stem-leaves with cordate, stem-clasping bases. Inflorescence capitate, with up to 50 large flowers. Calyx 1.4–1.8 cm, hairy. Petal-blade *c.* 1.5 cm, bifid, bright scarlet (or white); coronal scales conspicuous, ciliate. *European USSR; widely cultivated.* H2. Summer.

Usually cultivated as the typical scarlet-flowered plant; both double and white-flowered variants are known.

2. L. coronaria (Linnaeus) Desrousseaux (*Agrostemma coronaria* Linnaeus; *Coronaria tomentosa* Braun). Illustration: Coste, Flore de la France 1: 182 (1901); Everard & Morley, Wild flowers of the world, pl. 32 (1970).
Densely whitish-hairy plant, often biennial, with erect, often branched stems to 80 cm, and ovate-lanceolate leaves. Flowers large, long-stalked, in few-flowered inflorescences. Calyx 1.5–2 cm. Petal-blade *c.* 1.2 cm, entire or shallowly 2-toothed, purplish red, pale purple or white; coronal scales lanceolate, acute. *SE Europe; widely cultivated and locally naturalised.* H3. Summer.

All three colour-variants are grown, and are readily established from self-sown seed.

3. L. flos-jovis (Linnaeus) Desrousseaux (*Agrostemma flos-jovis* Linnaeus; *Coronaria flos-jovis* (Linnaeus) Braun). Illustration: Coste, Flore de la France 1: 183 (1901); Parey's Blumengärtnerei, 591 (1958).
Densely whitish-hairy plant with erect, usually unbranched, stems to 80 cm, and lanceolate-spathulate leaves. Inflorescence more or less capitate, with 4–10 flowers. Calyx 1–1.2 cm. Petal-blade *c.* 8 mm, bifid with broad, often cut lobes, red or white; coronal scales lanceolate, acute. *C Alps; widely cultivated and locally naturalised.* H3. Summer.

The bright red-flowered plant is usually grown; the white variant is less attractive.

4. L. × haageana Regel. Illustration: Parey's Blumengärtnerei, 592 (1958); Botanical Magazine, n.s., 314 (1958–9); Everard & Morley, Wild flowers of the world, pl. 105 (1970).
Stems to 60 cm, with downwardly pointing hairs; leaves lanceolate, glandular-hairy. Inflorescence a few-flowered dichasium, glandular-hairy. Calyx *c.* 1.8 cm. Petal-blade *c.* 2 cm, broadly obovate, bifid and usually with a narrow tooth on each side and variably developed teeth on margins of lobes; coronal scales toothed. Flower-colour bright scarlet or rich orange-red. *Garden origin.* H3. Summer.

A group of garden hybrids of somewhat uncertain parentage but clearly involving far-eastern species (especially *L. fulgens* Sims) not themselves cultivated in European gardens. The original *L. × haageana* is not particularly easy to overwinter, and in recent years has been replaced with 'Arkwrightii' hybrids (*L. chalcedonica × L. × haageana*), especially 'Vesuvius' with dark foliage and very large orange-scarlet flowers.

For a discussion of the parentage of these hybrids, see Turrill in the notes to the Botanical Magazine plate cited above.

5. L. flos-cuculi Linnaeus. Illustration: Garrard & Streeter, Wild flowers of the British Isles, 47 (1983).
Slightly hairy perennial with decumbent non-flowering shoots and flowering stems to 75 cm. Leaves hairless, oblanceolate and stalked on non-flowering shoots, oblong-lanceolate and stalkless on flowering stems. Inflorescence a loose, few-flowered dichasium; flowers large on slender stalks. Calyx 5–6 mm, hairless; petal-blades 1.2–1.5 cm, deeply 4-fid with narrow, unequal, spreading segments; coronal scales subulate. *Throughout much of Europe, USSR (Caucasus, Siberia).* H2. Summer.

White-flowered and double variants are cultivated. A dwarf variant (var. **congesta** Ascherson & Graebner) is also sometimes seen in gardens.

6. L. viscaria Linnaeus (*Viscaria vulgaris* Bernhardi). Illustration: Garrard & Streeter, Wild flowers of the British Isles, 47 (1983).
Almost hairless, tufted perennial with short, erect non-flowering shoots; flowering shoots to 60 cm, erect, very sticky beneath each node. Leaves elliptic-lanceolate to oblong-lanceolate, with ciliate margins. Inflorescence a narrow, interrupted, spike-like panicle. Calyx *c.* 1.2 cm, purplish; petal-blade 8–10 mm, entire or shallowly bifid; coronal scales lanceolate, conspicuous. *Europe & W Asia.* H2. Early summer.

Several cultivars are known in gardens, of which the commonest is a double-flowered variant, 'Plena', with the purplish red flowers of the typical wild plant; illustration: Parey's Blumengärtnerei, 594 (1958).

7. L. alpina Linnaeus (*Viscaria alpina* (Linnaeus) G. Don). Illustration: Garrard & Streeter, Wild flowers of the British Isles, 47 (1983).
Like *L. viscaria*, but only up to 15 cm; stem not sticky; inflorescence dense, corymbose to almost capitate; petal-blade more deeply bifid. *Subarctic regions and mountains of the northern hemisphere.* H2. Early summer.

Useful as a mat-forming plant in rock gardens on sunny, silica-rich soils.

3. PETROCOPTIS Braun
S.M. Walters

Like *Silene* but capsule opening with 5 teeth, and seeds with a tuft of hairs at the hilum. All are perennial, with 5 styles.

A small genus of 7 closely related species, endemic to the Pyrenees and the mountains of N Spain. Two species are cultivated to a limited extent in rock-gardens.
Literature: Rothmaler, W., Petrocoptis in Tutin, T.G. et al., *Flora Europaea* 1: 157–8 (1964).

1a. Petals white or pale pink; bracts green with scarious margins **1. pyrenaica**
  b. Petals pinkish purple; bracts entirely scarious **2. glaucifolia**

1. P. pyrenaica (Bergeret) Braun (*Lychnis pyrenaica* Bergeret). Illustration: Coste, Flore de la France 1: 184 (1901).
Loosely tussock-forming, hairless plant with woody base, and thin, fragile flowering stems to 15 cm from basal leaf-rosettes. Rosette leaves ovate-lanceolate, upper stem-leaves broadly ovate. Inflorescence a loose, few-flowered cyme; bracts green with scarious margins. Calyx 5–8 mm; petal-blade *c.* 10 mm, entire or slightly indented, white or pale pink; coronal scales small. *W Pyrenees.* H3. Summer.

2. P. glaucifolia (Lagasca) Boissier (*P. lagascae* (Willkomm) Willkomm).
Like *P. pyrenaica* but without basal leaf-rosettes; bracts scarious; and petal-blade 1.2–1.5 cm, pinkish purple. *N Spanish mountains.* H3. Summer.

4. AGROSTEMMA Linnaeus
S.M. Walters

Annual herbs with erect stems and narrow leaves. Flowers large, solitary or in few-flowered dichasia. Calyx with long, linear teeth. Petals with long claw, slightly notched blade and no coronal scales. Styles 5. Capsule opening with 5 teeth, not stalked.

As defined here, a genus of only 3 species, native of S Europe and W Asia but introduced into N America.

1. A. githago Linnaeus (*Lychnis githago* (Linnaeus) Scopoli). Illustration: Coste, Flore de la France 1: 182 (1901); Garrard & Streeter, Wild flowers of the British Isles, 47 (1983).
Flowering stems to 1 m, simple or sparingly branched, covered with adpressed, whitish hairs. Leaves linear-lanceolate, acute. Flowers usually solitary, to 5 cm in diameter, on long, hairy stalks. Calyx-teeth 3–5 cm, spreading. Petals pale purplish red. *Formerly widespread over much of Europe as a cornfield weed, now rare and decreasing in many parts through agricultural improvement; probably native only to the E Mediterranean & SW Asia.* H2. Summer.

Increasingly grown as an interesting and decorative garden plant. A white-flowered variant is known.

5. CUCUBALUS Linnaeus
S.M. Walters

Like *Silene* but fruit a berry.

A genus of a single species from Eurasia and N Africa.

1. C. baccifer Linnaeus. Illustration: Coste, Flore de la France 1: 169 (1901); Garrard & Streeter, Wild flowers of the British Isles, 47 (1983).
Hairy perennial to 80 cm, with brittle, widely branching stems and ovate, acute, short-stalked leaves. Flowers in a loose dichotomous cyme, drooping on short stalks. Calyx 8–12 mm, broadly bell-shaped, with 5 long, obtuse teeth turned back in the fruiting stage. Petals 6–10 mm, with deeply bifid, spreading, greenish white blades. Fruit a round, black berry, borne on a short carpophore inside the persistent, green calyx. *Eurasia, N Africa.* H3. Summer.

Formerly cultivated as a medicinal plant and therefore well-known in earlier times; now rarely grown, except in botanic gardens.

6. GYPSOPHILA Linnaeus
S.M. Walters

Annual or perennial with somewhat woody stock. Flowers usually numerous and small in large spreading panicles; but some species with relatively large or solitary flowers (*G. aretioides*). Calyx joined, 5-toothed, with 1 vein to each tooth and a band of membranous tissue between each vein. Petals 5, entire or notched, usually with no clear differentiation into blade and claw; coronal scales absent. Stamens 10. Styles 2. Fruit a capsule opening with 4 teeth. Ovary with or without a short carpophore.

More than 100 species, mostly in SE Europe and Asia. Familiar to gardeners by the use of the inflorescences of *G. paniculata* and *G. elegans* as cut sprays for floral decorations. Dwarf species are grown in rock gardens and specialised alpine collections.

1a. Stems more than 30 cm, erect, with numerous flowers in wide, spreading panicles   **2**
 b. Stems less than 30 cm, with more or less spreading, usually few-flowered inflorescences, or flowers solitary   **4**
2a. Annual; flower-stalks many times longer than calyx   **8. elegans**
 b. Perennial; flower-stalks up to 3 times longer than calyx   **3**
3a. Leaves linear-lanceolate; petals less than 5 mm   **6. paniculata**
 b. Leaves broadly lanceolate; petals more than 5 mm   **7. pacifica**
4a. Dense cushion plant; flowers often solitary, stalkless   **1. aretioides**
 b. Mat-forming or tufted plants, with usually few-flowered inflorescences   **5**
5a. Inflorescence capitate   **2. petraea**
 b. Inflorescence not capitate   **6**
6a. Leaves spathulate   **3. cerastioides**
 b. Leaves linear or linear-lanceolate   **7**
7a. Tufted plant with leaves to 10 cm; calyx-teeth triangular-ovate   **5. tenuifolia**
 b. Loosely mat-forming plant with leaves not more than 3 cm; calyx-teeth oblong   **4. repens**

1. G. aretioides Boissier (*G. imbricata* Ruprecht). Illustration: Bulletin of the Alpine Garden Society 51: 116 (1983).
Dense cushion-forming plant, usually with solitary, stalkless flowers, but sometimes with few-flowered inflorescences to 4 cm. Leaves very small, fleshy, oblong, triangular in section, blunt. Petals white, entire. *Mountains of N Iran and USSR (Caucasus).* H2. Summer.

Needs full sun in rock crevices or tufa holes, and even then difficult to bring to flower. 'Caucasica' is a very dense, grey-green variant.

2. G. petraea (Baumgarten) Reichenbach. Illustration: Šavulescu, Flora Republicii Populare Romane 2: 199 (1953).
Densely tufted plant with linear, hairless leaves to 5 cm. Stems 4–20 cm, simple, hairless below and downy above, with a more or less capitate inflorescence surrounded by large, triangular-ovate bracts. Calyx 2–4 mm, with acute teeth. Petals 3–6 mm, white, narrowly ovate, entire. *Bulgaria, Romania (Rhodope mountains, E & S Carpathians).* H2. Summer.

3. G. cerastioides D. Don. Illustration: Wehrhahn, Die Gartenstauden 1: 320 (1931); Huxley, Garden perennials and water plants, t. 124 (1970).
Loose, mat-forming plant, greyish hairy, with the appearance of a *Cerastium*. Leaves spathulate, the lower long-stalked, the upper almost stalkless and obovate. Flowers to 1.8 cm across, in loose corymbs; petals white or lilac with pink veins. *Himalaya.* H2. Summer.

4. G. repens Linnaeus. Illustration: Coste, Flore de la France 1: 187 (1901); Beckett, Concise encyclopaedia of garden plants, 170 (1983).
Mat-forming, blue-green, hairless plant with non-flowering leafy shoots and ascending flowering stalks to 20 cm. Leaves 1.5–3 cm, linear-lanceolate. Inflorescence with up to 25 flowers in a loose, somewhat corymbose panicle. Calyx 3–4 mm, with oblong, obtuse teeth. Petals *c.* 7 mm, narrowly ovate, entire or slightly notched, white, pink or lilac. *Mountains of C & S Europe from NW Spain to the Carpathians.* H2. Summer.

Several named cultivars are grown.

5. G. tenuifolia Bieberstein. Illustration: Kolakovski, Flora Abkhazii 2: t. 8 (1939).
Tufted, hairless plant with numerous narrowly linear basal leaves and flowering stems to 20 cm. Inflorescence a few-flowered, almost corymbose panicle. Calyx 4–5 mm, teeth triangular-ovate, obtuse or with a small point. Petals 8–10 mm, pink, obovate, notched. *USSR (Caucasus).* H2. Summer.

6. G. paniculata Linnaeus. Illustration: Parey's Blumengärtnerei, 597 (1958); Beckett, Concise encyclopaedia of garden plants, 170 (1983) – 'Bristol Fairy'.
Tall hairless plant with thick rootstock, and much-branched stems forming an intricate bush of flowering branches. Leaves to 7 cm, linear-lanceolate, stalkless. Inflorescence a many-flowered, loose panicle. Calyx 3–4 mm, teeth triangular-ovate, obtuse,

with a short point. Petals 2–4 mm, white or pink. *Native to C & E Europe, but widely naturalised from gardens.* H1. Summer.

Numerous named cultivars exist, several with semi-double or double flowers. Some of these are important in nurseries producing cut flowers for decoration. 'Rosenschleier' is of hybrid origin, *G. paniculata × G. repens* 'Rosea'; see Jelitto et al., Die Freiland Schmuckstauden, 271 (1986) for a detailed account.

7. G. pacifica Komarov. Illustration: Komarov & Klobukava-Alisova, Opredelitel' rastenii dlanevostochnogo kraya **1**: t. 151 (1931); Gardeners' Chronicle **96**: 295 (1934).
Like *G. paniculata* but less freely and delicately branched; leaves broadly lanceolate; petals 5–8 mm, pink. *USSR (E Siberia), China (Manchuria).* H1. Late summer.

Unusually for the genus, tolerant of acid soil.

8. G. elegans Bieberstein. Illustration: Pignatti, Flora d'Italia **1**: 126 (1982).
Hairless annual with stems to 50 cm, branching above. Leaves oblong-lanceolate to linear-lanceolate. Inflorescence a loosely branched panicle with relatively large flowers on very long, slender stalks. Calyx 3–5 mm. Petals 8–15 mm, white with pink or purplish veins. *USSR (S Ukraine, Caucasus) & Turkey; widely cultivated and sometimes naturalised.* H1. Summer.

Several colour-varieties are in cultivation, especially for cut flowers.

7. SAPONARIA Linnaeus
S.M. Walters

Perennial (rarely annual) herbs, sometimes with rather woody stocks; indistinguishable from *Silene* or *Lychnis* except by the possession of 2 styles (not 3 or 5) and the opening of the capsule by 4 (not 6 or 10) teeth. *Saponaria pumilio*, which has 2 or 3 styles, was earlier included in the genus *Silene*.

About 20 species, mostly on mountains in S Europe and SW Asia. Horticultural requirements are diverse (see individual species accounts).
Literature: Simmler, G., Monographie der Gattung Saponaria, *Denkschrift der Akademie der Wissenschaft, Wien* **85**: 433–509 (1910).

1a.	Annual, with numerous small flowers in a loose, spreading inflorescence	**6. calabrica**
b.	Perennial; flowers solitary or in relatively few-flowered inflorescences, often large	2
2a.	Petals yellow	3
b.	Petals reddish or purplish (rarely white)	4
3a.	Basal leaves spathulate; filaments yellow	**1. bellidifolia**
b.	Basal leaves linear-lanceolate; filaments purple	**2. lutea**
4a.	Flowers solitary or paired	5
b.	Inflorescences with more than 2 flowers	7
5a.	Petal-blade bifid	**5. sicula**
b.	Petal-blade entire or slightly notched	6
6a.	Calyx inflated; styles often 3	**4. pumilio**
b.	Calyx not inflated; styles 2	**3. caespitosa**
7a.	Stem robust, erect; leaves with 3 almost parallel veins	**8. officinalis**
b.	Stem relatively slender or short, often prostrate; leaves 1-veined or with single mid-vein and spreading lateral veins	8
8a.	Petal-blade bifid	**5. sicula**
b.	Petal-blade entire or slightly notched	9
9a.	Basal leaves ovate-spathulate	**7. ocymoides**
b.	Basal leaves linear-lanceolate	**3. caespitosa**

1. S. bellidifolia Smith. Illustration: Coste, Flore de la France **1**: 186 (1901).
Loosely tufted, almost hairless perennial with woody stock. Basal leaves spathulate, 1-veined. Flowering stems 10–50 cm, erect, with more or less capitate inflorescence. Calyx 6–7 mm; petal-blade 4–5 mm, pale yellow, narrowly spathulate, notched; filaments yellow, usually long-projecting. *European mountains from the Pyrenees to the Balkans.* H2. Early summer.

2. S. lutea Linnaeus. Illustration: Coste, Flore de la France **1**: 187 (1901).
Like *S. bellidifolia* but with shorter, hairy stems, linear-lanceolate basal leaves, deeper yellow, entire petals and deep purple filaments. *SW & C Alps, restricted to limestone.* H2. Early summer.

3. S. caespitosa de Candolle. Illustration: Coste, Flore de la France **1**: 187 (1901); Parey's Blumengärtnerei, 606 (1958).
Rather densely tufted and slightly hairy perennial with woody stock. Basal leaves linear, thick in texture, keeled. Flowering stems to 15 cm, erect, with few, almost stalkless flowers. Petal-blade 4–7 mm, obovate, entire, pinkish purple, with conspicuous coronal scales. *C Pyrenees.* H3. Late summer.

Species **1–3** are grown in specialist rock-gardens, and can be propagated from cuttings.

4. S. pumilio (Linnaeus) Braun (*S. pumila* (St. Lager) Janchen; *Silene pumilio* Wulf). Illustration, Jelitto et al., Die Freiland Schmuckstauden, edn 2, 557, 558 (1985) – both S. pumilio & S. × olivana; Griffith, Collins guide to alpines, edn 2, pl 26 (1985) - as 'Bressingham'.
Densely tufted, almost hairless perennial with woody stock and crowded linear, somewhat fleshy basal leaves. Flowers large, solitary or paired on very short stalks, mostly produced around the edge of the tussock. Petal-blade 7–10 mm, ovate, entire to notched, pinkish purple (rarely white), with conspicuous coronal scales. Styles often 3. *E Alps & SE Carpathians.* H3. Summer.

Intolerant of lime and not easy to cultivate, but can be represented in rock-garden collections by S. × olivana Wocke, hybrids involving *S. caespitosa* and *S. ocymoides* ('Bressingham') which are much easier to grow and very decorative.

5. S. sicula Rafinesque subsp. **intermedia** (Simmler) Chater (*S. haussknechtii* Simmler).
Tufted perennial with woody stock and hairless, narrowly obovate basal leaves. Flowering stems 5–50 cm with 1–30 flowers. Calyx 2–2.5 cm, cylindric; petal-blade 5–8 mm, deeply bifid, red. *Mountains in N Greece, Albania & SW Yugoslavia.* H2. Summer.

Dwarf plants, often with solitary flowers, to which the name *S. haussknechtii* applies, are sometimes cultivated in rock-gardens. A hybrid with *S. cypria* Boissier (Cyprus, not in general cultivation), S. × lempergii Anon. is also grown, sometimes as the cultivar 'Max Frei'.

6. S. calabrica Gussone. Illustration: Fiori & Paoletti, Iconographia Florae Italicae, edn 3, **1**: 148 (1933) – as a variety of S. ocymoides.
Annual, with widely branching stems hairless in the lower part and glandular-hairy in the many-flowered inflorescence. Basal leaves more or less spathulate. Petal-blade 3–5 mm, pale purplish. Flower-stalks often deflexed in fruit. *E Mediterranean area.* ?H4. Spring.

7. S. ocymoides Linnaeus. Illustration: Coste, Flore de la France **1**: 186 (1901); Jelitto et al., Die Freiland Schmuckstauden, edn 2, 558 (1985) – 'Rubra Compacta'.
Much-branched, loosely carpeting perennial with prostrate or ascending hairy

stems to 20 cm. Lower leaves ovate-lanceolate, upper narrower. Inflorescence loose, spreading, with numerous flowers in small clusters often covering the whole creeping mat. Petal-blade 3–5 mm, red, pink or white. *Mountains of SW & SC Europe, from Spain to Yugoslavia.* H2.

Lime-loving. Very commonly cultivated as an easy and decorative carpeting plant for rock-gardens, walls, etc., often effectively naturalised by self-sown seed. Named cultivars exist, e.g. 'Rubra Compacta' with neater habit and deep red flowers.

8. S. officinalis Linnaeus. Figure 27(1, 4), p. 178. Illustration: Coste, Flore de la France 1: 186 (1901); Garrard & Streeter, Wild flowers of the British Isles, 47 (1983). Robust, almost hairless perennial to 60 cm with thick underground stolons giving rapid vegetative spread. Basal and stem leaves ovate with 3 prominent, more or less parallel veins. Inflorescence a condensed dichasium. Flowers large; calyx *c.* 2 cm; petal-blade *c.* 1 cm, narrowly obovate, almost entire, pink, red or white; coronal scales linear. *Europe; widely naturalised and formerly much cultivated as a useful herb.* H1. Late summer.

Often represented by the pink-flowered double variant; white and red-flowered cultivars also exist in both single and double-flowered variants.

8. VACCARIA Medicus
S.M. Walters
Like *Saponaria*, but calyx-tube winged and without coronal scales.

A genus of 3 species native to the Mediterranean area.

1. V. hispanica (Miller) Rauschert (*V. pyramidata* Medicus; *Saponaria vaccaria* Linnaeus). Illustration: Coste, Flore de la France 1: 186 (1901).
Hairless, glaucous annual, with erect branching stems to 60 cm, and oblong-lanceolate, stalkless stem-leaves. Calyx inflated, with 5 green, winged angles and short triangular-acute teeth. Petal-blade 3–8 mm, wedge-shaped, toothed and sometimes slightly bifid, pinkish purple. *S & C Europe; introduced in N America.* H4. Summer.

Sometimes grown in gardens; in N Europe not infrequently arising from bird-seed and then preserved as an interesting plant. Also occasionally grown as a cut-flower crop.

9. DIANTHUS Linnaeus
R.F.L. Hamilton & S.M. Walters
Herbs, often hairless and with more or less woody stocks; rarely small shrubs. Leaves opposite, usually linear, entire, often glaucous. Flowering stems often swollen and brittle at nodes. Flowers conspicuous, solitary or in few- to many-flowered inflorescences, sometimes capitate with distinct involucral bracts surrounding the base of the head. Calyx usually cylindric, enclosed in 1–3 pairs of epicalyx bracts (see Figure 27(5), p. 178). Petals free, with narrow claw and wide spreading blade, often somewhat cut or toothed but not bifid. Coronal scales absent but petals often with a tuft of hairs ('bearded') in mouth of flower. Stamens 10. Styles 2. Fruit a capsule opening apically with 4 teeth. A short carpophore often present between calyx and capsule. Seeds numerous, concave on 1 side.

A large genus, easily recognised by the decorative, often scented flowers having 2 styles and epicalyx bracts. About 300 species in Europe and Asia, and extending to arctic N America (1 species) and to the African mountains; a few species long-cultivated for their attractive form and perfume.

A concise account of this genus in gardens can not satisfy even the general reader, and will certainly irritate the specialist. It is easy to recognise a *Dianthus*, whether the flower is single or double; but the long and complex history of garden 'Carnations' and 'Pinks' means that most garden representatives are not assignable to any particular species and can at best only be given a cultivar name (*The International Dianthus Register* (edn 2, 1983) and its annual supplement list more than 30 000 cultivar names.) Most 'species *Dianthus*' are grown by specialist alpine gardeners, and even here, the wild plant is often superseded by an 'improved' cultivar (as in the cases of *D. deltoides* and *D. gratiano-politanus*).

Hybridisation is frequent and surprisingly successful even between widely differing species, both in nature and, much more obviously, in cultivation (see Pax & Hoffmann, pp. 360–1, cited below, for further information).

In this account, both *D. caryophyllus* and *D. plumarius* are described and keyed, although the 'unimproved' wild species are hardly ever grown except in botanic gardens. It was felt that their intrinsic interest justifies this decision. Nearly all garden Carnations (*D. caryophyllus*) are

double-flowered, but many cultivated 'Pinks' (*D. plumarius* and other species) are single-flowered.

So-called 'annual' *Dianthus* cultivars, commonly grown from seed every year in gardens, involve mainly either *D. barbatus* or *D. chinensis* (which in the wild are short-lived perennials or biennials). Apart from these, nearly all *Dianthus* are truly perennial; most are at least slightly woody at the base, and characteristically the basal leaves are evergreen. They can all readily be propagated by layering or cuttings, and with few exceptions prefer a calcareous, well-drained soil. Many are therefore ideal rock-garden plants.

The classification and order of species adopted here is essentially that of Pax & Hoffmann (cited below), which is itself based on the work of Williams (also cited below). The definition of the species generally follows Tutin in *Flora Europaea* 1: 188–204 (1964).
Literature: Williams, F.N., Enumeratio Specierum Varietatumque Generis Dianthus, *Journal of Botany* 22: 340–349 (1885); Monograph of the Genus Dianthus, *Journal of the Linnean Society (Botany)*, 29: 346–378 (1893); Pax, F. & Hoffmann, K., Dianthus, *Die Natürlichen Pflanzenfamilien*, edn 2, 16c: 356–361 (1934) – especially for infrageneric classification and literature references; Whitehead, S.B., *Carnations today* (1956) – deals with the whole genus in gardens: particularly clear on history; Leslie, A.C., *International Dianthus Register*, edn 2 (1983) – especially the introductory 'explanatory notes'.

Habit. Small shrubs: **9,10.**
Leaves. Lanceolate-elliptic with prominent midrib and 2 more or less parallel lateral veins: **8.**
Inflorescence. Capitate, with several flowers in each head: **1–8,11.**
Calyx. Hairy: **1–5,22,24,28.**
Petals. Deeply cut: **24, 36–41,44.** Yellow: **2, 31** (cultivars), **36** (cultivars).

1a. Shrubs   2
  b. Herbs, sometimes more or less woody at the base   3
2a. Leaves linear, round in section **9. arboreus**
  b. Leaves oblong-lanceolate, flat **10. fruticosus**
3a. Flowers more or less stalkless, in heads surrounded by involucral bracts   4
  b. Flowers stalked, solitary or in a loose inflorescence, with no involucral bracts   12

4a. Flowers yellow   **2. knappii**
  b. Flowers purple, pink or white   5
5a. Leaves lanceolate-elliptic   **8. barbatus**
  b. Leaves narrowly oblong to linear   6
6a. Annual or biennial, obviously hairy
                1. armeria
  b. Perennial, hairless or only slightly
    hairy   7
7a. Stems 4-angled   8
  b. Stems more or less round in section
                10
8a. Epicalyx-bracts with terminal point
    less than half as long as calyx
                4. giganteus
  b. Epicalyx-bracts with terminal point
    more than half as long as calyx   9
9a. Petal-blades 5–8 mm   **3. cruentus**
  b. Petal-blades 1–1.5 cm
         7. carthusianorum
10a. Robust plant to 60 cm; leaves
    gradually narrowed from base
                5. capitatus
  b. Slender, compact plants to 40 cm;
    leaves more or less parallel-sided to
    near tip   11
11a. Dense cushion plant with bristle-like
    leaves   **11. pinifolius**
  b. Loosely tufted plant with rather soft
    leaves   **6. biflorus**
12a. Petal-blades deeply cut to at least one-
    third, usually bearded   13
  b. Petal-blades entire, toothed or cut to
    not more than one-quarter, bearded
    or beardless   22
13a. At least some leaves more than 4 cm
                14
  b. Leaves not longer than 3.5 cm   17
14a. Basal leaves more or less withered at
    flowering time   15
  b. Basal leaves persistent in a rosette at
    flowering time   16
15a. Calyx about 3 times as long as wide;
    epicalyx-bracts about half as long as
    calyx, with long terminal point
                24. chinensis
  b. Calyx at least 4 times as long as wide;
    epicalyx-bracts one-quarter to one-
    third as long as calyx, with abrupt,
    short terminal point   **40. superbus**
16a. Epicalyx-bracts less than one-third as
    long as calyx, abruptly pointed or
    almost truncate, adpressed
                36. plumarius
  b. Epicalyx-bracts at least one-third as
    long as calyx, gradually tapering to
    tip, somewhat spreading, especially at
    tip   **41. monspessulanus**
17a. Epicalyx-bracts at least half as long as
    calyx   **33. furcatus**
  b. Epicalyx-bracts not more than one-
    third as long as calyx   18

18a. At least some leaves blunt   19
  b. Leaves sharp-pointed   20
19a. Petal-blades cut to about one-third;
    plant downy below with minute hairs
                39. gallicus
  b. Petal-blades cut to more than half;
    plant hairless   **37. arenarius**
20a. Leaves spine-like, recurved, not
    crowded at base of stem
                38. squarrosus
  b. Leaves neither spine-like nor
    recurved, mostly basal   21
21a. Petal-blades more than 10 mm
                44. spiculifolius
  b. Petal-blades 4–10 mm   **43. petraeus**
22a. At least some leaves blunt   23
  b. Leaves sharp-pointed   36
23a. Epicalyx-bracts more or less covering
    calyx   24
  b. Epicalyx-bracts covering not more
    than three-quarters of calyx   29
24a. Flowers 2–6 (rarely 1) in a loose
    inflorescence   25
  b. Flowers solitary   26
25a. Calyx 1–1.2 cm   **13. nitidus**
  b. Calyx more than 1.2 cm
                23. haematocalyx
26a. Flowering stems with at least 5 pairs
    of leaves   **20. callizonus**
  b. Flowering stems with 4 or fewer pairs
    of leaves   27
27a. Basal leaves at least 3 mm wide; open
    flower flat   **12. alpinus**
  b. Basal leaves not more than 2.5 mm
    wide; open flower concave   28
28a. Leaves soft; calyx 1.2–1.6 cm; petals
    purplish red   **15. glacialis**
  b. Leaves rigid; calyx 9–10 mm; petals
    pink   **16. freynii**
29a. Leaves elliptic, those on basal shoots
    less than 5 mm   **29. myrtinervius**
  b. Leaves linear to oblanceolate, usually
    more than 10 mm   30
30a. Stems and leaves more or less hairy
                28. deltoides
  b. Plant hairless, though sometimes
    leaves rough on margin   31
31a. Calyx at least 1.3 cm   32
  b. Calyx less than 1.3 cm   33
32a.   Flowers solitary **42. gratianopolitanus**
  b. Flowers usually 2–4 together
                25. seguieri
33a. Epicalyx-bracts more than half as
    long as calyx   **13. nitidus**
  b. Epicalyx-bracts not more than half as
    long as calyx   34
34a. Epicalyx-bracts 2, separated from
    calyx by a short length of stalk
                17. microlepis
  b. Epicalyx-bracts 2 or 4, immediately
    below calyx and overlapping it   35

35a. Epicalyx-bracts 2, about half as long
    as calyx   **14. scardicus**
  b. Epicalyx-bracts 4, about one-third as
    long as calyx   **34. subacaulis**
36a. Epicalyx-bracts about equalling calyx
                37
  b. Epicalyx-bracts not more than three-
    quarters as long as calyx   39
37a. Calyx 1.2–1.5 cm   **21. pavonius**
  b. Calyx at least 1.6 cm   38
38a. Flowers usually solitary, bright pink
    with darker 'eye'   **20. callizonus**
  b. Flowers 2–5 (rarely 1), deep pinkish
    purple   **23. haematocalyx**
39a. Petal-blades entirely hairless   40
  b. Petal-blades bearded or otherwise
    hairy (sometimes minutely so)   45
40a. Epicalyx-bracts gradually tapering to
    a point   41
  b. At least the inner epicalyx-bracts
    truncate to abruptly apiculate   42
41a. Leaves c. 0.5 mm wide   **35. pungens**
  b. Leaves 1–3 mm wide   **33. furcatus**
42a. Leaves not more than 1 cm   43
  b. Leaves at least 2.5 cm   44
43a. Petal-blades not more than 3.5 mm,
    white; leaves acuminate
                32. anatolicus
  b. Petal-blades at least 5 mm, often pink
    or reddish; leaves acute or obtuse
                34. subacaulis
44a. Basal leaves twisted, wiry, usually less
    than 1 mm wide   **30. sylvestris**
  b. Basal leaves straight, rather soft, more
    than 1 mm wide   **31. caryophyllus**
45a. Sheaths of stem-leaves at least 3 times
    as long as the stem is wide   46
  b. Sheaths of stem-leaves not more than
    twice as long as the stem is wide   47
46a. Petals with short glandular hairs
    scattered over upper surface
                6. biflorus
  b. Petals with long non-glandular hairs
    on upper surface   **27. gracilis**
47a. Petals yellowish beneath   48
  b. Petals sometimes paler beneath but
    not yellowish   50
48a. Petal-blades c. 1.2 cm   **18. zonatus**
  b. Petal-blades not more than 7 mm   49
49a. Plant cushion-forming, to 12 cm
                19. brevicaulis
  b. Plant not cushion-forming, to 40 cm
                22. campestris
50a. Calyx-teeth obtuse to somewhat acute
                42. gratianopolitanus
  b. Calyx-teeth acute or acuminate   51
51a. Leaves bristle-like   **26. nardiformis**
  b. Leaves not bristle-like   52
52a. Calyx hairy   **28. deltoides**
  b. Calyx hairless   53

53a. Epicalyx-bracts at least two-thirds as
     long as calyx   **25. seguieri**
  b. Epicalyx-bracts less than two-thirds as
     long as calyx   **54**
54a. Plant hairless; petal-blades usually
     white   **43. petraeus**
  b. Plant somewhat hairy below; petal-
     blades reddish purple **45. graniticus**

Subgenus **Armeriastrum** Seringe. Flowers
in heads surrounded at the base by several
involucral bracts.

1. D. armeria Linnaeus. Illustration: Coste,
Flore de la France **1**: 190 (1901); Garrard &
Streeter, Wild flowers of the British Isles, 46
(1983).
Hairy annual or biennial to 40 cm. Basal
leaves in a rosette, narrowly oblong,
obtuse; stem-leaves linear, acute. Flowers
stalkless or nearly so, in heads surrounded
by linear, leaf-like involucral bracts
equalling or slightly overtopping the
flowers. Calyx 1.5–2 cm, woolly, narrowed
above the middle, with lanceolate, acute
teeth; epicalyx bracts 2, lanceolate-
subulate, equalling calyx. Petal-blade
4–5 mm, narrowly ovate, toothed, bearded,
bright pink with pale dots; claw narrow,
white. Capsule cylindric, equalling calyx.
Most of Europe but rare in the N; W Asia. H2.
Summer.

2. D. knappii (Pantocsek) Borbás.
Illustration: Addisonia **29**: pl. 610 (1935).
Hairy perennial to 40 cm. Leaves linear-
lanceolate with prominent midrib. Flowers
stalkless, in heads which are usually paired,
shortly stalked; outer involucral bracts leaf-
like, inner membranous. Calyx *c.* 1.5 cm
with ovate, acuminate teeth; epicalyx-
bracts 4 or 5, ovate with long, green points
nearly equalling the calyx. Petal-blade
c. 7 mm, yellow with purplish spot at base.
W Yugoslavia. H3. Summer.

Unique among wild *Dianthus* in having
yellow flowers and sometimes cultivated for
that reason.

3. D. cruentus Grisebach (*D. calocephalus*
Baldacci not Boissier; *D. lateritius* Halacsy;
D. holzmannianus Heldreich & Hauss-
knecht). Illustration: Flore des Serres **5**:
488 (1849).
Loosely tufted, somewhat glaucous
perennial with 4-angled stems to 60 cm
(rarely more). Leaves to 15 cm, linear.
Flowers in dense heads with ovate, hairy
involucral bracts with long terminal points.
Calyx *c.* 2 cm, hairy, usually reddish
purple, with acuminate teeth. Epicalyx-
bracts similar to involucral bracts but
smaller. Petal-blade 5–8 mm, toothed,

usually bearded, crimson or deep purple.
Balkan Peninsula. H3. Summer.

4. D. giganteus D'Urville. Illustration:
Sweet, British flower garden **3**: t. 288
(1829); Polunin, Flowers of Greece and the
Balkans, pl. 4, t. 187b (1980).
Robust perennial to 1 m, like *D. cruentus*,
but larger; involucral and epicalyx-bracts
with only very short terminal points. The
flowers are purple. *Balkan peninsula, north to
Romania.* H3. Summer.

5. D. capitatus de Candolle. Illustration:
Reichenbach, Icones Florae Germanicae et
Helveticae **6**: t. 249 (1844).
Like *D. cruentus* but stems not 4-angled,
usually with a single, dense head of purple
flowers, and leaves somewhat wider,
gradually narrowed from base. *SE Europe,
extending to S Urals.* H3. Summer.

6. D. biflorus Sibthorp & Smith. Illustration:
Sibthorp & Smith, Flora Graeca **4**: t. 393
(1825).
Loosely tufted, hairless perennial to 40 cm.
Leaves to 7 cm, linear, acuminate,
3–5-veined, rather soft in texture, evenly
spaced up stem; sheaths of stem-leaves
several times as long as stem is wide.
Flowers sometimes solitary, but more
usually 2–4 (rarely to 8) in a head
subtended by 2–4 long, pointed involucral
bracts nearly as long as flower(s). Calyx
2–2.5 cm × *c.* 5 mm, pale green, tapering
upwards from the middle. Epicalyx-bracts 4
or 6, obovate, each tapered to a very
narrow, bristle-like apex, leathery and pale,
one-third to half as long as calyx. Petal-
blade 8–16 mm, obovate, toothed, often
with short glandular hairs scattered over
the whole surface, reddish purple above,
yellowish beneath. *Mountains of Greece.* H3.
Summer.

7. D. carthusianorum Linnaeus (*D.
atrorubens* Allioni; *D. sanguineus* Visiani; *D.
tenuifolius* Schur). Illustration:
Reichenbach, Icones Florae Germanicae et
Helveticae **6**: t. 250–2 (1844) – showing
range of variation; Coste, Flore de la France
1: 191 (1901).
Usually hairless perennial with 4-angled
stems to 60 cm. Leaves linear, to 2 cm ×
5 mm, flat, with very long sheaths. Flowers
in more or less dense heads with lanceolate
or oblong involucral bracts. Calyx 1–2 cm;
epicalyx-bracts obovate or obcordate, with
short terminal points, more or less
membranous, shorter than calyx. Petal-
blade 1–1.5 cm, obovate, toothed, bearded,
purple, pink or (rarely) white. *S, W & C
Europe.* H3. Summer.

A very variable species. Deep reddish-
purple variants (*D. atrorubens* Allioni) are
sometimes cultivated.

8. D. barbatus Linnaeus. Illustration: Coste,
Flore de la France **1**: 190 (1901); Polunin
& Smythies, Flowers of southwest Europe,
pl. 4 (1973); Beckett, The concise
encyclopaedia of garden plants, 123 (1983)
– cultivars.
Hairless or sparsely hairy, short-lived
perennial with smooth, robust, ascending
stems to 70 cm and lanceolate-elliptic
leaves to 10 × 2 cm, with prominent
midribs and 2 more or less parallel lateral
veins. Flowers stalkless in large dense
heads, with leaf-like, pointed involucral
bracts usually as long as the flowers. Calyx
1.5–1.8 cm, hairless, with narrow, pointed
teeth; epicalyx-bracts 4, ovate, with
terminal point, equalling or exceeding
calyx. Petal-blade *c.* 1 cm, bearded, reddish
purple. *S Europe, from Pyrenees to
Carpathians and Balkans, long-cultivated and
widely naturalised.* H3. Summer.

A very familiar garden plant ('Sweet
William') often grown as an annual, and
represented by many colour and double
flowered variants. The relatively broad
leaves distinguish it easily from all other
species. For an account of its hybrids with
other species and resulting cultivars see
Whitehead's book cited on p. 185.

9. D. arboreus Linnaeus. Illustration:
Loddige's Botanical Cabinet **5**: t. 459
(1820); Gardeners' Chronicle **43**: 52 (1908).
Shrub to more than 1 m (in gardens), old
plants developing a thick trunk. Leaves
3–5 mm wide, linear-oblong, rather thick
and fleshy, greyish green. Flowers
numerous, large, scented, shortly stalked,
in heads of 2 or 3. Epicalyx-bracts 10–20,
obovate, tapering abruptly to a point. Calyx
c. 2 cm × 2–3 mm. Petal-blade *c.* 1 cm,
obovate, toothed, bearded, pink. *Crete and
the S Aegean; very rare on mainland Greece.*
H5–G1. Summer.

Cultivated in the 19th century, and
possibly involved in early hybridisation of
the 'perpetual' Carnation. Runemark,
Botaniska Notiser **133**: 475–490 (1980)
considers *D. arboreus* to be part of
D. fruticosus.

10. D. fruticosus Linnaeus. Illustration:
Sibthorp & Smith, Flora Graeca **5**: t. 407
(1825).
Like *D. arboreus* but with flat, elliptic-
oblanceolate leaves 4–8 mm wide, and
petals red. The flowers are said to be
scentless. *Aegean Isles (Cyclades).* H5–G1.
Summer.

11. D. pinifolius Sibthorp & Smith.
Illustration: Polunin, Flowers of Greece and
the Balkans, pl. 4, t. 194g (1980).
Dense cushion-forming perennial with
flowering stems to 40 cm. Leaves less than
1 mm wide, bristle-like, with a sheath
2–3 mm. Flowers several in dense heads on
stalks to 8 mm, and with ovate, pointed,
leathery involucral bracts. Calyx 1–2 cm;
epicalyx-bracts 6, obovate, with a terminal
point 2–4 mm, half as long as calyx. Petal-
blade 5–10 mm, toothed, sometimes
bearded, purple, pink or lilac. *Balkan
mountains.* H3. Summer.

Subgenus **Caryophyllum** Seringe. Flowers
solitary or in groups of 2 or 3 (rarely
more), each flower stalked. Involucral
bracts absent.

Section **Barbulatum** Williams. Petal-
blade bearded, entire or toothed but not
deeply divided.

12. D. alpinus Linnaeus. Illustration:
Wehrhahn, Die Gartenstauden 1: 341
(1931); Hay & Synge, Dictionary of garden
plants in colour, pl. 49 (1971).
Tufted, hairless perennial to 15 cm; leaves
to 2.5 cm × 5 mm, linear-lanceolate to
oblong-lanceolate, obtuse, shiny, with
prominent midrib. Flowers large, solitary
on stems usually 5–10 cm, with 2–3 pairs
of stem-leaves. Open flower flat. Calyx
1.5–2 cm × 6–8 mm, widened upwards,
dull purple, striped, hairless; teeth acute,
slightly hairy. Epicalyx-bracts usually 4
(rarely 2 or 6), ovate with subulate apex,
green with a membranous margin, from
half as long to as long as calyx. Petal-blade
1.5–2 cm, obovate, finely toothed, bearded,
deep pink with purplish spots on a white
ground in centre. *E Alps.* H3. Summer.

A handsome rock-garden plant. Several
cultivars are grown, including a white-
flowered 'Albus', and crosses to *D.* ×
'Allwoodii' have given a range of plants
usually referred to as 'Allwoodii Alpinus'
(for an account and illustration of these,
see Whitehead's book cited on p. 185).

13. D. nitidus Waldstein & Kitaibel.
Illustration: Reichenbach, Icones Florae
Germanicae et Helveticae 6: t. 261 (1844).
Like *D. alpinus* but to 30 cm, more loosely
tufted, with often branching stems with
2–5 flowers; calyx 1–1.2 cm and petal-
blade 8–10 mm, pink, sharply toothed.
W Carpathian mountains. H3. Summer.

14. D. scardicus Wettstein (*D. nitidus*
misapplied). Illustration: Polunin, Flowers
of Greece and the Balkans, pl. 4, t. 194f
(1980).

Like *D. alpinus* but more densely tufted;
epicalyx-bracts 2, with spreading teeth,
about half as long as calyx, and petal-blade
7–8 mm, pink. *Yugoslavia (Sar Planina).*
H3. Summer.

15. D. glacialis Haenke (*D. gelidus* Schott,
Nyman & Kotschy). Illustration:
Wehrhahn, Die Gartenstauden 1: 342
(1931); Parey's Blumengärtnerei, 599
(1958).
Densely tufted perennial, usually hairless,
to 10 cm. Leaves more or less linear, but
widest above middle, obtuse, soft in texture,
sometimes with ciliate margins. Flowers
usually solitary on stems with only 1–2
pairs of leaves. Calyx 1.2–1.6 cm ×
3–7 mm, widening upwards. Epicalyx-
bracts 2 or 4, ovate-lanceolate with long
terminal point, nearly equalling calyx.
Petal-blade 8–12 mm, purplish pink with a
white centre, yellowish beneath, finely
toothed, bearded. Petals ascending, so that
open flower is concave, not flat. *E Alps &
Carpathians.* H3. Summer.

16. D. freynii Vandas. Illustration:
Wehrhahn, Die Gartenstauden 1: 343
(1931); Bulletin of the Alpine Garden
Society 38: 392 (1970), 41: 46 (1973).
Like *D. glacialis* but leaves very narrow and
rigid; calyx 8–10 mm and petal-blade
5–9 mm. *Mountains of C Yugoslavia and
S Bulgaria.* H3. Summer.

17. D. microlepis Boissier. Illustration:
Wehrhahn, Die Gartenstauden 1: 340
(1931); Bulletin of the Alpine Garden
Society 41: 54 (1973); Polunin, Flowers of
Greece and the Balkans, pl. 4, t. 194g
(1980) – flower-colour wrong.
Dense cushion-forming perennial with very
short flowering stems usually only 1–2 cm
and bearing 1–2 (rarely 0) pairs of scale-
like stem-leaves. Basal leaves 1–2 cm ×
1.5 mm, linear, obtuse, with prominent
midrib. Flowers solitary. Calyx 9–11 mm,
widening upwards with short, acute teeth.
Epicalyx-bracts 2, ovate, membranous, not
more than half as long as calyx and not
closely adpressed to it. Petal-blade 6–7 mm,
irregularly toothed, bearded, purple.
Mountains of Bulgaria. H3. Summer.

Nos. **14–17** are choice 'alpines' needing
protection from winter wet.

18. D. zonatus Fenzl. Illustration: Davis
(ed.), Flora of Turkey 2: 109 (1967) –
epicalyx & calyx.
Tufted perennial to 30 cm. Leaves to 8 cm,
c. 2 mm wide, linear, acuminate. Flowers
solitary, calyx *c.* 2 cm × 5 mm, pale green
tinged with purple, teeth long, acuminate.

Epicalyx-bracts 4–8, long-acuminate, half
to three-quarters as long as calyx. Petal-
blade *c.* 1.2 cm, deep pink, yellowish
beneath. *Turkey, W Syria.* H3. Summer.

19. D. brevicaulis Fenzl. Illustration: Davis
(ed.), Flora of Turkey 2: 109 (1967) –
epicalyx & calyx.
Cushion-forming, glaucous perennial to
12 cm. Leaves 6–20 × 0.5–1.5 mm,
toothed, with very short point; often only 1
pair of stem-leaves close under the solitary
flower. Epicalyx-bracts 4 or 6, purplish,
ovate, sharp-pointed, up to half as long as
calyx. Petal-blade 4–7 mm, toothed,
bearded (sometimes minutely so), deep
pink, yellowish beneath. *Turkey.* H3.
Summer.

20. D. callizonus Schott & Kotschy.
Illustration: Farrer, English rock garden 1:
pl. 27 (1919); Wehrhahn, Die
Gartenstauden 1: 344 (1931); Šavulescu,
Republicii Populare Romane 2: 274, pl. 41
(1953).
Loosely mat-forming, hairless perennial to
20 cm. Leaves linear-lanceolate, the basal
obtuse, shorter than stem-leaves. Stem-
leaves at least 5 pairs, upper acute or
acuminate. Flowers large, solitary. Calyx
c. 1.6 cm, widening upwards, dark red with
ovate-lanceolate, acuminate teeth.
Epicalyx-bracts 2 or 4, ovate with long
subulate green tips, about equalling calyx.
Petal-blade 1–1.5 cm, more or less
triangular, deeply toothed, bright pink with
a basal zone of deeper colour composed of
dark spots. *S Carpathian mountains.* H3.
Summer.

A very handsome plant. The artificial
hybrid with *D. alpinus* (**D.** × **calalpinus**
invalid) is even better.

21. D. pavonius Tausch (*D. neglectus*
Loiseleur; *D. roysii* of gardens). Illustration:
Coste, Flore de la France 1: 193 (1901);
Wehrhahn, Die Gartenstauden 1: 345
(1931); Bulletin of the Alpine Garden
Society 41: 53 (1970).
Densely mat-forming, hairless perennial to
15 cm, usually greyish green. Leaves to
4 cm, narrowly linear, acuminate. Flowers
usually solitary but occasionally 2 or 3
together. Calyx 1.2–1.5 cm × 4–5 mm,
narrowing upwards, purplish with
membranous, acute teeth. Epicalyx-bracts
4, ovate-subulate, the outer pair often
longer than the calyx. Petal-blade
1–1.5 cm, toothed, bearded, crimson to
pale pink, buff beneath. *SW Alps (France,
Italy).* H3. Summer.

Unusually, this species has a reputation of
preferring neutral to acid soils in gardens.

22. D. campestris Bieberstein. Illustration: Flora SSSR **6**: 632 (1936); Šavulescu, Flora Republicii Populare Romane **2**: 274, pl. 41 (1953).
Perennial, usually hairy, with stout stock and numerous stems to 40 cm. Leaves linear, acute, the lower often withered at flowering-time. Flowers solitary or in pairs at tips of branches. Calyx 1.5–1.8 cm × 3–4 mm. Epicalyx-bracts 4 or 6, ovate-acuminate, one-third as long as calyx. Petal-blade 6–7 mm, toothed, bearded, pink or purplish, yellow-green beneath. *SE Europe, USSR (Caucasus, Siberia).* H2. Summer.

23. D. haematocalyx Boissier & Heldreich (*D. pindicola* Vierhapper). Illustration: Huxley & Taylor, Flowers of Greece and the Aegean, 77 (1977); Polunin, Flowers of Greece and the Balkans, pl. 4, t. 193e (1980).
Tufted, hairless perennial to 30 cm. Leaves linear, acute, with thickened margins, sheaths not more than twice as long as stem is wide. Flowers solitary or in clusters of 2–4, borne above the clusters of leafy shoots. Calyx 1.6–2.4 cm × 4–6 mm, narrowed upwards, usually reddish purple, with long, acuminate teeth. Epicalyx-bracts 4 or 6, half as long to as long as calyx. Petal-blade 6–12 mm, obovate, toothed, sparsely bearded, deep pinkish purple, greenish yellow beneath. *Balkan mountains.* H3. Summer.

24. D. chinensis Linnaeus. Illustration: Beckett, The concise encyclopaedia of garden plants, 123 (1983).
Biennial or short-lived perennial, usually somewhat hairy, to 70 cm. Basal leaves withering before flowering; stem-leaves to 8 cm × 6 mm, lanceolate, acute. Inflorescence a loose cluster of up to 15 large flowers. Calyx *c.* 2 cm, teeth shortly acuminate. Epicalyx-bracts 4 or 6, abruptly contracted to a long point, *c.* half as long as calyx. Petal-blade *c.* 1.5 cm, obovate, deeply toothed or cut to nearly half, pinkish lilac with purplish eye. *China, long cultivated there.* H4. Summer.

Represented in gardens by a range of showy cultivars usually raised from seed and treated as annuals. 'Heddewigii' is a name applied to a range of compact, free-flowering cultivars; illustration: Parey's Blumengärtnerei, 604 (1958).

25. D. seguieri Villars (*D. sylvaticus* Hoppe). Illustration: Coste, Flore de la France **1**: 193 (1901); Hegi, Illustrierte Flora von Mitteleuropa **3**: 341, t. 103 (1909);

Polunin & Smythies, Flowers of southwest Europe, pl. 4 (1973).
Loosely tufted, hairless, green perennial to 60 cm. Leaves linear-lanceolate, acute or obtuse; stem-leaves usually more than 4 pairs, with short sheaths. Inflorescence 1–few-flowered. Calyx 1.4–2 cm, with acute teeth. Epicalyx-bracts 2–6, ovate-subulate, often nearly as long as calyx. Petal-blade 7–17 mm, obovate, deeply toothed, bearded, pink with a pale purplish spotted band near base. *SW Europe, from Spain to Italy.* H3. Summer.

A variable species, represented in gardens mainly by its hybrids (see *D. monspessulanus*, p. 190).

26. D. nardiformis Janka. Illustration: Šavulescu, Flora Republicii Populare Romane **2**: 269 (1953).
Tufted perennial to 10 cm with slender, branched, somewhat woody stems. Basal leaves *c.* 1 cm, bristle-like; stem-leaves 6–10 pairs, longer than internodes. Flowers usually solitary; calyx 1.5–1.8 cm; epicalyx-bracts usually 6, about half as long as calyx, ovate, long-pointed. Petal-blade *c.* 5 mm, ovate, toothed, bearded, pink. *Bulgaria, Romania (lower Danube valley).* H3. Summer.

27. D. gracilis Sibthorp & Smith. Illustration: Sibthorp & Smith, Flora Graeca **5**: t. 404 (1825).
Loosely tufted, hairless perennial to 40 cm with obvious woody stock. Basal leaves to 4 cm, narrowly linear, usually withering at time of flowering; stem-leaves 4–6 pairs, shorter than internodes, with sheaths *c.* 3 times as long as stem is wide. Flowers usually solitary, occasionally short-stalked in small clusters. Calyx 1.3–1.8 cm, cylindric, with triangular, acuminate teeth. Epicalyx-bracts 4–6, obovate, apex abruptly tapered into a sharp point, one-third to two-thirds as long as calyx. Petal-blade 5–10 mm, obovate, toothed, bearded, deep pink above, yellow or purplish beneath. *Mountains of Balkan peninsula.* H3. Summer.

For possible confusion over the application of this name, and the identity of the plants illustrated by Sibthorp & Smith see Strid, Mountain Flora of Greece **1**: 186 (1986).

28. D. deltoides Linnaeus. Figure 27(5), p. 178. Illustration: Coste, Flore de la France **1**: 195 (1901); Wehrhahn, Die Gartenstauden **1**: 346 (1931); Garrard & Streeter, Wild flowers of the British Isles, 47 (1983).
Loosely tufted, somewhat hairy perennial

to 45 cm, green or glaucous, with more or less prostrate non-flowering shoots and erect flowering shoots. Leaves on non-flowering shoots 1–1.6 cm, narrowly oblanceolate, on flowering shoots 4–10 pairs, linear, acute. Flowers solitary (rarely 2–3) on ends of main branches. Calyx 1.2–1.8 cm, hairy, usually purplish, with acute teeth. Epicalyx-bracts 2 or 4, ovate-subulate, green with membranous margins, about half as long as calyx. Petal-blade *c.* 8 mm, obovate, irregularly toothed, bearded, pale to deep pink with dark basal band and pale spots. *Europe and temperate Asia.* H2. Summer.

Easily cultivated; a number of cultivars exists, with a range of flower-colour from white to scarlet.

29. D. myrtinervius Grisebach. Illustration: Annales Musei Goulandris **4**: 219 (1978).
Densely leafy, mat-forming hairless perennial to 5 cm. Leaves 3–8 mm, elliptic to oblong, obtuse, longer than internodes. Flowers solitary on very short stalks. Calyx 5–9 mm, bell-shaped, green. Epicalyx-bracts 2 or 4, outer pair usually leaf-like, two-thirds to almost as long as calyx. Petal-blade 4–6 mm, obovate, sparsely bearded, toothed, pink with purplish dots near base. *Mountains of N Greece and S Yugoslavia.* H3. Summer.

Section **Caryophyllum**. Like section *Barbulatum* (p. 188) but petal-blade not bearded.

30. D. sylvestris Wulfen. Illustration: Hegi, Illustrierte Flora von Mitteleuropa **3**: 323, t. 102 (1909); Thompson, Sub-alpine plants, 97, pl. 8 (1912); Farrer, English rock garden **1**: 292, p. 28 (1919).
Densely tufted, nearly hairless perennial with short, woody stock, to 40 cm. Basal leaves 2–10 cm × 0.5–1 mm, wiry, long-pointed, usually green; stem-leaves erect or spreading, decreasing in size upwards. Flowers 1–3 cm on erect stems held clear of basal tuft, scentless or nearly so. Calyx 1.2–3 cm, nearly cylindric. Epicalyx-bracts 2–8, broadly obovate, usually abruptly tapered into a long point, hard in texture, *c.* one-quarter as long as calyx. Petal-blade 6–15 mm, entire or toothed, not bearded, usually pink. *S Europe.* H3. Summer.

A variable species, divisible into several subspecies. The robust, large-flowered subsp. **siculus** (Presl) Tutin is handsome and decorative.

31. D. caryophyllus Linnaeus. Illustration: Boswell, (ed.), English botany, edn 3, **2**: 49, t. 214 (1886).

Loosely tufted, hairless perennial with branched woody stock, to 80 cm. Leaves 8–15 × 2–4 mm, linear, flat and soft, acuminate, somewhat glaucous, with conspicuous sheaths. Flowers 1–5 in loose cymes on stiff, ascending stems, strongly fragrant. Calyx 2.5–3 cm × 5–7 mm, teeth c. 0.5 mm, acute to more or less obtuse. Epicalyx-bracts 4 or 6, very broad, abruptly tapered to a long point, less than one-quarter as long as calyx. Petal-blade 1–1.5 cm, irregularly toothed, not bearded, bright pinkish purple. *Native region uncertain but probably from the Mediterranean area.* H3. Summer.

The above description refers to the 'wild' plant, the presumed parent of garden Carnations, which show an enormous range of flower-colour and shape in double-flowered cultivars. (See note under generic description, p. 185).

32. D. anatolicus Boissier. Illustration: Davis (ed.), Flora of Turkey **2**: 105 (1967) – epicalyx & calyx.
Loosely mat-forming perennial to 35 cm. Leaves linear, to 4.5 cm, long-acuminate. Flowers solitary or in groups of 2 or 3, their stalks more than 5 mm. Calyx 8–11 × 2–3.5 mm, contracted above, teeth 2 mm, often with a small point. Epicalyx-bracts usually 6, narrowly ovate, abruptly pointed, with membranous margins, one-third to half as long as calyx. Petal-blade 2.5–3.5 mm, linear-oblong, almost entire, not bearded, white. *Turkey.* H3. Summer.

33. D. furcatus Balbis. Illustration: Coste, Flore de la France **1**: 193 (1901).
Loosely tufted, glaucous, hairless perennial to 30 cm. Leaves 1–3 cm × 1–3 mm, linear-lanceolate, mostly basal. Flowers usually solitary, slightly pendent, on long stalks, fragrant. Calyx 1–1.5 cm, with lanceolate, acuminate teeth. Epicalyx-bracts 4, ovate-lanceolate, acute, usually c. half as long as calyx. Petal-blade 5–10 mm, entire or irregularly toothed, not bearded, pink or white. *Mountains of SW Europe.* H3. Summer.

A variable species divisible into several subspecies.

34. D. subacaulis Villars (*D. brachyanthus* Boissier). Illustration: Wehrhahn, Die Gartenstauden **1**: 347 (1931); Everard & Morley, Wild flowers of the world, pl. 32 (1970); Polunin & Smythies, Flowers of southwest Europe, pl. 4 (1973).
Like *D. furcatus* but dwarf, to 10 cm, leaves crowded, somewhat recurved, epicalyx-bracts one-third as long as calyx; calyx

with ovate, obtuse teeth. *Mountains of SW Europe.* H3. Summer.

35. D. pungens Linnaeus (*D. serratus* Lapeyrouse). Illustration: Coste, Flore de la France **1**: 193 (1901).
Like *D. furcatus* but leaves c. 2 cm × 0.5 mm, rigid, rough, acuminate; epicalyx-bracts ovate, abruptly long-acuminate, somewhat spreading from calyx. *E Pyrenees.* H3. Summer.

Nos. **32–35** are relatively small-flowered 'alpines' of limited appeal in the garden. *D. subacaulis* is much the most attractive of the group.

Section **Plumaria** Opiz. Petal-blade deeply divided, bearded or not.

36. D. plumarius Linnaeus. Illustration: Everard & Morley, Wild flowers of the world, pl. 13 (1970); Garrard & Streeter, Wild flowers of the British Isles, 46 (1983).
Loosely tufted, hairless, somewhat glaucous perennial to 40 cm. Leaves to 10 cm × 2 mm, narrowing towards the more or less acute tip, keeled. Flowers usually solitary at the ends of branches of a loose irregular cyme, very fragrant. Calyx 1.7–3 cm × 4–6 mm, green or purplish with ovate, usually obtuse teeth. Epicalyx-bracts usually 4, obovate, abruptly acuminate, about one-quarter as long as calyx. Petal-blade 1.2–1.8 cm, triangular-obovate, divided to about the middle into narrow lobes, usually bearded, bright pink or white, often with a darker centre. *E Central Europe; cultivated since the 17th century and naturalised on old walls, etc., in many parts of Europe.* H2. Summer.

The single-flowered wild plant is rarely grown today, but very many cultivars of the species and its hybrids with other *Dianthus* species are widely grown. Many of these are double-flowered variants.

D. 'Allwoodii' is a range of hybrids produced by crossing cultivars of this species with 'Perpetual-flowering Carnations' derived from *D. caryophyllus*.

37. D. arenarius Linnaeus. Illustration: Hegi, Illustrierte Flora von Mitteleuropa **3**: 340 (1909); Farrer, English rock garden **1**: 292, pl. 29 (1919).
Like *D. plumarius* but leaves not exceeding 4 cm, usually green; petal-blade divided to below the middle into narrow lobes, white with greenish spots and reddish purple centre. *N & E Europe.* H2. Summer.

More shade-tolerant in gardens than most *Dianthus*.

38. D. squarrosus Bieberstein. Illustration: Bieberstein, Centuria plantarum rariorum Rossiae meridionalis, t. 33 (1810); Bulletin of the Alpine Garden Society **41**: 59 (1970).
Like *D. plumarius* but leaves not exceeding 3 cm, recurved and rather rigid, acuminate; calyx tapering from base upwards; petal-blade c. 8 mm, deeply divided, pure white. *European and Asiatic USSR.* H2. Summer.

39. D. gallicus Persoon. Illustration: Coste, Flore de la France **1**: 192 (1901).
Loosely tufted perennial to 50 cm, glaucous, somewhat hairy towards base. Leaves 1–1.5 cm × 1.5–3 mm, obtuse or more or less acute; stem-leaves usually 6–10 pairs. Flowers 1–3 together at ends of stems, fragrant. Calyx 2–2.5 cm × 3–4 mm, slightly tapering upwards. Epicalyx-bracts 4, broadly ovate, abruptly tapered to a long point, mostly green, c. one-quarter as long as calyx. Petal-blade 1–1.5 cm, divided to about one-third into narrow segments, bearded, pink. *Atlantic coast of Europe from Portugal to NW France.* H3. Summer.

40. D. superbus Linnaeus. Illustration: Coste, Flore de la France **1**: 191 (1901); Die Natürlichen Pflanzenfamilien edn 2, **16c**: 359 (1934); Takeda, Alpine flora of Japan, pl. 56 (1963).
Robust, hairless perennial to 80 cm; stems decumbent below and branched above, with prominent nodes. Basal leaves mostly withered by flowering time, to 8 cm, linear-lanceolate, 3-veined, margins finely toothed. Flowers to 5 cm across, fragrant, usually solitary or paired at the ends of panicle branches. Calyx 1.5–3 cm × 3–6 mm, narrowed upwards, with teeth to 6 mm, acute. Epicalyx-bracts 2 or 4, broadly ovate, abruptly pointed. Petal-blades 1–3 cm, relatively narrow, not touching below, very deeply divided more than halfway into almost hair-like lobes, bearded, pink or purplish. *Europe except the extreme west and south; Asia.* H1. Summer.

A very handsome species, surprisingly rarely grown in gardens though involved in the hybrid origin of some cultivars.

41. D. monspessulanus Linnaeus (*D. sternbergii* Sieber). Illustration: Coste, Flore de la France **1**: 191 (1901); The Garden **101**: 555 (1976).
Loosely tufted, hairless perennial to 60 cm. Stems slender, branched above. Leaves 5–8 cm, linear or linear-lanceolate, to 3 mm wide, soft, acuminate. Flowers in groups of 2–7 (rarely solitary) on short

stalks, fragrant. Calyx 1.8–2.5 cm ×
3–5 mm, somewhat narrowed above, with
narrowly ovate, acute teeth. Epicalyx-
bracts 4, lanceolate, tapering into a narrow
green tip, one-third to more than half as
long as calyx. Petal-blade 1.2–2 cm,
divided to about halfway into narrow
segments, pink. *Mountains of S & C Europe.*
H3. Summer.

A variable species. Subsp. **sternbergii**
(Sieber) Hegi, from the E Alps, is more
compact, with glaucous foliage and larger,
solitary flowers, and makes an attractive
rock-garden plant. The hybrid with *D.
seguieri* (**D. × arvernensis** Rouy & Foucaud),
native to the Auvergne region of C France,
is grown (as several cultivars) in rock-
gardens. It has a compact cushion-form
and the leaves are grey-green. In most
horticultural works it has been treated as a
variant of *D. gratianopolitanus.*

42. D. gratianopolitanus Villars (*D. caesius*
Smith). Illustration: Coste, Flore de la
France 1: 195 (1901); Hay & Synge,
Dictionary of garden plants, pl. 51 (1952);
Garrard & Streeter, Wild flowers of the
British Isles, 46 (1983).
Compact, mat-forming, hairless, glaucous
perennial with woody stock, to 20 cm.
Leaves to 4.5 cm × 2 mm on non-flowering
shoots, linear, acute or obtuse, almost flat;
margins rough with minute teeth. Flowers
usually solitary, strongly fragrant, on erect
stems bearing 2 or 3 pairs of leaves. Calyx
1.3–2 cm, cylindric, with ovate, obtuse or
almost acute teeth. Epicalyx-bracts 4, ovate
with short point, about one-third as long as
calyx. Petal-blade 1–1.2 cm, obovate, with
irregular teeth to 3 mm, pink or red. *W &
C Europe from SW England to USSR (W
Ukraine).* H2. Summer.

An attractive rock-garden plant
represented in gardens by several double
and colour variants. It has been involved in
hybridisation, but many cultivars referred
to '*D. caesius*' are hybrids of other species in
this section (see note under *D. monspessulanus*).

43. D. petraeus Waldstein & Kitaibel (*D.
strictus* misapplied). Illustration:
Wehrhahn, Die Gartenstauden 1: 350
(1931); Bulletin of the Alpine Garden
Society **41**: 60 (1970); Strid, Mountain
Flora of Greece 1: 181 (1986).
Mat-forming or loosely tufted, hairless
perennial to 30 cm. Leaves to 2.5 cm,
linear, tapering from near middle to a long
point, sometimes rigid. Stem-leaves to 5
pairs. Flowers solitary or in loose clusters,
fragrant. Calyx 1.2–2.7 cm × 2–3 mm,

narrowing upwards from near base, teeth
long-acuminate. Epicalyx-bracts usually 4,
elliptic to ovate, abruptly narrowed to a long
point or acuminate, one-fifth to one-quarter
as long as calyx. Petal-blade 4–10 mm,
sometimes bearded, divided, toothed or
almost entire, white (rarely pink). *Mountains
of Balkan peninsula.* H3. Summer.

A variable species, especially with respect
to the degree of cutting of the petal-blade.

44. D. spiculifolius Schur. Illustration:
Šavulescu, Flora Republicii Populare
Romane 2: 264, p. 39 (1953).
Like *D. petraeus* but petal-blade pink,
bearded and deeply divided. *E Carpathians.*
H3. Summer.

45. D. graniticus Jordan. Illustration: Coste,
Flore de la France 1: 194 (1901); Bonnier,
Flore complète 2: pl. 82 (1912).
Like *D. petraeus* but stems somewhat hairy
below; calyx 1–1.5 cm, cylindric; epicalyx-
bracts *c.* half as long as calyx; petal-blade
toothed, bearded, reddish purple. *Southern C
France.* H3. Summer.

10. PETRORHAGIA (de Candolle) Link
S.M. Walters
Like *Dianthus* but with a membranous band
of tissue between the veins of the calyx.

A genus of about 25 species in Europe,
Asia and N Africa. Only 1 species is
cultivated. The distinction given applies to
this species, not to the genus as a whole.

1. P. saxifraga (Linnaeus) Link (*Dianthus
saxifragus* Linnaeus; *Tunica saxifraga*
(Linnaeus) Scopoli; *Kohlrauschia saxifraga*
(Linnaeus) Dandy). Illustration: Coste, Flore
de la France 1: 190 (1901); Wehrhahn, Die
Gartenstauden 1: 334 (1931).
Mat-forming perennial to 40 cm, usually
nearly hairless. Leaves linear, keeled.
Inflorescence a loose-flowered cyme.
Epicalyx bracts 4, membranous, pointed,
half as long as calyx. Calyx 3–6 mm, with
5 green angles. Petals 5–10 mm, white or
pink, indented, gradually narrowed into a
short claw. Capsule ovoid. *S & C Europe.*
H3. Summer.

Occasionally grown in rock gardens.
Several colour and double-flowered
variants have been in cultivation, but are
rarely seen today.

11. CERASTIUM Linnaeus
S.M. Walters
Perennial (ours) or annual herbs of varied
habit (the cultivated species tufted or mat-
forming), sometimes woody at base, with
stalkless leaves. Flowers in cymes or

solitary. Sepals usually 5, free; petals
usually 5, white, bifid or more or less
deeply indented; stamens usually 10. Styles
usually 5, on the same radii as the sepals.
Fruit a many-seeded capsule, usually more
or less cylindric and somewhat curved,
opening by 10 short teeth.

A genus of 60–100 species, mostly in
north temperate and arctic regions. The
classification of many species is difficult and
uncertain. In the following treatment a
conservative view is adopted.

1a. Leaves elliptic to obovate; leaf-axils
   with few or no tufts of leaves
                                3. alpinum
 b. Leaves linear to lanceolate-elliptic;
   leaf-axils bearing tufts of leaves   2
2a. Plant covered with a thick felt of
   irregular, wavy or twisted hairs
   giving a whitish appearance
                         2. tomentosum group
 b. Plant covered with straight, short
   hairs giving a grey-green appearance
                                1. arvense

1. C. arvense Linnaeus. Illustration: Fiori &
Paoletti, Iconographia Florae Italicae, edn
3, 141 (1933); Garrard & Streeter, Wild
flowers of the British Isles, 50 (1983).
Loose, mat-forming, glandular-hairy
perennial with many prostrate, leafy shoots
rooting at the nodes and with axillary tufts
of leaves. Flowering-stems ascending, to
30 cm. Leaves to 2 cm, linear to lanceolate-
elliptic. Inflorescence a loose, few-flowered
dichasium, with long, glandular-hairy
flower-stalks and large white flowers.
Sepals 5–8 mm, oblong-lanceolate. Petals
1–1.5 cm, obovate, bifid. Capsules
c. 10 mm, protruding from persistent calyx.
Temperate N hemisphere. H2. Summer.

A variable species, useful for carpeting on
dry, poor soils but not often cultivated (but
see No. **2**, below).

2. C. tomentosum group (**C. tomentosum**
Linnaeus, **C. biebersteinii** de Candolle).
Figure 27(2), p. 178. Illustration:
Wehrhahn, Die Gartenstauden 1: 363–4
(1931); Fiori & Paoletti, Iconographia
Florae Italicae, edn 3, 141 (1933).
Like *C. arvense* but covered in long, thick,
wavy, whitish hairs giving the plants a
very characteristic appearance.

Wild plants belonging to this group
occur in most of the mountains of S & C
Europe and W Asia, from Italy eastwards.
Their classification is much disputed. *C.
tomentosum* Linnaeus was early introduced
into European gardens from Italy, but later
introductions from, e.g. Crimea (*C.*

biebersteinii de Candolle) and the Caucasus confused the picture in the 19th century. It is not possible to elucidate the complex hybridisation which seems to have taken place in cultivated plants, for which the original Linnaean name seems best used in a collective sense. To complicate the picture further there is evidence of much polyploidy, and also of hybridisation with *C. arvense*; and individual clones are apparently self-incompatible, so that plants often appear sterile.

The following papers should be consulted for further information on this group: Favarger, C., Contribution à l'étude cytotaxonomique de la flore des Apennins II, Le group du Cerastium tomentosum L., *Saussurea* **3**: 65–71 (1972) and Nilsson, A., Spontaneous hybrids between Cerastium arvense and C. tomentosum, *Svensk Botanisk Tidskrift* **71**: 263–272 (1977). Both give extensive lists of references. '*C. tomentosum*' is one of the commonest garden plants, and compact variants (such as 'Columnae') are unrivalled as vigorous, free-flowering summer rock-plants. Their freely-rooting, invasive habit, however, reduces their attraction for the more expert gardener.

3. C. alpinum Linnaeus (*C. lanatum* Lamarck). Illustration: Fiori & Paoletti, Iconographia Florae Italicae, edn 3, 141 (1933); Garrard & Streeter, Wild flowers of the British Isles, 50 (1983).
Mat-forming perennial, with few or no axillary tufts of leaves, more or less densely covered with wavy, whitish hairs. Leaves to 1.5 cm, obovate to elliptic. Inflorescence a dichasium of 1–5 flowers, strongly hairy and sometimes more or less glandular. Flowers large, white; sepals 7–10 mm, oblong-lanceolate; petals 1.2–1.8 cm, obovate, shallowly bifid. Capsule *c.* 10 mm, protruding from persistent calyx. *Arctic regions and most of the European mountains.* H2. Summer.

A variable species, not much cultivated; much less aggressive in rock-gardens than '*C. tomentosum*'.

12. ARENARIA Linnaeus
S.M. Walters
Low-growing herbs, sometimes with woody stocks. Leaves very variable in shape, from linear-subulate to circular. Flowers usually in few-flowered cymes, sometimes solitary. Petals more or less entire, usually 5 and white; stamens 10; styles usually 3. Capsule opening with twice as many teeth as styles.

A genus of about 160 species, chiefly in temperate and arctic regions of the northern hemisphere. Only a few species are of horticultural importance and these are mat-forming, cushion-forming or rarely tufted perennials.

Because of the difficulties in discriminating the genera without ripe fruit and seed, species of *Minuartia* (p. 193) and *Moehringia* (p. 193) are included in the key below.

1a. Prostrate or ascending, loosely mat-
    forming plants   2
  b. Tufted or cushion-forming plants with
    more or less erect flowering stems, or
    solitary flowers   5
2a. Petals usually pale purplish; stems
    with downturned hairs in upper part
                    2. purpurascens
  b. Petals white; stems not as above   3
3a. Leaves broadly ovate to circular
                        7. balearica
  b. Leaves oblong-lanceolate to linear   4
4a. Robust hairy plant with large
    5-petalled flowers   **6. montana**
  b. Delicate hairless plant with small,
    usually 4-petalled flowers
          Moehringia muscosa (p. 194)
5a. Dense cushion-forming plants with
    solitary flowers   6
  b. Tufted or loosely cushion-forming
    plants with few-flowered
    inflorescences (rarely reduced to a
    single flower)   9
6a. Petals not longer than sepals,
    yellowish   **Minuartia sedoides**
                        (p. 193)
  b. Petals longer than sepals, white   7
7a. Petals 5; leaves lanceolate
                     4. aggregata
  b. Petals 4 (rarely 5); leaves ovate or
    oblong-elliptic   8
8a. Leaves ovate   **3. tetraquetra**
  b. Leaves oblong-elliptic
      Minuartia cherlerioides (p. 193)
9a. Petals pale purplish   **2. purpurascens**
  b. Petals white   10
10a. Leaves broadly ovate   **8. ciliata**
  b. Leaves lanceolate to linear   11
11a. Basal leaves 4–20 cm, grass-like
                     1. procera
  b. Leaves not more than 4 cm, not
    grass-like   12
12a. Plant hairless throughout
               5. ledebouriana
  b. Plant hairy, at least in part   13
13a. Sepals with 5 or 7 veins   14
  b. Sepals with 1 or 3 veins   15
14a. Leaves sickle-shaped, to 1 cm
      Minuartia recurva (p. 193)

  b. Leaves linear-lanceolate, to 4 cm
      Minuartia graminifolia (p. 193)
15a. Sepals acute, spreading in flower,
    usually glandular
          Minuartia verna (p. 193)
  b. Sepals obtuse, erect in flower,
    sometimes hairy but not usually
    glandular   **Minuartia laricifolia**
                        (p. 193)

1. A. procera Sprengel (*A. graminifolia* Schrader).
Flowering stems to 40 cm, erect, rigid, simple from a branching, woody stock, hairless below, variably hairy in inflorescence. Basal leaves 4–20 cm, linear, grass-like. Inflorescence a loose, few-flowered panicle. Sepals 2–5 mm, ovate, obtuse; petals 4–10 mm, white, obovate and usually slightly indented. *Temperate Asia, extending to E & C Europe.* H1. Summer.

2. A. purpurascens de Candolle.
Illustration: Botanical Magazine, 5836 (1870); Coste, Flore de la France 1: 207 (1901).
Stems to 10 cm, diffuse, ascending, branching and leafy above, hairless below and with downturned hairs above. Leaves to 10 mm, elliptic-lanceolate, hairless except for marginal hairs at base. Flowers usually 2–4 in dense clusters, rarely solitary. Sepals *c.* 5 mm, lanceolate, acute; petals 7–10 mm, oblong-obovate, pale purplish (rarely white). Capsule cylindric, much exceeding calyx. *Pyrenees and N Spanish mountains.* H3. Summer.

Unusual in the genus in having purplish flowers.

3. A. tetraquetra Linnaeus. Illustration: Coste, Flore de la France 1: 208 (1901); Wehrhahn, Die Gartenstauden 1: 358 (1931).
Dense cushion-forming plant with tightly packed, 4-ranked, ovate, obtuse, hairless leaves to 4 mm. Flowers solitary on stems to 1 cm. Sepals 4 or 5, 4–6 mm, lanceolate; petals 4 (rarely 5), slightly longer than sepals, lanceolate-spathulate. *C & E Pyrenees & mountains of E Spain.* H3. Summer.

4. A. aggregata (Linnaeus) Loiseleur (*A. erinacea* Boissier; *A. capitata* Lamarck). Illustration: Coste, Flore de la France 1: 208 (1901).
Like *A. tetraquetra* but petals 5 and leaves lanceolate, acute. *Mountains of SW Europe, from Portugal to Italy.* H3. Summer.

Nos. **3** & **4** are sometimes cultivated in troughs; they need light, well-drained soils.

5. A. ledebouriana Fenzl. Illustration: Boissier, Annales des Sciences naturelles, 257 (1854).

Tufted, hairless plant with short (to 10 mm), closely packed, rigid, subulate leaves, vegetative shoots and branched flowering stems to 15 cm. Inflorescence a few-flowered cyme. Sepals 3–4 mm, ovate-lanceolate, very acute; petals *c.* 8 mm, oblong-lanceolate. *Mountains of Turkey.* H3. Summer.

6. A. montana Linnaeus. Illustration: Coste, Flore de la France 1: 210 (1901); Wehrhahn, Die Gartenstauden 1: 359 (1931).

Robust, grey-green, creeping, hairy plant with prostrate vegetative shoots and ascending flowering stems. Leaves to 4 cm, oblong-lanceolate to linear. Flowers large, solitary or in few-flowered cymes. Sepals to 8 mm, ovate, acute; petals to 2 cm, obovate. *SW Europe.* H3. Spring.

A handsome rock-plant with the appearance of a *Cerastium*, but easily distinguished by its entire petals. Easily propagated by cuttings.

7. A. balearica Linnaeus. Illustration: Garrard & Streeter, Wild flowers of the British Isles, 54 (1983); Castroviejo et al., Flora Iberica 2: 34 (1987).

Plant with slender, prostrate, branching stems, forming a dense mat. Leaves 2–4 mm, broadly ovate to circular, rather shining but shortly hairy. Flowers solitary on slender stalks to 6 cm. Sepals *c.* 3 mm, ovate, more or less acute; petals *c.* 5 mm, obovate. *W Mediterranean islands; widely naturalised from gardens in England.* H4. Summer.

A very attractive carpeting plant, particularly effective on damp, shady rocks.

8. A. ciliata Linnaeus. Illustration: Coste, Flore de la France 1: 210 (1901); Garrard & Streeter, Wild flowers of the British Isles, 54 (1983).

Loosely tufted perennial to 7 cm. Leaves broadly ovate, more or less hairless except for marginal hairs at base, dense on short, prostrate, vegetative shoots. Flowering shoots with 1–6 flowers on slender stalks longer than calyx. Sepals *c.* 3 mm, ovate-lanceolate; petals 4–5 mm, oblong-ovate. *Mountains of C Europe; also N Europe.* H2. Summer.

The wild plant is widespread and variable. Variants are sometimes grown in rock-gardens and may be the closely related *A. norvegica* Gunnerus. The above description covers both species.

13. MINUARTIA Linnaeus
S.M. Walters

Differs from *Arenaria* chiefly in the capsule, which opens with as many teeth as styles. Most species have linear leaves.

About 100 species in the temperate and arctic regions of the northern hemisphere. Like *Arenaria*, of little horticultural importance, though some are grown in rock-gardens. All the garden species are mat-forming perennials.

Since it is difficult to distinguish *Minuartia* from *Arenaria* without ripe capsules, a combined key is provided under *Arenaria* (p. 192).

1. M. recurva (Allioni) Schinz & Thellung (*Alsine recurva* (Allioni) Wahlenberg). Illustration: Coste, Flore de la France 1: 204 (1901).

Densely tufted, glandular-hairy plant with somewhat woody base and flowering stems to 15 cm. Leaves to 10 mm, sickle-shaped. Inflorescence a 1–8-flowered cyme. Sepals 3–6 mm, 5–7-veined, ovate-lanceolate; petals 4–8 mm, more or less ovate. *Mountains of S & C Europe.* H3. Summer.

2. M. graminifolia (Arduino) Jávorka (*Alsine graminifolia* Vitman). Illustration: Fiori & Paoletti, Iconographia Florae Italicae, edn 3, 136 (1933).

Glandular-hairy cushion plant with unbranched flowering stems to 14 cm. Leaves to 4 cm, linear-lanceolate, rigid. Inflorescence a 2–7-flowered cyme. Sepals 6–10 mm, 5–7-veined, lanceolate; petals 8–10 mm, narrowly obovate. *Mountains of S Europe from Sicily to Romania.* H3. Summer.

3. M. cherlerioides (Hoppe) Becerer (*M. aretioides* (Sommerauer) Schinz & Thellung; *Alsine aretioides* (Sommerauer) Mertens & Koch; *A. octandra* (Sieber) Kerner). Illustration: Fiori & Paoletti, Iconographia Florae Italicae, edn 3, 137 (1933).

Dense cushion-plant with small (to 3 × 1 mm), oblong-elliptic, fleshy and usually hairless leaves, closely packed on the branching stems. Flowers solitary on very short stalks. Sepals 4, 2–4 mm, 3-veined, lanceolate, acute. Petals 4, 2–4 mm, lanceolate (sometimes absent). Stamens 8; styles usually 3. *C & E Alps.* H2. Summer.

4. M. verna (Linnaeus) Hiern (*Arenaria verna* Linnaeus; *Alsine verna* (Linnaeus) Wahlenberg). Illustration: Fiori & Paoletti, Iconographia Florae Italicae, edn 3, 136 (1933); Raven & Walters, Mountain flowers, pl. 6 (1956); Garrard & Streeter, Wild flowers of the British Isles, 53 (1983).

Loosely tufted perennials, usually somewhat glandular-hairy, with flowering stems to 15 cm. Leaves to 2 cm, more or less linear-subulate. Inflorescence a loose, few-flowered cyme. Sepals 3–5 mm, 3-veined, ovate-lanceolate, spreading in flower; petals 4–6 mm, obovate. *Widespread in Europe, mostly on mountains in the south.* H2. Summer.

A variable species not commonly grown in gardens, and then mainly as subsp. **verna** (to which the description applies). A double flowered variant is known. *M. verna* 'Aurea' of many nursery catalogues seems to be an error for *Sagina subulata* 'Aurea' (p. 194).

5. M. laricifolia (Linnaeus) Schinz & Thellung (*Alsine laricifolia* (Linnaeus) Crantz). Illustration: Fiori & Paoletti, Iconographia Florae Italicae, edn 3, 136 (1933).

Loosely tufted perennial with a somewhat woody stock. Flowering stems to 30 cm and hairless with linear, rigid and often sickle-shaped leaves to 10 mm. Inflorescence a few-flowered cyme, hairy on flower-stalks and sepals. Sepals 4–7 mm, 3-veined, oblong-lanceolate, erect in flower; petals 6–10 mm, obovate. *Mountains of S & C Europe from Spain to the Carpathians.* H3. Summer.

6. M. sedoides (Linnaeus) Hiern (*Cherleria sedoides* Linnaeus; *Alsine sedoides* (Linnaeus) Kittel; *A. cherleri* Fenzl). Illustration: Fiori & Paoletti, Iconographia Florae Italicae, edn 3, 137 (1933); Garrard & Streeter, Wild flowers of the British Isles, 53 (1983). Hairless, cushion-forming perennial. Leaves 5–15 mm, densely packed, somewhat fleshy, linear-lanceolate, channelled above. Flowers solitary, almost stalkless on the cushion. Sepals 5, 2–5 mm, 3-veined, ovate. Petals small, not exceeding sepals, yellowish, often absent in wild plants. *European mountains from Pyrenees to Carpathians, also in Scotland.* H2. Summer.

Often dioecious. Female flowers are usually without petals; male plants are therefore showier and more rewarding in cultivation.

14. MOEHRINGIA Linnaeus
S.M. Walters

Differs from *Arenaria* only in the presence of an appendage ('strophiole') on the seed. Most species are 4-petalled (a rare condition in *Arenaria*).

About 20 species in the temperate and arctic regions of the northern hemisphere. Since it is virtually impossible to distinguish *Moehringia* from *Arenaria* without seed, the single species here described is included in the combined key under the genus *Arenaria* (p. 192).

1. M. muscosa Linnaeus. Illustration: Coste, Flore de la France 1: 205 (1901). Delicate, hairless, loose mat-forming plant with thin, interlacing stems and linear, thread-like leaves to 4 cm. Flowers on long slender stalks in a very loose cyme. Sepals usually 4, *c.* 3 mm, lanceolate, acute; petals usually 4, 4–5 mm, narrowly ovate; styles usually 2. *Mountains of C & S Europe.* H3. Summer.

Occasionally grown in damp corners or rock-gardens as a small carpeting plant.

15. SAGINA Linnaeus
S.M. Walters

Small herbs with the general appearance of *Arenaria* and *Minuartia*, but always with linear-subulate leaves slightly joined at the base round the slender stem. Flowers small, solitary or in few-flowered cymes, spherical in bud. Sepals 4 or 5; petals 4 or 5, entire, white (sometimes very small or absent). Styles 4 or 5; fruit a more or less spherical capsule splitting to the base into 4 or 5 flaps.

About 20 species, chiefly in temperate regions of the northern hemisphere. Only 1 species has any real garden merit, though some (especially *S. procumbens* Linnaeus) occur frequently as weeds on paths and in rock-gardens, and may be tolerated as ground-cover.

1a. Loose mat-forming plant with straight leaves; flowers with petals
                                                1. subulata
  b. Dense cushion plant with recurved leaves; flowers more or less without petals   **2. boydii**

1. S. subulata (Swartz) Presl (*S. caespitosa*, *S. glabra*, *S. pilifera* all misapplied). Illustration: Wehrhahn, Die Gartenstauden 1: 368, 369 (1931); Garrard & Streeter, Wild flowers of the British Isles, 54 (1983); Beckett, The concise encyclopaedia of garden plants, 362 (1983) – 'Aurea'. Mat-forming perennial with numerous much-branched slender stems. Leaves to 1.2 cm, ending in a distinct white point. Flowers usually solitary on slender stalks to 4 cm. Sepals 5, *c.* 2 mm, ovate, with membranous margins; petals 5, ovate, more or less equalling sepals. Sepals persistent and closely enclosing the ripe capsule. *W & C Europe.* H3. Summer.

The typical wild plant is somewhat glandular-hairy, especially on the flower-

stalks, but hairless variants also occur in the wild. The plant seen in gardens is usually the yellow-green variant 'Aurea', which is quite hairless. It is popular as a carpeting plant for paved areas and rock-gardens. There is some doubt about the classification of this plant; it may be a hybrid between *S. subulata* and the very common perennial weed *S. procumbens*, which has 4-partite flowers that are usually more or less without petals.

In many nursery catalogues *S. subulata* 'Aurea' appears as *Minuartia verna* 'Aurea', a serious misidentification (see p. 193). Several other names in *Arenaria* and *Sagina* are also commonly given to it. Only the misapplied names in *Sagina* are given in the synonymy above.

2. S. boydii Buchanan-White. Illustration: Journal of Botany 30: 226, pl. 326b (1892); Moss, Cambridge British Flora 3: pl. 27 (1920); Ross-Craig, Drawings of British plants 5: pl. 56 (1951). Dense, slow-growing, hairless cushion-plant with crowded, rigid, somewhat recurved leaves. Flowers solitary on short stalks. Sepals 4 or 5, ovate, not opening widely. Petals very small or absent. *Native origin doubtful; said to have been discovered wild in Scotland, but never refound.* H2. Summer.

A horticultural curiosity, suitable only for specialist alpine gardens. Until recently the plant was thought never to set seed, but S.J. & V.G. Heyward recently claimed to have raised plants from seed (see *Watsonia* 15: 167, 1984).

Species of **Colobanthus** Bartling, a southern hemisphere genus with the characters of *Sagina*, but strictly without petals, are sometimes grown by alpine specialists, but have not yet achieved much popularity. **C. muscoides** J.D. Hooker, a New Zealand endemic of dense cushion-forming habit, has been longest in cultivation.

16. SCLERANTHUS Linnaeus
S.M. Walters

Low-growing annual to perennial herbs with opposite, subulate or linear leaves slightly joined at base. Flowers small, solitary, paired or in branched cymes. Sepals 4 or 5, situated on the rim of a perigynous zone, persistent in fruit; petals 0; stamens 1–10; styles 2. Fruit a usually 1-seeded nutlet enclosed in the perigynous zone.

A genus of *c.* 10 species, widely distributed in temperate regions of both hemispheres. The only species of

horticultural interest is from the southern hemisphere.

1. S. biflorus (Forster & Forster) J.D. Hooker.
An intricately branched, mat-forming yellowish perennial. Flowers usually in stalkless pairs, sometimes solitary, at the tip of a stalk borne in the axil of leaves; each flower has at the base a pair of small bracts which persist even after the fruit falls. Sepals 4 (rarely 5); stamen 1. *New Zealand; recorded also from Australia (Tasmania) and S America.* H4. Summer.

The above description covers 2 other New Zealand species (*S. uniflorus* Williamson & *S. brockiei* Williamson); for a detailed treatment, see Allan, *Flora of New Zealand* 1: 217–8 (1961). The plant is sometimes grown in alpine collections.

17. HERNIARIA Linnaeus
S.M. Walters

Low-growing annual to perennial herbs, sometimes with woody stocks. Leaves lanceolate to ovate or obovate, opposite near base but sometimes alternate above, with small, greenish stipules. Flowers small, greenish, perigynous, in dense axillary cymes. Sepals 5; petals 5, shorter than sepals; style 1, more or less stalkless, with bifid stigma. Fruit a nutlet inside the persistent sepals and perigynous zone, with a single black shining seed.

A genus of *c.* 15 species in Europe, Africa and Asia. Only 1 species has much claim to be in regular cultivation.

1. H. glabra Linnaeus. Illustration: Fiori & Paoletti, Iconographia Florae Italicae, edn 3, 134 (1933); Garrard & Streeter, Wild flowers of the British Isles, 55 (1983). *Common in much of Europe, extending to N Africa and Asia.* H3. Summer.

Can be used very effectively as a mat-forming ground-cover where bulbs are grown; it has the advantage of retaining well its green colour in dry soils and seasons.

18. PARONYCHIA Miller
S.M. Walters

Like *Herniaria*, but with silvery stipules, conspicuous, often silvery bracts more or less concealing the flowers, and usually 2 styles free from each other or partly joined. The fruit, though 1-seeded, is partly dehiscent.

A cosmopolitan genus of about 50 species, with limited horticultural value. Literature: Wehrhahn, H.R., *Die Gartenstauden* 1: 370–1 (1931).

1a. Sepals awned and with membranous
     margins   **1. argentea**
  b. Sepals not awned, entirely green   2
2a. Sepals more or less equal in length
                                  2. kapela
  b. Sepals very unequal in length
                                  3. capitata

1. P. argentea Lamarck. Illustration: Fiori &
Paoletti, Igonographia Florae Italicae, edn
3, 134 (1933); Beckett, Concise
encyclopaedia of garden plants, 289
(1983).
Mat-forming perennial with much-
branched stems. Leaves 4–10 mm, ovate to
lanceolate. Flowers in more or less compact
heads with conspicuous, silvery, broadly
ovate bracts partly covering the flowers.
Sepals *c.* 2 mm, awned and hooked. *S
Europe, N Africa, SW Asia.* H4. Summer.

An attractive silvery carpeting plant,
useful as ground-cover for warm, dry sandy
soils especially where bulbs are grown.

2. P. kapela (Hacquet) Kerner (*P.
serpyllifolia* de Candolle; *P. capitata*
misapplied). Illustration: Fiori & Paoletti,
Iconographia Florae Italicae, edn 3, 134
(1933).
Mat-forming perennial with very distinct
silvery heads of flowers and ovate-
lanceolate to elliptic leaves. Sepals
1.5–3 mm, more or less equal, green,
somewhat hooded at tip but without an
awn. *Mediterranean area.* H4. Summer.

3. P. capitata (Linnaeus) Lamarck (*P. nivea*
de Candolle). Illustration: Fiori & Paoletti,
Iconographia Florae Italicae, edn 3, 134
(1933).
Like *P. kapela* but leaves lanceolate to
linear-lanceolate, and sepals very unequal
in length. *Mediterranean area.* H4. Summer.

Nos. **2** & **3** have more attractive heads of
flowers than *P. argentea*, but are less
tolerant of winter wet conditions in
cultivation.

LV. CHENOPODIACEAE

Annual or perennial herbs or shrubs, rarely
becoming tree-like, frequently succulent
and with bladder-like hairs which give the
plants a mealy appearence. Leaves
alternate, rarely opposite, simple, lacking
stipules. Flowers with bracteoles, bisexual
or unisexual, radially symmetric. Perianth
with 1–5 segments, sometimes absent in
female flowers. Stamens 1–5, opposite the
perianth segments; anthers 2-celled,
opening lengthwise. Ovary superior, or

rarely partly inferior, 1-celled; stigmas 1–5;
ovule solitary, basal. Fruit an achene with
horizontally or vertically flattened seeds.

A cosmopolitan family of about 100
genera and 1400 species. Most are found in
arid salty places. Relatively few have any
ornamental value though several have
been consumed as vegetables for many
centuries.

1a. Stems segmented, with opposite
     branches; leaves apparently absent or
     rudimentary   2
  b. Stems not segmented; leaves present
                                          3
2a. Stamens 3–5   **10. Anabasis**
  b. Stamens 1–2   **9. Salicornia**
3a. Flowers mostly unisexual, female
     flowers usually without perianth, but
     with 2 bracteoles which become
     enlarged in fruit   4
  b. Flowers mostly bisexual, or if mostly
     unisexual, then female flowers with a
     perianth of 3 or more segments   5
4a. Bracteoles free in their upper half or
     more   **4. Atriplex**
  b. Bracteoles united almost to the apex
                              3. Spinacia
5a. Each perianth-segment (or at least
     some) bearing a spine on the back in
     fruit   **6. Bassia**
  b. Perianth-segments without spines in
     fruit, sometimes winged   6
6a. Ovary partly inferior, the lower part
     united with the thickened perianth in
     fruit   **1. Beta**
  b. Ovary completely superior, not united
     with perianth in fruit   7
7a. Plant smelling of camphor
                            5. Camphorosma
  b. Plant not smelling of camphor   8
8a. Leaves semi-cylindric in section,
     glaucous   **8. Suaeda**
  b. Leaves flat, not usually glaucous   9
9a. Perianth downy, segments winged in
     fruit   **7. Kochia**
  b. Perianth not downy, segments not
     winged in fruit   **2. Chenopodium**

1. BETA Linnaeus
E. H. Hamlet
Annual, biennial or perennial herbs. Leaves
alternate, entire, the basal long-stalked.
Flowers bisexual, 1–few in axillary clusters
in long spikes. Bracts 2, small. Perianth-
segments 5, usually green, entire or
laciniate; segments thickening in fruit.
Stamens 5. Stigmas 2–3. Ovary half-
inferior, united with the base of the
thickened perianth in fruit. Fruits often
held together by the swollen perianth and

receptacle. Seed horizontal, lens- or kidney-
shaped, glossy.

A genus of 6 species from the Mediterranean region. They prefer a well-drained,
moist soil and full sun. Propagation is by
seed.

1. B. vulgaris Linnaeus. Illustration: Maire,
Flore de l'Afrique du Nord 8: 14 (1962);
Masefield et al., The Oxford book of food
plants 161, 171 (1969); Täckholm,
Student's Flora of Egypt, edn 2, 103 (1974);
Jafri & El-Gadi, Flora of Libya 58: 6 (1978).
Annual, biennial or perennial herb, to 2 m,
hairless to stiffly hairy, decumbent to erect,
branched and leafy. Leaves usually to 12 ×
6 cm, stalked, dark green to reddish violet.
Basal leaves ovate to wedge-shaped, mostly
forming a rosette; stem leaves oblong to
linear-lanceolate. Flowers green or
purplish, in dense clusters of 1–4 forming
branched, long spikes. The lower flower-
clusters are each subtended by a narrow
leaf, the upper leafless. Perianth-segments
2–5 mm, often incurved and almost keeled
in fruit. *Mediterranean area.* H3.
Summer–autumn.

Many cultivated variants are used as root
and leaf vegetables, others, with brightly
coloured leaves, such as 'Cruenta',
'Dracaenifolia' and 'Victoria', are grown as
ornamentals.

2. CHENOPODIUM Linnaeus
J.R. Akeroyd
Annual to perennial herbs, mealy or with
mostly glandular hairs. Stems leafy. Leaves
mostly alternate, simple or pinnatifid.
Flowers bisexual (rarely female), greenish,
in spikes or panicles of cyme-like clusters.
Bracteoles absent. Perianth-segments 3–5,
papery or red and fleshy in fruit. Stamens
1–5. Stigmas 2 or 3. Ovary superior; fruit
an achene.

A genus of some 150 species, many of
them weeds of temperate and subtropical
regions. Few are cultivated in Europe, on
account of their inconspicuous flowers and
reputation as invasive weeds. However,
several species have been used for food,
usually as leaf-vegetables. Propagation is by
seed.
Literature: Beaugé, A., *Chenopodium album
et espèces affines: étude historique et statistique*
(1974).

1a. Perennial with woody rootstock;
     flowers in dense, tapering spikes
                            1. bonus-henricus
  b. Annual (rarely short-lived perennial)
     without woody rootstock; flowers in
     loose spikes or panicles   2

2a. Plant glandular-hairy, sticky, aromatic **3**

b. Plant hairless or mealy, not sticky, sometimes foetid but never aromatic **4**

3a. Leaves pinnatifid or deeply lobed; panicles more or less leafless **2. botrys**

b. Leaves entire or toothed; panicles usually leafy **3. ambrosioides**

4a. Perianth-segments 3–5, red and fleshy in fruit; stamen 1 **5**

b. Perianth-segments 5, papery in fruit; stamens usually 5 **6**

5a. Leaves with a few teeth; upper flower-clusters each without leaf at base **4. capitatum**

b. Leaves with many coarse teeth; all flower-clusters each with a leaf at base **5. foliosum**

6a. Largest leaves not more than 8 cm; stems rarely more than 1.5 m **7**

b. Largest leaves usually more than 8 cm and often to 15 cm; stems 1–3 m **8**

7a. Leaves ovate to lanceolate or diamond-shaped; plant green, rarely young leaves tinged with purple **6. album**

b. Leaves triangular, with conspicuous purple coloration **7. purpurascens**

8a. Leaves triangular to diamond-shaped; flowers all bisexual **8. giganteum**

b. Leaves ovate-triangular; flowers bisexual or female **9. quinoa**

1. C. bonus-henricus Linnaeus. Illustration: Hess, Landolt & Hirzel, Flora der Schweiz 1: 750 (1967); Ross-Craig, Drawings of British plants 25: pl. 20 (1968); Polunin, Flowers of Europe, pl. 10 (1969); Simpson, Flora of Suffolk, 509 (1982).
Robust perennial herb with stout, woody rootstock, densely mealy when young. Stems 20–80 cm, ascending to erect. Leaf-blade 4–10 × 2–9 cm, spear-shaped to triangular, usually acute, the margin more or less entire, wavy; stalk to twice length of blade. Flowers in dense tapering spikes to 25 cm, the lowest clusters remote and leafy. Perianth-segments 5, very weakly keeled. Stamens 5. Stigmas 2, to 1.5 mm. Seeds 1.5–2 mm, dark brown, shiny. *Mountains of C & S Europe; naturalised in the lowlands and in N Europe.* H1. Summer.

Cultivated since at least mediaeval times as a leaf-vegetable. Plants can be propagated by division of the rootstock.

2. C. botrys Linnaeus. Illustration: Coste, Flore de la France 3: 183 (1906); Bonnier, Flore complète 9: pl. 514 (1927); Hess, Landolt & Hirzel, Flora der Schweiz 1: 750 (1967); Pignatti, Flora d'Italia 1: 160 (1982).
Erect annual, 20–80 cm; whole plant glandular-hairy, sticky, aromatic. Leaves falling early, 1–5 cm, pinnatifid or deeply lobed, with 2–6 lobes on each margin; blade longer than stalk. Flowers yellowish, in narrow, cylindric, more or less leafless panicles of spreading cymes. Perianth-segments 5, not keeled. Seeds 0.5–0.8 mm, shiny. *S Europe to C Asia.* H3. Summer.

Grown for its inflorescences which are used in arrangements of dried flowers.

3. C. ambrosioides Linnaeus (*C. anthelminticum* misapplied). Illustration: Die natürlichen Pflanzenfamilien 3(1A): 58 (1893); Coste, Flore de la France 3: 183 (1906); Bonnier, Flore complète 9: pl. 514 (1927); Pignatti, Flore d'Italia 1: 160 (1982).
Differs from *C. botrys* by the less erect, more untidy habit (sometimes a short-lived perennial), stems to 1.2 m, entire or toothed leaves, and leafy panicles of flowers. *Tropical and subtropical S America; naturalised elsewhere.* H5–G1.

Formerly cultivated as a vermifuge.

4. C. capitatum (Linnaeus) Ascherson (*Blitum capitatum* Linnaeus). Illustration: Die natürlichen Pflanzenfamilien 3(1A): 60 (1893); Coste, Flore de la France 3: 187(1906).
Erect or ascending, hairless annual, 20–60 cm. Leaves to 10 cm, with a few teeth, rather fleshy; lower leaves triangular to arrow-shaped, upper leaves narrow, lanceolate. Flowers in dense clusters forming loose spikes towards ends of the branches; spikes leafless, except near base. Perianth-segments 3–5, weakly keeled, red and fleshy in fruit. Stamen 1. Fruiting clusters to *c.* 1.2 cm across. Seeds 0.8–1.2 mm, reddish brown, acutely keeled. *Europe & N America, perhaps of garden origin.* H1. Summer.

5. C. foliosum Ascherson (*Blitum virgatum* Linnaeus). Illustration: Coste, Flore de la France 3: 181 (1906); Hess, Landolt & Hirzel, Flore der Schweiz 1 : 753 (1967); Polunin, Flowers of Europe, pl. 10 (1969). Differs from *C. capitatum* by the densely and coarsely toothed leaves, more numerous and slightly smaller flower-clusters in leafy spikes, and seeds not acutely keeled. *Mountains of Asia, S Europe & NW Africa.* H1. Summer.

C. capitatum and *C. foliosum* have both been used as leaf-vegetables.

6. C. album Linnaeus. Illustration: Ross-Craig, Drawings of British plants 25: pl. 15 (1968); Polunin, Flowers of Europe, pl. 10 (1969).
Greyish, mealy erect annual, 10–150 cm, usually much-branched. Leaves to 8 × 5 cm, diamond-shaped, ovate-triangular or lanceolate, entire or toothed, acute to obtuse, sometimes purplish when young. Flowers in loose spike- or cyme-like panicles. Perianth-segments 5, weakly keeled. Stamens 5. Seeds 1.2–1.8 mm, black or brown, shiny. *Cosmopolitan.* H1. Summer–early autumn.

A very variable species. Probably never truly cultivated, but a weed of disturbed ground that has long been used as a leaf-vegetable and grown for its seeds (now only rarely).

7. C. purpurascens Jacquin (*C. atriplicis* Linnaeus filius). Illustration: Botanical Magazine, 5231 (1861).
Differs from *C. album* by the ovate-triangular, obtuse leaves which are conspicuously reddish or purple in colour, and the dense pyramidal panicle of flowers. *China.* H3. Summer–early autumn.

Very similar to variants of *C. album*, and often confused with *C. giganteum* on account of the purple colouration.

Nos. **6–8** belong to a difficult group of weedy species, the variation of which has been confused by migration, hybridisation and human selection, and requires further study.

8. C. giganteum D. Don (*C. amaranticolor* (Coste & Reynier) Coste & Reynier; *C. purpurascens*, misapplied). Illustration: Hay & Synge, Dictionary of garden plants in colour, 263 (1969) – as C. amaranticolor.
Differs from *C. album* by the conspicuous reddish or purplish young shoots and by the greater size, the stems 2–3 m, and the triangular to diamond-shaped, coarsely toothed leaves to 15 cm. *N India.* H3. Summer–early autumn.

Grown both as a leaf-vegetable and for the ornamental purple foliage.

9. C. quinoa Willdenow. Illustration: Botanical Magazine, 3641 (1838); Die natürlichen Pflanzenfamilien 3(1A): 59 (1893); Economic Botany 19: 225, 232 (1965); Advances in Applied Biology 10: 157 (1984).
Robust, erect, greyish annual 50–250 cm, usually with few branches, mealy when young, sometimes purplish. Leaves ovate-triangular, wedge-shaped at base, acute, usually with coarsely rounded teeth or

lobes. Flowers bisexual and female, in leafy panicles. Perianth-segments 5, keeled. Seeds 1.8–2.5 mm, black or dark brown, shiny. *Andes (mainly Peru)*. H1. Summer–early autumn.

An important food crop developed in pre-Colombian times; only known from cultivation. The plant is still cultivated in Peru for its seeds, which are made into flour. It is cold- and drought-tolerant and the seeds have a high protein content, and it has hence attracted much interest from plant breeders in recent years.

3. SPINACIA Linnaeus
E.H. Hamlet

Erect, annual or biennial, dioecious herbs. Leaves alternate, stalked. Flowers unisexual; male flowers with parts in 4s or 5s, in dense spike-like inflorescences; female flowers axillary, without perianth, but with 2 (more rarely 4) persistent bracteoles which become enlarged, united almost to the apex and hardened in fruit. Stamens 4–5. Stigmas 4–5. Seeds vertical.

A genus of 4 species from the E Mediterranean area to C Asia and Afghanistan. They grow best in fertile, slightly acid to neutral soil, which is kept moist. Propagation is by seed.

1. S. oleracea Linnaeus. Illustration: Polunin, Flowers of Europe, 73 (1969); Jafri & El-Gadi, Flora of Libya **58**: 32 (1978); Heukels & Van der Meijen, Flora van Nederland, 102 (1983).
Erect annual or biennial herbs to 1 m or more. Leaves ovate to triangular-hastate, entire or toothed; lower long-stalked, usually entire. Bracteoles in fruit circular to obovate, usually wider than long, free, with or without divergent spines at apex. *Origin unknown, possibly W Asia; occurring in most of Europe as an escape from cultivation*. H1. Summer.

Used as a leaf-vegetable.

4. ATRIPLEX Linnaeus
E.H.Hamlet

Annual or perennial herbs, hairless to mealy or scurfy. Male and female flowers on same or separate plants. Leaves alternate or opposite. Flowers unisexual, solitary or clustered, axillary or in terminal spikes or panicles. Male flowers with 3–5 perianth-lobes; stamens 3–5. Female flowers usually without perianth, but with 2 large persistent bracteoles, rarely some with 1–5, small scales or 3–5 perianth-lobes. Seed erect or inverted, rarely horizontal.

A genus of about 200 species from temperate and subtropical regions, grown as bedding plants or used as vegetables. They do well in ordinary soil in a sunny position. Propagation is by seed or cuttings.

Plant. Annual: **3,5**; perennial: **1,2,4**.
Leaves. Opposite: **3,5**; alternate: **1–5**.
Flowers. Male and female on the same plant: **1–5**; on separate plants: **1–3**.

1a. Annual   **2**
 b. Perennial   **3**
2a. Fruiting bracteoles circular to heart-shaped   **5. hortensis**
 b. Fruiting bracteoles broadly triangular with tapered bases   **3. patula**
3a. Fruiting bracteoles prominently 4-winged   **1. canescens**
 b. Fruiting bracteoles not winged   **4**
4a. Bracteoles hard in fruit   **2. nummularia**
 b. Bracteoles not hard in fruit   **4. halimus**

1. A. canescens (Pursh) Nuttall. Illustration: Hitchcock et al., Vascular plants of the Pacific Northwest **2**: 187 (1964); Wiggins, Flora of Baja California, 102 (1980); Benson & Darrow, Trees and shrubs of the southwestern Deserts, 165 (1981); Stubbendieck et al., North American range plants, edn 3, 290 (1986). Erect perennial shrub. Male and female flowers on same or separate plants. Leaves 1–5 cm × 2–12 mm, alternate, linear-spathulate to narrowly oblong, grey-scurfy, becoming hairless. Male flowers with 5 perianth-lobes, clustered in dense spikes in terminal panicles. Stamens 5. Female flowers each with 2 bracteoles, in short axillary spikes or panicles. Fruit 4–25 mm, conspicuously 4-winged. Seed brown, ovate. *Western N America*. H2. Flowering time unknown.

Grown as a hedging plant.

2. A. nummularia Lindley. Illustration: Rotherham et al., Flowers and plants of New South Wales and southern Queensland, 138 (1975); Costermans, Native Trees and shrubs of south-eastern Australia, 175 (1981); Elliot & Jones, Encyclopaedia of Australian plants **2**: 257 (1982).
Shrub to 3 m, grey-scurfy. Male and female flowers on same or separate plants. Leaves alternate; stalks 5–10 mm; blades 2–4 cm, grey-green, ovate to spathulate, obtuse, entire or toothed. Flowers in dense terminal panicles. Fruiting bracteoles 5–15 × 5–11 mm, spherical or fan-shaped, thick and corky, entire or toothed. *Australia,*

naturalised in USA (S California). H2. Flowering time unknown.

3. A. patula Linnaeus. Illustration: Hitchcock et al., Vascular Plants of the Pacific Northwest **2**: 193 (1964); Schauer, A field guide to the wild Flowers of Britain and Europe, 61 (1978); Keble Martin, The new concise British Flora, pl.72 (1982); Nicholson et al., The Oxford book of wild flowers, 55 (1982).
Annual herb to 1.5 m, smooth or slightly mealy, much-branched, erect or prostrate. Male and female flowers on same or separate plants. Stem strongly ridged. Leaves 2–14 × 1–6 cm, opposite, alternate or both, linear to broadly ovate, rounded to markedly hastate at base. Flowers clustered in axillary and terminal, simple to panicle-like spikes. Male flowers mixed with the female, or above them, with 3–5 perianth-lobes; female flowers without perianth but with 2 bracteoles. Fruiting bracteoles 3–12 mm, broadly triangular with tapered bases, entire or finely toothed. Seed 1.5–2.5 mm. *Europe, N Africa, W Asia & N America*. H1. Summer–autumn.

4. A. halimus Linnaeus. Illustration: Huxley & Taylor, Flowers of Greece and the Aegean, 74 (1977); Jafri & El-Gadi, Flora of Libya **58**: 34 (1978); Kunkel & Kunkel, Flore de Gran Canaria **2**: 43 (1978).
Erect perennial shrub to 3 m, silvery grey, much-branched. Male and female flowers on same plant. Leaves alternate, stalks 3–12 mm; blades to 6 × 4 cm, ovate to diamond-shaped or broadly triangular, leathery, entire or rarely toothed. Flowers clustered in dense spikes in terminal panicles. Male flowers at top of clusters, very small, with 5 perianth-lobes; female at base, with 2 bracteoles. Fruiting bracteoles 5 × 5 mm, rounded-cordate to broadly kidney-shaped, somewhat spongy. Seed 1–2 mm across, dark brown, vertical, lens-shaped. *Mediterranean area, N Africa & W Asia*. H1. Autumn.

5. A. hortensis Linnaeus. Illustration: Hitchcock et al., Vascular Plants of the Pacific Northwest, 189 (1964); Masefield et al., The Oxford book of food plants, 161 (1969); Polunin, Flowers of Europe, 71 (1969).
Annual herb to 2.5 m, erect or decumbent, branched. Male and female flowers on same plant. Leaves stalked, opposite near base, alternate above; blades 5–20 × 3–10 cm, ovate-triangular to broadly lanceolate, obtuse, entire or toothed, mealy at first, later smooth. Flowers in terminal or

axillary panicle-like spikes. Male flowers mixed with the female, or above them, with 3–5 perianth-lobes; stamens 5. Female flowers of 2 kinds, some lacking bracteoles, but with 3–5 perianth-lobes and horizontal seeds; more commonly without perianth, but with 2 bracteoles and vertical seeds. Fruiting bracteoles 5–15 mm, circular to heart-shaped, net-veined. *Asia, naturalised in C & S Europe.* H1. Spring–autumn.

Cultivated as a vegetable. 'Cupreatorosea' has leaves and stems red with copper tinge. 'Rubra' (var. *atrosanguinea* Bailey; var. *rubra* Linnaeus) has blood-red leaves.

5. CAMPHOROSMA Linnaeus
S.G. Knees
Annual or perennial herbs or small shrubs, smelling strongly of camphor (ours). Stems and leaves softly hairy. Leaves alternate, linear or awl-shaped. Flowers bisexual or female, solitary or in crowded ovoid cymes. Perianth-segments 4 or 5, 2 larger than the rest. Stamens 4 or 5, stigmas 2 or 3. Seeds vertically flattened.

A genus of about 11 species in Mediterranean Europe and C Asia. One species is regularly grown as a scree or trough plant. It requires well-drained soil and a position in full sun.

1. C. monspeliaca Linnaeus. Illustration: Reichenbach, Icones florae Germanicae et Helveticae **24**: t. 274 (1850).
Tufted perennial with short woody branches arising from the base. Leaves 2–10 mm, linear, stiff, often in bundles. Flowering stem 10–60 cm, smelling strongly of camphor. Perianth-segments 2–3.5 mm, hairless below, slightly hairy near apices. Anthers yellow, stigmas reddish. *S & E Europe.* H4–G1. Summer.

6. BASSIA Allioni
E.H. Hamlet
Annual or perennial downy herbs. Leaves alternate, oblong to linear, flat or cylindric, entire and fleshy. Flowers bisexual or female, solitary or in cymes arranged in a panicle. Perianth segments 5, becoming enlarged in fruit, stamens 5, stigmas 2–3. Fruiting perianth papery, smooth or with up to 5 horizontal spines (ours) developing on the backs of the segments.

A genus of *c.* 90 species; 10 are from C Europe, the Mediterranean and C Asia, the rest from Australia.

1a. Flowers solitary   **1. diacantha**
  b. Flowers in clusters of 2 or 3
                         2. hyssopifolia

1. B. diacantha (Nées) Mueller (*Sclerolaena diacantha* (Nées) Bentham). Illustration: Cunningham et al., Plants of western New South Wales, 251 (1981).
Rounded perennial to 30 cm; branches with densely felted hairs. Leaves 5–22 mm, narrowly to broadly linear, acute. Flowers solitary in the leaf axils, perianth hairy; stamens 5. Fruiting perianth 2.5–4 × 2–3 mm, depressed-spherical to pear-shaped, smooth or sparsely to densely hairy, 2-spined. Spines 2–3 mm long, divergent. Seed horizontal. *Australia.*

2. B. hyssopifolia (Pallas) Kuntze.
Erect annual to 1 m; branches slender with cottony hairs when young. Leaves linear to oblanceolate, 5–10 mm. Flowers in axillary clusters of 2 or 3. Fruiting perianth depressed-spherical, hairy, 5-lobed with 5 hooked, incurved spines arising from the base of the lobes. Seed horizontal. *Eurasia.*

7. KOCHIA Roth
E.H. Hamlet
Similar to *Bassia* but flowers neither hidden in hairs nor spiny, sometimes with a short flat wing, rarely a tubercle, on the back of the fruiting perianth-segments.

A genus of 90 species from Europe, temperate Asia, N & S Africa and Australia grown for the ornamental foliage as bedding plants or as pot-plants in the cool greenhouse. They require a light, well-drained soil and full sun. Propagation is by seed.

1a. Annual herb; fruiting perianth
      3–4 mm   **1. scoparia**
  b. Shrub; fruiting perianth 1.2–1.6 cm
                         2. rohrlachii

1. K. scoparia (Linnaeus) Schrader (*Bassia scoparia* (Linnaeus) Scott). Illustration: Maire, Flore de l'Afrique du Nord **8**: 48 (1962); Gartenpraxis for 1985: 46, 47.
Erect annual 20–150 cm, usually much-branched, columnar to spherical in habit. Leaves to 5 cm, flat, linear to lanceolate, 3-veined, turning purplish red in autumn. Fruiting perianth 3–4 mm, downy, with ovate to obtuse segments; wing very short or reduced to a tubercle. *Eurasia.* Summer–autumn.

'Childsii' (var. *childsii* Kraus) has green foliage. 'Trichophylla' (var. *trichophylla* (Voss) Boom) of dense habit, has leaves green turning to red in autumn. Illustration: Everett, The New York Botanical Gardens illustrated encyclopedia of horticulture **6**: 1899, 1900 (1981).

2. K. rohrlachii P.G. Wilson (*Maireana rohrlachii* (P.G. Wilson) P.G. Wilson.
Shrub to 1 m; branches with tufts of wool in leaf axils. Leaves 2–8 mm, alternate, obovoid to cylindric, frequently deciduous. Flowers solitary, bisexual, hairless. Fruiting perianth 1.2–1.6 cm across, hairless, pale brown when dry. Wing simple, horizontal and wavy. *Australia (S Australia, E Victoria).*

8. SUAEDA Scopoli
S.G. Knees
Annual or perennial herbs or small shrubs. Stems naked or sometimes with floury scales. Leaves alternate, fleshy, semi-cylindric or flat. Flowers bisexual or female, solitary or in cymes of 2–5. Bracteoles 2, very small; perianth-segments 5, fleshy, sometimes tuberculate; stamens 5; stigmas 2–5. Fruit with thin papery pericarp, seeds horizontally or vertically flattened; embryo in a flat spiral. Two seed types are often found in the same species, the earlier are usually black, shiny and smooth, the later are larger, brown or green, dull and net-patterned.

A cosmopolitan genus of about 110 species usually growing in coastal situations or in salt steppes. Only one is regularly cultivated and can be propagated from seed or cuttings of semi-ripened wood.

1. S. vera Gmelin (*S. fruticosa* invalid). Illustration: Pratt, Flowering plants, grasses, sedges & ferns of Great Britain **3**: pl. 180 (1873); Keble Martin, New concise British Flora, pl. 73 (1982).
Small, much-branched, hairless shrub, 40–120 cm. Stems erect, ascending, underground stems rooting freely. Leaves 5–18 × 1–1.5 mm, rounded or almost apiculate at apex, glaucous, stalkless, evergreen. Flowers in axillary cymes of 1–3. Stigmas 3. Seed 1.7–1.8 mm, ovoid, beaked, usually vertical. *Europe, Africa & SW Asia eastwards to India.* H1. Summer.

9. SALICORNIA Linnaeus
S.G. Knees
Succulent annual herbs with jointed stems. Leaves opposite, scale-like, inconspicuous, their bases joined and clasping the stem to form a segment. Flowers bisexual in groups of 3, sunken in the bracts of the segments, forming terminal spikes. Perianth with 3–4 lobes. Stamens 1 or 2, projecting. Ovary 1-celled with 2–4 style branches. Fruit bladder-like, surrounded by the perianth which hardens with age; seed coat thin, membranous.

A genus of about 35 species, all

occurring in saline habitats. They are distributed throughout the temperate and tropical world and the young shoots are often eaten as a vegetable.

1. S. europaea Linnaeus. Illustration: Phillips, Wild flowers of Britain, 16 (1977); Heywood, Flowering plants of the world, 7 (1978).
Herb 20–60 cm, stems simple or well-branched, glaucous or bright green, colouring yellow or red in autumn. Stem-joints longer than wide. Flowering spikes 1–5 cm with lateral flowers smaller than central flowers. Fertile segments 2.5–4.5 mm; upper edge with a narrow chaffy margin. *Europe, Asia & N America.* H1. Late summer.

10. ANABASIS Linnaeus
E.H. Hamlet
Perennial herbs or shrubs, succulent, with jointed stems. Leaves opposite, joined and clasping the stem, inconspicuous, scale-like. Flowers bisexual or female, solitary or several in the upper leaf axils. Female flowers with 2 bracteoles. Perianth-segments 5, each usually developing a transverse wing on the back in fruit. Stamens 5, alternating with 5 staminodes. Stigmas 2. Seeds vertical.

A genus of 30 species from the Mediterranean and C Asia.

1. A. aphylla Linnaeus. Illustration: Maire, Flore de l'Afrique du Nord 8: 171 (1962).
Stems 25–75 cm, woody below, branched from the base. Leaves reduced to obtuse, broadly triangular scales, joined in pairs into short sheaths, hairy in the axils. Flowers solitary, terminal, in the axils of bracteoles and forming a spike-like inflorescence. Perianth-segments concave, circular to ovate; 3 segments winged, 2 narrower segments wingless. Stigmas short and thick. Fruit succulent. *SW to C Asia.* Flowering time unknown.

LVI. AMARANTHACEAE
Annual or perennial herbs or shrubs. Leaves alternate or opposite, usually entire, without stipules, often coloured in cultivated species. Flowers small and inconspicuous, in complex inflorescences made up of very condensed cymes or racemes which are normally aggregated into conspicuous heads or spikes, more rarely in panicles, each flower usually subtended by a bract and 2 bracteoles. Perianth of usually 5 segments (rarely fewer), these free or united below, usually scarious. Stamens as many as perianth-segments, their filaments usually united, at least below, often with staminodes in between them. Ovary superior, 1-celled, usually with a single ovule, more rarely with several; styles 1–3. Fruit a capsule opening by a lid or indehiscent, rarely a berry.

A family of about 60 genera and 900 species, mostly from the tropics and subtropics. Cultivated species have either conspicuous bracts or inflorescences (the flowers being usually small and insignificant) or brightly coloured leaves.

1a. Evergreen shrub; fruit a berry
                                      2. Bosea
  b. Annual or perennial herb, if a shrub then not evergreen; fruit dry   2
2a. Leaves opposite or in whorls   3
  b. Leaves alternate   7
3a. Flowers rarely seen in cultivation; most of the leaves, or at least the midrib and stalk, red or crimson, the rest greenish or yellow   **9. Iresine**
  b. Plant not as above; if leaves reddish then flowers usually produced   4
4a. Some flowers in each inflorescence sterile, the perianth modified into a hard, woody column branching at the top into a number of hooks
                                      4. Pupalia
  b. Flowers not as above   5
5a. Flowers in a terminal head with conspicuous coloured bracts; plant annual or perennial   **8. Gomphrena**
  b. Flowers in small heads aggregated into axillary spikes or panicles; plants perennial   6
6a. 3 perianth-segments larger than the other 2   **7. Alternanthera**
  b. Perianth-segments all more or less similar in size   **5. Aerva**
7a. Plants annual (sometimes very large)
                                      8
  b. Plants perennial   9
8a. Filaments united below; ovary containing several ovules   **1. Celosia**
  b. Filaments free; ovary containing a single ovule   **3. Amaranthus**
9a. Basal (rosette) leaves much larger than the stem-leaves; perianth tubular below, the tips free, hairless and coloured   **6. Ptilotus**
  b. Rosette leaves absent; perianth not as above   **5. Aerva**

1. CELOSIA Linnaeus
S.G. Knees
Annual or perennial herbs, sometimes becoming woody at the base, occasionally ascending. Leaves alternate, simple, entire or variously lobed. Flowers bisexual, in compound spikes, with bracts and bracteoles. Perianth-segments 5, free, more or less equal. Stamens 5; filaments joined below, the free parts triangular above or swollen with teeth projecting on each side of the anther, anthers 4-celled. Ovary with few–many ovules, style long or occasionally almost absent, stigmas 2–3. Capsule with many black shiny seeds, seeds often strongly compressed with net-like, grooved or warty ornamentation.

A genus of about 50 species native to the warmer regions of America and Africa, though widely grown throughout the tropics and subtropics. They thrive in any fertile soil and are always grown from seed. They are not hardy but are frequently used for summer bedding in the cooler areas. The cultivated celosias are nearly all derived from *C. cristata* but are frequently sold under the name *C. argentea*.
Literature: Grant, W.F., Speciation and nomenclature in the genus Celosia, *Canadian Journal of Botany* 40: 1355–1363 (1962); Bose, R.B., The correct identity of 'cockscomb', *Bulletin of Botanical Survey of India* 18: 218–9 (1976).

1. C. cristata Linnaeus (*C. argentea* Linnaeus var. *cristata* Kuntze). Illustration: Everett, New York Botanical Gardens encyclopedia of horticulture 2: 671 (1981); Graf, Exotica, series 4 1: 83 (1985).
Hairless annual herb to 1.5 m, simple or with many ascending branches. Leaves stalked, 4–15 × 1–2 cm, linear to ovate-lanceolate, acute to obtuse, shortly mucronate. Flowers in terminal cylindric spikes 2.5–20 × 1.5–2.5 cm, white, yellow or purple. Bracts and bracteoles 3–5 mm, lanceolate or triangular, awned, papery. Perianth-segments 6–10 mm, narrowly elliptic-oblong. Stamens with long slender filaments, anthers cream or magenta. Stigmas 2–3, style 5–7 mm, thread-like. Ovary with 4–8 ovules. Capsule 3–4 mm, ovoid or spherical. Seeds 1.2–1.5 mm, black, shiny, finely net-patterned. *Tropics.* H5. Summer.

Two groups of cultivars have been recognised: those with plumose inflorescences are included in the 'Childsii' group (sometimes known as *C. thompsonii*) and those normally referred to as cockscombs are the 'Cristata' group.

2. BOSEA Linnaeus
J. Cullen & H.S. Maxwell
Upright evergreen shrub. Leaves alternate. Flowers in axillary panicles containing

many stalked flowers. Flowers with 2–4 bracteoles. Perianth-segments 5. Stamens 5, filaments united at the base and forming a disc around the ovary. Ovary ovoid, with 3 stigmas. Fruit a berry.

A small genus of an uncertain number of species, with a remarkable distribution: Canary Islands, Cyprus and E India. The species from the Canary Islands is grown as a cool glasshouse shrub for its freely produced berries. It is most easily propagated by seed.

1. B. yervamora Linnaeus. Illustration: Kunkel, Flora de Gran Canaria **1**: pl. 11 (1974).
Shrub to 3 m. Leaves stalked, ovate, entire, broadly tapered to the base, tapered to the apex, to 8.5 cm. Flower with 2 bracteoles, 2–4 mm in diameter. Stamens projecting beyond the perianth. Berries at first red, ultimately black. *Canary Islands.* H5–G1. Spring–autumn.

3. AMARANTHUS Linnaeus
J. Cullen & H.S. Maxwell
Large, monoecious, annual herbs. Leaves alternate, stalked. Flowers borne in very condensed cymes which are aggregated into spherical or cylindric spikes, themselves aggregated into compound series of spikes; spikes mostly axillary but in several species aggregated near the stem-apex with subtending leaves reduced or absent. Flowers unisexual with a bract and 2 bracteoles which resemble the perianth-segments. Perianth-segments mostly 3–5 (rarely 1–2), scarious. Stamens 1–5, free. Ovary 1-celled with 3 styles. Fruit a 1-seeded capsule opening by a lid or rarely indehiscent.

A genus of about 50 species from the warmer parts of the world. Various species have been cultivated for a long time for their edible seeds, as well as for ornament, and are now widely distributed as escapes from cultivation. Their classification is difficult, made so by difficulties of interpretation of old specimens (particularly Linnaeus's originals) by hybridisation and selection, and by the contradictory and often confusing application of names. The ornamental species are grown for their brightly coloured spikes or their brightly coloured leaves. They should be given sunny, open positions with plenty of space, and the coloured-leaved variants should be neither shaded nor over-fed, as the leaves tend to revert to green under these conditions. Propagation is by seed. Literature: there is a very large literature

on the genus, much of it contradictory; for the few ornamental species the most useful accounts are: Brenan, J.P.M., Amaranthus in Britain, *Watsonia* **4**(6): 261–280 (1961); Sauer, J.D., The grain amaranths and their relatives: a revised taxonomic and geographic survey, *Annals of the Missouri Botanic Garden* **54**: 103–137 (1967).

1a. Leaves linear to linear-lanceolate, many times longer than broad and drooping for about half their length
              4. tricolor var. **salicifolius**
  b. Leaves not as above   2
2a. Inflorescence (at least the long, terminal spike) drooping from near its base   **1. caudatus**
  b. Inflorescence not as above   3
3a. Perianth-segments of the female flowers 1–4; spikes all more or less axillary   **4. tricolor**
  b. Perianth-segments of female flowers mostly 5; upper spikes more or less leafless   4
4a. Perianth-segments of female flowers all (or at least the outer ones) tapering, acute, lanceolate to narrowly elliptic   **2. hybridus**
  b. Perianth-segments of the female flowers oblong or spathulate, all wide at the obtuse apex **3. retroflexus**

1. A. caudatus Linnaeus. Illustration: Jávorka & Csapody, Iconographia Florae Hungaricae, 130 (1930).
Large annual herb to 1 m or more. Leaves ovate-lanceolate to ovate-elliptic, stalked. Inflorescence a complex series of spikes, red, yellow or white, apical spike leafless, long and drooping almost from its base. Bracts shorter than the styles of the female flowers. Perianth-segments 5, recurved, the inner spathulate, obtuse or notched. *Andes.* H2. Summer–autumn.

2. A. hybridus Linnaeus (*A. chlorostachys* Linnaeus). Illustration: Jávorka & Csapody, Iconographia Florae Hungaricae, 130 (1930).
Erect herb to 1 m or more. Leaves ovate or narrowly ovate, stalked. Inflorescence a series of complex, erect spikes, greenish or reddish, the terminal spike more or less leafless. Bracts exceeding the style-branches of the female flowers. Perianth-segments 5, all (or at least the outer) tapering, acute, narrowly elliptic or lanceolate. *E USA to northern S America.* H5–G1. Summer–autumn.

A variable species, much confused with others in the past. In gardens the leaves are often tinged with red, and the spikes are usually red.

A. hypochondriacus Linnaeus. Bracts of female flowers shorter than the styles; style-branches thickened at the base; spikes usually red-purple. *N America.*

Often included in *A. hybridus* but treated as a quite separate species by Sauer (see reference above).

A. cruentus Linnaeus (*A. hybridus* subsp. *incurvatus* var. *cruentus* (Linnaeus) Mansfeld; *A. paniculatus* Linnaeus). Bracts of female flowers shorter than style-branches; style-branches not thickened at the base; spikes loose, usually red, sometimes yellowish. *S Mexico or Guatemala.* H5–G1. Summer.

3. A. retroflexus Linnaeus. Illustration: Jávorka & Csapody, Iconographia Florae Hungaricae, 130 (1930).
Stems erect to 1 m or more. Leaves ovate or narrowly ovate, stalked. Spikes congested, greenish. Bracts longer than the usually 5 perianth-segments in the female flowers. All perianth-segments of female flowers oblong, obtuse or rounded at the apex. *Tropical and subtropical America.* H5–G1. Summer–autumn.

4. A. tricolor Linnaeus (*A. melancholicus* Linnaeus; *A. gangeticus* Linnaeus).
Stem erect, much branched, to 1 m or more. Leaves ovate or narrowly ovate (but see below), stalked. Spikes mostly distinct in the axils of the leaves, greenish or purplish, congested terminal spikes few or absent. Perianth-segments of female flowers 1–4. *Tropical Asia.* H5–G1.

Normally grown for the variously coloured leaves. Var. **salicifolius** (Veitch) Aellen (*A. salicifolius* Veitch). Illustration: Flore de Serres, ser. 2, **9**: t. 1929, 1930 (1873); Brüggeman, Tropical plants, pl. 29 (1957); The Green Scene, March/April 1987, p. 25. Plant tall, pyramidal; leaves very long and narrow, drooping for about half their length, bright red in 'Splendens'.

4. PUPALIA Jussieu
J. Cullen & H.S. Maxwell
Herbs or shrubs, sometimes scrambling. Leaves opposite. Flowers in heads borne in spikes or panicles; some flowers in each head sterile, with a bract and the perianth converted into a column which is branched above into a number of hard, woody hooks which, when mature, spread in a star-like manner. Fertile flowers bisexual, also with a bract and with a scarious, hairy perianth of 5 segments. Stamens 5, filaments united into a cup below; staminodes absent. Fruit dry.

A genus of an uncertain number of

species from Africa and Asia; 1 is occasionally grown as a curiosity.

1. P. atropurpurea Moquin.
Herb, sometimes scrambling. Leaves long-stalked, ovate or ovate-lanceolate, entire. Flower-heads more or less spherical, 6–13 mm in diameter. *Tropical Africa and Asia.* G1.

5. AERVA Forsskål
J. Cullen & H.S. Maxwell
Herbs, low shrubs or shrubs. Leaves alternate, opposite or in whorls. Flowers small, borne in head-like inflorescences which are aggregated in spikes. Flowers bisexual or sometimes unisexual. Perianth-segments 4 or 5, translucent and scarious, all similar in size, all, or the inner 3, hairy with simple hairs. Stamens 4 or 5, filaments joined at the base to form a cup, usually with staminodes between the fertile stamens. Fruit indehiscent or opening by a lid, 1-seeded.

A genus of about 10 species in the tropics and subtropics of the Old World. Only 1 species is occasionally grown as a tropical ornamental.

1. A. scandens (Roxburgh) Wallich (*A. sanguinolenta* Blume; *A. sanguinea* Anon.). Illustration: Li et al., Flora of Taiwan 2: 371 (1976).
Slender perennial herb to 1 m, often woody towards the base. Leaves alternate or opposite, ovate, elliptic or lanceolate, tapered to the base and to the entire or acuminate apex, hairy, usually reddish in cultivated plants (forma **sanguinea** Standley). Inflorescences white or rarely reddish, spherical or cylindric, solitary in the axils, or in axillary spikes. Perianth-segments *c.* 2 mm, all hairy. *Old World tropics.* G2.

6. PTILOTUS R. Brown
J. Cullen & H.S. Maxwell
Herbs or shrubs. Leaves alternate on the flowering stems, larger and forming a rosette at the base (ours). Inflorescences dense, head-like, spherical or shortly cylindric. Bracts conspicuous. Perianth-segments 5, united below, the tips hardened and scarious, reddish or purplish at first, becoming yellow, densely hairy with jointed hairs below (ours). Stamens 5, filaments united below, staminodes, if present, small. Style excentric. Fruit dry.

A genus of about 100 species from Australia, of which 1 is grown as a glasshouse ornamental.

1. P. manglesii (Lindley) Mueller (*Trichinium manglesii* Lindley). Illustration: Botanical Magazine, 5448 (1864); Perry, Flowers of the world, 22 (1970).
Rosette leaves obovate, very gradually tapered to the long or short stalk, rather abruptly tapered to the obtuse apex, 3–8.5 cm, margin wavy or slightly toothed or entire. Flowering stems 11–30 cm. Stem-leaves small, oblanceolate, almost stalkless. Flower-heads spherical to cylindric, 2–5 × 3–6.5 cm, bracts brown, conspicuous against the white-hairy perianths. *W Australia.* G2.

The remarkable jointed hairs borne on the perianth are interesting objects at a magnification of 10–20 times.

7. ALTERNANTHERA Forsskål
E.H. Hamlet
Perennial herbs or shrubs, erect or prostrate. Leaves opposite, ovate to obovate, entire. Flowers bisexual, subtended by bracteoles, in small dense axillary heads. Perianth-segments unequal, usually 5. Stamens 3–5, united into a tube for most of their length, usually with staminode-like appendages between them. Style short or long. Stigma capitate. Fruit bladdery, ovoid to oblong, 1-seeded, indehiscent.

A genus of *c.* 200 species from tropical and subtropical regions, grown as bedding plants or pot-plants in the cool greenhouse. They prefer a well-drained soil and to be grown in full sun. Propagation is by division or cuttings.

1. A. ficoidea (Linnaeus) Roemer & Schultes. Illustration: Reitz (ed.), Flora Ilustrada Catarinense, Amara, 58 (1972).
Herbs with stems erect or prostrate, often finely downy. Leaf-stalks to 7 mm; blades 3–7 × 1–2 cm, elliptic to obovate, mucronate. Flower heads 5–10 mm, whitish, stalkless, axillary. Sepals 3–3.5 mm, unequal, lanceolate; outer 3 broad, hairy, 3-nerved, the inner 2 narrower. Stamens 5; anthers oblong, appendages longer than filaments. Style long. Seed dark, shining. *Mexico to Argentina.* H5–G1. Summer.

'Amoena' (*A. amoena* (Lemaire) Voss), of dwarf habit, has leaves blotched and veined with red and orange. Illustration: Graf, Tropica, 50 (1978).

'Bettzickiana' (*A. bettzickiana* (Regel) Nicholson), of erect habit, has leaves coloured and blotched in shades of cream or yellow to red. Illustration: Graf, Tropica 50 (1978).

'Versicolor' (*A. versicolor* (Lemaire) Seubert), of bushy habit, has green, copper or blood-red leaves with purplish pink veining.

8. GOMPHRENA Linnaeus
E.H. Hamlet
Annual or perennial hairy herbs, erect or prostrate. Leaves opposite, undivided. Flowers bisexual, in dense cyme-like heads, everlasting. Bracts concave, keeled, conspicuous. Perianth-segments 5, chaffy. Stamens 5, sometimes 4; filaments united into a tube. Fruit bladdery, ovoid-oblong, 1-seeded, indehiscent.

A genus of 100 species from tropical America, Australia and SE Asia, 2 grown as bedding plants or pot-plants in a cool greenhouse. They grow best in a well-drained soil and sunny position. Propagation is by seed sown under glass in February or March.

1a. Annual; free part of filaments oblong
       1. globosa
  b. Perennial with tapering root; free part of filaments long and narrow
       2. haageana

1. G. globosa Linnaeus. Illustration: Graf, Exotica edn 2, 58 (1959); Graf, Tropica edn 1, 52 (1978); Everett, New York botanical gardens illustrated encyclopedia of horticulture 5: 1509 (1981).
Annual herb with stems to 1 m, swollen at the nodes, often purplish. Leaf-stalks 5–20 mm; blades 2–10 cm, hairy, oblong or elliptic; base narrowed or obtuse; apex acute. Flower-heads subtended by 2 or 3 leafy bracts; heads 2–3 cm across, spherical or shortly oblong, dense, red-violet, (garden varieties may be white to orange, pink or red). Perianth-segments lanceolate to needle-like, densely woolly. Style slender; stigmas linear. Seed ovoid, yellowish. *India, widely cultivated in tropical S America.* H5–G1. Summer–autumn.

2. G. haageana Klotsch (*G. coccinea* Decaisne). Illustration: Revue Horticole 4(3): 161 (1854).
Perennial herb, but grown as an annual. Stems 20–70 cm. Leaf-stalks to 2 cm; blades 2–10 cm × 3–10 mm, oblanceolate to oblong-linear, tapered to the base; apex acute or acuminate. Flower-heads 2–2.8 cm across, yellow, spherical to shortly cylindric, subtended by 2 leafy bracts. Perianth-segments lanceolate to linear, acuminate. Style long; stigmas short. Seed brownish red, shining. *USA (Texas) to Mexico.* H5–G1. Summer.

9. IRESINE Browne
E.H. Hamlet
Perennial herbs or dwarf shrubs, mostly erect, sometimes climbing. Leaves opposite, usually ovate, simple, stalked. Flowers inconspicuous, in axillary and terminal compound panicles, bisexual or unisexual, white or greenish; bracts usually 3 to each flower; perianth 5-parted, in the fertile flowers usually woolly. Stamens usually 5, filaments united at base. Stigmas 2 or 3. Fruit bladdery, 1-seeded, indehiscent.

A genus of 80 species probably originating in S America, but widely cultivated in tropical and subtropical regions. Plants flower rarely in cultivation, but are popular for their ornate foliage colouring, grown as house plants or used for summer bedding purposes. They prefer a well-drained soil and the colourful foliage is enhanced in full sun. Propagation is by cuttings taken in late summer for house-plants, and February to March for summer bedding purposes.

1a. Leaves broadly ovate to almost
      circular, notched at apex   **1. herbstii**
  b. Leaves ovate or oblong-lanceolate,
      pointed   **2. lindenii**

1. I. herbstii Hooker. Illustration: Graf, Exotic Plant Manual, edn 4, 407 (1974); Wickham, The Houseplant Book, 131 (1977); Graf, Tropica, 50 (1978); Everett, New York botanical gardens illustrated encyclopedia of horticulture **6**: 1807 (1981).
Herb with stems 30–50 cm, bright carmine. Leaves 2–7.5 cm, broadly ovate to nearly spherical, concave, puckered, deeply notched at apex; midrib and main veins broadly margined carmine, deep crimson beneath. Flowers whitish, in panicles 5–10 cm long. *Brazil*. H5–G1. Flowering time unknown.

'Aureo-reticulata' has green or greenish red leaves with yellow veining; illustration: Graf, Tropica, 50 (1978). 'Brilliantissima' has crimson leaves and 'Wallisii' has leaves which are reddish purple.

2. I. lindenii Van Houtte. Illustration: Graf, Exotic Plant Manual, edn 4, 407 (1974); Everett, New York botanical gardens illustrated encyclopedia of horticulture **6**: 1807 (1981).
Herb with stems to 60 cm. Stems and leaf-stalks usually deep red in cultivation, sometimes green. Leaves 2.5–7.5 cm, ovate or oblong-lanceolate, acuminate, dark red with a central red band. *Ecuador*. H5–G1. Flowering time unknown.

'Formosa' has broader leaves, yellow with green between the crimson veins; illustration: Graf, Tropica, 50 (1978).

LVII. CACTACEAE

Highly specialised perennials of diverse habit; roots fibrous or tuberous; stems terete, spherical, tubercled, ribbed, winged or flattened, often segmented, mostly leafless and variously spiny; spines, new growth and flowers always arising from cushion-like areoles. Flowers solitary or rarely clustered, appearing stalkless (except *Pereskia*), nearly always bisexual, usually radially symmetric, receptacle enclosing, and more or less produced beyond, the zone around the ovary ('pericarpel') and between ovary and perianth ('tube'), naked or invested with bract-like scales and areoles, areolar trichomes (see below), hairs and/or spines; perianth-segments numerous in a graded series; stamens often very numerous; anthers 2-celled, splitting longitudinally; ovary inferior (except some *Pereskia* spp.), l-celled, placentation parietal, ovules numerous; style single, stigma lobed, variously papillate. Fruit juicy or dry, naked, scaly, hairy, bristly or spiny, indehiscent or variously dehiscent. Seeds numerous, sometimes with a caruncle or aril, coat variously patterned; embryo curved, usually strongly so, or nearly straight; cotyledons reduced or vestigial, rarely leaf-like.

An exclusively American family, apart from one species of *Rhipsalis* in Africa, Madagascar, the Mascarenes and Ceylon, comprising at least 1500 species. Amongst the most distinctive of flowering plants, the desert species have long attracted general curiosity, and the smaller-growing, at least, a worldwide clientele of specialist growers and collectors. Owing to the fleshy and spiny nature of the plants, they are poorly represented in herbaria. Few genera are covered by orthodox taxonomic treatments but horticultural literature is extensive and there are several societies and journals catering for the hobby.

A very high proportion of the known species of the more popular genera is obtainable in Europe. The status of many species given in reference works has yet to be evaluated and there are many names listed in nursery catalogues, etc., of no botanical standing. Many putative species are probably local variants worthy of varietal rank at most.

Cultivation. The subject is extensively covered in handbooks and general indications are all that can be given here. The majority of the smaller-growing terrestrial genera beloved of enthusiasts are native, not to the waterless deserts of popular imagination, but to upland temperate and subtropical areas of N and S America with annual rainfall of up to 800 mm, mainly falling in the summer growing season, followed by cool dry winters when the plants are dormant or nearly so. This seasonal cycle suits the plants for cultivation in Europe in cool greenhouses or frames with a minimum temperature in the range 3–10 °C, although, kept almost dry through the winter, many species will tolerate minimum temperatures of 0 °C or a few degrees lower. Few species (mainly *Opuntia* spp.) are reliably hardy, but interest in hardy species is increasing, particularly in continental Europe, as energy costs rise. For the purpose of this account, all species other than those which are reliably hardy are regarded as greenhouse subjects (hardiness code G1 or G2).

There are several groups of cacti which require warmer conditions, with a minimum of 10–15 °C recommended. These include some of the columnar or 'cereoid' genera with ribbed stems, *Melocactus*, *Discocactus*, some species in other terrestrial genera, and most of the epiphytic genera. For the last-named, orchid-house conditions are desirable, with shade in summer, good ventilation, and day temperatures not exceeding about 25 °C. Good ventilation is indeed desirable for all cacti grown under glass, and if ventilation is poor many of the less heavily-spined terrestrial species will need protection from full sun in spring and summer to prevent scorching.

Cacti succeed in a variety of composts, both loam- and peat-based, provided that they are well-drained and, for most species, lime-free. No such generalisation is possible in respect of watering (although rain-water is preferable to tap-water in some regions), but the natural growth-rate and the extent and vigour of the root-system of individual species are a good guide. Liquid feeding (high potash/low nitrogen, plus trace elements) is beneficial for all species when in active growth.

Pests such as mealy bug, sciarid or peat fly, and red spider-mite are liable to appear if hygiene is lax, and are not always easy to eradicate. Usual precautions such as sterilisation of compost, regular prophylactic treatment with systemic insecticides, and the isolation of any plant found to be infested, should be taken, and careful inspection, if not quarantine, of plants entering a collection is essential. Plants weakened by pests like mealy bug are prone to potentially fatal bacterial and fungal infections which are otherwise associated with bad cultivation.

Many species propagate readily from offsets or cuttings, and seed of numerous species is available through commercial suppliers. Grafting is recommended for many of the slower and more difficult species.

Literature: As noted above, the literature on cacti is very extensive. The principal reference works published in the past century are: Schumann, K., *Gesamt-beschreibung der Kakteen* (1897–99); edn 2, with Nachtrage 1898–1902 (1903); Britton, N.L. & Rose, J.N., *The Cactaceae* (1919–23); Bravo, H., *Las Cactàceas de Mexico* (1937); edn. 2, with H. Sanchez-Mejorada (1978–present); Backeberg, C., *Die Cactaceae*, 6 vols. (1958–62); Buxbaum, F., The phylogenetic division of the subfamily Cereoideae, Cactaceae, *Madroño* **14**: 177–216 (1958); Krainz, H. (ed), *Die Kakteen* (1956–75), 63 parts issued; Ritter, F., *Kakteen in Südamerika*, 4 vols (1979–81); Benson, L., *The Cacti of the United States and Canada* (1982).

Amongst useful works for collectors and growers are: Borg, J., *Cacti* (1937); edn 2 (1951); Backeberg, C., *Das Kakteenlexikon* (1966); edn 2 (1970); edn 3, Appendix by W. Haage (1976); English edn (*Cactus Lexicon*), translated by L. Glass (1977); Barthlott, W., *Kakteen* (1977); English edn (*Cacti*), translated by L. Glass (1979); Rowley, G. D., *Illustrated Encyclopedia of Succulents (including Cacti)* (1978); Hecht, H., *BLV Handbuch der Kakteen* (1982); Haustein, E., *Der Kosmos-Kakteenführer* (1983); English edn (1987); Cullmann, W., Götz, E. & Gröner, G., *Kakteen*, edn 5, (1984); English edn (1986).

Periodicals devoted to cacti and other succulent plants and including articles of taxonomic and horticultural interest include: *Monatsschrift für Kakteenkunde*, 1891, continued under various titles, and since 1937 as *Kakteen und andere Sukkulenten* (German); *Succulenta*, 1919, and continuing (Dutch); *Cactus & Succulent Journal of the Cactus and Succulent Society of America*, 1929 and continuing (U.S.A.); *The Cactus and Succulent Journal of Great Britain*, 1932–1982, and *The National Cactus and Succulent Journal*, 1946–1982,

amalgamated as *British Cactus & Succulent Journal*, 1983 and continuing (British); *Cactáceas y Suculentas Mexicanas*, 1955 and continuing (Mexican); *Sukkulentenkunde*, 1947–63 (Swiss); *Bradleya*, 1983 and continuing (British).

The generic classification adopted in this account is based on that of Hunt, D.R., in Hutchinson, J., *The Genera of Flowering Plants*, **2** (1967), modified in the light of subsequent research, and taking into account the recommendations of a working party of the International Organization for Succulent Plant Study (Hunt, D.R. & Taylor, N.P. (editors), The genera of the Cactaceae: towards a new consensus, *Bradleya* **4**: 65–78, 1986).

Hybrids. Although hybrids, both interspecific and intergeneric, are quite easily produced in the Cactaceae, there have been few notable hybridists, and interest has been largely confined to the larger-flowered epiphytic cacti (sometimes called 'orchid cacti'), mainly of Mexican origin, where several genera have been used and there are probably hundreds of named cultivars.

The only hybrids produced in commercial quantities, however, have been the interspecific crosses in *Schlumbergera* and *Hatiora* from SE Brazil, known as the 'Christmas Cactus' (*S.* × *buckleyi*) and 'Easter Cacti' (*H. gaertneri* and *H.* × *graeseri*) respectively. *Echinopsis* hybrids (notably those of *E. chamaecereus*) have been marketed in some quantity and variety, and there is specialist interest in *Astrophytum* hybrids, or curiosities such as × *Ferobergia* (*Ferocactus* × *Leuchtenbergia*). Several natural intergeneric hybrids are also on record. For further information, see Rowley, G.D., Zur Genealogie der 'Phyllohybriden' (Epicacti), in Backeberg, C., *Die Cactaceae* **6**: 3545–3572 (1962); Rowley, G.D., Intergeneric Hybrids in Succulents, *National Cactus & Succulent Journal* **37**: 2–6, 45–49, 76–80 (1982).

As cacti are generally so different in appearance from other plants, and are as widely grown for the form of their stems and spines as for their flowers, some introductory notes on the characters used in identification and in the key to genera are given below.

Growth habit. This is a useful starting point for identification of genera. The great majority of cacti are leafless, terrestrial, stem-succulents, but there are exceptions. *Pereskia* is a genus of leafy shrubs and trees, quite un-cactus like in appearance, apart from the presence of areoles, the felted (i.e.

trichome-bearing), cushion-like structures, which are the hallmark of the family. In many species of *Opuntia* the stem is flattened and ovate, and carries out the functions of the leaf while the leaves themselves are small and soon deciduous. The true nature of the stem is, however, evident from the presence of areoles, some of which can produce futher stem-segments and flowers. *Pereskia* and *Opuntia* and other smaller leafy genera between them comprise Group **A**. Flattened stems are also found in *Epiphyllum* and several other epiphytic genera (Group **B**) but their stems can be still more leaflike (cladodes), because, apart from the midrib, they are completely flat and thin and the areoles are confined to the margins. These flattened stems are often mistaken for leaves by laymen.

When cacti first reached Europe in the sixteenth century, they mostly became known as 'torch thistles', hence *Cereus* (a pre-Linnaean genus) or 'melon thistles' ('Echinomelocactus' etc.) according to whether they were tall and thin or short and fat. The terms 'cereoid' and 'cactoid' are still sometimes used to differentiate the cylindric or columnar kinds from the globular. Most genera are exclusively of one habit or the other, though some like *Echinopsis* (taken in a wide sense) exhibit the whole gamut. The second main feature of the cactus is that it is generally not smooth, but ribbed (Groups **C–E**) or tubercled (Groups **F–G**). Ribbing confers physical strength and the ability to expand and shrink and is characteristic of nearly all terrestrial cacti which attain more than 30 cm in height. Tubercled cacti are also capable of great shrinkage and expansion, and may grow to 1 m or more in length, but their habit soon becomes creeping or hanging or else the stems cluster and support one another to form dense mounds, as in some *Mammillaria* species; fleshy tubercles, unless decurrent (as in some *Opuntia* spp.) give no structural rigidity. They are commonly a feature of the seedling and kept until flowering in several early-maturing genera. Usually, the tubercles are inflated and teat-like in shape (Group **G**), but exceptionally, as in *Ariocarpus*, they resemble the leaves of succulent Liliaceae or Crassulaceae (Group **F**).

Spines, glochids and hairs (see Figures 28, 29, pp. 204, 205). The cactus areole is a highly modified axillary bud. It always bears abundant trichomes (i.e. hairs formed of a single chain of cells), giving a 'felted' or

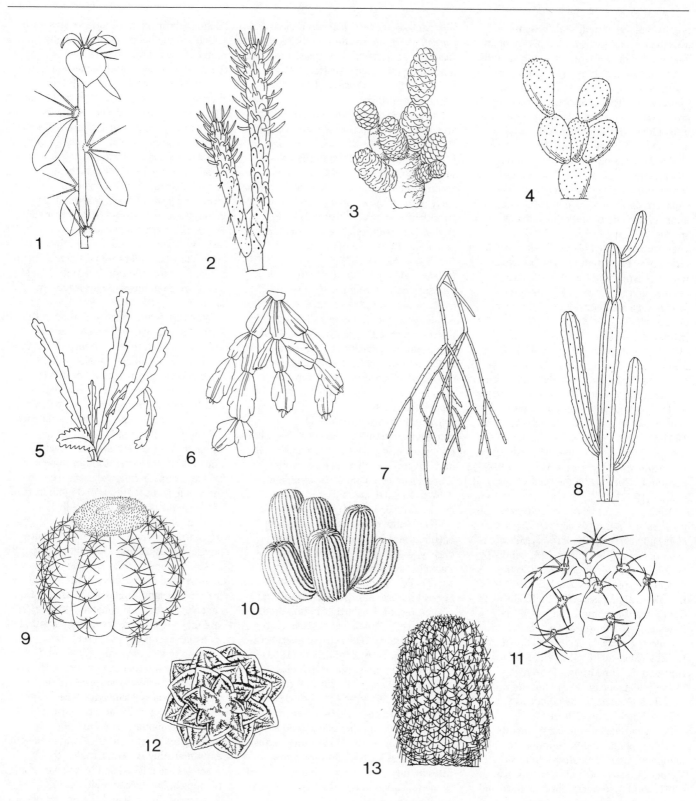

Figure 28. Diagrams of growth habit and stem form in Cactaceae. 1, *Pereskia sacharosa*, with felted areoles in the axils of the persistent leaves. 2, *Opuntia subulata*, stems cylindric, unsegmented, leaves deciduous, terete. 3, *O.* sp., stems cylindric, unsegmented, leaves small (fallen), spines absent. 4, *O. microdasys*, stems flattened, segmented, spines absent. 5, *Epiphyllum crenatum*, stems 2-winged, unsegmented, areoles confined to margins. 6, *Schlumbergera* × *buckleyi*, stems flat, segmented. 7, *Rhipsalis* sp., stems slender, cylindric. 8, *Myrtillocactus geometrizans*, stems long, cylindric. 9, *Melocactus matanzanus*, stem spherical with cap-like cephalium. 10, *Copiapoa cinerea*, stems spherical to shortly cylindric, clustering. 11, *Gymnocalycium buenekeri*, stem flattened-spherical, 5-ribbed. 12, *Ariocarpus fissuratus*, stem tubercled, resembling a rosette of leaves. 13, *Mammillaria polythele* stems with teat-like tubercles. (Not to scale)

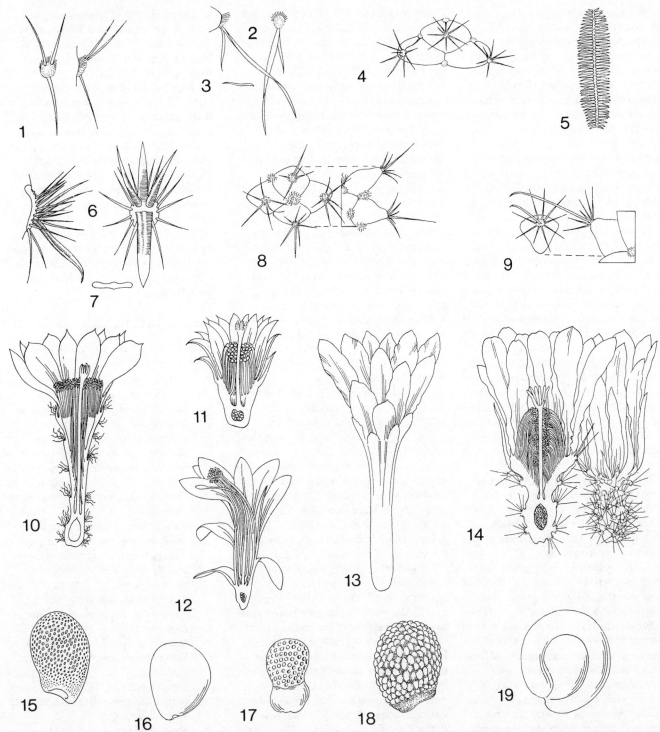

Figure 29. Diagrams of spines and hairs, flower-structure and seeds in Cactaceae. 1, *Opuntia stricta*, areole with 3 spines and a tuft of glochids. 2, *O.* sp., areole, glochids and papery spines. 3, *O.* sp., section of spine. 4, *Coryphantha clavata*, tubercles with the apical (spine-bearing) and axillary parts of the areoles connected by a groove. 5, *Pelecyphora aselliformis*, comb-like arrangement of spines. 6, *Ferocactus latispinus*, areole with 4 large, flattened, transversely striate central spines and numerous needle-like, radial spines. 7, *F. latispinus*, section of principal central spine. 8, *Mammillaria* sp., tubercles with no tubercular groove. 9, *M. poselgeri*, tubercle with the principal spine hooked. 10, Vertical section of tubular-funnel-shaped flower of *Echinopsis* *ancistrophora*, scales of pericarpel and tube with axillary hairs. 11, Vertical section of campanulate flower of *Mammillaria karwinskiana* with naked pericarpel and tube. 12, Vertical section of tubular flower of *Schlumbergera* × *buckleyi*, with naked pericarpel. 13, tubular-funnel-shaped flower of *Cereus* sp. 14, bowl-shaped flower of *Echinocereus pectinatus* with spiny pericarpel and tube. 15, seed of *Ferocactus emoryi* with pitted seed-coat. 16, seed of *Pereskia* sp. with smooth, black seed-coat. 17, seed of *Mammillaria leucantha* with pitted coat and hilum with corky appendage. 18, seed of *Echinocereus reichenbachii* with strongly tuberculate seed-coat. 19, seed of *Opuntia monacantha*, covered by a bony aril. (Not to scale)

'woolly' appearance and often the appearance of a 'pin-cushion' from which the spines protrude. Sometimes this 'wool' forms a dense mat covering the whole stem apex, and can be up to 4 cm thick in species of *Echinocactus*. As well as having trichomes, the areole usually bears spines, which are anatomically more complex and are modified leaf-primordia. Their arrangement, number, size, shape, strength, colour, persistence and other features vary from species to species and provide many useful diagnostic features. Usually, a distinction is made between central and radial (lateral) spines, according to their position and orientation in the areole, though the degree to which the two kinds differ varies and they can completely intergrade. Numbers of both kinds, and more especially the radials, often vary from areole to areole in an individual plant, but are constant within wide or narrow limits for each species. In the keys and descriptions which follow the spine-numbers given are for a single areole, and if not described in other terms, central spines are assumed to be hard and needle-like, and radial spines somewhat finer and less rigid. This is the most common pattern, but the family as a whole exhibits a wide variety of specialised spines and spine-formations, ranging from the broadly flattened, corrugated and fiercely hooked central spines of some *Ferocactus* species, and the papery, grasslike spines of *Leuchtenbergia* and some dwarf *Opuntia* species, to the fine and flexible bristle-spines ('bristles') of the 'Grenadier's Cap' (*Pachycereus militaris*), the very long fine hairs (more correctly 'hair-spines') wrapped around the stem of *Espostoa* species, and the minute comb-like spines of *Pelecyphora aselliformis*, which give the areoles a fancied resemblance to wood-lice. The spines of *Opuntia* and allied genera (Nos. **76–78**) are distinct from those of other cacti, and highly diagnostic of this group, being microscopically barbed and gripping the flesh after puncturing the skin. In some species of *Opuntia* these spines are covered at first with a papery sheath. In addition, all these genera bear 'glochids' – small, fine (rarely hair-like) bristles which are readily detached from the areole and easily lodge in human skin. They are very difficult to see and to remove and, though not poisonous, can cause considerable irritation. Species of *Opuntia*, even when spineless (like the attractive and deceptively innocuous-looking *O. microdasys*), should always be kept out of reach of small children.

Flowers (see Figures 28, 29, pp. 204, 205). In general the cactus flower is uniform throughout the family, though very diverse in size and colour. In all but a few *Pereskia* species the ovary is immersed in the receptacle, which, being stem-tissue, may bear areoles with wool, hairs, bristles or spines, more or less resembling those of the stem. The term 'pericarpel' is used throughout this account for the receptacular zone surrounding and below the ovary (no distinction being made between the pericarpel and the 'pedicel' zone beneath the ovary). In cactus flowers, the receptacle is often prolonged, sometimes greatly so, and usually more or less flared, between the ovary and the insertion of the numerous perianth-segments and stamens. This receptacle-tube is referred to in this account simply as the 'tube', and its proportions and covering of bract-like scales with (or without) areoles with wool, hairs, bristles or spines, provide a number of useful diagnostic characters, as do those of the pericarpel. The term 'limb' is used to denote the perianth proper.

Cactus-flowers may be classified in four principal categories, according to their pollinators, i.e. diurnal and nocturnal insects, birds (mainly humming-birds) and bats. The evolution of each of the flower-types has evidently occurred independently in genera not closely related. Superficially similar flowers, all adapted to pollination by humming-birds are found, for instance, in genera as diverse as *Cleistocactus*, *Echinocereus*, *Mammillaria*, *Neoporteria*, *Opuntia*, *Pereskia* and *Schlumbergera*.

Flowers arise, usually singly, at or immediately above an areole, or in a groove above the areole. In the exceptional case of *Mammillaria*, the spine-bearing and flower-bearing parts of the areole separate at an early stage of development, spines being largely confined to the tips of the tubercles, whilst flowers (and offsets) arise in the 'axils' between the tubercles, sometimes accompanied by bristle-like spines. In some species of *Echinocereus* and *Rhipsalis* the flower-buds burst through the stem-surface near an areole.

Areoles in the fertile zone often bear additional hairs, bristles or spines, and this zone, if sufficiently distinct, is termed the 'cephalium'. This, as the term implies, is usually terminal, as in *Melocactus*, and normal vegetative growth ceases once the cephalium develops. In several other genera, the cephalium is lateral, and growth in height can continue. Even when a cephalium is not developed, there may be an abrupt change in spine-development between the juvenile and flowering stems. *Fruit and seed*. As the outer layers of the fruit are derived from stem-tissue, rather than the ovary itself, the cactus fruit is, like the apple, a false fruit. It may be dry or fleshy, both types sometimes occurring in a single genus, and naked or bearing scales, hairs or spines which, in the absence of flowers, may allow identification to genus level. The mode of dehiscence (if any) is also important. As in other related families, the seeds are of considerable use, many genera and even species being recognisable by their seeds alone. The basic characters such as size, shape and colour, which can be seen with the naked eye or a hand-lens, are sometimes employed in the keys which follow, but microscopic details are not considered.

Key to Groups (vegetative characters only)

1a. True leaves produced   **Group A**
 b. True leaves vestigial or absent   2
2a. Stem flat (leaf-like) or 2-winged, at least in part; or if 3–5-ribbed or cylindric then less than 12 mm in diameter and not continuously ribbed; usually spineless or with fine bristles only   **Group B** (p.207)
 b. Stem 3-or more winged or ribbed, or tubercled   3
3a. Stem at least 6 times longer than thick, scrambling, trailing or pendent, often emitting aerial roots; spines usually weak and bristly or very short, or none   **Group C** (p.207)
 b. Stem erect or ascending, not scrambling, not producing aerial roots; spines often well-developed   4
4a. Stem long, cylindric or columnar, at least twice as long as thick before flowering   **Group D** (p.207)
 b. Stem globular or depressed, less than twice as long as tall (above ground) when flowering   5
5a. Stem ribbed, or if tubercled the tubercles disposed in more or less vertical rows   **Group E** (p.209)
 b. Stem tubercled, the tubercles spirally disposed, or very rarely both ribs and tubercles absent   6
6a. Tubercles somewhat leaf-like or scale-like, arranged as if in a rosette or cone   **Group F** (p.210)
 b. Tubercles not leaf-like, usually teat-like, inflated or pyramidal or absent   **Group G** (p.210)

Group A

1a. Leaves broad and flat, more or less persistent 2
 b. Leaves terete, mostly small and falling early 3
2a. Glochids absent; seeds without an aril, shiny black **1. Pereskia**
 b. Glochids present; seeds encased in whitish aril **76. Pereskiopsis**
3a. Glochids absent; dwarf shrub with persistent leaves up to 1 cm and short cylindric stem-segments; seeds shiny black **2. Maihuenia**
 b. Glochids present; habit various; seeds encased in a whitish aril 4
4a. Seeds not appreciably winged **77. Opuntia**
 b. Seeds winged all round **78. Pterocactus**

Group B

1a. Flowers large, usually more than 4.5 cm across, with a well-developed tube; stem not divided into short segments 2
 b. Flowers usually less than 4 cm across, tube very short or none, rarely to 5 cm or the stem divided into short segments less than 8 cm 3
2a. Flowers diurnal, pink or red **9. Nopalxochia**
 b. Flowers nocturnal, white **8. Epiphyllum**
3a. Flowers less than 2 cm long, creamy white, arising at the edges of stem-segments 4
 b. Flowers more than 2 cm long, brightly coloured (at least the style), arising at or near the tips of stem-segments 8
4a. Stem flat (2-winged, at least in part) 5
 b. Stem slender-cylindric or angled 6
5a. Adult stems of 2 forms, the main stem terete at least basally, the laterals flat, scolloped; seeds usually 1.5–2 mm, black **10. Disocactus**
 b. Adult stems all more or less similar, or if of 2 forms then the flat segments coarsely toothed; seeds less than 1.5 mm, black or brown 7
6a. Stem-segments mainly arising singly from the sides of older segments, never in apical clusters **16. Lepismium**
 b. Stem-segments mainly arising at the apices of older segments, often in clusters, or main stems producing

short lateral segments only 7–15 mm long **17. Rhipsalis**
7a. Stem-segments mainly arising singly from the sides of older segments, never in apical clusters **16. Lepismium**
 b. Stem-segments mainly arising at the apices of older segments, sometimes 2 or more together **17. Rhipsalis**
8a. Flowers radially symmetric, tube 5 mm or less; lowermost stamens not forming a ring round the style-base **18. Hatiora**
 b. Flowers more or less radially symmetric, tube 8 mm or more; lowermost stamens forming a ring round the style-base **19. Schlumbergera**

Group C

1a. Pericarpel with conspicuous triangular scales, enlarging in fruit; stems generally 3-winged and with very short spines **5. Hylocereus**
 b. Pericarpel without conspicuous triangular scales; habit various 2
2a. Flowers diurnal, bright red, orange, pink or purple 3
 b. Flowers nocturnal, white or pale pinkish 5
3a. Offsets and flower-buds bursting through the stem-surface **15. Echinocereus**
 b. Offsets and flowers not rupturing the stem-surface 4
4a. Stem 3–5-winged, -angled or -ribbed **11. Heliocereus**
 b. Stem 6–12-angled or -ribbed **12. Aporocactus**
5a. Stem climbing, trailing or pendent, often rooting aerially; spines very short or weak and bristly 6
 b. Stem erect, arching or clambering, not rooting aerially, except sometimes at the tip; spines often strong 7
6a. Flowers shortly funnel-shaped, the tube less than 5 cm **6. Weberocereus**
 b. Flowers elongate funnel-shaped, the tube more than 5 cm **7. Selenicereus**
7a. Scales of pericarpel, tube and fruit felted and hairy or bristly in their axils 8
 b. Scales of pericarpel, tube and fruit naked or nearly so in their axils **21. Monvillea**
8a. Roots tuberous; stem slender, often minutely hairy, dark-coloured and striate or speckled, not rooting

aerially; spines generally very short, appressed **14. Peniocereus**
 b. Roots not tuberous; stem various, not striate or speckled, sometimes rooting aerially; spines well-developed or short and weak 9
9a. Stems 3–4-angled or with 5–12 low ribs; tube relatively slender, not rigid or markedly flared, with numerous scales hairy in their axils **4. Harrisia**
 b. Stems sharply 3–5-winged or angled; tube stout, rigid, markedly flared, with scattered scales felted and sometimes spiny but not hairy in their axils **13. Acanthocereus**

Group D

This group includes the larger shrubby or tree-like cacti with ribbed stems, most of which grow too large for greenhouse cultivation except (or even?) in botanical gardens and professional collections. Many are relatively easily grown from seed but are usually only seen as non-flowering seedlings or juvenile plants. Genera which will rarely be seen in flower in general collections include: *Armatocereus*, *Austrocephalocereus*, *Browningia*, *Calymmanthium*, *Carnegiea*, *Cephalocereus*, *Cereus*, *Coleocephalocereus*, *Corryocactus*, *Espostoa*, *Eulychnia*, *Myrtillocactus*, *Neoraimondia*, *Pachycereus*, *Pilosocereus*, *Stenocereus* and *Stetsonia*. To construct a key to such genera using vegetative characters only is scarcely practicable, and the identification of young specimens must largely depend on familiarity with the plants or reference to illustrated guides such as those cited under Literature above.

Species of *Echinocereus* are also included here. They are readily diagnosed from the offsets and flower-buds which rupture the stem-surface as they arise; the flowers are diurnal, spiny, and mostly short-tubed with green stigmas.

1a. Flowering areoles similar to the non-flowering 2
 b. Flowering areoles different from the non-flowering 4
2a. Pericarpel and tube variously spiny, bristly, hairy or scaly 3
 b. Pericarpel and tube nearly naked, or with hairless scales only 21
3a. Pericarpel, tube and fruit spiny 5
 b. Pericarpel, tube and fruit with bristles, hairs, wool and/or scales 11
4a. Flowering areoles spineless or less strongly spiny than the non-flowering 24

b. Flowering areoles with longer spines, bristles or hairs and/or more conspicuous wool, than the non-flowering   27

5a. Stigmas green   **15. Echinocereus**

b. Stigmas not green   6

6a. Perianth concealed during bud-development; stem 3–4-ribbed, light green   **3. Calymmanthium**

b. Perianth visible during bud-development; stem various   7

7a. Flowers bell-shaped, the tube shorter than the pericarpel   8

b. Flowers tubular or funnel-shaped, the tube longer than the pericarpel   10

8a. Areoles long, proliferous, densely felted; ribs 4–8   **35. Neoraimondia**

b. Areoles cushion-like, not proliferous; ribs 4–17   9

9a. Areoles not hairy; ribs 4–12
             36. Corryocactus

b. Areoles with long white hairs; ribs 9–17   **37. Eulychnia**

10a. Stem with abrupt annual constrictions; flower-limb narrow
             33. Armatocereus

b. Stem without abrupt constrictions; flower-limb broad   **32. Stenocereus**

11a. Flower-tube very short   12

b. Flower-tube well-developed to long
                13

12a. Ribs 5–9; flowers very small, to 9 per areole; pericarpel scales slightly woolly (without bristles)
             31. Myrtillocactus

b. Ribs 9–20; flowers larger, solitary; pericarpel scales with conspicuous hairs   **37. Eulychnia**

13a. Tube at least twice as long as broad (at midpoint)   14

b. Tube less than twice as long as broad (at midpoint)   17

14a. Stamens borne in the tube in a single series; flowers 4–11 cm long
             38. Haageocereus

b. Stamens borne in 2 series or flowers c. 15 cm long   15

15a. Limb narrow, often somewhat bilaterally symmetric; flowers diurnal, mostly red   16

b. Limb broad, radially symmetric, flowers nocturnal, white
     45. Echinopsis (see also 4. Harrisia)

16a. Areoles without long white hairs; fruit 1–4 cm in diameter, pulpy, indehiscent or bursting at one side
             40. Cleistocactus

b. Areoles usually with long white hairs (hair-spines); fruit hollow, seeds escaping at base, or if pulpy then 4–5 cm in diameter   **42. Oreocereus**

17a. Areoles greatly enlarged and felted
             35. Neoraimondia

b. Areoles not greatly enlarged   18

18a. Flowers diurnal; tube slender; small shrubby plants   **46. Mila**

b. Flowers nocturnal; tube stout; trees or large shrubs   19

19a. Ribs 3–11   20

b. Ribs 12–50   **29. Carnegiea**

20a. Pericarpel and tube with bristles, hairs and wool   **28. Pachycereus**

b. Pericarpel and tube merely felted or woolly   **32. Stenocereus**

21a. Pericarpel and tube more or less scaly
                22

b. Pericarpel and tube naked or nearly so   23

22a. Pericarpel and tube densely covered with overlapping scales; withered perianth persistent   **34. Browningia**

b. Scales of tube distant, not overlapping; withered perianth falling early   **22. Stetsonia**

23a. Shrub with relatively slender stems; withered perianth and style persistent or both falling together
             21. Monvillea

b. Tree-like, with stout stems; withered perianth deciduous, usually leaving the persistent style   **20. Cereus**

24a. Flowering areoles elongating, bearing flowers repeatedly   **35. Neoraimondia**

b. Flowering areoles not elongating, bearing flowers once only   25

25a. Pericarpel and tube with overlapping scales nearly hairless in the axils
             34. Browningia

b. Pericarpel and tube with small, distant scales felted and/or bristly in the axils   26

26a. Flowering areoles separate
             29. Carnegiea

b. Flowering areoles touching or connected by a woolly groove
             28. Pachycereus

27a. Fertile zone discontinuous, alternating with vegetative growth and marked by intermittent collars of bristles
             25. Arrojadoa

b. Fertile zone apical or lateral, continuous   28

28a. Fertile zone covering the whole apex
                29

b. Fertile zone restricted to one side   31

29a. Cephalium composed of long or very dense bristle-spines   **28. Pachycereus**

b. Cephalium composed of wool and hairs or bristles   30

30a. Pericarpel and tube scaly and hairy; fruit yellowish green, opening at base; stems club-shaped   **42. Oreocereus**

b. Pericarpel and tube naked; fruit not yellowish green, indehiscent; stems short and cylindric to long and conical   **27. Melocactus**

31a. Fertile zone clearly defined by massive development of matted wool and/or bristles; ribs 12–30 or more   32

b. Fertile zone less clearly defined, but the flowers immersed in long woolly hairs; ribs 4–12 (rarely 19)
             23. Pilosocereus

32a. Pericarpel and fruit scaly, the scales conspicuous or minute, hairy in their axils, if hairless then the flowers red
                33

b. Pericarpel and fruit naked, or if with 1–2 scales these hairless in their axils and flowers not red   35

33a. Flowers produced only on stems c. 5 m; pericarpel and tube with minute, almost hairless scales
             30. Cephalocereus

b. Flowers produced on stems 2 m or less; pericarpel and tube with conspicuous scales or hairs   34

34a. Fertile zone densely woolly and sometimes also bristly   **39. Espostoa**

b. Fertile zone with bristles only
             40. Cleistocactus

35a. Fruit depressed-spherical to broadly ovoid, depressed at the point of attachment of the dried perianth, not red   36

b. Fruit obovoid-club-shaped to elliptic or spherical, not depressed at apex, red   39

36a. Fertile-zone forming a mat or fleece-like cephalium on one side of the stem only   37

b. Fertile zone with more or less conspicuous hair-tufts, but not forming a mat as above
             23. Pilosocereus

37a. Pericarpel and tube with minute, almost hairless scales; cephalium superficial   38

b. Flowers sparsely scaly, funnel-shaped; cephalium borne in a groove
             26. Coleocephalocereus

38a. Central spines stiffly protruding; flowers c. 4 × 4 cm
             24. Austrocephalocereus

b. Central spines masked by long white hair-spines; flowers c. 5–9 × 6 cm
             30. Cephalocereus

39a. Flowers numerous, less than 3 cm long, tube reddish outside
             25. Arrojadoa

b. Flowers few, more than 3 cm long, tube whitish or greenish yellow outside   **26. Coleocephalocereus**

Group E

1a. Flowers arising from a terminal cephalium   2
 b. Cephalium not developed, although the stem-apex sometimes densely woolly   3
2a. Flowers diurnal, usually less than 3 cm long, limb narrow, mostly pink or purplish, not scented
     27. Melocactus
 b. Flowers nocturnal, 3–8 cm long, funnel-shaped, white, fragrant
     50. Discocactus
3a. Pericarpel and tube bearing areoles with wool, hairs, bristles or spines   4
 b. Pericarpel and tube naked or with hairless scales only   20
4a. Pericarpel and tube bristly or spiny; flower-buds bursting through the stem-surface as they appear and/or stigmas green and spines neither hooked nor flattened and papery
     15. Echinocereus
 b. Pericarpel and tube not spiny, or the tube with some bristles; flower-buds not bursting the stem-surface; stigmas not green, or if green then some spines either hooked or flattened and papery   5
5a. Flowers regular, with distinct staminal throat-circle   6
 b. Flowers regular or bilaterally symmetric, lacking a staminal throat-circle   8
6a. Flowers mostly lateral, more than 4 cm long, variously coloured
     45. Echinopsis
 b. Flowers apical to almost so, 2.5–4 (rarely to 6) cm long, yellow- to reddish orange   7
7a. Stems cylindric, freely branched, covered with bristle-like spines
     46. Mila
 b. Stems spherical or slightly longer, simple or sparsely branched, not densely spiny   **47. Rebutia**
8a. Flowers at the shoulder of the stem or below   **47. Rebutia**
 b. Flowers near the stem-apex   9
9a. Scales of tube (at least the uppermost) with wool and bristles in the axils   10
 b. Scales of tube with hairs or wool but not bristles in the axils   13
10a. Fruit club-shaped to oblong, expanding and becoming partly hollow at maturity, opening at the base; seeds with relatively small hilum and no appendage, the seed-coat often folded and ridged   **52. Neoporteria**

 b. Fruit spherical to ovoid or cylindric, indehiscent, splitting vertically or disintegrating irregularly; or if club-shaped and expanding then the seeds with a broad hilum (equalling the diameter of the seed) and a caruncle   11
11a. Stem dwarf, 1–5 cm in diameter, or dark reddish brown, sometimes obscured by waxy whitish scales; spines not hooked; flowers and stigmas yellowish; seeds with depressed hilum and no appendage   12
 b. Stem more than 5 cm in diameter, light to dark green or glaucous, or some spines hooked or flowers and stigmas not yellowish and seeds not as above   **53. Parodia**
12a. Stem 1–5 cm in diameter; flowers either not opening or more than half the diameter of the stem   **55. Frailea**
 b. Stem more than 5 cm in diameter; flowers always opening, less than half the diameter of the stem
     54. Uebelmannia
13a. Scales of pericarpel and tube, and apex of outermost perianth-segments, narrow, sharply pointed, light to dark brown; flowers various, regular, not tubular or scarlet   14
 b. Scales and outermost perianth-segments not as above; flowers various, sometimes bilaterally symmetric and/or scarlet   18
14a. Stem-surface more or less spotted or covered with small whitish wool-scales; seeds cap-shaped, with deeply sunken hilum; flowers yellow, sometimes red in the throat
     59. Astrophytum
 b. Stem-surface and seeds not as above; flowers various   15
15a. Lower central spine recurved at tip, brownish, very stout, transversely ridged; fruit bright red; stem green
     61. Homalocephala
 b. Central spines nearly straight, or not as above; fruit not red; stem green or glaucous   16
16a. Flowers yellow; stem eventually massive, more than 30 cm in diameter, many-ribbed
     56. Echinocactus
 b. Flowers purple, pink or white; stems smaller   17
17a. Ribs 7–13, broad; flowers purplish pink; pericarpel densely woolly, with scales at the apex only
     56. Echinocactus

 b. Ribs c. 15, acute; flowers pink or white; pericarpel scaly throughout
     45. Echinopsis
18a. Flowers tubular, the limb not expanded, the stamens and style protruding; fruit splitting laterally when ripe, fleshy inside   **41. Denmoza**
 b. Flowers not as above; fruit opening at or splitting from the base, dry inside   19
19a. Flowers regular; fruit opening at base
     44. Oroya
 b. Flowers usually bilaterally symmetric; fruit splitting laterally or from the base   **43. Matucana**
20a. Stem ridged and wrinkled between the ribs; seeds minute, 0.5–0.7 mm, with an appendage   **68. Aztekium**
 b. Stem and seeds not as above   21
21a. Plant spineless, areoles tufted with woolly hairs only   **64. Lophophora**
 b. Plant more or less spiny   22
22a. Flowers regular   23
 b. Flowers bilaterally symmetric, the tube curved   **43. Matucana**
23a. Fruit dry, opening at apex, the base of the dried perianth detaching like a lid; flowers yellow with very short tube
     51. Copiapoa
 b. Fruit dry or juicy, opening basally or laterally or disintegrating irregularly, or indehiscent   24
24a. Ribs very thin, more or less wavy, or if few then the areoles with 1–3 large, flattened spines above and 1–2 much smaller terete spines below
     72. Stenocactus
 b. Ribs not thin and wavy; spines not as above   25
25a. Fruit fleshy, bursting laterally; ribs usually 'chinned' beneath the areoles; areolar glands absent
     48. Gymnocalycium
 b. Fruit dry, or juicy but indehiscent; ribs not 'chinned'; areolar glands often present   26
26a. Mature stem large, more than 20 cm in diameter, or smaller and the seeds pitted; ribs well-defined
     70. Ferocactus
 b. Mature stem 4–20 cm in diameter; seeds smooth, reticulate or tuberculate; ribs poorly-defined, strongly tubercled   27
27a. Spines obscuring the stem, or the central spines strongly recurved to hooked; flowers never yellow with red throat; stigmas sometimes green; hilum lateral or oblique to main axis of seed   **60. Sclerocactus**

b. Spines not as above, or flowers yellow with red throat; stigmas never green; hilum terminating long axis of seed **57. Thelocactus**

Group F

1a. Tubercles very long (to 15 cm) and thin, 3-angled, with long papery spines **71. Leuchtenbergia**
b. Tubercles leaf-like, flat, spineless or the spines falling early or bristle-like 2
2a. Tubercles rather thin, closely overlapping and adpressed; flowers purplish pink **69. Pelecyphora**
b. Tubercles thick, not closely overlapping; flowers purple or pale yellow 3
3a. Flowers purplish pink; areoles not apical on the upper side of the tubercles, sometimes in a groove; spines 0 (except occasional bristle-like spines on young tubercles) **63. Ariocarpus**
b. Flowers pale yellow; areoles apical, spiny at first **65. Obregonia**

Group G

1a. Flowers arising at the tips of the tubercles, or at the base of a short groove extending from the areole, or tubercles absent 2
b. Flowers arising in the axils between ungrooved tubercles 16
2a. Flowers borne at or near stem-apex 3
b. Flowers borne on the shoulder of the stem or below **47. Rebutia** (see also **Escobaria chihuahensis**) p. 285
3a. Pericarpel and fruit bearing scales, wool and sometimes bristles; fruit dry inside, indehiscent, disintegrating irregularly or partly hollow 4
b. Pericarpel and fruit naked or bearing hairless scales only; fruit juicy, indehiscent, or dry, opening at base, apex or laterally 6
4a. Flowers yellow, but often not opening, the perianth then scarcely developing; seeds 1-3 mm, shiny, with sunken hilum **55. Frailea**
b. Flowers yellow, red, pink or whitish, always opening; seeds c. 0.5-1.2 mm, shiny or dull, the hilum not sunken 5
5a. Fruit more or less spherical, 5-10 mm; seeds 0.5-1 mm, appendaged **53. Parodia**

b. Fruit club-shaped, 1-2.5 cm; seeds c. 1-1.2 mm, not appendaged **52. Neoporteria**
6a. Fruit opening near apex, the dried perianth-base partly detaching like a hinged lid 7
b. Fruit indehiscent or opening laterally or below, the dried perianth-base or its scar remaining firmly attached on all sides to the fruit-wall 8
7a. Roots fibrous; seeds 1.5-5 mm **58. Pediocactus**
b. Roots tuberous; seeds 1-1.5 mm **51. Copiapoa**
8a. Pericarpel and fruit always visible, the latter dry, opening laterally or basally, dull green or brownish 9
b. Pericarpel hidden between the tubercles or by spines and wool at flowering-time; or fruit juicy, indehiscent, bright green, pink or red, or dry and remaining hidden when ripe 12
9a. Spines flattened, resembling dried grass leaves; stem club-shaped-cylindric; stigmas green **60. Sclerocactus**
b. Spines not as above or stem depressed-spherical; stigmas not green 10
10a. Pericarpel and fruit naked, rarely with 1 or 2 scales; flowers to 3.5 × 4 cm; seeds 1-2 mm, black, not appendaged **62. Neolloydia**
b. Pericarpel and fruit with 2 or more scales; flowers to 6 × 8 cm or seeds only 0.5mm, brown, appendaged 11
11a. Fruit opening by a basal pore; seeds 1.5-2.5 mm, black, not appendaged; spines strong or dense **57. Thelocactus**
b. Fruit opening by 1-3 vertical splits; seeds 0.5 mm, brown, appendaged; spines few, weak, falling early **67. Strombocactus**
12a. Plant spineless, areoles tufted with woolly hairs only **64. Lophophora**
b. Plant spiny 13
13a. Tubercles conspicuous, more than 1 mm; areolar groove short or extending to the base 14
b. Tubercles minute, c. 1 mm, not grooved **66. Epithelantha**
14a. Fruit juicy, often brightly coloured; seed-coat smooth or pitted, black or brown 15
b. Fruit dry, dull-coloured; seeds tuberculate, black **62. Neolloydia**
15a. Outer perianth-segments entire, or areoles with glands **73. Coryphantha**

b. Outer perianth-segments fringed; areolar glands absent **74. Escobaria**
16a. Apex of tubercles with a spiny areole; axils woolly, bristly or naked 17
b. Apex of tubercles naked, the areole in the axil above **49. Neowerdermannia**
17a. Tubercles hatchet-shaped; spines comb-like; flowers borne near stem-apex **69. Pelecyphora**
b. Tubercles cylindric, conic or unequally swollen; spines various, if comb-like then the flowers lateral **75. Mammillaria**

1. PERESKIA Miller
D.R. Hunt & B.E. Leuenberger
Leafy trees, shrubs or scramblers; leaves broad flat, thin but somewhat fleshy, deciduous. Flowers solitary, clustered, or in corymbs or panicles, sometimes proliferous, stalked or stalkless perianth-segments free; stamens numerous, borne at base of perianth; ovary a cavity at the style-base or inferior. Fruit nearly spherical to pear-shaped, soft to tough-fleshy; seeds glossy black.

A genus of 16 species, native from Mexico and the West Indies to S America, and unusual among cacti in having normal leaves.

Literature: Leuenberger, B.E., Pereskia (Cactaceae), *Memoirs of the New York Botanical Garden* 41: 1-141 (1987).

1a. Scrambler; flowers white in open panicles; short, recurved, usually paired spines arising in the leaf-axils **1. aculeata**
b. Tree-like or shrubby; flowers pink or purplish, clustered; paired, recurved spines absent **2. grandifolia**

1. P. aculeata Miller. Illustration: Botanical Magazine, 7147 (1890); Hecht, BLV Handbuch der Kakteen, 194 (1982); Haustein, Der Kosmos-Kakteenführer, 53 (1983).
Scrambler to 10 m, with thin branches. Leaves shortly stalked, lanceolate to elliptic or ovate, to 11 × 4 cm. Spines on young growth short, recurved, usually paired, sometimes 1 or 3, arising in the leaf-axils; 'normal' spines several, straight, developing at areoles on older growth only. Flowers numerous, in panicles, white or nearly so, 2.5-5 cm in diameter, perfumed; ovary a hollow at the style-base (i.e. almost 'superior') Fruit c. 2 cm in diameter, pale yellow to orange, sometimes spiny, fleshy. *Tropical America*. G2. Autumn.

The cultivar 'Godseffiana', with variegated yellow to peach-coloured leaves, purplish-tinted beneath, is grown as a house-plant.

2. P. grandifolia Haworth. Illustration: Graf, Tropica, 241 (1978).
Shrub or small tree 2–5 or rarely to 10 m. Leaves narrowly elliptic ovate or obovate-lanceolate 9-23 × 4-6 cm, rather thin, lateral nerves *c.* 7–13. Spines 1–several, brownish black, straight, slender. Flowers few to many in proliferous, corymbose clusters 3–5 or rarely to 7 cm in diameter, pink to purplish pink; bracts with short wool in the axils; ovary inferior. Fruit pear-shaped, irregularly angled, 5–10 × 3–7 cm fleshy. *Brazil (widely cultivated in tropical countries)*. G2. Spring to autumn.

P. **sacharosa** Grisebach. Figure 28(1), p. 204. Shrub or small tree to 7 m. Leaves obovate to oblanceolate to 12 × 7 cm, rather fleshy, lateral veins *c.* 4–6. Spines 1–several, usually brown, very stout. Flowers solitary or in clusters of 2–4, to 7 cm in diameter, pink with white centre; axils of the upper receptacular bracts with coarse, longer hairs; ovary half-inferior. Fruit pear-shaped, 2.5–4 cm in diameter, tough, fleshy. *Argentina, W Paraguay, Bolivia, E Brazil*. G2. Spring–autumn.

P. *grandifolia* is probably more widespread in cultivation than *P. sacharosa*. Both species are frequently misidentified. Many plants named *P. sacharosa* are in reality *P. nemorosa* (= *P. amapola*) which is distinct with narrower and larger leaves, larger flowers and seeds, and tufts of hairs between inner perianth-segments and stamens. The true *P. sacharosa* is often cultivated under the synonym *P. sparsiflora* Ritter. Illustrations are rare.

2. MAIHUENIA (Weber) Schumann
D.R. Hunt
Low, clustering *Opuntia*-like plants with small spherical or short cylindric stem-segments; areoles woolly, without glochids. Leaves small, terete, persistent. Spines usually 3, the central long, the laterals short. Flowers terminal. Fruit obovoid or oblong, somewhat fleshy, with small, persistent, leaf-like scales; seeds glossy black.

A genus of probably no more than 2 species native to Patagonia (S Argentina and S Chile). They are winter-hardy in parts of Europe, given a well-drained soil.

1. M. poeppigii (Pfeiffer) Schumann. Illustration: Haustein, Der Kosmos-Kakteenführer, 53 (1983); Cullmann et al., Kakteen edn 5, 202 (1984).
Dwarf shrub, eventually forming low

mounds; stem-segments short-cylindric, to 6 × 1 cm; leaves *c.* 5 mm. Central spine 1.5–2 cm, whitish, laterals *c.* 4 mm. Flowers *c.* 3 × 3 cm, yellow, stigmas green. Fruit obovoid, 5 × 3 cm. *S Chile*. H3.

3. CALYMMANTHIUM Ritter
D.R. Hunt
Shrubby or tree-like with segmented, 3–4-winged, spiny stems. Flowers nocturnal, narrowly tubular-campanulate, the limb projecting beyond a sheath-like upgrowth of the outer layers of the tube from a point beneath the insertion of the perianth and stamens; tube enclosing the perianth in bud, the visible part bearing small scales, felted areoles and bristle-like spines, the invaginated part, scales; perianth-limb narrow, spreading; stamens borne in the throat and tube. Fruit fleshy, indehiscent, spineless or nearly so.

A genus of 1–2 species endemic to Peru and remarkable for the concealed development of the perianth. As in several genera, the old stem-areoles continue to develop spines, and the stems eventually become very fiercely spiny.

1. C. substerile Ritter. Illustration: Hecht, BLV Handbuch der Kakteen, 23 (1982); Haustein, Der Kosmos-Kakteenführer, 73 (1983).
Shrub or tree to 8 m; stem-segments to 1 m × 4–8 cm, 3–4-winged, wings scolloped, light green. Spines pale yellow or whitish, centrals to 6, 1–5 cm, radials 3–8, *c.* 1 cm. Flowers 9.5–11 × 3.5 cm, inner perianth-segments white to reddish brown. Fruit to 15 × 6 cm, 4–5-angled, light green; seeds 2.5 mm, greyish black. *Peru*. G2.

Seedlings are fast-growing and the 3–4-winged, light green stem with pale yellow spines is distinctive.

4. HARRISIA Britton
D.R. Hunt & B.E. Leuenberger
Tree-like, shrubby or climbing, stems usually slender, with 4–12 angles or ribs, often strongly spined, not rooting aerially. Flowers nocturnal, funnel-shaped, large, the tube as long as or longer than the limb; pericarpel and tube with scales felted and sometimes hairy in the axils; perianth white; stamens numerous in a single series. Fruit fleshy, with areoles and scales and/or spines, yellow or orange and not splitting (subgenus **Harrisia**) or red and usually splitting (subgenus **Eriocereus**); seeds dull black, thimble-shaped, coarsely tuberculate at the top, hilum deeply sunken.

A genus of about 20 poorly-defined

species in two disjunct groups, one in Florida and the West Indies, the other in Brazil, Bolivia, Paraguay and Argentina. Several members of the second group, subgenus **Eriocereus** Riccobono, are vigorous and commonly grown as grafting stocks.

1a.	Stem with 8–12 ribs	**1. gracilis**
b.	Stem with 3–8 angles or ribs	2
2a.	Stem with 4–8 deep, rounded ribs; pericarpel and lower part of tube with scales less than 1.5 cm; fruit 2–5 cm in diameter	3
b.	Stem with 3–4 (rarely 5) angles; pericarpel and lower part of tube with scales *c.* 2 cm; fruit 4–7 cm in diameter, with the large scales persisting	**5. guelichii**
3a.	Spines very unequal, the central spine more than 3 times as long as the average radials; pericarpel and lower part of tube with scales less than 3 mm	**4. martinii**
b.	Spines not very unequal, radials about half as long as the centrals; pericarpel and lower part of tube with conspicuous scales to *c.* 5–10 mm	4
4a.	Stem 6–10 cm in diameter; ribs usually 8; scales of pericarpel with dense curled hairs	**6. tetracantha**
b.	Stem 2–8 cm in diameter; ribs 4–7; scales of pericarpel felted and with sparse hairs and/or an occasional spine	5
5a.	Ribs 4–5, separated by nearly straight grooves; fruit smooth, with scales, spineless	**2. bonplandii**
b.	Ribs 6–7, separated by zig-zag grooves; fruit tubercled, usually with some spines	**3. tortuosa**

1. H. gracilis (Miller) Britton (*H. aboriginum* Small; *H. simpsonii* Small). Illustration: Haustein, Der Kosmos-Kakteenführer, 93 (1983).
Stems erect or sprawling, 1–5 m × 2.5–4 cm, not segmented, ribs 9–11, rounded; areoles *c.* 1.2 cm apart; spines 7–16, to 2.5 cm, rigid. Flowers 15–20 × 10–12 cm, scales of tube with white or brown axillary hairs 6–8 mm; inner perianth-segments irregularly toothed. Fruit depressed-spherical, 3–4 × 3–6 cm, orange-red or yellow; seeds 2 mm. *Jamaica, SE USA (Florida)*. G2. Summer.

H. **eriophora** (Pfeiffer) Britton (*H. fragrans* Small). Illustration: Benson, Cacti of the United States and Canada, pl. 67, f. 586–589 (1982). Ribs 9–12; spines 8–15, the longest 1.2–4.4 cm. Flowers 15–17.5 × 7.5–10 cm; scales of tube with white axillary hairs 1–1.5 cm; inner perianth-segments entire. *Cuba, SE United States (Florida)*. G2. Summer.

2. H. bonplandii (Pfeiffer) Britton & Rose (*H. pomanensis* (Weber) Britton & Rose; *Eriocereus pomanensis* (Weber) Riccobono). Illustration: Britton & Rose, The Cactaceae 2: f. 225 (1920); Haustein, Der Kosmos-Kakteenführer, 95 (1983).
Stems erect at first, then sprawling or clambering to 3 m or more, 3–8 cm in diameter, 4-(rarely 5) angled, jointed, glaucous; areoles 1.2–1.4 cm apart. Spines rigid, pinkish at first, later whitish, tipped blackish; central spine 1, 1–2.5 cm; radial spines 5–8, *c.* 1 cm. Flowers *c.* 15 cm; tube with acute scales 1 cm or more, hairy in the axils. Fruit spherical, 3–5 cm, red, splitting laterally to expose the spongy white funicles and black seeds. *S Brazil and Paraguay to N Argentina*. G2. Summer.

H. jusbertii (Schumann) Borg (*Eriocereus jusbertii* (Schumann) Riccobono). Illustration: Borg, Cacti, ed. 2, pl. 16b (1951). Probably a variant of *H. bonplandii*, but ribs 4–6 and spines very short, less than 5 mm. Known only in cultivation, where it is commonly used as a grafting stock.

3. H. tortuosa (Forbes) Britton & Rose. Illustration: Britton & Rose, The Cactaceae 2: pl. 21 (1920); Haustein, Der Kosmos-Kakteenführer, 95 (1983).
Intermediate between *H. bonplandii* and *H. martinii*; stems 2–4 cm in diameter, 6–7-ribbed, the ribs somewhat tuberculate and the grooves between slightly zig-zag. Central spine 1, 3–4 cm; radial spines 5–10, to 2 cm. Flowers 12–15 cm, scales of pericarpel short, with felted areoles and an occasional short spine, those of the tube acute, *c.* 1 cm, with axillary hairs; inner perianth-segments white to pinkish. Fruit spherical, 3–4 cm, red, tubercled and with a few short spines. *Argentina*. G2. Summer.

4. H. martinii (Labouret) Britton & Rose (*Eriocereus martinii* (Labouret) Riccobono). Illustration: Britton & Rose, The Cactaceae 2: pl. 20, f. 3 & pl. 21, f. 2 (1920).
Stem sprawling or clambering to 2 m or more × 1.5–2.5 cm; ribs 4– 5; areoles seated on broad tubercles, *c.* 1.5–2.5 cm apart. Central spine 1, 2–4 cm, stout; radial spines 1–3, to 3 mm. Flowers 15–22 × 15–17 cm; pericarpel and tube with small subulate scales 2–3 mm, felted and more or less hairy in the axils; inner perianth-segments white or pinkish. Fruit spherical, *c.* 3 cm, red, splitting down one

side, tuberculate and very shortly spiny. *Argentina*. G2. Summer.

The status of *H. regelii* (Weingart) Borg is uncertain. It may be a variety of *H. martinii*.

5. H. guelichii (Spegazzini) Britton & Rose. Illustration: Britton & Rose, The Cactaceae 2: f. 228 (1920).
Stems clambering, reputedly to 25 m × 3–5 cm, segmented, 3–4 angled. Spines *c.* 6, 1 stouter and longer than the rest. Flowers large; scales of pericarpel and tube *c.* 2 cm, nearly naked in the axils. Fruit 4–7 cm in diameter, strongly tuberculate, with persistent scales, spineless. *N Argentina*. G2. Summer.

6. H. tetracantha (Labouret) Hunt (*Roseocereus tetracanthus* (Labouret) Backeberg – as 'tephracanthus'). Illustration: Backeberg, Cactus Lexicon, f. 377 (1977).
Shrub or small tree to 4 m; stems 6–10 cm in diameter, bluish or greyish green; ribs 7–9, somewhat tubercled below the areoles. Central spine 1; radial spines 4–7. Flowers 18–22 cm; pericarpel scales acute, with dense curled hairs in the axils. Fruit ovoid. *Bolivia*. G2.

5. HYLOCEREUS (Berger) Britton & Rose
D.R. Hunt
Epiphytic, climbing or scrambling, often to 5 m or more; stems usually 3-winged or angled, the margins often horny, often producing aerial roots. Areoles with short spines or rarely spineless. Flowers usually very large, nocturnal, funnel-shaped, white or rarely red; pericarpel and tube stout with broad triangular scales, naked or nearly so in the axils; stamens numerous in a continuous series; style thick, stigma-lobes sometimes bifid. Fruit large, spherical or ovoid, fleshy, with broad scales; seeds kidney-shaped, black.

A genus of about 25 species, mostly ill-defined, in C America, the West Indies, Colombia and Venezuela.

1. H. undatus (Haworth) Britton & Rose (*H. triangularis* of some authors, not *Cactus triangularis* Linnaeus). Illustration: Botanical Magazine, 1884 (1817) – as Cactus triangularis; Hecht BLV Handbuch der Kakteen, 210 (1982).
Stems epiphytic or climbing to 5 m or more, usually jointed at intervals, 3-winged, 4–7.5 cm in overall diameter, margins horny. Areoles usually 4–5 cm apart with 1–3 short, conical, brownish grey spines to 3 mm. Flowers 25–27 ×

15–25 cm, white. Fruit spherical-oblong, 5–12.5 × 4–9 cm, scales long-pointed, to 2.5 cm. *Tropical America*. G2. Summer.

Along with *Selenicereus grandiflorus*, this is one of the most widely cultivated climbing cacti. According to Britton & Rose (The Cactaceae 2: 187, 1920), the true *H. triangularis* is a closely related but more slender-stemmed plant native to Jamaica with the stem-wings not horny and the flowers smaller.

6. WEBEROCEREUS Britton & Rose
D.R. Hunt
Epiphytes; stems terete, angled or flattened, sometimes rooting aerially; areoles spineless or with short bristle-like spines. Flowers shortly funnel-shaped; tube rather stout, bearing scales and areoles with hairs or bristles; perianth-segments pink, greenish or yellowish white; stamens numerous in 2 groups, one above the nectar chamber and one forming a throat-circle. Fruit fleshy, bristly or naked; seeds ovoid, black.

A genus of about 9 species occurring in C America and Ecuador and intergrading with *Selenicereus* and *Nopalxochia*.

1. W. tonduzii (Weber) Rowley (*Werckleocereus tonduzii* Britton & Rose). Illustration: Kew Magazine, pl. 43 (1985).
Stems climbing, creeping or pendent, segmented, the segments 3- or rarely 4-angled or winged, 10–40 × 2–3 (rarely to 5 cm) light green, rooting aerially. Areoles 1–3 cm apart, raised on low scollops, with short dark wool, usually spineless or with 1–2 brownish spines to 2 mm. Flowers 6–8 × 4–5 cm; pericarpel and tube 3.5–6 × 1.4 cm, green, with numerous, small, dark-felted spiny areoles; outer perianth-segments pale greenish or brownish yellow, inner creamy white. Fruit almost spherical, *c.* 4 cm in diameter, spiny, yellow; seeds obovoid, 1.5–2.2 × 1–1.5 mm, glossy brownish black. *Costa Rica*. G2. Winter.

7. SELENICEREUS (Berger) Britton & Rose
D.R. Hunt
Usually epiphytic or growing on rocks; stems climbing, clambering, trailing, often to 5 m or more, or pendent, slender, ribbed, more rarely angled or even flattened, often bearing aerial roots. Areoles with short or bristle-like spines or hairs, rarely spineless. Flowers solitary, lateral, funnel-shaped, regular, 12–40 × 10–20 cm, nocturnal; tube long, pericarpel and tube bearing small scales and areoles with hairs, bristles or spines; perianth-segments spreading,

white or the outer yellowish or pinkish; stamens very numerous in 2 series. Fruit large, spherical or ovoid, fleshy, pale yellowish red, with tufts of hairs, bristles or spines; seeds pear-shaped, black.

A genus of up to 20 mostly ill-defined species in the West Indies, Mexico and C America, extending to Venezuela and sometimes including species of uncertain affinity from elsewhere in South America.

1a. Areoles of pericarpel and tube bearing long hairs   2
  b. Areoles of pericarpel and tube not hairy, though usually with bristles or spines   4
2a. Areoles seated on a knob or spur   **3. hamatus**
  b. Areoles not seated on a knob or spur   3
3a. Spines 6–18, 4.5–15 mm, bristle-like; stems 1.2–2.5 cm in diameter   **1. grandiflorus**
  b. Spines 1–5, 1–3 mm, conical; stems 2.5–5 cm in diameter   **2. pteranthus**
4a. Stem 1–2 cm in diameter, with 4–6 or more ribs   **4. spinulosus**
  b. Stem usually 2 cm or more in diameter and 3-winged   **5. setaceus**

1. S. grandiflorus (Linnaeus) Britton & Rose. Illustration: Botanical Magazine, 3381 (1835) – as Cereus grandiflorus; Hecht, BLV Handbuch der Kakteen, 216 (1982).
Stems 1.2–2.5 cm in diameter, with 5–8 (rarely to 10) low ribs. Areoles 5–20 mm apart, with 6–18 bristle-like spines 4.5–15 mm, and (on young growth) whitish, yellowish or brown hairs, these later deciduous. Flowers 17.5–30 × 12.5–17.5 cm, outer segments pale yellow to brownish, inner white. *Tropical America.* G2. Summer.

Accurate classification of this species and its close allies, is hindered by doubt about its place of origin (it is widely cultivated and naturalised) and by hybridisation. Many species and varieties have been named which may belong here or to the next species.

2. S. pteranthus (Link & Otto) Britton & Rose. Illustration: Britton & Rose, The Cactaceae 2: pl. 38, f. 1 (1920); Hecht, BLV Handbuch der Kakteen, 217 (1982); Cullmann et al., Kakteen, edn 5, 292 (1984).
Like *S. grandiflorus* but stems stouter, 2.5–5 cm in diameter, 4–6-angled. Areoles 2–2.5 cm apart, with 1–5 short, hard, conical spines 1–3 mm. Fruit red. *E Mexico.* G2. Summer.

3. S. hamatus (Scheidweiler) Britton & Rose. Illustration: Guerke & Vaupel, Blühende Kakteen 3:161/162 (1914) – as Cereus hamatus.
Stems slender, *c.* 1.5 cm in diameter, 3–4-angled, the angles with prominent spurs to 1 cm long beneath the areoles; spines short and weak. Flowers 20–35 cm, white. *E Mexico.* G2. Summer.

In **S. macdonaldiae** (Hooker) Britton & Rose (Illustration: Botanical Magazine 4707, 1853 – as Cereus macdonaldiae), reputedly from Honduras, the areoles are seated on tubercles 2–3 mm high.

4. S. spinulosus (de Candolle) Britton & Rose. Illustration: Britton & Rose, The Cactaceae 2: pl. 38, f. 2 (1920).
Stem 1–2 cm in diameter, with 4–5 ribs. Areoles 1.5–2.5 cm apart with 6–8 spines only 1 mm. Flowers 10–12.5 × 7–8.5 cm; pericarpel spiny but without silky hairs; outer perianth-segments brownish green, inner white to pinkish. *E Mexico, USA (SE Texas).* G2. Spring–summer.

5. S. setaceus (de Candolle) Werdermann (*Mediocactus coccineus* (de Candolle) Britton & Rose). Illustration: Pfeiffer & Otto, Abbildung und Beschreibung bluehende Cacteen 1: 16 (1840) – as Cereus setaceus.
Stems usually 3-but sometimes 4–5-angled, 2–4 (rarely to 8) cm in diameter. Areoles 2–3 cm apart, with 1–2 conical brown spines 1–2 mm. Flowers 25–30 cm; pericarpel with felted and spiny areoles, tube with scales naked in their axils. Fruit ovoid, tuberculate and bristly, red. *Brazil to N Argentina.* G2.

S. megalanthus (Vaupel) Moran. Stems often only 1.5 cm in diameter; flowers 38 cm. *Peru.* G2.

8. EPIPHYLLUM Haworth
D.R. Hunt
Mostly epiphytes; old stems often terete and somewhat woody at base, rarely spiny; younger branches flattened, margins scolloped, toothed, lobed or pinnatifid between the areoles, spineless. Flowers mostly nocturnal, large; pericarpel with small scales or rarely hairs or bristles; tube long, with scales naked in the axils, abruptly dilated at the throat; limb broad, the outer perianth-segments whitish, yellowish or pinkish, the inner pale yellow or white; stamens numerous, inserted in the throat. Fruit ovoid or oblong, with small scales and areoles, these sometimes terminating low ridges; seeds kidney-shaped, black.

A genus of about 10 species in C

America and the West Indies, one species extending to Brazil.
Literature: Kimnach, M., Icones Plantarum Succulentarum, 11, Epiphyllum cartagense (Weber) Britton & Rose, *Cactus and Succulent Journal (US)* **30**: 23–26 (1958); Epiphyllum phyllanthus, **36**: 105–115 (1964); Epiphyllum grandilobum, **37**: 15–20 (1965); Epiphyllum thomasianum, **37**: 162–168 (1965); Bravo, H., *Las Cactáceas de Mexico* edn 2, **1**: 485–508 (1978).

1a. Margins of flattened stems scolloped or toothed   2
  b. Margins of flattened stems deeply lobed or pinnatifid   4
2a. Flattened stem-segments acuminate or tailed, thin, margins scolloped and more or less wavy   **1. oxypetalum**
  b. Flattened stem-segments acute or obtuse, relatively stiff, margins more or less deeply scolloped or scolloped-toothed   3
3a. Perianth-segments about equalling the tube, pale yellow; flowers remaining expanded for several days   **2. crenatum**
  b. Perianth-segments much shorter than the tube, white; flowers nocturnal   **3. phyllanthus**
4a. Stem-segments deeply lobed; areoles of pericarpel without bristles   **4. anguliger**
  b. Stem-segments pinnatisect (cut to the midrib); areoles of pericarpel bristly   **5. chrysocardium**

1. E. oxypetalum (de Candolle) Haworth. Illustration: Botanical Magazine, 3813 (1840) – as Cereus latifrons.
Primary stems terete, to 2–3 m, laterals flattened, elliptic-acuminate, to 30 × 10–12 cm, thin, margins scolloped and more or less wavy. Flowers 25–30 × 12–17 cm; tube 13–20 × 1 cm; outermost perianth-segments narrow, reddish, inner to 2.5 cm broad, whitish. *Mexico, Guatemala, widely introduced elsewhere in the tropics.* G2.

E. pumilum (Vaupel) Britton & Rose. Habit very similar but the flowers only 10–15 cm. *Guatemala, S Mexico.* G2.

2. E. crenatum (Lindley) G. Don. Figure 28(5), p. 204. Illustration: Edwards's Botanical Register **30**: pl. 31 (1844) – as Cereus crenatus.
Stems flattened for most of their length, terete or 3-angled and becoming woody at base, acute or obtuse at apex, to 50 × 3.5 cm overall, rather thick, margins

obliquely scolloped. Flowers 20–29 × 10–20 cm; outer perianth-segments pale yellow, inner white. *Mexico to Honduras.* G2.

A parent species of numerous intra- and inter-generic hybrids.

3. E. phyllanthus (Linnaeus) Haworth. Illustration: Botanical Magazine, 2692 (1826) – as Cactus phyllanthus; Britton & Rose, The Cactaceae 4: pl. 19 (1923) – as E. hookeri; Cactus & Succulent Journal (US) **36**: 107 (1964); Backeberg, Cactus Lexicon, fig. 101 (1977) – reproduction of the Britton & Rose plate cited.
Primary stems reaching 1–2 m, cylindric or 3-angled towards base, flattened above, laterals flattened except at base, linear, obtuse or acute, 25–30 × 3–10 cm, rather stiff, margins scolloped, the scollops nearly symmetric to strongly oblique, midrib prominent. Flowers variable in size, 7.5–30 × 4.5–23 cm, but the tube usually more than twice as long as the perianth-segments; outer segments tinged green or red; inner 3–10 mm broad, white. *Tropical America.* G2.

4. E. anguliger (Lemaire) Don (*E. darrahii* (Schumann) Britton & Rose). Illustration: Botanical Magazine, 5100 (1859) – as Phyllocactus anguliger; Haustein, Der Kosmos-Kakteenführer, 99 (1983).
Primary stems cylindric or 3-angled towards the base, flattened above, laterals flattened except at base, linear or narrowly lanceolate, obtuse or acute, to 100 × 4–8 cm, rather stiff, deeply lobed, the lobes more or less triangular. Flowers 15–18 × 10–13 cm, scented; outer segments lemon or golden yellow, inner white. *S Mexico.* G2. Late autumn.

5. E. chrysocardium Alexander (*Marniera chrysocardium* (Alexander) Backeberg). Illustration: Cactus & Succulent Journal (US) **28**: 3 (1956); Haustein, Der Kosmos-Kakteenführer, 99 (1983).
Vigorous scrambler, eventually developing a woody, spiny, cylindric trunk 3–4 cm in diameter, stems of current year to 1 m, pinnatifid and reminiscent of cycad fronds, elliptic overall, the lobes oblong, curved towards apex, acute, 13–15 × 4 cm, thinly leathery; areoles between the lobes, adjacent to midrib, naked or with 2–3 bristles. Flowers *c.* 32 × 20 cm; tube 16 × 1.25 cm, pale green below, dull purplish above, with green scales; outer perianth-segments dull purplish, inner white; filaments deep yellow, anthers brown. *Mexico (Chiapas).* G2. Winter.

9. NOPALXOCHIA Britton & Rose
D.R. Hunt
Epiphytes; stems terete at base, flattened above, margins scolloped, spineless except for the young growth. Areoles with sparse wool. Flowers solitary, diurnal, funnel-shaped, pink or red; tube shorter than or equalling the limb, bearing scales with short bristles or naked in the axils. Fruit spherical or ellipsoid, with areoles and with or without bristles. Seeds dark brown.

A group of 4 species native to S Mexico with affinities to *Aporocactus*, *Disocactus*, *Heliocereus* and *Weberocereus*. Further research may result in reorganisation of these genera.

1. N. phyllanthoides (de Candolle) Britton & Rose. Illustration: Botanical Magazine, 2092 (1819) – as Cactus phyllanthoides; Haustein, Der Kosmos-Kakteenführer, 97 (1983).
Primary stems to 40 cm × 6 mm, terete below, flattened at apex, laterals lanceolate, flat, with terete, stalk-like base, 15–30 × 2.5–4 cm overall, scolloped. Flowers campanulate funnel-shaped, 8–10 × 7–9 cm; tube 2.5–5 cm × 7–10 mm, greenish, scales naked in the axils; limb pink, outer perianth-segments opening irregularly before flowering, then spreading widely; inner remaining more or less erect; stamens, style and stigmas whitish. Fruit ellipsoid, 3–4 cm, with low ribs, green at first, later red. *S Mexico.* G2. Spring.

A beautiful and free-flowering species much used in hybridisation.

10. DISOCACTUS Lindley
D.R. Hunt
Epiphytic or growing on rocks; stems all flattened, or of 2 forms, the main stems terete and the lateral flattened, linear to narrowly lanceolate, margins scolloped or toothed, with small spineless areoles. Flowers usually solitary, diverse in size and form; tube 1–4 cm or very small, usually slender, with few, small scales; limb erect to spreading or disc-shaped, the inner segments 5–75 mm, whitish to purplish pink; stamens in 2 groups, one forming a ring in the throat, the remainder inserted further down the tube. Fruit berry-like, to 1.5 cm in diameter, with few small scales or almost naked; seeds ovoid to somewhat kidney-shaped, mostly 1.5–2 mm, black.

A genus of 10 species most numerous in C America but also in the West Indies and tropical S America. Many of the species are found only in specialist collections but 1, previously thought to be a *Rhipsalis*, is widespread.

1. D. ramulosus (Salm-Dyck) Kimnach (*Rhipsalis coriacea* Polakowsky). Illustration: Cactus & Succulent Journal (US) **33**: 13 (1961); Haustein, Der Kosmos-Kakteenführer, 101 (1983).
Stems of 2 forms, the main stems terete, to 70 cm × 2–5 mm, usually flattened and broadened in the apical 10–20 cm, lateral stems linear or narrowly lanceolate, flat, to 23 × 3 cm, 5 mm thick, with terete, stalk-like base, scolloped or obtusely toothed, often suffused purplish. Flowers disc-shaped, 7–12 × 10–14 mm, yellowish white, tinged pink or green; tube 4–5 mm. Fruit spherical, 4–8 × 4–8 mm, white to pinkish white, with small broadly ovate scales; seed 1.25–1.5 × 1 mm, black. *Tropical America.* G2. Spring.

11. HELIOCEREUS (Berger) Britton & Rose
D.R. Hunt
Epiphytic or growing on rocks; stems ascending, scrambling, procumbent or pendent, acutely 3–7-angled or ribbed, rarely flat. Areoles with short or bristle-like spines. Flowers solitary, diurnal, remaining open for several days, funnel-shaped, moderately large, mostly bright red, the tube usually shorter than or equalling the limb, with scales and areoles with bristle-like spines; stamens and style weakly deflected downwards. Fruit spherical or ovoid, fleshy, usually spiny; seeds large, kidney-shaped, black.

A genus of 3 or more species likely to be merged with *Aporocactus* or *Disocactus* after further research.

1. H. speciosus (Cavanilles) Britton & Rose. Illustration: Botanical Magazine, 2306 (1822) – as Cactus speciosissimus; Haustein, Der Kosmos-Kakteenführer, 97 (1983).
Variable; stems to 1 m × 1.5–2.5 cm, with 3–5 (rarely to 7) acute ribs. Areoles 1–3 cm apart, with 5–8 or more spines to 1 (rarely to 1.5) cm, yellowish at first, passing to brownish. Flowers 11–17 × 8–13 cm; tube *c.* 8 cm, green; perianth-segments oblanceolate, apiculate, scarlet or vermilion (white in var. **amecamensis** (Heese) Weingart). *C & S Mexico, Guatemala.* G2. Summer.

Extensively hybridised with species of allied genera.

12. APOROCACTUS Lemaire
D.R. Hunt
Creeping or trailing over rocks, or epiphytic; stems slender, cylindric, with 7–12 low ribs, sometimes producing aerial

roots. Areoles close-set; spines bristle-like. Flowers diurnal, narrowly tubular-funnel-shaped, bilaterally symmetric, brightly coloured; pericarpel and tube with scales and fine bristles; limb more or less oblique; stamens projecting. Fruit small, fleshy, reddish; seed brown.

A genus of 2 species, endemic to Mexico.

1a. Flowers purplish pink, the tube upcurved above the pericarpel, the limb oblique **1. flagelliformis**
  b. Flowers scarlet, the tube almost straight and the limb only slightly oblique **2. martianus**

1. A. flagelliformis (Linnaeus) Lemaire (*A. flagriformis* (Pfeiffer) Britton & Rose; *A. leptophis* (de Candolle) Britton & Rose). Illustration: Botanical Magazine, 17 (1787) – as Cactus flagelliformis; Haustein, Der Kosmos-Kakteenführer, 94 (1983). Stems slender, ascending at first, soon prostrate or pendent, to 1 m or more × *c*. 1 cm; ribs 7–12; areoles 4–8 mm apart. Spines *c*. 9–14, 4–5 mm, bristle-like. Flowers usually 5–8 × 2.5–4 cm, purplish pink, the limb bilaterally symmetric. *C Mexico (Hidalgo).* G1. Spring.

2. A. martianus (Zuccarini) Britton & Rose (*A. conzattii* Britton & Rose). Illustration: Bravo, Las Cactáceas de Mexico, edn 2, fig. 257 – as A. leptophis, and figs. 261, 262 – as A. conzattii (1978); Haustein, Der Kosmos-Kakteenführer, 94 (1983). Stems ascending at first, soon prostrate or pendent, to 2–5 m × 1.2–2.5 cm, with 8–10 low ribs. Spines 8–20, 5–12 mm, bristle-like. Flowers 6–9 × 5–6 cm, scarlet, the limb nearly regular. *S Mexico (Oaxaca).* G2. Summer.

13. ACANTHOCEREUS (Berger) Britton & Rose
D.R. Hunt
Clambering, arching or trailing with 3–5 (rarely to 7)-angled, segmented, strongly spined stems sometimes rooting at the tips. Flowers large, nocturnal, funnel-shaped, white or pale pink; receptacle-tube long, rather stiff, with small scales subtending conspicuous areoles and short stiff spines; stamens numerous in a single series. Fruit spherical or ovoid, dark red, fleshy and spiny; seeds black.

A genus of about 8 weakly defined species in tropical America.

1. A. pentagonus (Linnaeus) Britton & Rose. Illustration: Benson, Cacti of the United States & Canada, f. 592–594 (1982).

Stems several metres long, the terminal segments often 30–100 × 2.5–5 cm, acutely 3–5-angled; areoles 2.5–4 cm apart, with several spines, the central longest, 4 cm. Flower 17.5–25 × *c*. 10 cm, stigmas 10–15, *c*. 1.2 cm, slender. Fruit usually 3–6 × 2.5–4 cm, seeds 3 mm, nearly smooth, shiny. *USA (Florida and Texas), West Indies, Mexico to Venezuela, widespread near coasts.* G2.

14. PENIOCEREUS (Berger) Britton & Rose
N.P. Taylor
Climbing, with slender, ribbed, sparingly branched stems, hairless or papillose-downy, light green or rather dark; roots thickened, spherical or divided. Spines conspicuous, or adpressed and short. Flower nocturnal or diurnal, occasionally terminal; receptacle-tube long and slender, bearing areoles with bristles or spines. Fruit narrowly ovoid, tapered at apex, fleshy, red, the spines or bristles more or less deciduous; seeds mostly large, black.

A genus of about 20 species in C America, N to NW Mexico and SW USA. Seldom seen and, except for *P. viperinus* and *P. serpentinus*, rarely flowering in cultivation.

1a. Stem green, hairless; ribs 10–17 **3. serpentinus**
  b. Stem dark grey-brown, papillose-downy; ribs 3–10 **2**
2a. Stems strongly angled with 3–6 (rarely to 8) prominent ribs; flowers 15–21 cm **1. greggii**
  b. Stems terete with 6–10 low ribs; flowers 3–15 cm **2. viperinus**

1. P. greggii (Engelmann) Britton & Rose (*Cereus greggii* Engelmann). Illustration: Benson, Cacti of the United States & Canada, f. 625–631 (1982). Stems mostly 30–60 × 1–2 cm from a large spherical root; ribs 3–6, areoles *c*. 4.5–6 mm apart; spines 10–13, to 3 mm. Flower white, to 7.5 cm in diameter. Seeds 2–3 mm, rough. *C, N & NW Mexico, S USA.* G1.

2. P. viperinus (Weber) Kreuzinger (*Cereus viperinus* Weber; *Wilcoxia viperina* (Weber) Britton & Rose). Illustration: Bravo, Las Cactáceas de Mexico, edn 2, f. 233, 234 (1978); Cullmann et al., Kakteen, edn 5, 314, 315 (1984) – as Wilcoxia striata. Stems to 3 m × 1–1.5 cm from divided tuberous roots; ribs 8–10, very low; areoles 1–3 cm apart. Spines 9–12, to 5 mm. Flowers to 9 × 4 cm, bright pink to red. Seeds 2.5–4 mm, smooth. *Central S Mexico.* G1. Summer.

P. striatus (Brandegee) Buxbaum (*Cereus striatus* Brandegee; *Wilcoxia striata* (Brandegee) Britton & Rose; *Neoevansia striata* (Brandegee) Sanchez-Mejorada). Illustration: Bravo, Las Cactáceas de Mexico, edn 2, f. 228, 229 (1978); Benson, Cacti of the United States & Canada, f. 634–637 (1982). Similar but stems only 5–8 mm in diameter; ribs 6–9; areoles 5–15 mm apart. Spines to 3 mm. Flower 7.5–15 × 5–7.5 cm, white to pink or purplish. Seeds 1–2 mm, slightly tuberculate. *NW Mexico, USA (S Arizona).* G1.

3. P. serpentinus (Lagasca & Rodriguez) N.P. Taylor (*Cactus serpentinus* Lagasca & Rodriguez; *Nyctocereus serpentinus* (Lagasca & Rodriguez) Britton & Rose). Illustration: Bravo, Las Cactáceas de Mexico, edn 2, f. 199, 200 (1978). Stems 2–3 m × 3–5 cm from thickened, more or less tuberous roots; ribs 10–17; areoles *c*. 10 mm apart. Spines 10–14, to 3 cm. Flower 15–20 × 8 cm, white, reddish outside. Seeds 5 mm, smooth. *Mexico, widely cultivated.* G1. Summer.

Also probably belonging to this genus is **Wilcoxia papillosa** Britton & Rose, a poorly understood species from W Mexico (Sinaloa), which could be in cultivation. It may be related to *P. viperinus*, but differs in its 3–5-ribbed stems and areoles with only 6–8 spines.

15. ECHINOCEREUS Engelmann
N.P. Taylor
Low-growing; stems spherical to cylindric, ribbed, simple or clustering, self-supporting or rarely climbing; roots fibrous or tuberous. Flower-buds developing at the upper edge of the stem, areoles often bursting through the surface above them (as do the lateral shoots); flowers funnel-shaped, the pericarpel and tube bearing areoles with spines, bristles and sometimes abundant wool; stigmas greenish, rarely white. Fruit dehiscent or indehiscent, juicy, fleshy or dry, bearing spiny areoles or these falling off at maturity. Seeds 0.8–2 mm, mostly black, strongly warty.

A genus of 45 species native to Mexico and the SW United States. Many can withstand freezing temperatures when dry and some may be hardy outdoors in sheltered districts. The following account treats only the more commonly cultivated species.
Literature: Taylor, N.P., *The Genus Echinocereus* (1985).

1a. Flowers remaining open at night, bright red, orange, salmon or pale

pink, without admixture of blue; tube only gradually flared, sometimes very long; stems 4–12-ribbed
4. triglochidiatus

b. Flowers closing each night, variously coloured (including yellow and especially purplish pink) but never bright red or orange; tube broadly funnel-shaped or stems with more than 12 ribs   2

2a. Inner perianth-segments primarily yellow, sometimes contrasted with red or green at base, more than 5 mm wide   3

b. Inner perianth-segments primarily pinkish, purplish or white, or less than 5 mm wide   5

3a. Inner perianth-segments orange-red at base   **2. papillosus**

b. Inner perianth-segments entirely yellow or green at base   4

4a. Bases of perianth-segments green, fleshy and well-separated
3. pectinatus var. **dasyacanthus**

b. Bases of perianth-segments yellow, not fleshy, densely inserted at apex of tube   **10. subinermis**

5a. Stem(s) 4–10-ribbed or flowers not more than 3.5 cm   6

b. Stem(s) with more than 10 ribs and flowers more than 3.5 cm   12

6a. Stems clambering, long and very slender, to 2 cm thick, 8–10-ribbed, arising from a divided, tuberous rootstock   **12. poselgeri**

b. Stem(s) not as above   7

7a. Pericarpel and tube to 2 cm; perianth-segments to 7 mm wide   8

b. Pericarpel and tube 2.5–5 cm; perianth-segments 1–2.5 cm wide   9

8a. Stem-areoles with 8 or more spines
8. viridiflorus

b. Stem-areoles with 1–4 spines (rarely0)
14. knippelianus

9a. Perianth broadly bowl-shaped, more deeply coloured in the throat, the segments very fleshy and widely separated at base   **1. fendleri**

b. Perianth funnel-shaped, darker or lighter in the throat, the segments not as above   10

10a. Perianth usually more deeply coloured, not lighter, in the throat
5. enneacanthus

b. Perianth colour lighter or changing to white, very pale yellow or green in the throat   11

11a. Receptacular areoles bearing conspicuous diffuse wool; stem-areoles usually with only 1 central spine
9. pentalophus

b. Receptacular areoles with only short areolar felt (besides the usual spines); stem-areoles with 3–6 central spines
6. cinerascens

12a. Flower-buds and receptacle bearing conspicuous cobwebby wool   13

b. Flower-buds and receptacle not woolly, the areoles with short felt (besides the usual spines)   14

13a. Tube very slender, gradually flared above ovary, the narrow nectar-chamber 8–12 mm; stigmas very pale green or whitish   **13. adustus**

b. Tube strongly flared above ovary, the nectar-chamber to 4 mm; stigmas mid-green or darker
11. reichenbachii

14a. Spines 15–35 per stem-areole, less than 1.5 cm, the radials comb-like
3. pectinatus

b. Spines 2–16 per stem-areole, not comb-like or some more than 1.5 cm
15

15a. Outer perianth-segments ovate to obovate-oblanceolate, fleshy only near base; inner segments to 2.5 cm wide near apex, not darker coloured towards base; stems branching to form large, compact hemispherical clumps   **7. stramineus**

b. Outer perianth-segments linear to narrowly oblong-oblanceolate, rather fleshy and stiff throughout, deeply pigmented; inner segments 6–15 mm wide near apex and/or more deeply pigmented towards base; stems simple or loosely branched, rarely forming compact hemispherical clumps
1. fendleri

1. E. fendleri (Engelmann) Ruempler (*Cereus fendleri* Engelmann). Illustration: Benson, Cacti of the United States & Canada, f. 665, 666 (1982); Weniger, Cacti of Texas and neighboring states, 68 (1984). Simple or clustering, stems 7.5–50 × 4–8 cm, ovoid to cylindric, erect; ribs 8–18 (rarely to 22), strongly to weakly tuberculate; areoles circular 1–2.5 cm apart. Spines 2–18, usually 5–16, terete. Flowers 5–11 × 5–12 cm; limb broadly bowl-shaped, perianth-segments 6–15 mm wide, purplish magenta to white, darker towards the sometimes greenish base. Fruit nearly spherical, 3 cm, reddish, juicy. *N Mexico, SW USA.* G1. Spring–summer.

A complex species with 7 rather diverse but intergrading varieties:

Var. **fendleri**. Stems 1–2 (rarely to 5), 7.5–15 (rarely to 25 cm); ribs 8–10. Spines 5–10, usually 1 central, to 3.8 cm, curved upwards. Flowers large. *S USA (Colorado,*

New Mexico, Arizona, W Texas), Mexico (Chihuahua).

Var. **kuenzleri** (Castetter et al.) Benson (*E. kuenzleri* Castetter et al.; *E. hempelii* Fobe). Illustration: Hecht, BLV Handbuch der Kakteen, 255 (1982); Weniger, Cacti of Texas and neighboring states, 70 (1984). Stems 1–4 (rarely to 8), to 30 cm; ribs 9–12. Spines 2–7, very stout, central usually absent, rarely 1, to 2.9 cm. Flowers very large. *USA (New Mexico), N Mexico (Chihuahua).*

Var. **rectispinus** (Peebles) Benson (*E. rectispinus* Peebles). Illustration: Benson, Cacti of the United States & Canada, pl. 90 & f. 667–670 (1982). Stems 1–10, to 25 cm; ribs 9–10. Spines 8–10, 1 central, to 3.8 cm. Flowers large. *S USA (SE Arizona, SW New Mexico, W Texas), NW Mexico (N Sonora).*

Var. **ledingii** (Peebles) N.P. Taylor (*E. ledingii* Peebles). Illustration: Benson, Cacti of the United States & Canada, pl. 96, 97 (1982). Stems 4–12, to 50 cm; ribs 12–16. Spines 10–16, all yellow, 1–3 (rarely to 5) central, the principal one to 2.5 cm, strongly down-curved from near base. *USA (SE Arizona).*

Var. **bonkerae** (Thornber & Bonker) Benson (*E. bonkerae* Thornber & Bonker; *E. fasciculatus* var. *bonkerae* (Thornber & Bonker) Benson). Illustration: Lamb & Lamb, Colourful cacti, t. 123, 124 (1974); Benson, Cacti of the United States & Canada, f. 675–677 & pl. 94, 95 (1982). Stems 5–35, to 20 cm; ribs 11–16. Spines 12–15, centrals 1–3, less than 1 cm. Flowers small. *USA (Arizona).*

Var. **boyce-thompsonii** (Orcutt) Benson (*E. boyce-thompsonii* Orcutt; *E. fasciculatus* var. *boyce-thompsonii* (Orcutt) Benson). Illustration: Lamb & Lamb, Colourful cacti, t. 98 (1974); Benson, Cacti of the United States & Canada, f. 673, 674 (1982). Stems 3–12, to 25 cm; ribs 12–22 (usually 14–18). Spines 13–17, 3 central, the lowermost to 5 (rarely to 10) cm, strongly deflexed, very slender. *USA (C Arizona).*

Plants cultivated under this name often prove to be the following variety.

Var. **fasciculatus** (Jackson) N.P. Taylor (*Mammillaria fasciculata* Jackson; *E. fasciculatus* (Jackson) Benson). Illustration: Benson, Cacti of the United States & Canada, f. 671, 672 & pl. 93 (1982); Taylor, Genus Echinocereus, t. 1 in part (1985). Stems 5–20, 17.5–45 cm; ribs 8–13. Spines 10–15, 1–3 central, the lowermost to 7.5 cm, straight to slightly deflexed. *USA (Arizona, SW New Mexico), NW Mexico (Sonora).*

E. engelmannii (Engelmann) Lemaire
(*Cereus engelmannii* Engelmann).
Illustration: Benson, Cacti of the United
States & Canada, f. 680–687 & pl. 98–103
(1982); Taylor, The genus Echinocereus,
t. 1 in part (1985). Clustering, stems
cylindric, 5–60 × 4–8.7 cm; ribs 10–13,
not obviously tuberculate; areoles circular,
c. 6 mm in diameter, 2–3 cm apart. Spines
8–10 (rarely to 21), 4–7 (rarely as few as
2) central, 2–7 cm, angular. Flowers
shortly funnel-shaped, to 9 × 9 cm, purplish
red to magenta or lavender. Fruit to 3 ×
2.5 cm, red, juicy. *NW Mexico, SW USA.*
G1. Summer.

2. E. papillosus Ruempler (*E. angusticeps*
Clover). Illustration: Gürke, Blühende
Kakteen 2: t. 115 (1909); Lamb & Lamb,
The illustrated reference on cacti and other
succulents 5: t. 303 (1978).
Sparingly clustering (more so in var.
angusticeps (Clover) Marshall), stems 4–10
(rarely to 30) × 2–7 cm, cylindric,
brownish green; ribs 6–10, strongly
tuberculate, the tubercles to 9 × 2–3 mm,
often papilla-like; areoles circular, small,
9–12 mm apart. Spines 7–12, to 2 cm.
Flowers 8–12 cm in diameter, perianth
broadly bowl-shaped, yellow, red at base.
Fruit spherical, to 1.5 cm, fleshy. *USA (S
Texas); ?adjacent Mexico.* G2. Spring.

Needs a winter minimum temperature of
5–10 °C for reliable flower production.

3. E. pectinatus (Scheidweiler) Engelmann
(*Echinocactus pectinatus* Scheidweiler;
Echinocereus dasyacanthus Engelmann).
Figure 29(14), p. 205. Illustration:
Botanical Magazine, 4l90 (1845); Taylor,
The genus Echinocereus, t. 2 in part
(1985).
Simple, or sparingly branched in age; stems
to 8–35 × 13 cm, spherical to cylindric; ribs
12–23, composed of low tubercles; areoles
elliptic, to 6 mm, *c.* 6 mm apart. Central
spines 1–5 or more, minute or to 1.5 or
even 2.5 cm; radial spines 12–30, comb-
like, 5–15 mm. Flowers 5–12 (rarely to
16) cm in diameter, perianth broadly bowl-
shaped, pink to lavender (often yellow or
whitish in var. **dasyacanthus** (Engelmann)
N.P. Taylor), white or maroon in the
throat, green at base. Fruit spherical to
ellipsoid, to 6 × 4.5 cm, green to purplish,
fleshy, *SW USA, N Mexico.* G1. Spring.

4. E. triglochidiatus Engelmann (*E.
coccineus* Engelmann; *E. rosei* Wooten &
Standley; *E. polyacanthus* misapplied in
part). Illustration: Benson, Cacti of the
United States & Canada, f. 638–659 & pl.
81–87 (1982).

Simple or clustering, sometimes forming
very large clumps, stems 5–40 × 5–15 cm,
ovoid to cylindric, extremely variable; ribs
5–12, high and acute to low and
tuberculate, sometimes spiralled; areoles
circular, 6–12 mm apart, wool persistent.
Spines 1–22, usually 3–16, 1–7 cm
(absent or minute in var. **melanacanthus**
(Engelmann) Benson 'Inermis'), terete or
angled, straight (curved or twisted in var.
mojavensis (Engelmann & Bigelow)
Benson). Flower-buds bluntly pointed to
rounded at apex, flower tubular-funnel-
shaped, 3–7 (rarely to 9) × 2.5–7 cm,
brilliant scarlet, rarely pinkish, paler in the
throat, remaining open day and night,
unscented. Fruit spherical to obovoid, to
2.5 × 1.5 cm, eventually pink to red, juicy.
SW USA, adjacent N Mexico. G1. Spring.

E. scheeri (Salm-Dyck) Scheer (*Cereus
scheeri* Salm-Dyck; *E. salm-dyckianus*
Scheer?; *E. gentryi* Clover; *E. cucumis*
Werdermann). Illustration: Schumann &
Gürke, Blühende Kakteen 1: t. 29 (1902);
Taylor, The genus Echinocereus, t. 3
(1985). Stem 10–60 (rarely to 100) ×
1.5–4 cm, cylindric, ribs 4–14; areoles
2–7 mm apart. Spines to 2 cm (absent or
minute in var. **gentryi** (Clover) N.P.
Taylor). Flower-buds sharply pointed;
flowers narrowly tubular-funnel-shaped, to
14 × 8 cm, scarlet, orange or pink,
sometimes closing during warmest part of
the day. Fruit ovoid, to 2.7 × 2.4 cm, green.
N Mexico. G1. Spring.

5. E. enneacanthus Engelmann (*Cereus
dubius* Engelmann; *Echinocereus merkeri*
Schumann). Illustration: Benson, Cacti of
the United States & Canada, f. 688, 689,
692, 693 & pl. 105 (1982).
Clustering, stems mostly cylindric and
partly prostrate, to 60 × 3.5–5 cm (var.
brevispinus (Moore) Benson) or 5–15 cm
(var. **enneacanthus**) in diameter; ribs 7–10,
scarcely tuberculate or tubercles large but
flat; areoles circular, 16–45 mm apart.
Spines 7–17, to 4–8 cm, terete or partly
flattened and angled. Flowers funnel-
shaped, to 8 × 10 cm, magenta deepening
to crimson in the throat. Fruit spherical to
ovoid, to 4 × 2.5 cm. *Central N & NE
Mexico, USA (S Texas, S New Mexico).* G1.
Spring–Summer.

E. berlandieri (Engelmann) Haage (*Cereus
berlandieri* Engelmann; *E. blan(c)kii*
misapplied, not *Cereus blanckii* Poselger).
Illustration: Engelmann, Cactaceae of the
boundary, t. 5 (1859); Haustein, Der
Kosmos-Kakteenführer, 222 (1983). Stems
prostrate, 1.5–2.5 cm in diameter; ribs
5–7; areoles 8–20 mm apart. Spines 7–12,

to 3 cm, more or less terete. Flowers
broadly funnel-shaped, purplish. Fruit
ovoid, 2–2.5 × 1.7 cm. *USA (S Texas), NE
Mexico.* G1. Summer.

6. E. cinerascens (de Candolle) Lemaire
(*Cereus cinerascens* de Candolle; *C.
ehrenbergii* Pfeiffer; *Echinocactus
chlorophthalmus* Hooker; *Echinocereus
tulensis* Bravo). Illustration: Botanical
Magazine, 4373 (1848); Succulenta **50**(5):
cover illustration (1980).
Clustering, stems cylindric, in sprawling
clumps or mounds, 10–30 (rarely to 60) ×
1.5–12 cm; ribs 5–12, tuberculate or even-
edged, areoles circular, 8–15 mm or more
apart. Spines 7–16, to 4.5 cm, terete.
Flowers broadly funnel-shaped, 6–10 ×
6–12 cm, pinkish magenta, paler to white
or green in the throat. Fruit nearly
spherical, 2–3 cm, green. *Central E & N
Mexico.* G1. Summer.

7. E. stramineus (Engelmann) Ruempler
(*Cereus stramineus* Engelmann; *E.
enneacanthus* var. *stramineus* (Engelmann)
Benson). Illustration: Lamb & Lamb,
Colourful cacti, t. 27–30 (1974); Benson,
Cacti of the United States & Canada, f. 690,
691 & pl. 104 (1982); Hecht, BLV
Handbuch der Kakteen, 257 (1982);
Taylor, The genus Echinocereus, t. 5
(1985).
Clustering, forming mound-like clumps,
stems to 45 × 8 cm, tapering gradually
towards apex; ribs 11–17, somewhat
tuberculate; areoles circular, 7–18 mm
apart. Spines 9–18, the largest 4–9 cm,
mostly glassy white, terete to somewhat
flattened. Flowers broadly funnel-shaped,
6–12.5 × 6–12.5 cm or more, bright
pinkish magenta. Fruit spherical, to 5 cm,
red. *Central N & NE Mexico, USA (W Texas,
S New Mexico).* G1. Summer.

8. E. viridiflorus Engelmann (*E. davisii*
Houghton). Illustration: Weniger, Cacti of
Texas and neighboring states, 17, 20
(1984); Cullmann et al., Kakteen, edn 5,
158, 159 (1984).
Simple, rarely branched, spherical, ovoid or
cylindric, erect, 2.5–12.5 × 2.5–5 cm (only
1.2–2.5 cm × 9–12 mm in the dwarf var.
davisii (Houghton) Marshall); ribs 6–14,
tuberculate; areoles circular to elliptic,
3–6 mm apart. Central spines 0–3 (rarely
to 4), radial spines 8–24. Flowers funnel-
shaped, opening widely, green to yellowish
green, lemon-scented. Fruit spherical,
6–9 mm. *C & S USA.* G1. Spring.

E. chloranthus (Engelmann) Haage
(*Cereus chloranthus* Engelmann; *E.
russanthus* Weniger; *E. viridiflorus* var.

cylindricus (Engelmann) Ruempler).
Illustration: Weniger, Cacti of Texas and
neighboring states, 19, 23–25, 27 (1984);
Taylor, The genus Echinocereus, t. 7
(1985). Simple (occasionally clustering in
var. **russanthus** (Weniger) Rowley),
cylindric, 7.5–25 × 5–7.5 cm; ribs
10–18. Central spines 3–12 (0–3 in var.
cylindricus (Engelmann) N.P. Taylor),
radial spines 12–45. Flowers to 3.9 cm,
often not opening widely, yellowish green,
brownish or reddish, not lemon-scented.
Fruit 9–12 mm. *Central N Mexico
(Chihuahua), USA (W Texas, S New Mexico).*
G1. Spring.

9. E. pentalophus (de Candolle) Lemaire
(*Cereus pentalophus* de Candolle; *C.
procumbens* Engelmann; *Echinocereus
leonensis* Mathsson). Illustration: Benson,
Cacti of the United States & Canada, f. 662,
663 & pl. 88, 89 (1982); Weniger, Cacti of
Texas and neighboring states, 75, 76 – as
E. berlandieri (1984).
Clustering, sometimes branching by means
of underground suckers, stems cylindric,
sprawling or erect, 20–60 (rarely to 200) ×
1–6 cm; ribs 4–5 (6–8 in var. **leonensis**
(Mathsson) N.P. Taylor); areoles circular,
mostly 1–2 cm apart. Spines 3–10, very
short or to 6 cm. Flowers broadly funnel-
shaped, 8–10 × 10–15 cm, bright pinkish
magenta, white or yellowish in the throat,
rarely white. Fruit ovoid, to 5.5 × 3.2 cm.
Central E & NE Mexico, USA (S Texas). G2.
Summer.

10. E. subinermis Scheer (*E. luteus* Britton
& Rose; *E. ochoterenae* Ortega). Illustration:
Schumann & Gürke, Blühende Kakteen 1:
t. 4 ('3') (1900); Kakteen und andere
Sukkulenten 35: 164, 165 (1984).
Simple or sparingly clustering, depressed
spherical to cylindric, erect, 4–32.5 ×
4–15 cm, grey to dark bluish green; ribs
5–11, high, acute, not tuberculate; areoles
circular to elliptic, minute or to 3 mm long,
6–20 mm apart. Central spines 0–1 (1–4
in var. **ochoterenae** (Ortega) Unger),
minute or absent, or to 2 cm; radial spines
0–10, 1–30 mm. Flowers funnel-shaped, to
10 × 13 cm (only 4.5–6.5 × 4–7.5 cm in
var. **ochoterenae**), yellow, nectar chamber
to 6 mm. Fruit obovoid, 1.8–4 ×
1.3–2.2 cm, dark green, with white pulp.
NW Mexico. G1. Summer.

E. stoloniferus Marshall (*E. tayopensis*
Marshall). Illustration: Backeberg, Die
Cactaceae 4: f. l920a, b (1960); Succulenta
64: 109 (1985); Taylor, The genus
Echinocereus, t. 8 (1985). Clustering,
branches arising as underground suckers,

ovoid to cylindric, 9–30 × 3–8 cm, deep
olive-green; ribs 11–16, low, narrow,
areoles 3–9 mm apart, circular, to 2 mm in
diameter. Central spines 1–5, to 2–2.5 cm;
radials 8–13, to 1.5 cm long. Flowers
short-funnel-shaped, to 10 × 15 cm; nectar-
chamber *c.* 1.5–2 mm. Fruit described as
reddish. *NW Mexico.* G1. Summer–
Autumn.

11. E. reichenbachii (Walpers) Haage
(*Echinocactus reichenbachii* Walpers; *Cereus
caespitosus* Engelmann; *Echinocereus
caespitosus* (Engelmann) Engelmann; *E.
purpureus* Lahman). Figure 29(18), p. 205.
Illustration: Benson, Cacti of the United
States & Canada, f. 703–705 & pl. 110,
111 (1982); Weniger, Cacti of Texas and
neighboring states, 29, 30, 33 (1984).
Simple, more rarely clustering, stem
spherical to cylindric, erect, to 30–40 ×
10 cm, surface more or less obscured by
spines; ribs 10–19, narrow with low
tubercles; areoles elliptic to linear, to 4 mm
long, mostly 2–4.5 mm apart. Spines
12–40 or more, mostly rather short and
comb-like. Flowers funnel-shaped, to
5–12 × 6–16 cm; receptacular areoles
bearing fine bristles and much wool;
perianth-segments very numerous, thin
and delicate, shades of pink to magenta.
Fruit spherical to ovoid, 1.5–3 cm, green.
NE Mexico, SW USA. G1. Spring–summer.

A complex species with 5 distinctive
varieties:

Var. **reichenbachii**. Central spine 0,
radials 18–30 or more, to *c.* 6 mm; throat
white or deep magenta. *NE Mexico, USA
(Texas, adjacent Oklahoma).*

Var. **armatus** (Poselger) N.P. Taylor
(*Cereus pectinatus* var. *armatus* Poselger;
Echinocereus armatus (Poselger) Berger).
Illustration: Kakteen und andere
Sukkulenten 34: 179, f. 1, 2, 198, f. 2, 4
(1983); Cactus (Belgium) 9: 11 (1985).
Central spines 1–2 (rarely 0), to 2 cm,
radials to 23, to 8 mm; perianth pale in
throat, green at base. *NE Mexico (Nuevo
Leon).*

Var. **fitchii** (Britton & Rose) Benson (*E.
fitchii* Britton & Rose; *E. melanocentrus*
Lowry, invalid). Illustration: Weniger, Cacti
of Texas and neighboring states 35, 36
(1984). Central spines 1–7, to 9 mm,
radials to 22, to 7.5 mm; perianth deep
crimson in the throat. *USA (S Texas), NE
Mexico (N Nuevo Leon, Tamaulipas).*

Var. **perbellus** (Britton & Rose) Benson
(*E. perbellus* Britton & Rose). Illustration:
Benson, Cacti of the United States &
Canada, f. 706 & pl. 112 (1982). Central

spines 0–1, to 1 mm, radials to 16–20, to
6 mm; perianth crimson at base. *USA (S
Colorado, E New Mexico, N & C Texas).*

Var. **baileyi** (Rose) N.P. Taylor (*E. baileyi*
Rose; *E. reichenbachii* var. *albispinus*
(Lahman) Benson). Illustration: Cullmann
et al., Kakteen, edn 5, 157 (1984);
Weniger, Cacti of Texas and neighboring
states, 38, 39 (1984). Central spines 1–3,
to 3 mm, radials to 14, to 12 (rarely to
25) mm; perianth as for var. *reichenbachii.*
USA (S Oklahoma, adjacent Texas).

E. bristolii Marshall (*E. pectinatus* var.
pectinatus misapplied in part). Illustration:
Benson, Cacti of Arizona edn 3, f. 3.22
(1969); Taylor, The genus Echinocereus,
119 (1985). Simple or clustering, stem
cylindric, to 20 × 5 cm, surface partially or
totally obscured by spines; ribs 15–19;
areoles elliptic, 5–7 mm apart. Central
spines 1–5, to 1–1.5 cm; radials 20–24
(only 12–15 in var. **pseudopectinatus** N.P.
Taylor) to 9–12 mm. Flowers funnel-
shaped, 4–11 × 6–13 cm, receptacular
areoles bearing few stout spines; perianth-
segments light magenta, paler to greenish
near base. Fruit depressed-spherical,
c. 1.4 cm, red. *NW Mexico (E Sonora), USA
(SE Arizona).* G1. Spring–summer.

E. rigidissimus (Engelmann) Haage
(*Cereus pectinatus* var. *rigidissimus*
Engelmann; *E. pectinatus* var. *rubispinus*
Frank & Lau). Illustration: Benson, Cacti of
the United States & Canada, f. 695, 696 &
pl. 107 (1982); Taylor, The genus
Echinocereus, t. 2 in part (1985). Simple,
very rarely branched, spherical to cylindric,
6–30 × 4–11 cm, surface obscured by
spines; ribs 15–26 (normally 18–23),
areoles linear-elliptic, 2.5–7 mm, close.
Central spines 0; radials 15–23 (or 30–35
in var. **rubispinus** (Frank & Lau) N. P.
Taylor), comb-like, adpressed. Flowers
5–7 × 6–9 cm; receptacular areoles bearing
numerous short, stiff spines and some wool;
perianth-segments brilliant pinkish red to
magenta, white in the throat and at base.
Fruit ovoid, 2.5–5 × 1.5–3 cm, greenish,
brownish or red. *NW Mexico, USA (SW
New Mexico, SE Arizona).* G1. Summer.

12. E. poselgeri Lemaire (*Wilcoxia poselgeri*
(Lemaire) Britton & Rose; *W. tamaulipensis*
Werdermann; *W. tuberosa* Kreuzinger,
illegitimate). Illustration: Haustein, Der
Kosmos-Kakteenführer, 226 (1983);
Taylor, Genus Echinocereus, t. 10 (1985).
Stems scrambling, cylindric, to 60 (rarely
to 120) × 1–2 cm, ribs 8–10, very low;
areoles circular, 2–4.5 mm apart. Spines
9–17, adpressed, to 9 mm. Flowers funnel-

shaped, to 7.5 × 7.5 cm, bright pinkish magenta, darker in the throat. Fruit ovoid, to 2 × 1 cm, juicy, dark green-brownish. *NE Mexico, USA (S Texas)*. G1. Summer.

13. E. adustus Engelmann (*E. radians* Engelmann; *E. schwarzii* Lau). Illustration: Kakteen und andere Sukkulenten **34**: 179, f. 4, 6, 7 (1983); Taylor, The genus Echinocereus, t. 11 (1985).
Simple, or occasionally sparingly clustering (var. **adustus**), depressed-spherical to shortly cylindric, to 19 × 5–12 cm, ribs 11–16, slightly tuberculate; areoles oval-elliptic, 5–7 mm apart. Central spines 0–1 (1–5 in var. **schwarzii** (Lau) N.P. Taylor), to 3.2 cm, radials 15–31, to 8–18 mm, mostly adpressed. Flowers slender funnel-shaped, 3–6 × 4–5 cm (to 8 × 7 cm in var. **schwarzii**), stigmas very pale green or whitish. Fruit ovoid, to 2 cm long, ripening quickly and almost dry at maturity. *NW Mexico (Chihuahua, Durango)*. G1. Spring.

E. laui Frank. Illustration: Kakteen und andere Sukkulenten **29**: 74–77 (1978), **34**: 179, f. 5, 182, f. 9, 10 (1983). Clustering, stems cylindric, to 10 × 4 cm, surface obscured by dense spines; ribs 14–16, finely tuberculate, low, narrow; areoles 4–5 mm apart. Central spines 4, to 3 cm, radials 18–21, 5–10 mm long, bristle-like, more or less straight-projecting. Flowers 3–5 × 3–6.5 cm, pink. Fruit spherical, 8–12 mm in diameter. *NW Mexico (E Sonora)*. G1. Spring.

14. E. knippelianus Liebner. Illustration: Schumann & Gürke, Blühende Kakteen **1**: t. 12 (1901); Cullmann et al., Kakteen, edn 5, 155 (1984).
Simple or clustering, stem 3–8 cm in diameter, depressed-spherical, passing into a fleshy rootstock, dark green; ribs 5–8, low and broad; areoles 5–8 mm apart, circular, to 2 mm in diameter. Spines 1–4 (rarely 0), to 1.5–6 cm. Flowers shortly funnel-shaped, 2.5–4 × 4–6.5 cm, arising from stem sides (or from the stem apex in var. **kruegeri** Glass & Foster), pinkish lavender, purplish or white. Fruit spherical, very small, few-seeded. *NE Mexico (SE Coahuila & S Nuevo Leon)*. G1. Spring.

16. LEPISMIUM Pfeiffer
D.R. Hunt
Epiphytic or growing on rocks, often of pendulous habit; stems cylindric, ribbed, angled, winged or flat, usually segmented, sometimes spiny, the younger segments arising singly from the sides or apices of older segments; rudimentary scale-leaves often clearly visible. Flowers mostly very

small for the family, usually disc-shaped; pericarpel often tuberculate and spiny or angled and with or without spines, rarely almost terete; perianth-tube very short or none, rarely well-developed. Fruit berry-like, sometimes spiny; seeds oblong or semi-ovate in outline, about 1 mm, brown or black.
A genus now thought to contain 14 species, mostly from Bolivia and Argentina, a few extending into Brazil. Literature: Barthlott, W., New names in Rhipsalidinae (Cactaceae), *Bradleya* **5**: 97–100 (1987).

1a. Stem deeply 3–5-winged or -angled 2
 b. Stem flattened, except at the base 3
2a. Areoles in fertile zone sunken, tufted with white hairs **1. cruciforme**
 b. Areoles in fertile zone not sunken, not hairy **2. warmingianum**
3a. Stems deeply saw-toothed, 1–5 cm broad, tapering to a stalk-like base **3. houlletianum**
 b. Stems scolloped or wavy **1. cruciforme**

1. L. cruciforme (Vellozo) Miquel (*Rhipsalis cruciformis* (Vellozo) Castellanos). Illustration: Haustein, Der Kosmos-Kakteenführer, 115 (1983).
Stem-segments variable, 3–5-angled or -winged, or flat, to 50 × 2 cm, usually suffused with purple, margins more or less deeply scolloped; areoles in the scollops, tufted with white hairs. Flowers lateral, 1–5 per areole, 1–1.3 cm, creamy white, outermost segments reddish or brownish; pericarpel somewhat sunken in stem. Fruits almost spherical, *c.* 6 mm in diameter, purplish red. *SE Brazil to N Argentina*. G2.

2. L. warmingianum (Schumann) Barthlott (*Rhipsalis warmingiana* Schumann). Illustration: Britton & Rose, The Cactaceae **4**: pl. 30, f. 2 & pl. 34, f. 3, 4 (1923).
Stems slender, long, 3–4-angled or flat, segments to 30 × 1 cm, margins scolloped, areoles hairless. Flowers lateral, solitary, campanulate, *c.* 2 cm, white, the pericarpel angled. Fruit spherical, 5–6 mm in diameter, dark purple or nearly black. *E Brazil*. G2.

3. L. houlletianum (Lemaire) Barthlott (*Rhipsalis houlletiana* Lemaire). Illustration: Botanical Magazine, 6089 (1874); Hecht, BLV Handbuch der Kakteen, 213 (1982); Haustein, Der Kosmos-Kakteenführer, 109 (1983).
Stems 1–2 m, slender and terete below, 2–4 mm in diameter, flat above and then 10–20 × 1–5 cm, margins deeply saw-

toothed; areoles without bristles or hairs. Flowers numerous, campanulate, 1.5–2 cm, pericarpel 4–5-angled, creamy white. Fruit spherical, 5–6 mm in diameter, red. *E Brazil*. G2.

17. RHIPSALIS Gaertner
D.R. Hunt
Epiphytic or growing on rocks, often of pendulous habit; stems cylindric, ribbed, angled, winged or flat, usually segmented and spineless, the younger segments often arising singly or in clusters (verticils) at the apices of older segments. Flowers mostly very small for the family, disc-shaped, whitish, the perianth-tube very short or none, pericarpel terete, usually naked, rarely with soft, bristle-like spines. Fruit small, berry-like, usually naked; seeds oblong or semi-ovate in outline, about 1 mm, brown or black.
A genus of about 50 species in tropical America, with one species extending to tropical Africa, Madagascar and Asia, the only cactus which occurs spontaneously outside America. Literature: Barthlott, W., New Names in Rhipsalidinae (Cactaceae), *Bradleya* **5**: 97–100 (1987).

1a. Stem cylindric, ribbed or angled, not flat 2
 b. Stem flat or 2-winged, at least in part 11
2a. Stem-segments slender, cylindric or club-shaped, terete or if obscurely angled then less than 4 mm in diameter 3
 b. Stem with 3–9 low ribs, 4–10 mm in diameter; areoles usually with bristle-like hairs 3–8 mm **9. dissimilis**
3a. Plant with numerous, stubby, mostly 1-segmented lateral branches *c.* 1 cm **1. mesembryanthoides**
 b. Lateral branches exceeding 1 cm 4
4a. Flowers terminal or apical 5
 b. Flowers lateral 8
5a. Ovary and fruit hairless; stem-segments with very stout bristles 1–2 mm or hairless 6
 b. Ovary and fruit finely bristly; stem-segments with bristles 4–6 mm **2. pilocarpa**
6a. Stem-segments of 2 types, with longer segments subtending clusters of short segments, obscurely angled and very shortly bristly **3. cereuscula**
 b. Stem-segments all similar, shortly cylindric-club-shaped to club-shaped, usually hairless 7
7a. Stem-segments cylindric-club-shaped; flowers white **4. clavata**

b. Stem-segments club-shaped, usually with a distinct basal 'neck' like an inverted bottle; flowers yellow or orange   see **Hatiora salicornioides** (p.221)

8a. Flowering areoles not depressed or woolly, the flower-buds not rupturing the surface   9

b. Flowering areoles depressed, woolly, the flower-buds rupturing the surface   10

9a. Stems slender, less than 8 mm in diameter; flowers 5 mm in diameter   **5. baccifera**

b. Stems 8–12 mm in diameter; flowers *c.* 2 cm in diameter   **6. grandiflora**

10a. Stem slender, less than 8 mm in diameter; flowers 1.5–2 cm in diameter   **7. floccosa**

b. Stems 8–20 mm in diameter; flowers 3–4 cm in diameter   **8. megalantha**

11a. Stem-segments variable, the younger flat or 3-winged, less than 2 cm broad, the older sometimes 4–5-angled or terete   **10. micrantha**

b. Stem-segments all more or less similar, flat, oblanceolate to obovate, to 12 × 7 cm   **11. crispata**

1. R. mesembryanthoides Haworth. Illustration: Botanical Magazine, 3078 (1831); Haustein, Der Kosmos-Kakteenführer, 113 (1983); Cullmann et al., Kakteen, edn 5, 289 (1984).
Stems of 2 forms, the longer cylindric, 10–20 cm × 1–2 mm, becoming woody, and giving off numerous terete-cylindric, stubby, lateral branches 7–15 × 2–4 mm, remaining green, areoles with a few finer bristles *c.* 2 mm. Flowers lateral on the short shoots (rarely on the long shoots), *c.* 8 × 15 mm, white. Fruit spherical, *c.* 5 mm in diameter, white or tinged red, with persistent perianth. *Brazil (Rio de Janeiro).* G2.

2. R. pilocarpa Loefgren. Illustration: Barthlott, Cacti, f. 99 (1979); Hecht, BLV Handbuch der Kakteen, 213 (1982); Haustein, Der Kosmos-Kakteenführer, 113 (1983).
Stem slender-cylindric, segments to 4 cm × 6 mm, with whorls of branches at the apex; areoles green, often tinged purple, numerous, with 3–10 fine white bristles 4–6 mm. Flowers terminal or near stem-apex, solitary or paired, 2.5–4 cm in diameter, white, pericarpel bristly. Fruit spherical, to 1.2 cm in diameter, wine-red, bristly. *Brazil (Sao Paulo).* G2.

3. R. cereuscula Haworth (*R. saglionis* (Lemaire) Otto). Illustration: Botanical Magazine, 4039 (1843); Haustein, Der Kosmos-Kakteenführer, 111 (1983). Stems of 2 types, the longer cylindric, 10–30 cm × 3–4 mm, almost terete, the shorter arising in apical clusters and branching widely at their apices, 1–3 cm × 3–4 mm, obscurely 4–5-angled, the areoles marginal with 2–4 short bristles 1–2 mm. Flowers terminal, bell-shaped, 8–15 mm × 1–2 cm, white; pericarpel hairless. Fruit obovoid, *c.* 5 mm in diameter, white. *S Brazil to Argentina (Entre Rios).* G2.

R. cribrata (Lemaire) Ruempler. Similar to *R. cereuscula*, but terminal segments terete; fruit pinkish. *E Brazil.* G2.

4. R. clavata Weber. Illustration: Cullmann et al., Kakteen, edn 5, 289 (1984). Stem segments all similar, cylindric-club-shaped, slender, truncate, to 5 cm × 2–3 mm, with areoles at the apex only, and giving rise there to whorls of 2–7 younger segments, green or purplish. Flowers terminal or apical, bell-shaped, *c.* 1.5 cm, white. Fruit globular, white. *Brazil (Rio de Janeiro).* G2.

5. R. baccifera (Miller) Stearn (*R. cassutha* Gaertner). Illustration: Botanical Magazine, 3080 (1831); Barthlott, Cacti, f. 92 (1979). Stems long, cylindric, soon pendent, reaching 1–4 m, slender, usually 4–6 mm in diameter; areoles sometimes bearing 1–2 minute stiff bristles 1 mm. Flowers lateral, 5–10 mm in diameter, whitish. Fruit spherical, 5–8 mm in diameter, translucent white or pink; seeds 1 mm, black. *Tropical America, Africa, Madagascar, Sri Lanka.* G2. Winter-spring.

R. capilliformis Weber. Illustration: Britton & Rose, The Cactaceae 4: pl. 26, f. 4 (1923). Stems very slender, of 2 types, the main branches to 10–15 cm × 2–3 mm, the branchlets arising in clusters, short, only 1–1.5 mm in diameter. Flowers lateral, 6–8 mm in diameter, white. Fruit spherical, 4–5 mm in diameter, white; seeds brown. *Brazil.* G2. Autumn.

6. R. grandiflora Haworth. Illustrations: Botanical Magazine, 2740 (1827), imperfect; Britton & Rose, The Cactaceae 4: pl. 31 f. 1, 3 (1923). Stem-segments long, cylindric, to 25 cm × 8–12 mm, often tinged purplish near the areoles; flowering areoles not sunken, without bristles. Flowers lateral, *c.* 1.2 × 2 cm, creamy white, profuse. Fruit spherical, 6–7 mm in diameter, white or tinged purple. *Brazil (Rio de Janeiro).* G2. Spring.

7. R. floccosa Pfeiffer. Stem-segments slender, cylindric, to 25 cm × 5–6 mm, often tinged purplish near the areoles; flowering areoles sunken, woolly-felted, without bristles. Flowers lateral, 1.5–2 cm in diameter, creamy white, the buds rupturing the surface and surrounded by woolly hairs. Fruit spherical, *c.* 5 mm in diameter, white or pink-tinged. *Brazil, Bolivia.* G2. Winter–spring.

The following are doubtfully distinct:

R. gibberula Weber. Branches whorled and the areoles not floccose. Fruit 7–8 × 8–10 mm, pure white. *E Brazil.* G2. Winter–spring.

R. pulvinigera Lindberg. Illustration: Botanical Magazine n.s., 553 (1969). Branches whorled; areoles not floccose, but subtending leaf-rudiments conspicuous, reddish. Flowers 2 cm in diameter. Fruit red. *SE Brazil, N Paraguay.* G2. Winter-spring.

R. puniceodiscus Lindberg. Areoles somewhat woolly. Flowers whitish with pinkish or orange stamens. Fruit red or orange-yellow. *Brazil.* G2. Winter–spring.

8. R. megalantha Loefgren. Illustrations: Barthlott, Cacti, f. 95 (1979); Haustein, Der Kosmos-Kakteenführer, 115 (1983). Stems relatively stout, to 35 cm × 8–10 mm; areoles prominent, usually purple-tinged, without bristles. Flowers 3–4 cm in diameter, white with pinkish or orange stamens. Fruit spherical, 1.2 cm in diameter, intense purple-red, surrounded by woolly hairs. *SE Brazil (Ilha Sao Sebastiao).* G2. Autumn–winter.

9. R. dissimilis (Lindberg) Schumann (*Lepismium marnieranum* Backeberg). Illustration: Botanical Magazine, 8031 (1905).
Stem-segments variable and diverse, usually with 5–9 low ribs, sometimes 3–5-angled or nearly terete, 5–14 cm × 4–10 mm; areoles, at least those of the ribbed segments with bristle-like hairs 3–8 mm; flowering areoles sunken, woolly. Flower-buds reddish; flowers 1–1.5 cm in diameter, pale yellow. Fruit almost spherical, 8–10 mm in diameter, pink or purplish. *E Brazil.* G2.

R. cereoides (Backeberg & Voll) Castellanos. Illustration: Haustein, Der Kosmos-Kakteenführer, 109 (1983). Stem-segments 3–4-angled, 4–10 × up to 1–7 cm, not scolloped; areoles not sunken, sometimes with 2–4 very short bristles. Flowers solitary or 2–4 per areole, disc-shaped, 2 cm in diameter, white. Fruit pale pink. *E Brazil (Rio de Janeiro).* G2

10. R. micrantha (Kunth) de Candolle.
Illustration: Backeberg, Die Cactaceae **2**:
f. 627 (1959).
Stem-segments flat or 3-angled, the older
stems becoming 4–5-angled, scollops very
shallow, areoles often with 1–4 small
bristles. Flower *c.* 7 mm in diameter. Fruit
8–10 mm, white to reddish. *Ecuador,
N Peru.* G2.

11. R. crispata (Haworth) Pfeiffer.
Illustration: Britton & Rose, The Cactaceae
4: pl. 35, f. 3, f. 232 (1923).
Branches divided into flat or sometimes
3-winged, oblanceolate, elliptic or obovate
segments 6–10 × 2–4 cm, relatively thin,
green, margins scolloped or winged; areoles
without bristle-hairs. Flowers solitary or
2–4 per areole, disc-shaped, 1–1.2 cm in
diameter, creamy white. Fruit spherical,
7 mm in diameter, white. *E Brazil.* G2.

    R. pachyptera Pfeiffer. More robust, the
segments broadly elliptic to circular, to
14 × 7 cm, thick, often purplish, margins
deeply scolloped. *SE Brazil.* G2.

    R. crispata and *R. pachyptera* are
connected by numerous intermediate
species whose classification is yet to be
sorted out.

18. HATIORA Britton & Rose
D.R. Hunt
Plants epiphytic or growing on rocks, of
pendulous or shrubby habit; stems
cylindric, angled, winged or flat, segmented
and spineless or with soft, bristle-like spines
only, the segments usually less than 5 cm;
younger segments arising singly or in
clusters from a composite areole at the apex
of an older segment. Flowers regular, bell-
shaped, intense yellow, pink or red;
pericarpel angled or terete, naked; perianth-
tube short. Fruit small, obovoid, naked;
seeds oblong or semi-ovate in outline,
about 1 mm, brown or black.
    A genus of 5 species (including
Rhipsalidopsis Britton & Rose), from Brazil.
Two species and their hybrid enjoy
popularity as house-plants.
Literature: Barthlott, W., New names in
Rhipsalidinae (Cactaceae), *Bradleya* **5**:
97–100 (1987).

1a.   Stem-segments club-shaped, usually
     with a distinct basal 'neck' like an
     inverted bottle; flowers yellow or
     orange   **1. salicornioides**
  b.   Stem-segments flat or 3–5-angled 2
2a.   Stem-segments to 1 cm broad, not
     becoming woody; flowers pink,
     3–4 cm in diameter   **2. rosea**

  b.   Stem-segments 2–2.5 cm broad,
     eventually becoming woody and more
     or less terete, 4–7 × 2–2.5 cm; flowers
     scarlet, 4–7 cm   **3. gaertneri**

1. H. salicornioides (Haworth) Britton &
Rose (*Rhipsalis salicornioides* Haworth).
Illustration: Botanical Magazine, 2461
(1824); Hecht, BLV Handbuch der Kakteen,
209 (1982); Haustein, Der Kosmos-
Kakteenführer, 115 (1983).
Much-branched, the main stems erect or
pendent, eventually woody, cylindric, to
1 cm in diameter, the smaller branches
consisting of numerous club-shaped
segments 1.5–5 cm × 3–5 mm, these
arising in whorls of 2–6, each usually with
a distinctive basal 'neck'. Flowers terminal
or apical, to 1.3 × 1 cm, yellow to orange.
Fruit spherical, *c.* 5 mm in diameter, white.
SE Brazil. G2.

2. H. rosea (Lagerheim) Barthlott (*Rhipsalis
rosea* Lagerheim; *Rhipsalidopsis rosea*
(Lagerheim) Britton & Rose). Illustration:
Hecht, BLV Handbuch der Kakteen, 212
(1982); Haustein, Der Kosmos-
Kakteenführer, 117 (1983).
Stem-segments flat or 3–5-angled, oblong
or narrowly oblanceolate, 2–4 × *c.* 1 cm,
the areoles marginal and apical, usually
with fine, brownish bristles; new segments
and flowers arising from apical areoles.
Flowers regular, broadly funnel-shaped,
3–4 × 3–4 cm, pink, with angled pericarpel
and very short tube; stamens borne at base
of tube. Fruit depressed-spherical,
yellowish, faintly angled; seeds brown.
SE Brazil (Parana). G2. Spring.

    H. × graeseri (Werdermann) Barthlott
(*H. rosea × H. gaertneri*). Flowers 4–6 cm in
diameter, intermediate in size between
those of its parents, and with a range of
flower colours. A popular house-plant.

3. H. gaertneri (Regel) Barthlott (*Rhipsalis
gaertneri* (Regel) Vaupel; *Schlumbergera
gaertneri* (Regel) Britton & Rose;
Rhipsalidopsis gaertneri (Regel) Moran).
Illustration: Botanical Magazine, 7201
(1891); Haustein, Der Kosmos-
Kakteenführer, 117 (1983).
Stem-segments mostly flat, oblong or
elliptic, truncate, 4–7 × 2–2.5 cm, margins
shallowly scolloped, areoles small, usually
with 1–few fine, brownish bristles, these
deciduous; new segments and flowers
arising from the apical areoles. Flowers
regular, funnel-shaped, 4–5 × 4–7.5 cm,
intense scarlet with angled pericarpel and
short tube; stamens borne at base of the
tube. Fruit oblong, 1.5 cm, dull red. *E
Brazil (Minas Gerais).* G2. Spring.

19. SCHLUMBERGERA Lemaire
D.R. Hunt
Epiphytic or growing on rocks, stems
segmented, the segments flattened,
compressed or terete, oblong or obovate,
usually truncate, new segments and
flowers arising from the apices; spines
bristle-like, short, or none. Flowers nearly
regular to strongly bilaterally symmetric
with a distinct tube bearing petal-like
scales, these and the perianth red, purplish,
pink or white; stamens borne on the tube,
the lowermost united at the base to form a
short tube around the style, stigmas erect,
close together. Fruit berry-like, spherical to
obconic, ribbed or terete, the perianth
deciduous; seeds more or less kidney-
shaped or semi-ovoid, dark brown or black.
    A genus of 5 or 6 species from SE Brazil.
The species, and hybrids derived from
them, are popular as houseplants, and
there are more than 200 cultivars.
Literature: Hunt, D.R., A synopsis of
Schlumbergera Lem. (Cactaceae), *Kew
Bulletin* **23**: 255–263 (1969); Barthlott, W.
& Rauh, W., Notes on the morphology,
palynology, and evolution of the genus
Schlumbergera Lemaire (Cactaceae), *Cactus
& Succulent Journal (US)* **50**: 31–34 (1978);
McMillan, A.J.S., *Christmas Cacti/
Weihnachtskakteen* (1985).

1a.   Younger stem-segments 2-winged;
     areoles confined to margins and apex
                         2
  b.   Younger stem-segments cylindric or
     compressed and *Opuntia*-like; areoles
     distributed generally over the surface
                 3. opuntioides
2a.   Stem-segments acutely toothed;
     flowers strongly bilaterally symmetric;
     pericarpel terete or angled
                   1. truncata
  b.   Stem-segments obliquely scolloped;
     flowers with oblique limb or regular;
     pericarpel ribbed or winged
                 2. × buckleyi

1. S. truncata (Haworth) Moran
(*Epiphyllum truncatum* Haworth; *Zygocactus
truncatus* (Haworth) Schumann).
Illustration: Barthlott, Cacti, f. 101, 102
(1979).
Epiphyte; branches of several flat, 2-winged
segments, to 30 cm, becoming pendulous,
individual segments oblong, truncate,
2.5–8 × 1–4 cm, margins with 2–4 sharp,
forward-pointing teeth each side, midrib
prominent. Flowers bilaterally symmetric,
7 × 4.5 cm, the axis abruptly angled
between the pericarpel and tube, tube
c. 4 cm, limb strongly oblique the segments

c. 3 × 1.5 cm, recurved, variable in colour; stamens and style long-projecting; pericarpel terete. *SE Brazil (Rio de Janeiro).* G2. Autumn.

Many illustrations, supposedly of this species, depict the hybrid *S.* × *buckleyi*, which has scolloped, not toothed stem-segments.

S. orssichiana Barthlott & McMillan. Illustration: Haustein, Der Kosmos-Kakteenführer, 117 (1983). Stem-segments *c.* 5 × 3 cm, margins curled as well as toothed. Flowers 9 × 9 cm, tube 1 cm, segments 4–5 × 2 cm, purplish pink towards tip, white below; pericarpel 5–6-angled. *SE Brazil.* G2. Late summer; winter.

2. S. × buckleyi (Moore) Tjaden (*S. bridgesii* (Lemaire) Loefgren, misapplied; *S. trunctata* misapplied). *S. truncata* × *S. russelliana.* Figure 28(6), p. 204 & figure 29(12), p. 205. Illustration: Botanical Magazine, n.s., 823 (1981); Hecht, BLV Handbuch der Kakteen, 215 (1982) – as *S. truncata.*
Stems to 50 cm or more, eventually somewhat woody at base; segments oblong or obovate, truncate, 2–4.5 × 1.3–2.5 cm, obliquely scolloped. Flowers slightly bilaterally symmetric, 5.5–6.5 × 3–4 cm, tube 3–4 cm, limb somewhat oblique, segments spreading to recurving, stamens and style long-projecting, stigmas purplish; pericarpel ribbed. *Garden origin.* G2. Winter.

This is the popular 'Christmas Cactus'.

S. russelliana (Hooker) Britton & Rose (*Epiphyllum russellianum* Hooker). Illustration: Botanical Magazine, 3717 (1839); Hecht, BLV Handbuch der Kakteen, 215 (1982). Stem-segments 2–3.5 × 5–15 mm, margins scolloped with 1–2 notches each side, the areoles often bearing 1–2 weak bristles. Flowers nearly regular, to 5 × 3 cm, tube 3 cm, limb not oblique, deep pink, stamens and style projecting, stigmas creamy white; pericarpel strongly ribbed or winged. *SE Brazil (Rio de Janeiro).* G2. Winter.

3. S. opuntioides (Loefgren & Dusen) D. Hunt. Illustration: Haustein, Der Kosmos-Kakteenführer, 117 (1983).
Opuntia-like shrub, epiphytic or growing on rocks, with erect or arching branches to 40 cm (reaching 1.2 m in nature from woody base 10 cm in diameter); stem-segments obovate to oblong, more or less flattened, 5–7 × 1.5–4, the scattered areoles tufted with few to numerous bristle-like spines to 5 mm, yellowish. Flowers strongly bilaterally symmetric, resembling

those of *S. truncata, c.* 6 × 4.5 cm, deep pink, pericarpel obscurely 5–7-angled, stigmas whitish. *SE Brazil (Rio de Janeiro).* G2. Spring.

S. obtusangula (Schumann) D. Hunt. Similar to *S. opuntioides* but stem-segments terete or obtusely angled; flowers nearly regular, 5 × 2.5 cm. *SE Brazil (Rio de Janeiro).* G2. Spring.

20. CEREUS Miller

D.R. Hunt

Shrubby or tree-like, usually much branched with erect or ascending, strongly ribbed, often glaucous stems. Flowers mostly large, funnel-shaped, nocturnal, usually white; pericarpel and tube long, thick, naked or nearly so below and with scattered small scales above; perianth-limb broad or moderately so. Fruit fleshy, ovoid or oblong, usually red, naked, splitting along one side when mature to reveal white pulp; withered perianth and stamens deciduous but the style usually persistent; seeds usually ovate-kidney-shaped, black.

Considered here in the restricted sense adopted by Britton & Rose (1920) and most later authors, a genus of about 25 species native to the West Indies and eastern South America. The larger-growing species are not yet well understood and many of the names commonly used in cultivation are disputable.

1a. Shrubby species with relatively slender, often bloomed stems; ribs 6–8, not deeper than broad   2
  b. Tree-like species with stout stems, sometimes glaucous but not bloomed; ribs 4–6 in young plants, sometimes more when older, deeper than broad
        1. jamacaru
2a. Stems 5–10 cm in diameter, not appreciably tapered; ribs 6
        2. chalybaeus
  b. Stems 3–6 cm in diameter, tapered to the apex; ribs 6–8   **3. azureus**

1. C. jamacaru de Candolle (*C. lividus* Pfeiffer). Illustration: Botanical Magazine, 5775 (1869); Britton & Rose, The Cactaceae **2**: f. 9 (1920).
Tree to 10 m with short woody trunk to 60 cm in diameter; stems to 15 cm diameter; ribs 4–5 or rarely 6, to 3.5 cm deep, glaucous when young; areoles 2–4 cm apart. Spines of seedlings and young growth 9–11, yellow, to 2 cm, often much longer and darker on older growth. Flowers very large, 25–30 × 18–20 cm; outer perianth-segments pale green. *NE Brazil.* G1.

Very commonly raised from seed and used as a grafting stock, but rarely reaching flowering size.

C. hildmannianus Schumann. Illustration: Martius, Flora Brasiliensis **4**(2): t.41, f. 1 (1890). Nearly spineless, otherwise like *C. jamacaru* and perhaps only a cultivar of it.

C. pachyrhizus Schumann. Thick-rooted, with stems to 1–3 m × 10 cm; ribs 6, strongly compressed; areoles 2.5–3 cm apart. Spines 10–13, subulate, weakly differentiated into central and radial, the longest 3 cm. Fruit 5 × 3–4 cm. *Paraguay.* G1.

C. uruguayanus Kiesling (*C. peruvianus* misapplied). Like *C. jamacaru,* but ribs usually 6–9, spines darker, the flower with a shorter, thicker tube and brownish or reddish outer perianth-segments. *SE Brazil to N Argentina.* G2.

Several monstrous cultivars traditionally placed in *C. jamacaru* and '*C. peruvianus*' are popular.

C. forbesii Salm-Dyck and *C. validus* Haworth are among numerous names with a confused history which apparently apply to Argentinian species close to *C. uruguayanus.*

2. C. chalybaeus Foerster. Illustration: Gürke & Vaupel, Blühende Kakteen **3**: t.135 (1912); Backeberg, Die Cactaceae **4**: f. 2229 (1960).
Shrub to 3 m with few ascending branches, 6–10 cm in diameter, young growth glaucous or purplish and bloomed; ribs 5–6, shallowly scolloped with the areoles at the tops of the scollops. Spines *c.* 10–13, 1–2 cm, all dark brown or blackish. Flowers *c.* 20 cm; tube reddish to dark purple, with small scales; inner perianth-segments white. *N Argentina, Uruguay.* G2.

3. C. azureus Pfeiffer. Illustration: Schumann, Gesamtbeschreibung der Kakteen, 119, f. 26 (1897).
Shrubby, branching near base, to 2–3 m; stems slender, 3–3.5 cm in diameter, tapered to apex, bluish-bloomed, ribs 6–7, deeply scolloped, the areoles prominent. Central spines 1–3, to 1.5 cm, dark brown or blackish; radial spines 8–12, 5–10 mm, whitish. Flowers variously described as 10–12 or 20–30 cm; tube pale green; outer perianth-segments brownish, inner white. Fruit ellipsoid, 4.5 × 3 cm, naked, pink or purplish red. *N Argentina.* G2.

C. aethiops Haworth (*C. coerulescens* Salm-Dyck). Illustration: Botanical Magazine, 3922 (1842); Britton & Rose, The Cactaceae **2**: f. 16, 17 (1920). Ribs

7–8, tuberculate. Radial spines *c.* 9, black at least at the tip and base. Flowers 22 × 12 cm; outer perianth-segments pink, inner white. Fruit ovoid, 6 cm, red. *N Argentina.* G2.

21. MONVILLEA Britton & Rose
D.R. Hunt

Shrubby, thicket-forming or clambering, with rather slender, erect, arching or decumbent, ribbed stems. Flowers mostly moderately large, funnel-shaped, nocturnal; pericarpel with small scales naked in their axils; tube usually long, with more distant scales; perianth-segments long or short, the inner white or rarely yellow. Fruit spherical to ellipsoid or oblong, fleshy, red, often grooved, naked or with sparse scales, the withered perianth or at least the style persistent; seeds black, obliquely obovate in outline.

A genus of 10–15 species, some of them weakly defined, in South America.

1a. Ribs 3–5   **1. spegazzinii**
 b. Ribs 6–10   2
2a. Flowers 10–13 × 10 cm with slender tube and broad limb   **2. cavendishii**
 b. Flowers 6–10 × 4–6 cm, with stout tube and short limb   **3. smithiana**

1. M. spegazzinii (Weber) Britton & Rose (*Cereus spegazzinii* Weber). Illustration: Britton & Rose, The Cactaceae 2: 23, f. 22 (1920); Haustein, Der Kosmos-Kakteenführer, 121 (1983); Cullmann et al., Kakteen, edn 5, 238 (1984).
Stems ascending and arching or nearly prostrate, to 2 m × 1.5 cm, later woody at base, 2 cm in diameter, young growth bluish green, marbled whitish; ribs 3–5, tuberculate-toothed, with areoles at the tips of the tubercles, 2–4 cm apart. Spines 2 or 3 on young growth, short, blackish, later to 6, the longest 1.5 cm. Flowers 10–13 × 7–9 cm, outer perianth-segments purplish, inner white, acute. Fruit ellipsoid, pink, bloomed. *NE Argentina, Paraguay.* G2. Early summer.

2. M. cavendishii (Monville) Britton & Rose (*M. rhodoleucantha* (Schumann) Berger). Illustration: Haustein, Der Kosmos-Kakteenführer, 121 (1983).
More or less erect or clambering, branching at base, stems 1–3 m × 2–3 cm, dark green; ribs 7–9, areoles 5–15 mm apart. Central spines 1–4, to 2 cm, dark brown or black; radial spines 6–10, 5–10 mm, bristle-like, yellowish brown. Flowers 10–13 × 6–10 cm, outer perianth-segments pinkish or greenish, inner white,

acute. Fruit spherical to ellipsoid, 5–7 × 4–5 cm, red. *S Brazil, Paraguay & N Argentina.* G2. Late spring–early autumn.

3. M. smithiana (Britton & Rose) Backeberg (*Cephalocereus smithianus* Britton & Rose; *Cereus smithianus* (Britton & Rose) Werdermann). Illustration: Britton & Rose, The Cactaceae 2: f. 43, 46, 47 (1920).
Erect or clambering, branched or unbranched, stems 4–8 cm in diameter, green; ribs 8–11; areoles *c.* 2 cm apart. Spines *c.* 13, the centrals to 3–4 cm, black at first, the radials *c.* 1 cm, white. Flowers 6–8 × 4 cm, inner perianth-segments short, rounded and mucronate, white. Fruit ovoid, 4–5 × 3–4 cm, red, minutely scaly, splitting on one side when mature to reveal white pulp. *Venezuela.* G2.

M. campinensis (Backeberg & Voll) Backeberg (*Praecereus campinensis* (Backeberg & Voll) Buxbaum). Illustration: Backeberg, Die Cactaceae 4: f. 2216, 2217 (1960). Large shrub to 5 m with erect stems; ribs 7–9. Flowers bell- to funnel-shaped, to 10 × 6 cm, the tube 2 cm in diameter with prominent decurrent scales, inner perianth-segments greenish white. *SE Brazil.* G2.

22. STETSONIA Britton & Rose
D.R. Hunt

Tree with short thick trunk and numerous erect or ascending, cylindric, ribbed, branches. Flowers large, funnel-shaped, nocturnal; pericarpel with numerous broad, overlapping, mucronate scales; tube long, with scattered scales; perianth limb broad, spreading to disc-shaped; stamens numerous, borne in the tube. Fruit oblong-ovoid, red, with numerous small scales.

A genus of one species, in NW Argentina and S Bolivia.

1. S. coryne (Salm-Dyck) Britton & Rose. Illustration: Britton & Rose, The Cactaceae 2: 65, f. 96 & t. 9 (1920); Hecht, BLV Handbuch der Kakteen, 362 (1982).
Tree 5–10 m with short trunk and massive crown of ascending, bluish green branches *c.* 15 cm in diameter; ribs 8–9, spines 7–9, unequal, the longest 5 cm, needle-like. Flowers 12–15 × 8–10 cm; inner perianth-segments white. Fruit 5–6 × 4–5 cm, edible. *NW Argentina, S Bolivia.* G2.

23. PILOSOCEREUS Byles & Rowley
D.R. Hunt

Shrubby or becoming tree-like, usually more or less branched from the base or trunk, to 10 m; ribs 4–12 (rarely to 19), often cross-furrowed. Areoles, at least the

flowering, with more or less abundant woolly hairs, sometimes as long as 5 cm and forming skeins covering the ribs. Flowers medium-sized, nocturnal, tubular-bell-shaped; pericarpel and tube fleshy, without scales, often brownish or purplish; limb rather narrow, whitish or pinkish; stamens very numerous. Fruit ovoid to depressed-spherical, smooth, fig-like, with red pulp, the perianth persistent; seeds black or dark brown.

A genus of more than 60 species in Mexico, the West Indies and E tropical S America, formerly known under the misapplied name *Pilocereus* (i.e. *Pilocereus* in the sense of Schumann, not of Lemaire) or included in *Cephalocereus* Pfeiffer.

1a. Ribs 4–6, rarely more; flowering areoles only sparsely hairy
      1. pentaedrophorus
 b. Ribs 7 or more; flowering areoles moderately to densely hairy or woolly
      2
2a. Ribs 7–10; tree-like, to 6 m   3
 b. Ribs 13–17; shrubby, to 1.5 m
      4. werdermannianus
3a. Young spines light golden yellow; stem light blue-green; non-flowering areoles with a few hairs only; flowers whitish   **2. glaucescens**
 b. Young spines dull yellow-brown or blackish; stem dark green or bluish green; non-flowering as well as flowering areoles bearded with woolly hairs; flowers pinkish   **3. palmeri**

1. P. pentaedrophorus (Labouret) Byles & Rowley (*Cephalocereus pentaedrophorus* (Labouret) Britton & Rose). Illustration: Britton & Rose, The Cactaceae 2: 31, f. 31–33 (1920).
Stem simple, or few-branched at base, erect or decumbent, slender, 2–5 m or more tall, only *c.* 3–7 cm in diameter, vivid bluish; ribs usually 4–6, notched above the areoles; areoles with short hairs and up to *c.* 12 unequal, yellow spines, the longest 4 cm. Flowers 4–6 cm, curved; pericarpel and tube hairless, green; inner perianth-segments white. Fruit depressed-spherical, 3 × 5 cm, bluish green with purplish red pulp. *E Brazil (Bahia).* G2.

2. P. glaucescens (Labouret) Byles & Rowley (*Cephalocereus glaucescens* (Labouret) Borg). Illustration: Backeberg, Die Cactaceae 4: 2417, f. 2295 (1960).
Tree to 6 m, more or less branched; stems to 10 cm in diameter, vivid pale blue, ribs 8–10; areoles *c.* 1.5 cm apart, the non-flowering with spines and a few whitish

hairs, the flowering with a dense tuft of white hairs also. Spines c. 18–25, 5–15 mm, golden yellow. Flowers 6–7 cm; pericarpel and tube bluish or brownish green; inner perianth-segments white. Fruit depressed-spherical, c. 6 cm, bluish; seeds 1.5 mm. *E Brazil*. G2.

P. royenii (Linnaeus) Byles & Rowley (*Cephalocereus royenii* (Linnaeus) Britton & Rose). Tree-like, to 8 m or more, branched, stem stout, dark green to bluish, 7–11-ribbed; spines variable, often only 1 cm, sometimes much longer. Flowers 5 cm; tube greenish yellow to purplish; inner perianth-segments white. Fruit 5 cm in diameter, reddish green. *West Indies*. G2.

3. P. palmeri (Rose) Byles & Rowley (*Cephalocereus palmeri* Rose). Illustration: Hecht, BLV Handbuch der Kakteen, 351 (1982); Cullmann et al., Kakteen, edn 5, 274 (1984).
Becoming a tree to 6 m, more or less branched; stems 5–10 cm in diameter, dark green or glaucous and bluish green at first; ribs 7–9; areoles c. 1–1.5 cm apart, the upper densely clothed with white hairs 2–6 cm. Spines brownish or yellow at first, 7–12 radial, c. 1 cm, 1–2 central, 2–3 cm, stronger. Flowers 6–8 cm; outer segments purplish brown; inner segments pinkish. Fruit spherical, 6 cm in diameter, purplish; seeds 2 mm. *E Mexico*. G2.

4. P. werdermannianus (Buining & Brederoo) Ritter (*Pseudopilocereus werdermannianus* Buining & Brederoo). Illustration: Kakteen und andere Sukkulenten **26**(4): 75, 76 (1975); Ritter, Kakteen in Südamerika **1**: f. 46–49 (1979). Shrubby, branching from the base, the stems to c. 1.5 m × 4–6 cm, green; ribs 13–17 (sometimes more in flowering-zone), 4 mm high; areoles 3–6 mm apart, copiously white-hairy and with bristles 1–3 cm. Spines finely needle-like, pale golden yellow, darker at base, the centrals 4–15, 5–30 mm, the radials similar, 8–16, 4–15 mm. Flowers nocturnal, 3.5–6 cm, the pericarpel and tube green or red-green, inner perianth-segments white. Fruit 1.2–4 × 2.5–5 cm, red, with red pulp; seeds 1.8 mm, glossy black. *E Brazil (Minas Gerais)*. G2.

Var. **densilanatus** Ritter (*P. densilanatus* invalid). Illustration: Kakteen und andere Sukkulenten **22**(4): 61 (1971). Areoles extremely hairy. Central spines 4–8. Flowers 3.5–5 cm. Fruit 1.2–1.6 × 2.5–5 cm, pulp white or red. *E Brazil (Minas Gerais)*. G2.

24. AUSTROCEPHALOCEREUS Backeberg
D.R. Hunt
Shrubby, few- to many-branched, to 5 m; ribs 20–30 or more, low, with close-set areoles and numerous bristle-like spines. Flowering areoles densely woolly, forming a cephalium or fleece on one side of the stem-apex. Flowers rather small, nocturnal, bell-shaped; pericarpel naked; tube with tiny scales less than 1 mm, naked in the axils or with a few axillary hairs; perianth white. Fruit rather small, broadly ovoid to depressed-spherical, virtually naked, pinkish, with colourless pulp; seeds black or brownish black.

A genus of about 5 species, endemic to E Brazil.

1. A. dybowskii (Roland-Gosselin) Backeberg (*Cephalocereus dybowskii* (Roland-Gosselin) Britton & Rose). Illustration: Cactus & Succulent Journal of Great Britain **29**: 50 (1967); Hecht, BLV Handbuch der Kakteen, 227 (1982).
Shrub 2–4 m, mainly branching near the base; stems c. 8 cm in diameter; ribs 22–28; areoles close-set, with white hairs covering the whole stem. Central spines 1–3, pale yellow or brownish, 1–3 cm, radial spines c. 10, 5–8 mm, concealed by the areolar hair. Cephalium white. Flowers c. 4 cm, perianth white. Fruit broadly ovoid, 2.5 cm, pinkish; seeds black. *E Brazil (Bahia)*. G2.

25. ARROJADOA Britton & Rose
D.R. Hunt
Low shrubs, with cylindric stems, sometimes segmented; areoles close-set; spines small or bristly. Flowering areoles different from the non-flowering, tufted with bristles and hairs, and forming either a lateral cephalium or intermittent, ring-like, almost terminal cephalia through which vegetative growth can continue after flowering, leaving a collar of bristles. Flowers solitary at the areoles, small, often numerous, tubular, red, pink, violet or yellow, diurnal or nocturnal, the pericarpel and tube naked, the limb very short, erect or spreading; stamens and style not projecting. Fruit small, berry-like, juicy; seeds black or dark brown.

A genus of about 10 species, endemic to E Brazil.

1a. Ribs 10–12; non-flowering areoles not hairy; cephalia intermittent, almost terminal at first, later collar-like   2

b. Ribs 14–20; non-flowering areoles with silky hairs to 1 cm; cephalium lateral   **3. polyantha**
2a. Flowering stems 2–5 cm in diameter, not broader at apex   **1. rhodantha**
b. Flowering stems 1–1.5 cm in diameter, broader at apex   **2. penicillata**

1. A. rhodantha (Guerke) Britton & Rose. Illustration: Haustein, Der Kosmos-Kakteenführer, 151 (1983); Cullmann et al., Kakteen, edn 5, 130 (1984).
Stems erect at first, later sprawling or clambering, 1–2 m × × 2–4 cm; ribs 10–13, low; areoles usually less than 1 cm apart. Spines numerous, brown or yellowish at first, later white, c. 20 radial and 5–6 central, the longest to 3 cm. Flowers 3–4 × 2–3 cm, deep pink, the limb erect. Fruit oblong to obovate, c. 2 cm. *E Brazil*. G2.

2. A. penicillata (Guerke) Britton & Rose. Illustration: Hecht, BLV Handbuch der Kakteen, 223 (1982); Cullmann et al., Kakteen, edn 5, 130 (1984).
Stems ascending or sprawling 1–2 m × 1–1.5 cm, the apical flowering-zone broader; ribs 10–12, low, rounded; areoles close-set. Central spines 1–2, rigid, 1–3 cm, radial spines 8–12, short, adpressed. Flowers c. 3 cm, pale to deep pink, the limb spreading. Fruit ellipsoid, 1.5 cm; seeds dark brown. *E Brazil (Bahia)*. G2.

3. A. polyantha (Werdermann) D. Hunt (*Cephalocereus polyanthus* Werdermann; *Micranthocereus polyanthus* (Werdermann) Backeberg). Illustration: Werdermann, Brasilien und seine Saulenkakteen, 57 (1933); Haustein, Der Kosmos-Kakteenführer, 151 (1983).
Shrub to 1.2 m, branching from base; stems 3.5–5 cm in diameter; ribs c. 15–20; non-flowering areoles c. 1 cm apart, with dense woolly hairs 1–2 cm covering the stem. Central spines c. 3–7, with 1–3 stronger, to 3 cm, golden yellow, light brown or occasionally reddish; radial spines c. 20–30, 5–12 mm, whitish to golden yellow. Cephalium lateral, white, the wool intermixed with long bristle-like spines. Flowers numerous, only 1.6–1.8 cm, tubular; tube pinkish red; perianth-segments erect, 2–3 mm, the outer pink, the inner creamy white. Fruit 5–7 mm, pinkish red; seeds blackish. *E Brazil (Bahia)*. G2.

Micranthocereus densiflorus Buining & Brederoo and *M. violaciflorus* Buining are among recently introduced, related species.

26. COLEOCEPHALOCEREUS Backeberg
D.R. Hunt

Shrubby, branching from the base or unbranched, stems slender cylindric to squatly columnar with few to many ribs; areoles close together; spines weak or strong; flowering areoles different from the non-flowering, confluent and bearded with wool and bristles, forming an apical and lateral cephalium on several reduced and depressed ribs. Flowers tubular to funnel-shaped or bell-shaped, small or rather small, diurnal or nocturnal; pericarpel and tube naked. Fruit berry-like, obovoid-club-shaped or spherical, red, indehiscent; seeds black.

A genus (including *Buiningia* Buxbaum) of up to 10 species in E Brazil. All require warm greenhouse conditions for successful cultivation in Europe.

1a. Stem spherical to shortly cylindric; cephalium almost apical; flowers *c.* 3 cm, creamy yellow **1. brevicylindricus**
 b. Stem cylindric; cephalium lateral; flowers *c.* 4–7 cm, pinkish white. **2. fluminensis**

1. C. brevicylindricus (Buining) Ritter
(*Buiningia brevicylindrica* Buining). Illustration: Backeberg, Cactus Lexicon, f. 469 (1977); Haustein, Der Kosmos-Kakteenführer, 129 (1983); Cullmann et al., Kakteen, edn 5, 138 (1984). Stem spherical to shortly cylindric, like a *Melocactus*, to 30 (rarely to 60) × 17 cm, sometimes offsetting at base, pale green; ribs to 18, broad; non-flowering areoles 1–1.5 cm apart. Spines strong, yellowish, the centrals *c.* 4, unequal, the lower longest, to 6 cm, the upper to 2.5 cm; the radial spines *c.* 7, to 3 cm. Cephalium, almost apical at first, eventually extending down one side, of dense white wool interspersed with yellow bristles. Flowers tubular, 3.2 × 1.5 cm, light yellowish green, the limb nearly erect. Fruit spherical, 1.7 × 1.7 cm. *Brazil (Minas Gerais).* G2.

2. C. fluminensis (Miquel) Backeberg
(*Cactus melocactus* misapplied; *Cephalocereus melocactus* Schumann; *Cephalocereus fluminensis* (Miquel) Britton & Rose). Illustration: Vellozo, Flora Fluminensis 5: t. 20 (1827); Martius, Flora Brasiliensis 4(2): t. 43 (1890); Britton & Rose, The Cactaceae 2: 29, f. 26, 29 (1920); Backeberg, Cactus Lexicon, f. 64 (1977). Shrubby, the stems ascending and decumbent, to 2 m × 10 cm, dark green; ribs 10–17, acute. Spines flexible,

yellowish; variable in number (3–14), the longest to 3 cm. Cephalium decurrent from the apex, unilateral, with dense white wool and yellow bristles. Flowers funnel-shaped-campanulate, 5.5–7 × 4–5 cm; perianth-segments widely spreading, outer pinkish, inner white; stamens and style projecting. Fruit obovoid, 2–3 cm, red to purple. *SE Brazil (Rio de Janeiro).* G2.

C. goebelianus (Vaupel) Buining (*C. pachystele* Ritter; *Cephalocereus purpureus* misapplied). Illustration: Britton & Rose, The Cactaceae 2: f. 25, 27, 28 (1920). Columnar, unbranched, to 3 m; ribs 12–15. Central spines 8–10, unequal, the longer to 5 cm, brown, radial spines 15–20, 1 cm or less, white. *E Brazil (Bahia).* G2.

27. MELOCACTUS Link & Otto
N.P. Taylor

Low-growing, rarely more than 1 m high, simple, or branching if damaged, with depressed-spherical to cylindric, strongly ribbed, spiny stems; apex at maturity forming a woolly, usually bristly cephalium, bearing the flowers and fruit. Flowers small, tubular, red to pink, more or less immersed in the cephalium; pericarpel and tube naked. Fruit a juicy berry, usually club-shaped; seeds small, 1–2 mm, ovoid, seed-coat black, smooth or tuberculate.

A genus of about 20 poorly defined species with a circum-Amazonian distribution, especially E Brazil, Peru, Venezuela, C America and the Caribbean. With the exception of *M. matanzanus* (see under No. **4**), the Brazilian species (Nos. **1–5**) are those most frequently seen in cultivation, the others generally being more difficult to cultivate (some requiring a minimum temperature of 15 °C throughout the year). In the key and descriptions which follow, stem measurements are exclusive of the cephalium. Flowering can occur at any time of the year when the plant is growing.

1a. Stem to 60 × 30 cm; ribs 15–24, rounded; spines 2–7 cm, more or less straight; fruit 1–2.4 cm × 5–10 mm **8. intortus**
 b. Stem much smaller or either ribs, spines or fruit not as above 2
2a. Fruit red, or pink and central spines 3–4, projecting, nearly straight 3
 b. Fruit white to rather pale pink, or pink and central spine 1, curved or directed upwards 6
3a. Radial spines straight or curved upwards, needle-like, mostly slender 4

 b. Radial spines curved downwards, subulate, often stout 5
4a. Stem ellipsoid to cylindric; fruit 1.2–2.2 cm **9. harlowii**
 b. Stem depressed-spherical to conic, or ellipsoid and lowermost radial spine greatly elongated; fruit 1.5–4.5 cm **5. oreas**
5a. Flowers 1–2.3 cm × 5–10 mm; fruit 1–2.5 cm × 5–10 mm **6. peruvianus**
 b. Flowers 2–4 × 1.5–2 cm; fruit 2–6 × 1–1.5 cm **7. curvispinus**
6a. Cephalium apparently with creamy white wool only, bristles absent or hidden; isolated plants rarely fruiting (self-sterile) **2. glaucescens**
 b. Cephalium with wool and conspicuous bristles; isolated plants with or without fruit 7
7a. Fruit white to rather pale pink or not produced on isolated plants (self-sterile); seeds smooth **1. azureus**
 b. Fruit pink to pinkish mauve, usually produced even on isolated plants; seeds strongly tuberculate at end opposite hilum 8
8a. Fruit pinkish mauve, cylindric, often remaining partly immersed in the cephalium and drying up **5. oreas**
 b. Fruit pink, obovoid to club-shaped, usually expelled from the cephalium, very juicy 9
9a. Stem 13–50 cm high; spines rather stout, the central to 2.5 cm **3. zehntneri**
 b. Stem 5–9 cm high or if larger then central spine slender, to 4.5 cm **4. violaceus**

1. M. azureus Buining & Brederoo
(*M. ferreophilus* Buining & Brederoo; *M. krainzianus* Buining & Brederoo; *M. pachyacanthus* Buining & Brederoo). Illustration: Haustein, Der Kosmos-Kakteenführer, 225 (1983); Kakteen und andere Sukkulenten 35: 159 (1984). Stem to 14–30 × 14–20 cm, green, grey-green or intensely glaucous; ribs 9–12, acute; areoles 1.5–2 cm apart. Central spines 1–3, 2–4.6 cm; radials 7–11 very stout, lowermost longest, to 5 cm. Cephalium to 3.5–12 × 7–10 cm, with white wool and conspicuous reddish bristles. Flowers 1.4–2 cm × 4.5–7 mm, pink. Fruit 1.5–2.9 cm × 3.5–10 mm, white to rather pale pink; seeds 1–1.7 × 0.9–1.5 mm, almost smooth. *E Brazil (C, N & S Bahia).* G2.

M. levitestatus Buining & Brederoo (*M. diersianus* Buining & Brederoo; *M. rubrispinus* Ritter; *M. securituberculatus*

Buining & Brederoo; *M. warasii* Esteves Pereira & Buenecker; *M. uebelmannii* Braun). Illustration: Kakteen und andere Sukkulenten 26: 169, 170 (1975); Lamb & Lamb, Illustrated reference on cacti & other succulents 5: 1323 & t. 348 (1978). Very like *M. azureus* but stem 15–30 (rarely to 50) × 14–22 cm; ribs 9–15; areoles 1.5–2.5 cm apart. Radial spines shorter, to 3 cm. Flowers to 2.5 cm × 9 mm, red. Fruit 1.2–2 cm × 7–12 mm; seeds 1.2–1.8 mm. *E Brazil (C S Bahia & N Minas Gerais)*. G2. Probably only a variety of *M. azureus*.

2. M. glaucescens Buining & Brederoo. Illustration: Hecht, BLV Handbuch der Kakteen, 316 (1982) – both photographs, one as M. caesius.
Spherical to slightly pyramidal, to 14 × 14 cm, grey-green to grey-blue; ribs *c.* 11, acute; areoles 1–1.8 cm apart. Central spine 1, ascending, to 1.5 cm; radials 5–7, the lowermost to 2.5 cm. Cephalium *c.* 2–2.5 × 7 cm, with short creamy white wool, bristles very short, hidden. Flowers 2.1 cm × 6 mm, pinkish. Fruit 12 × 6 mm; seeds 1–1.3 × 0.8–0.9 mm, strongly tuberculate at end opposite hilum. *E Brazil (C Bahia)*. G2.

3. M. zehntneri (Britton & Rose) Luetzelburg (*Cactus zehntneri* Britton & Rose; *M. helvolilanatus* Buining & Brederoo; *M. macrodiscus* Werdermann; *M. canescens* Ritter). Illustration: Succulenta 56: 113, 117–119 (1977); Haustein, Der Kosmos-Kakteenführer, 225 (1983).
Depressed-spherical to shortly cylindric, 13–30 × 15–25 cm, blue-green; ribs 10–16 (rarely to 19), acute; areoles 1.2–2 cm apart. Central spine 1 (rarely absent), 1.5–2.5 cm, ascending; radials 6–10, to 3 cm, rather stout. Cephalium 4–12 × 6–9 cm, with white wool and reddish bristles. Flowers 1.5–2.5 cm × 4–10 mm, pink. Fruit 1.2–1.8 cm × 4.5–8 mm, pink; seeds 1.2–1.3 × 1–1.1 mm, strongly tuberculate at end opposite hilum. *E Brazil (Minas Gerais to Piaui & Pernambuco)*. G2.

M. curvicornis Buining & Brederoo (*M. giganteus* Buining & Brederoo). Illustrations: Kakteen und andere Sukkulenten 23: 34, 35 (1972); Cactus & Succulent Journal (US) 45: 227–230 (1973); Haustein, Der Kosmos-Kakteenführer, 225 (1983). Very like the above but stem 25–50 × 10–20 cm. Spines very stout. Cephalium to 30 cm. *E Brazil (Bahia)*. G2.
Probably only a variety of *M. zehntneri*.

4. M. violaceus Pfeiffer (*M. depressus* Hooker). Illustration: Cactus & Succulent

Journal (US) 47: 252 (1975); Lamb & Lamb, Illustrated reference on cacti & other succulents 5: 1328 & t. 352 (1978).
Depressed-spherical to almost pyramidal, 5–9 × 8–16 cm; ribs 9–12, acute; areoles 6–15 mm apart. Central spine absent or 1, ascending, shorter than the longest of the radials; radials 5–9 (rarely to 11), to 1–2 cm. Cephalium 1.5–7 × 4–6 cm, with wool and bristles. Flowers to 2.3 × 1.3 cm, pinkish red. Fruit 1.4–2.1 cm × 5–9 mm, pink; seeds 1.5–2 × 1.2–1.5 mm, tuberculate at end opposite hilum. *E Brazil (coastal region from Rio de Janeiro northwards)*. G2.

M. concinnus Buining & Brederoo (*M. axiniphorus* Buining & Brederoo; *M. lanssensianus* Braun; *M. robustispinus* Buining et al.; *M. saxicola* Diers & Esteves Pereira). Illustration: Cullmann et al., Kakteen, edn 5, 237 (1984); Kakteen und andere Sukkulenten 35: 196-199 (1984). Spherical or almost so, 6–12 × 10–16 cm; ribs 8–13, acute; areoles 1–2 cm apart. Central spine 1, 1.5–4.5 cm, ascending, curved; radials 5–9, to 2–4.5 cm. Cephalium 2.5–4.5 × 6–7.5 cm, with white wool and reddish bristles. Flowers 1.5–1.9 cm × 6.5-9 mm, pink. Fruit 1.4–1.8 cm × 4–9 mm, pink, seeds 1.1–1.4 × 0.9–1.15 mm, strongly tuberculate at end opposite hilum. *E Brazil (Bahia)*. G2.
Intermediate between *M. violaceus* and *M. zehntneri*.

M. matanzanus Leon (*M. actinacanthus* Areces Mallea). Figure 28(9), p. 204. Illustration: Hirao, Colour encyclopaedia of cacti, f. 600 (1979); Cullmann et al., Kakteen, edn 5, 237 (1984). Like the preceding but areoles sometimes only 8 mm apart. Spines to 1.5 (rarely to 2.4) cm. Cephalium to 4 × 5–6.5 cm, with dense orange-red bristles and less wool. Seeds *c.* 1 mm. *N Cuba (Matanzas & Las Villas Provinces)*. G2.
One of the smallest and most popular species.

The name *M. melocactoides* (Hoffmannsegg) de Candolle, based on *Cactus melocactoides* Hoffmannsegg, has been used for the plant called *M. violaceus* Pfeiffer here. Its naming is problematical (see Taylor, N.P., in Cactus & Succulent Journal of Great Britain 42: 67, 1980).

5. M. oreas Miquel (*M. ernestii* Vaupel; *M. rubrisaetosus* Buining & Brederoo; *M. longispinus* Buining & Brederoo; *M. interpositus* Ritter; *M. cremnophilus* Buining & Brederoo; *M. erythracanthus*

Buining & Brederoo; *M. florschuetzianus* Buining & Brederoo; *M. azulensis* Buining & Brederoo; *M. nitidus* Ritter). Illustration: Cactus & Succulent Journal of Great Britain 42: 67, 68 (1980).
Depressed-conic to ellipsoid, 8–25 × 12–22 cm; ribs 10–16, somewhat rounded to obtuse; areoles 1–2 cm apart. Central spines mostly 4, 2–9 cm; radials mostly 6–11, slender, the lowermost greatly elongate, to 14 cm. Cephalium 1.5–13 × 4–8 cm, with wool and bristles. Flowers 1.7–1.8 cm × 7–16 mm. Fruit club-shaped, to 2.5 (rarely to 4.5) cm × 6–10 mm, bright red or crimson-red, often somewhat flattened, especially below apex; seeds 1–1.3 × 0.8–1 mm, slightly tuberculate. *E Brazil*. G2.

M. albicephalus Buining & Brederoo. Illustration: Haustein, Der Kosmos-Kakteenführer, 225 (1983). Like the above but ribs more acute. Cephalium with white wool, bristles very short and hidden. *E Brazil (Bahia)*. G2.
Found growing with *M. oreas* and *M. glaucescens* and perhaps a hybrid.

M. brederooianus Buining (*M. amethystinus* Buining & Brederoo; *M. inconcinnus* Buining & Brederoo; *M. lensselinkianus* Buining & Brederoo; *M. griseoloviridis* Buining & Brederoo; *M. bahiensis* misapplied in part, not *Cactus bahiensis* Britton & Rose). Illustration: Lamb & Lamb, Illustrated reference on cacti & other succulents 5: 1318 (1978); Cactus & Succulent Journal of Great Britain 44: 90 (1982). Depressed-spherical to conic, 8–14 × 12–27 cm; ribs 10–14, acute; areoles 1–2.5 cm apart. Central spines 1–3, 1.5–4 cm, straight-projecting; radials 7–11, 1.5–4.5 cm , mostly straight, the lowermost only a little longer than the others. Cephalium 1–8 × 5.5–10 cm, with white wool and bristles. Flowers 1.8–2 cm × 9–12 mm. Fruit club-shaped, to 3.5 × 1.3 cm, red to crimson; seeds 1–1.3 × 1 mm, somewhat tuberculate. *E Brazil*. G2.

M. salvadorensis Werdermann (*M. bahiensis* misapplied in part, not *Cactus bahiensis* Britton & Rose). Illustration: Krainz, Die Kakteen, Lfg. 54 (1973). Like *M. brederooianus* but fruit rather narrowly cylindric, only 1.5–2.5 cm × 4–6.5 mm, pale pinkish mauve. *E Brazil*. G2.

6. M. peruvianus Vaupel (*M. jansenianus* Backeberg; *M. trujilloensis* Rauh & Backeberg; *M. fortalezensis* Rauh & Backeberg; *M. amstutziae* Rauh & Backeberg; *M. unguispinus* Backeberg;

M. huallancaensis Rauh & Backeberg).
Illustration: Backeberg, Die Cactaceae 4:
f. 2454, 2457–2463, 2467, 2476–2482,
pl. 204–210 (1960).
Globular to pyramidal or cylindric, to 20
(rarely to 40) × 20 cm; ribs 8–14, acute to
rounded and tuberculate, low. Central
spines absent or 1–4, to 6 cm; radials 4–14
(usually 6–10), to 6 cm, straight to
strongly curved. Cephalium to 20 × 8 cm,
with wool and bristles. Flowers to 2.3 ×
1 cm, purplish red. Fruit 1–2.5 cm ×
5–10 mm, red; seeds finely tuberculate.
N & C Peru, S Ecuador. G2.

7. M. curvispinus Pfeiffer (*M. delessertianus*
Lemaire; *M. salvador* Murillo; *M. maxonii*
(Rose) Guerke; *Cactus maxonii* Rose;
M. oaxacensis (Britton & Rose) Backeberg;
Cactus oaxacensis Britton & Rose).
Illustration: Lamb & Lamb, Illustrated
reference on cacti & other succulents **5**:
1327 (1978); Cullmann et al., Kakteen, edn
5, 237 (1984).
Depressed-spherical, spherical or ovoid,
10–40 cm in diameter; ribs l0–15, acute to
rounded. Central spines 1–2 (rarely to 3),
to 3.5 cm, ascending; radials 7–12,
1–2.8 cm , more or less curved. Cephalium
2–6 × 3–10 cm, usually rather small, with
white wool and reddish bristles. Flowers
2–4 × 1.5–2 cm, perianth-segments widely
spread. Fruit club-shaped, rather large,
2–6 × 1–1.5 cm, bright red; seeds
c. 1.5 mm, coat tuberculate. *C America (S &
E Mexico, Guatemala).* G2.

8. M. intortus (Miller) Urban (*Cactus
intortus* Miller; *C. coronatus* Lamarck, not
Melocactus coronatus Backeberg; *C.
melocactus* (var.) *communis* Aiton; *M.
communis* (Aiton) Link & Otto). Illustration:
Lamb & Lamb, Illustrated reference on cacti
& other succulents **5**: 1322 & t. 347
(1978).
Spherical to tapered-cylindric, to 60 (rarely
to 90) × 30 cm; ribs 12–24, rounded, thick.
Central spines 1–3; radials 10–14, 2–7 cm,
stout, more or less straight. Cephalium
cylindric, eventually to 50 cm or more
high, with white wool and many reddish
bristles. Flowers 1.5–2 cm × 7–10 mm.
Fruit 1–2.5 cm × 5–10 mm, pink to red;
seeds with strongly tuberculate coat. *West
Indies (Turks & Caicos southwards to
Martinique).* G2.

9. M. harlowii (Britton & Rose) Vaupel
(*Cactus harlowii* Britton & Rose; *Melocactus
acunai* Leon; *M. borhidii* Meszaros; *M. evae*
Meszaros; *M. radoczii* Meszaros; *M. nagyi*
Meszaros). Illustration: Riha & Subik,

Illustrated encyclopedia of cacti &
other succulents, 159, 160, f. 170, 171
(1981).
Ellipsoid to cylindric, to 30 × 6–16 cm,
light green, ribs 10–13; areoles closely set,
usually less than or *c.* 1 cm apart. Central
spines 3–4 (rarely 1), to 5 cm; radials
c. 12, 1–3 cm, slender. Cephalium 1–15 ×
3–7.5 cm with wool and bristles. Flowers
to 2 (rarely to 3) × 1–2 cm. Fruit obovoid,
1.2–2.2 cm × 5–11 mm. *SE Cuba.* G2.

The invalid names *Melocactus albescens*,
M. morrochapensis and *M. ocujalii* are of
uncertain application.

28. PACHYCEREUS (Berger) Britton & Rose
D.R. Hunt
Tree-like or shrubby, often massive, with
stout, erect, columnar stems. Flowering
areoles more or less different from the non-
flowering, usually confluent, or connected
by a groove, densely felted, spineless, or
with numerous long, bristle-like spines.
Flowers small to medium-sized, diurnal or
nocturnal, shortly tubular, funnel-shaped
or bell-shaped; tube with scales, these
usually areolate and sometimes woolly
and/or bristly in the axils. Fruit more or
less spherical, soon dry, with generally
numerous bristles, sometimes very long,
sometimes falling early or the bristles
confined to the apex; seeds relatively large,
shiny black.
As here understood, a genus of 9 species,
all native to Mexico, and including some of
the largest of all cacti.

1a. Flowering stems with a dense apical
    cap or cephalium of bristly spines   4
 b. Flowering cephalium lacking or only
    weakly developed   2
2a. Ribs 5–7; non-flowering areoles
    adjacent or confluent, weakly spined
                                3. marginatus
 b. Ribs 8–17; non-flowering areoles
    discrete, fiercely spined   3
3a. Ribs usually 8–10; spines *c.* 10 per
    areole on young growth   **1. weberi**
 b. Ribs usually 11–15; spines *c.* 20 per
    areole   **2. pringlei**
4a. Spines of cephalium greyish or tinged
    pink; flowers pink or white, not
    woolly   **4. schottii**
 b. Spines of cephalium golden yellow,
    later dark brown or blackish; flowers
    greenish yellow, woolly   **5. militaris**

1. P. weberi (Coulter) Backeberg
(*Lemaireocereus weberi* (Coulter) Britton &
Rose). Illustration: Backeberg, Die
Cactaceae 4: f. 2031 (1960).

Massive tree to 10 m with short thick trunk
and dense crown of numerous vertical
stems *c.* 20 cm in diameter, glaucous
green; ribs *c.* 10 ; areoles 2–3 cm apart,
greyish white felted, in adult stems
connected by a narrow groove. Central
spine 1, to 10 cm, stout, somewhat
flattened; radial spines 6–12, 1–2 cm.
Flowers funnel-shaped, 8–10 cm; pericarpel
and tube with brown hairs and sparse
bristles; inner perianth-segments white.
Fruit 6–7 cm in diameter, densely covered
with yellow bristle-like spines, splitting
when ripe to expose the red pulp and black
seeds. *S Mexico.* G2.

2. P. pringlei (Watson) Britton & Rose
(*Cereus pringlei* Watson). Illustration:
Backeberg, Die Cactaceae 4: f. 2013–2020
(1960); Hecht, BLV Handbuch der Kakteen,
340 (1982); Haustein, Der Kosmos-
Kakteenführer, 81 (1983).
Massive tree to 15 m, with short trunk to
1 m or more in diameter and few to many
erect branches 25–50 cm in diameter,
bluish green; ribs usually 11–15 ; areoles
discrete on young stems, 1–2 cm apart and
heavily spined, later becoming confluent or
connected by a felted groove and less spiny.
Spines 20 or more on young stems, 1–3 cm
or sometimes the central longer, stout.
Flowers 6–8 cm; pericarpel and tube
densely woolly in the axils of the scales;
inner perianth-segments white. Fruit
5–7 cm in diameter, brown-felted and
densely bristly. *NW Mexico (Baja California
and Sonora).* G2.

P. pecten-aboriginum (Engelmann)
Britton & Rose. Similar to *P. pringlei*, but
differing in having fewer (10–11) ribs and
spines (8–12), and less woolly flowers.
W Mexico. G2.

3. P. marginatus (de Candolle) Britton &
Rose (*Marginatocereus marginatus* (de
Candolle) Backeberg; *Stenocereus marginatus*
(de Candolle) Berger & Buxbaum.
Illustration: Backeberg, Die Cactaceae 4:
f. 2105–2115 (1960); Hecht, BLV
Handbuch der Kakteen, 341 (1982);
Haustein, Der Kosmos-Kakteenführer, 81
(1983).
Stem simple or little-branched, columnar,
to 3–7 m × 8–15 cm, dark green; ribs 4–7;
areoles adjacent or confluent. Central
spines 1–2, 1–1.5 cm, radials about 7,
usually only 2–4 mm; spines more
numerous and bristly in the flowering zone.
Flowers 1–2 per areole, tubular, 3–5 ×
3 cm; pericarpel and tube with woolly
areoles and sometimes small spines; inner
perianth-segments greenish white or pink.

Fruit 4 cm in diameter, yellowish red. *C & S Mexico.* G2.

4. P. schottii (Engelmann) D. Hunt (*Cereus schottii* Engelmann; *Lophocereus schottii* (Engelmann) Britton & Rose). Illustration: Benson, Cacti of the United States and Canada, pl. 76, 79, 80, f. 620–624 (1982); Hecht, BLV Handbuch der Kakteen, 295 (1982); Haustein, Der Kosmos-Kakteenführer, 83 (1983); Cullmann et al., Kakteen, edn 5, 201 (1984).
Large shrub, 3–7 m, branching at the base and with numerous erect stems 10–15 cm in diameter; ribs usually 5–7; areoles 2–2.5 cm apart on lower part of stem, with *c.* 8–10 spines to 1.2 cm; areoles on upper part of adult stems longer and only 5–6 mm apart, with very numerous bristly spines, weak, somewhat flattened and twisted, 3–7.5 cm. Flowers 1–several per areole, funnel-shaped, nocturnal, *c.* 4 × 3 cm with disagreeable smell; scales of pericarpel and tube sparsely woolly or naked in their axils; inner perianth-segments pink or white. Fruit spherical to ovoid, 2.5–3 × 2–2.5 cm, red, fleshy, spineless, splitting to reveal red pulp. *NW Mexico (just extending into Arizona).* G2.
'Monstrosus' is a peculiar spineless variant with irregular, broken ribs.

5. P. militaris (Audot) D. Hunt (*Backebergia militaris* (Audot) Sanchez-Mejorada). Illustrations: Haustein, Der Kosmos-Kakteenführer, 87 (1983); Cullmann et al., Kakteen, edn 5, 135 (1984).
Tree-like, to 5–6 m, with numerous erect branches *c.* 12 cm in diameter, dark greyish green; ribs 5–7 on young branches, later to 9–11; non-flowering areoles 5–10 mm apart. Spines 8–14, to 1 cm, weak. Flowering areoles with numerous dense golden bristles, *c.* 5.5 cm, the flowering zone (cephalium) 25–30 × 18–20 cm overall, 6–7 × 3.5–4 cm; pericarpel and tube greenish yellow, the numerous triangular scales with dense white, woolly hairs and a few bristles in the axils; inner perianth-segments pale green. Fruit oblong, 3.5 × 2 cm, red at first, soon dry, bristly at the apex only. *SW Mexico.* G2.
In the 1970s numerous top-cuts of stems with the remarkable 'Grenadier's Cap' cephalium were imported to Europe and rooted, but few have been coaxed into vegetative growth.

29. CARNEGIEA Britton & Rose
D.R. Hunt
Massive columnar cacti, sometimes unbranched, or with few to many side-branches from the principal trunk; ribs usually numerous, rarely less than 13; areoles also numerous, closely spaced, those in the flowering zone usually more densely felted and/or with fewer spines or more numerous bristle-like spines than the non-flowering. Flowers nocturnal or remaining open the following morning, small to medium-sized, tubular to bell-shaped or funnel-shaped; pericarpel and tube with numerous prominent decurrent scales, at least the lower areolate and sometimes bristly in the axils, and in some species nectar-bearing at the apex; perianth red, greenish red or white; stamens very numerous. Fruit green or red, more or less tuberculate, areolate, and occasionally bristly, splitting irregularly to reveal red or white pulp; seeds dark brown or black.
As here understood (including *Neobuxbaumia* Backeberg) a genus of 8 species in Mexico and the SW United States.

1a. Spines on lower part of stem 15–30, strong, brownish; flowers white
1. gigantea
b. Spines 7–9, bristle-like, golden yellow; flowers purplish red **2. polylopha**

1. C. gigantea (Engelmann) Britton & Rose (*Cereus giganteus* Engelmann). Illustration: Benson, Cacti of the United States and Canada, pl. 62–64, f. 568–585 (1982); Hecht, BLV Handbuch der Kakteen, 235 (1982); Haustein, Der Kosmos-Kakteenführer, 83 (1983).
Tree to 16 m, with trunk 30–75 cm in diameter, unbranched for many years but eventually with 1 or more lateral branches spreading and then erect; ribs 12–30; non-flowering areoles *c.* 1.2 cm apart, the flowering nearly contiguous and more densely felted. Spines 15–30, brownish or grey, stronger and more unequal on young and immature growth, the longest there to *c.* 7 cm, deflexed, thinner and without well-developed centrals at flowering areoles, the longest *c.* 3.5 cm. Flowers near stem-apex, funnel-shaped to bell-shaped, 9–12 × 5–6 cm; inner perianth-segments white. Fruit obovoid, 5–7.5 × 2.5–4.5 cm, red or greenish, scaly, pulp red; seeds black, 2 mm. *SW USA (Arizona), N Mexico (Sonora).* G2.

2. C. polylopha (de Candolle) D. Hunt (*Neobuxbaumia polylopha* (de Candolle) Backeberg). Illustration: Hecht, BLV Handbuch der Kakteen, 321 (1982); Haustein, Der Kosmos-Kakteenführer, 87 (1983).
Columnar, to 15 m × 50 cm, normally unbranched; ribs 22–36 or more; areoles 8–11 mm apart, the non-flowering and flowering similar; spines 7–9, 1–3 cm, bristle-like, yellow. Flowers shortly bell-shaped to funnel-shaped, 5–8 × 3.5–4.5 cm; scales of pericarpel and tube glandular, almost naked in their axils, brownish red; inner perianth-segments pinkish red. Fruit ovoid, 2.5–4 × 2–3.5 cm, scaly and with a few bristles, olive green, pulp white; seeds dark brown, 2.5–3.5 mm. *C Mexico.* G2.

30. CEPHALOCEREUS Pfeiffer
D.R. Hunt
Columnar, with many-ribbed stems to 10–12 m; flowering areoles different from the non-flowering, sometimes confluent, and/or bearded with long hairs or bristles. Flowers medium-sized, nocturnal, tubular-bell-shaped; pericarpel and tube with minute scales, sometimes woolly in their axils; stamens numerous, borne on the tube and throat, the lowermost arising from a rim nearly covering the nectar-chamber. Fruit ovoid, with very small scales, withered perianth persistent, seeds black or dark brown, shiny.
As understood here in a restricted sense, a genus of 2 species endemic to Mexico.

1. C. senilis (Haworth) Schumann. Illustration: Hecht, BLV Handbuch der Kakteen, 236 (1982); Haustein, Der Kosmos-Kakteenführer, 89 (1983); Cullmann et al., Kakteen, edn 5, 139 (1984).
Columnar, simple or branched near the base in wild plants, to 12 m × 40 cm; ribs 12 or more in young plants, later 25–30; areoles closely set, with 1–5 slender yellowish spines 1–2 cm and 20–30 hair-like, white bristles 6–12 cm, entirely covering the stem, eventually deciduous. Flowering areoles with abundant wool and bristles, forming a dense fleece (cephalium) at the apex and down one side of the stem. Flowers 5–9 × 6 cm; pericarpel and tube with numerous small scales hairy in the axils; inner perianth-segments whitish with pink midrib. Fruit 3 × 2–2.5 cm, red at first, becoming dry and brownish; seeds 2.5 mm. *C Mexico.* G2.
The Old Man cactus, a favourite with collectors, but not flowering until about 6 m tall.

31. MYRTILLOCACTUS Console
D.R. Hunt
Tree-like or shrubby with numerous ascending, few-ribbed, spiny stems. Flowers diurnal, small, up to 9 per areole; pericarpel

with small scales slightly woolly in the axils; tube very short; perianth disc-shaped; stamens relatively few. Fruit small, globular, berry-like, fleshy; seeds small, dull black, papillose.

A genus of 4 species in Mexico and Guatemala, closely allied and possibly subspecies within a single species. The edible fruits recall those of *Vaccinium myrtillus*, hence the generic name.

1. M. geometrizans (Martius) Console. Figure 28(8), p. 204. Illustration: Hecht, BLV Handbuch der Kakteen, 318 (1982); Haustein, Der Kosmos-Kakteenführer, 85 (1983); Cullmann et al., Kakteen, edn 5, 239 (1984).
Tree-like, 4–5 m with a short trunk and numerous upcurving, branching, bluish stems 6–10 cm in diameter; ribs 5–6; areoles 1.5–3 cm apart. Central spine 1, 1–7 cm, dagger-like and sometimes 6 mm broad at the base, blackish; radial spines 5–9, 2–10 mm, reddish at first, fading to grey. Flowers about 2 × 2.5 cm, creamy white. Fruit 1–2 cm in diameter, dark reddish or bluish purple. *Mexico*. G2.

32. STENOCEREUS Riccobono
D.R. Hunt
Tree-like, shrubby, thicket-forming or very rarely creeping, with thick, ribbed stems, often heavily spined; flowering areoles usually discrete, rarely confluent. Flowers diurnal or nocturnal, funnel-shaped or bell-shaped; pericarpel with numerous areoles and usually small spines, tube flared or not, with decurrent scales usually naked in the axils. Fruit spherical or ovoid, fleshy, often with deciduous spines; seeds shiny black, smooth or somewhat warty.

A genus of about 25 species native to Mexico, central America, the West Indies, Venezuela and Colombia, formerly known as *Lemaireocereus* Britton & Rose.

1a.	Young stems with conspicuous waxy bloom	2
b.	Young stems green or if glaucous then not conspicuously waxy	3
2a.	Spines 1–3; ribs usually 7–9	**1. beneckei**
b.	Spines 6 or more; ribs usually 5–6	**2. pruinosus**
3a.	Stem prostrate, creeping; principal spine strongly flattened	**3. eruca**
b.	Stem erect or arching; spines not strongly flattened	4
4a.	Areoles clearly separate, 6–50 mm apart	5
b.	Areoles adjacent, less than 5 mm apart, or confluent; ribs 5–7	**4. dumortieri**
5a.	Areoles white or whitish, without glandular hairs	**5. stellatus**
b.	Areoles dark brown, with glandular hairs	6
6a.	Ribs 6–8	**6. queretaroensis**
b.	Ribs 12–19	**7. thurberi**

1. S. beneckei (Ehrenberg) Buxbaum (*Lemaireocereus beneckei* (Ehrenberg) Britton & Rose). Illustration: Hecht, BLV Handbuch der Kakteen, 286 (1982); Haustein, Der Kosmos-Kakteenführer, 79 (1983).
Stems erect or decumbent, sparsely branched 1–2 m × 5–7 cm, bloomed towards the apex; ribs 7–9, tuberculate; areoles 2–5 cm apart. Spines usually 1–3, black at first, the uppermost longer, to 4 cm, the lower 3–15 mm. Flowers nocturnal, narrowly funnel-shaped, 6.5–8 × 4–5 cm; pericarpel and tube with small felted areoles and short spines, or none; inner perianth-segments greenish white or tinged pink. Fruit 4 cm in diameter, red; seeds 3.5 mm, shiny black. *W Mexico*. G2.

2. S. pruinosus (Pfeiffer) Buxbaum (*Lemaireocereus pruinosus* (Pfeiffer) Britton & Rose). Illustration: Hecht, BLV Handbuch der Kakteen, 287 (1982) – as 'Polaskia chichipe'.
Large shrub or small tree to 4–5 m, much-branched; stems 8–10 cm in diameter, white-bloomed near apex, 5–6 (rarely to 8)-ribbed; areoles to 3–4 cm apart. Central spines 1–4, 2–3 cm, grey; radial spines 5–8, 1–2 cm. Flowers funnel-shaped, c. 9 × 7 cm; inner perianth-segments white or pinkish. Fruit ovoid, 5–8 cm, varying from orange-green to purple, seeds 2–2.5 mm, black, warty. *S Mexico*. G2.

3. S. eruca (Brandegee) Gibson & Horak (*Lemaireocereus eruca* (Brandegee) Britton & Rose; *Machaerocereus eruca* (Brandegee) Britton & Rose). Illustration: Hecht, BLV Handbuch der Kakteen, 286 (1982); Haustein, Der Kosmos-Kakteenführer, 85 (1983).
Stems prostrate, creeping and branching, to 1–5 m × 4–10 cm, the apical 20–30 cm ascending, the older stem rooting adventitiously and eventually dying back from the base; ribs 10–12; areoles c. 2 cm apart. Spines c. 20, yellowish or whitish, the principal one to 3.5 cm, flattened and directed backwards, dagger-like, the remainder ascending and radiating, 1–2.5 cm. Flowers tubular-funnel-shaped, 10–14 × 7–8 cm; tube long, with few, distant scales; inner perianth-segments white, creamy yellow or tinged pink. Fruit

spherical, 3–4 cm; seeds dull black. *NW Mexico (Baja California)*. G2.

4. S. dumortieri (Scheidweiler) Buxbaum (*Cereus dumortieri* Scheidweiler; *Isolatocereus dumortieri* (Scheidweiler) Backeberg). Illustration: Backeberg, Die Cactaceae 4: f. 2122–2124 (1960); Hecht, BLV Handbuch der Kakteen, 286 (1982).
Tree 5–6 (rarely to 10) m, with short trunk and numerous erect glaucous branches 5–10 cm in diameter; ribs 5–7; areoles adjacent or sometimes confluent. Spines variable, relatively slender, centrals 1–4, to 4 cm, radials 9–11 or more, to 1 cm. Flowers opening at night, remaining open the following morning, 5 × 3 cm, tubular-funnel-shaped; outer perianth-segments brownish red, inner white. Fruit oblong 3–3.5 cm, orange-red; seeds 1.5 mm, black. *C & S Mexico*. G2.

5. S. stellatus (Pfeiffer) Riccobono (*Lemaireocereus stellatus* (Pfeiffer) Britton & Rose). Illustration: Backeberg, Die Cactaceae 4: f. 2116–2119 (1960); Haustein, Der Kosmos-Kakteenführer, 81 (1983).
Mostly branching from the base, stems to 4 m × 6–9 cm, dull green; ribs 8–12, obtuse, notched, the areoles seated in the notches, 1–2 cm apart, whitish. Central spines to 2.5 cm, slender, dark brown or blackish at first, fading to grey; radial spines c. 8–13, to 1.2 cm, spreading star-like. Flowers diurnal, tubular-bell-shaped, 5–6 × 3–4 cm; receptacle spiny; tube scarcely flared; perianth clear pink. Fruit spherical, 3 cm in diameter red, with spines falling early. *S Mexico*. G2.

6. S. queretaroensis (Weber) Buxbaum (*Lemaireocereus queretaroensis* (Weber) Safford; *Ritterocereus queretaroensis* (Weber) Backeberg). Illustration: Backeberg, Die Cactaceae 4: f. 2072–2074 (1960).
Tree 5–6 m or more with short trunk and numerous ascending green or somewhat reddish stems to 15 cm in diameter; ribs 6–8; areoles c. 1 cm apart, dark brown or blackish, with glandular hairs. Central spines 2–4, to 4 cm, stout; radial spines 6–9, unequal, the lower to 3 cm. Flowers funnel-shaped, 10–12 cm; inner perianth-segments white, tinged pink. Fruit spherical, 6 × 6 cm, reddish; seeds 2.5 mm, black, warty. *C Mexico*. G2.

7. S. thurberi (Engelmann) Buxbaum (*Cereus thurberi* Engelmann; *Lemaireocereus thurberi* (Engelmann) Britton & Rose). Illustration: Benson, Cacti of the United States and Canada, pl. 71, 72, f. 609–616

(1982); Hecht, BLV Handbuch der Kakteen, 287 (1982).

Massive shrub with numerous erect, cylindric stems arising near ground level to 3–7 m × 10–20 cm; ribs 12–19; areoles 1–3 cm apart, dark brown or blackish with glandular hairs. Spines 11–19, more or less equal, the longer 1.2–2.5 (rarely to 5) cm, blackish or brown. Flowers opening at night, often remaining open the following morning, funnel-shaped, 6–7.5 × *c*. 5 cm; inner perianth-segments clear purplish or pink. Fruit spherical, 3–7.5 cm in diameter, red; seeds 2–2.5 mm, shining black. *SW USA and NW Mexico (Sonoran Desert)*. G2.

33. ARMATOCEREUS Backeberg
D.R. Hunt

Shrubby or tree-like, branches stout, erect or ascending, cylindric, ribbed, constricted annually and thus more or less segmented. Flowers nocturnal, tubular, the tube not markedly flared, with felted, often spiny areoles; limb short, the perianth-segments erect to spreading, white or rarely red; stamens borne on the throat and upper part of tube. Fruit spherical to ovoid, fleshy, spiny; seeds black.

A genus of more than 10 species in Peru and Ecuador, probably very closely related to *Stenocereus*.

1. A. laetus (Kunth) Backeberg
(*Lemaireocereus laetus* (Kunth) Britton & Rose). Illustration: Britton & Rose, The Cactaceae 2: 99, f. 145 (1920); Backeberg, Die Cactaceae 2: f. 826 (1959).
Tall shrub to 6 m with short trunk and erect branches; branches to 15 cm in diameter, 6–9-ribbed, pale greyish green; areoles 2–4 cm apart. Central spines brownish black at first, 2–5, 7–30 mm, the radial spines 7–10, 5–10 mm. Flowers to 12 × 6 cm; tube spiny; perianth-segments white, red-tipped. *N Peru*. G2.

A. cartwrightianus (Britton & Rose) Backeberg. Illustration: Backeberg, Die Cactaceae 2: f. 810–812 (1959). Small tree 3–5 m with woody trunk; branch-segments 15–60 × 8–15 cm, typically 7–8-ribbed, green; areoles large, brown-felted; spines *c*. 20, 1–2 cm. Flowers narrowly tubular, 7–9 × 2–3 cm, the tube with scattered, spiny areoles, not tuberculate; perianth-segments short, spreading, white, tipped red. Fruit spherical to oblong, 8–9 cm, red with white pith. *Ecuador*. G2.

A. matucanensis Backeberg. Illustration: Backeberg, Die Cactaceae 2: f. 828–831 (1959). Broad-crowned, densely branched shrub to 5 m with short trunk; branches

4–8-ribbed, dark green. Spines 7 or more, very unequal, the longest to 10 cm. Flowers 10 × 8 cm; tube closely tuberculate, spiny; perianth white. *C Peru*. G2.

The name *Armatocereus giganteus* has not been traced.

34. BROWNINGIA Britton & Rose
D.R. Hunt

Tree-like with erect trunk and crown of erect, ascending or spreading, cylindric, ribbed branches; flowering areoles different from the non-flowering. Flowers solitary at the areoles, funnel-shaped, nocturnal; pericarpel and tube bearing broad acute scales naked or nearly so in their axils; perianth-segments spreading, rather short; stamens borne in the throat and tube. Fruit globular to ovoid, somewhat fleshy, covered with deciduous scales; seeds small to relatively large, black.

A genus of about 7 species, occurring in the Andes of N Chile, Peru and Bolivia.

1a. Stem green, with 30–34 ribs; spines of young and sterile plants 20–50; flowering branches with bristles only **1. candelaris**
  b. Stem vivid light blue, with *c*. 18 ribs; spines of young plants *c*. 5–8 increasing to *c*. 30 on flowering stems **2. hertlingiana**

1. B. candelaris (Meyen) Britton & Rose. Illustration: Haustein, Der Kosmos-Kakteenführer, 77 (1983).
Tree-like, to 3–5 m, with tapered trunk to 30 cm in diameter, branched at the top, the branches spreading or drooping, green; ribs 30–34, low; areoles *c*. 1 cm apart. Spines of trunk formidable, 20–50, 6–15 cm long, brown at first, spines of flowering branches weak, yellow, sometimes lacking. Flowers 8–12 × 6–8 cm; pericarpel and tube somewhat curved, with numerous overlapping scales; inner perianth-segments white or tinged pink. Fruit to 7 cm. *N Chile to S Peru*. G2.

2. B. hertlingiana (Backeberg) Buxbaum (*Azureocereus hertlingianus* (Backeberg) Backeberg). Illustration: Hecht, BLV Handbuch der Kakteen, 232 (1982); Haustein, Der Kosmos-Kakteenführer, 75 (1983).
Tree-like, to 8 m, main stem and side branches erect, to 30 cm in diameter, vivid light blue; ribs to 18 or more; areoles raised on tubercles. Spines of juvenile stems *c*. 5–8, unequal, longest to 8 cm, all yellow, tipped brown; upper part of older

stems with up to 30 more or less equal, more flexible spines. Flower 5 cm in diameter; tube curved, dark purplish brown, with ciliate scales almost black at maturity. *Peru*. G2.

35. NEORAIMONDIA Britton & Rose
D.R. Hunt

Shrubby or tree-like, much-branched from the base or with a definite trunk; stems erect, few-ribbed; non-flowering areoles large, brown-felted, usually spiny and one or more of the spines very long; flowering areoles enlarged, felted, nearly spineless, gradually prolonged into a spur, flowering annually. Flowers 1–2 per areole, small, the pericarpel and tube bearing small scales and felted areoles with or without bristles; limb short; stamens borne on the upper part of the tube. Fruit spherical to oblong, felted and more or less spiny; seeds black or dark brown.

A genus of 2 species, one from the pacific coastal belt of Peru and N Chile, the other native to Bolivia.

1. N. arequipensis (Meyen) Backeberg (*N. macrostibas* (Schumann) Backeberg; *N. roseiflora* (Werdermann & Backeberg) Backeberg). Illustration: Haustein, Der Kosmos-Kakteenführer, 73 (1983).
Robust shrub, branching from the base, the stems columnar, to 10 m × 20–40 cm; ribs 4–10; areoles 1–4 cm apart, large, the flowering areoles enlarging over several years to become a spherical or cylindric felted spur to 10 cm long; spines variable, to 12 or more, very unequal, the longest to 25 cm. Flowers 2.5–4 × 2–4 cm, pericarpel and tube felted; perianth disc-shaped, pink, purplish red or greenish white. Fruit to 7 cm in diameter, purplish, shedding the areoles when ripe; seeds dull black. *Peru, N Chile*. G2.

N. herzogiana (Backeberg) Buxbaum (*Neocardenasia herzogiana* Backeberg). Eventually a tree 7–10 m, with a distinct trunk 1–2 m × 50 cm; stems 15–20 cm in diameter; ribs sometimes 5 at first, later 6–7. Spines as in *N. arequipensis*. Flowers 7–7.5 × 5–6 cm; pericarpel and tube with dense bristle-like spines 1–2 cm; perianth-segments purple with paler or white margins. Fruit *c*. 5 cm in diameter, reddish yellow, the spiny areoles falling early; seeds somewhat smaller than in *N. arequipensis* and dark brown. *Bolivia*. G2.

36. CORRYOCACTUS Britton & Rose
D.R. Hunt

Shrubby, or rarely tree-like, mostly branching from base, with cylindric, 4–

10-ribbed, stout to slender, erect, ascending or prostrate stems, often strongly spined. Flowers bell-shaped; tube short, bearing numerous small scales felted and spiny in their axils; petal-blades spreading, the segments short, yellow, orange or purplish red; stamens borne in the broad throat. Fruit spherical, fleshy and spiny, the withered perianth falling; seeds curved-obovate in outline, black or brownish, warty and rough.

A genus of more than 20 species from Peru, Bolivia and Chile, several of which are occasionally cultivated.

1a. Stems 1–3 cm in diameter
1. squarrosus
b. Stems 3–15 cm in diameter 2
2a. Flowers purplish red
2. melanotrichus
b. Flowers orange or yellow
3. brevistylus

1. C. squarrosus (Vaupel) Hutchison (*Erdisia squarrosa* (Vaupel) Britton & Rose). Illustration: Backeberg, Die Cactaceae **2**: f. 775 (1959).
Stems prostrate or almost erect to 50 cm, branches *c.* 25 × 2–2.5 cm; roots tuberous; ribs 7–8, to 8 mm high, compressed; areoles to 2.5 cm apart. Central spine 1, strong; radial spines to 10, unequal. Flowers 4–4.5 cm, yellowish to reddish. Fruit 2.5 × 1.7 cm; seeds 2 mm, black. *C Peru (Junín)*. G2.

Britton & Rose's broad interpretation of this species included plants later considered to be: **C. erectus** (Backeberg) Ritter (*Erdisia erecta* Backeberg). Erect, to 1 m; ribs 5–6. Spines *c.* 10 (rarely to 18). Flowers scarlet-carmine or fiery red. *S Peru (Cuzco)*.

2. C. melanotrichus (Schumann) Britton & Rose. Illustration: Haustein, Der Kosmos-Kakteenführer, 191 (1983).
Stems erect, branching and clustering from the base, 1–2 m × 3–6 cm; ribs 7–9; areoles 2 cm apart. Spines 10–11, unequal, longest to 3 cm. Flowers 5 × 6 cm, purplish red. Fruit 4–8 cm in diameter, seed black. *Bolivia*. G2.

3. C. brevistylus (Schumann) Britton & Rose. Illustration: Britton & Rose, The Cactaceae **2**: f. 99, 101 (1920).
Stems robust, branching near the base, to 2–3 m × 15 cm; ribs 6 –7, areoles 3 cm apart. Spines *c.* 15, very unequal, the longest exceeding 20 cm in the wild. Flowers 9 × 10 cm, yellow, the throat 4.5 cm broad at the top. *S Peru*. G2.

C. brachypetalus (Vaupel) Britton & Rose. Illustration: Britton & Rose, The Cactaceae **2**: f. 99, 102, 103 (1920); Haustein, Der Kosmos-Kakteenführer, 191 (1983). Stems 2–4 × 6–10 cm; ribs 7–8. Flowers 6 × 5 cm, deep orange. Fruit 6–7 cm in diameter, green. *S Peru*. G2.

C. pulquinensis Cardenas. Stems to 4 m × 4 cm, sometimes clambering, 4–5-ribbed; areoles 3–4 cm apart. Spines 3–7, 0.5–2 cm. Flowers to 7.5 cm, orange-yellow. *Bolivia*. G2.

37. EULYCHNIA Philippi
D.R. Hunt
Shrubby or tree-like with ascending or prostrate, 9–16-ribbed stems, often fiercely spiny, even as seedlings. Flowers borne near tips of branches, small, broadly bell-shaped; receptacle-tube short, bearing numerous scales with woolly hairs and bristly spines in their axils; limb erect or somewhat spreading, the segments short, pink or white; stamens borne in the broad throat; style short and thick. Fruit spherical, crowned by the persistent perianth, fleshy and scaly or hairy but usually devoid of spines; seeds brownish or black.

A genus of 8 species, 7 in Chile and 1 in Peru.
Literature: Ritter, F., Kakteen in Südamerika **3**: 892–903 (1980).

1a. Pericarpel and tube with bristly spines
1. castanea
b. Pericarpel and tube without bristles 2
2a. Pericarpel and tube with very short sparse wool only **2. acida**
b. Pericarpel and tube with long, dense wool **3. breviflora**

1. E. castanea Philippi (*Philippicereus castaneus* (Philippi) Backeberg). Illustration: Britton & Rose, The Cactaceae **2**: f. 124 (1920); Backeberg, Die Cactaceae **2**: f. 1044–1046 (1959).
Shrubby, the stems erect at first, later spreading or decumbent, to 50–100 × 6–8 cm; ribs 8–13, low; areoles *c.* 1 cm apart. Spines 6–10 radiating, unequal 5–20 mm, 1–2 central, 3–10 cm. Flowers 5–5.5 × 4–5 cm; pericarpel tuberculate, scales minute, the areoles with short brown wool and slender bristly spines 1–1.5 cm; inner perianth-segments 1–1.5 cm, white or pinkish. Fruit 5 cm diameter, yellow-green, devoid of bristles except near top; seeds 1.5 mm, black. *Chile*. G2.

2. E. acida Philippi. Illustration: Britton & Rose, The Cactaceae **2**: f. 123 (1920); Backeberg, Die Cactaceae **2**: f. 1122–1124 (1959).

Shrubby or tree-like, stems to 1–4 m or more × 9–12 cm; ribs 10–16, broad; areoles 7–15 mm apart. Spines variable, to *c.* 12, radiating, 1 cm or more, 1–2 central, to 20 cm. Flowers 5–7 × 4–6 cm; pericarpel with small black-tipped scales and sparse grey-black hairs; perianth-segments white, usually with pink midstripe. Fruit 5–6 × 5 cm, brownish yellow-green; seeds 1.5 mm, dull black. *Chile*. G2.

3. E. breviflora Philippi (*E. spinibarbis* misapplied). Illustration: Hecht, BLV Handbuch der Kakteen, 268 (1982); Cullmann et al., Kakteen, edn 5, 172 (1984).
Shrubby or tree-like, much-branched, to 5 m, stems typically 6–10 cm in diameter; ribs 10–13, rounded; areoles 5–15 mm apart. Spines 10–22, radiating, unequal, longest to 3 cm, 3–6 central, not strongly differentiated, but the longest 5–15 cm. Flowers 7–8 × 5–6 cm; pericarpel with small green scales and long, dense, woolly, golden yellow to brown hairs; perianth-segments white, often pink above. Fruit *c.* 6 cm in diameter, green, covered with woolly hairs; seeds 2 mm, dull black or brownish black. *Chile*. G2.

E. ritteri Cullmann. Illustration: Cullmann et al., Kakteen, edn 5, 172 (1984). Areoles crowded, densely felted. Flowers small, 2 × 1.5 cm, pink. Fruit 3 cm in diameter, greenish orange. *S Peru*. G2.

E. saint-pieana Ritter. Illustration: Haustein, Der Kosmos-Kakteenführer, 195 (1983). Areoles of young plants densely white-woolly, those of mature plants crowded, densely felted. Flowers 6–7.5 × 5–7.5 cm; pericarpel densely woolly with brown to white hairs; perianth-segments white, usually with pink midstripe. Fruit *c.* 8 cm, pear-shaped; seeds 1.4 mm, greyish black, slightly rough. *N Chile*. G2.

Attractive as a seedling, on account of the strong spines and abundant snowy hair. **E. morromorenoensis** Ritter (*E. floresiana* invalid) is similar.

The name *E. marksiana* has not been traced.

38. HAAGEOCEREUS Backeberg
D.R. Hunt
Shrubby or tree-like or with erect, ascending, decumbent, or creeping stems, usually with numerous ribs, relatively large close-set areoles, and fierce spines; additional bristles sometimes present at flowering areoles. Flowers nocturnal

(though often remaining open the following morning), regular or nearly so, tubular-funnel-shaped, mostly white or dull red; tube stout, fleshy, with numerous small scales weakly to densely hairy in the axils; limb spreading, relatively narrow; stamens in a single series, style projecting. Fruit spherical to ovoid, fleshy, sparsely scaly and hairy, withered perianth persistent; seeds black.

A genus in which more than 50 species have been named (of which probably no more than 5–10 are distinct), native to the deserts of Peru and N Chile.

1a. Stems erect or ascending   2
  b. Stems decumbent or creeping
                                4. decumbens
2a. Stems usually 5 cm or less in diameter; flowers white   **3. versicolor**
  b. Stems 6–15 cm in diameter; flowers variously coloured   3
3a. Flowers regular, the tube straight   4
  b. Flowers slightly asymmetric, the tube slightly curved or S-shaped
                                5. weberbaueri
4a. Ribs 12–14; flowers greenish white
                                1. limensis
  b. Ribs 18–26; flowers dull purplish red, or greenish red to greenish white
                                2. multangularis

1. H. limensis (Salm-Dyck) Ritter (*H. acranthus* (Vaupel) Backeberg; *H. olowinskianus* Backeberg; *H. zonatus* Rauh & Backeberg). Illustration: Ritter, Kakteen in Südamerika 4: f. 1270 (1981).
Stems erect or ascending, branching from base, to 2 m × 7–10 cm; ribs 12–14, somewhat tuberculate. Principal central spines 1–3, 2–6 cm, very stout; other lesser spines in 2–3 series, to 30–40 in all, the outermost *c.* 1 cm. Flowers 6–8 × 5–6 cm; tube flared in upper half, greenish; perianth-segments white or pale pink. *C Peru.* G2.

2. H. multangularis (Willdenow) Ritter (*H. chosicensis* (Werdermann & Backeberg) Backeberg; *H. albispinus* (Akers) Backeberg; *H. pseudomelanostele* (Werdermann & Backeberg) Backeberg). Illustration: Haustein, Der Kosmos-Kakteenführer, 134 (1983); Cullmann et al., Kakteen, edn 5, 184, 185 (1984).
Stems erect or ascending, branching from base, to 1.5 m × 4–10 cm; ribs 18–26. Principal central spines 1–4, to 8 cm, very stout; radial spines 25–50 or more, intermixed with hair-like bristles. Flowers 4–8 × 2.5–4 cm; tube slightly flared in upper half, dull red, brownish or

greenish; perianth-segments red, pink or white. *C Peru.* G2.

Closely allied 'species' include *H. acanthocladus* Rauh & Backeberg, *H. chrysacanthus* (Akers) Backeberg, *H. dichromus* Rauh & Backeberg, *H. divaricatispinus* Rauh & Backeberg, *H. pacalaensis* Backeberg, *H. pachystele* Rauh & Backeberg, *H. pseudoversicolor* Rauh & Backeberg, *H. salmonoideus* (Akers) Backeberg, *H. seticeps* Rauh & Backeberg, *H. setosus* (Akers) Backeberg and *H. tenuispinus* Rauh & Backeberg, all from C Peru.

3. H. versicolor (Werdermann & Backeberg) Backeberg (*H. elegans* invalid; *H. icosagonoides* Rauh & Backeberg). Illustration: Backeberg, Die Cactaceae 2: f. 1148–1154 (1959); Haustein, Der Kosmos-Kakteenführer, 134 (1983).
Stems slender, erect, branching from the base to 1.5 m × 5 cm; ribs 16–22 ; areoles relatively small. Central spines 1–2, 1–4 cm, sometimes little different from the radials, variable in colour; radial spines 25–40, to 5 mm, reddish or yellowish brown to white, the colours varying in bands or zones toward the stem-apex. Flowers 6–8 × 4–6 cm; tube flared, light green; perianth-segments white. *N Peru.* G2.

4. H. decumbens (Vaupel) Backeberg (*H. ambiguus* Rauh & Backeberg; *H. australis* Backeberg; *H. litoralis* Rauh & Backeberg; *H. multi-colorispinus* Buining; *H. platinospinus* (Werdermann & Backeberg) Backeberg & Knuth). Illustration: Backeberg, Die Cactaceae 2: f. 1203 (1959); Haustein, Der Kosmos-Kakteenführer, 136 (1983).
Stems decumbent or pendent, to about 1 m × 4–8 cm; ribs 12–22; areoles 5–10 mm apart. Principal central spines 1–5, 1–5 cm, very stout, dark; other lesser spines in several series, 30–40, the outer to 5–8 mm, thin, pale. Flowers 6–8 × 5–6 cm; tube slightly flared in upper half, brownish or reddish green; perianth-segments white. *S Peru.* G2.

5. H. weberbaueri (Vaupel) D. Hunt (*Weberbauerocereus weberbaueri* (Vaupel) Backeberg; *W. albus* Ritter; *W. fascicularis* misapplied; *W. horridispinus* Rauh & Backeberg; *W. johnsonii* Ritter; *W. rauhii* Backeberg; *W. seyboldianus* Rauh & Backeberg). Illustration: Haustein, Der Kosmos-Kakteenführer, 132 (1983).
Large shrub or small tree, variable, to 6 m, much-branched below; stems 6–15 cm in

diameter, with 15–35 ribs; areoles usually very large. Central spines to 8, to 10 cm, fierce; radial spines 20–60, needle-like to hair-like. Flowers 6–11 × 2–4.5 cm; tube somewhat curved or S-shaped, dull reddish, brownish or greenish; perianth-segments relatively short, not widely spreading, pink to brownish or greenish white. Fruit spherical, 4 cm, orange-yellow, with short woolly hairs; seeds black. *S Peru.* G2.

Names not traced: *H. maritimus, H. pectinatus.*

39. ESPOSTOA Britton & Rose
D.R. Hunt
Shrubby or tree-like with cylindric or columnar, many-ribbed stems. Flowering areoles different from the non-flowering, with woolly hairs and often bristly spines forming a dense fleece (cephalium) over several ribs towards the stem-apex, the ribs more or less reduced and modified, and sometimes depressed in a groove. Flowers tubular-bell-shaped, usually nocturnal, in some species rather small; tube short, with small acute scales hairy in their axils; perianth-segments short, spreading to recurved; stamens borne in the throat and tube, the lowermost on a collar which partially closes the nectar chamber. Fruit spherical to obovoid, more or less naked or with tufted hairs; seeds small, black or brown.

A genus of about 10 species in Bolivia, Peru and S Ecuador.

1a. Ribs of the cephalium strongly reduced and depressed in a groove
                                1. lanata
  b. Ribs of the cephalium not depressed in a groove   2
2a. Cephalium with long dark brown or black bristles   **2. blossfeldiorum**
  b. Cephalium with no dark bristles
                                3. melanostele

1. E. lanata (Kunth) Britton & Rose (*E. sericata* (Backeberg) Backeberg).
Illustration: Britton & Rose, The Cactaceae 2: f. 87–91 (1920); Hecht, BLV Handbuch der Kakteen, 267 (1982); Haustein, Der Kosmos-Kakteenführer, 139 (1983); Cullmann et al., Kakteen, edn 5, 172 (1984).
Variable in habit, usually becoming shrubby or tree-like, to 4–7 m; stems to 10–15 cm in diameter; ribs 20–30, densely areolate. Non-flowering areoles with long silky hairs, 1–2 yellow, brown or black central spines 2–5 cm, and numerous yellowish radial spines 4–7 mm. Flowering

areoles with dense woolly hair varying from white to yellowish or brown. Flowers 4–6 cm; tube with numerous small triangular-lanceolate acute scales, at least the upper hairy in the axils; perianth-segments more or less spreading, whitish. Fruit nearly spherical, 3–4 cm, red; seeds matt black. *Peru, S Ecuador*. G2.

Recently introduced and closely related are several Peruvian 'species' probably representing different populations of *E. lanata*, all of which make attractive plants as seedlings. They include: *E. hylaea* Ritter, *E. mirabilis* Ritter, *E. ritteri* Buining and *E. superba* Ritter. *E. baumannii* Knize is said to be natural hybrid of another such species with a species of *Cleistocactus*.

2. E. blossfeldiorum (Werdermann) Buxbaum (*Thrixanthocereus blossfeldiorum* (Werdermann) Backeberg. Illustration: Backeberg, Die Cactaceae **4**: f. 2359–2362 (1960); Haustein, Der Kosmos-Kakteenführer, 141 (1983).
Simple or occasionally branched below, to 2–3 m; stems to 5–7 cm in diameter; ribs 18–25; non-flowering areoles numerous, *c*. 5 mm apart. Central spines 1–4, to 3 cm, brown or nearly black; radial spines 20–25, 5–10 mm, glassy white. Cephalium extending over 4–8 ribs, with dense pale yellowish wool and numerous dark brown or black bristles. Flowers 6–7 × 5 cm; tube with numerous acute, olive green scales, hairy in the axils; inner perianth-segments creamy yellow. Fruit *c*. 3 × 2.5 cm, dark green; seeds pale brown. *N Peru*. G2.

E. senilis (Ritter) N.P. Taylor (*Thrixanthocereus senilis* Ritter). Illustration: Backeberg, Die Cactaceae **4**: f. 2363 (1960); Cactus & Succulent Journal of Great Britain **40**: 54 (1978); Ritter, Kakteen in Südamerika **4**: f. 1358, 1361 (1981). Spines 60–80, mostly *c*. 1 cm, white, sometimes with 1–2 longer, darker, bristle-like centrals. Cephalium covering 6–12 ribs. Flowers 4.5–6 × 3–4 cm, purple. Fruit spherical, 2 × 2 cm, green; seeds shiny black. *Peru (Ancash)*. G2.

3. E. melanostele (Vaupel) Borg (*Pseudoespostoa melanostele* (Vaupel) Backeberg; *E. nana* Ritter). Illustration: Backeberg, Die *Cactaceae* **4**: f. 2364–2371 (1960); Hecht, BLV Handbuch der Kakteen, 267 (1982); Haustein, Der Kosmos-Kakteenführer, 139 (1983).
Shrub to 2 m, branching from the base; stems to 10 cm in diameter; ribs *c*. 25; areoles very numerous, the non-flowering with dense white or brownish hairs to 1 cm clothing the whole stem. Spines numerous,

at first yellow, later blackish; central spines 1–3, the longest to 4–10 cm; radial spines 40–50, 5–10 mm. Cephalium whitish, yellow or brown. Flowers 5–6 × 5 cm; scales of the pericarpel tiny, those of the tube larger, hairy in the axils; perianth white. Fruit 5 cm, yellowish white to reddish; seeds black. *C Peru*. G2.

40. CLEISTOCACTUS Lemaire
D.R. Hunt
Mostly shrubby, with erect, ascending, decumbent, prostrate or pendent, cylindric, usually slender, several- to many-ribbed stems. Non-flowering and flowering areoles similar, or rarely the flowering with more wool and/or bristles. Flowers narrowly tubular, regular or somewhat asymmetric, diurnal, red, orange, yellowish or green; tube straight or kinked beyond the ovary and more or less S-shaped, covered with numerous overlapping scales woolly or hairy in the axils; limb unexpanded, the segments scarcely larger than the scales of the tube and not or only slightly spreading, or if expanded then more or less strongly oblique, with the upper segments more or less erect and the lower spreading or recurved; stamens projecting or not, borne on the throat and tube in 2 series, the lowermost on a collar partially closing the nectar-chamber; style projecting. Fruit small, spherical, fleshy with funicular pulp, tufted with sparse hairs or almost hairless, the withered perianth usually persistent, indehiscent or bursting open when ripe; seeds small, black, shiny.

As here understood (including *Borzicactus* Riccobono), a genus of up to 50 poorly distinguished species distributed in South America from C Peru to Bolivia and N Argentina, Paraguay and Uruguay. Numerous new 'species' proposed and introduced in recent decades by authors like Backeberg, Cardenas and Ritter have yet to be evaluated.

1a. Flowering areoles similar to the non-flowering, or with a few additional bristles; flowers not lemon-yellow   2
 b. Flowering areoles with conspicuous additional bristles to 3 cm; flowers lemon-yellow   **10. ritteri**
2a. Limb of perianth not or only slightly expanded; stamens projecting or not   3
 b. Limb of perianth expanded; stamens projecting   8
3a. Ribs 7–14 (rarely to 16)   4
 b. Ribs 18–30   12
4a. Flower straight or downcurved   5

 b. Flower bent sharply upwards beyond the ovary   **1. baumannii**
5a. Flowers red, at least the tube   6
 b. Flowers white   **7. morawetzianus**
6a. Limb expanded, outer segments yellow, inner green   **6. smaragdiflorus**
 b. Limb not or slightly expanded, red   7
7a. Flowering areoles without extra bristles; flowers straight-tubed, limb *c*. 3 cm in diameter   **2. sepium**
 b. Flowering areoles developing extra bristles; flowers somewhat curved, limb *c*. 2 cm in diameter   **3. roezlii**
8a. Ribs usually 8–11, spines relatively strong   **2. sepium**
 b. Ribs usually 14 or more; spines relatively weak or bristle-like   9
9a. Innermost perianth-segments short, incurved against the filaments and closing the throat   **12. winteri**
 b. Innermost perianth-segments erect or spreading   10
10a. Flower strongly S-shaped, the tube laterally compressed   **13. samaipatanus**
 b. Flower nearly straight, the tube terete   11
11a. Spines 25–30, with some longer bristles at flowering areoles; flowers 7–8 cm   **4. icosagonus**
 b. Spines *c*. 20–25, extra bristles not developed at flowering areoles; flowers 4–5 cm   **5. acanthurus**
12a. Flower straight or nearly so   13
 b. Flower bent sharply upwards beyond the ovary   **11. brookei**
13a. Flowers 6–8 cm, dark red   **8. strausii**
 b. Flowers 3.5–4 cm, scarlet   **9. jujuyensis**

1. C. baumannii (Lemaire) Lemaire (*Cereus tweediei* Hooker). Illustration: Botanical Magazine, 4498 (1850); Britton & Rose, The Cactaceae **2**: t. 27, f. 2 (1920). Branching at base, erect or arching, to 2 m or more; stems 2.5–3.5 cm in diameter; ribs 12–16; areoles closely set, brown or black-felted; spines 15–20, the longest to 4 cm, needle-like. Flowers 5–7 × 1 cm, orange or scarlet, sharply upcurved and then narrowly S-shaped beyond the pericarpel; limb oblique; stamens and style shortly projecting. Fruit spherical, 1–1.5 cm in diameter, red with white pulp. *NE Argentina, Paraguay, Uruguay*. G2.

C. anguinus (Guerke) Britton & Rose. Decumbent; ribs 10–11. Flowers yellow-orange. *Paraguay*. G2.

C. santacruzensis Backeberg. Illustration: Backeberg, Cactus Lexicon, fig. 60 (1966).

Ribs 8–11. Flowers red. *SE Bolivia*. G2. Perhaps the same as *C. anguinus*.

2. C. sepium (Kunth) Roland-Gosselin (*Borzicactus sepium* (Kunth) Britton & Rose). Illustration: Britton & Rose, The Cactaceae 2: f. 229 (1920); Backeberg, Die Cactaceae 2: f. 894 (1959); Haustein, Der Kosmos-Kakteenführer, 143 (1983).
Stem simple, or branched near base, to 1.5 m × 4 cm, with *c*. 8–11 ribs; areoles 1.5–2 cm apart, seated on low tubercles. Central spines 1–3, to 4 cm, stout; radial spines 8–10, to 1 cm, thinner. Flowers to 7.5 × 3 cm, bright red, straight; tube *c*. 1.5 cm in diameter, scales with dark brown axillary hairs; limb nearly regular, segments spreading. Fruit spherical, 2 cm in diameter, fleshy, with white pulp. *Ecuador*. G2.

C. fieldianus (Britton & Rose) D. Hunt (*Borzicactus fieldianus* Britton & Rose; *B. tessellatus* Akers & Buining). Illustration: Britton & Rose, The Cactaceae 4: f. 250–254 (1923). Larger, forming thickets 3–6 m tall; ribs 5–7. *Peru*. G2.

C. serpens (Kunth) Roland-Gosselin (*Borzicactus serpens* (Kunth) Kimnach; *Haageocereus serpens* invalid). Illustration: Cactus & Succulent Journal (US) **32**: 94 (1960). Stems decumbent or prostrate, 2–3 m × 5–20 mm, with woody rootstock; ribs *c*. 10; areoles 4–10 mm apart. Central spines 1 (rarely to 4), 5–40 mm; radial spines *c*. 10, short. *Peru (La Libertad)*. G2.

3. C. roezlii (Haage) Backeberg (*Borzicactus roezlii* (Haage) Marshall; *Cereus roezlii* Haage; *Seticereus roezlii* (Haage) Backeberg). Illustration: Backeberg, Die Cactaceae 2: f. 914 (1959); Ritter, Kakteen in Südamerika 4: f. 1262 (1981); Haustein, Der Kosmos-Kakteenführer, 143 (1983).
Shrub or small tree 1–3 m, with stems 4.5–6 cm in diameter; ribs 7–14; areoles 8–15 mm apart. Central spines usually 1, 2–4 (rarely to 6) cm, sticking out or pointing downwards; radial spines 9–14 or more, to 1 cm, pale brown. Flowering areoles, at least on older stems, developing bristles. Flowers 6–7 × 2 cm, red, somewhat curved above; limb narrow, oblique, with the segments scarcely spreading. Fruit 2–4 × 2.5–4 cm, yellow-or reddish orange. *N Peru*. G2.

Three little-known species, perhaps related, are *C. areolatus* (Muehlenpfordt) Riccobono, *C. laniceps* (Schumann) Roland-Gosselin, and *Seticereus chlorocarpus* (Kunth) Backeberg (*Borzicactus chlorocarpus* invalid).

4. C. icosagonus (Kunth) Roland-Gosselin (*Borzicactus icosagonus* (Kunth) Britton & Rose; *B. aurivillus* (Schumann) Britton & Rose; *Seticereus icosagonus* (Schumann) Backeberg). Illustration: Backeberg, Die Cactaceae 2: f. 902–906 (1959); Cullmann et al., Kakteen, edn 5, 136 (1984).
Stems prostrate and ascending, 20–60 × 3–5 cm, with 12–20 low ribs; areoles close-set. Spines 25–50, to 1–1.5 cm, bristle-like, golden yellow, with some longer and finer bristles at flowering areoles. Buds covered with white hairs. Flowers 7–8 cm, orange to scarlet or pinkish, nearly straight, scales small, with few long or short hairs; limb oblique; anthers yellow. Fruit *c*. 2.5 cm in diameter, yellow. *S Ecuador, N Peru*. G1.

5. C. acanthurus (Vaupel) D. Hunt (*Borzicactus acanthurus* (Vaupel) Britton & Rose (*Haageocereus paradoxus* Rauh & Backeberg; *Loxanthocereus cantaensis* Rauh & Backeberg; *Haageocereus cantaensis* invalid). Illustration: Haustein, Der Kosmos-Kakteenführer, 145 (1983).
Stems spreading, decumbent or pendent, to 30 × 2–5 cm, ribs 15–18, low, rounded and indistinctly divided into tubercles, areoles 5–6 mm apart. Spines *c*. 20–25, short, yellowish, the centrals somewhat thicker and longer, to 1.5 cm. Buds clothed with silky white hairs. Flowers 4–7 × *c*. 2.5 cm, straight or slightly curved, scarlet, the lower segments recurved, the upper almost erect; stamens in a single series, anthers yellow. Fruit 2–2.5 cm in diameter. *C Peru*. G2.

6. C. smaragdiflorus (Weber) Britton & Rose. Illustration: Britton & Rose, The Cactaceae 2: f. 248 (1920); Haustein, Der Kosmos-Kakteenführer, 148 (1983); Cullmann et al., Kakteen, edn 5, 142 (1984).
Arching or decumbent, to 1 m; stems 2–3 cm in diameter; ribs 12–14; areoles pale brown-felted. Central spines 4–6, 1.5–3.5 cm, yellowish or brown; radial spines 10–14, to 10 mm. Flowers 4–5 cm, straight, slightly constricted above the pericarpel, the scales red, mucronate, with white hairs in the axils; outer perianth-segments yellow, inner vivid emerald-green; stamens not projecting, style slightly projecting. Fruit 1.5 cm in diameter. *N Argentina*. G1.

C. candellila Cardenas. Illustration: Haustein, Der Kosmos-Kakteenführer, 148 (1983); Cullmann et al., Kakteen, edn 5, 141 (1984). Ribs 11–12. Central spines

3–4, 1–2.5 cm, slightly flattened, yellowish brown; radial spines 13–15, 2–5 mm. Flowers 3–3.5 cm; scales of tube red with few or no axillary hairs; outer perianth-segments yellow, inner purplish, bordered white. *SE Bolivia*. G2.

C. fusiflorus Cardenas. Ribs 13–14. Central spines 1, 1–2 cm; radial spines 8–9, 3–7 mm. Flowers 3.5 cm; tube pale yellow at first, later reddish, the scales hairy in the axils; outer perianth-segments pink below, brownish green above, inner purple. *SE Bolivia*. G1.

Perhaps not distinct from *C. smaragdiflorus*. *C. rojoi* Cardenas and *C. villamontesii* Cardenas are among other introductions from SE Bolivia which may also not be distinct.

7. C. morawetzianus Backeberg. Illustration: Backeberg, Die Cactaceae 2: f. 938–940 & t. 66 (1959).
Much-branched shrub to 2 m; stems *c*. 5 cm in diameter, grey-green; ribs 12–14; areoles *c*. 1 cm apart. Spines to 14, unequal, the longest 1.5–5 cm, golden yellow at first, becoming grey. Flowers *c*. 5.5 cm × 9 mm, white, or tinged pale greenish or pinkish; style long-projecting. *C Peru*. G2.

8. C. strausii (Heese) Backeberg. Illustration: Haustein, Der Kosmos-Kakteenführer, 148 (1983); Cullmann et al., Kakteen, edn 5, 142 (1984).
Erect and branching from the base, 1–3 m; stems 4–8 cm in diameter; ribs 25–30; areoles close-set. Central spines *c*. 4, to 2 cm, pale yellow; radial spines 30–40, 1.5–5 cm, hair-like, pure white. Flowers 8–9 cm, dark red; tube with silky hairs. *S Bolivia*. G1.

One of the most attractive of all cacti when well grown.

C. tupizensis (Vaupel) Backeberg (*C. buchtienii* Backeberg; *C. ressinianus* Cardenas; *C. sucrensis* Cardenas). Illustration: Backeberg, Die Cactaceae 2: f. 944, 945 (1959); Krainz, Die Kakteen, Lfg 28/29 (1964). Shrubby, to 2–3 m; stems 4–5 cm in diameter; ribs 18–24; areoles *c*. 7 mm apart. Spines reddish brown to yellowish or whitish, unequal, the central 2–4 to 6 cm, the radials to 20, 4–20 mm. Flowers 6–8 cm, straight or somewhat upcurved, purplish or wine-red. *C & S Bolivia*. G1.

Related species include *C. hildegardiae* Ritter (*C. hildewinterae* invalid; *C. flavirufus* invalid), *C. luribayensis* Cardenas and *C. reae* Cardenas.

9. C. jujuyensis (Backeberg) Backeberg (*C. tarijensis* Cardenas). Illustration: Backeberg, Die Cactaceae **2**: f. 944–945 (1959).
Shrubby, to 1 m, branching from the base; stems 4–6 cm in diameter; ribs *c.* 20; areoles 5–8 mm apart. Spines 20–30, unequal, the longest to 3 cm or more, yellowish or brown. Flowers 3.5–4 cm, pale red outside, the inner perianth-segments purplish pink. *N Argentina, S Bolivia*. G1.

Perhaps the same as *C. hyalacanthus* (Schumann) Roland-Gosselin.

10. C. ritteri Backeberg (*Cephalocleistocactus ritteri* (Backeberg) Backeberg). Illustration: Backeberg, Die Cactaceae **6**: f. 3352, 3353 (1962); Barthlott, Cacti, f. 18 (1979).
Shrubby, branching at base; stems to 1 m or more × *c.* 3 cm; ribs 12–14 or more; areoles 5 mm apart. Central and radial spines not sharply differentiated, the centrals 5, 1 cm, yellowish, the radial *c.* 30, shorter, white; additional bristle-like white spines 3–4 cm developing in the flowering zone. Flower *c.* 4 cm × 5 mm, lemon-yellow, nearly hairless. *C Bolivia*. G2.

11. C. brookei Cardenas (*C. wendlandiorum* Backeberg; *C. flavescens* Haage, invalid). Illustration: Cactus & Succulent Journal (US) **24**: f. 89–91 (1952); Backeberg, Die Cactaceae **2**: f. 948, 949 (1959); Haustein, Der Kosmos-Kakteenführer, 148 (1983); Cullmann et al., Kakteen, edn 5, 141 (1984).
Stem usually unbranched, to 50 × 3–4.5 cm; ribs 22–24; areoles close-set, 3–4 mm apart. Spines 25–40, *c.* 1 cm, not differentiated into central and radial, fine, bristle-like, pale yellow. Flowers *c.* 5 cm × 8 mm, red or orange, sharply upcurved and somewhat inflated beyond the pericarpel, then S-shaped and somewhat flattened laterally; limb oblique; stamens and style shortly projecting. Fruit 8–10 mm in diameter, purple. *S Bolivia*. G1.

C. vulpis-cauda Ritter & Cullmann. Illustration: Cullmann et al., Kakteen, edn 5, 141 (1984). Prostrate, to 1.5 m; ribs 18–22. Spines to 50, 1–2 cm, white to reddish brown. Flowers as in *C. brookei*. *S Bolivia*. G1. Flowering nearly all year, according to Cullmann.

12. C. winteri D. Hunt (*Borzicactus aureispinus* (Ritter) Hutchison & Kimnach; *Hildewintera aureispina* (Ritter) Ritter; *Winterocereus aureispinus* (Ritter) Backeberg). Illustration: Haustein, Der

Kosmos-Kakteenführer, 143 (1983); Cullmann et al., Kakteen, edn 5, 188 (1984).
Stems prostrate or pendent, to 1.5 m × 2.5 cm, with 16–17 ribs. Areoles 3–5 mm apart. Spines *c.* 50, 5–10 mm, bright golden yellow, often somewhat longer on older, flowering stems. Flowers 4–6 × 5 cm, orange-red, the limb nearly regular, the outer segments linear, acute, 2–3.5 cm × 2.5–5 mm, spreading-recurved, the innermost only 3–8 × 3–5 mm, erect or incurved against the filaments, white or pale pink; stamens and style projecting, anthers violet, stigmas greenish yellow. Fruit 7–10 mm in diameter, green or reddish green. *Bolivia*. G1.

13. C. samaipatanus (Cardenas) D. Hunt (*Bolivicereus samaipatanus* Cardenas; *Borzicactus samaipatanus* (Cardenas) Kimnach). Illustration: Cactus & Succulent Journal (US) **23**: 92–94, figs. 40–44 (1951); Haustein, Der Kosmos-Kakteenführer, 141 (1983).
Erect, branching from base, stems to 1.5 m × 3.5–4 cm; ribs 14–16, low, rounded; areoles 3–4 mm apart. Spines 13–22 unequal, the shortest 4 mm, others 1–3 cm, straw-yellow or brownish to grey or white. Flowers 3.5 × 2 cm, bright red, the tube more or less S-shaped, the limb strongly oblique with narrow, acute segments; stamens and style projecting, filaments and anthers purplish, stigmas greenish yellow; staminodial hairs present. Fruit nearly spherical, 9–11 × 7–9 mm, with reddish acute scales, and covered with whitish or brown hairs, withered perianth persisting. *Bolivia*. G1.

The name *C. pallidus* has not been traced.

41. DENMOZA Britton & Rose
D.R. Hunt
Stem simple, spherical to stout-cylindric, eventually many-ribbed. Flowering areoles with or without extra bristles. Flowers tubular, slightly asymmetric, scarlet; tube slightly curved and dilated above the pericarpel, with scales hairy in their axils; perianth unexpanded, the segments short, erect; stamens borne in the throat and tube; filaments and style long-projecting; nectar-chamber plugged with a staminodial collar and hairs. Fruit spherical, bearing tufts of short hairs, drying and splitting when ripe to reveal white pulp; seeds black, warty.

A genus of 2 closely allied species endemic to NW Argentina, very weakly similar to *Cleistocactus*.

1. D. rhodacantha (Salm-Dyck) Britton & Rose (*Cleistocactus rhodacanthus* (Salm-Dyck) Lemaire). Illustration: Cullmann et al., Kakteen, edn 5, 149 (1984).
Stem spherical at first, with relatively few ribs, but eventually shortly columnar, to 30 × 15 cm or more, with 15–30 ribs; areoles 2–2.5 cm apart. Spines 8–10, stout, spreading and recurved, dark reddish brown, the lowermost longest, to 3 cm; sometimes 1 stronger central spine produced. Flowers arising near the apex, 6–7 × 1.5–2 cm, the flowering areoles without extra bristles. *NW Argentina*. G1.

D. erythrocephala (Schumann) Berger. Illustration: Haustein, Der Kosmos-Kakteenführer, 150 (1983). Eventually to 1.5 m × 30 cm, and with 40–50 ribs in very old plants. Spines 30 or more, to 6 cm, grading from needle-like to hair like, more or less curved, and with additional bristles at the flowering areoles. *NW Argentina*. G1.

Probably a subspecies or variety of *D. rhodacantha*.

42. OREOCEREUS (Berger) Riccobono
D.R. Hunt
Mostly shrubby, with erect or ascending, thickish cylindric stems; areoles, especially the flowering, often developing long white hairs, and densely spiny. Flowers diurnal, usually orange or red or purplish, more or less bilaterally symmetric, tubular-funnel-shaped, straight to somewhat curved; pericarpel and tube with numerous scales more or less hairy in the axils; limb narrow, oblique, with the upper segments almost erect and the lower spreading or recurved; stamens numerous, borne in the throat and tube, the lowermost filaments coalescent at the base to form a diaphragm over the nectar chamber, sometimes the diaphragm invested with staminodial hairs; style and stamens projecting. Fruit spherical to ovoid, hollow, dehiscing at the base, the pericarp fleshy but pulp usually not developed; seeds relatively large, black, dull, tuberculate.

A genus of about 5–7 species in W South America, in the Andes of S Peru, S Bolivia, N Chile and N Argentina.

1a. Shrubby, stems becoming long-cylindric, densely hairy   **2**
  b. Usually simple, spherical to short-cylindric, not conspicuously hairy
       3. hempelianus
2a. Flowers arising from unmodified areoles in upper part of stem
       1. celsianus
  b. Flowers arising in an apical tuft of hairs and bristles   **2. doelzianus**

1. O. celsianus (Salm-Dyck) Riccobono (*O. maximus* Backeberg; *O. neocelsianus* Backeberg). Illustration: Haustein, Der Kosmos-Kakteenführer, 145 (1983). Shrub to 1–3 m, branching from the base and above; stems to 12–20 cm in diameter; ribs 10–17, tuberculate, the sinuses relatively deep; areoles 1–1.8 cm apart, large, with woolly hairs to 5 cm long. Central spines 1–4, to 8 cm, stout, straw yellow to dark brown; radial spines *c.* 9, to 2 cm. Flowers 7–9 × 3 cm, dull pink, the tube lightly curved, limb oblique with rounded segments; anthers violet; stigmas long-projecting, yellowish green. Fruit spherical, yellowish green, opening basally. *Mountains of NW Argentina & Bolivia.* G1.

O. fossulatus (Labouret) Backeberg (*Cleistocactus fossulatus* Mottram). Illustrations: Backeberg, Die Cactaceae **2**: f. 968, 972 (1959); The Chileans **12**(42): 139–141 (1984). Similar to *O. celsianus*; ribs more prominently tubercled; seeds embedded in white pulp, *c.* 1.7 mm, shiny black. *Bolivia.*

According to Mottram (The Chileans **12**(42): 138–142, 1984 & **13**(43): 29–31, 1986), the name *O. fossulatus* was misapplied by Backeberg. Mottram formally renamed Backeberg's plant *Cleistocactus fossulatus*, on the basis of the pulpy fruit and smooth seeds, but in other respects it is a typical *Oreocereus*.

O. hendriksenianus Backeberg. Shrub to 1–1.4 m, branching from base, stems to 10 cm in diameter, with *c.* 10 broad ribs; flowers densely hairy. *S Peru, N Chile.* G1.

O. trollii (Kupper) Backeberg. Lower-growing, to 1 m, stems branching from base, *c.* 10 cm in diameter; flowers said to be only 4 cm. *S Bolivia, N Argentina.* G1.

2. O. doelzianus (Backeberg) Borg (*Morawetzia doelziana* Backeberg). Illustration: Backeberg, Die Cactaceae **2**: f. 983–989 (1959); Haustein, Der Kosmos-Kakteenführer, 147 (1983). Shrubby, branching from the base, stems to 1 m × 6–8 cm; ribs 10–11; areoles 1.5 cm apart, densely to sparsely hairy, or hairs absent. Spines variable, up to 20, to 3 cm, yellow to dark brown, 4 longer centrals sometimes developing later. Flowers arising in an apical tuft of bristles and hairs, to *c.* 10 × 3 cm, deep purplish pink, scales of tube with numerous loose axillary hairs, limb oblique, with rounded segments. Fruit ovoid-spherical, yellowish green, opening basally. *C Peru.* G1.

3. O. hempelianus (Guerke) D. Hunt (*Arequipa hempeliana* (Guerke) Oehme;

A. rettigii (Quehl) Oehme; *A. spinosissima* Ritter; *A. weingartiana* Backeberg; *Borzicactus hempelianus* (Guerke) Donald; *O. rettigii* (Quehl) Buxbaum). Illustration: Haustein, Der Kosmos-Kakteenführer, 147 (1983); Cullmann et al., Kakteen, edn 5, 128 (1984).
Stem simple or branched from the base, spherical at first but eventually short-cylindric, erect or decumbent, to 60 × 10–15 cm, usually greyish or glaucous green; ribs 10–20; areoles large, 5–15 mm distant. Spines very variable; centrals 3–10, to 5 cm, often curved, whitish, brown, or nearly black; radials 8–30, 1–3 cm, needle-like to finely bristle-like, yellowish to glassy white. Flowers arising near apex, 5–7.5 × 2.5–3.5 cm, scarlet to purplish red; tube slightly curved, the scales densely and softly hairy in the axils; limb more or less oblique, the segments acute. Fruit spherical to ovoid, *c.* 2 cm in diameter, yellowish, thin-walled, opening basally. *Mountains of S Peru, N Chile.* G1.

The following invalid *Oreocereus* names have been mentioned in catalogues: *O. churinensis, O. culpinensis, O. giganteus, O. lecoriensis, O. luribayensis, O. magnificus, O. urmiriensis.*

43. MATUCANA Britton & Rose
D.R. Hunt
Simple or clustering, spherical to short-cylindric; ribs few to numerous, broad, low; spines numerous, fine, to few or absent. Flowers narrowly tubular-funnel-shaped, somewhat bilaterally symmetric, scales of tube naked or hairy in the axils, limb usually narrow, red, orange or yellow. Fruit hollow, splitting longitudinally from the base, the pericarpel somewhat fleshy at first, later dry; seeds relatively large, dull black, warty.

A group of several low-growing species, native to Peru, and very similar to *Oreocereus.*

1a. Ribs 9–12; spines few, sometimes deciduous   **1. madisoniorum**
  b. Ribs *c.* 16–30, spines numerous, persistent   2
2a. Flowers regular, bright golden yellow   **4. aureiflora**
  b. Flowers slightly to strongly bilaterally symmetric, orange, scarlet or deeper red   3
3a. Ribs 25–30; scales of tube naked in the axils   **3. haynei**
  b. Ribs 16–26; scales of tube usually more or less hairy in the axils, rarely naked   4

4a. Stem short-cylindric, to 60 cm tall
          see **Oreocereus hempelianus**
  b. Stem spherical, usually less than 15 cm tall   **2. aurantiaca**

1. M. madisoniorum (Hutchison) Rowley (*Borzicactus madisoniorum* Hutchison). Illustration: Cactus & Succulent Journal (US) **35**: l67–172, f. 1–7 (1963); Haustein, Der Kosmos-Kakteenführer, 153 (1983). Stem simple, spherical to shortly columnar, to 10 (rarely to 30) × 8–15 cm, surface papillose, grey-green, sticky; ribs 7–12, obscure; areoles 2–2.5 cm apart. Spines absent or 1–3 on mature plants, 4–5 on cultivated seedlings, equal, 5–6 cm, curved or twisted irregularly, dark brown to blackish at first. Flowers apical, erect, almost regular, orange-red, 8–10 × 4.5–5 cm; tube straight or slightly twisted, scales distant, with dark brown axillary hairs; limb spreading, not or only slightly oblique; stamens not forming a diaphragm over the nectar-chamber; anthers and stigmas yellow. Fruit spherical, *c.* 2 cm in diameter, splitting longitudinally from base; seeds *c.* 2 × 1 mm, seed-coat expanded around the hilum, brown. *N Peru (Amazonas).* G2.

M. paucicostata Ritter (*Borzicactus paucicostatus* (Ritter) Donald). Illustration: Cullmann et al., Kakteen, edn 5, 235 (1984). Spines 4–9, persistent, flowers *c.* 5.5 cm. *Peru (Ancash).* G1.

2. M. aurantiaca (Vaupel) Buxbaum (*Borzicactus aurantiacus* (Vaupel) Kimnach & Hutchison). Illustration: Cactus & Succulent Journal (US) **29**: f. 30, 31 (1957); Haustein, Der Kosmos-Kakteenführer, 153 (1983).
Simple or clustering, with flattened-spherical to spherical or short-cylindric stems to 15 × 15 cm, surface dark shiny green; ribs usually about 16, often somewhat spiralled, tuberculate; areoles elliptic, to 8–15 mm, ending in a groove to the base of the tubercle above. Spines *c.* 20–25, intergrading, the centrals 3–7, to 4.5 cm, almost erect or horizontal, the radials *c.* 16–18, 5–25 mm, almost comb-like and recurved, all spines reddish brown below and yellowish above at first. Flowers borne near apex from upper edge of areoles, tubular-funnel-shaped, 7–9 × 5–7 cm, orange-red, the tube slightly curved, relatively thick, 1.2–1.7 cm in diameter, with numerous scales densely hairy with dark brown hairs to nearly naked in the axils; limb slightly to strongly oblique; filaments of lower stamens coalescent at base to form a diaphragm over the nectar-

chamber, and with a plug of staminodial hairs; anthers yellow; stigmas yellow-green. Fruit spherical, 1–2 cm in diameter, yellowish red with brownish red scales, splitting longitudinally; seeds 1.5 mm, dull brownish black or black, warty-tuberculate. *N Peru*. G1.

M. aurantiaca is variable and many populations have been given separate species names, e.g.: **M. ritteri** Buining. Ribs 12–22. Spines fewer (9–19) than in *M. aurantiaca*, black or reddish black. Flowers relatively narrow-tubed (tube 5–6 mm in diameter), reminiscent of *M. madisoniorum*. *Peru (La Libertad)*. G2.

Other variants in cultivation, allied to *M. ritteri*, are *M. formosa* Ritter; *Borzicactus huagalensis* Donald & Lau (*M. huagalensis* invalid); *M. intertexta* Ritter (*B. intertextus* (Ritter) Donald); *M. celendinensis* Ritter (*B. intertextus* var. *celendinensis* (Ritter) Donald; *B. celendinensis* invalid).

M. weberbaueri (Vaupel) Backeberg (*M. myriacantha* (Vaupel) Buxbaum). An eastern variant of *M. aurantiaca* with more numerous spines (25–40) and small flowers. *Peru (Amazonas)*. G2.

3. M. haynei (Otto) Britton & Rose (*M. crinifera* Ritter; *M. herzogiana* Backeberg; *M. multicolor* Rauh & Backeberg; *M. yanganucensis* Rauh & Backeberg). Illustration: Haustein, Der Kosmos-Kakteenführer, 153 (1983). Usually simple but some variants clustering, spherical to short-cylindric, to 60 × 10 cm; ribs 25–30, tuberculate; areoles small, close-set, woolly at first, eventually hairless. Central spines usually 3, 3.5–5 cm, dark-tipped, developed on plants of flowering-size; radial spines *c*. 30, to 2 cm, bristly, glassy white to yellowish brown. Flowers apical, 6–7 × 3.5 cm, nearly regular, scarlet to purplish crimson; scales of tube small, naked in their axils; limb scarcely oblique, the segments acute; anthers yellow; stigmas green. Fruit spherical, small. *N Peru*. G1.

4. M. aureiflora Ritter. Illustration: Cullmann et al., Kakteen, edn 5, 234 (1984).
Simple, flattened-spherical, to 13 cm in diameter; ribs 11–27, tuberculate, shiny grey-green; areoles oval, 5–8 × *c*. 2 mm, 7–11 mm apart. Spines *c*. 10–12, 7–18 mm, almost comb-like, yellow or yellow-brown above, darker below, 1–4 more central developed on old plants only, to 2.5 cm. Flowers regular, funnel-shaped, 3–4.5 × 4–5 cm, bright golden yellow; tube slender, flared, with scales sparsely hairy in

the axils; perianth-segments broadly spreading; lower stamens forming a diaphragm as in allied species; stigmas pale green or whitish. Fruit ovoid, *c*. 1.4 × 1 cm, greenish red or purplish, drying and splitting longitudinally; seeds 2 × 1.3 mm, black, finely tuberculate. *Peru (Cajamarca)*. G1.

Names not traced: *M. decajamarca* (probably just the locality of collection, 'de Cajamarca' (Peru), whence several species have been described); *M. lutea*.

44. OROYA Britton & Rose
D.R. Hunt
Stem simple or offsetting, flattened-spherical to very short cylindric, many-ribbed; areoles long; spines comb-like. Flowers borne near the apex, from the upper edge of the areoles, yellow to red, shortly funnel-shaped or bell-shaped, regular; tube very short, bearing small scales sparsely woolly in their axils; outer perianth-segments spreading, inner almost erect; stamens borne on the throat and tube, the lowermost on a collar partly enclosing the nectar-chamber; filaments and style not projecting beyond the perianth. Fruit obovoid, slightly fleshy with small scales, opening by a basal pore; seeds black, warty.

A genus of 2 species endemic to Peru.

1. O. peruviana (Schumann) Britton & Rose (*O. gibbosa* Ritter; *O. laxiareolata* Rauh & Backeberg; *O. neoperuviana* Backeberg; *O. subocculta* Rauh & Backeberg). Variable, stems simple or clustering, flattened spherical to very short cylindric, to 40 × 20 cm, with up to 35 ribs; areoles to 1.5 cm long. Central spines 0–5, to *c*. 2 cm, straight-projecting; radial spines comb-like, *c*. 10–30, to 1.5 cm, pale yellow to reddish brown. Flowers 1.5–3 × 1.5–2.5 cm, pale to deep pinkish red, usually yellow within. *Mountains of C Peru*. G1.

O. borchersii (Boedeker) Backeberg. Illustration: Cullmann et al., Kakteen, edn 5, 265 (1984). Flowers wholly greenish yellow to yellow. *N Peru*. G1.

Name not traced: *O. subgibbosa*.

45. ECHINOPSIS Zuccarini
D.R. Hunt
Habit very varied, stems erect, ascending, prostrate or very low, simple, branched or clustering, slender-cylindric to columnar, spherical to flattened-spherical, sometimes massive, usually distinctly ribbed, fiercely-

spined to nearly spineless; ribs few to numerous, often somewhat tuberculate beneath or between the areoles. Flowers nocturnal or diurnal, often large, funnel-shaped to almost bell-shaped, arising from areoles at the side or near the apex; pericarpel and tube bearing relatively narrow scales, often numerous, rarely bristle-tipped, more or less densely hairy and very rarely bristly in their axils; stamens numerous, variously borne on the throat and tube, usually the uppermost displaced to the mouth and forming a separate ring (throat-circle), sometimes their filaments contrasting in colour with the perianth and their bases fused into a distinct membranous ring (hymen); nectar-chamber rarely plugged with a tuft of staminodial hairs. Fruit spherical to narrowly ovoid or oblong, fleshy to dry; seeds almost globular to obovoid, seed-coat dullish black, more or less warty.

A large genus native to central South America, including (in this account) several genera usually separated in popular horticultural works. Many hundreds of species have been named, but it is doubtful whether there are really more than 50–60, falling into 3 principal and 3 or 4 subsidiary groups according to habit and whether day- or night-flowering. The 'true' *Echinopsis* species have relatively small, ribbed stems, and long-tubed nocturnal flowers; the larger-growing night-blooming species have been placed in *Trichocereus* Riccobono (including some Peruvian and Chilean species resembling *Eulychnia*) and the small, globular to shortcylindric, short-tubed, day-blooming species in *Lobivia* Britton & Rose. Groups which straddle these otherwise convenient boundaries include *Pseudolobivia* Backeberg, *Helianthocereus* Britton & Rose and *Soehrensia* (Backeberg) Backeberg; individual aberrant species are included in *Chamaecereus* Britton & Rose and *Setiechinopsis* (Backeberg) De Haas. *Acanthocalycium* Backeberg seems to be a mixture of species, including the very distinct *Echinopsis spiniflora*.

Numerous artificial hybrids have been raised, both within and between several of the groups just mentioned. Natural hybrids may also account for some of the problems.

1a. Receptacular scales ending in a spiny point   2
  b. Receptacular scales not spine-tipped; staminodes absent   4
2a. Flowers less than 6 cm long, diurnal; staminodes present at base of tube   3

b. Flowers 6–12 cm long, with very slender tube, nocturnal; staminodes absent **3. mirabilis**
3a. Ribs 16–20, acute; flowers mauve, pink or white **1. spiniflora**
b. Ribs 9–15, rounded; flowers yellow or white **2. thionantha**
4a. Flowers usually 10–20 cm long, mostly nocturnal, white, cream or pale pink 5
b. Flowers to 10 cm long, usually diurnal, brightly coloured 20
5a. Stems cylindric, even when young, erect or ascending, 30 cm–12 m tall; flowers with broad tube 6
b. Stems generally globular to short-cylindric, rarely reaching 30 cm, and flowering when 10 cm or less; flowers with narrow tube 13
6a. Flowers averaging 10 cm, remaining open several days **4. chiloensis**
b. Flowers averaging 18 cm, open one night and the following day only 7
7a. Ribs few, 4–8 **5. pachanoi**
b. Ribs more numerous, 8–20 or more 8
8a. Stems stout, columnar, to 3 m × 15 cm or more, eventually branching well above ground **6. terscheckii**
b. Stems more slender, usually less than 9 cm diameter, branching at ground level only 9
9a. Ribs continuous 10
b. Ribs tuberculate or transversely grooved **11. thelegona**
10a. Stems to 15 cm diameter; ribs 9–11; areoles 2–5 cm distant **7. candicans**
b. Stems c. 6 cm diameter; ribs 9–18; areoles c. 1 cm apart 11
11a. Stems erect, to 1–2 m, remaining simple until quite tall; ribs 10–15 **8. spachiana**
b. Stems ascending, 60 cm or less, soon clustering; ribs 14–18 12
12a. Stems to 60 cm, densely spiny; spines 1–5 cm **9. strigosa**
b. Stems to 25 cm; spines 5–10 mm **10. schickendantzii**
13a. Ribs not scolloped or tuberculate between the areoles; flowers mostly white or pale pink, rarely yellow 14
b. Ribs scolloped or tuberculate between the areoles; flowers red, orange or yellow, more rarely white 17
14a. Spines straight 15
b. Spines curved 16
15a. Flowers white **13. eyriesii**
b. Flowers pink **14. oxygona**
16a. Spherical to short-columnar; areoles 1.5–2.5 cm apart; spines 8 or fewer, to 2.5 cm **15. rhodotricha**

b. Becoming cylindric; areoles 1–1.5 cm apart; spines 8 or more, to 5 cm or more **16. leucantha**
17a. Plants large, hemispherical, to over 20 cm in diameter, with strong spines; ribs acute; flowers 7–10 cm **20. ferox**
b. Plants relatively small, depressed-spherical, to 15 cm in diameter, with weak spines 18
18a. Ribs with scolloped tubercles 19
b. Ribs with low, rounded tubercles **18. mamillosa**
19a. Ribs 13–16; spines curved **17. obrepanda**
b. Ribs 15–30; one or more central spines, if present, hooked **19. ancistrophora**
20a. Stem slender-cylindric, 30–100 × 4–5 cm in diameter, 12–18-ribbed **12. huascha**
b. Stem spherical to short-cylindric or columnar, large or small, or if slender-cylindric, less than 2 cm in diameter 21
21a. Stem spherical to columnar, often exceeding 15 cm in diameter 22
b. Stem spherical or rarely short cylindric, mostly less than 7 cm in diameter 23
22a. Stem columnar; spines very numerous; flowers more than 10 cm **21. tarijensis**
b. Stem spherical or short-cylindric; spines 15 or fewer; flowers less than 10 cm **22. bruchii**
23a. Stems slender cylindric, creeping **31. chamaecereus**
b. Stem depressed-spherical to short cylindric 24
24a. Stamens bunched round the style **24. maximiliana**
b. Stamens not bunched round the style 25
25a. Flowers c. 4 cm; perianth-limb shorter than the tube; fruit spiny **23. tegeleriana**
b. Flowers exceeding 4 cm, or the perianth-limb equalling or exceeding the tube; fruit not spiny 26
26a. Ribs scolloped-tubercled between the areoles, the tubercles acute, hatchet-shaped, often more or less offset 27
b. Ribs with low rounded tubercles 28
27a. Ribs apparent; flowers relatively long-tubed, often with white throat **25. pentlandii**
b. Ribs divided into acute tubercles; flowers relatively short-tubed, throat various **26. cinnabarina**

28a. Flowers with dark throat **28. chrysantha**
b. Flowers not dark-throated 29
29a. Pericarpel more than 5 mm diameter; scales on tube numerous, densely hairy 30
b. Pericarpel less than 5 mm diameter; scales on tube distant, sparsely hairy **30. saltensis**
30a. Stem short-cylindric **27. aurea**
b. Stem depressed-spherical to ovoid **29. kuehnrichii**

1. E. spiniflora (Schumann) Berger (*Acanthocalycium spiniflorum* (Schumann) Backeberg; *A. klimpelianum* (Weidlich & Werdermann) Backeberg; *A. peitscherianum* Backeberg; *A. violaceum* (Werdermann) Backeberg; *Lobivia violacea* (Werdermann) Berger).
Illustration: Hecht, BLV Handbuch der Kakteen, 218 (1982); Haustein, Der Kosmos-Kakteenführer, 165 (1983).
Spherical to short cylindric, eventually to 60 × 15 cm, pale to dark green; ribs c. 16–20, acute; areoles c. 1.5–2 cm apart. Spines 10–20, to 4 cm, needle-like, flexible, pale yellowish to brown, often darker tipped. Flowers borne close to the stem apex, erect, tubular-bell-shaped, 4–5 × 4 cm, pericarpel and tube green, with dense, narrow, spine-tipped brownish scales to c. 1 cm, sparsely hairy in their axils; limb pale mauve, pink or white; stamens in several series, incurved, with no separate throat-circle; staminodial plug whitish-woolly, 3–4 mm above the ovary; stigmas 6 mm, pale green. Fruit spherical, c. 10 mm, hard, with persistent scales, splitting vertically when ripe; seeds ovoid, dull brown to black, finely papillate, with small sunken basal hilum. *W Argentina (Córdoba).* G2.

2. E. thionantha (Spegazzini) Werdermann (*Acanthocalycium thionanthum* (Spegazzini) Backeberg; *A. chionanthum* (Spegazzini) Backeberg; *A. aurantiacum* Rausch; *A. glaucum* Ritter; *A. variiflorum* Backeberg).
Illustration: Hecht, BLV Handbuch der Kakteen, 218 (1982); Cullmann et al., Kakteen, edn 5, 124 (1984).
Spherical to short-cylindric, eventually 12 × 10 cm; ribs 9–15, rounded, dark green to grey- or blue-green. Central spines absent or 1–4, 5–20 mm, subulate, at first brown or black, later grey or horn-coloured; radial spines 5–10. Flowers to 4.5 × 4.5 cm, pericarpel and tube with dark, spine-tipped scales with brown to white hairs and bristles in their axils; limb bright yellow to red or white; stamens

erect, the uppermost forming a separate throat-circle; staminodial plug brownish; style pale yellow, stigmas red. Fruit and seed as in *E. spiniflora*. *NW Argentina (Salta)*. G2.

3. E. mirabilis Spegazzini (*Arthrocereus mirabilis* (Spegazzini) Marshall; *Setiechinopsis mirabilis* (Spegazzini) De Haas). Illustration: Backeberg, Cactus Lexicon, f. 386, 387 (1977); Hecht, BLV Handbuch der Kakteen, 361 (1982); Haustein, Der Kosmos-Kakteenführer, 189 (1983); Cullmann et al., Kakteen, edn 5, 292 (1984).

Usually simple, cylindric, to 15×2.5 cm, dark brownish green; ribs *c.* 11, straight, areoles small, *c.* 4–5 mm apart. Radial spines 9–14, 3–5 mm, bristly, whitish; central spine 1, 1–1.5 cm, stout, straight-projecting, dark brown or blackish, minutely velvety-downy. Flowers borne near apex, opening at night, tubular-funnel-shaped (salver-shaped), $6–12 \times 3–5$ cm, pericarpel and tube greenish with narrow, bristle-tipped scales, these with woolly hairs and sometimes bristles in their axils; perianth-segments narrow, acuminate, white. Fruit spindle-shaped, to $3–4$ cm $\times 6$ mm, dark green, becoming dry and splitting laterally; seeds spherical, 1.5 mm, dark brown. *Argentina*. G2.

4. E. chiloensis (Colla) Friedrich & Rowley (*Trichocereus chiloensis* (Colla) Britton & Rose; *E. litoralis* (Johow) Friedrich & Rowley). Illustration: Britton & Rose, The Cactaceae 2: f. 198–200 (1920); Haustein, Der Kosmos-Kakteenführer, 133 (1983). Stems stout-cylindric, to 8 m $\times 10–12$ cm, branching near base; ribs usually 16–17, low and broad, tuberculate. Central spine 1, 4–7 (rarely to 12) cm, straight-projecting; radial spines 8–12, 1–2 (rarely to 4) cm, pale yellow at first, brown-tipped, later grey. Flowers 14 cm, white. *Chile*. G2.

E. coquimbana (Molina) Friedrich & Rowley (*Trichocereus coquimbanus* (Molina) Britton & Rose). Shrubby, stems to *c.* 1 m $\times$ 7–8 cm; ribs 12–13. Spines *c.* 20, formidable, often 7–8 cm. Flowers *c.* 10 cm; scales of pericarpel and tube with black axillary hairs. *Chile (Coquimbo)*. G2.

E. deserticola (Werdermann) Friedrich & Rowley (*E. fulvilana* (Ritter) Friedrich & Rowley; *Trichocereus fulvilanus* Ritter). Illustration: Cullmann et al., Kakteen, edn 5, 307 (1984). Stems 1–1.5 m $\times$ 4–7 cm, branching below; ribs 8–12, deeply incised, furrowed above the areoles; areoles to 1.5 cm apart. Central spines 1–4, to 12 (rarely to 18) cm; radial spines irregular,

9–25, 1–3 cm, dark brown at first, later grey. Flowers near apex, $7–12 \times 7–9$ cm, fragrant, pure white, with dark wool outside. Fruit spherical, *c.* 4 cm in diameter; seeds matt black, 1.2 mm. *Chile (near Taltal)*. G2.

Perhaps not distinct from *E. coquimbana*.

5. E. pachanoi (Britton & Rose) Friedrich & Rowley (*Trichocereus pachanoi* Britton & Rose). Illustration: Backeberg, Die Cactaceae 2: f. 1072–1074 (1959). Plants tall, 3–6 m, with numerous erect branches, slightly glaucous when young; ribs 6–8, broad, rounded, with a deep horizontal depression above each areole. Spines few, sometimes absent, 3–7, the longest 1–2 cm, dark yellow to brown. Flowers borne near tips of branches, 19–23 cm, nocturnal, fragrant, white; pericarpel and tube with black hairs. *Ecuador*. G2.

E. lageniformis (Foerster) Friedrich & Rowley. Tall shrub to 5 m, stems to 15 cm in diameter; ribs 4–8. Spines 2–6, unequal, the longest to 10 cm. Flowers 18 cm. *Bolivia*. G2.

E. macrogona (Salm-Dyck) Friedrich & Rowley (*Trichocereus macrogonus* (Salm-Dyck) Riccobono). Illustration: Haustein, Der Kosmos-Kakteenführer, 131 (1983). Stems cylindric, to 3 m $\times$ 9 cm, bluish; ribs 6–9, usually 7, low and rounded; areoles large, 1.5–2 cm apart. Spines several, brown, the centrals 1–3, to 5 cm, the radials 6–9, to *c.* 2 cm. Flowers 18 cm, white. *Origin unknown*. G2. Modern descriptions (quoted here) deviate from the original, and the name may now be misapplied. As now understood, *E. macrogona* is said to be a good grafting stock.

E. peruvianus (Britton & Rose) Friedrich & Rowley. Large shrub to 4 m; stems 15–20 cm in diameter, glaucous when young; ribs 6–8; areoles 2–2.5 cm apart. Spines *c.* 10, unequal, to 4 cm, brown. *Peru (Matucana)*. G2.

6. E. terscheckii (Pfeiffer) Friedrich & Rowley (*Trichocereus terscheckii* (Pfeiffer) Britton & Rose; *E. gigantea* Meyer). Illustration: Britton & Rose, The Cactaceae 2: f. 203, 204 (1920). Stems columnar, reaching 10–12 m and with a trunk to 45 cm in diameter, branching well above ground level; branches 10–20 cm in diameter; ribs 8–18, obtuse, 2–4 cm high; areoles 1.5–3 cm apart, large, 1–1.5 cm in diameter, with dense chestnut-brown wool. Spines 8–15, more numerous at old areoles, 1–7 cm or

more, yellowish or brownish. Flowers bell-shaped to funnel-shaped, $15–20 \times 12.5$ cm; pericarpel and tube green, scales 5 mm with dense white or brownish axillary hairs; outer perianth-segments dark red to greenish, inner white. Fruit spherical, 3–5 cm, green, with acute scales woolly in the axils; seeds $1.2–1.5 \times 1–1.1$ mm, dark brown or almost black, somewhat warty-tessellate. *N Argentina*. G2.

The name is nowadays used in the sense adopted by Britton & Rose, but is probably misapplied.

E. pasacana (Ruempler) Friedrich & Rowley (*Helianthocereus pasacana* (Ruempler) Backeberg; *Trichocereus pasacana* (Ruempler) Britton & Rose). Less branched; ribs more than 20. Spines more numerous and more flexible. *N Argentina*. G2. Perhaps only a higher altitude variant of *E. terscheckii*.

Related Bolivian species include *E. tacaquirensis* (Vaupel) Friedrich & Rowley (*Trichocereus tacaquirensis* (Vaupel) Cardenas), *E. taquimbalensis* (Cardenas) Friedrich & Rowley (*T. taquimbalensis* Cardenas) and *E. werdermannianus* (Backeberg) Friedrich & Rowley (*T. werdermannianus* Backeberg).

7. E. candicans (Salm-Dyck) D. Hunt (*Trichocereus candicans* (Salm-Dyck) Britton & Rose; *T. courantii* (Schumann) Backeberg). Illustration: Backeberg, Die Cactaceae 2: f. 1085–1088, 1090, 1091 (1959). Clustering to form large clumps 1–3 m in diameter, individual stems to 60 $\times$ 14 cm, light green; ribs 9–11, low, obtuse; areoles 2–5 cm apart, $1.2–1.5 \times 1$ cm, white-felted. Spines subulate, brownish yellow, the centrals usually 4, 3–11 cm, the radials 9–12, 2–6 cm, spreading. Flowers tubular-funnel-shaped, $18–23 \times 11–19$ cm, white, fragrant. Fruit spherical to ellipsoid, *c.* 5 cm diameter, splitting down one side; seeds black. *W Argentina*. G1.

The name *E. lamprochlorus* (Lemaire) Friedrich & Glaetzle (*Trichocereus lamprochlorus* (Lemaire) Britton & Rose), based on *Cereus lamprochlorus* Lemaire, has been much confused and is best discarded. The plant usually cultivated as *E. lamprochlorus* is similar to *E. candicans*.

8. E. spachiana (Lemaire) Friedrich & Rowley (*Cereus spachianus* Lemaire; *Trichocereus spachianus* (Lemaire) Riccobono). Illustration: Hecht, BLV Handbuch der Kakteen, 368 (1982); Haustein, Der Kosmos-Kakteenführer, 133 (1983).

Stems to 2 m × *c.* 6 cm, branching from the base; ribs 10–15, low; areoles *c.* 1 cm apart, *c.* 3 mm. Spines straight, reddish yellow at first, later whitish, the central 1, 1–2 cm, the radials 8–10, 4–10 mm. Flowers 18–20 × 15 cm, white, tube very hairy. *Argentina.* G1.

Kiesling (in Darwiniana **21**: 291, 1978) unites *T. spachianus* with *T. santiaguensis* (Spegazzini) Backeberg, a tree-like species with a trunk 4–7 m × 16–20 cm, but the latter is not the plant well-known in cultivation as *T. santiaguensis* and popular as a grafting stock.

9. E. strigosa (Salm-Dyck) Friedrich & Rowley (*Trichocereus strigosus* (Salm-Dyck) Britton & Rose). Illustration: Britton & Rose, The Cactaceae **2**: f. 211 (1920); Backeberg, Die Cactaceae **2**: f. 1095 (1959).
Shrubby, stems to 60 × 5–6 cm, branching from the base to form clumps 1 m across; ribs 15–18, low; areoles *c.* 5 mm apart. Spines numerous, all similar, whitish to yellow or reddish brown and darker-tipped, the centrals 4, to 7 cm, radials 9–16, 1–5 cm. Flowers opening at night, funnel-shaped, *c.* 20 × 5 cm, white to salmon-pink. Fruit spherical, 4–6.5 cm, yellow or orange, fleshy; seeds 1.5 × 1 mm, glossy black. *W Argentina.* G1.

10. E. schickendantzii Weber (*Trichocereus schickendantzii* (Weber) Britton & Rose). Illustration: Marshall & Bock, Cactaceae, f. 43 (1941).
Stems oblong to cylindric, 15–25 × up to 6 cm, light green, offsetting at base to form several-headed clumps; ribs 14–18, low, somewhat acute; areoles close or touching. Spines 5–10 mm, slender, yellowish, radials 9, centrals 4, more numerous later. Flowers tubular-funnel-shaped, 20–22 cm, white, not fragrant, scales on tube with dense blackish axillary hairs. Fruit edible. *W Argentina (Tucuman).* G1.

A popular grafting-stock.

11. E. thelegona (Schumann) Friedrich & Rowley (*Trichocereus thelegonus* (Schumann) Britton & Rose). Illustration: Backeberg, Die Cactaceae **2**: f. 1096, 1097 (1959); Haustein, Der Kosmos-Kakteenführer, 131 (1983).
Stems prostrate or decumbent, to 2 m × 7–8 cm, little branched; ribs *c.* 13, low, divided into almost hexagonal tubercles; areoles at the apices of the tubercles, *c.* 8–10 mm apart, 2 mm in diameter. Spines pale honey-yellow at first, tipped brown; central spine 1, 2–4 cm, protruding; radial spines 6–7, lowest

longest, to 1.2 cm. Flowers funnel-shaped, *c.* 20 × 15 cm, scales of tube with reddish axillary bristles and hairs. Fruit spherical to ovoid, 5 cm diameter, tuberculate, yellowish to red, splitting laterally; seeds black. *W Argentina (Tucuman).* G1.

E. thelegonoides (Spegazzini) Friedrich & Rowley (*Trichocereus thelegonoides* (Spegazzini) Britton & Rose). Similar to *E. thelegona*, but stems thicker, more branched; ribs 15–16, not tuberculate but furrowed between the areoles. Central spines 4, 7–15 mm; radials *c.* 11, 5–7 mm. *N Argentina (Jujuy).* G1.

12. E. huascha (Weber) Friedrich & Rowley (*Lobivia huascha* (Weber) Marshall; *L. andalgalensis* (Schumann) Britton & Rose; *Trichocereus andalgalensis* (Schumann) Hosseus; *E. andalgalensis* invalid; *L. grandiflora* Britton & Rose; *Helianthocereus grandiflorus* (Britton & Rose) Backeberg; *Trichocereus grandiflorus* invalid; *E. pecheretianus* (Backeberg) Friedrich & Rowley; *E. rowleyi* Friedrich). Illustration: Haustein, Der Kosmos-Kakteenführer, 133 (1983); Cullmann et al., Kakteen, edn 5, 307 (1984).
Stems to 1 m × 4–5 (rarely to 10) cm, ascending or decumbent, branching from near the base to form clumps; ribs usually 14–17, low; areoles *c.* 1 cm apart. Central spines 1–3, 2–7 cm, somewhat thicker than the 9–11 radials, 1.5 cm, dark yellow to brown. Flowers funnel-shaped-bell-shaped, *c.* 10 × 6–7 cm; pericarp and tube with brownish green acute scales densely hairy in the axils; perianth-segments red, orange or yellow. Fruit spherical or ovoid, *c.* 3 cm in diameter, yellowish green or reddish; seeds almost kidney-shaped, 1.4 × 1 mm, black. *Argentina (Catamarca, La Rioja).* G1.

13. E. eyriesii (Turpin) Pfeiffer & Otto (*Echinocactus eyriesii* Turpin; *Echinopsis turbinata* (Pfeiffer) Pfeiffer & Otto). Illustration: Annales de L'Institut Royal Horticole de Fromont **2**: t. 2 (1830); Botanical Magazine, 3411 (1835); Haustein, Der Kosmos-Kakteenführer, 155 (1983).
Spherical at first, eventually short-cylindric, clustering from near the base, main stem to 15–30 × 10–15 cm; ribs 11–18, areoles *c.* 1 cm apart. Spines 12–15, very short, hard, blackish. Flowers 20–25 × 5–10 cm, white. *S Brazil to N Argentina (Entre Rios).* G1. Summer.

E. calochlora Schumann. Illustration: Haustein, Der Kosmos-Kakteenführer, 155 (1983). Smaller than the above, shiny pale

green; ribs *c.* 13. Central spines 3–4, slightly thicker and darker than the radials; radial spines 14–20, 5–10 mm, straw yellow. Flowers 16 × 10 cm, white. *Brazil (Goias).* G2. Summer.

E. tubiflora (Pfeiffer) Schumann (*E. grandiflora* Linke). Illustration: Haustein, Der Kosmos-Kakteenführer, 157 (1983). Dark green; ribs 10–12, prominent. Central spines 1–3, 1–2 cm, blackish; radial spines 7–9, shorter and weaker. Flowers 20–24 × 10 cm, white. *Argentina (Tucuman, Catamarca & Salta).* G1. Summer.

E. silvestrii Spegazzini. Like *E. tubiflora* but smaller and with greyish spines and broader perianth-segments. *Argentina (Tucuman, Salta).* G1.

14. E. oxygona (Link) Pfeiffer & Otto (*Echinocactus oxygonus* Link; *Cereus multiplex* Pfeiffer; *E. multiplex* (Pfeiffer) Pfeiffer & Otto). Illustration: Botanical Magazine, 3789 (1840), 4162 (1845); Britton & Rose, The Cactaceae **3**: pl. 6 f. 2 (1922).
Clustering, stems spherical, eventually 25–30 × 12–15 cm; ribs 13–15. Spines 13–15, to 2.5 cm, yellowish brown. Flowers *c.* 25 × 10 cm, tube reddish green, limb pink. *S Brazil, N Argentina.* G1. Summer.

15. E. rhodotricha Schumann. Illustration: Guerke, Bluehende Kakteen **2**: pl. 76 (1905).
Becoming short-columnar; ribs 8–13, areoles 1.5–2.5 cm apart. Central spine 0–1, to 2.5 cm; radial spines 4–7, *c.* 2 cm, yellowish, tipped brown. Flowers 15 cm, white. Stamens numerous with no separate throat-circle. *Paraguay, NE Argentina.* G1. Summer.

16. E. leucantha (Salm-Dyck) Walpers (*E. campylacantha* Pfeiffer; *E. shaferi* Britton & Rose; *E. spegazziniana* Britton & Rose). Illustration: Botanical Magazine, 4567 (1851); Britton & Rose, The Cactaceae **3**: f. 88 & pl. 7 f. 2 (1922).
Spherical at first, becoming cylindric and eventually 30–150 × 10–15 cm, usually simple; ribs 10–14. Central spine 1, to 10 cm, curved and sometimes almost hooked, dark brown; radial spines 7–8, to 2 cm, curved, brownish. Flowers 15–17 × 10 cm, white or pale pink. Stamens numerous with no separate throat-circle. *Argentina (Salta to Buenos Aires and Rio Negro).* G1.

E. comarapana Cardenas (*E. pereziensis* Cardenas; *E. ayopayana* Ritter & Rausch). Illustration: National Cactus & Succulent Journal **12**: 61 (1957). Smaller than

E. leucantha; short-cylindric, 10–15 × 4–8 cm; ribs 8–12; areoles 5–14 mm apart. Central spines absent or 1–4, 1.5–2 (rarely to 5) cm; radial spines *c.* 9–17, 5–11 mm. Flowers 13–15 cm, white. Fruit spherical, *c.* 3 cm. *Bolivia*. G1.

17. E. obrepanda (Salm-Dyck) Schumann (*E. cristata* Salm-Dyck). Illustration: Botanical Magazine, 4521 (1850), 4687 (1852).

Depressed-spherical, to 20 cm in diameter, later offsetting; ribs *c.* 13–16, acute, areoles *c.* 2 cm apart. Central spines 1–3, to 4–5 cm, bent at tip; radial spines 9–11, to 1 cm, mostly comb-like. Flowers 10–20 cm, white, parsley-scented, tube somewhat curved, outer segments spreading, inner upcurved, broadly obtuse and often mucronate. Fruit spherical, semi-dry; seeds spherical to ellipsoid, black. *Bolivia (Cochabamba)*. G1. Summer.

Var. **fiebrigii** (Guerke) Friedrich (*E. fiebriigii* Guerke) has flowers pleasantly scented (not of parsley) and the tube straight.

E. calorubra Cardenas. To 6–7 × 14 cm; ribs 16. Central spine 1, to 2.5 cm; radial spines 9–13, somewhat curved. Flowers to 15 cm, inner segments orange-red above, bluish pink towards base. *Bolivia (Valle Grande)*.

E. toralapana Cardenas. To 4 × 16 cm; ribs *c.* 13. Spines 6–10, to 5 cm, comb-like, curved. Flowers to 14 cm, red. *Bolivia (Arani)*.

Pseudolobivia carmineoflora Hoffmann & Backeberg invalid, may belong here. The plant described was depressed-spherical, to 4 × 7 cm, with 14 ribs. Central spines 2–4, to 3 cm; radial spines 10–12, to 2.6 cm. Flowers *c.* 7.5–10 × 6 cm, carmine, throat-membrane green. *Bolivia (Cochabamba)*.

E. frankii and *E. millarensis* are other invalid names which may belong under *E. obrepanda*.

18. E. mamillosa Guerke. Illustration: Backeberg, Die Cactaceae 2: f. 1254 (1959); Rausch, Lobivia 2: 89–91 (1976). Simple, rarely clustering, broadly spherical, to 12 × 10–25 cm; ribs straight, 17–32, tubercles rounded, areoles to 1.2–2 cm apart. Central spines 1–4, to 1 cm, yellowish, brown above; radial spines 8–10 (rarely to 12), 5–10 mm. Flowers near crown, 13–18 × to 8 cm, white, sometimes tinged pink. Fruit spherical, *c.* 2 cm, semi-dry; seed elongated-spherical, *c.* 1 mm, dull black, hilum with caruncle. *S Bolivia (Tarija)*. G1.

Var. **ritteri** (Boedeker) Ritter. Stem 10–25 cm in diameter; ribs 18–32. Central spines 3–8; radial spines 12–15, 1–2.5 cm. *Bolivia (Tarija)*.

Var. **kermesina** (Krainz) Friedrich (*E. kermesina* Krainz). Hemispherical, to over 15 cm in diameter; ribs 15–23. Central spines 4–6, to 2.5 cm; radial spines 11–16, to 1.2 cm. Flowers to *c.* 18 × 9 cm, carmine or paler, unscented. *N Argentina (Jujuy)*.

19. E. ancistrophora Spegazzini (*E. leucorhodantha* Backeberg; *E. pelecyrhachis* Backeberg). Figure 29(10), p. 205. Illustration: Haustein, Der Kosmos-Kakteenführer, 157 (1983). Simple or offsetting, flattened spherical, to 8 cm in diameter; ribs 15–20, straight, divided into numerous small tubercles. Central spines 0–1 (rarely to 4), 1–2 cm, often 1 or more bent or hooked; radial spines 3–10, 5–15 mm, recurved and spreading. Flowers lateral, 10–16 cm, white, sometimes tinged pink outside, unscented, tube slender. Fruit elliptic-ovoid, 16 × 8 mm, dirty greenish purple, semi-dry; seeds spherical to ellipsoid, slightly constricted above the hilum, black, with no aril. *W Argentina*. G1.

E. hamatacantha Backeberg. To 7 × 15 cm, ribs to 27. Spines 8–15, 4–12 mm, 1 curving to crown and bent or hooked. Flowers 20 cm, white, scented. *Argentina (Salta)*.

E. kratochviliana Backeberg. Spines 15–18, with 1–2 (rarely to 4) centrals, to 5 cm. Flower at most 5 cm, white, with rather plentiful blackish hairs. *Argentina (Salta)*.

E. polyancistra Backeberg. Ribs 17–30. Spines numerous, bristle-like. Flowers to 10 cm, white, scented. *Argentina (Salta)*.

E. arachnacantha (Buining & Ritter) Friedrich. Illustration: Cullmann et al., Kakteen, edn 5, 164 (1984). Flattened spherical, to 2 × 4 cm; ribs 14. Central spine 1, 5 mm, black, upcurved; radial spines 9–15, 5 mm, pale brown at first, later whitish, resembling spiders' legs. Flowers slender-tubed, yellow to orange. *Bolivia (Samaipata)*.

E. torrecillasensis Cardenas. Flat, 1–2 cm, with thick tap-root; ribs 16. Central spine 1, to 1 cm; radial spines 6–7, to 1 cm, all spines curved. Flowers 8 cm, red to salmon-red. *Bolivia (Santa Cruz)*.

20. E. ferox (Britton & Rose) Backeberg (*Lobivia ferox* Britton & Rose). Illustration: Britton & Rose, The Cactaceae 3: f. 63 (1922); Cullmann et al., Kakteen, edn 5, 165 (1984).

Simple, to 20 × more than 30 cm; ribs to 30, spiralled; areoles *c.* 3 cm apart. Central spines 3–4, to 15 cm, upcurving; radial spines 10–12, to 6 cm. Flowers 7–10 cm, white or rarely pink. Fruit spherical, relatively thin-skinned, splitting irregularly, with watery sap; seeds pear-shaped, brown-black. *Bolivia to N Argentina*. G1.

E. longispina (Britton & Rose) Backeberg (*E. ducis-pauli* misapplied; *Lobivia ducis-pauli* misapplied; *E. lecoriensis* Cardenas). Spherical to short-cylindric, to 25 cm in diameter; ribs 25–50. Spines to 15, over 8 cm, hooked on new growth, yellowish to brown. Flowers to 10 cm, white, yellow, orange, pink or red. *N Argentina (Jujuy)*.

E. nigra Backeberg becomes columnar, to 30 cm, with 12–14 spines, dark to blackish and is only a variant of *E. longispina*.

21. E. tarijensis (Vaupel) Friedrich & Rowley (*Trichocereus tarijensis* (Vaupel) Werdermann; *T. poco* Backeberg; *Helianthocereus poco* (Backeberg) Backeberg). Illustration: Hecht, BLV Handbuch der Kakteen, 368 (1982); Cullmann et al., Kakteen, edn 5, 308 (1984).

Stems columnar, to 5 m × 40 cm, simple or sometimes branched; ribs about 25; areoles large, close-set, densely felted. Spines 50 or more, 1–8 cm, pale brown to whitish. Flowers arising near the stem-apex, funnel-shaped, *c.* 12 × 9 cm; pericarpel and tube *c.* 8 cm, with acute scales *c.* 1 cm, with dense white or brown axillary hairs; outer perianth-segments brownish green, inner usually red, sometimes pink or creamy white. Fruit ovoid, 3.5–5 × 2–3 cm, green, covered with scales and hairs; pulp white; seeds almost kidney-shaped, 1.3–1.5 mm, black. *S Bolivia, N Argentina (Jujuy)*. G1.

22. E. bruchii (Britton & Rose) Friedrich & Glaetzle (*Lobivia bruchii* Britton & Rose; *Soehrensia bruchii* (Britton & Rose) Backeberg; *L. grandis* Britton & Rose; *S. grandis* (Britton & Rose) Backeberg; *L. kieslingii* Rausch; *E. korethroides* Werdermann). Illustration: Hecht, BLV Handbuch der Kakteen, 294 (1982); Haustein, Der Kosmos-Kakteenführer, 165 (1983).

Usually simple, depressed-spherical, reaching 50 cm in diameter, glossy dark green; ribs to 50 or more, somewhat inflated between the areoles. Spines *c.* 12–15, to 3 cm, yellowish or brown. Flowers 4–5 × 5 cm, deep red; tube short, axils of scales filled with woolly hairs. *Argentina (Tucuman)*. G1. Spring–Summer.

E. formosa (Pfeiffer) Salm-Dyck (*Soehrensia formosa* (Pfeiffer) Backeberg). Eventually 2 m × 50 cm; ribs 15–35, pale greyish green. Central spines 2–4, to 4 cm; radial spines 8–10. Flowers 6–8 cm, light to golden yellow. *Argentina (Mendoza)*.

E. walteri (Kiesling) Friedrich & Glaetzle (*Lobivia walteri* Kiesling). Stems to 16 × 16 cm, clustering; ribs *c.* 11. Spines to 15 or more, 1–2.5 cm, yellow. Flowers bell-shaped-funnel-shaped, 7–9.5 cm, yellow. Fruit spherical, 2–2.5 cm, yellow-green. *Argentina (Salta)*.

23. E. tegeleriana (Backeberg) D. Hunt (*Lobivia tegeleriana* Backeberg; *Acantholobivia tegeleriana* (Backeberg) Backeberg). Illustration: Rausch, Lobivia 1: 9, 10 (1975).
Spherical, rarely offsetting, with pronounced tap-root; ribs *c.* 16, divided into tubercles; areoles to 1.7 cm apart, oblong. Spines 12, 6 mm, mostly more or less curved, horn-coloured, tipped darker, the longest sometimes hooked. Flowers 4 cm, throat pinkish orange, perianth short, varying from red to orange or yellow. Fruit 2.5 cm in diameter, usually spiny, green, pulp watery at first, later sticky. *C Peru*. G1.

24. E. maximiliana Heyder (*Lobivia maximiliana* (Heyder) Backeberg; *L. westii* Hutchison). Illustration: Hecht, BLV Handbuch der Kakteen, 290 (1982) – as Lobivia chrysochete, 294 (1982); Haustein, Der Kosmos-Kakteenführer, 159 (1983); Cullmann et al., Kakteen, edn 5, 198 (1984).
Clustering, depressed-spherical to obovoid, *c.* 7.5 × 5 cm; ribs *c.* 17, acute, with hatchet-shaped tubercles between the areoles; areoles *c.* 2 cm apart. Spines very variable, central absent or 1, to 7 cm, upcurved, radials 4–12, to 3–5 cm, curved, unequal, brownish yellow. Flowers 5–8 cm, scarlet with orange-yellow throat, sometimes the inner perianth-segments tipped darker; innermost perianth-segments somewhat reduced, more or less erect; stamens bunched round the style. Fruit 1.2 cm in diameter, reddish green, hairy. *S Peru, N Bolivia*. G1.

25. E. pentlandii (Hooker) Salm-Dyck (*Echinocactus pentlandii* Hooker; *Lobivia pentlandii* (Hooker) Britton & Rose; *L. higginsiana* Backeberg; *L. larae* Cardenas). Illustration: Botanical Magazine, 4124 (1844); Haustein, Der Kosmos-Kakteenführer, 159 (1983); Cullmann et al., Kakteen, edn 5, 191 (1984).

Freely clustering, stems spherical to obovoid; ribs *c.* 12–15, deeply scolloped between the areoles, broken into long acute hatchet-shaped tubercles; areoles *c.* 2 cm apart. Spines very variable, centrals 0–1, 3–9 cm, radials 5–15, to 4 cm, yellowish brown, somewhat recurved. Flowers usually 4–6 cm; tube relatively stout, *c.* 1 cm in diameter; limb very variable in colour, purplish pink, red, orange or yellow, often with paler throat. Fruit nearly spherical, 1–1.2 cm in diameter. *S Peru, N Bolivia*. G1. Spring-Summer.

E. backebergii Werdermann (*Lobivia backebergii* (Werdermann) Backeberg; *L. wrightiana* Backeberg; *L. winteriana* Ritter). Illustration: Cullmann et al., Kakteen, edn 5, 201 (1984). Simple or clustering, spherical to obovoid, *c.* 4–5 cm in diameter; ribs *c.* 15, incised above the areoles, 3–4 mm high, areoles *c.* 1–1.5 cm distant. Spines all radial, 3–7 (usually 5), at first reddish brown, later grey, 5–50 mm, thin, unequal, sometimes hooked at tip. Flowers 4–9 cm; tube relatively slender, *c.* 6–7 mm in diameter; limb pale to dark carmine-red or violet, usually with whitish throat. Fruit semi-dry, splitting horizontally; seeds relatively large, finely rough. *E Bolivia, S Peru*. G1.

E. hertrichiana (Backeberg) D. Hunt (*Lobivia binghamiana* Backeberg; *L. allegraiana* Backeberg; *L. incaica* Backeberg; *L. planiceps* Backeberg; *Neolobivia divaricata* Ritter; *L. divaricata* invalid). Illustration: Backeberg, Die Cactaceae 3: f. 1366 (1959); Cullmann et al., Kakteen, edn 5, 196 (1984). Simple at first, later clustering, individual stems to 10 cm in diameter; ribs 11–22, straight or oblique, rounded to acute. Spines variable, *c.* 7–20, to 3 cm. Flowers *c.* 5 × 4–6 cm, bright red. *SE Peru*. G1.

E. pugionacantha Rose & Boedeker (*Lobivia pugionacantha* (Rose & Boedeker) Backeberg; *L. culpinensis* Ritter; *E. culpinensis* invalid). Illustration: Haustein, Der Kosmos-Kakteenführer, 161 (1983); Cullmann et al., Kakteen, edn 5, 199 (1984). Simple or clustering, with a long tap-root; ribs 15–30. Spines variable, typically 4–7, to 2.5 cm or more, flattened, dagger-like. Flowers 4.5–6 cm, reddish yellow. *Bolivia*. G1.

26. E. cinnabarina (Hooker) Labouret (*Echinocactus cinnabarinus* Hooker; *Lobivia cinnabarina* (Hooker) Britton & Rose; *L. oligotricha* Cardenas). Illustration: Botanical Magazine, 4326 (1847); Haustein, Der Kosmos-Kakteenführer, 159 (1983).

Simple, depressed-spherical, to 15 cm in diameter; ribs *c.* 20, irregular and oblique, divided into acute tubercles. Central spines 2–3, somewhat curved; radial spines 8–12, 6–12 mm, slender, more or less recurved. Flowers from crown of plant, 4 cm in diameter, scarlet. *Bolivia*. G1. Spring–Summer.

27. E. aurea Britton & Rose (*Pseudolobivia aurea* (Britton & Rose) Backeberg; *Lobivia cylindrica* Backeberg). Illustration: Britton & Rose, The Cactaceae 3: pl. 10 f. 1 (1922); Haustein, Der Kosmos-Kakteenführer, 155 (1983).
Simple or clustering, spherical to ellipsoid, to 10 cm tall; ribs straight, continuous, 11–15. Central spine 1 (rarely to 4), to 3 cm; radial spines 7–10, to 1 cm. Flowers 5.5–9 × 4–8 cm, lemon-yellow, deeper yellow inside. *Argentina (Córdoba)*. G1.

Lobivia shaferi Britton & Rose (not *Echinopsis shaferi* Britton & Rose) may belong here. It was described as clustering, becoming cylindric, 7–15 × 2.5–4 cm, with *c.* 10 ribs; central spines several, to 3 cm; radial spines 10–15, to 1 cm, white or brown; buds very hairy; flowers 4–6 × 3–4 cm, bright yellow, the tube stout, scales with long white hairs; style greenish white; stigmas cream. *Argentina (Catamarca)*. G1.

28. E. chrysantha Werdermann (*Lobivia chrysantha* (Werdermann) Backeberg; *L. vatteri* Krainz; *L. glauca* Rausch). Illustration: Hecht, BLV Handbuch der Kakteen, 292 (1982); Haustein, Der Kosmos-Kakteenführer, 163 (1983).
Simple, depressed-spherical to short-cylindric, *c.* 4–6 × 6–7 cm, from a carrot-like tap-root; ribs 8–26, continuous, straight or slightly oblique, 6–7 mm high; areoles *c.* 1.5 cm apart. Spines all radial, 5–7 (rarely 3), to 2 cm, reddish brown at first, later dingy grey, spreading laterally. Flowers to 5 cm; pericarpel *c.* 5 mm in diameter; perianth yellow to orange with darker throat; style and stigmas purple. *Argentina (Salta, Jujuy)*. G1.

E. marsoneri Werdermann (*Lobivia jajoiana* Backeberg; *L. muhriae* Backeberg). Illustration: Cullmann et al., Kakteen, edn 5, 197 (1984). Central spines 2–5, to 5 cm; radial spines 8–12, *c.* 3 cm. Flowers *c.* 5.5 × 6 cm, very variable, typically golden yellow with orange-brown to dark purplish throat, also pink, orange or red in other variants. *N Argentina (Jujuy)*. G1.

Probably not distinct from *E. chrysantha*.

29. E. kuehnrichii (Fric) Friedrich & Glaetzle (*Lobivia kuehnrichii* Fric; *E. densispina* Werdermann; *L. densispina* (Werdermann) Wessner; *L. nigrispina* Backeberg). Illustration: Cullmann et al., Kakteen, edn 5, 192, 193, 195 (1984). Simple, ovoid, 8 × 5.5 cm; ribs *c.* 17, 4–7 mm high, continuous but grooved between the areoles; areoles *c.* 5 mm apart. Central spines *c.* 4–7, to 1.5–2 cm, dark, tipped black; radial spines *c.* 16–22, horizontal, to 8 mm. Flowers *c.* 8.5 cm, inner perianth-segments *c.* 4.5 × 1 cm, typically golden yellow, sometimes orange, apex truncate-mucronate; stigmas green. *N Argentina (Jujuy)*.

Very variable. *E. densispina* was confused by Backeberg with *Rebutia famatimensis* which is superficially similar but does not have the staminal throat-circle characteristic of *Echinopsis*. *E. leucomalla* (Wessner) Friedrich is also similar to *E. kuehnrichii*, but is regarded as a close ally or variant of *E. aurea*.

E. schieliana (Backeberg) D. Hunt (*Lobivia schieliana* Backeberg). Illustration: Backeberg, Die Cactaceae 2: f. 1438 (1959); Ritter, Kakteen in Südamerika 2: f. 489–491, 586 (1980). Spherical, 3–6 cm in diameter, or becoming oblong-cylindric, dark or blackish green; ribs 13–21, straight or twisted, 3–5 mm high; areoles 3–6 mm apart. Spines white to yellowish or brown, centrals absent or 1–4, similar to the radials but often darker, radials 8–12, 2–15 mm. Flowers 3.5–4 cm; pericarpel 3–6 × 3–6 mm, tube 1.6–1.8 cm, stamens violet-red, perianth-segments 1.7–1.9 cm, violet-red with orange or brownish orange midstripe. Fruit spherical, 1 cm, blackish green, pulp juicy, white; seeds 1.5 × 1 mm, black. *Bolivia (La Paz)*. G1.

30. E. saltensis Spegazzini (*Lobivia saltensis* (Spegazzini) Britton & Rose; *E. cachensis* Spegazzini). Illustration: Britton & Rose, The Cactaceae 3: f. 70 (1922). Simple at first, later clustering, light green; ribs 17–18, low, shallowly tuberculate. Central spines 1–4, 1–1.2 cm; radial spines 12–14, to 6 mm, thinner than the centrals. Flowers lateral, 4 cm, red; scales of tube described as naked in the axils. *Argentina (Tucuman, Salta)*. G1.

E. schreiteri (Castellanos) Werdermann (*Lobivia schreiteri* Castellanos). Densely clustering to form cushions *c.* 30 cm in diameter; individual stems 1.5–3 cm in diameter, dark green; ribs 9–14, continuous. Central spines 0 or rarely 1; radial spines 6–8, to 5–10 mm. Flowers to 3 cm; pericarpel 5 mm, tube 1.5 cm, bright green outside with loosely hairy scales; perianth-segments purple, style and stigmas yellowish green. *Argentina (Tucuman)*.

31. E. chamaecereus Friedrich & Rowley (*Cereus silvestrii* Spegazzini; *Chamaecereus silvestrii* (Spegazzini) Britton & Rose; *Lobivia silvestrii* (Spegazzini) Rowley). Illustration: Botanical Magazine, 8426 (1912); Britton & Rose, The Cactaceae 3: f. 61 (1922); Hecht, BLV Handbuch der Kakteen, 238 (1982); Haustein, Der Kosmos-Kakteenführer, 167 and endpapers (1983). Creeping and branching to form mats or clumps, stems slender-cylindric, to 30 × 1–1.5 cm; ribs 6–9, very low; spines several, short, bristly, brownish or white. Flowers *c.* 7 cm, orange-scarlet, axils of scales bearing long brownish or whitish hairs. *Argentina (Tucuman)*. G1. Summer.

Very common in cultivation, rooting readily from offsets, and also a parent of various interspecific hybrids.

The following names have not been traced: *Echinopsis winteri*, *E. volcanensis*; *Lobivia soehrensiana*, *L. yamparezii*; *Trichocereus defatima*, *T. winterianus*.

46. MILA Britton & Rose
D.R. Hunt

Low, clustering plants with short-cylindric, densely spiny, ribbed stems and stout roots. Flowers small, funnel-shaped-bell-shaped, arising near the stem-apex; tube short, with small scales sparsely hairy in the axils. Fruit small, spherical, fleshy, almost naked; seeds small, semi-elliptic, black, warty.

A single variable species, native to C Peru, perhaps better included in *Echinopsis*. Literature: Donald, J.D., Occasional generic reviews No. 3, Mila Britt. & Rose, *Ashingtonia* 3: 31–35, 38–62 (1978).

1. M. caespitosa Britton & Rose (*M. pugionifera* Rauh & Backeberg; *M. fortalezensis* Rauh & Backeberg; *M. sublanata* Rauh & Backeberg). Illustration: Hecht, BLV Handbuch der Kakteen, 317 (1982); Haustein, Der Kosmos-Kakteenführer, 187 (1983); Cullmann et al., Kakteen, edn 5, 238 (1984). Stems 30 × 1–4 cm; ribs 10–13, 3–5 mm high; areoles 2–5 mm apart, at first densely brown-felted. Spines variable in number, length, texture and colour, usually more than 20, centrals 1–8, 1–3 cm; radials 10–40, 5–20 mm. Flowers 2–3.5 × 2–3 cm, yellow; pericarpel and tube with narrow scales with woolly or bristle-like axillary hairs. Fruit to 1.2–1.5 cm, reddish, with scales and woolly hairs; seeds oval. *C Peru*. G1.

47. REBUTIA Schumann
D.R. Hunt

Plants small, stems spherical to shortly cylindric, simple or clustering, tuberculate or weakly ribbed; areoles circular or oval to elliptic-linear; spines relatively weak, often hardly differentiated into radial and central. Flowers diurnal, freely produced, usually arising towards the stem-base, funnel-shaped, comparatively small (usually less than 5 cm), variously coloured; pericarpel and tube with small scales naked or with hairs and sometimes bristles in their axils; tube short or long, often slender, often curved, sometimes solid ('fused with the style'); stamens borne on the throat and tube, throat-membrane lacking (except in Nos. 1, 2). Fruit small, almost spherical, the pericarp juicy at first, drying papery, withered perianth persistent. Seeds numerous, cap-shaped, 1–1.5 mm; hilum basal, truncate, broad, often with a caruncle; seed-coat shiny black, somewhat warty or wrinkled, and often papillose towards the apex.

A popular genus of 20–30 species, some of them very variable, from the eastern cordilleras of the Andes from Bolivia (Cochabamba to Tarija) to NW Argentina (Jujuy to Tucuman), 1500–4000 m. Literature: Donald, J.D., Occasional generic review no. 6. Weingartia Werd., *Ashingtonia* 3: 87–139 (1982); Pilbeam, J., *Sulcorebutia and Weingartia, a Collector's Guide* (1985).

1a. Throat-membrane present; flowers at stem apex   2
 b. Throat-membrane absent; flowers at crown, sides or base   3
2a. Ribs *c.* 10–12, tubercles prominent; spines *c.* 12–13   **1. fidaiana**
 b. Ribs *c.* 14, tubercles low; spines *c.* 6–7   **2. neumanniana**
3a. Flowers at upper areoles; scales of tube naked in the axils   **3. neocumingii**
 b. Flowers at lateral or lower areoles, and/or the tube with hairs and sometimes bristles   4
4a. Areoles elliptic or linear   5
 b. Areoles round or slightly oval   14
5a. Flowers uniformly red or purple   6
 b. Flowers orange or yellow, or bicoloured   7
6a. Flowers arising at or above middle of stem; tube short, throat broad; tuberous roots present   **4. vizcarrae**
 b. Flowers arising below middle of plant; tube relatively long; roots not tuberous (stem-base sometimes thickened)   12

7a. Stem cylindric; flowers arising above
    middle   **5. cylindrica**
  b. Stem spherical; flowers arising below
    middle   8
8a. Flowers uniformly yellow   9
  b. Flowers bicoloured   11
9a. Spines more than 25, bristle-like,
    ascending   **6. glomeriseta**
  b. Spines less than 25   10
10a. Spines more than 4 mm, bristle-like
                7. candiae
  b. Spines less than 4 mm, stiff
               8. arenacea
11a. Flowers red with yellow throat
            9. canigueralii
  b. Flowers purple with white throat
            10. rauschii
12a. Spines less than 4 mm, comb-like
           11. taratensis
  b. Spines more than 4 mm   13
13a. Spines 7–9, needle-like
           12. steinbachii
  b. Spines 15–20 or more, bristle-like
            13. mentosa
14a. Scales of tube with axillary hairs
    and/or bristles   15
  b. Scales of tube naked or scarcely hairy,
    never bristly, in the axils 23
15a. Tube hollow to base (style free) or
    solid in the basal 2–5 mm only 16
  b. Tube solid for 5 mm or more above
    the ovary (style 'fused')   20
16a. Tube hollow to base   17
  b. Tube solid in basal 2–5 mm 19
17a. Stem with 18–24 low, tuberculate
    ribs   **14. famatinensis**
  b. Stem ribs indistinct   18
18a. Stem slender-cylindric; spines 10–12,
    2–3 mm, adpressed **15. einsteinii**
  b. Stem spherical; spines 15 or more,
    bristle-like, ascending **16. aureiflora**
19a. Stems ovoid to short-cylindric, 1–3 ×
    1–2 cm; spines 2–3 mm, adpressed
               17. pygmaea
  b. Stems depressed-spherical to oblong,
    2–3 × 1.5–2.5 cm; spines to 4 mm,
    bristle-like   **18. steinmannii**
20a. Flowers white or pale pink
              19. albiflora
  b. Flowers orange or red   21
21a. Spines short, comb-like, adpressed
            20. heliosa
  b. Spines bristle-like, ascending 22
22a. Spines less than 20, 3–5 mm; tube
    3–3.5 mm in diameter **21. deminuta**
  b. Spines 30–40, to 5–30 mm; tube
    2–3 mm in diameter **22. fiebrigii**
23a. Tube hollow; individual stems to 5 cm
    in diameter   **23. minuscula**
  b. Tube solid at the base; individual
    stems to 8 cm in diameter 24

24a. Flowers red   **24. wessneriana**
  b. Flowers yellow or orange yellow
             25. marsoneri

1. R. fidaiana (Backeberg) D. Hunt
(*Weingartia fidaiana* (Backeberg)
Werdermann; *W. cintiensis* Cardenas).
Illustration: Backeberg, Die Cactaceae 3:
1789, f. 1716, 1717 (1959); Haustein, Der
Kosmos-Kakteenführer, 177 (1983).
Stem usually simple, spherical to short-
columnar, tuberculate, eventually 30 ×
15 cm, from a thin-necked tuberous root,
grey-green; areoles large, raised, round,
felted. Spines subulate, upcurved, straw-
coloured to violet-black, the central 3–4, to
5 cm, the radials *c*. 9, to 3 cm. Flowers
arising near the apex, one per areole, to
3 × 3 cm, tube short, slender, perianth-
segments yellow; staminal throat-ring
distinct. Fruit small, oblong-spherical,
dehiscing basally; seeds few, cap-shaped,
nearly as broad as long, black, finely
areolate-papillate. *Bolivia*. G2. Spring.

2. R. neumanniana (Backeberg) D. Hunt
(*Weingartia neumanniana* (Backeberg)
Werdermann). Illustration: Backeberg, Die
Cactaceae 3: 1790, f. 1718, 1719 (1959).
Haustein, Der Kosmos-Kakteenführer, 177
(1983).
Simple, almost spherical, to 7 × 5 cm, from
an often branched tap-root, constricted at
neck, grey-green, velvety; ribs *c*. 14, with
low indistinctly hexagonal tubercles;
areoles *c*. 1 cm apart. Spines 1–3 cm, dark
brown to reddish black, rigid, the central
usually 1, the radials 6–8, somewhat
shorter. Flowers apical, *c*. 2.5 × 2.5 cm,
yellow to reddish orange; staminal throat-
circle present. Fruit spherical to ovoid,
brownish; seeds 1.5 × 1 mm, dark brown to
black. *NW Argentina (Jujuy)*. G2.

  The very similar *Weingartia kargliana*
Rausch came from an isolated population
in Bolivia (Potosi).

3. R. neocumingii (Backeberg) D. Hunt
(*Weingartia neocumingii* Backeberg; *W.
cumingii* misapplied; *W. corroana* Cardenas;
W. pulquinensis Cardenas; *W. lanata* Ritter;
W. longigibba Ritter; *W. riograndensis* Ritter;
W. sucrensis Ritter; *W. hajekyana* invalid).
Illustration: Backeberg, Die Cactaceae 3:
1791, f. 1720 (1959); Haustein, Der
Kosmos-Kakteenführer, 177 (1983);
Cullmann et al., Kakteen, edn 5, 313
(1984).
Stem simple, depressed-spherical to
spherical, to 10 cm diameter, light to mid-
green, tuberculate, the tubercles in 16–18
spirals; areoles slightly depressed, oval.

Spines *c*. 1 cm, projecting and radiating,
pale yellow, tipped darker, the centrals
c. 10, the radials *c*. 16, more slender.
Flowers arising around the crown (i.e. at
the shoulder), 1–several per areole,
c. 2.5 cm, orange or yellow, staminal
throat-membrane not developed. Seeds
1.4 × 1 mm, matt dark brown to black,
tessellate. *Bolivia*. G2.

4. R. vizcarrae Cardenas (*Sulcorebutia
vizcarrae* (Cardenas) Donald; *Weingartia
purpurea* Donald & Lau; *W. torotorensis*
Cardenas). Illustration: Cactus & Succulent
Journal (US) **42**: 186 (1970); Pilbeam,
Sulcorebutia and Weingartia, 100 and
colour plate (1985).
Spherical, 3–3.5 × 4–5 cm, grey-green,
sometimes becoming purplish, from *Dahlia*-
like tuberous roots; ribs *c*. 18, *c*. 4 × 5 mm;
areoles 1 cm apart, 4–6 mm long, grey
felted. Spines whitish yellow to brownish,
the centrals 2–5, 8–11 mm, spreading, the
radials *c*. 10–17, 4–8 mm, spreading,
comb-like. Flowers funnel-shaped, 3.5 cm,
clove-scented, tube short, 5 mm, perianth-
segments magenta. *Bolivia (Cochabamba and
Potosi)*. G2.

5. R. cylindrica (Donald) D. Hunt
(*Sulcorebutia cylindrica* Donald). Illustration:
Ashingtonia **1**: 55 (1974); Haustein, Der
Kosmos-Kakteenführer, 173 (1983);
Pilbeam, Sulcorebutia and Weingartia, 49
and colour plate (1985).
Cylindric, to 12 × 4.5 cm, from *Dahlia*-like
tuberous roots, offsetting below; areoles to
5 mm long, 10 mm apart. Central spines 4,
to 1.5 cm, white or yellow, tipped reddish
brown; radial spines 10–12, 5–10 mm,
white or pale yellow. Flowers 3 ×
3.5–4 cm, deep yellow, with musty scent.
Bolivia (Cochabamba). G2.

6. R. glomeriseta Cardenas (*Sulcorebutia
glomeriseta* (Cardenas) Ritter). Illustration:
Cactus & Succulent Journal (US) **23**: 95
(1951); Cullmann et al., Kakteen, edn 5,
297 (1984); Pilbeam, Sulcorebutia and
Weingartia, 52 and colour plate (1985).
Clustering, stems spherical, *c*. 5–6 cm in
diameter; tubercles in *c*. 20 spirals, 5 mm;
areoles 3–4 mm apart, 2–3 mm in
diameter. Spines more than 50, 2–3 cm,
bristle-like, interlacing, brownish at first.
Flowers 2.5 × 1.5 cm, golden yellow,
stamens in 2 series. *Bolivia (Cochabamba)*.
G2.

7. R. candiae Cardenas (*Sulcorebutia candiae*
(Cardenas) Buining & Donald). Illustration:
Cactus & Succulent Journal (US) **33**: 112
(1961); Cullmann et al., Kakteen, edn 5,

297 (1984); Pilbeam, Sulcorebutia and Weingartia, 42 and colour plate (1985). Clustering, stems spherical 2–3 × 4–5 cm, dark green; tubercles in *c.* 15–20 spirals, 5 × 10 mm; areoles 6–8 mm apart, elliptic, 5 mm. Spines all radial, 15–20, to 7 mm, comb-like, yellow to brown. Flowers 2.5–3 × 3–3.5 cm, yellow. Fruit spherical, brownish, 5 mm diameter; seeds 1–1.4 mm, brownish black. *Bolivia (Cochabamba).* G2.

This, and the closely allied **R. menesesii** Cardenas (spines 10–12, 3–25 mm, white or pinkish, downy), from the same area, link Nos. **6** and **8**, and they may all be the same species.

8. R. arenacea Cardenas (*Sulcorebutia arenacea* (Cardenas) Ritter). Illustration: Cactus & Succulent Journal (US) **23**: 94 (1951); Haustein, Der Kosmos-Kakteenführer, 175 (1983); Cullmann et al., Kakteen, edn 5, 295 (1984); Pilbeam, Sulcorebutia and Weingartia, 38 and colour plate (1985).
Simple or clustering, depressed-spherical, apex umbilicate, 2–3.5 × 2.5–5 cm; tubercles in *c.* 30 spirals, 3 mm broad; areoles 3 mm apart, oblong, 2.5 mm. Spines all radial, 14–16, *c.* 5 mm, comb-like, white to whitish yellow or brownish. Flowers 3 × 3 cm, inner perianth-segments orange-yellow. *Bolivia (Cochabamba).* G2.

R. krugeri (Cardenas) Backeberg. Spines bristle-like, usually brown. *Bolivia (Cochabamba).*

Variants with bicolored flowers have been given the name *Sulcorebutia hoffmanniana* (Backeberg) Backeberg.

9. R. canigueralii Cardenas (*Sulcorebutia canigueralii* (Cardenas) Buining & Donald; *S. tarabucoensis* Rausch). Illustration: Cactus & Succulent Journal (US) **36**: 26 (1964); Cullmann et al., Kakteen, edn 5, 297 (1984); Pilbeam, Sulcorebutia and Weingartia, 43 and colour plate (1985). Clustering, stems spherical, 1 × 2 cm, slate grey; tubercles in 13 series, 3–4 mm diameter; areoles 3 mm apart, elliptic, 3 mm. Central spines 0–2, erect; radial spines 11–14, 1.5–2 mm, bristle-like. Flowers 3–4 × 3–4 cm, tube upcurved, throat yellow, limb orange-red. *Bolivia (Chuquisaca).* G2.

10. R. rauschii (Frank) D. Hunt (*Sulcorebutia rauschii* Frank; *R. pulchra* Cardenas). Illustration: Kakteen und andere Sukkulenten **20**: 238 (1969); Haustein, Der Kosmos-Kakteenführer, 173 (1983); Cullmann et al., Kakteen, edn 5, 298

(1984); Pilbeam, Sulcorebutia and Weingartia, 72–75 and colour plate (1985).
Clustering, roots almost spherical, stems 1.5 × 3 cm, dark green to violet; tubercles in *c.* 16 spirals, *c.* 5 × 5 mm; areoles oblong, *c.* 2 mm. Spines all radial, to 11, 1–3 mm, subulate, adpressed, yellowish to black. Flowers 3 × 3 cm, magenta-pink, usually with paler or white throat. Fruit spherical, 4 × 4 mm, brownish; seeds 1.5 mm, grey-brown, minutely warty. *Bolivia (Chuquisaca).* G2.

Variable, and including forms with uniformly coloured flowers.

11. R. taratensis Cardenas (*Sulcorebutia taratensis* (Cardenas) Buining & Donald). Illustration: Cactus & Succulent Journal (US) **36**: 26 (1964); Cullmann et al., Kakteen, edn 5, 298 (1984); Pilbeam Sulcorebutia and Weingartia, 83 and colour plate (1985).
Clustering, from carrot-like tap-roots 5–8 cm, stems spherical 2–2.5 × 2–3.5 cm, dark green to purplish; tubercles in *c.* 16 spirals, 4–5 mm in diameter; areoles 6 mm apart, narrowly elliptic, 5 mm. Spines all radial, 13–16, 3–4 mm, white, tipped dark brown. Flowers 4 × 3.5 cm, magenta-red. *Bolivia (Cochabamba).* G2.

Variants of *R. rauschii* with uniformly coloured flowers will key out here, but have fewer, darker spines.

12. R. steinbachii Werdermann (*Sulcorebutia steinbachii* (Werdermann) Backeberg). Illustration: Cullmann et al., Kakteen, 299 (1984); Pilbeam, Sulcorebutia and Weingartia, 77 and colour plate (1985).
Clustering, depressed-spherical; tubercles in *c.* 13 spirals. Central spines 1–3, to *c.* 2 cm, blackish; radial spines 6–8, to 2.5 cm, blackish, ascending. Flowers *c.* 3.5 × 3.5 cm, the tube well-developed, limb scarlet to magenta, rarely yellow, musty-scented, stigma white. *Bolivia.* G2.

Very variable, and intergrading with the next species and others listed here. Variants with bicoloured flowers have been given the name *Rebutia tunariensis* Cardenas.

13. R. mentosa (Ritter) D. Hunt (*Sulcorebutia mentosa* Ritter; *S. flavissima* Rausch). Illustration: Pilbeam, Sulcorebutia and Weingartia, 63 and colour plate (1985).
Simple at first, eventually offsetting, to 9–10 × 6–7 cm; areoles elongate, 5–7 mm. Central spines 2–4 or more, 5–35 mm; radial spines 14–18, 5–25 mm, comb-like

and ascending, brown or yellow. Flowers 3 × 3.5 cm, purple or magenta. Fruit dark reddish brown with pale green scales. *Bolivia (Cochabamba).* G2.

R. tiraquensis Cardenas (*R. totorensis* Cardenas; *Sulcorebutia lepida* Ritter; *R. lepida* invalid; *S. bicolorispina* invalid). Central spines reddish brown, contrasting with the glassy white radials. *Bolivia (Cochabamba).* G2.

14. R. famatinensis (Spegazzini) Spegazzini (*Lobivia famatimensis* (Spegazzini) Britton & Rose). Illustration: Cullmann et al., Kakteen, edn 5, 195 (1984).
Usually simple, almost cylindric, umbilicate at apex, 3–3.5 × 2.5–2.8 cm, with tuberous roots; ribs *c.* 24, vertical, straight; tubercles hemispherical, 1.5 × 3–4 mm, separated by little grooves. Spines *c.* 12, 1.5–2 mm, weak, white or almost transparent, comb-like, adpressed. Flowers lateral, tubular-bell-shaped, 3 cm; pericarpel and tube densely clothed with woolly hairs; outer segments purple, inner orange above, yellow below; stamens in several series (throat-membrane not developed); style yellowish white, stigmas 8–12, 4 mm, cream. *W Argentina (La Rioja and San Juan).* G2.

Confused by Backeberg with *Echinopsis kuehnrichii* Fric (*Lobivia densispina* Werdermann) and re-named *Reicheocactus pseudoreicheanus* Backeberg. The name is usually given as 'famatimensis', and was originally spelt thus, but it appears to have been a typographical error, since the source was Famatina.

15. R. einsteinii Fric. Illustration: Ashingtonia **2**(6): 108,109, 112 (1976); Haustein, Der Kosmos-Kakteenführer, 167 (1983).
Clustering, stems slender-cylindric, to 8 × 3.5 cm, green or often bronze-tinted when growing in full sun; tubercles in 13–16 nearly vertical or spiral rows. Spines all radial, 8–12, 2–5 mm, bristle-like, brown to whitish. Flowers self-sterile, funnel-shaped to bell-shaped, 3 × 3.5 cm, golden yellow to orange, scales of tube with hairs and occasionally bristles; style free. Fruit flattened-spherical, naked; seeds small, bell-shaped to oblique, seed-coat brown, warty. *NW Argentina (Volcan Chani).* G2.

16. R. aureiflora Backeberg (*Lobivia euanthema* Backeberg; *R. euanthema* invalid; *Mediolobivia kesselringiana* Cullmann). Illustration: Ashingtonia **2**(5): 84, 85, 87 (1976).
Clustering, leaf-green, often red-tinged;

tubercles *c.* 6 mm. Spines *c.* 15–20, to 6 mm, among them 3–4 centrals to 1 cm, bristle-like. Flowers self-sterile, broadly funnel-shaped, *c.* 4 cm in diameter, typically yellow with white throat, but sometimes orange, red or purplish (forma **kesselringiana** (Cullmann) Donald); tube pale, scales tipped greenish red, with hairs and an occasional bristle in the axils; style free. Fruit flattened-spherical, reddish; seed bell- to cap-shaped, coat warty, dark brown. *NW Argentina.* G2.

Perhaps not distinct from *R. einsteinii.*

17. R. pygmaea (Fries) Britton & Rose. Illustration: Cullmann et al., Kakteen, edn 5, 286 (1984).
Simple, or with many short joints from thickened roots, ovoid to short-cylindric, 1–3 × 1.2–2 cm; tubercles in 8–12 spirals; areoles narrow. Spines all radial, 9–11, 2–3 mm, adpressed. Flowers from lower part of plant, curved-erect, 1.8–2.5 cm, rose-purple; scales of pericarpel and tube hairy in the axils; tube *c.* 3 mm in diameter, solid to 4 mm above the ovary. Fruit spherical, 6 mm. *NW Argentina.* G2.

18. R. steinmannii (Solms-Laubach) Britton & Rose (*R. costata* Werdermann; *R. eucaliptana* (Backeberg) Buining & Donald). Illustration: Haustein, Der Kosmos-Kakteenführer, 167 (1983).
Clustering, stems oblong, 2–3 × 1–2.2 cm; ribs low, often spiralled, tuberculate. Spines 6–10, to 4 mm, bristly, whitish. Flowers lateral, 3.5 × 2.5 cm; scales of tube with bristles and hairs; tube *c.* 3 mm in diameter, solid to *c.* 4 mm above ovary; outer perianth-segments oblong, mucronate, inner rounded, scarlet. *Bolivia (Chuquisaca).* G2.

19. R. albiflora Ritter & Buining. Illustration: Ashingtonia 3(5/6): f.27 (1980).
Freely clustering, individual stems spherical to short-cylindric, to 2.5 × 1.8–2.5 cm, bright green; tubercles in 12 or more rows, 1.5 mm; areoles *c.* 0.5 mm. Radial spines *c.* 15, 3–5 mm, hairlike, white; central spines *c.* 5, to 1.5 cm, erect, white with brownish base. Flowers 2.5 × 2.5 cm, white to pinkish white, tube very slender, to 7 × 1.75 mm; scales of pericarpel and tube with bristles and hairs; perianth-segments white or rose-white with deeper rose midstripe; filaments in 2 series. Fruit 2.5–3 mm in diameter, deep rose or olive green; seeds black, 0.3–0.4 mm. *Bolivia (Tarija).* G2.

20. R. heliosa Rausch. Illustration: Ashingtonia 3(5/6): f. 19–21 (1980);

Cullmann et al., Kakteen, edn 5, 282 (1984).
Simple at first, later clustering, stems flattened-spherical to short-cylindric, to 2 × 1.5–2.5 cm; tubercles in 35–40 spirals, *c.* 1 mm; areoles narrow, oval, 1–2 × 0.5 mm. Spines 24–26, 1 mm, comb-like, white. Flowers 4.5–5.5 × 4 cm; tube to 3 cm × 2–3 mm, deep rose to orange-pink, inner perianth-segments, orange with lilac midstripe; tube solid for more than half its length. Fruit 4 mm in diameter, deep crimson to purple; seeds hemispheric to helmet-shaped, black, papillate. *Bolivia.* G2.

R. albopectinata Rausch. Stems to 2–3 × 2–3.5 cm; radial spines to 13, to 3 mm, comb-like; central spines absent or 2, 1–2 mm. Flowers 2.5–5 × 3–4.5 cm, tube 3–4 mm in diameter; perianth red to crimson. *Bolivia.* G2.

21. R. deminuta (Weber) Berger (*R. pseudominuscula* (Spegazzini) Donald & Buining). Illustration: Ashingtonia 2(7): 144 (1976).
Simple or clustering, stems almost cylindric, 5–6 × 3.5 cm, dark shiny green; tubercles in 11–13 spirals, 5 × 7–8 mm; areoles 7–8 mm apart. Central spines 1–4, radials 7–14, 3–5 mm, bristle-like, at first ochre or dark pinkish, later white with rusty brown tip, finally almost transparent white. Flowers 2.5–3 × 3 cm; pericarp 6 mm in diameter, bristle-like, hairless, with 7–8 scales; tube 1.5 cm, solid, scales sparse, with axillary hairs; perianth-segments oblanceolate-spathulate, obtuse, 5–18 × 5–7 mm, red to orange or dark purple; filaments white, anthers yellow, style pink, stigmas 6. *NW Argentina.* G2.

R. pseudodeminuta Backeberg (*R. robustispina* Ritter). Stems to 7 × 5–6 cm; spines more numerous, to 3 cm, interlacing, white to yellow, orange or brown, often the centrals tipped red-brown or dark brown and the radials white. Flowers larger, to 3.5 × 4 cm, tube somewhat thicker, limb less disc-shaped, pure orange through scarlet to light crimson. *NW Argentina, Bolivia.* G2.

R. kupperiana Boedeker. Stems purplish; spines strongly projecting, chocolate-brown. Flowers larger, dark crimson. *Bolivia (Tarija).* G2.

R. spegazziniana Backeberg. Ribs well-defined with conic tubercles. Flowers dark red, perianth disc-shaped, filaments erect, projecting. Seeds larger with strongly warty coat and large hilum. *NW Argentina.* G2.

22. R. fiebrigii (Guerke) Britton & Rose (*R. albipilosa* Ritter; *R. muscula* Ritter &

Thiele; *R. cajasensis* Ritter; *R. flavistyla* Ritter). Illustration: Ashingtonia 2(9): 180 (1977).
Variable; spherical, depressed at apex, 5 cm high, tuberculate; areoles elliptic. Spines 30–40, not clearly differentiated into radial and central, but the more central to 2 cm, needle-like, straight-projecting, brown-tipped, the others shorter, white, or all spines bristle-like, white or pale yellow. Flowers funnel-shaped, 2–3 cm long, tube slender, upcurved, bright orange to red; scales of pericarpel small, woolly and bristly. Fruit small, purple. *Bolivia, NW Argentina.* G2.

R. spinosissima Backeberg. Spines bristle-like, white, the most central brown-tipped. *N Argentina (Salta).*

23. R. minuscula Schumann (*R. grandiflora* Backeberg; *R. scarlatea* Fric; *R. violaciflora* Backeberg). Illustration: Ashingtonia 2(3): 52, 53 (1975); Cullmann et al., Kakteen, edn 5, 283, 284 (1984).
Soon clustering, depressed-spherical to spherical, to *c.* 5 cm in diameter; tubercles in 16–20 spirals, light green. Spines 25–30, 2–3 mm, bristly. Flowers self-fertile, to 4 cm, pericarpel pale red, scales naked, perianth colour variable; style free to base, tube hollow. Fruit 3 mm in diameter, scarlet. *N Argentina (Tucuman).* G2.

R. margarethae Rausch. Simple, to 4 × 6 cm, roots spherical. Central spine 1, to 3 cm; radial spines 7–11, 1.5–2 cm, all spines dark brown with yellow base. Flowers *c.* 4 × 3.5 cm, orange-red, tube hollow. Fruit flattened-spherical, *c.* 4 mm diameter, brownish yellow with pink to green naked scales naked in their axils. *N Argentina (Salta).* G2.

R. senilis Backeberg (*R. chrysacantha* Backeberg; *R. kesselringiana* Bewerunge). Stem to 8 × 7 cm, deep green. Spines *c.* 25, to 3 cm, chalky white or yellowish. Flowers self-fertile, 3.5 cm, carmine with white throat or citron-yellow (forma **kesselringiana** (Bewerunge) Buining & Donald). *N. Argentina (Salta).* G2.

R. xanthocarpa Backeberg (*R. dasyphrissa* Werdermann; *R. graciliflora* Backeberg). Almost spherical, to 4.5 × *c.* 5 cm, offsetting. Spines 15–20, *c.* 4 ascending, to *c.* 7 mm, stouter, slightly yellowish, others 1–2 mm. Flowers only 2 cm in diameter, self-fertile, scarlet to carmine or salmon-pink, lighter inside. Fruit yellowish. *N Argentina (Salta).* G2.

24. R. wessneriana Bewerunge (*R. calliantha* Bewerunge). Illustration: Ashingtonia 2(4): 68, 72 (1976).

Stem depressed-spherical, to 7 (rarely to 15) × 8 cm, green, large examples freely offsetting; tubercles 2 mm, areoles 5 mm apart, 2.5 × 1 mm. Spines *c.* 15–25, the lowest to *c.* 2 cm, white, brown-tipped. Flowers 4.5–5.5 cm in diameter, bright red; style solid in basal 3–5 mm. *N Argentina.* G2.

R. krainziana Kesselring. Depressed-spherical, *c.* 4–5 × 3–4 cm, dull green; areoles *c.* 3–4 mm apart, oval, *c.* 2 × 1.5 mm. Spines 8–12, *c.* 1–2 mm, bristle-like, snow-white. Flowers self-sterile, *c.* 3 × 4 cm, bright red with violet sheen and yellowish in the throat. Scales of pericarpel violet-brown, hairless. G2.

Said to have originated in Bolivia, but probably only a cultivar of garden origin.

25. R. marsoneri Werdermann. Illustration: Ashingtonia 2(4): 72 (1976); Haustein, Der Kosmos-Kakteenführer, 171 (1983); Cullmann et al., Kakteen, edn 5, 283 (1984).
Usually simple, to 3 × 4.5 cm, pale green; tubercles *c.* 3.5–4.5 mm apart, 2 mm; areoles small, round. Spines *c.* 30–35, the lower 20 *c.* 3–5 mm, white, the upper 9–15, 8–15 mm, somewhat thicker, reddish brown or white with brown tip. Flowers 3.5–4.5 cm, closed, pericarpel *c.* 4 mm in diameter, tube 2–2.4 cm × 5 mm, perianth-segments yellow to orange-yellow; style *c.* 2 cm, stigmas 2 mm, whitish. *N Argentina (Jujuy).* G2.

Names not traced: *Mediolobivia arachnantha* (error for *Echinopsis arachnacantha* ?); *Rebutia grasshevii* (*graessneri* ?).

48. GYMNOCALYCIUM Pfeiffer
G.J. Swales & D.R. Hunt
Stem simple or clustering, spherical to depressed; ribs 4–15 or more, sometimes spiralling, variously tuberculate, spiny. Flowers usually arising near apex, funnel-shaped to bell-shaped, diurnal, usually white or pinkish, more rarely red or yellow; pericarpel and tube bearing broad, obtuse scales, their margins characteristically transparent, their axils naked; limb variously spreading or almost erect; stamens often in 2 groups, the lowermost surrounding the style immediately above the nectar-chamber, the remainder borne nearer the mouth of the tube. Fruit oblong-obovoid to spherical, dry or fleshy, splitting, cracking or deliquescent when ripe; seeds more or less globular to obovoid or somewhat flattened, often conspicuously

appendaged; seed-coat brown or black, shiny or matt, warty, smooth or with minute spicules.

A genus of perhaps 50 species (nearly 100 named) from Brazil, Paraguay, Bolivia, Uruguay and Argentina, the majority of which are available as European-grown seedlings.

Although some are slow-growing, all species will generally thrive in a well-drained, slightly acid compost. If kept dry from November to March, they require a minimum temperature of only 8 ° C during the winter, though some are said to be frost hardy (e.g. No. **8**). During the growing season, they respond to moderate watering and feeding with a low-nitrogen fertilizer. Slight shading may be necessary during the summer months in glasshouses exposed to full sun, as in nature a number of species grow in the partial shade of low shrubs and herbaceous vegetation. The flowering season is spring to early summer, though a few species may produce a few flowers well into the autumn (e.g. No. **33**). Others may produce buds late in the season which develop only slowly and open in the following spring (**11, 12, 15**).

Flowers. Yellow: **1, 6**; red: **9, 22, 24**; white without red throat: **2, 3, 4, 10, 17, 30, 31, 32**; pale pink to almost white without red throat: **11, 13, 23, 25, 28, 29**; white with red throat: **14, 15, 18, 19, 20, 21**; pale pink with deeper midstripe to petals, without red throat: **8, 12, 33**; pale pink with deeper midstripe petals and red throat: **5, 7, 26, 27**; pale peach fading to almost white: **34**; brownish green to yellow-green: **33**; pale lilac-pink with red throat: **16**.
Seeds. Pale brown to fawn, matt, 0.7–1.0 mm: **30, 31, 32, 33**; coat very dark brown to black, matt, but colour usually masked by a much folded and/or fragmented pale brown surface-layer, 1–1.8 mm: **2–9**; glossy chestnut brown, less than 0.8 mm: **34**; glossy chestnut brown, large conspicuous hilum more than 0.8 mm: **18–21**; matt black coat, 1.5–2 mm: **1, 10, 12, 13**; very dark brown to black, matt, 0.8-1.5 mm: **11, 22–27, 29**; very dark brown to black, lustrous, pale hilum, 1–1.8 mm: **14–17, 28**.

1a. Flowers pure yellow   2
 b. Flowers otherwise   3
2a. Spines rarely more than 3–5, strong, at first yellow, becoming whitish with age, centrals absent   **1. uruguayense**
 b. Spines usually 9–13, radials 8–10,

slender, dull white with brown base, centrals 1–3, dark brown   **6. andreae**
3a. Flowers deep pink to crimson, or red with a purple tinge   4
 b. Flowers otherwise   6
4a. Spines bristle-like, pale-tipped, darker below, later brown   **9. baldianum**
 b. Spines strong   5
5a. Spines light grey tinged pink   **22. oenanthemum**
 b. Spines pale brown, dark brown at base and tip   **24. mostii**
6a. Perianth-segments pale pink, pale peach, pale lilac-pink or almost white   7
 b. Perianth-segments otherwise   21
7a. Inner perianth-segments without darker midstripe   8
 b. Inner perianth-segments with dark midstripe   15
8a. Flowers without red throat   9
 b. Flowers red-throated   14
9a. Tube very short, the fully open flower scarcely exceeding the spines   **29. saglionis**
 b. Tube longer, flowers opening above the spines   10
10a. Ribs 5–6   **11. buenekeri**
 b. Ribs 10–17   11
11a. Stem yellowish green, lustrous   12
 b. Stem grey-, brown- or blue-green, matt   13
12a. Flowers funnel-shaped, to 8 cm   **13. monvillei**
 b. Flowers shortly bell-shaped, to 4 cm   **25. multiflorum**
13a. Spines pinkish grey   **23. mazanense**
 b. Spines whitish with dark tips   **28. castellanosii**
14a. Flowers pale peach, fading to almost white, tube very short   **34. pflanzii**
 b. Flowers very pale lilac-pink, tube longer, reddish brown externally   **16. chiquitanum**
15a. Flowers without a wine-red throat   16
 b. Flowers red-throated   18
16a. Stems small, often clustering, densely covered with rather weak spines   **8. bruchii**
 b. Stem otherwise   17
17a. Ribs 8, narrow, acute, dark olive- or bronze-green with characteristic lighter and darker horizontal cross-banding   **33. mihanovichii**
 b. Ribs 5–6, broad, rounded, lustrous yellowish green   **12. horstii**
18a. Spines all radial, short, slender, rough, greyish white; young spines faintly pink if moistened   **7. calochlorum**
 b. Spines otherwise   19

19a. Spines all radial, straight, pale horn-coloured, later becoming grey, to 2.5 cm   **5. capillaense**

   b. Spines 3–6 cm, more or less curved   20

20a. Ribs *c.* 8; central spines 1–2 on old plants only; radial spines 3–6   **26. cardenasianum**

   b. Ribs 10–15 or more; central spines 0–1, radials 5–7   **27. spegazzinii**

21a. Flowers white   22

   b. Flowers brownish green to yellowish green, inner perianth-segments sometimes tinged red; stem with characteristic light and darker horizontal cross-banding   **33. mihanovichii**

22a. Flowers with wine-red throat   23

   b. Flowers lacking wine-red throat, but outer segments sometimes tinged pinkish and/or with pale reddish midstripe   28

23a. Tube long and slender; stem greyish brown; spines minute   **21. ragonesei**

   b. Tube of moderate length, more or less equalling the width of the limb   24

24a. Stem dark green, matt, tending to bronze in full sun   25

   b. Stem otherwise   27

25a. Spines less than 5 mm, adpressed, fine, almost bristly   **18. quehlianum**

   b. Spines more than 5 mm, stronger   26

26a. Spines 1–3 rarely 5, strong, to 2 cm or more   **20. vatteri**

   b. Spines 3–5, moderately strong, to 1 cm   **19. stellatum**

27a. Stem deep green, lustrous   **15. fleischerianum**

   b. Stem pale green, dull, tending to bronze in full sun   **14. paraguayense**

28a. Tube long and slender, to 6–7 cm   29

   b. Tube shorter and more robust   30

29a. Spines 5–7, the lowest longest   **4. schroederianum**

   b. Spines 9–12, all bristle-like   **17. mesopotamicum**

30a. Ribs often only 5, broad and flattened; spines sparse, usually adpressed   **10. denudatum**

   b. Ribs 7–10 or more; spines not as above   31

31a. Radial spines 0.5–1.5 cm   **3. platense**

   b. Radial spines 2–3 cm   32

32a. Spines brown, becoming grey with age   **2. gibbosum**

   b. Spines otherwise   33

33a. Spines reddish grey or horn-coloured   **30. schickendantzii**

   b. Spines brownish towards tip, paler towards base   34

34a. Stem matt grey-green   **31. marsoneri**

   b. Stem leaf-green   **32. anisitsii**

1. G. uruguayense (Arechavaleta) Britton & Rose (*Echinocactus uruguayensis* Arechavaleta; *G. artigas* Herter). Illustration: Arechavaleta, Flora Uruguaya 2: 219, t.14 (1905); Krainz, Die Kakteen, Lfg. 40/41 (1969).
Stem depressed-spherical, 3–3.5 × 7–9 cm, dark green; ribs 12–14, obtuse, flattish, prominently tubercled. Central spines 0; radial spines 3 (rarely 5), 1.5–2 cm, adpressed, yellow at first, later pale dingy brown to almost white. Flowers few at any one time, bell-shaped, 4 × 5.5–6 cm, pale citron-yellow. Fruit green; seeds 2 mm, matt black. *Uruguay.* G1.

G. uruguayense 'Roseiflorum' (var. *roseiflorum* Ito) has a pale lilac-pink flower with yellow throat.

2. G. gibbosum (Haworth) Pfeiffer (*G. schlumbergeri* invalid). Illustration: Hecht, BLV Handbuch der Kakteen, 276 (1982).
At first simple, spherical, somewhat glaucous green, later club-shaped-cylindric, to 20 × 10–15 cm, brownish green, rarely offsetting; ribs 12–19, convex, with small, prominent, tubercles. Spines variable, rigid, straight or slightly curved, mostly brown, later greyish; central spines usually 1–3, 3 cm, often absent in young plants; radial spines 7–14, 2–3.5 cm. Flowers abundant, 6–6.5 × 3–5 cm, pure white, or outer segments with pale pink midstripe outside. Fruit green; seeds *c.* 1.5 × 1.25 mm, matt black. *Argentina (lat. 37–48° S).* G1.

3. G. platense (Spegazzini) Britton & Rose. Illustration: Britton & Rose, The Cactaceae 3: 164, f. 177 (1922).
Stem spherical, to 10 × 12 cm, dark olive green to purplish, never greyish; ribs 10–18, usually 13. Central spines 0; radial spines 5–7, 5–15 mm, subulate, rigid, base swollen, dark, remainder powdery white to greyish purple or blackish. Flowers several together, 5–6.5 × 5–6.5 cm, outer segments white with pinkish midstripe, inner snow-white. Fruit almost spherical, 1–2.5 × 1–2 cm, pink; seeds 1.2 × 1.2 mm, matt black. *Argentina (Buenos Aires).* G1.

Plants sold under this name but having flowers with a deep red throat and glossy brown seeds are probably *G. quehlianum* or a related species.

4. G. schroederianum Osten. Illustration: Backeberg, Die Cactaceae 3: t. 130, (1959) – reproduction of t. 50 from Osten, Notas Sobre Cactàceas, in Anales del Museo de Historia Natural de Montevideo, ser. 2, 5(1): t. 49, 50 (1941).
Stem depressed-spherical, to 7 × 14 cm, dark grey-green; ribs 24, broad, obtuse, with very prominent, broad tubercles. Central spines 0; radial spines usually 7, pale yellow at first, soon ash-grey, red at base, unequal, uppermost pair shortest, lateral pairs intermediate, lowermost spine longest, usually reaching the areole below. Flowers 7 × 5.5 cm, the tube slender, with conspicuous, olive green, white-bordered, kidney-shaped scales; perianth-segments pale greenish white, with pale green base. Fruit narrowly pear-shaped, *c.* 2.5 cm, pale grey-green; seeds 1.2 × 1.2 mm, matt black. *Uruguay (Rio Negro).* G1.

Sometimes united with the little-known *G. hyptiacanthum* (Lemaire) Britton & Rose. The spine arrangement is unique in the genus and has a fancied resemblance to a resting dragonfly. The flower-tube is pink at the base inside, but this is not externally visible.

G. leptanthum (Spegazzini) Spegazzini. Illustration: Backeberg, Die Cactaceae 3: 1713, f. 1640 (1959). Stem depressed-spherical, to 5 × 7 cm, glaucous green; ribs 8–11, robust and tuberculate. Central spines 0; radial spines usually 7, 7–10 mm, closely adpressed, straight or recurved. Flowers slender, 6–6.5 cm, segments white, throat reddish. *Argentina (Córdoba).* G1.

Often confused with *G. platense*, despite the different flower.

5. G. capillaense (Schick) Backeberg (*G. sigelianum* (Schick) Berger; *G. sutterianum* (Schick) Berger). Illustration: Backeberg, Die Cactaceae 3: f. 1647 (1959); Krainz, Die Kakteen Lfg. 42/43 (1969).
Stem depressed-spherical, dull bluish green, to 4–5 × 10 cm; ribs 8–11, broken into blunt tubercles with chin-like outgrowths below the areoles; areoles elliptic with yellowish wool in youth only. Central spines 0; radial spines 3–7, usually 5, 1–2.5 cm, radiating or somewhat adpressed, at first pale horn-coloured, later grey. Flowers to 10 × 6–7 cm; pericarpel and tube pale green, bearing broad, whitish green to pale pink semicircular scales; outer perianth-segments fleshy, ivory-white with greenish midstripe; inner perianth-segments whitish to pale pink with pink midstripe; throat wine-red. Fruit spindle-shaped, 4 × 1.5 cm, bluish green, bearing persistent floral remains; seeds *c.* 1 mm in diameter, coat matt black, hilum basal, tending to be triangular, depressed and often darkly pigmented. *Argentina (Córdoba).* G1.

6. G. andreae (Boedeker) Backeberg. Illustration: Cullmann et al., Kakteen, edn 5, 179 (1984).

Clustering, stems spherical, to 4.5 cm diameter, lustrous dark bluish or blackish green; ribs *c.* 8, flattish. Central spines 1–3, to 8 mm, slender, needle-like, often somewhat curved, rough, dark brown; radial spines 6 lateral and 1 descending, similar to the centrals but dull white, brown at base. Flowers funnel-shaped, 3 × 4.5 cm, outer segments pale greenish yellow with darker green midstripe outside, inner segments clear sulphur-yellow. Fruit spherical, 1.2 cm, bluish green; seeds *c.* 1 mm, matt black. *N Argentina (Córdoba).* G1.

7. G. calochlorum (Boedeker) Ito. Illustration: Krainz, Die Kakteen Lfg. 58 (1974).
Stem depressed-spherical, 4 × 6 cm, rarely offsetting, shining pale green; ribs 9–11, broad, flat. Central spines 0; radial spines directed sideways and downwards, to 9, *c.* 9 mm, often interlacing, slender, rough, pinkish at first, later greyish white. Flowers usually solitary, 5–6 × 5–6 cm, tube robust, *c.* 1 cm diameter, dark shining leaf-green, with rounded, mucronate, white scales 4 mm wide, throat rose-carmine, outer segments pink with grey-green tip and pale margins, inner segments satiny, pale pink with darker midstripe and greyish pink tip. Fruit ovoid, 6–12 mm diameter; seeds nearly spherical, *c.* 1 mm, matt black. *N Argentina (Córdoba).* G1.

8. G. bruchii (Spegazzini) Hosseus (*Frailea bruchii* Spegazzini; *G. lafaldense* Vaupel). Illustration: Anales de la Sociedad Científica Argentina **96**: 72 (repr. 16) (1923); Backeberg, Die Cactaceae **3**: f. 1631 (1959).
Offsetting from an early stage to form clumps or cushions 10–15 cm in diameter containing 10–30 stems; stems almost spherical, 1–2 × 1–2 cm, dull green, largely obscured by the spines; ribs *c.* 10, low, flat, with indistinct tubercles. Central spines 0–3, 2–5 mm, whitish to pale brown; radial spines 12–17, 2–5 mm, slender, recurved and adpressed, white, sometimes brownish at base. Flowers few or solitary, bell-shaped, 1.5–2 (rarely to 4) × 1.5–2 cm, sometimes faintly scented, perianth-segments pale pink with darker midstripe. Fruit spherical to ellipsoidal, 5–10 mm diameter, brown when ripe; seeds more or less spherical, 1–1.2 mm diameter, matt black. *N Argentina (Córdoba).* G1.

Often grown as *G. lafaldense.* Some plants appear to be male-sterile.

9. G. baldianum (Spegazzini) Spegazzini (*G. venturianum* (Fric) Backeberg;

G. sanguiniflorum (Werdermann) Werdermann). Illustration: Cullmann et al., Kakteen, edn 5, 179 (1984).
Stem simple, depressed-spherical, 2.5–4 × 4–7 cm, dark bluish to greyish green; ribs 9–11, obtuse, with small tubercles. Central spines 0; radial spines 3–7, 7–12 mm, slender, bristle-like, spreading, adpressed, straight or somewhat curved, pale-tipped, darker at base, later becoming brown. Flowers solitary or 2–3 together, 3.5–4 × 5–5.5 cm, perianth-segments deep pink to crimson or purplish red. Fruit not described; seeds 1 mm in diameter, matt black. *NW Argentina (Catamarca, La Rioja, Tucuman and Salta).* G1.

10. G. denudatum (Link & Otto) Pfeiffer (*Echinocactus denudatus* Link & Otto). Illustration: Link & Otto, Icones Plantarum Rariorum, t. 9 (1828).
Stem almost spherical, 5–10 × 5–15 cm, dark lustrous green; ribs 5–8, rounded, with narrow transverse furrows. Central spines 0; radial spines 5–8, 8–17 mm, bristly, sinuous, spreading, adpressed, one directed downwards, at first yellowish, later white. Flowers 5–7.5 × 6 cm, sometimes scented, outer segments white with greyish tinge, inner white, sometimes tinged green outside. Fruit ovoid, green, with scattered scales; seeds 1.75 × 2 mm, dark brown to matt black. *S Brazil (Rio Grande do Sul).* G1.

'Jan Suba' is said to be *G. denudatum* × *G. baldianum.* It has the habit of the former and the red or deep pink flower-colour of the latter. Illustration: Kakteen und andere Sukkulenten **32**: 254 (1981).

11. G. buenekeri Swales. Figure 28(11), p. 204. Illustration: Cactus & Succulent Journal of Great Britain **40**: 97, 98, f. 1, 2 (1978).
Stem hemispheric at first, becoming short-cylindric, to 10 cm or more in diameter, matt mid-green; ribs 5–6, rounded and broadly triangular, the tubercles indistinct. Central spines 0; radial spines 3 (rarely 5), to 2.5 cm or more on old plants, usually shorter, somewhat curved and outstanding, pale yellow at first, darkening to pale brown. Flowers 4.5 × 6.5 cm, rose-pink, throat somewhat deeper pink. Fruit (when fertile) ovoid, 5–6 × 3–4 cm, green, bloomed, scented ('melon and straw-berries'); seeds 1.2 × 1.3 mm, black. *S Brazil (Rio Grande do Sul).* G1.

Some plants appear to be male-sterile. Formerly confused with the following species, which differs conspicuously in its lustrous surface.

12. G. horstii Buining. Illustration: Cactus & Succulent Journal of Great Britain **40**: 99, f. 3 (1978).
Stem simple, depressed-spherical, to 7 × 11 cm, fresh glossy green; ribs 5–6, very broad, rounded, with indistinct tubercles. Central spines 0; radial spines usually 5, to 3 cm, rigid, straight, standing out obliquely, not adpressed, pale yellow to almost white. Flowers to 11 × 11 cm, often smaller in cultivation, outer segments deep pink, inner lilac-pink to almost white with deeper pink midstripe. Fruit almost spherical, 3.5 cm, green; seeds 1.3 × 1.2 mm, black. *S Brazil (Rio Grande do Sul).*

13. G. monvillei (Lemaire) Britton & Rose (*Echinocactus monvillei* Lemaire). Illustration: Lemaire, Cactearum Aliquot Novarum (1838), excluding details of flower and stigma.
Stem usually simple, depressed-spherical to shortly columnar, to 22 cm diameter, glossy yellowish green; ribs 13–17, deeply divided into pentagonal tubercles with pronounced 'chin' below. Central spines 0; radial spines 12–13, to 4 cm, bright yellow, reddish brown at base. Flowers broadly funnel-shaped, to 8 × 8 cm, outer segments white with broad green midstripe outside, inner whitish tinged pink, not lustrous. Fruit not described. *Paraguay.* G1.

Sometimes confused with *G. multiflorum,* which has smaller flowers and less well-developed tubercular 'chins'.

14. G. paraguayense (Schumann) Schuetz. Illustration: Friciana **8**: no. 47: f. 1, 3, 4 (1972).
Stem spherical, to 4–5 × 6 cm, eventually becoming columnar, pale to mid-green, tending to bronze in full sun; ribs 8–9 at first, eventually to 12 or more, straight, angular, divided into tubercles by well-defined transverse furrows. Central spines 0; radial spines usually 5, unequal, upper *c.* 1 cm, remainder to 2 cm, spreading, somewhat curved and adpressed, honey-yellow at first, fading to white. Flowers 4–5 × 4–5 cm, white with wine-red throat. Fruit not described; seeds *c.* 1.5 mm diameter, dark brown to black, with conspicuous pale hilum. *S Paraguay.* G1.

15. G. fleischerianum Backeberg. Illustration: Backeberg, Die Cactaceae **3**: f. 1636 (1959).
Variable; stem simple or offsetting, spherical to more or less cylindric, to 6–7 × 10 cm, glossy green; ribs to 8, rounded, tubercles not clearly defined. Spines to *c.* 20, to 2.5 cm, not differentiated into

centrals and radials, bristle-like, yellowish at first, later greyish. Flowers funnel-shaped, to 4 × 3.5 cm, white with deep pink throat. Fruit almost spherical, 1.5 cm, softening but not splitting when ripe; seeds 1.5 mm, black, with conspicuous pale hilum. *N Paraguay*. G1.

Sometimes incorrectly labelled as a variety of *G. denudatum*, under various names.

16. G. chiquitanum Cardenas (*G. hammerschmidii* Backeberg). Illustration: Krainz, Die Kakteen, Lfg. 38/39 (1968); Backeberg, Cactus Lexicon, f. 128, 134 (1977).
Simple, stem depressed-spherical, 2–4 × 6–9 cm, dull pale green, tending to bronze in full sun; ribs 6–7, rounded, divided into tubercles by transverse furrows. Central spines 0–1, 1.8 cm, greyish with darker brown tip; radial spines to 6, 1.5–2.3 cm, adpressed, somewhat curved. Flowers funnel-shaped, to 6 cm, very pale lilac-pink, with magenta throat. Fruit ellipsoid, 2 × 1 cm, purple; seeds 1.5 × 1.25 mm, glossy black. *S Bolivia*. G1.

17. G. mesopotamicum Kiesling. Illustration: Cactus & Succulent Journal of Great Britain **42**: 39, 40 (1980).
Usually simple, depressed-spherical, to 2.5 × 4 cm, dark green; ribs 7–9, low and obtuse, tubercles rather indistinct. Central spines 0; radial spines 9–12, unequal, 2–9 mm, bristle-like, adpressed, reddish brown at first, later greyish or pink-tinged. Flowers 5.5 × 6.5 cm; outer segments pinkish, inner white, throat reddish. Fruit club-shaped, 2–3 × 0.7–0.8 mm, dull green; seeds 1.7–1.8 mm diameter, dark brown to black, with conspicuous hilum. *NE Argentina (Corrientes)*. G1.

18. G. quehlianum (Haage) Berger (*G. bodenbenderianum* (Hosseus) Berger; *G. moserianum* Schuetz; *G. parvulum* (Spegazzini) Spegazzini). Illustration: Schumann, Gesamtbeschreibung der Kakteen, Nachtrag, f. 28 (1903); Cullmann et al., Kakteen, edn 5, 183 (1984).
Very variable; stem depressed-spherical, to 3.5 × 7 cm, dark greyish green, becoming bronze in full sun; ribs 8–11, strongly tuberculate. Central spines 0; radial spines usually 5, to 5 mm, adpressed, almost bristle-like, straight or curved, horn-coloured, reddish brown towards base. Flowers 3–4 × 3–4 cm, segments white, throat red. Fruit not described; seed helmet-shaped, *c.* 1 × 0.9 mm, glossy chestnut brown with large, pale hilum. *N Argentina (Córdoba)*. G1.

19. G. stellatum Spegazzini (*G. asterium* Ito). Illustration: Backeberg, Die Cactaceae **3**: f. 1662 (1959); Krainz, Die Kakteen Lfg. 61 (1975).
Very variable; stem depressed-spherical, to 2.5–5 × 10 cm, matt green, becoming reddish brown when grown in full sun; ribs 7–11, indistinct, divided into small tubercles. Central spines 0; radial spines 3–5, 5–10 mm, more or less projecting, dark grey-brown. Flowers broadly funnel-shaped, 4–5 cm, segments white, throat red. Fruit cylindric, bloomed; seed helmet-shaped, 1–1.2 × 0.9–1 mm, glossy chestnut-brown with large pale hilum. *Argentina (Córdoba)*. G1.

20. G. vatteri Buining. Illustrations: Krainz, Die Kakteen, Lfg. 6 (1957); Backeberg, Cactus Lexicon, f. 146 (1977).
Stem depressed-spherical, to 4 × 9 cm, matt dark olive-green, bronzing in full sun; ribs *c.* 11, large, tuberculate. Spines 1–3 (rarely to 5), 1–2 (rarely to 3) cm, strong, adpressed or projecting, horn-coloured or darker brown. Flowers 5 × 4 cm, segments white, throat red. Fruit 3 × 1 cm, deep bluish green; seeds helmet-shaped, 1–1.2 × 0.9 × 1 mm, glossy chestnut-brown with large pale hilum. *N Argentina (Córdoba)*. G1.

21. G. ragonesei Castellanos. Illustration: Krainz, Die Kakteen Lfg. 59 (1974).
Stem small, depressed-spherical, 1.5–2.5 × 3–5 cm, purplish brown to almost brick-red; ribs *c.* 10, flat, with low tubercles separated by shallow transverse grooves. Central spines 0: radial spines 5–7, usually 6, small, bristle-like, adpressed, spidery, whitish. Flowers 2–3 at a time, funnel-shaped, 3.5–4 cm, tube slender, outer segments greyish with white margins, inner white, often with light grey midstripe, throat dull red. Fruit slender, 3–4 cm, grey, with pale-edged scales; seed 1.5 × 1 mm, glossy chestnut-brown, hilum large, pale, margins spongy. *N Argentina (Catamarca)*. G1.

22. G. oenanthemum Backeberg (*G. tillianum* Rausch). Illustration: Krainz, Die Kakteen Lfg. 3 (1957).
Stem depressed-spherical, to 10–12 cm diameter, matt grey-green to bluish green; ribs 10–13, broad, obtuse, prominently tuberculate. Central spines 0; radial spines 5–7, to 1.5–2 cm, stout, slightly curved, pinkish at first, later grey. Flowers to 5 × 4.5 cm, wine-red or deep salmon-pink. Fruit ovoid, green; seed *c.* 0.8 mm diameter, dark brown to black, tuberculate, with small pale hilum. *N Argentina (Mendoza, Córdoba)*. G1.

23. G. mazanense (Backeberg) Backeberg (*G. glaucum* Ritter; *G. weissianum* Backeberg). Illustration: Backeberg, Die Cactaceae **3**: f. 1702 (1959).
Variable; stem depressed-spherical, matt grey-green, occasionally with brownish tinge. Ribs 10–14, fairly low, rounded, with well-defined tubercles. Areoles oval, felted. Spines more or less curved, pinkish grey at first, later grey; centrals at first 0, later 1, to 3.5 cm, radials usually 7, one directed downwards, the remainder spreading laterally, to 3 cm. Flowers *c.* 4 cm in diameter; tube short, dark or brownish green with pale-bordered scales; perianth-segments pale pink to almost white, darker in throat. Fruit almost spherical, dark green with bluish waxy bloom; seeds 0.8–1 mm in diameter, very dark reddish brown to black, matt, with conspicuous pale hilum. *N Argentina (La Rioja)*. G1.

24. G. mostii (Guerke) Britton & Rose (*Echinocactus mostii* Guerke). Illustration: Schumann & Guerke, Bluehende Kakteen **2**: pl. 93 (1907).
Stem depressed-spherical, dark to bluish green; ribs 11–14, vertical, later becoming somewhat spiralled, divided into well-defined tubercles; areoles elliptic to almost circular, 6–8 mm in diameter, with yellowish white wool at first, later naked. Spines strong, slightly curved towards the stem, pale horn-coloured with glossy brown base and tip, central 1, 1.8–2 cm, radials 7, 6–22 mm. Flowers 7–8 cm in diameter, tube short and thick, shiny bluish green, bearing more or less semicircular, pale-bordered scales; perianth-segments rose-pink with somewhat darker base. Fruit almost spherical, large, green; seeds more or less spherical, 0.7–0.9 mm, very dark reddish brown to black, matt, with whitish hilum. *N Argentina (Córdoba)*. G1.

25. G. multiflorum (Hooker) Britton & Rose (*Echinocactus multiflorus* Hooker). Illustration: Botanical Magazine, 4181 (1845); Cullmann et al., Kakteen, edn 5, 182 (1984).
Stem simple at first, almost spherical, to 9 × 12 cm, later short-cylindric and offsetting from the base to form clumps, lustrous yellowish green; ribs 10–15, with well-defined tubercles. Central spines 0; radial spines 7–10, to 3 cm, well-developed, almost comb-like but with one directed downwards, golden yellow, reddish brown at base. Flowers numerous (at least on old plants), shortly bell-shaped, to 4 cm, tube short, segments pink or whitish. Fruit

ellipsoid, *c.* 1 cm diameter, brown, shrivelling and splitting; seeds 1.1 × 0.9 mm, dark brown to black, tuberculate. *N Argentina (Córdoba).* G1.

Rather difficult to distinguish from *G. monvillei* when not in flower.

G. hybopleurum (Schumann) Backeberg (*G. pugionacanthum* Backeberg). Illustration: Backeberg, Die Cactaceae 3: 1751, fig. 1680 (1959). Broadly spherical, dull bluish to greyish green. Central spines 0; radial spines usually 9, to 3 cm, interlacing, reddish brown at first, later grey. Flowers *c.* 4 cm, outer segments white with conspicuous green midstripe, inner greenish white, throat greenish pink. Fruit green; seeds almost spherical, *c.* 0.9 mm diameter, dark reddish brown to black, hilum medium-sized, mid-brown. *N Argentina (Córdoba).* G1.

26. G. cardenasianum Ritter. Illustration: Cactus & Succulent Journal (US) **43**: 185, f. 11 (1971); Ritter, Kakteen in Südamerika **2**: 821, f. 643 (1980).
Stem depressed-spherical, 5–20 × 12–23 cm, grey-green; ribs *c.* 8, broad and flattened, straight, with indistinct tubercles. Spines variable in colour, very dark to pale brown, later greyish; centrals 1–2 on older plants only, to 8 cm; radials 3–6, 3–6 cm, strong and curved. Flowers 5 × 8–9 cm, tube rather short, segments pale purplish, fading to white at tip, with rust-red midstripe, throat dull reddish. Fruit *c.* 2 × 1.7 cm, greyish green, splitting and exuding the seeds with their fleshy funicles; seeds 1.2 × 0.8 mm, dark reddish brown, tuberculate, with small pale hilum. *S Bolivia.* G1.

27. G. spegazzinii Britton & Rose. Illustration: Britton & Rose, The Cactaceae **3**: 155, f. 162 (1922); Cactus & Succulent Journal (US) **43**: 153, f. 3 (1971).
Stem simple, almost spherical, to 20 × 18 cm, bluish or greyish green to brownish; ribs 10–15 or more, broad, flattish, with indistinct tubercles. Central spines 1, straight-projecting, or absent; radial spines 5–7, to 5.5 cm but usually shorter, strong, curved, more or less adpressed, dark brown to blackish, fading to grey. Flowers to 7 × 5 cm, segments whitish to pale pink, often with broad red midstripe, throat purplish. Fruit ellipsoid, grey-green, bloomed; seeds almost spherical, 0.9 mm, dark reddish brown to black, hilum inconspicuous. *N Argentina (Salta).* G1.

28. G. castellanosii Backeberg. Illustrations: Backeberg, Die Cactaceae **3**: f. 1695

(1959); Kakteen und andere Sukkulenten **21**(7): front cover (1970); Ito, The Cactaceae, 397, f. 177, 7, 4 (1981).
Stem simple, spherical to short-cylindric, to 15 × 10 cm, velvety matt bluish green; ribs 10–12, fairly broad, with rounded tubercles. Central spines 0–1; radial spines 5–7, to 2.5 cm, robust, slightly curved, white, tipped darker. Flowers bell-shaped to funnel-shaped, *c.* 4.5 cm diameter, white with pink sheen. Fruit almost spherical, green, somewhat bloomed; seed almost spherical, 0.8 mm, dark chestnut-brown, hilum inconspicuous. *N Argentina (Córdoba).* G1.

29. G. saglionis (Cels) Britton & Rose (*Brachycalycium tilcarense* (Backeberg) Backeberg; *G. tilcarense* (Backeberg) Schuetz). Illustration: Cactus & Succulent Journal (US) **43**: 153, f. 1 (1971); Hecht, BLV Handbuch der Kakteen, 279 (1982).
Stems simple, depressed-spherical, to 30 cm in diameter, bluish green; ribs 13–32, spiralling, often indistinct, divided into large tubercles. Spines more or less curved, reddish brown to black; centrals 1 or more; radials 8–10 at first, later to 15 or more, to 4 cm. Flowers in a ring round the apex, broadly funnel-shaped, to 3.5 cm, greenish outside, whitish to pale pink inside. Fruit almost spherical, 2 cm, reddish, fleshy; seeds almost spherical, 0.8 mm, dark brown or black, hilum inconspicuous. *NW Argentina (Salta, Tucuman, Catamarca).* G1.

Superficially like *G. pflanzii*, but the flower lacks the red throat of that species and the seed is of a different type.

30. G. schickendantzii (Weber) Britton & Rose. Illustration: Cullmann et al., Kakteen, edn 5, 183 (1984).
Stem simple, broadly depressed-spherical, to 10–15 × 30 cm, dark blue-green, bronzing in full sun; ribs 7–14 on older plants, somewhat spiralled, divided into moderately large tubercles. Central spines 0; radial spines 6–7, 2–3 cm, more or less flattened and recurved, reddish grey to horn-coloured, often darker-tipped. Flowers often forming a ring round the apex, bell-shaped to funnel-shaped, to 5 cm, tube bluish green, outer segments greenish or reddish, inner white tinged with reddish pink. Fruit ellipsoid, *c.* 2.5 cm, bright red; seeds 0.8–1 mm, light brown, including the hilum. *N Argentina (Córdoba, Catamarca, Tucuman).* G1.

Var. **delaetii** (Schumann) Backeberg. Flowers arising from older areoles at the sides of the plant.

31. G. marsoneri (Fric) Ito (*G. tudae* Ito; *G. megatae* Ito; *G. onychacanthum* Ito; *G. pseudomalacocarpus* Backeberg). Illustration: Backeberg, Cactus Lexicon, 623, f. 140 (1977).
Stem simple, depressed-spherical, matt greyish green, tending to bronze in full sun; ribs 10–15, divided into large spirally arranged tubercles. Central spines 0; radial spines *c.* 7, 2–3 cm, brownish above, lighter below at first, later darker. Flowers bell-shaped to funnel-shaped, 3–3.5 × 3–4.5 cm, pale greenish white to white. Fruit not known; seeds almost spherical, 1 mm, pale brown, including the inconspicuous hilum. *Paraguay?* G1.

32. G. anisitsii (Schumann) Britton & Rose (*G. damsii* (Schumann) Britton & Rose). Illustration: Schumann, Bluehende Kakteen **1**: t. 3 ['4'] (1900); Backeberg, Die Cactaceae **3**: f. 1711 (1959); Krainz, Die Kakteen, Lfg. 2 (1956).
Stems spherical at first, 5.5–8 × 7.5–10 cm, becoming short-cylindric, leaf-green, bronzing in full sun; ribs 8–11, acute, strongly tubercled. Spines 5–7 (rarely to 9), usually all radial and *c.* 2.5 cm, slender, tortuous, somewhat angled, nearly white, darker-tipped. Flowers profuse, funnel-shaped, 4 × 4 cm, outer segments greenish white, edged with reddish, inner pure white. Fruit 2.5 × 1 cm, red; seeds almost spherical, 1 mm, pale brown, hilum inconspicuous. *Paraguay.* G1.

33. G. mihanovichii (Fric & Guerke) Britton & Rose (*G. friedrichii* (Werdermann) Pazout; *G. chlorostictum* invalid). Illustration: Hecht, BLV Handbuch der Kakteen, 277 (1982); Cullmann et al., Kakteen, edn 5, 181, 182 (1984).
Very variable; stem depressed-spherical at first, eventually short-cylindric, to 8 cm or more in diameter, dark olive-green with characteristic lighter and darker horizontal cross-banding; ribs 8, prominent, acute, scarcely tuberculate. Spines all radial, 5–6, to 1 cm, sometimes breaking or dropping off, greyish yellow at first, tipped brown, later grey. Flowers bell-shaped to funnel-shaped, 4–4.5 cm, outer segments yellowish green tipped with reddish, inner green (var. **mihanovichii**) or deep pink (var. **friedrichii** Werdermann). Fruit club-shaped, 2 × 0.8 cm, red; seeds 0.9 × 0.8 mm, pale brown, hilum inconspicuous. *N Paraguay.* G1.

Several curious cultivars lacking chlorophyll, such as 'Red Cap', 'Pink Cap' and 'Gold Cap' (with bright red, pink and yellow stems respectively), mostly imported from Japan, have been marketed on a large scale in Europe. To survive, these plants must be grafted on normal green stock, vigorous genera such as *Harrisia* usually being preferred.

34. G. pflanzii (Vaupel) Werdermann (*G. izogzogsii* Cardenas; *G. lagunillasense* Cardenas; *G. millaresii* Cardenas; *G. zegarrae* Cardenas). Illustration: Ritter, Kakteen in Südamerika **2**: 821, f. 641 (1980).
Stem simple, depressed-spherical, to 50 cm but usually much less in cultivation, matt, slightly yellowish mid-green; ribs indistinct, spiralling, divided into large tubercles. Spines to 2.5 cm, smooth, reddish brown with darker tip at first, later rough, pinkish grey, dark-tipped; central 1, erect; radials 6–9, spreading, somewhat curved. Flowers *c.* 4 × 4 cm, variable in colour, lasting 3–4 days and tending to fade from pale peach through pale pink to white, with wine-red throat. Fruit almost spherical, red; seeds numerous, 0.7 × 0.5 mm, glossy chestnut-brown, hilum inconspicuous. *SE Bolivia (Tarija).* G1.

Names not traced: *G. herrisii* (error for *G. hennissii* invalid, a supposed hybrid of *G. platense* × *G. quehlianum*?); *G. pirattense* (error for *G. platense*?).

49. NEOWERDERMANNIA Backeberg
D.R. Hunt & G.J. Swales
Plants more or less spherical, with thick rootstock; ribs indistinct, *c.* 16, spiralling, deeply divided into triangular tubercles; areoles at base of upper side of tubercles. Flowers funnel-shaped with spreading limb, white or lilac-pink.

A genus of 2 closely allied species in southern S America, perhaps better included in *Gymnocalycium*.

1. N. vorwerkii (Fric) Backeberg.
Illustrations: Backeberg, Die Cactaceae **3**: f. 1723 (1959); Hecht, BLV Handbuch der Kakteen, 280 (1982).
Broadly flattened-spherical from well-developed tap-root; tubercles spirally arranged, bluntly conic, 3-sided, flattened above, keeled below, with the areoles situated in the depressions between. Spines up to 10, lowest directed downwards, almost hooked at tip, to 4 cm, blackish at first, remainder to 1.5 cm, whitish. Flowers *c.* 2 cm in diameter; perianth-segments white with light lilac-pink midstripe, or pale lilac. Fruit deep-seated between the tubercles, tearing open apically when ripe; seeds 2 × 1.5 mm, brown, rough, hilum inconspicuous. *N Argentina, S Bolivia, Peru, N Chile.* G2.

N. chilensis Backeberg. Flowers creamy white. *N Chile (Iquique) to border with Argentina.* G2.

50. DISCOCACTUS Pfeiffer
N.P. Taylor
Low-growing, with mostly solitary, depressed-spherical to spherical, spiny, ribbed stems; ribs distinct or strongly tuberculate. Flowers arising from a small and often depressed apical woolly cephalium, developing very rapidly, tubular-salver-shaped, 3–8 cm long, with a slender, scaly tube, white, nocturnal. Fruit a naked, spherical-club-shaped to oblong, slightly fleshy berry; seeds almost spherical, black, seed-coat strongly tuberculate.

About 5–7 species, native of the drier parts of Brazil, E Bolivia and N Paraguay. Uncommon in cultivation as they require a minimum temperature of at least 10° C in winter.
Literature: Taylor, N.P., Reconsolidation of Discocatus Pfeiff., *Cactus & Succulent Journal of Great Britain* **43**: 37–40 (1981).

1a. Fruit pinkish-red; spines more than 10, terete, needle-like, more than 2 cm, interlaced   **1. zehntneri**
 b. Fruit whitish, rarely pinkish, brownish or greenish; spines fewer than 10 or less than 2 cm, terete or flattened, mostly subulate, not interlaced though sometimes spreading over the adjacent rib   2
2a. Stem to 7 cm in diameter; ribs 15–22, straight and high, of even width not thickened at the areoles; spines adpressed to stem, to only 7.5 mm   **2. horstii**
 b. Stem more than 7 cm in diameter; ribs strongly tuberculate or fewer than 16 and broadening towards base; spines usually more than 10 mm   **3. placentiformis**

1. D. zehntneri Britton & Rose (*D. albispinus* Buining & Brederoo; *D. araneispinus* Buining & Brederoo; *D. boomianus* Buining & Brederoo). Illustrations: Cactus & Succulent Journal of Great Britain **43**: 39 (1981); Cullmann et al., Kakteen, edn 5, 150 (1984).
Stem to 10 × 6 cm, conspicuously and densely spiny; ribs to 20, usually spiralled, divided into tubercles. *E Brazil (N Bahia).* G2. Summer.

2. D. horstii Buining & Brederoo (*D. woutersianus* Brederoo & Vandenbroeck). Illustration: Cullmann et al., Kakteen, edn 5, 150 (1984).
Stem 2 × 4.5–7 cm, purplish brown with

inconspicuous, adpressed spines; ribs high and narrow, straight, not tuberculate. *E Brazil (N Minas Gerais).* G2. Summer.

Often grafted (see illustration cited), since it is difficult to cultivate on its own roots.

3. D. placentiformis (Lehmann) Schumann (*Cactus placentiformis* Lehmann; *Discocactus bahiensis* Britton & Rose; *D. latispinus* Buining & Brederoo; *D. tricornis* Pfeiffer). Illustration: Cactus & Succulent Journal of Great Britain **43**: 39 (1981).
Stem to 10 × 6–22 cm light, dark or bluish green; ribs mostly 10–16, acute, not or only slightly tuberculate, without acute sinuses between adjacent areoles on the same rib. Spines conspicuous, sometimes more than 2 cm long or 2 mm thick, often flattened. *E Brazil (E Bahia & E Minas Gerais).* G2. Summer.

D. heptacanthus (Rodrigues) Britton & Rose (*Malacocarpus heptacanthus* Rodrigues; *D. boliviensis* invalid; *D. silvaticus* Buining & Brederoo; *D. silicola* Buining & Brederoo; *D. catingicola* Buining & Brederoo). Illustration: Cactus & Succulent Journal of Great Britain **43**: 39 (1981). Like the above but ribs rounded and mostly formed from large tubercles often with deep sinuses between adjacent areoles on the same rib. *E Bolivia, Brazil (Mato Grosso, Goias, W Minas Gerais & Bahia).* G2. Summer.

D. hartmannii (Schumann) Britton & Rose (*Echinocactus hartmannii* Schumann; *D. magnimammus* Buining & Brederoo; *D. mamillosus* Buining & Brederoo; *D. patulifolius* Buining & Brederoo). Illustration: Cullmann et al., Kakteen, edn 5, 150 (1984). Like the above but ribs 15–22, completely divided into small tubercles. Spines less than 2 cm, to 1.5 mm thick, terete. *N Paraguay, Brazil (S Mato Grosso).* G2. Summer.

51. COPIAPOA Britton & Rose
N.P. Taylor
Low-growing or forming mounds of stems to 1 m or more high; rootstock fibrous or tap-root greatly enlarged and then sometimes connected to stem base via a slender neck; stems strongly to weakly ribbed or ribs divided into tubercles, rather spiny to almost spineless, apex often immersed in a conspicuous patch of dense areolar wool. Flowers shortly funnel-shaped to bell-shaped, small, yellow (in one, rarely cultivated species sometimes red); pericarpel and short tube almost naked or with few small scales naked in their axils (or with long hairs in subgenus **Pilocopiapoa** (Ritter) Ritter: *C. solaris* only); nectar-chamber short but broad, enclosed

above by the bases of the lowermost stamens. Fruit small, spherical to top-shaped, dehiscing at the truncate apex, naked or with some scales near apex; seeds 0.8–2 mm, obovoid, seed-coat black, smooth to tuberculate, shiny.

A difficult genus with an uncertain number (10–20) of species from the coastal desert region of N Chile. Literature: the only up-to-date account of the genus including descriptions is that of Ritter, F., *Kakteen in Südamerika* 3: 1044–1107, figs. 963–1058 (1980). This work, however, recognises 46 weakly defined species and lacks a key.

1a. Tap-root woody, not swollen and fleshy; stem very hard and rigid; radial spines stout and straight, or curved and more than 7 per areole   2
b. Tap-root swollen, tuberous, usually connected to stem base via a slender neck (tap-root may be absent in vegetatively propagated specimens), or radial spines 5–7 some strongly curved; stem hard or soft   3
2a. Stem green to brownish, never with a powdery white to grey coating, apical meristem with white, grey or brownish wool; spines 6–15, rather stout; ribs 8–17   **1. marginata**
b. Stem white or grey, with a powdery coating of wax, or green and apex with bright orange wool, or spines slender to hair-like, 12–30 or more; ribs 12–47   **5. cinerea**
3a. Radial and central spines very short, only 0.5-4 mm (or rarely to 5 mm, or one central rarely to 1.5 cm), those of adjacent areoles widely separated   **3. hypogaea**
b. Radial spines 5–50 mm, centrals to 6 cm, or spines from adjacent areoles almost touching or interwoven   4
4a. Stem light grey-green and/or spines subulate, stout, never all adpressed; ribs with broad low tubercles   **2. cinerascens**
b. Stem brown to blackish or olive-green; spines needle-like, slender or all short, adpressed; ribs often divided into spirally arranged, conic tubercles   **4. humilis**

1. C. marginata (Salm-Dyck) Britton & Rose (*Echinocactus marginatus* Salm-Dyck; *E. columnaris* Pfeiffer; *E. streptocaulon* Hooker). Illustration: Pfeiffer & Otto, Abbildung und Beschreibung Bluehender Kakteen 2: t. 30 (1850); Botanical Magazine, 4562 (1851); Cactus & Succulent Journal of Great Britain 43: 57 (1981).

Simple or branched, arising from a short, woody tap-root; stem 20–50 × 7–10 cm, cylindric, but often slightly tapered towards apex, grass-green; ribs 10–14, broad and obtuse, scarcely tuberculate; areoles close. Central spines 1–3, 2.5–4 cm, stoutly subulate, straight-projecting; radial spines 5–10, 1–1.5 cm, stout, nearly straight. Flowers 2.5–3.5 cm. Fruit *c.* 1 cm in diameter, green or tinged red; seeds *c.* 1 × 0.7 mm. *N Chile (Atacama)* . G2. Summer.

C. bridgesii (Pfeiffer) Backeberg (*Echinocactus bridgesii* Pfeiffer). Illustration: Cactus & Succulent Journal of Great Britain 43: 57 (1981). Very similar to the above but stem only 5–8 cm in diameter, with more wool at apex; ribs 8–12. Central spines more conspicuous, very stout, slightly curved upwards. Flowers 3–3.8 cm. Seeds *c.* 0.8 × 0.7 mm. *N Chile (N Atacama).* G2. Summer.

C. echinoides (Salm-Dyck) Britton & Rose (*Echinocactus echinoides* Salm-Dyck; *C. dura* Ritter; *C. cuprea* Ritter; *Echinocactus cupreatus* Hildmann?). Illustration: Pfeiffer & Otto, Abbildung und Beschreibung Bluehender Kakteen 2: t. 29 (1850); Cactus & Succulent Journal of Great Britain 43: 51 (1981). Like *C. marginata* but stem depressed-spherical to spherical, 7–18 cm in diameter, sometimes brownish; ribs 11–17; areoles 2–10 mm apart. Central spines straight or somewhat curved, to 5 cm; radial spines to 2.5 cm. Flowers 3.5–4 cm. Fruit *c.* 1.5 cm in diameter; seeds 1.7–2 × 1.3–1.5 mm. *N Chile (S Atacama).* G2. Summer.

Both the above are probably only varieties of *C. marginata*.

C. solaris (Ritter) Ritter (*Pilocopiapoa solaris* Ritter; *C. ferox* invalid). Illustration: Backeberg, Die Cactaceae 3: t. 160 (1959); Cactus & Succulent Journal of Great Britain 43: 56 (1981). Clustering (forming large hemispherical mounds of stems in the wild), stems 8–12 cm in diameter, cylindric, grey-green; ribs 8–12, to 3.5 cm high, not tuberculate; areoles very large, 1–1.75 cm in diameter, to 5 mm apart or close-set. Central spines 2–5, 2–6 cm; radial spines *c.* 7–10, 1.5–5 cm, often somewhat curved, rather stout. Flowers 2.5–3 cm, unique in the genus in having the tube woolly. Fruit 1.5 cm in diameter; seeds *c.* 2 × 1.3 mm. *N Chile (Antofagasta).* G2. Rarely flowering in Europe.

2. C. cinerascens (Salm-Dyck) Britton & Rose (*Echinocactus cinerascens* Salm-Dyck; *C. applanata* invalid). Illustration: Ritter, Kakteen in Südamerika 3: f. 1015–1018

(1980); Cactus & Succulent Journal of Great Britain 43: 58 (1981). Clustering (but remaining solitary for some time in cultivation), arising from a large swollen tap-root connected to the stem-base by a narrow neck; stems depressed-spherical, *c.* 8–15 cm in diameter, grey-green; ribs 15–20 or more, obtuse, tuberculate above the areoles; areoles 5–8 mm in diameter, 3–7 mm apart. Central spines 1–4, 1–2 cm, subulate, straight or slightly curved; radial spines 7–9, 5–15 mm, stoutly needle-like, straight, becoming grey or whitish. Flowers 2.7–3.7 cm. Fruit 1–1.2 cm in diameter, reddish; seeds *c.* 1.2 × 0.8 mm. *N Chile (N Atacama).* G2. Summer.

C. calderana Ritter (*C. lembckei* invalid). Illustration: Backeberg, Die Cactaceae 3: t. 160 (1959); Ritter, Kakteen in Südamerika 3: f. 1011–1013 (1980). Like the above but less tufted, stems spherical to short-cylindric, 5–10 cm in diameter; ribs 10–17, less tuberculate. Spines (at least in var. **spinosior** Ritter) 1–4 cm. Seeds to 1.5 × 1 mm. *N Chile (N Atacama).* G2. Summer.

C. grandiflora Ritter. Illustration: Ritter, Kakteen in Südamerika 3: f. 1014 (1980); Cactus & Succulent Journal of Great Britain 43: 58 (1981). Very like *C. cinerascens* but stem 6–l0 cm in diameter; ribs 12–19; areoles 2.5–4 mm in diameter, *c.* l0 mm apart. Central spines to 5 cm, needle-like; radial spines 7–10, to 3 cm, rather slender. Flowers 3–5.5 cm. Fruit to 1.5 cm in diameter; seeds *c.* 1 × 0.8 mm. *N Chile (S Antofagasta).*

Both the above are probably only varieties of *C. cinerascens*.

C. megarhiza Britton & Rose (*C. echinata* Ritter; *C. totoralensis* Ritter). Illustration: Ritter, Kakteen in Südamerika 3: f. 997, 1010 (1980); Cactus & Succulent Journal of Great Britain 43: 58 (1981). Simple or somewhat clustering, arising from a large tuberous tap-root, stems depressed-spherical to spherical, 5–10 cm in diameter, grey-green, very spiny; ribs 10–21, slightly tuberculate; areoles 4–7 mm in diameter, 5–10 mm apart. Central spines 1–10, 1.5–4 cm, mostly straight; radial spines 7–12, 5–25 mm, straight or somewhat curved, yellow, brown or black, later grey. Flowers 2.5–4 cm. Fruit *c.* 1 cm in diameter, green; seeds *c.* 1.4–1.8 × 0.9–1.4 mm. *N Chile (Atacama).* G2. Summer.

C. coquimbana (Ruempler) Britton & Rose (*Echinocactus coquimbanus* Ruempler; *C. coquimbana* var. *wagenknechtii* Ritter;

C. wagenknechtii invalid). Illustration: Ritter, Kakteen in Südamerika **3**: f. 1000–1002 (1980); Cactus & Succulent Journal of Great Britain **43**: 58 (1981). Clustering, stems 7–15 cm in diameter, green to bluish green; ribs 13–18, strongly tuberculate, especially in young specimens; areoles 6–10 mm in diameter, 5–15 mm apart. Central spine 0–1, to 6 cm; radial spines 5–7, 1–5 cm, stout, curved, blackish then grey. Flowers 3.5–5.5 cm. Fruit *c.* 1.5 cm in diameter; seeds *c.* 2 × 1.5 mm. *N Chile (N Coquimbo).* G2. Summer.

The name *C. coquimbana* (*Echinocactus coquimbanus* Ruempler) is of somewhat uncertain application; here it is used in the sense of Ritter's *C. coquimbana* var. *wagenknechtii* (see Ritter, Kakteen in Südamerika **3**: 1074–1075, 1980).

C. montana Ritter (*C. mollicula* Ritter; *C. olivana* Ritter). Illustration: Ritter, Kakteen in Südamerika **3**: f. 1024–1027 (1980); Cactus & Succulent Journal of Great Britain **43**: 58 (1981); Hecht, BLV Handbuch der Kakteen, 242 (1982). Simple or clustering, arising from a large tuberous tap-root connected to the stem base via a slender neck; stem 4–20 × 4–10 cm, depressed-spherical to oblong, grey-green to brownish; ribs 10–17, tuberculate; areoles 3–10 mm in diameter, 5–15 mm apart. Central spines 0–3, 1–3 cm, straight; radial spines 4–9, 5–20 mm, straight or slightly curved, stoutly needle-like, black, later grey. Flowers 2.5–4 cm. Fruit 1–1.3 cm in diameter, reddish; seeds 1.2–2 × 0.9–1.2 mm. *N Chile (S Antofagasta, N Atacama).* G2. Summer.

Intermediate between *C. cinerascens* and its immediate allies, and the following.

3. C. hypogaea Ritter (*C. laui* Diers; *C. hypogaea* var. *barquitensis* Ritter; *C. barquitensis* invalid). Illustration: Cactus & Succulent Journal of Great Britain **41**: 13 (1979); Kakteen und andere Sukkulenten **31**: 362 (1980); Ritter, Kakteen in Südamerika **3**: f. 1019–1023 (1980); Cullmann et al., Kakteen, edn 5, 144 (1984).
Simple or clustering, arising from a large tuberous tap-root connected to the stem base via a slender neck; stem depressed-spherical, 3–6.5 cm (or in the dwarf variant known as *C. laui* Diers, 1–3 cm) in diameter, grey or more commonly brownish or reddish; ribs more or less dissolved into low, often spiralled tubercles, 10–16; areoles only 0.7–3 mm in diameter, 2–10 mm apart. Central spine absent or 1, minute (or in var. **barquitensis**

Ritter occasionally to 1.5 cm), blackish; radial spines absent or 1–10, only 0.5–5 mm. Flowers 1.5–2.2 cm. Fruit small, *c.* 0.5 cm in diameter; seeds 1–1.15 × 0.7–0.9 mm. *N Chile (S Antofagasta, N Atacama).* G2. Summer.

4. C. humilis (Philippi) Hutchison (*Echinocactus humilis* Philippi). Illustration: Ritter, Kakteen in Südamerika **3**: f. 983–985 (1980); Cactus & Succulent Journal of Great Britain **43**: 57 (1981); Riha & Subik, Illustrated Encyclopedia of Cacti & other Succulents, 121, f. 126 (1981).
Simple or more commonly clustering, arising from a large tuberous tap-root connected to the stem base via a long slender neck; stems spherical or almost so, 3–9 cm in diameter, olive-green or darker in full sun; ribs 8–14, often divided into pronounced tubercles 5–15 mm high; areoles 2–4 mm in diameter, 8–15 mm apart. Central spines 1–4 or absent on young stems, 1–3.5 cm, rather slender, straight or curved; radial spines 7–13, 8–25 mm, more adpressed and only 2–5 mm in young plants, needle-like. Flowers 3–4 cm. Fruit *c.* 0.8 cm in diameter; seeds *c.* 1.4 × 1.2 mm. *N Chile (S Antofagasta).* G2. Summer.

A closely related plant which keys out here is '*C. tenuissima*' Ritter (invalid). Illustration: Cactus & Succulent Journal of Great Britain **43**: 57 (1981); Cullmann et al., Kakteen, edn 5, 144 (1984). From *C. humilis* it differs in its smaller stems, 2–5 cm in diameter; ribs 13–16, completely divided into small spirally-arranged tubercles. Central spine mostly absent; radial spines 8–14, only 3–6 mm. Flowers only 2–2.6 cm. *N Chile (Antofagasta).* G2. Summer.

It is not clear whether this should be treated as a species or as some lower rank.

5. C. cinerea (Philippi) Britton & Rose (*Echinocactus cinereus* Philippi; *C. longistaminea* Ritter; *C. tenebrosa* Ritter). Figure 28(10), p. 204. Illustration: Ritter, Kakteen in Südamerika **3**: f. 1037, 1038, 1045, 1058 (1980); Cullmann et al., Kakteen, edn 5, 144 (1984).
Simple or clustering and mound-forming, lacking a swollen tap-root; stem spherical to cylindric, 10–70 (rarely to 140) × 6–25 cm, more or less powdery white or grey, apex with much white, yellow or grey wool (but pale green, apex with orange-brown wool in var. **gigantea**); ribs 12–47, mostly low and rounded, somewhat tuberculate; areoles 2–10 mm in diameter,

close-set or to 1.8 cm apart. Central spines 0–4, 0.5–6 cm; radial spines 0–12, to 4 cm, needle-like to subulate, straight or slightly curved. Flowers 2.5–4.5 cm. Fruit 0.7–1.5 cm in diameter; seed 1.2–1.5 × 0.7–1 mm. *N Chile (Antofagasta).* G2. Summer.

A complex species with the following distinctive varieties:

Var. **gigantea** (Backeberg) N.P. Taylor (*C. gigantea* Backeberg; *C. haseltoniana* Backeberg; *C. cinerea* var. *haseltoniana* (Backeberg) N.P. Taylor; *C. eremophila* Ritter). Illustration: Cactus & Succulent Journal of Great Britain **43**: 51 (1981). Stem to 25 cm in diameter; ribs 14–37, perfectly rounded; areoles 5–10 mm in diameter, 4–18 mm apart. Spines 3–16, 1–4 cm, yellowish brown, slender. *N Chile.*

Var. **albispina** Ritter. Illustration: Cactus & Succulent Journal of Great Britain **43**: 59 (1981); Cullmann et al., Kakteen, edn 5, 143 (1984). Stem to *c.* 15 cm in diameter; ribs 12–21; areoles 4–5 mm in diameter, 1–15 mm apart. Spines 1–9, to 5 cm long, whitish to light brownish yellow, slender. *N Chile.*

Var. **cinerea**. Stem to 20 cm in diameter; ribs 12–30; areoles 4–5 mm in diameter, 1–15 mm apart. Spines 1–11, 0.5–4 cm long, blackish, stout. *N Chile.*

Var. **columna-alba** (Ritter) Backeberg (*C. columna-alba* Ritter; *C. melanohystrix* Ritter). Illustration: Ritter, Kakteen in Südamerika **3**: f. 1039–1041, 1043 (1980); Cactus & Succulent Journal of Great Britain **43**: 59 (1981). Stem to 20 cm in diameter; ribs 26–47, tuberculate; areoles 2–7 mm in diameter, 4–8 mm apart. Spines 0–12, 0.5–2.5 cm, black to yellowish brown, needle-like to subulate. *N Chile.*

C. malletiana (Salm-Dyck) Backeberg (*Echinocactus malletianus* Salm-Dyck; *C. dealbata* Ritter; *C. carrizalensis* Ritter). Illustration: Cactus & Succulent Journal of Great Britain **43**: 59 (1981); Riha & Subik, Illustrated Encyclopedia of Cacti & other Succulents, 123, f. 128 (1981). Like *C. cinerea* var. *cinerea*, but stem to 16 cm in diameter; ribs 15–33, higher and more acute, areoles 5–9 mm in diameter, 1–2.5 cm apart. Spines 1–8 (rarely 0), 1–6 cm, black to light brown, stout. *N Chile (Atacama).* G2. Summer.

C. krainziana Ritter. Illustration: Cactus & Succulent Journal of Great Britain **43**: 59 (1981). Clustering, lacking a tuberous tap-root; stem spherical to cylindric, 6–20 cm in diameter, pale greenish grey or somewhat powdery; ribs 13–24, acute to rounded; areoles 3–6 mm in diameter, to

7 mm apart. Spines 20–30 or more (only 12–20 in var. **scopulina** Ritter), 1–3.5 cm, needle-like to hair-like, straight or curved and tangled, white, grey or blackish. Flowers 2.5–3.5 cm. Fruit whitish to pink; seeds *c.* 1.6 × 1.3 mm. *N Chile (Antofagasta).* G2. Summer.

Probably only a more finely spined variety of *C. cinerea.* Seldom flowers when cultivated in N Europe.

The name *C. aureispina* has not been traced.

52. NEOPORTERIA Britton & Rose
D.R. Hunt

Mostly simple, rarely clustering, spherical to short-cylindric, ribbed, the ribs usually divided into prominent tubercles; areoles centrally placed, oval, depressed, felted. Flowers arising at the apex or crown, tubular-funnel-shaped, broadly funnel-shaped or bell-shaped; tube distinct or very short, bearing small scales variously felted and hairy or bristly in the axils, the uppermost areoles more strongly developed; perianth-segments spreading, or the outermost spreading and the inner erect, concealing the stamens; stamens borne on the throat and tube in one series. Fruit spherical to ovoid or elongated, sometimes balloon-like (*N. islayensis*), pericarp initially fleshy but the interior usually dry when mature, releasing the seeds through a basal opening or rarely disintegrating; seeds almost spherical to semi-oval, black, minutely warty and rough.

A genus of probably no more than 20–30 species, in Chile, S Peru and W Argentina, though many hundreds of names (mostly based on single plants or populations) have been published in *Neoporteria* itself and/or in *Horridocactus* Backeberg, *Islaya* Backeberg, *Neochilenia* Backeberg, *Pyrrhocactus* (Berger) Backeberg & Knuth, *Thelocephala* Ito, and others. In horticultural lists, many names originally published in these latter genera are listed under *Neoporteria*, though they have never been formally transferred.
Literature: Donald, J.D. & Rowley, G.D., Reunion of the genus Neoporteria, *Cactus & Succulent Journal of Great Britain* 28: 54–58, 74–77, 83 (1966).

1a. Fruit spherical, not elongating, green, eventually disintegrating
                  11. bulbocalyx
  b. Fruit elongating, pinkish, dehiscing by basal pore   2
2a. Flowers tubular-funnel-shaped, with narrow throat and almost erect inner

perianth-segments, purplish pink with purple or orange stigmas; fruit initially fleshy   4
  b. Flowers broadly funnel-shaped or bell-shaped, with broad throat and spreading inner perianth-segments, yellow to rose with cream to pink stigmas   3
3a. Flowers small, 2–2.5 cm, yellow, with very short, woolly and bristly tube and yellow style; stem-apex densely woolly; fruit inflated, balloon-like
                10. islayensis
  b. Flowers usually larger, various colours, but the style usually red; tube variously woolly or nearly naked; stem-apex woolly or not; fruit sometimes elongated, but not balloon-like   6
4a. Spines strongly curved or hairlike and tangled   **1. nidus**
  b. Spines more or less straight, not covering the body   5
5a. Radial spines needle-like, 16–24; flowers 3–6 cm, rarely as small as 1 cm   **2. subgibbosa**
  b. Radial spines hair-like, 12–20; flowers 2–2.5 cm   **3. villosa**
6a. Pericarpel and tube nearly naked, the scales only slightly felted in the axils
                5. horrida
  b. Pericarpel and tube more or less covered with woolly hairs   7
7a. Spines stout   8
  b. Spines less than 1 cm, or absent   10
8a. Stem pale or bright green; spines numerous, lustrous yellow or whitish
                4. chilensis
  b. Stem dark dull green, or tinged reddish brown; spines dull grey to blackish   9
9a. Radial spines *c.*18–20; flowers small, *c.* 3 × 3 cm   **6. taltalensis**
  b. Radial spines to about 10; flowers larger   **7. jussieui**
10a. Tube sparsely bristly   **8. napina**
  b. Tube with conspicuous fine bristles
                9. odieri

1. N. nidus Britton & Rose (*N. gerocephala* Ito; *N. multicolor* Ritter). Illustration: Hecht, BLV Handbuch der Kakteen, 326 (1982); Haustein, Der Kosmos-Kakteenführer, 195 (1983); Cullmann et al., Kakteen, edn 5, 245 (1984).
Simple, spherical to short-cylindric, to 30 × 5–9 cm; ribs 16–18, rounded, tuberculate, hidden by the spines. Spines *c.* 30, to 3–5 cm, weak and tortuous, interlaced and covering the apex, brown (at least the central), yellow or whitish. Flowers

tubular-funnel-shaped, 4–6 × 2.5 cm, pink, inner segments more or less erect, narrow, acute. *Chile.* G2. Autumn.

N. coimasensis Ritter appears to be intermediate between Nos. **1** and **2**.

2. N. subgibbosa (Haworth) Britton & Rose (*N. litoralis* Ritter; *N. mammillarioides* misapplied; *N. nigrihorrida* (Backeberg & Knuth) Backeberg; *N. robusta* Ritter). Illustration: Hecht, BLV Handbuch der Kakteen, 326 (1982); Haustein, Der Kosmos-Kakteenführer, 195 (1983). Variable; spherical to short cylindric at first, eventually to 1 m × 10 cm, green to grey-green; ribs 16–20, more or less tuberculate; areoles large. Central spines 4–8, to 4 cm, strong, yellow, brown or blackish at first; radial spines 16–24, 1–3 cm, amber-yellow. Flowers 3–6 cm, pink, whitish towards the throat, inner segments erect, acute, concealing the stamens. Fruit 1.5–2 × 1 cm, red or reddish green. *Chile.* G2.

N. clavata (Soehrens) Werdermann. Ribs 10–13, central spine 1, to 3 cm, radial spines 4–10, 1.5–3 cm, strong. *Chile.* G2.

N. wagenknechtii Ritter (*N. microsperma* Ritter; *N. rapifera* Ritter). Illustration: Haustein, Der Kosmos-Kakteenführer, 199 (1983). To 30 × 11 cm, grey-green; ribs 11–17, very obtuse, with chin-like projections; areoles 3–10 mm apart, 6–13 × 7 mm. Central spines 3–6, 2–3 cm, blackish at first, later grey-brown; radial spines 10–20, 1.5–2.5 cm, straight, dark grey. Flowers 2.2 cm, purplish. Fruit barrel-shaped, green or reddish; seeds brown. *Chile (La Serena).* G2.

3. N. villosa (Monville) Berger. Illustration: Hecht, BLV Handbuch der Kakteen, 326 (1982).
Short-cylindric, to 15 × 8 cm, greyish green, becoming tinged blackish purple; ribs 13–15, divided into prominent tubercles with large, felted areoles. Central spines 4, to 3 cm, bristle-like, dark; radial spines numerous, grading from bristle-to hair-like, pale brown or whitish. Flowers 2–2.5 cm, pink with whitish throat. *Chile (Huasco).* G2.

4. N. chilensis (Schumann) Britton & Rose (*Echinocactus chilensis* Schumann; *Neochilenia chilensis* (Schumann) Backeberg). Illustration: Gürke & Vaupel, Blühende Kakteen 3: t. 138 (1912). Simple or clustering from the base, spherical to short-columnar, woolly at apex; ribs 20–21, scolloped, pale green. Central spines 6–8, 2 cm; radial spines *c.* 20, 1 cm, glassy white. Flowers broadly funnel-shaped, 5 cm broad; scales of pericarpel and tube with wool and long white hairs; perianth-segments pink. *Chile.* G2.

5. N. horrida (Gay) D. Hunt (*N. tuberisulcata* (Jacobi) Donald & Rowley; *N. choapensis* (Ritter) Donald & Rowley; *N. nigricans* (Linke) Britton & Rose; *N. tuberisulcata* var. *froehlichiana* (Schumann) Donald & Rowley; *N. froelichiana* invalid; *Horridocactus tuberisulcatus* (Jacobi) Ito).
Illustration: Backeberg, Cactus Lexicon, f. 170, 171 (1977). Simple, short-cylindric, to 10 cm in diameter, dark green to bluish green; ribs 14–20, thickened and strongly tuberculate around the areoles; areoles white, to 1.5 cm. Central spines 1–5, to 2.5 cm, strong, subulate, black or brownish and yellow below; radial spines 8–12, later more, subulate. Flowers 4.5–5 × 3.5–4 cm; pericarpel and tube green with minute scales woolly in the axils; perianth white or dingy yellow, outer segments with reddish brown midstripe; stigmas reddish or purplish. Fruit almost hairless. *Chile (Santiago area).* G2.

N. curvispina (Bertero) Donald & Rowley (*N. kunzei* (Foerster) Backeberg; *Horridocactus kesselringianus* Doelz; *N. kesselringiana* invalid; *Pyrrhocactus curvispinus* (Bertero) Backeberg & Knuth; *P. vallenarensis* Ritter; *N. vallenarensis* invalid). Illustration: Ritter, Kakteen in Südamerika 3: f. 803-808 (1980). Spherical, to over 15 cm in diameter; ribs 16, 3 cm high, transversely grooved; areoles 1.5 cm. Spines stout, yellowish at first, darker above, more or less upcurved especially the 2–4 centrals; radials 6–10. Flowers 3.5 cm, straw yellow. Fruit almost hairless. *Chile (Santiago area).* G2.

6. N. taltalensis Hutchison. Illustration: Cactus & Succulent Journal of America 27: 182 (1955).
Simple, spherical, to 8 cm in diameter, dull dark green; ribs 10–16, with chin-like tubercles; areoles with pale yellowish brown felt at first. Central spines to *c.* 6, to 3 cm, dark greyish brown to blackish; radial spines merging with the centrals, *c.* 6–20, 3–20 mm, curving to twisted, brownish, later white. Flowers 3 × 2.5 cm or larger, fuchsia-purple, yellow or white. *Chile.* G2.

Of this affinity are: *Neoporteria intermedia* (Ritter) Donald & Rowley, *N. pilispina* (Ritter) Donald & Rowley, *N. scoparia* (Ritter) Donald & Rowley and *Neochilenia pygmaea* (Ritter) Backeberg.

7. N. jussieui (Monville) Britton & Rose (*N. dimorpha* (Ritter) Backeberg; *N. setosiflora* (Ritter) Donald & Rowley).
Illustration: Gürke, Blühende Kakteen 2: t. 67 (1905).
Simple, spherical or short-cylindric, dark or grey-green to almost black; ribs 12–17, rather stout, divided into prominent tubercles. Central spines 1–2, 2.5 cm; radial spines 7 or perhaps more, dark brown. Flowers 3–3.5 cm, pinkish or yellow with red stigmas; pericarpel with scales woolly in the axils. *Chile.* G2.

Names of doubtful application which may be belong here, or to the following species, include *N. fusca* (Muehlenpfordt) Britton & Rose and *N. hankeana* (Foerster) Donald & Rowley.

N. paucicostata (Ritter) Donald & Rowley (*Pyrrhocactus paucicostatus* (Ritter) Ritter).
Illustration: Ritter, Kakteen in Südamerika 3: f. 854, 855 (1980); Haustein, Der Kosmos-Kakteenführer, 197 (1983).
Simple, hemispheric at first, later cylindric, *c.* 15–30 × 6–8 cm, pale grey-green; ribs 8–13, 1–2 cm high, with chin-like tubercles; areoles 1–1.5 cm apart, 5–8 × 3–5 mm, white-felted, not sunken. Central spines 1–4, to 4 cm, greyish, tipped black; radial spines 5–8, 1.5–3 cm, somewhat recurved, greyish white. Flowers 3–5 × 3–5 cm; tube with small scales woolly in the axils; outer segments pinkish, inner white, style red. Fruit 1.5–2 × 1–1.5 cm, red or greenish red; seeds dull brownish black, somewhat tuberculate. *Chile.* G2.

Of this affinity are *Pyrrhocactus floccosus* Ritter (*N. floccosa* invalid), *P. neohankeanus* Ritter (*N. neohankeana* invalid), *P. neohankeanus* var. *flaviflorus* Ritter (*N. flaviflora* invalid), *P. pulchellus* Ritter (*N. pulchella* invalid), *N. cachytayensis* invalid. According to Ritter, *Delaetia woutersiana* Backeberg (invalid name) is a freakish variant of *P. neohankeanus* Ritter.

8. N. napina (Philippi) Backeberg (*Chileorebutia glabrescens* Ritter; *N. glabrescens* invalid; *Neochilenia mebbesii* (Schumann) Backeberg; *Neoporteria mebbesii* invalid; *Neochilenia mitis* (Philippi) Backeberg; *Neoporteria mitis* invalid; *Neochilenia napina* (Philippi) Backeberg).
Illustration: Backeberg, Cactus Lexicon, f. 258, 260 (1977); Haustein, Der Kosmos-Kakteenführer, 197 (1983); Cullmann et al., Kakteen, edn 5, 245 (1984).
Variable; spherical to ovoid, to 10 × 5 cm, greyish green or tinged red; ribs *c.* 14, more or less distinctly tubercled; areoles slightly felted or hairless. Central spines 0–1, blackish; radial spines usually 3–9, to

3 mm, black. Flowers 3–3.5 cm, tube with long hairs and dark curly bristles above; perianth pale yellow or the outer segments pinkish; style purplish, stigmas yellow. Fruit spherical to ovoid, moderately woolly. *Chile (Huasco).* G2.

N. reichei (Schumann) Backeberg (*Echinocactus reichei* Schumann).
Illustration: Schumann & Gürke, Blühende Kakteen 1: t. 42 (1903). Depressed-spherical, greyish green; ribs divided into spirally-arranged tubercles 4–5 mm in diameter; areoles elliptic, 2 mm. Spines 7–9, to 3 mm, equal, spreading, transparent or white at first, later grey. Flowers bell-shaped, 2.5–3.3 × 4 cm; scales of pericarpel and tube with woolly axillary hairs and bristles; perianth-segments yellow; style red; stigmas 10–12, red. *Chile.* G2.

Schumann's concept of this species is disputed, and modern usage of the name may be incorrect.

9. N. odieri (Lemaire) Backeberg (*Neochilenia aerocarpa* (Ritter) Backeberg; *Neoporteria aerocarpa* invalid; *Neoporteria esmeraldana* (Ritter) Donald & Rowley; *N. occulta* (Philippi) Britton & Rose; *N. krausii* (Ritter) Buxbaum; *Neochilenia krausii* (Ritter) Backeberg). Illustration: Haustein, Der Kosmos-Kakteenführer, 199 (1983); Cullmann et al., Kakteen, edn 5, 244, 246 (1984).
Simple or clustering, depressed-spherical, to 6 cm in diameter, dark reddish brown or blackish; ribs 8–13, conspicuously tubercled; areoles with scant wool. Central spines 0–1, to 1.5 cm or more; radial spines 6–10, to 5 mm, thin, blackish or reddish brown. Flowers 2.5–5 cm, white, yellow or pinkish, the tube with white wool and conspicuous fine bristles to 1 cm. *Chile (Copiapo to Huasco).*

Similar to *N. napina*, but the flower-tube with conspicuous bristles to 1 cm.

10. N. islayensis (Foerster) Donald & Rowley (*N. bicolor* (Akers & Buining) Donald & Rowley; *N. krainziana* (Ritter) Donald & Rowley; *N. lindleyi* (Foerster) Donald & Rowley; *Islaya bicolor* Akers & Buining; *I. copiapoides* Rauh & Backeberg; *I. divaricatiflora* Ritter; *N. divaricatiflora* invalid; *I. flavida* Ritter; *I. grandiflorens* Rauh & Backeberg; *I. maritima* Ritter, invalid; *N. maritima* invalid; *I. minor* Backeberg; *I. mollendensis* (Vaupel) Backeberg; *I. paucispina* Rauh & Backeberg; *I. paucispinosa* Rauh & Backeberg). Illustrations: Haustein, Der Kosmos-Kakteenführer, 201 (1983).

Simple, spherical to short-cylindric, usually 10–15 (rarely to 75) × 10 cm, apex densely woolly; ribs 12–21, more or less strongly tuberculate; areoles conspicuously white-woolly. Spines numerous, variable, usually scarcely differentiated into central and radial. Flowers 1.5–2.5 × 1.5–2.5 cm (to 4 × 4 cm in the variant known as *I. grandiflorens* Rauh & Backeberg), scales of pericarpel and tube with dense woolly hairs, perianth and style yellow. Fruit when ripe inflated, balloon-like, to 3–4 cm, pink or red. *S Peru, N Chile.* G2.

A complex species, to which many names have been given.

N. aricensis (Ritter) Donald & Rowley (*N. residua* (Ritter) Donald & Rowley; *Pyrrhocactus aricensis* Ritter; *P. residuus* Ritter). Illustration: Ritter, Kakteen in Südamerika 3: f. 785–787 (1980). Spherical at first, eventually cylindric and decumbent, to 55 × 10 cm; ribs 13–21, 5–10 mm high. Spines yellow-brown or rarely blackish brown, becoming grey; centrals 5–12, 1–3 cm, upcurved, radials 10–16, *c.* 8–15 mm. Flowers *c.* 2 × 2 cm; scales of pericarpel and tube with dense woolly hairs, tube very short; perianth pale yellow, style and stigmas yellow. Fruit 1–2 × 1–1.5 cm, dark red or reddish brown. *Chile (S Arica).* G2.

Pyrrhocactus vexatus Ritter (*Neoporteria vexata* invalid) appears to be a dwarf, weakly spined ally of the above. *Chile (Antofagasta).*

11. N. bulbocalyx (Werdermann) Donald & Rowley (*Pyrrhocactus bulbocalyx* (Werdermann) Backeberg). Illustration: Hecht, BLV Handbuch der Kakteen, 353 (1982); Haustein, Der Kosmos Kakteen-führer, 193 (1983). Simple, spherical to short-cylindric, eventually 50 × 12 cm, dull greyish green; ribs 12–17, swollen beneath the large areoles. Central spines usually 4, *c.* 2 cm, stout, upcurved, yellowish or dark brown to grey; radial spines *c.* 7–12, 1.5–2 cm, similar to the centrals or paler. Flowers urn-shaped, 4–4.5 cm; scales with dense wool and several bristles; perianth yellow with red throat. Fruit small, spherical, dry, green, eventually disintegrating. Seeds dull black. *N Argentina (Catamarca, La Rioja).* G2.

Name not traced: *Islaya longicarpa.*

53. PARODIA Spegazzini
N.P. Taylor
Simple or clustering, weakly to densely spiny, with mostly small, spherical to shortly cylindric, ribbed or tuberculate

stems (except·Nos. 1, 9 & 10). Flowers from the stem apex, diurnal, brightly coloured, shortly funnel-shaped; pericarpel and tube with fine scales, hairy and bristly in their axils, or the latter restricted to the uppermost axils of the tube (entirely absent in No. 1). Fruit spherical to club-shaped-cylindric, woolly and/or bristly, dry or nearly so, mostly thin-walled and disintegrating at or near base, or thick-walled and splitting laterally, or pink and fleshy at first, later hollow and dry; seeds bell- to helmet-shaped, spherical or ovoid, small to minute (0.3–1.5 mm), hilum mostly large, often bearing a corky caruncle, seed-coat smooth or tuberculate, rarely with spiny or hair-like projections, reddish brown to black.

About 35–50 species, from S Brazil, Uruguay, NE Argentina, S Paraguay and the E Andes of Bolivia and NW Argentina. Very popular for their freely borne, brightly coloured flowers and generally compact habit, only a few reaching any size. A great many ill-defined species have been named. Literature: There is no up-to-date treatment of the genus in either its present broad sense, or of its component groups, formerly recognised as genera. The latter may be distinguished as follows:
Blossfeldia Werdermann (species No. 1, below). Stem tiny, lacking ribs, tubercles and spines; flowers small (6–15 mm), whitish, pericarpel and tube with few minute scales and a little wool only; seeds minute, caruncle relatively large, seed-coat minutely hairy. *Andes.*
Parodia in the strict sense (species Nos. 2–8). Stem medium-sized, with tubercles or tuberculate ribs; spines hooked or straight; seeds minute or to 1.25 mm, spherical to ovoid, with a large or small caruncle, seed-coat smooth or weakly tuberculate. *Andes.*
Eriocactus Backeberg (*Notocactus* Fric: species Nos. 9–11). Stems medium-sized to very large (to 1.8 m), cylindric (eventually) with numerous or very high, well-defined ribs bearing many closely placed areoles; spines never hooked; flowers broad, yellow; seeds bell-shaped, expanded at the hilum or club-shaped, caruncle absent, seed-coat with low tubercles or spiny. *S Paraguay to S Brazil.*
Brasilicactus Backeberg (species Nos. 12 & 13). Stems finely tuberculate, hidden in dense straight spines; flowers small, with a narrow tube bearing conspicuous bristles but sparse wool; seeds cap-shaped, caruncle absent, seed-coat with large tubercles. *S Brazil.*
Brasiliparodia Ritter (No. 14). Like the

Andean parodias (some spines hooked) but seeds lacking a caruncle. *S Brazil.*
Notocactus in the sense of Backeberg (nos. 15–25). Stems medium-sized, weakly to strongly tuberculate-ribbed; spines straight (hooked in No. 18); flowers small to rather large (to 8 cm), stigma-lobes pink, red, purplish or orange (rarely yellow); seeds cap- to bell-shaped, without a caruncle or the hilum conspicuous and corky, seed-coat strongly to obscurely tuberculate, sometimes with a much wrinkled cuticle. *S Brazil, Uruguay, NE Argentina, S Paraguay.*
Wigginsia Porter (*Malacocarpus* Salm-Dyck invalid; No. 26). Like the preceding but fruit partly immersed in the densely woolly stem apex, fleshy at first, later drying and becoming hollow, often bright pink and partly naked. *S Brazil, Uruguay, NE Argentina.*

Parodia species need a well-drained acidic compost and should not be given alkaline (tap) water. Few species tolerate temperatures below freezing and the majority fare better with a minimum of 5–10 °C and a dry winter rest of only 2–4 months. Prolonged cold dry periods may result in root loss. All species tolerate full sun providing adequate ventilation is given. Propagation is by offsets or seeds, the species with minute seeds requiring particular care during the first year of the seedling phase.

1a. Stigma-lobes yellow, pale yellow or whitish   2
 b. Stigma-lobes orange, red, pink, crimson or purple   16
2a. Stems minute, 0.5–3.8 cm in diameter, lacking ribs and tubercles, spineless; seeds minute, with a very large white caruncle, seed-coat brown, hairy; flowers whitish
                                          1. liliputana
 b. Stems more than 3 cm in diameter, ribbed or tuberculate, spiny; seeds not as above or seed-coat hairless; flowers not whitish   3
3a. Ribs well-defined, vertical, with many closely placed areoles; flowers 4.5–6 cm in diameter, pale lemon to golden yellow; seeds bell-shaped, expanded at the hilum or club-shaped and spiny-tuberculate; spines bristle-like to hair-like, straight or slightly curved, never curved to hooked at apex   4
 b. Ribs poorly defined, spiralled or completely divided into tubercles or flowers less than 4.5 cm in diameter; seeds not as above; spines various, sometimes stout or curved to hooked at apex   6

4a. Ribs more than 20   5
 b. Ribs 11–16   **11. magnifica**
5a. Spines 6–11 per areole
    9. schumanniana
 b. Spines *c.* 18–24 or more per areole
    10. leninghausii
6a. Pericarpel, tube and fruit with
   conspicuous bristles and sparse
   whitish wool   7
 b. Pericarpel, tube and fruit with some
   bristles but these less obvious than the
   dense whitish to brownish wool   8
7a. Flowers red, innermost perianth-
   segments at first tightly clustered
   around style and hiding the stamens,
   others spreading   **12. haselbergii**
 b. Flowers yellowish green, inner
   perianth-segments erect or spreading,
   not hiding the stamens
    13. graessneri
8a. Bristles present in the axils of most
   scales on the pericarpel and tube;
   seeds smooth, reddish brown or
   tuberculate, blackish   9
 b. Bristles restricted to the axils of
   uppermost scales on the tube; seeds
   tuberculate, dark brown to black   12
9a. Ribs more or less vertical; usually
   some spines curved or hooked at
   apex; seeds black, strongly
   tuberculate, without a swollen
   whitish caruncle at the hilum
    14. buenekeri
 b. Ribs spiralled-tuberculate or
   completely divided into tubercles;
   spines straight, curved or hooked at
   apex; seeds black or brown, smooth or
   weakly tuberculate, with a
   conspicuous whitish caruncle at the
   hilum   10
10a. Apex of some central spines slightly
   curved to strongly hooked.
    2. microsperma
 b. Apex of all central spines perfectly
   straight   11
11a. Flowers red   **3. nivosa**
 b. Flowers yellow, sometimes reddish
   outside   **4. chrysacanthion**
12a. Apex of some central spines slightly
   curved to hooked   13
 b. Apex of all central spines perfectly
   straight   14
13a. Flowers 2.5–5 cm; stem spherical to
   cylindric, 7–25 cm in diameter
    7. maassii
 b. Flowers 1–3 cm; stem depressed-
   spherical, or cylindric and not
   exceeding 8 cm in diameter
    8. schwebsiana
14a. Fruit elongating when ripe, 2–4 cm ×
   4–10 mm, seeds only in the lower
   part, upper part hollow **6. ayopayana**

 b. Fruit not elongating, spherical or
   barrel-shaped, 5–10 × 5–8 mm, filled
   with seeds   15
15a. Ribs 5–10 mm high or less; seeds
   ovoid, *c.* 0.7–1 × 0.5 mm, glossy
    5. comarapana
 b. Ribs 7–20 mm high, seeds spherical,
   1 × 0.8–1 mm, dull   **19. horstii**
16a. Ribs low, to *c.* 5 mm high, rounded;
   fruits spherical to ovoid, not
   elongating at base when ripe; flowers
   uniformly coloured   17
 b. Ribs 5–25 mm high, more or less
   acute, or fruits club-shaped,
   elongating at base and becoming
   thin-walled or hollow and inflated;
   flowers sometimes bicoloured   20
17a. Stem depressed-spherical; ribs 10–16;
   spines 6–15 per areole; flowers
   reddish purple, whitish or yellow
    20. crassigibba
 b. Stem spherical to cylindric or with
   more than 16 ribs, or spines more
   than 15 per areole; flowers pale to
   deep yellow   18
18a. Flowers 5–8 × 5–8 cm; stem simple,
   usually depressed-spherical
    16. concinna
 b. Flowers 2–4.5 × 3–5 cm; stem simple
   or clustering, spherical to cylindric
    19
19a. Stem 4–8 × 2–4 cm, sometimes
   branching by means of underground
   suckers; ribs 11–22; central spines
   straight, curved or hooked at apex
    18. caespitosa
 b. Stem 5–50 × 2.5–11 cm, simple or
   branched above ground; ribs 18–40;
   central spines straight at apex
    17. scopa
20a. Flowers bicoloured, pale to deep
   purplish pink with a pale or yellowish
   throat, *c.* 5–8 cm in diameter; stem
   with 18–30 ribs   21
 b. Flowers not bicoloured, or less than
   5 cm in diameter; stem with 6–30
   ribs   22
21a. Stem *c.* 5 cm in diameter, ribs less
   than 5 mm high   **21. rutilans**
 b. Stem 10–15 cm or more in diameter,
   ribs *c.* 10 mm high   **22. herteri**
22a. Pericarpel and tube very short, partly
   immersed in the densely woolly stem
   apex; perianth-segments extremely
   glossy; fruit long, maturing the year
   after flowering, often pink, eventually
   dry and hollow, at first hidden in the
   woolly stem-apex   **26. erinacea**
 b. Flowers not as above; fruit maturing a
   few months after flowering, never
   pink and hollow, fully exposed   23

23a. Spines hair-like to slender needle-like,
   flexible, straight, curved or twisted,
   5–40 mm; stem simple or clustering,
   with 6–19 ribs   24
 b. Spines stout, stiff, straight, mostly
   c. 5–20 mm; stem simple, with
   13–24 ribs   25
24a. Flowers pale to golden yellow,
   2.5–6 × 2.5–7.5 cm; seeds glossy
    15. ottonis
 b. Flowers yellowish orange, reddish or
   purplish, 3–4 × 3–4 cm; seeds dull
    19. horstii
25a. Spines 8–30 or more per areole,
   usually one flattened; stem with
   13–24 ribs   **23. mammulosa**
 b. Spines *c.* 5–7 per areole, terete; stem
   with *c.* 15–16 ribs   26
26a. Stem dark green; flowers *c.* 5 cm in
   diameter   **24. allosiphon**
 b. Stem pale grey- to glaucous-green;
   flowers to 8 cm in diameter
    25. buiningii

1. P. liliputana (Werdermann) N.P. Taylor
(*Blossfeldia liliputana* Werdermann;
B. atroviridis Ritter; *B. campaniflora*
Backeberg; *B. fechseri* Backeberg; *B.
pedicellata* Ritter; *B. minima* Ritter).
Illustration: Krainz, Die Kakteen, Lfg.
28–29, f. 1–5 (1964), Lfg. 63 (1975);
Hirao, Colour Encyclopaedia of Cacti, 123,
f. 481 (1979); Hecht, BLV Handbuch der
Kakteen, 229 (1982); Cullmann et al.,
Cacti, 135 (1986).
Simple or clustering, stems depressed-
spherical, very small, only 5–38 mm in
diameter, very dark green to brownish or
greyish, lacking ribs or tubercles, the raised
or sunken areoles arranged in 13 to more
than 30 spirals, with whitish wool but
spineless. Flowers 6–15 mm, whitish;
pericarpel and tube scarcely 1.5–5 mm,
bearing minute scales woolly in their axils
but no bristles; perianth-segments few.
Fruit almost spherical to ellipsoid, *c.* 5 mm
in diameter, naked except for some tiny
scales with tufts of wool; seeds minute,
c. 0.5 × 0.2 mm, caruncle as large or larger
than remainder of seed; seed-coat brown,
glossy, covered in long, fine hair-like
papillae. *N Argentina (La Rioja, Catamarca,
Salta, Jujuy), S Bolivia (Tarija, Chuquisaca,
Potosi).* G2. Summer.
A miniature, cliff-dwelling species with
greatly reduced stems and flowers.
Commonly seen as a highly clustering,
grafted plant in cultivation, as it is very
difficult to grow on its own roots, which
are tuberous. Its seed-morphology allies it
with the following.

2. P. microsperma (Weber) Spegazzini
(*Echinocactus microspermus* Spegazzini;
Parodia catamarcensis Backeberg; *P.
sanagasta* Weingart; *P. sanguiniflora*
Backeberg; *P. setifera* Backeberg).
Illustration: Krainz, Die Kakteen, Lfg. 34,
f. 1, 5 (1966); Hecht, BLV Handbuch der
Kakteen, 345 (below) (1982); Haustein,
Der Kosmos-Kakteenführer, 211 (1983);
Rayzer, Cactus in fiore, 135 (1984).
Solitary, rarely offsetting, depressed-
spherical to spherical, later sometimes
becoming elongated, 5–20 × 5–10 cm; ribs
c. 15–21, more or less divided into spiralled
tubercles. Central spines 3–4, red,
brownish or darker, the lowermost hooked
at apex, 5–20 mm, rarely to 50 mm; radial
spines 7–30, bristle-like, white, c. 4–8 mm.
Flowers 3–3.5 × 4–5 cm, yellow or red;
pericarpel and tube with wool and bristles
throughout. Fruit nearly spherical,
c. 4–5 mm in diameter; seeds minute, to
only 0.5 mm long including the large white
caruncle, seed-coat smooth, brown, glossy.
N Argentina (Catamarca, Tucuman, Salta).
G2. Spring–Summer.

An extremely variable species of which
the following may be only variants:
P. mutabilis Backeberg (*P. aureispina*
Backeberg; *P. aurihamata* invalid).
Illustration: Hirao, Colour Encyclopaedia of
Cacti, 120, f. 466, 467 (1979); Hecht, BLV
Handbuch der Kakteen, 343 (below)
(1982); Haustein, Der Kosmos-
Kakteenführer, 211 (1983); Cullmann et
al., Cacti, 269 (1986). Central spines 4–10,
yellow, reddish or brownish; radial spines
20–50 or more, very fine, white. Flowers
3–5 cm in diameter, yellow. *N Argentina
(Salta)*. G2. Spring–Summer.
P. erythrantha (Spegazzini) Backeberg
(*Echinocactus microspermus* var. *erythranthus*
Spegazzini). Illustration: Krainz, Die
Kakteen, Lfg. 34, f. 3 (1966); Backeberg,
Cactus Lexicon, f. 303 (1977). Stem small,
to c. 4 cm in diameter. Flowers slender, to
3 cm in diameter, red or yellow; pericarpel
more or less naked; tube with wool and
bristles throughout. *N Argentina (Salta)*. G2.
Spring–Summer.
P. rigidispina Krainz. Illustration: Krainz,
Neue und seltene Sukkulenten, 7, 8 (1947);
Krainz, Die Kakteen, Lfg. 34, f. 6 (1966).
Central spines 4, all straight or one
somewhat curved at apex (but not hooked),
reddish to brownish, to 7 mm, stiff; radial
spines 6–11, c. 5 mm. Flowers slender,
3.7 cm, yellow, smelling of iodine.
N Argentina. G2.
P. heteracantha Weskamp is very
similar, differing in its central spines to

2.2 cm and flowers with the pericarpel
nearly naked. *N Argentina (Salta)*.

3. P. nivosa Backeberg. Illustration: Hirao,
Colour Encyclopaedia of Cacti, 119, f. 463
(1979); Hecht, BLV Handbuch der Kakteen,
344 (above) (1982); Haustein, Der Kosmos-
Kakteenführer, 211 (1983).
Simple, spherical or eventually somewhat
longer, to 15 × 8 cm; ribs divided into
spiralled tubercles. Central spines 4, white
or one dark at base, to 2 cm, all straight,
bristle-like; radial spines c. 18, white, like
the centrals but finer. Flowers borne 1–7
together, c. 3 × 2.5–5 cm, fiery red;
pericarpel and tube with wool and bristles
throughout. Seeds to 0.8 × 0.5 mm smooth,
brown, glossy. *N Argentina (Salta)*. G1.
Spring.
P. penicillata Fechser & van der Steeg.
Illustration: Krainz, Die Kakteen, Lfg. 6l
(1975); Riha & Subik, Encyclopedia of Cacti
& Other Succulents, 173, f. 189 (1981);
Hecht, BLV Handbuch der Kakteen, 344
(below) (1982). Spherical, then cylindric, to
30 × 7–12 cm (to 70 cm long in the wild);
ribs 17–20, spiralled, tuberculate. Central
and radial spines poorly differentiated, the
former to 10 or 20, straight and needle-
like, to 5 cm, whitish, yellowish or more
rarely brownish, the latter to c. 40 shorter,
glassy-whitish; spines developing early,
forming a dense tuft at stem apex. Flowers
borne 1–9 together, to 5 × 6 cm, orange to
blood-red. Seeds 0.5–0.75 mm; caruncle
large; seed-coat smooth, brown, glossy.
N Argentina (Salta). G1. Summer.

4. P. chrysacanthion (Schumann)
Backeberg (*Echinocactus chrysacanthion*
Schumann). Illustration: Krainz, Die
Kakteen, Lfg. 34, f. 7 (1966) & 35, f. 20
(1967); Hecht, BLV Handbuch der Kakteen,
342 (1982); Haustein, Der Kosmos-
Kakteenführer, 213 (1983); Cullmann et
al., Cacti, 268 (1986).
Solitary, depressed-spherical but eventually
elongating, to 12 × 10 cm, rarely larger;
stem apex depressed, very woolly,
surrounded by a ring of erect spines; ribs
divided into spiralled tubercles. Central and
radial spines similar, 30–40, golden yellow
or some more whitish, to 3 cm, all straight,
needle-like to bristle-like. Flowers borne
1–11 together, rather small, c. 2 cm,
yellow; pericarpel more or less naked, the
tube with wool and bristles throughout.
Seeds 0.75–1 mm, caruncle often large,
long, seed-coat tuberculate, dark brown to
blackish. *N Argentina (Jujuy)*. G1. Spring.
P. faustiana Backeberg. Illustration:
Backeberg, Die Cactaceae 3: 16l0 (1959);

Ritter, Kakteen in Südamerika 2: f. 273
(1980). Spherical, to 6 cm in diameter; ribs
spiralled, tuberculate. Central spines 4,
stout, to more than 2.5 cm, brown or
darker; radial spines c. 20, to 10 mm,
glassy white. Flowers yellow, red without.
Seeds c. 0.7 mm, caruncle large, deeply
notched, seed-coat weakly tuberculate.
N Argentina (Salta). G2.

Var. **tenuispina** Backeberg has red
flowers and is said to be intermediate
between this and *P. nivosa*.
P. formosa Ritter (*P. cardenasii* Ritter;
P. purpureoaurea Ritter; *P. chaetocarpa* Ritter;
P. setispina Ritter). Illustration: Krainz, Die
Kakteen, Lfg. 50–51 (1972); Ritter,
Kakteen in Südamerika 2: f. 416–420
(1980). Spherical, rarely elongating in age,
3–8 cm in diameter; ribs 16–26, more or
less divided into spiralled tubercles. Central
spines 1–12, finely needle-like, 3–25 mm,
all straight, reddish brown; radial spines
8–30, like the centrals, 3–12 mm, all white
or brown-tipped. Flowers few, 1.6–4 ×
3.5–4.5 cm, yellow. Fruit spherical, 8 mm
in diameter; seeds 0.5 × 0.2 mm or slightly
larger, caruncle very large, seed-coat
smooth, glossy brownish red. *S Bolivia
(Tarija)*. G2. Spring–Summer.
P. gibbulosa Ritter (*P. gibbulosoides*
Brandt?). Illustration: Ritter, Kakteen in
Südamerika 2: f. 421 (1980); Kakteen und
andere Sukkulenten 38: 42–45 (1987).
Like *P. formosa*, but flowers numerous, 5 or
more borne together, c. 1.8 cm; fruit only
2–2.5 mm in diameter; seeds 0.3 ×
0.2 mm. *Bolivia (Chuquisaca)*. G2.
Spring–Summer.

5. P. comarapana Cardenas (*P. neglecta*
Brandt; *P. neglectoides* Brandt; *P. saint-
pieana* Backeberg?). Illustration: Krainz, Die
Kakteen, Lfg. 34, f. 9, 18A (1966);
Backeberg, Cactus Lexicon, 8l8, f. 510, 511
(1977); Ritter, Kakteen in Südamerika 2:
f. 410, 411 (1980); Haustein, Der Kosmos-
Kakteenführer, 213 (1983).
Simple or clustering, stems 5–8 × 7–8 cm;
ribs 12–21, straight or spiralled, to 10 mm
high, tuberculate. Central and radial spines
poorly differentiated, the former c. 4–8,
whitish to yellowish, brownish-tipped,
1–2 cm, all straight and needle-like, the
latter c. 18–35, like the centrals or entirely
white, 3–10 mm. Flowers to 2.5 × 3 cm,
yellow to orange; scales on pericarpel and
tube very narrow and hairlike at apex.
Fruit spherical, 8 mm; seeds c. 1 mm,
caruncle small, seed-coat dark brown to
blackish, tuberculate, glossy. *C Bolivia*. G2.
Spring–Summer.

P. ocampoi Cardenas (*P. echinopsoides* Brandt). Illustration: Kaktus (Denmark) **11**(2): 40–41 (1976); Ritter, Kakteen in Südamerika **2**: f. 408 (1980). Clustering, stems shortly cylindric, 7–20 × 6–11 cm; ribs 13–17, straight, well-defined, 5–10 mm high. Central spine 1, straight, reddish brown, 5–25 mm; radial spines 8–9, reddish brown, 1–2.5 cm. Flowers *c.* 3 × 5 cm, yellow. Fruit obovoid, *c.* 5 mm; seeds ovoid, 0.7 × 0.5 mm, caruncle small, notched, seed-coat tuberculate, black. *C Bolivia (Cochabamba)*. G2. Spring–Summer.

P. gracilis Ritter. Illustration: Hirao, Colour Encyclopaedia of Cacti, 120, f. 465 (1979); Ritter, Kakteen in Südamerika **2**: f. 403 (1980). Solitary, spherical or somewhat longer, 5–10 cm in diameter; ribs 13–19, more or less vertical, well-defined, 5–8 mm high, tuberculate. Central spines 4–10, needle-like, brown, reddish brown or nearly white, all straight or rarely curved, not hooked except in very young plants, 7–10 mm; radial spines 14–20, like the centrals, needle-like to hair-like, 5–20 mm. Flowers 3–3.3 cm, golden to orange-yellow. Fruit almost spherical, *c.* 5 mm; seeds 0.7 × 0.5 mm, tuberculate, black. *S Bolivia*. G2. Spring–Summer.

6. P. ayopayana Cardenas. Illustration: Backeberg, Die Cactaceae **3**: 1606 (1959); Krainz, Die Kakteen, Lfg. 12 (1959) & 34, f. 14 & 15 (1966); Hirao, Colour Encyclopaedia of Cacti, 121, f. 474 (1979). Simple or clustering, stems spherical, 6–8 × 6–10 cm (or in var. **elata** Ritter to 60 × 12 cm); ribs *c.* 11, well-defined, *c.* 2 cm high, vertical; areoles large, very woolly, *c.* 1.2 cm apart. Central spines to 4, subulate, all straight, light brown to whitish, 3–3.5 cm; radial spines *c.* 6–11, needle-like, whitish, 1.2–2 cm. Flowers 3 cm, orange-yellow; pericarpel and tube densely clothed in white to orange wool, tube stout and bearing inconspicuous bristles at apex only. Fruit long, red, seeds only in the lower part; seeds spherical, 0.7–1.0 × 0.6–1.0 mm, caruncle flattened, rather small, seed-coat tuberculate, matt black. *C Bolivia (W Cochabamba)*. G2. Spring.

P. miguillensis Cardenas (*P. comosa* Ritter; *P. borealis* Ritter; *P. echinus* Ritter). Illustration: Ritter, Kakteen in Südamerika **2**: f. 409, 414, 415 (1980); Riha & Subik, Illustrated Encyclopedia of Cacti & other Succulents, 174, f. 192 (1981); Rayzer, Cactus in fiore, 134 (1984). Solitary, depressed-spherical to cylindric, 6–30 × 3–8 cm; ribs 8–16, well-defined, 7–10 mm high, vertical; areoles 1–5 mm apart. Central spines 4–9, needle-like or finer, all straight or somewhat curved, yellow-brown to dark brown, 1–2.5 cm; radial spines 12–18, hair-like, whitish to yellowish brown, 2–20 mm, intergrading with the centrals. Flowers 1.5–2.5 × 0.7–2 cm, light to golden yellow. Fruit 2–3 cm × 4–10 mm; seeds spherical, 0.7–0.8 mm, caruncle small, seed-coat tuberculate, matt black. *C W Bolivia (La Paz)*. G2. Summer.

Closely allied to *P. ayopayana* but more spiny.

7. P. maassii (Heese) Berger (*P. camargensis* Buining & Ritter; *P. roseoalba* Ritter; *P. ritteri* Buining; *P. tarabucina* Cardenas; *P. suprema* Ritter; *P. maxima* Ritter; *P. commutans* Ritter; *P. rubida* Ritter; *P. castanea* Ritter; *P. fulvispina* Ritter; *P. obtusa* Ritter). Illustration: Krainz, Die Kakteen, Lfg. 34, f. 8 & 17 (1966); Ritter, Kakteen in Südamerika **2**: pl. 17, f. 369–382, 385–389 (1980); Haustein, Der Kosmos-Kakteenführer, 213 (1983). Simple, seldom clustering, stem 10–50 × 7–25 cm; ribs *c.* 10–21, well-defined, straight or spiralled. Central spines 1–6, the lowermost longest and much stouter, strongly curved to hooked at apex, rarely almost straight, variously coloured, 2–7 cm; radial spines 6–18 (18–28 in forma **maxima** (Ritter) Krainz), straight or slightly curved, mostly needle-like, paler than the centrals, mostly 1–4 cm. Flowers *c.* 3–4.5 cm, red to yellow; tube woolly, bearing bristles on the uppermost part only. Fruit depressed-spherical or slightly longer, 5–10 mm in diameter (elongating to 2–5 cm in var. **commutans** (Ritter) Krainz); seeds spherical to somewhat ovoid, 0.7–1.2 mm, caruncle small or conspicuous, seed-coat tuberculate, blackish. *S Bolivia, N Argentina (Jujuy)*. G1. Spring–Summer.

A very variable species with many more synonyms than are given here.

P. aureicentra Backeberg (*P. varicolor* Ritter; *P. rauschii* Backeberg, invalid; *P. uhligiana* Backeberg, invalid). Illustration: Backeberg, Cactus Lexicon, f. 313 & 323 (1977); Ritter, Kakteen in Südamerika **2**: f. 278–281 (1980). Simple or clustering, stem spherical or longer, *c.* 15–40 × 8–15 cm; ribs *c.* 13–20, more or less spiralled, tuberculate. Central spines 4–12, some curved to hooked at apex, variously coloured, to 7 cm; radial spines 20–40, bristle-to hair-like, white to yellowish, to 1.2 cm. Flowers 3.5–5 cm, red. Fruit nearly spherical, *c.* 10 mm in diameter, seeds *c.* 0.7–0.8 × 0.3–0.5 mm, caruncle small, seed-coat tuberculate, black. *N Argentina (Salta)*. G1. Spring–Summer.

P. stuemeri (Werdermann) Backeberg (*Echinocactus stuemeri* Werdermann; *P. gigantea* Krainz; *P. friciana* Brandt; *P. tilcarensis* (Werdermann & Backeberg) Backeberg; *P. rubricentra* Backeberg; *P. pseudostuemeri* Backeberg, invalid). Illustration: Backeberg, Cactus Lexicon, f. 312 (below), 320, 321 (left), 503 (1977); Hirao, Colour Encyclopaedia of Cacti, 120, f. 464 (1979); Riha & Subik, Illustrated Encyclopedia of Cacti and other Succulents, 173, f. 190 (1981). Simple, spherical then cylindric, 15–25 × 7–12 cm; ribs *c.* 15–22, straight, rarely spiralled, tuberculate. Central spines 4–8, all straight or one or more slightly to strongly curved (not hooked) at apex, yellowish to brownish, 1.1–2.5 cm; radial spines 9–35, bristle-like or stouter, whitish to brownish, to 2 cm. Flowers 2.5–4 × 2.5–5 cm, reddish, orange or yellowish. Seeds to *c.* 0.8 mm, caruncle very small or pointed, seed-coat finely tuberculate, glossy black. *N Argentina (Salta, Jujuy)*. G1. Spring–Summer.

8. P. schwebsiana (Werdermann) Backeberg (*Echinocactus schwebsianus* Werdermann; *P. applanata* Brandt). Illustration: Krainz, Die Kakteen, Lfg. 24 (1963) & 34, f. 11 (1966); Haustein, Der Kosmos-Kakteenführer, 213 (1983); Cullmann et al., Cacti, 269 (1986). Usually simple, depressed-spherical to shortly cylindric, 2.3–12 × 8 cm, apex very woolly; ribs 13–20, low, somewhat spiralled, tuberculate. Central spines 1–4, the lowermost directed downwards, hooked, reddish to pale brown, 1–2 cm; radial spines *c.* 5–10, reddish or yellowish then grey, 5–12 mm. Flowers 2–3 × 2–2.5 cm, dark red; tube abruptly constricted above the pericarpel, woolly, bearing bristles only in the upper part. Fruit spherical, very small, 3–4 mm; seeds less than 1 mm, caruncle small, pointed, seed-coat tuberculate, glossy black. *C Bolivia (Cochabamba)*. G1. Summer.

This and the following are popular in cultivation.

P. mairanana Cardenas. Illustration: Krainz, Die Kakteen, Lfg. 34, f. 13 (1966) & Lfg. 40–41 (1969); Hecht, BLV Handbuch der Kakteen, 343 (1982); Haustein, Der Kosmos-Kakteenführer, 215 (1983); Cullmann et al., Cacti, 269 (1986). Simple, later clustering, depressed-spherical, *c.* 3–4 × 4–5.5 cm, dark green, not very woolly at apex; ribs *c.* 13–14, somewhat spiralled, scarcely tuberculate. Central spines 1–3, straight to hooked, light brown to blackish, 8–20 mm; radial spines *c.* 8–14, subulate, whitish to yellowish, 3–12 mm. Flowers 1–3.5 × 2–3.5 cm, reddish orange to golden yellow, tube gradually flared above the pericarpel. Fruit ellipsoid, to 8 mm; seeds ovoid, *c.* 1 mm, caruncle conspicuous, seed-coat tuberculate, dark brown. *C Bolivia (Santa Cruz).* G2. Spring.

P. tuberculata Cardenas (*P. otuyensis* Ritter). Illustration: Hirao, Colour Encyclopaedia of Cacti, 120, f. 469 (1979); Ritter, Kakteen in Südamerika **2**: f. 394, 396, 397 (1980). Simple, rarely clustering, depressed-spherical, 7–11 cm in diameter, apex not very woolly; ribs 13–20, spiralled, strongly tuberculate at first. Central spines 1–4, one hooked, brown to blackish or grey, 1.5–2.5 cm; radial spines 7–11, to 10 mm. Flowers 1.8–2.7 × 3 cm, yellowish red. Fruit spherical, 6–7 mm; seeds oblong 0.8–1.25 mm, caruncle conspicuous, seed-coat tuberculate, black. *Bolivia (Potosi, Chuquisaca).* G1. Spring–summer.

P. procera Ritter (*P. challamarcana* Brandt; *P. pseudoprocera* Brandt). Illustration: Backeberg, Cactus Lexicon, f. 501, 513 (1977); Hecht, BLV Handbuch der Kakteen, 345 (1982). Simple, spherical, club-shaped or cylindric, to 30 × 3–8 cm; ribs *c.* 13, high, somewhat spiralled, scarcely tuberculate. Central spines 4, straight or one curved to hooked at apex, brown, 1.5–3.5 cm, radial spines 7–10, hair-like, white to brownish, 7–15 mm. Flowers to 3 × 2.5–4 cm, yellow. Fruit spherical, 5–8 mm; seeds *c.* 0.5 mm, caruncle small, seed-coat finely tuberculate, black. *Bolivia (Chuquisaca).* G1. Spring–summer.

P. schuetziana Jajo. Illustration: Krainz, Die Kakteen, Lfg. 34, f. 12 (1966); Ritter, Kakteen in Südamerika **2**: f. 275 (1980); Riha & Subik, Illustrated Encyclopedia of Cacti & other Succulents, 174, f. 191 (1981); Haustein, Der Kosmos-Kakteenführer, 215 (1983). Simple, depressed-spherical or somewhat longer, to *c.* 11 cm in diameter; ribs *c.* 21, vertical or somewhat spiralled, tuberculate. Central spines 1–4, the lowermost hooked, brownish; radial spines *c.* 15, finely hair-like, interlaced, white. Flowers bell-shaped, *c.* 2 cm, dark red. *N Argentina (Jujuy).* G1. Spring–summer.

9. P. schumanniana (Nicolai) Brandt (*Echinocactus schumannianus* Nicolai; *Notocactus schumannianus* (Nicolai) Fric;

Eriocactus schumannianus (Nicolai) Backeberg; *E. ampliocostatus* Ritter; *E. grossei* (Schumann) Backeberg; *Notocactus grossei* (Schumann) Fric). Illustration: Krainz, Die Kakteen, Lfg. 35, f. 6E & 11 (1967); Hirao, Colour Encyclopaedia of Cacti, 118, f. 459 (1979); Haustein, Der Kosmos-Kakteenführer, 207 (1983); Pizzetti, Piante grasse le cactacee, no. 208 (1985). Usually simple, spherical then cylindric, green, to 1.8 m × 30 cm (flowering when much smaller); ribs 21–48 (fewer in non-flowering juvenile plants), straight, acute, well-defined. Spines golden yellow, brown or reddish, later grey, bristle-like, straight or slightly curved; centrals 3–4, 1–3 cm; radials *c.* 4, 7–50 mm. Flowers 4–4.5 × 4.5–6.5 cm, lemon to golden yellow; pericarpel and tube together *c.* 2–2.5 cm, densely clothed in wool and bristles throughout. Fruit spherical to ovoid, 1–1.5 cm; seeds bell-shaped, widest at hilum, 0.7 × 0.7 mm, reddish brown. *S Paraguay, NE Argentina.* G2. Summer.

P. claviceps (Ritter) Brandt (*Eriocactus claviceps* Ritter; *Notocactus claviceps* (Ritter) Krainz). Illustration: Ritter, Kakteen in Südamerika **1**: f. 89 (1979). Simple or clustering, stem depressed-spherical to shortly cylindric, 10–50 × 8–20 cm or larger; ribs 23–30. Spines pale yellow, needle-like, to 2–5 cm, the centrals 1–3, the radials 5–8. Flowers to 5.5 × 6 cm; pericarpel and tube *c.* 3.2 cm. Seeds bell-shaped, 1–1.2 × 0.6–0.8 mm, nearly black. *S Brazil (Rio Grande do Sul).* G2. Summer.

10. P. leninghausii (Schumann) Brandt (*Echinocactus leninghausii* Schumann; *Notocactus leninghausii* (Schumann) Berger; *Eriocactus leninghausii* (Schumann) Backeberg). Illustration: Hirao, Colour Encyclopaedia of Cacti, 117 (1979); Haustein, Der Kosmos-Kakteenführer, 209 (1983); Cullmann et al., Cacti, 254 (1986). Simple or more often clustering, stem cylindric, to 60 × 7–10 cm or taller, green, apex usually slanted; ribs 30–35, straight, the areoles close together. Spines pale to deep yellow or some brownish, straight to slightly curved, hair-like; centrals *c.* 3–4, 2–5 cm; radials 15–20 or more, 5–10 mm. Flowers *c.* 5 × 6 cm, lemon-yellow; pericarpel and tube densely clothed in brown wool and bristles throughout. Fruit spherical, *c.* 2 cm; seeds bell-shaped, widest at the hilum, *c.* 1 × 0.5 mm, brownish red. *S Brazil (Rio Grande do Sul).* G2. Summer.

A very popular, easily-grown species frequently seen for sale as a juvenile plant only 5–7 cm high.

11. P. magnifica (Ritter) Brandt (*Eriocactus magnificus* Ritter; *Notocactus magnificus* (Ritter) N.P. Taylor). Illustration: Hirao, Colour Encyclopaedia of Cacti, 118, f. 458 (1979); Hecht, BLV Handbuch der Kakteen, 335 (1982). Simple, rarely clustering, glaucous, spherical or eventually longer, 7–15 cm in diameter; ribs 11–15, straight, acute, areoles close together or almost contiguous. Spines 12–15 or more, bristle-like, golden yellow, 8–20 mm. Flowers 4.5–5.5 × 4.5–5.5, sulphur-yellow. Fruit spherical, 10 mm; seeds obovoid to club-shaped, 1 × 0.6 mm, reddish brown, spiny-tuberculate. *S Brazil (Rio Grande do Sul).* G2. Summer.

P. warasii (Ritter) Brandt (*Eriocactus warasii* Ritter; *Notocactus warasii* (Ritter) Hewitt & Donald). Illustration: Succulenta **58**: 313 (1979). Simple, rarely branched, to *c.* 50 × 10–15 cm, green; ribs 15–16, areoles 4–6 mm apart. Spines *c.* 15–20, needle-like, yellowish brown to light brown, 1–4 cm. Flowers 5–6 cm in diameter, golden to lemon yellow. Seeds bell-shaped, widest at the hilum, *c.* 1 × 0.7 mm, shiny, black, not spiny. *S Brazil (Rio Grande do Sul).* G2.

12. P. haselbergii (Ruempler) Brandt (*Echinocactus scopa* forma *haselbergii* Ruempler; *Notocactus haselbergii* (Ruempler) Berger; *Brasilicactus haselbergii* (Ruempler) Backeberg). Illustration: Krainz, Die Kakteen, Lfg. 35, f. 6D & 14 (1967); Hirao, Colour Encyclopaedia of Cacti, 118 (1979); Hecht, BLV Handbuch der Kakteen, 334 (1982); Haustein, Der Kosmos-Kakteenführer, 209 (1983). Simple, depressed-spherical to spherical, 4–15 cm in diameter, apex depressed and sometimes distorted or slanted in old plants; ribs 30–60 or more, very indistinct, divided into small tubercles. Spines *c.* 25–60, bristle-like to needle-like, straight, glassy white or the centrals somewhat yellowish, densely covering the stem, to 10 mm or more. Flowers long-lived, clustered at the stem apex, very numerous on large plants, rather variable in size and form, *c.* 1.5 cm × 9–11 mm, brilliant orange-red (orange-yellow in a rare cultivar); tube relatively long, bearing well-separated clusters of bristles but sparse wool; inner perianth-segments remaining erect, closely surrounding style and hiding the stamens at first, outer segments spreading. Fruit spherical to short-oblong, *c.* 10 mm, bristly, not woolly, yellowish to whitish; seeds cap-shaped, 0.8–1.2 × 0.5–0.7 mm, glossy, black, strongly tuberculate. *S Brazil (E Rio Grande do Sul)* G2. (Winter–)Spring.

Very common and free-flowering in cultivation; self-fertile and readily raised from seed.

13. P. graessneri (Schumann) Brandt (*Echinocactus graessneri* Schumann; *Notocactus graessneri* (Schumann) Berger; *Brasilicactus graessneri* (Schumann) Backeberg). Illustration: Hirao, Colour Encyclopaedia of Cacti, 119 (1979); Hecht, BLV Handbuch der Kakteen, 334 (1982); Haustein, Der Kosmos-Kakteenführer, 209 (1983).
Like No. **12** but spines pale to golden yellow, some brownish or nearly all white, to 2 cm or more. Flowers yellowish green, to *c.* 2.5 cm; perianth-segments half-erect or somewhat spreading but none closely surrounding style, the stamens visible. *S Brazil (E Rio Grande do Sul).* G2. Spring.

Self-sterile; the white-spined variants can be confused with *P. haselbergii* when not in flower.

14. P. buenekeri Buining (*Notocactus buenekeri* (Buining) Buxbaum; *Brasiliparodia buenekeri* (Buining) Ritter). Illustration: Backeberg, Cactus Lexicon, f. 297 (1977); Ritter, Kakteen in Südamerika 1: f. 86, 87 (1979); Hecht, BLV Handbuch der Kakteen, 329 (1982).
Simple or clustering, globular or longer, to 8 cm or more in diameter; ribs *c.* 15–29, more or less vertical, tuberculate but well-defined. Spines bristle-like, glassy-white, brownish or orange, later grey, often forming a brush-like tuft around the stem-apex; centrals 4–6, straight or one somewhat bent or hooked at apex, 5–50 mm; radials *c.* 14–20, laterals interlaced with those of adjacent ribs, 4–23 mm. Flowers 2.6–4 × 3.5–4 cm, golden yellow, the tube partly covered with tufts of brown wool and a few bristles. Fruit 8 × 8–10 mm, thinly covered in wool, seeds helmet-shaped, *c.* 1 mm, broadest at the hilum, black, strongly tuberculate. *S Brazil (SE Santa Catarina, NE Rio Grande do Sul).* G2. Spring.

Doubtfully distinct from the following but of easier cultivation, coming from higher elevations in the wild.

P. brevihamata Backeberg (*Notocactus brevihamatus* (Backeberg) Buxbaum; *Brasiliparodia brevihamata* (Backeberg) Ritter). Illustration: Krainz, Die Kakteen, Lfg. 35, f. 5C, 5D, 5E (1967) & 59 (1974); Ritter, Kakteen in Südamerika 1: f. 233 (1979); Hirao, Colour Encyclopaedia of Cacti; 117, f. 455 (1979) – as Notocactus

tenuicylindricus. Like the preceding but stem usually smaller (at least in cultivation). Central spines only 2–10 mm, 1 or more hooked. Flowers lemon yellow. *S Brazil (Rio Grande do Sul).* G2. Spring.

Both of the above are very similar to the following.

P. alacriportana Backeberg & Voll (*Notocactus alacriportanus* (Backeberg & Voll) Buxbaum; *Brasiliparodia alacriportana* (Backeberg & Voll) Ritter). Illustration: Backeberg, Die Cactaceae 3: f. 1591 (1959). Like *P. buenekeri* but stem usually smaller; ribs 17–31. Central spines 5–25 mm, one always hooked, radials whiter. *S Brazil (E Rio Grande do Sul).* G2. Spring.

15. P. ottonis (Lehmann) N.P. Taylor (*Cactus ottonis* Lehmann; *Notocactus ottonis* (Lehmann) Berger; *Echinocactus tenuispinus* Link & Otto; *E. arechavaletai* Spegazzini, not *E. arechavaletai* (Spegazzini) Arechavaleta; *Notocactus arechavaletai* (Spegazzini) Herter, not *N. arechavaletai* (Spegazzini) Krainz; *N. laetivirens* Ritter; *N. securituberculatus* Ritter; *N. acutus* Ritter; *N. oxycostatus* Buining & Brederoo; *N. glaucinus* Ritter; *N. globularis* Ritter; *N. ibicuiensis* Prestle). Illustration: Krainz, Die Kakteen, Lfg. 35, f. 1A, 1B, 6A (1967) & Lfg 50–51 (1972); Ritter, Kakteen in Südamerika 1: f. 96–111 (1979); Haustein, Der Kosmos-Kakteenführer, 205 (1983); Pizzetti, Piante grasse le cactacee, no. 206 (1985).
Simple at first, later usually clustering, stem more or less spherical, tapered at base, 3–15 cm in diameter, light, dark or bluish green or purplish; ribs 6–15, rarely more, well-defined, rounded or acute. Spines hair-like, straight, curved or twisted; centrals 1–6, light to dark brown, reddish brown or yellowish, 8–40 mm; radials 4–15, whitish to yellow or brown, 5–30 mm. Flowers 2.5–6 cm, yellow (in one rare variant orange-red); tube with brownish wool and bristles; stigma-lobes usually red or purplish, rarely orange to yellow. Fruit ovoid to short oblong, *c.* 9–12 mm in diameter, not elongating at maturity, thick-walled, splitting length-wise to expose the seeds and white pulp; seeds 15–100 per fruit, bell-shaped, 1.2–1.4 × 0.7–1.2 mm, black, glossy, strongly tuberculate. *S Brazil, Uruguay, NE Argentina, S Paraguay.* G2. Summer.

A common and very variable species with numerous synonyms for its various regional forms. The following varieties are recognised.

Var. **ottonis**. Stems mostly 3–10 cm in

diameter; ribs 6–12, rarely more. Flowers *c.* 4–6 cm, few. *S Brazil (Rio Grande do Sul), Uruguay, NE Argentina, S Paraguay.*

Var. **tortuosus** (Link & Otto) N.P. Taylor (*Echinocactus tortuosus* Link & Otto; *E. ottonis* [var.] *tortuosus* (Link & Otto) Haage; *Cactus linkii* Lehmann; *Notocactus linkii* (Lehmann) Herter; *N. villa-velhensis* (Backeberg & Voll) Slaba; *N. carambeiensis* Buining & Brederoo; *N. megapotamicus* (Osten) Herter; *Echinocactus megapotamicus* Osten). Illustration: Succulenta 52(9): front cover (1973); Backeberg, Cactus Lexicon, f. 494 (1977); Ritter, Kakteen in Südamerika 1: f. 112–117, 236 (1979).
Stems to 15 cm in diameter; ribs 10–15, occasionally more. Flowers *c.* 2.5–4 cm, many produced together. *S Brazil (Parana, Santa Catarina, Rio Grande do Sul).*

16. P. concinna (Monville) N.P. Taylor (*Echinocactus concinnus* Monville; *Notocactus concinnus* (Monville) Berger; *N. tabularis* (Ruempler) Berger; *Echinocactus concinnus* var. *tabularis* Ruempler; *E. apricus* Arechavaleta; *N. apricus* (Arechavaleta) Berger; *N. agnetae* Vliet; *N. blaauwianus* Vliet; *N. vanvlietii* var. *gracilis* Rausch; *N. multicostatus* Buining & Brederoo; *N. eremiticus* Ritter; *N. muricatus* (Pfeiffer) Berger, misapplied?). Illustration: Krainz, Die Kakteen, Lfg. 34 (1966), 35, f. 1C, 1D, 1H, 1I (1967) & 55–56 (1973); Hirao, Colour Encyclopaedia of Cacti, 113–114, f. 438, 438-1, 439 (1979); Hecht, BLV Handbuch der Kakteen, 332 (1982); Internoto 8(1): 3 (1987).
Simple, depressed-spherical, or somewhat longer in age, 3–10 × 4–10 cm, depressed at apex, dark green; ribs 15–32, low, with conspicuous chin-like tubercles between the areoles. Spines hair-like to bristle-like, some more or less curved to twisted, brown, reddish or partly whitish to yellowish; centrals 4–6 or more, the longest *c.* 1–2.5 cm, often poorly differentiated from the 9–25, shorter, adpressed and interlaced radials. Flowers rather large, commonly 5–8 × 5–8 cm, *c.* 1–5 produced together, lemon-yellow; pericarpel and tube elongate, equal to or longer than the perianth-segments. Fruit ovoid to spherical, *c.* 1.5 cm, thin-walled, not elongating but splitting or disintegrating at maturity; seeds often more than 100 per fruit, bell-shaped, expanded and broadest at the hilum, *c.* 0.8–1 mm, tuberculate, shiny, black. *S Brazil (Rio Grande do Sul), Uruguay.* G2. Spring.

A very free-flowering and popular species.

P. werdermanniana (Herter) N.P. Taylor (*Notocactus werdermannianus* Herter). Illustration: Osten, Notas sobre Cactàceas, pl. 37, 38 (1941); Mace, Notocactus, 76 (1975). Like the preceding but stem usually larger, to 13 × 10 cm; ribs to 40. Spines bristle- to needle-like, straight, yellow to white, radials not adpressed. Flowers *c.* 6 × 7 cm, sulphur-yellow, 6 or more produced together in a ring around the stem apex. *Uruguay.* G2. Spring.

17. P. scopa (Sprengel) N.P. Taylor (*Cactus scopa* Sprengel; *Notocactus scopa* (Sprengel) Berger). Illustration: Krainz, Die Kakteen, Lfg. 18 (1961) & 35, f. 1E, 1F (1967); Hirao, Colour Encyclopaedia of Cacti, f. 450–453 (1979); Hecht BLV Handbuch der Kakteen, 332 (1982); Pizzetti, Piante grasse le cactaceae, no. 209 (1985). Simple or clustering, spherical to cylindric, 5–50 × 6–10 cm, dark green but stem more or less obscured by the dense spines; ribs 25–40, low, finely tuberculate, the areoles only 3–8 mm apart. Central spines 3–4, brown, red or white, needle-like, 6–12 mm; radial spines *c.* 35–40 or more, finely bristle-like, glassy white or yellowish, 5–7 mm. Flowers 2–4 × 3.5–4.5 cm, bright yellow, many produced together in a ring around the stem apex. Fruit nearly spherical, *c.* 7 mm in diameter, its base remaining attached to stem after dehiscence; seeds more than 100 per fruit, *c.* 1 mm, tuberculate, dull black. *S Brazil (Rio Grande do Sul), Uruguay.* G2. Summer.

P. succinea (Ritter) N.P. Taylor (*Notocactus succineus* Ritter ('sucineus'); *N. neobuenekeri* Ritter; *N. fuscus* Ritter?). Illustration: Ritter, Kakteen in Südamerika 1: f. 121, 125, 126 (1979); Hirao, Colour Encyclopaedia of Cacti, f. 454 (1979); Riha & Subik, Illustrated Encyclopedia of Cacti & other Succulents, 169, f. 185 (1981); Pizzetti, Piante grasse le cactacee, no. 210 (1985). Like the preceding but stem smaller, 2.5–7 cm in diameter; ribs 18–26. Central spines 4–12, yellow, brownish or violet-grey, to 2.5 cm; radial spines 12–40, whitish, yellowish or brownish, 3–10 mm. Flowers 3–3.6 × 3–4 cm. Seeds 0.6–0.9 × 0.5–1 mm. *S Brazil (Rio Grande do Sul).* G2. Summer.

A little-known species that may be related to the above is **Notocactus schlosseri** Vliet. Illustration: Succulenta 53: 11 (1974). Stem large, to 18 × 11 cm; ribs 22. Central spines 4, to 2.4 cm; radial spines *c.* 34, to 1.1 cm. Flowers 4.5 × 5 cm. Fruit with *c.* 400 seeds. *Uruguay.* G2.

18. P. caespitosa (Spegazzini) N.P. Taylor (*Echinocactus caespitosus* Spegazzini; *Frailea caespitosa* (Spegazzini) Britton & Rose; *Notocactus caespitosus* (Spegazzini) Backeberg; *N. minimus* Fric & Kreuzinger; *N. tenuicylindricus* Ritter). Illustration: Krainz, Die Kakteen, Lfg. 19 (1962) & 35, f. 4A, 4B (1967); Ritter, Kakteen in Südamerika 1: f. 128 (1979); Hirao, Colour Encyclopaedia of Cacti, f. 456 (1979); Haustein, Der Kosmos-Kakteenführer, 205 (1983). Simple or branching by means of underground suckers, stem cylindric, 4–8 × 2–4 cm; ribs 11–22, only 1–4 mm high; areoles 1.5–4 mm apart. Central spines 1–4, brown or reddish brown, straight or some curved to hooked at apex, 3–15 mm; radial spines 9–17, bristle-like, white or yellowish, 3–6 mm. Flowers rather large for the size of the plant, 2.7–4.2 cm, sulphur- or lemon-yellow, only 1–4 produced together. Fruit ovoid, *c.* 10 mm in diameter. Seeds bell-shaped, 0.7–1.2 × 0.9–1.2 mm, expanded and broadest at the hilum, tuberculate, dull brown to black. *S Brazil (Rio Grande do Sul), Uruguay.* G2. Summer.

19. P. horstii (Ritter) N.P. Taylor (*Notocactus horstii* Ritter; *N. purpureus* Ritter). Illustration: Hirao, Colour Encyclopaedia of Cacti, 113, f. 437 (1979); Ritter, Kakteen in Südamerika 1: pl. 8, f. 129–131, 238 (1979); Pizzetti, Piante grasse le cactacee, nos. 203, 207 (1985). Simple or sparingly clustering, spherical, later elongating, to 14 cm in diameter, green but soon becoming corky from the base upwards; ribs 12–19, well-defined, 7–20 mm high; areoles in shallow notches between low tubercles, 5–9 mm apart. Central spines 1–6, needle-like, yellow to brown, straight, curved or twisted, 8–30 mm or more; radials *c.* 10–15, finer, white to pale brown, 6–30 mm. Flowers 3–4 cm, yellowish orange, reddish or purplish, arising very close to the stem apex; stigma-lobes orange-yellow, pinkish, purplish or almost white. Fruit spherical to barrel-shaped, 7–10 × 6–8 mm; seeds 1 × 0.8–1 mm, tuberculate, matt black. *S Brazil (Rio Grande do Sul).* G2. Summer–autumn, occasionally spring.

20. P. crassigibba (Ritter) N.P. Taylor (*Notocactus uebelmannianus* Buining; *N. crassigibbus* Ritter; *N. arachnites* Ritter). Illustration: Backeberg, Cactus Lexicon, f. 496 (1977); Hirao, Colour Encyclopaedia of Cacti, f. 434, 434-1, 435, 435-1, (1979); Hecht, BLV Handbuch der Kakteen, 329,

333 (1982); Cullmann et al., Cacti, 255 (1986). Simple, depressed-spherical 4–17 cm in diameter, shining dark green; ribs 10–16, very low, rounded, with broad chin-like tubercles between the areoles. Spines all more or less adpressed, mostly curved, whitish to grey or pale brown, 5–30 mm; central spine absent or 1; radials 6–14. Flowers rather variable in size, form and colour, 3.5–6 × 4.5–6 cm, nearly white, yellow or reddish purple. Fruit barrel-shaped, 5–10 mm; seeds to 1.4 × 1.4 mm, hilum corky, seed-coat minutely roughened under a lens, matt black. *S Brazil (Rio Grande do Sul).* G2. Spring–summer.

21. P. rutilans (Daeniker & Krainz) N.P. Taylor (*Notocactus rutilans* Daeniker & Krainz). Illustration: Krainz, Die Kakteen, Lfg. 2 (1956) & 35, f. 6B (1967); Hecht, BLV Handbuch der Kakteen, 33l (1982); Haustein, Der Kosmos-Kakteenführer, 207 (1983); Cullmann et al., Cacti, 255 (1986). Simple, spherical or somewhat elongated, *c.* 5 cm in diameter or slightly larger, dark green; ribs 18–24, vertical or weakly spiralled, low, with chin-like tubercles between the areoles. Central spines 2, the lower one to 7 mm, reddish brown, straight, stiff, radials *c.* 14–16, slender, needle-like, whitish, darker-tipped, to 5 mm. Flowers 3–4 × 6 cm, pink, becoming paler to yellowish in the throat. Fruit elongating at base, *c.* 1.5 cm, thin-walled; seeds *c.* 1 mm, dull black, finely tuberculate. *N Uruguay.* G2. Summer.

A justly popular species for its compact habit and relatively large bicoloured flowers.

22. P. herteri (Werdermann) N.P. Taylor (*Echinocactus herteri* Werdermann; *Notocactus herteri* (Werdermann) Buining & Kreuzinger). Illustration: Krainz, Die Kakteen, Lfg. 25 (1963) & 35, f. 1G (1967); Hecht, BLV Handbuch der Kakteen, 330 (1982); Cullmann et al., Cacti, 254 (1986). Simple, spherical or slightly elongated, 10–15 cm in diameter, dark green, becoming corky at base; ribs *c.* 20–30, high and straight, the areoles set in notches with chin-like tubercles between. Central spines 4–6, subulate, brown, to 2 cm; radial spines *c.* 8–17, needle-like, entirely white or brown-tipped, to 1.2 cm. Flowers *c.* 4 × 5 cm, brilliant pink or darker, pale to whitish in the throat, many produced together on large plants, in a ring around stem apex. Fruit spherical; seeds *c.* 1 mm, hilum corky, seed-coat finely tuberculate,

shiny black. *S Brazil (S Rio Grande do Sul),
Uruguay.* G2. Summer.

A very beautiful species which does not
normally flower until the stem is at least
10 cm in diameter.

Notocactus roseoluteus Vliet is probably
only a variant of the next species, with
stems to 18 cm in diameter, which keys to
here on account of its pinkish flowers to
8 cm in diameter. *Uruguay.* G2.

23. P. mammulosa (Lemaire) N.P. Taylor
(*Echinocactus mammulosus* Lemaire;
Notocactus mammulosus (Lemaire) Berger;
N. submammulosus (Lemaire) Backeberg;
Echinocactus submammulosus Lemaire;
E. floricomus Arechavaleta; *Notocactus
floricomus* (Arechavaleta) Berger;
N. pampeanus (Spegazzini) Backeberg).
Illustration: Krainz, Die Kakteen, Lfg. 35,
f. 9A, 9B (1967); Hirao, Colour
Encyclopaedia of Cacti, f. 446–448 (1979);
Haustein, Der Kosmos-Kakteenführer, 209
(1983); Cullmann et al., Cacti, 255 (1986).
Simple, spherical or somewhat elongated,
5–13 cm in diameter, dark green; ribs
13–21, rarely more, vertical, well-defined,
with large, pointed, chin-like tubercles
between the areoles. Central spines 2–4 or
more and then less easily distinguished
from the radials, straight, rather stout and
stiff, white to grey or brownish, usually one
strongly flattened, to c. 2 cm; radials
c. 6–25, needle-like or stouter, whitish to
brownish, 5–10 mm. Flowers
c. 3.5–5.5 cm, pale to golden yellow;
pericarpel and tube short and broad, with
very dense pale wool and few dark bristles.
Fruit spherical at first, then elongating
below and eventually releasing seeds at
base, thin-walled; seeds helmet- to bell-
shaped, c. 1.2 mm, hilum conspicuous,
corky, seed-coat finely tuberculate, matt
brownish. *S Brazil (Rio Grande do Sul),
Uruguay, NE Argentina.* G2. Summer.

P. mueller-melchersii (Backeberg) N.P.
Taylor (*Notocactus mueller-melchersii*
Backeberg). Illustration: Krainz, Die
Kakteen, Lfg. 19 (1962) & 35, f. 9C, 9D
(1967); Hirao, Colour Encyclopaedia of
Cacti, 114, f. 443 (1979). Simple, 5–8 ×
5–6 cm, dark green; ribs 21–24, low, with
small rounded tubercles. Central spines
1–3, straight, not flattened, subulate to
needle-like, pale yellow, darker at base and
apex, 4–20 mm; radials 14–18 or more,
slender, needle-like, whitish, 2–8 mm.
Flowers c. 3 × 4.5–5 cm, pale golden
yellow. Fruit elongating at base, thin-
walled, c. 7 mm in diameter; seeds
c. 1.2 mm, dull black, tuberculate.
Uruguay. G2. Summer.

The plant known as *Notocactus mueller-
moelleri* Fleischer & Schuetz appears to be
somewhat intermediate between *P.
mammulosa* and *P. mueller-melchersii*. It is
said to resemble the former in its flowers
and fruits, but its stem is smaller and the
reddish to brownish, terete central spines
are only 5 mm. *Uruguay.* G2.

24. P. allosiphon (Marchesi) N.P. Taylor
(*Notocactus allosiphon* Marchesi).
Illustration: Ashingtonia 1: 67 (1974);
Mace, Notocactus, 24 (1975).
Simple, spherical, 8–12 × 11–13 cm, dark
green; ribs c. 15–16, very high, straight
and well-defined, with low tubercles below
the areoles. Central spines 4, terete, stiff
and sharp-pointed, dark red to black, then
grey, 8–20 mm; radials c. 2 (more in
young plants), thinner, laterally directed.
Flowers 5.5 × 5 cm, pale yellow; pericarpel
and tube much shorter than the perianth-
segments. Fruit elongating below, to 3 cm,
not splitting, seeds escaping at base; seeds
1–1.1 mm, black, tuberculate. *Uruguay.*
G2.

25. P. buiningii (Buxbaum) N.P. Taylor
(*Notocactus buiningii* Buxbaum). Illustration:
Hirao, Colour Encyclopaedia of Cacti, f. 449
(1979); Hecht, BLV Handbuch der Kakteen,
336 (1982).
Simple, depressed-spherical to spherical, to
8 × 12 cm, pale grey- or glaucous-green;
ribs c. 16, straight, thin and sharply acute,
with narrow blade-like tubercles between
the areoles. Central spines 3–4, straight,
stiff, yellowish, dark brown at base,
c. 2–3 cm; radial spines c. 2–3, similar to
the centrals but smaller. Flowers to 7 ×
8 cm, yellow; pericarpel and tube with
thick dark to light brown wool. Fruit
elongating at base, to c. 3 cm; seeds
helmet-shaped, to 1.4 mm, hilum
conspicuous, corky, seed-coat finely
tuberculate, matt black. *N Uruguay, S Brazil
(S Rio Grande do Sul).* G2. Summer.

A very distinctive and beautiful species,
but not the easiest to grow, requiring very
well-drained compost, a warm sunny
position and even watering throughout the
year except in the depths of winter.

26. P. erinacea (Haworth) N.P. Taylor
(*Cactus erinaceus* Haworth; *Malacocarpus
erinaceus* (Haworth) Ruempler; *Wigginsia
erinacea* (Haworth) Porter; *Notocactus
erinaceus* (Haworth) Krainz; *Malacocarpus
corynodes* (Pfeiffer) Salm-Dyck; *Wigginsia
corynodes* (Pfeiffer) Porter; *Notocactus
corynodes* (Pfeiffer) Krainz; *Malacocarpus
tephracanthus* (Link & Otto) Schumann;
Wigginsia tephracantha (Link & Otto) Porter;

Notocactus tephracanthus (Link & Otto)
Krainz; *Malacocarpus sellowii* (Link & Otto)
Schumann; *Wigginsia sellowii* (Link & Otto)
Ritter; *Malacocarpus sessiliflorus* (Pfeiffer)
Backeberg; *Wigginsia sessiliflora* (Pfeiffer)
Porter; *Notocactus sessiliflorus* (Pfeiffer)
Krainz; *Malacocarpus fricii* (Arechavaleta)
Berger; *Wigginsia fricii* (Arechavaleta)
Porter; *Notocactus fricii* (Arechavaleta)
Krainz; *Malacocarpus arechavaletai*
(Spegazzini) Berger; *Wigginsia arechavaletai*
(Spegazzini) Porter; *Notocactus arechavaletai*
misapplied; *Echinocactus acuatus* var.
arechavaletai Spegazzini; *Malacocarpus
vorwerkianus* (Werdermann) Backeberg;
Wigginsia vorwerkiana (Werdermann)
Porter; *Malacocarpus pauciareolatus*
(Arechavaleta) Berger; *Notocactus
pauciareolatus* (Arechavaleta) Krainz).
Illustration: Krainz, Die Kakteen, Lfg. 35,
f. 6C, 7 (1967) & Lfg. 53–54 (1973); Hirao,
Colour Encyclopaedia of Cacti, f. 422–425
(1979); Hecht, BLV Handbuch der Kakteen,
336 (1982); Cullmann et al., Cacti, 253
(1986).
Simple, depressed-spherical, spherical, or
short-cylindric when old, 6–30 cm in
diameter, light to dark green; apex very
woolly in old plants; ribs 12–30, sharply
acute, well-defined, areoles situated in
notches. Spines whitish, grey or brown,
straight to strongly curved, subulate, to
2 cm; central spine absent or 1; radial
spines 2–12, mostly adpressed to stem.
Flowers 3–5 × 4–7 cm, glossy yellow,
borne singly or 2–3 arising together from
the stem apex; pericarpel and tube short,
covered in dense brownish wool, often
partly hidden in the wool of the stem apex.
Fruit elongating when ripe, club-shaped,
pink or reddish, partly naked with sticky
pulp at first, later drying and appearing
hollow, to 4 cm; seeds bell-shaped,
c. 1 mm, black, finely roughened. *S Brazil
(Rio Grande do Sul), Uruguay, NE Argentina.*
G2. Summer.

An extremely variable species, the young
plants generally very different from the
more woolly, older individuals.

P. neohorstii (Theunissen) N.P. Taylor
(*Notocactus neohorstii* Theunissen; *Wigginsia
horstii* Ritter). Illustration: Ritter, Kakteen
in Südamerika 1: f. 150, 151 (1979);
Hirao, Colour Encyclopaedia of Cacti,
f. 427–428 (1979). Simple, spherical,
3–9 cm; ribs 18–26. Central spines 1–6,
straight, needle-like, pale below, dark
brown to black at apex, 1–3 cm; radial
spines 14–24, adpressed, more slender,
whitish, 3–7 mm. Flower 2.5–4 ×
2.5–3.5 cm, shiny yellow. Fruit 8 × 4 mm,

hollow, hidden in the wool of the stem apex; seeds *c.* 1 × 1 mm, shiny, black. *S Brazil (Rio Grande do Sul). G2. Summer.*

The names '*Notocactus incomptus*' and '*N. muegelianus*' are of uncertain application and do not appear to be valid at present. The exact identity of *N. muricatus* (Pfeiffer) Berger (*Echinocactus muricatus* Pfeiffer) is debatable: according to Mace (Notocactus, 50, 1975), some plants in cultivation are close to *P. concinna*, but the name may refer to a variant of *P. ottonis* or some other species.

54. UEBELMANNIA Buining
N.P. Taylor

Simple, spiny, with mostly small spherical to cylindric, ribbed stems; surface smooth or finely tuberculate, sometimes covered in waxy scales (cf. *Astrophytum*); small to large cells filled with gum-like sap present beneath or amongst the green stem tissue. Flowers from stem apex, diurnal, small, shortly funnel-shaped, yellow; pericarpel and tube with minute scales, much wool and few inconspicuous bristles. Fruit spherical to cylindric, reddish, naked below, woolly and bristly at apex, thin-walled and dry at maturity, disintegrating; seeds small or to 2.3 mm, spherical, ovoid or cap-shaped, hilum somewhat depressed, seed-coat smooth or weakly tuberculate, reddish brown to black.

A genus of 3–5 species from the mountains of E Brazil (Minas Gerais). They are difficult to grow unless grafted, requiring a minimum temperature of *c.* 15 °C. When they are grown on their own roots, a warm sunny position, well-drained acidic compost and very careful watering, coupled with spraying, are essential.

1a. Ribs lacking tubercles; areoles 0–3 mm apart **1. pectinifera**
 b. Ribs tuberculate; areoles 3 mm or more apart **2. gummifera**

1. U. pectinifera Buining. Illustration:
Krainz, Die Kakteen, Lfg. 55–56 (1973), 62 (1975); Hirao, Colour Encyclopaedia of Cacti, figs. 502, 503 (1979); Haustein, Der Kosmos-Kakteenführer, 219 (1983); Cullmann et al., Cacti, 311 (1986). Spherical to cylindric, to 85 × 10–17 cm, but usually much smaller in cultivation, light grey-green to reddish brown, more or less covered in minute whitish scales (these almost absent in some cultivated plants); ribs *c.* 15–20, vertical, acute, not tuberculate; areoles contiguous or 1–3 mm apart. Spines 3–6 per areole, more or less

erect and comb-like in arrangement, light grey, dark brown or blackish, straight, needle-like, 5–15 mm. Flower 1.4–1.8 cm × 8–10 mm. Fruit 15–25 × 8 mm; seeds *c.* 10, 2–2.3 × 1.6–1.8 mm. *Brazil (Minas Gerais). G2. Summer.*

2. U. gummifera (Backeberg & Voll) Buining (*Parodia gummifera* Backeberg & Voll). Illustration: Hirao, Colour Encyclopaedia of Cacti, fig. 504 (1979). Spherical to slightly elongated, to 12 × 9 cm, grey-green; ribs *c.* 32, strongly tuberculate; areoles 3 mm or more apart. Spines *c.* 4–6 , more or less erect, subulate, light to dark grey, brownish tipped, 3–15 mm. Flower *c.* 2 × 1.5 cm. Fruit 1.8 cm × 7–8 mm. *E Brazil (Minas Gerais). G2. Summer.*

Rarely cultivated, as are the following, which are closely related and may not be distinct.

U. buiningii Donald. Illustration: Hirao, Colour Encyclopaedia of Cacti, fig. 505 (1979); Haustein, Der Kosmos-Kakteenführer, 219 (1983); Cullmann et al., Cacti, 311 (1986). Like No. **2** but stem greenish red-brown to deep chocolate brown; ribs *c.* 18. Spines 4–8, some incurved, yellow-brown, black-tipped, later whitish. Flowers 2.7 × 2 cm. *E Brazil (Minas Gerais). G2. Summer.*

U. meninensis Buining. Illustrations: Krainz, Die Kakteen, Lfg. 55–56 (1973); Pizzetti, Piante grasse le cactacee, no. 297 (1985). Very similar to *U. gummifera* but stem light to dark green or reddish, to 50 × 10 cm (usually much smaller in cultivation); ribs to 40. Spines 2–4, to 2 cm. Flower 2.2–3.5 × 2–3 cm. Fruit 8 × 6 mm, yellowish green; seeds numerous, 1.3 × 0.8 mm. *E Brazil (Minas Gerais). G2. Summer.*

55. FRAILEA Britton & Rose
N.P. Taylor

Low-growing with dwarf, depressed-spherical to cylindric, mostly weakly ribbed or tuberculate stems, simple or clustering. Flowers arising near stem-apex, shortly funnel-shaped, yellow, diurnal, but opening only briefly, or cleistogamous; pericarpel and tube clothed with small scales, their axils with much wool and bristles. Fruit thin-walled, dry, densely packed with seeds, the withered perianth and/or receptacular wool and bristles persistent, indehiscent or rupturing irregularly at maturity. Seeds 1–3 mm or more in diameter, often broadest near the hilum, cap or helmet-shaped; hilum depressed, often large and

long; seed-coat brown or blackish brown, glossy, often with minute, hair-like projections.

A genus of about 10–15 species, native of E Bolivia, Paraguay, S Brazil, and adjacent parts of E Argentina, with 1 doubtfully native in Colombia. They flower unpredictably during summer and autumn, but in cultivation flowers are often cleistogamous, the young flower-buds becoming converted into fruits almost before the perianth has begun to develop. Most species can be raised to maturity from seed in 1.5–3 years. They need regular watering during the warmer months of the year and a minimum temperature of 5–10 °C in winter.

1a. Stem cylindric; central spines much darker than the whitish radials, projecting **1. gracillima**
 b. Stem depressed-spherical to spherical or if cylindric then spines not as above 2
2a. Ribs conspicuously tuberculate or divided into large, more or less spiralled tubercles 3
 b. Ribs nearly flat and tubercles minute or very low 7
3a. Clustering, forming small mounds of stems in time; spines pale yellowish white, sometimes dark-tipped **2. grahliana**
 b. Simple or sparingly clustering in age, or if forming mounds then spines brown to blackish 4
4a. Spines dark, short, not interlaced, scarcely or not exceeding the adjacent rib sinuses or tubercle margins; stem usually offsetting **4. schilinzkyana**
 b. Spines light or dark, more or less interlaced, exceeding the adjacent rib sinuses, or stem mostly simple or offsetting only when old 5
5a. Tubercles with reddish purple crescent-shaped markings beneath and/or stigma-lobes to 5.5 mm, very slender; hilum of seed narrowly ovate-oblong; central spines absent or inconspicuous 6
 b. Tubercles lacking crescent-shaped markings beneath; stigma-lobes 1.5–3.5 mm, stout; hilum of seed broadly ovate-elliptic; central spines conspicuous, rarely absent **3. pumila**
6a. Areoles large, with persistent white wool; spines mostly dark brownish, later grey, to 3 mm **5. chiquitana**
 b. Areoles small, the wool soon disappearing; spines yellow, to 5 mm **6. mammifera**

7a. Stem depressed-spherical to hemispheric, or with very short, tightly adpressed spines   8
 b. Stem spherical to shortly cylindric; spines interlaced, not all tightly adpressed   9
8a. Spines radiating; ribs 15–26
                                    9. cataphracta
 b. Spines directed downwards; ribs 8–15
                                    10. castanea
9a. Ribs 13–24 (see also No. 6)
                                    7. pygmaea
 b. Ribs *c.* 32   **8. curvispina**

1. F. gracillima (Lemaire) Britton & Rose (*Echinocactus gracillimus* Lemaire; *Frailea alacriportana* Backeberg & Voll; *F. albifusca* Ritter; *F. itapuensis* invalid). Illustration: Ritter, Kakteen in Südamerika **1**: f. 152, 155 (1979); Haustein, Der Kosmos-Kakteenführer, 217 (1983); Kakteen/ Sukkulenten (DDR) **19** (3/4): front cover (1984).
Simple, cylindric, 6–10 × 1.5–3 cm, erect, grey-green; ribs 14–22, tuberculate. Central spines 2–5, 4–10 mm, dark reddish to blackish brown, projecting; radial spines 8–13, 2–4 mm, white. Flowers to 3.5–4 × 3.5–4 cm; tube very densely clothed in wool and bristles; inner segments oblanceolate, gradually tapered towards apex, pale yellow; stigma-lobes *c.* 9–10, 4–6 mm. Fruit *c.* 9 mm in diameter; seeds 1.3–1.7 × 1.5–2.7 mm, brown to blackish, smooth. *Central N Uruguay, S Brazil (Rio Grande do Sul).* G2.

F. horstii Ritter. Illustration: Ritter, Kakteen in Südamerika **1**: f. 153 (1979); Hirao, Colour Encyclopaedia of Cacti, 122, f. 480 (1979). Stem to 18 × 2–2.5 cm, eventually sprawling and branching; ribs 20–33. Central spines 3–6, yellowish brown to reddish brown; radial spines 15–20. Flowers to 4.3 × 5 cm; seeds 1.3 × 1.5–1.8 mm. *S Brazil (Rio Grande do Sul).* G2. Perhaps only a variety of *F. gracillima*.

F. lepida Buining & Brederoo. Illustration: Krainz, Die Kakteen, Lfg. 54 (1973). Stem to 5 × 1–1.5 cm, erect, grey to blackish green; ribs sometimes purplish beneath the tubercles. Radial spines 9–20, 1.5–2 mm. Flowers 1.5–2 × 2.2 cm; stigma-lobes 5–6, *c.* 3 mm. Fruit 7–8 mm in diameter; seeds to 2.4 × 1.5–1.7 mm. *S Brazil (Rio Grande do Sul).* G2.

2. F. grahliana (Schumann) Britton & Rose (*Echinocactus grahlianus* Schumann). Illustration: Succulenta 1971: 49 (1971); Haustein, Der Kosmos-Kakteenführer, 217 (1983) – as F. colombiana; Pizzetti, Piante grasse le cactaceae, no. 115 (1985).

Clustering, forming mounds of depressed-spherical stems; individual heads to 1.75–3 × 2.5–4 cm, dark green to reddish or bronze-coloured; ribs to 13, conspicuously tuberculate; areoles with white to yellowish wool. Spines scarcely interlaced, to 3.5–5 mm, pale yellow then grey, centrals mostly absent, radials 7–11. Flowers variable in size, to 3.5 × 4 cm; perianth-segments rounded to acuminate, canary-yellow; stigma-lobes 7–8, *c.* 2.5 mm. Fruit 6 mm in diameter; seeds 1.3–1.5 × 1.2 mm, shiny brown, hilum broadly elliptic. *S Paraguay, ?Argentina (Misiones).* G2.

F. moseriana Buining & Brederoo (*F. grahliana* misapplied?). Illustration: Krainz, Die Kakteen, Lfg. 50–51 (1972), 59 (1974); Herbel, Alles ueber Kakteen, 16 (1978). Clustering, forming mounds of depressed-spherical to spherical stems; individual heads 2.5–4 × 3–4 cm, light to dark green; ribs 13–19, strongly tuberculate; areoles large and woolly. Spines partly interlaced, to 5 mm, yellowish white, darker tipped, centrals 1–2, more or less straight-projecting, radials 9–11 (rarely to 13). Flowers 2.4–3 × 2.1–3.5 cm; perianth-segments rounded to acuminate-mucronate, pale or sulphur-yellow; stigma-lobes 4–7, 2.5–3 mm. Fruit 6–7 mm in diameter; seeds 1.3–1.7 × 1.2–1.5 mm, hilum ovate. *S Paraguay.* G2.

3. F. pumila (Lemaire) Britton & Rose (*Echinocactus pumilus* Lemaire; *F. friedrichii* Buining & Moser; *F. albiareolata* Buining & Brederoo; *F. carminifilamentosa* Backeberg, invalid). Illustration: Krainz, Die Kakteen, Lfg. 50–51 (1972), Lfg. 53 (1973); Backeberg, Cactus Lexicon, f. 117 (1977); Ritter, Kakteen in Südamerika **1**: f. 163 (1979).
Simple or sparingly clustering in age, stem depressed-spherical to spherical, to 3 × 3–5 cm, dark green to reddish; ribs 17–20, conspicuously tuberculate. Spines more or less interlaced, to 5 mm, yellowish to brown or reddish, centrals 1–4, directed outwards, conspicuous, radials 12–16, more or less curved. Flowers 1.8–2.7 × 2–3.5 cm; perianth-segments tapering to acuminate, lemon or golden yellow; stigma-lobes 7–11, 2.5–3.5 mm. Fruit 6–10 mm in diameter; seeds 1.2–2 × 1.2–1.7 mm, brown, hilum ovate-elliptic, small. *S Paraguay, S Brazil (W Rio Grande do Sul), NW Uruguay.* G2.

The following are very similar and may not be distinct from the above:

F. ignacionensis Buining & Moser (*F.*

ybatensis Buining & Moser). Illustration: Krainz, Die Kakteen, Lfg. 48/49 (1972). Simple, stem depressed-spherical, 2.5–3 × 4–4.5 cm, light to dark green; ribs *c.* 18–24, more or less divided into conspicuous, spiralled tubercles. Spines mostly interlaced, to 4–5 mm, whitish to pale brown, centrals absent or 1, radials 12–13. Flowers 2.4–3 × 2–4.5 cm; perianth-segments acute, deep or lemon yellow; stigma-lobes 5, *c.* 3 mm. Fruit 5–7 mm in diameter; seeds 1–1.4 × 1.5–1.9 mm, dark or blackish brown, hilum broadly ovate-elliptic. *S Paraguay.* G2.

F. knippeliana (Quehl) Britton & Rose (*Echinocactus knippelianus* Quehl). Illustration: Krainz, Die Kakteen, Lfg. 59 (1974); National Cactus & Succulent Journal **32**: 4 (1977). Simple or branching when old, stem shortly cylindric, *c.* 4 × 2 cm, grass green; ribs 14–15, conspicuously tuberculate. Spines interlaced, to 5 mm, yellowish, centrals *c.* 2, radials 12–14. Flowers to 2.5 × 4.5 cm; perianth-segments bluntly acute to rounded, canary-yellow; stigma-lobes 6, 1.5 mm. Fruit *c.* 5.5 mm in diameter; seeds 0.9–1 × 0.9–1 mm, brown, hilum broadly ovate. *S Paraguay.* G2.

F. colombiana (Werdermann) Backeberg (*Echinocactus colombianus* Werdermann; *F. pumila* misapplied; *F. chrysantha* Hrabe?). Illustration: Krainz, Die Kakteen, Lfg. 19 (1962). Stems depressed-spherical to shortly cylindric, to 6 × 4 cm, dark to light green; ribs 17–30, tuberculate. Spines interlaced, to 6 mm, brown, yellow or yellow tipped brown, very dense, centrals 2–6, stoutly straight-projecting, radials 16–20. Flowers to 2.5 × 2–2.5 cm; perianth-segments acuminate; stigma-lobes *c.* 7. Fruit *c.* 5 mm in diameter; seeds *c.* 1–2 × 1 mm, brown, hilum broadly elliptic, small. *Colombia (introduced?), S Brazil & Uruguay.* G2.

4. F. schilinzkyana (Schumann) Britton & Rose (*Echinocactus schilinzkyanus* Schumann). Illustration: Herbel, Alles ueber Kakteen, 16 (1978); Haustein, Der Kosmos-Kakteenführer, 217 (1983). Clustering, rarely remaining simple, stems depressed-spherical to spherical, 2–4 × 2–4 cm, bright green; ribs ill-defined, 10–20, or completely divided into low but conspicuous, spirally arranged tubercles. Spines not interlaced, to 3 mm, brown to blackish, centrals absent or 1, inconspicuous, radials 10–14. Flowers 2–3.5 × 2.5–3.5 cm; perianth-segments abruptly acuminate, sulphur-yellow; stigma-lobes 5–7, long and slender. Fruit *c.* 5 mm in diameter; seeds 1.8–2 mm in diameter, shiny brown, smooth, hilum elongate-ovate. *S Paraguay, Argentina (Misiones).* G2.

5. F. chiquitana Cardenas (*F. pullispina* Backeberg, invalid). Illustration: Succulenta **49**(9): front cover (1970); Backeberg, Cactus Lexicon, f. 119–121, 123.3 (1977); Cullmann et al., Kakteen, edn 5, 176 (1984) – as Frailea sp.
Simple, rarely offsetting, depressed-spherical to spherical, rarely short-cylindric, 2–3 × 2.5–3.5 cm, yellowish or bluish green or tinged with pink to red; ribs *c.* 18–24, straight and well-defined but broken up into conspicuous, blunt tubercles 2–3 mm in diameter, often with pronounced reddish markings beneath; areoles large with persistent white wool. Spines not or slightly interlaced, to 3 mm, whitish or more often dark brown to blackish, centrals absent or 1–3, radials 8–12, adpressed. Flowers 1.7–3.2 × 2.2–4 cm; perianth-segments attenuate-acute, yellow; stigma-lobes 5–7, to 5.5 mm, very slender. Fruit to 10 mm in diameter; seeds 2 mm in diameter, shiny dark brown to blackish, smooth, hilum elongate-ovate-oblong. *E Bolivia.* G2.

6. F. mammifera Buining & Brederoo. Illustration: Krainz, Die Kakteen, Lfg. 50/51 (1972).
Simple, rarely branching, spherical or somewhat elongated, *c.* 3 × 2.5 cm, shiny, dark green; ribs *c.* 15–19, broken up into conspicuous, blunt tubercles, *c.* 2.5 mm in diameter, with pronounced reddish purple markings beneath; areoles minute, wool soon disappearing. Spines only slightly interlaced, to 5 mm, golden yellow, later whitish, centrals absent, radials 6–8, not adpressed. Flowers 2.2–2.5 × 2.6 cm; perianth-segments light yellow; stigma-lobes 5–8, *c.* 3.5 mm. Fruit 10 mm in diameter; seeds 2 mm in diameter, seed-coat shiny, brown, hilum narrowly ovate-oblong. *S Brazil (S Rio Grande do Sul), E Argentina.* G2.

Closely related to the above, but lacking a valid name, is 'F. magnifica', represented in cultivation by the collection Horst-Uebelmann 64. Illustration: Cactus & Succulent Journal (US) **44**: 36, f. 5 (1972); Hirao, H., Colour Encyclopaedia of Cacti, f. 479 (1979). From *F. mammifera* it differs in its eventually cylindric habit (to 6 × 1–2 cm), much smaller tubercles, and more numerous (*c.* 12–15), densely arranged

radial spines, only 1.5–2 mm. *S Brazil (Rio Grande do Sul).* G2.

7. F. pygmaea (Spegazzini) Britton & Rose (*Echinocactus pygmaeus* Spegazzini; *E. pulcherrimus* Arechavaleta; *Frailea pulcherrima* (Arechavaleta) Spegazzini; *F. aurea* Backeberg; *F. dadakii* Berger; *F. asperispina* Ritter; *F. fulviseta* Buining & Brederoo; *F. aureispina* Ritter; *F. aureinitens* Buining & Brederoo). Illustration: Ritter, Kakteen in Südamerika 1: f. 156–161, 168, 169, 243 (1979).
Simple, sometimes clustering when old, stem spherical to shortly cylindric, 1–7 × 1–2.5 cm, light to dark or grey-green; ribs 13–23, composed of minute, flattish and generally inconspicuous tubercles, often with dark reddish to purplish, crescent- or V-shaped markings beneath (especially conspicuous in young plants); areoles minute, with white, grey or brownish wool. Spines not or partly interlaced, 1–4 mm, bristle-like, white, yellow or brownish at first, usually more or less adpressed, centrals absent or 1–3, often inconspicuous, radials 6–19. Flowers large, commonly 3.5–5 × 3.5–5 cm; perianth-segments attenuate-acute to acuminate, mostly pale or very pale yellow; stigma-lobes 7–12, to 6.5 mm. Fruit 6–10 mm in diameter; seeds 1.2–1.6 × 1.5–2 mm, shiny, dark brown to black, hilum rather narrowly oblong. *Argentina (Entre Rios), S Brazil (Rio Grande do Sul), Uruguay.* G2.

The following are very similar and probably only varieties of the above.

F. albicolumnaris Ritter. Illustration: Ritter, Kakteen in Südamerika 1: f. 167 (1979). Stem shortly cylindric, 4–6 × 2–2.6 cm; ribs 21–24, finely tuberculate, without purple markings beneath the minute areoles. Spines partly interlaced, 2–5 mm, whitish, centrals 2–4, conspicuous, radials 14–18. Flowers *c.* 4 × 5 cm, sulphur yellow; stigma-lobes 11, 5 mm long. Seeds 2 × 2.5 mm, shiny, brownish black, smooth. *S Brazil (Rio Grande do Sul).* G2.

F. perumbilicata Ritter. Illustration: Ritter, Kakteen in Südamerika 1: f. 164–166 (1979). Simple, stem spherical, 1.5–3 cm in diameter, dark or somewhat grey-green; ribs 14–19, flat and scarcely tuberculate, without markings beneath the minute areoles. Spines not or partly interlaced, 2–3 mm, light to dark brown, centrals absent or 1–3, radials 6–11. Flowers 3.5–5.3 × 4–5.5 cm; perianth-segments abruptly acuminate, very pale to sulphur-yellow; stigma-

lobes 7–11, 5–7 mm. Fruit to 10 mm in diameter; seeds 2 × 2.5–3 mm, shiny, dark brown, smooth, hilum very narrowly oblong. *S Brazil (Rio Grande do Sul).* G2.

8. F. curvispina Buining & Brederoo. Illustration: Krainz, Die Kakteen, Lfg. 50/51 (1972); Cactus & Succulent Journal (US) **44**: 37, f. 7 (1972).
Simple or clustering, stem to 5 × 3 cm, grey-green; ribs *c.* 32, finely tuberculate. Spines 4–6 mm, glassy white, whitish or somewhat yellowish, curved and twisted, interlaced and obscuring the stem, central 1, radials *c.* 14. Flowers *c.* 3 × 2.6 cm; tube slender, densely clothed in wool and bristles; stigma-lobes 8, 4.5 mm. Fruit 1.2 cm in diameter; seeds 1.5 × 1.5 mm, chestnut-brown, smooth, hilum narrowly oval. *S Brazil (Rio Grande do Sul).* G2.

9. F. cataphracta (Dams) Britton & Rose (*Echinocactus cataphractus* Dams; *Frailea cataphracta* var. *tuyensis* Buining & Moser). Illustration: Krainz, Die Kakteen, Lfg. 48/49 (1972); Haustein, Der Kosmos-Kakteenführer, 219 (1983).
Simple or sometimes clustering, stem depressed-spherical, 1–3.5 × 2.5–4 cm, dull deep green, grey-green or brownish to reddish bronze; ribs to 15–25, very low with almost flat tubercles; areoles white to brownish, commonly with dark crescent-shaped markings beneath. Spines not interlaced, to 4 mm, mostly adpressed or partly deciduous, centrals absent or 1–2; radials 5–11. Flowers 2–3.8 × 2–4 cm; perianth-segments attenuate-acute, pale yellow; stigma-lobes 5–9, 3–4 mm. Fruit 4–8 mm in diameter; seeds 1.5–2 × 2–2.5 mm, shiny brown to black, smooth or minutely hairy, hilum very narrowly oblong. *E Bolivia, Brazil (S Mato Grosso), S & E Paraguay.* G2.

Var. **cataphracta**. Areoles with brownish wool or almost naked. Spines to 2 mm, persistent, pale yellow or glassy-white, brown at base, all strongly adpressed, inconspicuous, centrals absent. *S Paraguay.*

Var. **duchii** Moser (*F. melitae* Buining & Brederoo; *F. matoana* Buining & Brederoo; *F. cataphractoides* and *F. uhligiana* Backeberg, invalid). Illustration: Cactus & Succulent Journal (US) **43**: 139 (1971); Kakteen und andere Sukkulenten 25(6): 121 (1974); National Cactus & Succulent Journal **32**: 83 (1977). Areoles with conspicuous whitish wool. Spines to 4 mm, deciduous or persistent, yellowish grey or reddish brown to almost black, centrals absent or 1–2, projecting, radials less adpressed. *E Bolivia, Brazil (S Mato Grosso), S & E Paraguay.*

F. phaeodisca (Spegazzini) Spegazzini (*Echinocactus pygmaeus* [var.] *phaeodiscus* Spegazzini; *Frailea pygmaea* var. *phaeodisca* (Spegazzini) Ito; *F. perbella* Prestle). Illustration: Ashingtonia 1(5): 54 (1974); Succulenta **59** (5): 116–119, 184–185 (1980); Andersohn, Cacti and Succulents, 212 (1983) – as *F. castanea*. Simple, 1.5–3 × 1.5–3.5 cm; roots long, tapered; stem pale to dark green or dark brown; ribs 22–26, completely flat and recognisable only by the straight, dark, vertical line marking each sinus, with dark crescent-shaped markings beneath the areoles in juvenile plants and seedlings; areoles violet-black. Spines not interlaced, to 2.5 mm, white, brownish black at base, completely adpressed, centrals absent, radials 10–14. Flowers 2.2–3.5 × 2.2–4 cm; perianth-segments acuminate, sulphur-yellow or paler; stigma-lobes 6–7, to 3.5 mm. Fruit 10 mm in diameter; seeds 1.5–2 × 2.3–3 mm, shiny, dark brown, smooth, hilum narrowly oblong-ovate. *S Brazil (Rio Grande do Sul), Uruguay.* G2.

10. F. castanea Backeberg (*F. asterioides* Werdermann). Illustration: Ashingtonia 1(5): 54–55 (1974); Backeberg, Cactus Lexicon, f. 118 (1977); Haustein, Der Kosmos-Kakteenführer, 219 (1983). Simple, depressed-spherical, 1–2 × 4.5 cm; rootstock spherical; stem chocolate-brown to dark green; ribs 8–15, flat to slightly convex but scarcely tuberculate, sinuses straight; areoles brown to whitish, conspicuous. Spines slightly or not interlaced, 0.5–5 mm, shiny brown then dull brownish to blackish, stout, usually all directed downwards, more or less adpressed, centrals usually absent, radials 3–11. Flowers 3–4 × 3–5 cm; perianth-segments rounded to broadly acuminate, golden yellow; stigma-lobes 8–11, 3.5 mm. Fruit to 1.3 cm in diameter; seeds 2 × 3–3.3 mm, shiny chestnut-brown, hilum narrowly oblong. *NE Argentina (Misiones), S Brazil (Rio Grande do Sul), N Uruguay.* G2.

The name *F. pseudopulcherrima* Ito is of uncertain application.

56. ECHINOCACTUS Link & Otto
N.P. Taylor
Simple (rarely clustering in cultivation) with massive spherical, barrel-shaped or short-cylindric, strongly ribbed stems; areoles large, long and more or less contiguous in adult plants, forming a broad woolly crown at the stem apex, bearing

stout spines; nectar-secreting glands absent (see *Ferocactus*, p. 279). Flowers arising from and sunken into the woolly stem apex, shortly funnel-shaped to bell-shaped; pericarpel and tube short with numerous, narrow, pointed scales above, with wool in their axils. Fruit top-shaped to oblong-club-shaped, pale yellow, bearing narrow scales and wool, interior with juicy pulp or nearly dry, indehiscent; seeds large, mussel-shaped, shiny brown to black, smooth, hilum small.

This description applies mainly to *E. platyacanthus* and *E. grusonii*, from central and northern Mexico. *E. horizonthalonius* (N Mexico, SW USA) has a much smaller, depressed stem like *Homalocephala* (p. 274) and flowers agreeing closely with those of *Astrophytum*, to either of which it may be more closely related. Two other, smaller-growing species, from N Mexico and SW United States, are also of uncertain position, but are seldom cultivated. Flowering in the typical species rarely occurs unless they are given a free root-run, a very sunny position and allowed to reach a considerable size and age.

1a. Stem spherical to barrel-shaped, eventually massive, green when adult; flowers yellow   2
  b. Stem depressed-spherical, rarely elongated, to 10–20 cm in diameter, blue grey-green; flowers pale pink to magenta   **3. horizonthalonius**
2a. Spines 4–12, dark to light brown, paler at apex, eventually grey or blackish   **1. platyacanthus**
  b. Spines *c.* 12–14, yellow to whitish, later darkening   **2. grusonii**

1. E. platyacanthus Link & Otto (*E. ingens* Zuccarini; *E. grandis* Rose; *E. palmeri* Rose; *E. visnaga* Hooker). Illustration: Hirao, Colour Encyclopaedia of Cacti, f. 4–6 (1979); Cullmann et al., Cacti, 152 (1986). Stem to 3 × 1 m, but much smaller in cultivation, green but glaucous and sometimes purple-banded when juvenile; ribs 5–20 in young plants, later increasing to 30–60 or more, at first strongly tuberculate, later high and even-edged; areoles round, but becoming elliptic to linear and continuous in mature plants. Spines brownish, *c.* 1–8 cm, oval to angular, centrals 1–4, radials 3–8. Flowers 3–6 × 3–8 cm; perianth-segments broad, yellow, the outer ones tapering to narrow brownish points. Fruit narrowly club-shaped, to *c.* 8 × 2.5 cm. *C & N Mexico.* G2. Summer, but rarely flowering in cultivation.

2. E. grusonii Hildmann. Illustrations: Hirao, Colour Encyclopaedia of Cacti, f. 1–3 (1979); Cullmann et al., Cacti, 151–152 (1986).
Stem to 1.3 m × 80 cm, spherical but eventually elongating and sometimes offsetting, green; ribs 5–40 or more, initially tuberculate, later high and even-edged; areoles round, later continuous. Spines golden to pale yellow, sometimes white, to 3–5 cm, angular, centrals 4, radials 8–10. Flowers to 6 × 3–5 cm; perianth-segments narrow, yellow, all gradually tapering to a sharp and usually brownish point. Fruit top-shaped, *c.* 2 cm, with white, juicy flesh. *Central E & NE Mexico.* G2. Summer.

The 'Golden Barrel Cactus' found in collections the world over, but rare and endangered in the wild.

3. E. horizonthalonius Lemaire. Illustration: Hirao, Colour Encyclopaedia of Cacti, 4–5, f. 9 & 9.1 (1979); Cullmann et al., Cacti, 152 (1986).
Stem to 10–15 × 10–20 cm, depressed-spherical to hemispheric (or eventually shortly cylindric in the rare var. **nicholii** Benson), blue grey-green; ribs 7–13, tuberculate or nearly even-edged, low, broadly convex; areoles round, separate, but becoming close through compression. Spines grey to black, to 3 cm, somewhat recurved towards the stem, conspicuously tranverse-ridged, centrals 1–5, flattened, radials 5–7. Flowers 5–6 × 5–7 cm; perianth-segments, broad, pale pink to magenta, darker near base, only the outermost sharply pointed. Fruit *c.* 2.5 × 1.2 cm, juicy then dry. *N Mexico and SW USA.* G2. Summer.

57. THELOCACTUS Britton & Rose
N.P. Taylor
Simple or clustering, weakly to densely spiny, with small depressed-spherical to shortly cylindric, ribbed or tuberculate stems; areoles with or without nectar-secreting glands, sometimes extending at the upper edge to form a short groove. Flowers from the stem apex, arising from newly developed areoles, small to large, funnel-shaped; pericarpel and tube bearing conspicuous scales naked in their axils. Fruit spherical, either dark green to brownish, dry and dehiscent by a basal pore, or (in No. 1) bright red, somewhat fleshy and indehiscent; seeds pear-shaped, small or to 2.3 mm, hilum large, seed-coat net-like to strongly tuberculate, black.

A genus of 11 highly variable species,

from central and northern Mexico, 2 extending into Texas; popular in cultivation for their varied stems and flowers, the latter large and strikingly coloured in some species.

Literature: Anderson, E. F., A revision of the genus Thelocactus B. & R. (Cactaceae), *Bradleya* 5: 49–76 (1987).

1a. Central spines hooked at apex, fruits bright red; flowers yellow, red in the throat **1. setispinus**
 b. Central spines not hooked at apex; fruits dark green, brownish or dull purplish; flowers not as above **2**
2a. Flowers distinctly bi- or tricoloured (pinkish magenta, and sometimes also white, throat red); stigma-lobes yellow or red **2. bicolor**
 b. Flowers uniformly coloured or at least without a sharply defined contrasting colour in the throat (perianth-segments sometimes with darker midstripes); stigma-lobes white to yellow **3**
3a. Stem clustering, ribs 7–14, vertical or spiralled; flowers yellow or magenta; areoles with nectar-secreting glands **3. leucacanthus**
 b. Stem simple, lacking well-defined ribs or these more than 14, or flowers whitish, areoles lacking nectar-secreting glands **4**
4a. Radial spines 9–25, white; stem lacking ribs, with numerous spiralled tubercles **4. conothelos**
 b. Radial spines less than 9 or brownish; stem lacking ribs or ribs present though poorly defined **5**
5a. Stem dark green, brownish or reddish brown; flowers whitish or magenta **5. tulensis**
 b. Stem pale yellow-green or pale blue grey-green; flowers white to pale pink **6. hexaedrophorus**

1. T. setispinus (Engelmann) Anderson (*Hamatocactus setispinus* (Engelmann) Britton & Rose; *Ferocactus setispinus* (Engelmann) Benson; *Echinocactus setispinus* Engelmann). Illustration: Hirao, Colour Encyclopaedia of Cacti, f. 49 (1979); Hecht, BLV Handbuch der Kakteen, 282 (1982); Haustein, Der Kosmos-Kakteenführer, 237 (1983); Cullmann et al., Cacti, 174 (1986). Simple, rarely clustering, spherical to shortly cylindric, *c.* 7–12 × 5–9 cm, yellowish green to dark green; ribs 12–15, well-defined, acute, not tuberculate but sometimes sinuous; areoles with nectar-secreting glands. Central spine 1, yellowish white to reddish, apex pale, hooked, to

2.7 cm; radial spines 9–17, needle-like, reddish or white, 9–24 mm. Flower *c.* 5 × 5 cm, yellow with a red throat; tube well-developed; stigma-lobes pale yellow to white. Fruit spherical, *c.* 10 mm in diameter, bright red with white pulp; seeds 1.3–1.7 × 0.5–0.8 mm, finely tuberculate. *NE Mexico (Tamaulipas), USA (S Texas).* G2. Summer–autumn.

2. T. bicolor (Pfeiffer) Britton & Rose (*Echinocactus bicolor* Pfeiffer; *E. rhodophthalmus* Hooker; *E. bolaensis* Runge ('bolansis'); *E. wagnerianus* Berger; *Thelocactus wagnerianus* Berger; *T. schwarzii* Backeberg; *T. flavidispinus* (Backeberg) Backeberg). Illustration: Botanical Magazine, 4486, 4634 (1850, 1852); Hirao, Colour Encyclopaedia of Cacti, f. 124–127 (1979); Hecht, BLV Handbuch der Kakteen, 365 (1982); Cullmann et al., Cacti, 301 (1986).
Simple, rarely clustering, depressed-spherical, ovoid or cylindric, yellowish green to deep green, 5–38 × 5–14 cm (to only 5 × 7 cm in var. **flavidispinus** Backeberg); ribs 8–13, tuberculate, well-defined or obscure, vertical or somewhat spiralled; areoles with nectar secreting glands. Spines reddish, yellowish or white, paler at apex and at base; centrals 1–4 (usually absent in var. **schwarzii** (Backeberg) Anderson), stout, stiff, with a straight, sharp point, to 3.5 cm, the lowermost straight-projecting, the upper ones somewhat flattened or terete, pointing towards stem apex; radials 8–17, needle-like, usually to 2.5 cm, rarely longer. Flowers 3.5–7 × 4–8 cm, pinkish magenta with a red throat, sometimes with a band of white in between (especially in var. **schwarzii**); tube well-developed; stigma-lobes yellow or red. Fruit spherical-oblong, *c.* 10 mm in diameter, dark green to brownish, dry, seeds escaping at base, 1.2–2.5 × 1.1–1.5 mm, tuberculate. *N & NE Mexico, USA (S Texas).* G1. Summer.

An extremely variable species, especially in stem-size and spines.

T. heterochromus (Weber) Oosten (*Echinocactus heterochromus* Weber; *Thelocactus pottsii* misapplied). Illustration: Hirao, Colour Encyclopaedia of Cacti, f. 118 (1979); Hecht, BLV Handbuch der Kakteen, 363 (1982). Depressed-spherical, 4–7 × 6–15 cm, green to bluish green; ribs 7–11, with very large rounded tubercles; areoles lacking nectar-secreting glands. Central spines 1–4, very stout and flattened, yellowish white and red, to 3 cm, the lowermost down-pointed and sometimes

grooved, the others curving towards stem apex; radial spines 6–9, subulate, to 2.8 cm, otherwise like the centrals. Flowers like those of *T. bicolor* but to 10 cm in diameter, throat orange-red. Fruit *c.* 1.5 cm in diameter; seeds 1.6–2.2 × 1–1.5 mm, tuberculate. *N Mexico (Chihuahua & Durango).* G1. Summer.

3. T. leucacanthus (Pfeiffer) Britton & Rose (*Echinocactus leucacanthus* Pfeiffer; *E. ehrenbergii* Pfeiffer; *Thelocactus ehrenbergii* (Pfeiffer) Knuth; *T. sanchezmejoradae* Meyran). Illustration: Backeberg, Die Cactaceae 5: 2814–2820 (1961); Haustein, Der Kosmos-Kakteenführer, 243 (1983); Bradleya 5: 67 (1987).
Simple then clustering freely, stem spherical to shortly cylindric, light green, 4.5–15 × 2.5–5 cm; ribs 7–14, tuberculate but well-defined, vertical or slightly spiralled; areoles with nectar-secreting glands. Central spines absent or 1, rarely 3, straight-projecting, yellowish white or reddish to blackish, 5–50 mm; radial spines 6–20, yellowish to reddish, to 7 mm or more. Flower to 5 × 5 cm or more, yellow or magenta; tube short. Fruit spherical, 6–8 mm in diameter, dark green; seeds 1.4–2 × 1–1.8 mm, seed-coat net-like. *Central E Mexico.* G2. Summer.

Extremely variable in the number and length of its spines.

4. T. conothelos (Regel & Klein) Knuth (*Echinocactus conothelos* Regel & Klein; *Gymnocactus conothelos* (Regel & Klein) Backeberg; *Thelocactus saussieri* (Weber) Berger). Illustration: Hirao, Colour Encyclopaedia of Cacti, 39, f. l28, l28-1, 129 (1979); Cullmann et al., Cacti, 301 (1986); Bradleya 5: 67 (1987).
Simple, spherical or somewhat elongated, 6–25 × 7–17 cm, yellowish green to pale green; ribs lacking, replaced by upward pointing spiralled tubercles; areoles normally lacking nectar-secreting glands. Central spines 1–4, straight-projecting or ascending, nearly straight, dark blackish brown to grey, 1–5 cm; radials 9–23, needle-like, white, 5–20 mm. Flowers *c.* 5 × 4 cm, magenta or orange-yellow; tube well-developed, narrow. Fruit spherical-oblong, to 9 mm in diameter; seeds 1.5–2.1 × 1.2–1.5 mm, with pointed tubercles. *NE Mexico.* G1. Spring.

T. macdowellii (Quehl) Glass (*Echinocactus macdowellii* Quehl; *Echinomastus macdowellii* (Quehl) Britton & Rose; *Neolloydia macdowellii* (Quehl) Moore). Illustration: Hirao, Colour Encyclopaedia of Cacti, f. 109 (1979); Haustein, Der Kosmos-

Kakteenführer, 239 (1983); Cullmann et al., Cacti, 241 (1986); Bradleya **5**: 67 (1987). Like *T. conothelos* but sometimes clustering, spherical to club-shaped, 4–15 × 5–12 cm. Spines all yellowish white or white; radials 15–25, finer. Flower to 5 cm or more in diameter, magenta; tube short. Seeds 2 × 1 mm, tubercles low and rounded. *NE Mexico, rare.* G1. Summer.

This species has been regarded as a variety of *T. conothelos* by some authors, but is amply distinct in its flowers and seeds.

5. T. tulensis (Poselger) Britton & Rose (*Echinocactus tulensis* Poselger; *Thelocactus buekii* (Klein) Britton & Rose; *T. matudae* Sanchez-Mejorada & Lau). Illustration: Hirao, Colour Encyclopaedia of Cacti, f. 119, 120 (1979); Hecht, BLV Handbuch der Kakteen, 363 (1982); Bradleya **5**: 70, 75 (1987).
Simple (sometimes clustering in var. **tulensis**), depressed-spherical to spherical, 5–25 × 6–18 cm dark green, purplish, brownish or reddish brown; ribs *c*. 10, strongly tuberculate and ill-defined (var. **tulensis**) or ribs absent, replaced by spiralled tubercles (var. **buekii** (Klein) Anderson); areoles without nectar-secreting glands. Central spines 1–7, the lowermost straight-projecting, brown then grey, 5–55 mm; radial spines 5–8, rarely more, needle-like, brownish to whitish, 7–15 mm. Flowers 2.5–5 × 3.5–8 cm, whitish (var. **tulensis**) or magenta (var. **buekii**); tube very short. Fruit spherical to oblong, to 10 mm in diameter; seeds 1.6–2.7 × 0.6–1.7 mm, net-patterned. *NE Mexico.* G2. Spring–Summer.

6. T. hexaedrophorus (Lemaire) Britton & Rose (*Echinocactus hexaedrophorus* Lemaire; *E. fossulatus* Scheidweiler; *Thelocactus fossulatus* (Scheidweiler) Britton & Rose; *T. lloydii* Britton & Rose). Illustration: Hirao, Colour Encyclopaedia of Cacti, f. 115–117 (1979); Haustein, Der Kosmos-Kakteenführer, 241; Cullmann et al., Cacti, 302 (1986); Bradleya **5**: 70 (1987).
Simple, depressed-spherical to spherical 3–8 × 8–15 cm, light green to grey-green; ribs 8–13, ill-defined, composed of large blunt tubercles; areoles without nectar-secreting glands. Spines 4–10, straight or curved, terete, stout, red, pink, yellow or grey, poorly differentiated into centrals and radials, the former absent or 1–3, to 2.5 cm, the latter 4–8, to 3.5 cm. Flowers 2.7–3.5 × 3.3–5.5 cm, white; tube very short. Fruit spherical, 8–12 mm in

diameter; seeds 1.5–2 × 1.2–1.5 mm, net-patterned. *N Mexico.* G1. Summer.

T. rinconensis (Poselger) Britton & Rose (*Echinocactus rinconensis* Poselger; *Thelocactus nidulans* (Quehl) Britton & Rose; *T. phymatothelos* (Ruempler) Britton & Rose; *T. lophothele* (Salm-Dyck) Britton & Rose; *Echinocactus lophothele* Salm-Dyck?) Illustration: Hirao, Colour Encyclopaedia of Cacti, f. 121–123 (1979); Hecht, BLV Handbuch der Kakteen, 364 (1982), – both figures (one as 'T. tulensis'); Haustein, Der Kosmos-Kakteenführer, 241 (1983); Cullmann et al., Cacti, 303 (1986). Simple, depressed-spherical to spherical, 4–15 × 8–20 cm, light grey-green, sometimes tinged with purple; ribs to 31, but poorly defined, tuberculate, tubercles conical, angled. Central spines absent or 1–4, straight, terete to angled, dark brown, yellowish or grey, pale at apex, very short or to 6 cm; radial spines absent or 1–5, similar to the centrals, 3–35 mm. Flowers 3–4 × 2.7–3 cm, white or pale pinkish, sometimes pale yellow in the throat; tube short. Fruit spherical to oblong, 7–9 mm in diameter; seeds 1.7–2 × 0.5–1 mm, net-patterned. *NE Mexico.* G2. Summer.

Like *T. hexaedrophorus*, this species is extremely variable in its spines. It differs from the former in its more numerous ribs and tubercles, the latter pointed not blunt, and in its smaller flowers.

58. PEDIOCACTUS Britton & Rose
N.P. Taylor
Low-growing with mostly dwarf, simple or clustered, tuberculate stems. Flowers from near the stem-apex, bell-shaped; pericarpel more or less naked; tube very short, scaly. Fruits mostly top-shaped, nearly naked, dull-coloured, the dried perianth partly deciduous leaving a cap which opens like a lid as the fruit-wall splits vertically; seeds small to large, obliquely obovoid, hilum lateral to the long axis, seed-coat tuberculate, sometimes with larger folds or wrinkles, grey to black.

A genus of about 6 species native to the USA (Colorado Plateau, Columbia River and Great Basins, and the Rocky Mountains). Rarely cultivated, except by specialist growers.
Literature: Heil, K., Armstrong, B. & Schleser, D., A Review of the genus Pediocactus, *Cactus & Succulent Journal (US)* **53**: 17–39 (1981).

With the exception of *P.simpsonii*, species of this genus are difficult to grow on their own roots and are usually grafted. Flowering is assisted by a cold winter rest

(below 0 °C), but failure can occur if water is not given once the buds have formed in early spring or late winter. *P. simpsonii* is hardy out-of-doors as far north as S England provided it is protected from damp.

1a. Stem 5–15 cm in diameter
              1. simpsonii
  b. Stem 1–5 cm in diameter
              2. knowltonii

1. P. simpsonii (Engelmann) Britton & Rose (*Echinocactus simpsonii* Engelmann). Illustration: Benson, Cacti of the United States and Canada, pl. 137–139 (1982); Cullmann et al., Cacti, 271 (1986). Simple or clustering, stem spherical to ovoid, 2.5–15 × 3–15 cm, with spirally arranged tubercles. Central spines 4–10, reddish brown, paler at base, 5–28 mm; radial spines usually 15–28 (rarely as few as 10 or as many as 35) white, 3–19 mm. Flower 1.2–3 × 1.5–2.5 cm, white, pink, magenta, yellow or yellow-green. Fruit 6–11 × 5–10 mm; seeds 2–3 × 1.5–2 mm. *W USA.* G1. Spring.

P. sileri (Engelmann) Benson (*Echinocactus sileri* Engelmann; *Utahia sileri* (Engelmann) Britton & Rose). Illustration: Cactus & Succulent Journal (US) **53**: 28 (1981); Benson, Cacti of the United States & Canada, pl. 143 (1982). Simple, depressed-ovoid to shortly cylindric, 5–15 (rarely to 25) × 6–11.5 cm. Central spines 3–5, brownish black, becoming grey, 1.3–3 cm; radial spines 11–15, 1.1–2.1 cm, white. Flowers to 2.2 × 2.5 cm, yellow; scales of tube and outer perianth-segments conspicuously ciliate. Fruit 1.2–1.5 cm × 6–9 mm; seeds 3.5–5 × 3–3.5 mm. *USA (Arizona).* G1. Spring.

2. P. knowltonii Benson. Illustration: Cactus & Succulent Journal (US) **53**: 31 (1981); Benson, Cacti of the United States & Canada, pl. 140, 141 (1982); Hecht, BLV Handbuch der Kakteen, 346 (1982); Cullmann et al., Cacti, 270 (1986). Simple or clustering, stem 0.7–5.5 × 1–3 cm; tubercles tiny, only 2–4 × 1–2 mm. Central spines absent; radial spines 18–26, pale brown, pink or white, directed back towards stem, 1–1.5 mm. Flowers 1–3.5 × 1–2.5 cm, pink. Fruit 4 × 3 mm; seeds 1.5 × 1–1.2 mm. *USA (Colorado, New Mexico).* G1. Spring.
Very rare in the wild.

P. peeblesianus (Croizat) Benson (*Navajoa peeblesiana* Croizat; *N. fickeisenii* Backeberg). Illustration: Haustein, Der Kosmos-Kakteenführer, 245 (1983); Cullmann et al., Cacti, 271 (1986). Obovoid, spherical or

depressed-spherical, 2.2–6 × 2–5.5 cm;
tubercles 3–7 × 4–6 mm. Central spines
absent or 1–2, 5–21 mm, white to pale
grey, corky; radials 3–7, 2–9 mm, like the
centrals. Flowers 1.5–2.5 cm in diameter,
cream, yellow or yellowish green. Fruit
7–11 × 6–11 mm; seeds 3 × 2 mm. *USA
(Arizona)*. G1. Spring.

P. paradinei Benson (*Pilocanthus paradinei*
(Benson) Benson & Backeberg). Illustration:
Hecht, BLV Handbuch der Kakteen, 347
(1982); Haustein, Der Kosmos-
Kakteenführer, 245 (1983). Simple,
3–7.5 × 2.5–3.8 (rarely to 8) cm; tubercles
to 5 × 3–5 mm. Central and radial spines
poorly differentiated, the centrals 3–6,
8–28 mm, the radials 13–22, 2–5 mm, all
hair-like and white to pale grey. Flowers to
2.2 × 1.9–2.5 cm, pale yellow or pink. Fruit
7–10 × 4.5–8 mm; seeds 2.5 × 2 mm. *USA
(Arizona)*. G1. Spring.

59. ASTROPHYTUM Lemaire
N.P. Taylor

Simple, with hemispheric, spherical or
shortly columnar, few-ribbed stems, the
surface green but hidden to varying degrees
by minute, whitish scales; areoles large,
distinct but sometimes close, spiny or
spineless. Flowers borne at stem apex,
shortly funnel-shaped, yellow or bicoloured
yellow and red; pericarpel and tube bearing
narrow, pointed scales with woolly axils.
Fruit spherical, scaly and slightly woolly,
with scant flesh, packed with numerous,
shiny, cap-shaped seeds, the coat expanded
and inrolled around the sunken hilum.

A genus of 4 distinct species, native of
central-eastern to north-eastern Mexico
and southernmost Texas. Numerous
varieties (meriting cultivar status only)
have been recognised in cultivation, based
on the number of ribs, form of the spines
and on the abundance or absence of the
epidermal scales. Hybrids between the
species have also arisen, and these combine
the specific characters in almost every
possible way. No attempt to account for
them is made here. All members of the
genus are justly popular for their
remarkable appearance and beautiful
flowers produced in summer.

1a. Stem spineless   2
  b. Stem spiny, the spines sometimes very
     few   3
2a. Stem hemispheric, with *c.* 8 nearly
     flat, low ribs   **4. asterias**
  b. Stem spherical to columnar with
     4–10 acute or high ribs
                   2. myriostigma

3a. Spines nearly straight, not or scarcely
     flattened; flowers entirely yellow; fruit
     opening at apex to expose the seeds
                   1. ornatum
  b. Spines curved or twisted, somewhat
     flattened; flowers red in the throat;
     fruit indehiscent or breaking open
     irregularly   **3. capricorne**

1. A. ornatum (de Candolle) Britton & Rose
(*Echinocactus ornatus* de Candolle; *E. mirbelii*
Lemaire). Illustration: Hecht, BLV
Handbuch der Kakteen, 226 (1982);
Haustein, Der Kosmos-Kakteenführer, 221
(1983).
Stem at first spherical, later columnar, to
20–30 cm in diameter, white scales more
abundant in young specimens, sometimes
almost absent later; ribs 6–8 ; areoles close
or to 5 cm apart. Spines 5–11, to 3 cm,
yellow or brownish, more or less straight.
Flowers to 10 cm in diameter, pale to deep
yellow. Fruit opening at apex. *Central E
Mexico*. G2.

2. A. myriostigma Lemaire (*A. coahuilense*
(Moeller) Kayser; *A. tulense* (Kayser)
Sadovsky & Schuetz). Illustration: Hirao,
Colour Encyclopaedia of Cacti, f. 72–83
(1979); Hecht, BLV Handbuch der Kakteen,
226 (1982).
Spherical or shortly columnar, to
10–20 cm in diameter, white scales usually
very dense giving the stem a chalky-white
appearance, but sometimes completely
absent (then known as 'Nudum'); ribs 3–5
or more (5–10 in var. **tulense** (Kayser)
Borg), acute to rounded; areoles
approximate or to *c.* 1 cm apart. Spines
absent, but present though minute in
seedlings. Flowers *c.* 4–6 cm in diameter,
entirely yellow (or with a red to orange
throat in var. **coahuilense** (Kayser) Borg).
Fruit opening at apex (except in var.
coahuilense). *NE Mexico*. G2.

3. A. capricorne (Dietrich) Britton & Rose
(*Echinocactus capricornis* Dietrich;
Astrophytum senile Fric; *A. niveum* (Kayser)
Haage & Sadovsky; *A. crassispinum*
(Moeller) Haage & Sadovsky). Illustration:
Hirao, Colour Encyclopaedia of Cacti,
f. 84–88.1 (1979); Hecht, BLV Handbuch
der Kakteen, 225 (1982); Cullmann et al.,
Cacti, 132 (1986).
Spherical or elongated, rarely columnar, to
15 cm in diameter, white scales dense,
sparse or absent; ribs 7–9 sharply acute;
areoles 1–3 cm apart. Spines 1–20, rarely
almost absent, to 7 cm or more, light to
dark brown, blackish or grey, variously
curved and twisted, flattened, very variable.

Flowers 6–10 cm in diameter, yellow
with a red throat. Fruit indehiscent or
rupturing irregularly at or near base.
N Mexico. G2.

4. A. asterias (Zuccarini) Lemaire
(*Echinocactus asterias* Zuccarini).
Illustration: Hecht, BLV Handbuch der
Kakteen, 224 (1982); Haustein, Der
Kosmos-Kakteenführer, 221 (1983).
Stem hemispheric, 4–15 cm in diameter;
ribs 6–10 flat or very slightly convex;
areoles *c.* 5 mm apart. Spines absent.
Flowers 4–6.5 cm in diameter, yellow with
an orange or red throat, rarely the throat
colour extending towards the tips of the
perianth-segments and giving the flower a
reddish tinge. Fruit indehiscent or breaking
irregularly near base. *NE Mexico, USA
(S Texas)*. G2.

Now very rare and endangered in the
wild.

60. SCLEROCACTUS Britton & Rose
N.P. Taylor

Low-growing, with small, depressed-
spherical to club-shaped or cylindric,
tuberculate-ribbed, spiny, usually simple
stems; areoles more or less extended above
the spine-bearing part, often with nectar-
secreting glands. Flowers from the stem
apex, shortly funnel-shaped or bell-shaped;
pericarpel bearing few to numerous scales
naked in their axils. Fruit ovoid to cylindric,
club- or barrel-shaped, usually scaly, with
persistent perianth remains, either fleshy to
juicy and indehiscent, or becoming dry and
opening irregularly at or towards base or
by means of 1–3 vertical splits in the fruit-
wall; seeds 1.5–4 mm, obliquely obovoid,
hilum more or less lateral to the long axis,
coat with broad, flattened tubercles and
net-like markings, brown to black.

A genus of about 15–18 species, native
of N Mexico and SW USA. Closely allied to,
and questionably distinct from *Pediocactus*;
differing in having weakly ribbed stems
(except No. **10**), a more or less scaly
pericarpel, persistent perianth remains, a
different fruit form or mode of dehiscence,
and in microscopic features of seed-
morphology. In addition, more than half
the species have hooked or strongly
recurved central spines and long areoles
with nectar-secreting glands.

Included here are species formerly
classified in *Toumeya*, *Echinomastus* and
Ancistrocactus by Britton & Rose (1922,
1923). *Toumeya papyracantha* (Engelmann)
Britton & Rose has been placed in
Pediocactus, with which it agrees in its

tuberculate stem and spine-characters, but on flower-, fruit- and seed-morphology it clearly belongs in *Sclerocactus*. *Echinomastus* differs from *Sclerocactus* only in its slightly longer areoles and characteristic spines, while *Ancistrocactus* (including *Glandulicactus* Backeberg), whose areoles are also rather long, has more or less fleshy to juicy, indehiscent fruits.

Both of these groups agree with *Sclerocactus* in flower- and seed-morphology, with minor points of difference.

Literature: there is no comprehensive, up-to-date account of the genus in either its present, broad sense, or of its component species groups formerly recognised as separate genera.

With the possible exceptions of *S. uncinatus* and *S. glaucus*, species of *Sclerocactus* are difficult to grow on their own roots in Europe and, even as grafted specimens, are seldom seen outside specialist collections. Unlike the species of *Pediocactus*, they will not withstand low temperatures for long periods. A well-drained, gritty compost, very careful watering and full sunlight, are essential.

1a. Lowermost 3 radial spines in each areole strongly recurved to hooked at apex, stout **1. uncinatus**

b. Lowermost radial spines not as above 2

2a. Perianth-segments predominantly greenish to brownish, or lowermost central spine strongly down-curved, fang-like 3

b. Perianth-segments predominantly whitish, yellow, pink or magenta; lowermost central spine more or less projecting or ascending, straight, hooked, or flattened and papery 4

3a. Lowermost central spine hooked, projecting at right angles to stem or slightly upward- to downward-directed; stigma-lobes spreading **2. scheeri**

b. Lowermost central spine only weakly recurved or strongly down-curved from near base, but not hooked; stigma-lobes remaining erect **3. unguispinus**

4a. Central spines all flattened, papery and grass-like; stem to 3.5 cm in diameter **10. papyracanthus**

b. Central spines not as above; stem more than 3.5 cm in diameter 5

5a. Central spines all straight or slightly curved at apex 6

b. One or more central spines hooked at apex 9

6a. Spines *c.* 28–36 per areole, mostly white, never reddish brown to pink, nearly or completely obscuring stem, not all adpressed, the centrals more than 5 mm; stigma-lobes green **4. mariposensis**

b. Spines 7–24 per areole and/or partly reddish brown to pink, not obscuring stem, or all adpressed and the projecting central spine only 1–5 mm; stigma-lobes green or pink 7

7a. Radial spines 16–25 per areole or flowers whitish; central spines 0.4–0.6 mm in diameter near base or the projecting one to only 5 mm **5. intertextus**

b. Radial spines 6–15; flowers never whitish; central spines 0.7–1.5 mm in diameter near base, 1.2–4 cm 8

8a. Areoles with 12–19 spines, nectar-secreting glands normally absent; stem green **6. johnsonii**

b. Areoles with 7–11 spines, nectar-secreting glands present, conspicuous; stem often somewhat glaucous **7. glaucus**

9a. Hooked central spines 1–3 per areole **8. whipplei**

b. Hooked central spines *c.* 4–8 per areole **9. polyancistrus**

1. S. uncinatus (Galeotti) N.P. Taylor (*Echinocactus uncinatus* Galeotti; *Ferocactus uncinatus* (Galeotti) Britton & Rose; *Hamatocactus uncinatus* (Galeotti) Orcutt; *Glandulicactus uncinatus* (Galeotti) Backeberg; *Ancistrocactus uncinatus* (Galeotti) Benson; *A. crassihamatus* (Weber) Benson; *Ferocactus crassihamatus* (Weber) Britton & Rose; *Glandulicactus crassihamatus* (Weber) Backeberg). Illustration: Hirao, Colour Encyclopaedia of Cacti, f. 110, 111 (1979); Benson, Cacti of the United States & Canada, pl. 162, f. 847, 848 (1982); Kakteen und andere Sukkulenten **33**: 246–247 (1982); Weniger, Cacti of Texas and neighboring states, 103 (1984).

Depressed-spherical to shortly cylindric, 7.5–20 × 5–7.5 cm (to 15 cm or more in diameter in var. **crassihamatus** (Weber) N. P. Taylor), glaucous green; ribs *c.* 13, well-defined, the areoles on large tubercles, elongated above and bearing active nectar-secreting glands for much of the year. Central spines 1–4, uppermost 3, when present, nearly straight or incurved, reddish brown or whitish, often flattened, lowermost 1 strongly hooked, ascending, reddish brown, 5–11 cm, slender (only 2–5 cm, very stout and stiff in var. **crassihamatus**); radial spines 7–11, to

5 cm, the upper and laterals nearly straight, pale at first, the lowermost 3 hooked, reddish brown. Flowers 2–4 × 2.5–3 cm; perianth-segments deep pinkish to brownish red (with broad white margins in var. **crassihamatus**); stigma-lobes *c.* 10, whitish, yellowish or pink, spreading. Fruit green, juicy, indehiscent, 3–4.4 × 1–1.8 cm (to 5 × 2 cm in var. **crassihamatus**, or red, drier, 1.5–2.5 × 1.5–2 cm in var. **wrightii** (Engelmann) N. P. Taylor); seeds *c.* 1.5 × 1 mm, black. *N Mexico (Guanajuato northwards), S USA (S Texas, S New Mexico)*. G2. Summer.

2. S. scheeri (Salm-Dyck) N.P. Taylor (*Echinocactus scheeri* Salm-Dyck; *Ancistrocactus scheeri* (Salm-Dyck) Britton & Rose; *A. brevihamatus* (Engelmann) Britton & Rose; *A. megarhizus* (Rose) Britton & Rose). Illustration: Hirao, Colour Encyclopaedia of Cacti, f. 112–114 (1979); Benson, Cacti of the United States and Canada, f. 840–846 (1982); Weniger, Cacti of Texas and neighboring states, 111 (1984); Cullmann et al., Cacti, 126 (1986).

Spherical to narrowly club-shaped, 2.5–15 × 2.5–7 cm, dark green; ribs *c.* 13, tuberculate; areoles elongated into a groove above and bearing very active nectar-secreting glands. Central spines 1–4, uppermost 1–3, when present, nearly straight, flattened, white to yellowish, lowermost 1 strongly hooked, directed upwards to downwards, pale yellowish to brownish, 1–4.5 cm; radial spines 12–20, to *c.* 1 cm, needle-like, white to pale yellow, all straight. Flowers 2.5–3 cm, yellowish green; stigma-lobes *c.* 10, pale greenish. Fruit green, club-shaped, juicy at first, indehiscent, 1.2–2.5 × 0.6–0.9 cm; seeds 1.5 × 2 mm, brown. *NE Mexico (Nuevo Leon, Tamaulipas), USA (S Texas)*. G2. Summer.

3. S. unguispinus (Engelmann) N.P. Taylor (*Echinocactus unguispinus* Engelmann; *Echinomastus unguispinus* (Engelmann) Britton & Rose; *E. laui* Frank & Zecher; *E. durangensis* (Runge) Britton & Rose; *E. mapimiensis* Backeberg). Illustration: Krainz, Die Kakteen, Lfg. 52 (1973); Cactus & Succulent Journal (US) **47**: 220–221, f. 3–6, 9–10 (1975); Kakteen und andere Sukkulenten **29**(2): front cover (1978); Hirao, Colour Encyclopaedia of Cacti, f. 108, 108-1 (1979).

Depressed-spherical to shortly cylindric, 7–14 × 7–15 cm; ribs 13–21, acute but strongly tuberculate; areoles slightly elongated above, nectar-secreting glands absent or inconspicuous. Central spines

3–9, pale yellow to grey with dark brownish to blackish tips, to 3.5 cm, the lowermost stout and strongly down-curved from near base, recurved at apex, others ascending, stout or finer (all much finer, lowermost scarcely recurved in var. **durangensis** (Runge) N.P. Taylor); radial spines 15–30, white to yellowish with darker tips, nearly straight, needle-like, 1–3.3 cm, all interlaced. Flowers to *c.* 3 cm; perianth-segments greenish brown or somewhat reddish brown, with pale margins; stigma-lobes *c.* 10–15, pale yellow to greenish yellow, remaining erect. Fruit green to brown, 1.3–2.5 × 0.8–1.2 cm; seeds 2 mm, black. *N Mexico (W San Luis Potosí, Zacatecas, E Durango, S Chihuahua).* G2. Summer.

4. S. mariposensis (Hester) N.P. Taylor (*Echinomastus mariposensis* Hester; *Neolloydia mariposensis* (Hester) Benson). Illustration: Cactus & Succulent Journal (US) **47**: 221–222, figs. 7–8, 11 (1975); Benson, Cacti of the United States & Canada, pl. 152 (1982); Weniger, Cacti of Texas and neighboring states, 133, 135 (1984).
Depressed-spherical to ovoid, 6–10 × 4–6 cm; ribs to *c.* 21, poorly defined, tuberculate. Central spines 2–4, pale brown with darker tips, the lowermost down-curved, 0.7–1.5 cm, the other(s) curving upwards, to 2 cm, all rather slender, to 0.6 mm in diameter near base; radials *c.* 26–32, whitish, 0.6 cm, interlaced and obscuring stem. Flowers comparatively large for the size of the plant, *c.* 2.5 × 4 cm; perianth-segments white to pale pink, with darker or greenish midstripes; stigma-lobes 5–8, green, erect. Fruit spherical to oblong, to 9 mm, yellowish green at first; seeds 1.3 × 1.5 mm, black. *N Mexico (Coahuila), USA (SW Texas).* G2. Summer.

5. S. intertextus (Engelmann) N.P. Taylor (*Echinocactus intertextus* Engelmann; *Neolloydia intertexta* (Engelmann) Benson; *Echinomastus intertextus* (Engelmann) Britton & Rose; *E. dasyacanthus* (Engelmann) Britton & Rose). Illustration: Krainz, Die Kakteen, Lfg. 22 (1962); Kakteen und andere Sukkulenten **32**(10): front cover (1981) – as *E. warnockii*; Benson, Cacti of the United States and Canada, pl. 154, f. 829–831 (1982); Weniger, Cacti of Texas and neighboring states, 127, 129, 130 (1984).
Spherical to shortly cylindric, 5–15 × 4–7.5 cm; ribs *c.* 13, tuberculate. Central spines 4, the lowermost projecting, but very short, 1–5 mm, uppermost 3 adpressed to

stem, to 1.2–1.5 cm (lowermost projecting to 2–4 cm and upper 3 less adpressed, to 2–4 cm in var. **dasyacanthus** (Engelmann) N.P. Taylor); radials 13–25, 9–15 mm, tightly adpressed; spines all pinkish or grey to yellowish with pink tips, to 0.6 mm in diameter near base. Flowers 2–3 × 2.5–3, pale pink or salmon to white; stigma-lobes 6–12, pink to crimson, erect. Fruit green then brown, 1.2 × 0.6 cm; seeds 1.5 × 2 mm, black. *N Mexico (Sonora, Chihuahua, Coahuila), SW USA (SE Arizona, New Mexico, SW Texas).* G2. Summer.

S. warnockii (Benson) N. P. Taylor (*Neolloydia warnockii* Benson; *Echinomastus warnockii* (Benson) Glass & Foster). Illustration: Cactus & Succulent Journal (US) **47**: 218, 223 (1975); Hirao, Colour Encyclopaedia of Cacti, f. 107 (1979); Weniger, Cacti of Texas and neighboring states, 132 (1984). Ovoid to elongated, 7–15 × 5–10 cm; ribs 13–21, rarely more, well-defined. Central spines 1–6 (depending on interpretation), lowermost projecting-ascending, 1.2–2.5 cm, others ascending or resembling radials, light yellow to pale brown with chalky blue tips; radials 12–14, shorter but otherwise like the centrals. Flowers *c.* 2.5 × 2.5 cm; perianth-segments white to pale pink, with pale green midstripes; stigma-lobes 5–10, light green, erect. Fruit green then brownish, spherical, 6 mm; seeds 1.5 × 2 mm, black. *USA (SW Texas).* G2. Summer.

6. S. johnsonii (Engelmann) N.P. Taylor (*Echinocactus johnsonii* Engelmann; *Echinomastus johnsonii* (Engelmann) Baxter; *Neolloydia johnsonii* (Engelmann) Benson). Illustration: Benson, Cacti of the United States & Canada, pl. 157–161 (1982). Ovoid to cylindric, 10–25 × 5–10 cm; ribs 17–21, acute. Central spines 4–9, all similar, variously directed, straight or curved, pink to reddish brown, to 3–4 cm, 1–1.5 mm in diameter near base; radial spines *c.* 9–10, paler, 1.2–2 cm. Flowers 5–6 × 5–7.5 cm, bicoloured, pinkish magenta with red in the throat, or yellow with deep green in the throat; stigma-lobes *c.* 10, light yellowish green to brownish green, erect. Fruit green at first, 1.2 × 1 cm, splitting vertically to release the 2 × 2.5 mm seeds. *SW USA (E California, S Nevada, W Arizona).* G2. Summer.

S. erectocentrus (Coulter) N. P. Taylor (*Echinocactus erectocentrus* Coulter; *Neolloydia erectocentra* (Coulter) Benson; *Echinomastus erectocentrus* (Coulter) Britton & Rose; *E. acunensis* Marshall). Illustration: Benson, The Cacti of Arizona, edn 3, 193

(1969); Krainz, Die Kakteen, Lfg. 53 (1973); Benson, Cacti of the United States & Canada, pl. 156, f. 833–838 (1982). Like No. **6**, but central spines 1, or 2–4 and the upper 1–3 like the radials, the lowermost 1 much longer, to 1.2–3.5 cm; radials 11–15, 1.2–2.5 cm. Flowers 4–5 × 3.8–5 cm, pale to orange-pink or rarely white. Fruit 1 × 0.75 cm; seeds 1.5 × 2 mm. *N Mexico (N Sonora), SW USA (S Arizona).*

7. S. glaucus (Schumann) Benson (*Echinocactus glaucus* Schumann; *S. franklinii* Evans). Illustration: Ashingtonia **3**(5/6): pl. 13, 14 (1979); Benson, Cacti of the United States and Canada, f. 762–764 (1982). Ovoid to spherical, 4–6 × 4–5 cm or larger, light green or often glaucous; ribs *c.* 12, well-defined; areoles with conspicuous nectar-secreting glands. Central spines 1–3, straight or slightly curved, rarely hooked, terete or flattened, to 2.5 cm, light to dark brown; radials 6–8, needle-like, to 2 cm, white or some brown. Flowers 3–3.8 × 4–5 cm, magenta; stigmas pinkish, erect. Fruit 9–12 × 9 mm; seeds *c.* 1.5 × 2.5 mm, black. *SW USA (E Utah, W Colorado).* G1. Summer.

8. S. whipplei (Engelmann & Bigelow) Britton & Rose (*S. intermedius* Peebles; *S. parviflorus* Clover & Jotter). Illustration: Benson, Cacti of the United States and Canada, pl. 133–135, f. 780–786 (1982); Weniger, Cacti of Texas and Neighboring States, 105 (1984).
Depressed-spherical to cylindric, 7.5–20 × 5–9 cm; ribs *c.* 13–15, well-defined; areoles with nectar-secreting glands. Central spines *c.* 4, the lowermost hooked, purplish pink to reddish, 2.5–4.5 cm, laterals similar but not hooked, uppermost flattened, white, to 5 cm; radial spines 7–11, to 2.5 cm, whitish. Flowers 1.5–5 × 2.5–5.5 (rarely to 7) cm, yellow, pink, purplish or rarely white; stigma-lobes 5–6, green, erect. Fruit 1.2–2.5 × 0.6–1 cm, dehiscing by means of a median horizontal split in the fruit wall; seeds 2 × 3 mm, black. *SW USA (Arizona, New Mexico, Colorado, Utah).* G2. Summer.

S. wrightiae Benson (*Pediocactus wrightiae* (Benson) Arp). Illustration: Benson, Cacti of the United States and Canada, f. 769, 770 (1982); Cullmann et al., Cacti, 271 (1986). Depressed-spherical to obovoid, 5–9 × 5–7.5 cm; ribs *c.* 13. Central spines 4, to 1.2 cm, lowermost hooked, dark brown; radials 8–10, to 0.6–1.2 cm, white. Flower 2–2.5 (rarely to 4) × 2–2.5 (rarely to 4) cm; perianth-segments whitish to pale pink, with brownish midribs; stigma-lobes 5–8, green. Fruit naked or with 1–2 scales,

spherical, 9–12 mm; seeds 2 × 3.5 mm.
USA (Utah). G2. Summer.

9. S. polyancistrus (Engelmann & Bigelow)
Britton & Rose (*Echinocactus polyancistrus*
Engelmann & Bigelow). Illustration: Britton
& Rose, The Cactaceae 3: f. 224 (1922);
Krainz, Die Kakteen, Lfg. 48/49 (1972);
Benson, Cacti of the United States and
Canada, pl. 136, f. 787 (1982).
Cylindric, 10–15 × 5–7.5 cm; ribs
c. 13–17, well-defined. Central spines
c. 9–11, the upper 3 white, flattened, the
remainder red or reddish brown, rarely
yellow, the majority (4–6 or more) hooked,
to 7.5–9 cm; radial spines 10–15, to 2 cm,
white. Flowers *c.* 5–6 × 5 cm, magenta;
stigma-lobes *c.* 10, pink. Fruit barrel-
shaped, *c.* 2.5 × 1.5–2 cm; seeds 3 ×
2.3 mm, black. *SW USA (S California, SW
Nevada)*. G2. Summer.

Very difficult to cultivate in Europe.

10. S. papyracanthus (Engelmann) N.P.
Taylor (*Mammillaria papyracantha*
Engelmann; *Toumeya papyracantha*
(Engelmann) Britton & Rose; *Pediocactus
papyracanthus* (Engelmann) Benson).
Illustration: Hirao, Colour Encyclopaedia of
Cacti, f. 153 (1979); Benson, Cacti of the
United States & Canada, pl. 146,
f. 808–811 (1982); Haustein, Der Kosmos-
Kakteenführer, 245 (1983); Cullmann et
al., Cacti, 271 (1986).
Simple (or clustering freely when grafted),
2.5–10 × 1.2–3.5 cm; ribs absent or
completely divided into small tubercles;
areoles with or without nectar-secreting
glands. Central spines 1–4, 1–5 cm,
brown, flattened, papery, flexible, often
curved but never with a rigid hook; radial
spines 5–9, 3–4 mm, white. Flowers to
2.5 × 2–2.5 cm; perianth-segments whitish
or the outer ones brownish; stigma-lobes
c. 5, pale green, erect. Fruit small, only
4.5–6 × 4.5 mm, with or without a few
scales, the wall splitting vertically down
one side to release the few seeds; seeds
2.5 × 3 mm, black. *SW USA (E Arizona,
New Mexico)*. G2. Summer.

Can be grown and flowered on its own
roots but liable to succumb to rot at stem
base; grows well grafted and quite common
in specialist collections.

61. HOMALOCEPHALA Britton & Rose
N.P. Taylor
Low-growing with depressed-spherical to
disc-shaped, simple, ribbed stems, to
12–20 × 30 cm, light to dark green; ribs
13–27, acute, thickened at the large,
woolly areoles. Spines very heavy, strongly

transverse-ridged, to 7.5 cm, central spine
1, much flattened, directed downwards and
strongly incurved at apex, brownish,
radials 5–7, brown, or the laterals whitish.
Flowers from near the stem apex, 5–6 ×
5–6 cm; pericarpel and tube bearing
narrow scales with tufts of wool in their
axils; perianth-segments numerous,
narrow, pinkish, to orange-red in the
throat, margins paler, conspicuously torn
or fringed, outermost segments narrowly
pointed and brownish at apex. Fruit *c.* 5 ×
2.5–3.8 cm, bright red, juicy, splitting
irregularly at maturity; seeds *c.* 2 ×
2.5 mm, black, net-patterned, hilum region
depressed.

A genus of a single species, allied to
Sclerocactus, but differing in the somewhat
woolly pericarpel and tube, and in having
large, flattened stems.

1. H. texensis (Hopffer) Britton & Rose
(*Echinocactus texensis* Hopffer). Illustration:
Benson, Cacti of the United States &
Canada, f. 759, pl. 130 (1982); Weniger,
Cacti of Texas and neighboring states, 99
(1984).
NE Mexico, Texas. G2. Summer.

62. NEOLLOYDIA Britton & Rose
N.P. Taylor
Low-growing with small or dwarf
depressed-spherical to short-cylindric,
tuberculate stems, simple or clustering, tap-
root sometimes spherical; spines various
but never hooked. Flowers mostly small,
shortly funnel-shaped, arising from near
the upper side of the spine-bearing areole
(or from the base of a tubercular groove in
N. conoidea); receptacle-tube narrow but
short, pericarpel naked or rarely with 1–2
small scales. Fruit spherical to top-shaped,
naked or almost so, dry or slightly fleshy,
splitting vertically or disintegrating to
release the small (1–2 mm), black, pear-
shaped, tuberculate seeds.

A genus of about 10–14 species, native
to E & NE Mexico and SW Texas. Closely
related to *Thelocactus*, but with smaller
stems and flowers, the receptacle-tube
(including pericarpel) usually naked, not
conspicuously scaly.
Literature: Anderson, E.F., A revision of the
genus Neolloydia B. & R. (Cactaceae),
Bradleya 4: 1–28, many illustrations (1986).

1a. Tubercles grooved above, the groove
    connecting the spine-bearing areole
    with a very woolly spineless flower-
    bearing areole in the axil of the
    tubercle; flowers 4–6 cm in diameter
    1. conoidea

 b. Tubercles not grooved, the flower-
    bearing part of the areole merely
    shortly extended above the spine-
    bearing part; flowers 1.5–4 cm in
    diameter   2
2a. Spines *c.* 25–50 per areole, all
    addressed to stem, feathery or comb-
    like, *c.* 1–2 mm   **8. valdeziana**
 b. Spines absent or 1–31 per areole,
    never all addressed, 1–42 mm   3
3a. Spines absent or 1–9 per areole, or
    stems 1–3 cm in diameter   4
 b. Spines 10–31 per areole; stems
    3–9 cm in diameter   5
4a. Stem pale green or light blue to grey-
    green; spines slender, needle-like,
    terete, straight or slightly curved,
    whitish or tipped with dark grey to
    black, 4–9 (rarely only 2) per areole
    6. gielsdorfiana
 b. Stem dark blue-green to grey or
    brownish; spines thickened, subulate,
    flattened and sometimes papery,
    straight, curved, twisted or tangled,
    light to dark brown, greyish or
    blackish, 0–23 per areole
    7. schmiedickeana
5a. Stem tapered at base into a neck
    connected to a large tuberous tap-root
    3. mandragora
 b. Stem not as above, though tap-root
    sometimes tuberous   6
6a. Stem depressed-spherical, flowers
    white or pink; central spines 1–2
    (rarely 3) per areole   **4. saueri**
 b. Stem spherical to short cylindric;
    flowers magenta, or if white then
    central spines 3–5 per areole   7
7a. Stem simple; perianth-segments
    magenta with conspicuously paler
    margins; style reddish   **2. smithii**
 b. Stem simple or clustering; perianth-
    segments magenta with scarcely paler
    margins, or white; style white
    5. horripila

1. N. conoidea (de Candolle) Britton & Rose
(*Mammillaria conoidea* de Candolle;
Neolloydia grandiflora (Pfeiffer) Knuth; *N.
ceratites* (Quehl) Britton & Rose; *N. texensis*
Britton & Rose; *N. matehualensis* Backeberg).
Illustration: Hecht, BLV Handbuch der
Kakteen, 324 (1982); Bradleya 4: 5, f. 2–7,
pl. 1 (1986).
Simple or clustering, 5–24 × 3–6 cm,
spherical-ovoid to shortly cylindric, grey to
bluish green or slightly yellowish green;
tubercles large, ascending, 3–10 × 6–10 ×
5–9 mm, with a woolly groove connecting
the apical spine-bearing and 'axillary'
flower-bearing areoles on the outer surface.

Central spines absent or 1–6, 5–30 mm, black to reddish brown, projecting, straight; radial spines 8–28, to 5–13 mm, white or whitish and dark tipped. Flowers 2–3 × 4–6 cm; perianth-segments to 3 cm × 9–10 mm, magenta. Seed-coat extended over part of the hilum making it V-shaped. *E & NE Mexico, SW USA (SW Texas).* G2. Summer.

An extremely variable species, especially in the number and arrangement of its tubercles and spines.

2. N. smithii (Muehlenpfordt) Kladiwa & Fittkau (*Echinocactus smithii* Muehlenpfordt; *E. beguinii* var. *senilis* Josefski; *N. beguinii* Britton & Rose; *Gymnocactus beguinii* Backeberg) Illustration: Hirao, Colour Encyclopaedia of Cacti, f. 136–137 (1979); Haustein, Der Kosmos-Kakteenführer, 247 (1983); Bradleya **4**: 18 (1986).
Simple, mostly 7–12 × 4–9 cm, spherical to short-cylindric, grey or blue-green; tubercles 3–5 × 3–4 × 2–3 mm. Central spines 1–4, to 1.2–3 cm, white to yellowish brown, darker towards apex, straight; radial spines 12–27, 3–18 mm, white, darker tipped, eventually yellowish. Flowers 1.8–3.5 × 1.2–4 cm; perianth-segments to 2.9 cm × 7.5 mm, magenta with paler margins. *NE Mexico (SE Coahuila, Nuevo Leon, N San Luis Potosii).* G1. Spring.

3. N. mandragora (Berger) Anderson (*Echinocactus mandragora* Berger; *Gymnocactus mandragora* (Berger) Backeberg; *Neolloydia subterranea* (Backeberg) Moore; *N. subterranea* var. *zaragosae* (Glass & Foster) Anderson). Illustration: Hirao, Colour Encyclopaedia of Cacti, f. 138 (1979); Hecht, BLV Handbuch der Kakteen, 273 (1982); Haustein, Der Kosmos-Kakteenführer, 247 (1983); Bradleya **4**: 26 (1986).
Simple, 3–7 × 3–6 cm, spherical to obovoid, green to grey-green, arising from a large spherical root narrowed at its junction with the stem; tubercles 3–4 × 3–4 × 3–5 mm. Central spines 1–2, 1–2.2 cm, whitish, darker at base or tip, straight; radial spines 8–25, 3–15 mm, white. Flowers 2–3 × 1.5–3.5 cm; perianth-segments to 2.5 cm × 5 mm, white with greenish, brownish or magenta midstripes. *NE Mexico (S Coahuila, S Nuevo Leon).* G2. Spring.

4. N. saueri (Boedeker) Knuth (*Echinocactus saueri* Boedeker; *Gymnocactus saueri* (Boedeker) Backeberg). Illustration: Hirao,

Colour Encyclopaedia of Cacti, f. 130 (1979); Bradleya **4**: 26 (1986).
Simple, 3–5 × 4–7.5 cm, depressed-spherical, grey- to blue-green; tubercles 9–10 × 7–9 × 2–5 mm. Central spines 1–3, 1–2 cm, greyish black, curving slightly upwards; radial spines 7–14, 5–15 mm, white. Flowers 1.5–2.3 × 2–2.5 cm; perianth-segments to l.2 cm × 3.5 mm, white with a pinkish midstripe. *NE Mexico (SW Tamaulipas).* G2.

N. knuthiana (Boedeker) Knuth (*Echinocactus knuthianus* Boedeker; *Gymnocactus knuthianus* (Boedeker) Backeberg; *G. ysabelae* (Schlange) Backeberg?). Illustration: Hirao, Colour Encyclopaedia of Cacti, f. 132 (1979) – as Gymnocactus spec.; Haustein, Der Kosmos-Kakteenführer, 243 (1983); Bradleya **4**: 23 (1986). Simple or clustering; tubercles 4–5 × 5–7 × 5–7 mm. Central spines 1–2, to 1–1.6 cm; radial spines l4–20, 6–8 mm. Flowers 2.3–3 × 1.8–2.5 cm; perianth-segments to 19 × 5 mm, pale pink with a darker midstripe. *E Mexico (San Luis Potosi).* G2.

5. N. horripila (Lemaire) Britton & Rose (*Mammillaria horripila* Lemaire; *Gymnocactus horripilus* (Lemaire) Backeberg; *Thelocactus goldii* Bravo). Illustration: Hirao, Colour Encyclopaedia of Cacti, f. 133 (1979); Hecht, BLV Handbuch der Kakteen, 274 (1982); Bradleya **4**: 23 (1986).
Simple or often clustering, 7–18 × 4–9 cm, spherical or elongate, yellowish to olive or blue-green; tubercles 7–9 × 5–7 × 5–7 mm. Central spines rarely absent or 3, usually 1, 1.2–4.2 cm, white to yellowish, with a dark tip, straight; radial spines 12–14, 9–41 mm, white to yellowish. Flowers 2.2–4 × 2.5–4 cm; perianth-segments to 2.3 cm × 6 mm, magenta, paler to white near base. *E Mexico (Hidalgo).* G2. Spring.

N. viereckii (Werdermann) Knuth (*Echinocactus viereckii* Werdermann; *Gymnocactus viereckii* (Werdermann) Backeberg; *G. viereckii* var. *major* Glass & Foster). Illustration: Cactus & Succulent Journal (US) **50**: 285 (1978); Hirao, Colour Encyclopaedia of Cacti, f. 134 (1979); Bradleya **4**: 23 (1986). Tubercles 4–6 × 8–15 × 5–6 mm; areoles with abundant wool. Central spines 3–5, 1.5–2 cm, dark brown towards apex; radial spines 13–22, 8–13 mm. Perianth-segments white or magenta and white near base. *NE Mexico (S Nuevo Leon, SW Tamaulipas).* G2.

6. N. gielsdorfiana (Werdermann) Knuth (*Echinocactus gielsdorfianus* Werdermann; *Gymnocactus gielsdorfianus* (Werdermann) Backeberg). Illustration: Hirao, Colour Encyclopaedia of Cacti, f. 135 (1979); Hecht, BLV Handbuch der Kakteen, 272 (1982); Bradleya **4**: 23 (1986).
Simple, rarely clustering, to 5–7 × 4.5–5 cm, spherical to ovoid or short cylindric, light blue to grey-green or somewhat yellowish green; tubercles 6–10 × 3–5 × 3–5 mm. Central spine absent or l, to 2 cm, white, dark-tipped or blackish, curved upwards; radial spines 5–7, to 2 cm, white with dark or blackish tips. Flowers 1.3–2.4 × 1.5–2 cm; perianth-segments to 1.3 cm × 4 mm, rather pale yellowish cream with faint midstripes; filaments white. *NE Mexico (SW Tamaulipas).* G2.

N. lophophoroides (Werdermann) Anderson (*Thelocactus lophophoroides* Werdermann; *Turbinicarpus lophophoroides* (Werdermann) Buxbaum and Backeberg). Illustration: Hirao, Colour Encyclopaedia of Cacti, f. 150 (1979); Hecht, BLV Handbuch der Kakteen, 370 (1982); Bradleya **4**: 23 (1986). Depressed-spherical to ovoid arising from a tuberous rootstock; tubercles 9–12 × 10–12 × 2–4 mm. Central spine absent or 1, to 11 mm; radial spines 2–5 (rarely to 6), 2–9 mm. Flowers to 3.5 cm in diameter; perianth-segments to 2 cm × 5 mm, silvery-white to pale pink. *E Mexico (San Luis Potosí).* G2. Summer.

N. laui (Glass & Foster) Anderson (*Turbinicarpus laui* Glass & Foster). Illustration: Bradleya **4**: 23 (1986). Depressed-spherical, 0.5–1.5 × 1.2–3.5 cm; tubercles 2–3 × 3–10 × 3–5 mm. Radial spines 6–8, 1.2–2.2 cm, brownish white. Flowers to 3.5 cm in diameter; perianth-segments to 1.9 cm × 5 mm, pale pink with darker mid stripes; filaments red. *E Mexico (San Luis Potosí).* G2. Summer.

7. N. schmiedickeana (Boedeker) Anderson (*Echinocactus schmiedickeanus* Boedeker; *Turbinicarpus schmiedickeanus* (Boedeker) Buxbaum & Backeberg). Illustration: Hecht, BLV Handbuch der Kakteen, 371 (1982); Bradleya **4**: 26 (1986).
Simple or clustering in cultivation, 1–3 × 1.5–5 cm, depressed-spherical to spherical or short cylindric, dark blue to grey-green; tubercles 1.5–7 × 3–18 × 2–8 mm. Spines absent or 1–8 or more, diverse, 1–30 mm, straight or curved, flattened and papery, or flexible and needle-like. Flowers 1.5–2.6 × 1–3.2 cm; perianth-segments to 2.7 cm × 7 mm, white, yellow or pink to magenta.

NE Mexico (S Nuevo Leon, SW Tamaulipas, San Luis Potosí). G1. Spring.

Var. **gracilis** (Glass & Foster) Anderson (*Turbinicarpus gracilis* Glass & Foster). Illustration: Cactus & Succulent Journal (US) **48**: 176–177 (1976), **49**: 167 (1977); Hirao, Colour Encyclopaedia of Cacti, f. 148 (1979). Stem grey-green; tubercles slender, 1.5–1.9 × 3–5 × 7–8 mm. Central spine 1, 1.8–2.3 cm × 1–2 mm, thin and papery, curved near apex, flexible; radial spines, 1–3, 2 mm, white. Flowers 2 × 1.5 cm; perianth-segments white, sometimes with faint pinkish midstripes; stigma-lobes whitish. *NE Mexico (S Nuevo Leon)*.

Var. **schmiedickeana**. Stem dull dark green; tubercles 3–4 × 5–7 × 4–5 mm. Spines 2–4, 1.5–2.2 cm × 1 mm, dark brown to grey, flattened, curved and sometimes twisted. Flowers 2–2.7 × 1.8–2.8 cm; perianth-segments white to pink with magenta midstripes; stigma-lobes white. *NE Mexico (SW Tamaulipas)*.

Var. **flaviflora** (Frank & Lau) Anderson (*Turbinicarpus flaviflorus* Frank & Lau). Illustration: Bradleya **4**: 26 (1986). Stem grey-green. Spines 4–6, to 3 cm, brown, flattened, curved towards stem apex. Flowers 1.5 × 1–1.5 cm; perianth-segments pale yellow; stigma-lobes whitish. *E Mexico (San Luis Potosí)*.

Var. **klinkeriana** (Backeberg & Jacobsen) Anderson (*Turbinicarpus klinkerianus* Backeberg & Jacobsen). Illustration: Hirao, Colour Encyclopaedia of Cacti, f. 146, 147 (1979). Stem blue-green to grey-green or brownish; tubercles 3–6 × 5–9 × 5–8 mm. Spines 1–3, 7–8 × 1 mm, dark brown then pale grey, cylindric, curved upward. Flowers 1.5–2.3 × 1–2.7 cm; perianth-segments white with magenta midstripe; stigma-lobes white. *E Mexico (San Luis Potosí)*.

Var. **schwarzii** (Shurly) Anderson (*Turbinicarpus schwarzii* (Shurly) Backeberg; *T. polaskii* Backeberg, invalid). Illustration: Cullmann et al., Kakteen, edn 5, 309 (1984); Bradleya **4**: 26 (1986). Stem grey-green; tubercles flattened, 5–7 × 6–8 × 2–3 mm. Spines absent or 1–3, to 1–1.4 cm, light brown, paler at base, darker at apex, flattened, curved towards stem apex. Flowers 2–2.5 × 2.5–3.2 cm; perianth-segments white with faint pinkish midstripes; stigma-lobes pink. *E Mexico (N San Luis Potosí)*.

Var. **macrochele** (Werdermann) Anderson (*Turbinicarpus macrochele* (Werdermann) Buxbaum & Backeberg). Illustration: Bradleya **4**: 26 (1986). Stem grey-green to yellow-green; tubercles

6–8 × 12–18 × 2–4 mm. Spines rarely absent, usually 4–6, to 2–2.7 cm × 1–1.5 mm, pale brown, strongly curved and twisted. Flowers 2–2.6 × 2.3–3.2 cm; perianth-segments white with faint brown to pink midstripe; stigma-lobes pink. *E Mexico (N San Luis Potosí)*.

Var. **dickisoniae** (Glass & Foster) Anderson (*Turbinicarpus schmiedickeanus* var. *dickisoniae* Glass & Foster). Illustration: Cactus & Succulent Journal (US) **54**: 74 (1982); Bradleya **4**: 26 (1986). Stem dark green. Central spines 1–3, 1.3–2.2 cm, brownish grey, cylindric, slightly curved; radial spines 18–20, *c*. 2.5 mm, white. Flowers 2 × 1.7 cm; perianth-segments white with pale reddish brown midstripe; stigma-lobes white. *NE Mexico (S Nuevo Leon)*.

N. pseudomacrochele (Backeberg) Anderson (*Turbinicarpus pseudomacrochele* (Backeberg) Buxbaum & Backeberg; *T. krainzianus* (Frank) Backeberg). Illustration: Bradleya **4**: 23 (1986). Simple, rarely clustering, 2–4 × 2.5–3.5 cm, spherical to short-cylindric, dark blue-green; tubercles 6–8 × 7–10 × 3–5 mm. Spines 5–8, to 1.5–3 cm, dirty white below, blackish brown to grey above, slender, bristly, curved and twisted. Flowers 2.5–3.2 × 3–3.5 cm; perianth-segments to 2.7 cm × 5 mm, creamy white or pink with darker midstripe; stigma-lobes white. *E C Mexico (Queretaro)*. G2. Summer.

8. N. valdeziana (Moeller) Anderson (*Pelecyphora valdeziana* Moeller; *Normanbokea valdeziana* (Moeller) Kladiwa & Buxbaum; *Gymnocactus valdezianus* (Moeller) Backeberg, invalid; *G. valdezianus* var. *albiflorus* (Pazout) Backeberg, invalid; *Turbinicarpus valdezianus* (Moeller) Glass & Foster). Illustration: Hirao, Colour Encyclopaedia of Cacti, f. 140, 141 (1979); Cullmann et al., Kakteen, edn 5, 310 (1984); Bradleya **4**: 18 (1986). Simple, 1–2.5 × 1.5–2.5 cm, depressed-spherical to ovoid, arising from a spherical rootstock, bright green but hidden by the spines; tubercles 2–3 × 1–2 × 2–3 mm; areoles circular. Central spines absent; radial spines *c*. 25–30, 1–2 mm, strongly adpressed, feathery, white. Flowers 2–2.5 × 2.2–2.5 cm; perianth-segments to 1.2 cm × 6 mm, white with pale pinkish midstripe, or magenta with paler margins. *NE Mexico (SE Coahuila, S Nuevo Leon)*. G1. Spring.

N. pseudopectinata (Backeberg) Anderson (*Pelecyphora pseudopectinata* Backeberg; *Normanbokea pseudopectinata*

(Backeberg) Kladiwa & Buxbaum; *Turbinicarpus pseudopectinatus* (Backeberg) Glass & Foster). Illustration: Cullmann et al., Kakteen, edn 5, 310 (1984); Bradleya **4**: 18 (1986). Simple, 2–3 × 2–3.5 cm; tubercles laterally compressed, 3–3.5 × 2–3 × 3 mm; areoles linear. Central spines absent; radial spines *c*. 50, 1–2 mm, white, comb-like, adpressed, not feathery. Flowers like those of *N. valdeziana* or slightly larger. *NE Mexico (SW Tamaulipas, S Nuevo Leon)*. G1.

63. ARIOCARPUS Scheidweiler
N.P. Taylor

Low-growing with depressed, mostly simple, tuberculate, virtually spineless stems and very thick rootstocks; tubercles arranged as in a rosette, triangular, short or long, with a woolly groove above, a circular areole below the tip or naked but immersed in wool at base. Flowers from the base of the tubercular groove, or from the areole on the tubercle or at its base, funnel-shaped; receptacle-tube (including pericarpel) well-developed but usually hidden by the tubercles and wool, naked; perianth-segments oblanceolate, spreading. Fruit spherical to oblong or club-shaped, fleshy, but eventually drying and disintegrating within the wool at the stem-apex or between the tubercles, inconspicuous; seeds ovoid, black, tuberculate, 1–1.5 mm.

A genus of 6 species native to N & E Mexico and S Texas. Very slow-growing and requiring many years to flower when propagated from seed. Formerly imported as mature field-collected specimens, but Nos. **4–6** are now regarded as threatened or endangered and trade in wild plants is prohibited.

Literature: Anderson, E.F., A revision of Ariocarpus (Cactaceae), *American Journal of Botany* **50**: 724–732 (1963), **51**: 144 (1964).

1a. Tubercles with a woolly groove above
                                                            2
  b. Tubercles not grooved, naked or bearing a circular areole above   3
2a. Tubercles convex above, often deeply fissured and wrinkled; stem 5–15 cm in diameter   **1. fissuratus**
  b. Tubercles flat above, slightly roughened but not fissured and wrinkled; stem 2–7 cm in diameter
                                     2. kotschoubeyanus
3a. Flowers pale yellow   **4. trigonus**
  b. Flowers white pink or magenta   4

4a. Tubercles light grey-, yellow- or blue-green, as long or slightly longer than broad; flowers white, rarely pinkish **3. retusus**

b. Tubercles dark green to brownish, usually more than twice as long as broad; flowers magenta 5

5a. Tubercles bearing a woolly areole above **6. agavoides**

b. Tubercles lacking an areole, woolly only around the base **5. scaphorostrus**

1. A. fissuratus (Engelmann) Schumann (*Mammillaria fissurata* Engelmann; *Roseocactus fissuratus* (Engelmann) Berger; *R. intermedius* Backeberg & Kilian, invalid; *Ariocarpus lloydii* Rose). Figure 28(12), p. 204. Illustration: Hirao, Colour Encyclopaedia of Cacti, f. 179, 180–1 (1979); Hecht, BLV Handbuch der Kakteen, 222 (1982); Haustein, Der Kosmos-Kakteenführer, 257 (1983).
Stem 5–10 cm in diameter (to 15 cm in var. **lloydii** (Rose) Marshall); tubercles 1–2 × 1.5–2.5 cm, grey-green to brownish, deeply fissured and wrinkled (less so in var. **lloydii**), with a woolly groove above. Flowers 1.5–3.5 × 2.5–4.5 cm, pale to deep magenta; stigma-lobes 5–10. Fruit 5–15 × 2–6 mm, whitish or greenish. *SW USA (W Texas), NE Mexico*. G1. Autumn.

2. A. kotschoubeyanus (Lemaire) Schumann (*Anhalonium kotschoubeyanum* Lemaire; *Roseocactus kotschoubeyanus* (Lemaire) Berger; *A. macdowellii* Marshall, invalid). Illustration: Hirao, Colour Encyclopaedia of Cacti, f. 182–184 (1979); Hecht, BLV Handbuch der Kakteen, 222 (1982); Cullmann et al., Kakteen, edn 5, 129 (1984).
Stem 2–7 cm in diameter, occasionally branched; tubercles 5–13 × 3–10 mm, dark green to brownish, only slightly roughened and with a woolly groove above. Flowers 1.8–2.5 × 1.5–2.5 cm, magenta or white; stigma-lobes 4–6. Fruit 5–18 × 1–3 mm, reddish to pinkish. *E & NE Mexico*. G1. Autumn–early winter.

3. A. retusus Scheidweiler (*A. furfuraceus* (Watson) Thompson). Illustration: Hirao, Colour Encyclopaedia of Cacti, f. 186–188 (1979); Hecht, BLV Handbuch der Kakteen, 221 (1982); Haustein, Der Kosmos-Kakteenführer, 257 (1983).
Stem 10–25 cm in diameter; tubercles 1.5–4 × 1–3.5 cm, outspread, light grey-, yellow- or blue-green, smooth or with a slightly wavy surface, with or without an areole near the tip, not grooved above.

Flowers 2–4.2 × 4–5 cm, white or pale pinkish; stigma-lobes 7–16. Fruit 1–2.5 cm × 3–10 mm, white to pinkish red. *NE Mexico*. G1. Autumn.

4. A. trigonus (Weber) Schumann (*Anhalonium trigonum* Weber; *A. elongatum* Salm-Dyck; *Ariocarpus elongatus* (Salm-Dyck) Lee).
Illustration: Hirao, Colour Encyclopaedia of Cacti, f. 189 (1979); Haustein, Der Kosmos-Kakteenführer, 257 (1983).
Stem 7–30 cm in diameter; tubercles 3.5–8 × 2–2.5 cm, erect or curving upwards or inwards, yellow-green or brownish grey-green, smooth, lacking an areole, not grooved above, apex sharply pointed. Flowers 2.5–4 × 3–5 cm, pale yellow; stigma-lobes 6–10. Fruit 1.5–2.5 cm × 4–12 mm, whitish. *NE Mexico*. G2. Autumn.

5. A. scaphorostrus Boedeker. Illustration: Hirao, Colour Encyclopaedia of Cacti, 57, f. 190 (1979); Cullmann et al., Kakteen, edn 5, 130 (1984).
Stem 4–9 cm in diameter; tubercles to 5 × 2 cm, almost erect or somewhat incurved, very dark green to brown, finely roughened, lacking an areole, not grooved, apex resembling the prow of a boat. Flowers to 4 × 4 cm, magenta; stigma-lobes 3–5. *NE Mexico (Nuevo Leon)*. G2. Autumn.

6. A. agavoides (Castaneda) Anderson (*Neogomesia agavoides* Castaneda).
Illustration: Hecht, BLV Handbuch der Kakteen, 221 (1982); Haustein, Der Kosmos-Kakteenführer, 257 (1983).
Stem (including tubercles) 5–8 cm in diameter, occasionally offsetting in cultivation; tubercles to *c.* 4 cm × 6 mm, dark green to brownish, dying off towards the tip when old, bearing a large woolly areole above with a few short spines. Flowers 4–5 × 4–5 cm, magenta; stigma-lobes *c.* 5–6. Fruit to 2.5 cm, club-shaped, red. *NE Mexico (Tamaulipas)*. G2. Early winter.

64. LOPHOPHORA Coulter
N.P. Taylor
Low-growing with depressed, usually clustering, weakly tuberculate-ribbed, spineless stems, arising from a spherical or tapered rootstock. Flowers arising out of the densely woolly stem apex, bell-shaped, self-fertile, receptacle-tube (including pericarpel) naked; stamens sensitive, closing around the style when touched. Fruit cylindric to club-shaped, pink, naked, juicy when ripe, but soon drying, withered

perianth deciduous, seeds ovoid, black, tuberculate, *c.* 1.5 mm.

A genus of 1 or 2 species, native to E & N Mexico and S Texas. Well-known as 'peyote' for its narcotic properties, though *L. diffusa* contains only trace amounts of the intoxicating alkaloids involved. Literature: Anderson, E. F., The biogeography, ecology and taxonomy of Lophophora (Cactaceae), *Brittonia* 21: 299–310 (1969); Anderson, E.F., *Peyote, The Divine Cactus*, University of Arizona Press (1980).

1a. Stem blue-green, usually with recognisable tuberculate ribs and sinuses; flowers pinkish, rarely whitish **1. williamsii**

b. Stem yellow-green, ribs absent or poorly defined, flat; flowers commonly whitish to yellowish white **2. diffusa**

1. L. williamsii (Salm-Dyck) Coulter (*Echinocactus williamsii* Salm-Dyck; *Lophophora lewinii* (Coulter) Rusby; *L. echinata* Croizat; *L. jourdaniana* Habermann; *L. fricii* Habermann). Illustration: Hirao, Colour Encyclopaedia of Cacti, f. 160–162 (1979); Haustein, Der Kosmos-Kakteenführer, 253 (1983).
Stems 2–6 × 4–11 cm, blue-green; ribs 4–14, more or less tuberculate; areoles on mature stems bearing tufts of wool, regularly spaced. Flowers to 2.4 × 2.2 cm; inner perianth-segments 2.5–4 mm wide, pink with pale to white margins. *N & NE Mexico, USA (S Texas)*. G1. Spring–autumn.

2. L. diffusa (Croizat) Bravo (*L. echinata* var. *diffusa* Croizat). Illustration: Hirao, Colour Encylopaedia of Cacti, 50, f. 163, 164 (1979) – f. 163 as L. lutea.
Stems 2–7 × 5–12 cm, yellow-green; ribs absent or very poorly defined, flat; areoles irregularly and widely spaced. Flowers to 2.4 × 2.2 cm; inner perianth-segments 2–2.5 mm wide, white, yellowish white or faintly pink. *E Mexico (Queretaro)*. G2. Summer.

Geographically isolated, but questionably distinct from the preceding. Rare in cultivation.

65. OBREGONIA Fric
N.P. Taylor
Low-growing with mostly simple, depressed, yellow-green to brownish green, tuberculate stems 5–20 (rarely to 30) cm in diameter, arising from a tapered tap-root; tubercles 5–15 × 7–15 mm, triangular, arranged as in a rosette, apex

acuminate, bearing a small areole with up to 5 weak and soon deciduous, pale spines to 1.5 cm. Flowers 2–3.6 × 2.5 cm, whitish, arising out of the densely woolly stem apex; pericarpel and tube well-developed but immersed in wool, naked; perianth-segments narrow, 1–1.5 mm wide, stamens sometimes sensitive as in *Lophophora*, filaments pink. Fruit 1.6–2.5 cm × 3–6 mm, club-shaped, white, fleshy then dry, withered perianth deciduous; seeds ovoid, black, tuberculate, 1.1–1.4 mm.

A genus of a single species endemic to NE Mexico, which may be related to *Lophophora*. Rare and vulnerable; the importation of wild-collected specimens is prohibited, but the species can be propagated easily, if slowly, from seed.

1. O. denegrii Fric. Illustration: Hirao, Colour Encyclopaedia of Cacti, f. 102 (1979); Hecht, BLV Handbuch der Kakteen, 337 (1982); Cullmann et al., Kakteen, edn 5, 256 (1984).
NE Mexico (Tamaulipas). G2. Summer.

66. EPITHELANTHA Britton & Rose
N.P. Taylor

Low-growing, with spherical to obovoid, simple or clustering, tuberculate, densely spined, stems 1–6 cm in diameter; tubercles minute, often obscured by the fine dense spines. Spines 19–38 per areole, 1–8 mm, white to pale grey or yellowish, the uppermost club-shaped, functioning as glands at first, the tip later deciduous. Flowers arising adjacent to the spine clusters at the woolly stem apex, 3–12 mm in diameter, bell-shaped, whitish or pale orange to pink; pericarpel and tube naked; perianth-segments and stamens few; stigma-lobes mostly 3–4. Fruit 3–18 mm, club-shaped-cylindric, red, rather juicy, few-seeded, indehiscent, the withered perianth deciduous; seeds ovoid, black, 1–1.5 mm.

A genus of one variable species divisible into 6 varieties.

Literature: Glass, C. & Foster, R., A revision of the genus Epithelantha, *Cactus & Succulent Journal (US)* 50: 184–187, with illustrations (1978).

1. E. micromeris (Engelmann) Britton & Rose (*Mammillaria micromeris* Engelmann). Illustration: Benson, The Cacti of the United States & Canada, 774–775 (1982).
SW USA, NE Mexico. G1. Summer.

Var. **micromeris**. Simple or clustering, stems 1.3–3.8 (rarely to 6) cm in diameter, more or less obscured by spines. Spines *c*. 20, 1–3 mm, the uppermost much

longer and club-shaped. Flowers *c*. 6 × 3–4.5 mm. Fruit to 1.2 cm. *USA (SE Arizona, S New Mexico, SW Texas and adjacent NE Mexico)*.

Var. **bokei** (Benson) Glass & Foster (*E. bokei* Benson). Illustration: Cactus & Succulent Journal (US) 50: 185, f. 4 (1978); Hirao, Colour Encyclopaedia of Cacti, f. 170 (1979); Benson, Cacti of the United States & Canada, pl. 148 (1982). Simple or clustering, 2.5–5 cm in diameter, completely obscured by spines. Spines 35–38, longest to 4.5 mm, all tightly adpressed to stem. Flowers large, to 1.2 × 1.2 cm. Fruit to 9 mm. *SW Texas to N Mexico (SE Coahuila)*.

Var. **greggii** (Engelmann) Borg (*Mammillaria micromeris* var. *greggii* Engelmann). Illustration: Cactus & Succulent Journal (US) 50: 184–185, f. 1–3 (1978). Simple or clustering, stems 5–7.5 cm in diameter, nearly obscured by spines. Spines 1–4 mm, the upper ones much longer, not strongly adpressed, giving the stem a bristly appearance. Fruit to 1.8 cm. *NE Mexico (S Coahuila)*.

Var. **unguispina** (Boedeker) Backeberg (*Mammillaria micromeris* var. *unguispina* Boedeker). Illustration: Cactus & Succulent Journal (US) 50: 186, f. 5 (1978); Hirao, Colour Encyclopaedia of Cacti, f. 168 (1979). Clustering, stems 3–5 cm in diameter, not obscured by spines. Spines *c*. 19, lowermost longest, to *c*. 5 mm, not adpressed to stem, others 1–2.5 mm. *NE Mexico (C Nuevo Leon)*.

Var. **pachyrhiza** Marshall. Illustration: Cactus & Succulent Journal (US) 50: 186, f. 6, 7 (1978). Stems sparingly dichotomously branched, 2.5 cm or more in diameter arising from branched, tuberous, tapered roots connected to the stem by a narrow neck. Spines 24–28, to 1.75 mm, strongly adpressed, comb-like. *NE Mexico (S Coahuila)*.

Var. **polycephala** (Backeberg) Glass & Foster (*E. polycephala* Backeberg). Illustration: Cactus & Succulent Journal (US) 50: 187, f. 8, 9 (1978); Hirao, Colour Encyclopaedia of Cacti, f. 171 (1979) – as E. sp.?. Clustering, stems slender, to 7 × 1.5–2 cm. Spines 22–27, 0.5–2 mm, the upper ones to 3.5 mm, all yellowish orange, darker tipped. Flowers 11 × 5 mm. *NE Mexico (SE Coahuila)*.

67. STROMBOCACTUS Britton & Rose
N.P. Taylor

Low-growing with depressed, simple, tuberculate, weakly spined stems, 2–8 × 3–17 cm; tubercles spiralled, irregularly

rhomboid, pale grey-green; areoles soon losing their wool; spines 1–5, to 1.5 cm, erect, dirty whitish, darker near apex, often early deciduous. Flowers from the stem apex, shortly funnel-shaped, to 3.2 × 3.2 cm, pale yellow to whitish, reddish in the throat; pericarpel and tube bearing few greenish to reddish scales; stigma-lobes 7–11. Fruit conspicuous, 7–10 × 6–7 mm, reddish brown to green, dehiscing by 2–4 longitudinal splits; seeds only 0.5 × 0.3 mm, brown, with a large caruncle.

A genus of a single species, of slow growth. Its only close relative appears to be *Aztekium*. Propagation (by seed) is difficult, the unusually minute seedlings readily being overwhelmed by moss or algae.

1. S. disciformis (de Candolle) Britton & Rose (*Mammillaria disciformis* de Candolle). Illustration: Hirao, Colour Encyclopaedia of Cacti, f. 101 (1979); Hecht, BLV Handbuch der Kakteen, 362 (1982); Haustein, Der Kosmos-Kakteenführer, 251 (1983). *E Mexico (Hidalgo, Queretaro), on vertical cliff-faces*. G1. Summer.

68. AZTEKIUM Boedeker
N.P. Taylor

Low-growing with depressed, simple or clustering, ribbed, weakly spined, pale grey-green stems to *c*. 3 × 5 cm; ribs 8–11, transversely wrinkled and with longitudinal rib-like ridges in between; areoles contiguous, bearing wool and 1–3 upward-curved and adpressed, flattened, yellowish to grey spines to *c*. 4 mm long. Flowers from the stem-apex, funnel-shaped, to 12 × 14 mm, white, reddish purple outside; pericarpel and tube naked, rather slender; stigma-lobes *c*. 4. Fruit hidden, poorly known, *c*. 3 mm; seeds like those of *Strombocactus* but *c*. 0.7 × 0.5 mm.

A rare genus of a single species allied to and questionably distinct from *Strombocactus*. Very slow-growing and threatened in the wild through illegal commercial collecting. As with *Strombocactus*, propagation by seed is difficult and very slow. The practice of grafting offsets results in rather lush, atypical growth.

1. A. ritteri (Boedeker) Boedeker (*Echinocactus ritteri* Boedeker). Illustration: Hirao, Colour Encyclopaedia of Cacti, f. 152 (1979); Hecht, BLV Handbuch der Kakteen, 228 (1982); Haustein, Der Kosmos-Kakteenführer, 251 (1983). *NE Mexico (Nuevo Leon), on vertical cliff faces*. G1. Spring–autumn.

69. PELECYPHORA Ehrenberg
D.R. Hunt

Simple or clustering, with spherical or obconic tubercled stems; tubercles hatchet-shaped and laterally compressed,or scale-like and flattened against stem and then acute, keeled and somewhat incurved; areoles with a vegetative, spine-bearing part at the tubercle-apex and a narrow tubercular groove or shallow corky ridge to the woolly, flower-bearing axil on the upper side. Flowers arising near the axils of young tubercles, shortly funnel-shaped or bell-shaped; tube naked; perianth purplish pink. Fruit small, dry; seeds kidney-shaped, coat brown, almost smooth, hilum small.

A genus of 2 species in NE Mexico, both popular with collectors, but slow-growing from seed. Commercial gathering of wild plants has diminished natural populations of both species, and is now prohibited. For two other species sometimes referred to this genus, *P. pseudopectinata* and *P. valdeziana*, see genus **62, Neolloydia** (p. 274).

1a. Tubercles hatchet-shaped, the narrowly elongate areole with numerous tiny spines in two comb-like rows   **1. aselliformis**
b. Tubercles scale-like, as if in a rosette or cone, the apical areole with a bunch of few, deciduous spines   **2. strobiliformis**

1. P. aselliformis Ehrenberg. Figure 29(5), p. 205. Illustration: Hecht, BLV Handbuch der Kakteen, 348 (1982); Haustein, Der Kosmos Kakteenführer, 255 (1983); Cullmann et al., Kakteen, edn 5, 272 (1984) – as P. asseliformis.
Simple or clustering, individual stems spherical or obconic, to *c.* 10 × 5 cm diameter, dull greyish green or tinged purplish brown; tubercles hatchet-shaped, compressed laterally, *c.* 5 mm high; areoles linear, to 8 mm, tubercular groove at first marked by fine hairs, later corky. Spines *c.* 40–60, to 4 mm, fused at the base in two comb-like rows, white or grey. Flowers arising in the youngest tubercular axils, bell- to funnel-shaped, *c.* 2 × 3 cm, purplish pink. Fruit spindle-shaped, soon dry after ripening and disintegrating in the stem apex; seeds *c.* 1 mm, dark brown. *E C Mexico (San Luis Potosí)*. G1. Summer.

2. P. strobiliformis (Werdermann) Kreuzinger (*Encephalocarpus strobiliformis* (Werdermann) Berger). Illustration: Hecht, BLV Handbuch der Kakteen, 264 (1982). Haustein, Der Kosmos Kakteenführer, 255 (1983); Cullmann et al., Kakteen, edn 5, 272 (1984).

Depressed-spherical to spherical or ovoid, to 6 cm in diameter, with thick tap-root; tubercles overlapping, scale-like, flattened against stem, acute, keeled and somewhat incurved; areoles apical. Spines *c.* 9–12, to 5 mm, whitish, deciduous. Flowers funnel-shaped *c.* 3 × 4 cm, tube slender, outer segments fringed, pale, inner purplish pink. Fruit dry, seeds small, brownish black. *NE Mexico (Nuevo Leon, Tamaulipas)*. G1. Summer.

70. FEROCACTUS Britton & Rose
N.P. Taylor

Simple or clustering, often fiercely spiny, with large or very large, depressed-spherical to cylindric, mostly strongly ribbed stems; areoles bearing nectar-secreting glands. Flowers funnel-shaped, often shortly so, or bell-shaped, pericarpel and tube scaly, the scales naked in their axils; perianth-segments and stamens separated by a ring of hairs. Fruit spherical to oblong, thick-walled, interior dry at maturity, the seeds escaping via a basal pore formed as the fruit becomes detached from the areole (section **Ferocactus**), or interior juicy, sweet, the fruit indehiscent or splitting irregularly to release the seeds in the liquid pulp (section **Bisnaga** (Orcutt) N.P. Taylor & J.Y. Clark); seeds mussel-shaped to ovoid, 1–3 mm, shiny or dull, brown to black, coat smooth, warty, net-patterned or finely and deeply pitted.

A genus of 23 well-defined species, native to Mexico and SW United States. Popular in cultivation for their bold stems and spines, but mostly flowering only after reaching a fair size. The key below uses vegetative characters as far as possible, and is designed to identify plants seen in cultivation; it will not work reliably for all wild forms.
Literature: Taylor, N.P., A review of Ferocactus Britton & Rose, *Bradleya* 2: 19–38 (1984).

Spines. Central all straight or only slightly curved: **1–4,8,13,14**; one central spine strongly recurved to hooked, if only in young plants: **5–7,9–12,15.**
Flowers. Produced freely when plants only 8–13 cm in diameter: **3,5,6,7,9,11,12**; cultivated specimens rarely flowering or mostly juvenile: **1,8,9,10,13,15.**
Fruit. Juicy when ripe: **1–7,13**; interior dry at maturity, seeds escaping basally: **8–12,14–15.**
Seeds. Smooth or with weakly net-patterned coat: **1–3,14**; finely pitted, the pits of small diameter: **4–7,9, 13–15.**

1a. Principal spines 6–9 (rarely to 12), per areole, red, stout, to 5 cm, straight or slightly curved (never hooked or strongly recurved near apex); other spines reduced to fine, hair-like whitish bristles or absent; flowers orange-red, the perianth-segments remaining erect   **13. pilosus**
b. Spines and flowers not as above   **2**
2a. Spines 1–13 per areole, straight or slightly curved, none strongly flattened above or recurved to hooked at apex.
b. Spines more than 13 per areole or at least one strongly flattened and/or recurved to hooked at apex   **9**
3a. Stem depressed-spherical to flattened and disc-shaped, surface light to dark blue-green; areoles sunken into notches in the ribs; perianth-segments purplish pink with paler or whitish margins   **5. macrodiscus**
b. Stem spherical to cylindric, or surface, areoles or perianth-segments not as above   **4**
4a. Ribs acute   **5**
b. Ribs rounded or obtuse and tuberculate   **8**
5a. Spines 1–6 (rarely to 8) to 2.5 cm, almost equal, the central not differentiated from the radials   **3. glaucescens**
b. Spines 7–13, to 10 cm, the central longest or at least well-differentiated   **6**
6a. Spines yellow or red, sometimes brown at base; seeds finely pitted   **7**
b. Spines mostly dull brown or grey; seeds smooth   **2. echidne**
7a. Central spines 1–4, the lowermost often slightly down-curved   **4. histrix**
b. Central spine 1, straight or slightly ascending   **14. pottsii**
8a. Radial spines 7–9, to 5–8 cm; central spine to 10–25 cm; young plants glaucous or light grey-green   **15. emoryi**
b. Radial spines 3–8, to 2–4.5 cm; central spine to 3–7.5 cm; young plants dark green   **14. pottsii**
9a. Ribs 7–9; stems offsetting freely   **8. robustus**
b. Ribs 11–13 or more; stems simple (offsetting in No. **1**)   **10**
10a. Perianth-segments and scales on pericarpel and tube linear-lanceolate, tapered toward apex, very narrow   **11**
b. Perianth-segments and scales not as above   **12**

11a. Lowermost central spine strongly recurved to hooked at apex; stem simple **6. latispinus**

b. Lowermost central spine straight or slightly curved throughout; stems offsetting at maturity **1. flavovirens**

12a. Stem light glaucous- to grey-green. **15. emoryi**

b. Stem dark blue-green to dark yellow-green **13**

13a. Flowers funnel-shaped, 6–10 × 6.5–7.5 cm, yellow; tube well-developed; fruit very juicy when ripe **7. hamatacanthus**

b. Flowers bell-shaped to shortly or broadly funnel-shaped, 3–7.5 × 2.5–6 cm, purplish, red or greenish, or the perianth-segments yellow striped with orange or red; tube very short; fruit interior dry at maturity, the seeds escaping via a basal pore **14**

14a. Flowers violet-purple to lilac, or greenish, borne on plants only 8–12 cm in diameter **15**

b. Flowers yellow, orange or reddish, if greenish then plants not flowering until at least 20 cm in diameter **16**

15a. Flowers purplish; seeds to 2.5 mm **12. fordii**

b. Flowers greenish; seeds c. 1.5 mm **11. viridescens**

16a. Spines clearly differentiated into stout dark centrals plus upper and lower radials, and finer whitish laterally directed radials, or the latter few or absent and the seeds lacking pits **9. wislizeni**

b. Spines of each areole intergrading in size and colour; seeds pitted or with a pronounced raised net-pattern **10. cylindraceus**

1. F. flavovirens (Scheidweiler) Britton & Rose (*Echinocactus flavovirens* Scheidweiler). Illustration: Hirao, Colour Encyclopaedia of Cacti, f. 47 (1979); Kakteen und andere Sukkulenten 34: 40–41 (1983).
Clustering, stems to 30–40 × 20 cm light green; ribs 13–15 (rarely 11); areoles widely spaced. Central spines 4–6, to 8 cm, light brown; radials 12–20, light brown to grey, some bristle-like. Flowers from centre of stem apex, to 3.5 cm, red; perianth-segments linear-lanceolate, only 2–3 mm wide. Fruit nearly 3 × 2 cm, red with long ciliate or bristly scales; seeds c. 1 mm, dark brown, with net-like markings. *S Mexico*. G2. Summer.

A relatively unattractive species which is seldom grown.

2. F. echidne (de Candolle) Britton & Rose (*Echinocactus echidne* de Candolle; *E. victoriensis* Rose; *E. rafaelensis* Purpus; *F. rafaelensis* (Purpus) Borg). Illustration: Hirao, Colour Encyclopaedia of Cacti, f. 39–41 (1979) – as F. rhodanthus.
Simple or clustering, almost spherical to cylindric or club-shaped, to 35 (rarely to 100) × 20 (rarely to 30) cm, dull or grey-green; ribs 13–21; areoles well separated. Central spine 1, 5–10 cm long; radials 7–9. Flowers 2–4.5 × 3–3.5 cm, yellow in cultivated plants. Fruit spherical to ovoid, 2 × 1.5 cm, light green to white, tinged pink or red; seeds 1–1.75 mm, dark red to black, very smooth. *E & NE Mexico*. G2. Summer.

3. F. glaucescens (de Candolle) Britton & Rose (*Echinocactus glaucescens* de Candolle). Illustration: Hirao, Colour Encyclopaedia of Cacti, f. 37 (1979); Cullmann et al., Kakteen, edn 5, 174 (1984).
Simple or clustering, spherical to cylindric, to 45 (rarely to 70) × 50 (rarely to 60) cm, usually rather glaucous, sometimes strikingly so; ribs 11–34 (rarely to 44); areoles close in older specimens. Spines 4–6 (rarely to 8), almost equal, to 2.5–3.5 cm, yellow. Flowers 2–4.5 × 2.5–3.5 cm, yellow. Fruit spherical to ovoid, 1.5–2.5 × 2 cm, whitish or yellowish tinged with red, with yellow scales; seeds to 1.5 mm, dark brown to almost black, very smooth. *Central E Mexico*. G2. Summer.

F. schwarzii Lindsay. Illustration: Hirao, Colour Encyclopaedia of Cacti, f. 43 (1979); Bradleya 2: 24 (1984). Simple, spherical to obovoid, to 80 × 50 cm, though mature at 10 cm in diameter, dark green; ribs 13–19; areoles confluent. Spines 1–4 (rarely 0 or 5), more numerous in young plants, 0.5–5.5 cm, yellow. Flowers to 5 × 4 cm, yellow. Seeds 1.5 mm, black, with net-like markings. *W Mexico (Sinaloa)*. G2. Summer.

4. F. histrix (de Candolle) Lindsay (*Echinocactus histrix* de Candolle; *E. melocactiformis* de Candolle?; *F. melocactiformis* Britton & Rose; *Echinocactus electracanthus* Lemaire?). Illustration: Backeberg, Die Cactaceae 5: f. 2595–2597 (1961); Hirao, Colour Encyclopaedia of Cacti, f. 38 (1979).
Simple, depressed-spherical to short-cylindric, to 1.1 m × 80 cm, but usually much smaller; ribs to 20–40 or more; areoles almost confluent in old plants. Spines yellow, red to brown at base; centrals 1–4, uppermost 2–3 to 3.5 cm, lowermost to 9 cm, projecting and often slightly down-curved, except in young plants; radials 6–9. Flowers 2–3.5 × 2.5–3.5 cm, yellow. Fruit 2–3 cm, pinkish; seeds c. 1 mm, dark brown, coat finely pitted. *N C Mexico*. G1. Summer.

F. haematacanthus (Salm-Dyck) Backeberg & Knuth (*Echinocactus electracanthus* [var.] *haematacanthus* Salm-Dyck, not *F. stainesii* var. *haematacanthus* Backeberg). Illustration: Hirao, Colour Encyclopaedia of Cacti, f. 42 (1979).
Simple, spherical, to cylindric, to 30–120 × 26–36 cm, glaucous when young; ribs 13–27; areoles confluent in old plants. Spines blood-red with yellow tips; centrals 4, 4–8 cm; radials 6–7. Flowers 6–7 × 5.8–6.8 cm, purplish pink. Fruit ovoid, 2.2–3.5 × 1.4–2.7 cm, purplish; seeds c. 1.8 mm, black, coat finely pitted. *E Mexico (Puebla-Veracruz border)*. G1. Summer.

5. F. macrodiscus (Martius) Britton & Rose (*Echinocactus macrodiscus* Martius; *E. macrodiscus* var. *multiflorus* Meyer). Illustration: Kakteen und andere Sukkulenten 29: 65–66, f. 1–5 (1978); Hirao, Colour Encyclopaedia of Cacti, f. 32, 33 (1979); Bradleya 2: 26 (1984).
Simple, depressed-spherical or flattened, to 10 × 30–40 cm; ribs 13–35 ; areoles sunken in notches in the ribs. Central spines 4, 3.5 cm; radials 6–8, c. 2–3 cm. Flower 3–4 × 3–4 cm, purplish pink. Fruit nearly spherical, to 4 × 3 cm, red; seeds to 2 mm, dark brown, finely pitted. *C & S Mexico (Guanajuato, Oaxaca)*. G1. Summer.

6. F. latispinus (Haworth) Britton & Rose (*Cactus latispinus* Haworth; *C. recurvus* Miller?; *C. nobilis* Linnaeus?; *Ferocactus recurvus* Borg; *F. nobilis* Britton & Rose; *Echinocactus spiralis* Pfeiffer). Figure 29(6, 7), p. 205. Illustration: Backeberg, Die Cactaceae 5: f. 2583 (1961) – var. spiralis; Hirao, Colour Encyclopaedia of Cacti, f. 34–36 (1979); Cullmann et al., Kakteen, edn 5, 175 (1984).
Usually simple, depressed-spherical (var. **latispinus**) or spherical to cylindric (var. **spiralis** (Pfeiffer) N.P. Taylor), to 10–40 (rarely to 100) × 16–40 cm; ribs 13–21, acute, sometimes spiralled. Central spines 4, the lowermost usually recurved at apex, very broad (4–9 mm) and flattened above (less so or more nearly terete in var. **greenwoodii** (Glass) N.P. Taylor); radial spines 5–7, stout (or 9–15 and finely needle-like in var. **latispinus**). Flowers 4–6 × 2.5–4 cm, purplish pink, whitish or yellow; perianth-segments linear-lanceolate. Fruit spherical to cylindric,

2.5–8 × 1.8–2.5 cm, clothed in long tapering scales; seeds 1.2–1.5 mm, dark brown, finely pitted. *Northern C & S Mexico.* G1. Autumn.

7. F. hamatacanthus (Muehlenpfordt) Britton & Rose (*Echinocactus hamatacanthus* Muehlenpfordt; *E. sinuatus* Dietrich; *F. sinuatus* invalid; *Hamatocactus sinuatus* (Dietrich) Orcutt; *Brittonia davisii* Armstrong). Illustration: Weniger, Cacti of Texas and neighboring states, 119, 121 (1984).
Simple, or sometimes clustering, hemispheric to cylindric, to 60 × 30 cm (but var. **sinuatus** (Dietrich) Benson mature when only 10 cm in diameter); ribs 12–17, rounded and strongly tuberculate (var. **hamatacanthus**) or more acute and better defined (var. **sinuatus**). Central spines 4–8, the lowermost strongly recurved to hooked at apex, terete (or flattened above in var. **sinuatus**), to 8 cm long; radials 8–20, 1.5–4 (rarely to 8) cm long. Flowers funnel-shaped, large, 6–10 × 6.5–7.5 cm, yellow. Fruit spherical to oblong, to 2.5–5 × 2.5 cm, pinkish red, brownish or greenish; seeds 1–1.6 mm, black, finely pitted. *N & NE Mexico, USA (SE New Mexico, S & W Texas).* G1. Summer–autumn.

Var. *sinuatus* is sometimes confused with *Thelocactus setispinus* but has pitted seeds and brown to greenish fruits. It is free-flowering.

8. F. robustus (Pfeiffer) Britton & Rose (*Echinocactus robustus* Pfeiffer). Illustration: Kakteen und andere Sukkulenten **33**: 170–171 (1982); Bradleya **2**: 28 (1984).
Clustering, stems cylindric; ribs 7–9, acute; areoles widely spaced. Central spines 4–7, to 6 cm, straight or slightly curved, angled or flattened; radials 10–14, mostly bristle-like, whitish. Flowers 3–4 × 3–4 cm, yellow. Fruit 2–3 × 2 cm, yellow; seeds *c.* 1.5 mm, black, with net-like markings, very finely warty. *S Mexico (SE Puebla).* G1.

9. F. wislizeni (Engelmann) Britton & Rose (*Echinocactus wislizeni* Engelmann; *Ferocactus herrerae* Ortega). Illustration: Hirao, Colour Encyclopaedia of Cacti, f. 26–29 (1979); Benson, Cacti of the United States & Canada, pl. 122 (1982).
Simple, spherical to cylindric, tapered towards apex, to 1.6 (rarely to 3) m × 45–80 cm; ribs to 20–30 (or 13 in var. **herrerae** (Ortega) N.P. Taylor), high and acute. Central spines 4 (rarely to 8), brownish to grey, uppermost 3 terete, lowermost flattened and usually recurved to hooked at apex; radials 12–30 (few or absent in var. **herrerae**), bristle-like to needle-like, whitish. Flowers 5–7.5 × 4.5–6 cm, yellow, orange or red. Fruit ovoid to 5–6 × 3 cm, yellow; seeds to 2.5 mm black, coat finely warty. *SW USA, NW Mexico (Sonora, Chihuahua, Durango, Sinaloa).* G2. Summer.

F. peninsulae (Weber) Britton & Rose (*Echinocactus peninsulae* Weber; *F. horridus* Britton & Rose; *F. townsendianus* Britton & Rose; *F. santa-maria* Britton & Rose). Illustration: Britton & Rose, The Cactaceae **3**: f. 133, 140 (1922); Cactus & Succulent Journal of Great Britain **41**: 112 (1979).
Simple, spherical, ovoid or club-shaped, to 70–250 × 50 cm (much smaller and flowering when *c.* 10–13 cm in diameter in var. **townsendianus** (Britton & Rose) N.P. Taylor and var. **santa-maria** (Britton & Rose) N.P. Taylor); ribs 12–20, acute. Spines and flowers very similar to those of *F. wislizeni.* Fruit nearly spherical, *c.* 3.5 cm long (only *c.* 2.5 × 2 cm in var. **townsendianus** and to 5 × 4 cm in var. **santa-maria**), yellow; seeds 1.2–2 mm, brown to black, with net-like markings. *NW Mexico (C & S Baja California).* G2. Spring–autumn.

Cultivated plants can be reliably distinguished from *F. wislizeni* var. *herrerae* only when seeds or details of wild source are available (see Taylor, N.P. & Clark, J.Y. in Bradleya **1**: 8, 9, 16, f. 2, 3, 5, 6, 7, 36, 1983). Some plants cultivated under the synonym *F. horridus* may be *F. wislizeni* var. *herrerae.*

F. gracilis Gates (*F. coloratus* Gates; *F. viscainensis* Gates; *F. gatesii* Lindsay?). Illustration: Pizzetti, Piante grasse le Cactacee, nos. 103, 107 (1985). Simple, eventually cylindric, to 1.5–3 m × 30 cm; ribs 16–32. Central spines 4–12, red, curved, twisted and tangled, the larger upper and lower ones both flattened, to 7 cm, seldom hooked; radials *c.* 8–12, slender, whitish. Flowers 4–6 × 3.5–5 cm, red. Fruit oblong-cylindric, 2.5–7.5 cm, yellow, sometimes tinged reddish; seeds 1.75–2.5 mm, black, finely pitted to strongly net-patterned. *NW Mexico (C Baja California).* G1. Rarely flowering in Europe.

With very attractive spines, but not of vigorous growth in cultivation.

10. F. cylindraceus (Engelmann) Orcutt (*Echinocactus viridescens* [var.] *cylindraceus* Engelmann; *E. lecontei* Engelmann; *F. lecontei* (Engelmann) Britton & Rose; *F. tortulispinus* Gates; *F. acanthodes* misapplied; *F. acanthodes* var. *eastwoodiae* Benson; *F. eastwoodiae* (Benson) Benson).

Illustration: Benson, Cacti of the United States & Canada, pl. 117–121 (1982).
Simple, eventually cylindric or barrel-shaped, to 3 m × 40–50 cm; ribs 18–30, tuberculate. Central spines 4–7, reddish, orange or yellow, variously terete, flattened, straight, curved or twisted, the largest 7–17 cm, sometimes hooked, often recurved; radials 15–25, fine and hair-like or some stout (only 12–14 and all stout in var. **eastwoodiae** (Benson) N.P. Taylor). Flowers 3–6 × 4–6 cm, green, yellow, or tinged with red. Fruit 3–4 × 1.5–2 cm, yellow; seeds to 2–3 mm, black, pitted to net-patterned. *NW Mexico (NE & E Baja California Norte, N Sonora), SW USA.* G2. Spring–summer.

F. chrysacanthus (Orcutt) Britton & Rose (*Echinocactus chrysacanthus* Orcutt). Illustration: Hirao, Colour Encyclopaedia of Cacti, f. 18 & 18-1 (1979). Simple, spherical to short-cylindric, to 100 × 30 cm; ribs *c.* 21, tuberculate. Central spines *c.* 10, yellow or red, flattened, curved and twisted, to *c.* 5 cm; radials finer, 12 or more. Flowers 4.5 × 4 cm, yellow to orange; perianth-segments with reddish midstripes. Fruit oblong-cylindric, 3 × 1.5 cm, yellow; seeds to 2.5 mm, black, net-patterned. *NW Mexico (Isla Cedros).* G2. Summer.

11. F. viridescens (Torrey & Gray) Britton & Rose (*Echinocactus viridescens* Torrey & Gray; *E. orcuttii* Engelmann; *F. orcuttii* (Engelmann) Britton & Rose). Illustration: Hirao, Colour Encyclopaedia of Cacti, f. 19, 20 (1979); Bradleya **2**: 31, 32 (1984).
Simple, rarely clustering, spherical to short-cylindric, to 30 (rarely to 130) × 18–40 cm, bright green; ribs 13–25 (21–34 in var. **littoralis** Lindsay), tuberculate, low. Central spines 4–9, to 5 cm, yellow or reddish, larger upper and lower ones flattened, curved, hooked only in juveniles (much more dense and mostly less flattened in var. **littoralis**); radials to 19 or more, some bristle-like, others stouter, terete. Flowers to 5 × 6 cm (only 3 × 2.5 cm in var. **littoralis**), greenish or the outer perianth-segments tinged with red. Fruit to 3.5 × 1.5–2.5 cm, yellow or occasionally red; seeds *c.* 1.5 cm, black, net-patterned. *NW Mexico (NW Baja California), USA (SW California, near San Diego).* G2. Spring–summer.

12. F. fordii (Orcutt) Britton & Rose (*Echinocactus fordii* Orcutt). Illustration: Schumann, Bluehende Kakteen **1**: t. 11 (1903); Hirao, Colour Encyclopaedia of Cacti, f. 22 (1979).

Simple, depressed-spherical to spherical, to 25 cm in diameter, dark blue-green; ribs to c. 21, tuberculate, low. Central spines 4–7, red or grey, lowermost to 4–7 cm, flattened, hooked, sometimes twisted, others more or less terete and straight; radials c. 15, to 3 cm, paler. Flowers c. 4 × 4 cm, lilac to purple. Fruit oblong, 3–4 cm, yellow or tinged reddish; seeds to 2.5 mm, black, net-patterned. *NW Mexico (W coast of C & N Baja California)*. G2. Spring–summer.

Probably the easiest species to bring into flower.

13. F. pilosus (Salm-Dyck) Werdermann (*Echinocactus pilosus* Salm-Dyck; *E. pilosus* [var.] *stainesii* Salm-Dyck; *F. stainesii* (Salm-Dyck) Britton & Rose; *Echinocactus pilosus* [var.] *pringlei* Coulter; *Ferocactus pringlei* (Coulter) Britton & Rose; *F. stainesii* var. *haematacanthus* Backeberg, not *F. haematacanthus* (Salm-Dyck) Backeberg & Knuth; *F. piliferus* (Ehrenberg) Unger; *Echinocactus piliferus* Ehrenberg). Illustration: Lamb & Lamb, Illustrated Reference on Cacti and other Succulents 5: f. 331, 332 (1978); Cullmann et al., Kakteen, edn 5, 175 (1984). Simple, rarely clustering (but in the wild sometimes forming massive clumps), eventually cylindric, to 3 m × 50 cm; ribs 13–20, acute in young plants, straight and even; areoles close on mature stems. Spines of two distinct kinds: straight, red principal spines and fine, white, hair-like radial spines, the latter sometimes very few or absent. Flowers to 4 × 2.5 cm, orange-red, perianth-segments not spreading. Fruit ovoid, 3–4 cm, yellow or partly red, not opening at base, remaining somewhat fleshy when ripe; seed c. 1.75 mm, black, finely pitted. *C N Mexico*. G2.

Rarely flowering in cultivation.

14. F. pottsii (Salm-Dyck) Backeberg (*Echinocactus pottsii* Salm- Dyck; *E. alamosanus* Britton & Rose; *F. alamosanus* (Britton & Rose) Britton & Rose; *F. alamosanus* var. *platygonus* Lindsay; *F. guirocobensis* Schwarz, invalid). Illustration: Britton & Rose, The Cactaceae 3: f. 145 (1922); Hirao, Colour Encyclopaedia of Cacti, f. 46 (1979). Simple, rarely clustering, spherical to short-cylindric, to 100 × 50 cm (spherical, 15–30 cm in diameter in var. **alamosanus** (Britton & Rose) Unger); ribs 13–25, rather broad and obtuse (acute and narrow in var. **alamosanus**). Central spine 1, to 3 cm, rarely to 7.5 cm, straight or slightly

upward-curved; radials 3–8, to 4.5 cm (often very short in var. **pottsii**, dense and yellow in var. **alamosanus**). Flowers to 4.5 × 3.5 cm, yellow. Fruit to 4 × 3 cm, yellow (red in var. **alamosanus**), interior dry at maturity; seeds to 3.2 mm, finely pitted. *NW Mexico (SE Sonora, SW Chihuahua, N Sinaloa)*. G2. Summer.

F. reppenhagenii Unger. Illustration: Kakteen und andere Sukkulenten 25: 50–54 (1974); Hirao, Colour Encyclopaedia of Cacti, f. 45 (1979) (young plant). Simple, spherical to short-cylindric, to 30 (rarely to 80) × 9–24 cm; ribs 12–18, the edge rounded; areoles confluent on old plants. Spines yellowish, central 1, 2.8–8 cm, straight; radials 7–9 (rarely 6 or 11), to 4 cm. Flowers 2–3 × 2–3 cm, yellow to orange. Fruit ovoid, 1.5–2.2 × 0.8–1.7 cm, bright to dark red, very juicy when ripe; seeds to 2.3 mm, reddish brown to black, very smooth. *SW Mexico*. G2. Summer.

Superficially like *F. pottsii* var. *alamosanus*, but differing markedly in its fruit and seeds; related to Nos. **2 & 3**.

15. F. emoryi (Engelmann) Orcutt (*Echinocactus emoryi* Engelmann; *E. emoryi* [var.] *rectispinus* Engelmann; *F. rectispinus* (Engelmann) Britton & Rose; *F. covillei* Britton & Rose). Figure 29(15), p. 205. Illustration: Benson, Cacti of the United States & Canada, pl. 123 (1982); Bradleya 2: 37 (1984). Simple, spherical to cylindric to 1.5 (rarely to 3) m × 45–60 (rarely to 100) cm; ribs 15–30 or more, strongly tuberculate, rounded. Central spine 1, reddish or later brownish, to 4–10 cm, hooked or straight (but to 25 cm, straight in var. **rectispinus** (Engelmann) N.P. Taylor); radials 7–9, to 8 cm, red or whitish. Flowers to 6–7.5 × 5–7 cm, yellow or red. Fruit to 5 × 3 cm (to 3.5 × 2.5 cm in var. **rectispinus**), yellow; seeds to 2–2.5 mm, brown to black, coat finely pitted. *NW Mexico, USA (SW Arizona)*. G2. Summer–autumn, but rarely flowering in Europe.

F. diguetii (Weber) Britton & Rose (*Echinocactus diguetii* Weber). Illustration: Rowley, Illustrated Encyclopaedia of Succulent Plants, f. 2.1 (1978); Hirao, Colour Encyclopaedia of Cacti, f. 23 & 24 (1979). Like *F. emoryi* but even larger, to 4 m × 60 cm; juvenile plants very similar but spines yellow. Flowers 4 × 4 cm, red. Fruit c. 3 × 2 cm; seeds 1.5–2 mm, pitted. *NW Mexico (islands in Gulf of California)*. G2.

This rare giant species has probably never flowered in cultivation, where it is seen only as a seedling or juvenile plant.

71. LEUCHTENBERGIA Hooker
N.P. Taylor

Low-growing, rarely to 70 cm, simple or occasionally clustering, with a spherical to short-cylindric stem clothed in 10–12 cm, glaucous, triangular tubercles; rootstock fleshy. Spines to 15 cm, papery, flattened and flexuous. Flowers borne at the ventral edge of the spine-bearing areoles of young tubercles near the stem apex, to 8 × 5–6 cm, yellow, otherwise like those of *Ferocactus*, but ring of hairs absent. Fruit as in *Ferocactus* section *Ferocactus*, ovoid-oblong, c. 3 cm, excluding perianth, grey-green; seeds ovoid, to 2.5 mm, black or dark brownish, weakly tuberculate.

A genus of a single species, endemic to N Mexico and closely allied to *Ferocactus*.

1. L. principis Hooker. Illustration: Britton & Rose, The Cactaceae 3: f. 117, 117a (1920); Cullmann et al., Kakteen, edn 5, 190 (1984).
Central N Mexico. G2. Summer–autumn.

72. STENOCACTUS (Schumann) Hill
N.P. Taylor

Low-growing with almost spherical to short-cylindric, usually many-ribbed stems, simple, less often clustering; ribs acute, very thin and often wavy (except in No. 1), the areoles usually widely spaced. Spines commonly differentiated into a large upper series (central spines) and small lower series (radials), the former often strongly flattened. Flowers small, shortly funnel-shaped or bell-shaped, pericarpel scaly but hairless. Fruit small, mostly spherical, greenish, scaly, dry, dehiscent at one side; seeds obovoid, black or dark grey, pitted to net-patterned.

A small genus with an uncertain number of highly variable species, endemic to Mexico. Correct use of many of the published names is problematical. The genus was formerly called *Echinofossulocactus* Lawrence.

1a. Ribs 10–15, straight, stout
      1. coptonogonus
  b. Ribs 20–120, straight or wavy, often very thin   **2**
2a. Inner perianth-segments white, pinkish purple or violet
      2. multicostatus
  b. Inner perianth-segments pale to sulphur-yellow, the darker midstripes dirty yellow to brownish   **3**
3a. Radial spines 2–7   **3. phyllacanthus**
  b. Radial spines 8–27   **4. vaupelianus**

1. S. coptonogonus (Lemaire) Hill
(*Echinocactus coptonogonus* Lemaire;
Echinofossulocactus coptonogonus (Lemaire)
Lawrence). Illustration: Schumann &
Gürke, Blühende Kakteen 1: t. 28 (1904);
Oudshoorn, 126 Cacti & Succulents in
Colour, 66 (1977).
Simple, depressed-spherical to spherical,
5–10 × 7–15 cm, grey to blue-green; ribs
10–15, acute, notched at the areoles.
Spines 3–5 (rarely to 7). Flowers to 3 ×
4 cm, tube very short; perianth-segments
striped white, pinkish purple or violet.
Northern C Mexico. G1. Spring.

2. S. multicostatus (Schumann) Hill
(*Echinocactus multicostatus* Schumann;
Echinofossulocactus multicostatus
(Schumann) Britton & Rose; *E. lloydii*
Britton & Rose?). Illustration: Britton &
Rose, The Cactaceae 3: f. 118 (1920);
Cullmann et al., Kakteen, edn 5, 161
(1984).
Simple or clustering, depressed-spherical to
spherical, 6–10 × 6–12 cm, pale green; ribs
up to 120, very thin. Spines 6–18. Flowers
to 2.5 cm, tube short; perianth-segments
white with faint pink to purplish midstripes,
or pinkish purple. *NE Mexico.* G1. Spring.

  S. obvallatus (de Candolle) Hill
(*Echinocactus obvallatus* de Candolle; *E.
pentacanthus* Lemaire; *E. violaciflorus* Quehl;
Echinofossulocactus violaciflorus (Quehl)
Britton & Rose; *Echinocactus arrigens*
Dietrich; *Echinofossulocactus arrigens*
(Dietrich) Britton & Rose; *E. caespitosus*
Backeberg). Illustration: Britton & Rose,
The Cactaceae 3: f. 121, 122 (1920);
Hecht, BLV Handbuch der Kakteen, 259
(1982). Like the above but stems blue-
green; ribs 20–50. Spines 5–12. *N &
Eastern C Mexico.* G1. Spring.

  S. crispatus (de Candolle) Hill
(*Echinocactus crispatus* de Candolle;
E. lamellosus Dietrich; *Echinofossulocactus
lamellosus* (Dietrich) Britton & Rose;
Echinocactus lancifer Dietrich; *E. guerraianus*
Backeberg). Like *S. multicostatus* but stems
dark or bluish green; ribs 25–60. Spines
6–10. Flowers to 4 cm, tube well-
developed. *C Mexico.* G1. Spring.

  S. ochoterenaus Tiegel
(*Echinofossulocactus ochoterenaus* (Tiegel)
Whitmore; *E. lexarzai* Croizat?; *E.
bustamentei* Croizat?; *Echinocactus
heteracanthus* Muehlenpfordt?). Illustration:
Moeller's Deutsche Gaertner-Zeitung **48**:
398 (1933); Kakteen und andere
Sukkulenten **21**: 161 (1970). Like *S.
multicostatus* but ribs 30–50. Spines 11–26
or more. Flowers to 4 cm, tube well-
developed. *NC Mexico.* G1. Spring.

3. S. phyllacanthus (Dietrich & Otto) Hill
(*Echinocactus phyllacanthus* Dietrich & Otto;
Echinofossulocactus phyllacanthus (Dietrich &
Otto) Lawrence; *Echinocactus tricuspidatus*
Scheidweiler). Illustration: Pfeiffer & Otto,
Abbildung und Beschreibung bluehender
Cacteen 1: t. 9 (1839); Cullmann et al.,
Kakteen, edn 5, 161 (1984).
Simple, rarely clustering, depressed-
spherical to short-cylindric, 4–10 cm in
diameter, dark green; ribs 26–60. Spines
2–7. Flowers 1–2.3 cm, yellow, borne
singly or 2–3 together, tube short. *Northern
C Mexico.* G1. Summer.

4. S. vaupelianus (Werdermann) Backeberg
& Knuth (*Echinocactus vaupelianus*
Werdermann; *Echinofossulocactus
vaupelianus* (Werdermann) Whitmore; *E.
albatus* misapplied, not *Echinocactus albatus*
Dietrich). Illustration: Hecht, BLV
Handbuch der Kakteen, 260 (1982).
Simple, depressed-spherical to spherical,
pale green, with much wool at apex; ribs
27–40 or more, very thin. Spines 14–27.
Flowers to 2.5 cm, yellow, 4 or more borne
together, tube short. *Eastern C Mexico (N
Hidalgo).* G1. Spring.

  S. sulphureus (Dietrich) Bravo
(*Echinocactus sulphureus* Dietrich;
Echinofossulocactus sulphureus (Dietrich)
Ito). Illustration: Cactàceas y Suculentas
Mexicanas **22**: f. 18, 19 (1977). Like the
above but spines 8–16; flowers to 3.5 cm,
tube well-developed. *Eastern C Mexico
(N Hidalgo).* G1.

  The following unclear names are
haphazardly applied (under *Echinocactus* or
Echinofossulocactus) to various members of
this genus in cultivation: *E. anfractuosus*
Pfeiffer, *E. dichroacanthus* Pfeiffer,
E. grandicornis Lemaire (*Echinofossulocactus
grandicornis* (Lemaire) Britton & Rose),
Echinocactus tetraxiphus Schumann,
illegitimate, *E. wippermannii* Muehlenpfordt,
E. xiphacanthus Miquel (*Echinofossulocactus
xiphacanthus* (Miquel) Backeberg).
E. gladiatus Link & Otto is also in use for a
Stenocactus, but probably does not apply to
a member of this genus.

73. CORYPHANTHA (Engelmann) Lemaire
D.R. Hunt
Simple or clustering, stems spherical to
cylindric, to 50 cm, tuberculate; tubercles
terete, cylindric, unequally swollen or
pyramidal, grooved above except in
seedlings and juvenile plants; areoles of
adult plants two-parted, consisting of an
apical spine-bearing but sterile part and an
'axillary' fertile part, connected by the
tubercular groove; areoles of juvenile plants
either apical only, later with a rudimentary
groove (subgenus **Coryphantha**) or two-
parted (apical and 'axillary') but not
connected by a groove (subgenus
Neocoryphantha Backeberg). Spines
normally developed, or in some species
some of the spines modified into coloured
capitate glands. Flowers funnel-shaped or
bell-shaped, to 6.5 × 10 cm; pericarpel
naked or with small, sparse scales only.
Fruit berry-like, spherical to ovoid, oblong
or club-shaped, naked, green or yellowish,
indehiscent, the withered perianth
persistent; seeds usually semi-ovoid, brown
(black in *C. odorata*).

  A genus of up to 45 species, native to the
SW United States and Mexico. There is no
reliable recent literature, and several early
names in use for cultivated species are now
thought to have been misapplied from the
time of Britton & Rose's account, published
in 1923 (A. Zimmerman, unpublished
data). For these reasons, only the broadest
treatment is attempted here.

1a.	Central spines 3–4, all hooked, radials 7–9; flowers small, 1.5 cm, whitish	**8. odorata**
b.	Central spines at most curved, or rarely 1 hooked; flowers 2 cm or more, of various colours	2
2a.	Tubercular groove of adult stage not extending fully to the axil; flowers lilac with ciliate perianth-segments	**1. macromeris**
b.	Tubercular groove of adult stage extending fully to the axil; flowers not as above	3
3a.	Axillary areoles present, even on juvenile (non-flowering) plants; stem cylindric	7
b.	Axillary areoles present only on adult (flowering) plants; stem spherical or depressed-spherical	4
4a.	Tubercles conic, somewhat compressed and ascending	5
b.	Tubercles obtuse and gibbous, often large and spreading	**5. elephantidens**
5a.	Radial spines 6–8	**2. poselgeriana**
b.	Radial spines 12–20 or more (rarely 7)	6
6a.	Spines all more or less similar, radiating and adpressed	**3. radians**
b.	Spines of two or more kinds	**4. cornifera**
7a.	Radial spines 8–14, amber-yellow; flowers yellow, 5 cm or more	**6. erecta**
b.	Radial spines 8–9, whitish, brown-tipped; flowers 2–4 cm, whitish, outer segments brownish	**7. clavata**

1. C. macromeris (Engelmann) Lemaire. Illustration: Benson, Cacti of the United States & Canada, pl. 163–165, f. 851–853 (1982) – including var. runyonii; Cullmann et al., Kakteen, edn 5, 147 (1984) – var. runyonii.

Simple or clustering, individual stems 5–15 × 5 cm; tubercles cylindric, flabby, to c. 1.5 cm × 6–9 mm, with the groove extended only about midway from the tip to the axil. Central spines to 4 (rarely to 6), 2.5–5 cm, blackish, brown or grey; radial spines 9–15, mostly 2–2.5 cm, thinner and paler than the centrals. Flowers 4.5–6 × 3–4.5 (rarely to 8) cm, pericarpel c. 7.5 × 4.5 mm, greenish, perianth segments pink to reddish purple, fringed. Fruit 1.6–2.5 cm × 6–9 mm, green; seeds 2 mm. *SW United States, N Mexico (Chihuahua to Zacatecas)*. G2.

Var. **runyonii** (Britton & Rose) Benson, a plant of lower elevations (*USA: Texas, Rio Grande Plain*), has smaller stems and tubercles, and fewer spines.

2. C. poselgeriana (Dietrich) Britton & Rose. Illustration: Cullmann et al., Kakteen, edn 5, 147 (1984).

Spherical, bluish green; tubercles large, conic-pyramidal, 2.5 cm broad at base. Central spine 1, 2–4 cm; radial spines 9–13, the 4–5 lower similar to the central spine and the 5–8 upper close, weak; glands usually present, pale red. Flowers to 6 cm, pale yellow to pink, with red throat. *N Mexico (Coahuila)*. G2.

C. durangensis (Schumann) Britton & Rose. Illustration: Britton & Rose, The Cactaceae 4: f. 40, 41 (1923). Becoming short-cylindric, to 10 cm; tubercles somewhat compressed, very woolly in the axils. Central spine 1, erect, blackish; radial spines 6–8, to 1 cm, spreading. Flowers small, c. 2 × 2.5–4 cm, outer segments purplish, inner pale yellow. Fruit spherical, 5–8 mm, greenish; seeds c. 1 mm broad. *NW Mexico (Durango)*. G2.

C. maiz-tablasensis Backeberg. Clustering, depressed-spherical, to 3 × 5.5 cm, bluish green; tubercles c. 1 cm. Spines all radiating, 6–7, to 1.2 cm, greyish white. Flowers yellowish. *E Mexico (San Luis Potosí)*. G2.

C. sulcata (Engelmann) Britton & Rose (*C. calcarata* (Engelmann) Lemaire). Clustering, stems almost spherical or ovoid, 4–13 × 4–13 cm, green; tubercles to 1.2 × 1 cm, lax. Central spines absent or 1–3, 9–12 mm; radial spines 8–14, 8–18 mm, white. Flowers to 6 × 6 cm, yellow with red throat. Fruit 3 × 2 cm; seeds 2 mm. *SW USA (Texas), NE Mexico*. G2.

3. C. radians (de Candolle) Britton & Rose. Illustration: Haustein, Der Kosmos-Kakteenführer, 263 (1983).

Simple, spherical, 7.5 cm in diameter. Spines all radial, 16–18, 1–1.2 cm, whitish or yellowish. Flowers large, to 7–10 cm broad, lemon-yellow, outer segments tinged red. *C Mexico*. G2.

4. C. cornifera (de Candolle) Lemaire. Illustration: Britton & Rose, The Cactaceae 4: pl. 2, f .4 (1923).

Simple, spherical or ovoid, to 12 cm, pale or glaucous green; tubercles diamond-shaped-conic, to 2.5 × 2.5 cm. Central spine 1 (rarely 0), c. 1.5 cm, curved downwards, dark; radial spines 16–17, 1–1.2 cm, greyish. Flowers to 6 × 7 cm, yellow. *C Mexico*. G2.

The earliest-known of a widespread group in which many poorly differentiated variants have been named.

C. pallida Britton & Rose. Illustration: Britton & Rose, The Cactaceae 4: f. 38 (1923). Central spines 3, radials 20 or more; flowers 7 × 6–7 cm, outer segments with reddish midstripe. Fruit 2 cm, greenish brown. *SE Mexico (Puebla)*. G2.

C. echinus (Engelmann) Britton & Rose (*C. cornifera* var. *echinus* (Engelmann) Benson; *C. pectinata* (Engelmann) Britton & Rose). Illustration: Benson, Cacti of the United States & Canada, pl. 175, f. 877–879 (1982). Central spines absent in juvenile plants, later 1 or 3–4, longest to 1.7 cm; radial spines 16–30, to 1.6–2.8 cm, interlacing. *SW United States (Texas)*. G2.

C. ramillosa Cutak. Simple, spherical to broadly obovoid, 6–7.5 × 6 cm; tubercles 1.2–1.5 × 1.2–1.5 cm. Central spines 4, longest to 4 cm; radial spines 9–20, to 2 cm, white. Flowers c. 6.5 × 5 cm, pink, purple or crimson, with white throat. *SW USA (Texas), N Mexico (NW Coahuila)*. G2.

C. difficilis (Quehl) Orcutt. Depressed spherical, to 6 × 8 cm, greyish or bluish green. Central spines 4, to c. 2 cm, reddish brown, darker-tipped; radial spines 12–14, 1–2 cm, whitish, often brown-tipped. Flowers 4–5 cm broad, yellow. *N Mexico (Coahuila)*. G2.

C. scolymoides (Scheidweiler) Berger. Spherical to ovoid, pale green; tubercles overlapping. Spines 3, to 3.5 cm or more, curved, darker than the radials; radials 14–20, to 2 cm, fewer at first, light, dark-tipped. Flowers over 5 cm broad, yellow with red throat, inner perianth-segments finely toothed. *C Mexico*. G2.

Regarded as part of *C. cornifera* by Britton & Rose.

5. C. elephantidens (Lemaire) Lemaire. Illustration: Hecht, BLV Handbuch der Kakteen, 246 (1982); Cullmann et al., Kakteen, 146 (1984).

Simple, depressed-spherical, to 14 × 19 cm; tubercles large, gibbous, to 4 × 6 cm, densely woolly in the axils; areoles elliptic. Spines c. 8, to 2 cm, subulate, spreading, yellowish, tipped brown. Flowers to 11 cm, deep purplish pink or whitish with red throat. *SW Mexico (Michoacan)*.

C. bumamma (Ehrenberg) Britton & Rose. Illustration: Britton & Rose, The Cactaceae 4: pl. 5, f. 6 (1923). Similar to *C. elephantidens*, and perhaps only a variant of it, but the flowers only 5–6 cm in diameter and pale yellow. *SW Mexico (Morelos, Guerrero)*.

C. sulcolanata (Lemaire) Lemaire. Depressed-spherical, clustering, stems c. 5 × 6.5 cm, shining green; tubercles gibbous; groove and axils woolly at first. Central spines absent; radial spines 9–10, 1.2–1.6 cm, unequal, upper and lower short, laterals longer, at first yellow-white, tipped purple, later dark, tipped blackish. *S Mexico (Oaxaca)*.

Misidentified by Britton & Rose as a species from C Mexico, perhaps *C. radians*.

C. retusa (Pfeiffer) Britton & Rose. Illustration: Britton & Rose, The Cactaceae 4: f. 36 (1923). Depressed-spherical, 5–10 cm in diameter, apex very woolly; tubercles large, unequally swollen, areoles long. Spines 6–12, subulate except the uppermost, yellowish or brownish. Flowers only 3 × 4 cm, yellow. *S Mexico (Oaxaca)*.

6. C. erecta (Pfeiffer) Lemaire. Illustration: Haustein, Der Kosmos-Kakteenführer, 265 (1983).

Simple at first, eventually clustering from the base; stems cylindric, to 30 (rarely to 50) × 5–8 cm; tubercles conic, c. 8 mm. Spines amber-yellow, central spines 2 (rarely to 4), to 2 cm, radials 8–14 (rarely to 18), to 1.2 cm; axillary areoles with small yellow glands. Flowers variable in size, usually 5–7 cm, yellow. Fruit cylindric, c. 2 × 1 cm, green. *C Mexico*. G2.

7. C. clavata (Scheidweiler) Backeberg. Figure 29(4), p. 205. Illustration: Haustein, Der Kosmos-Kakteenführer, 265 (1983). Cylindric to club-shaped, to 7 cm in diameter, dark bluish green. Tubercles obliquely conic, to 2 cm; glands 1–2, reddish. Central spines 0–1, yellowish to brown; radial spines 8–9, 8–15 mm, brownish, darker tipped. Flowers variable in size, to 5 × 4 cm, pale yellow. Fruit 2 cm, pale green. *C Mexico*. G2.

Seedlings slender-stemmed, with characteristic chalky-white spines which are unusually broad for their length.

8. C. odorata Boedeker (*Neolloydia odorata* (Boedeker) Backeberg; *Cumarinia odorata* (Boedeker) Buxbaum). Illustration: Hecht, BLV Handbuch der Kakteen, 325 (1982); Haustein, Der Kosmos-Kakteenführer, 247 (1983).

Clustering; stems ovoid, to 6 × 3 cm; tubercles cylindric, *c.* 10 × 4 mm, flabby. Central spines 3–4, 2–2.5 cm, hooked, slender, dark brown; radial spines 7–9, 8–10 mm, pale brown or whitish. Flowers small, narrowly funnel-shaped, 1.5–2 × 1 cm, creamy white, outer segments tinged pinkish brown. Fruit slender-cylindric, to 1.5 cm × 3 mm, dull purplish green; seeds black. *NE Mexico (Tamaulipas, San Luis Potosí).*

A species of uncertain affinity, perhaps deserving a separate genus, as proposed by Buxbaum.

74. ESCOBARIA Britton & Rose
N.P. Taylor

Low-growing with depressed-spherical to cylindric, tubercled stems, simple or clustering; tubercles grooved above (or groove lacking in section **Acharagma** N.P. Taylor), extra-floral nectary-glands absent, apex bearing an areole with straight spines. Flowers mostly small, arising singly in 'axils' between the tubercles (or from upper edge of the areole in section **Acharagma**); pericarpel naked, tube short; outer perianth-segments usually ciliate-margined. Fruit berry-like, quite naked or with one or more small scales near apex, ovoid, shortly club-shaped or cylindric, green, pink or red; seeds black or brown, minutely pitted (shallowly so in section **Pleurantha** N.P. Taylor).

A genus of about 16 recognisable species, native to N Mexico, SW United States, 1 species ranging into S Canada, and 1 in Cuba. Details of the stigma-lobes, fruits and seeds are essential for accurate identification.

Literature: Taylor, N.P., Die Arten der Gattung Escobaria Britton & Rose., *Kakteen und andere Sukkulenten* **34**: 76–79, 120–123, 136–140, 154–158, 184–188 (1983). Taylor, N.P., The Identification of Escobarias (Cactaceae), *British Cactus & Succulent Journal* **4**: 36–44 (1986).

1a. Flowers arising from the axils of grooved tubercles; outer perianth-segments ciliate   **2**
  b. Flowers arising adjacent to the spine-bearing part of areole, tubercular

groove absent; perianth-segments not ciliate (section **Acharagma** N.P. Taylor)   **8. roseana**
2a. Flowers from the sides of the stem, just below its apex; seeds with broad shallow pits, brown (section **Pleurantha** N.P. Taylor)   **1. chihuahuensis**
  b. Flowers from the centre of the stem-apex; seeds with deep pits confined to the central part of the surface   **3**
3a. Stigma-lobes whitish, pink or violet; seeds brown   **4**
  b. Stigma-lobes green, greenish yellow or brownish; seeds black or dark brown (lighter brown in No. **4**)   **5**
4a. Fruit deep pink to dull red when ripe, not scaly at apex   **2. tuberculosa**
  b. Fruit green when ripe, often scaly near apex   **3. vivipara**
5a. Seeds light brown   **4. emskoetteriana**
  b. Seeds dark brown to black (section **Neobesseya** (Britton & Rose) N.P. Taylor)   **6**
6a. Stems cylindric and/or simple; spines 16–75, smooth   **7**
  b. Stems depressed-spherical, rarely spherical, clustering; spines 8–21, finely hairy   **7. missouriensis**
7a. Withered perianth falling off leaving a very small white scar at apex of ripe fruit   **6. zilziana**
  b. Withered perianth persisting at fruit apex   **5. dasyacantha**

1. E. chihuahuensis Britton & Rose. Illustration: Kakteen und andere Sukkulenten **34**: 77, 120 (1983).

Simple or clustering, stems spherical to cylindric, to 8 cm in diameter, very variable in size and shape, rootstock tuberous. Spines mostly somewhat adpressed to stem. Flowers to 2 × 2 cm, pale pink to purplish; stigma-lobes, 5–6 white. Fruit green. *N Mexico (C Chihuahua).* G1. Spring.

E. henricksonii Glass & Foster. Illustration: Cactus & Succulent Journal (US) **49**: 195–196 (1977); British Cactus & Succulent Journal **4**: 40 (1986). Clustering, stems cylindric, to 3 cm in diameter. Spines very dense and tightly adpressed to stem. Flowers to 2.5 cm in diameter; stigma-lobes 4–5. Fruit 8 × 3 mm, reddish. *N Mexico (S Chihuahua & NE Durango).* G1. Spring.

Probably only a more slender southern variety of the preceding.

2. E. tuberculosa (Engelmann) Britton & Rose (*Mammillaria tuberculosa* Engelmann; *Escobaria strobiliformis* misapplied); *Coryphantha (Escobaria) varicolor* Tiegel). Illustration: Kakteen und andere

Sukkulenten **34**: 78 (1983); Weniger, Cacti of Texas and neighboring states, 193, 194 (in part), 201 (1984).

Simple or clustering, stems ovoid to cylindric, 5–20 × 2.5–7.5 cm, very variable, some of the spines projecting. Flowers to 2.5 × 3.5 cm, pale pink to whitish, sweetly scented; stigma-lobes 4–7, white. Fruit club-shaped to cylindric, to 2 cm, deep pink to dull red, rarely green tinged with red. *N Mexico, USA (S New Mexico, W Texas).* G1. Summer.

3. E. vivipara (Nuttall) Buxbaum (*Cactus viviparus* Nuttall; *Coryphantha vivipara* (Nuttall) Britton & Rose; *Mammillaria vivipara* (Nuttall) Haworth; *M. radiosa* Engelmann; *M. arizonica* Engelmann; *M. deserti* Engelmann; *Coryphantha deserti* (Engelmann) Britton & Rose; *M. vivipara* subvar. *neomexicana* Engelmann; *Coryphantha neomexicana* (Engelmann) Britton & Rose; *C. alversonii* (Zeissold) Orcutt; *C. bisbeeana* Orcutt; *C. rosea* Clokey). Illustration: Benson, Cacti of the United States & Canada, pl. 167–169, 171–172 (1982); Weniger, Cacti of Texas and neighboring states, 181, 183–185, 187, 189 (1984).

Clustering or simple, stems depressed-spherical to ovoid (or cylindric in var. **deserti** (Engelmann) D. Hunt and var. **neomexicana** (Engelmann) Buxbaum). Central spines 2–12; radial spines 12–40. Flowers 2.5–5 × 2.5–6 cm, shades of pink, magenta or purple, rarely white (greenish yellow, yellow-orange, or brownish in var. **deserti**); stigma-lobes 5–10, white, pink or violet. Fruit 1.2–2.5 cm × 6–15 mm, green; seeds 1–2.4 mm. *Canada (Alberta, Saskatchewan & Manitoba), USA (Great Plains and southwards), N Mexico (N Sonora & N Chihuahua).* G1. Spring.

E. hesteri (Wright) Buxbaum (*Coryphantha hesteri* Wright). Illustration: Benson, Cacti of the United States & Canada, pl. 173 (1982); Kakteen und andere Sukkulenten **34**: 154 (1983). Clustering, stems spherical, 2.5–4 cm. Central spines absent or 1–4; radials 12–20. Flowers to 2.3 × 2.5 cm, light purple; stigma-lobes 3–4, white. Fruit 6–7 × 3–6 mm; seeds 0.75–1 mm. *USA (W Texas).* G1. Summer.

4. E. emskoetteriana (Quehl) Borg (*Mammillaria emskoetteriana* Quehl; *Escobaria runyonii* Britton & Rose; *E. muehlbaueriana* (Boedeker) Knuth; *E. bella* Britton & Rose?). Illustration: Kakteen und andere Sukkulenten **34**: 155, 156 (1983);

Weniger, Cacti of Texas and neighboring states, 205 (1984) – as Mammillaria roberti.
Clustering, stems spherical to short-cylindric, to 5 × 2–4 cm, very variable. Central spines 6–8; radials 15–30. Flowers 1.5–3 × 2–3 cm, off-white, greenish yellow or pale purple, the perianth-segments with darker or brownish midstripes; stigma-lobes greenish, sometimes tipped with brown. Fruit spherical to elliptic, to 9 mm, pink, red or purplish, rarely remaining green; seeds brown. *NE Mexico, USA (S Texas)*. G2. Summer.

5. E. dasyacantha (Engelmann) Britton & Rose (*Mammillaria dasyacantha* Engelmann; *Escobesseya duncanii* Hester; *Coryphantha duncanii* (Hester) Benson; *C. chaffeyi* (Britton & Rose) Fosberg; *Escobaria chaffeyi* Britton & Rose). Illustration: Benson, Cacti of the United States & Canada, pl. 178 (1982); Weniger, Cacti of Texas and neighboring states, 194 (in part), 196–198 (1984). Simple, rarely 2–3-branched, almost spherical to cylindric, 7.5–17.5 × 5–7.5 cm (only 2.5–6 × 2.5–3.5 cm in var. **duncanii** (Hester) N.P. Taylor). Spines 25–50 (to 75 in var. **duncanii**), of uneven length and thickness (finer and more uniform in var. **chaffeyi** (Britton & Rose) N.P. Taylor). Flowers 1–3 × 1–2.5 cm, pinkish to brownish. Fruit club-shaped to cylindric, 1–2 cm, bright red to deep pink, withered perianth persistent. *USA (S New Mexico, W Texas), N Mexico*. G1. Spring.

6. E. zilziana (Boedeker) Backeberg (*Coryphantha zilziana* Boedeker; *Neobesseya zilziana* (Boedeker) Boedeker; *E. strobiliformis* misapplied; *E. tuberculosa* misapplied; *E. muehlenbaueriana* misapplied). Illustrations: Krainz, Die Kakteen, Lfg. 17 (1961) – as E. muehlbaueriana; Cactus & Succulent Journal of Great Britain 40: 36 (1978) – as E. strobiliformis; British Cactus & Succulent Journal 4: 42 (1986). Clustering, stems mostly cylindric, to 10 × 3 cm. Central spines 0–1 or more; radials 16–22. Flowers to 2.5 × 3 cm, the perianth-segments pale yellow, olive-green or whitish with pink midstripe. Fruit narrowly club-shaped-cylindric, to 2 cm, bright pink or red; seeds black. *N Mexico (C & S Coahuila)*. G1. Summer

Not uncommon in cultivation under the misapplied names given above.

7. E. missouriensis (Sweet) D. Hunt (*Mammillaria missouriensis* Sweet; *Neobesseya missouriensis* (Sweet) Britton &

Rose; *Mammillaria similis* Engelmann; *Coryphantha asperispina* Boedeker; *C. marstonii* Clover). Illustration: Benson, Cacti of the United States & Canada, pl. 179, 180 (1982); Kakteen und andere Sukkulenten **34**: 184, 186 (1983); Weniger, Cacti of Texas and neighboring states, 177, 178 (1984).
Clustering, rarely simple, stems depressed-spherical, rarely spherical, to 10 cm in diameter. Spines 8–21, centrals 0–1. Flowers opening widely, to 4–5 cm in diameter (to only 2.5 cm in var. **missouriensis**, to 6 cm in var. **similis** (Engelmann) N.P. Taylor); perianth-segments greenish, yellow or pink (cream with brownish pink midstripes in var. **asperispina** (Boedeker) N.P. Taylor). Fruit spherical to obovoid, to 2 × 2 cm, bright red, withered perianth sometimes deciduous. *USA (Montana, North Dakota & W Minnesota, S to Arizona, E Texas & W Louisiana), NE Mexico (SE Coahuila & S Nuevo Leon)*. G1. Spring.

8. E. roseana (Boedeker) Buxbaum (*Echinocactus roseanus* Boedeker; *Gymnocactus roseanus* (Boedeker) Glass & Foster). Illustration: Kakteen und andere Sukkulenten **34**: 187 (1983); Cullmann et al., Kakteen, edn 5, 171 (1984). Clustering, rarely simple, stems spherical to cylindric, 3–5 cm in diameter, bright green; tubercles small, *c.* 3 mm, groove lacking. Spines *c.* 20. Flowers to 2 × 2 cm; perianth-segments creamy yellow with darker reddish yellow midstripes; stigma-lobes 6–7, cream. Fruit club-shaped, 1–1.5 cm, greenish; seeds brown. *NE Mexico (SE Coahuila & S Nuevo Leon)*. G1. Summer.

E. aguirreana (Glass & Foster) N.P. Taylor (*Gymnocactus aguirreanus* Glass & Foster). Illustration: Cactus & Succulent Journal (US) **44**: 81 (1972), **50**: 219, f. 8 (1978); British Cactus & Succulent Journal **4**: 42 (1986). Usually simple, depressed-spherical to spherical, to 7 cm in diameter, dark greenish bronze. Flowers yellowish to reddish yellow; stigma-lobes 5–6 yellow. Fruit 1.2 cm, tinged bronze; seed dark purplish red to black. *NE Mexico (Coahuila)*. G1. Summer.

75. MAMMILLARIA Haworth
D.R. Hunt
Low-growing with flattened, spherical or cylindric, tubercled stems, simple, clustering or mound-forming, some with latex; tubercles not grooved or glandular, diverse in shape, apex spiny but non-

flowering. Flowers mostly small, arising singly in 'axils' between the tubercles; ovary naked, tube usually short. Fruit berry-like, oblong or club-shaped, often bright red, rarely immersed in stem; seeds black or brown, minutely pitted or wrinkled (smooth and black in No. 1).

About 150 recognisable species, mostly native to Mexico, with several in the SW United States and a few in the West Indies, Central America and Northern S America. Long the most popularly grown and collected of all cacti, with virtually all the known species and varieties cultivated and more or less readily available. The key below, which requires flowering material, is designed to identify about 70 species, representing the principal groups. Other species are briefly described under those to which they are allied.
Literature: Craig, R. T., *The Mammillaria Handbook* (1945); Hunt, D.R., Schumann and Buxbaum reconciled, *Cactus & Succulent Journal of Great Britain* 33: 53–72 (1971); *A new review of Mammillaria names*, reprinted from *Bradleya* 1: 105–128 (1983), 2: 65–96 (1984), 3: 53–66 (1985), 4: 39–64 (1986), 5: 17–48 (1987); Pilbeam, J.W., *Mammillaria, A Collector's Guide* (1981).

Hooked spines.
  2,3,6–9,12–20,22–29,48,58.
Latex in tubercles. 51–70.
Seeds. Black: 1,2,7–14,16–24,26–37,43; brown: 3,6,38–41, 44–70.

1a. Flowers bright yellow, orange or
   scarlet   2
  b. Flowers white, creamy yellow, pink or
   purple   8
2a. Flowers yellow or orange, with or
   without distinct tube   3
  b. Flowers scarlet, with distinct tube   7
3a. One or more spines hooked   4
  b. No spines hooked   5
4a. Flowers orange, salver-shaped
                                  2. beneckei
  b. Flowers bright yellow, funnel-shaped
                                  3. surculosa
5a. Flowers more than 2 cm   6
  b. Flowers less than 2 cm
                                  64. marksiana
6a. Spines up to 15, bristle-like
                                  4. longimamma
  b. Spines more than 30, hair-like
                                  5. baumii
7a. Limb of flower spreading, regular;
   radial spines 30 or more   **7. senilis**
  b. Limb of flower 2-lipped, irregular;
   radial spines fewer than 25
                                  8. poselgeri

8a. Ovary and fruit immersed in stem; flowers more than 2 cm, the withered remains persistent   **9**

  b. Ovary and fruit not, or only slightly, immersed in stem; flowers less or more than 2 cm   **11**

9a. One or more spines hooked   **9. longiflora**

  b. No spines hooked   **10**

10a. Flowering stem 1–2 cm in diameter; flowers with conspicuous slender tube   **10. saboae**

  b. Flowering stems 3 cm or more in diameter; flowers with short tube   **11. napina**

11a. Flowers narrowly funnel-shaped, *c.* 2.5 cm, perianth-segments whitish, striped pink; tube with distinct solid portion between ovary and base of style   **6. carretii**

  b. Flowers various; flower-tube (if evident) hollow   **12**

12a. One or more spines hooked; tubercles without latex   **13**

  b. No spines hooked (except *M. uncinata*); tubercles with or without latex   **30**

13a. Flowers averaging more than 2 cm   **14**

  b. Flowers averaging less than 2 cm   **22**

14a. Radial spines more than 35   **15**

  b. Radial spines fewer than 35   **16**

15a. Radial spines bristle-like; stem ovoid or cylindric   **12. tetrancistra**

  b. Radial spines hair-like; stem spherical   **13. guelzowiana**

16a. Tubercles slender, terete, up to 2.5 cm; flowers 3–4 cm, white, tinged pink or yellow   **14. zephyranthoides**

  b. Tubercles rounded, cylindric or terete, rarely reaching 1.5 cm; flowers various   **17**

17a. Flowers near apex, more or less erect; outer segments distinctly lacerated-ciliate; filaments twisted round style   **15. wrightii**

  b. Flowers lateral or from periphery of apex, not erect; outer segments at most shortly ciliate; filaments not twisted round style   **18**

18a. Stigmas bright purple   **16. thornberi**

  b. Stigma-colour various, not purple   **19**

19a. Tubercle-axils with bristles   **17. dioica**

  b. Tubercle-axils without bristles   **20**

20a. Flowers whitish   **18. hutchisoniana**

  b. Flowers pink, lavender or purple   **21**

21a. Radial spines usually *c.* 12; stem grey-green; fruit breaking off near the base   **19. schumannii**

  b. Radial spines usually 15–35; stem green or grey-green; fruit detaching intact   **20. grahamii**

22a. Stems slender, more than twice as long as broad; stigmas more than 4 mm   **22. occidentalis**

  b. Stems spherical to cylindric but less than twice as long as broad; stigmas less than 4 mm   **23**

23a. Flowers pink, purple or clear dark red   **24**

  b. Flowers creamy yellow, pale pinkish or white   **27**

24a. Radial spines 30 or more   **23. bombycina**

  b. Radial spines less than 30   **25**

25a. Stems spherical, to 5 cm tall, tubercles cylindric, soft   **26**

  b. Stems cylindric, to 15 cm or more tall, tubercles conic or pyramidal, hard   **48. nunezii**

26a. Tubercle-axils without bristles; stigmas yellowish   **24. zeilmanniana**

  b. Tubercle-axils with bristles; stigmas purplish pink   **25. erythrosperma**

27a. Tubercle-axils slightly woolly or naked   **28**

  b. Tubercle-axils with bristles or long hairs   **29**

28a. Radial spines feathery, even to the naked eye   **26. pennispinosa**

  b. Radial spines smooth or at most minutely downy   **27. mercadensis**

29a. Radial spines 25–50, hair-like   **28. bocasana**

  b. Radial spines 25 or fewer, bristle-like   **29. wildii**

30a. Spines more than 50, not sharply differentiated into central and radial   **31**

  b. Spines fewer than 50, all similar, or more or less sharply differentiated into central and radial   **35**

31a. Flowers deep pink or purple   **30. humboldtii**

  b. Flowers not deep pink or purple   **32**

32a. Flowers 2 cm or more; spines bristle-like   **33**

  b. Flowers less than 2 cm; spines hair-like   **34**

33a. Stem-diameter soon exceeding 3 cm; tubercle-axils bristly   **1. candida**

  b. Stems rarely exceeding 3 cm diameter; tubercle-axils naked   **31. herrerae**

34a. Spines feathery, white   **33. plumosa**

  b. Spines at most downy, usually yellow   **34. schiedeana**

35a. Spines 1.5–2 mm, arranged like a comb   **32. pectinifera**

  b. Spines not arranged like a comb   **36**

36a. Flowers bright purple, to 3–4 × 3 cm   **21. mazatlanensis**

  b. Flowers various, not exceeding *c.* 2.5 × 2 cm   **37**

37a. Some or all spines very weak and hair-like; latex absent   **38**

  b. Spines flexible (bristle-like) to rigid; latex present or absent   **39**

38a. Stem simple or offsetting sparingly   **35. picta**

  b. Stem offsetting abundantly   **36. prolifera**

39a. Stem offsetting very freely even near the apex, the offsets often very lightly attached; latex absent   **37. gracilis**

  b. Offsets, if any, produced at the stem base or sides and firmly attached; latex present or absent   **40**

40a. Tubercles 1–2 cm or more, slender; latex absent; spines 10 or fewer   **38. decipiens**

  b. Tubercles not as above; latex present or absent; spines various (see also variants of Nos. 17, 18, 20)   **41**

41a. Flowering stems slender, at least twice as long as broad and not exceeding 4.5 cm in diameter (excluding spines); latex absent   **42**

  b. Flowering stems flattish, spherical or club-shaped, if cylindric then usually exceeding 4.5 cm in diameter; latex present or absent   **47**

42a. Spines smooth, glassy or chalky   **43**

  b. Spines downy   **46**

43a. Flowers creamy yellow   **44**

  b. Flowers purplish   **45**

44a. Radial spines less than 20; stems not exceeding 3 cm in diameter   **39. elongata**

  b. Radial spines 30 or more; stem 3–4.5 cm in diameter   **40. microhelia**

45a. Central spines several; radial spines up to 45   **42. pottsii**

  b. Central and radial spines not differentiated, or central spine 1; other spines less than 30   **43. sphacelata**

46a. Flowers and central spines reddish   **47. spinosissima**

  b. Flowers and central spines yellowish   **49. eriacantha**

47a. Tubercles not exuding milky latex when pricked   **48**

  b. Tubercles exuding latex when pricked   **54**

48a. Flowers pale yellow or whitish, or with a tinge of pink or red   **49**

  b. Flowers pink to deep purplish pink   **50**

49a. Stem becoming cylindric, base not tuberous; central and radial spines not strongly differentiated   **41. densispina**

b. Stem usually depressed-spherical, thickened at base; central and radial spines strongly differentiated **44. discolor**

50a. Central spines strongly differentiated from the radial, or the radials rudimentary or absent (spines less than 7) 51

b. Central spines little stronger than the radials (total spines at least 9) 52

51a. Radial spines present 53

b. Radial spines absent or rudimentary; centrals 2–8 **46. polythele**

52a. Tubercle-axils without bristles **47. spinosissima**

b. Tubercle-axils with bristles **48. nunezii**

53a. Tubercle-axils usually with bristles but scant wool; central spines usually more than 2; flowers averaging 1.5–2 cm **45. rhodantha**

b. Tubercle-axils without bristles, but often densely woolly; central spines usually 2; flowers averaging 1–1.5 cm **50. haageana**

54a. Stem largely hidden by the numerous white radial spines (and/or axillary bristle-hairs); tubercles small, 5–10 mm; flowers small, averaging 1.2 cm 55

b. Stem not hidden by the spines; tubercles medium-sized to large, usually 8–15 mm; flowers medium-sized, usually exceeding 1.5 mm 60

55a. Flowers deep purplish pink 56

b. Flowers clear pink to white or pale yellow 59

56a. Uppermost central spine, longest, to 4 cm; radial spines 16–20; stem offsetting, not dividing apically **51. geminispina**

b. Central spines short, usually less than 8 mm, or the lowest longest, to 3.5 cm; radial spines 12–50; stem simple or dividing apically 57

57a. Individual stems small, to 5 cm in diameter; axillary bristle-hairs inconspicuous or scarcely exceeding the tubercles; radial spines 12–22 **52. perbella**

b. Individual stems 5–15 cm in diameter; axillary bristle-hairs conspicuous, often twice or more as long as the tubercles; radial spines 20–50 (sometimes rudimentary) 58

58a. Central spines dark brown to black, at least towards the tip **53. hahniana**

b. Central spines pale yellow to golden yellow **54. muehlenpfordtii**

59a. Radial spines usually 30–35; lowest central spine 1–3.5 cm **55. parkinsonii**

b. Radial spines less than 30; central spines usually 4–8 mm **56. formosa**

60a. Hooked spine(s) present **58. uncinata**

b. No spines hooked 61

61a. Radial spines 40–50; central spines golden or straw yellow **54. muehlenpfordtii**

b. Radial spines less than 30 62

62a. Tubercle-axils without bristles 63

b. Tubercle-axils with bristles 72

63a. Spines 6 or fewer 64

b. Spines 7 or more 68

64a. Flowers pale pink or creamy white 65

b. Flowers purplish pink or red 67

65a. Spines 2 or 4, very short, up to 4 mm long, straight **57. sempervivi**

b. Spines 4 or 5, longer than 4 mm, often greatly so, straight or curved 66

66a. Flowers creamy yellow or tinged brown; tubercles not sharply 4-angled **66. magnimamma**

b. Flowers flesh pink; tubercles 4-angled **68. carnea**

67a. Tubercles bluntly and unequally swollen, broader than long, bluish green **67. compressa**

b. Tubercles conic-pyramidal, as long or longer than broad, dark green **66. magnimamma**

68a. Flowers yellowish or whitish 69

b. Flowers pink or red **61. melanocentra**

69a. Flowers less than 1.2 cm, not exceeding the tubercles **59. mammillaris**

b. Flowers 1.5–3 cm, projecting beyond the tubercles 70

70a. Central spines 4, one or more of them exceeding 1.5 cm; radial spines to 12, stem spherical, eventually 15 cm or more in diameter **62. gigantea**

b. Central spines 0–2, usually less than 1.5 cm (rarely 4–6 and then accompanied by 10–18 radials); radial spines 7–25; stem various 71

71a. Stem flat-topped to flattish-spherical, not becoming cylindric unless more than 15 cm diameter; flowers 2–3 cm, whitish or pale pink **60. heyderi**

b. Stem usually spherical to short-cylindric, rarely flattish; flowers 1.5–2 cm, pale yellow, outer segments tinged brown or red **65. johnstonii**

72a. Spines 10 or more 73

b. Spines fewer than 10 74

73a. Flowering axils densely woolly **63. standleyi**

b. Flowering axils with scant wool **69. mystax**

74a. Flowers red or purple **67. compressa**

b. Flowers creamy yellow or pale pink **70. karwinskiana**

1. M. candida Scheidweiler (*Mammilloydia candida* (Scheidweiler) Buxbaum). Illustration: Pilbeam, Mammillaria, 44, opp. 96 (1981); Graf, Exotica, edn 4, **1**: 712 (1982); Haustein, Der Kosmos-Kakteenführer, 271 (1983).
Simple or clustering, stems spherical, to 14 cm in diameter, eventually short-cylindric. Spines very numerous, *c.* 10 mm, pure white or tipped pink or brown. Flowers near apex, 2–3 cm, pale pink or tinged creamy brown, stigmas purplish. Fruit oblong, pinkish, the perianth not persistent; seeds black, shiny, not pitted. *NE Mexico*. G1. Spring.

Probably not closely related to the rest of the genus, in view of the different seeds.

2. M. beneckei Ehrenberg (*M. balsasoides* Craig). Illustration: Craig, Mammillaria Handbook 158, f. 139 (1945); Pilbeam, Mammillaria, 38 (1981); Haustein, Der Kosmos-Kakteenführer, 270 (1983).
Simple or clustering, stems spherical to short-cylindric, to *c.* 10 × 7 cm, apex often oblique, tubercles soft-textured. Central spines 2–6, 8–12 mm, 1 or 2 longer, hooked, dark brown; radial spines 8–15, 6–8 mm, pale. Flowers salver-shaped, 2–2.5 × 3–4 cm, orange-yellow, tube solid below insertion of stamens. Seeds largest in genus, 2.5-3 mm, black, wrinkled. *W Mexico*. G2. Summer (rarely flowering in Europe).

3. M. surculosa Boedeker (*Dolichothele surculosa* (Boedeker) Buxbaum). Illustration: Craig, Mammillaria Handbook 177, fig. 159 (1945); Pilbeam, Mammillaria, 131 (1981); Haustein, Der Kosmos-Kakteenführer, 273 (1983).
Freely clustering, stems *c.* 3 × 2 cm with tuberous rootstock, tubercles cylindric, *c.* 8 × 4 mm. Central spine 1, to 2 cm, slender, hooked, amber-yellow, darker tipped; radial spines *c.* 15, 8–10 mm, bristle-like, pale yellow. Flowers funnel-shaped, *c.* 2.5 × 2 cm, bright yellow. Fruit oblong, greenish brown; seeds light brown. *NE Mexico*. G1. Spring.

4. M. longimamma de Candolle (*Dolichothele longimamma* (de Candolle) Britton & Rose; *D. uberiformis* (Zuccarini) Britton & Rose). Illustration: Pilbeam, Mammillaria, 84, 85 (1981); Graf, Exotica edn 4, **1**: 708, 711 (1982); Haustein, Der Kosmos-Kakteenführer, 272 (1983).

Simple or clustering, short-stemmed from tuberous rootstock, to 10–12 cm in diameter, tubercles oblong-terete, the largest in the genus, 2–5 cm × 6–12 mm. Central spines 0–3, usually 1, bristle-like, protruding, to 2.5 cm; radial spines 6–10, 1.2–1.8 cm, all spines whitish. Flowers large, 3–6 × 3–6 cm, bright yellow. Fruit ovoid, yellowish or purplish green; seeds dark brown. *Eastern C Mexico*. G1. Summer.

M. sphaerica Dietrich (*Dolichothele sphaerica* (Dietrich) Britton & Rose). Very similar, but tubercles generally smaller; radial spines *c.* 12–14; seeds black. *NE Mexico, USA (SE Texas)*. G1. Summer.

5. M. baumii Boedeker (*Dolichothele baumii* (Boedeker) Werdermann & Buxbaum). Illustration: Pilbeam, Mammillaria, 36 (1981); Bradleya **1**: 113 (1983). Freely clustering, stems ovoid, to 6–7 × 5–6 cm; tubercles cylindric 8–10 × 5 mm, largely hidden by the numerous bristle-and hair-like, pale yellow to white spines *c.* 1.5 cm. Flowers 2.5 × 3 cm, bright yellow. Fruit greenish; seeds dark brown. *NE Mexico*. G1. Spring.

6. M. carretii Schumann. Illustration: Craig, Mammillaria Handbook, 176, f. 158 (1945); Pilbeam, Mammillaria, 46 (1981); Haustein, Der Kosmos-Kakteenführer, 273 (1983). Simple or clustering, stems of cultivated variant flattened, spherical, tubercles 7–10 × 7 mm. Central spine *c.* 1.5 cm, hooked, brown; radial spines 12–14, slightly shorter, pale yellow. Flowers 2.5 × 1.5 cm, creamy white, inner segments with pink midstripe. Fruit green; seeds brown. *NE Mexico*. G1. Spring.

7. M. senilis Salm-Dyck (*Mamillopsis senilis* (Salm-Dyck) Britton & Rose). Illustration: Pilbeam, Mammillaria, 125 and colour plate (1981); Kakteen und andere Sukkulenten **33**(2): front cover (1982); Haustein, Der Kosmos-Kakteenführer, 275 (1983). Eventually forming large clusters, stems *c.* 6 cm in diameter, densely covered by the numerous, bristle-like, pure white or yellowish-tipped spines, some hooked. Flowers 5–7 × 2–2.5 cm with straight scaly tube and spreading limb, bright scarlet, stamens and style projecting. Fruit red; seeds black. *Mountains of NW Mexico*. G1. Spring.

8. M. poselgeri Hildmann (*Cochemiea poselgeri* (Hildmann) Britton & Rose). Figure 29(9), p. 205. Illustration: Pilbeam,

Mammillaria, 113 and colour plate (1981); Cactus & Succulent Journal of Great Britain **42**(1): front cover (1980); Haustein, Der Kosmos-Kakteenführer, 275 (1983). Stems long, to 60 cm or more × 4 cm in diameter, branching from base and developing a tuberous rootstock; tubercles distinctly upswept, conic-pyramidal, *c.* 1 cm. Central spine 1, 1.5–2 cm, hooked; radial spines 7–9, *c.* 10 mm, all spines brownish at first, eventually whitish. Flowers near apex, 3 cm or more long with curved tube and oblique limb with double ring of recurving segments, bright scarlet, stamens and style well-projecting. Fruit obovoid, red; seeds black. *NW Mexico (S Baja California)*. G1. Summer.

M. pondii Greene (*Cochemiea pondii* (Greene) Walton), from *Mexico (Cedros Island, Baja California)* (and related species from the peninsula mainland) have more numerous spines. G1. Summer.

9. M. longiflora (Britton & Rose) Berger (*Neomammillaria longiflora* Britton & Rose). Illustration: Pilbeam, Mammillaria, 83, 84 and colour plate (1981); Haustein, Der Kosmos-Kakteenführer, 277 (1983). Usually simple, stems spherical, 3–9 cm in diameter, tubercles *c.* 1 cm long, plump. Central spines 4, the upper 3 straight, the lowermost strongly hooked, 1.1–2.5 cm, dark brown to yellow or whitish; radial spines 25–30, 1–1.3 cm, bristle-like, whitish. Flowers funnel-shaped, 2–3.5 × 3 cm, pink or purplish pink, with distinct tube, the ovary and fruit immersed in the stem. Seeds black. *Mountains of NW Mexico*. G1. Spring.

10. M. saboae Glass (*M. goldii* Glass & Foster; *M. haudeana* Lau & Wagner). Illustration: Pilbeam, Mammillaria, 120 and colour plate (1981). Freely clustering or remaining simple (*M. goldii*), stems ovoid, 1–2 × 1–3 cm from fleshy roots, tubercles short, rounded. Spines all radial (rarely 1 very short central spine present), 17–45, 2 mm, whitish. Flowers 3.5–4.5 × 3 cm, violet or purplish pink, the slender tube 2–3 cm long. Fruit and black seeds ripening immersed in the stem, the withered perianth persistent. *Mountains of NW Mexico*. G1. Spring.

M. theresae Cutak. Very similar but stems to 4 cm, often tinged purple, radial spines feathery. *Mountains of NW Mexico*. G1. Spring.

11. M. napina Purpus. Illustration: Pilbeam, Mammillaria, 101 (1981); Cactus & Succulent Journal of Great Britain **43**: 95 (1981).

Simple, spherical, to 5 cm in diameter, with tuberous tap-root. Spines about 12, normally all radial, 8–9 mm, glassy white. Flowers 4 cm in diameter, petals pale pink with deeper midstripe. Fruit and black seeds ripening almost immersed in stem. *SW Mexico*. G1. Summer.

M. deherdtiana Farwig and **M. dodsonii** Bravo lack the tuberous tap-root and have more numerous spines, usually including 1 or more straightish, reddish brown centrals. *SW Mexico*. G1. Summer.

12. M. tetrancistra Engelmann (*M. phellosperma* Engelmann). Illustration: Craig, Mammillaria Handbook, 196, f. 177 (1945); Pilbeam, Mammillaria, 133 and colour plate (1981); Haustein, Der Kosmos-Kakteenführer, 277 (1983). Simple and clustering from stout rootstock, stems ovoid to cylindric, to 15 (rarely to 25) × 7 cm; tubercle-axils with bristles. Central spines 3–4, dark brown, longest to 1.8 cm, one or more hooked; radial spines 30–60, to *c.* 10 mm, bristle-like, white or dark-tipped. Flowers 2.5 × 2.5–3.5 cm, lavender-pink, stigmas cream. Fruit club- or sausage-shaped to 2.5 cm, red, perianth not persistent; seeds black, wrinkled, with corky appendage larger than the seed itself. *SW United States, NW Mexico*. G2. Summer.

13. M. guelzowiana Werdermann. Illustration: Cactus & Succulent Journal of Great Britain **41**: 12 (1979); Pilbeam, Mammillaria, 67 and colour plate (1981); Haustein, Der Kosmos-Kakteenführer, 277 (1983); Cullmann et al., Kakteen, edn 5, 218 (1984). Simple or eventually clustering, stems flattened-spherical, to 7 cm or more in diameter, roots fibrous, tubercles plump. Central spine 1, 8–16 mm, hooked, reddish or yellowish; radial spines 60–80, hair-like, white. Flowers amongst largest in genus, to 5 × 6 cm, bright purplish pink with green stigmas. Fruit spherical-cylindric, pinkish or yellowish, not exceeding the tubercles; seeds black, each with a corky appendage. *NW Mexico*. G1. Summer.

14. M. zephyranthoides Scheidweiler. Illustration: Pilbeam, Mammillaria, 143 (1981); Cullmann et al., Kakteen, edn 5, 233 (1984). Simple, flattened-spherical, to 8 × 10–15 cm, with stout roots, tubercles to 2.5 cm, slender, flabby. Central spines 1–2, to 1.4 cm, hooked, reddish brown or paler; radial spines 12–18, 8–10 mm, hair-like, white. Flowers to 3 × 4 cm, tube hollow above ovary, perianth-segments white with pink midstripe; stigmas green. Fruit red, ovoid; seeds black. *C Mexico*. G1. Summer.

M. heidiae Krainz. Illustration: Cactus & Succulent Journal of Great Britain **42**: 73 (1980); Cullmann et al., Kakteen, edn 5, 220 (1984). Later clustering; radial spines 16–24; flowers 3 × 2.5 cm, yellow. *Mexico (Puebla)*. G1. Summer.

15. M. wrightii Engelmann. Illustration: Cactus & Succulent Journal (US) **49**: 25 (1977); Pilbeam, Mammillaria, 140 (1981). Usually simple, flattened-spherical to short cylindric, to 6 cm in diameter, tubercles terete, averaging 1.3 cm, with naked axils. Central spines 1–several, 1 or more hooked, 5–21 mm, brown; radial spines bristle-like, 8–20 (averaging 13), white. Flowers 2.5–5 × 2.5–7.5 cm, purple (white in one variant). Fruit grape-like, to 2.8 × 2.6 cm (averaging 1.9 × 1.5 cm), dull purplish; seeds dark brown or black. *SW USA, N Mexico*. G1. Spring.

M. viridiflora (Britton & Rose) Boedeker (*Neomammillaria viridiflora* Britton & Rose) has smaller, paler flowers and more numerous spines. *SW USA, N Mexico*. G1. Spring.

The group to which these belong has been extensively studied in the field by Zimmerman & Zimmerman (see Cactus & Succulent Journal (US) **49**: 23–34, 51–62, 1977).

16. M. thornberi Orcutt (*M. fasciculata* misapplied). Illustration: Cactus & Succulent Journal of Great Britain **33**(3): 61 (1971); Pilbeam, Mammillaria, 134 (1981). Clustering, stems slender-cylindric, to 8 × 2 cm or taller, tubercle-axils naked. Central spine usually 1, to 1.8 cm, hooked, brownish; radial spines 13–20, 5–7 mm, bristle-like, whitish, dark tipped. Flowers funnel-shaped, about 2 × 2 cm, pink, with purple stigmas to 1.1 cm. Fruits scarlet; seeds black. *SW USA (Arizona), N Mexico*. G1. Summer.

M. yaquensis Craig. Illustration: Haustein, Der Kosmos-Kakteenführer, 279 (1983). Stems only 1–1.5 cm in diameter. *N Mexico*. G1. Summer.

M. mainiae Brandegee. Stems spherical, to 10–12 × 6–7 cm, radial spines 10–15. *SW USA (Arizona), N Mexico*. G2. Summer.

M. fraileana (Britton & Rose) Boedeker (*Neomammillaria fraileana* Britton & Rose). Stems cylindric, to 15 × 3 cm or taller, radial spines usually 11–12. *NW Mexico (S Baja California)*. G1. Summer.

All these species have long, purple stigma-lobes.

17. M. dioica Brandegee. Illustration: Pilbeam, Mammillaria, 53 and colour plate (1981).
Simple or clustering, stems cylindric to columnar, to 10–15 × 3–8 cm, tubercle-axils with bristles. Central spines 3–4, one larger, to 1.5 cm, hooked, dark brown; radial spines 11–22, bristle-like, whitish. Flowers creamy yellow, bisexual or female, the bisexual flowers usually larger, to 2.5 × 2 cm, the female only 1–1.5 × 1–1.5 cm, stigmas yellowish or brownish green. Fruit scarlet; seeds black. *NW Mexico (Baja California), SW USA (SW California)*. G1. Summer.

M. armillata Brandegee. To 30 × 4–5 cm, radial spines 9–15, flowers bisexual, creamy yellow to pinkish. *NW Mexico (Cape region of Baja California)*. G1. Summer.

A mainland plant of uncertain status is **M. swinglei** (Britton & Rose) Boedeker (*Neomammillaria swinglei* Britton & Rose). Illustration: Haustein, Der Kosmos-Kakteenführer, 281 (1983). Stems cylindric, 10–20 × 3–5 cm; axils of tubercles more or less bristly. Central spines 4, ascending, dark brown or black, the lowest longest, 1–1.5 cm, hooked or sometimes straight; radial spines 11–18, dull white with dark tips. Outer perianth-segments greenish or sometimes pinkish; inner perianth-segments nearly white with brown midstripe; style pink, twice as long as the pink filaments; stigma-lobes 8, linear, pointed, green. Fruit dark red, club-shaped, 1.4–1.8 cm; seeds 1 mm, black. *NW Mexico (Sonora)*. G1. Summer.

18. M. hutchisoniana (Gates) Boedeker (*Neomammillaria hutchisoniana* Gates; *M. louisiae* Lindsay). Illustration: Pilbeam, Mammillaria, 75 (1981).
Clustering; stems cylindric to 15 × 4–6 cm, tubercle-axils without bristles. Central spines 4, lowermost longest, about 1 cm, hooked, dark-tipped; radial spines 10–20, bristle-like, pale brown or whitish. Flowers about 2.5 × 2.5 cm, pale pink or creamy white, outer segments striped dull maroon or brown outside, stigmas green or olive green. Fruit scarlet; seeds black. *NW Mexico (Baja California)*. G1. Summer.

The earlier-named *M. goodridgei* Salm-Dyck (*Mexico, Cedros Island*) may be the same.

M. capensis (Gates) Craig (*Neomammillaria capensis* Gates). Stems to 25 cm, central spine 1, radial spines about 13, rigid. *NW Mexico (Cape region of Baja California)*. G1. Summer.

M. multidigitata Lindsay. Freely clustering, central spines normally straight,

radial spines 15–25, flowers nearly white. *NW Mexico (Isla San Pedro Nolasco)*. G1. Summer.

19. M. schumannii Hildmann (*Bartschella schumannii* (Hildmann) Britton & Rose). Illustration: Pilbeam, Mammillaria, 123 (1981); Graf, Exotica, edn 4, **1**: 710 (1982); Haustein, Der Kosmos-Kakteenführer, 281 (1983).
Clustering, from thickish roots, stems ovoid to short-cylindric, to 6–8 × 3–4 cm; tubercles short, grey-green; axils soon naked. Central spine usually 1, 1–1.5 cm, hooked, dark brown to white with dark tip; radial spines 9–15, 6–12 mm. Flowers *c.* 2.5 × 2.5–4 cm, lavender-pink, stigmas greenish or creamy yellow. Fruit to 2 cm, orange-scarlet, partly embedded in the stem and tending to break off where it emerges; seeds black. *NW Mexico (S Baja California)*. G1. Summer.

20. M. grahamii Engelmann. Illustration: Benson, Cacti of the United States & Canada, f. 933–938 (1982).
Simple or branched at base, spherical to ovoid, 7.5–10 × 7.5–11 cm, grey-green, from thickened roots; tubercles ovoid-cylindric, 6–12 × 4–5 mm; axils naked. Central spines 1–3, the longest hooked, to 1.8 (rarely to 2.5) cm, dark brown, the hook 1.5 mm across, the others when present straight, shorter, paler; radial spines 20–35, 6–12 mm, white. Flowers 2.5–4.4 cm in diameter, pink, outer segments fringed; stigmas greenish, to 8 mm. Fruit almost spherical to barrel-shaped, 1.2–2.5 cm × 6 mm, red; seeds 0.8–1.0 mm, black. *SW USA*. G1. Summer.

M. milleri (Britton & Rose) Boedeker (*M. microcarpa* Engelmann, provisional name). Illustration: Pilbeam, Mammillaria, 94 (1981); Haustein, Der Kosmos-Kakteenführer, 279 (1983). Simple, later clustering, stems cylindric, to *c.* 15 × 4–8 cm, green. Hook of principal central spine 3 mm across; radial spines 18–28. Flowers about 2 × 2–3 cm, lavender-pink with light green or pale brownish stigmas. Fruits of 2 forms, either club-shaped, 2–2.5 cm long, scarlet, or small, ovoid, green; seeds black. *SW USA, N Mexico*. G1. Summer.

M. blossfeldiana Boedeker (*M. shurliana* (Gates) Gates). Spherical to short-cylindric, green; roots tuberous or not; radial spines 13–20. *NW Mexico (C Baja California)*. G1. Summer.

M. boolii Lindsay. Illustration: Haustein, Der Kosmos-Kakteenführer, 281 (1983). Spherical, to 5 cm, grey-green, with tapered roots. Central spine 1, hooked, yellow or horn-coloured; radial spines *c.* 20. Fruit orange. *NW Mexico (coast of Sonora).* G1. Summer.

M. insularis Gates. Illustration: Haustein, Der Kosmos-Kakteenführer, 281 (1983). Depressed-spherical, bluish green, with tuberous roots. Central spine 1, blackish; radial spines 20–30, white. *NW Mexico (Los Angeles Bay area of Baja California).* G1. Summer.

M. sheldonii (Britton & Rose) Boedeker (*Neomammillaria sheldonii* Britton & Rose). Illustration: Britton & Rose, The Cactaceae 4: f. 175 (1923). Stems slender-cylindric, *c.* 8 cm. Central spine 1, hooked; radial spines 20–24, pale with dark tips, the 3 or 4 upper ones darker, a little stouter and 1 or 2 of them more or less central. Perianth-segments light purple with very pale margins; filaments and style light purple; stigma-lobes 6, green. Fruit club-shaped, 2.5–3 cm, pale scarlet. *NW Mexico (Sonora).* G1. Summer.

Variants with straight central spines occur with the above and have been given species names (*M. oliviae* Orcutt, *M. inaiae* Craig, *M. gueldemanniana* Backeberg).

21. M. mazatlanensis Schumann. Illustration: Pilbeam, Mammillaria, 91 (1981); Graf, Exotica, edn 4, 1: 725 (1982); Haustein, Der Kosmos-Kakteenführer, 279 (1983).
Clustering, slender-cylindric, to 15 × 2–4 cm, tubercle-axils usually naked. Central spines 3–4, the lowermost typically or predominantly straight, sometimes more or less strongly hooked, reddish brown; radial spines 13–15 bristly, white. Flowers to 3–4 × 3 cm, bright purple with green stigmas. Fruit reddish; seeds black. *Coast of NW Mexico (Sonora to Nayarit).* G1. Summer.

22. M. occidentalis (Britton & Rose) Boedeker (*Neomammillaria occidentalis* Britton & Rose). Illustration: Pilbeam, Mammillaria, 103 (1981).
Like the preceding, but the lowermost central spine hooked, flowers less than 2 × 2 cm, pale pink or whitish, with long pinkish orange stigmas; fruit red. *Coast of W Mexico (Colima).* G1. Summer.

23. M. bombycina Quehl. Illustration: Pilbeam, Mammillaria, 40 (1981); Graf, Exotica edn 4, 1: 718 (1982); Cullmann et al., Kakteen, edn 5, 204 (1984).
Clustering, stems short-cylindric, to 20 × 7–8 cm, tubercle-axils white-woolly. Central spines 4, lowermost to 2 cm,

hooked, yellow to reddish brown; radial spines 30–40, stiff, glassy white. Flowers 1.5 × 1.5 cm, light purple. Fruit whitish; seeds black. *Origin uncertain, presumably Northern C Mexico.* G1. Spring.

24. M. zeilmanniana Boedeker. Illustration: Pilbeam, Mammillaria, 143 (1981); Haustein, Der Kosmos-Kakteenführer, 285 (1983); Cullmann et al., Kakteen, edn 5, 232 (1984).
Soon clustering, stems to 6 × 4.5 cm, tubercle-axils naked. Central spines 4, the lowermost hooked, about 10 mm, reddish brown; radial spines about 15–18, finely bristle-like, white. Flowers to 2 cm, varying from purple to pink (white variants also occur), with yellowish stigmas. Fruit pale pink or whitish; seeds black. *C Mexico (Guanajuato).* G1. Spring.

25. M. erythrosperma Boedeker. Illustration: Pilbeam, Mammillaria, 58 and colour plate (1981).
Like the preceding, and confused with it, but stems usually narrower, tubercle-axils with bristles, flowers clear dark red (rarely white), fruit red, seeds brown. *Eastern C Mexico (San Luis Potosí).* G1. Spring.

26. M. pennispinosa Krainz. Illustration: Pilbeam, Mammillaria, l07 (1981); Haustein, Der Kosmos-Kakteenführer, 283 (1983); Cullmann et al., Kakteen, edn 5, 226 (1984).
Usually simple, spherical, to about 5 × 5 cm, from stout root-stock; tubercle-axils woolly at first. Central spine 1, 1–1.2 cm, hooked, yellow below, brownish red above; radial spines 16–20, feathery, greyish white. Flowers *c.* 1.5 × 1.5 cm, segments whitish with pale pink midstripe; stigmas yellowish. Fruit to 2 cm, red; seeds black, with corky appendage. *N Mexico (Coahuila).* G1. Spring.

27. M. mercadensis Patoni. Illustration: Pilbeam, Mammillaria, 92 (1981); Haustein, Der Kosmos-Kakteenführer, 283 (1983).
Simple or clustering, stems usually more or less spherical, 5–6 cm in diameter, tubercle-axils naked. Central spines 4 or more, usually only one hooked, 1.5–2.5 cm, varying from pale straw yellow to dark reddish brown; radial spines 25–30, stiffly bristle-like, white or pale yellowish. Flowers about 1.5 × 1.5 cm, pale pink, with creamy white stigmas. Fruit red; seeds black. *Mexico (E flank of Sierra Madre Occidental).* G1. Spring.

Probably the same are *M. sinistrohamata* Boedeker, *M. jaliscana* (Britton & Rose)

Boedeker (*Neomammillaria jaliscana* Britton & Rose) and some other species from the same region.

28. M. bocasana Poselger. Illustration: Pilbeam, Mammillaria, 40 (1981); Graf, Exotica, edn 4, 1: 717 (1982); Haustein, Der Kosmos-Kakteenführer, 285 (1983).
Freely clustering, stems more or less spherical, usually almost hidden by the spines. Central spines 1 or more, 1–2 hooked, 5–10 mm, reddish or brown; radial spines 25–50, hair-like, white. Flowers 1.3–2.2 × up to 1.5 cm, creamy white or pale pink. Fruit slender-cylindric, to 4 cm, red; seeds brown or black. *C Mexico (San Luis Potosí).* G1. Spring.

Distinctive in its best cultivated variants, some of which have cultivar names, but variable and often seen under other, mostly problematical, names such as *M. erectohamata* Boedeker, *M. hirsuta* Boedeker, *M. icamolensis* Boedeker, *M. kunzeana* Boedeker & Quehl, *M. longicoma* (Britton & Rose) Berger and *M. schelhasii* Pfeiffer.

Since about 1910, numerous 'species' related to this and the next species have been named, nearly all by European amateurs, notably F. Boedeker, from individually imported live specimens of imprecise or uncertain origin. Material has rarely been preserved, descriptions are often incomplete, and many of the names seem to have been subsequently mixed up by nurserymen and collectors. For example, a host of names (including *M. nana* Backeberg and *M. trichacantha* Schumann) has been applied to a range of plants of this affinity with downy spines.

29. M. wildii Dietrich. Illustration: Pilbeam, Mammillaria, 139 (1981); Graf, Exotica, edn 4, 1: 726 (1982); Haustein, Der Kosmos-Kakteenführer, 285 (1983).
Clustering, stems spherical to cylindric, to about 12 × 5 cm, tubercle-axils with bristle-like hairs. Central spines 4, 8–10 mm, lowermost hooked, straw yellow to brown; radial spines 8–10, bristle-like, pale yellow to white. Flowers small, whitish. Fruit red; seeds black. *C Mexico (Hidalgo).* G1. Spring.

Amongst numerous related species all from EC Mexico, are: **M. aurihamata** Boedeker (*M. boedekeriana* misapplied; *M. crinita* misapplied), depressed-spherical, radial spines 15–20; **M. painteri** Rose, tubercle-axils without bristles, central spines dark brown; **M. pygmaea** (Britton & Rose) Berger (*Neomammillaria pygmaea* Britton & Rose; *M. knebeliana* Boedeker; *M. mollihamata* Shurly, etc.), radial spines about 15, finely bristle-like; **M. leucantha**

Boedeker (*M. sanluisensis* misapplied), stem depressed-spherical, often markedly tuberous, radial spines about 18 (Figure 29(17), p. 205). The name **M. glochidiata** has been misapplied to a pink-flowered variant with 12–15 radial spines.

30. M. humboldtii Ehrenberg. Illustration: Pilbeam, Mammillaria, 74 and colour plate (1981); Haustein, Der Kosmos-Kakteenführer, 289 (1983).
Simple or clustering, stems spherical, tubercle-axils with bristles. Spines very numerous (80 or more), 4–6 mm, bristle-like, pure white. Flowers about 1.5 × 1.5 cm, bright purplish red. Fruit red; seeds black. *E C Mexico*. G1. Spring.

M. laui D. Hunt. Illustration: Cactus & Succulent Journal of Great Britain **41**: 103 (1979); Pilbeam, Mammillaria, 80, 81, colour plate (1981). Clustering, tubercle-axils without bristles. Spines variable, all more or less similar or graded from central to radial, white, yellow or the central brown-tipped. Flowers pink. Fruit whitish or pale pink; seeds black. *Eastern C Mexico*. G1. Spring.

31. M. herrerae Werdermann. Illustration: Pilbeam, Mammillaria, 71 and colour plate (1981); Hecht, BLV Handbuch der Kakteen, 301 (1982).
Simple or clustering from base, stems spherical to cylindric, usually 2–3 cm in diameter, tubercular axils naked. Spines very numerous (60–100 or more), 1–5 mm, bristle-like, white. Flowers 2–3.5 × 2.5–3 cm, lavender-pink or whitish with green stigmas. Fruit almost spherical, 6 mm in diameter, whitish; seeds black. *Eastern C Mexico*. G1. Spring.

32. M. pectinifera Weber (*Solisia pectinata* (Stein) Britton & Rose). Illustration: Pilbeam, Mammillaria, 106 and colour plate (1981); Haustein, Der Kosmos-Kakteenführer, 289 (1983); Cullmann et al., Kakteen, edn 5, 227 (1984).
Simple, stems 1–3 cm in diameter, spherical to shortly cylindric, with tubercle-axils naked, areoles narrowed vertically, with 20–40 comb-like spines, 1.5–2 mm, bristle-like, adpressed chalky white. Flowers about 2.5 × 2.5 cm, segments creamy yellow or pinkish white with brownish mid-stripe. Fruit small, oblong, greenish; seeds black. *Southeastern C Mexico*. G1. Spring.

33. M. plumosa Weber. Illustration: Pilbeam, Mammillaria, lll (1981); Graf, Exotica, edn 4, **1**: 722 (1982); Haustein, Der Kosmos-Kakteenführer, 287 (1983).
Clustering to form mounds to 40 cm or

more in diameter, individual stems spherical to 6–7 cm diameter, hidden by the spines. Spines up to 40, 3–7 mm, feathery, white. Flowers small, about 1.5 × 1.2 cm, creamy white or tinged brownish pink. Fruit pinkish to dull red; seeds black. *NE Mexico*. G1. Winter.

34. M. schiedeana Ehrenberg. Illustration: Pilbeam, Mammillaria, 122 (1981); Graf, Exotica, edn 4, **1**: 724 (1982); Haustein, Der Kosmos-Kakteenführer, 289 (1983).
Clustering, visible part of individual stems flattened-spherical, to 6 cm diameter, becoming tuberous at base, tubercles slender, the axils with white hairs. Spines very numerous (*c.* 80) 2–5 mm, adpressed, hair-like or finely bristle-like, yellow or golden yellow below, whitish at the tip. Flowers *c.* 1–1.5 × 1 cm, whitish. Fruit slender-cylindric. bright red; seeds black *Eastern C Mexico*. G1. Autumn.

M. carmenae Castaneda. Illustration: Cactus & Succulent Journal of Great Britain **41**: 102 (1979); Pilbeam, Mammillaria, 45, colour plate (1981); Haustein, Der Kosmos-Kakteenführer, 289 (1983). Stems spherical-ovoid, spines somewhat upstanding, finely bristly. Fruit short, greenish; seeds black. *Eastern C Mexico*. G1. Spring.

35. M. picta Meinshausen. Illustration: Pilbeam, Mammillaria, 109 (1981).
Usually simple, becoming tuberous at base, spherical or ovoid, often *c.* 5 × 5 cm, tubercles cylindric-terete, their axils with fine bristles. Spines 11–15, *c.* 10 mm, 1–2 protruding, bristle-like, dark brown, 7–8 radiating, bristle-like, tipped brown, paler below, 3–5 lowermost hair-like, white. Flowers *c.* 2 × 1.5 cm, pale greenish white. Fruit red; seeds black. *Northeastern C Mexico*. G1. Spring.

This species intergrades with the preceding by way of *M. viereckii* Boedeker and *M. dumetorum* Purpus, and with the next (*M. prolifera*) by way of:
M. albicoma Boedeker. Central spines 0 or 3–4, 4–5 mm, white, tipped darker; radial spines 30–40, 8–10 mm, hair-like, white. *Northeastern C Mexico*. G1. Spring.

M. pilispina Purpus (*M. sanluisensis* Shurly). Illustration: Haustein, Der Kosmos-Kakteenführer, 287 (1983). Central spines 5–8, 6–10 mm, brown in upper half, white or yellow below; radial spines *c.* 40, 7–10 mm. *Eastern C Mexico*. G1. Spring.

36. M. prolifera (Miller) Haworth (*Cactus proliferus* Miller; *M. multiceps* Salm-Dyck). Illustration: Pilbeam, Mammillaria, 114

(1981); Graf, Exotica, edn 4, **1**: 708 (1982); Haustein, Der Kosmos-Kakteenführer, 287 (1983).
Offsetting very freely and forming dense clumps, stems to 9 × 4.5 cm, but often smaller, tubercle-axils nearly naked or with fine white hairs. Central spines 5–12, 4–9 mm, white to yellow or reddish, sometimes dark-tipped; radial spines 25–40, 3–12 mm, intergrading with the centrals, bristle- to hair-like, white. Flowers *c.* 1–1.8 × 1.2 cm, creamy or pinkish yellow, outer segments brown-striped. Fruit to 2 cm, red; seeds black. *NE Mexico, SW USA, Cuba, Hispaniola*. G1. Spring.

37. M. gracilis Pfeiffer. Illustration: Pilbeam, Mammillaria, 65 (1981); Graf, Exotica, edn 4, **1**: 716 (1982); Hecht, BLV Handbuch der Kakteen, 301 (1982).
Offsetting very freely, the offsets easily detached, principal stems cylindric, to 13 × 3 cm, tubercles to 6 mm, their axils naked. Central spines none or to 2–5 on mature stems, 1–1.2 cm, bristle-like, white or dark brown; radial spines 11–17, 3–8 mm, chalky white. Flowers small, about 1.2 cm × 8 mm, creamy white. Fruit red; seeds black. *Eastern C Mexico*. G1. Spring.

M. vetula Martius (*M. magneticola* Meyran; *M. kuentziana* invalid). Offsets numerous but more firmly attached than in *M. gracilis*, main stem to 4 cm in diameter; tubercles to 8 mm. Central spines 1–7; radial spines 18–45, 4–12 mm. Flowers 1–1.5 cm, pale yellow. *Eastern C Mexico*. G1. Autumn.

38. M. decipiens Scheidweiler. Illustration: Pilbeam, Mammillaria 51 (1981).
Clustering, stems 4–7 cm in diameter, tubercles cylindric 1–2.2 cm × 5–7 mm, with a few fine axillary bristles. Central spines 1–2, rarely none, 1.8–2.7 cm, bristle-like, straight, brown; radial spines 5–11, 7–15 mm, bristle-like, straight, yellowish or white. Flowers *c.* 1.5 × 1 cm, white, slightly scented. Fruit reddish green; seeds brown. *Eastern C Mexico*. G1. Autumn.

M. camptotricha Dams. Illustration: Haustein, Der Kosmos-Kakteenführer, 273 (1983); Cullmann et al., Kakteen, edn 5, 215 (1984). Central spines 0; radial spines usually 4–5, to 3 cm, almost straight to strongly curved or twisted, yellow to golden brown. *Eastern C Mexico*. G1. Autumn.

M. albescens Tiegel. Central spines 0; radial spines shorter than in *M. camptotricha*, white. *Eastern C Mexico*. G1. Autumn.

39. M. elongata de Candolle (*M. echinaria* de Candolle). Illustration: Pilbeam, Mammillaria, 57 and coloured plate (1981); Graf, Exotica, edn 4, **1**: 708, 714 (1982); Hecht, BLV Handbuch der Kakteen, 299 (1982); Haustein, Der Kosmos-Kakteenführer, 291 (1983).
Clustering, stems cylindric, long, 1–3 cm in diameter, tubercles short, their axils naked. Central spines 0–3, to 1.5 cm, pale yellow to dark brown; radial spines 14–25, 4–9 mm, whitish to golden yellow, regularly radiating. Flowers about 10 × 10 mm, pale yellow or tinged pink. Fruit dull pink or reddish; seeds brown. *C Mexico*. G1. Spring.

40. M. microhelia Werdermann (*M. microheliopsis* Werdermann). Illustration: Pilbeam, Mammillaria, 95 (1981); Graf, Exotica, edn 4, **1**: 723 (1983); Haustein, Der Kosmos-Kakteenführer, 291 (1983).
The typical variant usually simple, but others clustering, stems cylindric, to 25 × 6 cm, tubercle-axils naked. Central spines to 8 or more, to about 1.1 cm, dark reddish brown, contrasting with the paler radial spines, these numerous (30–50), 4–6 mm, regularly radiating and often slightly recurved. Flowers about 1.5 × 1.5 cm, whitish cream to purplish red. Fruit pale greenish or pinkish; seeds brown. *C Mexico (Queretaro)*. G1. Spring.

41. M. densispina (Coulter) Orcutt (*Cactus densispinus* Coulter). Illustration: Pilbeam, Mammillaria, 52 (1981); Hecht, BLV Handbuch der Kakteen, 298 (1982); Haustein, Der Kosmos-Kakteenführer, 291 (1983).
Usually simple, stems spherical to short-cylindric, reaching 10 cm in diameter, tubercle-axils without bristles. Central spines 5–6, 1–1.2 cm, typically longer, more rigid and darker than the 20–25 yellowish radials. Flowers about 1 × 1.5 cm, pale yellow. Fruit greenish pink; seeds brown. *C Mexico*. G1. Spring.

42. M. pottsii Salm-Dyck (*M. leona* Poselger). Illustration: Pilbeam, Mammillaria, 113 (1981); Graf, Exotica, edn 4, **1**: 710 (1982); Hecht, BLV Handbuch der Kakteen, 303 (1982).
Stems slender-cylindric, to 20 × 3 cm, offsetting near the base, tubercle-axils without bristles. Central spines about 7, 7–12 mm, uppermost longest, tipped dark brown, becoming greyish white below; radial spines 40–45, about 6 mm, bristle-like, chalky white. Flowers about 1.2 × 1.2 cm, deep red. Fruit reddish; seeds dark brown or black. *N Mexico, extending into SW USA (SW Texas)*. G1. Spring.
Red-flowered plants of *M. microhelia* also key out here. *M. pottsii* is distinguished by the more slender stem and recurved, erect upper central spine.

43. M. sphacelata Martius (*M. viperina* Purpus). Illustration: Pilbeam, Mammillaria, 128 (1981); Haustein, Der Kosmos-Kakteenführer, 291 (1983).
Freely clustering and eventually forming mounds or colonies 50 cm or more across, stems cylindric, erect or decumbent, 20 or more × 1.5–3 cm, tubercles short, tubercle-axils usually without bristles. Central and radial spines not strongly differentiated, 12–30, 4–8 mm, chalky white, or commonly reddish brown tipped, or brownish throughout. Flowers purplish red, 1.5 × 1 cm. Fruit red; seeds black. *S Mexico (Puebla)*. G1. Spring.

M. kraehenbuehlii (Krainz) Krainz (*Pseudomammillaria kraehenbuehlii* Krainz). Illustration: Pilbeam, Mammillaria, 79 (1981). Densely clustering, the stems to 3.5 cm in diameter, more softly fleshy than *M. sphacelata*; central spine 1, white, tipped brown or absent; radial spines 18–24, very thin. Flowers 1.8 cm, lilac-pink. *S Mexico (Oaxaca)*. G1. Spring.

44. M. discolor Haworth. Illustration: Pilbeam, Mammillaria, 54 (1981); Haustein, Der Kosmos-Kakteenführer, 293 (1983).
Simple, flattened-spherical to short-columnar, some variants eventually 15 × 10 cm or larger, tubercle-axils without bristles. Central spines 4–8, 1 cm or longer, amber-yellow to dark brown; radial spines 16–28, glossy white or pale yellow. Flowers 2–2.5 × 1–1.5 cm, varying from cream-yellow through pale brownish pink to bright pink. Fruit red; seeds brown. *Mountains of C Mexico*. G1. Spring.

M. wiesingeri Boedeker. Roots thickened; flowers smaller, 1.2 × 1 cm, reddish pink. *Eastern C Mexico (Hidalgo)*. G1.

45. M. rhodantha Link & Otto. Illustration: Pilbeam, Mammillaria, 118 (1981); Cullmann et al., Kakteen, edn 5, 228 (1984).
Usually simple or dividing apically, columnar, to 30 × 10 cm, the tubercle-axils with fine bristles. Central spines usually 4–6, 1–1.5 cm, straight or slightly curved, typically reddish brown, in some variants straw or golden yellow; radial spines 16–24, 3–10 mm, glossy white to yellowish. Flowers 2–2.3 × 1.5–1.8 cm, smaller in some variants, intense purplish pink. Fruit dull purplish pink; seeds brown. *Mountains of C Mexico*. G1. Summer.

M. fera-rubra Craig. Very similar but radial spines 12–16, flowers smaller, *c.* 1.5 × 1.2 cm. *Mountains of Western C Mexico*. G1.

M. moellendorffiana Shurly. Very similar, but more slender, to 6–7 cm diameter, radial spines 23–28, flowers smaller, 1–1.5 × 1 cm. *C Mexico (Hidalgo)*. G1. Summer.

M. mundtii Schumann. Illustration: Kakteen und andere Sukkulenten **31**: 1 (1980). Depressed-spherical, central spines 2–4 reddish brown, radial spines 11–12. *Mountains of C Mexico (near Toluca)*. G1. Summer.

M. pringlei (Coulter) Brandegee (*Cactus pringlei* Coulter). One of several yellow- or golden-spined variants of *M. rhodantha. C Mexico*. G1. Summer.

46. M. polythele Martius (*M. tetracantha* Pfeiffer; *M. obconella* Scheidweiler; *M. pyrrhocephala* misapplied; *M. hidalgensis* Purpus). Figure 28(13), p. 204. Illustration: Pilbeam, Mammillaria 112 (1981); Bradleya **4**: 52 (1986).
Simple, cylindric, 30 (rarely to 80) × 8–12 (rarely to 17) cm (but flowering when 5 × 5 cm), tubercles pyramidal-conic, 9–14 × 8 mm, dark bluish green, without axillary bristles. Central spines 2–4, 8–17 (rarely to 25) mm, usually dark brown, in some variants yellowish; radials spines 0 or rudimentary. Flowers pinkish purple, 1.5–1.9 × 1–1.4 cm. Fruit dull purplish; seeds brown. *C Mexico (Hidalgo)*. G1. Summer.

M. kewensis Salm-Dyck (*M. durispina* Boedeker; *M. kelleriana* Craig). Illustration: Haustein, Der Kosmos-Kakteenführer, 293 (1983). Very similar, but spines 6–8. *C Mexico (Queretaro, Hidalgo)*. G1. Summer.

47. M. spinosissima Lemaire. Illustration: Pilbeam, Mammillaria, 129 and colour plate (1981); Haustein, Der Kosmos-Kakteenführer, 295 (1983); Cullmann et al., Kakteen, edn 5, 229 (1984).
Simple at first, later often clustering, stems cylindric, to 30 × 8 cm, tubercle-axils slightly woolly but lacking bristles. Central spines 4–9 or more, 1–1.5 cm, bristle-like, reddish or pinkish brown, or yellowish; radial spines 14–26, 4–10 mm, bristle-like, whitish. Flowers 1.5–2 × 1.5 cm, purplish pink. Fruit greenish to dull purple; seeds brown. *C Mexico*. G1. Spring.

M. backebergiana Buchenau. More slender than the above, 4–6 cm diameter, and with fewer spines; centrals 1–3 radials 8–12. *C Mexico (State of Mexico)*. G1. Spring.

M. matudae Bravo. Illustration: Haustein, Der Kosmos-Kakteenführer, 293 (1983). Slender, 3.5–5 cm in diameter; central spine 1, pointing upwards, to 5 mm, reddish brown; radial spines 18–20, 2–3 mm. *C Mexico (State of Mexico)*. G1. Spring.

M. meyranii Bravo. Slender, to 55 × 5 cm, central spines 2, to 10 mm, orange-yellow, tipped light brown; radial spines 17–19, 3–6 mm. *C Mexico (State of Mexico)*. G1. Spring.

48. M. nunezii (Britton & Rose) Orcutt (*Neomammillaria nunezii* Britton & Rose; *N. solisii* Britton & Rose; *Mammillaria solisii* (Britton & Rose) Boedeker; *M. bella* Backeberg; *M.deliusiana* Shurly). Illustration: Pilbeam, Mammillaria, 103 (1981).
Simple or later clustering, stems spherical to cylindric, to 15 × 8 cm, tubercle-axils with bristles. Central spines 2–6, 1–1.2 cm, all straight, or lowermost longer, hooked, brown, or glassy white with brown tip; radial spines 10–30, 5–8 mm, bristle-like, white. Flowers 2 × 1.5–1.8 cm, purplish pink. Fruit white or greenish, later tinged pink; seeds brown. *Southwestern C Mexico (Guerrero)*. G1. Spring.

M. duoformis Craig & Dawson. Very similar, but stem more slender, elongated, 3–4 cm in diameter, flowers somewhat smaller. *Southern C Mexico (Puebla)*. G1. Spring.

M. hamata Pfeiffer (*Cactus cylindricus* Ortega). Clustering from the base, main stem cylindric, to 60 × 10 cm or more, tubercle-axils without bristles. Central spines 3–4, lowest longest, usually hooked, to 3 cm, brown; radial spines 15–20, white. *Southern C Mexico (Puebla)*. G1. Spring.

M. magnifica Buchenau. Illustration: Pilbeam, Mammillaria, 86 (1981). Stems to 40 × 5–9 cm, tubercle-axils with bristles. Central spines 4–5 or more, the lowest 1.5–5.5 cm, hooked, remainder shorter, straight, all clear yellowish brown; radial spines 17–24, 3–8 mm, glassy white or yellowish. Flowers 1.7–2 × 1.1–1.5 cm, pinkish red. *Southern C Mexico (Morelos-Puebla border)*. G1. Spring.

M. pilcayensis Bravo (as 'pitcayensis'). Stems slender cylindric, to 50 × 4 cm. Central and radial spines all similar, about 30, 5–6 mm, bristle-like, pale glassy yellow. *Southwestern C Mexico (Guerrero, Barranca de Pilcaya)*. G1. Spring.

M. rekoi (Britton & Rose) Vaupel (*Neomammillaria rekoi* Britton & Rose; *M. pullihamata* invalid). Very similar to *M. nunezii*, but spines typically stronger and one central consistently hooked. *S Mexico (Oaxaca)*. G1. Spring.

49. M. eriacantha Pfeiffer. Illustration: Pilbeam, Mammillaria, 58 (1981); Haustein, Der Kosmos-Kakteenführer, 295 (1983).
Simple, some clustering later, stem slender cylindric to 30 (rarely to 50) × 5 cm; tubercles 7 × 6 mm, their axils without bristles but often woolly in flowering zone. Central spines 2, 8–10 mm, golden yellow, minutely hairy; radial spines 20–24, 4–6 mm, paler golden yellow, also minutely hairy. Flowers small, 1–1.2 × 1.2–1.4 cm, greenish yellow. Fruit purplish; seeds brown. *E Mexico (Veracruz)*. G1. Spring.

50. M. haageana Pfeiffer (*M. elegans* de Candolle, misapplied; *M. collina* Purpus; *M. vaupelii* Tiegel). Illustration: Pilbeam, Mammillaria, 68, 69 (1981); Haustein, Der Kosmos-Kakteenführer, 295 (1983) – as *M. collina*; Cullmann et al., Kakteen, edn 5, 219 (1984).
Usually simple (clustering in var. **schmollii** (Craig) D. Hunt), very variable, reaching 15 × 5–10 cm in larger variants; tubercles small, crowded, often woolly in their axils and sometimes with bristles. Central spines usually 2, sometimes 1 or 4, to 1.5 cm, usually brown or dark brown; radial spines 15–25, 3–6 mm, white. Flowers small, 1.2 × 1 cm, deep purplish pink. Fruit red; seeds light brown. *Southeastern C Mexico*. G1. Spring.

M. albilanata Backeberg (*M. fuauxiana* Backeberg). Densely woolly in the tubercular axils. Central spines 2 or 4, very short or up to 7 mm, white or pale yellow, often brown-tipped. Flowers very small, 7–8 mm. *SW Mexico*. G1. Spring.

M. columbiana Salm-Dyck (*M. yucatanensis* (Britton & Rose) Orcutt; *Neomammillaria yucatanensis* Britton & Rose; *M. bogotensis* Werdermann). Tubercle-axils densely woolly. Central spines 3–7, 6–8 mm, pale yellow to brown; radial spines 18–30, 4–6 mm, white. Flowers small, 7–8 mm, pink. *SE Mexico (Yucatan), Guatemala, Jamaica, Northern S America*. G1. Spring.

M. dixanthocentron Mottram (*Neomammillaria celsiana* Britton & Rose, not *M. celsiana* Lemaire). Illustration: Cullmann et al., Kakteen, edn 5, 217 (1984). Stout, cylindric, to 20 × 7–8 cm,

tubercle-axils woolly. Lower central spine to 1.5 cm, pale yellow to brown. Flowers 8–10 × 8–10 mm, pale clear red. *S Mexico (Oaxaca)*. G1. Spring.

M. supertexta Martius (*Neomammillaria lanata* Britton & Rose; *M. lanata* (Britton & Rose) Orcutt). Tubercles small, 2–4 mm, axils woolly. Central spines 0–2, usually very short, white, tipped dark brown. Flowers lateral (rather than near apex), small, 6–7 mm, deep reddish pink. *S Mexico (Puebla, Oaxaca)*. G1. Spring.

51. M. geminispina Haworth. Illustration: Pilbeam, Mammillaria, 62 and coloured plate (1981); Graf, Exotica, edn 4, **1**: 722 (1982), Cullmann et al., Kakteen, edn 5, 213 (1984).
Soon clustering and forming mounds, stems becoming cylindric, 6–8 cm in diameter, tubercle-axils with wool and short bristles. Central spines 2 or 4, the uppermost 1.5–4 cm, others 7–15 mm, chalky white, tipped brown or black; radial spines 16– 20, 5–7 mm, chalky white. Flowers 1.5 × 1.5 cm, deep pink. Fruit red; seeds brown. *C Mexico*. G1. Summer–autumn.

M. leucocentra Berg. Simple or clustering, spherical to short-columnar, to *c.* 12 × 11 cm; tubercles terete-conic, *c.* 8 × 9 mm, bluish green, axils with white bristles. Central spines 4–6, to 1.2 cm, white with dark tip; radial spines 30–35, *c.* 6 mm, chalky white. Flowers as in *M. geminispina*. *C Mexico*. G2. Summer.

52. M. perbella Schumann (*M. aljibensis* invalid). Illustration: Pilbeam, Mammillaria, 108 (1981); Bradleya **1**: 120 (1983).
Forming low mounds by repeated apical division of stem, heads depressed-spherical, *c.* 5 cm diameter; tubercles small, 5–7 × 3–4 mm, axils with short bristly hairs. Central spines usually 2, sometimes 1 or 0, 1–4 mm, brown or white with reddish to black tip; radial spines 12–22, 1.5–4 mm. Flowers *c.* 10 × 10 mm purplish pink. Fruit red; seeds brown. *C Mexico*. G1. Spring.
For *M. pseudoperbella* Quehl, see No. **56**.

53. M. hahniana Werdermann.
Illustration: Pilbeam, Mammillaria, 70 and colour plate (1981); Graf, Exotica, edn 4, **1**: 723 (1982); Hecht, BLV Handbuch der Kakteen, 301 (1982); Cullmann et al., Kakteen, edn 5, 220 (1984).
Simple or clustering, stems to 20 × 12 cm; tubercle-axils with hair-like bristles to 1.5–4 cm. Central spine 1 or sometimes 2–4, to 4 mm, white with reddish brown tip; radial spines 20–30, 5–15 mm, hair-like, white. Flowers *c.* 1.2 × 1.2–1.5 cm,

deep purplish pink. Fruit red; seeds brown. *C Mexico*. G1. Spring.

M. mendeliana (Bravo) Werdermann (*Neomammillaria mendeliana* Bravo; *M. woodsii* Craig). Very similar to *M. hahniana*, but axillary bristle-hairs 1.5–2.5 cm, lower central spine to 2 cm, radial spines poorly developed or lacking. *C Mexico (E Guanajuato)*. G1. Spring.

M. morganiana Tiegel. Axillary bristle-hairs to 2 cm. Central spines 4–16, 10 mm; radial spines 40–50, to 1.2 cm. *C Mexico*. G1. Spring.

54. M. muehlenpfordtii Foerster (*M. celsiana* Lemaire, misapplied; *M. neopotosina* Craig). Illustration: Pilbeam, Mammillaria, 98 (1981).
Simple at first, often dividing apically later, heads depressed spherical, 10–15 cm in diameter; tubercle-axils with fine bristles; young areoles with pale yellow wool. Central spines 4–7, the lowermost longest, 5–35 mm, often strong and curved, remainder 4–14 mm, pale to dark yellow, often tipped brown; radial spines 40–50, 2–8 mm, glassy, whitish. Flowers 1.5 × 1 cm, deep purplish pink. Fruit red; seeds brown. *C Mexico*. G1. Summer.

55. M. parkinsonii Ehrenberg (*M. rosensis* Craig). Illustration: Pilbeam, Mammillaria, 106 (1981); Cullmann et al., Kakteen, edn 5, 226 (1984).
Simple at first, later dividing apically, individual stems 7–15 cm in diameter; tubercles 8–10 × 4–6 mm, axils with wool and bristles. Central spines 2–4, rarely 5, upper 6–8 mm, lowermost variable, up to 3.5 cm, often subulate, curved, white or reddish brown, tipped dark brown; radial spines 30–35, white. Flowers 1.2–1.5 × 1.2–1.5 cm, creamy yellow tinged brownish or pink. Fruit red; seeds brown. *C Mexico*. G1. Spring.

56. M. formosa Scheidweiler. Illustration: Pilbeam, Mammillaria, 60 (1981).
Simple, depressed-spherical to spherical or short-cylindric, to 15 cm in diameter; tubercle-axils white-woolly but without bristles. Central spines usually 6, sometimes 4 (or 7), to 8 mm, pinkish brown, darker tipped, later greyish; radial spines 20–25, 3–6 mm, white. Flowers 1–1.8 × 1–1.5 cm, pale pink or whitish. Fruit red; seeds brown. *Northeastern C Mexico*. G1. Spring.

M. chionocephala Purpus. Tubercle-axils with bristle-like hairs to 2 cm. Central spines 2–4, rarely 5–6. Flowers white. *NE Mexico*. G1. Spring.

M. pseudoperbella Quehl (? misapplied). Illustration: Pilbeam, Mammillaria, 115 (1981). Cylindric, to 10 × 8 cm; tubercle-axils with hair-like bristles somewhat exceeding the 20–30 radial spines. Flowers pale to mid-pink. *Unknown in the wild*. G1. Summer.

57. M. sempervivi de Candolle (*M. pseudocrucigera* Craig). Illustration: Pilbeam, Mammillaria, 125 (1981); Graf, Exotica, edn 4, **1**: 725 (1982); Haustein, Der Kosmos-Kakteenführer, 297 (1983).
Simple at first, sometimes later dividing apically or offsetting, depressed-spherical, individual stems to 10 cm in diameter; tubercles angled-conic, to 10 × 7 mm, dark bluish green; axils woolly. Central spines usually 2, sometimes 4, to 4 mm, brown or black, later grey; radial spines rudimentary, on juvenile stems only. Flowers 1–1.2 × 1–1.2 cm, whitish or pale pink. Fruit red; seeds brown. *Eastern C Mexico*. G1. Spring.

58. M. uncinata Zuccarini. Illustration: Backeberg, Die Cactaceae 5: 3142, f. 2935 (1961); Cullmann et al., Kakteen, edn 5, 231 (1984).
Usually simple, depressed-spherical, to 12 cm or more in diameter; tubercles conic-pyramidal, obtuse, *c.* 10 × 10 mm, axils naked or slightly woolly. Central spines 1–2, upper, if present, 4–7 mm, straight or semi-hooked, lower, 6–10 mm, stronger, hooked, reddish brown; radial spines 3–7, 4–6 mm, white, tipped darker. Flowers 1.5–2 × 1.5–2 cm, whitish, outer segments striped brown or brownish red. Fruit red; seeds brown. *C Mexican plateau*. G1. Spring.

59. M. mammillaris (Linnaeus) Karsten (*Cactus mammillaris* Linnaeus; *M. simplex* Haworth). Illustration: Pilbeam, Mammillaria, 88 (1981).
Simple at first, later clustering, stems spherical or short-cylindric to 6 × 7 cm or larger; tubercles terete-conic, 5–7 mm, axils slightly woolly. Central spines 3–5, 7–10 mm, reddish brown, darker-tipped; radial spines 10–16, 5–8 mm, tinged reddish brown. Flowers small, 8–12 × 6–10 mm, pale yellow. Fruit red; seeds brown. *Dutch West Indies, Grenadines, Venezuela*. G2. Summer.

60. M. heyderi Muehlenpfordt (*M. gummifera* Engelmann; *M. hemisphaerica* Engelmann; *M. meiacantha* Engelmann; *M. macdougalii* Rose). Illustration: Pilbeam, Mammillaria, 72, 73 (1981).
Stem usually simple, flattish to almost spherical, 7.5–15 cm in diameter; tubercles 9–12 mm; axils naked or woolly, without

bristles. Central spines 0–2, 3–10 mm, brown; radial spines 6–22, 6–14 mm, brown or whitish. Flowers usually 2–3 × 1.5–3 cm, white, pale yellow or pale pink. Fruit obovoid, red; seeds brown. *SW USA, N Mexico*. G1. Spring.

M. coahuilensis (Boedeker) Moran (*Porfiria coahuilensis* Boedeker; *M. albiarmata* Boedeker). Illustration: Haustein, Der Kosmos-Kakteenführer, 299 (1983). Stem to 5 cm in diameter. Central spines 0–1, 6 mm; radial spines 16–25. Seeds distinctly pitted. *N Mexico (Coahuila)*. G1. Spring.

M. gaumeri (Britton & Rose) Orcutt (*Neomammillaria gaumeri* Britton & Rose). Illustration: Pilbeam, Mammillaria, 62 (1981). Clustering, spherical. Radial spines usually 10–12. Flowers 1–1.5 cm. *SE Mexico (Yucatan)*. G1. Spring.

M. grusonii Runge. Simple, cylindric, massive, to 50 × 25 cm. Central spines 2–3; radial spines 12–14. Flowers 2.5 cm, yellowish. *N Mexico (Coahuila)*. G1. Spring.

61. M. melanocentra Poselger. Illustration: Pilbeam, Mammillaria, 92 (1981).
Simple, depressed-spherical, to 15 cm in diameter, tubercles large, pyramidal, to *c.* 1.4 × 1.4 cm, their axils woolly at first, without bristles. Central spine 1, to 3 cm, black, becoming brownish grey later; radial spines 6–9, 6–22 mm, the lowermost longest. Flowers *c.* 2 × 2 cm, bright deep pink. Fruit pink or red; seeds brown. *N Mexico (Nuevo Leon)*. G1. Spring.

M. petterssonii Hildmann. Simple, spherical, to 30 cm in diameter, tubercles dark green. Central spines 5–7, one to 4.5 cm, straight-projecting, the others shorter, brown at first, tipped darker; radial spines *c.* 10, 2–10 mm. *C Mexico (Guanajuato)*. G1. Spring.

M. scrippsiana (Britton & Rose) Orcutt (*Neomammillaria scrippsiana* Britton & Rose). Simple at first, later clustering, individual stems to 8 cm in diameter; tubercles bluish green, with woolly axils. Central spines 2, 8–10 mm reddish brown; radial spines 8–10, 6–8 mm, whitish. *Western C Mexico (Jalisco)*. G1. Spring.

M. sonorensis Craig (*M. tesopacensis* Craig). Illustration: Cactus & Succulent Journal (US) **12**: 155 (1940). Simple and later clustering, depressed-spherical to stout-cylindric; tubercles 8–15 × 8–18 mm, dull bluish green; axils with wool, usually without bristles. Spines variable; central spines 1–4, 5–45 mm; radial spines 8–15, variable, 1–20 mm. Flowers *c.* 2 × 2 cm, deep pink (inner perianth-segments pale yellow with pink midstripe in one variant); style and stigmas olive-green. Fruit club-shaped, 1.2–1.8 × 1 cm, scarlet; seeds 1 × 0.6 mm, light brown. *NW Mexico.* G1. Spring.

62. M. gigantea Schumann. Illustration: Pilbeam, Mammillaria, 62 (1981); Cullmann et al., Kakteen, edn 5, 210 (1984).
Simple, spherical, to 30 cm in diameter, tubercles dark green, with woolly axils. Central spines 4–6, robust, lowermost longest, to 2 cm or more, dark, yellowish or brown at first; radial spines up to 12, less than 5 mm, white. Flowers 1.5 × 1.5 cm, pale greenish yellow. Fruit purplish red; seeds brown. *C Mexico.* G1. Spring.

M. obscura Hildmann may be the correct name for this species, but its application is uncertain.

M. zeyeriana Schumann. Illustration: Haustein, Der Kosmos-Kakteenführer, 299 (1983). Simple, shortly cylindric, to 20 × 20 cm. Central spines 4, uppermost 2 cm or more, lower shorter; radial spines *c.* 10, to 10 mm. Flowers *c.* 2.5 × 2–2.5 cm, white or pale yellow, tinged brownish pink. *N Mexico.* G1. Spring.

63. M. standleyi (Britton & Rose) Orcutt (*Neomammillaria standleyi* Britton & Rose). Illustration: Britton & Rose, The Cactaceae **4**: f. 93 (1923); Pilbeam, Mammillaria, 130 (1981).
Simple or clustering, depressed-spherical, to 9 × 12 cm; tubercles obtuse, 8 × 12 mm, pale blue-green, axils with wool and bristles. Central spines 4–5, 5–9 mm, white, tipped reddish brown; radial spines 13–19, 4–8 mm, white. Flowers *c.* 1.2–1.8 × 1.2 cm, purplish pink. Fruit red; seeds brown. *NW Mexico (Sonora).* G1. Spring.

M. canelensis Craig. Simple, spherical; tubercle-axils with dense wool and bristles. Central spines 2–4, 3 cm, straight or curved, yellow to orange-brown; radial spines 22–25, 5–15 mm, white. Flowers red. *NW Mexico.* G1. Summer.

64. M. marksiana Krainz. Illustration: Pilbeam, Mammillaria, 89 (1981).
Simple at first, later clustering, flat-spherical, to 12 cm in diameter; tubercles unequally swollen-pyramidal, bright green, flowering axils densely woolly, without bristles. Spines usually 9–11, 8–11 mm, golden yellow, sometimes 2–13 additional bristle-like, rudimentary spines present. Flower *c.* 1.5 × 1.5 cm, yellow. Fruit short, purplish; seeds brown. *NW Mexico.* G1. Summer.

65. M. johnstonii (Britton & Rose) Orcutt (*Neomammillaria johnstonii* Britton & Rose). Illustration: Pilbeam, Mammillaria, 76, 77 (1981); Haustein, Der Kosmos-Kakteenführer, 301 (1983).
Usually simple, spherical to short-cylindric, eventually 15–20 × 10 cm; tubercles bluish, their axils with some wool but no bristles. Central spines usually 2, sometimes 4 or 6, 1–2.5 cm, purplish brown to black; radial spines 10–18, 6–9 mm, white, tipped brown. Flowers 2 × 2 cm, whitish. Fruit red; seeds brown. *NW Mexico (Sonora).* G1. Summer.

M. baxteriana (Gates) Backeberg & Knuth (*Neomammillaria baxteriana* Gates; *N. marshalliana* Gates; *N. pacifica* Gates; *Mammillaria marshalliana* (Gates) Backeberg & Knuth; *M. pacifica* (Gates) Backeberg & Knuth). Central spine usually 1, white with brown tip; radial spines 7–13, 1–1.5 cm. Flowers 1.5 × 2 cm, pale yellow. Fruit red. *NW Mexico (S Baja California).* G1. Summer.

M. brandegei (Coulter) Brandegee (*Cactus brandegei* Coulter). Flowers pale brownish yellow. Fruit dull pink or purplish, slow-ripening. *NW Mexico (deserts of central Baja California).* G1. Summer.

66. M. magnimamma Haworth (*M. zuccariniana* Martius; *M. centricirrha* Lemaire). Illustration: Pilbeam, Mammillaria, 87 (1981); Graf, Exotica, edn 4, **1**: 721 (1982); Cullmann et al., Kakteen, edn 5, 222 (1984).
Variable, but usually clustering and forming mounds 50 cm or more in diameter, heads to 10–12 cm in diameter; tubercles dark green, pyramidal-conic but not sharply angled, *c.* 10 × 10 mm, the flowering axils woolly. Spines 3–6, usually 1 longer, to 5 cm, stronger than the others and more or less curved. Flowers *c.* 2 × 2 cm, pale brownish yellow to deep purplish pink. Fruit red; seeds brown. *C Mexico.* G1. Spring.

M. lloydii (Britton & Rose) Orcutt (*Neomammillaria lloydii* Britton & Rose). Flat-spherical, 6–7 cm in diameter. Spines 3–4, 2–5 mm, uppermost red or dark brown, others white. Flowers 1.5 × 1.5 cm, petals white, striped red. *NC Mexico (Zacatecas).* G1. Spring.

Perhaps a variant of *M. uncinata*, with which it is said to grow.

67. M. compressa de Candolle. Illustration: Pilbeam, Mammillaria, 49 (1981); Graf, Exotica, edn 4, **1**: 717 (1982); Bradleya **1**: 126 (1983).
Clustering to form mounds 60 cm or more

in diameter, heads 5–8 cm in diameter; tubercles bluntly and unequally swollen, angled, 4–6 × 8–15 mm, light bluish green, axils with wool and bristles, the latter sometimes inconspicuous or lacking. Spines 4–6, unequal, upper short, lowermost to 1.5–7 cm, white to reddish with darker tip. Flowers 1–1.5 × 1–1.5 cm, deep purplish pink. Fruit red; seeds brown. *C Mexico.* G1. Spring.

68. M. carnea Zuccarini. Illustration: Pilbeam, Mammillaria, 46 (1981); Graf, Exotica edn 4, **1**: 717 (1982).
Simple or more often clustering, stems spherical to cylindric, to 8.5 cm in diameter; tubercles pyramidal, 4-angled to 1.3 cm × 8–10 mm, mid-green, often brownish red towards tip, axils with wool but no bristles. Spines usually 4, variable, all 6–15 mm, or the uppermost longest, to 2 cm, or the lowermost longest, to 5 cm, pinkish brown, tipped black. Flowers 1.5–2 × 1.2–1.5 cm, flesh pink. Fruit red; seeds brown. *S Mexico.* G1. Spring.

69. M. mystax Martius. Illustration: Pilbeam, Mammillaria, 99 (1981); Hecht, BLV Handbuch der Kakteen, 304 (1982).
Simple, spherical to cylindric, to 15 × 7–10 cm, eventually dividing apically; tubercles pyramidal, angled and keeled, 1–1.5 cm × 8 mm, axils with bristles. Central spines 3–4, 1.5–2 cm often one in addition straight-projecting, to 7 cm, twisted, purplish brown at first, tipped darker, later grey; radial spines 3–10, 4–8 mm, white, brown-tipped. Flowers 2.5 × 2 cm, deep purplish pink. Fruit red; seeds brown. *S Mexico.* G1. Spring.

70. M. karwinskiana Martius (*M. fischeri* Pfeiffer; *M. praelii* Muehlenpfordt; *M. multiseta* Ehrenberg; *M. confusa* (Britton & Rose) Orcutt; *Neomammillaria confusa* Britton & Rose). Figure 29(11), p. 205. Illustration: Pilbeam, Mammillaria, 77 (1981).
Simple at first, later dividing apically and offsetting, individual heads depressed-spherical to short-cylindric, to 15 × 10 cm, tubercles pyramidal-conic, obscurely angled, 10 × 8–9 mm, axils with bristles. Spines 4–7, more or less equal, 5–9 mm, or the upper and lower longer, 1–1.2 cm, occasionally one in addition straight-projecting, to 2.5 cm, all spines reddish brown at first, greyish, tipped darker later. Flowers 2–2.5 × 1.5–2 cm, very pale yellow, often striped red on outer petals. Fruit red; seeds brown. *S Mexico.* G1. Spring or autumn.

M. nejapensis Craig & Dawson. Illustration: Pilbeam, Mammillaria, 101 (1981). Very similar to *M. karwinskiana* but distinctive in cultivation by virtue of the abundant long white bristles in the tubercle-axils. *SW Mexico*. G1. Spring or Autumn.

M. voburnensis Scheer (*M. collinsii* (Britton & Rose) Orcutt; *Neomammillaria collinsii* Britton & Rose). Clustering by offsets rather than apical division. Central spines 1–2, to 1.2 cm; radial spines 6–9. *S Mexico, Guatemala*. G1. Autumn.

Names of doubtful application in cultivation: *M. longispina*; *M. pseudosupertexta*.

76. PERESKIOPSIS Britton & Rose
D.R. Hunt
Leafy shrubs or scramblers; leaves elliptic, obovate or circular, fleshy; areoles with wool, hairs, glochids and normally 1–several spines. Flowers usually lateral, solitary, stalkless, *Opuntia*-like. Fruit fleshy; seeds few, each covered with a bony aril and matted hairs.

A genus of about 8 species in Mexico and Guatemala.

1. P. diguetii (Weber) Britton & Rose (*P. velutina* Rose).
Shrub to 1 m or more, young stems minutely downy. Leaves elliptic to ovate, abruptly acuminate, tapered, 2–6 × 1.5–2.5 cm, minutely velvety-downy. Spines 1–4, usually short and weak in cultivated material, (to 7 cm and almost black in wild specimens). Flowers stalkless on second year growth, *c.* 5 cm in diameter, yellow, outer segments tinged red; pericarpel with reduced leaves. Fruit 3 cm, red, minutely downy. *Mexico*. G2.
Widely used as a grafting-stock.

77. OPUNTIA Miller
D.R. Hunt
Small to large shrubs or trees with fleshy, cylindric, club-shaped, almost spherical, flattened or very rarely ribbed, segmented branches; areoles often raised on more or less prominent tubercles, with glochids and usually 1–many spines, sometimes sheathed and barbed, occasionally lacking. Leaves terete or subulate, usually small, falling early. Flowers solitary, stalkless, lateral or almost terminal, rarely terminal; pericarpel with leaves, areoles, glochids and often spines; perianth disc-shaped or spreading, rarely erect, without tube; stamens numerous, sometimes touch-sensitive; ovary inferior. Fruit fleshy or dry,

with a depression at the top; seeds encased in a bony aril.

A genus of *c.* 200 species, extending throughout the range of the family (53° N to 50° S) in America. A few species are grown for forage or edible fruits and some of those introduced to Australia and South Africa have become noxious pests. Various species, mostly small-growing, are in horticulture. Many are best not touched, and should be kept out of reach of small children, as the minutely barbed glochids are easily detached from the plant but less easily removed from the skin.

1a. Stem-segments spherical, cylindric, ribbed or tubercled, not appreciably flattened   2
  b. At least some of the stem-segments disc-like or appreciably flattened   15
2a. Spines covered by a papery sheath   3
  b. Spines not sheathed   4
3a. Spine solitary; stem-segments only 3–5 mm in diameter   **1. leptocaulis**
  b. Spines several; stem-segments mostly 1.5–3 cm in diameter   **2. imbricata**
4a. Stem-segments cylindric, 8–15 mm in diameter, areoles on low tubercles 15 × 2–3 mm; flowers strictly terminal, 6–8 cm in diameter; roots tuberous   **3. marenae**
  b. Stem-segments various; flowers not normally terminal; roots usually fibrous   5
5a. Stem-segments spherical, club-shaped or short-cylindric   6
  b. Stem-segments cylindric, long   10
6a. Stem-segments obovoid to club-shaped, strongly tubercled; spines sheathed at apex   **4. invicta**
  b. Stem-segments globular to oblong, smooth or weakly tubercled; spines not sheathed   7
7a. Areoles with abundant glochids but no spines or hairs   **5. articulata**
  b. Areoles with spines and/or hairs   8
8a. Plant more or less covered with hair-like spines   **7. floccosa**
  b. Plants lacking hair-like spines   9
9a. Spines somewhat flattened, papery or ribbon-like   **5. articulata**
  b. Spines terete, needle-like   **6. pentlandii**
10a. Plant (or at least the younger segments) clothed with long hairs   **8. vestita**
  b. Plant not hairy   11
11a. Stem-segments distinctly tubercled   12
  b. Stem-segments smooth, not tubercled   14

12a. Stem-segments 1–1.5 cm in diameter   **9. verschaffeltii**
  b. Stem-segments 2.5 cm more in diameter   13
13a. Leaves usually 2 cm or longer, persistent   **10. subulata**
  b. Leaves about 1 cm, falling early   **11. cylindrica**
14a. Stem-segments club-shaped or crested; spines small, adpressed   **12. clavarioides**
  b. Stem-segments slender, long; spines bristle-like, spreading   **13. salmiana**
15a. Plant developing an erect, unjointed trunk with flat branches   **14. brasiliensis**
  b. Plant not forming an erect, unjointed trunk   16
16a. Stem-segments not strongly flattened, to 15 × 1.5 × 1 cm, readily detached   **15. aurantiaca**
  b. Stem-segments mostly strongly flattened, not readily detached   17
17a. Stem-segments small, obliquely oval, to 5 × 2.5 × 1.5 cm, not strongly flattened   **16. corrugata**
  b. Stem-segments larger, mostly strongly flattened   18
18a. Surface of stem-segments minutely velvety   19
  b. Surface of stem-segments smooth, hairless   21
19a. Spines, especially those on older growth, long, weak and bristle- or hair-like   **17. leucotricha**
  b. Spine 1, short or lacking, but glochids conspicuous   20
20a. Stem-segments bluish grey, often purple tinged around the edges and areoles; flowers purplish red; glochids reddish brown, deciduous   **18. basilaris**
  b. Stem-segments green; glochids yellow, white or reddish brown, persistent; flowers yellow   **19. microdasys**
21a. Spines mixed with long white or yellowish hairs   **20. scheeri**
  b. Spines not mixed with long hairs   22
22a. Stem-segments relatively small, *c.* 5–13 × 2.5–10 cm; plants mostly prostrate, spreading   23
  b. Stem-segments usually larger, *c.* 10–50 × 10–30 cm; plants mostly shrubby or ascending   24
23a. Spines numerous, largely covering the segments   **21. polyacantha**
  b. Spines 1–2 per areole or lacking   **22. compressa**
24a. Perianth-segments spreading, yellow or orange; stamens and style not projecting   25

b. Perianth-segments erect, red; stamens and style projecting
28. cochenillifera
25a. Spines brown or yellow, at least when young 26
  b. Spines white 28
26a. Spines brown 27
  b. Spines yellow **23. stricta**
27a. Stem-segments ovate to almost circular, thick, not tapered below; spines 1–4 or more **24. phaeacantha**
  b. Stem-segments oblong to obovate, rather thin, tapered below; spines 1–2 **25. monacantha**
28a. Stem-segments obovate to elliptic-oblong, green or bluish green **26. ficus-indica**
  b. Stem-segments nearly circular, bluish-bloomed **27. robusta**

1. O. leptocaulis de Candolle. Illustration: Benson, Cacti of the United States & Canada, pl. 17, f. 331–336 (1982). Brittle, thin-stemmed bush to 50 cm or more, sometimes developing a thin trunk; stem-segments slender, often 5–15 cm × 3.5 mm, hardly tubercled, branching more or less at right-angles, segments easily detached. Leaves to 1.2 cm, falling early. Spine usually 1, 1–5 cm, slender, sheathed, the sheath yellowish brown to whitish. Flowers 2 cm, greenish or yellowish. Fruit globular to obovoid, 1–1.8 cm, orange, red or yellow, often proliferous. *Mexico, SW USA.*

2. O. imbricata (Haworth) de Candolle (*O. arborescens* Engelmann). Illustration: Benson, Cacti of the United States & Canada, f. 294–297 (1982); Haustein, Kosmos-Kakteenführer, 59 (1983). Eventually tree-like, to 3 m; larger stem-segments 12–38 × 2–3 cm, strongly tubercled. Leaves to 1.5 cm, falling early. Spines 8–30, to 3 cm, sheath dull brown. Flowers *c.* 6 × 6 cm, purple. Fruit nearly globular, 3 cm, yellow, spineless. *Mexico, SW USA.*

O. tunicata (Lehmann) Link & Otto. Illustration: Benson, Cacti of the United States & Canada, f. 275–278 (1982); Hecht, BLV Handbuch der Kakteen, 197 (1982); Haustein, Der Kosmos-Kakteenführer, 59 (1983). Shrubby, to 60 cm, densely branched; spines 6–10, to 5 cm, sheath conspicuous, yellow or whitish. Flowers yellow, 3 × 3 cm. *Mexico, introduced to S America.*

3. O. marenae Parsons. Illustration: Haustein, Der Kosmos-Kakteenführer, 61 (1983).

Low shrub 15–60 cm with tuberous roots; stem-segments to 20 cm × 8–15 mm, faintly tubercled. Leaves 5–10 mm, falling early. Spines several, adpressed, 3–10 mm, fine, white, 1–2 or more stronger, projecting downwards, to 2 cm. Flowers terminal, disc-shaped, 6–8 cm in diameter, satiny white, stamens sensitive. Fruit not externally visible until tip of stem splits. *Mexico (Sonora).* G2.

4. O. invicta Brandegee. Illustration: Britton & Rose, The Cactaceae 1: pl. 16, f. 2 (1919). Low shrub 20–50 cm eventually forming colonies 2 m in diameter; stem-segments obovoid to club-shaped, 6–10 × 2–6 cm, strongly tubercled. Leaves subulate, 8–14 mm, reddish, deciduous. Spines formidable, *c.* 16–22, to 3.5 cm, flattened, sheathed at apex, reddish at first, becoming dull brown. Flowers 5 cm in diameter, yellow. *Mexico (Baja California).* G2.

5. O. articulata (Pfeiffer) D. Hunt (*Cereus articulatus* Pfeiffer). Illustration: Haustein, Der Kosmos-Kakteenführer, 63 (1983); Cullmann et al., Kakteen, edn 5, 259 (1984). Dwarf shrub with brittle erect branches to 20–30 cm; stem-segments spherical to oblong, usually 2.5–5 × 2.5–5 cm, easily detached. Spines lacking or 1–4, to 5 cm or more × 7 mm, flat, paper- or raffia-like, brownish or white. Flowers 3–4 cm in diameter, white or pinkish. *Argentina.* G1.

O. molinensis Spegazzini. Similar to spineless forms of *O. articulata*, but with conspicuous tufts of reddish brown glochids. *N Argentina.* G1.

O. glomerata Haworth. Forming dense mounds, stem-segments conical, to 4 × 2 cm. Spines always present, more or less flattened. Flowers yellow (rarely seen). *N Argentina.* H5.

6. O. pentlandii Salm-Dyck (*O. subinermis* Backeberg; *Tephrocactus pentlandii* (Salm-Dyck) Backeberg). Illustration: Backeberg, Die Cactaceae 1: f. 300–304 (1958); Succulenta 45: 105 (1966). Dwarf shrub forming mounds eventually 1 m in diameter; stem-segments spherical to cylindric or obovoid, 2–5 × *c.* 4 cm. Spines variable in number and length, needle-like (not flattened), more or less deflexed, sometimes absent. Flowers yellow, orange or red, rarely seen in cultivation. *Bolivia.* H5–G1.

O. sphaerica Foerster. Stem-segments spherical, *c.* 2.5 cm in diameter with numerous shortly woolly and variably spiny areoles. *Peru.* G1.

7. O. floccosa Salm-Dyck. Illustration: Haustein, Der Kosmos-Kakteenführer, 63 (1983). Forming mounds eventually 2 m in diameter; stem-segments to 10 × 3 cm, densely clothed with long white areolar hair. Leaves terete, 8–13 × 2–3 mm, persisting. Spines 1–3, 1–3 cm, pale yellow. Flowers *c.* 3 × 3.5 cm, yellow or orange. *Peru, Bolivia.* H5?

8. O. vestita Salm-Dyck. Illustration: Cullmann et al., Kakteen, edn 5, 259 (1984). Low, fragile-stemmed shrub; stem-segments to 20 × 1–2 cm, largely hidden by soft white areolar hair. Leaves *c.* 1 cm. Spines 4–8, usually short and weak but sometimes to 1.5 cm. Flowers 3.5 × 3 cm, violet-red. *Bolivia.* G1.

A monstrous form is commonly grown.

9. O. verschaffeltii Cels. Illustration: Haustein, Der Kosmos-Kakteenführer, 57 (1983). Forming low, dense clumps; stem segments usually long in cultivation, 6–20 × 1–1.5 cm, with low tubercles. Leaves terete, to 3 cm, persistent. Spines 1–3 or more, 1–3 cm, bristly, or absent. Flowers orange to deep red. *Bolivia, N Argentina.* G1.

10. O. subulata (Muehlenpfordt) Engelmann. Figure 28(2), p. 204. Illustration: Hecht, BLV Handbuch der Kakteen, 196 (1982); Haustein, Der Kosmos-Kakteenführer, 57 (1983); Cullmann et al., Kakteen, edn 5, 258 (1984). Becoming a small tree, 2–4 m, or with several branches near the base; stems cylindric, unsegmented, 5–7 cm in diameter, with low tubercles, green. Leaves terete, acute, 5 cm or more, persisting more than a year. Spines 1–2 (or more on older growth), pale yellow. Flowers reddish, 7 cm. *S Peru.*

O. exaltata Berger. Tree to 6 m; stems somewhat glaucous. Leaves 1–7 cm. Spines brownish. *Peru(?)* G1.

Perhaps only a form of *O. subulata.*

11. O. cylindrica (Lamarck) de Candolle (*Austrocylindropuntia cylindrica* (Lamarck) Backeberg). Illustration: Britton & Rose, The Cactaceae 1; pl. 14, f. 2 (1919); Backeberg, Die Cactaceae 1: f. 78 (1958). Becoming a small tree 3–4 m, stems cylindric, unsegmented, 3–5 cm in diameter, with low tubercles, green. Leaves terete, acute, 1–1.3 cm, deciduous. Spines 2–6, to 1 cm, whitish, or lacking, usually

mixed with fine areolar hairs. Flowers
c. 2.5 cm in diameter, red. *S Ecuador.* G1.
A crested variant is also cultivated.

12. O. clavarioides Pfeiffer. Illustration:
Hecht, BLV Handbuch der Kakteen, 196
(1982); Haustein, Der Kosmos-
Kakteenführer, 57 (1983).
Low shrublet, much-branched from
tuberous roots; stem-segments obconic,
truncate or concave, often crested at apex,
to c. 2 × 1.5 cm, not tubercled. Leaves
1.5 mm, reddish, falling early. Spines
4–10, minute, adpressed, white. Flowers
c. 5 cm in diameter, brownish, rarely seen
in cultivation. *Argentina.* G1.
A curious species, usually grafted on
other species of *Opuntia*, or *Cereus* or
Harrisia.

13. O. salmiana Pfeiffer
(*Austrocylindropuntia salmiana* (Pfeiffer)
Backeberg). Illustration: Britton & Rose,
The Cactaceae 1: f. 88, 89 (1919);
Backeberg, Die Cactaceae 1: f. 95–97
(1958).
Shrub 30–50 cm or more, much branched;
stem-segments slender-cylindric to 25 ×
1 cm, not tubercled, often tinged red.
Leaves very small, 1–2 mm, purplish,
falling early. Spines 3–5, to 1.5 cm, or
lacking. Flowers produced rather freely,
2–3.5 cm in diameter, pale yellow, with
sensitive stamens. Fruits oblong-ellipsoid,
c. 1 cm in diameter, red, barren in
cultivated plants, but proliferous. *S Brazil to
Argentina.* G1.

14. O. brasiliensis (Willdenow) Haworth.
Illustration: Botanical Magazine, 3293
(1834); Benson, Cacti of the United States &
Canada, f. 557, 558 (1982); Haustein, Der
Kosmos-Kakteenführer, 69 (1983).
Tree-like, to 6–9 m, with cylindric,
unjointed trunk and branches; ultimate
stem-segments flat and somewhat leaf-like,
obovate to oblong-lanceolate, to 15 × 6 cm,
4–6 mm thick, eventually deciduous.
Leaves small, subulate, falling early. Spines
1–3, to 1.5 cm, on young growth, more
numerous on the trunk, or lacking. Flowers
c. 5 × 5 cm, pale yellow, staminodial hairs
present. Fruit spherical, 2.5–4 cm in
diameter. *Central S America.* G2.

O. spinosissima (Martyn) Miller. Tree to
5 m with densely spiny trunk to 20 cm in
diameter; stem-segments flattened,
narrowly oblong, 12–40 × 5–10 cm,
6–9 mm thick. Areoles with conspicuous
glochids and several spines, the longest to
8 cm, directed downwards; flowering
segments usually spineless. Flowers with

short, orange perianth c. 2 cm in diameter,
pericarpel 4–5 cm. *Jamaica.* G2.

15. O. aurantiaca Lindley. Illustration:
Edwards's Botanical Register, t. 1606
(1833).
Low spreading shrub to 30 cm; stem-
segments almost terete at base, somewhat
compressed above, to 15 × 1.5 × 1 cm, not
tubercled, easily detached. Spines usually
2–3, 1–3 cm, brownish. Flowers 2.5–4 cm
in diameter, orange-yellow. Fruit to 3 cm,
purplish red, spiny. *Uruguay and adjacent
Argentina.* G2.

O. fragilis (Nuttall) Haworth. Illustration:
Benson, Cacti of the United States &
Canada, f. 389–394 (1982). Forming
clumps 10 × 30 cm; stem-segments variable
in shape, 2–4.5 × 1.2–2.5 (rarely to 3.8) ×
1.2–2.5 cm. Spines 1–6 (rarely to 9). *USA,
Canada.* H3.

O. pestifer Britton & Rose. Brittle
prostrate shrub; stem-segments nearly
terete, 2–8 × 1–3 cm, or when young
flattened and 2–3 cm broad. Spines
2–5. *Ecuador, Peru.* G2.

16. O. corrugata Salm-Dyck (*Platyopuntia
corrugata* (Salm-Dyck) Ritter). Illustration:
Ritter, Kakteen in Südamerika 2: f. 266
(1980).
Creeping shrublet, to 10 cm tall; stem-
segments obliquely ovate or elliptic to
circular, flattened but not strongly so, to
5 × 2.5 × 1.5 cm, with low tubercles,
somewhat pale bluish green.
Leaves subulate, 2 mm, reddish. Spines
several, 8–12 mm, often with 1–2 much
longer. Flowers reddish, 6 × 5 cm, stigma
green. *NW Argentina.* G1.

O. picardoi Marnier-Lapostolle. Stem-
segments to 7 × 3.5 cm, green; spines very
short. *N Argentina.* G1.

17. O. leucotricha de Candolle. Illustration:
Benson, Cacti of the United States &
Canada, f. 552–555 (1982).
Becoming a small tree 3–4 m; terminal
stem-segments oblong to broadly ovate, to
25 × 12 cm, about 1 cm thick, velvety.
Leaves small, subulate, falling early. Spines
becoming longer and more numerous on
older segments, 1–6, to 7.5 cm, bristle- or
hair-like, white, almost covering the stem.
Flowers c. 5 × 5 cm, yellow. *Mexico.* G1.

18. O. basilaris Engelmann & Bigelow.
Illustration: Benson, Cacti of the United
States & Canada, pl. 33–35, f. 416–426
(1982).
Forming clumps 30–60 cm or more tall;
stem-segments obovate to nearly circular,
often truncate to notched at apex, usually

8–20 × 6–15 cm, bluish grey and often
tinged purplish around the edges and
areoles, velvety. Glochids reddish brown,
deciduous. Spines usually 0, or 1 or rarely
5, short. Flowers 5–7.5 × 5–7.5 cm,
usually deep purplish red. Fruit dry,
spherical to obovoid, with a depression at
the top. *SW USA, NW Mexico (Sonora).* G1.

19. O. microdasys (Lehmann) Pfeiffer.
Figure 28(4), p. 204. Illustrations: Hecht,
BLV Handbuch der Kakteen, 200, 201
(1982); Haustein, Der Kosmos-
Kakteenführer, 65 (1983); Cullmann et al.,
Kakteen, edn 5, 262 (1984).
Shrub, forming thickets 40–60 cm or more
tall; stem-segments oblong, obovate or
almost circular, 6–15 × 6–12 cm, green,
velvety. Glochids usually yellow, also white
or reddish brown. Spines 0, rarely 1, very
short. Flowers c. 4 × 4 cm, yellow, outer
segments often tinged red. Fruit nearly
spherical, c. 3 cm in diameter, fleshy,
purplish red. *C & N Mexico.* G1.
Decorative and popular, despite the
glochids which easily detach in numbers
and lodge in the skin if the plant is touched.
The variant with white glochids is
'Albispina'; that with brown glochids is var.
rufida (Engelmann) Schumann, which
ranges into SW USA (Texas) and is
sometimes treated as a distinct species.

20. O. scheeri Weber. Illustration: Haustein
Der Kosmos-Kakteenführer, 65 (1983).
Shrubby, to 1 m; stem-segments oblong to
circular, 15–30 × 9–22 cm, bluish green.
Spines 8–12, to 1 cm, yellow, mixed with
long white or yellow hairs. Flowers large,
to 10 cm in diameter, yellow at first, fading
to salmon-pink. *C Mexico.* G1.

O. pailana Weingart. Spines white at
first, later brown. *N Mexico.* G1.

21. O. polyacantha Haworth. Illustration:
Benson, Cacti of the United States &
Canada, pl. 21, 22, f. 374–387 (1982).
Forming mats or clumps 15 cm ×
30–100 cm or more; stem-segments
circular to broadly obovate, 5–10 ×
4–10 cm, bluish green. Spines 5–10
unequal, the longest to 5 cm, not markedly
flattened in section, mostly deflexed and
largely covering the segments. Flowers
4.5–8 × 4.5–6 cm yellow. Fruit dry, spiny.
Canada, USA, N Mexico. H3.

O. erinacea Engelmann & Bigelow
(*O. hystricina* Engelmann & Bigelow).
Illustration: Benson, Cacti of the United
States & Canada, pl. 24–27, f. 398–401
(1982); Hecht, BLV Handbuch der Kakteen,
199 (1982) – as O. hystricina. Similar to

O. polyacantha, but with at least some of the spines flattened basally. *SW USA.* H3–G1?

Benson treats several popularly cultivated plants as varieties of *O. erinacea*, notably the following:

Var. **ursina** (Weber) Parish. Illustration: Benson, Cacti of the United States & Canada, f. 406, 407 (1982); Haustein, Der Kosmos-Kakteenführer, 67 (1983) – as O. hystricina. Stem-segments oblong or oblong-elliptic, 5–10 × 2.5–6 cm; spines numerous, from all areoles, to 10 cm, very slender and flexuous. Flowers orange or pink.

Var. **utahensis** (Engelmann) Benson (not *O. utahensis* Purpus; *O. rhodantha* Schumann). Illustration: Benson, Cacti of the United States & Canada, pl. 31, f. 409–412 (1982); Haustein, Der Kosmos-Kakteenführer, 67 (1983) – as O. rhodantha; Cullmann et al., Kakteen, edn 5, 261 (1984) – as O. hystricina. Stem-segments obovate, 5–8.5 × 5–7.5 cm; areolar areas often purplish brown; spines from upper areoles only, 3 (rarely to 10) cm. Flowers 7–8 × 7–8 cm, purplish red or pink.

22. O. compressa (Salisbury) Macbride (*O. humifusa* (Rafinesque) Rafinesque; *O. rafinesquei* Engelmann; *O. vulgaris* invalid). Illustration: Benson, Cacti of the United States & Canada, pl. 38, f. 438–439 (1982); Cullmann et al., Kakteen, edn 5, 261 (1984).

Forming clumps or mats 10–30 cm tall; stem-segments elliptic to obovate or circular, 5–12.5 × 4–10 cm, green, often purple tinged. Leaves subulate, 4–7 mm, falling early. Spines usually 0, sometimes 1–2, especially on margined areoles, to 2.5 cm, terete, not flattened. Flowers 4–6 × 4–6 cm, yellow, often with reddish centre. Fruit obovoid, 2.5–4 × 2–3 cm, fleshy, purplish or red. *USA.* H5.

Naturalised in Switzerland.

23. O. stricta Haworth (*O. dillenii* (Ker Gawler) Haworth; *O. inermis* (de Candolle) de Candolle). Figure 29(1), p. 205. Illustration: Benson, Cacti of the United States & Canada, pl. 49, f. 509–513 (1982).

Sprawling or erect shrub, 50–200 cm; stem-segments obovate to oblong, 10–40 × 7.5–25 cm, bluish green. Spines few or lacking in var. **stricta**, to 11 in var. **dillenii** (Ker Gawler) Benson, usually 1.5–4 cm, stout, straight or commonly curved, flattened, yellow or with brown bands. Flowers *c.* 5–6 × 5–6 cm, yellow. Fruit spherical to pear-shaped 4–6 × 2.5–3 cm,

fleshy, purple. *SE USA to N Venezuela, naturalised in various tropical countries.* G2.

O. paraguayensis Schumann. Stem-segments oblanceolate or narrowly elliptic, to 30 × 8 × 2 cm, dark glossy green. Spines absent or 1, to 1 cm, yellowish. Flowers *c.* 10 × 6.5 cm, orange. *Paraguay.* G2.

24. O. phaeacantha Engelmann (*O. engelmannii* misapplied, not Salm-Dyck). Illustration: Benson, Cacti of the United States & Canada, pl. 46, 48, f. 477–496 (1982).

Sprawling and spreading shrub 30–100 cm tall; stem-segments obovate or circular, 10–40 × 7.5–22.5 cm, bluish green, sometimes purple-tinged. Spines 1–8 or more, to 6 cm, flattened or elliptic in section, brown or reddish brown, sometimes confined to upper part of segment, rarely lacking. Flowers 6–8 × 6–7.5 cm, yellow, sometimes red-tinged within. Fruit pear-shaped, 4–8 × 2–4 cm, purplish. *SW USA, N Mexico.* G1.

O. violacea Engelmann (*O. gosseliniana* Weber). Stem-segments persistently purple-tinged; spines absent or 1 (rarely 3), to 10 (rarely to 17) cm, not flattened at base. *SW USA, N Mexico.* G1.

25. O. monacantha (Willdenow) Haworth (*O. vulgaris* misapplied). Figure 29(19), p. 205. Illustration: Benson, Cacti of the United States & Canada, f. 435–437 (1982).

Shrubby or tree-like, to 2 m; stem-segments oblong to obovate, tapered towards the base, 10–30 × 7.5–12.5 cm. Spines 1 or 2, unequal, longer to 4 cm, brown towards tip and base, greyish between, spines more numerous on trunk. Flowers 5–7.5 × 7.5–10 cm, yellow or orange-yellow, outer perianth-segments reddish. Fruit pear-shaped, 5–7.5 × 4–5 cm, reddish purple. *S Brazil to Argentina, naturalised in various warm countries.* G2.

26. O. ficus-indica (Linnaeus) Miller (*O. engelmannii* Salm-Dyck; *O. megacantha* Salm-Dyck). Illustration: Benson, Cacti of the United States & Canada, pl. 54–57, f. 528–533 (1982).

Large shrub or small tree to 5 m with a trunk sometimes 1 m in diameter; stem-segments obovate to oblong 20–60 × 10–40 cm, green or bluish green. Spines variable, 1–2 or more, the larger to 2.5 cm, white or whitish, or lacking. Flowers 6–7 × 5–7 cm, yellow. Fruit 5–10 × 4–9 cm, yellow, orange, red or purple in different cultivars. *Mexico.* G1.

Long cultivated in warmer regions for the fruits, and widely naturalised.

O. lanceolata Haworth. Perhaps only a form of *O. ficus-indica*. Stem-segments diamond-shaped or lanceolate, 20–30 × 6–8 cm. *Origin uncertain.*

27. O. robusta Pfeiffer. Illustration: Britton & Rose, The Cactaceae 1: pl. 34, f. 4 (1919); Backeberg, Die Cactaceae 1: f. 520 (1958).

Shrub to 2 m; stem-segments circular or nearly so, massive, to 40 × 40 cm or more, bluish bloomed. Spines 2–12, unequal, longest to 5 cm, white, brownish or yellowish below, terete. Flowers 5 × 5–7 cm, yellow. Fruit spherical to ellipsoid, 7–8 cm, deep red. *C Mexico.* G1.

28. O. cochenillifera (Linnaeus) Miller (*Nopalea cochenillifera* (Linnaeus) Salm-Dyck). Illustration: Benson, Cacti of the United States & Canada, pl. 50, 51, f. 515 (1982).

Shrub or tree to 4 m or more; stem-segments elliptic to obovate, 8–25 × 5–12 cm, dark green, with wide-spaced (2.5 cm) areoles. Spines usually lacking, rarely 1–3, less than 1 cm. Flowers 5–6 × 1.2–1.5 cm, perianth erect, bright red, stamens and style exserted, pink. Fruit ellipsoid, 2.5–3.8 × 2.5–3 cm, fleshy, red. *Mexico.* G2.

Long cultivated in tropical America and elsewhere, formerly as a host for the cochineal insect. Commonly confused (since it rarely flowers in cultivation) with *O. paraguayensis* Schumann or an allied species.

O. auberi Pfeiffer (*Nopalea auberi* (Pfeiffer) Salm-Dyck). Illustration: Britton & Rose, The Cactaceae 1: pl. 5 (1919); Haustein, Der Kosmos-Kakteenführer, 69 (1983). Stem-segments narrower than in *O. cochenillifera*; spines usually present, 2–3, to 3 cm, greyish, tipped brown. Flowers *c.* 9 cm, dull red; stamens pinkish; stigma-lobes green. *S Mexico.* G2.

78. PTEROCACTUS Schumann
D.R. Hunt

Dwarf tuberous-rooted shrubs with spherical, cylindric or club-shaped stems, often brownish or reddish and usually papillate. Leaves small, subulate, falling early. Glochids numerous to few or absent. Spines needle-like, subulate or papery. Flowers immersed in the apex of stem-segments, disc-shaped, stamens touch-sensitive. Fruit dry, with a depression at the top, dehiscent by splitting transversely near the top; seeds surrounded by a broad papery wing formed from the funicle.

A genus of 9 closely allied species endemic to Argentina.

Literature: Kiesling, R., The genus Pterocactus, *Cactus & Succulent Journal of Great Britain* **44**: 51–56 (1982).

1. P. tuberosus (Pfeiffer) Britton & Rose (*P. kuntzei* Schumann). Illustration: Britton & Rose, The Cactaceae **1**: f. 37, 38 (1919); Haustein, Der Kosmos-Kakteenführer, 55 (1983).
Aerial stem-segments 7–20 × 8–15 mm, brown or greenish brown, with a vertical violet line below the areoles. Spines 8–12, 5–10 mm, whitish. Flowers 3–5 cm in diameter, yellowish, brownish or coppery. Seeds (including the wing) 1–1.2 cm in diameter. *Argentina*. G1.

According to Kiesling, the name *Opuntia tuberosa* Pfeiffer (on which the name of this species is based) is of doubtful application.

LVIII. DIDIEREACEAE

Trees or shrubs, often with succulent spiny stems. Leaves simple, linear, elliptic or circular, slightly fleshy, deciduous. Flowers unisexual, borne in cymes or densely branched panicles, occasionally in umbels borne directly on the stems. Calyx of 2 persistent sepals, more or less decurrent on the flower-stalks. Petals 4, in 2 pairs, the 2 outer often larger than the inner. Stamens 8–10 (rarely 6 or up to 14), anthers dorsifixed, reduced to staminodes in female flowers. Ovary ovoid or pyramidal, 1-celled, superior, rudimentary in male flowers. Style long or short with 3 or 4 lobes. Fruit dry, indehiscent.

A small family of 4 genera and 11 species, resembling some cacti and euphorbias, but differing from them and other succulents in the arrangement of leaves in relation to spines. The family is endemic to Madagascar. All species are cultivated for their succulent and spiny stems rather than their flowers, which are rarely produced in cultivation. Because of this, flowering times are not given for many of the species included here.

1a. Branches strongly zig-zag **1. Decarya**
 b. Branches straight, never zig-zag 2
2a. Branched shrubs with slender woody shoots **2. Alluaudiopsis**
 b. Shrubs or trees with stout branches or stems 3
3a. Spines in pairs or solitary (minute in 1 species) **3. Alluaudia**
 b. Spines in clusters of 4 (rarely 5) **4. Didierea**

1. DECARYA Choux
S.G. Knees
Small trees or shrubs to 6 m with markedly zig-zag branching, more pronounced in youngest shoots. Leaves shortly stalked, 5–10 mm, oval, flattened, mucronate, downy, solitary or in pairs at branch angles, borne between two spines. Flowers bisexual, in cymes. Sepals 4–7 mm, leaf-like, greenish yellow. Petals 3–6 mm, white. Stamens 6 or 8, filaments 3–4 mm. Ovary stalkless, tapering into a shortly 3-lobed style. Fruit 2–3 × 1.5 mm.

A genus of a single species grown for its curious branching habit. Very distinctive and somewhat dissimilar from other genera in the family. Warmth and careful cultivation are necessary. A freely drained compost and a position in full sun are also essential. Propagation is mainly from seed.

1. D. madagascariensis Choux. Illustration: Humbert, Flore de Madagascar **121**: 33 (1963); Ashingtonia **2**(1): 12 (1975); Graf, Exotica, International series 4, **1**: 950 (1985).
SW Madagascar. G1. Spring.

2. ALLUAUDIOPSIS Humbert & Choux
S.G. Knees
Shrubs 2–4 m in the wild, very branched at the base, bearing paired or solitary spines for much of their length. Paired leaves borne above the spines. Flowers unisexual, relatively large, in cymes of 2, 3 or 5 on short branches. Male and female flowers borne on separate plants.

A genus of 2 species with similar cultural requirements to *Decarya*. Propagation by seed or from cuttings.

1a. Spines solitary; flowers whitish yellow **1. fiherenensis**
 b. Spines paired; flowers reddish pink **2. marnieriana**

1. A. fiherenensis Humbert & Choux. Illustration: Bulletin de la Société Botanique de France **82**: pl. 2–4 (1935); Humbert, Flore de Madagascar **121**: 11 (1963); Ashingtonia **2**(1): 5 (1975); Heywood, Flowering plants of the world, 74 (1978).
Shrubs 1.5–2 m, with stems 3–6 cm in diameter. Young branches brown fading to grey with age. Leaves shortly stalked, fleshy, 1–4 cm × × 2–5 mm (often longer in cultivation), cylindric, oblong-elliptic, shortly mucronate. Spines solitary, 7–20 mm, spirally arranged along branches. Inflorescences borne at the ends of branches, 7–14 cm. Female flowers on stalks 5–7 mm. Sepals unequal, 1–1.5 cm × 5–8 mm, broadly triangular,

membranous, olive green, decurrent on stalks, joined at base. Petals 4, the 2 outer 1–1.2 × 5 mm, pale yellowish green, the 2 inner 8–10 × 3 mm. Staminodes 7 or 8, filaments red. Ovary stalkless, 1.5–2 mm. Style 6–8 mm, 2-or 3-lobed. Male flowers smaller than female, on stalks 3–5 mm, stamens 7–9, filaments red, 7–8 mm; ovary very reduced. Fruit ovoid, 3–4 mm, enveloped by sepals. *SW Madagascar*. G2. Summer.

2. A. marnieriana Rauh. Illustration: Cactus **7**: 47–50 (1962); Humbert, Flore de Madagascar **121**: 7 (1963).
Branched shrub 1.5–4 m, stems to 10 cm in diameter near the base. Leaves in pairs, 7–15 × 2–4 mm on male plants, 5 × 3 mm on female plants, oblong-lanceolate, mucronate. Spines paired, 5–10 mm. Flowers in groups of 2–5. Male flowers 1–1.5 cm across; sepals unequal, 5–8 × 6–7 mm, membranous, yellowish, decurrent on stalks; petals 4, the outer 2 to 1.2 cm × 6 mm, the inner 2 to 1.2 cm × 4 mm, bright reddish pink; stamens 9–10, filaments red, to 8 mm. Female flowers 2 cm or more across; sepals unequal, 1–1.5 cm × 3–6 mm; petals 1–1.5 cm × 3–7 mm; staminodes 8–11; ovary stalkless, 2–4 mm; styles 8–12 mm, 3-lobed. Fruit ovoid, 3–4 mm, enveloped by persistent sepals. Seeds ovoid, 3 mm, with white aril. *SW Madagascar*. G2. Summer.

3. ALLUAUDIA Drake
S.G. Knees
Succulents with ascending or spreading branches reaching 10–15 m. Spines solitary or paired, occasionally absent. Branches usually leafless or sometimes with leaves in pairs beneath the spines. Leaves fleshy, circular, almost stalkless. Flowers unisexual in umbels or clusters. Sepals persistent, eventually enveloping the fruit. Corolla constricted, 7–8 mm. Fruit 3–6 mm, oblong, enclosed by sepals; seeds often with arils.

A genus of 6 species requiring careful cultivation. Propagation by seed or cuttings.

1a. Spines absent or not more than 2 mm; plants usually leafless **3. dumosa**
 b. Spines at least 1.5 cm; plants leafy, at least during the growing season 2
2a. Branches of equal height, producing flat-topped, funnel-shaped bushes; flowers in umbels **2. comosa**
 b. Branch height variable; flowers in branched clusters 3

3a. Flowering clusters 30 cm or more   4
 b. Flowering clusters 15 cm or less   5
4a. Plant branched; leaves rounded, with fine hairs   **6. procera**
 b. Plant unbranched; leaves notched at tip, hairless   **5. montagnacii**
5a. Branches curved in mature plants; leaves circular, notched at the tip   **1. ascendens**
 b. Branches erect in mature plants; leaves obovate with scolloped margins   **4. humbertii**

1. A. ascendens (Drake) Drake (*Didierea ascendens* Drake). Illustration: Graf, Exotica International series 4, **1**: 949 (1985); Graf, Tropica, edn 3, 389, 391 (1986).
Plant 5–12 m, with few branches, trunk to 30 cm across. Spines 1.5–2 cm, conical, arranged irregularly along the stems. Leaves 1.3–2.5 cm, fleshy, circular or reverse heart-shaped, notched at the apex, dark green. Flowers inconspicuous (rarely seen in cultivation), sepals keeled. *S Madagascar*. G2.

2. A. comosa Drake. Illustration: Jacobsen, Lexicon of succulent plants, t. 10 (1974); Graf, Exotica International series 4, **1**: 950 (1985); Graf, Tropica, edn 3, 389 (1986).
Plant 1–10 m, branches 4–6, erect. Twigs short, numerous, irregularly arranged, giving the plants a characteristic flat-topped, funnel-shaped appearance, with thorns 1.5–3.5 cm. Leaves solitary, 1–2.2 cm, ovate or almost circular, occasionally notched at apex. Flowers in umbel-like clusters to 3 cm. *S & SW Madagascar*. G2.

3. A. dumosa Drake. Illustration: Jacobsen, Lexicon of succulent plants, t. 10 (1974); Graf, Exotica International series 4, **1**: 950 (1985).
Plant 2–8 m, with erect, succulent branches. Twigs with few isolated spines 2 × 1 mm or spines absent. Leaves *c.* 8 mm, cylindric, falling early. Flowers borne directly on the branches. Male flowers spherical, to 8 mm, female flowers cylindric, to 6 mm. *S Madagascar*. G2.

4. A. humbertii Choux. Illustration: Humbert, Flore de Madagascar **121**: 23 (1963).
Plant 6–7 m with slender, woody branches. Spines isolated, 5–23 mm, slender. Leaves in pairs, usually beneath the spines, 5–16 × 5–10 mm, ovate-reverse-heart-shaped, fleshy. Male flowers 8–10 mm across, female flowers 4–6 mm. *C, S & SW Madagascar*. G2.

5. A. montagnacii Rauh. Illustration: Humbert, Flore de Madagascar **121**: 17 (1963); Jacobsen, Lexicon of succulent plants, t. 10 (1974); Graf, Tropica, edn 3, 391 (1986).
Plant to 8 m, rarely branched. Stems columnar, with ring-like constrictions. Spines to 2.5 cm on new growth, silvery, conical. Leaves to 1.5 cm, circular or broadly ovate, deeply notched at apex, shortly stalked. Flowers in well-branched clusters to 30 cm. *SW Madagascar*. G2.

6. A. procera (Drake) Drake (*Didierea procera* Drake). Illustration: Jacobsen, Lexicon of succulent plants, t. 10 (1974); Graf, Exotica International series 4, **1**: 949 (1985).
Woody plant eventually to *c.* 20 m. Branches few, ascending or spreading. Spines 2–2.5 cm, broadly conical, densely arranged in spirals throughout the length of the stem. Leaves 7–25 × 4–12 mm, ovate or ovate-oblong, fleshy, borne in pairs. Flowers numerous in well-branched cymes 12–30 cm across. *S & SW Madagascar*. G2.

4. DIDIEREA Baillon
S.G. Knees
Deciduous succulent shrubs, becoming tree-like. Stems with vertical and horizontal branches covered in warty shoots. Leaves fleshy, in rosettes, arising from the centres of groups of 4–5 spines. Flowers shortly stalked, in cymes. Sepals persistent, eventually enclosing the fruit. Corolla flattened, of 4 or 5 petals.
A genus of 2 species endemic to SW Madagascar. Propagation is normally by seed or cuttings. *Didierea* has also been successfully grafted on to members of the Cactaceae.
Literature: Rauh, W., Bemerkenswerte Sukkulente aus Madagascar, *Kakteen und andere Sukkulenten* **12**(6): 82–88 (1961).

1a. Juvenile branches prostrate, mature stems with horizontal branches; leaves 1.5–3 cm   **1. trollii**
 b. All branches more or less erect; leaves 5–10 cm   **2. madagascariensis**

1. D. trollii Capuron & Rauh. Illustration: Kakteen und andere Sukkulenten **12**: 83–88 (1961); Humbert, Flore de Madagascar **121**: 31 (1963); Jacobsen, Lexicon of succulent plants, pl. 51 (1974); Graf, Tropica, edn 3, 389 (1986).
Branched shrub with prostrate juvenile stems and eventually erect branches with short shoots. Spines 2–4 cm, in groups of

5. Leaves 1–2 cm × 3–8 mm, ovate-elliptic to oblong, 2–5 per rosette. Male flowers on stalks 5–10 mm with triangular, mucronate sepals 3–4 × 2–3 mm; petals 7–9 × 3–4 mm, greenish yellow; stamens 8, filaments red, hairy. Female flowers in groups of 4–10 on stalks 5–20 mm; sepals 8–12 × 8–9 mm; petals 6–10 × 2–3 mm, white or greenish yellow; staminodes 8; ovary 3.5 mm, stalkless, with club-shaped style. Fruit 5–7 mm, enveloped by the sepals. *S Madagascar*. G2.

2. D. madagascariensis Baillon (*D. mirabilis* Baillon). Illustration: Grandidier, Histoire physique, naturelle et politique de Madagascar, Plantes volume **5**, Atlas **3**: pl. 261, 262 (1894); Humbert, Flore de Madagascar **121**: 33 (1963); Ashingtonia **2**(1): 5 (1975); Graf, Exotica International series 4, **1**: 949, 950 (1985).
Erect branched shrubs 4–6 m with woody stems to 40 cm in diameter, often curved at the apex and densely covered with spiny short shoots. Spines 4–10 cm, in groups of 4, up to 12 on each shoot. Leaves 7–15 cm × 5–10 mm, linear, mucronate, 3–10 per rosette. Flowers numerous in clusters arising from between the spines. Male flowers on stalks 5–10 mm, sepals yellowish green, 4–5 × 2–3 mm; petals pinkish yellow, oblong, 8–18 × 3–4 mm; stamens 8, filaments grey, anthers red. Female flowers with greenish yellow sepals 1.1–1.6 cm × 8–10 mm; petals greenish yellow with pinkish veins, oblong, 3.5 × 2 mm; staminodes 8; ovary stalkless, 1.5 mm, 3-sided, style 3-lobed, reddish pink, 3–5 mm. Seeds ovoid-oblong, to 3 mm, with white aril. *S Madagascar*. G2.

LIX. MAGNOLIACEAE
Evergreen or deciduous trees or shrubs, the buds enclosed by stipules and the branchlets with conspicuous ring-like stipule-scars at the nodes. Leaves alternate, sometimes crowded into false whorls, with free or attached stipules, the blades simple, sometimes lobed. Flowers solitary, terminal or axillary, bisexual; perianth-segments 6–9, or numerous, free, all more or less similar, or of 2 forms; stamens numerous, inwardly or laterally dehiscent. The numerous carpels spirally arranged on the long floral axis, each with 2–6 ovules. Fruits cone-like heads of persistent follicles or deciduous samaras.
About 12 genera and 200 species distributed in temperate and tropical E & SE

Asia and in the western hemisphere in eastern N America, from southern Canada southwards through the West Indies and C America to E Brazil.

Literature: Dandy, J.E., The genera of Magnoliaceae, *Kew Bulletin* for 1927: 257–264; Dandy, J.E., A survey of the genus Magnolia together with Michelia and Manglietia, *Camellias and Magnolias (RHS)*, 64–81 (1950); Spongberg, S.A., Magnoliaceae hardy in temperate N America, *Journal of the Arnold Arboretum* **57**: 250–312 (1976); Noteboom, H.P., Notes on Magnoliaceae with a revision of Pachylarnax and Elmerillia and the Malesian species of Manglietia and Michelia, *Blumea* **31**: 65–121 (1985).

1a. Fruits cone-like heads of closely overlapping 2-seeded samaras, the samaras falling at maturity leaving the persistent, spindle-shaped floral axis; plants deciduous, the leaves variously lobed   **4. Liriodendron**
  b. Fruits cone-like heads of follicles, the follicles persistent on the floral axis; plants evergreen or deciduous, the leaves simple, occasionally cordate or auriculate at base or notched at apex   **2**
2a. Flowers axillary; ovary stalked; plants evergreen   **3. Michelia**
  b. Flowers terminal; ovary stalkless or rarely shortly stalked; plants evergreen or deciduous   **3**
3a. Carpels with 2 ovules, the follicles with 2 seeds (sometimes 1 or 0 through abortion); plants evergreen or deciduous   **1. Magnolia**
  b. Carpels with 4 or more ovules, the follicles with 4 or more seeds; plants evergreen   **2. Manglietia**

1. MAGNOLIA Linnaeus
S.A. Spongberg

Evergreen or deciduous trees or shrubs with smooth, rough or deeply furrowed bark; branchlets with septate or continuous pith, the buds enclosed by stipule-scales. Leaves alternate, sometimes in false whorls, stalked, with free or attached stipules which fall early, the blades thickly leathery to membranous, entire, sometimes cordate to auriculate at base or notched at the apex. Flowers fragrant, appearing before or with the leaves, terminal, solitary. Perianth of 6–9 or numerous (to 33) segments, white or pink to purple, occasionally greenish or pure yellow, in whorls of 3 (occasionally more), all more or less similar or sometimes those of the outer whorl

reduced in size and sepal-like. Stamens numerous, spirally arranged, the linear pollen-sacs inwardly or laterally dehiscent. Carpels many, each with 2 ovules, spirally arranged on a short to elongated floral axis. Fruits more or less spherical to cylindric cone-like heads of free, longitudinally-dehiscent follicles, the 1 or 2 seeds suspended on thin threads, with orange, red or pink aril-like seed-coats.

Approximately 80 species distributed in temperate and tropical regions in E Asia from Manchuria to Java and in the western hemisphere from southern Canada through the eastern USA and southwards into the West Indies, Mexico, C America to SE Venezuela. All that have been introduced into cultivation are highly desirable ornamentals and the genus is one of the most important groups of ornamental woody plants. Numerous hybrids have been produced, many still rare in cultivation and a multitude of cultivars has been selected and named. *Magnolia*, journal of the Magnolia Society and its predecessor, *Newsletter of the American Magnolia Society* should be consulted for a wealth of information on the culture, propagation, hybridisation, history, classification, nomenclature and cultivars of this popular and widely cultivated group.

Two species, *M. biondii* Pampanini and *M. zenii* Cheng, both from China, have only recently been introduced into cultivation in Europe and N America. Because they are poorly known botanically and of extremely limited distribution in cultivation, they have not been accounted for here.

Literature: Johnstone, G.H., *Asiatic Magnolias in cultivation* (1955); Fogg, J.M. & McDaniel, J.C. (eds), *Checklist of the cultivated Magnolias* (1975); Treseder, N.G., *Magnolias* (1978); Treseder, N.G. & Blamey, M., *The book of Magnolias* (1981); Ueda, K., Nomenclatural revision of the Japanese Magnolia species, together with two long-cultivated species, II, *M. tomentosa* and *M. praecocissima*, *Taxon* **35**: 344–346 (1986); Meyer, F.G. & McClintock, E., Rejection of the names Magnolia heptapeta and M. quinquepeta, *Taxon* **36**: 590–600 (1987).

Habit. Evergreen: **1–5**; deciduous: **5–32**.
Branchlets. With septate or incompletely septate pith: **1–6**.
Leaf-stalks. Lacking stipule-scars: **1,3**.
Leaves. In false whorls: **6–14, 18**. With notched apices: **13,24**. With deeply cordate to auriculate bases: **9,10**.
Flowers. Appearing before the leaves: **19–30**; appearing with the leaves: **1–8, 11–15,31,32**. Pendent: **15–17,24,25**.

Green to clear yellow: **32**. With outermost perianth-whorl sepal-like: **26–32**. With ovary shortly stalked: **3**.

1a. Anthers dehiscing inwardly; flowers neither appearing before the leaves, nor with an outer whorl of sepal-like perianth-segments; plant evergreen or deciduous   **2**
  b. Anthers dehiscing laterally or more or less so; flowers appearing before the leaves and/or with the outermost whorl of perianth-segments sepal-like; plant deciduous   **23**
2a. Stipules free, not leaving scars on the leaf-stalks; plant evergreen   **3**
  b. Stipules attached to the leaf-stalks, soon falling, leaving scars; plant evergreen or deciduous   **5**
3a. Ovary shortly stalked, with stamens or stamen-scars separated by a short gap from carpels or follicles   **3. nitida**
  b. Ovary stalkless, stamens or stamen scars continuous with the lowermost carpels or follicles   **4**
4a. Perianth-segments 7.5–12 cm; stamens 1.3–2.1 cm, yellowish; fruiting heads ovoid, 5.5–7.5 cm   **1. grandiflora**
  b. Perianth-segments 5.5–9 cm; stamens *c*. 1.2 cm, rosy purple; fruits *c*. 4 cm, rarely developed   **2. grandiflora × virginiana**
5a. Plant evergreen   **6**
  b. Plant deciduous   **8**
6a. Stipule-scars large, extending almost the entire length of the leaf-stalk to the base of the blade   **7**
  b. Stipule-scars very small, at the base of the leaf-stalk, often obscure   **2. grandiflora × virginiana**
7a. Leaf-stalks 3.5–8.7 cm; blades thickly leathery, pale green or yellowish beneath; fruiting heads 6.7–9.5 cm   **4. delavayi**
  b. Leaf-stalks 1.5–3 cm; blades thinly leathery to membranous, glaucous or rarely green beneath; fruiting heads 2–5 cm   **5. virginiana**
8a. Leaves crowded into false whorls at the ends of the branches, distinctly alternate on new shoots; blades 10–60 cm or more   **9**
  b. Leaves distinctly alternate, not crowded into false whorls at the ends of the branchlets; blades not exceeding 23 cm   **17**
9a. Leaf-blades cordate to auriculate at the base   **10**
  b. Leaf-blades wedge-shaped to rounded at base   **13**

10a. Lower surfaces of the leaves, stipules, carpels and follicles downy · 11

b. Lower surface of the leaves, stipules, carpels and follicles hairless · 12

11a. Tree; stamens 1.5–1.8 cm; fruiting heads more or less spherical · **7. macrophylla**

b. Large shrub; stamens 1.1–1.3 cm; fruiting head more or less cylindric to ovoid · **8. ashei**

12a. Stamens 8–15 mm; ovary *c.* 2.5 cm at flowering; fruiting heads 6.5–11 cm · **9. fraseri**

b. Stamens 4–6 mm; ovary *c.* 1.5 cm at flowering; fruiting heads 3.5–6 cm · **10. pyramidata**

13a. Hairs on the lower surfaces of the leaves usually rust-coloured · 14

b. Hairs on the lower surfaces of the leaves not rust-coloured · 15

14a. Leaves in loose false whorls, the blades with 10–15 pairs of lateral veins · **18. × wieseneri**

b. Leaves in distinct false whorls, the blades with *c.* 30 or more pairs of lateral veins · **14. rostrata**

15a. Flowers with an offensive odour; fruiting heads 7–10 cm; small tree of open habit · **11. tripetala**

b. Flowers pleasantly fragrant; fruiting heads 13.5–20 cm, rarely shorter; large, widely branching tree · 16

16a. Leaf-blades mostly oblong-obovate, with acute to rounded apices; young branchlets purplish or silvery; lowermost follicles concave, decurrent along the axis · **12. hypoleuca**

b. Leaf-blades mostly elliptic to obovate, sometimes deeply notched at apex; young branchlets yellowish or yellowish grey; lowermost follicles convex, not as above · **13. officinalis**

17a. Flowers and fruiting heads more or less erect · 18

b. Flowers and fruiting heads pendent · 20

18a. Anthers rosy crimson, rounded or blunt at the apices, the connectives only rarely produced into apiculate appendages · **18. × wieseneri**

b. Anthers yellow, the connectives with short, acute appendages · 19

19a. Pith septate; perianth-segments 6–15, usually 3–5 cm; leaf-blades 6.5–10.5 cm, generally narrowly oblong to elliptic or ovate · **5. virginiana**

b. Pith incompletely septate; perianth-segments 9, 7–11 cm; leaf-blades 12–17.5 cm, broadly elliptic to oblanceolate · **6. × thompsoniana**

20a. Branchlets becoming dark reddish or purplish brown · 21

b. Branchlets becoming light brown or silvery tan · 22

21a. Leaf-blades elliptic, broadly ovate or usually obovate, 7–9 cm wide, generally reddish-downy beneath; stamens 1.2–1.7 cm · **15. globosa**

b. Leaf-blades elliptic or usually lanceolate to oblong-ovate, 3–7 cm wide, generally yellowish- or silvery-downy beneath; stamens 9–12 mm · **16. wilsonii**

22a. Some or all hairs on the lower surfaces of the leaves rust coloured, the hairs essentially straight; leaf-blades usually broadly ovate · **17. sieboldii**

b. Hairs on the lower surfaces of the leaves clear, not rust-coloured, more or less crinkled; leaf-blades usually broadly elliptic to obovate · **17. sieboldii** subsp. **sinensis**

23a. Perianth-segments more or less equal, the outer whorl petal-, not sepal-like; flowers before the leaves, sometimes continuing as the leaves expand, perianth-segments white to purple · 24

b. Perianth-segments very unequal, the outer whorl sepal-like, simulating a calyx, sometimes falling early; flowers appearing before or with the leaves; petal-like perianth-segments white, purplish, greenish or yellow · 33

24a. Perianth-segments 12–33, narrowly oblong or strap-shaped, 6–17 mm wide · **29. stellata**

b. Perianth-segments 6–16, spathulate to obovate or oblanceolate, 1.4–8 cm wide · 25

25a. Perianth-segments of the innermost whorl erect, enclosing the stamens and carpels, the outer whorls more or less horizontal or occasionally reflexed; leaf-blades generally elliptic to oblong-ovate with 12 or more pairs of major lateral veins · 26

b. Perianth-segments of the innermost whorl not as above, the outer whorls more or less horizontal or drooping; leaf-blades generally obovate to oblanceolate with 12 or fewer pairs of major lateral veins · 27

26a. Perianth-segments 12–16; fruiting heads 11–20 cm; leaf-blades hairless above · **19. campbellii**

b. Perianth-segments 7–10; fruiting heads 7–10 cm; leaf-blades with short hairs along the veins above · **20. × veitchii**

27a. Flowers more or less erect; leaves usually obovate to oblanceolate, acute or abruptly short acuminate at the apex · 28

b. Flowers more or less horizontal; leaves usually obovate to oblanceolate with rounded or sometimes notched apices · 31

28a. Perianth-segments (or their scars) 9, the segments tapering to broad bases · 29

b. Perianth-segments (or their scars) 12 or more, the segments tapering to relatively narrow, more or less clawed bases · 30

29a. Flowers before the leaves; perianth-segments white, the outer whorl more or less equal to the inner · **21. denudata**

b. Flowers before the leaves, sometimes continuing after the leaves have expanded; perianth-segments rarely white throughout, the outer surfaces usually flushed with pink to dark purple, the outer whorl often about half as long as the inner whorl · **22. × soulangiana**

30a. Perianth-segments rosy red on outer surfaces; leaves generally less than twice as long as broad · **23. sprengeri**

b. Perianth-segments white on outer surfaces; leaves generally more than twice as long as broad · **23. sprengeri** var. **elongata**

31a. Perianth-segments 9–12, pale pink on the outer surfaces; leaf-blades with a strong network of veins, the lower surfaces hairless or with few, scattered hairs along the veins · **25. dawsoniana**

b. Perianth-segments 10–16, purplish pink on the outer surfaces; leaf-blades with a fine network of veins, the lower surfaces finely downy over the surface or adjacent to the midveins · 32

32a. Perianth-segments 12–14, 6.4–9 cm; leaf-blades generally less than twice as long as broad · **24. sargentiana**

b. Perianth-segments 10–16, 8.4–12.5 cm; leaf-blades generally more than twice as long as broad · **24. sargentiana** var. **robusta**

33a. Flowers before the leaves; petal-like perianth-segments white, sometimes flushed pink to purple at the base or all pink to purple · 34

b. Flowers appearing with the expanded leaves; perianth-segments purplish, greenish or yellow · 40

34a. Flowers often continuing after the leaves have expanded; petal-like perianth-segments flushed pink to purple on the outer surfaces, white, pink or purplish within **35**

b. Flowers not continuing after the leaves have expanded; petal-like perianth-segments white on both surfaces, sometimes tinged pink to purple basally, rarely with outer surfaces flushed pink throughout **36**

35a. Shrub; flowers continuing to appear with the leaves; stipule-scars extending almost to the base of the leaf-blades, the blades often decurrent along the stalk, and with hairs along the midvein, or hairless beneath **31. liliflora**

b. Large shrub or usually a tree; flowers primarily before the leaves, some appearing with the leaves; stipule-scars small, basal, the leaf-blade not decurrent, usually finely hairy over the lower surface **22. × soulangiana**

36a. Petal-like perianth-segments 12–33, narrowly oblong or strap-shaped, 6–17 mm wide, the outer whorl of sepal-like segments often soon falling or sometimes abortive; young branchlets finely silky **29. stellata**

b. Petal-like inner perianth-segments 6–16, spathulate, obovate or oblanceolate, 2–4.5 cm wide. the outer whorl of sepal-like segments persistent at flowering; young branchlets hairless or finely silky **37**

37a. Lower surfaces of the leaves finely hairy over the entire surface or at least adjacent to the midvein **38**

b. Lower surface of the leaves hairless or with long hairs scattered along the veins and/or in the axils of the major lateral veins **39**

38a. Flowers more or less erect; petal-like perianth-segments 6, more or less erect; branchlets finely silky **26. cylindrica**

b. Flowers more or less horizontal; petal-like perianth-segments 6 (rarely to 12), more or less drooping; branchlets hairless **27. salicifolia**

39a. Petal-like perianth-segments 5–7 **28. kobus**

b. Petal-like perianth-segments 11–16 **30. × loebneri**

40a. Large shrub; petal-like perianth-segments purple **31. liliflora**

b. Large tree or rarely a shrub; petal-like perianth-segments greenish to golden yellow **41**

41a. Branches of the current and previous year hairless; flower-stalks hairy or rarely velvety; perianth-segments greenish yellow throughout or sometimes golden yellow within **32. acuminata**

b. Branches of the current and previous year with short hairs or roughened by hair-bases; flower-stalks velvety or rarely hairless; perianth-segments light yellow on the outer surfaces, golden yellow within, rarely greenish yellow throughout **32. acuminata var. subcordata**

1. M. grandiflora Linnaeus. Figure 30 (1–4), p. 306. Illustration: Botanical Magazine, 1952 (1818); Journal of the Arnold Arboretum **57**: 261 (1976); Treseder, Magnolias, pl. 2 (1978); Treseder & Blamey, The book of Magnolias, 11 (1981).

Evergreen tree, often pyramidal, to 25–30 m, the branchlets with septate pith. Leaf-stalks 9–27 mm, lacking stipule-scars, leaf-blades stiffly leathery, 8–20 × 6–9 cm, usually broadly ovate to almost circular, brilliantly glossy green above, downy beneath. Flowers large, cup-shaped, the 9–12 (occasionally more) perianth-segments creamy white, thick and fleshy. Fruiting heads ovoid, 5.5–7.5 cm, follicles finely downy. *Coastal plain of SE USA*. H4. Late spring–early summer.

Over 150 cultivars have been described, of which 'Exmouth' (*M. grandiflora* var. *exoniensis* Loudon) is an old clone of narrow, pyramidal habit with oblong-elliptic to oblong-lanceolate leaves, finely downy beneath.

2. M. grandiflora × M. virginiana (*M. freemannii* invalid). Illustration: The National Horticultural Magazine **16**: 161 (1937).

Evergreen or semi-evergreen tree to 10 m, the branchlets with sepatate pith. Leaf-stalks 1.5–2 cm, the minute stipule-scars often obscured by hairs; leaf-blades leathery, 8.8–18 × 4–8.5 cm, elliptic to broadly ovate, densely downy or somewhat glaucous beneath. Flowers cup-shaped, the 9–12 perianth-segments white. *Garden origin*. H4. Late spring.

The result of intentional hybridisation by O.M. Freeman in 1930 and 1931, best known in cultivation by the cultivars 'Freeman' and 'Maryland'.

3. M. nitida W.W. Smith. Illustration: Botanical Magazine, n.s., 16 (1948); Johnstone, Asiatic Magnolias in cultivation, pl. 14, f 20 (1955); Treseder, Magnolias, pl.

12 (1978); Treseder & Blamey, The book of Magnolias, 34 (1981).

Hairless evergreen tree or large shrub to *c.* 15 m, the branchlets with septate pith and conspicuous encircling stipule-scars. Leaf-stalks 1.5–2.5 cm, lacking stipule-scars, the blades stiffly leathery, 8–11.5 × 3–5.5 cm, elliptic to oblong-lanceolate, dark, glossy green above. Flowers saucer-shaped, the 9–12 perianth-segments fleshy, creamy white, the outer 3 flushed purplish. *SW China (Yunnan, SE Xizang), Burma*. H5. Spring.

4. M. delavayi Franchet. Illustration: Botanical Magazine, 8282 (1909); Treseder & Blamey, The book of Magnolias, 27 (1981).

Large evergreen shrub or tree to *c.* 10 m, the branchlets with continuous pith. Leaf-stalks 3.5–6.7 cm, with conspicuous, long stipule-scars, blades thickly leathery, 13–30 × 5–16 cm, elliptic to ovate or oblong-ovate, dark, dull green above, pale green beneath. Flowers cup-shaped, the 3 outer perianth-segments greenish white, reflexed, the inner 6 creamy white. Fruiting heads 6.5–9.5 cm. *W China (Yunnan, Sichuan)*. H4. Late spring–early summer.

5. M. virginiana Linnaeus (*M. glauca* Linnaeus). Illustration: Botanical Magazine, n.s., 457 (1964); Treseder & Blamey, The book of Magnolias, 31 (1981).

Deciduous, semi-evergreen or evergreen tree to 30 m or large, multiple-trunked shrub of open habit, the branchlets with septate pith. Leaf-stalks 1.5–2 cm, with conspicuous, long stipule-scars, blades 6.5–10.5 × 3.5–4.5 cm, narrowly oblong to elliptic or ovate, shining green above, silvery glaucous or rarely green beneath. Flowers creamy white, spherical to cup-shaped, perianth-segments 6–15 or sometimes more. Fruiting heads ovoid to almost spherical, 3–4.5 cm. *E USA*. H3. Late spring.

6. M. × thompsoniana (Loudon) de Vos (*M. glauca* var. *major* Sims; *M. glauca* var. *thompsoniana* Loudon). Illustration: Botanical Magazine, 2164 (1820); Treseder & Blamey, The book of Magnolias, 69 (1981).

Deciduous or rarely semi-evergreen shrub or small tree, like *M. virginiana* in habit, the branchlets with incompletely septate pith. Leaf-stalks 1.8–2.5 cm, with basal stipule-scars, the leaves alternate or in loose, false whorls, blades 12–17.5 × 5.5–7.5 cm, elliptic to oblanceolate, glaucous and finely downy beneath. Flowers vase-shaped, the 9

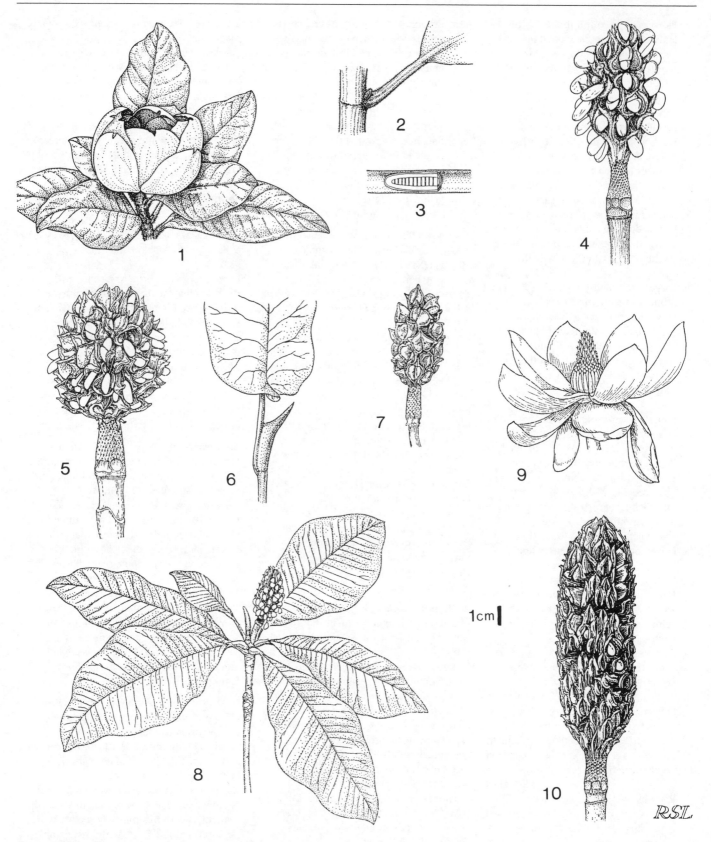

Figure 30. Diagnostic details of Magnoliaceae. 1–4, *Magnolia grandiflora*: 1, flower; 2, leaf-base; 3, septate pith; 4, fruiting head. 5, *M. macrophylla*, fruiting head. 6,7, *M. ashei*: 6, leaf-base; 7, fruiting head. 8, *M.tripetala*, habit and fruiting head. 9, 10, *M. hypoleuca*: 9, flower, 10, fruiting head.

perianth-segments greenish white. *Garden origin.* H3. Late spring.

A chance hybrid between *M. virginiana* and *M. tripetala.*

7. M. macrophylla Michaux. Figure 30(5), p. 306. Illustration: Botanical Magazine, 2189 (1820); Journal of the Arnold Arboretum **57**: 267 (1976).
Deciduous tree to 15 or rarely to 20 m, usually of upright habit, with smooth, light grey bark, the stout branchlets at first green and thickly hairy with silvery hairs, becoming hairless and reddish brown. Leaves in false whorls, the stalks 5.5–13 cm, with long stipule-scars, the blades 24–60 × 11–32 cm, oblanceolate, with cordate to auriculate bases, chalky white to silvery glaucous beneath. Flowers large, cup-shaped, the 3 outer perianth-segments greenish white, the inner 6 creamy white, often purplish-spotted near the base inside. Stamens 1.5–1.8 cm. Fruiting heads 5.5–8 cm, almost spherical, the lowermost follicle dehiscing downwards. *SE USA.* H3. Late spring.

8. M. ashei Weatherby (*M. macrophylla* subsp. *ashei* (Weatherby) Spongberg). Figure 30(6, 7), p. 306. Illustration: Journal of the Arnold Arboretum **57**: 267 (1976).
Widely spreading shrub to 10 m, similar to *M. macrophylla* but the branchlets initially strongly glaucous and finely downy with silvery hairs. Leaves in false whorls, the stalks with long stipule-scars, the blades oblanceolate with cordate to auriculate bases. Flowers large, cup-shaped, the 9 perianth-segments creamy white, sometimes purplish-spotted near the base inside. Stamens 1.1–1.3 cm. Fruiting heads 5–7 cm, more or less cylindric to ovoid, the lowermost follicle dehiscing outwards. *S USA (Florida).* H4. Late spring.

9. M. fraseri Walter. Illustration: Botanical Magazine, 1206 (1809).
Upright deciduous tree to *c.* 14 m with smooth bark. Leaves in false whorls, stalks 3–6 cm, with long stipule-scars, blades 14.5–26.5 × 8.5–18 cm, oblanceolate with cordate to auriculate bases, becoming glaucous beneath. Flowers vase-shaped, becoming saucer-shaped, the 9 perianth-segments yellowish or creamy white. Stamens 8–15 mm. Ovary 2–2.8 cm. Fruiting heads 6.5–11 cm, ellipsoid. *SE USA.* H3. Spring.

10. M. pyramidata Bartram (*M. fraseri* var. *pyramidata* (Bartram) Pampanini).
Small, deciduous tree to 16 m, of pyramidal habit, closely resembling *M. fraseri.* Leaves in false whorls, the stalks with long stipule-scars, the blades fiddle-or diamond-shaped in outline with auriculate bases. Flowers vase-shaped, the 9 perianth-segments creamy white. Stamens 4–6 mm. Ovary *c.* 1.5 cm. Fruiting heads 3.5–6 cm. *SE USA.* H4. Late spring.

11. M. tripetala Linnaeus. Figure 30(8), p. 306. Illustration: Treseder & Blamey, The book of Magnolias, 63 (1981).
Deciduous tree to 12 m, sometimes shrub-like, with smooth, greyish bark. Leaves in false whorls, the stalks 2–3.5 cm, with long stipule-scars, the blades 22–38 × 10–23 cm, obovate-lanceolate with wedge-shaped bases, pale green and finely downy beneath. Flowers of unpleasant odour, vase-shaped, erect, the 9 or 12 perianth-segments light greenish white. Fruiting heads 7–10 cm, ovoid-cylindric. *E USA.* H3. Late spring.

Hybrids between *M. tripetala* and *M. sieboldii* are represented in cultivation by 'Charles Coats', a clone with a *tripetala*-like habit and *sieboldii*-like flowers borne upright on the branches.

12. M. hypoleuca Siebold & Zuccarini (*M. obovata* Thunberg). Figure 30(9, 10), p. 306. Illustration: Botanical Magazine, 8077 (1906); Kurata, Illustrated important forest trees of Japan **1**: pl. 65 (1971); Newsletter of the American Magnolia Society **14**(1): 4 (1978); Treseder & Blamey, The book of Magnolias, 19 (1981).
Widely branching deciduous tree to 30 m, with silvery grey bark. Leaves in false whorls, stalks 2.5–3.5 cm, with prominent stipule-scars, blades 16–45 × 9–20 cm, oblong-obovate, with wedge-shaped bases, silvery or greyish green and finely downy beneath. Flowers cup- or saucer-shaped, the 9–12 perianth-segments creamy white. Fruiting heads 13.5–20 cm, oblong-cylindric. *Japan, Kurile & Ryukyu islands.* H3. Late spring.

13. M. officinalis Rehder & Wilson. Illustration: Newsletter of the American Magnolia Society **14**(1): 4, 6 (1978); Treseder & Blamey, The book of Magnolias, 55 (1981).
Deciduous tree to 20 m, with ash-grey bark. Leaves in false whorls, stalks 2.5–3.8 cm, with prominent stipule-scars, blades 22–40 × 11–20 cm, oblong-obovate with rounded or generally wedge-shaped bases, finely downy and distinctly glaucous beneath. Flowers cup- or saucer-shaped, the 9–12 perianth-segments pale greenish

white. Fruiting heads 9–15 cm, oblong-ovoid, the lowermost follicle convex. *Known only in cultivation; perhaps from C China.* H3. Late spring.

Var. **biloba** Rehder & Wilson has reddish-downy bud scales and some or all of the leaves with deeply bilobed apices. *Known only in cultivation; perhaps from China.*

Perhaps a distinct species.

14. M. rostrata W.W. Smith. Illustration: Johnstone, Asiatic Magnolias in cultivation, pl. 13, f. 19 (1955).
Deciduous tree to 24 m with smooth, grey bark. Leaves in false whorls, the stalks 3.5–7.5 cm with long stipule-scars, the blades 19–50 × 12.5–22.5 cm, obovate, glaucous and finely reddish-downy beneath. Flowers cup-shaped, the 9–11 perianth-segments white. Fruiting heads *c.* 14 cm, oblong, with long-beaked follicles. *SW China (Yunnan, SE Xizang), adjacent Burma.* H5. Late spring.

15. M. globosa Hooker & Thomson. Figure 31(1), p. 308. Illustration: Botanical Magazine, 9467 (1936); Treseder & Blamey, The book of Magnolias, 45 (1981).
Widely spreading deciduous shrub to *c.* 7 m, the branchlets initially covered with reddish hairs. Leaf-stalks with long stipule-scars, blades 10–21 × 7–9 cm, elliptic to obovate with rounded or almost cordate bases, glaucous and downy beneath, initially with dense reddish, yellowish or silvery hairs. Flowers cup-shaped or almost spherical, pendent, the 9–12 perianth-segments creamy white. Fruiting heads 6–8 cm, cylindric. *Himalaya, from Nepal to N Burma, China (Yunnan).* H4. Early summer.

16. M. wilsonii (Finet & Gagnepain) Rehder (*M. × highdownensis* Dandy). Figure 31(2, 3), p. 308. Illustration: Botanical Magazine, 9004 (1924); Treseder & Blamey, The book of Magnolias, 53 (1981).
Slender deciduous shrub or small tree to *c.* 8 m, the branchlets initially downy with reddish hairs, eventually hairless and grayish purple. Leaf-stalks with long stipule-scars, blades 6.5–15.5 × 3–7 cm, elliptic or more usually lanceolate, downy beneath with yellowish and silvery hairs. Flowers cup- to saucer-shaped, pendent, the 9–12 perianth-segments pure white, the stamens rosy purple. Fruiting heads 6–10 cm, cylindric. *W China.* H3. Early summer.

17. M. sieboldii K. Koch (*M. parviflora* Siebold & Zuccarini). Figure 31(4–7), p. 308. Illustration: Botanical Magazine, 7411 (1895); Hay & Synge, The dictionary

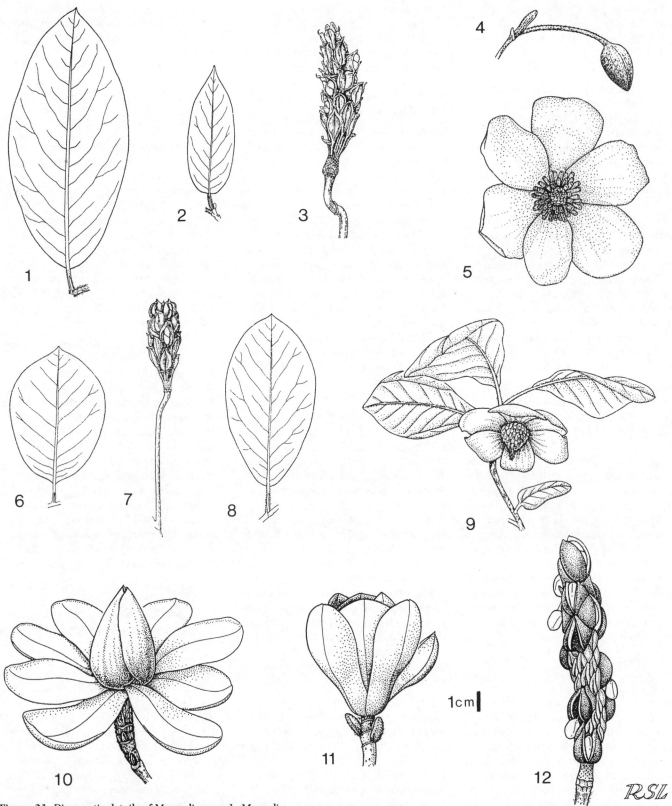

Figure 31. Diagnostic details of Magnoliaceae. 1, *Magnolia
globosa*, leaf. 2, 3, *M. wilsonii*: 2, leaf; 3, fruiting head.
4–7, *M. sieboldii*: 4, bud; 5, flower; 6, leaf; 7, fruiting head.
8, 9, *M. sieboldii* var. *sinensis*: 8, leaf; 9, habit and flower.
10, *M. campbellii*, habit and flower. 11, 12, *M. denudata*:
11, flower, 12, fruiting head.

of garden plants in colour, 212 (1969);
Treseder & Blamey, The book of Magnolias,
39 (1981).
Small deciduous tree to 10 m or often a
large shrub, the branchlets initially downy,
becoming silvery tan. Leaf-stalks
1.3–4.5 cm, with long stipule-scars, blades
9–12 × 5.5–10 cm, usually obovate and
generally downy beneath. Flowers cup- to
saucer-shaped, pendent, the 9–12 perianth-
segments pure white, the stamens reddish
purple. Fruiting heads 2–7 cm, cylindric.
E Asia. H2. Early summer.

Subsp. **sinensis** (Rehder & Wilson)
Spongberg (*M. sinensis* Rehder & Wilson).
Figure 31(8, 9), p. 308. Illustration:
Journal of the Arnold Arboretum 57: 276
(1976); Treseder & Blamey, The book of
Magnolias, 23 (1981). Young branches
with yellowish or reddish hairs; leaf-stalks
to 6.4 cm, with stipule-scars for more than
half their length, leaf-blades to 21.5 ×
15.9 cm, sometimes almost circular with
crinkled, clear, adpressed hairs beneath.
W China (Sichuan). H3. Early summer.

18. M. × wieseneri Carrière (*M. × watsonii*
J.D. Hooker). Illustration: Botanical
Magazine, 7157 (1891); Arnoldia 36: 137
(1976); Treseder & Blamey, The book of
Magnolias, 69 (1981).
Small, deciduous, bushy tree or a large
shrub to 7 m. Leaves alternate or in loose,
false whorls, stalks 2.4–3 cm, with
prominent stipule-scars, blades 13.8–19 ×
7.4–10.5 cm, broadly elliptic to obovate,
glaucous and finely downy beneath.
Flowers erect, cup-shaped, the 9 perianth-
segments white to cream, stamens rosy
crimson. *Garden origin*. H3. Late spring.

Thought to be a hybrid between *M.
hypoleuca* and *M. sieboldii* which originated
in Japan.

19. M. campbellii Hooker & Thomson.
Figure 31(10), p. 308. Illustration:
Botanical Magazine, 6793 (1885);
Johnstone, Asiatic Magnolias in cultivation,
pl. 2, 3 (1955); Hay & Synge, Dictionary of
garden plants in colour, 210 (1969);
Treseder & Blamey, The book of Magnolias,
15 (1981).
Ultimately large tree to 35 m or large
shrub, bark greyish tan. Leaf-stalks 1–5 cm
with small, basal stipule-scars, blades
leathery, variable, 11.5–25 × 4.6–14.5 cm,
elliptic to oblong-oblanceolate, with acute
to shortly acuminate apices and wedge-
shaped to rounded, often unequal bases,
hairless to finely adpressed-silky-hairy
beneath. Flowers large, before the leaves,
the 12–16 perianth-segments pure white to

pink or crimson, the outer whorls held
horizontally to reflexed and drooping, the
innermost whorl held erect, enclosing the
stamens and ovary. *Himalaya (Nepal) to
China (Yunnan)*. H4. Early spring.

Subsp. **mollicomata** (W.W. Smith)
Johnstone. Illustration: Johnstone, Asiatic
Magnolias in cultivation, pl. 4, 5 (1955);
Treseder & Blamey, The book of Magnolias,
33, 47 (1981). Differing in its densely
yellow-hairy flower-stalks, branchlets and
its long buds. *SW China (Yunnan), adjacent
Burma*. H4. Early spring.

Plants of this subspecies tend to be
hardier in gardens and begin flowering at
an early age.

20. M. × veitchii Bean. Illustration:
Treseder & Blamey, The book of Magnolias,
69 (1981).
Vigorous deciduous tree to 30 m, when in
flower similar in appearance to *M. ×
soulangiana*. Leaf-stalks 2.5–3 cm, with
basal stipule-scars, blades 14–22 ×
8–11.5 cm, obovate to oblong, with a fine
network of veins beneath. Flowers before
the leaves, vase-shaped, the 7–10 perianth-
segments white flushed pink to purple.
Garden origin. H4. Early spring.

Hybrids between *M. campbellii* and *M.
denudata*, first obtained by P.C.M. Veitch in
1907; 'Peter Veitch' is the most widely
grown selection.

21. M. denudata Desrousseaux (*M.
conspicua* Salisbury; *M. heptapeta* (Buc'hoz)
Dandy; *M. yulan* Desfontaines). Figure
31(11,12), p. 308. Illustration: Botanical
Magazine, 1621 (1814); Johnstone, Asiatic
Magnolias in cultivation, f. 1, 2 (1955);
Hay & Synge, Dictionary of garden plants
in colour, 211 (1969); Treseder & Blamey,
The book of Magnolias, 29 (1981).
Twiggy, deciduous tree to *c.* 15 m. Leaf-
stalks 1.3–2 cm, with small, basal stipule-
scars, blades 5.5–13.5 × 4.6–8.5 cm,
usually obovate to oblong-obovate, with
abruptly short-acute apices and wedge-
shaped bases. Flowers lemon-scented,
before the leaves, more or less vase-shaped,
the 9 (or 12) perianth-segments pure white
to ivory. *E China*. H3. Spring.

22. M. × soulangiana Soulange-Bodin.
Illustration: Hay & Synge, Dictionary of
garden plants in colour, 213 (1969);
Treseder & Blamey, The book of Magnolias,
51, 69 (1981).
Spreading, deciduous tree of twiggy habit,
to *c.* 15 m. Leaf-stalks 7–30 mm, with
basal stipule-scars, blades 7.5–16.5 ×
3–12.5 cm, broadly elliptic to oblanceolate

or broadly obovate, usually abruptly short-
acuminate at the apices, often shortly
downy beneath. Flowers before the leaves,
variable, the 9 perianth-segments white to
dark purple, often somewhat 2-coloured.
Garden origin. H3. Spring.

A variable group of hybrids between
M. denudata and *M. liliflora*, first created in
France in 1820. Perhaps the most widely
and frequently cultivated Magnolias
represent selections from among these
hybrids and multitudes of cultivars have
been selected and named.

23. M. sprengeri Pampanini. Figure 32 (1),
p. 310. Illustration: Botanical Magazine,
9116 (1926); Johnstone, Asiatic Magnolias
in cultivation, pl. 9 (1955); Treseder &
Blamey, The book of Magnolias, 61 (1981).
Deciduous tree to 20 m, often of slender,
pyramidal habit, in old plants the smooth
bark flaking. Leaf-stalks *c.* 1.5 cm with
small, basal stipule-scars, blades 7–15 ×
3.8–10 cm, generally less than twice as
long as broad, obovate with abruptly short-
acuminate apices and wedge-shaped bases.
Flowers before the leaves, saucer-shaped,
the 12–14 perianth-segments all more or
less similar, rosy red to pale pink. Fruiting
heads 6–13 cm, cylindric. *China*. H4. Early
spring.

Best-known in cultivation is 'Diva' (*M.
diva* Millais; *M. sprengeri* var. *diva* (Millais)
Johnstone).

Var. **elongata** (Rehder & Wilson)
Johnstone. Figure 32(2), p. 310.
Illustration: Johnstone, Asiatic Magnolias
in cultivation, pl. 10, f. 13 (1955). Large
shrub or small tree with the leaves
10.7–14.5 × 4–5.8 cm, oblong-
oblanceolate, usually more than twice as
long as broad. Flowers slightly smaller than
var. *sprengeri*; perianth-segments white,
sometimes tinged pink. *China*. H4.

24. M. sargentiana Rehder & Wilson.
Figure 32(3), p. 310. Illustration:
Johnstone, Asiatic Magnolias in cultivation,
pl. 7 (1955).
Upright, deciduous tree to 25 m, with tan-
grey bark. Leaf-stalks 1.5–4.5 cm, with
small, basal stipule-scars, blades somewhat
leathery, 11–18 × 5–10.5 cm, obovate to
oblanceolate or broadly elliptic, sometimes
notched at apex. Flowers before the leaves,
pendent, the 12–14 perianth-segments all
more or less similar, white to pale pink or
purplish pink. Fruiting heads 9–11 cm,
cylindric. *SW China*. H4. Early spring.

Var. **robusta** Rehder & Wilson.
Illustration: Johnstone, Asiatic Magnolias
in cultivation, frontispiece, pl. 8 & f. 12

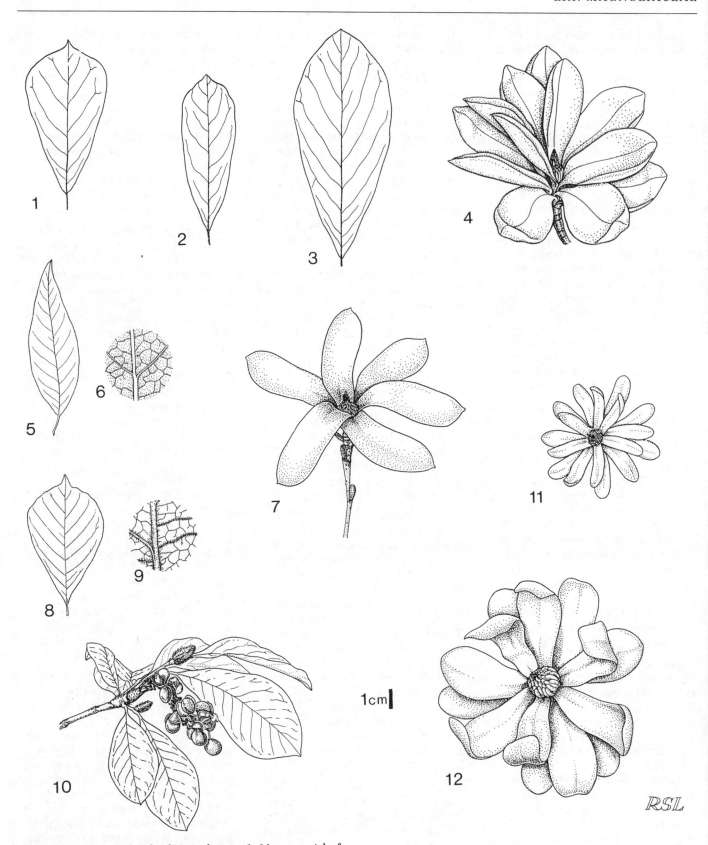

Figure 32. Diagnostic details of Magnoliaceae. 1, *M. sprengeri*. leaf.
2, *M. sprengeri* var. *elongata*, leaf. 3, *M. sargentiana*, leaf.
4, *M. dawsoniana*, flower. 5, 6, *M. salicifolia*: 5, leaf; 6, leaf surface.
7–10, *M. kobus*: 7, habit and flower; 8, leaf; 9, leaf surface;
10, habit and fruiting head. 11, *M. stellata*, flower. 12, *M.* × *loebneri*
'Merrill', flower.

(1955); Treseder & Blamey, The book of Magnolias, 25 (1981). Spreading bushy tree, usually branched from near the base, the oblanceolate leaves 14.5–20.5 × 5.9–8.7 cm, more than twice as long as broad, generally notched at apex. Flowers larger than in var. *sargentiana*, with 12–16 perianth-segments. Fruiting heads 15–19.5 cm, cylindric. *W China*. H5. Early spring.

25. M. dawsoniana Rehder & Wilson. Figure 32(4), p. 310. Illustration: Botanical Magazine, 9678, 9679 (1948); Johnstone, Asiatic Magnolias in cultivation, pl. 6, f. 9 (1955).
Deciduous tree to 12 m or usually a large, twiggy shrub. Leaf-stalks 1.3–2.9 cm with small, basal stipule-scars, blades 7.5–13.5 × 5–7.8 cm, broadly elliptic or usually obovate. Flowers before the leaves, the 9–12 perianth-segments all more or less similar, white to pale pink. Fruiting heads 7–10 cm, cylindric. *W China*. H4. Early spring.

26. M. cylindrica Wilson. Illustration: Treseder & Blamey, The book of Magnolias, 43 (1981).
Small deciduous tree to 9 m, with silvery grey or tan bark. Leaf-stalks 1.2–2 cm, with small, basal stipule-scars, blades 8–13.5 × 1.6–4.6 cm, elliptic to oblong-elliptic. Flowers before the leaves, vase-shaped, erect, the outer 3 perianth-segments sepal-like, the inner 6 petal-like, white, sometimes tinged pink. Fruiting heads 5–7.5 cm, cylindric. *E China*. H3. Early spring.

27. M. salicifolia (Siebold & Zuccarini) Maximowicz (*Talauma salicifolia* Siebold & Zuccarini var. *concolor* Miquel; *M. × proctoriana* Rehder; *M. × slavinii* Harkness; *M. × kewensis* Pearce). Figure 32(5, 6), p. 310. Illustration: Botanical Magazine, 8483 (1913); Hay & Synge, Dictionary of garden plants in colour, 211 (1969); Kurata, Illustrated important forest trees of Japan 3: pl. 9 (1971).
Large deciduous shrub or small, often pyramidal tree to 12 m, the twigs usually with a lemon- or anise-like odour when bruised. Leaf-stalks 1.2–2 cm, with small, basal stipule-scars, blades 6.5–12 × 2–5.5 cm, usually lanceolate or oblong-lanceolate, very finely and inconspicuously downy and often glaucous beneath. Flowers before the leaves, the outer 3 perianth-segments sepal-like, the inner 6 (to 12) petal-like, white. Fruiting heads 4.5–6 cm, cylindric. *Japan*. H3. Early spring.

28. M. kobus de Candolle (*M. kobus* var. *borealis* Sargent). Figure 32(7–10), p. 310. Illustration: Botanical Magazine, 8428 (1912); Kurata, Illustrated important forest trees of Japan 1: pl. 64 (1971).
Deciduous shrub or usually tree to 20 m. Leaf-stalks 1–2.5 cm, with basal stipule-scars, the blades 4.5–14.5 × 4–8.3 cm, usually obovate to oblanceolate, hairless or with long hairs scattered along the veins. Flowers before the leaves, the outer 3 perianth-segments sepal-like, the inner 6 petal-like, white, usually pinkish at base. Fruiting heads 3.5–11 cm, cylindric. *Japan*. H2. Early spring.
According to Ueda (reference on p. 303) the correct name for this species is *M. praecocissima* Koidzumi.

29. M. stellata (Siebold & Zuccarini) Maximowicz (*M. kobus* var. *stellata* (Siebold & Zuccarini) Blackburn. Figure 32(11), p. 310. Illustration: Botanical Magazine, 6370 (1878); Hay & Synge, Dictionary of garden plants in colour, 212 (1969); Kurata, Illustrated important forest trees of Japan 3: pl. 10 (1971).
Low, twiggy shrub or small tree to 3 m. Leaf-stalks with basal stipule-scars, the blades 6–13.2 × 2.5–6.5 cm, elliptic to obovate or oblanceolate. Flowers before the leaves, the outer 3 perianth-segments sepal-like, falling early, the 12–33 inner segments petal-like, strap-shaped, 4–6.5 cm × 6–14 mm, white to pink. *Japan*. H3. Early spring.
According to Ueda the correct name for this species is *M. tomentosa* Thunberg.

30. M. × loebneri Kache (*M. kobus* var. *loebneri* (Kache) Spongberg). Illustration: Treseder & Blamey, The book of Magnolias, 17 (1981).
Deciduous shrub or usually a small tree to 3.5 m. Leaf-stalks with basal stipule-scars, blades 6.5–14.3 × 2.3–6 cm, oblanceolate to obovate or broadly elliptic. Flowers before the leaves, the outer 3 perianth-segments sepal-like, the 11–16 inner segments petal-like, narrowly to broadly spathulate, white to pink. *Garden origin*. H2. Early spring.
Hybrids between *M. kobus* and *M. stellata*, first obtained in Germany. This cross has been repeated often and several popular cultivars have been selected and named. 'Leonard Messel' has pink perianth-segments, and 'Merrill' (Figure 32(12), p. 310; illustration: Hay & Synge, Dictionary of garden plants in colour, 211, 1969) has white, broadly strap-shaped

perianth-segments; these are the most widely cultivated.

31. M. liliiflora Desrousseaux (*M. quinquepeta* (Buc'hoz) Dandy; *M. discolor* Ventenat; *M. purpurea* Curtis). Illustration: Botanical Magazine, 390 (1797); Johnstone, Asiatic Magnolias in cultivation, pl. 11 (1955); Treseder & Blamey, The book of Magnolias, 21 (1981).
Deciduous shrub to *c.* 3 m. Leaf-stalks 1–2 cm, with long, downy stipule-scars, blades 13–20 × 6.5–10 cm, broadly elliptic to oblanceolate. Flowers sometimes appearing before the leaves, continuing after the leaves have expanded, vase-shaped, the outer 3 perianth-segments sepal-like, falling early, the 6 inner segments petal-like, whitish to deep purple, often 2-coloured. Fruiting heads 3.5–5.5 cm, cylindric to almost spherical. *China*. H3. Late spring.
Hybrids between this and *M. acuminata* have been given the name **M. × brooklynensis** Kalmbacher and are represented by 'Evamaria' and 'Woodsman'.

32. M. acuminata (Linnaeus) Linnaeus. Illustration: Botanical Magazine, 2427 (1823); Treseder & Blamey, The book of Magnolias, 49 (1981).
Large, deciduous tree to 30 m, bark becoming scaly in age. Leaf-stalks 2.5–3.5 cm, with basal stipule-scars, blades 10–24 cm, elliptic to broadly ovate or oblong-ovate, usually somewhat glaucous beneath. Flowers appearing with the leaves, cup-shaped, the outer 3 perianth-segments sepal-like, the inner 6 petal-like, green or greenish yellow, sometimes purplish-tinged and glaucous. Fruiting heads 3.5–7 cm, ovoid to oblong-cylindric. *Eastern N America*. H2. Late spring.
Var. **subcordata** (Spach) Dandy (*M. cordata* Michaux). Tree or sometimes large shrub with the branchlets densely and closely silky with silvery hairs. Leaves densely hairy beneath. Flowers borne on densely velvety (rarely hairless) stalks, perianth-segments yellowish green to clear yellow. *SE USA*. H3. Late spring.

2. MANGLIETIA Blume
S.A. Spongberg
Evergreen trees or large shrubs, very similar to *Magnolia*. Flowers terminal, carpels with 4 or more ovules.
About 25 species from tropical and subtropical Asia. Keng (reference below) has suggested that *Manglietia* be incorporated into *Magnolia*.

Literature: Treseder, N.G., *Magnolias* (1978); Keng, H., The delimitation of the genus Magnolia, *The Garden's Bulletin Singapore* **31**: 127–131 (1978).

1. M. insignis (Wallich) Blume (*M. hookeri* misapplied; *M. forrestii* misapplied). Illustration: Botanical Magazine, n.s., 443 (1964); Treseder & Blamey, The book of Magnolias, 65 (1981).
Tree to 12 m or usually a large, erect and compact, multiple-stemmed shrub to 9 m. Leaf-stalks 1.3–2.5 cm, blades leathery, oblong-elliptic, 10–20 × 4–6.5 cm, somewhat bluish beneath. Flowers with 12 perianth-segments, all more or less similar, ranging in colour from white to yellowish to rose-pink. *C Himalaya to W China and N Vietnam.* H5. Spring.

3. MICHELIA Linnaeus
S.A. Spongberg
Evergreen trees or shrubs similar to *Magnolia*. Leaves stalkless to distinctly stalked, stipules free or attached to the leaf-stalk, falling early, when detached leaving obscure to conspicuous scars. Flowers usually fragrant, axillary, the 6–21 perianth-segments all more or less similar. Stamens numerous, filaments hardly differentiated. Ovary conspicuously stalked, of numerous carpels each with 2 or more ovules. Fruits in cone-like heads of free, loose follicles.

About 45 species from tropical and warm-temperate climates of SE Asia, India and Sri Lanka.

1a. Leaves stalkless or very shortly stalked **1. figo**
  b. Leaves distinctly stalked   2
2a. Leaf-stalks with long stipule-scars extending almost their entire length **2. champaca**
  b. Leaf-stalks with small, basal stipule-scars or scars absent   3
3a. Leaf-blades 5–10 cm   **3. compressa**
  b. Leaf-blades 10–18 cm   **4. doltsopa**

1. M. figo (Loureiro) Sprengel (*M. fuscata* (Andrews) Wallich). Illustration: Botanical Magazine, 1008 (1807).
Bushy shrub to *c.* 6 m, the new growth and buds coppery-downy. Leaves stalkless or stalks at most to 4 mm, blade 3–8 × 2–4 cm, elliptic to oblanceolate, wedge-shaped at base. Flowers *c.* 3 cm across, with the 6 more or less similar perianth-segments creamy white or yellowish, often tinged purplish brown. *SE China.* H5–G1. Spring.

2. M. champaca Linnaeus.
Small tree to *c.* 10 m. Leaf-stalks 1–2.6 cm, finely downy, with long stipule-scars, blades 10–15.5 × 3.5–8.5 cm, ovate to lanceolate, with acute to acuminate apices. Flowers *c.* 4 cm across, the 12 perianth-segments creamy white to pale yellow. *S China.* H5–G1. Spring.

3. M. compressa (Maximowicz) Sargent. Illustration: Kurata, Illustrated important forest trees of Japan 3: pl. 11 (1971).
Erect tree to 20 m, the young branchlets and buds densely brown-downy. Leaf-stalks 2–3 cm, lacking stipule-scars, blades leathery, 5–10 × 2–4 cm, obovate-oblong to oblanceolate, glaucous beneath. Flowers 3–4 cm across, the 12 perianth-segments white, purplish towards the base. *Japan, Ryukyu islands.* H5–G1. Spring.

4. M. doltsopa de Candolle. Illustration: Botanical Magazine, 9645 (1943); Treseder & Blamey, The book of Magnolias, 67 (1981).
Large shrub or tree to 15 m. Leaf-stalks 1.5–3 cm, lacking stipule-scars, blade leathery, dark green but generally without lustre above, bluish beneath, 10–18 × 3.5–8 cm, elliptic to oblong. Flowers usually 7.5–10 cm across, occasionally smaller, variable, freely produced, the 12–16 perianth-segments pale yellow to white. *W China & E Himalaya.* H5–G1. Spring.

4. LIRIODENDRON Linnaeus
S.A. Spongberg
Deciduous trees with the bark becoming fissured into longitudinal plates; young branchlets with septate pith; buds enclosed by fused stipules. Leaves alternate, long-stalked, the large, free stipules early deciduous; blades with truncate or widely notched apices and 1 or 2 lateral lobes on each side towards the base. Flowers odourless, appearing with the leaves, terminal. Perianth of 9 segments, the outer 3 reflexed and sepal-like, the inner 6 petal-like, in 2 whorls, simulating a tulip or cup-shaped corolla. Stamens numerous. Carpels many, each with 2 ovules, densely overlapping and spirally arranged. Fruits in a spindle-shaped, cone-like head of free, tightly overlapping, 2-seeded samaras which fall at maturity, the axis persistent.

Two species, 1 from eastern N America, the second from China and northern Indo-China. Hybrids between the 2 species have been produced experimentally and may prove to be of horticultural significance in the future.

Literature: Santamour, F.S., Interspecific hybrids in Liriodendron and their chemical verification, *Forest Science* **18**: 233–236 (1972); Parks, C.R., Miller, N.G., Wendel, J.F. & McDougal, K.M., Genetic divergence within the genus Liriodendron, *Annals of the Missouri Botanical Garden* **70**: 658–666 (1983).

1a. Inner perianth-segments 4–6 cm, each with an irregular orange band near the base on both surfaces; ovary included within the perianth **1. tulipifera**
  b. Inner perianth-segments 2–4 cm, green with yellow veins on the outer surface; ovary protruding from the perianth **2. chinensis**

1. L. tulipifera Linnaeus. Figure 33(1–5), p. 313. Illustration: Botanical Magazine, 275 (1794); Mitchell, A field guide to the trees of Britain and northern Europe, pl. 24 (1974); Journal of the Arnold Arboretum **57**: 310 (1976).
Tree to 50 (rarely to 60) m, the bark becoming deeply furrowed. Leaf-stalks 5–10 cm, the large stipules to 3.5 cm; blades variable in outline, 7–12 cm, usually saddle-shaped with 2 pairs of lateral lobes near the base and 2 lobes creating a truncate or shallowly notched apex, surfaces pale green, the lower becoming faintly glaucous. Flowers tulip-shaped, the 3 outer perianth-segments sepal-like, greenish, strongly reflexed, the inner 6 erect, 4–6 × 1.8–3 cm, pale green each with an orange band near the base on both surfaces. Fruiting heads to 7 cm, the samaras to 4.5 cm, with acute apices, the lowermost samaras often persistent on the axis through the winter. *Eastern N America.* H2. Late spring.

Numerous variants, most based either on habit or the shape and coloration of the leaves, have been named.

2. L. chinensis (Hemsley) Sargent (*L. tulipifera* var. *chinensis* Hemsley). Figure 33(6, 7), p. 313. Illustration: Journal of the Arnold Arboretum **57**: 310 (1976).
Upright tree to *c.* 60 m, similar to *L. tulipifera* but not as tall. Leaf-stalks 7–18 cm, the stipules to 3.5 cm. Leaf-blades variable in outline, usually saddle-shaped, with 1 pair of lobes near the base and another towards the apex, creating a truncate to shallowly notched apex, the lobes usually drawn out and apiculate; lower surface becoming glaucous. Flowers cup-shaped, the 3 outer perianth-segments sepal-like, greenish, reflexed, the 6 inner

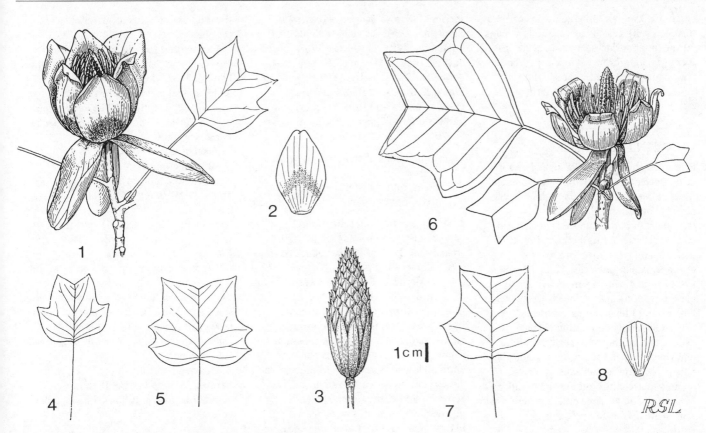

Figure 33. Diagnostic details of Magnoliaceae. 1–5, *Liriodendron tulipiferum*: 1, habit and flower; 2, petal; 3, fruit; 4 & 5, leaves. 6, 7, *L. chinense*: 6, habit and flower; 7, leaf.

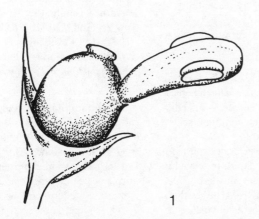

Figure 34. Floral units of Chloranthaceae. 1, *Sarcandra glabra*. 2, *Chloranthus spicatus*.

erect, 3–4 cm, becoming glaucous green on the outer surfaces, green with yellow veins on the inner surfaces. Fruiting heads to 9 cm, the light brown samaras to 4 cm, with acute or rounded apices. *China & northern Indo-China*. H4. Late spring.

LX. WINTERACEAE

Evergreen trees or shrubs, some dioecious or monoecious. Leaves aromatic, simple, entire, alternate, without stipules. Flowers bisexual or unisexual, in axillary or terminal compound inflorescences, floral parts indefinite in number; sepals 2–6; petals 2–many; stamens numerous, anthers opening inwards. Carpels 1–many, free or united, with 1–many ovules; style absent. Fruit a capsule, follicle or berry. Embryo minute, endosperm absent.

A small family containing about 6 genera and perhaps 80 species, confined almost entirely to the southern hemisphere, but there is considerable disagreement about generic limits and species numbers.

In European gardens only 2 genera are represented. They are propagated by cuttings and seed.

1a. Leaves entirely green, or with red stalks; inflorescences terminal; calyx deciduous; stamens with terete filaments **1. Drimys**
 b. Leaves coloured, not entirely green; inflorescences axillary; calyx not deciduous; stamens with flattened filaments **2. Pseudowintera**

1. DRIMYS Forster & Forster
E.C. Nelson

Trees or shrubs, sometimes dioecious. Leaves hairless. Flowers bisexual or unisexual, terminal, solitary or in umbels. Calyx enclosing the bud, splitting into 2–3 lobes, deciduous. Petals few–numerous. Stamens with terete filaments. Fruit a berry.

There are perhaps about 30 species in this southern hemisphere genus but the generic limits are disputed, as is the number of species. Plants cultivated in European gardens are generally assigned to 2 species, 1 S American, the other Austral-asian, which some botanists separate into the genus *Tasmannia* R. Brown.

All cultivated *Drimys* have uniformly green leaf-blades; plants with highly coloured foliage belong to the related genus *Pseudowintera*.

Literature: Smith, A.C., The American

species of Drimys, *Journal of the Arnold Arboretum* **24**: 1–33 (1942); Smith, A.C., Taxonomic notes on the Old World species of Winteraceae, *Journal of the Arnold Arboretum* **24**: 119–164 (1942); Vink, W., The Winteraceae of the Old World I, Pseudowintera and Drimys – morphology and taxonomy, *Blumea* **18**: 225–354 (1970).

1a. Stems and leaf-stalks red; dioecious shrubs **1. lanceolata**
 b. Stems and leaf-stalks green; shrubs or trees, not dioecious **2. winteri**

1. D. lanceolata (Poiret) Baillon (*D. aromatica* (R. Brown) Mueller; *Tasmannia aromatica* R. Brown). Illustration: Cochrane et al., Flowers and plants of Victoria, 129 (1968); Morley & Toelken, Flowering plants in Australia, 37 (1983).

Dioecious shrub or small tree to 5 m, young stems and leaf-stalks crimson. Leaves leathery, narrowly lanceolate to elliptic, to 13 × 1–4 cm, apex acute or obtuse. Flowers apparently in umbels, each flower actually solitary, stalk to 1.5 cm. Sepals to 5 mm, deciduous. Petals 2–8, greenish white. Male flowers with 20–25 buff stamens, ovary rudimentary. Fruit a black berry. *Australia (New South Wales, Victoria, Tasmania)*. H4. Summer.

An attractive shrub with fragrant foliage and richly coloured stems and leaf-stalks. The fruits have been used as a substitute for pepper. When this is regarded as a species of *Tasmannia* the correct name is *T. aromatica*.

2. D. winteri Forster & Forster (*D. andina* invalid; *D. aromatica* Murray; *D. latifolia* invalid). Illustration: Botanical Magazine, 4800 (1854); Heywood, Flowering plants of the world, 29 (1978).

Tree to 20 m, or shrub. Leaves leathery, oblong-elliptic to oblanceolate, 2.5–20 × 1–8 cm, apex obtuse, bright green, shiny, hairless, glaucous to almost white on undersurface.
Flowers white, in umbel-like inflorescences (sometimes solitary in var. *andina*). Sepals to 6 × 4–12 mm, red. Petals 1.5–2.5 cm × 5 mm, lanceolate. Fruit a shining black berry. *Chile, Argentina*.

Some botanists place all American plants in this species, but A.C. Smith restricted the name to populations from southern Chile and Argentina (ranging from the southern and western parts of Tierra del Fuego northwards to 30° S). For horticultural purposes this is a sensible arrangement. The species displays substantial variation in

the wild and in gardens; 3 varieties are recognised:

Var. **winteri** (*D. winteri* var. *punctata* (Lamarck) de Candolle). Illustration: Moore, Flora of Tierra del Fuego, 67 (1983). Tree to 17 m, leaves with 7–11 pairs of veins; umbels with less than 10 flowers, petals 5–7. *S Chile, Argentina, including Tierra del Fuego*. H4. Summer.

Perhaps not in cultivation in Europe.

Var. **chilense** (de Candolle) Gray (*D. chilense* de Candolle; *D. winteri* var. *latifolia* Miers). Illustration: Botanical Magazine, n.s., 200 (1953). Shrub or tree 3–15 m; leaves with 7–15 pairs of veins; umbels with 15–40 flowers, petals 6–14. *C Chile (30–44° S)*. H5. Summer.

The most commonly cultivated variant, growing into a tall, multi-stemmed tree.

Var. **andina** Reiche. Shrub, rarely more than 1 m; leaves with 5–7 pairs of veins; flowers in umbels with 2–4 flowers or solitary, petals 4–9. *C Chile, W Argentina*. H4. Summer.

This dwarf shrubby variety is the hardiest variant, but not the most commonly grown. It flowers freely when young.

2. PSEUDOWINTERA Dandy
E.C.Nelson

Trees or shrubs. Leaves hairless, with obvious glands. Flowers in fascicles or solitary, bisexual. Calyx cup-shaped, not deciduous, not covering the petals in bud. Petals 5–6. Stamens to 15, filaments flattened. Fruit a berry with 2–6 seeds.

Pseudowintera (formerly included in *Drimys*) is a genus of 3 species all endemic to New Zealand; only 1 is widely cultivated in Europe, although the identity of the cultivated plants has not been studied, and hybrids may be grown.

Literature: Dandy, J.E., The Winteraceae of New Zealand, *Journal of Botany* **71**: 119–122 (1933); Sampson, F.B., Natural hybridism in Pseudowintera (Winteraceae), *New Zealand Journal of Botany* **18**: 43–51 (1980).

1. P. colorata (Raoul) Dandy (*Drimys colorata* Raoul; *Wintera colorata* (Raoul) van Tieghem). Illustration: Salmon, The native trees of New Zealand, 100, 101 (1980). Leaves green, spotted or margined with red-purple or yellow-orange and red above, cream-grey to blue-grey beneath, elliptic, 2–6 × 1–3 cm. Flowers 2–5 (rarely to 10) in a fascicle. Calyx-lobes shallow or absent. Petals yellow-green. Stamens 5–12. Fruit black. *New Zealand*. H5. Spring.

A remarkable and handsome shrub with most unusual foliage colours, but tender, thriving only in moist, frost-free, sheltered gardens where the soil is lime-free and peaty. The leaves colour best in full sun.

P. axillaris (Forster & Forster) Dandy (*Drimys axillaris* Forster & Forster). Illustration: Salmon, The native trees of New Zealand, 102 (1980). Differs from *P. colorata* in having green leaves and red fruits, but not commonly cultivated in Europe. *New Zealand*. H5. Spring.

LXI. ANNONACEAE

Trees, shrubs and climbers, containing oil passages. Wood aromatic. Leaves alternate, arranged in 2 ranks, entire, without stipules, with pinnately arranged veins, aromatic. Flowers bisexual, usually solitary but sometimes in compound inflorescences, often fragrant. Sepals and petals often indistinguishable from each other, in whorls of 3. Stamens many, rarely few, usually spirally arranged. Carpels many, usually separate, ovules 1 to many, basal or parietal, anatropous. Fruit a berry, or more commonly, an aggregate of berries. Seeds large with hard shiny testas, often with arils.

A family containing 120 genera and about 2100 species, found mainly in the tropics of the Old World but with some representatives in the New World.

1a. Fruit an aggregate of berries   2
  b. Fruit a cluster of distinct carpels   3
2a. Outer petals 6–10 cm   **6. Monodora**
  b. Outer petals not more than 3.5 cm
                     1. Annona
3a. Plant a climber   **3. Artabotrys**
  b. Plant not a climber   4
4a. Petals not more than twice as long as broad   **2. Asimina**
  b. Petals at least 4 times longer than broad   5
5a. Fruits oblong with seeds in 2 rows
                     4. Cananga
  b. Fruits with seeds in 1 row   **5. Xylopia**

1. ANNONA Linnaeus
S.G. Knees
Evergreen or briefly deciduous trees and shrubs, 1–10 m in cultivation. Leaves stalked, elliptic to oblong-elliptic, hairless or hairy, fragrant. Flowers bisexual, usually in small clusters. Sepals 3. Petals 6, the inner 3 often greatly reduced. Fruit formed by the fusion of carpels with the receptacle,

surface smooth, spiny or with warty tubercles.

Leaves. Hairless: **1,2**; with at least some hairs on veins beneath: **3,5**; with dense felted hairs beneath: **4**.
Flowers. With 6 petals of similar size: **1,2**; with 3 outer petals much larger than the inner 3: **3–5**.
Fruits. With spiny surface: **2**; with warty tubercles: **3**; with smooth surface: **1,4**. With net-like veins: **5**.

1a. Leaves with densely felted hairs covering undersurface   **4. cherimola**
  b. Leaves hairless or with few hairs confined to veins beneath   2
2a. Fruit with long curved spines in longitudinal rows   **2. muricata**
  b. Fruit without spines, but sometimes with bulging tubercles   3
3a. Fruit with bulging tubercles
                     3. squamosa
  b. Fruit smooth though sometimes with net-like veining   4
4a. Fruit with net-like veining, reddish green when ripe; petals greenish
                     5. reticulata
  b. Fruit smooth, yellow when ripe; petals cream and crimson   **1. glabra**

1. A. glabra Linnaeus. Illustration: Botanical Magazine, 4226 (1846).
Tree 3–12 m. Young shoots smooth. Leaf-stalks curved, blades 7–15 × 6 cm, oblong-elliptic, acute or shortly acuminate, hairless. Flowers fragrant. Petals 6, the 3 outer 2.5–3 cm long, cream with a crimson spot near the base of inner face, the 3 inner 2–2.5 cm, white without, crimson within. Stigmas sticky, quickly deciduous. Fruit of variable shape, usually spherical or ovoid, 12 × 8 cm, yellow when ripe, pulp pinkish orange, strongly aromatic. Seeds 1.5 × 1 cm, light brown. *Coasts of tropical America and W Africa*. G2. Spring–summer.

2. A. muricata Linnaeus. Illustration: Masefield et al., Oxford book of food plants, 97 (1969); Graf, Exotica International series 4, 1: 133 (1985); Lötschert & Beese, Collins' guide to tropical plants, f. 242 (1983); Graf, Tropica, edn 3, 72 (1986).
Tree 5–8 m. Leaf-stalks 5–10 mm, curved, blades 10–25 × 3.5–8 cm, hairless. Flowers pendent. Petals 6, the 3 outer 3–3.5 × 2–3 cm, 2 mm thick, greenish yellow, the 3 inner slightly shorter and thinner. Fruit 15–20 × 10 cm, obliquely ovoid with large curved spines in longitudinal rows, green when ripe, pulp white. Seeds 1.5 × 1 cm,

brown or black. *C America and the West Indies*. G1. Summer–autumn.

3. A. squamosa Linnaeus. Illustration: Botanical Magazine, 3095 (1831); Heywood, Flowering plants of the world, 30 (1978); Graf, Exotica International series 4, 1: 133 (1985); Graf, Tropica, edn 3, 72 (1986).
Semi-deciduous tree or shrub, 5–8 m. Young shoots downy at first, becoming smooth with maturity. Leaves 7–12 × 3–4 cm, oblong to narrowly elliptic, glaucous, usually hairless. Flowers in clusters of 3–4. Petals 6, the 3 outer 1.5–2.5 cm, narrowly oblong, light green without, yellow or white with a reddish purple spot near base within, the 3 inner very reduced. Fruit 6–10 cm in diameter, spherical with bulging segments, glaucous green when ripe, pulp white. Seeds dark brown or black. *C America & the West Indies*. G1. Summer–autumn.

4. A. cherimola Miller. Illustration: Botanical Magazine, 2011 (1818); Masefield et al., Oxford book of food plants, 97 (1969); Graf, Exotica International series 4, 1: 133 (1985); Graf, Tropica, edn 3, 72 (1986).
Tree or shrub, 3–7 m. Young shoots covered with yellowish hairs. Leaves 10–15 × 5–10 cm, ovate-lanceolate, acuminate, with densely felted hairs beneath. Flowers fragrant. Petals 6, covered with a rusty down, the 3 outer 2–2.5 cm, yellowish white with a purplish spot near base of inner face, the 3 inner very small. Fruit 20 × 10 cm, broadly conical with a smooth skin, green when ripe, pulp white. Seeds 1.5 cm × 8 mm, black. *Andes, now widely cultivated throughout the tropics*. G1. Summer.

The cultivar 'Booth' has larger than average fruits weighing up to 500 gm with very sweet flesh.

5. A. reticulata Linnaeus. Illustration: Botanical Magazine, 2911–12 (1829); Lötschert & Beese, Collins' guide to tropical plants, f. 243 (1983).
Semi-deciduous trees 5–10 m. Young shoots covered with reddish brown hairs. Leaves 10–15 × 2–5 cm, elliptic-oblong, almost hairless beneath. Flowers pendent in axillary clusters. Petals 6, the 3 outer 2–3 cm, concave, greenish yellow with a purplish basal blotch, the 3 inner very reduced. Fruit 10–12 cm in diameter, variable in shape but usually spherical, skin smooth but with net-like veining over whole surface, reddish green when ripe,

pulp granular, yellowish. Seeds 1.5 cm, dark brown or black. *Tropical America, from USA (S Florida) through Mexico to Peru and Brazil, West Indies.* G1. Late summer.

2. ASIMINA Adanson
S.G. Knees

Evergreen or deciduous shrubs or small trees. Leaves alternate, simple, entire. Flowers usually axillary, solitary or in small clusters on pendulous stalks. Sepals 3, deciduous; petals 6; stamens numerous on short filaments. Carpels 3–15, separate, 1-celled. Fruit a berry, seeds many.

A genus of 8 species from eastern North America, only 1 of which is regularly cultivated. All species are readily propagated from seed and a rich loamy soil is required for their successful cultivation.

1. A. triloba (Linnaeus) Dunal. Illustration: Botanical Magazine, 5854 (1870); Journal of the Royal Horticultural Society, xli (1915); Bean, Trees and shrubs hardy in the British Isles, edn 8, **1**: 346 (1970); Heywood, Flowering plants of the world, 30 (1978).
Deciduous tree to 10 m. Shoots downy when young, becoming hairless at maturity. Leaf-stalks 5–10 mm, blades 10–23 cm obovate. Flowers solitary, axillary. Sepals 1.2–1.8 cm, ovate, green, falling early. Petals 6, the 3 outer 2–2.5 cm, dullish purple, the 3 inner smaller, marked with horizontal bands of yellow. Fruit ovoid, 7.5–16 cm, yellowish, containing several seeds, within a fleshy endocarp. *Southeastern N America.* H5–G1. Summer.

3. ARTABOTRYS R. Brown
S.G. Knees

Climbing or prostrate shrubs to 2 or 3 m. Leaves glossy, oblong-lanceolate, acuminate. Flowers fragrant, solitary or clustered on hard stalks. Sepals 3, joined at the base. Petals 6, free. Stamens many. Ovary with 2 basal ovules. Fruit a berry. Seeds oblong.

A genus of about 100 species native to tropical Africa and Indomalesia. Some have edible fruits and several species are grown in other parts of the tropics. Soil with a high organic content is necessary for the successful cultivation of these species. Most species can be propagated from cuttings of ripened wood, and rooted in sand with bottom heat, during the cooler months. Alternatively seed can be sown as soon as the fruits are ripe.

1. A. hexapetalus (Linnaeus) Bhandari (*A. uncinatus* (Lamarck) Merrill). Illustration: Bailey, Standard cyclopedia of horticulture, 398 (1939); Herklots, Flowering tropical climbers, 35 (1976). Climbing shrub to 4 m. Young shoots slightly hairy becoming hairless with age. Leaves stalked, 12–25 × 5–8 cm, lanceolate or oblong-lanceolate, acuminate. Flower-stalks hooked. Sepals reflexed, yellowish. The 3 outer petals 2.5–3 cm, brownish red with basal hook on midrib of outer face, the 3 inner 2–2.5 cm. Ovaries 8–12. Fruits several, 3–4 cm, rounded, ovoid with abruptly pointed apices, in clusters to 30 cm long, on the hardened receptacle. *Southern India & Sri Lanka.* G2. Summer.

Widely cultivated in the Old World tropics.

4. CANANGA (de Candolle) Hooker & Thomson
S.G. Knees

Evergreen trees to 30 m. Leaves alternate, simple, oblong-ovate, long-acuminate. Flowers drooping in axillary clusters. Sepals 3. Petals 6 in 2 series. Stamens numerous. Carpels numerous. Fruits oblong, many, containing seeds in two rows.

A genus of 2 species in tropical Asia and Australasia, only one of which is cultivated.

1. C. odorata Hooker & Thomson. Illustration: Bailey, Standard cyclopedia of horticulture, 652 (1914); Graf, Exotica International Series 4, **1**: 132 (1985); Graf, Tropica, 72 (1986).
Tree to 25 m. Leaves 15–20 cm. Flowers greenish yellow, fragrant. Petals 4–5 cm, narrowly lanceolate, all of similar length. Fruit stalked, 2–2.5 cm, oblong, containing 6–12 seeds. *India, Indonesia & Philippine Islands.* G2. Summer.

Cultivated for its flowers which yield the perfume Ylang-ylang or Macassar oil.

5. XYLOPIA Linnaeus
S.G. Knees

Evergreen trees or shrubs to 45 m, rarely exceeding 15 m in our species. Leaves leathery, shortly stalked. Flowers solitary or in small clusters in the leaf axils, stalkless or very shortly stalked. Sepals 3, thickened and joined at the base. Petals 6, in 2 series. Stamens numerous. Carpels 3–35, styles short, ovules numerous in 1 or 2 series. Fruit a schizocarp, cylindric or obovoid, stalked or stalkless. Seeds 1–8, with arils.

A genus of between 100 and 150 species, mostly native to Africa. The fruits are sometimes used as peppers. Propagation of

well-ripened woody cuttings will succeed as long as bottom heat is provided. Once established the plants will flourish in a compost of peat, charcoal, sand and organic manure.

1a. Branches with lenticels; flowers 4–5 cm **1. aethiopica**
  b. Branches smooth; flowers 1–2 cm **2. quintasii**

1. X. aethiopica (Dunal) Richard. Illustration: Aubréville & Leroy (eds), Flore du Gabon 16: 167 (1969); Berhaut, Flore illustrée du Sénégal 1: 340 (1971).
Tree 5–10 m in cultivation. Branches with lenticels, brown. Leaf-stalks 2–6 mm, blades 8–20 × 3–8 cm, oblong-elliptic, cuneate at base, undersurface slightly hairy. Flowers in clusters of 2–6, borne in leaf axils, on stalks 5–10 mm. Sepals 2.5–3 mm, triangular, green. Petals 4–5 cm, linear, cream or greenish white. Styles 3 mm, carpels 30–32, ovaries cylindric, ovules 6–8 in 1 series. Fruit 5–6 cm, cylindric. Seeds 2–3 mm including the orange aril. *Tropical W Africa.* G2. Summer.

2. X. quintasii Engler & Diels (*X. striata* Engler). Illustration: Aubréville & Leroy (eds), Flore du Gabon 16: 157 (1969); Berhaut, Flore Illustrée du Sénégal 1: 348 (1971).
Tree to 8 m in cultivation. Branches reddish brown. Young leaves pink at first, stalks 5–6 mm, blades 5.5–10 × 2.5–5 cm, oblong to oblong-obovate. Flowers perfumed, solitary or in cymes of 3–11, stalks 3–8 mm. Sepals triangular-ovate, 2–3 mm, hairy on outer face, hairless on inner. Outer petals 1–2 cm × 2.5–3 mm, linear-oblong, obtuse at apex, inner petals slightly shorter and narrower. Styles very short, 0.5–0.7 mm, carpels 2–4, ovules 4, in 1 series. Fruit 3–5 cm, splitting transversely into 1-seeded segments. *Tropical W Africa.* G2. Summer.

6. MONODORA Dunal
S.G. Knees

Evergreen trees and shrubs. Leaves alternate, entire, stalked. Flowers usually solitary, large, fragrant, pendulous on long stalks. Sepals distinctly thickened. Petals united at the base, with wavy margins. Stamens short, many, the connective dilated at apex. Ovary 1-celled, ovules numerous. Fruit formed by the fusion of the previously free carpels. Seeds with woody testa.

A genus of 20 species native of tropical

Africa and Madagascar. Plants can be propagated from seed or by cuttings rooted in sand with bottom heat and a high humidity.

1. M. myristica (Gaertner) Dunal. Illustration: Botanical Magazine, 3059 (1831); Paxton's Magazine of Botany, 137 (1849); Heywood, Flowering plants of the world, 30 (1978); Graf, Tropica, 72 (1986). Tree to 8 m. Leaf-stalks 8–20 mm, blades 10–40 × 8–20 cm, obovate or oblong-elliptic, shortly acuminate, heart-shaped at base, glossy, pale green, with prominent veins beneath. Flower-stalks 10–25 cm. Sepals 1.5–3 cm, margins wavy. Outer petals 6–10 cm, broadly elliptic with wavy margins, yellow with purplish markings. Inner petals 2.5–4 cm, obovate, whitish without, covered in soft hairs, yellowish within with crimson spots, glossy. Carpels united. Fruit spherical, 9–15 cm in diameter, containing many seeds. *Tropical W Africa, from Sierra Leone to Cameroon and south to Angola.* G2. Summer.

The seeds contain a high percentage of aromatic oil and are used locally as a substitute for nutmeg.

Although *Oxandra lanceolata* Baillon is listed in some works, there is considerable confusion regarding the identity of this species, which is doubtfully in cultivation.

LXII. MYRISTICACEAE

Large, dioecious trees. Leaves usually evergreen, entire, alternate, without stipules. Flowers in racemes, cymes or clusters, stalked or not, unisexual, radially symmetric. Perianth usually 3-lobed (rarely 2–5-lobed) at the apex. Stamens 2–10 or more; filaments united into a column at the top of which the anthers are borne. Carpel 1, free, containing a single basal ovule, style very short. Fruit usually large with a fleshy or leathery pericarp which splits into 2 segments to reveal the single large seed which has a conspicuous, large, often divided aril and convoluted endosperm.

A tropical family of mainly large trees in 15 genera and *c.* 250 species. Mostly they are plants of the rain forests and so are not easy to grow in Europe.

1a. At least the male flowers subtended by a bracteole; anthers 6 or more **1. Myristica**
 b. Bracteoles completely absent; anthers 2–4 **2. Pycnanthus**

1. MYRISTICA Gronovius
J. Cullen
Large trees. Flowers in racemes or clusters, the individual flowers stalked and at least the male flowers each subtended by a bracteole. Perianth 3-lobed. Anthers 6 or more. Fruits large.

A genus of 70–100 species from tropical areas of the Old World (though cultivated throughout the tropics). Perhaps 2 species may be seen in gardens with large glasshouses, grown mainly for their interest as economic plants: *M. fragrans* provides the mace (aril) and nutmeg (seed) of commerce. Literature: Sinclair, J., Flora Malesiana Praecursores – XLII: the genus Myristica in Malesia and outside Malesia, *Gardens' Bulletin, Singapore* **23**: 1–540 (1968).

1. M. fragrans Houttuyn (*M. officinalis* Linnaeus filius). Illustration: Bentley & Trimen, Medicinal plants **3**: pl. 218 (1880); Masefield et al., The Oxford book of food plants, pl. 131 (1969); Horticulture **49**(3): 50 (1971); Der Palmengarten, special issue, 36 (1987).
Large trees; young shoots smooth, with distant, scattered lenticels. Leaves silvery with a covering of small stellate scales when young, these soon falling, the silver colour soon lost. Flowers less than 10 mm. Fruits to 3 cm. Aril pink. *Native in Indonesia (Moluccas) but widely cultivated elsewhere in the tropics.* G2.

M. argentea Warburg. Illustration: Gardens' Bulletin, Singapore **23**: 236 (1968). Very similar to *M. fragrans* but young twigs rough with crowded lenticels, leaves persistently silver-scaly beneath. *W New Guinea.* G2.

2. PYCNANTHUS Warburg
J. Cullen
Large trees. Flowers stalkless in heads borne in panicles; bracteoles absent. Perianth usually 3-lobed. Anthers 2–4.

A genus of about 7 species from tropical Africa, of which 1 may be occasionally grown.

1. P. angolensis (Welwitsch) Warburg (*P. kombo* (Baillon) Warburg). Illustration: Thonner, Flowering plants of Africa, pl. 49 (1915); Hutchinson & Dalziel, Flora of west tropical Africa, edn 2, **1**: 60 (1954). Large trees, bark exuding sticky orange sap when cut. Leaves cordate at base, brown-hairy beneath with branched hairs. Clusters of male flowers *c.* 5 mm in diameter. *Tropical Africa.* G2.

The wood is used commercially as timber (Kombo or Illoma).

LXIII. CANELLACEAE

Small trees or shrubs with aromatic bark. Leaves alternate, simple, entire, without stipules. Flowers in cymes or corymbs, rarely solitary, bisexual, radially symmetric. Sepals 3. Petals 4 or 5. Stamens more than 10, the filaments united in a tube around the ovary, anthers opening by slits. Ovary superior, 1-celled, with parietal placentas bearing 2 or more ovules; style short, stigma 2–6-lobed. Fruit a berry with 2 or more smooth, hard seeds embedded in pulp.

A family of 6 genera and about 20 species from tropical and subtropical Africa and America.

1. CANELLA P. Brown
J. Cullen
Trees or shrubs. Leaves evergreen. Flowers red, purple or violet, in terminal corymbs. Petals 5. Stamens 10–20, anthers touching. Ovary with 2–3 parietal placentas, each bearing 2 ovules. Berry spherical, containing few seeds.

A genus of perhaps 2 species from areas around the Caribbean.

1. C. winterana (Linnaeus) Gaertner (*C. alba* Murray). Illustration: Bentley & Trimen, Medicinal plants **1**: t. 26 (1876); Fournet, Flore de Guadeloupe et de Martinique, 511 (1978); Proctor, Flora of the Cayman Islands, 286 (1984). Shrub or small tree to 15 m. Leaves 3–12 cm, gland-dotted. Petals deep red, to 6 mm. Berries deep crimson, 8–10 cm in diameter. *USA (Florida) to West Indies; ?Venezuela.* G1–2.

The bark of this species was formerly used as a medicine.

LXIV. SCHISANDRACEAE

Monoecious or dioecious woody climbers. Leaves alternate, simple, without stipules. Flowers unisexual, axillary, radially symmetric. Perianth made up of a single whorl of 5–20 free segments, usually all more or less of the same size, sometimes a few (outer) smaller. Stamens 4–80; the filaments united at least at the base, sometimes for the greater part of their length or completely united into a more or less spherical mass on which the anthers are borne (occasionally on very short stalks). Carpels numerous, free, each with 2–3 ovules. Fruit a group of fleshy 'berries' borne in a head or on a long, spike-like receptacle.

A family of 2 genera mainly from SE Asia with a single species from SE USA.

1a. Fruits forming a compact head **2. Kadsura**
  b. Fruits borne along a long spike-like receptacle **1. Schisandra**

1. SCHISANDRA Michaux

J. Cullen & J. Howe

Deciduous or evergreen, monoecious or dioecious. Leaves entire or finely and distantly toothed. Perianth cup-shaped, segments 5–20. Stamens 4–60, filaments variously united. Fruit a group of 'berries' borne on a long, spike-like receptacle.

A genus of 25 species, mainly from E Asia, but 1 from SE USA. A few species are grown as ornamental climbers with interesting, if rather small flowers and unusual fruits (not often produced, at least in northern Europe). They are easily grown in good soil in a sunny position and are propagated by seed, stem- or root-cuttings, suckers or layers. The genus name is often incorrectly spelled 'Schizandra'.

Literature: Smith, A.C., The families Illiciaceae and Schisandraceae, *Sargentia* 7: 1–224 (1947).

1a. Filaments completely united in a more or less spherical head in which the anthers are sunk **6. propinqua**
  b. Filaments variously united, but not as above, some part of the filament always free **2**
2a. Leaves conspicuously glaucous beneath **3. glaucescens**
  b. Leaves not glaucous beneath **3**
3a. Free part of filament longer than the anther **4**
  b. Free part of filament shorter than the anther **5**
4a. Perianth whitish, cream or pinkish **1. grandiflora**
  b. Perianth bright to deep red **2. rubriflora**
5a. Stamens 4, 5 or 6 **5. chinensis**
  b. Stamens numerous **4. sphenanthera**

1. S. grandiflora Hooker & Thomson. Hairless climber with round stems. Leaves lanceolate to narrowly elliptic or oblanceolate, tapered to the base, tapered to the acute or acuminate apex, 6–15 × 2–7 cm, obscurely toothed or almost entire, green beneath. Flowers fragrant, 2.5–3 cm in diameter, whitish, cream or pinkish. Perianth-segments 7–8. Stamens 33–60, filaments united only at the extreme base. Fruit scarlet. *N India, Bhutan, Nepal.* H5. Summer.

The name *S. grandiflora* has been frequently mistakenly applied to the next species as well as to several others not in general cultivation.

2. S. rubriflora Rehder & Wilson (*S. grandiflora* var. *rubriflora* (Rehder & Wilson) Schneider). Illustration: Botanical Magazine, 9146 (1928); Perry, Flowers of the world, 283 (1970); Grey-Wilson & Matthews, Gardening on walls, pl. 1 (1983).

Very similar to *S. grandiflora*, differing only in the bright to deep red perianth-segments. *W China, adjacent India and Burma.* H5. Summer.

3. S. glaucescens Diels. Hairless dioecious climber with round stems. Leaves oblong-elliptic to elliptic or obovate, tapered to the base and to the acuminate apex, 5–10 × 1.5–5.5 cm, very finely and obscurely distantly toothed, glaucous beneath. Flowers orange-red, 1.5–2 cm in diameter. Perianth-segments 6–7. Stamens 18–25, filaments united for most of their length into an almost spherical mass, the free parts shorter than the anthers. Fruits scarlet. *C & W China.* H5. Summer.

4. S. sphenanthera Rehder & Wilson. Illustration: Botanical Magazine, 8921 (1938).

Very similar to *S. glaucescens* but leaves broadly obovate to elliptic, 3–11 × 3–7 cm, not glaucous beneath, stamens 10–15. *S & W China.* H5. Summer.

5. S. chinensis (Turczaninow) Baillon. Illustration: Kitamura & Okamoto, Coloured illustrations of Japanese trees and shrubs, pl. 31 (1977).

Usually a dioecious climber, often large (to 8 m or more). Leaves elliptic, elliptic-ovate or elliptic-obovate, tapered to the base, and more or less abruptly to the acute apex, 3–14 × 2–9 cm, finely toothed, green and sometimes with short brown hairs on the veins beneath. Flowers to 1.5 cm in diameter, cream to pinkish. Perianth-segments 6–8. Stamens usually 5, rarely 4 or 6, filaments united almost to the top into a column. Fruit scarlet or pink. *E Asia.* H4? Summer.

6. S. propinqua (Wallich) Baillon. Illustration: Botanical Magazine, 4614 (1851); Hooker's Icones Plantarum 18: t. 1715 (1887); Menninger, Flowering vines of the world, f. 176 (1970).

Hairless, usually monoecious climber with angled branches. Leaves lanceolate, narrowly ovate or almost elliptic, rather rounded at the base, tapering to the shortly acuminate apex, 4–16 cm × 8–50 mm, remotely toothed, green beneath. Flowers orange, c. 1.5 cm in diameter. Perianth-segments 6–10. Stamens 6–16; filaments completely united into a more or less spherical mass in whose surface the anthers are sunk. Fruit scarlet. *Himalaya, C & W China.* H5. Summer.

A rather variable species, divided into 3 varieties by Smith (cited above); it is uncertain which of these are in cultivation.

2. KADSURA Jussieu

J. Cullen & J. Howe

Evergreen, monoecious, woody climbers. Leaves entire or slightly toothed. Flowers unisexual. Perianth of 7–24 segments which are overlapping in several series, the inner larger than the outer. Stamens 20–80, completely covering the surface of a fleshy column formed from the filaments. Carpels forming a head. Fruit a compact head of numerous 'berries'.

A genus of about 20 species from E & SE Asia. Only 1 species is generally grown as a wall-covering. It is easily grown in a good soil and is propagated by cuttings.

Literature: Smith, A.C., The families Illiciaceae and Schisandraceae, *Sargentia* 7: 1–224 (1947).

1. K. japonica (Linnaeus) Dunal. Illustration: Kitamura & Okamoto, Coloured illustrations of Japanese trees and shrubs, pl. 31 (1977).

Hairless climber to 3 m. Leaves elliptic to ovate-lanceolate, 4–11 × 2.5–6.5 cm, usually toothed, sometimes obscurely so. Flowers axillary, solitary or a few together, stalks 1–3 cm. Perianth of 9–17 rather fleshy segments, yellowish white. Fruit scarlet, a more or less spherical head. *Japan, Korea.* H5. Autumn.

'Variegata' has the margins of the leaves yellowish or cream.

LXV. ILLICIACEAE

Hairless shrubs or small trees with aromatic or fragrant bark. Leaves more or less evergreen, alternate or sometimes clustered in false whorls, without stipules. Flowers solitary or in clusters of 2–3 in the leaf-axils, bisexual, radially symmetric. Perianth of numerous petal-like segments which often differ among themselves in size, white, yellow or red. Stamens 5–20, sometimes very fleshy and incurved. Carpels 5–many in a single whorl, free, each with a short style and a single, more or less basal ovule. Fruit a collection of follicles in a single whorl, each containing a single seed.

A family of a single genus.

1. ILLICIUM Linnaeus
J. Cullen & J. Howe
Description as for family.

A genus of about 40 species from SE Asia and SE USA, Mexico and parts of the Caribbean area. A few are cultivated as greenhouse shrubs (some perhaps hardy in the mildest parts of Europe). They are easily grown in good soil in a semi-shaded position and may be propagated by seed or by cuttings of half-ripe wood.
Literature: Smith, A.C., The families Illiciaceae and Schisandraceae, *Sargentia* **7**: 1–224 (1947); Hopkins, H., Illicium: an old plant with new promise, *Journal of the Royal Horticultural Society* **97**: 525–530 (1972).

1a. Inner perianth-segments oblong or lanceolate, not forming a cup 2
  b. Inner perianth-segments ovate to almost circular, forming a cup 3
2a. Perianth-segments white to yellowish; stamens 17–25; carpels 8–10
    1. anisatum
  b. Perianth-segments deep red to purple; stamens 30 or more; carpels 10–20
    2. floridanum
3a. Carpels 11–14   **3. parviflorum**
  b. Carpels 5–9   4
4a. Perianth-segments dark red
                    4. henryi
  b. Perianth-segments yellow **5. verum**

1. I. anisatum Linnaeus (*I. religiosum* Siebold & Zuccarini). Illustration: Botanical Magazine, 3965 (1843); Perry, Flowers of the world, 142 (1972); Journal of the Royal Horticultural Society **97**: f. 259 (1972).
Shrub or small tree to 8 m. Leaves elliptic or narrowly obovate, 4–12 × 1.5–5 cm, tapered to the base, and somewhat abruptly to the ultimately rounded apex. Flowers axillary on stalks 4–30 mm. Perianth-segments 17–24, white to yellowish, the longer (inner) thin, oblong to lanceolate, 1–2.5 cm, all spreading. Stamens 17–25. Carpels 8–10. *Japan, Korea.* H5–G1. Spring–summer.

2. I. floridanum Ellis. Illustration: Botanical Magazine, 439 (1799); Journal of the Royal Horticultural Society **97**: f. 261, 262 (1972); Dean et al., Wildflowers of Alabama, 67 (1973).
Shrub or small tree to 3 m. Leaves lanceolate, elliptic or obovate, 5–15 × 1.5–6 cm, tapered to the base, and to the more or less acute apex. Flowers axillary on

stalks 2–10 cm. Perianth-segments 21–33, deep red to purple, the longer (inner) thin, oblong to very narrowly obovate, 1.5–3 cm, all spreading. Stamens 30 or more. Carpels 10–20. *SE USA.* Spring.

3. I. parviflorum Ventenat. Illustration: Journal of the Royal Horticultural Society **97**: f. 260 (1972).
Large spreading shrub or small tree to 7 m or more. Leaves alternate or in false whorls, narrowly elliptic to obovate, 6–18 × 2–5 cm, tapered to the base, and rather abruptly to the blunt apex. Flowers axillary, solitary or 2 or 3 together, on stalks 1–3 cm. Perianth-segments 13–15, yellow, the inner (larger) 8–9, circular to oblong-obovate, 5–7 mm, forming a cup. Stamens 5–7, fleshy, incurved, borne at the bottom of the perianth-cup. Carpels 11–14. *SE USA (Florida).* G1. Summer.

4. I. henryi Diels. Illustration: Journal of the Royal Horticultural Society **97**: f. 263 (1972).
Shrub to 7 m. Leaves alternate or in false whorls, lanceolate or oblanceolate, 6–15 × 1–5 cm, tapered to the base, and rather abruptly to the usually acuminate apex. Flowers axillary, solitary or 2 or 3 together, on stalks 1–5 cm. Perianth-segments 10–14, dark red, the largest broadly elliptic, 6–10 mm, forming a cup. Stamens 11–28, fleshy, incurved, borne at the bottom of the perianth-cup. Carpels 7–8, style 2–3 mm, clearly longer than the carpel at flowering. *C & W China.* H5–G1. Summer?

5. I. verum Hooker. Illustration: Botanical Magazine, 7005 (1888); de Wit, Plants of the world **1**: pl. 19 (1965).
Small tree to 20 m or a smaller shrub. Leaves alternate or in false whorls, oblong-elliptic to obovate-elliptic or oblanceolate, 5–16 × 1.5–5.5 cm, tapered to the base, and rather abruptly to the acute apex. Flowers axillary, solitary, stalks 1–4 cm. Perianth-segments 17–18, yellow, the largest broadly elliptic, 6–10 mm, forming a cup. Stamens 8–20, fleshy, incurved, borne at the bottom of the perianth-cup. Carpels 8–10, style 1–2 mm, shorter than to as long as the carpel at flowering. *SE China, Vietnam.* H5–G1. Summer–autumn.

LXVI. MONIMIACEAE
Trees or shrubs with aromatic bark, wood and leaves. Leaves usually evergreen, opposite, without stipules. Flowers solitary

or in cymes or racemes, usually unisexual, radially symmetric. Perianth of 4–many segments, often in 2 or more series, but not clearly differentiated into calyx and corolla. Stamens generally numerous, usually each with a pair of glands or cup-like appendage at the base of the filament, anthers opening by slits or from the base upwards by flaps. Perianth and stamens borne on the rim of a perigynous zone. Ovary of usually several free carpels, each containing a single ovule. Fruit a group of achenes or drupes borne within the persistent perigynous zone.

A family of 34 genera and about 450 species, mainly from the tropics and subtropics (extending south to Chile and New Zealand); only a few are cultivated. Literature: Perkins, J., Monimiaceae, *Das Pflanzenreich* **4** (1901).

1a. Anthers opening by longitudinal slits; leaves yellowish-hairy beneath, at least when young   **1. Peumus**
  b. Anthers opening by flaps; leaves hairless   2
2a. Flowers solitary, each stalked and subtended by an involucre formed from 2 bracts which are edge-to-edge in bud   **2. Atherosperma**
  b. Flowers in axillary racemes or cymes, without an involucre   **3. Laurelia**

1. PEUMUS Molina
J. Cullen
Evergreen, dioecious tree or shrub. Leaves entire, leathery, dark green, yellowish-hairy beneath, at least when young. Flowers in axillary or terminal cymes, 5–6 mm in diameter, white. Perianth-segments 10–12 in 2 series, those of the outer series broader than those of the inner. Stamens numerous; filaments short with 2 small glands at or near the base; anthers opening by longitudinal slits. Carpels 3–5, hairy, with thread-like styles; a few staminodes occur in the female flowers. Fruit of 3–5 drupes.

A genus of a single species from Chile.

1. P. boldus Molina. Illustration: Botanical Magazine, 7024 (1888) – as *P. fragrans*; Das Pflanzenreich **4**: 16 (1901); Rodriguez, Matthei & Quezada, Flora arbórea de Chile, pl. 61 (1983).
Chile. G1. Late summer.

2. ATHEROSPERMA Labillardière
J. Cullen
Dioecious trees. Leaves evergreen, hairless, thick, often toothed, at least in the upper half. Flowers solitary, stalked, each subtended by an involucre of 2 bracts

which are edge-to-edge in bud. Perianth-segments 8–10, more or less equal, overlapping in 2 series. Stamens 10–18, each with 2 long appendages at the base of the filament; anthers opening by flaps. Staminodes numerous in female flowers. Carpels many, each tapering into a long, hairy style. Fruit made up of the more or less spherical perigynous zone, containing many achenes.

A genus of 2 species from Australia; only 1 is cultivated as a handsome, usually small (in cultivation) evergreen tree which flowers early in the year. It requires a protected, semi-shaded situation to grow outdoors in Europe. Propagation is by cuttings which are slow to root and establish.

1. A. moschatum Labillardière. Illustration: Das Pflanzenreich **4**: 78 (1901); Botanical Magazine, n.s., 43 (1948); Galbraith, Collins' field guide to the wild flowers of south-east Australia, pl. 67 (1977); Morley & Toelken, Flowering plants in Australia, 40 (1983).
Ultimately large trees (to 50 m in the wild, usually much less in gardens). Leaves oblong to linear-lanceolate, entire or toothed in the upper half. Flowers white, *c.* 1 cm in diameter. *Australia (Victoria, New South Wales, Tasmania).* H5. Late winter.

3. LAURELIA Jussieu
J. Cullen
Tall, aromatic, usually dioecious trees. Leaves leathery, hairless, entire or toothed. Flowers in axillary racemes (ours) or cymes. Perianth-segments 6–12 in 2–3 series, all more or less equal. Flowers mostly unisexual, some bisexual flowers found on basically female plants. Male flowers with an almost flat perigynous zone and with 5–12 stamens, each with a cup-like gland at the base; anthers opening by flaps. Female flowers with a deep perigynous zone; carpels numerous, very hairy, with long, hairy styles; staminodes often present. Fruit a group of achenes.

A genus of 3 species from Chile and New Zealand, 1 occasionally grown as an interesting evergreen.

1. L. sempervirens (Ruiz & Pavon) Tulasne (*L. serrata* Bertero). Illustration: Das Pflanzenreich **4**: 76 (1901); Botanical Magazine, 8279 (1909); Rodriguez, Matthei & Quezada, Flora arbórea de Chile, pl. 38 (1983).
Tall trees in the wild. Leaves pale green, ovate-oblong or lanceolate, entire or toothed, 6–9 × 2–3 cm. Flowers 5–10 mm in diameter, greenish. *Chile.* H5–G1.

LXVII. CALYCANTHACEAE

Deciduous (ours) or evergreen shrubs with aromatic bark and wood. Leaves opposite, entire, without stipules. Flowers solitary, axillary on short shoots or terminal, subtended by bracts, fragrant, sometimes borne before the leaves. Perianth of numerous free segments. Fertile stamens 5–30, some stamens sterile. Carpels many, free, borne within a cup-shaped perigynous zone on whose rim the perianth-segments and stamens are borne. Ovules 1–2 per carpel, placentation marginal. Fruit a group of achenes borne within the persistent, enlarged perigynous zone.

A family of 2 genera from N America, Australia, Japan and China, both grown for their scented flowers.

1a. Flowers borne before the leaves; perianth-segments mostly yellow
                                        2. Chimonanthus
 b. Flowers borne with the leaves; perianth-segments reddish brown to greenish purple   **1. Calycanthus**

1. CALYCANTHUS Linnaeus
J. Cullen & J. Howe
Deciduous shrubs. Axillary buds hidden by the expanded bases of the leaf-stalks or exposed. Flowers fragrant, borne with the leaves, terminal on the branches or on short lateral branches, appearing axillary. Perianth-segments numerous, reddish brown, brownish or greenish purple. Fertile stamens 10–30, brown or yellowish brown. Fruiting perigynous zone bell-shaped or contracted at the apex.

A genus of about 5 species from N America and Australia. Two species are commonly grown in shrubberies or as specimen plants. They are easily grown and propagated by seed, division or layering.

1a. Buds hidden within the expanded leaf-stalk bases; fruiting perigynous zone contracted towards the mouth
                                        1. floridus
 b. Buds exposed; fruiting perigynous zone not contracted at the mouth
                                        2. occidentalis

1. C. floridus Linnaeus.
Spreading shrub to 3 m. Leaves ovate to narrowly elliptic or elliptic-oblong, 5–15 cm, acute or acuminate, tapering or rounded at the base, hairy at least on the veins, or hairless beneath; axillary buds hidden within the expanded bases of the leaf-stalks. Flowers terminal on short shoots, appearing axillary, dark reddish

brown to greenish purple, *c.* 5 cm in diameter, perianth-segments hairy outside. Fruiting perigynous zone contracted towards the mouth. *Eastern N America.* H2. Summer–autumn.

A variable species in which a number of varieties (or other ranks) has been described. Only 2 seem to be of general significance.

Var. **floridus**. Illustration: Botanical Magazine, 503 (1800); Justice & Bell, Wild flowers of North Carolina, 72 (1968); Bärtels, Gartengehölze, 123 (1973); Gartenpraxis for 1978: 411. Leaves green and hairy (at least on the veins) beneath; perianth-segments reddish brown.

Var. **laevigatus** (Willdenow) Torrey & Gray (*C. laevigatus* Willdenow; *C. glaucus* Willdenow; *C. fertilis* Walter). Illustration: Edwards's Botanical Register **5**: t. 404 (1819); Bärtels, Gartengehölze, 123 (1973). Leaves more or less glaucous and hairless beneath; perianth-segments greenish purple.

2. C. occidentalis Hooker & Arnott. Illustration: Botanical Magazine, 4808 (1854); Heywood, Flowering plants of the world, 35 (1978); Csapody & Toth, A colour atlas of flowering trees and shrubs, pl. 31 (1982).
Shrub to 3 m or more. Leaves ovate to ovate-oblong or elliptic, acute at the apex, broadly rounded or slightly cordate at the base, rough above, hairless or very slightly hairy beneath; axillary buds exposed. Flowers 5–7 cm in diameter, mostly terminal on the long shoots, perianth-segments hairy outside, mostly brownish purple, the tips of some of the segments often pale brown. Fruiting perigynous zone bell-shaped. *Western USA (California).* H4. Summer.

2. CHIMONANTHUS Lindley
J. Cullen & J. Howe
Deciduous shrubs (ours). Axillary buds exposed. Flowers subtended by numerous bracts, borne before the leaves (ours) on leafless branches, almost stalkless. Perianth-segments numerous, the outer yellow, the inner yellow or yellow marked with purplish brown. Fertile stamens 5 or 6. Fruiting perigynous zone narrowed towards the mouth.

A genus of 4 species from China and Japan, 1 of them widely cultivated for its early, fragrant flowers. It is easily grown in good soil and is most easily propagated by layering. At least 1 of the species not cultivated is evergreen.

1. C. praecox (Linnaeus) Link (*Calycanthus praecox* Linnaeus; *Chimonanthus fragrans* (Loiseleur) Lindley; *Meratia praecox* (Linnaeus) Rehder & Wilson). Illustration: Botanical Magazine, n.s., 184 (1952); Gartenpraxis for 1977: 98; Heywood, Flowering plants of the world, pl. 31 (1978); Csapody & Toth, A colour atlas of flowering trees and shrubs, pl. 31 (1982). Shrub to 3 m. Leaves 7–20 cm, elliptic-ovate to ovate-lanceolate, acuminate at the apex, tapered or rounded at the base, rough above, hairless or with a few hairs on the veins beneath and on the margins; axillary buds exposed. Flowers developing before the leaves, almost stalkless, subtended by numerous bracts, fragrant, to 3.5 cm in diameter. Outer perianth-segments yellow, inner yellow or yellow marked with brownish purple. Fruiting perigynous zone ellipsoid. *China, Japan*. H5. Late winter.

Several varieties are described by Turrill in the text to the Botanical Magazine plate cited above, but it is not known which of these are in cultivation.

LXVIII. LAURACEAE

Trees or shrubs with hard, often foetid wood. Leaves alternate or almost opposite, leathery, often gland-dotted; veins pinnate or 3–5. Flowers bisexual or unisexual, usually small and inconspicuous and borne in umbels, cymes or panicles. Bracts often falling early or totally absent. Perianth of 4 or 6 segments in 2 series, sometimes enlarging in fruit. Stamens or staminodes usually twice as many as perianth-segments, in 4 or more whorls; anthers opening by flaps. Ovary stalkless, of 1 carpel, 1-celled. Style terminal, simple, stigma entire or irregularly lobed. Ovule solitary, pendulous, anatropous. Fruit a fleshy berry, a drupe or dry and indehiscent, sometimes partly covered by the persistent perianth. Seed pendulous, without endosperm.

A large, homogeneous family of about 47 genera and about 2000 species, distributed in the tropics and subtropics. Many species are economically important.
Literature: Stern, W.L., Comparative anatomy of xylem and phylogeny of Lauraceae, *Tropical Woods* **100**: 1–75 (1954); Wood, C.E., Citation of some genera of Lauraceae, *Journal of the Arnold Arboretum* **39**: 213–215 (1958).

1a.	Stamens 6 or 12; perianth-segments four	2
b.	Stamens 9; perianth-segments six	3
2a.	Stamens 6	**6. Neolitsea**
b.	Stamens 12	**8. Laurus**
3a.	Perianth persistent and hardening in fruit, enclosing the base of the fruit	**2. Phoebe**
b.	Perianth deciduous, or if persistent then not hardening in fruit	4
4a.	Leaves prominently 3-veined, almost opposite	**4. Cinnamomum**
b.	Leaves pinnately veined, rarely 3–5-veined, alternate	5
5a.	Leaves evergreen, always entire	6
b.	Leaves deciduous, sometimes lobed	7
6a.	Stigma peltate	**5. Umbellularia**
b.	Stigma not peltate	**3. Persea**
7a.	Tree to 15 m or more; fruit a drupe	**1. Sassafras**
b.	Shrubs or small trees rarely exceeding 9 m; fruit a berry	**7. Lindera**

1. SASSAFRAS Trew
S.G. Knees

Deciduous aromatic trees to 15 m or more (to 30 m in the wild). Young branches smooth and green, bark deeply furrowed. Leaves alternate, entire or with 1–3 lobes, with prominent midvein and 2 lateral veins giving a 3-veined appearance. Flowers unisexual (when males and females usually on separate plants) or bisexual, borne in small racemes and appearing before the leaves. Perianth of 6 sepal-like segments. Male flowers with 9 stamens, female flowers with 6 staminodes. Style slender. Fruit-stalk thick, fleshy; fruit an ovoid drupe.

A genus of 2 or 3 species, 1 from N America and 1 or 2 from China. The bark yields oil of sassafras, an important flavouring. Plants are usually raised from imported seed, although cuttings from roots or suckering stems, taken early in the year are equally successful.

1. S. albidum (Nuttall) Nees (*Laurus albida* Nuttall; *S. officinale* Nees & Ebermaier var. *albidum* (Nuttall) Blake). Illustration: Marshall Cavendish encyclopaedia of gardening **6**: 1977 (1969); Krüssmann, Manual of cultivated broad-leaved trees and shrubs **3**: 309 (1986).
Tree to 15 m, suckering when mature. Leaves 7–16 × 5–10 cm, ovate, often lobed on one or both sides, apparently 3-veined, softly hairy beneath when young, colouring well in autumn. Flowers greenish yellow, in panicles 2–5 cm. Fruits 1.2–1.5 cm, dark blue, borne on red stalks. *Eastern N America from Ontario to Florida, west to Kansas and Texas*. H2. Spring.

2. PHOEBE Nees
S.G. Knees

Trees or shrubs. Leaf-buds small with few scales. Leaves alternate, pinnately veined. Flowers bisexual, in panicles or corymbs. Perianth of 6 persistent, sepal-like segments, hardening in fruit. Stamens in 3 series, anthers 4-celled. Fruit immersed in the hardened calyx.

About 70 species in Indomalaysia, eastern Asia, tropical America and the West Indies. Propagation is by seed or cuttings.

1. P. formosana (Hayata) Hayata (*P. sheareri* (Hemsley) Gamble). Illustration: Li et al., Flora of Taiwan **2**: 466 (1976).
Evergreen tree to 5 m (in cultivation). Leaves 12–18 cm, lanceolate-ovate, apex acute, hairless, leathery, with 10 pairs of veins. Flowers in axillary panicles to 10–15 cm. Perianth small, to 4 mm across. Stamens 9. Fruit *c*. 1 cm, dark purple, oblong, sunk in the persistent perianth. *Taiwan*. H4. Spring–summer.

3. PERSEA Miller
S.G. Knees

Evergreen trees or shrubs. Leaves alternate, entire, pinnately veined. Flowers bisexual or unisexual, in axillary or terminal panicles. Perianth-segments 6. Stamens 9 in male flowers, anthers 4-celled. Ovary stalkless, style slender, terminating in a small, flattened stigma. Fruit a spherical or oblong berry.

A genus of about 150 species, distributed mainly in eastern Asia and tropical and subtropical America. Well-known for its edible fruit, the avocado, otherwise a genus containing several ornamental species; most of these are grown only in the tropics.

1a.	Flowers in terminal panicles; leaves elliptic or ovate	**1. americana**
b.	Flowers in axillary panicles; leaves lanceolate	**2. indica**

1. P. americana Miller. Illustration: Botanical Magazine, 4580 (1851); Masefield et al., The Oxford book of food plants, 115 (1969).
Tree or shrub, 5–20 m. Leaves 10–20 cm, elliptic or ovate, acuminate, acute or shortly pointed. Flowers in terminal panicles. Perianth-segments greenish, softly hairy. Fruit in cultivated specimens 8–15 cm, ovoid or spherical, skin leathery, glossy, green, yellowish or purple, often pitted. Seed solitary, 2–4 cm. *C America, widely cultivated elsewhere*. H5–G1. Flowering season variable.

Three races of avocados have been selected: Mexican, Guatemalan and West Indian; numerous varieties have been selected and named.

2. P. indica (Linnaeus) Sprengel. Illustration: Bramwell & Bramwell, Wild flowers of the Canary Islands, 54 (1974). Tree to 20 m with broad rounded crown and stout twigs; shoots with silky hairs when young. Leaves 8–23 × 3–8 cm, lanceolate, obtuse or acute at the apex, hairless. Perianth inconspicuous, to 1.1 cm across, greenish yellow. Fruit 1.8–2.1 cm, ovoid-ellipsoid, bluish black when ripe. *Atlantic Islands: Azores, Canary Islands.* H3. Spring.

4. CINNAMOMUM Schaeffer
S.G. Knees
Evergreen, aromatic trees or shrubs. Buds with scales or sometimes naked. Leaves almost opposite, occasionally alternate, leathery and usually 3-veined. Flowers bisexual, very inconspicuous, in terminal or axillary panicles. Perianth-tube short, bell-shaped, lobes 6, almost equal. Stamens 9 in 3 whorls, the inner whorl with glandular filaments, anthers 4-celled, Ovary stalkless, 1-celled, narrowed into the style, stigmas disc-like. Fruit a berry borne within the enlarged perianth, the lobes falling early.

A genus of about 250 species from tropical and subtropical Asia, Australia and the Pacific Islands. Two species have been economically important for centuries; *C. zeylanicum*, for its bark which yields the spice cinnamon and *C. camphora*, for its wood from which camphor is distilled. Propagation is readily achieved from cuttings of slender shoots taken in spring. Bottom heat and a rich peaty compost are recommended. Their ornamental value is mainly in the young leaves which are often bright red.

1a. Buds with overlapping scales   2
 b. Buds naked   3
2a. Bud-scales many in several series,
    buds hairy   **1. camphora**
 b. Bud-scales few in 1 series, buds
    hairless   **2. japonicum**
3a. Young branches black; fruits black
                           3. micranthum
 b. Young branches pale brown; fruits
    purplish   **4. zeylanicum**

1. C. camphora (Linnaeus) Siebold (*Laurus camphora* Linnaeus). Illustration: Botanical Magazine, 2658 (1826); Makino, Illustrated Flora of Japan, 1092 (1925); Li et al., Flora of Taiwan **2**: 414 (1976); Graf, Tropica, 534 (1979).
Large evergreen tree with blackish, hairy branches and buds. Bud-scales many in several series. Leaf-stalks 1–1.5 cm, blades 7.5–10 × 3–4 cm, elliptic, acuminate, papery, 3-veined at the base with 2 glands in the vein-axils. Flowers many in axillary panicles. Fruits 6–10 mm across, blackish when ripe. *Tropical Asia, Malaysia, Taiwan, Japan.* G2.

2. C. japonicum Nees (*C. pedunculatum* Nees).
Evergreen tree with slender hairless branches. Bud-scales few, in 1 series. Leaves alternate or almost opposite, stalks 1.5 cm, flattened, blades 6–7 × 2.5–3.5 cm, ovate-oblong to oblong-lanceolate, apex acute, base obtuse, 3-veined, leathery. Flowers few, in cymose umbels, 4–5 cm across, on stalks 6–7 cm. Fruits 10 × 7 mm, oblong. *Japan, Korea, Taiwan.* G2.

3. C. micranthum (Hayata) Hayata (*Machilus micranthum* Hayata; *Machilus kanehirai* Hayata). Illustration: Li, Woody Flora of Taiwan, 203 (1963).
Large evergreen tree to 30 m, with hairless blackish branches. Leaf-stalks 1.5–2.5 cm, stout, blades 9.5–10.5 × 4–5 cm, broadly elliptic to obovate, acute, leathery, 3-veined at the base with 3–4 additional pairs of veins diverging above the base. Flowers terminal or axillary in cymose panicles, fragrant. Fruits 1–1.2 cm across, ellipsoid or almost spherical, blackish. *Taiwan.* G1–G2.

4. C. zeylanicum Blume. Illustration: Botanical Magazine, 2028 (1826); Graf, Tropica, 534 (1979).
Small tree, 10–13 m with pale brown papery bark. Leaf-stalks to 2 cm, blades 7–18 cm, reddish when young, becoming dark green with age and very leathery, strongly 3-veined, veins whitish, conspicuous. Flowers yellowish white in terminal or axillary racemes. Fruits 9–12 mm, ovoid, purplish. *S India, Sri Lanka, widely cultivated elsewhere.* G2.

5. UMBELLULARIA Nuttall
S.G. Knees
Strongly aromatic evergreen tree reaching 15–20 m. Leaves alternate, 5–12 cm, narrowly oblong-elliptic, leathery and glossy, entire and tapered at each end. Flowers bisexual in shortly stalked, many-flowered umbels to 2 cm across. Perianth with short tube and 6 equal lobes. Stamens 9, the 3 innermost each with 2 basal glands. Fruit a pear-shaped berry to 2.5 cm, green at first, becoming purple when ripe.

A genus of a single species occurring in western N America. Propagation is by seed sown as soon as ripe. A temperature of 13–15 °C must be maintained until germination. Alternatively, plants can be propagated by layering in spring. Although hardy, protection from north-east winds is advised. These trees do not thrive in alkaline soils.

1. U. californica (Hooker & Arnott) Nuttall. Illustration: Phillips, Trees in Britain, 214 (1978); Everett, New York Botanical Gardens illustrated encyclopedia of horticulture **10**: 3446 (1982); Krüssmann, Manual of cultivated broad-leaved trees and shrubs **3**: 289 (1986).
W USA (Oregon, California). H4. Spring.
Standing near this tree for too long in warm weather can cause headaches.

6. NEOLITSEA (Bentham) Merrill
S.G. Knees
Evergreen, dioecious trees. Leaves alternate, stalked, entire, normally 3-veined, rarely pinnately veined, softly hairy when young. Flowers unisexual, perianth-segments 4, soon falling. Stamens 6 (rarely 8) in male flowers, the inner 2 each with 2 basal glands, the outer 4 without. Fruit a berry.

A genus of about 80 species in E & SE Asia.

1. N. sericea (Blume) Koidzumi (*Laurus sericea* Blume; *N. glauca* misapplied; *N. latifolia* misapplied). Illustration: Kew Magazine **1**(3): 138 (1984).
Small tree or shrub to 7 m. Young branches softly hairy. Leaves 5–7 × 2–2.5 cm, ovate-elliptic, usually clustered near the tips of the branches, golden-hairy beneath. Flowers in umbels of 5–10. Fruit 1.1–1.3 cm, ellipsoid, red, with remains of perianth at base. *China, Korea, Japan, Taiwan.* H2. Autumn.

7. LINDERA Thunberg
S.G. Knees
Small trees or shrubs, mostly deciduous. Leaves pinnately 3-or 5-veined, alternate. Flowers dioecious, borne in false umbels and surrounded by 4 conspicuous persistent bracts. Perianth segments 6. Male flowers with 9 (rarely 12) stamens. Female flowers with 9 staminodes; stigma peltate, conspicuous. Fruit fleshy or dry, splitting at maturity, containing a single stone and seated in a more or less

developed cup or disc, sometimes surrounded by the persistent perianth.

A genus of about 100 species native of eastern Asia from the Himalayas south to Malaysia, north to China and Japan, some species also in North America. Propagation by seed, layers or cuttings of green wood.

1a. Leaves distinctly 3-veined at base, entire or 3-lobed   2
  b. Leaves pinnately veined, or only weakly 3-veined, always entire   3
2a. Leaves acuminate, narrowly ovate, entire   **7. strychnifolia**
  b. Leaves acute or obtuse, broadly ovate, sometimes 3-lobed near apex   **6. obtusiloba**
3a. Leaves evergreen, 10–20 cm   **3. megaphylla**
  b. Leaves deciduous, less than 12 cm   4
4a. Flowers opening before the leaves   5
  b. Flowers opening with or after the leaves   6
5a. Leaf-stalk 5–15 mm; fruit red   **1. benzoin**
  b. Leaf-stalk 1.5–2.5 cm; fruit yellowish brown   **2. praecox**
6a. Leaves 6–12 cm   **4. umbellata**
  b. Leaves 3–5 cm   **5. glauca**

1. L. benzoin (Linnaeus) Blume (*Laurus benzoin* Linnaeus; *Benzoin aestivale* Nees; *B. odoriferum* Nees). Illustration: Everett, New York Botanical Gardens illustrated encyclopedia of horticulture **6**: 2031 (1981); Krüssmann, Manual of cultivated Broad-leaved trees and shrubs **2**: f. 148, pl. 88 (1986).

Deciduous, highly aromatic shrub, forming a low broad bush 2–4 m tall. Branches greyish brown, buds with 2–3 outer scales. Leaf-stalks 5–15 mm, blades 7–13 × 2.5–7 cm, broadly elliptic, pinnately veined, hairless when young, becoming hairy with age. Flowers 5 mm across, stalkless, in clusters of 2–5, pale yellow, borne on the previous year's growth and usually opening with the leaves. Fruits 1 cm, ellipsoid, bright red. *SE USA (Carolinas to Missouri)*. H2. Spring.

2. L. praecox (Siebold & Zuccarini) Blume (*Benzoin praecox* Siebold & Zuccarini; *Parabenzoin praecox* (Siebold & Zuccarini) Nakai). Illustration: Useful Plants of Japan, t. 316 (1895); Makino, Illustrated Flora of Japan, 1096 (1925).

Deciduous shrub or small tree to 7 m. Young branches brown with white lenticels. Leaf-stalks 1.5–2.5 cm, blades 4–9 × 3–5 cm, ovate-elliptic, acuminate, pinnately veined, glaucous beneath. Flowers 5–6 mm across, in small clusters of 3–6, greenish yellow. Fruits 1.5–2 cm, spherical, yellowish or brown. *China, Japan*. H2. Spring.

3. L. megaphylla Hemsley. Illustration: Yung, Illustrated manual of Chinese trees and shrubs, 353 (1937); Li et al., Flora of Taiwan **2**: 431 (1976).

Evergreen shrub or small tree to 8 m in cultivation (to 20 m in the wild). Branches dark purplish brown, with whitish lenticels when young. Leaf-stalks 1–2 cm, blades 10–20 × 2–6 cm, oblanceolate, acuminate, shiny dark green above, glaucous beneath, pinnately veined. Flowers many in shortly stalked umbels to 3 cm in diameter, yellowish. Fruits 1.5 cm, spherical–ovoid, black. *C & W China*. H2. Spring.

4. L. umbellata Thunberg. Illustration: Thunberg, Flora Japonica, t. 21 (1794); Makino, Illustrated Flora of Japan, 1094 (1925); Yung, Illustrated manual of Chinese trees and shrubs, 355 (1937); Krüssmann, Manual of cultivated broad-leaved trees & shrubs **2**: pl. 88 (1986).

Deciduous shrub to 3 m. Branches dark red, without lenticels. Leaf-stalks 1–1.8 cm, blades 6–12 cm, obovate-elliptic, acute or acuminate, pinnately veined. Flowers on stalks 6–10 mm opening with the leaves, in umbels 2–2.5 cm across, yellow. Fruits to 8 mm, almost spherical, black. *Japan*. H2. Spring.

5. L. glauca (Siebold & Zuccarini) Blume (*Benzoin glaucum* Siebold & Zuccarini). Illustration: Makino, Illustrated Flora of Japan, 1095 (1925); Yung, Illustrated manual of Chinese trees and shrubs, 353 (1937).

Deciduous tree to 10 m. Young branches softly hairy. Leaf-stalks to 2 mm, blades 3–5 × 2–3 cm, oblong, acute, pinnately veined with 5–7 pairs of veins. Flowers in umbels of 5–10, on stalks 5–7 mm, yellow, opening with or slightly after the leaves. Fruits 6–8 mm, spherical. *China*. H2. Spring.

6. L. obtusiloba Blume. Illustration: Makino, Illustrated Flora of Japan, 1096 (1925); Yung, Illustrated manual of Chinese trees and shrubs, 359 (1937); Krüssmann, Manual of cultivated broad-leaved trees & shrubs **2**: pl. 88 (1986).

Deciduous shrub to 10 m. Branches yellowish grey, occasionally purplish, with lenticels. Leaf-stalks 1–2 cm, blades 6–12 × 3–10 cm, broadly ovate, acute or obtuse and often with 3 lobes near apex, base of leaf distinctly 3-veined. Flowers yellow, in umbels 1–1.5 cm across, produced before the leaves on stalks 3–4 mm. Fruits 7–8 mm, spherical, shiny, black. *Korea, China & Japan*. H2. Spring.

7. L. strychnifolia (Siebold & Zuccarini) Vilmorin (*Daphnidium strychnifolium* Siebold & Zuccarini). Illustration: Makino, Illustrated Flora of Japan, 1097 (1925); Yung, Illustrated manual of Chinese trees and shrubs, 357 (1976); Li et al., Flora of Taiwan **2**: 431 (1976).

Evergreen tree or shrub to 12 m. Branches brown, softly hairy when young. Leaf-stalks 8–10 mm, blades 4–6 × 2–3 cm, ovate to ovate-oblong, leathery, distinctly 3-veined. Flowers in clusters of 3–8 with densely woolly stalks to 2 mm. Fruits broadly ovoid, hairy. *China, Taiwan, Philippine Islands and SE Asia*. H5–G1. Spring–summer.

8. LAURUS Linnaeus
S.G. Knees

Aromatic evergreen trees or shrubs. Leaves alternate, simple, pinnately veined. Flowers in axillary umbels, bisexual or unisexual. Perianth-segments 4. Male flowers with 12 (or more) stamens, anthers opening by 2 flaps. Fruit a berry.

A genus of 2 species native to the Mediterranean region and Atlantic Islands, including the sweet bay, which is used as a condiment and in veterinary medicine. They are usually propagated by cuttings of ripened wood, less commonly by seed.

1a. Young branches hairless; leaves narrowly lanceolate   **1. nobilis**
  b. Young branches softly hairy; leaves circular or broadly lanceolate   **2. azorica**

1. L. nobilis Linnaeus. Illustration: Graf, Tropica, edn 3, 534, 535 (1986).

Slender small tree or shrub, 3–15 m, with slender, hairless twigs. Leaves 5–10 × 2–4 cm, oblong-lanceolate, acute, glossy dark green above, hairless beneath. Male flowers with 12 stamens, most of them with 2 basal glands. Female flowers with 2–4 staminodes. Fruit 1–1.5 cm, on stalks 3–4 mm, ovoid, shiny black when ripe. *Mediterranean area, widely cultivated elsewhere*. H4. Spring.

Two variants regularly available are 'Angustifolia' with very narrow leaves (less than 1 cm) and 'Crispa' with leaves with wavy margins.

2. L. azorica (Seubert) Franco (*L. canariensis* Webb & Berthelot). Illustration: Bramwell & Bramwell, Wild flowers of the Canary Islands, 53 (1974).

Tree to 10 m with stout branches covered in soft hair when young. Leaves 5–12 × 3–8 cm, almost circular to broadly lanceolate, hairy beneath. Flowers like those of *L. nobilis*. Fruit to 1.2 cm on stalks to 6 mm, ovoid, black. *Atlantic Islands: Azores, Canary Islands*. H5. Spring.

LXIX. TETRACENTRACEAE

Small to medium-sized trees with slender ascending primary branches, each marked with the closely packed ring-scars of fallen leaves. Buds long, slightly curved, pointed. Leaves stalked, alternate, ovate or heart-shaped, cordate at base, margins saw-toothed. Inflorescence pendulous, with 80–120 flowers, borne on a short spur. Flowers bisexual, very small, yellowish, parts in 4s, petals absent. Ovary superior. Fruit a 4-celled capsule.

A family closely allied to the Trochodendraceae; both have primitive characteristics in their wood anatomy (lack of vessels). Only a single genus is known, containing a single species from the Himalayas and China.

1. TETRACENTRON Oliver
S.G. Knees
Deciduous trees 17–30 m in the wild, rarely exceeding 15 m in cultivation. Buds 9–12 mm, light orange-brown. Leaves with stipules, stalks 1–3 cm, blades 4.5–7 cm, with 5 prominent veins radiating from a central point at the leaf-base, colouring a rich bluish red in autumn. Sepals 4, overlapping, petals absent. Stamens 4. Stigmas 4, arising from the base of the ovary. Carpels 4, united, each with up to 6 ovules. Fruit a capsule with a spur-like appendage on each cell. Seeds oily, linear-oblong.

The single species, although hardy, may be damaged by spring frosts. It will thrive in most soil types. Propagation is by cuttings of the current year's growth, taken in summer, or by seed, though this is rarely available. Literature: Smith, A.C., A taxonomic review of Trochodendron and Tetracentron, *Journal of the Arnold Arboretum* **26**: 123–142 (1945).

1. T. sinense Oliver. Illustration: Phillips, Trees in Britain, Europe and North America, 205 (1978); Bean, Trees and shrubs hardy in the British Isles, edn 8, 4: pl. 80 (1980).
NE India, Nepal, Burma, SW China. H3. Summer.

LXX. TROCHODENDRACEAE

Evergreen trees or shrubs with stalked, saw-toothed, almost whorled leaves. Flowers bisexual, regular or slightly asymmetric, in terminal, raceme-like clusters. Perianth absent. Stamens and carpels numerous; carpels free, in a ring. Fruit a ring of coalesced, many-seeded follicles. Seeds with oily endosperm.

A family of a single genus and a single species from E Asia.

1. TROCHODENDRON Siebold & Zuccarini
S.G. Knees
Medium-sized trees or shrubs to about 10 m in cultivation, 20–25 m in the wild. Terminal bud to 2.1 cm × 8 mm, with papery scales. Leaves glossy above, leathery, broadly ovate to elliptic, 5–12 × 3–7 cm. Inflorescence 5–13 cm long, with 10–20 or more flowers. Stamens 40–70, 3.5–7 mm, spreading or reflexed, falling early. Carpels 6–11, the whole ovary obovoid, 2–2.5 mm, ovules 16–24 in each carpel. Style 0.5–2 mm. Fruits 7–10 mm in diameter, with 7–12 seeds per follicle.

Propagation is usually by seed or semi-hardwood cuttings taken in autumn. Literature: Smith, A.C., A taxonomic review of Trochodendron and Tetracentron, *Journal of the Arnold Arboretum* **26**: 123–142 (1945).

1. T. aralioides Siebold & Zuccarini. Illustration: Botanical Magazine, 7375 (1894); Li et al., Woody Flora of Taiwan, 154 (1963); Li, Flora of Taiwan, 474 (1976).
Korea (Quelpaert Island). Japan (Honshu), Ryukyu Islands, Taiwan. H3. Summer.

LXXI. EUPTELEACEAE

Small, erect to spreading, branched trees or shrubs. Leaves deciduous, in (false) whorls at the ends of short shoots, long-stalked, without stipules, irregularly and unequally toothed. Flowers bisexual, radially symmetric, borne in the leaf-axils. Perianth absent or forming a minute cup. Stamens 8–18. Ovary of 8–18 free, long-stalked carpels, each containing 1–3 ovules with marginal placentation; stigmatic area conspicuous. Fruit a group of stalked, 1–3-seeded samaras.

A family of a single genus from the eastern Himalayas, China and Japan.

1. EUPTELEA Siebold & Zuccarini
J. Cullen & J. Howe
Description as for the family.

A genus of 2 very similar species, occasionally grown as interesting specimen trees with reddish brown young foliage. They can be propagated by seed or by cuttings.
Literature: Smith, A.C., A taxonomic review of Euptelea, *Journal of the Arnold Arboretum* **27**: 175–185 (1946).

1a. Larger teeth of the leaf more than twice the size of the smaller; leaf very abruptly narrowed at the apex into an entire point which is 1–4 cm long
1. polyandra
  b. Larger teeth of the leaf scarcely larger than the smaller; leaf gradually tapering at the apex to an entire point which is up to 1.5 cm long
2. pleiosperma

1. E. polyandra Siebold & Zuccarini. Illustration: Schneider, Illustriertes Handbuch der Laubholzkunde 1: 270 (1904); Kurata, Illustrated important forest trees of Japan 2: pl. 21 (1968).
Trees 5–15 m with rough, greyish bark. Leaves circular, sometimes broader than long, 6–15 × 5–16 cm, usually hairless beneath except on the veins and in the vein-axils, deeply and irregularly toothed, the larger teeth more than twice as large as the smaller, the whole leaf abruptly tapered to an entire point 1–4 cm long; base broadly tapered, rounded or truncate. Samaras with 1–2 seeds. *Japan*. H2. Spring.

2. E. pleiosperma Hooker & Thomson (*E. franchetii* van Tieghem). Illustration: Journal of the Linnaean Society, Botany 7: t. 2 (1864).
Very similar to *E. polyandra* but leaves narrower, somewhat glaucous and hairless beneath, base broadly to narrowly tapered, apex gradually tapered to an entire point to 1.5 cm long, margins unequally toothed but with the larger teeth little larger than the smaller. Samaras with 1–3 seeds. *E Himalaya, China*. H2. Spring.

LXXII. CERCIDIPHYLLACEAE

Large deciduous, dioecious trees with pendulous branches and spirally twisted trunks. Leaves stalked, opposite on long shoots and alternate on short shoots. Flowers unisexual, the male almost stalkless, the female stalked. Male with 4 free perianth-segments and 15–20 stamens. Female with 4 free perianth-segments and 4–6 free carpels. Fruit a cluster of dehiscent follicles with woody endocarp. Seeds compressed, winged and 4-sided.

An interesting family whose relations to other families are still uncertain. It contains a single genus with a single species.

1. CERCIDIPHYLLUM Siebold & Zuccarini
S.G. Knees
Tree 20–30 m with trunk solitary or branched from the base into 3–5 smaller boles. Leaves 5–10 cm, broadly ovate, cordate at base, margins scolloped, stalks 2–3 cm. Male flowers 1.8–2.2 cm, borne in the leaf-axils, either solitary or in bundles. Female flowers 5–8 mm, with the 4 sepals green and fringed; carpels 4–6, styles thread-like. Fruits pod-like, 1.2–2 cm, in clusters of 2–4 on a stalk to 5 mm.

Although hardy in most parts of Europe, the trees come into leaf and flower early and are often checked by late frosts. Moist soil conditions are generally preferred since die-back of the current year's growth may occur when plants are grown in light, sandy soils in areas of low rainfall. The foliage colours well in autumn, especially on trees grown in lime-free soils. Propagation is normally by seed, which should be sown in March; the seedlings should be transplanted 2 or 3 years later. Literature: Swamy, B.G.L. & Bailey, I.W., The morphology and relationships of Cercidiphyllum, *Journal of the Arnold Arboretum* 30: 187–210 (1949).

1. C. japonicum Siebold & Zuccarini. Illustration: Journal of the Arnold Arboretum 30: 188 (1949); Bean, Trees and shrubs hardy in the British Isles, edn 8, 1: pl. 38 (1976); Phillips, Trees in Britain, Europe and North America, 100 (1978). *China, Japan*. H4. Spring.

Var. **magnificum** Nakai has smooth bark until well into maturity, is usually smaller, but has larger leaves which are more rounded in outline. It is commercially available and is probably superior as an ornamental, but demands very moist conditions. *Japan*. H4. Spring.

LXXIII. RANUNCULACEAE
Herbs or rarely woody and/or climbing plants. Leaves mostly alternate or all basal, rarely opposite, usually without stipules, often deeply divided into numerous segments. Flowers solitary and terminal or in racemes or panicles. Perianth of a single, usually petal-like whorl or of 2 whorls, calyx and corolla both present. Sepals or perianth-segments 3–many, free. Petals 3 or more, sometimes flat and petal-like, sometimes small, variously shaped, almost always with a nectar-secreting area on the surface, at the base or at the apex. Stamens usually numerous, more rarely as few as 5, free; anthers opening by longitudinal slits. Carpels usually numerous (rarely 1–few), usually free. Ovules 1–many per carpel, placentation marginal. Fruit usually a group of follicles or achenes, more rarely a single follicle or berry-like.

A family of about 50 genera and 2000 species, mostly from temperate regions, a few from mountains in the tropics. There are difficulties and controversies concerning the interpretation of the various whorls of the perianth in this family, particularly in those genera where there are 2 whorls. In some genera, e.g. *Ranunculus*, the segments of the inner whorl are clearly petal-like, but they have a small nectary (often covered by a flap) on the upper surface near the base. In other genera (e.g. *Helleborus*, *Eranthis*) the nectar-bearing organs are relatively small and not very petal-like, though they are in the position normally occupied by petals; for this reason, some authorities (especially the German school, following Engler) refer to them as honey-leaves (Honigblätter). They are referred to throughout this account as petals. A further complication arises with the genus *Hepatica*; the flowers of this genus have 3 greenish segments borne immediately below a whorl of petal-like segments. These 3 greenish segments are referred to here as sepals, even though they probably represent the same structure as the involucre of 3 bracts found some distance below the flower in the related genus *Anemone* (in which the perianth is considered to consist of a single whorl).

1a. Leaves opposite or rarely absent; plants usually woody and/or climbing **26. Clematis**
  b. Leaves alternate or all basal (sometimes with a whorl of bracts below the flowers which can be mistaken for leaves); plants usually herbaceous and usually not climbing 2

2a. Plant a shrub with yellow wood; flowers small, brown-purple in drooping racemes; stamens 5–10 **3. Xanthorhiza**
  b. Combination of characters not as above 3

3a. Fruit 1 or more follicles 4
  b. Fruit a group of achenes, or berry-like 22
4a. Flowers bilaterally symmetric 5
  b. Flowers radially symmetric 7
5a. Ovary of a single carpel; follicle 1 **20. Consolida**
  b. Ovary of 2 or more carpels; follicles 2 or more 6
6a. Upper perianth-segment hooded or helmet-like **18. Aconitum**
  b. Upper perianth-segment spurred **19. Delphinium**
7a. Perianth of a single whorl of segments 8
  b. Perianth of 2 whorls of segments 10
8a. Perianth-segments 4, mauve or rarely white **2. Glaucidium**
  b. Perianth-segments 5 or more, yellow or white 9
9a. Leaves 2–3-pinnate **9. Isopyrum**
  b. Leaves simple **15. Caltha**
10a. Petals as large as or larger than the sepals 11
  b. Petals smaller than the sepals (if as long or longer, then narrower) 12
11a. Each petal with a backwardly-pointing spur (except in a few spurless cultivars) **11. Aquilegia**
  b. Petals not spurred but slightly pouched at base **10. Semiaquilegia**
12a. Flowering stem leafless but with a whorl of bracts just below the solitary, terminal flower 13
  b. Flowering stems without such a whorl of bracts 14
13a. Annual; flowers blue; leaves very finely divided **17. Nigella**
  b. Perennial; flowers white or yellow; leaves not finely divided **8. Eranthis**
14a. Outer perianth-segments persisting around the group of follicles **7. Helleborus**
  b. Outer perianth-segments deciduous in fruit 15
15a. Carpels (and follicles) united, at least towards the base **17. Nigella**
  b. Carpels (and follicles) completely free 16
16a. Sepals 3 **14. Anemonopsis**
  b. Sepals 4 or more 17
17a. Flowers very numerous, in long racemes or panicles **5. Cimicifuga**
  b. Flowers 1–4, solitary or in short racemes 18
18a. Leaves evergreen, all basal; carpels and follicles stalked **4. Coptis**
  b. Leaves deciduous; carpels and follicles not stalked 19
19a. Sepals mauve **12. Paraquilegia**
  b. Sepals yellow, orange or white 20

20a. Annual herb; sepals white, 4–5 mm
                      13. Leptopyrum

  b. Perennial herb; sepals larger, yellow
     or orange or more rarely white   21

21a. Leaves 2–3-pinnate   **9. Isopyrum**

  b. Leaves palmately divided **16. Trollius**

22a. Fruit berry-like   23

  b. Fruit dry, not berry-like   24

23a. Leaves 3, simple, palmately lobed, 1
     basal, 2 on the stem; flower solitary,
     terminal   **1. Hydrastis**

  b. Leaves several, pinnate; flowers in
     dense racemes   **6. Actaea**

24a. Perianth of a single whorl   25

  b. Perianth of 2 whorls   28

25a. Flowers subtended by a whorl of 3
     bracts   26

  b. Flowers not subtended by a whorl of
     bracts   **21. Thalictrum**

26a. Styles elongating and becoming
     feathery in fruit; nectar-secreting
     staminodes present   **24. Pulsatilla**

  b. Styles neither elongating nor
     becoming feathery in fruit; nectar-
     secreting staminodes absent   27

27a. Style absent, stigma broad and
     depressed, borne directly on top of the
     carpel; achenes strongly 8–10-ribbed
                    22. Anemonella

  b. Style present, stigma borne on its
     inner side; achenes not strongly
     ribbed   **23. Anemone**

28a. Sepals with backwardly directed spurs
     adpressed to the flower-stalk
                    28. Myosurus

  b. Sepals without backwardly-directed
     spurs   29

29a. Leaves all basal, main lobes 3–5,
     broad; sepals 3, green   **25. Hepatica**

  b. Combination of characters not as
     above   30

30a. Petals without nectaries **29. Adonis**

  b. Petals with nectaries   31

31a. Petals white or pinkish, each with an
     orange spot at the base; leaves
     pinnately divided **30. Callianthemum**

  b. Petals yellow, white, pink or red,
     without orange spots, if petals white
     or pink, leaves not pinnately divided
                   27. Ranunculus

1. HYDRASTIS Linnaeus
A.C. Whiteley

Perennial rhizomatous herbs with usually a single basal leaf and a simple, hairy stem bearing a solitary flower and 2 stem-leaves. Sepals 3, soon falling. Petals absent. Stamens numerous. Carpels 12 or more, forming a head of scarlet, usually 1-seeded berries in fruit.

A genus of 2 species, 1 from NE Asia, the other from northeastern N America; only the American species is commonly cultivated. It is easily grown in a shady position. Propagation is by division of the rhizome or by seed.

1. H. canadensis Linnaeus. Illustration: Botanical Magazine, 3019 (1830), 3232 (1833); Stary, Poisonous plants, 112–113 (1983).

Rootstock thick, yellow. Basal leaf with blade 12–20 cm wide and long, palmately 5–9-lobed, the lobes broad, with sharp, forwardly-pointing teeth. Leaf-stalk hairy, to 15 cm. Stem to 30 cm with leaves near the top. Flower to 1.2 cm across. Sepals greenish white or pinkish. Stamens to 5 mm. Carpels with short, curved styles. Head of fruit *c.* 1.7 cm. *Northeastern N America.* H2. Spring.

2. GLAUCIDIUM Siebold & Zuccarini
P.G. Barnes

Herbaceous perennial with short, stout rhizomes. Basal leaves several, thin, the blade 10–20 cm long and wide, kidney-shaped or rounded, palmately lobed with deeply toothed lobes, the base cordate; stalks 10–15 cm. Flowering stems usually solitary, erect, 20–40 cm, with 2 alternate, stalked leaves in the upper half. Flowers 5–10 cm wide, solitary, shortly stalked, borne in the axil of stalkless, leaf-like bracts. Sepals 4, petal-like, 3–5 × 2–4 cm, rounded or obovate, mauve; petals absent. Stamens numerous, 5–10 mm. Carpels 2, ovoid or oblong, 5 mm, united at the base and spreading widely, stigmas almost capitate. Fruit of 2 oblong, compressed follicles united at the base. Seeds numerous, flat, obovate, winged, *c.* 1.5 × 1 cm. A genus of a single species, requiring a cool, humus-rich soil and light shade. Propagation is by seeds, which should be sown fresh; the seedlings should be kept in a cold frame; they grow slowly and take several years to reach flowering size. Established plants may be divided in early spring.

1. G. palmatum Siebold & Zuccarini. Illustration: Botanical Magazine, 9432 (1936); Evans, The peat garden and its plants, pl. 14 (1974); Bulletin of the Alpine Garden Society **46**: 123 (1978). *Japan.* H3. Spring.

A white-flowered variant, var. **leucanthum** Makino is occasionally grown, but does not breed completely true from seed. Illustration: Hay & Synge, Dictionary of garden plants in colour, f. 80 (1969).

3. XANTHORHIZA Marshall
D.M. Miller

Deciduous shrubs with creeping rootstock and yellow wood. Stems 50–100 cm, erect, with a few branches from the upper parts. Leaves clustered at stem-apex, pinnate with 3–5 leaflets, each to 10 × 8 cm, stalkless, deeply lobed and irregularly toothed; leaf-stalk to 18 cm. Flowers numerous, 6–8 mm across, unisexual or bisexual, brown-purple, in simple or branched, drooping racemes to 15 cm, borne at the tops of the stems. Sepals 5, *c.* 3 × 2 mm, petal-like, ovate, spreading; petals 5, bilobed, very small. Stamens 5–10. Carpels 5–10. Fruit a single-seeded follicle.

A genus of a single species originally published as *Zanthorhiza* L'Heritier but generally recognised as *Xanthorhiza* Marshall. It spreads by suckers and makes an effective ground-cover in semi-shaded or open positions in moist soils. Propagation is by division in spring, or by seed if available.

1. X. simplicissima Marshall (*Zanthorhiza apiifolia* L'Héritier). Illustration: Botanical Magazine, 1736 (1815); Bean, Trees and shrubs hardy in the British Isles, edn 8, **4**: 760 (1980). *Eastern USA (New York to Florida).* H3. Spring.

4. COPTIS Salisbury
P.G. Barnes

Low-growing perennials with creeping yellow rhizomes. Leaves persistent, long-stalked, palmate or bipinnate, or divided into 3 leaflets. Flowering stems erect with 1–4 bisexual or unisexual flowers. Sepals 5–7, petal-like, often falling early. Petals 5–7, small, each with a narrow claw, either club-shaped or hooded with a nectary at the apex, or linear and with a nectary near the base. Stamens 10–25. Carpels 5–10, stalked. Fruit an umbel-like cluster of several stalked, oblong follicles, each with 4–8 seeds.

A genus of about 10 species native to temperate N America and Asia, requiring cool, moist, acid soil and light shade. Propagation is by seed, sown fresh and kept in a cold frame, or by division in spring.

1a. Leaves with 3 stalkless leaflets or
     palmate with 5 leaflets; stems
     1-flowered; sepals obovate   2

  b. Leaves divided into 3 leaflets or
     bipinnate, leaflets clearly stalked; stem
     usually 2–4-flowered; sepals
     lanceolate or linear   3

2a. Leaflets 3; follicles with long styles
                    1. trifolia

  b. Leaflets 5; follicles with very short
     styles   **2. quinquefolia**

3a. Leaves with 3 primary leaflets, these pinnatifid or pinnate; petals club-shaped, with a terminal nectary **3. japonica**

b. Leaves bipinnate; petals with a linear blade, nectary at the base **4. asplenifolia**

1. C. trifolia (Linnaeus) Salisbury. Illustration: Loddiges' Botanical Cabinet 2: pl. 173 (1821); Hitchcock et al., Vascular plants of the Pacific Northwest 2: 347 (1971); Clark, Wildflowers of the Pacific Northwest, 122 (1976).
Plant 5–10 cm, leaves divided into 3 leaflets. Leaflets 1–2.5 × 1–2 cm, stalkless, broadly obovate, slightly lobed and toothed. Stems 1-flowered. Sepals 5–10 mm, elliptic or narrowly obovate, white. Petals 3–5 mm, club-shaped, with a nectary at the apex. Follicles 3–5 mm, oblong, long-stalked, with styles 2–4 mm. *Western N America and NE Asia.* H3. Spring.

Subsp. **groenlandica** (Oeder) Hultén, from Greenland, differs in its shortly stalked leaflets, smaller sepals and longer follicles. Illustration: Gleason, Illustrated Flora of the northeastern United States and adjacent Canada 2: 162 (1952).

2. C. quinquefolia Miquel. Illustration: Terasaki, Nippon shokubutsu zufu, pl. 759 (1933); Makino, Illustrated Flora of Nippon, pl. 1712 (1942); Li et al., Flora of Taiwan 2: pl. 386 (1976).
Plant 10–20 cm, with palmate leaves. Leaflets 5, obovate, 1–2.5 cm × 5–20 mm, toothed and slightly lobed. Stems 1-flowered. Sepals 5–10 mm, elliptic or obovate, white. Petals *c.* 4 mm, club-shaped, each with a nectary at the hooded apex. Follicles short-styled. *Japan, Taiwan.* H4. Spring.

3. C. japonica (Thunberg) Makino. Illustration: Terasaki, Nippon shokubutsu zufu, pl. 4 (1933); Makino, Illustrated Flora of Nippon, pl. 1709 (1942).
Plant 10–25 cm, leaves divided into 3 leaflets or bipinnate. Leaflets 2–4 cm, stalked, broadly ovate, pinnatisect or pinnate, toothed. Stems 1–3-flowered. Sepals *c.* 8 × 2 mm, white. Petals club-shaped, *c.* 4 mm, each with a nectary at the hooded apex. Follicles with very short styles. *Japan.* H4. Spring.

A variable species. Var. **dissecta** (Yatabe) Nakai has bipinnate leaves and var. **major** (Miquel) Satake (*C. brachypetala* Siebold & Zuccarini) has tripinnate leaves.

4. C. asplenifolia Salisbury. Illustration: Hitchcock et al., Vascular plants of the Pacific Northwest 2: 347 (1971); Clark, Wildflowers of the Pacific Northwest, 167 (1976).
Plant 10–25 cm. Leaves long-stalked, glossy, triangular, bipinnate with 3–5 stalked primary leaflets. Leaflets toothed or pinnatifid. Stems 2–3-flowered; flower-stalks 2–4 cm. Sepals 6–10 mm, linear-lanceolate or thread-like, greenish white. Petals linear, each with a narrow blade above the nectary. Follicles 6–9 mm, styles short. *Northwest N America.* H3. Spring

5. CIMICIFUGA Linnaeus
J. Cullen
Large perennial herbs. Leaves all basal, or basal and on the stem, the lower long-stalked; blades 1–4 times divided into 3 segments, the ultimate segments toothed or lobed. Flowers radially symmetric in long racemes or spikes (sometimes branched below), each flower usually with a bract and 2 bracteoles (these sometimes inconspicuous or absent). Sepals 4–5 (rarely fewer), falling early, petal-like, usually whitish, sometimes brownish purple in bud. Petals to 8, (sometimes absent), usually bilobed or 2-fid, each bearing a nectary (the petals sometimes appear transitional to stamens). Stamens numerous. Carpels 1–8, free, shortly stalked or stalkless. Fruit a single follicle or a group of follicles, follicles stalked or not. Seeds smooth or scaly.

A genus of about 10 species from north temperate regions, grown for their long spikes or racemes of late, whitish flowers. They require moist soil and some shade; because of their size they tend to be planted at the backs of borders. Propagation is by division in spring or by seed, which should be sown as soon as it is ripe.
Literature: Thomas, G.S., Bugbane by any other name, *The Garden* **100**: 230–233 (1976).

1a. Carpel 1 (rarely 2) 2
b. Carpels 3 or more (rarely 2) 4
2a. Stem-leaves present; flowers shortly stalked **7. racemosa**
b. Stem-leaves absent or very small; flowers stalkless 3
3a. Ultimate segments of leaves acute or obtuse, hairy above, at least on the veins **5. japonica**
b. Ultimate segments of leaves acuminate, hairless above except on the margins **6. acerina**
4a. Carpels and follicles not stalked **4. dahurica**
b. Carpels and follicles stalked 5

5a. Bracts and bracteoles minute, inconspicuous or absent **3. simplex**
b. Bracts and bracteoles present, conspicuous though small 6
6a. Flowers white; bracteoles, or at least 1 of them, borne on the flower-stalk; follicle-stalk about as long as follicle **2. americana**
b. Flowers greenish white; bracteoles both borne at the base of the flower-stalk; follicle-stalk much shorter than the follicle **1. foetida**

1. C. foetida Linnaeus.
Plant to 2 m. Leaves divided into 3 segments which are themselves 2-pinnate or further divided into 3s, the ultimate segments deeply toothed, the terminal often 3-lobed. Bract and bracteoles borne at the base of the flower-stalk. Sepals greenish white. Carpels 4–8, stalked, the stalk much shorter than the follicle in fruit. *USSR (Siberia), E Asia.* H3. Summer–autumn.

2. C. americana Michaux. Illustration: Britton & Brown, Illustrated Flora of the northern United States and Canada 2: 57 (1897), edn 2 2: 92 (1913); Rickett, Wild flowers of the United States 1: pl. 34 (1966).
Plant to 2 m. Leaves 2–3 times divided into 3 segments, these segments pinnate with 3–5, ovate or oblong, deeply toothed ultimate segments. Racemes erect, often branched at the base. At least 1 of the bracteoles borne on the flower-stalk some distance above its base. Carpels 3–8, borne on short stalks; stigma minute; stalk in fruit about as long as the follicle. *Eastern USA.* H3. Summer–autumn.

3. C. simplex de Candolle. Illustration: Addisonia 2: t. 58 (1917); The Garden 100: 233 (1975); Kitamura & Murata, Coloured illustrations of herbaceous plants of Japan, pl. 50 (1977).
Plant to 1 m. Leaves 2–3 times divided into 3 segments, ultimate segments ovate or narrowly ovate, sometimes 3-lobed. Racemes often branched at the base, the branches arching; bracts and bracteoles minute, inconspicuous or absent. Carpels 3–7 (rarely 2), stalked, the stalk elongating in fruit. *Japan, E USSR (Sakhalin, Kamtchatka).* H2. Summer–autumn.

'Elstead' has brownish purple flower-buds.

4. C. dahurica (Turczaninow) Maximowicz. Illustration: Huxley, Garden perennials and water plants, pl. 85 (1971).
Plant to 2 m. Leaves 2–3 times divided into 3 segments, these segments pinnatifid or

lobed, the terminal segment cordate at base. Spikes numerous, often branched, the branches arching. Carpels usually 3 (more rarely 2), not stalked. Follicles not stalked. *E USSR, Korea.* H2. Summer–autumn.

5. C. japonica (Thunberg) Sprengel. Illustration: Flore des serres, ser. 2, **12**: t. 2363 (1874); Kitamura & Murata, Coloured illustrations of herbaceous plants of Japan, pl. 50 (1977).
Plant to 1 m. Leaves basal, stem-leaves absent or inconspicuous, 1–2 times divided into 3 segments, the ultimate segments ovate or broadly ovate, toothed or palmately lobed, acute or obtuse at apex, hairy on the veins above, hairy beneath. Spikes often branched at the base. Carpel 1 (rarely 2), not stalked. Follicle not stalked. *Japan.* H3. Summer–autumn.

6. C. acerina (Siebold & Zuccarini) Tanaka (*C. japonica* var. *acerina* (Siebold & Zuccarini) Huth). Illustration: Kitamura & Murata, Coloured illustrations of herbaceous plants of Japan, t. 50 (1977).
Like *C. japonica* but taller (to 1.3 m), leaflets acuminate, hairless above, except on the margins. *Japan.* H3. Summer–autumn.

7. C. racemosa (Linnaeus) Nuttall.
Very similar to *C. americana* but stems to 3 m, ultimate leaf-segments tapered to cordate at base; carpel 1 (rarely 2), not stalked; stigma broad and flat. *Eastern USA.* H3. Summer–autumn.

Variable; 2 varieties may be found in cultivation. Var. **racemosa**. Illustration: Bentley & Trimen, Medicinal plants **1**: t. 8 (1872); Britton & Brown, Illustrated Flora of the northern United States and Canada **2**: 56 (1897), ed. 2 **2**: 91 (1913); The Garden **100**: 233 (1975). Ultimate leaf-segments numerous, their bases tapered or rounded. Var. **cordifolia** (Pursh) Gray (*C. cordifolia* Pursh). Illustration: Botanical Magazine, 2069 (1819); Britton & Brown, Illustrated Flora of the northern United States and Canada **2**: 57 (1897), edn 2 **2**: 91 (1913). Ultimate leaf-segments fewer, their bases cordate.

6. ACTAEA Linnaeus
D.M. Miller
Perennial herbs with short rhizomes. Leaves from the base and on the stem, alternate, bi- or tripinnate; stem-leaves similar to but smaller than the basal. Inflorescence a dense raceme, usually terminal, with small, persistent bracts, lengthening in fruit. Flowers many, small, white. Sepals 3–5, petal-like, deciduous. Petals 4–10 (rarely 0),

spathulate. Stamens numerous, longer than petals. Carpel 1 with a 2-lobed stigma. Fruit a many-seeded berry.

A genus of about 8 species from north temperate regions. About 6 are in general cultivation for the glossy, brightly coloured berries, which are poisonous if eaten. They grow best in light shade but will tolerate most garden conditions in a moisture-retentive soil. Propagation is by division or by fresh seed. The species are difficult to distinguish without fruit.

1a.	Fruit-stalk thick and fleshy	2
b.	Fruit-stalk not thick and fleshy	3
2a.	Fruit white or occasionally red	
		1. alba
b.	Fruit black	**2. asiatica**
3a.	Fruit red or white	4
b.	Fruit black	5
4a.	Fruit more than 10 mm in diameter	
		3. rubra
b.	Fruit less than 10 mm in diameter	
		4. erythrocarpa
5a.	Leaflets with long acuminate tips, more or less hairless	**6. acuminata**
b.	Leaflets without long acuminate tips, sparsely hairy beneath	**5. spicata**

1. A. alba (Linnaeus) Milller (*A. spicata* Linnaeus var. *alba* Linnaeus; *A. pachypoda* Elliot). Illustration: Gleason, Illustrated Flora of northeastern United States and adjacent Canada **2**: 159 (1974).
Plant to 90 cm. Basal leaves to 60 cm, tripinnate, more or less hairless. Leaflets 3–8 × 2–5 cm, ovate, deeply toothed and incised, with an acuminate tip. Raceme *c.* 3 × 3 xm on a stalk 6–10 cm, sparsely hairy. Flowers to 10 mm on short stalks. Fruiting raceme to 20 cm on a stalk to 50 cm; berries to 8 mm across, spherical to ovoid, usually white, on a thickened, red, horizontal stalk to 3.5 cm. *Eastern & Central N America.* H2. Spring–Summer.

Forma **rubrocarpa** (Killip) Fernald, with red fruits, may be found occasionally.

2. A. asiatica Hara. Illustration: Iconographia cormophytorum Sinicorum **1**: 664 (1972).
Plant to 70 cm. Basal leaves to 50 cm, bi-to quadripinnate, with a few scattered hairs at first. Leaflets 4–10 × 2–6 cm, ovate, sharply toothed each with an acuminate tip. Raceme 3–5 cm, ovoid, with short hairs. Flowers to 6 mm across, on short stalks. Fruiting raceme to 10 cm; berries to 6 mm across, more or less spherical, black, on a thickened, dark red, horizontal stalk to 1.5 cm. *China, Korea, Japan.* H2. Spring–summer.

3. A. rubra (Aiton) Willdenow (*A. spicata* Linnaeus var. *rubra* Aiton). Illustration: Rickett, Wild flowers of the United States **5**(1): pl. 36 (1971); Hitchcock et al., Vascular plants of the Pacific Northwest **2**: 326 (1971); Clark, Wild flowers of the Pacific Northwest, pl. 158 (1976).
Plant to 80 cm. Basal leaves 40 cm or more, hairy at first. Leaflets 3–9 cm, ovate, toothed and lobed. Raceme 3–5 cm, ovoid, with short hairs. Flowers *c.* 8 mm across; flower-stalk to 10 mm. Fruiting raceme to 10 cm; berries to 1.4 cm across, spherical to ellipsoid, bright red, on a slender green stalk to 1.5 cm. *N America.* H2. Spring–summer.

Forma **neglecta** (Gillman) Robinson (*A. neglecta* Gillman; *A. alba* Mackenzie & Rydberg; *A. eburnea* Rydberg) is taller and has berries to 1.2 cm across, white, on slender stalks. Subsp. **arguta** (Nuttall) Hultén (*A. arguta* Nuttall) is a variant found in the western USA which is said to have smaller berries; it is usually included in *A. rubra*.

4. A. erythrocarpa Fischer.
Similar to *A. rubra* but fruit-stalks slender, deflexed; berries 8–9 mm across, red. *NE Europe to Japan, Siberia, Korea.* H1. Spring–summer.

Plants in cultivation under this name are probably *A. rubra*.

5. A. spicata Linnaeus (*A. nigra* Gaertner). Illustration: Ross-Craig, Drawings of British Plants **1**: t. 44 (1948).
Plant to 80 cm. Basal leaves to 60 cm, sparsely hairy beneath. Leaflets 6–10 × 3–6 cm, ovate, sharply toothed and incised. Raceme 2–6 cm, ovoid, on a stalk to 10 cm. Flowers 8–10 mm across on a stalk to 5 mm. Fruiting raceme 10 cm or more; berries 6–10 mm across, ovoid, black, on a slender stalk to 1.5 cm. *Europe to W Asia.* H1. Spring–summer.

6. A. acuminata Royle (*A. spicata* var. *acuminata* (Royle) Hara). Illustration: Polunin & Stainton, Flowers of the Himalayas, pl. 3 (1985).
Like *A. spicata* but leaflets with long acuminate tips, more or less hairless. *Afghanistan to China (Xizang).* H1. Summer.

Although occasionally listed, the genuine species may not be in cultivation.

7. HELLEBORUS Linnaeus
B. Mathew
Rhizomatous perennials, herbaceous or with rather woody overwintering stems. Leaves basal or on the stem, pedate,

palmate or with 3 leaflets. Flowers solitary or in cymes, with 5 large, persistent perianth-segments (sepals) and an inner whorl of 5–15 small, tubular or funnel-shaped nectaries (petals). Stamens numerous. Carpels usually 3–8, free or partly united. Fruit of 3–5 follicles, each containing several seeds.

A genus of 15 species mainly from Europe and western Asia. They are popular for their winter or early spring flowers, *H. niger* being one of the earliest. The hybrids of *H. orientalis*, are the most well-known, and there is a wide range of named cultivars. Hellebores are suitable for semi-shaded situations in good soil, well supplied with humus. Propagation is by seed or division of clumps in autumn or early spring; the divisions may take 1–2 years to become established.

Literature: Schiffner, V., Die Gattung Helleborus, *Botanische Jahrbücher* 11: 97–122 (1889); Ulbrich, E., Die Arten der Gattung Helleborus, *Blatter fur Staudenkunde* (1938); Merxmuller H. & Podlech, D., Über die europaischen Vertreter von Helleborus sect. Helleborus, *Feddes Repertorium* 64: 1–8 (1961); Mathew, B., A gardener's guide to hellebores, Bulletin of the Alpine Garden Society 35: 1–32 (1967).

1a. Plant dying down in summer; follicles greatly inflated   **1. vesicarius**
  b. Plant remaining leafy in summer; follicles not greatly inflated   2
2a. Basal leaves absent, the leaves and flowers carried on the same stem   3
  b. Basal leaves present, on stems separate from those of the flowers   5
3a. Leaves pedate with 7–11 segments   **2. foetidus**
  b. Leaves with 3 leaflets   4
4a. Leaflets spiny-toothed; plant to 1.2 m   **3. argutifolius**
  b. Leaflets entire or with a few widely spaced teeth; plant to 45 cm   **4. lividus**
5a. Bracts entire; flowers white, 4.5–11 cm in diameter   **5. niger**
  b. Bracts divided and toothed; flowers variously coloured, to 7 cm in diameter   6
6a. Follicles free at base, very shortly stalked   7
  b. Follicles shortly fused at base   8
7a. Flowers clear green; leaves densely downy beneath when young, not overwintering   **7. cyclophyllus**
  b. Flowers creamy white, purplish or greenish; leaves hairless or sparsely

hairy beneath, overwintering   **6. orientalis**
8a. Flowers purple or violet, at least outside   9
  b. Flowers green or yellowish green inside and out   11
9a. Leaves hairless beneath, strongly pedate with the central leaflet undivided   **12. atrorubens**
  b. Leaves downy beneath, especially when young, pedate or palmate, usually all leaflets divided   10
10a. Leaves palmate, more or less circular in outline; flowers 5–7 cm in diameter   **14. purpurascens**
  b. Leaves pedate, more or less kidney-shaped in outline; flowers usually 3–5 cm in diameter   **13. torquatus**
11a. Flowers usually 2.5–3.5 cm in diameter; leaves strongly pedate with the central leaflet undivided, hairless beneath   **11. dumetorum**
  b. Flowers at least 3.5 cm in diameter; leaves weakly pedate or with much-divided leaflets, or, if strongly pedate, then leaflets downy beneath   12
12a. Leaves overwintering, densely downy beneath, strongly pedate with the central leaflet undivided   **8. odorus**
  b. Leaves not overwintering, hairless or downy, but if densely downy then all leaflets much-divided   13
13a. Leaves with all leaflets divided into many narrow segments   **10. multifidus**
  b. Leaves with only some leaflets subdivided   **9. viridis**

1. H. vesicarius Aucher. Illustration: Botanical Magazine, n.s., 116 (1950); Journal of the Royal Horticultural Society 91: f. 185 (1966); Bulletin of the Alpine Garden Society 35: 12 (1967).
Herbaceous perennial with stems to 60 cm, dying down in summer after fruiting. Basal leaves with 3 coarsely toothed leaflets, the lateral ones deeply divided, all hairless; stem-leaves similar but smaller. Inflorescence a loose cyme; bracts divided into 3 and coarsely toothed; flower-stalks slender, 1–2.5 cm. Flowers bell-shaped, 1.5–2 × 1.5–1.7 cm, erect at first, becoming pendent; perianth-segments green with a purple or brownish stain near the apex. Carpels 3, fused at the base; follicles much-inflated when mature, forming a 3-winged, yellowish green fruit to 7.5 cm. *S Turkey*. H5. Spring.

2. H. foetidus Linnaeus. Illustration: Keble Martin, The concise British Flora in colour, pl. 4 (1965); Polunin, Flowers of Europe, pl.

19, f. 199 (1969); Fitter, New generation guide to the wild flowers of Britain and northwest Europe, 40 (1987); Gilmour & Walters, Wild flowers, pl. III (1954).
Evergreen perennial with rather woody stems to 80 cm which are replaced annually after fruiting. Basal leaves absent. Stem-leaves leathery, dark green, pedate with 7–10 segments; central leaflet undivided; segments narrowly lanceolate with toothed margins. Inflorescence a terminal, many-flowered cyme; bracts elliptic or ovate, the lower ones with leaf-like divisions at the apex, the upper ones entire. Flowers bell-shaped, 1.5–2 × 1.5–2.5 cm, pendent; segments green, often suffused purple-brown towards the apex. Carpels usually 3, fused at the base. *W Europe*. H2. Winter–spring.

3. H. argutifolius Viviani (*H. lividus* subsp. *corsicus* (Briquet) Yeo). Illustration: Bulletin of the Alpine Garden Society 35: 20 (1967); Botanical Magazine, n.s., 578 (1970).
Evergreen perennial with woody stems to 1.2 m, which are replaced annually after fruiting. Basal leaves absent. Stem-leaves with 3 leaflets, leathery, pale or mid-green, leaflets spiny-toothed, unequal, the central one regularly elliptic, the laterals irregular, unequal-sided. Inflorescence a many-flowered terminal cyme; bracts ovate or elliptic, the lower ones with leaf-like tips, the upper ones entire. Flowers cup-shaped, 2.5–5 cm in diameter, pendent at first, becoming horizontal or more or less erect; segments pale green. Carpels 3–5, fused at the base. *Corsica, Sardinia*. H4. Late winter–spring.

Hybrids between *H. argutifolius* and *H. lividus*, with intermediate characters, have been given the name **H. × sternii** Turrill. Hybrids between *H. argutifolius* and *H. niger* have been named **H. × nigercors** Wallich.

4. H. lividus Aiton. Illustration: Bulletin of the Alpine Garden Society 20: 23 (1952).
Evergreen perennial with woody stems to 45 cm, which are replaced annually after fruiting. Basal leaves absent. Stem-leaves with 3 leaflets, leathery, deep green, the upper surface with conspicuous pale veins, the lower surface suffused with pinkish purple; leaflets entire or with a few shallow teeth, the central elliptic, the laterals irregular, unequal-sided. Inflorescence a few-flowered terminal cyme; lower bracts 3-lobed and leaf-like, upper entire, ovate or elliptic. Flowers bowl-shaped to flattish, 3–5 cm in diameter, pendent at first, becoming horizontal or more or less erect;

segments creamy green suffused with pinkish purple. Carpels usually 5, fused at the base. *Spain (Majorca)*. H5. Winter–spring.

5. H. niger Linnaeus. Illustration: Bulletin of the Alpine Garden Society **35**: 19 (1967); Polunin, Flowers of Europe, pl. 19, f. 201 (1969).

Herbaceous perennial. Basal leaves overwintering, leathery, dark green, pedate with 7–9 segments; central leaflet undivided; segments oblong or oblanceolate, wedge-shaped at the base, toothed at the apex, hairless. Inflorescence often 1-flowered but sometimes with 2 or 3 flowers; inflorescence-stalk stout, 5–20 cm; bracts ovate or obovate, entire. Flowers flattish, 4.5–8 cm in diameter, usually facing outwards or slightly pendent; segments white, usually greenish at the base and often changing to pinkish purple with age. Carpels 5–8, fused at the base. *European Alps*. H1. Winter–spring.

A variable species. Subsp. **macranthus** (Freyn) Schiffner has bluish green, broadly lanceolate leaf-segments and larger flowers 8–11 cm in diameter. *Italy & N Yugoslavia*.

6. H. orientalis Lamarck (*H. olympicus* Lindley; *H. abchasicus* Braun; *H. antiquorum* Braun; *H. caucasicus* Braun; *H. guttatus* Braun & Sauer; *H. kochii* Schiffner). Illustration: Bulletin of the Alpine Garden Society **35**: 14 (1967).

Herbaceous perennial. Basal leaves overwintering, leathery, pedate with 5–11 segments; central leaflet usually undivided; segments broadly lanceolate or oblanceolate, coarsely toothed, hairless or sparsely downy beneath. Inflorescence a few-flowered cyme; inflorescence-stalk stout, to 60 cm; bracts with leaf-like divisions, toothed. Flowers flattish or cup-shaped, unscented, 5–7 cm in diameter, pendent or facing outwards, segments creamy white, green at the base, sometimes internally spotted with reddish purple and sometimes wholly suffused with purplish red, becoming greenish with age. Carpels 3–7, free at the base and each very shortly stalked. *Turkey, USSR (Caucasus)*. H3. Spring.

Very variable and sometimes divided into 3 species, largely based on flower-colour: *H. orientalis*, with a wide distribution, has creamy white flowers; *H. abchasicus* from the Caucasus has purplish red flowers; *H. guttatus*, also from the Caucasus, has white flowers spotted reddish purple inside. Many hybrids have been raised between these variants and a wide range of cultivars

is available; these do not breed true, so that named clones must be vegetatively propagated

7. H. cyclophyllus Boissier. Illustration: Polunin & Huxley, Flowers of the Mediterranean, pl. 19 (1965); Polunin, Flowers of Europe, pl. 18, f. 200 (1969). Herbaceous perennial. Basal leaves usually not overwintering, weakly pedate with 5–9 segments; central leaflet undivided; segments broadly lanceolate, toothed, densely downy beneath, especially when young. Inflorescence a few-flowered cyme; inflorescence-stalk stout, to 30 cm; bracts with leaf-like divisions, toothed. Flowers flattish or saucer-shaped, scented, 5–7 cm in diameter, pendent or facing outwards; segments clear green inside and out. Carpels 3–7, free at base, each shortly stalked. *S Yugoslavia, Greece, Bulgaria, Albania*. H4. Spring.

8. H. odorus Waldstein & Kitaibel. Illustration: Polunin, Flowers of Greece and the Balkans, pl. 5, f. 200c (1980). Herbaceous perennial. Basal leaves usually overwintering, strongly pedate with 7–11 segments; central leaflet often undivided but sometimes shallowly lobed; segments elliptic or oblanceolate, coarsely toothed, densely downy beneath. Inflorescence a few-flowered cyme; inflorescence-stalk 10–20 cm when the first flowers open, elongating during flowering; bracts with leaf-like divisions, toothed. Flowers saucer-shaped, scented, 5–6 (rarely to 7) cm in diameter, facing outwards or slightly pendent; segments clear green inside and out. Carpels 3–6, shortly fused at base. *C & N Yugoslavia, S Hungary, S Romania*. H3. Spring.

9. H. viridis Linnaeus. Illustration: Keble Martin, Concise British Flora in colour, pl. 4 (1965); Ary & Gregory, The Oxford book of wild flowers, pl. 51 No. 3 (1980); Fitter, New generation guide to the wild flowers of Britain and northwest Europe, 40 (1987).

Herbaceous perennial. Basal leaves not overwintering, weakly pedate with 7–13 segments; central leaflet usually undivided; segments oblanceolate, coarsely toothed, hairless or sparsely downy beneath. Inflorescence a few-flowered cyme; inflorescence-stalk 10–20 cm; bracts with leaf-like divisions, toothed, often coarsely so. Flowers cup-shaped, not scented, 3.5–5 cm in diameter, more or less pendent; segments green inside and out or occasionally with a purple blotch inside

near the base. Carpels 3–5, shortly fused at base. *W & C Europe*. H3. Spring.

The western and northern European plants including those from Britain are sometimes distinguished as subsp. **occidentalis** (Reuter) Schiffner; they have more coarsely toothed leaf-margins and smaller flowers and are less showy as garden plants than subsp. **viridis** from central Europe.

10. H. multifidus Visiani (*H. bocconei* Tenore; *H. siculus* Schiffner). Illustration: Bulletin of the Alpine Garden Society **35**: 12 (1967); Botanical Magazine, n.s., 698 (1975).

Herbaceous perennial. Basal leaves not overwintering, weakly pedate but the arrangement obscured because each leaflet, including the central, is divided into many narrow segments, giving a total of 30 or more per leaf; segments linear-lanceolate, toothed, downy beneath. Inflorescence a few-flowered cyme; inflorescence-stalk slender, 15–25 cm; bracts with leaf-like divisions, toothed. Flowers cup-shaped or somewhat conical, scented, pendent, 3.5–4.5 cm in diameter; segments yellowish green inside and out, or occasionally with a faint purplish suffusion on the outside. Carpels 5–7, shortly fused at base. *W Yugoslavia*. H4. Spring.

Plants with rather few leaf-divisions, from NW Yugoslavia, known as subsp. **istriacus** (Schiffner) Merxmuller & Podlech, are occasionally cultivated.

11. H. dumetorum Waldstein & Kitaibel (*H. pallidus* Host; *H. viridis* subsp. *dumetorum* (Waldstein & Kitaibel) Hayek). Illustration: Polunin, Flowers of the Greece and the Balkans, pl. 5, f. 200e (1980). Herbaceous perennial. Basal leaves not overwintering, strongly pedate, usually with 9–11 segments; central leaflet undivided; segments narrowly oblong-lanceolate or narrowly elliptic, hairless or with few hairs beneath. Inflorescence a few-flowered cyme; inflorescence-stalk slender, 15–25 cm, elongating slightly in fruit; bracts with leaf-like divisions, finely toothed. Flowers cup-shaped, not scented, usually 2.5–3.5 cm in diameter, pendent; segments green inside and out. Carpels 2–5, shortly fused at base. *E Europe*. H2. Spring.

12. H. atrorubens Waldstein & Kitaibel (*H. atropurpureus* Schultes; *H. cupreus* Host; *H. dumetorum* subsp. *atrorubens* (Waldstein & Kitaibel) Merxmuller & Podlech). Illustration: Waldstein & Kitaibel, Icones

Plantarum Rariorum Hungariae **3**: pl. 271 (1812); Reichenbach, Icones Florae Germanicae et Helveticae, pl. 160 (1838–39).

Herbaceous perennial. Basal leaves not overwintering, strongly pedate with 7–11 segments; central leaflet undivided; segments narrowly elliptic, hairless. Inflorescence a few-flowered cyme; inflorescence-stalk 15–30 cm; bracts with leaf-like divisions, toothed. Flowers flattish when fully open, not scented, usually 4–5 cm in diameter, facing outwards or slightly pendent; segments deep violet outside, violet or greenish inside. Carpels 5–6, shortly fused at base. *NW Yugoslavia.* H2. Spring.

A plant which has been in cultivation under the name *H. intermedius* Host is probably a variable natural hybrid between *H. atrorubens* and a green-flowered species, most likely *H. multifidus.* The name *H. atrorubens* is often misapplied to a frequently cultivated plant with large purple flowers produced very early in the year in the winter. It is almost certainly a hybrid, possibly between *H. orientalis* and *H. purpurascens.*

13. H. torquatus Archer-Hind (*H. serbicus* Adamovič; *H. multifidus* subsp. *serbicus* (Adamovič) Merxmuller & Podlech). Illustration: Journal of the Royal Horticultural Society **91**: f. 186 (1966) – as *H. dumetorum* subsp. *atrorubens*; Polunin, Flowers of Greece and the Balkans, pl. 5, f. 200d (1980).

Herbaceous perennial. Basal leaves not overwintering, pedate with up to c. 30 segments, all leaflets, including the central, divided; segments narrowly lanceolate, downy beneath. Inflorescence a few-flowered cyme; inflorescence-stalk 15–25 cm at first, elongating in fruit; bracts with leaf-like divisions, toothed. Flowers cup-shaped, not scented, usually 3–5 cm in diameter, pendent or facing outwards; segments deep purple outside, purple, green or purple-streaked inside. Carpels 3–5, shortly fused at base. *C & S Yugoslavia.* H3. Spring.

H. torquatus has been hybridised with cultivars of *H. orientalis*, giving rise to very dark purple-flowered plants such as the well-known 'Ballard's Black'.

14. H. purpurascens Waldstein & Kitaibel. Illustration: Bulletin of the Alpine Garden Society **29**(1): 37 (1961).

Herbaceous perennial. Basal leaves not overwintering, palmate, usually with 5 primary divisions and each of these divided into 2–6 segments, rarely some undivided; segments lanceolate, downy beneath. Inflorescence a few-flowered cyme; inflorescence-stalk stout, 5–15 cm at first, elongating to 15–30 cm in fruit; bracts with leaf-like divisions, toothed. Flowers cup-shaped, not scented or sometimes unpleasantly scented, 5–7 cm in diameter, pendent or facing outwards; segments purple and glaucous outside, purple or greenish suffused with purple inside. Carpels 5–7, shortly fused at base. *E to C Europe.* H3. Spring.

8. ERANTHIS Salisbury
F. McIntosh

Low perennial herbs with tuberous rootstocks. Leaves mostly basal, palmately or pinnately divided, stalked; stem-leaves similar but stalkless, arranged in a whorl, forming an involucre just beneath the solitary, terminal flower. Sepals 5–8, yellow or white. Petals modified into 2-lipped, tubular nectaries. Stamens many. Carpels few to many, fruit a group of follicles.

A genus of abaout 8 species from Eurasia.

1a. Leaves pinnately divided; sepals white
           1. pinnatifida
  b. Leaves palmately divided; sepals yellow   2
2a. Plants 5–15 cm tall   **2. hyemalis**
  b. Plants less than 5 cm tall   **3. sibirica**

1. E. pinnatifida Maximowicz.
Plants 10–15 cm tall. Basal and stem leaves pinnately divided. Flowers stalked above the involucre. Sepals white. Carpels scarcely stalked. *Japan.* H1 or 2.

2. E. hyemalis (Linnaeus) Salisbury. Illustration: Hay & Synge, Dictionary of garden plants, f. 729 (1971); Journal of the Royal Horticultural Society **97**: 44 (1972); Mathew, Dwarf bulbs, pl. 39 (1973); The Garden **101**: 351 (1976).

Plants 5–15 cm tall. Basal leaves long-stalked, generally palmately 3–5-lobed almost to the base, the lobes further cut into segments. Involucral leaves similar, the lobes divided to various depths. Flowers 5–6 cm in diameter, opening wide. Sepals yellow. Follicles to 1.5 cm, brownish. *Europe, SW Asia as far east as Afghanistan.* H1. Winter–spring.

A rather variable and widely distributed species. Plants from SE Turkey tend to have leaves with more numerous leaflets and smaller flowers and have been called *E. cilicicus* Schott & Kotschy; Illustration: Hay & Synge, Dictionary of garden plants, f. 728 (1971); Gartenpraxis **1**: 40 (1984). The variation between these and the more usual type is, however, quite continuous (see Blakelock, Kew Bulletin for 1948: 378, 1949) and the two are not really distinguishable in the wild, though, by selection, they can be in gardens. The situation is compounded by the existence of a hybrid between *E. hyemalis* and '*E. cilicicus*', known in gardens as *E.* × *tubergenii* Hoog (illustration: Hay & Synge, Dictionary of garden plants, f. 730 (1971)); Gartenpraxis **1**: 40, 1984), which tends to a rather more clumped growth than the other variants.

3. E. sibirica de Candolle. Very similar to *E. hyemalis*, but much smaller, plants never more than 5 cm tall. *Eastern USSR (Siberia).* H1.

9. ISOPYRUM Linnaeus
A.C. Whiteley

Perennial herbs with rhizomatous or tuberous rootstocks. Stems slender, branching, hairless. Leaves bi- or tripinnate, the divisions all 3-parted, ultimate leaflets 2–3-lobed. Flowers axillary and terminal, white or pinkish. Sepals 5, petal-like, deciduous. Petals absent or reduced to small nectaries. Stamens 10–40, shorter than sepals. Carpels 2–6, occasionally more, styles pointed. Fruit a group of oblong to ovoid follicles, each containing 2 or more seeds.

A genus of about 20 species distributed in the northern hemisphere; American species have been placed in the genus *Enemion* Rafinesque. They are suitable for a shady position in moist, rich soil. Propagation is by division of the rootstock or by seed.

Literature: Drummond, J.R. & Hutchinson, J., A revision of Isopyrum and its nearer allies, *Kew Bulletin* for 1920: 145–169.

1a. Follicles 2; rootstock rhizomatous
           1. thalictroides
  b. Follicles 3–6; rootstock tuberous
           2. biternatum

1. I. thalictroides Linnaeus. Illustration: Hegi, Illustrierte Flora von Mitteleuropa **3**: 479 (1909); Kew Bulletin for 1920, 162 (1920).

Stems slender, 10–30 cm, often glaucous, from a creeping rhizome. Basal leaves stalked, ultimate leaflets 3-lobed. Stem-leaves with stipules, stalkless, the upper often simply 3-lobed. Flowers 1–2 cm across, white. Petals 1–1.5 mm. Follicles 2, ovoid, flattened. *C Europe.* H2. Spring.

2. I. biternatum (Rafinesque) Torrey & Gray (*Enemion biternatum* Rafinesque). Illustration: Gleason, Illustrated flora of the northeastern United States and adjacent Canada **2**: 168 (1952); Kew Bulletin for 1920, 160.

Stems erect, to 30 cm or more. Roots fibrous with small tubers. Basal leaves long-stalked, stem-leaves shortly stalked or stalkless, 1–2 times divided or simply 3-lobed. Flowers 1–2 cm across, white. Petals absent. Follicles usually 4, spreading. *Northwestern N America.* H2. Spring.

10. SEMIAQUILEGIA Makino
P.G. Barnes

Tufted herbaceous perennials with long-stalked basal leaves and leafy stems. Leaves pinnate or bipinnate with 3 main divisions, the leaflets usually long-stalked, the segments lobed. Flowering stems bearing several flowers in a loose panicle. Sepals 5, petal-like, somewhat spreading. Petals 5, erect, concave or pouched at the base. Stamens numerous with several small, membranous, lanceolate staminodes. Carpels 5–10, long-styled. Fruit of 3–5 erect, many-seeded follicles.

A genus of about 6 species native to E Asia. They are easily grown in any well-drained soil and an open situation. Propagation is by seed, sown fresh, or by division.
Literature: Drummond, J.R. & Hutchinson, H., A revision of Isopyrum and its nearer allies, *Kew Bulletin* for 1920: 145–169; Ulbrich, E., Ranunculaceae novae vel criticae VII, *Notitzblatt des Königlichen botanischen Gartens und Museums zu Berlin* **9**: 209–228 (1925).

1. S. ecalcarata (Maximowicz) Sprague & Hutchinson (*S. simulatrix* Drummond & Hutchinson; *Aquilegia ecalcarata* Maximowicz).
Illustration: Botanical Magazine, 9382 (1935); Bulletin of the Alpine Garden Society 50: 303 (1982).
Leaves bipinnate with stalked primary divisions. Leaflets 1.5–2.5 cm, obovate, tapered to the base, lobed, slightly glaucous. Stem 20–40 cm, flowers pendent, stalks 3–7 cm. Sepals 1.3–1.6 cm × 4–6 mm, lanceolate, acute. Petals 1.5–1.8 cm × 5–7 mm, oblong, truncate and notched at apex, pouched at the base. *China.* H3. Early summer.

11. AQUILEGIA Linnaeus
D.M. Miller & P.G. Barnes
Perennial herbs. Basal leaves tufted, usually twice or more divided into 3 segments, on long stalks with expanded bases. Stem-leaves usually present but less divided than basal leaves. Stems often branched, with leaf-like bracts and few to many, terminal, erect or pendent flowers. Sepals 5, petal-like, usually spreading and smaller than petals. Petals 5 each with a flat, broad blade and a backwardly-projecting, hollow, nectar-producing spur (absent in a few cultivars). Stamens many, the innermost reduced to membranous staminodes. Carpels usually 5, free, stalkless. Fruit a group of erect, many-seeded follicles.

A genus of about 70 species from north temperate areas, grown for their showy flowers. The smaller species may be grown in the rock garden, the larger in a border, preferably with a light, sandy soil. They may be propagated by seed or by division in spring. Many hybrid groups are grown, most of which are derived from *A. vulgaris*, *A. chrysantha* and *A. caerulea*.
Literature: Munz, P.A., Aquilegia, the wild and cultivated columbines, *Gentes Herbarum* **7**: 1–150 (1946).

1a. Petal-spurs absent
Some cultivated varieties
  b. Petal-spurs present   2
2a. Spurs strongly hooked   3
  b. Spurs straight or curved but not hooked   11
3a. Flowers of one colour   4
  b. Flowers distinctly bicoloured   6
4a. Spur of petal distinctly longer than blade   **1. atrata**
  b. Spur of petal as long as or shorter than blade   5
5a. Follicles hairless   **2. sibirica**
  b. Follicles glandular-hairy   **3. vulgaris**
6a. Flowers deep red to purple, and white   **4. oxysepala**
  b. Flowers not as above   7
7a. Follicles hairy; blade of petals whitish   8
  b. Follicles hairless; blade of petals cream or yellow   10
8a. Flowers more or less erect   **5. glandulosa**
  b. Flowers arching over or drooping   9
9a. Stem-leaves absent or very small   **6. amaliae**
  b. Stem-leaves like the basal leaves but smaller   **7. olympica**
10a. Sepals 2 cm or more   **8. flabellata**
  b. Sepals to 1.3 cm   **9. saximontana**
11a. Flowers blue or blue and white   12
  b. Flowers neither blue nor blue and white   23
12a. Flowers distinctly bicoloured   13
  b. Flowers of shades of one colour   16

13a. Spurs of petals much longer than blades   14
  b. Spurs of petals as long as or shorter than blades   16
14a. Plant 20–80 cm, branched   **10. caerulea**
  b. Plant 5–20 cm, densely tufted   **11. scopulorum**
15a. Stem glandular-hairy; flowers many, pendent   **15. guarensis**
  b. Stem more or less hairless; flowers few, almost erect   **14. discolor**
16a. Follicles hairless   **12. jonesii**
  b. Follicles hairy or sticky   17
17a. Stamens projecting beyond petals by more than 2 mm   18
  b. Stamens not projecting, or projecting less than 1 mm   19
18a. Flowers purplish blue   **16. grata**
  b. Flowers blue, sepals with greenish tips   **17. nevadensis**
19a. Spurs of petals longer than blades   **18. alpina**
  b. Spurs of petals not longer than blades   20
20a. Leaflets hairless beneath or almost so   21
  b. Leaflets distinctly hairy beneath   22
21a. Spurs 1–1.6 cm; sepals 2–3.5 cm   **13. pyrenaica**
  b. Spurs 7–10 mm; sepals 1.5–1.9 cm   **19. einseleana**
22a. Sepals 9–14 mm wide; blades of petals truncate   **21. bertolonii**
  b. Sepals 7–8 mm wide; blades of petals rounded   **20. thalictrifolia**
23a. Flowers yellow or yellow and red   24
  b. Flowers not yellow or yellow and red   29
24a. Flowers yellow and red   25
  b. Flowers yellow (or yellow and cream)   27
25a. Sepals widely spreading to reflexed   **22. formosa**
  b. Sepals not spreading   26
26a. Sepals about 2 × as long as petal-blades; spurs 2–2.5 cm   **23. canadensis**
  b. Sepals 1–1.5 × as long as petal-blades; spurs 1.6–2 cm   **24. elegantula**
27a. Flowers pale yellow; spurs 9–15 cm   **25. longissima**
  b. Flowers bright yellow or yellow and cream; spurs less than 8 cm   28
28a. Flowers bright yellow, erect   **26. chrysantha**
  b. Flowers yellow and cream, pendent   **27. flavescens**
29a. Flowers purple and cream   30
  b. Flowers not purple and cream   31

30a. Flowers not fragrant; sepals to 2.5 cm
                     28. buergeriana
  b. Flowers fragrant; sepals 2.5–3 cm
                         29. fragrans
31a. Flowers cream to white; sepals widely
    spreading to reflexed   **30. lactiflora**
  b. Flowers not cream or white; sepals
    not widely spreading   32
32a. Flowers fragrant, green to purplish
    green   **31. viridiflora**
  b. Flowers not fragrant, greenish orange
    to red   **32. skinneri**

1. A. atrata Koch (*A. atroviolacea* Beck).
Illustration: Reichenbach, Icones florae
Germanicae et Helveticae 1: t. 115
(1838–39); Bonnier, Flore complète 1: pl.
18 (1911); Gentes Herbarum 7: f. 17
(1946).
Plant to 80 cm. Leaflets of basal leaves to
3 cm, obovate, 2–3-lobed, each lobe deeply
toothed, hairless, glaucous beneath. Stem
hairy, especially in the upper part. Stem-
leaves smaller than basal leaves. Flowers
few to many, pendent, dark purple. Sepals
1.5–2.5 cm × 8–9 mm. Blades of petals
8–12 × 7–9 mm; spurs longer than blades,
hooked, hairy. Stamens much longer than
petal-blade. Follicles usually 5, glandular-
hairy, 1.5–2 cm, styles 8–10 mm. *Alps
(Italy, Germany, Switzerland, Austria).* H1.
Early summer.

2. A. sibirica Lamarck. Illustration: Sweet,
British flower garden 4: pl. 90 (1831);
Gentes Herbarum 7: f. 6 (1946).
Plant 30–70 cm. Leaflets of basal leaves
1–4 cm, rounded or obovate, deeply
3-lobed, glaucous and hairless beneath.
Stems leafless but for 1–2 small, 3-lobed,
stalkless bracts near the top. Flowers 1–3,
pendent. Sepals 2–3 cm, broadly elliptic,
blue or purple. Blades of petals 1–1.3 cm,
rounded, blue. Spurs 5–15 mm, strongly
hooked. Follicles 2–2.5 cm, hairless. *USSR
(Siberia).* H2. Early summer.

3. A. vulgaris Linnaeus. Illustration: Gentes
Herbarum 7: f. 13 (1946); Ross-Craig,
Drawings of British plants 1: pl. 41 (1948);
Keble Martin, The concise British Flora in
colour, pl. 4 (1965).
Plant to 60 cm. Leaflets of basal leaves
1.5–4 cm, rounded to ovate or obovate,
deeply 2–3-lobed, each lobe scolloped,
glaucous and sparingly hairy beneath.
Stems hairy especially in the upper part.
Stem-leaves less divided than the basal.
Flowers few to many, pendent, usually blue
to purple, sometimes pinkish or white.
Sepals 1.8–2.5 × 1–1.2 cm. Blades of petals
1–1.3 cm × 9–10 mm, with rounded tip;

spurs about as long as blades, strongly
hooked, hairy. Stamens about as long as
the petal-blades. Follicles usually 5, densely
glandular-hairy. 1.5–2.5 cm; styles to
7 mm, hairless. *Most of Europe.* H3. Early
summer.

    This is an extremely variable species
which has been in cultivation for hundreds
of years and from which many cultivated
groups raised from seed or propagated
vegetatively have been developed. Colour
variants (e.g. 'Nivea') and double-flowered
plants ('Flore Plena') of various colours,
with spurs, may be grown. Plants with
double, spurless flowers may be found
under the name 'var. *stellata*', or, if the
flowers are very large, under '*A.
clemataquila*'. Plants with semi-double, red
and white, spurred flowers are sometimes
listed as 'var. *caryophylloides*'. The name
'*A. baicalensis*' is sometimes applied to a
dwarf, short-spurred variant of *A. vulgaris.*

4. A. oxysepala Trautvetter & Meyer.
Illustration: Gentes Herbarum 7: f. 15
(1946).
Plant 50–100 cm. Leaflets of basal leaves
3–6 cm, broadly obovate, deeply 3-lobed
and toothed, slightly glaucous beneath.
Stem-leaves several, the upper stalkless and
1–3-lobed. Flowers several, pendent. Sepals
2–3 cm, narrowly ovate, acute, deep red to
violet. Blades of petals *c.* 1.2 cm, oblong,
truncate, cream. Spurs 1.5–2 cm, strongly
hooked, dark red or purple. Stamens not
projecting. Follicles 2.5–3 cm, glandular-
hairy. *USSR (Siberia).* H2. Early summer.

5. A. glandulosa Link. Illustration: Sweet,
British flower garden 4: pl. 55 (1830);
Wehrhahn, Die Gartenstauden, 1: 428
(1931); Gentes Herbarum 7: f. 10 (1946).
Plant 10–40 cm. Leaflets of basal leaves
1–4 cm, rounded, deeply 3-lobed, glaucous
on both surfaces. Stem almost leafless,
bracts few. Flowers 1–5, nearly erect.
Sepals 2–4.5 × 1.5–2.5 cm, broadly ovate,
spreading widely, light blue. Blades of
petals 1.5–2.5 × 1–1.5 cm, rounded, violet-
blue to white. Spurs 5–12 mm, blue,
strongly hooked. Follicles 6–12, glandular-
hairy, 2–3 cm. *USSR (Siberia).* H2. Early
summer.

6. A. amaliae Boissier. Illustration: Gentes
Herbarum 7: f. 16 (1946); Bulletin of the
Alpine Garden Society 37: 263 (1969);
Polunin, Flowers of Greece and the
Balkans, 126 (1980).
Plant to 30 cm. Leaflets of basal leaves to
3 cm, obovate, deeply 2–3-lobed, glaucous
and with long hairs beneath. Stem

glandular-hairy; stem-leaves none or very
small. Flowers 1–3, pendent, bicoloured.
Sepals 1.8 cm × 8–9 mm, pale bluish
purple, spreading. Blades of petals
1.3–1.4 cm × 8 mm, tips rounded, white;
spurs as long as blades, pale purple to
white, hooked, hairless. Stamens as long as
blades. Follicles 5, glandular-hairy. *Balkan
peninsula (Greece, Albania, Yugoslavia).* H4.
Early summer.

7. A. olympica Boissier. Illustration: Revue
Horticole for 1896, 108; Gentes Herbarum
7: f. 14 (1946).
Plant to 60 cm. Leaflets of basal leaves
ovate to obovate, deeply 2–3-lobed, each
lobe scolloped, sparsely hairy beneath. Stem
glandular-hairy in the upper part; stem-
leaves similar to the basal but smaller.
Flowers several, pendent, bicoloured. Sepals
2–4.5 × 1.3–2 cm, pale bluish purple.
Blades of petals 1.4–2 cm, whitish. Spurs
about as long as blades, bluish, hooked.
Stamens as long as blades. Follicles 5–8,
glandular-hairy, 2–3 cm; styles 4–5 mm.
USSR (S Russia), Turkey, Iran. H4. Early
summer.

8. A. flabellata Siebold & Zuccarini
(*A. japonica* Nakai & Hara; *A. amurensis*
misapplied). Illustration: Gentes Herbarum
7: f. 6 (1946); Suzuki, Alpine plants in
eastern Japan, 118 (1982).
Plant 15–45 cm. Leaflets of basal leaves
1.5–4 cm, broadly obovate, slightly
glaucous or purple-tinged above, glaucous
beneath. Stem-leaves few, the upper small,
stalkless and with narrower leaflets.
Flowers 1–3, pendent. Sepals broadly
elliptic, somewhat spreading, violet-blue.
Blades of petals *c.* 1.5 cm, cream, blue at
the base. Spurs 8–20 mm, strongly hooked
or incurved, blue. Stamens not projecting.
Follicles usually 5, hairless. *Japan.* H3. Early
summer.

    Var. **pumila** Kudo (*A. akitensis* Huth) is a
commonly cultivated dwarf variant, usually
less than 30 cm tall, with 1–2 flowers and
smaller leaflets. Illustration: Readers' Digest
encyclopedia of garden plants and flowers,
51 (1973). 'Nana Alba' is a white-flowered
cultivar.

9. A. saximontana Rydberg. Illustration:
Gentes Herbarum 7: f. 5 (1946); Rickett,
Wild flowers of the United States 6: pl. 51
(1973).
Compact, tufted plant to 25 cm. Leaves
mostly basal. Leaflets of basal leaves
1–1.5 cm, broadly obovate, mostly 3-lobed,
hairless. Stems 1-flowered, the flower
pendent. Sepals 9–12 mm, ovate-elliptic,

blue. Blades of petals 7–8 mm, rounded, yellow. Spurs 3–7 mm, hooked, blue. Follicles *c.* 1 cm, hairless; styles 5 mm. *Midwestern USA.* H2. Spring.

10. A. caerulea James. Illustration: Botanical Magazine, 5477 (1864); The Garden 16: 264 (1879); Gentes Herbarum 7: f. 35 (1946); Rickett, Wild flowers of the United States 6: pl. 52 (1973).
Plant 20–80 cm. Leaflets of basal leaves deeply lobed and toothed, usually hairy beneath. Stem-leaves similar but smaller, becoming bract-like above. Stem loosely branched with several erect flowers. Sepals 2–4 cm, ovate, blue, spreading widely. Blades of petals 1.5–2.5 cm. oblong, almost truncate, white. Spurs 3–4.5 cm, slender, nearly straight, blue. Stamens shorter than blades. Follicles 5–10, hairless, 2–3 cm; styles *c.* 1 mm, hairy. *Midwestern USA.* H2. Early summer.

A variable species and a parent of many garden hybrids in a wide range of colours.

11. A. scopulorum Tidestrom. Illustration: Bulletin of the Alpine Garden Society 10: 222 (1942); Gentes Herbarum 7: f. 7 (1946).
Densely tufted plant, 5–20 cm. Leaves glaucous, leaflets of basal leaves 5–15 mm, rounded, deeply 3-lobed and overlapping; stem-leaves few, the upper 3-lobed and bract-like. Flowers 1 or few, erect. Sepals 1.5–2 cm, ovate, blue or white (rarely red), widely spreading. Blades of petals *c.* 1 cm, oblong, rounded, white, cream, blue or rarely red. Spurs 2.5–3.5 cm, straight, usually blue or yellowish. Stamens not or little projecting. Follicles 1.5–1.8 cm, glandular-hairy; styles *c.* 8 mm. *Midwestern USA.* H2. Early summer.

12. A. jonesii Parry. Illustration: Bulletin of the Alpine Garden Society 8: 146 (1940); Gentes Herbarum 7: f. 7 (1946); Rickett, Wild flowers of the United States 6: pl. 51 (1973); Everett, New York Botanical Garden illustrated encyclopedia of horticulture 1: opposite p. 184 (1980).
Compactly tufted plant 5–10 cm. Leaves all basal; leaflets rounded, deeply 3-lobed, glaucous, hairy. Stems 1-flowered, hairy, flower erect. Sepals *c.* 2 cm × 9 mm, ovate, blue or purple. Blades of petals oblong, rounded. Spurs 8–15 mm, nearly straight. Follicles 5–6, hairless, 1.4–2.2 cm; styles *c.* 1.2 cm. *Western N America (Rocky mountains).* H2. Spring.

13. A. pyrenaica de Candolle (*A. beata* Rapaics). Illustration: Bonnier, Flore complète 1: pl. 18 (1911); Gentes Herbarum 7: f. 20 (1946); Botanical Magazine, n.s., 435 (1963).
Plant to 30 cm. Leaflets of basal leaves 5–15 mm, obovate, deeply 2–3-lobed with each lobe itself lobed, stalkless, glaucous and hairless beneath. Stem glandular-hairy; stem-leaves usually undivided, linear. Flowers 1–3, pendent, bright blue. Sepals 2–3.5 × 1–1.5 cm, spreading. Blades of petals 1.2–1.5 cm × 8–10 mm, with rounded tips. Spurs more or less equal to petal-blades, slender, straight or curved, hairy. Stamens shorter than to as long as petal-blades. Follicles 5, glandular-hairy, 1.3–1.7 cm; style 8 mm. *Spain, France (Pyrenees).* H2. Early summer.

14. A. discolor Levier & Leresche. Illustration: Bulletin of the Alpine Garden Society 3: 49 (1939).
Similar to *A. pyrenaica* and sometimes included in it, but with smaller, nearly erect flowers with whitish petal-blades and more or less hairless, few-flowered stems. *N Spain.* H3. Early summer.

15. A. guarensis Losa. Similar to *A. discolor* but stems glandular-hairy, branched and with many flowers; leaflets with short, down-like hairs. *NE Spain.* H3. Early summer.

16. A. grata Zimmeter. Illustration: Botanical Magazine, 9405 (1935); Gentes Herbarum 7: f. 18 (1946).
Plant to 45 cm. Leaflets of basal leaves to 4 cm, obovate, 3-lobed, toothed, glandular-hairy above and beneath, greyish green beneath. Stem glandular-hairy; stem-leaves with 3 linear lobes. Flowers 3–5, becoming erect, reddish blue. Sepals to 2.9 × 1.2 cm; blades of petals to 1 cm; spurs longer than blades, straight. Stamens longer than blades. Follicles to 1.5 cm, glandular-hairy. *Yugoslavia.* H4. Early summer.

17. A. nevadensis Boissier & Reuter. Plant to 60 cm. Leaflets of basal leaves obovate to circular, 3-lobed, glandular-hairy above and beneath. Stems glandular-hairy; stem-leaves divided into 3 lobes. Flowers pendent, pale blue. Sepals narrow, spreading, each with a greenish tip. Blades of petals with rounded tip. Spurs longer than blades, straight to slightly curved. Stamens longer than the blades, Follicles 5, sticky. *S Spain.* H5. Early summer.

18. A. alpina Linnaeus. Illustration: Botanical Magazine, 8303 (1910); Bonnier, Flore complète 1: pl. 18 (1911); Gentes Herbarum 7: f. 20 (1946).
Plant to 80 cm. Leaflets of basal leaves to 3 cm, obovate to rounded, deeply 2–3-lobed, more or less hairless and somewhat glaucous beneath. Stem hairy especially in the upper part; stem-leaves similar to the basal but smaller. Flowers 2–3, pendent, bright blue. Sepals 3–4.5 × 1.4–2 cm, spreading. Petals with blades 1.4–1.7 cm × 8–11 mm, tips rounded. Spurs longer than blades, straight to curved. Stamens shorter than blades. Follicles 5–7, hairy, to 2.8 cm, styles 6–7 mm. *Alps (France, Switzerland, Italy).* H2. Early summer.

Variants with different flower-colours (e.g. 'Alba') are cultivated. 'Atroviolacea', with dark reddish purple flowers, may be of hybrid origin.

19. A. einseleana Schultz. Illustration: Hegi, Illustrierte Flora von Mitteleuropa 3: 484 (1909); Gentes Herbarum 7: f. 19 (1946); Kohlhoupt, I fiori della Dolomiti, pl. 7 (1984).
Plant to 45 cm. Basal leaves few; leaflets to 2 cm, obovate, 2–3-lobed, stalkless, somewhat leathery, usually hairless and glaucous beneath. Stem sparingly glandular-hairy in the upper part; stem-leaves small with 1–3 linear lobes. Flowers 1–3, pendent, bluish purple. Sepals 1.5–1.9 cm × 7–8 mm, spreading. Blades of petals 8–10 × 6–9 mm, tips rounded. Spurs more or less equal to blades, almost straight, hairy. Stamens shorter than blades. Follicles 5, glandular-hairy, 9–10 mm; styles 6–7 mm. *Alps (Italy, Austria, Germany).* H2. Early summer.

20. A. thalictrifolia Schott & Kotschy. Illustration: Hegi, Illustrierte Flora von Mitteleuropa 3: 484 (1909); Gentes Herbarum 7: f. 18 (1946).
Similar to *A. einseleana* but with slightly larger flowers and with glandular hairs on stems and leaves. *N Italy.* H3. Early summer.

21. A. bertolonii Schott (*A. reuteri* Boissier). Illustration: Coste, Flore de la France 1: 54 (1901); Gentes Herbarum 7: f. 19 (1946).
Plant to 30 cm. Leaflets of basal leaves to 2 cm, obovate, 2–3-lobed, stalkless, glaucous and hairy beneath. Stem glandular-hairy in upper part; stem-leaves linear or with 3 linear lobes. Flowers 1–3, pendent, bluish purple. Sepals 1.8–3.3 cm × 9–14 mm, spreading. Blades of petals 1–1.4 cm × 6–8 mm, tips truncate. Spurs more or less equal to the blades, straight to slightly curved, hairy. Stamens shorter than the blades. Follicles 5, hairy, to 1.2 cm; styles to 8 mm,

glandular-hairy. *S France, Italy*. H4. Early summer.

22. A. formosa Fischer. Illustration: Botanical Magazine, 6552A (1881); Clark, Wild flowers of the Pacific Northwest, 170 (1976); Rickett, Wild flowers of the United States 6: pl. 51 (1973).
Plant 50–100 cm. Basal leaves with leaflets 2–4 cm, obovate or rounded, deeply lobed and toothed, usually hairy beneath. Lower stem-leaves similar, the upper stalkless, with 1–3 narrow lobes. Stems branched, flowers numerous, pendent. Sepals 1.5–2.5 cm × 5–10 mm, narrowly ovate, widely spreading or reflexed, red. Blades of petals 4–5 mm, truncate or rounded, yellow. Spurs 1–2 cm, straight, red. Stamens projecting *c.* 1 cm. Follicles glandular-hairy, 1.5–2.5 cm, styles 1–1.5 cm. *Western N America*. H3. Early summer.

23. A. canadensis Linnaeus. Illustration: Botanical Magazine, 246 (1793); Gentes Herbarum 7: f. 32 (1946); Gleason, Illustrated Flora of the eastern United States and adjacent Canada 2: 165 (1974); Everett, New York Botanical Garden illustrated encyclopedia of horticulture 1: 213 (1980).
Plant 15–40 cm. Basal leaves with leaflets 1–2 cm, obovate, lobed, somewhat glaucous and usually hairy beneath. Stem leaves several, similar to basal leaves, the upper less divided. Stems branched, flowers several, pendent. Sepals 1–1.4 cm, ovate, not spreading widely, red. Blades of petals 6–8 mm, with a small abrupt point, yellow. Spurs 2–2.5 cm, straight, rather stout in the basal half, red. Stamens projecting. Follicles glandular-hairy, 1.5–2.5 cm, styles 1–2 cm. *Eastern USA*. H2. Early summer.

24. A. elegantula Greene. Illustration: Gentes Herbarum 7: f. 31 (1946).
Plant 10–40 cm. Leaflets of basal leaves 1–3 cm, broadly obovate, deeply lobed, glaucous beneath. Stem-leaves smaller, the upper much reduced. Flowers usually 1–4, pendent. Sepals 7–11 mm, erect, ovate, red (sometimes yellow at the tip). Blades of petals 6 mm, rounded, yellow. Spurs *c.* 2 cm, red, straight. Stamens projecting. Follicles 1.3–2 cm, glandular-hairy, styles *c.* 1.5 cm. *Southern USA and Mexico*. H4. Early summer.

25. A. longissima Watson. Illustration: Gardeners' Chronicle 95: 385 (1934); Bulletin of the Alpine Garden Society 5: 162 (1937); Gentes Herbarum 7: f. 37 (1946); Rickett, Wild flowers of the United States 4: pl. 43 (1970).
Plant 50–90 cm. Leaflets of basal leaves obovate or wedge-shaped, deeply lobed. Stem-leaves several, the lower resembling the basal. Flowers several, erect, pale yellow. Sepals 2.5–3 cm, lanceolate, spreading. Blades of petals obovate, rounded. Spurs 9–15 mm, slender, straight. Stamens projecting *c.* 1 cm. Follicles to 2.5 cm, glandular-hairy, styles 1.5–2.5 cm. *Southern USA and Mexico*. H4. Summer.

26. A. chrysantha Gray. Illustration: Botanical Magazine, 6073 (1873); The Garden 16: 264 (1879); Gentes Herbarum 7: f. 36 (1946); Rickett, Wild flowers of the United States 6: pl. 51 (1973).
Plant 40–100 cm. Leaflets of basal leaves 1–3 cm, wedge-shaped or obovate, 3-lobed, hairy or hairless beneath. Stem-leaves numerous, the upper small and bract-like. Stem branched with several erect, bright yellow flowers. Sepals 2–3.5 cm, narrowly ovate. Blades of petals 8–16 mm, oblong, rounded, rather spreading. Spurs 4–7 cm, diverging, slender. Stamens projecting by 1 cm. Follicles 2–3 cm, glandular-hairy, styles *c.* 1.6 cm. *Midwestern USA*. H2. Early summer.
    A parent of various long-spurred garden hybrids.

27. A. flavescens Watson (*A. formosa* var. *flavescens* Hooker). Illustration: Botanical Magazine, 6552B (1881); Hitchcock et al., Vascular plants of the Pacific Northwest 2: 334 (1971); Clark, Wild flowers of the Pacific Northwest, 170 (1976).
Plant 20–70 cm. Leaflets of the basal leaves 1–4 cm, wedge-shaped or obovate, sometimes hairy beneath. Stem-leaves mostly small and bract-like. Stem branched, flowers numerous. Sepals 1.5–2.2 cm, narrowly ovate, yellow, widely spreading or reflexed. Blades of petals 6–10 mm, rounded, cream. Spurs 6–18 mm, pale yellow, curved near the tips. Follicles 1.5–2.2 cm, glandular-hairy, styles 8–10 cm. *Western N America*. H2. Early summer.

28. A. buergeriana Siebold & Zuccarini. Illustration: Makino, Illustrated Flora of Nippon, f. 1702 (1942); Gentes Herbarum 7: f. 23 (1946).
Plant 50–80 cm. Leaflets of basal leaves 1–4 cm, obovate, deeply 2-or 3-lobed, glaucous beneath. Stem leafy, branched, flowers several, pendent. Sepals 1.5–2.5 cm, narrowly ovate, spreading, purple (rarely pale yellow). Blades of petals 1–1.5 cm, truncate, cream. Spurs 1.5–2 cm, slightly curved, purple. Stamens shorter than petal-blades. Follicles 5–7, densely hairy, 2–2.5 cm. *Japan*. H3. Early summer.

29. A. fragrans Bentham (*A. glauca* Lindley; *A. suaveolens* Durand & Jackson). Illustration: Maund, Botanist 4: pl. 181 (1840); Gentes Herbarum 7: f. 21 (1946); Polunin & Stainton, Flowers of the Himalaya, pl. 1 (1985).
Plant 40–80 cm. Leaflets of lower leaves 2–4 cm, obovate, deeply 3-lobed and toothed, glaucous beneath. Stem-leaves several, the upper smaller and less divided. Flowers several, horizontal or pendent, scented. Sepals 2.5–3 cm, narrowly ovate, spreading, pale purple. Petals white, the blades *c.* 1.8 × 1.4 cm. Spurs 1.5–1.8 cm, curved. Stamens equalling the blades of the petals. Follicles 6–9, densely hairy, to 2 cm, styles *c.* 8 mm. *Himalaya*. H4. Summer.

30. A. lactiflora Karelin & Kirilow. Illustration: Gentes Herbarum 7: f. 25 (1946).
Plant 40–80 cm. Leaflets of basal leaves 1–2 cm, deeply 3-lobed, hairy beneath. Upper stem-leaves bract-like, simple or 3-lobed. Flowers 1–several, somewhat pendent, white or cream. Sepals *c.* 1.6 cm, narrowly ovate, widely spreading or reflexed. Blades of petals 7–10 mm. Spurs 1.5–2.5 cm, straight. Follicles 5–6, glandular-hairy, 1.3–1.5 cm, styles *c.* 6 mm. *Southern USSR*. H2. Spring.

31. A. viridiflora Pallas. Illustration: The Garden 80: 252 (1916); Gentes Herbarum 7: f. 22 (1946); Iconographia cormophytorum sinicorum 1: f. 1338 (1972).
Plant 30–50 cm. Leaflets of basal leaves 1–3 cm, wedge-shaped or obovate, glaucous and slightly hairy beneath. Stem-leaves several. Flowers several, somewhat pendent, scented. Sepals 1–1.5 cm, ovate, not spreading widely. Petals brownish or purplish green, the blades obovate, truncate. Spurs 1–1.8 cm, straight. Stamens slightly projecting. Follicles glandular-hairy, 1–1.5 cm. *N China*. H2. Early summer.

32. A. skinneri Hooker. Illustration: Botanical Magazine, 3919 (1842); Wehrhahn, Die Gartenstauden, 1: 429 (1931); Gentes Herbarum 7: f. 33 (1946).
Plant 60–100 cm. Leaflets of basal leaves 1.5–4.5 cm, broadly obovate, deeply lobed, slightly hairy beneath. Stem-leaves similar, the upper linear, bract-like. Flowers several,

pendent. Sepals 1.8–2.8 cm, narrowly ovate, slightly spreading, greenish yellow. Blades of petals rounded, *c.* 8 mm, greenish. Spurs 3.5–5 cm, straight, red. Stamens projecting by 1–2 cm. Follicles 2–3 cm, glandular-hairy, styles *c.* 2 cm. *N Mexico.* H4. Early summer.

12. PARAQUILEGIA Drummond & Hutchinson
P.G. Barnes

Densely tufted perennials with the remains of the leaf-stalks persisting and crowded at the top of the rootstock. Leaves long-stalked, pinnate or bipinnate (each into 3 divisions), with long-stalked primary leaflets and deeply dissected ultimate segments. Flowering stems several, 1-flowered, with 2 small bracts above the middle. Sepals 5, petal-like. Petals 5, small, oblong or rounded, notched, slightly concave at the base. Stamens numerous. Fruit of 5–7, usually erect follicles with shining, slightly keeled seeds.

A genus of about 6 species native to mountainous areas from C Asia to W China. Although hardy in Europe, they are not easily cultivated, requiring cool conditions and very free drainage. They are usually grown in scree beds, alpine houses or crevices in tufa blocks. Propagation is by seed, sown when fresh and kept in a cold frame. Stem-cuttings can be taken in early summer.
Literature: Drummond, J.R. & Hutchinson, J., A revision of Isopyrum and its nearer allies, *Kew Bulletin* for 1920, 145–169; Ulbrich, E., Ranunculaceae novae vel criticae VII, *Notitzblatt des Königlichen botanischen Gartens und Museums zu Berlin* 9: 209–228 (1925).

1. P. microphylla (Royle) Drummond & Hutchinson.
Illustration: Kew Bulletin for 1920, f. 2; Bulletin of the Alpine Garden Society 46: 290 (1978); Polunin & Stainton, Flowers of the Himalaya, pl. 2, f. 4 (1984).
Plant 5–10 cm. Leaves mostly bipinnate, hairless and glaucous. Flowers 2–3 cm wide. Sepals 1.5–2.5 × 1–2 cm, broadly elliptic, obtuse or acute, pale mauve. Petals *c.* 6 mm. Follicles erect, slightly compressed, 1.2 cm × 4 mm, style persistent, 2 mm. Seeds hairless. *Bhutan, India (Sikkim), W China.* H3. Spring.

Plants from the eastern end of the range are said to be the most satisfactory.
P. anemonoides (Willdenow) Ulbrich (*P. grandiflora* (de Candolle) Drummond & Hutchinson is distinguished by its minutely hairy seeds.

13. LEPTOPYRUM Reichenbach
A.C. Whiteley

Annual, tufted herbs with slender stems, erect or spreading, 15–20 cm. Basal and stem leaves divided into numerous fine segments. Stem-leaves often whorled. Flowers solitary at the tips of the shoots, to 1 cm across. Sepals 4–5, ovate, white, to 5 mm. Petals 4–5, yellow, minute. Stamens numerous. Carpels 12–20, narrow, pointed; follicles with 1–several seeds.

A genus of a single species from eastern Asia, easily grown in a shady position in rich soil.
Literature: Drummond, J.R. & Hutchinson, J., A revision of Isopyrum and its nearer allies, *Kew Bulletin* for 1920, 145–169.

1. L. fumarioides (Linnaeus) Reichenbach (*Isopyrum fumarioides* Linnaeus).
Illustration: Reichenbach, Icones florae Germanicae et Helveticae 1: t. 113 (1839); Kew Bulletin for 1920, 159. *E Asia, naturalised in parts of Europe.* H2. Summer.

14. ANEMONOPSIS Siebold & Zuccarini
P.G. Barnes

Herbaceous perennial, 60–100 cm. Leaves hairless, mostly basal, 3 times pinnate with ovate, acute or acuminate, sharply toothed or lobed segments; stalk and axis blackish, blade light green, paler beneath; stalk expanded at the base and clasping the stem. Stem erect, almost black, hairless or sparsely woolly near the nodes, bearing smaller leaves above. Inflorescence a loose panicle or raceme, the branches with small leaf-like or undivided, oblong bracts at the base. Flowers *c.* 3 cm across, pendent. Sepals 3, oblong, *c.* 1.5 cm × 7 mm, pale mauve. Petals 8–10, almost square, *c.* 5 mm, deep violet-purple, white at the base. Stamens numerous, the filaments flattened below the anthers, greenish. Carpels 2–4, styles slender with small terminal stigmas. Fruit of 2–4 many-seeded follicles.

A genus of a single species resembling *Anemone × hybrida* and flowering in late summer and autumn. It requires an acid or neutral soil rich in humus and not subject to drought. A position in light shade is suitable, but a more open situation may be tolerated in cooler areas. Propagation is by seed, which should be sown fresh and exposed to frost. Careful division of established plants may also be attempted.

1. A. macrophylla Siebold & Zuccarini.
Illustration: Botanical Magazine, 6413 (1879); Revue Horticole for 1909, 510; Wehrhahn, Die Gartenstauden 1: 395

(1931); Thomas, Perennial garden plants, edn 2, pl. 2A (1982). *Japan.* H3. Late summer–autumn.

15. CALTHA Linnaeus
F. McIntosh

Low, fleshy, perennial herbs generally growing in damp places. Leaves alternate, stalked, entire or toothed. Flowers axillary or terminal, stalked. Sepals 5–9 or more, petal-like, yellow, white or rarely pink. Petals absent. Stamens numerous. Ovary of 4–many carpels. Fruit of 4–many follicles.

A genus of about 20 species from both north and south temperate zones, generally growing in damp places at the edges of ponds.

1a.	Plants to 60 cm; leaf-blades 5–20 cm wide; sepals usually 5, bright yellow **1. palustris**
b.	Plants to 30 cm; leaf-blades 1–5 cm wide; sepals 6–12, white 2
2a.	Leaves longer than broad; follicles scarcely stalked at maturity **2. leptosepala**
b.	Leaves broader than long; follicles distinctly stalked at maturity **3. biflora**

1. C. palustris Linnaeus.
Illustration: Rickett, Wild flowers of the United States 1: pl. 178 (1964); The Garden 101: 204 (1976); Perry, The water garden, 72 (1981).
Stems to 60 cm, hollow, branched above. Basal leaves long-stalked, stalks of the upper leaves progressively shorter, blades kidney-shaped, 5–20 cm wide. Flowers to 5 cm in diameter. Sepals usually 5, bright yellow (white in var. **alba** Anon.). Follicles shortly stalked. *North temperate regions.* H1. Spring.

2. C. leptosepala de Candolle.
Illustration: Clements, Plant physiology and ecology, 166 (1907); Rickett, Wild flowers of the United States 4: pl. 46 (1970).
Plant to 30 cm. Leaves longer than broad, the blades 1–5 cm wide. Flowers 1–many, borne on stalks from ground-level. Sepals 6–12, white. Follicles scarcely stalked at maturity. *Western N America.* H1. Late spring.

3. C. biflora de Candolle.
Illustration: Rickett, Wild flowers of the United States 5: pl. 127 (1971).
Like *C. leptosepala* but leaves broader than long, sepals 6–9, follicles distinctly stalked at maturity. *Western N America.* H1.

16. TROLLIUS Linnaeus
F. McIntosh

Herbaceous perennials. Roots thickened, fibrous. Stems erect. Leaves basal and borne on the stem, palmately lobed or divided, the segments usually further divided or toothed. Flowers terminal, usually solitary, large. Outer perianth-segments (sepals) petal-like, 5–20 or more, yellow, white or purplish. Inner perianth-segments (petals) 5–15, longer or shorter than the outer, each with a nectary pit at the base. Stamens numerous. Carpels 5–20 or more. Fruit a group of follicles.

A genus of perhaps 20 species from Eurasia.

Plant. Dwarf with conspicuous sheathing leaf-bases: **7**.

Leaves. Mostly in a basal rosette: **5**.

Outer perianth-segments. Conspicuously incurved, producing a globe-shaped flower: **4**. Orange inside, purplish outside: **5**.

Inner perianth-segments. Conspicuous: **1,2**. With spoon-shaped tips: **3–5**. Much shorter than the stamens: **6**.

Style. Conspicuous, to 5 mm long: **8**.

1a. Inner perianth-segments 2–3 × as long as stamens   **2**
  b. Inner perianth-segments shorter than or equal to the stamens   **3**
2a. Inner perianth-segments to 1.5 cm; follicles (including the style) to 8 mm   **1. asiaticus**
  b. Inner perianth-segments more than 2.5 cm; follicles 9–12 mm   **2. chinensis**
3a. Styles to 1 mm; follicles (including styles) to 1 cm   **4**
  b. Styles 2–5 mm; follicles 1.1–2 cm   **5**
4a. Stamens to 8 mm; outer perianth-segments spreading   **3. yunnanensis**
  b. Stamens 9–11 mm; outer perianth-segments strongly incurved, the flower globe-shaped   **4. europaeus**
5a. Outer perianth-segments orange inside, deep purplish crimson outside; most leaves in a basal rosette   **5. pumilus**
  b. Outer perianth-segments yellow inside and out; leaves arranged along the stem, or, if mostly basal, with conspicuous sheathing bases   **6**
6a. Inner perianth-segments darker in colour than the outer; follicles to 2 cm, styles 2–5 mm   **7**
  b. Inner and outer perianth-segments of the same colour; follicles to 1.3 cm, styles to 2 mm   **7. laxus**
7a. Plant to 25 cm leaf-bases conspicuous, sheathing   **6. acaulis**
  b. Plant to 55 cm; leaf bases not conspicuously sheathing   **8. ranunculinus**

1. T. asiaticus Linnaeus (*T. giganteus* invalid). Illustration: Robinson, The English flower garden, 282 (1883).
Plant to 60 cm. Leaves basal and on the stem, mostly 5-parted, the segments finely divided and toothed. Flowers *c.* 4.5 cm in diameter. Outer perianth-segments 10–15 or more, spreading, golden yellow, inner perianth-segments 2–3 × as long as the stamens, to 1.5 cm. Carpels usually numerous. Follicles (including the styles) to 8 mm. *USSR (Turkestan, Siberia).* H1. Spring–summer.

The name *T. giganteus* has been applied to large variants of this species. 'Fortunei' has doubled outer perianth-segments.

2. T. chinensis Bunge (*T. ledebourii* Reichenbach). Illustration: Botanical Magazine, 8565 (1914); Parey's Blumengärtnerei, 631 (1954); Hay & Synge, Dictionary of garden plants, 176 (1969); Flora reipublicae popularis sinicae 27: 84 (1979).
Plant to 85 cm. Leaves basal and on the stem, mostly with 5 leaflets which are finely divided and toothed. Flowers to 5 cm in diameter. Outer perianth-segments spreading, golden yellow, inner perianth-segments 2–3 times as long as the stamens, to 2.5 cm, protruding beyond the outer perianth-segments. Stamens numerous. Carpels 11–17. Follicles to 1.2 cm. *NE China.* H1. Summer.

3. T. yunnanensis (Franchet) Ulbrich. Illustration: Botanical Magazine, 9143 (1927–1928); Thomas, Perennial garden plants, pl. X, 1 (1976); Flora reipublicae popularis sinicae 27: 76 (1979); Everett, New York Botanical Gardens illustrated encyclopedia of horticulture, 3412 (1982).
Plant to 70 cm. Leaves basal and on the stem, 3-lobed or with the 2 outer lobes deeply divided and appearing 5-lobed, lobes divided and toothed. Flowers 2–6 cm in diameter. Outer perianth-segments spreading, yellow, inner perianth-segments to 7 mm, almost as long as the stamens. Stamens numerous, to 8 mm. Carpels 11–20 or more. Follicles to 1 cm, styles to 1 mm. *SW China (Yunnan).* H5. Summer.

4. T. europaeus Linnaeus. Illustration: Robinson, The English flower garden, 715 (1893); Perrin & Boulger, British flowers 2: pl. LXX (1914); Ary & Gregory, The Oxford book of wild flowers, 4 (1962); Everett, The New York Botanical Gardens illustrated encyclopedia of horticulture, 3411 (1982).
Plant to 80 cm. Leaves basal and also on the stem, 5-lobed, the lobes divided and toothed. Flowers to 5 cm in diameter. Outer perianth-segments yellow, strongly incurved producing a globe-shaped flower, inner perianth-segments shorter than the stamens. Stamens 9–11 mm. Carpels more than 20. Follicles to 1 cm, styles to 1 mm. *Most of Europe, USSR (Caucasus), northern N America.* H1. Summer.

A variable species. The invalid name *T. giganteus* has been applied to large variants (see above under *T. asiaticus*). Hybrids between this species and *T. asiaticus* and *T. chinensis* are widely cultivated and are referred to as **T. × cultorum** Bergmans (*T. × hybridus* invalid).

5. T. pumilus D. Don. Illustration: Flora reipublicae popularis sinicae 27: 76 (1979); Everett, The New York Botanical Gardens illustrated encyclopedia of horticulture, 3412 (1982).
Dwarf plants to 30 cm. Leaves usually all basal, 5-lobed with divided lobes. Flowers *c.* 3 cm in diameter. Outer perianth-segments spreading, usually orange within and red- or purple-crimson outside. Inner perianth-segments shorter than stamens. Carpels 6–20, usually 14–18. Follicles 1.1–2 cm, styles 2–5 mm. *Himalaya.* H5. Summer.

'Farreri' (*T. farreri* Stapf) is a somewhat taller variant.

6. T. acaulis Lindley. Illustration: Everett, The New York Botanical Gardens illustrated encyclopedia of horticulture, 3412 (1982).
Very similar to *T. pumilus*, but with leaves on the stem as well as in a basal rosette, and the outer perianth segments entirely yellow. *Himalaya.* H5. Summer.

7. T. laxus Salisbury. Illustration: Abrams & Ferris, Illustrated flora of the Pacific States, 177 (1944); Rickett, Wild flowers of the United States 1: pl. 35 (1969), 5: pl. 141 (1971), 6: pl. 57 (1973).
Plant to 55 cm. Leaves basal and on the stems, divided into 5 leaflets which are divided and toothed. Flowers 2.5–5 cm in diameter. Outer perianth-segments spreading, inner perianth-segments shorter than the stamens, all of the same yellow. Carpels usually 11 or 12. Follicles to 1.3 cm, style to 2 mm. *Northern part of N America.* H1. Spring–summer.

'Albiflorus' (*T. albiflorus* Rydberg) has white perianth-segments.

8. T. ranunculinus (Smith) Stearn
(*T. patulus* Salisbury; *T. caucasicus* Steven).
Plant to 55 cm, conspicuously leafy. Leaves
basal and on the stems, 5-lobed, the lobes
finely divided and toothed. Flowers
2.5–4 cm in diameter. Outer perianth-
segments spreading, yellow. Inner perianth-
segments about as long as the stamens,
darker yellow than the outer perianth-
segments. Carpels usually 14 or more.
Follicles to 2 cm, styles 2–5 mm. *Turkey,
USSR (Caucasus), Iran.* H5.
Spring–summer.

17. NIGELLA Linnaeus
A.C. Whiteley
Annual, simple or branching herbs to
80 cm. Leaves alternate, 1–3 times
pinnatisect, the ultimate divisions often
thread-like. Flowers terminal and axillary.
Sepals 5, often petal-like, blue, yellow or
white, falling in fruit. Petals 5–10, much
smaller than sepals, nectar-producing,
2-lipped, the outer lip bifid. Stamens
numerous. Carpels usually 5, united, at
least at the base, the ripened follicles
forming a capsule with long, persistent
styles. Seeds numerous.

A genus of about 22 species native to the
Mediterranean area and extending
northwards to Germany and eastwards to
Iran. They are easily grown by sowing
seeds in spring in the place where the
plants are to flower.

1a. Involucre present; fruit inflated
                                3. damascena
  b. Involucre absent; fruit not inflated   2
2a. Flowers *c.* 6 cm across, deep blue;
      stamens deep red   **1. hispanica**
  b. Flowers *c.* 4 cm across, bluish white;
      stamens pale   **2. sativa**

1. N. hispanica Linnaeus. Illustration:
Botanical Magazine, 1265 (1810); Marshall
Cavendish Encyclopedia of gardening, 1328
(1968–70); Polunin & Smythies, Flowers of
southwest Europe, 188 (1973).
Stems branched, 35–45 cm. Flowers
4–7 cm across, deep blue with red stamens.
Involucre absent. Carpels reddish, fully
united, usually densely glandular, the
glands persisting in fruit. *Spain, S France.*
H3. Summer.

Cultivars 'Alba' and 'Atropurpurea' are
occasionally seen.

2. N. sativa Linnaeus. Illustration: Polunin,
Flowers of Europe, pl. 18 (1969).
Stems erect, branched, hairy, to 30 cm.
Flowers 3.5–4.5 cm across, bluish white,
the sepals ovate. Involucre absent. Carpels

fully united, narrow, tuberculate. Styles as
long as capsules in fruit. *SE Europe, SW
Asia, widely naturalised elsewhere.* H3.
Summer.

Widely cultivated in S Europe for the
aromatic seeds used to flavour bread.

3. N. damascena Linnaeus. Illustration:
Botanical Magazine, 22 (1787); Marshall
Cavendish encyclopedia of gardening,
1328, 1329 (1968–70); Polunin, Flowers
of Europe, pl. 18 (1969).
Stems erect, simple or branched,
20–50 cm. Flowers 3.5–4.5 cm across,
blue, with a stiff involucre of segments
which resemble the leaves. Carpels fully
united, ripening to a smooth, inflated
capsule with 10 chambers, the inner 5
containing the seeds. Styles shorter than
capsule in fruit. *S Europe.* H2. Summer.

Widely cultivated and naturalised. Most
cultivars have double or semi-double
flowers, and colours vary from white or
pink to blue and purple. Common cultivars
are 'Miss Jekyll' (blue) and 'Persian Jewels'
(double, mixed colours).

18. ACONITUM Linnaeus
J. Cullen
Perennial herbs. Roots usually tuberous,
sometimes long and clustered. Stems erect,
or scrambling or climbing, often very leafy.
Leaves ovate to almost circular in outline,
base often cordate, the blade deeply
palmately lobed or divided into 3–7 lobes
or separate leaflets, the lobes or leaflets
themselves toothed or shallowly to deeply
lobed. Inflorescence a raceme or panicle;
flower-stalks long or short, stiff or flexuous,
each bearing a bracteole. Flowers showy,
bilaterally symmetric. Sepals 5, petal-like
(the conspicuous parts of the flower), the
uppermost forming a large, erect,
hemispheric to cylindric hood and known
as the helmet. Petals 2–10, small, hidden
within the sepals, the 2 uppermost with
long, nectar-secreting spurs which project
into the helmet. Stamens numerous.
Carpels 3–5. free. Fruit a group of 3–5
follicles. Seeds angled or winged, sometimes
with transverse plates or folds.

A genus of perhaps 300 species from
north temperate regions. Its classification is
very difficult and the recognition of species
is not easy. Fortunately only a small
number of species persist in general
cultivation, though many more have been
introduced over the past 2 centuries. All
are poisonous, some containing medically
active compounds which have led to their
cultivation over long periods. They are

easily grown in any good garden soil and
propagation is generally by seed.
Literature: Munz, P.A., The Cultivated
Aconites, *Gentes Herbarum* 6: 463–506
(1945).

Roots. Long, clustered: **1**; tuberous: **2–11**.
Stems. Erect: **1–7**; twining or scrambling, at
    least above: **8–11**.
Leaves. Divided to the base into distinct
    leaflets: **3–5,9,10**; divided to two-thirds
    or more of their diameter but not to the
    base: **1,6–8,11**; divided into linear
    segments: **2**.
Flower-stalks. Covered with long, spreading
    or slightly deflexed, sometimes glandular,
    hairs: **4,8,10**; hairless or with sparse to
    dense crisped hairs: **1–3,5–7,9,11**.
Helmet: At least 3 × longer than wide,
    cylindric or sac-like: **1**; less than 3 ×
    longer than wide, hemispherical,
    hemispherical-conical or hemispherical-
    cylindric: **2–11**.

1a. Helmet 3 or more times longer than
      wide, cylindric or sac-like; nectary-
      spurs usually somewhat spirally
      curved   **1. lycoctonum**
  b. Helmet not more than 2 times longer
      than wide, rounded and more or less
      hemispherical; nectary-spurs straight
                                            2
2a. Sepals yellowish, usually persistent
      around the follicles; leaves deeply
      divided into linear segments
                                2. anthora
  b. Sepals usually purplish or blue, falling
      early; ultimate segments of the leaves
      broader   3
3a. Stems flexuous, scrambling or
      twining, at least above   4
  b. Stems stiffly erect, not scrambling or
      twining   7
4a. Flower-stalks covered with long,
      spreading or slightly deflexed hairs 5
  b. Flower-stalks covered with sparse
      crisped hairs, especially towards the
      apex, or hairless   6
5a. Leaf-segments broad, toothed but not
      themselves deeply divided
                                8. uncinatum
  b. Leaf-segments narrow, themselves
      deeply divided   **10. volubile**
6a. Leaves divided into 3 distinct,
      shortly-stalked leaflets   **9. henryi**
  b. Leaves divided into 3–5 lobes to two-
      thirds or more of their diameter (not
      divided to base)   **11. hemsleyanum**
7a. Leaves divided to two-thirds or
      slightly more of their diameter into
      3–5 lobes   8

b. Leaves divided to the base or almost
  so into 3–7 leaflets   9
8a. Upper leaves clasping the stem;
  flower-stalks hairless, long and
  flexuous, spreading; sepals hairless
                          7. amplexicaule
b. Upper leaves not clasping the stem;
  flower-stalks crisply hairy, short,
  straight or almost so, erect; sepals
  crisply hairy   **6. carmichaelii**
9a. Flower-stalks with long, straight or
  slightly deflexed, often glandular hairs
                          4. paniculatum
b. Flower-stalks hairless or with crisped
  hairs   10
10a. Helmet usually higher than wide;
  seeds winged on 1 angle and with
  transverse plates or folds on the sides
                          5. variegatum
b. Helmet usually wider than high; seeds
  winged on 3 angles, without
  transverse plates or folds   **3. napellus**

1. A. lycoctonum Linnaeus.
Tall, erect perennial herbs with long roots.
Leaves more or less circular in outline or
somewhat broader than long, deeply
5–7-lobed, the lobes themselves variously
toothed or lobed, the basal long-stalked;
hairless or hairy above, usually hairy along
the veins beneath. Inflorescence a loose to
dense, few–many-flowered panicle. Flower-
stalks hairy. Helmet cylindric or sac-like,
3 or more times longer than wide, usually
hairy outside; spurs of upper petals spirally
coiled. Follicles usually 3. Seeds brownish
black, obtusely 4-angled. *Most of Europe,
N Africa.*

A very variable species which here
includes, as subspecies, several units often
recognised as individual species. These are
difficult to distinguish as they tend to
intergrade. The names applied to them are
also very confused and liable to change;
those used here follow the revision for the
2nd edition of *Flora Europaea.*
  Subsp. **lycoctonum** (*A. septentrionale*
Koelle). Figure 35(1), p. 340. Illustration:
Botanical Magazine, 2196 (1820);
Lindman, Nordens Flora 1: pl. 225 (1964).
Flowers dark violet, helmet with a wide
base, tapering abruptly from it into the
long, sac-like upper part. *N Europe.* H1.
Summer.
  Subsp. **vulparia** (Reichenbach) Schinz &
Keller (*A. vulparia* Reichenbach). Figure
35(2), p. 340 Illustration: Gentes
Herbarum 6: 462 (1945); Huxley,
Mountain flowers in colour, f. 170 (1967).
Leaf-segments toothed or divided as far as
the middle; terminal raceme small and

rather few-flowered; flowers yellowish;
helmet not conspicuously wider at the base
and not abruptly tapered. *C & S Europe.*
H2. Summer.
  Subsp. **moldavicum** (Hacquet) Jalas
(*A. moldavicum* Hacquet) from E Europe is
similar but has bluish flowers and may
occasionally be found in gardens. The
name *A. orientale* is sometimes misused in
gardens for plants of subsp. *vulparia.* This,
however, is incorrect; genuine *A. orientale*
Miller, from Turkey and the Caucasus is
similar, but has dark blue flowers and
hairless flower-stalks and helmets; it is
apparently not cultivated.
  Subsp. **neapolitanum** (Tenore) Nyman
(*A. lamarckii* Reichenbach; *A. pyrenaicum*
Linnaeus, in part). Figure 35(3), p. 340.
Illustration: Huxley, Mountain flowers in
colour, f. 171 (1967). Leaf-segments deeply
divided beyond the middle; terminal raceme
large and many-flowered; flowers
yellowish; helmet not conspicuously
widened at the base. *Mountains of S Europe,
Morocco.* H3. Summer.

2. A. anthora Linnaeus. Figure 35(4),
p. 340. Illustration: Botanical Magazine,
2654 (1826); Bonnier, Flore complète 1: pl.
20 (1911); Gentes Herbarum 6: 475
(1945); Huxley, Mountain flowers in
colour, f. 172 (1967).
Roots tuberous. Stems erect, little-
branched. Leaves more or less circular in
outline, deeply divided into numerous
linear segments, completely hairless or with
sparse, crisped hairs on the margins and
beneath. Inflorescence a compact, many-
flowered raceme or with a few branches
from the base. Flower-stalks and helmet
with dense, crisped hairs. Flowers yellow,
helmet more or less hemispherical, the
perianth persistent around the developing
follicles. Spurs of inner petals straight.
Follicles usually 5. Seeds black, acutely
4-angled. *S Europe, W & C Asia.* H3.
Summer.

3. A. napellus Linnaeus (*A. pyramidale*
Miller; *A. tauricum* Wulfen; *A. firmum*
Reichenbach). Figure 35(5), p. 340.
Illustration: Botanical Magazine, 8152
(1907); Bonnier, Flore complète 1: pl. 21
(1911); Gentes Herbarum 6: 481 (1945);
Huxley, Mountain flowers in colour, f. 175
(1967).
Roots tuberous. Stems erect, very leafy.
Leaves more or less circular in outline,
divided to the base or almost so into 5–7
leaflets which are themselves toothed or
lobed, hairless or sparsely hairy above and
beneath. Inflorescence a dense raceme,

occasionally with short lateral branches
from the base. Flower-stalks with sparse to
dense crisped hairs or hairless. Flowers blue
or purplish. Helmet hemispherical, usually
wider than high. Spurs of inner petals
straight. Filaments hairy or not. Follicles
usually 3. Seeds with 3 narrowly winged
angles. *Most of Europe.* H1. Summer.

An extremely variable species within
which several variants have been described
(as distinct species or as subspecies). It is
uncertain which of these are in cultivation.

4. A. paniculatum Lamarck. Figure 35(6),
p. 340.
Illustration: Bonnier, Flore complète 1: pl.
21 (1911); Huxley, Mountain flowers in
colour, f. 173 (1967).
Roots tuberous. Stems erect, often
branched. Leaves more or less circular in
outline, divided to the base or almost so
into 5–7 leaflets, the leaflets themselves
lobed or toothed. Inflorescence an open
panicle. Flower-stalks spreading, covered
with long, straight or slightly deflexed,
glandular hairs. Flowers blue or blue-violet.
Helmet hemispherical, often with long hairs
on the outside. Spurs of inner petals
straight. Filaments hairless. Follicles
usually 3–5. Seeds black, winged on 1
angle, with numerous transverse plates or
folds. *Southern part of C Europe.* H3.
Summer.

5. A. variegatum Linnaeus. Figure 35(7),
p. 340. Illustration: Bonnier, Flore complète
1: pl. 21 (1911); Gentes Herbarum 6: 483
(1945); Huxley, Mountain flowers in
colour, f. 174 (1967).
Similar to both *A. paniculatum* and *A.
napellus,* differing from the former in the
flower-stalks being hairless or with sparse
to dense, crisped hairs, and from the latter
in its helmet usually higher than wide and
the seeds with numerous transverse plates
or folds. *C & S Europe, Turkey.* H3. Summer.

A variable species, divided into
subspecies by some authors. A. ×
cammarum Linnaeus (*A. bicolor* Schultes;
A. × *stoerckianum* Reichenbach) is the name
applied to plants which seem to be hybrids
between *A. variegatum* and *A. napellus.* It is
intermediate between the parents but has a
helmet which is mainly whitish but with
purplish margins. It has been cultivated for
many years (illustration: Gentes Herbarum
6: 485, 486, 1945).

6. A. carmichaelii Debeaux (*A. fischeri*
Forbes & Hemsley not Reichenbach). Figure
35(8), p. 340. Illustration: Botanical
Magazine, 7130 (1890); Gentes Herbarum
6: 489, 490 (1945).

Figure 35. Leaves of *Aconitum* species. 1, *A. lycoctonum* subsp.
lycoctonum. 2, *A. lycoctonum* subsp. *vulparia*. 3, *A. lycoctonum*
subsp. *neapolitanum*. 4, *A. anthora*. 5, *A. napellus*.
6, *A. paniculatum*. 7, *A. variegatum*. 8, *A. carmichaelii*. 9, *A. amplexicaule*.
10, *A. uncinatum*. 11, *A. henryi*. 12, *A. volubile*.
13, *A. hemsleyanum*. Scale = 1 cm.

Roots tuberous. Stems erect, often very tall (to 2 m). Leaves more or less ovate in outline, divided to about two-thirds of their diameter into 3–5 lobes which are themselves sparsely toothed or lobed, leathery, dark green above, paler beneath, hairless or with crisped hairs on the veins above and beneath. Inflorescence a dense panicle with upright branches; flower-stalks short, covered with crisped hairs. Flowers deep purple outside, paler or whitish within. Helmet hemispherical-cylindric, sparsely crisped-hairy outside. Follicles usually 3. Seeds covered with clear, transverse plates or folds. *C & W China.* H3? Summer.

Often found in gardens under the incorrect name *A. fischeri.* The genuine *A. fischeri* Reichenbach, from eastern Asia, is not cultivated; its upper stems twine or scramble and the leaves are thin and bright green.

7. A. amplexicaule Lauener. Figure 35(9), p. 340.
Roots tuberous. Stems erect, often purplish. Leaves more or less circular in outline, divided to two-thirds of their diameter into 3–5 toothed lobes, hairless or sparsely hairy along the veins above, the upper leaves clasping the stem at their bases. Inflorescence a widely spreading panicle with long, flexuous branches. Flower-stalks long, hairless. Flowers blue-purple. Helmet hemispherical, hairless. Follicles usually 3. *Nepal.* H3? Summer.

8. A. uncinatum Linnaeus. Figure 35(10), p. 340. Illustration: Botanical Magazine, 1119 (1808); Gentes Herbarum 6: 495 (1945); Justice & Bell, Wild flowers of North Carolina, 62 (1968).
Roots tuberous. Stems thin, weak, scrambling or twining. Leaves ovate in outline, deeply 3–5-lobed, the lobes broad and sparsely toothed, dark green above, paler beneath, hairless. Inflorescence few-flowered. Flower-stalks covered with long, spreading, straight or slightly deflexed hairs. Flowers deep blue, helmet hemispherical-conical. Follicles usually 3. *Eastern USA.* H1. Summer.

9. A. henryi Pritzel (*A. californicum* misapplied). Figure 35(11), p. 340. Illustration: Bulletin de la Société Botanique de France **51**: t. 6, fig. 32 (1904); Gentes Herbarum 6: 499, 500 (1945).
Roots tuberous. Stems erect below, scrambling or twining above, branched. Leaves ovate in outline, divided into 3 (rarely 5) distinct, shortly stalked

(especially conspicuous in the terminal leaflet) leaflets which are narrow, regularly toothed or lobed and tapered towards their bases, light green, sparsely hairy on the veins or hairless. Raceme few-flowered. Flower-stalks hairless or with sparse, crisped hairs. Flowers deep bluish purple. Helmet hemispherical. Follicles 3–5, hairy. *C & W China.* H3. Summer.

A. vilmorinianum Komarov is similar, but has hairless follicles and broader leaflets. *W China.* H5. Summer.

10. A. volubile Pallas. Figure 35(12), p. 340. Illustration: Gentes Herbarum 6: 504 (1945).
Roots tuberous. Stems thin, weak, twining or scrambling. Leaves almost circular in outline, divided almost to the base into 3 narrow lobes which are themselves deeply lobed, hairy. Racemes few-flowered. Flower-stalks covered with long, spreading or slightly deflexed hairs. Flowers purple and green or bluish and green. Helmet hemispherical-conical. Follicles 3–5. *E Asia.* H2. Summer.

Doubtfully in cultivation; the name *A. volubile* was applied indiscriminately in the past to all the climbing species introduced from eastern Asia and China.

11. A. hemsleyanum Pritzel (*A. volubile* misapplied). Figure 35(13), p. 340. Illustration: Journal of the Royal Horticultural Society **28**: 58 (1903); Gardeners' Chronicle **66**: 150 (1919).
Roots tuberous. Stems twining or scrambling. Leaves ovate in outline, divided to two-thirds or more of their diameter into 3–5 lobes which are themselves broadly toothed, dark green above, noticeably paler beneath, hairless or sparsely hairy. Inflorescence a collection of widely spreading, long-stalked racemes. Flower-stalks hairless or with sparse, crisped hairs, especially towards the apex. Flowers dark purplish blue. Helmet hemispherical-cylindric, hairy outside. Follicles 3–5. *C & W China.* H3. Summer.

Often cultivated under the name *A. volubile.*

19. DELPHINIUM Linnaeus
J. Cullen & H.S. Maxwell
Annual, biennial or perennial herbs with woody, fibrous or tuberous roots. Leaves palmately lobed, the lobes usually themselves toothed or lobed, sometimes the leaves made up of very narrow segments. Inflorescence a raceme or panicle, each flower stalked and subtended by a bract and 2 bracteoles. Flowers bilaterally

symmetric. Sepals 5, the uppermost with a backwardly-pointing spur. Petals 4, sometimes paler than the sepals, sometimes darker to black or dark grey, the 2 uppermost with nectar-secreting projections which extend into the sepal-spur. Stamens numerous. Ovary of 3–5 free carpels. Fruit a group of 3–5 follicles.

A large genus of perhaps 100 or more species from north temperate regions, extending south to the equator on the mountains of East Africa. Identification is difficult as the species are variable and no worldwide study has been produced since 1895 (see below). The relationships of the species are also problematic and confused and this is reflected in the order in which they are presented here.

Most of the commonly cultivated Delphiniums are hybrids, many with double flowers. Most of these have a long central raceme with shorter side branches and have been developed from crosses involving *D. elatum* and other species. The 'Belladonna' Delphiniums are distinguished by having no central raceme and are derived from *D. elatum* × *D. grandiflorum.* Red- and pink-flowered hybrids have been developed by crossing *D. elatum* and its hybrids with red-flowered north American species.

Delphiniums are easily grown, though prone to various diseases. They require a sunny site with good soil and are propagated by seed or division. The hardiness of many of the species is uncertain, and the codes given here are speculative.
Literature: Huth, E., Monographie der Gattung Delphinium, *Botanische Jahrbücher* **20**: 322–499 (1895); Ewan, J.A., A synopsis of the North American species of Delphinium, *University of Colorado Studies* D **2**(2): 55–244 (1945); Munz, P.A., A synopsis of African species of Delphinium and Consolida, *Journal of the Arnold Arboretum* **48**: 30–55 (1967); Munz, P.A., A synopsis of the Asian species of Delphinium sensu stricto, *Journal of the Arnold Arboretum* **48**: 249–302, 476–545 (1967), **49**: 73–166 (1968). There are many books on Delphinium cultivation: Edwards, C., *Delphiniums* (1981) is one of the most recent.

1a. Plants annual or biennial; petals
  hairless   2
 b. Plants perennial; petals hairy   3
2a. Spur 4–6 mm; bracteoles borne above
  the base of the flower-stalk
                                                         26. requienii

b. Spur up to 4 mm; bracteoles borne at the base of the flower-stalk **25. staphisagria**

3a. Flowers red 4

b. Flowers blue, purple, yellow or white 5

4a. Leaves (at least those on the upper part of the stem) divided into linear or narrowly lanceolate segments; flowers usually scarlet **17. cardinale**

b. Leaves divided into broad primary divisions; flowers orange-red or dull red **16. nudicaule**

5a. Roots tuberous; sepals to 1 cm; flower-stalks to 1 cm, raceme long, dense, many-flowered, parallel-sided 6

b. Combination of characters not as above 8

6a. Flowers cream or yellow, sometimes with orange on the petals; leaves glaucous, very finely divided into thread-like segments **11. semibarbatum**

b. Flowers white, pale blue or blue; leaves not glaucous, divided into oblong-linear ultimate segments 7

7a. Follicles hairy; hairs on inflorescence spreading and glandular; flowers white or very pale blue **18. virescens**

b. Follicles hairless; hairs on inflorescence crisply deflexed; flowers blue **13. fissum**

8a. Sepals 2 cm or more, strongly veined, persistent and papery; spur narrowly conical, 4–10 mm in diameter at base 9

b. Sepals usually smaller, not as above; spur narrow, less than 4 mm in diameter at the base 11

9a. Spur as long as to longer than the rest of the upper sepal **10. caucasicum**

b. Spur much shorter than the rest of the upper sepal 10

10a. Many of the hairs in the inflorescence with swollen, yellowish bases, or glandular heads **1. brunonianum**

b. No hairs in the inflorescence as above **2. cashmerianum**

11a. Bracteoles, or at least 1 of them, borne close below the flower and overlapping it 12

b. Bracteoles distant from the flower, or, if rather close, then not overlapping it 17

12a. Spur 1.7 cm or more 13

b. Spur less than 1.7 cm 14

13a. Petals black or dark grey; raceme long, compact, many-flowered **4. elatum**

b. Petals blue; raceme loose, few-flowered **19. trolliifolium**

14a. Spur less than 1.2 cm; racemes many-flowered, compact, flower-stalks shorter than to as long as sepals **22. nuttallii**

b. Spur 1.3–1.7 cm; racemes loose, few-flowered, flower-stalks mostly longer than sepals 15

15a. Follicles spreading horizontally, claw-like, forming a 3-armed star-shape **15. tricorne**

b. Follicles erect, strict 16

16a. Lower petals deep blue, shallowly notched **21. bicolor**

b. Lower petals white or pale blue, deeply notched **23. nuttallianum**

17a. Spur 1.7 cm or more 18

b. Spur up to 1.7 cm 21

18a. Spur 3.3–4.3 cm **24. leroyi**

b. Spur not more than 3 cm 19

19a. Some hairs in the inflorescence with swollen, yellowish bases, some others with glandular tips; leaves with broad, obovate, toothed primary segments **14. delavayi**

b. No hairs as above; leaves with finely divided segments 20

20a. Sepals 2.2–2.8 × 1.2–1.6 cm **9. pylzowii**

b. Sepals 1.4–2.6 × up to 1.1 cm **20. grandiflorum**

21a. Petals black or dark grey 22

b. Petals blue, purple or pale 23

22a. Inflorescence hairless **6. huetianum**

b. Inflorescence hairy, some hairs with swollen, yellowish bases and some with glandular tips **5. maackianum**

23a. Plant tufted, less than 15 cm high **3. muscosum**

b. Plant not tufted, much taller 24

24a. Flowers bright pale blue or mauve **12. denudatum**

b. Flowers dark blue or purplish 25

25a. Spur 1.3–1.5 cm; inflorescence densely hairy **8. brachycentrum**

b. Spur 1.5–1.7 cm; inflorescence sparsely hairy **7. cheilanthum**

1. D. brunonianum Royle (*D. jacquemontianum* Cambessedes). Figure 36(1), p. 343. Illustration: Botanical Magazine, 5461 (1864); Flora SSSR 7: pl. IX (1937); Polunin & Stainton, Flowers of the Himalaya, pl. 4 (1984).
Perennial herb with a musky smell when fresh; roots slender. Stems to 1 m but often much less, hairy, the hairs in the inflorescence mixed, some simple, some glandular, some with swollen, yellowish bases. Lower leaves divided almost to the base into 3 lobes which are themselves coarsely toothed or shallowly lobed. Inflorescence a raceme; flower-stalks more than 1 cm, bracteoles distant from the flower. Sepals 2 cm or more, strongly veined, persistent and papery after flowering; spur narrowly conical, 4–10 mm in diameter at the base, shorter than the rest of the upper sepal. Petals black or dark grey. Follicles hairy. *Himalaya (Afghanistan to China), USSR (C Asia).* H3. Summer.

A handsome but variable species.

2. D. cashmerianum Royle (*D. aitchisonii* Huth). Figure 36(2), p. 343. Illustration: Botanical Magazine, 6189 (1875); Garden **18**: 568 (1880); Polunin & Stainton, Flowers of the Himalaya, pl. 2 (1984).
Very similar to *D. brunonianum* but inflorescence often paniculate, the hairs in it all simple, without glands or swollen yellowish bases. *N India, China (Xizang).* H3. Summer–autumn.

3. D. muscosum Exell & Hillcoat. Figure 36(3), p. 343. Illustration: Journal of the Royal Horticultural Society **87**: f. 6 (1962).
Tufted perennial less than 15 cm high, roots slender. Stems 1–few-flowered, hairy, hairs mostly deflexed and crisped, spreading just below the flowers. Leaves mostly basal, more or less circular in outline, very finely divided into linear or almost thread-like segments. Bracteoles borne some distance from the flower and not overlapping it. Sepals to 2 cm, dark blue or purple; spur to 1.4 cm, all yellow-hairy outside. Petals paler than sepals. Follicles hairy. *Bhutan.* H5. Summer.

4. D. elatum Linnaeus. Figure 36(4), p. 343. Illustration: Bonnier, Flore complète 1: pl. 20 (1911); Huxley, Mountain flowers in colour, f. 169 (1967).
Perennial herb to 2 m, with a woody rhizome. Stems mostly hairless, but with sparse hairs with swollen, yellowish bases in the inflorescence. Leaves numerous, deeply palmately divided into 5–7 coarsely toothed lobes, hairless or hairy. Inflorescence a long raceme; flower-stalks exceeding 1 cm, bracteoles borne closely beneath the flower and overlapping it. Sepals pale to dark blue, to 1.6 cm, spurs 1.7 cm or more. Petals almost black. Follicles hairless. *C Europe to USSR (Siberia).* H1. Summer.

5. D. maackianum Regel. Figure 36(5), p. 343. Illustration: Gartenflora **10**: t. 344 (1861).
Perennial herb with a woody rhizome.

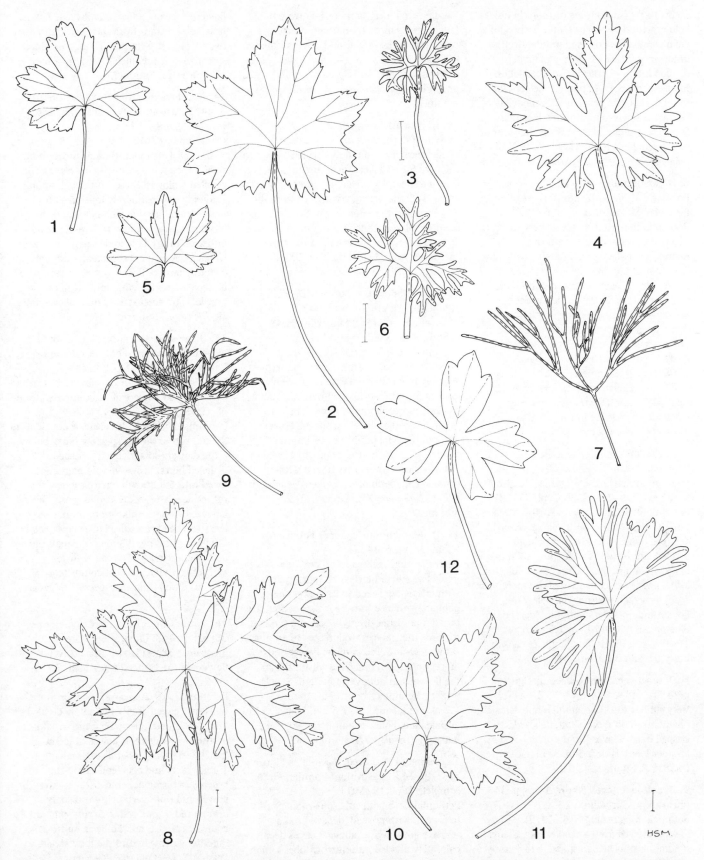

Figure 36. Leaves of *Delphinium* species. 1, *D. brunonianum*. 2, *D. cashmerianum*. 3, *D. muscosum*. 4, *D. elatum*. 5, *D. maackianum*. 6, *D. pylzowii*. 7, *D. semibarbatum*. 8, *D. denudatum*. 9, *D. fissum*. 10, *D. delavayi*. 11, *D. tricorne*. 12, *D. nudicaule*. Scale in bottom right-hand corner = 1 cm, applies to all except 3, 6 & 8, which have their own 1 cm scales.

Stems to 80 cm, hairless or sparsely hairy below, with dense hairs, some with swollen yellowish bases, others glandular, in the inflorescence. Leaves deeply palmately divided into 5–7 coarsely toothed lobes. Inflorescence a loose panicle; flower-stalks more than 1 cm, bracteoles borne well below the flower and not overlapping it. Sepals to 1.6 cm, blue, sparsely hairy; spur to 1.7 cm. Follicles hairless. *USSR, N China (Manchuria), Korea.* H1. Summer.

6. D. huetianum Meikle (*D. formosum* Boissier & Huet not Beaton). Illustration: Botanical Magazine, n.s., 566 (1970). Perennial herb to 2 m, stems very sparsely hairy or hairless even within the inflorescence. Leaves palmately divided into 5–7 coarsely toothed lobes. Inflorescence a dense, broad raceme; flower-stalks more than 1 cm, bracteoles borne distant from the flower and not overlapping it. Sepals to 2 cm, deep violet-blue, hairless or sparsely hairy outside; spur to 1.5 cm. Petals black or dark grey. Follicles usually hairless. *N Turkey.* H4. Summer.

The nomenclatural confusion involving this species and *D. formosum* has been elucidated by Meikle in the notes to the illustration cited above.

7. D. cheilanthum de Candolle (*?D. formosum* Beaton). Figure 37(7), p. 345. Illustration: Edwards's Botanical Register 6: t. 473 (1820); Flora SSSR 7: pl. X (1937). Perennial herb from a woody root. Stems to 1 m, hairless below, with sparse, crisped, deflexed hairs above. Leaves palmately divided almost to the base into 5–7 narrow lobes which are themselves lobed. Raceme loose, few-flowered; flower-stalks more than 1 cm; bracteoles borne well below the flower and not overlapping it. Sepals deep blue or rarely whitish, to 1.6 cm, spur 1.5–1.7 cm. Petals blue. Follicles densely hairy. *USSR (E Siberia).* ?H3. Summer.

8. D. brachycentrum Ledebour. Figure 37(4), p. 345.
Very similar to *D. cheilanthum* but stems hairy throughout with soft, deflexed, crisped hairs, leaves less deeply lobed with broader lobes, spur 1.3–1.5 cm. *Eastern USSR.* H2. Summer.

9. D. pylzowii Regel. Figure 36(6), p. 343. Illustration: Gartenflora 25: t. 879 (1876); Botanical Magazine, 8613 (1919). Perennial herb from a woody root. Stems to 60 cm, hairless below, sparsely to densely spreading-hairy above. Leaves deeply palmately 5–7-lobed into narrow lobes which are themselves deeply lobed. Raceme

few-flowered (occasionally 1-flowered), flower-stalks more than 1 cm, bracteoles distant from the flower and not overlapping it. Sepals 2.2–2.8 × 1.2–1.6 cm, deep blue or bluish purple, densely hairy; spur 1.7 cm or more. Follicles hairy. *C & W China.* H3. Summer.

10. D. caucasicum Meyer.
Perennial herb, stem to 40 cm. Leaves rather leathery, palmately divided almost to the base into 3 lobes, these further lobed. Raceme short, loose, few-flowered, flower-stalks 4–6 cm, bracteoles distant from the flower and not overlapping it. Sepals 2 cm or more, blue, rather sparsely hairy outside; spur as long as or longer than the rest of the upper sepal. Follicles hairy. *USSR (Caucasus).* H3. Summer.

11. D. semibarbatum Boissier (*D. zalil* Aitchison). Figure 36(7), p. 343. Illustration: Botanical Magazine, 7049 (1889); Flora SSSR 7: pl. XI (1937). Perennial herb from a thickened root. Stems to 80 cm, hairless below, sometimes sparsely hairy above. Leaves very finely divided into thread-like segments, glaucous. Raceme loose, elongate; flower-stalks to 1 cm, bracteoles distant from the flower and not overlapping it. Sepals bright yellow, to 1 cm, hairless; spur to 1 cm. Petals yellow, sometimes tinged with orange. Follicles hairless. *Iran, Afghanistan, USSR (Turkestan, Transcaspia).* ?H3. Summer.

12. D. denudatum Hooker & Thomson. Figure 36(8), p. 343.
Perennial herb. Stems to 80 cm, usually freely branched, hairless or sparsely hairy with crisped, deflexed hairs. Leaves deeply palmately divided into 5–7 lobes which are broad and rather deeply lobed and toothed towards their apices. Inflorescence an open, rather few-flowered panicle; flower-stalks more than 1 cm, bracteoles distant from the flower and not overlapping it. Sepals bright pale blue or bright pale mauve, 1.2–1.5 cm, spur 1.4–1.5 cm. Petals white or pale blue. Follicles hairless or sparsely hairy. *W Himalaya.* ?H3. Summer.

13. D. fissum Waldstein & Kitaibel. Figure 36(9), p. 343. Illustration: Bonnier, Flore complète 1: pl. 19 (1911).
Perennial herb with tuberous roots. Stems to 60 cm, with crisped, deflexed hairs, sparse below, denser above. Leaves deeply palmately divided into narrow lobes which are themselves deeply lobed. Inflorescence a narrow, parallel-sided, dense, many-

flowered raceme; flower-stalks to 1 cm, bracteoles distant from the flower and not overlapping it. Sepals to 1 cm, pale blue; spur to 1.5 cm. Follicles hairless. *S Europe, N Turkey.* H3. Summer.

14. D. delavayi Franchet. Figure 36(10), p. 343. Illustration: Botanical Magazine, n.s., 68 (1949).
Perennial herb from a woody root. Stems to 1 m, densely spreading-hairy throughout, some of the hairs in the inflorescence with swollen yellowish bases, others glandular. Leaves deeply palmately lobed into 5–7 lobes which are themselves lobed or bluntly toothed, densely hairy. Inflorescence a loose or dense, many-flowered raceme; flower-stalks 1 cm or more, bracteoles not very distant from the flower but not overlapping it. Sepals 1.4–1.6 cm, bluish violet to deep purple; spur at least 1.7 cm. Follicles hairy. *W China.* H4. Summer.

15. D. tricorne Michaux. Figure 36(11), p. 343. Illustration: Loddige's Botanical Cabinet 4: t. 306 (1819); Justice & Bell, Wild flowers of North Carolina, 61 (1968); Rickett, Wild flowers of the United States 6: pl. 46 (1973).
Perennial herb with tuberous roots. Stem to 30 cm, rather fleshy, sparsely hairy below, more densely so above with crisped, deflexed hairs. Leaves deeply palmately lobed into 5 lobes which are themselves rather deeply lobed. Racemes loose, several-flowered; flower-stalks more than 1 cm, at least 1 of the bracteoles borne close below the flower and overlapping it. Sepals blue or violet; spur 1.3–1.7 cm. Follicles spreading widely, almost completely hairless when mature. *Eastern USA.* H3. Summer.

16. D. nudicaule Torrey & Gray. Figure 36(12), p. 343. Illustration: Botanical Magazine, 5819 (1870); Rickett, Wild flowers of the United States 5: pl. 36 (1971).
Perennial herb, stems to 30 cm or more, hairless. Leaves deeply palmately divided into 3–5 broad lobes which are irregularly toothed towards their apices. Raceme loose, few-flowered; flower-stalks more than 1 cm, bracts and bracteoles reddish, bracteoles borne at some distance from the flower and not overlapping it. Sepals orange-red or dull red, *c.* 1 cm; spur rather conical, 1.4–1.8 cm. Follicles hairless, tapering smoothly into the long styles. *Western USA (California, Oregon).* H4. Summer.

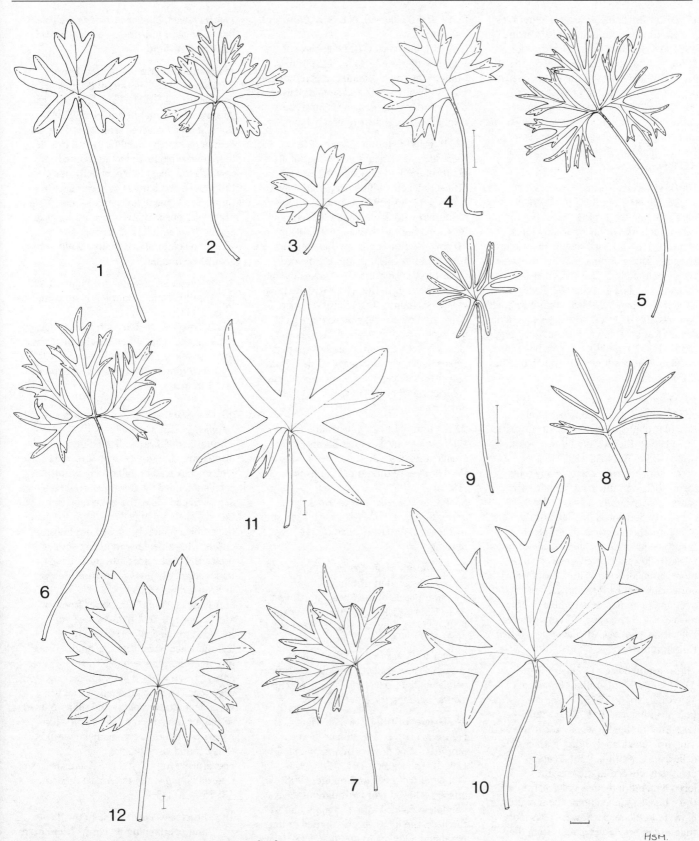

Figure 37. Leaves of *Delphinium* species. 1, *D. cardinale*.
2, *D. virescens*. 3, *D. trolliifolium*. 4, *D. brachycentron*. 5, *D. grandiflorum*.
6, *D. tatsienense*. 7, *D. cheilanthum*. 8, *D. bicolor*.
9, *D. nuttallianum*. 10, *D. leroyi*. 11, *D. requienii*. 12, *D. staphisagria*.
Horizontal scale in bottom right-hand corner = 1 cm, applying to 1–3,
5–7; the others have individual 1 cm scales.

17. D. cardinale J.D. Hooker. Figure 37(1), p. 345. Illustration: Botanical Magazine, 4887 (1855); Illustration Horticole for 1856: t. 92; Gartenflora **6**: t. 208 (1857); Rickett, Wild flowers of the United States **4**: pl. 40 (1970), **5**: pl. 36 (1971).
Like *D. nudicaule* but stems often hairy above, at least the upper leaves divided into numerous narrow segments, sepals scarlet. *USA (California), Mexico (Baja California).* H5. Summer.

18. D. virescens Nuttall. Figure 37(2), p. 345. Illustration: Rickett, Wild flowers of the United States **4**: pl. 39 (1970).
Perennial herb with deep woody roots. Stems to 1.75 m, with crisped hairs below and fine, soft, glandular hairs above. Leaves deeply palmately 5–7-lobed, the lobes themselves lobed and toothed. Raceme long, dense, parallel-sided, many-flowered; flower-stalks to 1 cm, bracteoles somewhat distant from the flower and not overlapping it. Sepals *c.* 1 cm, white or pale blue; spur 1–2 cm. Follicles hairy. *Central N America.* H3. Summer.

19. D. trolliifolium Gray. Figure 37(3), p. 345. Illustration: Hitchcock et al., Vascular plants of the Pacific northwest **2**: 365 (1964); Rickett, Wild flowers of the United States **5**: pl. 34 (1971).
Perennial herb with deep, woody roots. Stems to 1.5 m, thick, leafy, hairless or sparsely crisped-hairy. Leaves deeply palmately 5–7-lobed, the lobes rather broad, toothed towards their apices. Inflorescence a rather short raceme with at most 30 flowers; flower-stalks 1 cm or more, bracteoles (or at least 1 of them) borne close beneath the flower and overlapping it. Sepals deep blue or blue-purple, *c.* 1.5 cm; spur 1.7 cm or more. Follicles hairy. *W USA (California, Oregon).* H3. Summer.

20. D. grandiflorum Linnaeus. Figure 37(5), p. 345. Illustration: Botanical Magazine, 1686 (1815); Edwards's Botanical Register **6**: t. 472 (1820).
Perennial herb with woody roots. Stems to 1 m, branched, finely hairy with crisped, deflexed hairs throughout. Leaves palmately divided to the base into 3–5 lobes which are themselves deeply lobed, the whole leaf made up of linear segments. Flowers in a panicle; flower-stalks more than 1 cm, bracteoles distant from the flower and not overlapping it. Sepals 1.4–2.6 × up to 1.1 cm, usually deep blue, rarely pale or almost white; spur 1.7 cm or more. Petals violet-blue. Follicles usually

hairy. *USSR (Siberia), N,C & W China.* H2. Summer.

A variable species. **D. tatsienense** Franchet (Figure 37(6), p. 345) from W China, is extremely similar; it is reputed to have less leafy stems and smaller flowers with relatively larger spurs, but these distinctions all appear to break down.

21. D. bicolor Nuttall. Figure 37(8), p. 345. Illustration: Hitchcock et al., Vascular plants of the Pacific northwest **2**: 352 (1964); Rickett, Wild flowers of the United States **6**: pl. 45 (1973); Porsild, Rocky mountain wild flowers, 179 (1974).
Perennial herb with deep roots. Stems to 50 cm with fine spreading hairs or hairless. Leaves rather thick, palmately divided almost to the base into 5–7 lobes, the lobes themselves lobed, the whole leaf made up of oblong segments. Raceme rather open, few-flowered; flower-stalks more than 1 cm, bracteoles (or at least 1 of them) borne close beneath the flower and overlapping it. Sepals violet, *c.* 1.2 cm, spur 1.3–1.7 cm. Lower petals deep blue, shallowly notched. Follicles hairy or hairless. *W USA.* H4. Summer.

22. D. nuttallii Gray. Illustration: Hitchcock et al., Vascular plants of the Pacific northwest **2**: 363 (1964); Rickett, Wild flowers of the United States **5**: pl. 34 (1971).
Similar to *D. bicolor* but racemes dense, many-flowered, flower-stalks shorter, spur to 1.2 cm. *Western N America.* H4. Summer.

23. D. nuttallianum Pritzel. Figure 37(9), p. 345. Illustration: Hitchcock et al., Vascular plants of the Pacific northwest **2**: 361 (1964); Rickett, Wild flowers of the United States **5**: pl. 34 (1973), **6**: pl. 44 (1973).
Similar to *D. bicolor* but with a short rhizome, racemes with 3–8 flowers, lower petals white or pale blue, deeply notched. *Western N America.* H4. Summer.

24. D. leroyi Huth (*D. wellbyi* Hemsley). Figure 37(10), p. 345. Illustration: Moriarty, Wild flowers of Malawi, pl. 33 (1975); The Garden **101**: 358 (1976).
Tall perennial herb to 2 m. Stem with crisped deflexed hairs throughout. Leaves palmately divided into 3–5 lobes almost to the base, the lobes deeply toothed or almost pinnatifid. Flowers in an open, few-flowered raceme or panicle; flower-stalks more than 1 cm, bracteoles distant from the flower and not overlapping it. Sepals white to deep blue, to 2 × 1.5 cm; spur 3.3–4.3 cm.

Petals clawed, blue or brownish. Follicles hairy. *Tropical E Africa, on mountains.* H5. Summer–autumn.

25. D. staphisagria Linnaeus. Figure 37(12), p. 345. Illustration: Bonnier, Flore complète **1**: pl. 20 (1911).
Annual or biennial herb to 1 m. Stems thick, densely covered with spreading hairs. Leaves palmately lobed for two-thirds or more of the blade, the lobes toothed towards the apex, hairy on both sides. Inflorescence a panicle or raceme with numerous flowers; flower-stalks 8–20 mm, bracteoles borne more or less at the base. Sepals deep blue, 1.3–2 cm; spur very short. Follicles inflated, hairy. *Mediterranean area.* H2. Summer.

26. D. requienii de Candolle. Figure 37(11), p. 345. Illustration: Bonnier, Flore complète **1**: pl. 20 (1911).
Like *D. staphisagria* but with longer, denser hairs on the stem; bracteoles borne above the base of the flower-stalk; flowers pale, spur 4–6 mm. *S France, Corsica; ?Sardinia.* H5. Summer.

20. CONSOLIDA (de Candolle) S.F. Gray
J. Cullen & H.S. Maxwell
Annual herbs. Leaves divided into numerous, thread-like segments, alternate. Flowers in racemes, bilaterally symmetric, each subtended by a bract and 2 bracteoles. Sepals 5, petal-like, the uppermost spurred. Corolla made up of perhaps 4 united petals, the whole corolla 3–5-lobed or almost entire, with a nectar-producing spur projecting backwards into the spur of the upper sepal. Stamens numerous. Carpel 1. Fruit a single follicle.

A genus of about 50 species from the Mediterranean area to C Asia, often included in *Delphinium* but easily distinguished from it by the single carpel (among other characters). The plants are easily grown in most soils and propagation is by seed. The 3 cultivated species are variable in flower colour and many named selections have been made, as they also have from hybrids, especially between *C. ajacis* and *C. orientalis*.
Literature: Munz, P.A., The Asian species of Consolida, *Journal of the Arnold Arboretum* **48**: 159–202 (1967).

1a. Bracteoles borne close to the flower and overlapping its base   **1. orientalis**
 b. Bracteoles borne some distance below the flower on the flower-stalk, not overlapping the base of the flower   2

2a. Follicles hairy; branches few, rather
    erect; spur shorter than to as long as
    the rest of the corolla   **2. ajacis**
  b. Follicles hairless or almost so;
    branches numerous, wide-angled;
    spur longer than the rest of the
    corolla   **3. regalis**

1. C. orientalis (Gay) Schrodinger
(*Delphinium orientalis* Gay; *Delphinium ajacis*
misapplied). Illustration: Botanical
Magazine, n.s., 186 (1952); Davis (ed),
Flora of Turkey 1: 129 (1965); Polunin &
Smythies, Flowers of SW Europe, 116
(1973).
Stems to 75 cm, erect, sparsely branched
with rather erect branches, hairy above,
many of the hairs with swollen, yellowish
bases and glandular tips. Leaves divided
into numerous thread-like segments.
Bracteoles borne close to the flower and
overlapping its base. Flowers pale to intense
violet, rarely pink or whitish. Spur shorter
than the rest of the sepal. Follicles hairy,
irregularly swollen, the stigma borne
slightly at one side of the apex and
incurved. *Mediterranean area to C Asia*. H1.
Summer.

2. C. ajacis (Linnaeus) Schur (*Delphinium
ajacis* Linnaeus; *C. ambigua* (Linnaeus) Ball
& Heywood). Illustration: Bonnier, Flore
complète 1: pl. 19, 85 (1912).
Stems to 75 cm, sparsely branched with
rather erect branches, hairy above, most of
the hairs deflexed and without swollen,
yellowish bases or glandular tips. Leaves
divided into numerous thread-like
segments. Bracteoles borne at some
distance below the flowers and not
overlapping their bases. Flowers blue,
purple, pink or white; spur shorter than the
rest of the sepal. Follicles hairy, the stigmas
borne at the apex and curved outwards.
Mediterranean area. H1. Summer.

The naming of this plant is complex,
turning on the interpretation of Linnaeus's
Delphinium ajacis, a name which has been
applied by others to more than one species.
Ball & Heywood in Flora Europaea 1: 217
(1965) rejected the name *C. ajacis* and
provided the name *C. ambigua* instead.
However, more recent opinion supports the
view that *C. ajacis* is the correct name.

3. C. regalis S.F. Gray (*Delphinium consolida*
Linnaeus). Illustration: Bonnier, Flore
complète 1: pl. 19, 84 (1912); Davis (ed),
Flora of Turkey 1: 129 (1965); Polunin,
Flowers of Europe, 20 (1969).
Stem to 40 cm, frequently branched, the
branches wide-angled, the upper parts with

deflexed hairs without swollen yellow bases
or glandular tips. Leaves divided into
numerous thread-like segments. Bracteoles
borne on the flower-stalk at some distance
from the flowers, not overlapping their
bases. Flowers mauve or violet-blue. Spur
narrow and considerably longer than the
rest of the sepal. Follicles usually hairless,
stigma borne at the apex, more or less
straight. *SE Europe, SW Asia*. H1. Summer.

A variable species with several varieties
named to cover the wild variation; the most
commonly grown is var. **paniculata** (Host)
Soó.

21. THALICTRUM Linnaeus
D.M. Miller
Perennial herbs. Leaves 2–4-pinnate,
sometimes divisions only 3, usually
alternate, with stipules; leaflets lobed or
toothed. Flowers small, usually bisexual in
axillary or terminal racemes, panicles or
corymbs. Sepals 4–5, petal-like, spreading,
usually deciduous if small and
inconspicuous. Petals absent. Stamens
numerous, often coloured and showy.
Carpels few. Fruiting achenes stalked or
stalkless, ribbed, angled or winged.

A genus of 130 or more species mainly
from north temperate regions, with
attractive foliage and numerous small
flowers, many conspicuous only because of
the numerous stamens. The smaller species
are suitable for the rock garden or alpine
house. Larger species may be planted in
any good garden soil but more care is
required for those from Asia, which need a
moist, rich but well-drained soil in light
shade and grow best in cooler regions with
somewhat higher humidity. All may be
propagated by seed or by division in spring.
Literature: Lecoyer, J.C., Monographie du
Genre Thalictrum, *Bulletin de la Société
Royale de Botanique de Belgique* 24(1):
78–325 (1885).

Plant. Under 20 cm: **1,6,12,13**; over 20 cm:
    2–11, 14–21. With long rhizomes:
    6,7,9,12; with tubers: **3**.
Flower. White, pink or purple: **2–4,10,11,
    13–20**; greenish or yellow: **1,5–9,12,21**.
Sepals. Longer than stamens: **2–4,15–20**;
    shorter than stamens: **1,5–14,21**.
Filaments. Thread-like: **1–9, 15–21**;
    thickened or club-shaped: **10–14**.
Achenes. Stalked: **10–13,15–20**; stalkless:
    1–9,14,21.

1a. Leaflets peltate   **13. ichangense**
  b. Leaflets not peltate   2
2a. Flowers in racemes   **1. alpinum**
  b. Flowers not in racemes   3

3a. Achenes stalkless or almost so   4
  b. Achenes distinctly stalked   13
4a. Sepals longer than stamens   5
  b. Sepals shorter than stamens   7
5a. Leaflets very narrow to cylindric
    2. foeniculaceum
  b. Leaflets ovate to rounded   6
6a. Rootstock with tubers   **3. tuberosum**
  b. Rootstock without tubers
    4. orientale
7a. Flowers unisexual   **21. fendleri**
  b. Flowers bisexual   8
8a. Filaments club-shaped
    14. javanicum
  b. Filaments thread-like   9
9a. Stamens drooping   10
  b. Stamens erect   12
10a. Plant densely glandular and foetid
    5. foetidum
  b. Plant hairless or with a few glandular
    hairs, not foetid   11
11a. Leaflets about as long as wide
    6. minus
  b. Leaflets much longer than wide
    9. simplex
12a. Rootstock with creeping rhizomes
    7. flavum
  b. Rootstock tufted, without rhizomes
    8. lucidum
13a. Sepals much shorter than stamens
    14
  b. Sepals as long as or longer than
    stamens   16
14a. Plant dwarf, under 20 cm
    12. kiusianum
  b. Plant taller, more than 20 cm   15
15a. Achenes pendent, 3-winged or
    -angled   **10. aquilegiifolium**
  b. Achenes erect, ribbed but not winged
    11. calabricum
16a. Plant glandular or hairy   17
  b. Plant completely hairless   18
17a. Leaflets very small, less than 1 cm
    wide   **15. diffusiflorum**
  b. Leaflets usually well over 1 cm wide
    20. reniforme
18a. Achenes with 2 narrow wings
    18. dipterocarpum
  b. Achenes with more than 2 ribs but
    not winged   19
19a. Achenes with stalks 1–1.5 mm
    16. rochebrunianum
  b. Achenes with much longer stalks   20
20a. Leaflets 3-lobed or entire
    17. delavayi
  b. Leaflets 7–13-lobed or -toothed
    19. chelidonii

1. T. alpinum Linnaeus. Illustration:
Botanical Magazine, 2237 (1821); Ross-
Craig, Drawings of British plants 1: t. 2

(1948); Keble Martin, Concise British Flora in colour, pl. 1 (1965).

Plant to 15 cm with short rhizomes. Leaves mostly basal, leaflets 3, each 3-segmented, hairless; segments 3–8 mm, almost circular, 3–5-lobed or -toothed, glaucous beneath. Raceme to 15 cm; flowers c. 5 mm, greenish purple. Sepals 4. Stamens c. 6 mm, much longer than sepals, drooping, filaments thread-like, purplish. Achenes 2–3, almost stalkless, ribbed. *North temperate regions*. H2. Early summer.

2. T. foeniculaceum Bunge (*T. psilotifolium* invalid). Illustration: Journal of the Royal Horticultural Society 70: pl. 75 (1945); Iconographia Cormophytorum Sinicorum 1: 664 (1972).

Plant to 45 cm, hairless. Leaves 3–5 times divided into 3 parts; leaflets very narrow to cylindric, c. 1 mm wide. Flowers few in a loose corymb. Sepals to 1.4 cm, white to pale pink. Stamens shorter than sepals, erect, filaments thread-like. Achenes 4–7, stalkless, ribbed. *E China*. H3. Late spring.

3. T. tuberosum Linnaeus. Illustration: Bonnier, Flore complète 1: pl. 4 (1911); Polunin & Smythies, Flowers of SW Europe, 116 & pl. 5 (1973).

Plant to 50 cm with short rhizome and tuberous roots. Leaves mostly basal, 2–3-pinnate, hairless; leaflets 8–10 mm, rounded, 3-lobed. Flowers few. Sepals 4–5, white or yellowish white, to 1.5 cm. Stamens shorter than sepals, erect, filaments thread-like. Achenes 5–11, stalkless or almost so, ribbed. *Spain to SW France*. H3. Early summer.

4. T. orientale Boissier. Illustration: Polunin, Flowers of Greece and the Balkans, 88 (1980).

Plant to 30 cm with short rhizome. Leaves mostly on the stem, twice divided into 3 segments, hairless. Leaflets to 2 cm, almost circular, 3-lobed and scolloped. Flowers few. Sepals 4, white to pale purple, to 1.2 cm. Stamens shorter than sepals, erect, filaments thread-like. Achenes 2–6, stalkless, ribbed. *Greece to SW Asia*. H4. Early summer.

5. T. foetidum Linnaeus. Illustration: Bonnier, Flore complète 1: pl. 2 (1911).
Plant to 60 cm with short rhizome, glandular-hairy, foetid. Leaves 3–4-pinnate, leaflets c. 5 mm, ovate to obovate or circular, lobed. Flowers numerous in loose, drooping panicles, greenish yellow. Sepals c. 3 mm. Stamens longer than sepals, drooping, filaments

thread-like. Achenes 8–10, stalkless, ribbed. *Europe to temperate Asia*. H3. Summer.

6. T. minus Linnaeus. Illustration: Ross-Craig, Drawings of British plants 1: pl. 3–5 (1948); Keble Martin, Concise British Flora in colour, pl. 1 (1965).

Plant 15–150 cm, tufted or with long rhizomes. Leaves hairless or somewhat glandular, green or glaucous, 3–4-pinnate; leaflets almost round to ovate, lobed. Flowers numerous in dense or loose panicles. Sepals 4–5, yellow or purplish green. Stamens longer than sepals, drooping, filaments thread-like. Achenes 3–15, stalkless, ribbed, erect. *Europe to temperate Asia*. H3. Summer.

An extremely variable species, especially in size and habit. Many variants may be found as species, subspecies or at other ranks. The following names are often used in gardens: *T. adiantifolium* Besser, *T. arenarium* Butcher, *T. elatum* Boissier, *T. kemense* Fries, *T. majus* Crantz, *T. pubescens* Schleicher, *T. purpureum* Shang, *T. saxatilis* de Candolle.

7. T. flavum Linnaeus. Illustration: Bonnier, Flore complète 1: pl. 2 (1911); Ross-Craig, Drawings of British plants 1: pl. 6 (1948); Keble Martin, Concise British Flora in colour, pl. 1 (1965).

Plant to 1 m with long rhizomes. Leaves 2–3-pinnate, usually hairless; leaflets obovate to oblong with 3–4 pointed lobes at apex. Flowers numerous in erect, narrow to ovoid panicles, fragrant. Sepals 4, yellow, c. 3 mm. Stamens longer than sepals, erect, filaments thread-like. Achenes c. 15, stalkless, ribbed. *Europe to the E Mediterranean area*. H3. Summer.

Subsp. **glaucum** (Desfontaines) Battandier (*T. speciosissimum* Linnaeus; *T. glaucum* Desfontaines; *T. rugosum* Aiton), from SW Europe and N Africa, has the stem and leaves glaucous (illustration: Reichenbach, Icones florae Germanicae et Helveticae 1: pl. 46 (1838–39).

8. T. lucidum Linnaeus (*T. angustifolium* misapplied). Illustration: Reichenbach, Icones florae Germanicae et Helveticae 1: pl. 38 (1838–39); Hay & Synge, Dictionary of garden plants in colour, 175 (1969).
Plant to 1 m, tufted, hairless. Leaves 2–3-pinnate, leaflets oblong to linear, much longer than broad, entire or lobed. Flowers numerous in erect panicles of dense clusters. Sepals 4, greenish yellow, c. 3 mm. Stamens longer than sepals, erect, filaments thread-like. Achenes 6–12,

stalkless, ribbed. *Europe to temperate Asia*. H3. Summer.

9. T. simplex Linnaeus (*T. angustifolium* misapplied; *T. gallicum* Rouy & Foucaud). Illustration: Reichenbach, Icones florae Germanicae et Helveticae 1: pl. 32 (1838–39); Bonnier, Flore complète 1: pl. 3 (1911).

Very similar to *T. lucidum* but with long rhizomes and flowers in loose panicles, with drooping stamens. *Europe to temperate Asia*. H3. Summer.

Variable and perhaps not in general cultivation.

10. T. aquilegiifolium Linnaeus.
Illustration: Bonnier, Flore complète 1: pl. 4 (1911); Hay & Synge, Dictionary of garden plants in colour, 175 (1969); Polunin, Flowers of Europe, 109 (1969).
Plant to 1 m with short rootstock. Leaves 2–3-pinnate, hairless, leaflets to 3 cm or more, obovate, scolloped and with stipels. Flowers sometimes unisexual, numerous in large panicles. Sepals 4–5, greenish white, c. 4 mm. Stamens longer than sepals, erect, filaments club-shaped, usually purple, sometimes pink or white. Achenes pendent, long-stalked, 3-winged. *Europe to temperate Asia*. H3. Early summer.

A range of selected variants (e.g. 'Album') is available.

11. T. calabricum Sprengel.
Like *T. aquilegiifolium* but achenes shortly stalked, erect, ribbed; filaments purple; flowers few. *Italy*. H5. Early summer.
Not common in cultivation.

12. T. kiusianum Nakai. Illustration: Bulletin of the Alpine Garden Society 1: 79 (1930–32); Everett, New York Botanical Gardens illustrated encyclopedia of horticulture 10: 3322 (1982).
Plant usually less than 15 cm, hairless, with creeping rootstocks. Leaves once or twice divided into 3 segments; leaflets to 1.5 mm ovate, 3–5-lobed. Flowers few in small corymbs. Sepals minute–3 mm, purple to white. Stamens longer than sepals, erect, pale purple, filaments club-shaped. Achenes stalked, ribbed. *Japan*. H3. Summer.

13. T. ichangense Oliver. Illustration: Iconographia cormophytorum sinicorum 1: 678 (1972).
Plant to 20 cm, hairless. Leaves twice divided into 3 segments, leaflets 1–4 cm wide, ovate to round, scolloped, all or most peltate. Flowers few in loose corymbs. Sepals c. 3 mm, pale purple or white.

Stamens longer than sepals, erect, filaments club-shaped. Achenes 5–10, stalked, ribbed. *N to E China*. H3. Summer.

Plants grown as *T. coreanum* Léveillé probably belong to this species.

14. T. javanicum Blume. Illustration: Collett, Flora Simlensis, 7 (1902).
Plant to 2 m, hairless. Leaves 3–4-pinnate, leaflets 5–25 mm, ovate to almost round, 3–7-toothed, more or less glaucous beneath. Flowers in panicles, white. Sepals 4, white or pale purple inside, *c.* 4 mm. Stamens longer than sepals, erect, filaments club-shaped. Achenes 8–30, more or less stalkless. ribbed, each with the persistent, hooked style. *Himalaya to India and W China, Indonesia*. H4. Summer.

15. T. diffusiflorum Marquand & Airy-Shaw.
Plant to 1 m, minutely glandular or hairy. Leaves 2–3-pinnate or basically divided into 3 segments, leaflets usually less than 5 mm across, almost circular, 3–5-toothed, grey-green. Flowers few to many in large panicles, to 4 cm across. Sepals to 2 cm, pale purple. Stamens much shorter than sepals, drooping, filaments thread-like. Achenes 10–20, stalked, flattened. *China (SE Xizang)*. H2. Summer.

16. T. rochebrunianum Franchet & Savatier.
Plant to 1 m, hairless, with short rootstock. Leaves 3–4 times divided into 3 segments or pinnate; leaflets 2–3 cm, obovate to elliptic, entire or lobed. Flowers in panicles. Sepals *c.* 7 mm, purple or white. Stamens about as long as sepals, filaments thread-like. Achenes 10–20, shortly stalked, ribbed. *Japan*. H3. Summer.

17. T. delavayi Franchet (*T. dipterocarpum* misapplied). Illustration: Botanical Magazine, 7152 (1890); Gardeners' Chronicle **38**: 450 (1904); Anderson et al., The Oxford book of garden flowers, 148 (1973).
Plant to 1.2 m, hairless, stems blackish. Leaves 2–3 times divided into 3 segments or pinnate; leaflets to 1.5 cm, 3-lobed or entire. Flowers numerous in loose panicles. Sepals 4, pale purple to white, *c.* 1.5 cm. Stamens shorter than sepals, drooping, filaments thread-like. Achenes 10–20, flattened but not winged, to 1 cm, long-stalked. *W China*. H3. Summer.

White-flowered variants ('Album') are sometimes found. A double, purple-flowered variant grown as *T.* 'Hewitt's Double' probably belongs to this species.

18. T. dipterocarpum Franchet. Illustration: Gardeners' Chronicle **45**: 216 (1909); Revue Horticole **14**: 568 (1915).
Like *T. delavayi* but achenes flattened, shortly stalked, to 6 mm, with 2 narrow wings; flowers somewhat smaller. *W China*. H3. Summer.

Plants grown under this name are usually *T. delavayi*; some botanists consider them both to belong to the same species.

19. T. chelidonii de Candolle. Illustration: Journal of the Royal Horticultural Society **75**: pl. 176 (1950); Zhang, Alpine plants of China, 124 (1982).
Plant 30–250 cm, hairless, sometimes with bulbils in the leaf-axils. Leaves 2–3-pinnate or divided into 3 segments. Leaflets 1–4 cm, ovate to round, with many lobes or teeth. Flowers few to many in loose panicles. Sepals 4, pink to pale purple, to 1.2 cm. Stamens shorter than sepals, drooping, filaments thread-like. Achenes 10–15, long-stalked, flattened. *Himalaya*. H3. Summer.

20. T. reniforme Wallich. Illustration: Botanical Magazine n.s., 592 (1970–72); Polunin & Stainton, Flowers of the Himalayas, 8 (1984).
Similar to and sometimes confused with *T. chelidonii*, but leaves and flower-stalks glandular-hairy and flowers usually larger. *Himalaya*. H3. Summer.

21. T. fendleri Engelmann. Illustration: Abrams & Ferris, Illustrated Flora of the Pacific States **2**: 213 (1944); Rickett, Wild flowers of the United States **4**: pl. 48 (1970); Hitchcock et al., Vascular plants of the Pacific northwest **2**: 408 (1971).
Plant to 2 m, minutely downy. Leaves 3–4 times divided into 3 segments, upper leaves almost stalkless; leaflets to 2 cm, round, 3-lobed, scolloped. Flowers numerous, in panicles, unisexual, greenish white. Sepals of male flowers to 6 mm, those of female flowers smaller. Stamens longer than sepals, filaments thread-like, yellow. Achenes 7–11, stalkless or almost so, ribbed. *Western N America*. H2. Summer.

22. ANEMONELLA Spach
D.M. Miller
Perennial, hairless herbs to 25 cm with tuberous roots. Leaves basal, 2–3 times pinnate, each leaflet divided into 3 parts, 1–2 cm, ovate to round, 3-lobed at the tip. Flowers in umbels of 2–5 or more, with an involucre of 2–3 stalkless leaves each divided into 3 stalked leaflets. Sepals 5–10, petal-like, 5–15 mm, ovate to elliptic,

spreading, white to pink; petals absent. Stamens numerous. Style absent, stigma depressed, borne directly on top of the carpel. Achenes 4–15, stalkless, 8–10-ribbed.

A genus of a single species which grows well, if slowly, on the rock garden or peat bank in moist soils in light shade. It is best left undisturbed but may be propagated by division or by seed sown in a cold frame immediately after collection.

1. A. thalictroides (Linnaeus) Spach (*Anemone thalictroides* Linnaeus; *Thalictrum anemonoides* Michaux). Illustration: Botanical Magazine, 866 (1805); Rickett, Wild flowers of the United States **6**: pl. 157 (1973); Everett, The New York Botanical Gardens illustrated encyclopedia of horticulture **1**: 165 (1980). *Eastern USA*. H4. Spring.

23. ANEMONE Linnaeus
C.D. Brickell
Perennial herbs, occasionally woody-based, with rhizomatous, fleshy, fibrous or woody rootstocks. Basal leaves stalked, usually lobed, dissected or compound, sometimes lacking. Stem-leaves shortly stalked or stalkless, often forming an involucre below the inflorescence, sometimes remote from it. Flowers radially symmetric, bisexual, solitary or in cymes, saucer-shaped to shallowly cup-shaped. Perianth-segments petal-like, 5–20 or more, white, yellow, blue, purple or red. Stamens numerous. Carpels free, numerous, each with a single pendulous ovule and a short, persistent style. Fruit an achene, hairless to densely woolly.

A genus of 120 species, cosmopolitan in distribution with the major representation in the northern hemisphere. It is closely related to *Pulsatilla*, but differs in the short-styled fruits and the absence of nectar-secreting staminodes.

The majority of species in cultivation are hardy and easily grown under ordinary garden conditions, although a number prefer woodland soils. More specialised growing conditions are required for species such as *A. coronaria* and *A. pavonina*, their close relatives and the St Brigid, De Caen and St Bavo groups of Anemones derived from them. In nature these occur in regions with hot, dry summers and in cultivation require warm, sunny, well-drained positions in the garden to thrive. A few species, including *A. biflora* and its allies, are best grown as alpine-house plants or in frames, where a dry, summer resting period can be provided without difficulty. Most species are readily propagated from seed; this can be made easier by rubbing the seed in sand to remove the down. The rhizome-bearing species can be increased by lifting the rhizomes (which, when congested, are referred to as 'tubers' in the trade) in the summer, dividing them and replanting in the autumn.

Literature: Ulbrich, E., Über die systematische Gliederung und geographische Verbreitung der Gattung Anemone, *Botanischer Jahrbücher* **37**: 172–334 (1906).

1a. Plant rhizomatous, the rhizomes long and slender spindle-shaped or congested and tuber-like; mainly spring-and early summer-flowering
   2
  b. Plant with fibrous or woody rootstock; summer- and autumn-flowering   21
2a. Rhizomes long and slender or thicker and spindle-shaped   3
  b. Rhizomes congested and tuber-like
   11
3a. Involucral leaves stalkless or virtually so   4
  b. Involucral leaves clearly stalked   6
4a. Stems 20–80 cm, with forking branching; achenes flattened, winged
   1. dichotoma
  b. Stems 5–30 cm, unbranched; achenes neither flattened nor winged   5
5a. Involucral leaves 3-parted; flowers 1–3; perianth-segments 5, white, occasionally pink-flushed   **2. flaccida**
  b. Involucral leaves pinnately divided; flowers solitary; perianth-segments 12–15, pale purple or nearly white
   3. keiskeana
6a. Leaves 3-parted, toothed, not further divided or lobed; anthers blue or occasionally white   **4. trifolia**
  b. Leaves with primary divisions further lobed; anthers yellow or white   7
7a. Anthers white or pale yellow; flowers blue (rarely white or pink)
   5. apennina
  b. Anthers bright yellow; flowers yellow or white (occasionally blue, purplish red or pink)   8
8a. Perianth-segments 5–8   9
  b. Perianth-segments 8–15   10
9a. Perianth-segments 5, yellow, downy outside   **6. ranunculoides**
  b. Perianth-segments usually more than 5, white (occasionally blue, purplish red or pink), hairless   **7. nemorosa**
10a. Basal and involucral leaves lobed and

toothed; stems and stalks of basal leaves hairless; perianth-segments 8–9 (rarely to 12)   **8. altaica**
  b. Basal and involucral leaves lobed but not toothed; stems and stalks of basal leaves with long hairs; perianth-segments 10–15   **9. raddeana**
11a. Basal leaves absent or 1 (rarely 2); involucral leaves stalked; perianth-segments hairless outside   **10. blanda**
  b. Basal leaves 3 or more; involucral leaves stalkless or almost so; perianth-segments downy outside   12
12a. Flowers solitary   13
  b. Flowers 2–3 together, rarely solitary
   16
13a. Flowers yellow; basal leaves almost circular, shallowly lobed   **11. palmata**
  b. Flowers scarlet, blue, violet-purple, pink or white; basal leaves deeply divided and lobed   14
14a. Involucral leaves much divided
   13. coronaria
  b. Involucral leaves undivided or only toothed at apex   15
15a. Flowers 2–4 (rarely to 6) cm in diameter, rosy mauve; perianth-segments 12–20   **12. hortensis**
  b. Flowers 3–8 cm in diameter, scarlet, blue, violet-purple, pink or white, sometimes white or yellowish white at base; perianth-segments 7–9 (rarely to 12)   **14. pavonina**
16a. Leaflets of basal leaves stalked or stalkless; involucral leaves stalkless but flattened and wing-like at base
   17
  b. Leaflets of basal leaves stalkless; involucral leaves stalkless but without a wing-like base   20
17a. Leaflets of basal leaves stalkless or almost so   **18. gortschakowii**
  b. Basal leaves with all leaflets or at least the central distinctly stalked   18
18a. Stalks of the leaflets of the basal leaves more or less equal in length; flowers yellow, reddish outside; anthers yellow   **15. petiolulosa**
  b. Stalk of the central leaflet of the basal leaves much longer than those of the almost stalkless lateral leaflets; flowers red (occasionally coppery or yellow); anthers purple or yellow   19
19a. Flowers red; anthers purple; leaflets of basal leaves very dissected
   16. bucharica
  b. Flowers red, occasionally coppery or yellow; anthers yellow; leaflets of basal leaves lobed, sometimes deeply, but never very dissected   **17. biflora**

20a. Flowers yellow; anthers yellow
   19. eranthoides
  b. Flowers white, sometimes suffused purplish pink; anthers purple or violet-purple   **20. tschernjaewii**
21a. Plant 60–150 cm or more; autumn-flowering   22
  b. Plant 10–60 (rarely to 75) cm; summer-flowering   26
22a. Leaves always entire, *Vitis*-like; blade deeply and variably lobed but never divided into 3 leaflets   **21. vitifolia**
  b. Leaves divided into 3 leaflets, occasionally entire and lobed on young shoots   23
23a. Leaves densely white-hairy beneath
   22. tomentosa
  b. Leaves only sparsely hairy beneath
   24
24a. Perianth-segments 5 or 6, more or less circular; plant to 60 cm
   23. hupehensis
  b. Perianth-segments 7–20 or more (rarely 6), narrower; plant 60–150 cm or more   25
25a. Perianth-segments numerous, 20 or more; pollen fertile; plants 60–90 cm
   23. hupehensis var. **japonica**
  b. Perianth-segments fewer, 6–11 (rarely more); pollen mostly sterile; plants 1.2–1.5 m or more
   24. × hybrida
26a. Achenes embedded in dense, woolly hairs   27
  b. Achenes not embedded in dense, woolly hairs, hairless, with adpressed bristles or thinly downy   32
27a. Flowers solitary (rarely 2 together) from each involucre; perianth 2.5–8 cm in diameter   28
  b. Flowers 2 or more together from each involucre (rarely solitary) perianth 5–20 mm in diameter   30
28a. Primary divisions of basal leaves 5; rootstock increasing by root-buds; plant 15–50 cm   **25. sylvestris**
  b. Primary divisions of basal leaves 3; rootstock not increasing by root-buds; plants 5–15 (rarely to 20) cm   29
29a. Flowers 2.5–4 cm in diameter; perianth-segments 8–10 (rarely 5); head of achenes ovoid-oblong
   26. baldensis
  b. Flowers 4–8 cm in diameter; perianth-segments 5, rarely 6; head of achenes spherical   **27. rupicola**
30a. Involucral leaves stalkless; basal leaves much-divided with linear segments; fruiting heads more or less spherical; flowers white, yellow or red
   28. multifida

b. Involucral leaves stalked; basal leaves divided, but lobes not linear; fruiting heads almost spherical to cylindric; flowers white or greenish white   31

31a. Fruiting heads cylindric, 2–4 cm; styles densely long-hairy; basal leaves divided into 3 or 5, the segments further cut to more than half their length   **29. cylindrica**

b. Fruiting heads ovoid, 1.5–3 cm; styles shortly downy; basal leaves deeply cut into 3–5 segments which are not further divided   **30. virginiana**

32a. Achenes strongly compressed, virtually flat and more or less winged   33

b. Achenes oblong-cylindric or ovoid-oblong, not or only slightly compressed   35

33a. Plant 40–75 cm; perianth-segments 4 (sometimes 5–7), white; style hooked; achenes *c.* 1 cm across; basal leaves deeply 5-lobed, 8–20 cm; lobes saw-toothed   **31. tetrasepala**

b. Plant 20–50 cm; perianth-segments 5 or 6, white, red or mauve; style straight or hooked; achenes 5–6 mm across; basal leaves usually smaller and lobed, lobes not saw-toothed   34

34a. Plant 20–30 (rarely to 40) cm; flowers white, sometimes purplish or blue flushed outside; style hooked   **32. narcissiflora**

b. Plant to 50 cm; flowers red, purple or white; style straight   **33. polyanthes**

35a. Plant 15–30 cm; leaves 2–5 cm across, deeply 3-lobed; flowers purple, blue, yellow or white; achenes coarsely hairy but not woolly   **34. obtusiloba**

b. Plant 30–90 cm; leaves 8–15 cm across, deeply 3-lobed; flowers white flushed with blue-violet outside; achenes hairless   **35. rivularis**

1. A. dichotoma Linnaeus.

Perennial herb 20–80 cm with a slender, blackish brown rhizome. Stems erect, repeatedly branched. Basal leaves few, deeply 5–7-lobed, long-stalked. Involucral leaves stalkless, deeply 2–3-lobed, in whorls of 3 in the lower part of the stem, in pairs higher up the stem. Flowers white, 2–3 cm across, borne singly on long stalks from the axils of the involucral leaves. Perianth-segments 5. Achenes strongly compressed, winged, hairless, 4–5 mm. *USSR (Siberia), Japan, China (Manchuria), Korea.* H1. Early summer.

2. A. flaccida Schmidt. Illustration:
Kitamura & Murata, Coloured illustrations of alpine plants of Japan, 454 (1978).

Perennial herb 5–20 (rarely to 30) cm with a thick, shortly creeping, black rhizome. Stem erect, unbranched, forming mounds of thick, rather fleshy foliage, bronze-green when young, darker green and white-marked at bases of lobes later. Basal leaves long-stalked, 3-parted, lateral segments further divided, central segment 3-cleft, diamond-shaped to obovate. Involucral leaves stalkless or virtually so, similar to but smaller than basal leaves. Flowers 1–3, white, occasionally pink-flushed, 1.5–3 cm across. Perianth-segments 5 (rarely to 7). Achenes ovoid, *c.* 2.5 mm, with short, white hairs. *Eastern USSR, China, Japan.* H1. Late spring.

3. A. keiskeana Ito. Illustration: Makino's
new Illustrated Flora of Japan, 177 (1963); Kitamura & Murata, Coloured illlustrations of the alpine plants of Japan, 453 (1978). Perennial herb to 15 cm with a slender, creeping rhizome. Stem erect, unbranched. Basal leaves long-stalked, divided into 3 triangular-ovate, stalkless leaflets. Involucral leaves stalkless, narrowly ovate or narrowly oblong, pinnately cut. Flowers borne singly, pale purple, sometimes white within, 1.5–3.5 cm across. Perianth-segments 12–15. Achenes ovoid *c.* 2 mm, ?hairless. *Japan.* H1. Early spring.

4. A. trifolia Linnaeus. Illustration:
Pignatti, Flora d'Italia, 293 (1982). Perennial herb to 15 (rarely 20) cm with a slender, creeping, brown rhizome. Basal leaves rarely produced, stalked, 3-parted, toothed but not lobed. Involucral leaves borne in whorls of 3, each stalked, 3-parted and toothed. Flowers borne singly on long stalks above involucral leaves, white (rarely slightly pink), *c.* 2 cm across. Perianth-segments 5–8. Anthers blue or white. Achenes ovoid, shortly bristly, *c.* 2 mm. *C Italy & Austria to Hungary and N Yugoslavia; NW Spain & Portugal.* H1. Spring.

Two subspecies are recognised and both are grown. The commoner, subsp. **trifolia**, which occurs throughout the range except for the Iberian peninsula, has blue anthers and the heads of achenes are held erect; subsp. **albida** (Matiz) Tutin, endemic to NW Spain and Portugal, has white anthers and the heads of achenes are drooping.

5. A. apennina Linnaeus. Illustration:
Journal of the Royal Horticultural Society **93** (April): cover (1968); Polunin, Flowers of Europe, t. 22 (1969); Gartenpraxis for April 1985: 36.

Perennial herb to 15 cm with a creeping, brown rhizome (often incorrectly described as a congested tuber). Basal leaves 3-parted, the segments further deeply lobed and cut; primary divisions stalked, acute; downy beneath. Involucral leaves stalked, 3-parted, usually further divided or deeply lobed. Flowers borne singly on stalks 2–3 cm above involucral leaves, blue, occasionally white (var. **albiflora** Strobl) or pink-tinged, *c.* 2.5–3.5 cm across. Perianth-segments 8–23. Anthers pale yellow or white. Achenes ovoid, *c.* 2.5 mm, in a more or less spherical, erect head. *S Europe.* H1. Early spring.

Often confused with *A. blanda* which differs in its hairless or almost hairless leaves, congested, tuber-like rhizome, and drooping head of achenes.

6. A. ranunculoides Linnaeus. Illustration:
Huxley, Garden perennials and water plants, pl. 35 (1970); Gartenpraxis for 1984: 69; The Garden **112**: pl. 65 (1987). Perennial herb to 15 cm with a creeping, brown rhizome. Basal leaves lacking or 1, stalked, 3-parted, the segments further deeply divided and lobed. Involucral leaves similar but with short stalks. Flowers borne singly 2–3 cm above the involucral leaves, deep yellow, 1.5–2 cm across. Perianth-segments 5 (occasionally more), downy outside. Anthers bright yellow. Achenes ovoid, *c.* 2 mm. *Widespread in most of Europe except the Mediterranean area.* H1. Early spring.

Two variants, 'Grandiflora', with flowers 2–2.5 cm and 'Pleniflora' ('Flore Pleno'), with 12 or more ragged perianth-segments, are grown.

7. A. nemorosa Linnaeus. Illustration:
Keble Martin, The concise British Flora in colour, pl. 1 (1965); Polunin, Flowers of Europe, pl. 22 (1969); Il Giardino Fiorito 48: 269 (1982); Gartenpraxis for April 1986: 53.

Perennial herb to 20 cm with a creeping, brown rhizome. Basal leaves stalked, 3-parted, the lateral segments further subdivided and lobed, the central segment deeply divided. Involucral leaves similar but with short, fairly broad stalks. Flowers borne singly, 2–3 cm above involucral leaves, white, sometimes suffused with purple or pink (variants with blue, purplish or pink flowers have also been found in the wild), 2–3 cm across. Perianth-segments 6–8 (rarely as few as 5 or as many as 12), hairless. Anthers bright yellow. Achenes *c.* 2 mm, downy, in drooping heads. *Widespread in Europe except for the Mediterranean area.* H1. Early spring.

Very variable in flower colour and degree

of doubling. Many garden variants have been selected including 'Vestal', which is double with a neat central boss of modified filaments; 'Alboplena', which is double (sometimes ragged), white; 'Robinsoniana' which is pale lavender-blue, greyish outside; 'Allenii' which is deep lavender-blue, deep lilac outside; 'Lismore Pink' which is pale pink; 'Leeds Variety' which has flowers 4–5 cm; and 'Bracteata' which is an oddity with a ruff of leaves immediately below the semi-double, ragged flower with sometimes greenish segments.

A. × **lipsiensis** Beck (*A.* × *intermedia* Winkler; *A. seemannii* Camus). *A. ranunculoides* × *A. nemorosa*. This occurs in some areas where the parents grow together. It is intermediate between them, with light creamy yellow flowers about 1.5–2 cm across.

8. A. altaica Meyer. Illustration: Makino's new illustrated Flora of Japan, t. 704 (1963).
Perennial herb to 20 cm with a cylindric, creeping, yellowish brown rhizome. Basal leaf stalked, stalk hairless, the segments acuminate and toothed except at the base, sometimes further divided. Stems hairless with an involucre of 3 leaves which are 3-parted, sometimes further divided and toothed like the basal leaves. Flowers borne singly above the involucre, 2–4 cm across, white, occasionally flushed with violet outside. Perianth-segments 8–9 (rarely to 12). Achenes narrowly ovoid, *c.* 2 mm, shortly hairy. *Arctic USSR (Siberia), Japan.* H1. Spring.

9. A. raddeana Regel. Illustration: Makino's new illustrated Flora of Japan, t. 705 (1963); Kitamura & Murata, Coloured illustrations of alpine plants of Japan, 457 (1978).
Perennial herb to 15 (rarely to 25) cm with a short, spindle-shaped rhizome. Basal leaf absent or 1, long-stalked, covered with long hairs, divided into 3 broad, ovate segments, each with a long stalk, the segments further lobed but not toothed. Stem covered with long hairs. Involucral leaves in whorls of 3, stalked, divided into 3 oblong or linear segments, each shallowly lobed at the apex. Flowers borne singly above the involucral leaves, 2–4 cm across, white with a purplish flush outside. Perianth-segments 10–15. Achenes narrowly ovoid, *c.* 2 mm, very shortly downy. *China (Manchuria), USSR (Sakhalin), Japan, Korea.* H1. Spring.

10. A. blanda Schott & Kotschy. Illustration: Polunin, Flowers of Europe, pl. 21 (1969); Journal of the Royal Horticultural Society **100** (March): cover (1975); Huxley & Taylor, Flowers of Greece and the Aegean, pl. 43 (1977).
Perennial herb to 18 cm with a congested, tuber-like rhizome. Basal leaves 0–1 (rarely 2), stalked, broadly triangular, divided into 3 almost stalkless or shortly stalked, irregularly lobed segments. Involucral leaves stalked, deeply 3-sect or 3-lobed, adpressed-hairy above. Flowers borne singly 2–5 cm above the involucre, 2–4 cm across, blue, mauve, white or pink. Perianth-segments 9–15, hairless outside. Achenes 1-2 mm, hairless or thinly hairy. *SE Europe, Cyprus, W Turkey, USSR (Caucasus).* H1. Spring.

The variability of *A. blanda* in flower colour in the wild has been exploited to produce a range of large-flowered cultivars including 'Atrocaerulea' and 'Ingramii', both deep blue; 'Charmer', deep pink; 'Radar', bright red with a white centre and 'White Splendour' with pink-flushed exterior. 'Scythinica', originally selected from the wild in Turkey, sometimes treated as a botanical variety, is slightly later-flowering than other cultivars, with perianth-segments sapphire-blue outside, white within.

A. **caucasica** Ruprecht (*A. blanda* var. *parvula* de Candolle). Similar but smaller, involucral leaves becoming hairless above, perianth-segments 8–11. *USSR (Caucasus), N Iran.* H3.

11. A. palmata Linnaeus. Illustration: Polunin, Flowers of Europe, pl. 21 (1969); Polunin & Smythies, Flowers of southwest Europe, pl. 6 (1973).
Perennial herb to 15 cm with a congested, tuber-like rhizome. Basal leaves stalked, more or less circular with 3–5 shallow, toothed lobes. Involucral leaves stalkless with 3–5 linear-lanceolate divisions joined at the base. Flowers solitary (? occasionally 2 together), 2.5–3.5 cm across, yellow, sometimes flushed with red outside. Perianth-segments 10–15. Achenes oblong-ovoid, *c.* 1.5 mm. *SW Europe.* H2. Spring.

12. A. hortensis Linnaeus (*A. stellata* Lamarck). Illustration: Polunin & Huxley, Flowers of the Mediterranean, pl. 23 (1965); Polunin, Flowers of Europe, pl. 21 (1969).
Perennial herb to 30 cm with a congested, tuber-like rhizome. Basal leaves stalked, divided completely or almost to the base into 3 irregularly lobed and finely dissected segments. Involucral leaves stalkless, undivided, usually toothed at the apex.

Flower solitary, 2–4 cm across, rosy mauve, sometimes paler at the base. Perianth-segments 12–20. Achenes oblong-ovoid, *c.* 2 mm, densely woolly. *C Mediterranean area.* H2. Spring.

A. **heldreichiana** (Boissier) Gandoger (*A. stellata* var. *heldreichii* Boissier) has perianth-segments white inside, blue-grey outside. *Crete.* H3. Spring.

13. A. coronaria Linnaeus. Illustration: Journal of the Royal Horticultural Society **82**: f. 44 (1957); Polunin & Huxley, Flowers of the Mediterranean, pl. 24, 26, 27 (1965); Polunin, Flowers of Europe, pl. 21 (1969).
Perennial herb to 30 cm with a congested, tuber-like rhizome. Basal leaves stalked, divided into 3 stalked, very dissected segments. Involucral leaves stalkless, undivided, with apex deeply cut. Flower solitary, 3–8 (rarely to 10) cm across, scarlet, blue, pink or white, sometimes parti-coloured. Perianth-segments 5–6. Achenes oblong-ovoid, *c.* 2 mm, densely woolly. *S Europe, Mediterranean area.* H2. Spring.

Many colour-variants have been treated as varieties. These include var. **coronaria** (var. *coccinea* (Gordon) Burnat) – scarlet, var. **cyanea** (Risso) Arduino – blue, var. **rosea** (Hanry) Rouy & Foucaud and var. **alba** (Goaty & Pons) Burnat – white.

The St Brigid and De Caen group of Anemones, grown for the cut-flower industry, derive from *A. coronaria*. The original St Brigid group, raised in Ireland during the late nineteenth century, was semi-double but the name is now applied to a range of semi-double, double and chrysanthemum-flowered types. The De Caen group, first developed in the mid-nineteenth century near Caen in France, was double-flowered but the name is now used mainly for single-flowered Anemones.

14. A. pavonina Lamarck. Illustration: Polunin, Flowers of Europe, pl. 21 (1969); Polunin & Smythies, Flowers of southwest Europe, pl. 5 (1973); Huxley & Taylor, Flowers of Greece and the Aegean, pl. 41, 42 (1977).
Perennial herb to 30 cm with a congested, tuber-like rhizome. Basal leaves stalked, deeply 3-parted, the segments further lobed or cut. Involucral leaves stalkless, undivided, with a few apical teeth. Flower solitary, 3–10 cm across, scarlet, violet, purple or pink, sometimes white or yellow, white at the base. Perianth-segments 7–9 (rarely to 12). Achenes oblong-ovoid,

c. 2 mm, densely woolly. *S Europe, Mediterranean area.* H2. Spring.

Various colour variants have been recognised, including var. **ocellata** (Moggridge) Bowles & Stearn – flowers scarlet with yellowish white centre and var. **purpureo-violacea** (Boissier) Halacsy – flowers violet or pinkish violet with white centres.

A. × fulgens Gay. *A. pavonina* × *A. hortensis.* Intermediate between its parents, occurring in southern France and widely cultivated. It is distinguished by its bright scarlet flowers usually with 15 or more perianth-segments, greater in number and narrower than those of *A. pavonina*, which in other characters it closely resembles. Two clones are offered: 'Annulata Grandiflora', with a yellowish white centre akin to that of *A. pavonina* var. *ocellata* and 'Multipetala' with semi-double flowers of 20 or more perianth-segments.

The St Bavo group of Anemones grown mainly for cut flowers, with flowers to 10 cm or more across in a wide variety of colours and varying from single to semi-double, is derived from *A. pavonina.*

15. A. petiolulosa Juzepczuk. Illustration: Rix & Phillips, The bulb book, 65 (1981); Mathew, The smaller bulbs, t. 4 (1987). Perennial herb to 18 cm with a congested, tuber-like rhizome. Basal leaves stalked, divided into 3 segments with more or less equal stalks; segments further subdivided into 3 and deeply cut and lobed. Involucral leaves with a flattened, wing-like base, cut into 3 lobes, the lobes also 3-fid and toothed. Flowers usually 2–4 together, rarely 1, 2–4.5 cm across, slightly drooping, yellow, reddish outside. Perianth-segments 5, densely adpressed hairy outside. Anthers yellow. Achenes oblong-ovoid, *c.* 3 mm, densely woolly. *USSR (C Asia).* H2. Spring.

16. A. bucharica Finet & Gagnepain. Illustration: Botanical Magazine, n.s., 622 (1972). Perennial herb to 20 (rarely to 30) cm with a congested, tuber-like rhizome. Basal leaves stalked, divided into 3 segments, the lateral segments more or less stalkless and bifid, the central segment with a long stalk, 3-fid, each segment further cut and lobed. Involucral leaves with a flattened, wing-like base, divided into 3 segments, each further 2–3-parted. Flowers usually 2 together, rarely 1 or 3, 3–4 cm across, erect, red or purplish red. Perianth-segments 5, densely adpressed-hairy outside. Anthers purple.

Achenes oblong-ovoid, *c.* 4 mm, densely woolly. *USSR (C Asia).* H2. Spring.

Sometimes considered to be the same species as *A. coronaria.*

17. A. biflora de Candolle. Illustration: Journal of the Royal Horticultural Society **88**: f. 72 (1963), **90**: f. 9 (1965); Mathew, Dwarf bulbs, pl. 2 (1973); Rix & Phillips, The bulb book, 67 (1981). Perennial herb to 12 cm with a congested, tuber-like rhizome. Basal leaves stalked, divided into 3 segments, the lateral segments more or less stalkless and further cut and lobed, the central segment with a short but distinct stalk and also lobed and cut. Involucral leaves with a flattened, wing-like base, 2–3-fid, the segments further deeply lobed. Flowers 2–3 together (rarely solitary), 2–5 cm across, erect, red, sometimes coppery or yellow. Perianth-segments 5, densely adpressed-hairy outside. Anthers yellow. Achenes ovoid-oblong, *c.* 2–3 mm, densely woolly. *Iran, Afghanistan, Pakistan, India (Kashmir).* H2. Spring.

18. A. gortschakowii Karelin & Kirilov. Perennial herb to 15 cm with a congested, tuber-like rhizome. Basal leaves stalked, divided into 3 stalkless or almost stalkless segments, the segments further bifid or 3-fid and lobed. Involucral leaves with a wing-like base, 3-parted, the lobes then further cut. Flowers usually 2 together (rarely solitary), 1–2 cm across, pale yellow, sometimes red-suffused with age. Perianth-segments 5, densely adpressed hairy outside. Anthers yellow. Achenes ellipsoid, *c.* 2–2.5 mm, densely woolly. *USSR (C Asia).* H2. Spring.

19. A. eranthoides Regel. Perennial herb to 12 cm with congested, tuber-like rhizome. Basal leaves stalked, divided into 3 or 5 stalkless leaflets, the leaflets shallowly 3-fid and toothed. Involucral leaves stalkless, narrowly obovate and shallowly to deeply 3-fid. Flowers usually 2 together, rarely solitary or 3, 1–2.5 cm across, golden yellow inside, greenish yellow outside. Perianth-segments 5 (rarely to 8), adpressed-hairy outside. Anthers yellow. Achenes ovoid-oblong, *c.* 2 mm, densely woolly. *USSR (C Asia).* H2. Spring.

20. A. tschernjaewii Regel. Illustration: Journal of the Royal Horticultural Society **90**: pl. 213 (1965); Mathew, Dwarf bulbs, pl. 13 (1973); Rix & Phillips, The bulb book, 65 (1981). Perennial herb to 30 cm with a congested,

tuber-like rhizome. Basal leaves stalked, divided into 3 stalkless segments which are shallowly 3-fid. Involucral leaves similar but smaller. Flowers 2–3 together (rarely solitary), 2–4.5 cm, white suffused with purple-pink at base. Perianth-segments 5, hairy outside (?). Anthers purple or violet-purple. Achenes ellipsoid, *c.* 2–2.5 mm, closely woolly. *USSR (C Asia).* H2. Spring.

21. A. vitifolia de Candolle. Robust perennial to 90 cm with a woody, fibrous rootstock. Basal leaves stalked, *Vitis*-like, shallowly 5-lobed (but never divided into 3 leaflets), the lobes conspicuously toothed, sparsely white-woolly beneath. Involucral leaves similar but smaller. Stems branched with terminal, fairly loose umbels of white flowers which are 3.5–5 cm across. Perianth-segments 5–6. Achenes in a more or less spherical, densely woolly cluster 5–8 mm across. *Afghanistan to W China and Burma.* H3. Late summer–early autumn.

22. A. tomentosa (Maximowicz) P'ei (*A. vitifolia* misapplied; *A. vitifolia* 'Robustissima'). Robust perennial to 1 m or more with a woody, fibrous rootstock, colonising from underground shoots. Leaves divided into 3, deeply toothed, *Vitis*-like leaflets, covered beneath with dense white hairs. Young leaves on young shoots occasionally entire. Involucral leaves similar to basal but smaller. Stems branched with terminal sprays of pale pink flowers which are 5–8 cm across. Perianth-segments 5–6. *N China.* H3. Late summer–early autumn.

23. A. hupehensis Lemoine. Illustration: Hay & Synge, Dictionary of garden plants in colour, pl. 964 (1969). Perennial to 60 cm with a woody, fibrous rootstock. Leaves with 3 leaflets which are deeply and sharply toothed, *Vitis*-like, sparsely hairy beneath. Young leaves on young shoots occasionally entire. Involucral leaves similar to basal but smaller. Stems branched, with umbels of pink flowers which are crimson-pink outside. Perianth-segments 5 or 6, more or less circular. *C China.* H1. Late summer–autumn.

Var. **japonica** (Thunberg) Bowles & Stearn (*A. nipponica* Merrill). Illustration: Kitamura & Murata, Coloured illustrations of alpine plants of Japan, 451 (1978). Differs in its more numerous (20 or more), narrower perianth-segments and its slightly more robust habit. *Naturalised in Japan and S China.* H1. Late summer–autumn.

Appears to differ from *A. hupehensis* only in its semi-double flowers with narrow perianth-segments.

24. A. × hybrida Paxton (*A. × elegans* Decaisne). Illustration: Huxley, Garden perennials and water plants, pl. 3 (1970); Perry, Flowers of the world, 252 (1972). Differs from *A. hupehensis* and its var. *japonica* in the lack of fertile pollen and more robust habit, usually reaching 1.2–1.5 m or more; and from var. *japonica* in having only 6–11 (rarely to 15) perianth-segments. H1. Late summer–autumn.

The common Japanese Anemone of gardens, a hybrid of *A. vitifolia* and *A. hupehensis* var. *japonica*. Numerous selections have been introduced of which the pure white 'Honorine Jobert' is very frequently grown.

25. A. sylvestris Linnaeus. Illustration: Lindman, Nordens Flora 4: pl. 227 (1964); Huxley, Garden perennials and water plants, pl. 36 (1970).
Perennial herb 15–50 cm with a woody, fibrous rootstock spreading rapidly by root-buds. Basal leaves stalked, deeply palmately divided into 5 obovate, deeply lobed segments. Involucral leaves similar but smaller. Flowers solitary (occasionally 2 together), 2.5–8 cm across, white. Perianth-segments 5 (occasionally more). Head of achenes more or less spherical, densely woolly. *C & E Europe, USSR (Caucasus).* H1. Early summer.

A large-flowered cultivar, 'Macrantha', with flowers about 8 cm across is sometimes grown and a double, 'Flore Pleno', has been cultivated occasionally.

26. A. baldensis Linnaeus. Illustration: Coste, Flore de la France 1: 43 (1901). Perennial herb to 20 cm with a fibrous rootstock. Basal leaves stalked, divided into 3 leaflets, each further cut into 3-lobed divisions. Involucral leaves similar but smaller. Flower solitary, 2.5–4 cm across, white, sometimes flushed with pink outside. Perianth-segments 8–10 (rarely as few as 5). Head of achenes ovoid-oblong, densely woolly. *Central S Europe.* H1. Summer.

27. A. rupicola Cambessedes. Illustration: Botanical Magazine, 9496 (1937); Polunin & Stainton, Flowers of the Himalaya, pl. 7 (1985).
Perennial herb to 15 cm with a somewhat woody, fibrous rootstock. Basal leaves stalked, 3-lobed to the base, the lobes deeply 3-lobed and sharply toothed. Involucral leaves stalkless, similar to basal

but smaller and less deeply cut. Flower solitary, 4–8 cm across, white, occasionally suffused with pink or blue outside. Perianth-segments 5 (rarely 6). Head of achenes spherical, densely woolly. *Afghanistan to SW China.* H1. Summer.

28. A. multifida Poiret (*A. globosa* Pritzel; *A. hudsoniana* Richardson). Illustration: Hitchcock et al., Vascular plants of the Pacific Northwest 2: 328 (1964); Clark, Wild flowers of British Columbia, 147 (1973).
Perennial herb 10–60 cm with a stout, fibrous rootstock. Basal leaves stalked, divided into 3 or 5 segments themselves further cut 2 or 3 times into linear segments. Involucral leaves stalkless, similar but smaller. Flowers 2–3 (rarely solitary) from each involucre, 5–20 mm across, white, yellowish or red. Perianth-segments 5–9. Head of achenes more or less spherical, densely woolly. *N America.* H1. Spring–early autumn.

Extremely variable in height, flower colour and density of hairiness. Hybridised with *A. sylvestris* it has given rise to **A. × lesseri** Wehrhahn with flowers 2–3 (rarely to 4) cm across, varying from purple, yellow and white to pink; the pink-flowered variant is sometimes available.

Plants grown as *A. magellanica* in gardens appear to be variants of *A. multifida*.

29. A. cylindrica Gray. Illustration: Hitchcock et al., Vascular plants of the Pacific Northwest 2: 326 (1964).
Perennial herb 20–60 (rarely to 80) cm with a stout, fibrous rootstock. Basal leaves divided into 3 (rarely 5) segments which are further cut to more than half their length, toothed at apex. Involucral leaves stalked, similar to basal but smaller. Flowers 2–6 together (rarely solitary) from each involucre, 5–20 mm across, white or greenish white. Perianth-segments 5 (rarely 6). Styles densely long-hairy. Fruiting head cylindric, 2–4 cm long, densely woolly. *Western N America.* H1. Summer.

30. A. virginiana Linnaeus. Illustration: House, Wild flowers, pl. 66 (1961). Perennial herb 30–60 (rarely to 80) cm with a stout, fibrous rootstock. Basal leaves stalked, deeply cut into 3–5 diamond-shaped-ovate segments, each sharply toothed and cut in the upper half. Involucral leaves stalked, similar to basal but smaller. Flowers 2–3 together (rarely solitary) from each involucre, 5–25 mm

across, white or greenish white. Perianth-segments 5 or 6. Styles shortly hairy. Fruiting head ovoid, 1.5–3 cm long, densely woolly. *Central & eastern N America.* H1. Summer.

A. canadensis Linnaeus. Similar but with stalkless involucral leaves and with a spherical head of flattened, winged achenes, adpressed-bristly but not densely woolly. *N America.* H1. Summer.

31. A. tetrasepala Royle. Illustration: The Garden **102**: 452 (1977).
Robust perennial herb to 75 cm with a stout, woody, fibrous rootstock. Basal leaves stalked, leathery, deeply lobed and sharply toothed. Involucral leaves stalkless, similar but smaller and usually with narrower lobes. Flowers 5 or more in umbels (often several umbels to each involucre), 3 cm across, white. Perianth-segments 4 (sometimes 5–7). Styles hooked. Achenes strongly compressed and winged, *c.* 1 cm across. *Afghanistan to India (Kashmir).* H3. Summer.

A. demissa Hooker & Thomson. Similar, but with smaller, rounded basal leaves cut into obovate, stalked segments; smaller narrower, 3-lobed involucral leaves and spreading umbels of 3–6 white, blue or purple flowers. *W Himalaya to SW China.*

32. A. narcissiflora Linnaeus (*A. fasciculata* Linnaeus). Illustration: Takeda, Alpine flora of Japan, pl. 51 (1963); Hay & Synge, Dictionary of garden plants in colour, pl. 966 (1969); Polunin, Flowers of Europe, pl. 22 (1969); Tosco, Mountain flowers, 111 (1978).
Perennial herb to 40 cm with a stout, woody, fibrous rootstock. Basal leaves stalked, deeply palmately divided into 5 and further cut into linear-lanceolate lobes. Involucral leaves stalkless, similar but smaller. Flowers 3–8 in umbels, 2–4 cm across, white, sometimes pink- or blue-flushed outside.
Perianth-segments 5–6. Styles hooked. Achenes strongly compressed and winged, 5–6 mm across. *C & S Europe, Turkey, USSR (Caucasus, Siberia), Japan, N America.* H1. Summer.

Extremely variable.

A. zephyra Nelson. Leaves less divided, flowers solitary, lemon-yellow. *Western N America (Rocky Mountains).*

Closely related and perhaps not a separate species.

A. biarniensis Juzepczuk. Primary leaf-divisions 3, stalked. *Northeast USSR.*

33. A. polyanthes D. Don. Illustration: Bulletin of the Alpine Garden Society **52**: 255 (1984); Polunin & Stainton, Flowers of the Himalayas, pl. 6 (1985). Similar to *A. tetrasepala* but smaller (to 50 cm), with rounded basal leaves shallowly lobed and with rounded, not sharply toothed segments. Flowers in umbels, white, red or purple-blue. Styles straight, achenes strongly compressed, winged, *c.* 5 mm across. *Pakistan to Bhutan.* H1. Late spring–summer.

34. A. obtusiloba D. Don. Illustration: Journal of the Royal Horticultural Society **93**: f. 127 (1968); Polunin & Stainton, Flowers of the Himalayas, pl. 7 (1985). Perennial herb to 15 cm with a fibrous rootstock. Basal leaves stalked, 2–5 cm across, with rounded, deeply 3-lobed leaves, the lobes further cut and toothed. Involucral leaves stalkless, 3-lobed at apex. Flowers borne singly, 1.5–5 cm across, purple, blue, white or yellow. Perianth-segments 5–7. Achenes oblong-cylindric, coarsely hairy but not woolly. *Pakistan to China (SE Xizang) and Burma.* H1. Late spring–summer.

A. trullifolia Hooker & Thomson . Illustration: Hara, Photo-album of plants of eastern Himalaya, pl. 223 (1968). Similar but basal leaves narrower, 3-lobed towards the apex, the lobes themselves shortly 3-lobed. *E Himalaya to SW China.* H3. Summer.

The plant currently in cultivation has white flowers.

35. A. rivularis de Candolle. Vigorous perennial herb to 90 cm with a woody, fibrous rootstock. Basal leaves stalked, 8–15 cm across, rounded, deeply 3-lobed, the lobes further cut and shallowly toothed. Involucral leaves with a broad base and divided into 3 lobed, toothed segments. Flowers 2–5 or more in an umbel, 1.5–3 cm across, white flushed with blue-violet outside. Perianth-segments 5–8. Achenes oblong-cylindric, hairless. *India, SW China.* H1. Summer.

24. PULSATILLA Miller
C.D. Brickell

Tufted, perennial herbs with woody, fibrous rootstocks. Leaves mostly basal, pinnately or palmately divided. Stem-leaves (involucral leaves) 3, usually stalkless and united at the base, occasionally shortly stalked. Flowers solitary, perianth spreading or narrowly campanulate to openly campanulate, erect to pendent. Perianth-segments 6 (occasionally 5, 7 or 8), frequently silky on the outer surface. Stamens numerous, surrounded by whorls of nectar-secreting staminodes. Carpels numerous in a head, free; styles long, becoming longer and feathery in fruit. Ovule 1.

A genus of some 30 species from north temperate areas, often from mountainous regions. It is closely related to *Anemone* and often united with it, but is distinguished by the styles, which become long and feathery in fruit and the presence of nectar-secreting staminodes.

Identification is complicated by the considerable variation in leaf size and dissection, degree of hairiness and flower colour which occur within populations of some species. Many of the names in botanical and horticultural literature represent only minor variants.

All the species in cultivation are hardy and, with the exception of *P. occidentalis*, not difficult to grow in sunny, open, well-drained sites, many preferring alkaline conditions but thriving also in acid soils. Propagation is normally from seed, although particularly attractive colour-forms have been successfully (if slowly) increased by division.

Literature: Aichele, D. & Schwegler, H.-W., Die Taxonomie der Gattung Pulsatilla, *Feddes Repertorium* **60**: 1–230 (1957); Krause, K., Zur taxonomischen Gliederung, Verbreitung und Genetik der Pulsatilla halleri (All.) Willd., *Botanische Jahrbücher* **78**: 1–68 (1958).

1a. Involucral leaves shortly stalked, resembling the basal leaves   **1. alpina**
  b. Involucral leaves stalkless, not resembling the basal leaves, divided into linear segments   **2**
2a. Basal leaves palmately divided   **7. patens**
  b. Basal leaves pinnately divided   **3**
3a. Basal leaves overwintering (evergreen), pinnate with 3–5 segments; flowers white, flushed pink or blue-violet externally   **2. vernalis**
  b. Basal leaves dying down in autumn, 2–4 times pinnate; flowers violet, purple, red-purple, occasionally yellow or white   **4**
4a. Flowers distinctly pendent at maturity   **5**
  b. Flowers erect or slightly pendent at maturity   **6**
5a. Perianth-segments not recurved at apex, at least twice the length of the stamens   **3. montana**
  b. Perianth-segments recurved at apex, not more than one and a half times the length of the stamens   **4. pratensis**
6a. Leaves well-developed at flowering, finely dissected and usually with more than 100 lobes; flower-stems and involucral leaves thinly hairy, later hairless or almost so; flowers somewhat pendent, usually narrowly campanulate   **5. vulgaris**
  b. Leaves not developed or only slightly so at flowering, coarsely dissected and usually with fewer than 100 lobes; flower-stems and involucral leaves densely and often persistently hairy; flowers erect, shallowly campanulate to campanulate   **6. halleri**

1. P. alpina (Linnaeus) Delarbre (*Anemone alpina* Linnaeus).
Perennial herb to 30 cm in flower, to 45 cm in fruit. Basal leaves dying down in autumn, long-stalked, downy, twice pinnate, lobes frequently recurved, the terminal segments not divided to the midrib. Involucral leaves similar to basal leaves, but with broad, short stalks. Flowers erect or almost so, shallowly campanulate, 4–6 cm across, white flushed blue-purple outside, or pale yellow. *Mountains of C & S Europe, USSR (Caucasus).* H1. Spring–early summer.

Two well-defined subspecies occur:

Subsp. **alpina**. Illustration: Bonnier, Flore complète **1**: pl. 7 (1911); Gartenpraxis for 1979: 120; Rasetti, I fiori delle Alpi, t. 37 (1980). Outer perianth-segments white or white flushed blue-purple outside, inner perianth-segments white; usually occurring on alkaline soils.

Subsp. **apiifolia** (Scopoli) Nyman (subsp. *sulphurea* (de Candolle) Ascherson & Graebner). Illustration: Bonnier, Flore complète **1**: pl. 7 (1911); Gartenpraxis for 1979: 121; Rasetti, I fiori delle Alpi, t. 37 (1980). Perianth-segments pale yellow; usually occurring on acid soils.

P. alba Reichenbach is similar but has leaves hairless or almost so, and terminal leaf-segments divided to the midrib; it is usually smaller than *P. alpina* and is never yellow-flowered. *C Europe.* H1. Spring–early summer.

P. occidentalis (Watson) Freyn (*Anemone occidentalis* Watson). Illustration: Abrams & Ferris, Illustrated Flora of the Pacific States **2**: f. 1810 (1944); Journal of the Royal Horticultural Society 92: f. 250 (1967). Similar to *P. alpina* but leaves divided into 3 segments, these then 2–3 times pinnate;

involucral leaves shortly stalked; flowers creamy white, sometimes flushed blue on the outside. *Western N America (mountains of British Columbia, Washington, Oregon, California, N Idaho, Montana).* H1. Late spring–late summer.

2. P. vernalis (Linnaeus) Miller (*Anemone vernalis* Linnaeus). Illustration: Bonnier, Flore complète 1: pl. 6 (1911); Gartenpraxis for 1979: 120; Rasetti, I fiori delle Alpi, t. 37 (1980); Bulletin of the Alpine Garden Society 54: 274 (1986).
Perennial herb to 15 cm in flower, to 45 cm in fruit. Basal leaves overwintering (evergreen), more or less hairless, pinnate with 3–5 deeply toothed segments. Involucral leaves stalkless, united below. Flowers pendent in bud, then erect, campanulate, 4–6 cm across, white-silky; outer perianth-segments strongly suffused with pink or blue-violet outside, the colour less pronounced on the inner segments. *Europe (mountains from Scandinavia to S Spain, eastwards to Bulgaria), USSR (Siberia).* H1. Spring–summer.

3. P. montana (Hoppe) Reichenbach (*Anemone montana* Hoppe). Illustration: Bonnier, Flore complète 1: pl. 6 (1911); Gartenpraxis for 1979: 120; Il Giardino Fiorito, 37 (1987).
Perennial herb to 15 cm in flower, to 45 cm in fruit. Basal leaves dying down in autumn, downy, 3 times pinnate, the segments deeply cut into many narrow lobes. Involucral leaves with about 25 lobes, stalkless, united below. Flowers pendent, tubular-campanulate, 3–4 cm across, dark or bluish violet, the perianth twice the length of the stamens, segments not recurved at apex. *C & E Europe (Switzerland & Italy to Romania & Bulgaria).* H1. Spring–early summer.

P. rubra (Lamarck) Delarbre. Illustration: Polunin & Smythies, Flowers of southwest Europe, pl. 8 (1973). Differs in its fewer-lobed involucral leaves and in its dark red-purple, blackish or brownish red flowers with perianth-segments 2–2.5 times the length of the stamens. *C & S France, C & E Spain.* H1. Spring–early summer.

4. P. pratensis (Linnaeus) Miller (*Anemone pratensis* Linnaeus; *P. nigricans* Störk). Illustration: Bonnier, Flore complète 1: pl. 6 (1911); Gartenpraxis for 1979: 121.
Like *P. montana* but involucral leaves with about 30 lobes, perianth-segments 1.5 times the length of the stamens and recurved at apex; flowers very variable in colour, populations with dark to pale

purple and yellowish or greyish violet occurring in various areas of its range. *C & E Europe, extending north to Denmark and southern Norway.* H1. Spring–early summer.

Plants with dark purple flowers are sometimes offered as *P. nigricans* Störk (*P. pratensis* subsp. *nigricans* (Störk) Zamels).

P. albana (Steven) Berchtold & Presl (*Anemone albana* Steven). Similar, but with yellow or occasionally bluish flowers. *USSR (Caucasus).*

P. armena (Boissier) Ruprecht (*Anemone armena* Boissier; *A. albana* subsp. *armena* (Boissier) Smirnov). Similar, but flowers violet blue. *USSR (Caucasus, Transcaucasia); Iran?*

Both these species are occasionally offered.

5. P. vulgaris Miller (*Anemone pulsatilla* Linnaeus).
Illustration: Bonnier, Flore complète 1: pl. 6 (1911); Journal of the Royal Horticultural Society 98: f. 131 (1973); The Garden 101: 554 (1976); Gartenpraxis for 1979: 120 & front cover of part 3.
Perennial herb to 15 cm in flower, to 45 cm in fruit. Basal leaves dying down in autumn, thinly hairy when young, later more or less hairless, pinnate, usually divided into 7–9 primary segments, each segment further 2–3 times pinnatisect, the lobes linear to linear-lanceolate. Involucral leaves stalkless, united below. Flowers appearing when the leaves fairly well developed, usually slightly pendent, narrowly campanulate (occasionally campanulate), 4–9 cm across, deep to pale purple, occasionally white. *Europe.* H1. Spring–early summer.

Extremely variable and occurring in numerous isolated populations in the wild. These have often been recognised as distinct species or as subspecies or varieties of *P. vulgaris*. They are, however, somewhat ill-defined, with considerable overlap of characters and numerous intermediates occurring.

This is the most widely grown species, and a number of colour-variants varying from white ('Alba') to pink ('Mrs Van der Elst', 'Rosea' and 'Barton's Pink') and deep red ('Coccinea', 'Rubra') have been raised from seed. While some individual colour-variants were formerly propagated by division, most are nowadays maintained by selection to a specific colour range. It is probable that some of the colour variation is due to hybridisation with *P. rubra* and other species.

6. P. halleri (Allioni) Willdenow. Illustration: Bonnier, Flore complète 1: pl. 6 (1911); Journal of the Royal Horticultural Society 88: f. 120 (1963); Gartenpraxis for 1979: 120; Rasetti, I fiori delle Alpi, t. 37 (1980).
Perennial herb to 15 cm in flower, to 45 cm in fruit. Basal leaves dying down in autumn, densely hairy at least when young, pinnate, usually with 3–5 primary segments, the terminal segments long-stalked, each segment pinnatifid with oblong-lanceolate lobes. Involucral leaves stalkless, united below. Flowers appearing before the leaves, more or less erect, shallowly to more deeply campanulate, 4–9 cm across, violet-purple to lavender-blue. *C & SE Europe, USSR (Crimea).* H1. Late spring–early summer.

The *P. vulgaris-P. halleri* complex has been subject to a variety of classifications, with the name *P. grandis* Wenderoth in particular being attached to both species at various ranks. Meikle (Botanical Magazine, n.s., 475, 1964) points out the clear distinctions between the western and eastern European representatives of this complex. These distinctions lie in the occurrence of flowers with the finely dissected leaves, thinly hairy, later hairless leaves and stems and slightly nodding, narrowly campanulate flowers of the *P. vulgaris* (western) group, compared with the flowers before the more coarsely dissected leaves, densely hairy stems and leaves (at least when young) and erect, shallowly campanulate flowers of the *P. halleri* (eastern) group.

Subsp. **grandis** (Wenderoth) Meikle (*P. grandis* Wenderoth). Illustration: Botanical Magazine, n.s., 475 (1964). Hairs on the stems and involucral leaves dense, silvery or tawny. *C & E Europe (S Germany, Austria, Czechoslovakia, Hungary).*

The plant known in gardens as *P. vulgaris, P. halleri* or *P. grandis* 'Budapest' is now considered to be a selected colour-variant of *P. halleri* subsp. *grandis* with lavender-blue flowers; it was originally raised from seed collected in the Budapest area (Hungary). A number of other subspecies of *P. halleri* are currently recognised, of which subsp. **halleri**, usually with 5 primary leaf-divisions, from the SW & C Alps, and subsp. **slavica** (Reuss) Zamels with 3 primary leaf-divisions, from the Carpathians, are available; both have dark violet flowers.

7. P. patens (Linnaeus) Miller (*Anemone patens* Linnaeus; *A. hirsutissima* (Britton) Macmillan; *A. ludoviciana* de Candolle;

A. nuttallii Nuttall; *A. nuttalliana* de
Candolle; *A. patens* subsp. *hirsutissima*
(Britton) Zamels; *A. patens* var. *nuttalliana*
(de Candolle) Gray; *A. wolfgangiana* Besser;
Pulsatilla hirsutissima Britton; *P. ludoviciana*
(Nuttall) Heller; *P. nuttalliana* (de Candolle)
Sprenger; *P. patens* var. *wolfgangiana*
(Besser) Regel). Illustration: Rickett, Wild
flowers of the United States **3**: pl. 29
(1969).
Perennial herb to 15 cm in flower, to
45 cm in fruit. Basal leaves dying down in
autumn, roughly hairy, palmately divided
into 3–7 primary divisions, these further
divided into 15–80 linear to linear-
lanceolate segments. Involucral leaves
stalkless, united below. Flowers erect with
spreading perianth-segments, 5–7 cm
across, blue-violet, lilac or occasionally
yellowish or yellowish white. *Europe, N Asia
(Siberia), N America.* H1. Late
spring–summer.

Very variable, with several subspecies
recognised on the basis of leaf division and
flower colour. Populations from N America
appear similar to those from Siberia, with
finely dissected leaves, and are sometimes
recognised as subsp. **multifida** (Pritzel)
Zamels. The N American populations are
sometimes regarded as a separate species,
when their correct name is *P. nuttalliana*
(de Candolle) Sprenger.

25. HEPATICA Miller
F. McIntosh
Small perennial herbs with rhizomes.
Leaves all basal, persistent, long-stalked,
3–5-lobed, the lobes broad, entire or
further divided or toothed. Flowers solitary
on long stalks arising from ground level.
Sepals 3, green. Petals several, white,
purplish, bluish or rose-coloured. Stamens
numerous. Ovary of several–many carpels.
Fruit a group of achenes.

A genus of 6 species from Europe, C & E
Asia and eastern N America, sometimes
included in *Anemone*. The 3 organs referred
to here as sepals are often interpreted as 3
bracts, as in *Anemone*: however, as they are
closely adpressed to the coloured floral
segments, and look like a calyx it seems
sensible here to regard them as sepals.

1a. Leaf-lobes divided or toothed
1. transsilvanica
 b. Leaf-lobes entire 2
2a. Leaf-lobes acute **2. acutiloba**
 b. Leaf-lobes obtuse or rounded 3
3a. Flowers mostly 2.5 cm or more in
 diameter; plant slightly hairy
3. nobilis

 b. Flowers to 2 cm in diameter; plant
 conspicuously hairy **4. americana**

1. H. transsilvanica Fuss (*Anemone
transsilvanica* (Fuss) Heuffel). Illustration:
Šavulescu, Flora Republicii socialiste
Romania **2**: pl. 71, f.2 (1953); The Garden
102: 470 (1977).
Rhizomes long. Plant 20–30 cm. Leaves
3-lobed (rarely 5-lobed), the lobes with 3–5
broad, ovate teeth, hairless beneath and
somewhat shining when mature. Flowers
to 5 cm in diameter, sepals light blue or
whitish. *Romania.* H2. Spring.

2. H. acutiloba de Candolle. Illustration:
Rickett, Wild flowers of the United States **1**:
pl. 69 (1967); House, Wild flowers, pl. 69
(1967); Justice & Bell, Wild flowers of North
Carolina, 66f (1968).
Plants to 25 cm. Leaves 3-lobed, lobes
entire, pointed. Flowers 1.5–2.5 cm in
diameter. Petals bluish or white. *Eastern
USA.* H1. Spring.

3. H. nobilis Miller (*H. triloba* Chaix;
Anemone hepatica Linnaeus). Illustration:
Polunin, Flowers of Europe, pl. 20 (1969).
Rhizome short. Plants to 15 cm (rarely
more), very slightly hairy. Leaves 3-lobed,
lobes entire, obtuse or rounded, purplish
beneath. Flowers 2.5–3.5 cm in diameter,
petals white, pink or purplish blue. *Eurasia.*
H1. Spring.

H. × media Simonkai is the hybrid
between *H. nobilis* and *H. transsilvanica*; it
occurs in the wild and may well be found
in gardens and is distinguished by its sterile
pollen.

4. H. americana (de Candolle) Ker
Gawler(*H. triloba* misapplied). Illustration:
Rickett, Wild flowers of the United States **1**:
pl. 126, 127 (1967), **2**: pl. 179, 180
(1968); Justice & Bell, Wild flowers of North
Carolina, 67 (1968).
Like *H. nobilis* but plants conspicuously
hairy, flowers to 2 cm in diameter. *Eastern
N America.* H1. Spring.

26. CLEMATIS Linnaeus
W.A. Brandenburg
Woody climbers, low shrubs or perennial
herbs. Leaves opposite or sometimes
alternate, simple or compound, when
divided into 3s or pinnate or bipinnate.
Inflorescences basically 1–many-flowered
cymes borne at the apex of young shoots or
axillary on young or old shoots. Flowers
with 4–8 (rarely many) free, petal-like
perianth-segments, sometimes unisexual,
when plants dioecious or monoecious.
Staminodes sometimes present between

perianth and stamens. Stamens numerous.
Ovary of usually many free carpels. Fruit a
group of achenes which sometimes have
long, feathery persistent styles. A
cosmopolitan genus of about 200 species
with its main distribution in northern
temperate zones. The greatest diversity of
species occurs in the far east. The plants
are generally easy to grow; many need
suitable, robust support. A certain amount
of lime in the soil is beneficial (mortar from
walls can provide this). Propagation is by
seed or cuttings.

The majority of the large-flowered
Clematis are cultivars of hybrid origin and
cannot be assigned to a particular species.
A brief outline of the cultivar groups is
presented at the end of this account
(p. 364).
Literature: Kuntze, O., Monographie der
Gattung Clematis L., *Verhandlungen der
Botanischer Verein Brandenburg* **26**: 83–202
(1885); Pringle, J.S., The cultivated taxa of
Clematis Sect. Atragene (Ranunculaceae),
Baileya **19**: 49–83 (1973); Tamura, M., A
classification of genus Clematis, *Acta
Phytotaxonomica et Geobotanica* **38**: 33–44
(1987).

1a. Perennial herb or low, non-climbing
 shrub 2
 b. Woody climber 10
2a. Leaves all simple 3
 b. At least some of the leaves compound
 4
3a. Leaves hairless or sparsely hairy;
 perianth-segments downy at the
 margins **20. integrifolia**
 b. Leaves woolly beneath; perianth-
 segments woolly throughout
22. ochroleuca
4a. Lower leaves simple, upper leaves
 compound 5
 b. All leaves compound 6
5a. Lower leaves stalkless, upper leaves
 simple to pinnate; perianth-segments
 purplish **15. addisonii**
 b. Lower leaves stalked, upper leaves
 simple to pinnate; perianth-segments
 deep mauve to violet **23. douglasii**
6a. Leaves divided one or many times into
 3 leaflets 7
 b. Leaves pinnate 9
7a. Leaves divided two or three times into
 3 leaflets; perianth-segments pale
 yellow to creamy white
3. aethusifolia
 b. Leaves simply divided into 3 leaflets;
 perianth-segments blue or blue-violet
 8

8a. Flowers in long-stalked cymes; perianth-segments spreading, violet-to pale blue   **36. × aromatica**
  b. Flowers shortly stalked, solitary in bract-axils; perianth-segments dark violet-blue to pale blue
                                      4. heracleifolia
9a. Flowers solitary, violet to grey-brown; perianth-segments brown-hairy outside   **18. fusca**
  b. Flowers numerous in panicle-like inflorescences, white; perianth-segments not brown-hairy outside
                                              33. recta
10a. Leaves all simple   11
  b. At least the lower leaves compound   14
11a. Leaves absent or at most to 1 cm, or reduced to leaf- and leaflet-stalks which act as tendrils   **40. afoliata**
  b. Leaves present, not as above   12
12a. Inflorescence-stalk bearing an involucre of fused bracts **30. cirrhosa**
  b. Inflorescence-stalk without an involucre of fused bracts   13
13a. Stems trailing; leaves ovate-lanceolate, 10–15 cm, not glaucous
                                          48. henryi
  b. Stems erect; leaves linear to lanceolate, 4–8 (rarely to 13) cm, glaucous   **37. songarica**
14a. Leaves divided into 3 leaflets (sometimes simple)   15
  b. Leaves all compound   17
15a. Leaves glaucous beneath, simple and divided into 3 on the same plant
                                      16. glaucophylla
  b. Leaves not glaucous beneath; upper leaves simple   16
16a. Flowers in 3-flowered cymes; perianth-segments 4–6, obovate
                                      44. × jackmanii
  b. Flowers solitary; perianth-segments 6–8, broadly ovate   **47. lanuginosa**
17a. Leaves divided into 3 leaflets (once or many times)   18
  b. Leaves pinnately divided (once or many times)   27
18a. All leaves divided into 3 leaflets   19
  b. Leaves variously divided into 3 leaflets once, twice or 3 times   23
19a. Leaves evergreen   **34. armandii**
  b. Leaves deciduous   20
20a. Inflorescences borne on wood of previous year   21
  b. Inflorescences borne on shoots of current year   22
21a. Perianth-segments brown-violet; filaments dilated above   **5. barbellata**
  b. Perianth-segments white or pink; filaments thread-like   **32. montana**

22a. Flowers unisexual, plants dioecious
                                          41. indivisa
  b. Flowers bisexual   **24. virginiana**
23a. Leaves once or twice divided into 3 leaflets   24
  b. Leaves always twice divided into 3 leaflets   25
24a. Flowers in panicle-like inflorescences; perianth-segments 4–5   **31. aristata**
  b. Flowers 1–few in cymes; perianth-segments 5–8   **42. petriei**
25a. Inflorescences mostly borne on wood of previous year   **6. alpina**
  b. Inflorescences all on shoots of the current year   26
26a. Flowers yellow   **9. serratifolia**
  b. Flowers white   **38. potaninii**
27a. Leaves pinnate   28
  b. Leaves pinnate, bipinnate or tripinnate on the same plant   44
28a. Flowers unisexual, plants dioecious
                                      25. ligusticifolia
  b. Flowers bisexual   29
29a. Inflorescences at first on the wood of the previous year, later on shoots of the current year   **46. patens**
  b. Inflorescences only on shoots of the current year   30
30a. Flowers solitary in the leaf-axils   31
  b. Flowers numerous in inflorescences   36
31a. Flowers with perianth-segments spreading   32
  b. Flowers bell-shaped to tubular, perianth-segments not spreading   33
32a. Perianth-segments 4, blue, purple, red or white   **43. viticella**
  b. Perianth-segments 5–7, white with purple veins   **39. phlebantha**
33a. Flowers cylindric (though perianth-segments spreading at apex)
                                          21. crispa
  b. Flowers not cylindric   34
34a. Flowers urn-shaped   **19. pitcheri**
  b. Flowers bell-shaped   35
35a. Perianth-segments very thick
                                          14. viorna
  b. Perianth-segments thin
                                      17. versicolor
36a. Flowers in axillary or terminal cymes
                                              37
  b. Flowers in panicle-like inflorescences
                                              41
37a. Flowers with 4–6 perianth segments on the same plant   **12. grewiiflora**
  b. Flowers all usually with 4 perianth-segments   38
38a. Flowers creamy or greenish white when first open   39
  b. Flowers not white or creamy white when first open   40

39a. Flowers greenish white; perianth-segments obtuse   **26. vitalba**
  b. Flowers creamy white; perianth-segments not obtuse   **29. grata**
40a. Flowers yellow; leaves glaucous
                                      8. orientalis
  b. Flowers pale blue fading to white; leaves not glaucous   **28. × jouiniana**
41a. Leaf-bases fused into a flat disc or cup
                                          1. connata
  b. Leaf-bases not fused as above   42
42a. Perianth-segments spreading
                                  27. brevicaudata
  b. Perianth-segments not spreading, flowers tubular or bell-shaped   43
43a. Flowers tubular; perianth-segments fleshy   **11. buchananiana**
  b. Flowers bell-shaped; perianth-segments not fleshy   **13. nutans**
44a. Leaves pinnate or bipinnate   45
  b. Leaves bipinnate or tripinnate   46
45a. Perianth-segments spreading
                                          45. florida
  b. Perianth-segments not spreading, flowers tubular or bell-shaped
                                      2. rehderiana
46a. Leaves bipinnate   **35. flammula**
  b. Leaves bipinnate and tripinnate on the same plant   47
47a. Leaves mostly bipinnate, sometimes tripinnate; flowers yellow, violet when faded   **10. graveolens**
  b. Leaves bipinnate and tripinnate on the same plant; flowers white, blue or violet   **7. macropetala**

1. C. connata de Candolle (*C. amplexicaulis* Edgeworth; *C. gracilis* Edgeworth; *C. velutina* Edgeworth; *C. venosa* Royle). Illustration: Garden and Forest 4: 235 (1891).

Woody climber with strong stems. Leaves pinnate, with 3–7 ovate-lanceolate, coarsely toothed, hairless leaflets 6–12 cm, cordate at base; leaf-bases fused into a flat disc or cup. Flowers fragrant, in panicle-like inflorescences. Perianth-segments 4, oblong, obtuse, woolly on both surfaces, 2–2.5 cm, very pale yellow to creamy white. Stamens with hairy filaments. Achenes downy, with persistent styles 4–5 cm. *Pakistan, Himalaya, SW China.* H5. Summer.

2. C. rehderiana Craib (*C. veitchiana* Craib; *C. nutans* var. *thyrsoidea* Rehder & Wilson). Illustration: Botanical Magazine, n.s., 523 (1966–68); Hay & Synge, Dictionary of garden plants in colour, pl. 1965 (1969); Grey-Wilson & Matthews, Gardening on walls, pl. 7 (1983).

Woody climber to 4 m. Leaves pinnate or

bipinnate, leaflets ovate-lanceolate, irregularly lobed or toothed, acute, cordate or rounded at base, to 7 cm. Flowers tubular or bell-shaped in axillary panicle-like inflorescences. Perianth-segments 4, ovate, hairless inside, downy outside, 1–2 cm, recurved at apex, pale yellow. Achenes compressed, with feathery styles. *China (Xizang)*. H2. Summer.

3. C. aethusifolia Turczaninow. Illustration: The Garden 6: 423 (1874); Botanical Magazine, 6542 (1881); Schneider, Illustriertes Handbuch der Laubholzkunde 1: 275, 282 (1904).
Erect or climbing perennial herb to 60 cm, the whole plant with sparse, adpressed hairs. Leaves once or twice divided into 3 variable leaflets varying from linear to ovate, broadly tapered at base, toothed, downy beneath. Flowers drooping, fragrant, bell-shaped, in axillary cymes. Perianth-segments 4, oblong, acute, margins white-downy, 1.5–2 cm, pale yellow to creamy white. Stamens with thread-like, downy filaments. Achenes ribbed, downy, with persistent styles 1.5–2 cm. *N China, USSR (E Siberia)*. H1. Summer.

4. C. heracleifolia de Candolle (*C. tubulosa* Turczaninow). Illustration: Decaisne, Les Clématites a fleurs tubuleuses, pl. 9 (1881). Much-branched perennial herb, variably woody at base, to 1.5 m, with erect, grooved stems. Leaves divided into 3 broadly ovate coarsely toothed leaflets with rounded bases and acuminate tips, 5–8 cm. Flowers tubular, shortly stalked in 3-flowered cymes in the axils of bracts which may be divided into 3 or simple. Perianth-segments 4, oblong, obtuse and recurved at apex, hairy outside, 1.5–2 cm, pale blue to dark violet-blue. Stamens with filaments as long as anthers, dilated above. Carpels with long hairs at the base. Achenes ovate, hairy, with persistent, feathery styles to 3.5 cm *NE China. Mongolia*. H1. Summer.

C. davidiana Decaisne. Illustration: Decaisne, Les Clématites a fleurs tubuleuses, pl. 10 (1881). Similar but flowers unisexual (plants dioecious), stalkless, perianth-segments oblong-ovate and spreading for at least half of their length. *NE China*. H1. Summer.

C. stans Siebold & Zuccarini. Illustration: Decaisne, Les Clématites a fleurs tubuleuses, pl. 12 (1881). Similar to *C. davidiana* but with many-flowered, panicle-like inflorescences. *Japan*. H2. Summer.

5. C. barbellata Edgeworth (*C. nepalensis* Royle). Illustration: Botanical Magazine, 4794 (1854).
Woody climber to 4 m. Leaves divided into 3, leaflets ovate-lanceolate, rounded at base, acute at tip, irregularly toothed, 4–8 cm. Flowers solitary or several together clustered at the nodes on growth of the previous year, bell-shaped. Perianth-segments 4, oblong-ovate, downy inside and out and with longer hairs at the margins, 2–4 cm, purple to brown-violet. Stamens with dilated, downy filaments, anthers with long hairs at the apex. Achenes diamond-shaped, hairless but with feathery styles. *Himalaya*. H3. Spring.

C. napaulensis de Candolle (*C. montana* D. Don not de Candolle). Similar but with 2 united bracteoles on the flower-stalks. *Himalaya*.

6. C alpina (Linnaeus) Miller (*C. sibirica* (Linnaeus) Miller). Illustration: Robert & Schroeter, Alpine flowers, pl. 3 (1938); Polunin, Flowers of Europe, pl. 23 (1969); Gartenpraxis for 1976: 322; The Garden 113: 30 (1988).
Woody climber to 2 m. Leaves twice divided into 3, leaflets ovate-lanceolate, toothed, to 5 cm. Flowers solitary, drooping, borne on wood of the previous year, more or less bell-shaped. Perianth-segments 4, ovate, acuminate, densely hairy outside, violet-blue or yellowish white, 3–5 cm. Staminodes about half as long as perianth-segments, yellowish white but fading violet-blue. Achenes diamond-shaped, ribbed, with persistent, feathery styles. *N & C Europe, Asia*. H1. Spring–summer.

C. columbiana Torrey & Gray. Illustration: Rickett, Wild flowers of the Unites States 6: pl. 53 (1973). Similar but a woody climber with leaves divided into 3 leaflets. *N America*. H1. Spring–summer.

C. ochotensis Poiret. Also similar, but a woody climber with obtuse perianth-segments. *Korea. Japan*. H1. Spring.

Probably more often used in Clematis breeding than is generally acknowledged.

7. C. macropetala Ledebour. Illustration: Hay & Synge, Dictionary of garden plants in colour, pl. 1962 (1969); Gartenpraxis for 1981: 200; Grey-Wilson & Matthews, Gardening on walls, pl. 6 (1983).
Woody climber to 5 m. Leaves bi- or tripinnate, leaflets ovate, sometimes cordate at base, deeply lobed and irregularly toothed, to 7 cm. Flowers solitary in axils on the wood of the previous year, drooping. Perianth-segments oblong-lanceolate,

acute, densely hairy throughout, violet-blue or white, 2.5–4 cm. Staminodes many, at least half as long as the perianth-segments. Achenes ovoid with persistent, feathery styles. *N China, Mongolia, USSR (E Siberia)*. H1. Spring–summer.

A direct or indirect parent of many hybrids.

8. C. orientalis Linnaeus (*C. glauca* Willdenow).
Woody climber to 8 m. Leaves very variable in shape and size, pinnate with 5–7 (rarely 3 or 9) oblong, lanceolate, elliptic or ovate to broadly ovate or broadly elliptic, greyish green, hairless or sparsely hairy, entire leaflets; the leaflets themselves may be divided almost to the base or further divided into 3, acute or acuminate, more or less abruptly narrowed at the base. Inflorescences terminal and axillary, 3–many-flowered cymes on the branchlets, the apical flower of each cyme sometimes aborted or absent. Perianth-segments 4, oblong or elliptic, spreading, recurved later, woolly at the margins, somewhat less so within, with sparse hairs outside, 1–2 cm, yellow, pale yellow or greenish yellow, sometimes light purplish brown inside or tinged with red-violet outside. Anthers yellow, to 5 mm, filaments dilated above, hairy towards the base, mostly dark red-purple. Achenes diamond-shaped, slightly ribbed at the margins, downy, dark brown, with persistent styles to 5.5 cm, covered with long, erect hairs. *Temperate Asia*. H2. Summer.

The name *C. orientalis* is very often wrongly applied in cultivation; genuine *C. orientalis* is often in cultivation under the name *C. glauca*.

9. C. serratifolia Rehder. Illustration: Gartenpraxis for 1981: 200.
Woody climber to 5 m, hairless. Leaves twice divided into 3, leaflets oblong-lanceolate to ovate-lanceolate, acute, toothed, 4–8 cm. Flowers more or less hanging in axillary 3-flowered cymes with small bracts. Perianth-segments 4, spreading, ovate-lanceolate, acuminate, hairless outside, sparsely hairy inside, woolly at margins, 2–2.5 cm, pale yellow tinged or veined with violet. Filaments hairy, dilated above. Achenes ovoid with persistent, feathery styles to 4 cm. *Korea*. H2. Summer.

C. koreana Komarov. Leaves once divided into 3 leaflets, and flowers solitary, yellow to violet. *Korea*.

This name is often wrongly applied in gardens to plants of *C. serratifolia*, and the occurrence of genuine *C. koreana* in cultivation is doubtful.

10. C. graveolens Lindley.
Woody climber to 4 m. Leaves variously pinnate, bipinnate or tripinnate; leaflets very variable in shape and size, irregularly lobed or toothed. Flowers 1–3 in axillary cymes. Perianth-segments 4, spreading, ovate-obovate, densely and finely hairy at the margins, 2–4 cm, yellow. Anthers linear, filaments hairy, dilated above. Achenes diamond-shaped with feathery styles. *Himalaya.* H2. Summer.

This species is part of a complex of which the following 3 variants are traditionally treated as species:

C. tangutica (Maximowicz) Korshinsky. Illustation: Botanical Magazine, 7710 (1900); Revue Horticole for 1902: 528; Grey-Wilson & Matthews, Gardening on walls, pl. 7 (1983). Woody climber to 5 m. Leaves pinnate with 5–7 oblong-lanceolate to lanceolate, irregularly toothed or slightly lobed leaflets 4–8 cm. Flowers solitary, axillary or terminal on young shoots, bell-shaped. Perianth-segments 4, ovate-lanceolate, strongly ribbed, acuminate, hairy at margins, 2–4 cm, bright yellow. Achenes ovoid, hairy with persistent styles to 7 cm. *USSR (Pamirs), NW China.* H2. Summer.

C. tibetana Kuntze. Like *C. tangutica* but leaflets linear, glaucous, perianth-segments with violet spots. *China (Xizang).* H2. Summer.

C. vernayi Fischer. Like *C. tangutica* but perianth-segments very thick and spreading. *Himalaya west to Nepal, W China (Xizang).* H2. Summer.

This variant was long-cultivated as *C. orientalis.*

11. C. buchananiana de Candolle. Illustration: Schneider, Illustriertes Handbuch der Laubholzkunde **1**: 275, 282 (1904).
Woody climber with hairy young shoots. Leaves pinnate with 5–7 broadly ovate leaflets, cordate at base, coarsely toothed or sometimes lobed, hairy, 5–10 cm; leaf-bases slightly fused around the stem. Flowers fragrant, tubular, in long, leafy panicle-like inflorescences. Perianth-segments 4, thick, linear-oblong, recurved at the apex, woolly outside, 2–2.5 cm, creamy white to pale yellow. Filaments densely hairy. Achenes woolly with persistent styles to 5 cm. *Himalaya from Pakistan to W China.* H5–G1. Summer–autumn.

12. C. grewiiflora de Candolle (*C. loasaefolia* de Candolle). Illustration: Botanical Magazine, 6369 (1878); Hara, Photo-album of plants of Eastern Himalaya, pl. 70 (1968).
Woody climber with woolly stems. Leaves pinnate with 3–5 broadly ovate, leaflets which are acute, tapered or cordate at the base, toothed and densely woolly, 8–10 cm. Flowers broadly bell-shaped. Perianth-segments 4, ovate, slightly recurved at apex, 1–3 cm, yellow. Filaments thread-like, downy. *Himalaya, Burma.* H5–G1. Summer.

13. C. nutans Royle.
Woody climber to 5 m. Leaves pinnate with 5–7, oblong, ovate or lanceolate, downy leaflets 3–8 cm, each usually with 3–5 acute or obtuse lobes. Flowers hanging in panicle-like inflorescences, bell-shaped. Perianth-segments 4, oblong, downy outside, 2–4 cm, pale yellow. Anthers short, filaments thread-like, downy towards the base. Achenes diamond-shaped, compressed, with long, feathery styles. *Himalaya.* H3. Summer.

14. C. viorna Linnaeus. Illustration: Justice & Bell, Wild flowers of North Carolina, 64 (1968); Journal of the Royal Horticultural Society **95**: f. 113 (1970); Grey-Wilson & Matthews, Gardening on walls, pl. 7 (1983). Woody climber. Leaves pinnate with 7–9 leaflets which are hairless, entire, lobed or further divided into 3 leaflets. Flowers solitary, bell-shaped. Perianth-segments 4, very thick, recurved at apex, violet-blue. Achenes diamond-shaped with brown feathery styles 3–5 cm. *Eastern N America (Pennsylvania to Georgia and Indiana).* H3. Spring–summer.

15. C. addisonii Vail. Illustration: Britton & Brown, Illustrated Flora of the northern United States and Canada **2**: 69 (1897).
Low shrub to 1 m. Leaves glaucous, the lower simple, stalkless, sometimes lobed, obtuse, 4–8 cm; the upper simple to pinnate with 2–4 leaflets, the uppermost leaflet transformed into a tendril. Flowers solitary, terminal and axillary, urn-shaped, pendulous. Perianth-segments 4, thick, lanceolate, recurved at the apex, purplish. Achenes diamond-shaped to almost spherical, ribbed, downy, with persistent styles 4–5 cm, covered with brown hairs. *Eastern N America (Virginia, N Carolina).* H3. Spring–summer.

16. C. glaucophylla Small.
Woody climber to 4 m, with conspicuous red stems. Leaves simple or divided into 3;

when simple, entire or lobed, ovate, acute, cordate at base, 5–10 cm; when divided, leaflets like the simple leaves but smaller. Flowers bell-shaped. Perianth-segments 4, thick, lanceolate, recurved at the apex, *c.* 2.5 cm, red-purple. Achenes almost spherical, with long, persistent styles. *Eastern USA (Kentucky, Alabama, Florida).* H3. Summer.

17. C. versicolor Rydberg.
Woody climber to 3.5 m. Leaves with slender stalks, pinnate, leaflets ovate, lanceolate or oblong, 2–8 cm, with a network of veins, glaucous beneath. Flowers solitary, bell-shaped, pendulous. Perianth-segments 4, thin, lanceolate, *c.* 2 cm, recurved at apex, purple or blue. Achenes diamond-shaped with long, persistent, white, feathery styles. *USA (Missouri, Arkansas).* H3. Summer.

18. C. fusca Turczaninow. Illustration: Lavallée, Les Clématites a grand fleurs, t. 20 (1884); Schneider, Illustriertes Handbuch der Laubholzkunde **1**: 275 (1904).
Climbing or erect perennial to 2 m with grooved stems, hairy at nodes. Leaves pinnate, sometimes with long stalks which act as tendrils; leaflets very variable, ovate-lanceolate, acute, entire, toothed or lobed, the veins beneath sparsely hairy. Flowers solitary, pendulous. Perianth-segments 4–6, oblong-ovate, red- or brown-hairy outside, margins white-downy, inside hairless, 2–2.5 cm, violet to grey-brown or dark red-brown. Anthers and filaments with grey-brown hairs. Achenes diamond-shaped, downy, with persistent, grey-brown, feathery styles 4–6 cm. *NW China, Japan.* H1. Summer.

19. C. pitcheri Torrey & Gray. Illustration: Revue Horticole for 1878, 10; Rickett, Wild flowers of the United States **2**: pl. 60 (1967).
Woody climber to 6 m, with hairy young shoots. Leaves pinnate with 5–7 (rarely 3 or 9) ovate, thick, acute net-veined leaflets; leaflets entire, lobed or sometimes futher divided into 3. Flowers solitary in the leaf-axils, urn-shaped. Perianth-segments 4, thick, ovate, downy and recurved at the apex, pale bluish purple. Achenes with downy styles to 3 cm. *Eastern USA (Indiana to Missouri, Nebraska and Texas).* H3. Summer.

C. texensis Buckley (*C. coccinea* Englemann). Illustration: Botanical Magazine, 6594 (1881); Addisonia **21**: pl. 675 (1939–42); Rickett, Wild flowers of the

United States 3: pl. 29 (1969); Grey-Wilson & Matthews, Gardening on walls, pl. 5 (1983). Low shrub similar to *C. pitcheri* but with broadly ovate, greyish green leaflets and scarlet-red flowers. *USA (Texas)*. H3. Summer.

A direct or indirect parent of some large-flowered hybrids.

20. C. integrifolia Linnaeus. Illustration: Botanical Magazine, 65 (1788); Reichenbach, Icones Florae Germanicae et Helveticae 4: t. 60 (1840); Polunin, Flowers of Europe, pl. 24 (1969).
Perennial with erect stem to 70 cm. Leaves simple, stalkless, ovate-lanceolate, entire, hairless or sparsely hairy along the veins beneath, 4–9 cm. Flowers solitary, terminal or occasionally axillary, pendulous, broadly bell-shaped. Perianth-segments 4, ovate, acute, downy at the margins, blue or rarely pink or white. Filaments downy, yellow, dilated. Styles in flower yellow-downy. Achenes diamond-shaped with feathery styles. *C Europe, Asia*. H2. Summer.

C. × durandii Kuntze. Illustration: The Garden 73: 302 (1909). Similar, but leaves simple, elliptic-ovate, entire. *Garden Origin*. H2. Summer.

This plant was originally thought to be a species, and is often found under the name above; however, it must be considered a cultivar of unknown origin, and should probably be called *C.* 'Durandii'.

21. C. crispa Linnaeus. Illustration: Botanical Magazine, 1892 (1817); Lavallée, Les Clématites a grand fleurs, t. 14 (1884); Rickett, Wild flowers of the United States 2: pl. 60 (1967).
Woody climber. Leaves pinnate, most of the divisions further divided into 3, the ultimate divisions entire or lobed, hairless. Flowers solitary, terminal or axillary, pendulous, perianth cylindric at the base but spreading towards the apex. Perianth-segments 4, margins wavy, bluish purple. Achenes with persistent silky styles to 4 cm. *Eastern USA (Pennsylvania to Missouri, Arkansas, Florida and Texas)*. H3. Summer.

22. C. ochroleuca Solander. Illustration: Loddige's Botanical Cabinet 7: t. 661 (1822); Rickett, Wild flowers of the United States 2: pl. 61 (1967).
Perennial to 60 cm. Leaves simple, stalkless, entire or sometimes lobed, obtuse, hairless above, woolly beneath. Flowers solitary, terminal and sometimes also axillary, urn-shaped, pendulous. Perianth-segments 4, thick, ovate, recurved at the apex, woolly throughout, bluish outside

yellowish inside. Achenes diamond-shaped with persistent, yellowish brown, feathery styles 3–6 cm; fruiting head erect. *Eastern USA (Pennsylvania to Georgia)*. H2. Spring–summer.

23. C. douglasii W.J. Hooker (*C. hirsutissima* Pursh). Illustration: Hooker, Flora Boreali-Americana, t. 1 (1833); Rickett, Wild flowers of the United States 4: pl. 60 (1970).
Perennial with shoots densely hairy when young, hairless when older, to 60 cm. Leaves stalked, the lower simple (entire or variously lobed), the upper pinnate or bipinnate; leaflets oblong, lanceolate or ovate, mostly entire, sometimes with a few teeth. Flowers solitary, mostly terminal but sometimes also axillary, bell-shaped, pendulous on long stalks. Perianth-segments 4, thick, ovate-lanceolate, *c.* 2.5 cm, deep mauve to violet. Stamens as long as or longer than perianth-segments. Achenes diamond-shaped with persistent, brown, feathery styles 3–5 cm. *Northwest N America*. H3. Summer.

24. C. virginiana Linnaeus. Illustration: House, Wild flowers, pl. 73 (1961); Rickett, Wild flowers of the United States 1: pl. 63 (1967); Justice & Bell, Wild flowers of North Carolina, 64 (1968).
Woody climber to 6 m. Leaves divided into 3 hairless, broadly ovate, toothed or lobed, acute leaflets which are sometimes slightly cordate at the base. Flowers many in axillary cymes. Perianth-segments 4 (rarely 5), spreading, white, 1–1.5 cm. Achenes with persistent, white, feathery styles 3–6 cm. *Eastern N America*. H1. Summer.

25. C. ligusticifolia Torrey & Gray. Illustration: Rickett, Wild flowers of the United States 6: pl. 53 (1973); Clark, Wild flowers of British Columbia, 163 (1973).
Dioecious woody climber to 6 m, almost hairless. Leaves pinnate with 5, oblong, lanceolate or ovate leaflets, acute at the apex, tapered at the base, coarsely toothed or lobed, 5–8 cm. Inflorescences many-flowered, leafy, axillary cymes. Perianth-segments 4 (rarely 5), spreading, white, 1–1.5 cm. Achenes with persistent, white, feathery styles 3–6 cm. *Eastern USA*. H1. Summer.

26. C. vitalba Linnaeus. Illustration: Ary & Gregory, The Oxford book of wild flowers, pl. 1 (1962); Polunin, Flowers of Europe, pl. 24 (1969); Polunin & Everard, Trees and bushes of Europe, 67 (1976); Bärtels, Gartengehölze, 131 (1981).
Very vigorous woody climber to 30 m.

Leaves pinnate; leaflets ovate or broadly ovate to lanceolate, more or less entire or coarsely lobed and toothed. Cymes many-flowered, terminal and axillary. Perianth-segments 4, spreading, obtuse and recurved at apex, downy, greenish white, 1–2 cm. Achenes diamond-shaped with short, feathery styles. *Eurasia*. H1. Summer.

27. C. brevicaudata de Candolle. Illustration: Schneider, Illustriertes Handbuch der Laubholzkunde 1: 280 (1904).
Woody, climbing, much-branched shrub. Leaves pinnate with the lower leaflets divided into 3, the upper simple; all leaflets ovate-lanceolate, acute, toothed, with conspicuous impressed veins above, hairless or slightly hairy. Flowers in axillary, panicle-like inflorescences. Perianth-segments 4, spreading, oblong-lanceolate, downy outside, white or creamy white, to 1.5 cm. Filaments hairless. Achenes downy with persistent styles to 3 cm. *Japan, N China, Mongolia*. H1. Summer.

28. C. × jouiniana Schneider. Illustration: Schneider, Illustriertes Handbuch der Laubholzkunde 1: 280 (1904); Gartenpraxis for 1981, 200.
Woody climber. Leaves pinnate with 5–7 ovate-lanceolate, irregularly lobed and toothed leaflets 4–8 cm. Cymes many-flowered, terminal and axillary. Perianth-segments 4, at first forming a tubular flower, later spreading and recurved at the base, obtuse, pale blue fading to white, 1–2 cm. Achenes (if present) diamond-shaped with short feathery styles. *Garden Origin*. H1. Summer.

Reputedly the hybrid between *C. heracleifolia* and *C. vitalba*.

29. C. grata Wallich.
Vigorous woody climber to 10 m, stems very grooved. Leaves pinnate with 5 ovate-lanceolate, acute, coarsely toothed or sometimes lobed leaflets 3–8 cm, hairy beneath. Flowers fragrant, numerous in axillary cymes. Perianth-segments 4, spreading, ovate-oblong, to 1 cm, woolly outside, creamy white. Achenes hairy with persistent styles 2.5–4 cm. *Afghanistan, Himalaya, China*. H2. Summer.

30. C. cirrhosa Linnaeus (*C. balearica* Richard; *C. calycina* Solander). Illustration: Botanical Magazine, 959 (1806), 1070 (1807); Polunin and Smythies, Flowers of southwest Europe, pl. 5 (1973); Polunin & Everard, Trees and bushes of Europe, 68 (1976); Huxley & Taylor, Flowers of Greece and the Aegean, pl. 35 (1977).

Evergreen woody climber. Leaves simple, ovate, toothed, sometimes lobed at the apex, 4–6 cm. Flowers solitary or in clusters in the axils of the previous year's growth, the flower-stalk bearing an involucre of fused bracts. Perianth-segments 4, ovate-obovate, obtuse, sometimes slightly toothed or lobed, 1.5–2.5 cm, creamy white. Achenes with feathery styles. *Mediterranean area*. H5–G1. Winter–spring.

31. C. aristata R. Brown. Illustration: Edwards's Botanical Register, t. 238 (1817); Loddige's Botanical Cabinet 7: t. 620 (1822); Cochrane et al., Flowers and plants of Victoria, f. 424 (1968); Rotherham et al., Flowers and plants of New South Wales and Queensland, pl. 260 (1975).
Woody climber, downy in young parts and inflorescence, otherwise hairless. Leaves long-stalked, once or twice divided into 3; when once divided, leaflets irregularly toothed, when twice divided, leaflets entire; leaflets thick and leathery. Flowers in axillary, panicle-like inforescences. Perianth-segments 4 (rarely 5), spreading, oblong or oblong-lanceolate, hairless or hairy, 2–2.5 cm, white or creamy white. Anthers oblong, each with a needle-like appendage at the apex. Achenes ovoid, hairless or downy with persistent, feathery styles to 5 cm. *Australia*. H4. Summer.

C. glycinoides de Candolle (*C. stenosepala* de Candolle). Similar, but leaves always divided into 3 thinner, broader, sparsely toothed leaflets, smaller flowers and anthers without appendages. *Australia*. H4. Summer.

C. microphylla de Candolle (*C. linearifolia* Steudel; *C. stenophylla* Fraser). Similar to *C. aristata*, leaves usually twice divided into 3, leaflets oblong to ovate-lanceolate, 1.5–3 cm, flowers small in short, panicle-like inflorescences, anthers without appendages, achenes with wrinked ribs and styles to 8 cm. *Australia*. H4. Summer.

32. C. montana de Candolle. Illustration: Edwards's Botanical Register 26: t. 53 (1840); Bärtels, Gartengehölze, 130 (1981); Grey-Wilson & Matthews, Gardening on walls, pl. 6 (1983).
Woody climber. Leaves divided into 3, leaflets ovate-lanceolate, acute, tapered or rounded at the base, usually coarsely toothed, slightly or somewhat deeply 3-lobed at the apex, rarely entire, 2.5–9 cm. Flowers solitary or several together clustered at the nodes on the old wood, mostly appearing with the leaves;

flower-stalks hairy, 2–21 cm. Perianth-segments 4 (rarely 5), spreading, oblong, ovate or obovate, obtuse or acute, tapered at base, hairy outside on the veins, margins downy, hairless or nearly so inside, 1–4 cm, white or pink. Achenes diamond-shaped with persistent, white feathery styles 1–1.5 cm. *Himalaya, W China (Xizang, Yunnan)*. H2. Spring–summer.

Many varieties and cultivars of this species have been named, of which var. **rubens** Wilson, with pink flowers, and var. **sericea** Franchet (*C. spooneri* Rehder & Wilson) which is hairy in all parts are among the best-known. Many cultivars have been selected, especially from var. *rubens*.

The name *C. chrysocoma* Franchet has been used for a non-climbing, erect variant with dense yellow down on the young growth.

33. C. recta Linnaeus. Illustration: Hay & Synge, Dictionary of garden plants in colour, pl. 1054 (1969).
Erect perennial to 1 m. Leaves pinnate with 5–7, ovate, entire leaflets 5–9 cm. Flowers in cymes grouped together in panicle-like inflorescences. Perianth-segments 4, spreading, linear to oblong, margins hairy, white, 1–2 cm. Achenes diamond-shaped, ribbed, with feathery styles. *S & C Europe, adjacent Asia*. H2. Summer.

C. terniflora de Candolle (*C. maximowicziana* Franchet & Savatier; *C. paniculata* Thunberg; *C. dioscoreifolia* Léveillé & Vaniot). Similar, but to 2 m, leaves with 3–5 ovate leaflets, tapered to cordate at base, sometimes lobed, 3–10 cm; flowering more profuse. *E China, Korea, Japan, Taiwan*. H2. Summer.

C. terniflora var. **mandshurica** (Ruprecht) Ohwi has somewhat woody stems and more or less bipinnate leaves.

34. C. armandii Franchet. Illustration: Botanical Magazine, 8587 (1914); Menninger, Flowering vines of the world, f. 175 (1970); Journal of the Royal Horticultural Society 91: 212 (1966); The Garden 101: 500 (1976).
Evergreen woody climber to 4 m, hairless throughout. Leaves leathery, divided into 3 leaflets, leaf-stalk to 8 cm; leaflets oblong-lanceolate, cordate or rounded at abse, shortly acuminate at apex, entire, with 5 distinct veins with a network between them running from base to apex, 8–12 cm. Flowers fragrant in axillary cymes with small bracts. Perianth-segments 5–7, spreading, obovate or oblong, white or pink, 2–3 cm. Filaments dilated above, the

inner shorter than the anthers, the outer longer. Achenes ovoid, adpressed-hairy with feathery styles to 3 cm. *SW China*. H5. Spring.

C. meyeniana Walpers (*C. oreophila* Hance). Illustration: Botanical Magazine, 7897 (1903). Similar, but young parts hairy, cymes longer. *Burma, S China, Indochina, Taiwan & the Philippine islands*. H5–G1.

35. C. flammula Linnaeus. Illustration: Polunin & Everard, Trees and bushes of Europe, 67 (1976).
Woody climber to 5 m. Leaves bipinnate, leaflets very variable in size and shape, linear-oblong to almost circular, entire or 2–3-lobed. Flowers in cymes aggregated in panicle-like inflorescences, fragrant. Perianth-segments 4, spreading, obtuse, downy on margins and outer surfaces, white, 1–2 cm. Achenes compressed, diamond-shaped, with feathery styles. *Mediterranean area*. H2. Summer.

36. C. × aromatica Lenné & Koch (*C. odorata* invalid; *C. caerulea odorata* invalid). Illustration: Lavallée, Les Clématites a grandes fleurs, t. 9 (1884).
Much-branched low shrub to 2 m, shoots with green and black stripes, shortly hairy throughout. Leaves divided into 3 or sometimes simple, basal leaflets stalkless, ovate or oblong-ovate, to 5 cm, rounded at the base and mucronate at the apex, the terminal leaflets sometimes 3-lobed. Flowers in terminal, long-stalked cymes with small bracts which are simple or 2–3-lobed. Perianth-segments 4, spreading, oblong-lanceolate, sometimes acuminate, recurved at the apex, colour fading from violet-blue to pale blue, 1–1.5 cm. Stamens yellowish white, filaments dilated above. Ovary hairless, style hairy. Achenes diamond-shaped, compressed, with long feathery styles. *Garden Origin*. H2. Summer.
A hybrid of unknown parentage.

37. C. songarica Bunge (*C. gebleriana* Bongard). Illustration: Flora SSSR 7: pl. 20 (1937).
Shrub with erect stems to 1 m. Leaves simple, entire or toothed, linear or lanceolate, glaucous, 4–8 (rarely to 13) cm. Flowers in panicle-like inflorescences. Perianth-segments 4, spreading, oblong-obovate, hairless inside, hairy outside, 1–2 cm, white or creamy white. Filaments thread-like, shorter than or as long as the anthers. Achenes downy with feathery styles to 3 cm. *USSR (C Asia)*. H2. Summer.

38. C. potaninii Maximowicz (*C. fargesii* Franchet). Illustration: Botanical Magazine, 8702 (1917).
Woody climber to 3 m. Leaves twice divided into 3, leaflets ovate, irregularly lobed or toothed, acuminate, tapered or rounded at the base, to 7 cm. Flowers solitary or in 2-flowered axillary cymes with small bracteoles. Perianth-segments 6, spreading, obovate, mucronate, downy outside, to 4 cm, white sometimes with a yellow flush. Filaments slightly dilated above, much longer than the yellow anthers. Achenes diamond-shaped, hairless, with feathery styles. *SW China*. H3. Summer.

Var. **souliei** Finet & Gagnepain has somewhat larger flowers.

39. C. phlebantha Williams. Illustration: Botanical Magazine, n.s., 574 (1970); Grey-Wilson & Matthews, Gardening on walls, pl. 5 (1983).
Woody climber to 1.5 m. Leaves pinnate with 5–9 broadly lanceolate leaflets, lobed at apex, 5–15 mm, green, more or less hairless above, white-woolly beneath. Flowers axillary, mostly solitary. Perianth-segments 5–7, spreading, elliptic, 1.5–2 cm, woolly outside, white with purple veins. *Nepal*. H4. Spring–summer.

40. C. afoliata Buchanan (*C. aphylla* Kuntze). Illustration: Botanical Magazine, 8686 (1916); Salmon, Field guide to the alpine plants of New Zealand, pl. 180, 181 (1968); Grey-Wilson & Matthews, Gardening on walls, 71 (1983).
Much-branched shrub, sprawling rather than climbing. Branches yellow-green, terete, finely grooved. Leaves simple, to 1 cm, entire, ovate, present only on young or shaded shoots, otherwise reduced to the leaf- and leaflet-stalks which act as tendrils. Cymes axillary, 1–5-flowered, stalk hairy, 1–2 cm; flowers unisexual. Perianth-segments 4 (rarely 5), greenish yellow, ovate-oblong, 1.5–2.5 cm in male flowers, smaller in female flowers. Achenes reddish brown, hairy, with persistent styles to 2 cm. *New Zealand*. H5. Summer.

41. C. indivisa Willdenow (*C. paniculata* Gmelin). Illustration: Journal of the Royal Horticultural Scoeity **91**: f. 20 (1966); The Garden **101**: 515 (1976).
Dioecious woody climber to 4 m in cultivation (more in the wild). Leaves divided into 3 broadly ovate, acute leaflets with cordate-truncate bases. Flowers numerous in axillary inflorescences with bracts like the upper leaves or simple. Male flowers with 4 ovate-oblong, hairless

perianth-segments, 2.5–5 cm; female flowers smaller, sometimes with a few abortive stamens. Achenes small, downy, with persistent feathery styles 3–5 cm. *New Zealand*. H5–G1. Summer.

C. forsteri Gmelin (*C. colensoi* J.D. Hooker; *C. hexapetala* Linnaeus; *C. hexasepala* de Candolle). Similar, but smaller in all parts, perianth-segments 5–8. *New Zealand*. H5–G1. Summer.

42. C. petriei Allan.
Woody climber to 4 m. Leaves once or twice divided into 3, leaflets ovate-oblong, obtuse, the larger truncate at the base, entire or with 1–2 blunt lobes, 1–4 cm. Flowers 1–few in axillary cymes with small, ovate bracts which are sometimes slightly united at their bases. Perianth-segments 5–8, ovate-oblong, downy, greenish yellow, in male flowers to 1.5 cm, smaller in female flowers. Achenes ovoid, dark red, with strong ribs, hairless when mature, persistent styles to 3 cm. *New Zealand*. H5. Summer.

43. C. viticella Linnaeus. Illustration: Botanical Magazine, 565 (1802); Horticulture **46**: 53 (1968); Polunin & Everard, Trees and bushes of Europe, 68 (1976); Grey-Wilson & Matthews, Gardening on walls, pl. 7 (1983).
Woody climber to 4 m with reddish brown stems. Leaves pinnate, leaflets very variable, ovate or lanceolate, often lobed or subdivided into 3, 2–7 cm. Flowers solitary, terminal or axillary, somewhat pendulous. Perianth-segments 4, spreading, obovate, obtuse, blue, purple, red or white, 2–4 cm. Filaments dilated. Styles hairless in flower. Achenes diamond-shaped, ribbed, with short, feathery styles. *Mediterranean area, adjacent Asia*. H2. Summer.

The direct or indirect parent of many large-flowered hybrids.

C. campaniflora Brotero. Illustration: Gartenpraxis for 1981: 200; Grey-Wilson & Matthews, Gardening on walls, pl. 5 (1983). Similar but more vigorous (to 7 m), stems green when not woody, flowers bell-shaped with pale violet perianth-segments to 2 cm, styles hairy in flower. *Portugal*. H2. Summer.

C. × eriostemon Decaisne is similar but of garden origin, and should probably be considered a cultivar – illustration: Revue Horticole, ser. 4 **1**: 341 (1852); Lavallée, Les Clématites a grandes fleurs, t. 12 (1884).

44. C. × jackmanii Moore. Illustration: Gardener's Chronicle for 1864: 825; Floral Magazine **4**: t. 226 (1865); Flore des Serres,

ser. 2, t. 1628, 1629 (1965–67); Grey-Wilson & Matthews, Gardening on walls, pl. 6 (1983).
Woody climber to 4 m. Upper leaves simple, the rest divided into 3 ovate-lanceolate, acute leaflets which are slightly cordate at the base, 4–8 cm. Cymes 3-flowered, borne at the tips of the young shoots. Perianth-segments 4 (rarely to 6), broadly obovate, with 3 conspicuous veins running their whole length, hairy outside, dark violet-blue. Filaments greenish white, anthers brownish. Achenes diamond-shaped, compressed, with persistent, long, feathery styles. *Garden Origin*. H2. Summer.

A hybrid (*C. viticella* × *C. lanuginosa*) raised by Jackman in 1858. It is the direct or indirect parent of many other hybrids, and serves as a standard for the Jackmanii group (see p. 364).

45. C. florida Thunberg. Illustration: Botanical Magazine, 834 (1805); Revue Horticole ser. 4 **5**: 41 (1856); Lavallée, Les Clématites a grand fleurs, t. 5, 6 (1884); Grey-Wilson & Matthews, Gardening on walls, pl. 6 (1983).
Woody climber to 4 m. Leaves pinnate or bipinnate, the major divisions simple, trilobed or divided into 3, the minor divisions ovate, rounded at the base, mucronate, entire or 2–3-lobed, or with a few teeth. Flowers solitary in the axils on the old wood or in axillary or terminal cymes; flower-stalks with 2 bracteoles halfway up. Perianth-segments usually 6 (more rarely 4–8), spreading, broadly ovate or obovate, acuminate, 5–8 cm, white or creamy white. Staminodes sometimes present, to 2.5 cm, dark violet, transitional to stamens. Filaments white, anthers dark violet. Carpels and styles dark violet. Achenes diamond-shaped, compressed, brown-violet with persistent, feathery styles. *E China, Japan*. H2. Spring–summer.

A direct or indirect parent of many hybrids.

46. C. patens Morren & Decaisne. Illustration: Revue Horticole, ser. 4 **5**: 261 (1856); Lavallée, Les Clématites a grandes fleurs, t. 1, 3 (1884).
Woody climber. Leaves pinnate with 3–5 ovate leaflets, 4–7 cm. Flowers on old and young shoots. Perianth-segments 6–8, scarcely overlapping, elliptic-obovate, 5–8 cm, creamy white or bright blue. Anthers purplish brown. Achenes with conspicuous ribs, dark brown with long persistent styles covered with yellowish hairs. *Japan, China (Shandong, Liaoning)*. H2. Spring–summer.

47. C. lanuginosa Lindley & Paxton.
Illustration: Paxton's Flower Garden **3**:
t. 107 (1852–3).
Woody climber. Leaves simple or divided
into 3; if simple, then ovate, acuminate,
cordate, 5–10 cm; if divided leaflets similar
in shape but smaller; all woolly beneath.
Flowers solitary in the axils or at the tips of
the young shoots; buds woolly. Perianth-
segments 6–8, broadly ovate, overlapping,
8–10 cm, sky-blue, woolly outside.
Achenes ribbed, with persistent feathery
styles. *Not known in the wild.* H2. Summer.

This species was named from Chinese
garden material. It is a parent of many
hybrids.

48. C. henryi Oliver. Illustration: Hooker's
Icones Plantarum, t. 1819 (1889).
Woody climber to 5 m. Leaves simple,
ovate-lanceolate, entire or irregularly
toothed, cordate at base, acute or
acuminate at apex, 10–15 cm. Flowers
solitary, axillary. Perianth-segments 4,
spreading, ovate, acute, hairy outisde,
white or creamy white, to 2 cm. Achenes
oblong, hairy, with very short feathery
styles. *C & S China, Vietnam, Taiwan.* G1.
Summer.

Cultivar-groups of large-flowered hybrids

Florida-group. Plants flowering on the old
or ripened wood, mostly with semi-double
or double flowers; woody climbers
flowering spring–summer.

Jackmanii-group. Plants flowering profusely
on the young shoots over a long period of
summer–autumn; woody climbers.

Lanuginosa-group. Plants flowering on short
side-shoots on the current year's growth;
flowers very large, spread over the whole
plant, summer–autumn; woody climbers.

Patens-group. Plants mostly flowering
spring–summer on old or ripened wood,
mostly with single flowers with pointed
perianth-segments; woody climbers.

Texensis-group. Plants flowering profusely
on young shoots over a long period of the
summer; flowers bell-shaped; plants low
shrubs.

Viticella-group. Plants flowering profusely
over a rather short period of
summer–autumn; woody climbers.

27. RANUNCULUS Linnaeus
A.C. Leslie
Annual or perennial herbs. Leaves
alternate, the basal usually stalked, lobed,
divided or toothed, sometimes simple and
entire. Flowers bisexual, solitary or in
cymose panicles, radially symmetric. Sepals
3–7, usually 5. Petals 0–16 (or more),
usually 5, yellow, white, orange, red or
purple, each with a nectar-producing scale
towards the base. Stamens numerous.
Carpels numerous, the style persistent as a
beak in fruit. Fruit a head of achenes.

A genus of about 400 species, widely
distributed but chiefly in the temperate and
colder regions. Their cultural requirements
are very varied. Some, such as *R. acris* and
R. lanuginosus are readily grown under a
wide range of conditions in the herbaceous
border. Others such as *R. montanus* or *R.
gramineus* are suitable for the rock garden,
but some of the high mountain species (e.g.
R. glacialis) need alpine house treatment,
with very sharp drainage. *R. aquatilis* and
R. circinatus, in contrast, are aquatics,
whilst *R. flammula* and *R. lingua* are also
suitable for waterside planting. Propagation
may be by division or seed, or in some
species by tubers or stolons.

Roots. Some or all tuberous: **8–13,17,21.**
Stolons present. **1,23.**
Stock swollen, corm-like. **7.**
Leaves. Some with thread-like segments:
33,34. Entire: **25–30.**
Petals. Red or purple: **10;** orange: **11.**

1a. Petals white, pink, red or purplish   2
  b. Petals yellow or orange   19
2a. Leaves entire   3
  b. Leaves bluntly angled or toothed or
    lobed   6
3a. Flowers 4–5 cm wide
      30. calandrinioides
  b. Flowers 1–3 cm wide   4
4a. Stem-leaves absent or not clasping the
    stem   **25. pyrenaeus**
  b. Stem-leaves clasping the stem   5
5a. Sepals hairy; achenes smooth
      28. parnassifolius
  b. Sepals hairless; achenes veined
      29. amplexicaulis
6a. Some leaves divided into thread-like
    segments; plants aquatic   7
  b. No leaves divided into thread-like
    segments; plants not aquatic   8
7a. Thread-like segments of the leaves all
    lying in 1 plane   **34. circinatus**
  b. Thread-like segments of the leaves not
    all lying in 1 plane   **33. aquatilis**
8a. Basal leaves lobed   9
  b. Basal leaves toothed but not lobed   17
9a. At least some roots tuberous   10
  b. Roots not tuberous   11
10a. Basal leaves hastate, greyish green
      21. acetosellifolius

  b. Basal leaves broadly ovate to circular
    in outline, green   **10. asiaticus**
11a. Sepals hairy   **20. glacialis**
  b. Sepals hairless   12
12a. Stems 4–20 cm   13
  b. Stems more than 20 cm   16
13a. Receptacle hairy   **19. seguieri**
  b. Receptacle hairless   14
14a. Basal leaves shallowly 3-lobed at
    apex; achenes glaucous   **16. crenatus**
  b. Basal leaves deeply 3–5-lobed;
    achenes not glaucous   15
15a. Basal leaves dark glossy green; stem-
    leaves usually 3-lobed   **14. alpestris**
  b. Basal leaves matt green; stem-leaves
    all simple   **15. traunfellneri**
16a. Middle lobe of basal leaves free to
    base; stems to 60 cm
      17. aconitifolius
  b. Middle lobe of basal leaves not free to
    base; stems to 1.3 m
      18. platanifolius
17a. Root-tubers present   **11. ficaria**
  b. Root-tubers absent   18
18a. Flowers 5–8 cm wide; achenes hairy
      31. lyallii
  b. Flowers 2–2.5 cm wide; achenes
    hairless   **16. crenatus**
19a. Some or all roots tuberous   20
  b. Roots fibrous or slightly fleshy, not
    tuberous   26
20a. Leaves lobed   21
  b. Leaves entire or toothed, not lobed
      23
21a. Stems 50–120 cm   **9. cortusifolius**
  b. Stems at most 30 cm   22
22a. Anthers purplish black   **10. asiaticus**
  b. Anthers yellow   **8. millefoliatus**
23a. Leaves linear-lanceolate
      27. abnormis
  b. Leaves ovate or kidney-shaped   24
24a. Stems hairy; receptacle hairless
      12. bullatus
  b. Stems hairless; receptacle hairy   25
25a. Achenes hairless; leaves kidney-
    shaped   **13. thora**
  b. Achenes hairy; leaves broadly ovate
      11. ficaria
26a. Receptacle hairy, at least above   27
  b. Receptacle hairless   32
27a. Stolons present   **1. repens**
  b. Stolons absent   28
28a. Rootstock swollen, corm-like; sepals
    reflexed   **7. bulbosus**
  b. Rootstock not swollen and corm-like;
    sepals spreading   29
29a. Basal leaves toothed or at most
    shallowly lobed   **32. insignis**
  b. Basal leaves deeply lobed   30

30a. Stems 30 cm or more; achenes keeled
and furrowed at margin
       5. carpaticus
  b. Stems to 30 cm; achenes keeled but
not furrowed at margin   **31**
31a. Apex of rhizome with numerous hairs
3–4 mm long   **6. gouanii**
  b. Apex of rhizome hairless or with a
few short hairs   **4. montanus**
32a. Basal leaves deeply lobed   **33**
  b. Basal leaves not lobed (often withered
by flowering time)   **35**
33a. Achenes 1–2 mm; stem often
unbranched   **22. eschscholtzii**
  b. Achenes 2–5 mm; stem usually
branched   **34**
34a. Achenes 2–3.5 mm; petals golden
yellow   **3. acris**
  b. Achenes 4–5 mm; petals orange-
yellow   **2. lanuginosus**
35a. Achenes conspicuously veined; leaves
glaucous   **26. gramineus**
  b. Achenes smooth; leaves not glaucous
    36
36a. Flower-stalks furrowed; achenes
c. 1.5 mm   **24. flammula**
  b. Flower-stalks not furrowed; achenes
c. 2.5 mm   **23. lingua**

1. R. repens Linnaeus. Illustration: Bonnier,
Flore complète 1: pl. 13 (1911); Ross-Craig,
Drawings of British plants 1: pl. 30 (1948);
Keble Martin, The concise British Flora in
colour, pl. 3 (1965).
Nearly hairless or hairy perennial, with
long, leafy stolons which produce fibrous
roots at the nodes. Stems 15–60 cm. Basal
leaves triangular-ovate, 3-lobed, the middle
lobe long-stalked; all lobes further divided
and toothed. Flowers few to numerous,
2–3 cm wide, stalk furrowed. Sepals
spreading. Petals yellow. Receptacle hairy.
Achenes 3 mm, bordered, with a short,
hooked beak. *Europe, Asia.* H1. Summer.
'Flore Pleno' has double flowers.

2. R. lanuginosus Linnaeus. Illustration:
Coste, Flore de la France 1: 29 (1901);
Grey-Wilson & Blamey, The alpine flowers
of Britain and Europe, 49 (1979); Pignatti,
Flora d'Italia 1: 307 (1982).
Very hairy perennial with a short stock and
fibrous roots. Stems usually 30–50 cm,
rarely more. Basal leaves mostly 3-lobed,
divided to approximately three-quarters of
the distance to the base, the lobes broadly
ovate, irregularly cut and toothed. Flowers
numerous, 1.5–4 cm wide, stalks not
furrowed. Sepals spreading. Petals deep
orange-yellow. Receptacle hairless.
Achenes 4–5 mm, bordered, with a short,

broad, strongly recurved beak. *C & S
Europe, Caucasus.* H1. Spring–summer.
'Pleniflorus' has double flowers.

3. R. acris Linnaeus. Illustration: Bonnier,
Flore complète 1: pl. 15 (1911); Ross-Craig,
Drawings of British plants 1: pl. 29 (1948);
Keble Martin, The concise British Flora in
colour, pl. 3 (1965).
Hairless or hairy perennial with a short
rhizome or vertical stock and fibrous roots.
Stems 15–100 cm. Basal leaves angled or
more or less circular, usually divided into
3–7 stalkless segments which are ovate-
cuneate, toothed and usually again divided.
Flowers few to many, 1.5–2.5 cm wide,
stalks not furrowed. Sepals spreading.
Petals golden yellow. Receptacle hairless.
Achenes 2–3.5 mm, bordered, with a
short, hooked beak. *Europe, Asia.* H1.
Summer.
'Flore Pleno' has tightly double flowers.

4. R. montanus Willdenow (*R. geraniifolius*
misapplied). Illustration: Coste, Flore de la
France 1: 30 (1901); Bonnier, Flore
complète 1: pl. 14 (1911); Huxley,
Mountain flowers in colour, 32 (1967).
Hairless or sparsely hairy perennial;
rhizome covered with persistent fibres.
Stems 5–25 cm. Basal leaves 3–5 (rarely to
7)-lobed, the lobes obovate, toothed or
entire; stem-leaves with narrower lobes
which are sometimes toothed, often partly
clasping the stem at the base. Flowers
1–few, 2–4 cm wide. Sepals finely hairy.
Petals yellow. Receptacle hairless at the
insertion of the stamens, hairy above.
Achenes 2.5–3.5 mm, strongly keeled, not
furrowed, with a hooked beak one quarter
to one-third as long as the achene. *Alps,
Jura, Black Forest.* H1. Summer.
  Very variable and part of a group of
closely similar species distributed in the
mountains of C, E & S Europe.

5. R. carpaticus Herbich. Illustration:
Botanical Magazine, 7266 (1892).
Like *R. montanus* but taller, with larger
flowers and the achene margin both
keeled and furrowed. *E Carpathians.* H1.
Summer.

6. R. gouanii Willdenow. Illustration: Grey-
Wilson & Blamey, The alpine flowers of
Britain and Europe, 49 (1979).
Like *R. montanus* but taller, more hairy and
with hairs 2–4 mm long at the apex of the
rhizome. *Pyrenees.* H1. Summer.
  A double-flowered plant grown as '*R.
speciosus plenus*' probably belongs to this
species.

7. R. bulbosus Linnaeus. Illustration:
Bonnier, Flore complète 1: pl. 13 (1911);
Ross-Craig, Drawings of British plants 1: pl.
30 (1948); Keble Martin, The concise
British Flora in colour, pl. 3 (1965).
Hairy perennial with a swollen, corm-like
stock and fibrous roots. Stems 15–50 cm.
Basal leaves ovate, 3-lobed, terminal lobe
usually distinctly stalked, sometimes
stalkless; all lobes further toothed and
lobed. Flowers 1–several, 1.5–3 cm wide,
stalks furrowed. Sepals reflexed. Petals
bright yellow. Receptacle hairy. Achenes
2–4 mm, margin keeled and grooved, with
a short, curved beak. *Europe, N Africa,
Caucasus.* H1. Spring.
  'Pleniflorus' has double flowers, whilst in
'F.M. Burton' the petals are primrose
yellow.

8. R. millefoliatus Vahl. Illustration:
Botanical Magazine, 3009 (1830); Polunin,
Flowers of Greece and the Balkans, 238
(1980); Pignatti, Flora d'Italia 1: 317
(1982).
Usually hairy perennial, with both fibrous
and tuberous roots. Stems 8–30 cm. Basal
leaves broadly triangular-ovate,
2–3-pinnatisect with linear-lanceolate,
acute lobes. Flowers 1 or 2, 1.5–3 cm wide,
on thick stalks. Sepals becoming hairless,
spreading. Petals yellow. Receptacle
hairless, longer in fruit. Achenes 3.5 mm,
broadly keeled, with a broad hooked beak
half as long as the achene. *South and eastern
C Europe, N Africa, W Asia.* H2. Spring.

9. R. cortusifolius Willdenow. Illustration:
Botanical Magazine, 4625 (1852);
Bramwell & Bramwell, Wild flowers of the
Canary Islands, f. 13 (1974); Sjøgren,
Açores Flores, f. 59 (1984).
Densely hairy perennial with both fibrous
and tuberous roots. Stems 50–120 cm.
Basal leaves circular to kidney-shaped, to
30 cm wide, shallowly 3–5-lobed, lobes
scolloped or shallowly lobed. Stem-leaves
more deeply lobed. Flowers numerous, to
5 cm wide, stalks not furrowed. Petals
yellow. Receptacle almost hairless, longer
in fruit. Achenes 3 mm, hairless, with a
short, recurved beak. *Azores, Madeira,
Canary Islands.* H5. Summer.

10. R. asiaticus Linnaeus. Illustration:
Polunin & Huxley, Flowers of the
Mediterranean, 30 (1965); Zohary, Flora
Palaestina 1: pl. 295 (1966); Polunin,
Flowers of Greece and the Balkans, pl. 7
(1980).
Hairy perennial with both fibrous and
tuberous roots. Stems 10–30 cm. Basal

leaves variable, broadly ovate to circular, usually 3-lobed, the lobes toothed, occasionally some or all of the leaves deeply further dissected into many narrow lobes. Stem-leaves 2–3-pinnatisect. Flowers 1–few, 3–6 cm wide. Petals white, yellow, red or purple. Anthers purplish black. Receptacle hairy, longer in fruit. Achenes 2–3 mm, keeled, tapered into a broad, hooked beak. *E Mediterranean area, SW Asia.* H3. Spring.

Very variable. At one time many cultivars were grown as garden plants or as florists' flowers, most with double or semi-double flowers; these have now largely been replaced by seed races of mixed colours.

11. R. ficaria Linnaeus. Illustration: Ross-Craig, Drawings of British plants 1: pl. 35 (1948); Keble Martin, The concise British Flora in colour, pl. 3 (1965); Polunin, Flowers of Europe, pl. 25 (1969). Hairless perennial with both fibrous and tuberous roots. Stems 5–30 cm, ascending. Basal leaves broadly ovate, bluntly angled, scolloped or toothed, often with silver or brownish markings. Flowers 1–few, 1.5–5 cm wide. Sepals usually 3, greenish, soon falling. Petals usually 7–13, usually yellow, fading to white. Receptacle hairy. Achenes *c.* 2.5 mm, keeled, hairy, with a short beak. *Europe, NW Africa, SW Asia.* H1. Spring.

Very variable. Three subspecies occur: subsp. **ficaria** lacks axillary bulbils and has flowers 2–3 cm wide, with broad, overlapping petals and good seed set. Subsp. **bulbifer** (Marsden-Jones) Lawalrée differs in having axillary bulbils, narrower petals and poor seed set. Both of these are widespread; the latter, in particular, can be a troublesome weed. Subsp. **ficariiformis** Rouy & Foucaud (*R. ficaria* var. *grandiflora* (Robert) Strobl) from the Mediterranean area, is a much more robust plant, with flowers 3–5 cm wide, with broad, overlapping petals and good seed set.

Most selected garden variants belong to subsp. **ficaria** and vary in petal colour from almost white through various shades of yellow to coppery orange; there is also a range of doubles, differing in form and colour. Leaf markings vary enormously, and in 'Brazen Hussy' the entire upper surface is suffused a dark greenish brown.

12. R. bullatus Linnaeus. Illustration: Coste, Flore de la France 1: 24 (1901); Polunin & Smythies, Flowers of SW Europe, 62 (1973); Pignatti, Flora d'Italia 1: 319 (1982).

Tufted perennial with tuberous roots. Stems 5–20 cm, hairy. Leaves all in a basal rosette, ovate, with rounded teeth, hairy and with a puckered surface. Flowers usually solitary, 2–2.5 cm wide, scented. Petals yellow. Receptacle hairless. Achenes 4 mm wide, narrowly bordered, with a short, curved beak. *Mediterranean area.* H2 or 3. Autumn–spring.

13. R. thora Linnaeus. Illustration: Coste, Flore de la France 1: 24 (1901); Bonnier, Flore complète 1: pl. 11 (1911); Huxley, Mountain flowers in colour, 32 (1967). Tufted perennial with tuberous roots. Stems 10–30 cm. Basal leaves appearing after flowering, kidney-shaped, toothed only above, glaucous and hairless. Lower stem-leaves similar, almost stalkless, the upper smaller, 3–5-lobed. Flowers 1–few, 1–2 cm wide. Sepals hairless. Petals yellow. Receptacle sparsely hairy. Achenes *c.* 4 mm, hairless, strongly veined, with a short, hooked beak. *C Pyrenees, NW Spain, Jura, Alps, Carpathians, Balkan peninsula.* H1. Summer.

14. R. alpestris Linnaeus. Illustration: Coste, Flore de la France 1: 24 (1901); Huxley, Mountain flowers in colour, 33 (1967); Grey-Wilson & Blamey, The alpine flowers of Britain and Europe, 51 (1979). Hairless tufted perennial with fibrous roots. Stems 3–12 cm. Basal leaves 3–5-lobed; lobes obovate with deep rounded teeth, dark green and glossy. Lower stem-leaves usually 3-lobed. Flowers 1–3, *c.* 2 cm wide. Sepals hairless. Petals sometimes more than 5, notched at the tip, white. Receptacle hairless. Achenes *c.* 2 mm, smooth, with a slender beak, curved at the tip. *Pyrenees, N Spain, Jura, Alps, Apennines, Carpathians.* H1. Spring–summer.

15. R. traunfellneri Hoppe. Illustration: Huxley, Mountain flowers in colour, 33 (1967); Grey-Wilson & Blamey, The alpine flowers of Britain and Europe, 52 (1979). Very similar to *R. alpestris* but smaller and not tufted. Basal leaves matt green, usually 3-lobed, the lobes narrower, more deeply divided. Stem-leaves all simple. Flowers *c.* 1.5 cm wide, usually solitary. *SE Alps.* H1. Spring–summer.

16. R. crenatus Waldstein & Kitaibel. Illustration: Thompson, Alpine plants of Europe, pl. 4 (1911); Hegi, Illustrierte Flora von Mitteleuropa edn 2, 3: f. 190 (1974); Polunin, Flowers of Greece and the Balkans, 168 (1980). Hairless perennial with fibrous roots. Stems 4–15 cm. Basal leaves almost circular, with

deep, usually rounded teeth, sometimes shallowly 3-lobed at the apex. Stem-leaves lanceolate to linear. Flowers 1–2, 2–2.5 cm wide. Petals shallowly notched, white. Achenes *c.* 2 mm, glaucous, smooth, with a slender beak hooked at apex. *E Alps, Apennines, E Carpathians, Balkan peninsula.* H1. Summer.

17. R. aconitifolius Linnaeus. Illustration: Huxley, Mountain flowers in colour, 33 (1967); Polunin, Flowers of Europe, t. 25 (1969); Grey-Wilson & Blamey, The alpine flowers of Britain and Europe, 50, 51 (1979). Perennial with fibrous roots. Stem to 60 cm. Basal leaves palmately 3–5-lobed, dark green; lobes obovate to lanceolate, toothed, the middle lobe free to the base, lateral lobes stalkless. Flower-stalks 1–2 times as long as subtending leaf, hairy above. Flowers numerous, 1–2 cm wide. Sepals reddish or purple below, hairless and soon falling. Petals white. Receptacle hairy. Achenes 3–5 mm, veined and keeled, usually with a slender beak which is curved only at the tip. *C Europe.* H1. Spring–summer.

'Flore Pleno' has tight double flowers.

18. R. platanifolius Linnaeus. Illustration: Grey-Wilson & Blamey, The alpine flowers of Britain and Europe, 51 (1979). Very similar to *R. alpestris* but taller (to 1.3 m) with 5–7-lobed leaves, the middle lobe not free to the base, and with relatively longer, less hairy flower-stalks. *C & S Europe.* H1. Summer.

19. R. seguieri Villars. Illustration: Coste, Flore de la France 1: 23 (1901); Huxley, Mountain flowers in colour, 33 (1967); Grey-Wilson & Blamey, The alpine flowers of Britain and Europe, 51 (1979). Tufted, hairy or almost hairless perennial with fibrous roots. Stems 8–20 cm. Basal leaves deeply 3–5-lobed, the lobes narrow, acute, further divided and toothed, hairy beneath. Flowers 1–few. 2–2.5 cm wide. Sepals hairless. Petals usually notched, white. Receptacle hairy. Achenes *c.* 4 mm, veined, hairy above, not keeled, with a slender curved beak. *Alps, NW Spain, Apennines, SW Yugoslavia.* H1. Summer.

20. R. glacialis Linnaeus. Illustration: Bonnier, Flore complète 1: pl. 13 (1911); Huxley, Mountain flowers in colour, 33 (1967); Grey-Wilson & Blamey, The alpine flowers of Britain and Europe, 53 (1979). Perennial with fibrous roots. Stems 4–25 cm, erect or ascending. Basal leaves with a metallic sheen, divided into 3 segments which are usually stalked and divided into elliptic or oblong lobes. Flowers 1–4, 2.5–4 cm wide. Sepals with numerous purple-brown hairs. Petals white becoming pink or purplish. Achenes 2.5 mm, winged on 2 sides, hairless; beak 1–1.5 mm, curved only at the tip. *Greenland, N Europe, high mountains of C Europe, Pyrenees, Spain (Sierra Nevada).* H1. Summer.

21. R. acetosellifolius Boissier. Illustration: Schimper, Plant-Geography, f. 471 (1903); Polunin & Smythies, Flowers of SW Europe, 77 & t. 5 (1973); Bulletin of the Alpine Garden Society **46**: 289 (1978).
Hairless perennial with a fibrous stock and tuberous roots. Stems 3–20 cm, ascending or prostrate but turning up at the ends. Leaves all basal, hastate, irregularly laciniate at base, with a long middle lobe, greyish green. Flowers usually solitary, 1.5–2.5 cm wide. Sepals purplish. Petals white to pale pink. Receptacle hairless. Achenes 2 mm, keeled on inner margin, veined, with a short, curved beak. *Spain (Sierra Nevada).* H1. Spring.

22. R. eschscholtzii Schlechtendahl (*R. suksdorfii* Gray; *R. adoneus* Gray; *R. trisectus* Eastwood). Illustration: Abrams & Ferris, Illustrated Flora of the Pacific States **2**: f. 1840 (1944); Hitchcock et al., Vascular plants of the Pacific northwest **2**: 386 (1964).
Perennial with short, vertical stock and fibrous roots; hairless or with yellowish hairs above. Stems 5–35 cm. Basal leaves kidney-shaped, ovate or broadly obovate, shallowly 3-lobed and toothed or dissected 2–3 times into linear segments. Flowers usually solitary. Petals yellow. Receptacle hairless, longer in fruit. Achenes 1–2 mm, with a slender, usually straight beak. *Mountains of western N America.* H1. Summer.

A variable species.

23. R. lingua Linnaeus. Illustration: Bonnier, Flore complète **1**: pl. 11 (1911); Ross-Craig, Drawings of British plants **1**: pl. 27 (1948); Keble Martin, The concise British Flora in colour, pl. 3 (1965).
Stoloniferous perennial with fibrous roots. Stems 50–200 cm, hollow, usually hairless. Basal leaves produced in autumn, ovate, long-stalked, soon withering. Stem-leaves oblong-lanceolate, more or less arranged in 2 rows, shortly stalked or stalkless, half clasping the stem, usually more or less toothed. Flower-stalks hairy, not furrowed. Flowers several, 2–5 cm wide. Petals

yellow. Receptacle hairless. Achenes *c.* 2.5 mm, hairless, bordered, with a short, broad, slightly curved beak. *Europe, Caucasus, Siberia and C Asia.* H1. Summer.

24. R. flammula Linnaeus. Illustration: Bonnier, Flore complète **1**: pl. 25 (1911); Ross-Craig, Drawings of British plants **1**: pl. 25 (1948); Keble Martin, The concise British Flora in colour, pl. 3 (1965).
Similar to *R. lingua* but generally smaller in all its parts. Stems 8–80 cm. Flowers 8–25 mm wide, on furrowed stalks. Achenes *c.* 1.3 mm. *Europe, Asia.* H1. Summer.

25. R. pyrenaeus Linnaeus. Illustration: Coste, Flore de la France **1**: 23 (1901); Huxley, Mountain flowers in colour, 33 (1967); Polunin & Smythies, Flowers of SW Europe, 149 (1973).
Tufted perennial with fibrous or slightly fleshy roots; stock covered with coarse fibres. Stems 5–15 cm, branched or simple. Basal leaves linear to broadly lanceolate, stalkless, grey-green, hairless or with a few hairs at base, entire. Stem-leaves 1–3. Flower-stalks hairy above. Flowers 1–4, 1–3 cm wide. Sepals hairless, whitish. Petals white, often imperfect or some lacking (rarely the flowers double). Receptacle hairy. Achenes 2 mm, veined and keeled, with a short curved beak. *Alps, Pyrenees, mountains of Spain, Corsica.* H1. Spring.

The description above is of subsp. **pyrenaeus**. Plants from the Alps with broader leaves and more robust stems have been distinguished as subsp. **plantagineus** (Allioni) Rouy & Foucaud, and those from Spain (Sierra Nevada), with usually unbranched stems, 0–1 stem-leaves and smaller flowers (1–1.3 cm wide) as subsp. **alismoides** (Bory) Bolos & Font-Quer.

26. R. gramineus Linnaeus (*R. graminifolius* Salisbury). Illustration: Coste, Flore de la France **1**: 25 (1901); Bonnier, Flore complète **1**: pl. 11 (1911); Polunin & Smythies, Flowers of SW Europe, pl. 7 (1973).
Hairless or rarely hairy perennial, with fibrous roots and a stout stock covered with coarse fibres. Leaves mostly basal, linear to lanceolate, glaucous, entire. Flowers 1–few, 2–3 cm wide. Sepals hairless, yellowish. Petals yellow. Receptacle hairless. Achenes 3 mm, hairless, conspicuously veined, keeled, with a very short, stout, slightly curved beak. *S Europe, N Africa.* H1. Spring–summer.

'Flore Pleno' has double flowers.

27. R. abnormis Cutanda & Willkomm. Illustration: Polunin & Smythies, Flowers of SW Europe, 111 & pl. 5 (1973); Bulletin of the Alpine Garden Society **43**: 214 (1975) & **45**: 324 (1977).
Perennial with a fibrous stock and tuberous roots. Stems 5–20 cm, sometimes hairy. Leaves mostly basal, linear-lanceolate, hooded at tip, entire, green. Flowers 1–4, 2–3 cm wide. Sepals more or less hairy, yellowish. Petals 8–10, yellow. Receptacle hairless. Achenes *c.* 1.5 mm, smooth, keeled, with a very short, curved beak. *W & C Spain, Portugal.* H3. Spring.

28. R. parnassifolius Linnaeus. Illustration: Botanical Magazine, 386 (1797); Huxley, Mountain flowers in colour, 33 (1967); Pignatti, Flora d'Italia **1**: 327 (1982).
Perennial with fibrous roots. Stems 4–20 cm. Basal leaves ovate or broadly lanceolate, dark green, entire, hairy at first especially at the margins and on veins beneath. Stem-leaves clasping the stem. Flower-stalks hairy above. Flowers 1–several, 2–2.5 cm wide. Sepals hairy. Petals white or reddish. Achenes 3 mm, smooth, with a short, hooked beak. *Alps, Pyrenees, N Spain.* H1. Summer.

29. R. amplexicaulis Linnaeus. Illustration: Botanical Magazine, 266 (1794); Coste, Flore de la France **1**: 22 (1901); Grey-Wilson & Blamey, The alpine flowers of Britain and Europe, 53 (1979).
Perennial with a fibrous stock and fibrous or slightly fleshy roots. Stems 4–30 cm, hairless. Basal leaves ovate-lanceolate, grey-green, entire, hairless or with a few long hairs. Stem-leaves lanceolate, clasping the stem. Flowers 1–several, 2–2.5 cm wide. Sepals hairless, greenish, soon falling. Petals sometimes more than 5, white. Achenes 2 mm, strongly veined, keeled, with a short, curved beak. *Pyrenees, N Spain.* H1. Spring–summer.

30. R. calandrinioides Oliver. Illustration: Botanical Magazine, n.s., 38 (1948); RHS Dictionary of gardening **4**: 1739 (1956); Bulletin of the Alpine Garden Society **29**: 67 (1961).
Hairless perennial with a thick fibrous stock and fleshy roots. Stems to 20 cm, flushed pink below. Basal leaves lanceolate to ovate-lanceolate, slightly fleshy, grey-green, with entire, wavy margins. Flowers 1–few, 4–5 cm wide. Sepals reddish. Petals white, sometimes flushed pink. Achenes 2–2.5 mm, veined and keeled, with a very short, curved beak. *Morocco.* H3. Winter–spring.

31. R. lyallii J.D. Hooker. Illustration: Botanical Magazine, 6888 (1886); Cheeseman, Illustrations of the New Zealand Flora 1: pl. 3 (1914); Fisher, The alpine Ranunculi of New Zealand, f. 109 (1965).

Rhizomatous perennial with fleshy roots. Stems 30–120 cm, hairy. Basal leaves circular, 10–40 cm wide, peltate, with fine, rounded teeth, almost hairless above, sparsely hairy beneath, edged with red; early leaves may be kidney-shaped and less peltate. Flowers 5–15, 5–8 cm wide. Petals 10–16 (or more), white. Sepals hairy. Receptacle hairy. Achenes *c.* 3 mm, very hairy, with a slender style *c.* 3 times as long as the body of the achene. *New Zealand (South Island).* H5. Spring–summer.

32. R. insignis J.D. Hooker. Illustration: Fisher, The alpine Ranunculi of New Zealand, f. 102 (1965); Everard & Morley, Wild flowers of the world, pl. 128 (1970); Bulletin of the Alpine Garden Society **41**: 294 (1973).

Usually brownish hairy, rarely hairless, perennial, with a short, stout stock covered by persistent leaf-bases and with slightly fleshy main roots. Stems 3–75 cm. Basal leaves ovate-lanceolate to heart-shaped, with 7–15 shallow lobes or teeth, hairy or hairless. Flowers to 20 or more, 2–5 cm wide. Sepals hairless or silky-hairy, often red at the tips. Petals 5–15 (or more), yellow. Receptacle hairy. Body of achene 1–3 mm, with a slender, almost straight beak to 3 mm. *New Zealand.* H1. Spring–summer.

Variable in size, leaf-shape and hairiness.

33. R. aquatilis Linnaeus (*R. heterophyllus* Weber). Illustration: Bonnier, Flore complète 1: pl. 9 (1911); Ross-Craig, Drawings of British plants 1: pl. 17 (1948); Keble Martin, The concise British Flora in colour, pl. 2 (1965).

Annual or perennial aquatic, usually with both flat and lobed leaves and leaves divided into thread-like segments. Stems 10–150 cm. Leaves with thread-like segments 3–8 cm, shorter than internodes, segments not all lying in 1 plane; flat leaves almost circular, more or less deeply divided into 3–7 wedge-shaped, usually straight-sided and often toothed segments. Flowers 1.2–1.8 cm wide. Petals white, each with a yellow base. Receptacle hairy. Achenes 1.5–2 mm, often hairy, transversely ridged and with a short beak. *Europe.* H1. Spring–summer.

34. R. circinatus Sibthorp (*R. divaricatus* misapplied). Illustration: Coste, Flore de la France 1: 22 (1901); Ross-Craig, Drawings of British plants 1: pl. 12 (1948); Keble Martin, The concise British Flora in colour, pl. 2 (1965).

Aquatic perennial. Stems 10–100 cm. Leaves 5–30 mm, all divided into thread-like segments, circular, shorter than internodes; segments rigid, all lying in 1 plane. Flowers 8–18 mm wide. Petals white, each with a yellow base. Receptacle hairy. Achenes 1.2–1.5 mm, transversely ridged, hairless or sparsely hairy above, with a short beak. *Most of Europe except the SW.* H1. Summer.

28. MYOSURUS Linnaeus
A.C. Whiteley

Small annual herbs with linear leaves in a basal rosette. Flowers small, solitary, on slender, upright stalks. Sepals 5 or more, each with a basal spur. Petals 5–7 or absent, tubular, secreting nectar. Stamens usually 5–10. Carpels numerous, ripening to achenes borne on a greatly elongate receptacle. A genus of about 7 species from the northern hemisphere, Chile and New Zealand. The one cultivated species is easily grown in moist soil by sowing seeds where the plants are to flower.

1. M. minimus Linnaeus. Illustration: Coste, Flore de la France 1: 32 (1901); Hegi, Illlustrierte Flora von Mitteleuropa 3: t. 117 (1909); Ross-Craig, Drawings of British plants 1: pl. 10 (1948); Keble Martin, The concise British Flora in colour, edn 2, pl. 1 (1969).

Leaves hairless and somewhat fleshy, 1–4 cm × *c.* 1 mm. Flower-stalks 5–12 cm. Flowers greenish yellow. Sepals 2–4 mm, the spur pointing downwards from the base of each. Petals 2–4 mm. Stamens 5–10. Receptacle long, to 1.5–4 cm with numerous, closely packed achenes in fruit. Achenes brown, *c.* 1.1 mm, each with a low ridge projecting into a short beak at the tip. *Europe, SW Asia, N Africa, naturalised elsewhere.* H2. Summer.

29. ADONIS Linnaeus
F. McIntosh

Annual or perennial herbs. Leaves divided 1–3 times with narrow, more or less linear segments. Flowers solitary, radially symmetric. Sepals 5–8, often coloured and somewhat petal-like. Petals 5–20, red, yellow or whitish, nectaries absent. Stamens numerous, anthers blackish or yellow. Carpels borne on a long or roundish head. Fruit a loose or crowded head of achenes. Achenes beaked, not laterally compressed, often with a tooth below the beak.

A genus of about 20 species from temperate Eurasia.

Plant. Hairy, sometimes so only at base: **1,5,8**.
Lower leaves. Scale-like: **5–7**.
Sepals. Hairy: **1,6,7**.
Petals. With a dark basal spot: **1,3,4**.
Anthers. Black: **1–4**; yellow: **5–9**.

1a. Annual; anthers black; fruiting heads long   2
  b. Perennial; anthers yellow; fruiting heads rounded   5
2a. Achenes not crowded; style indigo; petals 4 or more times longer than broad; sepals hairy   **1. flammea**
  b. Achenes crowded; style green; petals 2–3 times longer than broad; sepals not hairy   3
3a. Flowers 4–5 cm in diameter; petals bright crimson, without dark basal spots   **2. aleppica**
  b. Flowers 1.5–2.5 cm in diameter; petals crimson, each with a dark basal spot   4
4a. Achenes 5–6 mm, each with a dorsal projection   **3. aestivalis**
  b. Achenes 3.5–5 mm, without dorsal projections   **4. annua**
5a. Lower leaves reduced to scales; achenes conspicuously hairy, even if only at the base   6
  b. Lower leaves developed; achenes not hairy or only very sparsely so   8
6a. Plant generally hairy, sepals hairless   **5. amurensis**
  b. Plant generally hairless, sepals hairy   7
7a. Flowers 3–3.5 cm in diameter; achenes hairy only at the base   **6. volgensis**
  b. Flowers 4–8 cm in diameter; achenes hairy all over the surface   **7. vernalis**
8a. Styles 2 mm, distinctly shorter than the achenes, which may be up to 6 mm; base of stems and leaf-sheaths with small hairs   **8. pyrenaica**
  b. Styles 2 mm, almost as long as the achenes, which are 2–3 mm; plant completely hairless   **9. chrysocyathus**

1. A. flammea Jacquin. Illustration: Jávorka & Csapody, Iconographia florae hungaricae, 177 (1930); Davis (ed), Flora of Turkey 1: 177 (1965).

Annual, 10–40 cm, sparingly branched. Leaf-segments linear. Flowers 2–3 cm in diameter. Sepals more or less hairy. Petals to 4 times longer than broad, deep scarlet (rarely yellow), each often with a black base. Anthers black. Style indigo, shorter than the achenes, achenes each with a rounded tooth. Fruiting-heads long, achenes not crowded. *Mediterranean area, SW Asia east to Iran and the Caucasus.* H? Summer.

2. A. aleppica Boissier. Illustration: Davis (ed), Flora of Turkey 1: 177 (1965). Annual, 20–40 cm, branched from the base. Leaf-segments linear. Flowers 4–5 cm in diameter. Sepals hairless. Petals to 2 times longer than broad, crimson, lacking a dark spot at the base. Anthers black. Style green, as long as the achenes, 5–7 mm. Achenes not toothed. Fruiting heads long, the achenes crowded. *Turkey, Syria, N Iraq.* H5. Late spring.

3. A. aestivalis Linnaeus. Illustration: Reichenbach, Iconographia Florae Germanicae 3: t. 24 (1838–1839). Annual, 10–40 cm, often widely branched. Leaf-segments linear. Flowers 1.5–2.5 cm in diameter. Sepals hairless. Petals to 2 times longer than broad, bright scarlet, each with a dark basal spot. Anthers black. Styles green. Achenes 5–6 mm, each with a tooth. Fruiting heads long, achenes crowded. *Europe, N Africa, SW Asia.* H1. Summer.

4. A. annua Linnaeus (*A. autumnalis* Linnaeus). Illustration: Keble Martin, The concise British flora in colour, pl. 1 (1965); Polunin & Huxley, Flowers of the Mediterranean, pl. 32 (1965); Polunin, Flowers of Europe, pl. 24 (1969); Huxley & Taylor, Flowers of Greece and the Aegean, pl. 45 (1977). Annual, 10–40 cm, often widely branched. Leaf-segments linear. Flowers 1.5–2.5 cm in diameter. Sepals hairless. Petals 2–3 times longer than broad, bright scarlet, each with a dark basal spot. Anthers black. Styles green, short. Achenes 3.5–5 mm, not toothed. Fruiting heads long, achenes crowded. *Europe, SW Asia.* H1 Spring–summer.

5. A. amurensis Regel & Radde. Illustration: Huxley, Garden perennials and water plants, pl. 25 (1971); Journal of the Royal Horticultural Society 97: pl. 75 (1972); The Garden 101: 149 (1976). Hairy perennial, 10–30 cm, often branched. Lower leaves reduced to scales, segments of the upper leaves lanceolate. Flowers 3–4 cm in diameter. Sepals hairless. Petals 2–3 times longer than

broad, yellow. Anthers yellow. Style short and recurved. Achenes 3–3.5 mm, hairy, not toothed. Fruiting heads rounded. *Japan, Korea, eastern USSR.* H1. Spring.

6. A. volgensis Steven. Illustration: Jávorka & Csapody, Iconographia florae hungaricae, 177 (1930). Like *A. amurensis* but the whole plant hairy, apart from the sepals. Flowers 3–3.5 cm in diameter, achenes hairy only at the base. *Hungary, European USSR, Caucasus (Armenia).* H? Spring.

7. A. vernalis Linnaeus. Illustration: Jávorka & Csapody, Iconographia florae hungaricae, 177 (1930); Journal of the Royal Horticultural Society 89: f. 118 (1964); Polunin, Flowers of Europe, pl. 24 (1969); The Garden 101: 149 (1976). Perennial, 10–40 cm. Lower leaves reduced to scales, upper leaves not or scarcely longer than wide, segments linear, hairless. Flowers 4–8 cm in diameter. Sepals hairy. Petals to 3 times longer than broad, yellow. Anthers yellow. Style short, achenes to 3.5 mm, not toothed, hairy all over. Fruiting heads rounded. *S, E & C Europe, USSR (Siberia, Caucasus).* H1. Spring.

8. A. pyrenaica de Candolle. Illustration: Huxley, Mountain flowers in colour, f. 126f (1967); Taylor, Wild flowers of the Pyrenees, 75 (1971); Polunin & Smythies, Flowers of SW Europe, pl. 4f (1973). Perennial, 25–40 cm, with small hairs on the base of the stem and the leaf-sheaths. Lower leaves developed, upper leaves broader than long. Flowers to 6 cm in diameter. Sepals hairless. Petals 2–3 times longer than broad, yellow. Anthers yellow. Style to 2 mm, achenes to 6 mm, not toothed, hairless or very sparsely hairy. Fruiting heads rounded. *Spain, France (Pyrenees and Alpes Maritimes).* H5. Summer.

9. A. chrysocyathus J. D. Hooker & Thomson. Illustration: Blatter, Beautiful flowers of Kashmir 1: 2 (1927); Journal of the Royal Horticultural Society 93: f. 129 (1968); The Garden 102: 454 (1977). Like *A. pyrenaica* but style *c.* 2 mm, almost as long as the achene, which is 2–3 mm, plant entirely hairless. *W Himalaya.* H5.

30. CALLIANTHEMUM Meyer
P.G. Barnes
Low-growing perennials with short rhizomes. Basal leaves long-stalked, pinnate with a terminal leaflet, the lobes pinnatifid. Flowering stems 1–3, leafless or bearing a few shortly stalked leaves. Flowers solitary

or 2–3 in a short raceme. Sepals 5, broadly ovate or rounded, greenish, sometimes falling early. Petals 5–20, white or pink, each with an orange nectary at the base. Stamens numerous. Carpels numerous with very short styles. Fruit a head of achenes.

About 10 species in the mountains of C Europe, C & E Asia. They are occasionally cultivated by alpine plant enthusiasts, usually with the protection of a cold frame or alpine house, in scree or peat beds. An open or lightly shaded situation and freely draining soil are essential. Propagation is normally by fresh seed (old or dry seeds may germinate poorly), but careful division of established plants in spring may also be attempted.
Literature: Witasek, J., Die Arten der Gattung Callianthemum, *Verhandlungen der Königliche-Kaiserliche zoologisch-botanische Gesellschaft in Wien* 49: 316–356 (1899).

1a. Flowering stems leafless; leaves with 3–6 pairs of leaflets **1. pimpinelloides**
  b. Flowering stems with 1–2 small leaves; basal leaves with 2–4 pairs of leaflets 2
2a. Basal leaves fully grown at flowering; petals broad, length *c.* 2 times width **2. coriandrifolium**
  b. Basal leaves developing after flowering; petals narrow, length *c.* 3 times width 3
3a. Lowest pair of leaflets distinctly stalked; fruit wrinkled **3. anemonoides**
  b. Lowest pair of leaflets scarcely stalked; fruit smooth **4. kernerianum**

1. C. pimpinelloides (D. Don) Hooker & Thomson (*C. cachemirianum* Cambessedes). Illustration: Blatter, Beautiful flowers of Kashmir 1: pl. 2 (1927); Polunin & Stainton, Flowers of the Himalaya, pl. 4, f. 30 (1984).
Plant 5–10 cm. Leaves all basal, long-stalked, pinnate with 3–6 pairs of deeply lobed and overlapping leaflets, the lowest almost stalkless. Flowering stems 4–8 cm. Sepals *c.* 5 × 3 mm, oblong or elliptic, apex rounded, whitish green or reddish. Petals 1–1.2 cm × 2–4 mm, oblong, white, reddish beneath. Achenes *c.* 3.5 × 2 mm, wrinkled. *Himalaya & W China.* H4. Spring.

2. C. coriandrifolium Reichenbach (*C. rutifolium* misapplied). Illustration: Bonnier, Flore complète 1: pl. 10, f. 43 (1911) – as Ranunculus rutaefolius; Huxley, Mountain flowers in colour, 151 (1967); Hegi, Illustrierte Flora von Mitteleuropa edn 2, 3: pl. 110, f. 5 (1974).

Plant 10–30 cm. Basal leaves with stalks 3–7 cm, the blades 2.5–5 cm, oblong, with 2–3 pairs of pinnae, the lowest pair shortly stalked; leaflets finely divided with narrow ultimate segments. Flowering stem usually solitary, bearing 1 or 2 leaves 1.5–2 cm long and 1 or occasionally 2–3 flowers each 2.5–3 cm in diameter. Sepals oblong, 6–10 × 2–4 mm. Petals 9–13, 9–12 × 5–6 mm, broadly obovate or elliptic, white. Achenes *c.* 3 × 2.5 mm, wrinkled. *S Europe.* H3. Spring.

3. C. anemonoides (Zahlbruckner) Heynhold (*C. rutifolium* misapplied). Illustration: Botanical Magazine, 7603 (1898); Schacht, Rock gardens and their plants, f. 42 (1963); Hegi, Illustrierte Flora von Mitteleuropa edn 2, 3: pl. 110, f. 4 (1974).
Plant 8–20 cm. Basal leaves not fully developed at flowering, pinnate with 2–3 pairs of leaflets, the lowest distinctly stalked and pinnately lobed. Flowering stem bearing a small, shortly stalked leaf. Flower usually solitary, 2.5–3 cm wide. Sepals rounded, pale green. Petals 12–16, narrowly oblong, *c.* 1.3 cm × 4 mm, white or pink. Fruit *c.* 4.5 × 2 mm, wrinkled. *Austrian Alps.* H3. Spring.

4. C. kernerianum Freyn. Illustration: Wehrhahn, Die Gartenstauden 1: 402 (1931); Bulletin of the Alpine Garden Society 1: 71 (1931) & 36: 12 (1968); Hegi, Illustrierte Flora von Mitteleuropa edn 2, 3: f. 61, a–c (1974).
Plant 2–10 cm. Basal leaves not fully developed at flowering, pinnate with 2–3 pairs of leaflets, the lowest pair shortly stalked and pinnatisect with narrow, oblong lobes. Flowering stem with 1–2 small, shortly stalked leaves, usually bearing a single flower, 2.5–3 cm wide. Sepals *c.* 10 × 6 mm, broadly ovate, obtuse or rounded, whitish. Petals 10–15, 1.3–1.6 cm × 4–6 mm, oblong, white or pink. Fruit *c.* 3.5 × 2.5 mm, smooth. *Italian Alps.* H4. Spring.

LXXIV. BERBERIDACEAE

Shrubs or perennial herbs with rhizomes or tubers. Leaves alternate or (in *Podophyllum* and *Ranzania*) opposite, simple or compound, usually divided into 3s, pinnately or palmately veined. Flowers bisexual, radially symmetric, solitary or in racemes, corymbs, panicles, spikes or clusters. Sepals and petals together in 2 or 3 whorls or (in *Achlys*) absent, each with 6 or 4 members, the outermost ('outer sepals') small, soon falling, the middle ('inner sepals') often petal-like, the innermost ('petals') often reduced and nectar-producing. Stamens as many as the petals and on the same radii as them, or (in *Podophyllum*) twice as many; anthers opening by longitudinal slits or by flaps up-rolling from the base and hinged at the top. Ovary superior, 1-celled; ovules basal and then 1, or lateral when 1–many; style long, short or absent; stigma small to large. Fruit a capsule or berry or (in *Achlys*) an achene.

The family is accepted here in its traditional sense, thus including *Nandinaceae*, *Diphylleiaceae*, *Leonticaceae* and *Podophyllaceae*.
Literature: Janchen, E., Die systematische Gliederung der Ranunculaceen und Berberidaceen, *Denkschriften der Akademie der Wissenschaft in Wien, Mathematisch-naturwissenschaftliche Klasse* **108**, No. 44: 1–82 (1949); Chatterjee, R., Studies in Indian Berberidaceae, *Records of Botanical Survey of India* **16**(2) (1953); Ernst, W.R., The genera of Berberidaceae, Lardizabalaceae and Menispermaceae in the southeastern United States, *Journal of the Arnold Arboretum* **45**: 1–20 (1964); Rix, M., The herbaceous Berberidaceae, *The Plantsman* **4**: 1–15 (1984); Terabayashi, S., The comparative floral anatomy and systematics of the Berberidaceae, *Acta Phytotaxonomica et Geobotanica* **36**: 1–13 (1985); Seedling morphology of the Berberidaceae, *Acta Phytotaxonomica et Geobotanica* **38**: 63–74 (1987).

Plant. Shrub: **1–4**; herb with tubers: **11–13**.
Leaves. Peltate: **14,15**; with 2 leaflets: **4,8**.
Flower-stems. Leafless: **6–9,13**.
Flowers. Solitary: **8,15**. Lavender-blue or light purple: **5,8**.
　　Sepals and petals absent: **9**.
Petals. With 2 rounded glands at base: **2,4,5**. With long, nectar-producing spurs: **7**.
Stamens. 4: **7**. Anthers opening by longitudinal slits: **1,15**.
Fruit. A many-seeded berry: **5,13**. Berry blue: **2,4,14**. A papery, inflated capsule: **11,12**.
Seeds. Ripening outside remains of fruit: **10,12**.

1a. Woody plants 60 cm or more high　　2
　b. Herbaceous plants not more than 1 m high　　5
2a. Leaves pinnately much divided, with more than 65, entire, spineless leaflets; flowers white; anthers opening by longitudinal slits
　　　　　　　　　　1. Nandina
　b. Leaves undivided or pinnate with 3–37 toothed or spiny leaflets; flowers yellow; anthers opening by 2 flaps up-rolling from base　　3
3a. Leaves pinnate with 3–37 leaflets
　　　　　　　　　　2. Mahonia
　b. Leaves undivided or sometimes also with 3 leaflets　　4
4a. Leaves always undivided; spines present on shoots　　**4. Berberis**
　b. Leaves on same plant variable, undivided or with 3 leaflets; spines absent on shoots　**3. × Mahoberberis**
5a. Flowers stalkless in a many-flowered, slender spike; sepals and petals absent; anthers as broad as long
　　　　　　　　　　9. Achlys
　b. Flowers stalked, solitary or in racemes, corymbs or panicles; sepals and petals present; anthers longer than broad　　6
6a. Flowers solitary on a basal, leafless stem; leaves all basal, with 2 stalkless, opposite leaflets or a single lobed blade　　**8. Jeffersonia**
　b. Flowers several or many or if solitary then at apex of stem between 2 leaves　　7
7a. Leaves undivided but sometimes lobed almost to stalk, peltate, major veins radiating palmately from top of stalk　　8
　b. Leaves divided into leaflets, not peltate, pinnately veined　　9
8a. Stem-leaves 2, opposite, with the solitary flower arising between them or leaf 1 and flowers several, clustered, drooping; anthers opening by longitudinal slits; fruit a large red or yellow berry　　**15. Podophyllum**
　b. Stem-leaves 2, well-separated; flowers in erect, many-flowered cymes; anthers opening by up-rolling flaps; fruit a small blue berry
　　　　　　　　　　14. Diphylleia
9a. Leaflets stalkless; rootstock a tuber　10
　b. Leaflets stalked; rootstock a rhizome　　12
10a. Leaves all basal, simply pinnate, with 5–8 pairs of leaflets; inner sepals small; petals flat, conspicuous
　　　　　　　　　　13. Bongardia
　b. Leaves basal and on stem, divided into 3 leaflets; inner sepals large, petal-like; petals small, nectar-producing　11

11a. Stems with several, much-divided leaves, bearing terminal and axillary racemes; fruit large, inflated, membranous **11. Leontice**

b. Stem with 1 leaf, not branched, bearing a terminal raceme only; fruit small, opening before ripening of seeds and exposing them **12. Gymnospermium**

12a. Leaflets deeply 3–5-lobed 13

b. Leaflets entire or only slightly 3-lobed 14

13a. Flowers greenish, yellowish or purplish, in an erect, terminal raceme **10. Caulophyllum**

b. Flowers lavender-violet on long, drooping stalks between 2 leaves **5. Ranzania**

14a. Petals and stamens 4; petals various but not narrowly oblong **6. Epimedium**

b. Petals and stamens 6; petals narrowly oblong with a flat or hooded nectar-producing tip **7. Vancouveria**

1. NANDINA Thunberg
W.T. Stearn

Shrub. Leaves alternate with numerous entire, spineless, pinnately-veined leaflets which are pinnately arranged or in 3s. Flowers very numerous in terminal panicles. Sepals numerous. Petals 6. Stamens 6, anthers almost stalkless, opening by longitudinal slits. Ovary with 1 lateral ovule; style short, stigma minute. Fruit a berry.

A genus of 1 species from China and Japan. This is generally hardy but preferably should be given a warm, sheltered position in northern Europe. Propagation is best by seed obtained from southern Europe or eastern Asia.

1. N. domestica Thunberg. Illustration: Botanical Magazine, 1109 (1808); Sim, Flowering trees and shrubs in South Africa, 114 (1919); Hutchinson, Families of flowering plants, edn 3, 506 (1973). Evergreen upright, hairless shrub to 3 m. Leaves to 50 cm; leaflets numerous (65–97), stalkless, lanceolate, acute, 3–10 cm. Panicle to 35 cm. Flowers white, *c.* 6 mm across. Anthers yellow. Berry spherical, normally red, but white in var. **leucocarpa** Yanagawa, nearly 1 cm across. *C China, Japan.* H4. Summer.

2. MAHONIA Nuttall
N.P. Taylor & S.G. Knees

Evergreen, hairless shrubs, occasionally tree-like. Stems spineless, with alternate leaves. Leaves with stipules, pinnate with a terminal leaflet often larger than the rest, sometimes only 3-lobed, leaflets with more or less wavy, spiny margins. Flowers bisexual, in clustered, erect, ascending or drooping panicles or racemes, borne in the leaf-axils but usually near the apices of lateral branches or terminal stems. Perianth-segments 15 in 2–3 series, the outer 3 often smaller, often greenish or purplish and sepal-like, the remainder concave, with a pair of glandular spots at the base, yellow in all but one of the species cultivated. Stamens 6, sensitive to touch, anthers opening by flaps. Ovary 1-celled, superior. Stigma circular, depressed. Fruit a bloomed berry, bluish black, purple or more rarely red (in one species pale yellow). Seeds ellipsoid to pear-shaped with a hard coat.

A genus of an uncertain number of species (30–100) of which about half occur in Asia from the Himalayas eastwards to Japan and Indonesia (Sumatra); the remainder are distributed throughout N & C America. Propagation of most species is easily achieved from seed sown as soon as possible after ripening, though some species hybridise readily and may not come true. Alternatively, cuttings of semi-ripened wood may be taken in autumn and winter; Nos. **1 & 2** are best propagated from leaf-bud cuttings, the remainder from basal cuttings, although some species cannot be propagated reliably by this method. Some of the blue-leaved species are often grafted on to stock of *Berberis thunbergii* (p. 384).

Literature: Fedde, F., Versuch einer Monographie der Gattung Mahonia, *Botanische Jahrbücher* **31**: 30–133 (1901); Ahrendt, L.W.A., Berberis and Mahonia, *Journal of the Linnean Society, Botany* **57**: 1–410 (1961).

1a. Plant to 1 m; stems underground and suckering freely, or or creeping, persistently scaly throughout 2

b. Plant erect, more than 1 m, or if a low spreading shrub then without underground stems; stems with widely spaced clusters of scales, or scales absent 5

2a. Leaflets 11–15 **6. nervosa**

b. Leaflets 3–9 3

3a. Leaflets shiny above, or green and lacking papillae beneath **8. aquifolium**

b. Leaflets dull above and white-waxy or grey-papillose beneath 4

4a. Leaflets broadly ovate, oblong or almost circular, not markedly tapered at apex; perianth-segments all yellow **7. repens**

b. Leaflets elliptic, finely tapered at apex; outer perianth-segments reddish purple **3. gracilipes**

5a. Stems not branched at base, or if branched near base then habit very upright and leaves with 17 or more leaflets 6

b. Stems branched at or near base forming spreading shrubs; leaflets 3–19 8

6a. Leaves with 17 or more leaflets; one-year-old stems *c.* 1–2 cm or more in diameter **2. × media**

b. Leaves with fewer than 17 leaflets; one-year-old stems less than 9 mm in diameter 7

7a. Leaves with 5–9 leaflets, terminal leaflet without a stalk **4. fortunei**

b. Leaves with 11–15 leaflets, terminal leaflet usually stalked **5. confusa**

8a. Leaves 30 cm or more, with 13–19 leaflets **1. japonica**

b. Leaves 25 cm or less, with 3–11 leaflets 9

9a. Fruit bright red or yellow; leaves mostly light blue-grey, dull 10

b. Fruit waxy blue or black; leaves dark blue-green or shiny 11

10a. Terminal leaflet with 2–12 broadly based marginal teeth **10. trifoliolata**

b. Terminal leaflets with 14–26 fine marginal spines **11. nevinii**

11a. Leaves with 3 and 5 leaflets on the same plant **9. trifolia**

b. Leaves all with 5 or more leaflets **8. aquifolium**

1. M. japonica (Thunberg) de Candolle. Illustration: Botanische Jahrbücher **31**: 128 (1901); Marshall Cavendish encyclopaedia of gardening **4**: 1177 (1969); Krüssmann, Handbuch der Laubgehölze **2**: t. 114, 117 (1977); The Plantsman **1**(1): 17 (1979). Erect, branched shrub to 3 m. Leaves 30–40 cm, with 13–19 leaflets; leaflets 5–11 cm, ovate-lanceolate or oblong-ovate, with acuminate apices and rounded bases. Flowers borne in apparently terminal clusters, racemes 6–10, loose, spreading, 15–25 cm. Fruits 6–7 mm, deep bluish purple. *China, widely cultivated elsewhere.* H1. Winter–spring.

A Chinese species originally named from plants cultivated in Japan. The name *M. bealei* (Fortune) Carrière has been applied to a Chinese variant of the species which probably merits only cultivar status. It is notable for its broader leaflets and erect, dense racemes. 'Hiemalis' is lower-growing with larger leaves (to 50 cm), narrower leaflets with longer spines and paler flowers in longer racemes (to 35 cm).

2. M. × **media** Brickell (*M. japonica* × *M. lomariifolia*). Illustration: The Plantsman 1(1): 17 (1979).

Erect, branched shrub to 4 m. Leaves 28–40 cm, with 17–25 leaflets; leaflets 9–11 cm, ovate-lanceolate, with acuminate apices and 5–11 marginal spines. Flowers in up to 20 ascending, loose racemes, 25–30 cm. Fruits 1.1–1.5 cm, bluish black and heavily bloomed with whitish wax, ovoid. *Garden Origin*. H2. Late autumn to early spring.

Numerous cultivars have been selected, including 'Charity', 'Winter Sun' and 'Lionel Fortescue'.

M. lomariifolia Takeda. Illustration: Bean, Trees and shrubs hardy in the British Isles, edn 8, **2**: t. 95 (1973). Less commonly cultivated and differing from *M.* × *media* and *M. japonica* in having leaves with 19–41 leaflets which are 1–2 cm wide and erect, compact racemes 10–20 cm. *W China, Burma*. H3. Winter.

M. napaulensis de Candolle (*M. acanthifolia* Don) and **M. siamensis** Takeda are closely related to *M. lomariifolia* but differ in their wider leaflets (3–4.5 cm) and loose racemes in *M. siamensis*. The cultivar 'Cantab', selected from a cross between *M. japonica* and *M. siamensis* combines the best features of both species.

3. M. gracilipes (Oliver) Fedde (*Berberis gracilipes* Oliver). Illustration: Hooker's Icones Plantarum **8**: t. 1754 (1887); Botanische Jahrbücher **31**: 128 (1901).

Erect, suckering shrub to 1 m. Leaves 30–60 cm, with 5–9 elliptic leaflets; leaflets 3.8–5.5 cm wide, with 5–11 marginal spines, margin smooth in lower third, all white-waxy beneath, eventually becoming pale green, dark green above. Flowers in 5–12 very loose racemes 25–50 cm, flower-stalks 1.1–1.8 cm. Perianth in 3 series, the outer 2–3 mm, sepal-like, reddish purple, the middle series to 4 mm, also reddish purple, the inner series pale yellowish cream. Fruits bluish black (rarely seen in cultivation). *China (Sichuan)*. G1. Autumn.

4. M. fortunei (Lindley) Fedde (*Berberis fortunei* Lindley). Illustration: Flore des serres ser. 1, **3**: t. 73 (1847); Botanische Jahrbücher **31**: 114 (1901).

Erect to 2 m, stems unbranched. Leaves 15–25 cm, with 5–9 lanceolate leaflets; leaflets 1.1–1.5 cm wide, with 11–19 marginal spines, lower third of leaflet entire; terminal leaflet without a stalk. Flowers in 4–7 compact, spreading racemes 3–5 cm; flower-stalks *c.* 1 mm. Fruits

bluish black, with white waxy bloom. *China*. H3. Late autumn.

5. M. confusa Sprague.

Erect, little-branched shrub to 1–1.5 m. Leaves 40–50 cm, with 11–15 (rarely to 17) lanceolate-elliptic leaflets; leaflets 2–3 cm wide with 5–11 marginal spines. Flowers in 5–9 loose racemes to 15 cm, flower-stalks 1–2 mm. Fruit *c.* 7 mm, bluish black. *China*. H3. Winter.

6. M. nervosa (Pursh) Nuttall (*Berberis nervosa* Pursh; *B. glumacea* Sprengel). Illustration: Botanical Magazine, 1426 (1831), 3949 (1842).

A low, creeping shrub to *c.* 50 cm, often spreading by suckers, stems persistently and densely scaly throughout. Leaves to 45 cm, with 11–15 leaflets; leaflets 3–8 cm, ovate-lanceolate with 10–18 marginal spines. Flowers in erect, loose racemes 17–25 cm. Fruit 7–9 mm, oblong-ovoid, bluish black. *Western N America*. H2. Spring.

7. M. repens (Lindley) Don (*Berberis repens* Lindley; *B. nana* Greene; *M. nana* (Greene) Fedde). Illustration: Loddige's Botanical Cabinet **19**: t. 1847 (1832); Clements & Clements, Rocky Mountain flowers, t. 4 (1914).

Dwarf, to *c.* 30 cm, with scarcely branched stems arising underground, Leaves to 20 cm, with 3–7 leaflets; leaflets to 9 cm, broadly ovate, oblong or almost circular, truncate or obliquely so at base, with *c.* 15–30 marginal spines or teeth, dull bluish green above, grey and papillose beneath. Flowers in dense, apparently terminal racemes to 8 cm, deep yellow. Fruits to 9 cm, spherical, waxy, blue-black. *Western N America*. H1. Spring.

'Rotundifolia' is taller, to 60 cm or more, with green, sometimes almost entire leaflets.

M. pumila (Greene) Fedde (*Berberis pumila* Greene) has leaves to 14 cm, with 5–9 leaflets; leaflets to 6 cm, dull grey-green above, glaucous beneath; racemes to 5 cm, flower-stalks with bracteoles. *W USA (California, Oregon)*. H2. Spring.

8. M. aquifolium (Pursh) Nuttall (*Berberis aquifolium* Pursh). Illustration: Marshall Cavendish encyclopaedia of gardening **4**: 1177, 1178 (1969); Everett, New York Botanical Gardens illustrated encyclopedia of horticulture **6**: 2103, 2104 (1981). Spreading shrub, 1–2 m, sometimes suckering. Leaves to 25 (rarely to 30) cm, with 5–9 (rarely 11) leaflets; leaflets to 9 cm, broadly to narrowly ovate, more or

less truncate at base, with *c.* 15–35 marginal spines, bright to dark purplish green, shiny. Flowers in erect, apparently terminal and axillary clusters, racemes to 8 cm, flowers golden-yellow. Fruits as in *M. repens*. *Western N America*. H1. Late winter–spring.

Rather variable through hybridisation with the preceding. 'Atropurpurea' has leaves dark reddish purple in winter.

M. dictyota (Jepson) Fedde and **M. piperiana** Abrams differ from *M. aquifolium* in having leaflets very papillose beneath, with only 6–20 marginal spines.

M. pinnata (Lagasca) Fedde (*Berberis pinnata* Lagasca; *B. toluacensis* Anon.; *M. moranensis* (Schultes) Johnston; *M. fascicularis* de Candolle). Illustration: Delessert, Icones Plantarum **2**: t. 3 (1823); Edwards's Botanical Register **9**: t. 702 (1823). Like *M. aquifolium*, but taller, to 3 m; leaves with 7–11 (rarely 13) smaller, ovate-lanceolate, less shiny, more wavy leaflets; flower-stalks bracteolate. *USA (S California), C Mexico*. H4. Spring.

Rarely cultivated.

M. × **wagneri** (Jouin) Rehder (*M. pinnata* misapplied). *M. aquifolium* × *M. pinnata*. This name is in use for a varied group of cultivars including 'Aldenhamensis', 'Undulata' and 'Vicaryi', which are more or less intermediate between the parents.

M. 'Heterophylla' (*M. toluacensis* misapplied) has 5–7 lanceolate to narrowly oblong leaflets, 2.5–8 cm × 6–18 mm, irregularly twisted or curled and variably stalked. *Garden Origin*.

9. M. trifolia Chamisso & Schlechtendahl (*Berberis trifolia* (Chamisso & Schlechtendahl) Schultes; *B. schiedeana* Schlechtendahl; *M. schiedeana* (Schlechtendahl) Fedde; *M. eutriphylla* misapplied). Illustration: Botanical Magazine, n.s., 832 (1981).

Low spreading shrub, 50–100 × 100 cm. Leaves to 11 cm with 3–5 leaflets; leaflets to 4 cm, ovate-elliptic to nearly circular, pale green or deep blue-green to purplish when mature, strongly wavy, with 6–8 sharp marginal spines, the terminal leaflet with or rarely without a stalk. Flowers densely clustered in very short, axillary and seemingly terminal, 2–4-branched racemes; flower-stalks *c.* 8 mm, minutely bracteolate. Fruit to 1 cm, spherical to ovoid, waxy bluish. *C Mexico*. H3. Spring.

10. M. trifoliolata (Moricand) Fedde (*Berberis trifoliolata* Moricand; *B. trifoliata* Lindley; *M. trifoliata* (Lindley) Lavallée). Illustration: Edwards's Botanical Register

31: t. 10 (1845); Flore des Serres, t. 56 (1845); Botanische Jahrbücher **31**: 81, f. 1N (1901); Krüssmann, Handbuch der Laubgehölze **2**: 293 (1977).

Freely branched shrub to 2 m or more with slender, ascending or spreading branches often bearing short, lateral, spur-like shoots. Leaflets 3, grey-blue to whitish, to *c.* 3–7 cm, unstalked, narrowly elliptic to lanceolate, margins strongly wavy, with 2–10, broadly based, marginal, spiny teeth. Flowers in 1–4 slender, loose clusters, 2–4 cm, at each leaf-axil, bracts minute. Fruit 8–10 mm, spherical, red when ripe. *S USA, N Mexico.* H4. Spring.

M. haematocarpa (Wooton) Fedde (*Berberis haematocarpa* Wooton). Illustration: Botanische Jahrbücher **31**: 81 (1901). Like *M. trifoliolata* but leaves with 3–9 leaflets 3–11 cm, the terminals often stalked, marginal teeth 4–12. *S USA, N Mexico.* H4. Spring.

Scarcely differing is **M. fremontii** (Torrey) Fedde (*Berberis fremontii* Torrey) with ovate or broadly oblong leaflets to 3.5 cm and light yellow or red fruits to 1.5 cm. *USA (Colorado, Utah, New Mexico, Arizona).* H4. Spring.

M. higginsiae (Munz) Ahrendt is very similar, differing in its smaller fruits (to 8 mm). *USA (S California), NW Mexico.* H4.

11. M. nevinii (Gray) Fedde (*Berberis nevinii* Gray). Illustration: Bean, Trees and shrubs hardy in the British Isles, edn 8 **2**: 686 (1973); Krüssmann, Handbuch der Laubgehölze **2**: t. 113 (1977).

Like *M. trifoliolata* but leaflets 5–7 (rarely to 9), the terminal stalked, to 5 cm, with *c.* 14–26 marginal spines. Flowers in solitary clusters to 6 cm, in leaf-axils of the short spur-shoots. Fruit 6–8 mm, red. *W USA (S California).* H3. Spring.

M. swaseyi (Buckland) Fedde is similar with flower-clusters bearing small, leaf-like bracts and fruits to 1.2 cm. *USA (Texas).*

3. × MAHOBERBERIS Schneider

A genus of hybrids between various species of *Mahonia* and *Berberis*, known only in gardens. They are distinguished from *Mahonia* by having leaves simple and with 3 leaflets on the same plant, and from *Berberis* by the same character and by the lack of spines on the shoots. The most frequently seen is × **M. neubertii** (Baumann) Schneider (*Berberis neubertii* Baumann), a hybrid between *M. aquifolium* and *B. vulgaris*, which is evergreen or partially so. A few others are known, but are rarely cultivated.

4. BERBERIS Linnaeus

D.F. Chamberlain & H.S. Maxwell

Spiny prostrate to erect shrubs; stems much-branched or arching, to 5 m; spines with 1–3 (rarely 7) prongs, occasionally leaf-like. Leaves persistent and leathery or deciduous and usually flexible, simple, entire or with spine-like teeth. Flowers borne in the axils of the leaves, solitary or in clusters, umbels, racemes or panicles. Flowers greenish to bright yellow or orange, 3–25 mm across. Sepals petal-like, in 2 (rarely 1 or 3) rows of 3, petals in 2 (rarely 1) rows of 3; stamens 6. Berries spherical to ellipsoid, fleshy, red or black, with or without a style; ovules 1–16.

Over 600 species mostly in the temperate parts of both the old and new worlds. About 100 species are commonly grown in Europe; most are reasonably hardy.

Literature: Ahrendt, L. W. A., Berberis and Mahonia, A Taxonomic Revision, *Journal of The Linnean Society (Botany)* **57**: 1–295 (1961).

The species of *Berberis* are difficult to identify and care will be needed in using the key; the following notes are intended to highlight some of the problems.

Stems. These may be grooved, angled or round in cross section. In some species the young shoots are hairy or warty. Specimens, especially those collected in the wild, often have stem parasites that look like hairs or warts. Careful examination with a lens will therefore be necessary. Where the stem does not have either hairs or warts it is described as being smooth.

Spines. These are either simple or branched. In some species they are leaf-like.

Leaves. In some species the leaf-shape is extremely variable depending on the position on the plant. The degree of toothing is also variable; leaves that are entire or with several pairs of coarse or fine teeth occur on the same plant. In 3 species the leaves are finely hairy; unless otherwise stated it is assumed that the leaves lack hairs.

Inflorescence. The arrangement of the flowers is important in the identification of *Berberis.* For illustration of this character see Figure 38, p. 379.

Berries. The ripe berries have a persistent stigma that may be borne on a conspicuous style or the style may be absent. The number of ovules in the ovary is cited for each species. Where there are several ovules only a proportion may mature into ripe seeds. The number of ovules will

therefore include those that have aborted. Again, care must be taken with this character.

1a. Flowers and berries solitary, rarely in pairs   2
  b. Flowers and berries in clusters, umbels, racemes, corymbs or panicles   19
2a. Berries black, with or without a bloom; leaves persistent, leathery, if deciduous (**96. chillanensis**), then flowers yellow with an orange centre   3
  b. Berries red, without a bloom; leaves deciduous and usually flexible, or persistent (**34. concinna** and **35. tsangpoensis**); flowers entirely yellow   9
3a. Berries ovoid to oblong-ovoid, 8–16 mm; leaves always toothed   4
  b. Berries spherical, 6–9 mm diameter; leaves entire   7
4a. Leaves 1.5–6 cm, with 5–12 pairs of teeth   5
  b. Leaves with 1–2.5 cm, with 2–5 pairs of teeth   6
5a. Flowers *c.* 2.5 cm across; leaves 1–2 cm broad   **1. calliantha**
  b. Flowers *c.* 1.5 cm across; leaves 6–10 mm broad   **3. chrysosphaera**
6a. Stems strongly papillose; flowers *c.* 2 cm across   **6. verruculosa**
  b. Stems finely warty; flowers *c.* 1.5 cm across   **7. candidula**
7a. Leaves linear-elliptic, 1–3 mm broad; flowers 1.2–1.4 cm across, golden-yellow   **92. empetrifolia**
  b. Leaves elliptic to broadly elliptic, 2–14 mm broad; flowers 1.4–1.8 cm across, orange   8
8a. Flower-stalks hairy, 5–10 mm   **96. chillanensis**
  b. Flower-stalks hairless, 1.4–1.8 cm   **97. buxifolia**
9a. Flowers 8–10 mm across; berries 6–9 mm; dwarf shrub, 20–65 cm   10
  b. Flowers 1.2–1.8 cm across; berries 1–1.8 cm (6–7 mm diameter in **43. approximata**, then shrub to *c.* 2 m)   11
10a. Leaves 1–2.5 cm × 6–14 mm, flexible   **44. sibirica**
  b. Leaves 6–15 × 2–4 mm, thick   **45. mucrifolia**
11a. Prostrate shrub 15–25 cm high; flower-stalks *c.* 3 mm   **35. tsangpoensis**
  b. Shrub 60–300 cm high; flower-stalks 5–50 mm   12

12a. Lower surface of leaves without a bloom, yellow-green and shining; berries ovoid to almost spherical, without a bloom, 1–1.3 times as long as broad **32. angulosa**

b. Lower surface with a grey or whitish bloom, sometimes papillose; berries ellipsoid, 1.5–2 times as long as broad 13

13a. Young shoots with a glaucous bloom; flowers pale yellow, stalks 5–15mm 14

b. Young shoots without a bloom; flowers bright to deep yellow, stalks 8–50mm 16

14a. Ovules 6–11; leaves 2–4.5 cm **36. temolaica**

b. Ovules 3–6; leaves 1–2.2 (rarely to 3) cm 15

15a. Berries ellipsoid, 1–1.3 cm × 6–8 mm **42. dictyophylla**

b. Berries spherical, 7–9 mm in diameter **43. approximata**

16a. Young shoots minutely hairy; flowers 1.5–1.8 cm across, stalks to 5 cm **33. ludlowii**

b. Young shoots smooth to slightly warty; flowers 1.2–1.5 cm across, stalks to 3 cm 17

17a. Leaves with under-surface white-papillose, semi-evergreen **34. concinna**

b. Leaves with a grey bloom beneath, deciduous 18

18a. Shrub 60–100 cm; young shoots yellowish; spines 1.2–2.5 2 cm **38. diaphana**

b. Shrub 1.5–3 m; young shoots purple; spines 6–15 mm **39. aemulans**

19a. Flowers in clusters or simple umbels 20

b. Flowers in simple umbel-like racemes, racemes, corymbs or panicles 64

20a. Leaves deciduous; berries red or black when mature 21

b. Leaves evergreen; berries black when mature 34

21a. Flowers in umbels 22

b. Flowers in clusters, sometimes with one or more branches compound 27

22a. Flowers 1.8 cm across; ripe berries black with a blue bloom **94. cabrerae**

b. Flowers 8–10 mm across; ripe berries red to dark red with or without a bloom 23

23a. Leaves c. 4 times as long as broad; berries almost spherical, 1.2–1.4 times as long as broad **51. amoena**

b. Leaves 1.2–3 times as long as broad; berries ovoid to ellipsoid, 1.4–2.5 times as long as broad 24

24a. Leaf-stalks 5–7 mm 25

b. Leaf-stalks to 1 mm 26

25a. Leaves 1–1.5 (rarely to 2) cm; berries 7–8 × c. 4 mm **57. thunbergii**

b. Leaves 1.3–3.5 cm; berries c. 9 × 5 mm **58. × ottawensis**

26a. Leaves 2.5–4.5 cm **31. umbellata**

b. Leaves 1.1–2.3 cm **50. tsarongensis**

27a. Flowers 6–10 mm across, in clusters of 4–20; berries 4–6.5 mm 28

b. Flowers 1–2 cm across, in clusters of 2–9; berries 6–16 mm 29

28a. Leaf margin with 2–6 pairs of teeth; ovules 2 **79. arido-calida**

b. Leaf margin entire; ovules 3–5 **80. wilsoniae**

29a. Flowers with at least the centre part orange; berries black 30

b. Flowers pale to bright yellow, never orange; berries red 31

30a. Flowers in clusters of 4–9, c. 1.2 cm across; berries with style absent **88. heteropoda**

b. Flowers in clusters of 2–3, 1.5–1.8 cm across; berries with style 2–3 mm **95. montana**

31a. Flowers 1.6–1.8 mm across; ovules 1–3 **52. yunnanensis**

b. Flowers 1.2–1.5 mm across; ovules 4–15 32

32a. Leaves oblanceolate, 3–4 times as long as broad; berries 8–9 mm long **37. morrisonensis**

b. Leaves obovate to oblong, 2–2.5 (rarely to 3) times as long as broad; berries 1.3–1.6 cm long 33

33a. Small shrub 60–100 cm, mature stems yellow **38. diaphana**

b. Shrub 1.5–3 m, mature stems purple **39. aemulans**

34a. Flowers c. 3 mm across; spines leaf-like with 3–4 spine-like teeth **91. crispa**

b. Flowers 5 mm or more across; spines not leaf-like 35

35a. Flowers orange-yellow to apricot; ripe berries with style to 3mm 36

b. Flowers pale or deep yellow; ripe berries with style absent or to 1.5mm 40

36a. Leaves 2–3 mm broad **93. × stenophylla**

b. Leaves 7–50 mm broad 37

37a. Flowers 1.8–2.2 cm across; flower-stalks 1.5–2 cm 38

b. Flowers 1.5–1.8 cm across; flower-stalks 2–15 mm 39

38a. Leaves entire, 3–3.5 times as long as broad **102. linearifolia**

b. Leaves with 2–6 pairs of teeth, 1.4–2.5 times as long as broad **101. ilicifolia**

39a. Flowers 8–12 in clusters or short umbels; flower-stalks to c. 1.5 cm **99. × lologensis**

b. Flowers 2–3 in clusters; flower-stalks 2–3 mm **100. comberi**

40a. Leaves 8–25 mm, if more than 2 cm then spines with up to 7 branches that are flattened at base 41

b. Leaves 2–22 cm, spines with 1–3 branches that are rounded at base 42

41a. Leaves 2–3 mm broad, entire; spines with 1–3 branches that are rounded at base **93. × stenophylla**

b. Leaves 5–10 (rarely to 15) mm broad, spines with up to branches which are flattened at base, with 3 pairs of marginal teeth **89. actinacantha**

42a. Flowers 1.5–2.5 cm across; berries 1.1–1.5 cm; leaves at least 3 times as long as broad 43

b. Flowers 8–15 mm across; berries 5–12 mm, if more than 10 mm long then leaves less than 2.5 times as long as broad 44

43a. Flowers 2.5 cm across; ovules 10–15 **1. calliantha**

b. Flowers 1.5–2 cm across; ovules 4–9 **2. hookeri**

44a. Leaves 1.9–2.5 times as long as broad 45

b. Leaves greater than 2.5 times as long as broad 52

45a. Leaves with a strong white or greyish bloom beneath 46

b. Leaves sometimes paler, but apparently lacking a bloom beneath 48

46a. Berries 1.1–1.2 cm; flowers in clusters of 3–6 **4. coxii**

b. Berries 6–9 mm; flowers in clusters of 6–18 (rarely 4) 47

47a. Leaves entire or with 1–6 pairs of teeth **19. pruinosa**

b. Leaves with 6–10 pairs of teeth **9. hypokerina**

48a. Flowers pale greenish yellow; leaves with 15–25 pairs of teeth, veins conspicuous **11. sargentiana**

b. Flowers lemon to rich yellow; leaves entire or with 1–20 pairs of teeth; veins obscure to conspicuous 49

49a. Young shoots deeply grooved; flower-stalks 4–6 (rarely to 12 mm) **10. kawakamii**

b. Young shoots rounded to angled; flower-stalks 7–20 mm 50

50a. Lower surface of leaf with conspicuous veins; ovules 4–5 **18. manipurana**

b. Lower surface of leaf with more or less obscure veins; ovules 1–3 51

51a. Ovules 2–3; leaf-margin entire or with 1–6 pairs of teeth **19. pruinosa**

b. Ovules 1–2; leaf-margin with 4–20 pairs of teeth **21. julianae**

52a. Leaves with a white bloom beneath, narrow, strongly recurved; margin entire or with up to 6 pairs of teeth 53

b. Leaves lacking a bloom beneath, broad or narrow, not strongly recurved; margin with 4–30 pairs of teeth 54

53a. Stems rounded, smooth; leaves dull green above; ovules 2 **16. replicata**

b. Stems grooved, warty; leaves shiny above; ovules 4 **17. taliensis**

54a. Leaves 2–3 times as long as broad 55

b. Leaves 3–12 times as long as broad 57

55a. Flowers greenish yellow, in clusters of 4–8; leaves with 15–25 pairs of teeth **11. sargentiana**

b. Flowers yellow, in clusters of 10–20; leaves with 4–20 pairs of teeth 56

56a. Leaves with veins more or less obscure beneath; ovules 1–2 **21. julianae**

b. Leaves with veins conspicuous beneath; ovules 4–5 **18. manipurana**

57a. Leaves 6–12 times as long as broad 58

b. Leaves 3–6 times as long as broad 59

58a. Flowers 1.2–1.5 cm across; berries 9–10 × c. 6 mm **5. gagnepainii**

b. Flowers c. 10 mm across; berries 6–8 × c. 3 mm **12. sanguinea**

59a. Leaves with 5–11 pairs of teeth 60

b. Leaves with 10–30 pairs of teeth 62

60a. Leaves 4.5–6 times as long as broad **22. atrocarpa**

b. Leaves 3–4 times as long as broad 61

61a. Flowers in clusters of 10–25 **14. bergmanniae**

b. Flowers in clusters of 3–6 **15. lempergiana**

62a. Flower-stalks 2–3.5 mm **13. veitchii**

b. Flower-stalks 5–20 mm 63

63a. Leaves dull beneath, up to 2 cm broad; ovules 1–2 **20. dumicola**

b. Leaves shining beneath, up to 5 cm broad; ovules 4–7 **8. insignis**

64a. Flowers in simple racemes or corymbs 65

b. Flowers in racemes that are branched below or in panicles 114

65a. Leaves persistent, leathery, thick; berries black 66

b. Leaves deciduous usually flexible; berries red to dark red (black in **25. glaucocarpa, 85. hispanica, 88. heteropoda**) 72

66a. Leaves broadly ovate to circular, about as long as broad, with almost cordate bases **90. congestiflora**

b. Leaves linear-elliptic to hexagonal at least 2 times as long as broad; leaf-base tapering 67

67a. Leaves 2.5–7 times as long as broad 68

b. Leaves up to 2.5 times as long as broad 69

68a. Flowers bright yellow, 15–30 in racemes 2–9 cm; leaves 2–6 cm **24. lycium**

b. Flowers golden-yellow or orange, sometimes reddish in bud, up to 14 in racemes to 2.5 cm; leaves 8–20 mm **93. × stenophylla**

69a. Leaves 1–2 cm, obovate-hexagonal **98. darwinii**

b. Leaves 2–8 cm, elliptic to obovate 70

70a. Berries red, without a bloom; leaves 2.5–3 cm; spines 2–4.5 cm **26. potaninii**

b. Berries black, with a bloom; leaves to 7.5 cm; spines 1–2.5 (rarely to 3.5 cm) 71

71a. Flowers yellow, c. 1.5 cm across; berries oblong-ovoid **23. asiatica**

b. Flowers orange, 6–8 mm across; berries spherical **103. valdiviana**

72a. Leaves hairy beneath 73

b. Leaves hairless beneath 75

73a. Petals shorter than inner sepals **69. mitifolia**

b. Petals slightly longer than inner sepals 74

74a. Leaves 1.3–2.9 (rarely to 3.3) cm across, 1.3–2.3 times as long as broad **67. brachypoda**

b. Leaves 3–15 mm across, 2.7–3.3 times as long as broad **68. gilgiana**

75a. Leaves 2.7–5 times as long as broad 76

b. Leaves 1.3–2.6 times as long as broad 83

76a. Leaves 8–16 × 2–4 mm **51. amoena**

b. Leaves 1.5–10 cm (rarely less) × 3–35 mm 77

77a. Leaves 2.7–3.5 times as long as broad 78

b. Leaves 4–5 times as long as broad 82

78a. Flowers 1.4–1.7 cm across; berries 1.2–1.3 cm **30. coriaria**

b. Flowers 3.5–12 mm across; berries 3.5–11 mm 79

79a. Flowers 3.5 mm across, stalks 1–3 mm; berries spherical 3.5–4.5 mm in diameter **66. vernae**

b. Flowers 5–12 mm across, stalks 3–23 mm; berries ovoid to oblong, 7–13 mm 80

80a. Leaves 3.5–10 × 1.5–3.5 cm **56. virgetorum**

b. Leaves 1.6–3.5 cm × 5–10 mm 81

81a. Berries with a short style **48. franchetiana** var. **macrobotrys**

b. Berries without a style **49. forrestii**

82a. Berries black with a bluish bloom, style short; racemes 4–8-flowered **65. lepidifolia**

b. Berries red, without a bloom, style absent; racemes 11–18-flowered **64. poiretii**

83a. Leaves with 10–150 pairs of teeth 84

b. Leaves entire or with up to 10 pairs of teeth 95

84a. Leaves with 50–150 pairs of teeth **53. sieboldii**

b. Leaves with 10–40 pairs of teeth 85

85a. Leaves 1.3–1.5 times as long as broad; flowers 4–5 mm across **54. zabeliana**

b. Leaves 1.5–3 times as long as broad; flowers 5–20 mm across 86

86a. Spines leaf-like, at least in upper parts of young shoots; berries spherical **71. koreana**

b. Spines not leaf-like; berries spherical to elliptic 87

87a. Leaves with a tapered apex; flowers c. 6 mm across, in racemes of 10–20 **75. dielsiana**

b. Leaves with a rounded to acute apex; flowers 5–12 mm across, if less than 8 mm then in racemes of 20–40 88

88a. Leaves with a grey bloom beneath 89

b. Leaves green, though sometimes paler beneath 93

89a. Flowers 5–7 mm across, in racemes
of 20–40　　**62. jamesiana**
  b. Flowers 8–12 mm across, in
racemes of 4–25　　90
90a. Flower-stalks 1.5–3.5 cm, rarely
shorter　　**40. tischleri**
  b. Flower-stalks to 10 mm　　91
91a. Leaves with 25–40 pairs of teeth;
berries 2.2–2.5 times as long as
broad　　**73. regeliana**
  b. Leaves with up to 22 pairs of teeth;
berries 1.4–1.7 times as long as
broad　　92
92a. Shrub 2–2.5 m; young shoots
brown; leaf-stalks to 2 mm
　　74. henryana
  b. Shrub 1–1.3 m; young shoots red;
leaf-stalks to 1.4 cm　**72. amurensis**
93a. Leaves 1.5–2 times as long as broad;
ovules 3–5　　**46. orthobotrys**
  b. Leaves 2–2.7 times as long as broad;
ovules 2　　94
94a. Flower-stalks 4–6 mm; berries
c. 7 mm　　**55. quelpaertensis**
  b. Flower-stalks 6–12 mm; berries
9–14 mm　　**70. vulgaris**
95a. Leaf-stalks 6–25 mm (rarely 5 mm)
　　96
  b. Leaf-stalks absent or to 5 (rarely to
7) mm　　102
96a. Flower-stalks 1–1.7 cm　　97
  b. Flower-stalks 3–10 mm　　99
97a. Flowers orange-yellow; berries black
　　88. heteropoda
  b. Flowers yellow; berries red　　98
98a. Leaves with upper surface dull,
lower surface with a grey bloom
　　59. silva-taroucana
  b. Leaves with upper surface shining,
lower surface lacking a bloom
　　60. mouillacana
99a. Racemes 20–40-flowered; leaves
c. 17 times as long as broad
　　62. jamesiana
  b. Racemes 5–20-flowered; leaves
2–2.4 times as long as broad　　100
100a. Berries almost spherical, 5–6 ×
c. 5 mm　　**61. fendleri**
  b. Berries oblong, 8–9 × 4–6 mm　101
101a. Flowers 5–10 (rarely to 15) in
racemes 2.5–3.5 cm
　　56. virgetorum
  b. Flowers 10–20 in racemes 5–7 cm
　　75. dielsiana
102a. Flowers and fruits 2–4; berries
c. 1.5 cm　　**81. × rubrostilla**
  b. Flowers and fruits, 3–25; berries
4–13 mm　　103
103a. Young shoots finely hairy; berries
4–5 mm　　**79. arido-calida**

  b. Young shoots smooth or warty;
berries 7–13 mm　　104
104a. Flower-stalks 1.5–3.5 cm (rarely
only 5 mm)　　**40. tischleri**
  b. Flower-stalks 8–15 mm (rarely to
2 cm)　　105
105a. Flowers 6–7 mm across; berries
black when ripe　　106
  b. Flowers 8–12 mm across; berries
red or black when ripe　　107
106a. Shrub to 1.5 m, stems remaining
red　　**85. hispanica**
  b. Dwarf shrub, 30–60 cm, stems
becoming yellow at maturity
　　86. aetnensis
107a. Leaves with a grey or white bloom
beneath　　108
  b. Leaves lacking a bloom, sometimes
dull or papillose　　111
108a. Leaves 1.7–1.9 × as long as broad;
berries 2.8–3.7 times as long as
broad　　**47. johannis**
  b. Leaves 1.9–3 × as long as broad;
berries 1.4–2 times as long as broad
　　109
109a. Flowers 10–20, in racemes 3–6 cm
　　74. henryana
  b. Flowers 4–9, in racemes
1.5–3.5 cm　　110
110a. Flowers greenish- to sulphur-yellow;
berries with a bloom, ovules 4–5
　　41. virescens
  b. Flowers bright yellow; berries
without a bloom, ovules 2–3
　　50. tsarongensis
111a. Flower-stalks 1–1.7 cm; flowers
orange-yellow　　**88. heteropoda**
  b. Flower-stalks 3–10 mm; flowers
bright to lemon-yellow　　112
112a. Leaves 2.2–2.5 times as long as
broad; berries 1.2–1.3 times as long
as broad　　**25. glaucocarpa**
  b. Leaves 1.8–2 times as long as broad;
berries at least 1.4 times as long as
broad　　113
113a. Leaves 1.5–2.5 (rarely to 3) cm;
racemes 2.8–3.5 cm
　　46. orthobotrys
  b. Leaves 2.5–5 cm; racemes 4–6 cm
　　28. floribunda
114a. Leaf-stalk 5–25 mm　　115
  b. Leaf-stalk absent or to 5 mm　117
115a. Berries black; leaves entire or rarely
with up to 2 pairs of teeth; flowers
in open panicles　　**87. oblonga**
  b. Berries red; leaves with up to 30
pairs of teeth; flowers in racemes
that are compound below　　116
116a. Leaves with a grey bloom beneath
　　62. jamesiana

  b. Leaves pale green beneath
　　63. francisci-ferdinandi
117a. Flowers 1.6–2 cm across; berries
1.2–1.4 cm × 4–6 mm　**29. chitria**
  b. Flowers 5–12 mm across; berries
3–12 mm, spherical to ellipsoid
　　118
118a. Leaves persistent; berries black
　　27. aristata
  b. Leaves deciduous, (semi-persistent in
82. × carminea); berries black or red
　　119
119a. Flowers *c.* 5 mm, up to 80 in
cylindrical panicles; berries 3–6 mm
　　78. prattii
  b. Flowers 6–10 mm across, 10–30, in
panicles; berries 6–12 mm　　120
120a. Flower-stalks 6–15 mm
　　76. beaniana
  b. Flower-stalks 1–5 (rarely to 7) mm
　　121
121a. Panicles 1–3 cm; berries almost
spherical　　**77. aggregata**
  b. Panicles 3–16 cm; berries oblong-
ovoid　　122
122a. Dwarf shrub 30–100 cm; berries
red, without a bloom; leaves semi-
persistent　　**82. × carminea**
  b. Shrub 2–3 m, berries black with a
bloom; leaves deciduous　　123
123a. Flower-stalks 1–2 mm, bracts as
long as flower-stalks　**83. gyalaica**
  b. Flower-stalks 2–5 mm, bracts half
as long as flower-stalks
　　84. sherriffii

1. B. calliantha Mulligan. Illustration:
Gardeners' Chronicle 97: f. 390, 391
(1935).
Compact or open shrub to 1 m; young
shoots grooved, smooth, soon brown;
spines 1–2 cm. Leaves leathery, persistent,
elliptic to oblong, 2–6 × 1–2 cm, base
tapered into a stalk 1–2 mm, margin with
5–10 pairs of coarse, spine-tipped teeth,
venation web-like, less marked beneath,
shiny above, whitish beneath. Flowers
solitary or in simple clusters of 2–3; flower-
stalks 2–4 cm, reddish; flowers hanging,
c. 2.5 cm, pale to lemon-yellow. Berries
ovoid, 1.1–1.4 cm × 6–9 mm, black with a
blue bloom, style absent; ovules
10–15. *China (SE Xizang)*. H2. Late spring.

2. B. hookeri Lemaire (*B. jamesonii*
misapplied; *B. wallichiana* misapplied).
Illustration: Botanical Magazine, 4656
(1852).
Shrub 1–1.5 m; young shoots grooved,
smooth, yellow; spines 1–2.5 cm. Leaves
leathery, persistent, elliptic to oblong,
3–6 cm × 8–20 mm, base tapered with

leaf-stalk more or less absent, margin with 7–15 pairs of coarsely spine-tipped teeth, venation web-like, shiny above, dull green, usually with a whitish bloom beneath. Flowers in simple clusters of 3–6; flower-stalks 1.5–2.5 cm, red. Flowers 1.5–2 cm across, yellowish green. Berries oblong, black with a slight bloom, 1.2–1.5 cm × 6–8 mm; style absent; ovules 6–9 (rarely 4–5). *Nepal, N India (Sikkim to Assam), Bhutan, China (S Xizang)*. H2. Early summer.

Plants with leaves green beneath have been called var. **viridis** Schneider.

3. B. chrysosphaera Mulligan. Illustration: Journal of the Royal Horticultural Society **65**: f.77, 78 (1940).
Dwarf shrub 30–60 cm; young shoots slightly grooved, more or less smooth, red; spines 1–2 cm. Leaves leathery, persistent, elliptic to ovate, 1.5–4 cm × 4–10 mm, base tapered with leaf-stalk more or less absent, margin strongly recurved, with 5–12 pairs of spine-tipped teeth, venation more or less obscure, green and shiny above, with a white bloom beneath. Flowers solitary; flower-stalks 1.8–2.5 cm. Flowers *c.* 1.5 cm across, yellow. Berries ovoid, blue-violet, *c.* 10 × 6 mm; ovules 9–12. *China (SE Xizang)*. H2. Late spring.

4. B. coxii Schneider.
Dense shrub *c.* 2 m; young shoots rounded, finely warty, yellowish brown; spines 1.5–2 cm. Leaves leathery, persistent, elliptic to elliptic-ovate, 3–6 × 1.3–2.6 cm, base tapered into a stalk 2–3 mm, margin with 6–13 pairs of spine-tipped teeth, venation open, dark green and shining above, with a greyish bloom beneath. Flowers in simple clusters of 3–6; flower-stalks 1.3–1.7 cm, red. Flowers *c.* 1.2 cm across, pale yellow. Berries oblong, bluish black, with a bloom, 1.1–1.2 cm × *c.* 5 mm, style absent; ovules 4. *NE Burma*. H2–3. Early summer.

Plants cultivated from the first collection (*Farrer* 1030) have broad leathery leaves with rounded tips. Some plants in cultivation, with narrower and less leathery leaves, have been wrongly called *B. coxii*.

5. B. gagnepainii Schneider. Illustration: Botanical Magazine, 8185 (1908) – as B. acuminata; Gardeners' Chronicle **54**: 335 (1913); Gartenflora **81**: 227 (1932).
Open shrub 2–2.5 m; young shoots rounded, smooth, yellowish brown; spines 1.2–2 cm. Leaves firm but not leathery, persistent, linear-lanceolate, 3–10 cm ×

5–8 mm, base tapered with leaf-stalk more or less absent, margin wavy, with 4–6 (rarely to 10) pairs of spine-tipped teeth, venation open above, more or less obscure beneath, dull green above, light green beneath. Flowers in simple clusters of 2–5; flower-stalks 1–1.7 cm. Flowers 1.2–1.5 cm across, bright yellow. Berries oblong-ovoid, blue-black, with a bloom, 9–10 × *c.* 6 mm; style absent; ovules 3–4. *China (Hubei, Sichuan)*. H2. Early summer.

Var. **lanceifolia** Ahrendt appears to be the most common variant of this species in cultivation.

B. wisleyensis Ahrendt (*B. triacanthophora* misapplied), which is almost certainly a hybrid of *B. gagnepainii*, differs in the grey bloom on the under-surface of the leaves and the longer flower-stalks, (up to 3 cm).

B. × hybridogagnepainii Ahrendt (*B. × wokingensis* Ahrendt), a hybrid between *B. gagnepainii* and *B. candidula*, differs in the densely warty stems, the shorter, broader leaves which have a slight bloom underneath and solitary flowers and berries that are borne on very short stalks.

B. × chenaultii Ahrendt (*B. gagnepainii × B. verruculosa*) differs from *B. gagnepainii* in its slow-growing habit with arching branches and in its glossy leaves.

6. B. verruculosa Hemsley & Wilson. Illustration: Botanical Magazine, 8454 (1912); Journal of the Royal Horticultural Society, **37**: 242, f. 151 (1912); Gartenflora **81**: 226 (1932).
Dense shrub *c.* 1 m; young shoots rounded, strongly papillose, olive green; spines 1–1.5 cm. Leaves leathery, persistent, obovate to elliptic, 1–2 cm × 5–10 mm, base tapered into a stalk 1–3 mm, margin with 2–4 pairs of fine spine-tipped teeth, venation open, upper surface shiny, lower surface with a white bloom. Flowers solitary; flower-stalks 4–9 mm, reddish. Flowers *c.* 2 cm across, yellow. Berries oblong-ovoid, 9–12 × 6–7 mm, dark purple, with a whitish bloom, style absent; ovules 4–5. *China (Sichuan)*. H2. Late spring.

The hybrid **B. × frikartii** Schneider (*B. verruculosa × B. candidula*) is sometimes cultivated.

B. × bristolensis Ahrendt (*?B. verruculosa × B. calliantha*) differs in its finely warty (not papillose) stems, larger leaves (2–3.5 cm × *c.* 1.8 cm) and flower-stalks (*c.* 1.5 cm). H4. Flowering period not known.

A garden hybrid, **B. × interposita**

Ahrendt (*?B. verruculosa × B. hookeri* var. *viridis*) differs from *B. verruculosa* in its sparsely and finely warty stems, slightly narrower and larger leaves, to 2.5 × 1.2 cm, longer flower-stalks (*c.* 1.5 cm), and apparently smaller flowers, *c.* 1.5 m across. *Garden origin*. H3. Early summer.

7. B. candidula Schneider.
Dwarf shrub 60–100 cm, forming dense mounds; young shoots rounded, finely warty, green; spines 6–13 mm. Leaves leathery, persistent, narrowly elliptic to ovate, 1.2–2.5 cm × 4–8 mm, base tapered with leaf-stalk more or less absent, margins strongly recurved, with 2–4 pairs of spine-tipped teeth, venation more or less obscure, dark green and shiny above, with a white bloom beneath. Flowers solitary; flower-stalks 2–8 mm. Flowers *c.* 1.5 cm across, yellow. Berries ovoid, 8–9 × 4–5 mm, blackish purple, with a slight bloom, style absent; ovules 4–5. *China (Hubei)*. H2. Late spring–early summer.

Fairly common in cultivation.

8. B. insignis Hooker & Thomson.
Open shrub 1–2 (rarely to 2.5) m; young shoots rounded, smooth, dark red; spines to 1 cm, though usually absent. Leaves thinly leathery, flexible, persistent, elliptic to lanceolate, 7–22 × 1.5–5 cm, acuminate, base tapering into a leaf-stalk 3–5 mm; margin with 10–20 (rarely to 30) pairs of coarse, spine-tipped teeth, venation open, upper surface green, shiny, lower surface paler, shiny. Flowers in simple clusters of 6–30; flower-stalks 5–20 mm. Flowers *c.* 1.5 cm across, yellow. Berries oblong to spherical, 5–9 × 5–6 mm, black, without a bloom, style short; ovules 4. *Himalaya (Sikkim) to China (SE Xizang) and NE Burma)*. H4. Spring.

Subsp. **insignis**. Flower-stalks 5–15 mm, thick; berries oblong, 8–9 × 5–6 mm, ovules 4.

Subsp. **incrassata** (Ahrendt) Chamberlain & Hu (*B. incrassata* Ahrendt). Illustration: Gardeners' Chronicle **115**: 259 (1944). Flower-stalks 1–2 cm, slender though thickened below berries; berries spherical, 5–6 mm in diameter; ovules 5–7.

A variant of subsp. *incrassata* with few (4–8) flowers in each cluster, is known as var. **bucahwangensis** Ahrendt. Plants apparently intermediate between the two subspecies occur in cultivation.

9. B. hypokerina Airy-Shaw.
Open, sparsely branched shrub 60–250 cm; young shoots rounded, finely warty, dark red. Leaves leathery, persistent, elliptic to

oblong-elliptic, 8–12 × 3–6.5 cm, base tapered with leaf-stalk very short, margin with 6–10 pairs of strong spine-tipped teeth, lateral veins indistinct, dark green above, silvery white beneath. Flowers in simple clusters of 6–12; flower-stalks 7–11 mm (to 1.8 cm in fruit). Flowers 1–1.5 cm across, pale yellow. Berries ellipsoid, c. 7 × 4 mm, black with a dense white bloom, style short; ovules 4. *NE Upper Burma*. H4. Late spring–early summer.

10. B. kawakamii Hayata (*B. formosana* Ahrendt). Illustration: Botanical Magazine, 9622 (1942).

Shrub c. 2 m; young shoots deeply grooved, finely warty, whitish yellow; spines 1–3 cm. Leaves leathery, persistent, elliptic to obovate, 2.2–4.5 (rarely to 6) × 1.1–2.7 cm, base tapered into a stalk 1–2 mm; margins with 4–7 (rarely to 12) pairs of spine-tipped teeth, venation open or web-like, mid-green and shiny above, paler beneath. Flowers in simple clusters of 10–15; flower-stalks 4–6 (rarely to 12) mm, greenish. Flowers 8–10 mm across, rich yellow. Berries ovate, c. 6 × 3–4 mm, blue-black, with a bloom, style short; ovules 4. *Taiwan*. H2. Spring.

11. B. sargentiana Schneider.

Open shrub c. 2 m; young shoots rounded, finely warty, red at maturity; spines stout, 2–6 cm. Leaves leathery, flexible, persistent, oblong to elliptic, 4–10 × 2–3.3 cm, acuminate, base tapered into a winged leaf-stalk c. 5 mm; margin with 15–25 pairs of fine spine-tipped teeth, upper surface green and slightly shiny, with prominent web-like lateral veins, lower surface dull, yellowish green. Flowers in simple clusters of 4–8; flower-stalks 1–2 cm, reddish. Flowers 1–1.2 cm across, pale greenish yellow. Berries oblong-ellipsoid, 6–8 × 4–6 mm, black without a bloom, style absent; ovules 4–5. *China (Sichuan, Hubei)*. H2. Late spring.

12. B. sanguinea Schneider (*B. panlanensis* Ahrendt).

Shrub 2–3 m; young shoots grooved, finely warty, greyish when mature; spines 1.7–4.3 cm. Leaves leathery, persistent, linear-lanceolate, 3.3–6.2 cm × 3–8 mm, base tapered with leaf-stalk more or less absent, margin with 6–10 (rarely to 14) pairs of spine-tipped teeth, venation faint or more or less obscure, mid-green and slightly glossy above, paler beneath. Flowers in simple clusters of 2–7 (rarely to 9); flower-stalks 6–15 (rarely to 20) mm,

red-green. Flowers c. 10 mm across, golden-yellow. Berries oblong, 6–8 × c. 3 mm, reddish at first, turning blue-black, apparently with a style; ovules 3–4 (rarely 2). *China (Sichuan)*. H2. Spring.

13. B. veitchii Schneider (*B. acuminata* misapplied).

Open, spreading shrub; young shoots rounded, finely warty, bright red at first; spines 1.5–2.5 cm. Leaves leathery, persistent, narrowly lanceolate, 5–11 × 1–2 cm, base tapered with leaf-stalk very short, margin wavy, with 10–24 pairs of fine spine-tipped teeth, dull grey-green above, paler and shiny beneath. Flowers in simple clusters of 4–10; flower-stalks 2–3.5 cm. Flowers c. 1.5 cm across, pale straw-coloured flushed brownish. Berries ovoid to ellipsoid, c. 9 × 6 mm, bluish black, with a bloom, style absent; ovules 2–3. *China (Hubei, Guizhou)*. H2. Late spring.

14. B. bergmanniae Schneider.

Shrub 1.6–2 m; young shoots angled, finely warty; spines 9–33 mm. Leaves very leathery, persistent, elliptic to oblong-elliptic, 2.6–4.5 (rarely to 10) cm × 6–15 (rarely to 30) mm, base tapered into a winged leaf-stalk 2–4 mm; margin with 5–10 pairs of coarse spine-tipped teeth, venation open, faint, more or less obscure beneath, upper surface green, shiny, lower surface paler and shiny. Flowers in simple clusters of 10–15; flower-stalks 1–1.5 cm. Flowers c. 1.2 cm across, yellow. Berries oblong-ellipsoid, c. 8 × 6 mm, dark blue, with a bloom, style short; ovules 3–4. *China (Sichuan)*. H2. Early summer.

Var. **acanthophylla** Schneider differs in leaves dull above, with fewer coarse spiny teeth. B. × **wintonensis** Ahrendt, a hybrid of *B. bergmanniae*, differs in its narrow leaves, up to 6 × as long as broad, with fine teeth, and shorter (5–10 mm) flower-stalks.

15. B. lempergiana Ahrendt (*B. cavaleriei* misapplied). Illustration: Botanical Magazine, n.s., 90 (1950).

Shrub c. 2 m; young shoots round, finely warty; spines 2–4 cm, yellowish grey. Leaves leathery, persistent, oblong-elliptic, 4–8 × 1.3–2 cm, base tapered with leaf-stalk very short, margin with 6–11 pairs of fine spine-tipped incurved teeth, dull grey-green above, yellowish green beneath, venation open, faint. Flowers in simple clusters of 3–6; flower-stalks 8–19 mm, red. Flowers 1–1.2 cm across, yellow. Berries oblong-ellipsoid, bluish black, with a strong white bloom, 7–10 × 3–5 mm,

style present; ovules 2–3. *E China (Zhejiang)*. H2. Spring.

16. B. replicata W.W. Smith. Illustration: Botanical Magazine, 9076 (1925); Journal of the Royal Horticultural Society **49**: f. 13 (1924).

Shrub 1–1.6 m; young shoots rounded, smooth, slender, yellowish; spines 7–15 mm. Leaves leathery, persistent, linear-oblong, 2–3.5 cm × 3–5 mm, base tapered with leaf-stalk more or less absent, margin strongly recurved, entire or with 1–3 (rarely to 6) pairs of spine-tipped teeth, more or less inrolled, dull green above, with a white bloom beneath. Flowers in simple clusters of 3–12; flower-stalks 8–12 mm. Flowers c. 9 mm across, bright yellow. Berries oblong, 6–8 × 3–5 mm, red, turning purplish black, without a bloom, style very short; ovules 2. *China (Yunnan)*. H2. Spring.

17. B. taliensis Schneider.

Dense, slow-growing shrub 30–160 cm; young shoots grooved, warty, stout, brownish; spines 1–1.5 cm. Leaves leathery, persistent, narrowly elliptic, 2.5–4 cm × 4–10 mm, base tapered with leaf-stalk more or less absent, margin strongly recurved, entire or with up to 6 pairs of spine-tipped teeth, venation more or less obscure, upper surface dark green and shining, lower surface with a white bloom. Flowers in simple clusters of 2–6; stalks 1–1.5 cm. Flowers c. 10 mm, yellow. Berries oblong, 8–9 × 3–4 mm, bluish black, with a bloom, style absent; ovules c. 4. *China (Yunnan)*. H2. Summer.

Probably not as common in cultivation as the closely allied *B. replicata*.

18. B. manipurana Ahrendt (*B. knightii* misapplied; *B. xanthoxylon* misapplied).

Open shrub 1.5–3 m; young shoots slightly angled, finely warty, brown; spines 1–2.5 cm. Leaves thinly leathery, persistent, oblong-elliptic, 3–7 × 1–3.5 cm, base tapered into a winged leaf-stalk 1–3 mm, margin with 6–12 pairs of fine spine-tipped adpressed teeth, green and shiny on both surfaces, venation web-like. Flowers in simple clusters of 12–15; stalks 7–20 mm, green. Flowers c. 1.2 cm across, bright yellow. Berries oblong, c. 10 × 6 mm, bluish black with a bloom, style absent; ovules 4–5. *NE India (Manipur)*. H2. Late spring.

19. B. pruinosa Franchet.

Shrub 2–3 m; young shoots rounded, finely warty, yellow; spines 1–3 cm, stout. Leaves leathery, stiff, persistent, elliptic to ovate,

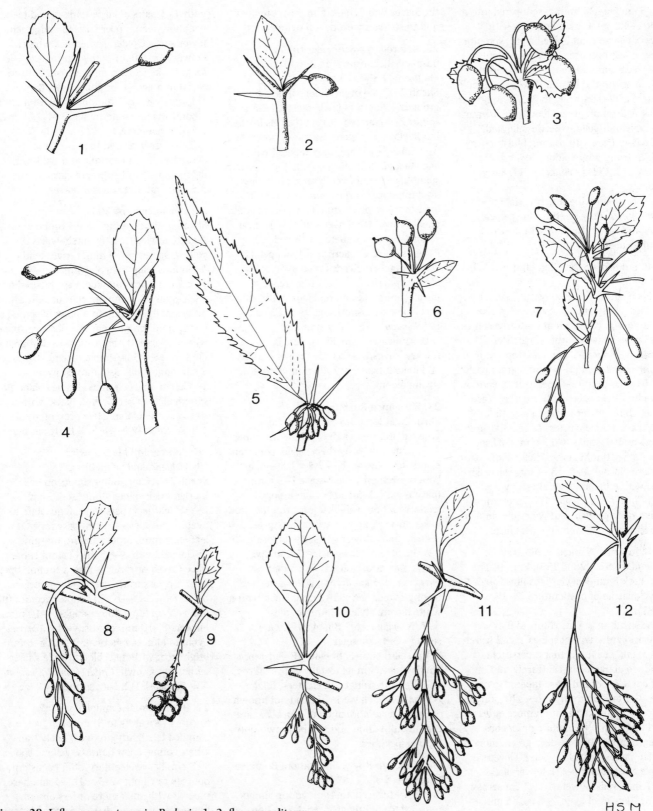

Figure 38. Inflorescence types in *Berberis*. 1–2, flowers solitary;
3–6, flowers in clusters; 7, flowers in umbel-like clusters;
8–9, flowers in simple racemes; 10, flowers in racemes which are
compound below; 11–12, flowers in panicles. 1, *B. ludlowii*.
2, *B. dictyophylla*. 3, *B. diaphana*. 4, *B. yunnanensis*. 5, *B. dumicola*.
6, *B. montana*. 7, *B. umbellata*. 8, *B. silva-taroucana*. 9, *B. aggregata*.
10, *B. jamesiana*. 11, *B. francisci-ferdinandi*. 12, *B. chitria*.

HSM

2.5–5 cm × 6–25 mm, base tapered into a stalk 1–3 mm; margin entire or with 1–6 pairs of coarse spine-tipped teeth, venation more or less obscure or open and faint, mid-green and dull above, paler beneath, with or without a bloom. Flowers in simple clusters of 4–18 (rarely to 25); flower-stalks 7–20 mm, green. Flowers *c.* 10 mm across, lemon-yellow. Berries ellipsoid, 6–9 × 3–5 (rarely to 6) mm, bluish purple, with a strong white bloom, style short; ovules 2–3. *China (Yunnan).* H2. Early Summer.

A species showing much variation in the leaves which may be entire to coarsely spiny and either green or with a strong bloom beneath.

20. B. dumicola Schneider. Figure 38(5), p. 379.

An open shrub to 1.5 m; young shoots rounded, finely and sparsely warty, red when young, yellowish at maturity; spines 6–15 mm. Leaves leathery, persistent, elliptic, 5–9.5 × 1–2 cm, base tapered into a winged leaf-stalk 3–5 mm; margin with 15–30 pairs of spine-tipped teeth, venation web-like, obvious beneath, upper surface green, shiny, lower surface paler, dull. Flowers in simple clusters of 15–30 (rarely as few as 10); stalks 8–15 mm, reddish. Flowers 8–10 mm across, rich yellow. Berries ellipsoid, 7–9 × 3–4 mm, blackish blue, with a bloom, style short; ovules 1–2. *China (W Yunnan).* H4. Late spring.

Like *B. sargentiana* but generally smaller and with more flowers in each cluster.

21. B. julianae Schneider. Illustration: Gartenflora **81**: 228 (1932); Everett, The New York Botanical Gardens illustrated encyclopaedia of horticulture **2**: 401 (1981).

Dense shrub to 2.5 m; young shoots angled, slightly warty; spines 1.2–2 (rarely to 3.5) cm. Leaves leathery, persistent, elliptic to obovate, 2–4.5 (rarely to 7.5) cm × 7–18 mm, base tapered into a stalk 2–3 mm; margin with 4–20 pairs of spine-tipped teeth, venation more or less open and faint above, more or less obscure beneath, upper surface deep green, more or less shiny, lower surface paler. Flowers in simple clusters of 10–20; flower-stalks 9–17 mm, red. Flowers 1–1.2 cm across, yellow. Berries oblong, 7–8 × 3–5 mm, blackish, with a dense white bloom, style short; ovules 1–2. *China (Sichuan, Hubei).* H2. Early summer.

Wild specimens have short leaves, rounded at the apex, and berries with solitary ovules. Cultivated plants differ in

the leaves which have a more acute apex and the berries with paired ovules.

22. B. atrocarpa Schneider (*B. levis* misapplied). Illustration: Botanical Magazine, 8857 (1920).

Shrub 2–3 m; young shoots angled to grooved, smooth to finely warty, yellow; spines 2–4 (rarely to 6) cm. Leaves leathery but flexible, persistent, narrowly lanceolate, 3–7 cm × 5–15 mm, base tapered with leaf-stalk absent, margin with 5–10 pairs of spine-tipped teeth, venation more or less obscure, upper surface mid-green, dull, lower surface paler, dull. Flowers in simple clusters of 4–10; flower-stalks 5–10 mm, yellowish green sometimes tinged red. Flowers *c.* 1.2 cm across, yellow, petals deeply notched. Berries broadly ovoid to almost spherical, *c.* 5 × 4 mm, reddish purple at first, becoming black, with little or no bloom, style short; ovules 2. *China (W Sichuan).* H2. Early summer.

B. soulieana Schneider differs in its thicker leaves, 10–20 flowers per cluster, and oblong fruits, *c.* 7 × 5 mm, with a strong bloom.

23. B. asiatica Roxburgh.

Shrub to at least 3.5 m; young shoots grooved, minutely hairy, yellowish; spines 1–1.5 cm. Leaves leathery, rigid, persistent, elliptic to obovate, 1.8–7.5 × 1.5–3 cm, base tapered into a leaf-stalk 1–5 mm; margin with 2–5 pairs of spine-tipped teeth, venation web-like, less prominent on lower surface, upper surface dark green, slightly shiny, lower surface papillose with a white bloom. Flowers 15–25, in open simple racemes; stalks 1.5–2.5 cm, red. Flowers *c.* 1.5 cm across, yellow. Berries oblong-ovoid, 5–8 × 3–7 mm, black with a white bloom, style *c.* 2 mm; ovules 5–7. *N India, Nepal, Bhutan.* H4. Late spring– early summer.

Var. **clarkeana** Schneider (*B. hypoleuca* Lindley) differs in its usually entire leaves and flowers arranged in clusters. *Bhutan, NE India.* H4. Flowering period not known.

Both var. *asiatica* and var. *clarkeana* are rare in cultivation and can be grown only in mild localities.

24. B. lycium Royle. Illustration: Botanical Magazine, 7075 (1889).

Shrub 1–3 (rarely to 4) m; young shoots rounded to slightly grooved, finely hairy at least at first, yellow, spines 6–20 mm. Leaves leathery, flexible, persistent, narrowly oblanceolate to elliptic, 2–6 cm × 8–11 (rarely to 15) mm, base tapered with stalk more or less absent, margin entire or

with 1–3 pairs of large spine-tipped teeth, venation open, upper surface mid-green, lower surface paler, sometimes whitish. Flowers 11–30, in simple racemes 2–9 cm long; stalks 5–12 mm, greenish. Flowers 8–10 mm across, bright yellow. Berries oblong, 7–8 × *c.* 5 mm, black, with a bluish bloom, style present; ovules 3–4. *NW India.* H2. Early summer.

The closely related **B. parkeriana** Schneider, from Kashmir and Pakistan, differs in its 8–11-flowered racemes to 3.5 cm, and in its thicker leaves.

25. B. glaucocarpa Stapf.

Shrub 2–4 m; young shoots rounded to slightly angled, usually finely warty, yellow; spines 5–10 mm. Leaves leathery, deciduous, elliptic to obovate, 2.5–5 × 1–2.2 cm, apex rounded, base tapered into a leaf-stalk 1–4 mm; margin usually entire, occasionally with up to 4 pairs of spine-like teeth, venation more or less obscure above, web-like beneath, both surfaces dull green. Flowers 15–25 in simple racemes up to 5 cm long; stalks 3–7 mm. Flowers 1–1.2 cm across, yellow. Berries oblong to spherical, 8–9 × *c.* 7 mm, black, with a strong white bloom, style present; ovules 4. *NW India, W Nepal.* H2. Early summer.

26. B. potaninii Maximowicz (*B. leichtensteinii* Schneider).

Shrub 1–1.5 m; young shoots angled, hairless, red; spines stout, 2–4.5 cm. Leaves leathery, persistent, lanceolate to ovate, 2.5–3 × 1–1.3 cm, base tapered with leaf-stalk more or less absent, margin thickened, with 2–6 pairs of stout spine-tipped teeth or rarely entire, venation open, faint, upper surface green, lower surface paler. Flowers 6–12, in simple racemes that are 3–5 cm long; flower-stalks 5–10 mm. Flowers 8–10 mm across, yellow. Berries ovoid to almost spherical, 7–8 × *c.* 6 mm, bright red, without a bloom, style short; ovules not known. *China (Gansu, Shaanxi, NW Sichuan).* H2. Late spring.

27. B. aristata de Candolle.

Shrub 1–2 (rarely to 3) m; young shoots rounded to slightly grooved, finely hairy when young, soon hairless, yellow; spines 1–3 cm. Leaves leathery, stiff, persistent, obovate to elliptic, 2–7 × 1.2–4 cm, base tapered with stalk more or less absent, margin entire or with 1–4 pairs of distant, large, spine-tipped teeth, both surfaces with prominent web-like venation, upper surface mid-green, lower surface white-papillose. Flowers 15–30, in racemes, 4–7 cm long, that are branched below; stalks 8–15 mm.

Flowers 1–1.2 cm across, yellow tinged with orange. Berries oblong-ovoid, 7–8 × c. 7 mm, black with a dense white bloom, style present; ovules 3–5. *N India, Nepal, Bhutan.* H4. Early summer.

This tender species is rare in cultivation. **B. sikkimensis** (Schneider) Ahrendt differs from *B. aristata* in its simple racemes and generally smaller leaves (to 3.8 cm) and from *B. floribunda* in its 3–12-flowered inflorescence.

28. B. floribunda G. Don (*B. aristata* misapplied).
Shrub 2–3 m; young shoots rounded, hairless, becoming pale yellow; spines absent or to 10 mm. Leaves flexible, deciduous, elliptic to ovate-elliptic, 2.5–5 × 1.2–2.4 cm, base tapered with leaf-stalk more or less absent, margins entire or with 5–7 pairs of spine-like teeth, those on new summer shoots sometimes coarsely toothed, both surfaces with a prominent web-like venation, upper surface dull green, lower surface yellow-green, slightly papillose. Flowers 10–25, in simple racemes to 4–6 cm; stalks 5–7 mm. Flowers lemon-yellow, c. 10 mm across. Berries broadly oblong-ellipsoid, 9–12 mm, dark red to purple with a bloom, style short; ovules 4–5. *Nepal.* H2. Late spring.

Closely allied to *B. aristata* and *B. chitria*.

29. B. chitria Ker Gawler. Figure 38(12), p. 379. Illustration: Botanical Magazine, 2549 (1825); Journal of the Royal Horticultural Society **59**: f. 167 (1934).
Shrub c. 4 m; young shoots angled, minutely hairy, reddish brown; spines 1–2 (rarely to 3) cm. Leaves flexible, deciduous, obovate to elliptic, 2–5 × 1–2 (rarely to 3) cm, base tapered with leaf-stalk more or less absent, margin entire or with 1–8 pairs of spine-like teeth, both surfaces with prominent web-like venation, bright green. Flowers 6–20, in panicles 3–5 (rarely to 11) cm long; stalks 8–15 mm, red. Flowers 1.6–2 cm across, pale yellow. Berries ellipsoid, 1.2–1.4 cm × 4–6 mm, dark red, with a slight bloom, style to 2 mm; ovules 4–5. *NW India, W Nepal.* H4. Early summer.

B. chitria is an invalid name but no other name is available. It may be the same as *B. aristata*, of which no authentic material has been seen. Garden plants with the name *B. aristata* belong either to this species or to *B. floribunda*.

B. × macrocantha Ahrendt, probably a hybrid between *B. chitria* and *B. vulgaris*, is cultivated in the United States and may also be available in Germany. It differs from

B. chitria in the leaf-margin with 5–18 pairs of teeth and its hairless young shoots.

30. B. coriaria Lindley. Illustration: Edwards's Botanical Register **27**: t. 46 (1841).
Shrub c. 3 m; young shoots rounded to finely grooved, hairless, pale yellow; spines 1–3 cm. Leaves leathery but flexible, deciduous, narrowly obovate, 2–5 cm × 7–18 mm, base tapered with leaf-stalk more or less absent, margin entire, upper surface dull green, lower surface green and shining, with fine, open, elevated venation. Flowers 10–24, in simple racemes, 4–5 cm; flower-stalks 3–5 mm. Flowers 1.4–1.7 cm across, yellow. Berries oblong, 1.2–1.3 cm × 6–8 mm, red, without a bloom, style present; ovules 4–5. *NW India, W Nepal.* H2. Summer.

Var. **patula** Ahrendt differs in its loose racemes 5–8 cm, and its flower-stalks 6–12 mm.

31. B. umbellata G. Don. Figure 38(7), p. 379. Illustration: Botanical Magazine, n.s., 145 (1951).
Shrub c. 2.5 m; young shoots grooved, warty, deep red; spines 7–17 mm. Leaves slightly leathery, deciduous, oblanceolate to elliptic, 2.5–4.5 × 1.2–1.6 cm, base tapered with leaf-stalk more or less absent, margin entire on young shoots, with 12 (rarely to 20) pairs of spine-tipped teeth on main stems, venation prominent, web-like, upper surface green and slightly shiny, lower surface grey. Flowers 3–6, in long-stalked umbels that are c. 4 cm long; flower-stalks 1.2–1.7 cm. Flowers c. 10 mm across, yellow. Berries oblong, 1–1.1 cm × 4–6 mm, dark red with a purplish bloom, style absent; ovules 2–3. *Nepal.* H3. Late spring.

32. B. angulosa Hooker & Thomson. Illustration: Botanical Magazine, 7071 (1886).
Shrub 1–1.5 (rarely to 3) m; young shoots deeply grooved, minutely hairy, dark brown; spines 5–15 mm. Leaves flexible, deciduous, obovate, 1–2 (rarely to 4) cm × 6–10 (rarely to 18) mm, base tapered into a winged leaf-stalk 1–2 mm, margin entire or with up to 4 pairs of spine-like teeth, venation prominent, open, upper surface shiny, lower surface yellow-green, shining. Flowers solitary; stalks 1–1.5 cm, minutely hairy, reddish. Flowers 1.2–1.5 cm across, yellow. Berries ovoid to almost spherical, 1–1.2 cm × 9–12 mm, bright red, without a bloom, style absent; ovules 6–10. *C & E Nepal, Sikkim, Bhutan.* H2. Late spring.

B. parisepala Ahrendt, which is supposed to differ in its shortly styled berries, with only 4 ovules, in the shorter (5–12 mm) flower-stalks, is almost certainly no more than a variant of *B. angulosa*. The plate of *B. parisepala* in the Botanical Magazine (n.s., 147, 1950) does not agree with Ahrendt's original description as the ovary has 7 ovules.

33. B. ludlowii Ahrendt (*B. capillaris* Ahrendt). Figure 38(1), p. 379.
Shrub 1.3–3 m; young shoots grooved, minutely hairy, purple; spines absent or to 2 cm. Leaves flexible, deciduous, obovate, 2–4 × 1–2 cm, base tapered into a stalk 2–3 mm, margin entire or with up to 6 pairs of spine-like teeth, venation open, prominent, upper surface dull, grey-green, lower surface grey. Flowers solitary, 1.5–1.8 cm across, rich yellow, stalks 2–5 cm. Berries ellipsoid, 1.4–1.8 cm × 8–9 mm, (rarely as little as 10 × 6 mm), dull red, without a bloom, style absent; ovules 5–7. *NE Burma, SW China (Xizang, Yunnan).* H3. Spring.

Doubtfully distinct from **B. macrosepala** Hooker & Thomson, from Sikkim, Bhutan and SE Xizang, a species that is rare in cultivation.

34. B. concinna Hooker & Thomson. Illustration: Botanical Magazine, 4744 (1853).
A low compact shrub c. 1 m; young shoots strongly grooved, more or less smooth, becoming dark red and shiny; spines 1–1.5 cm. Leaves semi-persistent, oblong to obovate, 1.5–2.5 cm × 7–15 mm, base tapered with leaf-stalk more or less absent, margin with 3–5 pairs of spine-tipped teeth, venation open, faint, upper surface deep green and shiny, with open veins, lower surface white and finely papillose. Flowers solitary, 1.2–1.5 cm across, deep yellow; stalks 1.5–2.5 cm. Berries oblong, 1.3–1.6 cm × 6–8 mm, dull red, without a bloom, style absent; ovules 6–8. *Nepal to India (Sikkim).* H3. Early summer.

35. B. tsangpoensis Ahrendt.
A prostrate spreading shrub 15–25 cm high; young shoots angled to grooved, smooth, yellow to deep brown; spines 1–2 cm. Leaves more or less persistent, leathery, obovate, 1.3–2 cm × 6–10 mm, base tapered, margin with 2–5 pairs of small spine-like teeth, upper and lower surfaces dull green with web-like veins. Flowers solitary, 1.5–1.8 cm across, yellow; stalks c. 3 mm, slender. Berries oblong, 1.2–1.3 cm × 9–11 mm, whitish at

first, later bright red; ovules 12–15. *China (SE Xizang), collected only once.* H2. Summer.

Closely resembles *B. concinna* but with a more prostrate habit and leaves with a web-like venation.

36. B. temolaica Ahrendt.
Shrub to 2.5 m with arching stems; young shoots rounded, smooth, dark purple, with a strong whitish bloom at first. Leaves flexible, deciduous, oblong to obovate, 2–4.5 cm × 8–20 mm, base tapered into a stalk 1–2 mm; margin with 3–9 pairs of spreading spine-like teeth, venation web-like, upper surface dull, blue-green, lower surface with a strong whitish bloom. Flowers solitary, *c.* 1.5 cm across, pale yellow; stalks 5–10 mm. Berries narrowly elliptic 1.1–1.4 cm × 6–7 mm, red with a whitish bloom at first, style more or less absent; ovules 6–11. *China (SE Xizang).* H2. Mid-summer.

The whitish bloom on the young stems and leaves makes this a most striking plant. Var. **artisepala** Ahrendt differs, in cultivation at least, in its slender flower-stalks 1–1.5 cm and its narrower leaves, about 3 times as long as broad.

37. B. morrisonensis Hayata. Illustration: Botanical Magazine, 9017 (1924).
Shrub to 2 m; young shoots deeply grooved, smooth, yellowish brown or reddish; spines 1–2 cm. Leaves deciduous, flexible, oblanceolate, 1–3 cm × 3–7 mm, base tapered into a stalk 1–2 mm, margin entire or with 1–3 (rarely to 7) pairs of spine-like teeth, venation web-like, prominent, upper surface yellow-green, shiny, lower surface with a grey bloom at first, later sometimes pale green. Flowers in simple clusters of 2–5; stalks 1.2–2.5 cm. Flowers *c.* 1.2 cm across, pale yellow. Berries spherical to broadly obovoid, 8–9 × 6–7 mm, bright red, translucent, with a slight bloom, style absent; ovules 4–8. *Taiwan.* H2–3. Early summer.

38. B. diaphana Maximowicz
(*B. yunnanensis* misapplied). Figure 38(3), p. 379. Illustration: Botanical Magazine, 8224 (1908) – as *B. yunnanensis*; Sargent, Trees & Shrubs **2**: t. 109 (1917).
Shrub 60–100 cm; young shoots rounded, smooth, yellowish at maturity; spines 1.2–2.5 cm. Leaves flexible, deciduous, obovate to oblong, 1.5–4 cm × 5–15 mm, base tapered into a winged stalk 1–3 mm; margin with 4–12 pairs of spine-like teeth, upper surface dull, grey-green, with elevated, web-like venation, lower surface

with a greyish bloom. Flowers rarely solitary, usually 2–5, in simple or compound clusters; stalks 1.2–2.2 cm. Flowers 1.2–1.5 cm across, yellow. Berries ellipsoid, 1.3–1.6 cm × 6–7 mm, bright red without a bloom, style short; ovules ?6–10. *China (Gansu).* H2. Late spring–early summer.

The plates cited above, the accompanying description and the specimens seen, indicate that the berries are 1.3–1.6 cm long, not 1–1.2 cm as described by Ahrendt.

At least some material in cultivation under the name *B. diaphana*, with dark red fruits and solitary flowers, represents var. **uniflora** Ahrendt.

Confused with the closely allied *B. circumserrata* Schneider, which is said to differ in its broader leaves, with 20–40 pairs of teeth. *B. circumserrata* var. **subarmata** Ahrendt (including var. *occidentalior* Ahrendt) is said to differ from *B. diaphana* in its berries with 3–5 ovules; this may be a variant of *B. diaphana*. *B. diaphana* is also closely allied to *B. aemulans*.

39. B. aemulans Schneider. Illustration: Botanical Magazine, n.s., 179 (1952).
Shrub with erect arching stems 1.5–3 m; young shoots rounded to slightly grooved, finely warty, becoming purple; spines 6–15 mm. Leaves flexible, deciduous, obovate, 1.5–4 × 1.1–1.9 cm, base tapered into a stalk 1–2 mm; margin with 5–12 pairs of spine-like teeth, venation web-like, upper surface dull green, lower surface with a grey bloom. Flowers solitary or 2–4, in simple or compound clusters; stalks 8–30 mm, reddish. Flowers 1.2–1.5 cm across, yellow. Berries ellipsoid, 1.3–1.6 cm × 6–8 mm, red, without a bloom, style short; ovules 7–15. *China (W & N Sichuan).* H3. Late spring.

Closely allied to *B. diaphana* but differing in its taller stature and purple stems.

40. B. tischleri Schneider.
Shrub 2.5–3.5 m; young shoots deeply grooved, more or less smooth, reddish when young, brown at maturity; spines 1–2.5 cm. Leaves deciduous, flexible, obovate to oblong, 2–3.5 (rarely to 5) × 1–1.7 (rarely to 2.4) cm, base tapered into a winged stalk 2–3 mm; margin with 4–12 pairs of small spine-like teeth, venation open, faint, upper surface dull, lower surface with a grey bloom. Flowers 4–15 in loose simple racemes up to 9 cm long; flower-stalks 5–35 mm (usually more than 1.5 cm) long. Flowers *c.* 10 mm across, yellow. Berries ellipsoid, 1–1.2 cm ×

5–6 mm, red with a glaucous bloom, style conspicuous; ovules 3–4. *China (W Sichuan).* H2. Late spring.

Closely allied to *B. aemulans* but with smaller flowers that are more numerous, arranged in a loose raceme; the berries are smaller, with a more obvious bloom.

41. B. virescens Hooker & Thomson (*B. ignorata* misapplied). Illustration: Botanical Magazine, 7116 (1880); Gardeners' Chronicle **93**: 59 (1933).
Shrub 2–3 m; young shoots ridged or grooved, smooth, reddish purple; spines 6–15 mm. Leaves deciduous, flexible, oblong to obovate, 1–3 cm × 4–15 mm, base tapering into a stalk 2–6 mm; margin entire or with up to 4 pairs of spine-tipped teeth, venation open, upper surface dull green, lower surface with a greyish bloom. Flowers 4–8, arranged in umbel-like racemes *c.* 2 cm long (elongating in fruit to 4 cm); flower-stalks 8–15 mm, reddish. Flowers 8–10 mm across, greenish yellow to sulphur. Berries narrowly oblong-ovoid, 8–11 × 3.5–6 mm, reddish with a bloom, style short; ovules 4–5. *E Nepal to W Bhutan.* H3. Early summer.

This species has been confused in cultivation with the black-fruited *B. paravirescens* Ahrendt and *B. ignorata* (see under **47**), with bright yellow flowers.

42. B. dictyophylla Franchet. Figure 38(2), p. 379.
Shrub *c.* 2 m; young shoots slightly grooved, smooth, red, usually with a whitish bloom, at least at first; spines 8–18 mm. Leaves flexible, deciduous, obovate to elliptic, 1–2.2 (rarely to 3) cm × 4–10 mm, base tapered into a stalk 1 mm; margin entire or with up to 5 pairs of obscure spine-tipped teeth, venation more or less open, upper surface dull green, lower surface with a marked whitish bloom. Flowers solitary or paired; flower-stalks 5–15 mm. Flowers *c.* 1.6 cm across, pale yellow. Berries ellipsoid, 1–1.3 cm × 6–8 mm, red, with a whitish bloom, style short; ovules 3–6. *China (Yunnan, W Sichuan).* H2. Late spring.

Plants lacking the characteristic bloom on the shoots and leaves, and apparently with bright yellow flowers, have been called var. **epruinosa** Schneider. Some plants in cultivation under this name are *B. approximata*.

43. B. approximata Sprague. Illustration: Botanical Magazine, 7833 (1902) – as *B. dictyophylla*.
Similar to *B. dictyophylla* but differing in its

more strongly toothed leaves and in its spherical berries, 7–9 mm in diameter, with style up to 1 mm. *China (E Sichuan).* H2. Flowering period not known.

44. B. sibirica Pallas. Illustration: Edwards's Botanical Register **6**: t. 487 (1820). Shrub 50–65 cm; young shoots strongly grooved, minutely hairy at first, often soon hairless, pale yellow-brown; spines 3–10 mm, strongly grooved, with up to 11 prongs. Leaves flexible, deciduous, obovate, 1–2.5 cm × 6–14 mm, base tapered with stalk more or less absent, margin with 4–7 pairs of spine-like teeth, both surfaces with elevated web-like venation, upper surface bright green, lower surface green and shiny. Flowers solitary or rarely in pairs; flower-stalks 7–10 mm. Flowers 8–10 mm across, bright yellow. Berries obovoid, 7–9 × 6–7 mm, dark red without a bloom, style present; ovules 3–5. *USSR (Siberia), N Mongolia.* H2. Flowering period not known.

May no longer be in cultivation.

45. B. mucrifolia Ahrendt. Illustration: Botanical Magazine, n.s., 643 (1972). Dwarf shrub 20–60 cm; young shoots strongly grooved, hairy when young, yellowish; spines 7–20 mm. Leaves thick, deciduous, narrowly elliptic to narrowly obovate, 6–15 × 2–4 mm, base tapered with leaf-stalk more or less absent, margin inrolled, entire or with 1 pair of spine-like teeth, apex spine-tipped, upper surface green, slightly shiny, lower surface dull, veins totally obscured. Flowers solitary or rarely paired, *c.* 9 mm across, bright yellow; stalks 5–10 mm. Berries spherical, often curved, 6–7 mm in diameter, bright red without a bloom, style conspicuous; ovules 4–5. *Nepal.* H2. Spring.

46. B. orthobotrys Aitchison. Shrub 1–1.2 m; branches rounded to slightly angled, reddish; spines 1–1.5 cm. Leaves flexible, deciduous, obovate, 1.5–2.5 (rarely to 3) cm × 7–15 (rarely to 17) mm, base tapered into a winged stalk 2–3 mm; margin with 5–22 pairs of spine-like teeth, both surfaces green, veins prominent, web-like. Flowers 5–15, in simple racemes 2.8–3.5 cm long; stalks 5–10 mm. Flowers *c.* 10 mm across, yellow. Berries oblong, 9–10 × 6–7 mm, dull red, without a bloom, style fairly short or absent; ovules 3–5. *Afghanistan, Pakistan, NW India and adjacent parts of W China.* H2. Early summer.

Var. **canescens** Ahrendt differs in its narrower leaves, 3–4 times as long as broad, that have a grey bloom beneath. Plants in cultivation under the name var. *canescens* as *Stainton, Sykes & Williams 8100* may be var. **rubicunda** Ahrendt.

47. B. johannis Ahrendt. Illustration: Botanical Magazine, n.s., 57 (1949). Shrub 1–2 m; young shoots slightly grooved, more or less smooth, greenish to red; spines 5–8 (rarely to 14) mm. Leaves flexible, deciduous, ovate, 1.5–2 cm × 8–12 mm, base tapered into a sometimes winged leaf-stalk 1–7 mm; margin entire or with up to 5 pairs of spine-like teeth, venation indistinctly open or more or less obscure, upper surface dull green, lower surface with a greyish bloom. Flowers 3–7, in simple racemes 2–3.5 cm long; flower-stalks 8–10 mm. Flowers *c.* 10 mm across, yellow. Berries narrowly oblong-ovoid, 1.1–1.3 cm × 3–4.5 mm, bright red, without a bloom, style absent; ovules 2–4. *China (SE Xizang).* H3. Summer.

B. ignorata, as known in cultivation, originates from Kingdon-Ward seed (5724) from SE Xizang. This plant resembles *B. johannis* but differs in its more dense inflorescence, with up to 12 flowers, and in its broader leaves, 1–1.3 cm × 5–7 mm; it may not belong to *B. ignorata* as described by Schneider, which is based on a plant from Sikkim.

48. B. franchetiana Schneider var. **macrobotrys** Ahrendt. Shrub 1.3–2 m; young shoots slightly grooved, smooth, reddish brown; spines absent or to 2 cm. Leaves flexible, deciduous, narrowly ovate, 1.6–3.5 cm × 5–10 mm, base tapered into a winged leaf-stalk 3–5 mm; margin entire or with 1–4 pairs of spine-like teeth, venation web-like, upper surface dull, lower surface with a white bloom. Flowers 8–14 (rarely to 17), in simple racemes 3–7 cm long; flower-stalks 5–15 (rarely to 23) mm. Flowers 1–1.2 cm across, yellow. Berries ovoid, 8–11 × 4–7 mm, red, probably with a bloom, style short; ovules 2. *China (Yunnan).* H2. Early summer.

Similar to *B. forrestii* and probably to *B. lecomtei*, differing chiefly in the short style on the ripe berries.

Var. **glabripes** Ahrendt, which comes from NW Yunnan, close to the Xizang border, differs in its 3–8-flowered racemes, 2–5 cm long. Var. *franchetiana* is not known in cultivation.

49. B. forrestii Ahrendt (*B. pallens* misapplied).
Differs from *B. franchetiana* var. *macrobotrys*

in its berries that lack a style. *China (W Yunnan).* H2. Late spring.

Doubtfully distinct from *B. franchetiana.*
B. lecomtei Schneider, with 4–8-flowered racemes, has been confused with *B. forrestii.*

50. B. tsarongensis Stapf. Illustration: Botanical Magazine, 9332 (1933). Shrub 60–200 cm; young shoots rounded, warty, red; spines 5–20 mm. Leaves flexible, deciduous, ovate to elliptic, 1.1–2.3 c × 5–8 mm, base tapered into a winged leaf-stalk *c.* 1 mm; margin entire or with up to 4 pairs of spine-like teeth, venation open, upper surface bright green, lower surface with a whitish bloom. Flowers 4–9, in simple umbels or umbel-like racemes that are 1.5–3.5 cm long; flower-stalks 8–15 (rarely to 20) mm. Flowers 8–10 mm across, yellow. Berries oblong to ellipsoid, 8–10 × 5–7 mm, red, without a bloom, style absent; ovules 2–3. *China (NW Yunnan, SE Xizang).* H2. Early summer.

Closely resembling *B. franchetiana* var. *glabripes*, which may be the same as this species. The plate cited above indicates that plants originating from a single batch of wild-collected seed (*Forrest 14290*), may have berries with or without styles.
B. stearnii Ahrendt, which differs in its dark red fruits and shorter petals, was raised from seed (*Forrest 29042*) that originated from a plant thought to be *B. tsarongensis*

51. B. amoena Dunn (*B. leptoclada* Diels). Shrub 60–200 cm; young shoots grooved, slightly hairy, red; spines 6–11 mm. Leaves flexible, deciduous, obovate, 8–16 × 2–4 mm, base tapering with leaf-stalk absent, margin entire or with 1–2 pairs of spine-like teeth, venation web-like, upper surface dull, lower surface with a white bloom, veins visible. Flowers 5–7, in simple umbels or racemes 2.5–4 cm; flower-stalks 5–18 mm. Flowers 8–9 mm across, yellow to orange-yellow. Berries almost spherical, 6–7 × *c.* 5 mm, red, without a bloom, style short; ovules 2–3. *China (Xizang, W Yunnan).* H2–3. Early summer.

Resembling *B. wilsoniae* but differing in its fewer ovules, etc. This species is rare in cultivation.

52. B. yunnanensis Franchet. Figure 38(4), p. 379.
Shrub 1.5–3 m; young shoots strongly grooved, finely warty, red; spines stout, 1.5–2.5 cm. Leaves flexible, deciduous, obovate, 2–4 cm × 6–14 mm, tapered

below into a winged stalk 1–3 mm; margin with 2–6 pairs of spine-like teeth, occasionally entire, venation open, upper surface dull green, lower surface greyish. Flowers 2–4, in clusters; flower-stalks 1.2–1.8 cm. Flowers 1.6–1.8 cm across, golden-yellow. Berries oblong-ovoid, 1–1.4 cm × 5–7 mm, red, without a bloom, style absent; ovules 1–3. *China (W Yunnan).* H3. Early summer.

Most of the material in cultivation under this name is either *B. forrestii* or *B. diaphana*.

53. B. sieboldii Miquel.

Compact shrub 60–100 cm; young shoots slightly angled, smooth, greenish at first, becoming reddish; spines 1–1.5 cm. Leaves flexible, deciduous, lanceolate to diamond-shaped, 2.5–7 × 1–3 cm, margin with 50–150 pairs of fine spine-like teeth, base tapered into a winged stalk 2–3 mm, venation web-like, upper surface bright green, lower surface pale green and shiny. Flowers 3–6, in umbel-like racemes 2–3 cm; flower-stalks 3–7 mm. Flowers *c.* 8 mm across, pale yellow. Berries 5–6 mm diameter, dark red, without a bloom, style short; ovules 2. *Japan.* H3. Early summer.

54. B. zabeliana Schneider.

Shrub 1.7–2 m; young shoots usually finely angled or grooved; spines absent or 3–12 mm. Leaves flexible, deciduous, 2–4 (rarely to 6) × 1.6–2.7 (rarely to 3.8) cm, base tapered into a leaf-stalk 5–22 mm, margin with 10–35 pairs of fine spine-tipped teeth, upper surface with web-like venation, light green, dull, lower surface with a grey bloom at first Flowers 10–20, in simple racemes 3–5 cm; flower-stalks 4–7 mm. Flowers 4–5 mm across, yellow. Berries oblong-obovoid, 9–11 × 4–5 mm, colour and styles not known; ovules 2. *Afghanistan, Pakistan.* H2. Early summer.

Rare in cultivation. Apparently allied to *B. sieboldii.*

55. B. quelpaertensis Nakai.

Shrub 1.6–1.9 m; young shoots grooved, slightly warty, red, becoming yellow; spines stout, 1–1.5 cm. Leaves thick, deciduous, oblong-ovate, 2–6.5 × 1–2.5 cm, margin with 20–35 fine spine-like teeth, base tapered into a stalk 2–6 mm; venation web-like, prominent, upper surface dull, lower surface paler and shiny. Flowers 8–12 (rarely to 15), in simple, racemes 2–4 cm; flower-stalks 4–6 mm. Flowers *c.* 10 mm across, yellow. Berries oblong, *c.* 7 × 4 mm, red, without a bloom, style

absent; ovules 2. *Korea (Cheju Do Island).* H3. Spring.

56. B. virgetorum Schneider.

Shrub *c.* 2 m; young shoots angled, slightly warty, pale yellow; spines solitary, 1–2.5 cm. Leaves flexible, deciduous, oblong-diamond-shaped, 3.5–10 × 1.5–3.5 cm, margin entire though wavy, base tapering into a stalk 5–18 mm; upper surface dull, yellow-green, lower surface green, veins not prominent. Flowers 5–10 (rarely to 15) in racemes 2–3.5 cm that are sometimes more or less umbel-like; flower-stalks 4–10 mm. Flowers *c.* 8 mm across, yellow. Berries oblong, *c.* 8 × 6 mm, red, without a bloom, style absent; ovules solitary. *S China (Jiangxi).* H2. Spring.

A distinctive species on account of its leaves.

57. B. thunbergii de Candolle. Illustration: Botanical Magazine, 6646 (1882); Everett, The New York Botanical Gardens illustrated encyclopaedia of horticulture **2**: 400 (1981).

A dense compact shrub, to *c.* 1 m; young shoots slightly grooved, finely warty or smooth, dark red; spines 5–10 mm. Leaves flexible, deciduous, blade diamond-shaped-ovate, 1–1.5 (rarely to 2) cm × 6–13 mm, base tapered into a winged stalk 5–7 mm; margin entire, venation more or less obscure, upper surface yellow-green, dull, lower surface grey-papillose. Flowers 2–5 (rarely to 12), in umbels 1–2 cm or rarely solitary; flower-stalks 5–10 mm. Flowers *c.* 10 mm across, yellow. Berries elliptic, 7–8 × *c.* 4 mm, red, shiny, without a bloom, style absent; ovules 1–2. *Japan.* H2. Late spring–early summer.

Widely grown, with several well-known cultivars. The following are obtainable in the trade: 'Atropurpurea', with leaves deep purple; 'Atropurpurea Nana', like 'Atropurpurea' but only 40–60 cm high; 'Aurea', leaves lemon- to golden-yellow. For a more complete list of cultivars see Krüssmann, *Handbuch der Laubgehölze* **1** (1976). **B. × media** Grottenberg (*B. × hybrido-gagnepainii × B. thunbergii*) and **B. 'Mentorensis'** (*B. thunbergii × julianae*) are sometimes cultivated.

58. B. × ottawensis Schneider (B. thunbergii × B. vulgaris).

Like *B. thunbergii* but a little larger, a shrub to *c.* 1.3 m; young shoots yellow, leaf-margin entire or with a few spine-like teeth; leaves 1.3–3.5 cm × 6–15 mm; berries *c.* 9 × 5 mm. *Garden Origin.*

Selected variants include 'Superba', a shrub to 2.5 m with bronze-red leaves.

59. B. silva-taroucana Schneider. Figure 38(8), p. 379.

Shrub *c.* 2 m; young shoots grooved, smooth, soon yellow; spines *c.* 1 cm. Leaves flexible, deciduous, oblong-obovate, blade 2.5–5 cm × 8–30 mm, margin entire or with up to 10 pairs of spine-like teeth, base tapered into a stalk 1.2–2 cm; upper surface dull, lower surface with a greyish bloom, venation web-like. Flowers 6–12, in loose, simple racemes 3–7 cm; flower-stalks slender, 1–1.7 cm. Flowers 7–10 mm across, yellow. Berries ovoid, *c.* 10 mm, scarlet, without a bloom, style absent; ovules 2. *China (W Sichuan).* H1. ?Late spring.

60. B. mouillacana Schneider.

Differs from *B. silva-taroucana* in its usually smaller leaves, 1–4.5 cm × 5–23 mm, green and shining above, and without a bloom beneath. *China (W Sichuan).* H1. Spring.

Rare in cultivation.

61. B. fendleri Gray.

Shrub 1–1.5 m; young shoots rounded, smooth, dark red and shiny; spines 3–7 mm. Leaves flexible, deciduous, oblong-obovate, 1.5–4.5 cm × 6–20 mm, base tapered into a winged stalk 4–12 mm; margin entire or with up to 10 pairs of coarse spine-tipped teeth, upper surface bright green, more or less shining, lower surface paler. Flowers 4–8, in simple umbel-like racemes 2–2.5 cm; flower-stalks 3–6 mm. Flowers *c.* 8 mm across, with outer segments orange, inner segments light yellow. Berries almost spherical, 5–6 × *c.* 5 mm, red, without a bloom, style absent; ovules 2–4. *SW USA.* H3–4. Late spring.

B. canadensis Miller, from NE USA, differs in its angled stems, leaves with a greyish bloom beneath, and ellipsoid berries, *c.* 8 × 6 mm. Both species are rare in cultivation.

62. B. jamesiana Forrest & W.W. Smith. Figure 38(10), p. 379. Illustration: Botanical Magazine, 9298 (1933).

Shrub 2.5–3.5 cm; young shoots rounded, not or only slightly warty, red; spines stout, 1–4 (rarely to 5) cm. Leaves thick, deciduous, blade broadly ovate to broadly elliptic, 1.7–5 × 1–3 cm, base tapered abruptly into a winged stalk 7–25 mm; margin entire or occasionally with up to 27 pairs of fine spine-like teeth, venation prominent, web-like, upper surface light green, dull, lower surface with a grey bloom. Flowers 20–40, in racemes that are sometimes compound below, 7–10 cm;

flower-stalks 6–10 mm. Flowers 5–7 mm across, golden-yellow. Berries spherical to ovoid, 6–11 × 4–9 mm, creamy white at first, finally light red, without a bloom, style absent; ovules 2. *China (NW Yunnan)*. H2. Early summer.

63. B. francisci-ferdinandi Schneider. Figure 38(11), p. 379. Illustration: Botanical Magazine, 9281 (1932).
Shrub 2.5–3.5 m; young shoots finely grooved, minutely warty, yellow becoming red; spines stout, 2–4 cm. Leaves flexible, deciduous, blade obovate to elliptic, 2–7 × 1–3.6 cm, base more or less narrowed into a winged stalk 5–15 mm; margin with 9–30 pairs of spine-tipped teeth, venation web-like, upper surface green, more or less shining, lower surface pale green. Flowers 20–35 (rarely to 40), in racemes 5–12 cm, compound below; stalks 5–14 mm. Flowers 6–8 mm across, yellow. Berries ovoid-ellipsoid, 7–10 × 3–5 mm (10–12 × 7–9 mm, according to Ahrendt), scarlet, without a bloom, style short; ovules 2. *China (W Sichuan)*. H1. Summer.

The specimens seen do not agree with Ahrendt's description with respect to the size of the fruit.

64. B. poiretii Schneider.
Shrub *c.* 2 m; young shoots grooved, warty, reddish brown; spines stout, 5–9 mm. Leaves flexible, deciduous, narrowly oblanceolate, 1.5–4 cm × 3–8 mm, margin entire, venation open, both surfaces dull green; leaf-stalks absent. Flowers 11–18, in narrow, spike-like, simple racemes; stalks 3–6 mm. Flowers 5–6 mm across, pale yellow. Berries oblong, 7–11 × 3–5 mm, bright red, without a bloom, style absent; ovules solitary. *NE China, Mongolia, Soviet E Asia*. H1. Early summer.

Rare in cultivation; plants grown as *B. chinensis* are sometimes this species.

65. B. lepidifolia Ahrendt.
Differs from *B. poiretii* in the 4–8-flowered, simple racemes and in the berries that are black with a bluish bloom, and with short styles. *China (NW Yunnan, SW Sichuan)*. H2. Late spring–early summer.

Wild-collected material has leaves somewhat broader than those of *B. poiretii* (4–5 × as long as broad). Plants in cultivation can have very narrow leaves, up to 10 times as long as broad. This species is apparently rare in cultivation.

66. B. vernae Schneider. Illustration: Botanical Magazine, 9089 (1926).
Shrub 1–1.3 m; young shoots grooved, smooth, becoming dark red; spines single, stout, 2–4 cm. Leaves flexible, deciduous, oblanceolate, sometimes narrowly so, 1.5–4 cm × 4–12 mm, base tapered into a leaf-stalk 2–6 mm, margin entire or with up to 6 pairs of coarse spine-tipped teeth, venation open, web-like, both surfaces green. Flowers 15–35, in spike-like, simple racemes 2–4 cm; stalks 1–3 mm. Flowers *c.* 3.5 mm across, golden yellow. Berries spherical, 3.5–4.5 mm in diameter, red, without a bloom, style absent; ovules 1–2. *China (Gansu, N Sichuan)*. H1. Late spring–early summer.

Similar to *B. poiretii*.

67. B. brachypoda Maximowicz.
Illustration: Maximowicz, Flora Tangutica, t. 7 (1889).
Shrub *c.* 1 m; young shoots grooved, finely hairy, soon becoming smooth and yellow-grey; spines 1–3 (rarely to 4) cm. Leaves flexible but firm, deciduous, blade oblong-elliptic, 1.6–4.5 (rarely to 7.5) × 1.3–2.9 (rarely to 3.3) cm, base tapered into a winged, hairy leaf-stalk 5–10 mm, margin with 25–40 pairs of spine-tipped teeth, venation web-like, prominent, upper surface light green, dull, hairy at first, lower surface green, with longer and more persistent hairs. Flowers 20–30, in spike-like, simple racemes 7–12 cm; stalks 2–4 mm, hairy. Flowers *c.* 8 mm in diameter, outer parts red, inner parts yellow, inner sepals *c.* 4.5 mm, petals *c.* 5 mm. Berries oblong, *c.* 9 × 5 mm, blood-red, shining, without a bloom, style short; ovules 1–2. *China (Gansu, Shaanxi)*. H2. Summer.

It is not certain that plants in cultivation under the name *B. brachypoda* actually belong to this species. Plants collected by Rock in Gansu differ significantly in the structure of the flowers (petals shorter than inner sepals) and the berry shape (almost spherical, *c.* 6 × 5 mm). Furthermore, *B. brachypoda* has been confused with *B. mitifolia*. This species is, however, rarely grown.

68. B. gilgiana Fedde.
Differs from *B. brachypoda* in its red-brown to purple mature shoots, smaller leaves, 1–4 cm × 3–15 mm, petals that are slightly longer than the sepals, and ovoid berries with a slight bloom. *China (Shaanxi)*. H2. Flowering period not known.

Closely allied to *B. brachypoda*, sharing with it the hairy leaves and flower-stalks; possibly confused with it in cultivation. Only rarely grown.

69. B. mitifolia Stapf. Illustration: Botanical Magazine, 9236 (1931).
Shrub to 2.5 m; young shoots finely grooved, persistently finely hairy, reddish, soon becoming yellow; spines stout, 1.5–3 cm. Leaves flexible, papery, deciduous, blade oblong-ovoid, 1.7–9.2 (rarely as little as 1) cm × 6–33 mm, base tapered into a slightly hairy leaf-stalk 1–1.5 cm, margins with 15–45 pairs of spine-tipped teeth, venation open, web-like, upper surface yellow-green, slightly shiny, more or less hairless, lower surface green, persistently hairy. Flowers 20–30, in narrow spike-like simple racemes that are 5–9 cm long; flower-stalks 3–4 (rarely to 6) mm, hairy. Flowers *c.* 10 mm across, outer parts red, inner parts yellow, petals shorter than inner sepals. Berries oblong, 1–1.2 cm × 3–7 mm, red, without a bloom, style absent; ovules 1–2. *China (W Hubei)*. H2. Early summer.

Plants in cultivation under the name *B. brachypoda* may be this species.

70. B. vulgaris Linnaeus (*B. lucida* Schrader). Illustration: Ross-Craig, Drawings of British plants 2: t. 1 (1948).
Shrub 1–2 m; young shoots angled or grooved, smooth, yellow. Spines 5–25 mm. Leaves flexible, deciduous, blade ovate to elliptic, 2.5–5 cm × 7–35 mm, base tapered into a stalk to 8 mm, margin with 9–30 (rarely to 36) pairs of fine spine-tipped teeth, venation web-like, both surfaces green, dull. Flowers 9–25, in simple racemes 4–6 cm; flower-stalks 6–12 mm. Flowers *c.* 10 mm across, yellow. Berries ellipsoid, 9–14 × 4–6 mm, red, dull, without a bloom, style absent; ovules 2. *Europe, Turkey, USSR (Caucasus)*. H1. Late spring–early summer.

Very common in cultivation. Several selected and named variants are available.

71. B. koreana Palibin. Illustration: Gardeners' Chronicle 86: 409 (1929).
Compact shrub 1–1.3 m; young shoots grooved, smooth, dark red; spines 8–15 mm, the upper often strongly flattened in the lower half, often more or less leaf-like. Leaves flexible, deciduous, blade oblong-ovate to oblong-elliptic, 2.8–6.5 × 1.2–3.3 cm, base tapered into a stalk up to 8 (rarely to 14) mm long, margin with 10–20 pairs of weak spine-like teeth, venation web-like, often red at first, upper surface light green, dull, lower surface with a grey bloom. Flowers 12–20, in simple racemes 4–6 cm; flower-stalks 4–6 mm. Flowers *c.* 10 mm across, yellow. Berries spherical, 7–8 mm in diameter, red,

shiny, style absent; ovules 1–2. *Korea*. H1.
Late spring–early summer.

72. B. amurensis Ruprecht.
Differs from *B. koreana* in the spines that
are not flattened, and oblong berries,
c. 10 × 6 mm, red with a slight bluish
bloom. *Soviet E Asia, N China, Korea*. H2.
Early summer.

Rare in cultivation.

73. B. regeliana Schneider (*B. japonica*
(Regel) Schneider).
Shrub *c.* 2 m; young shoots grooved,
probably smooth, pale yellow-grey; spines
1–2 cm. Leaves flexible, deciduous, blade
ovate, 3–6 × 1.5–4 cm, base tapered into a
stalk 7–15 mm, margin with 25–40 pairs
of small spine-tipped teeth, venation finely
web-like, upper surface dull green, lower
surface with a grey bloom. Flowers 10–18
(rarely 5), in open, simple racemes 3–6 cm;
flower-stalks 4–9 mm. Flowers 8–10 mm
across, yellow. Berries oblong, 1–1.1 cm ×
4–5 mm, red with a slight bluish bloom at
base, style absent; ovules 2. *Japan*. H2.
Summer.

Plants in cultivation under the names *B.
koreana* and *B. sieboldii* are sometimes this
species.

74. B. henryana Schneider.
Shrub 2–2.5 m; young shoots finely
grooved, smooth, brown; spines *c.* 7 mm
(on mature shoots up to 3 cm). Leaves
flexible, deciduous, elliptic-obovate,
1.5–4.2 cm × 8–18 mm, tapered into a
minute leaf-stalk, margin more or less
entire or with up to 20 pairs of hooked
spine-like teeth, venation web-like, upper
surface light green, dull, lower surface with
a greyish bloom. Flowers 10–20, in loose,
simple racemes 3–6 cm; flower-stalks
5–10 mm. Flowers 1–1.2 cm across,
yellow. Berries ellipsoid, *c.* 9 × 6 mm, red,
with a slight bloom, style short; ovules 2.
*China (W Hubei, adjacent parts of Sichuan,
Shaanxi)*. H2. Late spring.

Rare in cultivation.

75. B. dielsiana Fedde.
Shrub 2–3 m; young shoots finely angled
to grooved, warty, becoming dark red;
spines 8–30 mm (rarely as little as 3 mm).
Leaves flexible, deciduous, blade lanceolate
to obovate-lanceolate, 1.5–5 (rarely to
9) cm × 7–22 mm, apex attenuate, base
tapered into a stalk *c.* 1 mm, margin entire
or with up to 19 pairs of small, incurved,
spine-tipped teeth, venation web-like, upper
surface deep green, dull, lower surface at
first with a grey bloom. Flowers 10–20, in
narrow simple racemes 5–7 cm; flower-

stalks 3–5 mm. Flowers *c.* 6 mm across,
yellow. Berries oblong, 8–9 × 4–5 mm, red,
without a bloom, style short; ovules 2.
China (Shaanxi, Henan). ?H1. Spring.

76. B. beaniana Schneider. Illustration:
Botanical Magazine, 8781 (1918).
Dense shrub *c.* 2 m; young shoots grooved,
finely warty, red-brown at first, soon
becoming grey; spines 1.5–2 cm. Leaves
flexible, deciduous, narrowly elliptic to
ovate-elliptic, 1.4–5 cm × 3–13 mm, base
tapered into a stalk 1–3 mm, margin with
5–12 pairs of spine-like teeth, both surfaces
with a web-like venation, at least when
dry; upper surface slightly shiny, deep
green, lower surface with a grey bloom.
Flowers 10–20, in loose panicles 2–4 cm;
flower-stalks 6-15 mm. Flowers 8–9 mm
across, deep yellow. Berries ovoid, 9–10 ×
5–6 mm, purplish red with a dense bloom,
style absent; ovules 3–4. *?China
(W Sichuan)*. H2. Summer.

This species is only known in cultivation
and was raised from seed collected in China
by Wilson.

77. B. aggregata Schneider. Figure 38(9),
p. 379. Illustration: Botanical Magazine,
8722 (1917); Gartenflora **84**: 267 (1935).
Dense shrub 1–1.5 m; young shoots
strongly grooved, minutely hairy, becoming
pale yellow; spines slender, 8–15 mm.
Leaves flexible, deciduous, oblong-ovate,
8–30 × 4–11 mm, base tapered with stalk
absent, margin with 1–5 pairs of minute
spine-like teeth, both surfaces with a
prominent web-like venation, upper surface
dull, lower surface with a grey bloom.
Flowers 10–30 in dense panicles 1–3 cm,
flower-stalks 2–4 mm. Flowers *c.* 7 mm
across, pale yellow. Berries almost
spherical, 6–7 mm in diameter, pale red
with a slight bloom, style short, ovules
2. *China (Gansu, Sichuan)* H2. Summer.
Similar to *B. prattii*.

78. B. prattii Schneider (*B. aggregata*
Schneider var. *prattii* (Schneider) Schneider;
B. polyantha Hemsley misapplied).
Illustration: Botanical Magazine, n.s., 286
(1957).
Shrub 2–3 m, young shoots grooved, finely
hairy, becoming pale yellow, spines
5–15 mm. Leaves slightly leathery,
deciduous, obovate, 1.4–3 cm × 6–16 mm,
base tapered into a stalk 2–5 mm, margin
with 3–8 pairs of spine-tipped teeth, both
surfaces with a fine web-like venation,
upper surface slightly shiny, lower surface
with a grey bloom. Flowers 20–80, in
cylindric panicles 7–8 (rarely to 20) cm;
stalks 2–5 mm. Flowers *c.* 5 mm across,

yellow. Berries spherical, 3–6 × 2–6 mm,
salmon-red, without a bloom, style
conspicuous; ovules 2. *W China (Sichuan)*.
H1. Summer.

Similar to *B. aggregata* which differs in its
panicles 3 cm, with 10–30 flowers.
Var.**recurvata** Schneider apparently differs
only in its curved fruit-stalks.

79. B. arido-calida Ahrendt.
Shrub 1.3–1.5 m; young shoots grooved,
finely hairy, reddish at first, becoming pale
yellow; spines slender, 8–13 mm. Leaves
flexible, deciduous, elliptic-obovate,
1–2 cm × 5–10 mm, base tapered with
stalk absent, margin with 2–6 pairs of
spine-tipped teeth, both surfaces with a
web-like venation, upper surface yellow-
green, dull, lower surface with a greyish
bloom. Flowers 6–10 (rarely to 14), in
clusters or simple contracted racemes
c. 1.5 cm; flower-stalks 2–4 mm. Flowers
c. 7 mm across, yellow. Berries oblong-
ellipsoid, 4–6.5 × 3–5 mm, red, style short;
ovules 2. *China (Gansu)*. H2. Summer.
Allied to *B. wilsoniae*.

80. B. wilsoniae Hemsley. Illustration:
Botanical Magazine, 8414 (1912); Journal
of the Royal Horticultural Society **33**:
f. 105 (1908).
Shrub 30–100 cm, to 2 m across; young
shoots grooved, minutely hairy, dark red,
spines slender 7–20 mm. Leaves flexible,
semi-persistent, obovate-spathulate,
1–2.3 cm × 3–6 mm, base tapered, stalk
absent, margin entire, both surfaces with a
web-like venation, upper surface dull, lower
surface with a greyish bloom. Flowers 4–7
(rarely to 20), in clusters (rarely in short
panicles); flower-stalks 2–7 mm. Flowers
6–10 mm across, light yellow. Berries
spherical, 4–6 mm in diameter, salmon-
pink without a bloom, style short; ovules
3–5. *China (W Sichuan, Yunnan)*. H2. Early
summer.

This species is often cultivated and
several varieties are recognised including
var. **parvifolia** (Sprague) Ahrendt (*B.
parvifolia* Sprague) with all its parts smaller,
leaves 4–8 × 1–2 mm, flowers *c.* 7 mm
across, berries *c.* 4 mm diameter; var.
subcaulialata (Schneider) Schneider, with
smooth stems and bluish green leaves; var.
stapfiana (Schneider) Schneider, with
leaves that have a more marked bloom on
the under-surface and ellipsoid berries.
B. coryi Veitch, which, in cultivation,
apparently differs in its flowers arranged in
panicles 3–5 cm (wild-collected specimens
have flowers in clusters), is doubtfully
distinct from *B. wilsoniae*.

81. B. × rubrostilla Chittenden. Illustration: The Garden **80**: 563 (1916).
Shrub 1–1.5 m, young shoots grooved, finely hairy, becoming dark red, spines 1.5–2.5 cm. Leaves flexible, deciduous, obovate, 1.6–3 cm × 8–12 mm, base tapered, stalk absent, margin with up to 6 pairs of spine-tipped teeth, venation web-like, upper surface bright green, dull, lower surface grey-blue. Flowers 2–4, in umbellate racemes 2–2.5 cm; flower-stalks 4–7 mm. Flowers 1.8–2 cm across, yellow. Berries ovoid, bright red, *c.* 1.5 cm × 9 mm, without a bloom, style absent; ovules 4–5. *Garden Origin.*

A hybrid raised in cultivation with *B. wilsoniae* as one parent. H2. Summer.

'Chealii' is a variant with dark purple berries, 1–1.1 cm long. 'Crawleyensis' has larger racemes 3–6 cm long, with berries 1.5–1.8 cm × 5–6 mm, bright red. 'Cherry Ripe', with berries that are creamy-white when young, becoming cherry-red, was raised from seed of 'Crawleyensis' and is grown for its spectacular berries.

82. B. × carminea Ahrendt (*B. wilsoniae × ?B. aggregata*).
Differs from *B. wilsoniae* in its relatively narrower leaves, 1.5–3.5 cm × 4–8 mm, the margins with 2–3 pairs of spine-like teeth; flowers 10–16, in the open panicles to 5 cm long, slender flower-stalks, to 5 (rarely to 7) mm, and ovoid berries, 8–9 × *c.* 7 mm. Several named cultivars of this hybrid have been grown. Of these 'Barbarossa' is the most commonly available.

83. B. gyalaica Ahrendt. Illustration: Botanical Magazine, n.s., 22 (1948).
Shrub 2–3 m; young shoots grooved, hairless to minutely hairy, brown; spines 5–13 mm. Leaves flexible, deciduous, obovate to elliptic, 8–25 × 4–10 mm, tapered into a stalk 1–3 mm, margin entire on new shoots (sometimes with up to 4–7 pairs of spine-like teeth on old shoots), both surfaces with open web-like venation, upper surface dull, lower surface with a greyish bloom. Flowers 20–30, in open panicles 3–7 (rarely to 16) cm; flower-stalks 1–2 mm, as long as bracts. Flowers *c.* 10 mm across, yellow. Berries oblong-ovoid, 9–12 × 4–5 mm, black with a bluish bloom, style absent; ovules 3–5. *China (SE Xizang).* H1. Summer.

The closely related **B. taylorii** Ahrendt differs in its leaves with a close web-like venation and berries with solitary ovules.

84. B. sherriffii Ahrendt.
Differs from *B. gyalaica* in its consistently hairless shoots and flower-stalks, 10–25-flowered panicles, bracts that are only half as long as the flower-stalks (which are 2–5 mm) and the smaller, berries (*c.* 7 × 3–4 mm) with 1–2 (or rarely 3) ovules. *China (SE Xizang).* H2. Early summer.

85. B. hispanica Boissier & Reuter.
Stiffly erect shrub to 1.5 m; young shoots grooved, with black pustules, dark red when mature; spines stout, 1–2 cm. Leaves flexible, deciduous, elliptic to obovate, 1–2.5 cm × 5–10 mm, base tapered into a stalk 1–2 mm, margin usually entire, occasionally with up to 6 pairs of spine-like teeth, both surfaces with open web-like venation, green, slightly shiny. Flowers 6–15, in contracted, simple racemes 1.5–2 cm; flower-stalks 4–8 mm. Flowers *c.* 6 mm across, orange-yellow. Berries ovoid, *c.* 9 × 5 mm, black, usually with a slight bluish bloom, style absent; ovules 2. *S Spain, Algeria, Morocco.* H2. Late spring.

86. B. aetnensis Presl.
Compact shrub 30–70 cm; young shoots grooved, smooth, reddish at first, becoming pale yellow; spines 1–2 cm. Leaves flexible, deciduous, oblong to obovate, 10–35 cm × 5–18 mm, stalk short, margin with 8–10 pairs of spine-like teeth towards the apex, venation prominent, web-like; upper surface grey-green, lower surface green, without a bloom. Flowers 10–20 (rarely as few as 6) in simple racemes mostly 2.4–4 cm (rarely 1.5–7.5 cm); flower-stalks 6–10 mm. Flowers 6–7 mm across, yellow. Berries oblong, 8–9 × *c.* 5 mm, red at first, becoming black, with or without a slight bloom, style absent; ovules 2. *S Italy, Sicily.* H2. Late spring–early summer.

87. B. oblonga (Regel) Schneider.
Loose, spreading shrub to 2 m; young shoots rounded, not or only slightly warty, green; spines stout, 5–15 mm. Leaves papery, deciduous, blade oblong-obovate, 3–6 × 1.3–2.3 cm, base tapering into a stalk 8–15 mm, margin entire or rarely with 1–2 pairs of spine-like teeth, both surfaces with open venation, upper surface grey-green, dull, lower surface with a grey bloom. Flowers 10–25 (rarely to 50) in open panicles 4–7 cm, flower-stalks 5–10 mm. Flowers *c.* 10 mm across, yellow, sepals in 1 series of 3. Berries oblong, 1–1.1 cm × 6–8 mm, with a bluish bloom, style *c.* 1 mm; ovules 3–4. *USSR*

(Pamirs, Tien-Shan). H2. Late spring–early summer.

88. B. heteropoda Schrenk.
Shrub *c.* 2 m; young shoots finely grooved at first, becoming rounded, smooth, dark red; spines 5–10 mm. Leaves flexible, deciduous, obovate to elliptic, 2–4 cm × 1–2.7 (rarely to 6 × 4) cm, base tapered into a stalk 3–10 mm, margin entire or with a few minute spines, both surfaces with a web-like venation, green and shiny, sometimes paler beneath. Flowers 4–9, in clusters or umbellate racemes 1.5–5 cm; flower-stalks 9–17 mm. Flowers *c.* 1.2 cm across, orange-yellow. Berries oblong to spherical, 10 mm in diameter, black with a slight bluish bloom, style absent; ovules 4–5 (rarely 6). *Soviet E Asia, Mongolia, China (Xinjiang).* H2.

89. B. actinacantha Martelli. Illustration: Edwards's Botanical Register **31**: t. 55 (1845).
Shrub *c.* 1 m, with spreading, arching branches; young shoots finely grooved to rounded, smooth, becoming yellow; spines stout, 5–15 mm, with up to 7 branches that are flattened at the base. Leaves usually persistent, more or less leathery, leaf-blade oblong, 1–2.5 cm × 5–10 (rarely to 15) mm, base rounded or tapered with stalk less than 1 mm, margin with 3–4 pairs of spine-tipped teeth, both surfaces with web-like venation, slightly shiny, the upper surface grey-green, the lower surface green. Flowers 2–6, in clusters, flower-stalks 7–10 mm. Flowers *c.* 10 mm across, deep yellow. Berries spherical, 6–7 mm diameter, black with a bluish bloom, style absent; ovules 5–7. *Chile.* H3. Late spring.

90. B. congestiflora Gay.
Shrub 1.6 m with slender arching branches; young shoots slightly grooved, minutely hairy at first, becoming hairless, pale yellow; spines leaf-like, encircling the stems, 5–10 mm broad, with 10–20 spine-tipped teeth. Leaves thinly leathery, persistent, blade ovate to circular, 1–2.5 cm in diameter, base slightly cordate, stalk 2–5 cm, minutely hairy, margin with 6–20 pairs of broadly triangular spine-tipped teeth, both surfaces with an open elevated venation, upper surface grey-green, dull, lower surface with a grey bloom. Flowers 8–10, in umbel-like corymbs 3–4 cm; flower-stalks 2–3 mm. Flowers 5–7 mm across, golden yellow. Berries spherical, 4–5 mm in diameter, black with a blue bloom, style absent; ovules 3–5. *Chile.* H2. Spring–early summer.

The related **B. hakeoides** (J.D. Hooker) Schneider, also from Chile, differs in its hairless stems, more leathery leaves, and in its 8–20-flowered clusters or racemes with flower-stalks 4–7 mm. The two species are remarkable for their more or less circular leaf-blades.

91. B. crispa Gay.
Shrub 60–90 cm; young shoots grooved, minutely hairy, becoming dark red-brown; spines leaf-like, with 5–7 spine-like teeth. Leaves thinly leathery, persistent, oblong to oblong-circular, 5–9 × 2.5–6 mm, base contracted into stalk *c.* 1 mm, margin with 3–4 pairs of spine-tipped teeth, both surfaces with a web-like venation, upper surface dull, lower surface with a grey bloom. Flowers 4–8, in clusters or umbels 6–10 mm; flower-stalks *c.* 2 mm. Flowers *c.* 3 mm across, yellow. Berries spherical, 3–4 mm in diameter, black with a blue bloom, style absent; ovule number not known. *Chile.* H3? Late spring.

Like *B. congestiflora*, but with smaller flowers, berries and leaves.

92. B. empetrifolia Lamarck. Illustration: Edwards's Botanical Register **26**: t. 27 (1840).
Prostrate shrub 15–30 cm; young shoots round, finely hairy, becoming hairless, soon dark red; spines 4–15 mm. Leaves leathery, persistent, linear-elliptic, 10–20 × 1–3 mm, stalk absent, margin strongly inrolled, entire, venation obscure, upper surface dark green, dull, lower surface with a grey bloom. Flowers 1–2; flower-stalks *c.* 8 mm. Flowers 1.2–1.4 cm across, golden yellow. Berries spherical, 6–7 mm in diameter, blackish with a blue bloom, style absent; ovules 8–12. *Chile, Argentina.* H3. Late spring.

93. B. × stenophylla Lindley
(*B. empetrifolia* × *B. darwinii*). Illustration: Revue Horticole, 526 (1913) – as *B.* 'Corallina'; and as *B.* × *irwinii*.
Dense shrub 2–3 m, with arching stems; young shoots rounded, hairy, red; spines 2–5 mm. Leaves persistent, narrowly- to linear-elliptic, 8–20 × 2–3 mm, leaf-stalk absent, margin entire, strongly inrolled, venation obscure, upper surface dark green, shining, lower surface pale green. Flowers to 14, in clusters or simple racemes to 2.5 cm long; flower-stalks 3–6 mm. Flowers to *c.* 1.2 cm across, golden yellow or orange, sometimes red in bud. Berries spherical, 6–7 mm in diameter, blackish with a blue bloom, style conspicuous; ovules 7–10. H2. Spring.

This hybrid is common in cultivation and has itself been used as a parent to produce a number of cultivars including: 'Autumnalis', with golden yellow flowers and notched petals that are longer than the inner sepals, and 'Corallina', with flowers that are reddish in bud, orange when open, and petals that are shorter than the inner sepals.

B. × irwinii Byharrer, which differs in its more compact habit, *c.* 1 m tall, with leaves 3–5 mm broad with 1 pair of spine-like teeth, flowers orange-yellow, is a distinct hybrid of the same parentage as *B.* × *stenophylla*.

94. B. cabrerae Job.
Shrub 1–3.5 cm; young shoots finely grooved, smooth, yellowish; spines 8–10 mm. Leaves flexible, deciduous, obovate, 1.5–3 cm × 8–18 mm, base gradually tapered with stalk absent, margin entire, venation open, upper surface green, dull, lower surface greyish. Flowers 4–6, in open, simple, umbel-like racemes 3–6 cm; flower-stalks 1–1.8 cm. Flowers *c.* 1.8 cm across, yellow to orange. Berries ellipsoid to spherical, 8–10 × 6–8 mm, black with a blue bloom; style conspicuous, up to 2 mm long; ovules *c.* 8. *Argentina.*

This species is apparently rare in cultivation.

95. B. montana Gay. Figure 38(6), p. 379.
Erect shrub 2–3.5 m; young shoots stout, angled to grooved, hairless, red; spines stout, 7–12 mm. Leaves flexible, deciduous, elliptic to obovate, 1–2 cm × 6–11 mm, venation open, faint, leaf-stalk absent, margin entire, both surfaces green, dull, lower surface papillose. Flowers 2–3, in clusters; flower-stalks 1.3–2 cm. Flowers 1.5–2 cm across, yellow and light orange. Berries spherical, 6–8 mm diameter, black with a blue bloom; style 2–3 mm; ovules 6–9. *Chile, Argentina.* H2. Late spring.

Closely allied to *B. cabrerae* but differing in the form of the inflorescence.

96. B. chillanensis (Schneider) Sprague var. **hirsutipes** Sprague. Illustration: Botanical Magazine, 9503 (1937).
Shrub 2–4 m; young shoots slightly angled, minutely hairy, at least at first, grey-brown; spines 4–10 (rarely to 20) mm. Leaves flexible, deciduous, elliptic, 5–20 × 2–7 mm, gradually tapered into a stalk *c.* 1 mm, margin entire, venation open, upper surface green, lower surface paler and dull. Flowers solitary; flower-stalks 5–10 mm, hairy. Flowers *c.* 1.8 cm across, yellow with a pale orange centre. Berries

spherical to ovoid, *c.* 8 mm diameter, black with a bluish bloom, style conspicuous; ovules 6–12. *Chile, Argentina.* H3. Late spring–early summer.

Var. **chillanensis**, which differs in its hairless flower-stalks and apiculate stamens, may also be in cultivation.

97. B. buxifolia Lamarck (*B. dulcis* Sweet). Illustration: Botanical Magazine, 6505 (1880).
Shrub 60–300 cm, compact to spreading; young shoots rounded to angled, usually hairy, dark brown; spines 2–7 (rarely to 15) mm. Leaves leathery, persistent, elliptic to broadly elliptic, 1.7–2.2 cm × 8–14 mm, tapered into a stalk 1–3 mm, margin entire, upper surface with open venation, mid- to dark green, slightly shiny, lower surface greyish, dull, venation more or less obscure. Flowers solitary; flower-stalks 1.4–1.8 cm. Flowers 1.4–1.6 cm across, deep orange. Berries almost spherical, 6–8 mm in diameter, blackish red with a blue bloom, style absent; ovules 6 or more. *Chile, Argentina.* H2. Late spring.

Var. **nana** Usteri is a compact shrub, *c.* 60 cm tall, while var. **buxifolia** usually reaches a height of 2 m.

B. × antoniana Ahrendt (*B. buxifolia* × *B. darwinii*) has leaves with a rigid texture and shiny surfaces as in *B. darwinii*, but margins more or less entire as in *B. buxifolia*. The solitary flowers, and berries with long styles and 8–10 ovules, are more reminiscent of *B. buxifolia*.

98. B. darwinii W.J. Hooker. Illustration: Botanical Magazine, 4590 (1851).
Dense shrub 1.5–2.5 m; young shoots round, densely hairy, becoming brown; spines 5–10 mm. Leaves leathery, thick, persistent, obovate-hexagonal, 1–2 cm × 5–10 mm, apex acute, spine-tipped, base tapered, stalk absent, margin with 1–2 pairs of spine-tipped teeth, at least the upper pair broadly triangular, upper surface dark green, shiny, with open venation, lower surface pale and more or less dull; Flowers 10–30, in simple racemes 3–4 cm; flower-stalks 5–10 mm. Flowers 1–1.2 cm across, orange. Berries spherical, 6–7 mm in diameter, black with a blue bloom, style *c.* 3 mm; ovules 4. *Chile, Argentina.* H3. Spring.

Several cultivars of this species are commonly grown.

99. B. × lologensis Sandwith (*B. darwinii* × *B. linearifolia*).
Intermediate between the parents. Young shoots slightly grooved, slightly hairy, grey-

brown. Leaves to 4 × 1 cm, otherwise similar to those of *B. darwinii*. Flowers 8–12, in clusters or short umbels to 2 cm; flower-stalks to 1.5 cm. Flowers *c.* 1.5 cm across, deep orange-yellow. Berries ovoid, *c.* 9 × 6 mm, black with a bluish bloom, style *c.* 3 mm long; ovules 3–4. *Argentina.* H2. Spring.

A wild-collected natural hybrid more like *B. darwinii* in its foliage but with larger and deeper-coloured flowers that are arranged in more contracted clusters.

100. B. comberi Sprague. Illustration: Kew Bulletin, 177 (1927).

Shrub 30–130 cm; young shoots stout, rounded, hairless, yellowish; spines absent. Leaves leathery, thick, persistent, obovoid-hexagonal, 1.7–3.5 × 1–2.5 cm, apex acute, spine-tipped, base abruptly tapered, stalk absent, margin with 2 pairs of spine-tipped teeth, venation obscure, upper surface pale green, dull, lower surface with a grey bloom. Flowers 2–3, in clusters; flower-stalks 2–3 mm. Flowers 1.4–1.8 cm across, orange-yellow, sepals, petals and stamens 5. Berries not known; ovules 8–10. *Argentina.* H4. Spring.

This species is unique in having 5 petals, sepals and stamens. It is apparently rare in cultivation, possibly due to its lack of hardiness.

101. B. ilicifolia Forster. Illustration: Botanical Magazine, 4308 (1847).

Straggling shrub to 1.5 m in cultivation; young shoots strongly grooved, hairless, brownish red; spines 1–1.5 cm. Leaves leathery, thick, persistent, oblong-elliptic, 1.7–5 × 1.2–2.1 cm, apex acute, spine-tipped, base tapered into a stalk 2–5 mm, margin with 2–5 (rarely to 6) pairs of broadly triangular spine-tipped teeth, venation open, upper surface grey-green, shiny, lower surface paler, shiny. Flowers 4–7, in clusters or contracted umbels that are up to 2 cm long; flower-stalks 1.5–2 cm. Flowers 1.8–2.2 cm across, orange-yellow. Berries obovoid, *c.* 8 × 6.5 mm, black with a bluish bloom, style 2–3 mm long; ovules 3–5. *Chile, Argentina.* H4. Spring.

A tender species that is rarely grown. Some plants in cultivation under the name *B. ilicifolia* are the bigeneric hybrid X **Mahoberberis neubertii** (p. 373).

102. B. linearifolia Philippi. Illustration: Botanical Magazine, 9526 (1938); Gardeners' Chronicle **89**: 333–335 (1931). Loosely branched shrub 1.3–2.5 cm; young shoots angled, smooth, pale brown; spines

7–10 mm. Leaves leathery, persistent, narrowly obovate to linear-elliptic, 2–3.5 cm × 7–10 mm, apex spine-tipped, base tapered, stalk more or less absent, margin slightly inrolled, entire, venation not strongly marked, upper surface green and dull, lower surface grey-green with a slight bloom. Flowers 2–4, in clusters; flower-stalks 1.5–1.8 cm, slender. Flowers *c.* 2 cm across, orange-red to apricot. Berries ellipsoid, 8–10 × 6–7 mm, black with a blue bloom, style 2–3 mm long; ovules *c.* 7. *Chile, Argentina.* H2. Late spring–early summer.

103. B. valdiviana Philippi. Illustration: Botanical Magazine, n.s., 139 (1951).

Sturdy shrub 3–5 m; young shoots rounded, smooth, reddish at first, becoming yellow; spines 1.4–2.5 (rarely to 3.5) cm. Leaves leathery, persistent, on young shoots to 8 × 3 cm, leaves elliptic to oblong-ovate 2–6.5 × 1–3.3 cm, base tapered into a stalk 2–6 mm, margin entire or with up to 4 pairs of minute spine-like teeth, venation more or less obscure, upper surface deep green, slightly shiny, lower surface yellow-green and finely papillate. Flowers 10–25, in narrowly cylindric simple racemes 4–6 cm; flower-stalks, 3–5 mm. Flowers 6–8 mm across, orange. Berries spherical, 6–7 mm diameter, black with a bluish bloom, style *c.* 2 mm; ovules 3–5. *Chile.* H4. Late spring–early summer.

Doubtful and Excluded Species

B. chinensis Poiret.

The application of the name *B. chinensis* as it applies to plants in cultivation is confused. Both *B. poiretii* and *B. crataegina* are represented under this name.

B. suberecta Ahrendt.

Only known in cultivation. Supposed to have been raised from seed collected in China (W Yunnan). Probably a chance hybrid. It is described as only an upright form of *B.* × *rubrostilla*.

B. validisepala Ahrendt.

This species, with var. **primaglauca** Ahrendt, is only known in cultivation and may be a hybrid of *B. yunnanensis*.

5. RANZANIA Ito
W.T. Stearn

Herbaceous perennials with slender rhizomes. Stems with 2 opposite leaves. Leaves divided into 3 shortly stalked, deeply lobed, palmately veined leaflets. Flowers 1–6 in a cluster between the 2 leaves. Outer sepals 3, small. Inner sepals 6, much

larger than outer. Petals smaller than inner sepals, flat, each with 2 nectar-producing glands at the base. Stamens 6, anthers opening from the base by 2 up-rolling flaps. Ovary with numerous lateral ovules and with a large, stalkless stigma. Fruit a berry.

A genus of a single woodland species from Japan. It grows best in a compost of leaf-mould and peat with a little sand, and is propagated by division of the rhizomes or by seed.

1. R. japonica (Ito) Ito (*Yatabea japonica* (Ito) Yatabe). Illustration: Botanischer Jahrbücher **31**: 87 (1902); Bulletin of the Alpine Garden Society **5**; 223 (1937); Botanical Magazine, n.s., 76 (1949); Satake et al., Wild flowers of Japan **2**: pl. 89 (1984).

Plant in flower 20–50 cm, hairless. Leaflets with 3–5 lobes, 8–12 cm long and wide. Flowers-stalks 4–8 cm; flowers drooping, pale purple (lavender-violet), 2–5 cm across. *Japan (Honshu).* H5. Spring.

6. EPIMEDIUM Linnaeus
W.T. Stearn

Herbaceous perennials with rhizomes. Stems leafless or with 1–6 leaves. Leaves with 2–60 leaflets, arranged in 3s or pinnately. Flowers in a raceme or panicle. Outer sepals 4, soon falling. Inner sepals 4, petal-like. Petals 4, flat or extended outwards into a nectar-producing pouch or spur, shorter than or longer than sepals. Stamens 4, anthers opening from the base by 2 up-rolling flaps. Ovary with numerous lateral ovules; style slender with a slightly swollen stigma. Fruit a capsule opening by 2 flaps. Seeds with arils.

A genus of about 25 woodland species in S Europe, NW Africa, SW Asia, the Himalayas, China and Japan. They grow in any good well-drained soil, preferably rich in humus and in partial shade. Propagation is by division of the rhizomes in autumn or late winter.

Individual plants of *Epimedium* and clones derived from them are self-sterile but produce seed when pollinated from other clones, which in gardens usually belong to other species, hence the many interspecific hybrids of garden origin, some deliberately raised, some spontaneous.

Literature: Stearn, W.T., Epimedium and Vancouveria (Berberidaceae), a monograph, *Journal of the Linnean Society, Botany* **51**: 409–535 (1938); Ying, T.-S., On the Chinese species of Epimedium L., *Acta Phytotaxonomica Sinica* **13**: 49–55 (1975); Weaver, R., In praise of

Epimediums, *Arnoldia* **39**: 51–66 (1979); Suzuki, K., Pollination system and its significance on isolation and hybridization in Japanese Epimedium, *Botanical Magazine, Tokyo* **97**: 381–396 (1984).

Flower-stem. Leafless: **1,2,3,5,10,12**.
Leaves. With 2 leaflets: **12,13**. Two on flowering stem: **15–18**.
Petals. Flat, without spur or pouch: **15–18**.

1a. Flowering stems bearing 1 or more leaves or leafy and leafless on the same plant   2
  b. Flowering stems all leafless   18
2a. Flowering stems bearing 1 or more leaves   3
  b. Flowering stems leafy or leafless on the same plant
      5. × warleyense & 10. × versicolor
3a. Flowering stem normally with only 1 leaf, with 2 or several leaflets   4
  b. Flowering stem with 2 (rarely 3) opposite leaves with 3 , 5 or 9 leaflets   14
4a. Flowers 3–5 cm across; spurs of petals fine and slender, much exceeding the inner sepals, to 2.5 cm   5
  b. Flowers 1–2.5 cm across; spurs of petals absent, or, if present, shorter than (or rarely sometimes a little longer than) inner sepals   6
5a. Leaves dying down in autumn
         8. grandiflorum
  b. Leaves evergreen   **11. sempervirens**
6a. Spurs of petals cylindric, blunt, yellowish or yellow; stamens protruding from, or just enclosed by the petals   7
  b. Spurs of petals absent or short and conical or slender and pointed; stamens not protruding, much shorter than petals   13
7a. Spurs of petals with blade reduced or absent; stamens protruding   8
  b. Spurs of petals with distinct blade; stamens not protruding   12
8a. Inner sepals dark crimson or yellowish white tinged with pink but not coppery, slightly longer than the slipper-like petals; filaments *c.* 1 mm; anthers yellow   9
  b. Inner sepals coppery red, twice as long as petals; filaments *c.* 2 mm; anthers greenish   **5. × warleyense**
9a. Flowers 1.8–2.5 cm across; inflorescence sparsely hairy or hairless
         9. × rubrum
  b. Flowers 9–13 mm across; inflorescence glandular   10
10a. Rhizome long, slender; mature leaflets sparsely hairy or hairless beneath,

dying in autumn; inflorescence shorter than the stem-leaf
         4. alpinum
  b. Rhizome compact, stout; mature leaflets finely hairy beneath, evergreen; inflorescence overtopping the stem-leaf   11
11a. Inflorescence 25–35 cm long from stem-leaf, rising much above the leaves; inner sepals rose-tinged or almost white   **6. pubigerum**
  b. Inflorescence 15–30 cm long from stem-leaf, not rising much above the leaves; inner sepals dull red
         7. × cantabrigiense
12a. Inflorescence simple; inner sepals broadly ovate, flat, pale yellow, copper or dull rose   **10. × versicolor**
  b. Inflorescence compound; inner sepals narrowly ovate, boat-shaped, bright crimson   **9. × rubrum**
13a. Petals spurless, flat, white; leaves normally with only 2 nearly spineless leaflets   **12. diphyllum**
  b. Petals spurless or spurred, obovate, white or rose, the spur varying even on one plant from short and conical to slender, sharp and more or less equalling the inner sepals; leaves variable with 2–6 or 9 leaflets
         13. × youngianum
14a. Flowers 2–4 cm across; spurs of petals 1–2.5 cm   15
  b. Flowers 8–10 mm across; spurs of petals 2–4 mm   16
15a. Inner sepals 8–12 mm; petals without basal blades, horn-shaped, 1.5–2.5 cm   **15. acuminatum**
  b. Inner sepals 4 mm; petals with basal blades forming a cup around the stamens, with spur 1–1.5 cm
         14. davidii
16a. Stem-leaves with 9 ovate or broadly ovate leaflets; inner sepals 10 mm
         16. brevicornu
  b. Stem-leaves with 3–9 narrowly ovate to lanceolate leaflets; inner sepals 5–7 mm   17
17a. Leaflets hairy beneath with numerous spreading or curled hairs
         18. pubescens
  b. Leaflets hairless or hairy beneath with short, adpressed hairs   **17. sagittatum**
18a. Petals much shorter than the petal-like, conspicuous inner sepals, blades almost absent, 2–3 mm high; spurs to 4 mm, brown or yellow; stamens protruding   19
  b. Petals conspicuous, often as long as inner sepals, blades distinct, petal-like, forming a cup 5 mm high around

stamens; spurs cylindric-tapering, 6–9 mm; stamens usually not protruding   **10. × versicolor**
19a. Inner sepals at first red, changing to copper; anthers greenish
         5. × warleyense
  b. Inner sepals yellow; anthers yellow   20
20a. Leaves with 3 leaflets (rarely 1), very spiny and wavy at the margins
         2. perralderianum
  b. Leaves with 3, 5 or 9 leaflets, often sparsely spiny or even spineless at the margin   21
21a. Leaves often with 9 leaflets, though sometimes with 5 or 3; petals to 2 mm, spurs scarcely 1 mm
      1. pinnatum subsp. **pinnatum**
  b. Leaves with 3 or 5 (never 9) leaflets; petals *c.* 4 mm, spurs 2 mm   22
22a. Leaflets sparsely spiny or sometimes even spineless (spines 0.1–0.9 mm, averaging 0.5 mm); spurs of petals straight, projecting at about 90° from the blade
      1. pinnatum subsp. **colchicum**
  b. Leaflets usually distinctly spiny at the margin (spines 0.2–2.5 mm), of firmer texture and more evident veining on the upper surface; spur of petal slightly upcurved
         3. × perralchicum

1. E. pinnatum de Candolle.
Plant in flower 20–45 cm. Rhizome long, *c.* 5 mm thick. Leaves all basal; leaflets 3–11, ovate to broadly ovate, tips acute, margins sparsely spiny-toothed, becoming almost hairless beneath, evergreen, to 15 × 11 cm. Flowering stem leafless, ending in a loose, glandular or hairless raceme of 15–30 flowers. Flowers *c.* 1.6–2 cm across. Outer sepals 3–5 mm; inner sepals rounded, yellow, the outer pair broadly ovate, *c.* 8 × 7 mm, the inner pair almost elliptic. Petals to 3.5 mm in total length, with toothed, reduced, yellow blades *c.* 2 mm deep; spurs straight or slightly upcurved, brown, *c.* 2 mm. Stamens protruding, 5–6 mm; anthers yellow.

Subsp. **pinnatum**. Illustration: Fischer & Meyer, Sertum Petropolitanum, t. 1 (1846). Leaves usually with 9 leaflets, sometimes 5 or 11. Petals *c.* 2 mm, spurs *c.* 1 mm. *N Iran, USSR (eastern Caucasus).* H3. Spring–summer.
Rare in cultivation, introduced about 1950.

Subsp. **colchicum** (Boissier) Busch. Illustration: Botanical Magazine, 4456 (1849). Leaves with 3 or 5 leaflets; petals to

3.5 mm. *USSR (western Caucasus), adjacent N Turkey.*

Common in cultivation, introduced in about 1842.

2. E. perralderianum Cosson. Illustration: Botanical Magazine, 6509 (1880).
Plant 15–30 cm. Rhizome creeping, 2–4 mm thick. Leaves all basal; leaflets 3 (rarely 1), ovate to broadly ovate, tips acute, margins wavy, rigid and very spiny-toothed, becoming almost hairless beneath, evergreen, to 10 × 7 cm. Flowering stems leafless, ending in a very glandular raceme of 9–25 flowers. Flowers 1.5–2.3 cm across. Outer sepals 4–5 mm; inner sepals obovate, rounded, yellow, 8–11 × 5–9 mm. Petals 2.5 mm in total length, with toothed yellow blades 2–3 mm deep; spurs brown, upcurved, 1–2 mm. Stamens protruding, 5 mm long; anthers yellow. *Algeria (Chaine de Babors).* H5. *Spring–summer.*

3. E. × perralchicum Stearn.
A group of hybrids between *E. pinnatum* subsp. *colchicum* and *E. perralderianum*, not exactly agreeing with either parent but combining their characteristics in various ways. *Garden Origin.* H3. *Spring–summer.*

4. E. alpinum Linnaeus. Illustration: Smith & Sowerby, English botany 7: 4.438 (1798); Moss, Cambridge British Flora 3: t. 165 (1920).
Plant to 30 cm. Rhizome creeping, 2–4 mm thick. Leaves basal and on the stems; leaflets usually 9, ovate, tips acute or acuminate, margins spiny, membranous, at first finely hairy beneath, but usually becoming almost hairless, variable in size, but to 13 × 8.5 cm. Flowering stem bearing 1 leaf; inflorescence compound, loose, glandular, usually 8–26-flowered, nearly always slightly or much shorter than stem-leaf. Flowers 9–13 mm across. Outer sepals 2.5–4 mm; inner sepals narrowly ovate, blunt or almost acute, boat-shaped, dull garnet red, 5–7 × 3 mm. Petals slightly shorter than inner sepals, slipper-like, cylindric, 4 mm, canary yellow with no basal blades. Stamens protruding, 3 mm long. *S Europe (Italy, Austria, Yugoslavia, Albania).* H4. *Spring–summer.*

5. E. × warleyense Stearn. Illustration: Stoker, A gardener's progress, t. 11 (1938); Journal of the Linnean Society, Botany 51: 520 (1938).
Plant 20–55 cm. Rhizome long, 4–8 mm thick. Leaves basal and on stem; leaflets usually 9 or 5, sometimes 3, ovate or broadly ovate, acute, sparsely spiny-toothed, finely hairy beneath, becoming somewhat leathery and often remaining green all winter, to 13 × 9.5 cm. Flowering stem leafless or with 1 leaf with 5 or 9 leaflets; inflorescence simple or compound, glandular, 10–30-flowered, sometimes with a large secondary raceme in the leaf-axil. Flowers to 1.5 cm across. Outer sepals 3–4 mm; inner sepals ovate-oblong, blunt, coppery red or orange-pink (the yellow ground-colour being conspicuously tinged and veined with red, especially in the opening flower), fading to a yellowish salmon, to 8 × 5 mm. Petals much shorter than inner sepals, yellow; spurs blunt, occasionally red-streaked, scarcely 4 mm long, and blades small, bilobed (not jaggedly toothed) *c.* 3 × 3 mm. Stamens protruding, to 4.5 mm long, anthers greenish. *Garden Origin.* H3. *Spring–summer.*

A hybrid between *E. alpinum* and *E. pinnatum* subsp. *colchicum*.

6. E. pubigerum (de Candolle) Morren & Decaisne. Illustration: Hooker's Icones Plantarum 32: t. 3116 (1927).
Plant 20–50 cm. Rhizome short, stout, compact, *c.* 5 mm or more thick. Leaves basal and on stem; leaflets usually 9, rarely 3, ovate to broadly ovate or almost circular, tips acute, margins sparsely spiny, firm, persistently hairy beneath with crowded, soft, white hairs, evergreen, *c.* 4.5 × 2.5 cm but to 9.5 × 6 cm occasionally. Flowering stem bearing 1 leaf; inflorescence compound, loose, very glandular, usually many-flowered, overtopping the stem-leaf. Flowers 8–12 mm across, pale. Outer sepals *c.* 3 mm; inner sepals narrowly ovate, blunt, boat-shaped, pale rose or nearly white, 5–7 × 2 mm. Petals slightly shorter than inner sepals, slipper-like, cylindric, blunt, 3.5–4 mm, canary yellow with no basal blades. Stamens protruding, 3 mm long. *SE Europe (Bulgaria, European Turkey), Turkey along Black Sea coast to USSR (Georgia).* H4. *Spring–summer.*

7. E. × cantabrigiense Stearn. Illustration: Plantsman 1: 189 (1979).
Plant 30–60 cm. Rhizomes short, to 5 mm thick. Leaves basal and on the stem; leaflets usually 9, sometimes to 17, ovate or broadly ovate, acute or shortly acuminate, margins sparsely spiny, persistently hairy beneath, to 10 × 7 cm. Inflorescence compound, loose, many-flowered, overtopping the stem-leaf, glandular with numerous red-tipped hairs. Flowers *c.* 1 cm across. Outer sepals *c.* 2 mm; inner sepals boat-shaped, blunt, dull red, *c.* 5.5 ×

3.5 mm. Petals slightly shorter than inner sepals, slipper-like, blunt, *c.* 5 mm, pale yellow, with no blades. Stamens protruding, *c.* 3 mm long. *Garden Origin.* H4. *Spring–summer.*

A hybrid between *E. alpinum* and *E. pubigerum* which arose between 1938 and 1950 in St John's College garden, Cambridge.

8. E. grandiflorum Morren (*E. macranthum* Morren & Decaisne).
Plant 12–35 cm. Rhizome long, 3–5 mm thick. Leaves basal and on stem; leaflets 9 or more, narrowly to broadly ovate, tips acute or acuminate, very spiny-toothed, membranous, becoming hairless or sparsely hairy beneath, not evergreen, 3–13 × 2–8 cm. Flowering stem bearing 1 leaf; inflorescence simple or compound, loose, hairless or rarely sparsely hairy, 4–16-flowered, usually overtopping the stem-leaf. Flower 2–4.5 cm across, white, pale yellow, deep rose or violet. Outer sepals 4–5 mm; inner sepals narrowly ovate to lanceolate, acute, flat, 8–18 × 3–6 mm. Petals usually much longer than inner sepals, with distinct, rounded blades 5–8 mm deep; spurs slender, tapering, needle-like, 1–2 cm. Stamens not protruding, 5 mm long. *Japan, N China, N Korea.* H3. *Spring–summer.*

This long-spurred species is variable in flower-colour and for horticultural purposes a number of forms have been distinguished.

Forma **grandiflorum**. Illustration: Edwards's Botanical Register 22: t. 1906 (1836); Maund, Botanist 2: t. 90 (1838); Paxton's Magazine of botany 5: 151 (1838); Satake et al., Wild flowers of Japan 2: t. 91 (1984). Inner sepals white tinged with violet; petals white.

Forma **violaceum** (Morren) Stearn. Illustration: Botanical Magazine, 3751 (1839); Edwards's Botanical Register 26: t. 43 (1840). Inner sepals and petals light violet.

Forma **flavescens** Stearn. Illustration: Satake et al., Wild flowers of Japan 2: t. 90 (1984). Inner sepals and petals clear yellow.

The names *E. koreanum* Nakai, *E. sulphurellum* Nakai, *E. cremeum* Nakai and *E. longifolium* Decaisne apparently belong here.

'Rose Queen'. Illustration: Satake et al., Wild flowers of Japan 2: t. 90 (1984). Inner sepals and petals crimson carmine.

9. E. × rubrum Morren (*E. alpinum* var. *rubrum* (Morren) J.D. Hooker). Illustration: Botanical Magazine, 5671 (1867).

Plant 25–35 cm. Rhizome long, *c.* 3 mm thick. Leaves basal and on stem; leaflets usually 9 or more, ovate or narrowly ovate, acuminate, sometimes bifid or trifid, spiny-toothed, at first hairy beneath, later almost hairless, membranous, often bright red when young, to 14 × 9 cm. Flowering stem bearing 1 (rarely 2) leaves usually with 9 leaflets; inflorescence compound, loose, hairless or sparsely hairy, 10–23-flowered, shorter than, equalling or overtopping the stem-leaf. Flowers 1.5–2.5 cm across. Outer sepals 3–4 × 1.5–3 mm; inner sepals narrowly oblong-ovate, blunt, somewhat boat-shaped, light crimson-carmine or carmine, 1–1.2 cm × 4–5 mm. Petals about as long as or shorter than inner sepals, slipper-like, cylindric, pale yellow or white, tinged with red, to 1 cm, the inflated, blunt tips curving upwards, with slight or distinct rounded basal blades to 4 mm high, tending to enclose the stamens. Stamens not or slightly protruding, to 4 mm long. *Garden Origin.* H3. *Spring–summer.*

A hybrid between *E. alpinum* and *E. grandiflorum.*

10. E. × versicolor Morren.
Plant 20–50 cm. Rhizome 3–5 mm thick. Leaves basal and on the stem; leaflets usually 9, sometimes more or 3 or 5, ovate or narrowly ovate, acute or acuminate, spiny-toothed, sparingly hairy beneath, usually dying in winter in 'Versicolor' (as in *E. grandiflorum*) but some remaining green all winter in 'Sulphureum' and 'Neosulphureum' (as in *E. pinnatum* subsp. *colchicum*). Flowering stem leafless or bearing 1 leaf with 9 leaflets on the same plant; inflorescence usually simple, 8–20-flowered. Flowers to 2 cm across, like those of *E. grandiflorum* but smaller and possessing shorter petal-spurs and broader inner sepals; inner sepals broadly ovate, almost acute or blunt, 1–1.4 cm × 5–8 mm. Petals shorter or slightly longer than inner sepals, yellow, with distinct blades 5–6 mm high; spurs cylindric, slightly upcurved and swollen at the tips. Stamens not usually projecting, to 4.5 mm long. *Garden Origin.*

A group of garden hybrids between *E. grandiflorum* and *E. pinnatum* subsp. *colchicum*; the following are widely grown:
'Versicolor' (*E. versicolor* Morren). Leaflets usually 9, conspicuously mottled or entirely red when young, later green, membranous, dying in autumn, to 10 × 6 cm; inflorescence glandular with 10–20 flowers; inner sepals old rose; petals yellow with red-tinged spurs.
'Sulphureum' (*E. sulphureum* Morren; *E.*

citrinum Baker; *E. pinnatum* var. *sulphureum* (Morren) Bergmans). Leaflets 5–11 (usually 9), green or sometimes red- or brown-mottled, almost leathery, sometimes remaining green all winter, to 8 × 6 cm; inflorescence usually hairless with 8–20 flowers; inner sepals pale yellow; petals usually about as long as inner sepals and brighter yellow.

Probably the most commonly cultivated member of the genus.
'Neosulphureum'. Illustration: Journal of the Linnean Society, Botany **51**: t. 31 (1938). Basal leaves usually with 3, less often with 5 or 9 leaflets; stem-leaves nearly always with 3, more rarely with 5 or 9 leaflets; leaflets narrowly ovate, to 9 × 7 cm; flowering stem sometimes leafless, often bearing 1 leaf with 3 or occasionally 5 leaflets; inflorescence simple, loose, almost hairless, 7–16-flowered; flowers to 2 cm across, pale yellow; inner sepals pale creamy yellow; petals 3–5 mm shorter than inner sepals, with pale lemon-yellow blades 5–6 mm high, 6 mm broad, enclosing the stamens; spurs brownish-tinged, slightly upcurved, 3–4 mm. Stamens 4 mm.

11. E. sempervirens Nakai.
Illustration: Satake et al., Wild flowers of Japan **2**: t. 91 (1984).
Very similar to *E. grandiflorum* but with firmer, evergreen leaflets more glaucous beneath. Flowers white or purplish. *Japan (Honshu).* H4.

12. E. diphyllum Graham (*Aceranthus diphyllus* Morren & Decaisne). Illustration: Loddiges' Botanical Cabinet **19**: t. 1858 (1832); Botanical Magazine, 3448 (1835); Satake et al., Wild flowers of Japan **2**: t. 90 (1984).
Plant 10–30 cm. Rhizome short, compact, 1–2 mm thick. Leaves basal and on the stem; leaflets 2, sometimes 6, usually narrowly ovate, blunt, margins usually almost spineless, membranous, hairy beneath, partly evergreen, 2–5 × 1–3 cm. Flowering stem bearing 1 leaf usually with 2 leaflets; inflorescence simple or the lower stalk 2-flowered, almost hairless, 4–15-flowered. Flowers bell-shaped, drooping, white. Outer sepals 3 mm; inner sepals narrowly ovate, bluntish, spreading horizontally, 6 × 2.5 mm. Petals slightly longer and broader than inner sepals, obovate, rounded, flat, spurless but with a slight central furrow, 7 × 3–4 mm. Stamens not protruding, 3 mm long. *S Japan (Shikoku, Honshu).* H4. Spring–summer.

13. E. × youngianum Fischer & Meyer (× *Bonstedtia youngiana* (Fischer & Meyer) Wehrhahn).
Plant 10–30 cm. Leaves basal and on stem; leaflets 2–9, ovate to narrowly ovate, thin, becoming more or less hairless beneath. Flowering stem with 1 leaf with 2–9 leaflets; inflorescence simple or with the lower stalks 2-flowered, almost hairless, 2–12-flowered. Flowers bell-shaped, drooping, white or rose, 1.6–2 cm across. Inner sepals narrowly ovate to lanceolate, blunt to almost acute, horizontally spreading, 8–11 × 3–5 mm. Petals broader and slightly shorter than inner sepals, obovate, rounded, close, 7–10 × 4–6 mm, spurless or with a short conical projection or a slender, incurving, sometimes subulate spur to 1 cm, frequently varying in length and form in the same inflorescence. Stamens not protruding, 3–4 mm long, anthers 2.5–3 mm.

A group of hybrids between *E. diphyllum* and *E. grandiflorum* which arose in Japan. They differ in habit, leaf and flower colour. In 'Youngianum' the leaf usually has 9 leaflets, but in 'Roseum' and 'Niveum' there are usually 6 leaflets, with, however, leaves with 7, 4, 3 or 2 leaflets sometimes being found on the same plant.
'Youngianum'. Illustration: Botanical Magazine, 3745 (1839); Maund, Botanic garden **12**: t. 281 (1848). Plant 15–30 cm; leaflets usually 9, sometimes 3, narrowly ovate or ovate, tips abruptly long-acuminate, margins sparsely or distinctly spiny-toothed, 2–8 × 1–5 cm; flowering stem with 1 leaf with 9 leaflets; inflorescence with 3–8 flowers, much shorter than the stem-leaf; flowers white with a greenish tinge.

An uncommon but historically important plant.
'Roseum' (*E. lilacinum* Donckelaar; *E. concinnum* Vatke; *E. roseum* Vilmorin-Andrieux; × *Bonstedtia lilacina* (Donckelaar) Wehrhahn). Illustration: Journal of Japanese Botany **13**: 811 (1937). Plant 10–30 cm; leaves very variable: leaflets 2, 6 or 7, very rarely 9, ovate, the lateral leaflets usually with very unequal basal lobes, tips blunt, margins sparsely spiny-toothed or entire, 2–5 × 1–2.5 cm; flowering stem with 1 leaf usually with 6 or 2 leaflets; inflorescence loose, 4–12-flowered, about as long as or longer than stem-leaf; flowers purplish mauve, varying in depth of colour.
'Niveum' (*E. musschianum* misapplied, not Morren & Decaisne; *E. niveum* Vilmorin-Andrieux). Illustration: Gartenflora **86**: 53

(1937). Like 'Roseum, but flowers white; spur, when present, shorter than inner sepal.

'Yenomoto'. Like 'Roseum' and 'Niveum' in leaf but flowers white with spur straight, slender, to 1.2 cm, longer than inner sepal.

E. trifoliolatobinatum (Koidzumi) Koidzumi, with twinned leaflet-stalks each bearing 3 acute leaflets and small white flowers bearing spreading, fine petal-spurs about as long as inner sepals, is considered by Suzuki to be a species derived from *E. diphyllum* and *E. grandiflorum* in southern Japan and stabilised by short-tongued pollinating bees.

14. E. davidii Franchet. Illustration: Franchet, Plantae Davidianae **2**: t. 6 (1885).
Plant 30–50 cm. Rhizome fairly long, 3 mm thick. Leaves basal and on the stem; leaflets 3 or 5, narrowly to broadly ovate, tips usually rounded and mucronate, very spiny-toothed, rather leathery, both sides with a distinct network of veins, sparingly hairy beneath with short adpressed hairs, usually less than 6 × 4.5 cm. Flowering stem normally with 2 opposite (rarely alternate) leaves with 3 or 5 leaflets; inflorescence usually compound below, simple above, loose, very glandular, 6–24-flowered. Flowers 2–3 cm across, yellowish. Outer sepals 2–4 mm; inner sepals narrowly ovate, almost acute, 4 × 1 mm. Petals much longer than inner sepals, with distinct rounded blades forming a cup 7–13 mm deep; spurs slender, curved, subulate, 1–1.5 cm. Stamens not protruding, *c.* 4 mm long. *W China (Sichuan)*. H5. Spring–summer.

15. E. acuminatum Franchet. Illustration: Journal of Japanese botany **57**: 316 (1982).
Plant 20–50 cm. Rhizome sometimes long, 2–5 mm thick. Leaves basal and on the stem; leaflets 3, narrowly ovate to lanceolate, long-acuminate, very spiny-toothed, thin and hairless when young, leathery when mature, then hairless or thickly or sparsely hairy beneath with short, adpressed, fairly stout bristles, 3–18 × 1.5–7 cm. Flowering stem normally with 2 opposite leaves with 3 leaflets; inflorescence compound with the lower stalks 2–5-flowered, loose, hairless or rarely sparsely glandular, the whole 10–55-flowered. Flowers 3–4 cm across, yellow, white, rose-purple or pale violet. Outer sepals 3–4.5 mm × 2–4 mm; inner sepals ovate-elliptic, acute, 8–12 × 3–7 mm. Petals much longer than inner sepals, horn-shaped, tapering from the

swollen but bladeless base, curving outwards, 1.5–2.5 cm. Stamens 3–4 mm. *W & C China*. H4.

Although very variable in the wild in flower colour, the only variant in cultivation (from Sichuan, Emei Shan) has purple flowers.

16. E. brevicornu Maximowicz. Illustration: Journal of the Linnean Society, Botany **51**: t. 2a (1938).
Plant 20–60 cm. Rhizome short, compact, 3 mm thick. Leaves basal and on the stem; leaflets usually 9, rarely 3 or 5, ovate to broadly ovate, acute or short-acuminate, spiny-toothed, firm and parchment-like when mature, almost hairless beneath, to 8 × 6.5 cm. Flowering stems usually with 2 opposite leaves usually with 9 leaflets; inflorescence compound, loose, glandular, 20–50-flowered. Flowers 1.5 cm across. Outer sepals 1–3 mm; inner sepals lanceolate, acute, white or yellowish, to 10 × 4 mm. Petals much shorter than inner sepals, with very slight blades; spurs narrow, conical, blunt, 2–3 mm. Stamens protruding, 3–4 mm long. *N China*. H3. Spring–summer.

17. E. sagittatum (Siebold & Zuccarini) Maximowicz. Illustration: Terasaki, Nippon Shokubutsu Zufu, 153 (1933).
Plant 25–50 cm. Rhizome short, 3–5 mm thick. Leaves basal and on the stem; leaflets usually 3, sometimes up to 9, narrowly ovate to lanceolate, acute or acuminate, very spiny, leathery when mature, at first quite hairless beneath, becoming sparsely or densely hairy with short, stout adpressed hairs, often to 5 × 3 cm, rarely to 19 × 8 cm. Flowering stem with 2 (rarely 3) opposite leaves with 3–9 leaflets; inflorescence compound, usually hairless, 20–60-flowered, rather narrow in outline, 10–20 (rarely to 30) × 2–4 cm. Flowers 8 mm or less across. Outer sepals 3.5–4.5 × 1.5–2 mm; inner sepals ovate-triangular, acute, white, to 4 × 2 mm. Petals minute, almost as long as inner sepals, sac-like, blunt, brownish yellow with slight lateral flanges at base, 2–4 mm. Stamens protruding, 5 mm long. *C China, long cultivated in China & Japan as a medicinal plant*. H4. Spring–summer.

18. E. pubescens Maximowicz. Illustration: Bulletin, Académie Impériale des Sciences de St. Pétersbourg **29**: t. 1 (1883).
Plant 20–60 cm. Rhizome sometimes long, 3–4 mm thick. Leaves basal and on the stem: leaflets normally 3, ovate, narrowly ovate or lanceolate, acuminate (sometimes

abnormally rounded), very spiny-toothed, leathery when mature, persistently hairy beneath with numerous fine spreading or curled grey hairs, 3–15 × 2–8 cm. Flowering stem with 2 (rarely 3) opposite leaves with 3 leaflets; inflorescence compound, loose, usually glandular, to 30-flowered, 10–20 × 5–6 cm (towards the base). Flowers to 1 cm across. Outer sepals 2–3 mm; inner sepals lanceolate or narrowly lanceolate, acute or acuminate, white, 5–7 × 1.5–3.5 mm. Petals minute, much shorter than inner sepals, sac-like, blunt, brownish, to 2 mm, with no basal blades. Stamens protruding, 4 mm long. *N, W & C China*. H3. Spring–summer.

7. VANCOUVERIA Morren & Decaisne
W.T. Stearn

Herbaceous perennials with long slender rhizomes. Stems leafless. Leaves divided into 3s, with numerous stalked leaflets lacking marginal spines. Flowers pendulous in a raceme or panicle. Outer sepals 6–9, soon falling. Inner sepals 6, petal-like, reflexed. Petals 6, smaller than the sepals, reflexed, with a hooded or flat tip, nectar-producing. Stamens 6; anthers opening by 2 up-rolling flaps. Ovary with lateral ovules; style slender with slightly swollen stigma. Fruit a capsule opening by 2 flaps. Seeds with arils.

A genus of 3 woodland species in coast ranges of the western USA from north Washington to central California. They are easily grown in half-shade in soil with much leaf-mould.

Literature: Stearn, W.T., Epimedium and Vancouveria (Berberidaceae), a monograph, *Journal of the Linnean Society, Botany* **51**: 409–535 (1938).

1a. Leaflets thin, not thickened at margin, dying in autumn; flower-stalks hairless   **1. hexandra**
 b. Leaflets leathery, thickened at margin, evergreen; flower-stalks glandular   2
2a. Inflorescence with 20–50 white flowers; stamens and ovary hairless   **3. planipetala**
 b. Inflorescence with 6–18 yellow flowers; stamens and ovary glandular   **2. chrysantha**

1. V. hexandra (W.J. Hooker) Morren & Decaisne. Illustration: Hooker, Flora Boreali-Americana **1**: t. 13 (1829); Journal of the Linnean Society, Botany **51**: 447 (1938); Abrams & Ferris, Illustrated Flora of the Pacific States **2**: 222 (1944); Rickett, Wild flowers of the United States **5**: 207 (1972).
Plant 10–40 cm. Rhizome creeping. Leaves

normally basal, with profuse white hairs when young; leaflets 9 or more, narrowly to broadly ovate or broader than long, often 3-lobed, tips rounded and indented, margins not thickened, always thin and membranous, dying down in autumn, becoming almost hairless, very variable in size and shape, to 7.5 × 7 cm. Flowering stem leafless (or abnormally bearing 1 leaf), ending in a hairless, loose inflorescence of 6–45 flowers; flower-stalks hairless. Flowers white, 1–1.3 cm. Outer sepals glandular; inner sepals spathulate, 8–9 mm, 3–4 mm broad towards the tip. Petals shorter than inner sepals, the narrowly oblong stalk to 5 × 1 mm, expanded and folded over at the tip to form a rounded, nectar-producing pocket. Stamens sparingly glandular. Ovary very glandular. *W USA.* H5. Spring.

2. V. chrysantha Greene. Illustration: Abrams & Ferris, Illustrated Flora of the Pacific States **2**: 222 (1944); Rickett, Wild flowers of the United States **5**: 207 (1972). Plant 20–40 cm. Rhizome creeping, 2 mm thick. Leaves basal, stalks and nodes with dense, slender, spreading hairs; leaflets 9 (rarely 3 or 5), ovate to much broader than long, obscurely 3-lobed, margins thickened and crisped, leathery in texture and almost evergreen, usually very downy beneath with numerous soft hairs, variable in size but to 4 × 4 cm. Flowering stem leafless, ending in a very glandular, loose inflorescence of 4–15 flowers; flower-stalks to 3 cm, glandular. Flowers yellow, 1–1.3 cm. outer sepals glandular; inner sepals spathulate, 7–8 mm, to 3.5 mm wide towards the tip. Petals shorter than inner sepals, the narrowly oblong stalk *c.* 4 mm, the tip expanded and folded over to form a rounded nectar-producing pocket. Stamens sparingly glandular. Ovary glandular. *W USA (S Oregon).* H5. Spring.

3. V. planipetala Calloni (*V. parviflora* Greene). Illustration: Malpighia **1**: t. 6 (1887); Journal of the Linnean Society, Botany **51**: 153 (1938); Abrams & Ferris, Illustrated Flora of the Pacific States **2**: 222 (1944); Rickett, Wild flowers of the United States **5**: 207 (1972). Plant 18–32 cm. Rhizome long, 2–4 mm thick. Leaves basal; leaflets 9 or more, broadly ovate, often broader than long and 3-lobed, margins thickened and slightly or very crisped, leathery and evergreen, with short sparse hairs beneath, very variable in size, to 6 × 5 cm. Flowering stems leafless, ending in a loose, glandular panicle of 20–50 flowers; flower-stalks glandular.

Flowers white or lavender-tinged, 6–8 mm (including stamens and reflexed sepals). Outer sepals hairless; inner sepals spathulate, 4 mm, to 2 mm broad towards the tip. Petals shorter than inner sepals, oblanceolate, flat and notched at the tip with a slight central lobe and large yellow lateral lobes, not pouched, 3 × 1 mm. Stamens hairless. Ovary hairless. *W USA.* H5. Spring.

8. JEFFERSONIA Barton
W.T. Stearn
Herbaceous, low-growing perennials with short rhizomes. Stems leafless, 1-flowered. Leaves basal, long-stalked; blade entire, lobed or divided into 2 stalkless, palmately-veined leaflets. Sepals 4, soon falling. Petals 8, flat. Stamens usually 8; anthers opening from the base by 2 up-rolling flaps. Ovary with many lateral ovules; style short with small, 2-lobed stigma. Capsule opening either by an almost horizontal, incomplete slit below the top (thus appearing to have a lid), or by an oblique, downward slit from the top. Seeds numerous, each with a small, lacerate aril.

A genus of 2 woodland species, 1 from eastern N America, the other from China (Manchuria) and Korea. They grow best in a sandy soil with abundant leaf-mould or peat, in partial shade. Propagation is by careful division or by seed.
Literature: Hutchinson, J., Jeffersonia and Plagiorhagma, *Kew Bulletin for 1920*, 242–245 (1920).

1a. Leaf-blade divided into 2 fan-shaped leaflets; flowers white   **1. diphylla**
  b. Leaf-blade not divided but somewhat kidney-shaped; flowers lavender-blue
                                                  2. dubia

1. J. diphylla (Linnaeus) Persoon. Illustration: Botanical Magazine, 1513 (1812); Loddiges' Botanical Cabinet **11**: t. 1036 (1825); House, Wild flowers of New York **1**: 116 (1918); Rickett, Wild flowers of the United States **1**: pl. 46 (1966). Plant 10–20 cm. Leaves with 2 entire or shallowly lobed, somewhat fan-shaped leaflets to 15 × 8 cm. Flower white, 2–3 cm across. Capsule pear-shaped, 1.5–2 cm, opening by an almost horizontal slit below the top. *Eastern N America (Ontario south to Alabama and Georgia).* H3. Spring.

2. J. dubia (Maximowicz) Baker & Moore (*Plagiorhegma dubium* Maximowicz). Illustration: Kew Bulletin for 1920, 242 (1920); Botanical Magazine, 9681 (1948); Plantsman **4**: 11 (1982). Plant 5–14 cm. Leaf-blade not divided, but

kidney-shaped with indented apex, to 10 cm wide. Flowers lavender-blue, 2–3.5 cm across. Capsule *c.* 1 cm, opening by an oblique slit from below the top. *China (Manchuria), Korea.* H2. Spring.

9. ACHLYS de Candolle
W.T. Stearn
Herbaceous perennials with rhizomes. Stems leafless. Leaves basal with 3 stalkless, fan-shaped, palmately veined leaflets. Flowers numerous in a long, slender spike. Sepals and petals absent. Stamens 6–12, usually 9, with long filaments; anthers opening by 2 up-rolling valves. Ovary with 1 basal ovule; stigma stalkless, broad. Fruit a small achene.

A genus of 3 very similar species in western N America and Japan. They should be given half-shade and soil rich in leaf-mould.
Literature: Takeda, H., On the genus Achlys, *Botanical Magazine Tokyo* **29**: 169–184 (1913); Fukuda, I. & Baker, H., Achlys californica (Berberidaceae), *Taxon* **19**: 341–344 (1970).

1. A. triphylla (Smith) de Candolle. Illustration: Armstrong, Field book of western wild flowers, 157 (1915); Abrams & Ferris, Illustrated Flora of the Pacific States **2**: 222 (1944); Rickett, Wild flowers of the United States **5**: 207 (1972). Plant hairless, 20–40 cm. Leaflets shallowly lobed, to 8 cm broad. Spike 2–5 cm. Stamens 3–4 mm, white. *Western N America (British Columbia to N California).* H5. Spring.

The related **A. californica** Fukuda & Baker is slightly larger in all parts.

10. CAULOPHYLLUM Michaux
W.T. Stearn & E.H. Hamlet
Herbaceous perennials with short rhizomes. Stem with 1 leaf. Leaves compound, leaflets in 3s, stalkless but having the 3 primary divisions long-stalked with numerous, 2–5-lobed, pinnately-veined leaflets. Flowers greenish, numerous in racemes or panicles. Outer sepals 3 or 4; inner sepals 6, petal-like. Petals 6, much shorter than inner sepals, thick, fan-shaped, hooded, nectar-producing. Stamens 6, anthers opening by 2 up-rolling flaps. Ovary with short style, stigma gland-like, minute. Fruit almost non-existent because the 2 developing seeds burst the ovary-wall early and ripen naked, blue and berry-like.

A genus of 2 very similar species, 1 from eastern N America, the other from Japan. They require a soil rich in leaf-mould in partial shade. Propagation is by division in spring or just after flowering.

1. C. thalictroides (Linnaeus) Michaux. Illustration: Rickett, Wild flowers of the United States **1**: pl. 46 (1966); Justice & Bell, Wild flowers of N Carolina, 68 (1968); Plantsman **4**: 13 (1982); Zichmanis & Hodgins, Flowers of the wild, 22, 23 (1982).
Erect, hairless, 30–80 cm. Leaves 5–8 cm, leaflets broadly obovate, 2–5-lobed. Flowers yellowish green or greenish purple, to 1 cm across. Seeds 5–8 mm, dark blue. *Eastern N America*. H2. Spring.

11. LEONTICE Linnaeus
W.T. Stearn
Perennial herbs with tubers. Leaves basal and on the stem, long-stalked, divided into 3s and/or pinnately, with numerous shortly stalked or stalkless, entire leaflets. Flowers numerous in axillary and terminal racemes often forming a panicle. Sepals 6, flat, petal-like. Petals small, nectar-producing. Stamens 6, anthers opening by 2 up-rolling flaps. Ovary with 2–4 basal ovules; style short, stigma truncate. Fruit much inflated, bladder-like, net-veined, breaking open irregularly at the top when dry.

About 3 species in N Africa, SE Europe and SW Asia. They are suitable only for cultivation in a cool greenhouse and should be given a dry resting period with limited water, but not allowed to dry out completely. Propagation is by seed.

1. L. leontopetalum Linnaeus. Illustration: Holmboe, Vegetation of Cyprus, 228 (1914); Zohary, Flora Palaestina **1**: t. 315 (1966); Ali & Jafri, Flora of Libya **3**: f. 1 (1976); Huxley & Taylor, Wild flowers of Greece and the Aegean, pl. 57, 58 (1977).
Plant hairless, 15–60 cm. Leaves much divided, 10–25 cm long, leaflets almost circular to broadly obovate, obtuse. Inflorescence loose, mostly broadly pyramidal, 15–30 cm across. Flowers yellow, *c.* 1.6 cm across. Fruit to 3 cm. *N Africa, east Mediterranean area to Iran.* H5. Spring.

12. GYMNOSPERMIUM Spach
W.T. Stearn
Herbaceous perennials with tubers. Leaves basal and 1–3 on the stems, palmate with 4–7 narrow, entire leaflets. Flowers in a short, unbranched raceme. Sepals 6, petal-like. Petals much shorter than sepals, semi-cylindric, nectar-producing. Stamens 6, anthers opening from the base by 2 up-

rolling flaps. Ovary with 2–4 basal ovules; style long, stigma minute. Fruit short, opening early by short lobes, the seeds ripening in an exposed state.

A genus of about 6 species ranging from the Balkan peninsula across C Asia to China, formerly included in *Leontice* until separated by Stearn & Webb in *Flora Europaea* **1**: 244 (1964). They are best cultivated in an alpine house, and can be propagated only by seed.
Literature: Takhtajan, A.L., On the genus Gymnospermium, *Botanicheskii Zhurnal* **55**: 1191–1193 (1970).

1a. Stem-leaf 1; flowers to 1.8 cm across; sepals inconspicuously veined; ovary stalkless   **1. altaicum**
  b. Stem leaves 1–3; flowers to 2.5 cm across; sepals conspicuously red-veined; ovary on a short stalk   **2. alberti**

1. G. altaicum (Pallas) Spach (*Leontice altaica* Pallas). Illustration: Botanical Magazine, 3245 (1833).
Plant 5–20 cm. Leaflets narrowly elliptic, 1.5–4 cm. Flowers to 1.8 cm across. Sepals yellow without conspicuous veining. Ovary stalkless. *Romania to USSR (Crimea to C Asia).* H5. Spring.

2. G. alberti (Regel) Takhtajan (*Leontice alberti* Regel). Illustration: Botanical Magazine, 6900 (1886); Plantsman **4**: 2 (1982).
Plant 18–25 cm. Leaflets broadly elliptic, to 6 cm. Flowers to 2.5 cm across. Sepals yellow, the outer reddish, all with conspicuous red veining. Ovary shortly stalked. *USSR (C Asia).* H2. Spring.

13. BONGARDIA Meyer
W.T. Stearn & E.H. Hamlet
Herbaceous perennials with tubers. Leaves all basal, pinnately divided, with paired or whorled, stalkless, lobed leaflets. Flowers numerous, long-stalked, in a loose, narrow panicle with ascending branches. Sepals 6, concave. Petals 6, flat, conspicuous, each with a small nectar-bearing pore at the base. Stamens 6, anthers opening from the base by 2 up-rolling valves. Ovary with 5–6 basal ovules; stigma almost stalkless, folded or lobed. Fruit an ellipsoid, slightly inflated, papery capsule opening by short, acute flaps at the top.

A genus of a single species in SW Asia, from Turkey to Pakistan. It is best grown in an alpine house and kept fairly but not completely dry in summer. Propagation is by offsets and seed.

1. B. chrysogonum (Linnaeus) Endlicher (*Leontice chrysogonum* Linnaeus; *B. rauwolfii* Meyer). Illustration: Botanical Magazine, 6244 (1876); Polunin, Flowers of Greece and the Balkans, pl. 8 (1980); Everett, New York Botanical Gardens illustrated encyclopedia of horticulture **2**: 447 (1981); Plantsman **4**: 3 (1982).
Plant 20–60 cm. Leaves 10–40 cm; leaflets 8–14, in pairs or sometimes whorls of 3 or 4, stalkless, obovate or wedge-shaped leaflets which are up to 3 cm long and 3–6-lobed. Flowers yellow, to 2 cm across. Fruit to 1.5 cm. Seeds black, bloomed. *E Mediterranean area to USSR (C Asia), Afghanistan, Pakistan.* H4. Spring.

14. DIPHYLLEIA Michaux
W.T. Stearn
Herbaceous perennials with short rhizomes. Stem with 2 well-separated leaves. Basal leaves with centrally peltate blade, stem-leaves with marginally peltate blades, the lower long-stalked, the upper stalkless; blades circular to kidney-shaped, lobed and toothed, palmately veined. Flowers in an umbel-like cyme. Outer sepals 6, soon falling. Petals 6, flat. Stamens 6, anthers opening by 2 up-rolling valves. Ovary with few lateral ovules; stigma almost stalkless. Fruit a spherical berry with 2–4 seeds.

A genus of 3 woodland species from eastern N America and Japan.
Literature: Li, H.L., The genus Diphylleia, *Journal of the Arnold Arboretum* **28**: 442–444 (1941).

1. D. cymosa Michaux. Illustration: Botanical Magazine, 1666 (1841); Britton & Brown, Illustrated flora of the northern United States and Canada, edn 2, **2**: 129 (1913).
Plant hairless, 60–100 cm. Blade of basal leaf to 20 cm broad. Flowers white, *c.* 1.5–2 cm across. Petals narrowly obovate, *c.* 1 cm. Berries blue, *c.* 1 cm long and wide. *Eastern N America (Virginia to Georgia).* H4. Spring.

15. PODOPHYLLUM Linnaeus
W.T. Stearn
Herbaceous perennials with rhizomes. Stem with 1 or 2 leaves. Leaves broad, peltate, deeply lobed to almost entire but then angled. Flower solitary between the 2 leaves or flowers several in a cluster. Sepals 6, soon falling. Petals 6 or 9, flat. Stamens 12–18, anthers opening by longitudinal slits. Ovary with numerous lateral ovules; stigma stalkless or almost so. Fruit a large berry with the seeds surrounded by pulp.

A genus of 9 species, 1 in eastern N America, 1 in the Himalayas and W China, 7 in China (some of the Chinese species sometimes separated off into the genera *Dysosma* and *Sinopodophyllum*). Cultivation is easy in good garden soil preferably enriched with leaf-mould. The rhizomes of *P. peltatum* and *P. hexandrum* collected in the wild have long been used as purgatives but are now the source of anti-cancer drugs and so, by excessive collecting, may become endangered species.

Literature: Woodson, R.E., Dysosma, a new genus of Berberidaceae, *Annals of Missouri Botanical Garden* **15**: 335–340 (1928); Chatterjee, R., Indian Podophyllum, *Economic Botany* **64**: 342–354 (1952); Or Ying, T.S., On Dysosma Woodson and Sinopodophyllum Ying, *Acta Phytotaxonomica Sinica* **17**: 15–22 (1979).

1a. Flowers solitary, white or rose   2
  b. Flowers several in a cluster, dark red
                                            3
2a. Leaf-margin coarsely toothed with 4–5 teeth per 2.5 cm of margin; ripe fruits normally yellow   **1. peltatum**
  b. Leaf-margin finely toothed with 6–12 teeth per 2.5 cm of margin; ripe fruits red   **2. hexandrum**
3a. Leaves divided to one-third of their radius; petals 4.5–6 cm
                                3. pleianthum
  b. Leaves divided to about half of radius; petals 2.5–3 cm   **4. versipelle**

1. P. peltatum Linnaeus. Illustration: Botanical Magazine, 1819 (1816); Britton & Brown, Illustrated Flora of the northern United States and Canada **2**: 92 (1897); Gleason, Illustrated Flora of the northern United States and adjacent Canada **2**: 189 (1952); Rickett, Wild flowers of the United States **1**: 157 (1966).

Plant 30–50 cm. Leaf-blade to 27 cm broad, divided almost to the base into 5–8 lobes which are themselves coarsely toothed and sometimes 2-cleft. Flower solitary, white, 3–5 cm across. Berry yellow, 3–5 cm. *Eastern N America (from Ontario south to Texas and Florida)*. H3. Summer.

2. P. hexandrum Royle (*P. emodi* Wallich; *Sinopodophyllum emodi* (Wallich) Ying). Illustration: Flore des serres **6**: t. 1659, 1660 (1866); Blatter, Beautiful flowers of Kashmir, t. 7 (1928); Economic Botany **64**: 3435, 346 (1952).

Plant 15–30 cm. Leaf-blade to 23 cm broad, deeply divided to below the middle or almost to the base into 3–5 finely toothed lobes. Flower solitary, pink or white, 2.5–3.5 cm across. Berry red, 2.5–10 cm. *Afghanistan, Himalaya from Kashmir to China*. H5. Summer.

At flowering the leaves usually droop and are not yet unfolded. A Chinese variant has been described as *P. emodi* var. *chinense* Sprague (illustration: Botanical Magazine, 8850, 1920).

3. P. pleianthum Hance (*Dysosma pleiantha* (Hance) Woodson). Illustration: The Garden **2**: 299 (1899); Botanical Magazine, 7098 (1890); Li et al., Flora of Taiwan **2**: 520 (1976).

Plant to 30 cm or more. Leaf-blade to 35 cm broad, shallowly divided (to *c*. one-third of the radius) into 6–10 broadly triangular, ciliate lobes which are broader than long. Flowers in a cluster, cup-shaped, dark reddish purple; petals 4.5–6 cm. *Taiwan, China*. H3. Summer.

4. P. versipelle misapplied, not Hance. Illustration: Botanical Magazine, 8154 (1907).

Plant to 60 cm. Leaf-blade to 35 cm broad, deeply divided to about half of the radius into 6–7 broadly ovate or oblong, ciliate lobes which are longer than broad. Flowers in a cluster, cup-shaped, dark purplish red. Petals 2.5–3 cm. *C China*. H3. Summer.

Plants grown under this name are not the same as Hance's original specimens and the correct name of this species is uncertain. It may be the same as the Chinese species named *Dysosma tsayuensis* Ying.

LXXV. LARDIZABALACEAE

Woody plants, monoecious or dioecious, usually twining. Leaves alternate, without stipules, palmate or trifoliolate, rarely pinnate. Flowers radially symmetric, bisexual or unisexual, usually in racemes, occasionally solitary. Sepals usually 6, occasionally 3, petal-like. Petals 6 or absent. Nectaries often present. Stamens 6, united or not; anthers opening towards the outside of the flower. Carpels 3–9, superior, free, with many ovules in vertical rows. Fruit dehiscent or not.

A family of 7–9 genera of which 6 are in cultivation. The unisexual flowers generally have rudimentary organs of the other sex.

Habit. Erect shrub: **1**. Deciduous: **1,5,6**; evergreen: **2–4,6**.
Leaves. Pinnate: **1**.
Flowers. Male and female on separate plants: **2–5**; male and female or bisexual on the same plant: **1,3,4,6**.
Sepals. Three: **6**. Greenish yellow: **1**; white or whitish, sometimes violet-tinged: **2,4,5**; purple-brown to red-purple: **3,6**.
Stamens. United: **1–3**; free: **4–6**.
Carpels. 3, rarely 6: **1–5**; 5–10: **6**.

1a. Erect, non-twining shrub; leaves pinnate; sepals greenish yellow
                                1. Decaisnea
  b. Climber with twining stems; leaves not pinnate; sepals white to purplish
                                              2
2a. Sepals 3; male and female flowers borne in the same inflorescence; female flowers with 5–10 carpels
                                6. Akebia
  b. Sepals 6; male and female flowers borne in separate inflorescences or, if in the same inflorescence, then sepals whitish; female flowers with 3, rarely 6 carpels   3
3a. Stamens united   4
  b. Stamens free   5
4a. Leaves palmate with 3–7 narrow leaflets; male and female flowers borne on separate plants; sepals white tinged with violet   **2. Stauntonia**
  b. Leaves once, twice or 3 times divided into 3 segments; male and female flowers borne on the same plant; sepals purplish brown   **3. Lardizabala**
5a. Plant evergreen; male and female flowers borne on the same plant
                                4. Holboellia
  b. Plant deciduous; male and female flowers borne on separate plants
                                5. Sinofranchetia

1. DECAISNEA Hooker & Thomson
V. A. Matthews

Deciduous shrubs. Flowers bisexual or functionally male, borne on the same plant, with 6 sepals in 2 whorls of 3, petals absent. Stamens united, sometimes shortly so. Fruit a many-seeded, cylindric berry.

A Himalayan and Chinese genus of two species, requiring a good, loamy soil; propagation is by seed.

1. D. fargesii Franchet. Illustration: Botanical Magazine, 7848 (1902); Hay & Synge, Dictionary of garden plants in colour, 195 (1969) – fruit; Huxley, Deciduous garden trees and shrubs, f. 65 (1973); Journal of the Arnold Arboretum **60**: 314 (1979).

Erect shrub to *c*. 3 m. Leaves pinnate with 13, 15 or 17 ovate to elliptic leaflets, 4.5–15 cm, which are finely downy beneath when young; leaf- and leaflet-stalks jointed at the base. Flowers borne in erect or drooping racemes (which may form loose panicles) at the ends of lateral branches; bisexual and male flowers are produced in separate racemes or on different panicle-branches. Sepals greenish yellow, lanceolate, minutely downy on the backs, 1.5–3 cm, the 3 outer *c.* 3 mm wide, the 3 inner 6–7 mm wide and strongly keeled. Bisexual flowers with filaments united at the base for *c.* 2 mm, and 3 cylindric carpels which protrude from the filament-tube, each with a conspicuous stigma. Male flowers filaments joined for most of their length (2.5–4 mm); carpels rudimentary and hidden within the filament-tube. Fruit 5–10 × *c.* 1.5 cm, more or less straight, cylindric, blue, thick-skinned, containing numerous black seeds embedded in white pulp. *W China.* H4. Spring.

2. STAUNTONIA de Candolle
V.A. Matthews

Evergreen climbers with twining stems. Flowers unisexual, the male and female borne on separate plants, with 6 fleshy sepals and no petals. Stamens united. Fruit an ellipsoid berry.

A genus of at least 6 and possibly 16 species from eastern Asia. It grows best in a shady, damp place in soil with plenty of humus, and can be grown up a wall, fence or other support. Propagation is best by cuttings of half-ripened wood. Fruit is sometimes produced when the male plant is absent, suggesting that the female flowers or plants may not be completely unisexual.

1. S. hexaphylla (Thunberg) Decaisne. Illustration: Gardeners' Chronicle **5**: 597 (1876); Kitamura & Okamoto, Coloured illustrations of trees and shrubs of Japan, f. 156 (1977).
Stems climbing to 13 m. Leaves with 3–7 leaflets arranged palmately; leaflets 5–13 cm, ovate to elliptic. Flowers in few-flowered racemes borne in the leaf-axils, white with a violet tinge, fragrant, *c.* 1.8 mm across, the male with 6 stamens united into a column, the female with 3 carpels. Fruit 2.5–5 cm, purple. *Japan, South Korea, Taiwan.* H5. Spring.

Sometimes this species occurs in catalogues under the invalid name *Holboellia hexaphylla.*

3. LARDIZABALA Ruiz & Pavon
V.A. Matthews

Evergreen climbers with slender, twining stems. Flowers functionally unisexual, male and female borne on the same plant, each with 6 fleshy sepals in 2 whorls of 3, and reduced petals. Fertile stamens united. Fruit a many-seeded, pulpy berry developed from 1 carpel.

A genus of 2 species from Chile. Any good loam is suitable for growth, but if grown out-of-doors the plants usually need the protection of a wall.

1. L. biternata Ruiz & Pavon. Illustration: Botanical Magazine, 4501 (1850); Paxton's Flower Garden **1**: 29 (1853); Gardeners' Chronicle **52**: 467 (1912); Grey-Wilson & Matthews, Gardening on walls, pl. 11 (1983).
Stems climbing 3–4 m. Leaves once, twice or 3 times divided into 3 segments, with dark green, leathery, usually ovate leaflets 5–10 cm with the centre one of each 3 the largest; margin shallowly scalloped with occasional teeth. Male flowers in drooping spikes 7.5–10 cm, arising in the leaf-axils, the flowers 8–12 mm with fleshy, broadly ovate, deep brown-purple sepals, narrow, white petals and 6 stamens joined together by their filaments; carpels rudimentary. Female flowers solitary in the leaf-axils, 1.5–1.8 cm with 3 or sometimes 6 carpels and 6 free, sterile stamens; stalks *c.* 2.5 cm. Only 1 carpel develops into the fruit which is dark purple and sausage-shaped. *Chile.* H5. Winter.

4. HOLBOELLIA Wallich
V.A. Matthews

Evergreen, twining shrubs with long-stalked leaves composed of palmately-arranged leaflets. Male and female flowers are produced in few-flowered racemes or corymbs on the same plant. Flowers with 6 rather fleshy sepals, petals reduced to nectaries. Stamens 6, free. Carpels 3. Fruit an indehiscent, fleshy pod containing many black seeds.

A genus of about 5 species from the Himalayas and mainland south-east Asia. They will grow in most soils, either in sun or shade. They do not fruit readily in more northern latitudes where hand-pollination may encourage the formation of fruit. Propagation is by seed, layers or cuttings.

1a. Leaflets 3; male flowers 1.1–1.3 cm, not fragrant, male and female flowers usually borne in separate inflorescences **1. coriacea**
  b. Leaflets 3–7; male flowers 1.5–1.8 cm, fragrant, male and female flowers usually borne in the same inflorescence **2. latifolia**

1. H. coriacea Diels. Illustration: Botanical Magazine, n.s., 447 (1964).
Stems to 7 m, the young shoots often purplish. Leaves with 3 leaflets 6.5–15 cm, with the central one larger than the laterals, ovate to obovate or lanceolate, entire, leathery. Male flowers 1.1–1.3 cm, borne in a group of corymbs which make a cluster *c.* 7.5 × 10 cm. Sepals whitish, stamens with mauve filaments. Female flowers larger than the male, sepals greenish white tinged with purple, carpels cylindric, *c.* 6 mm. Fruit 4.5–6 × *c.* 2.5 cm, purple. *C China.* H3. Spring.

2. H. latifolia Wallich (*Stauntonia latifolia* (Wallich) Wallich). Illustration: Edwards's Botanical Register **32**: t. 49 (1846); The Garden **14**: 369 (1878); Revue Horticole for 1890: 348; Grey-Wilson & Matthews, Gardening on walls, pl. 9 (1983).
Differs from *H. coriacea* in having 3–7 leaflets, male flowers 1.5–1.8 cm, fragrant, usually in the same inflorescence as the females, and the fruit 5–7.5 cm. *Himalaya.* H5. Spring.

5. SINOFRANCHETIA (Diels) Henry
V.A. Matthews

Deciduous climber with twining stems to 15 m. Leaves with 3 leaflets 6–14 cm, ovate, entire. Flowers unisexual, white, to 8 mm across, each with 6 sepals and 6 nectaries, in drooping racemes, male and female flowers produced on separate plants. Male flowers with 6 free stamens. Female flowers with 3 carpels. Fruit a blue-purple berry containing many black seeds; the berries are borne alternately along a stalk 20 cm or more.

A genus containing only a single species from China. It is a very vigorous plant, needing plenty of room and is not for the small garden. Any soil or position suits it; it is grown mainly for its foliage and fruit, the flowers being insignificant and often apparently male. Propagation is by seed.

1. S. chinensis (Franchet) Hemsley. Illustration: Hooker's Icones Plantarum **29**: t. 2842 (1907); Schneider, Illustriertes Handbuch der Laubholzkunde **2**: 913 (1912); Botanical Magazine, 8720 (1917). *C & W China.* H4. Late spring.

6. AKEBIA Decaisne
V.A. Matthews

Deciduous or evergreen climbers with twining stems. Flowers unisexual, male and female borne in the same pendent raceme, the males smaller, numerous, the female few and larger and produced at the base of the raceme. Sepals 3 (occasionally 4), petals absent. Male flowers with 6–8 free stamens. Female flowers with 5–10 carpels. Fruit a fleshy follicle containing numerous seeds.

A genus of 4 or 5 species from eastern Asia. They need a rich, loamy soil. Fruits are, unfortunately, not produced abundantly in cultivation. Propagation is best by layers, or stem- or root-cuttings. Once established, plants should not be moved as they dislike root disturbance. Literature: Spongberg, S. A. & Burch, I.H., Lardizabalaceae hardy in temperate North America, *Journal of the Arnold Arboretum* **60**: 305–312 (1979).

1a.	Leaflets widest above the middle; flowers fragrant; female flowers 2–3.5 cm across	**2. quinata**
b.	Leaflets widest at or below the middle; flowers not or only slightly fragrant; female flowers 1.4–2 cm across	2
2a.	Leaflets usually 3; flowers not fragrant; male flowers appearing stalkless at flowering, the sepals 2.5–3 mm	**1. trifoliata**
b.	Leaflets usually 4 or 5; flowers slightly fragrant; male flowers stalked at flowering, the sepals 3–4.5 mm	**3. × pentaphylla**

1. A. trifoliata (Thunberg) Koidzumi (*A. lobata* Decaisne). Illustration: Botanical Magazine, 7485 (1896); Kitamura & Okamoto, Coloured illustrations of trees and shrubs of Japan, f. 155 (1977); Journal of the Arnold Arboretum **60**: 308 (1979). Deciduous climber. Leaflets usually 3, 3.5–10 cm, widest below the middle, margin irregularly wavy to shallowly lobed. Flowers dark reddish purple, unscented. Male flowers 7–20, sepals 2.5–3 mm, reflexed; stamens 6–8. Female flowers 2 or 3, sepals 6–10 mm, spreading or reflexed; carpels 5–10, 3–6 mm. Fruit 5–13 × 3.5–6.5 cm, sausage-shaped, pale violet, splitting when ripe to reveal black or dark brown seeds embedded in white pulp. *China, Japan.* H4. Spring.

2. A. quinata (Houttuyn) Decaisne. Illustration: Everard & Morley, Wild flowers of the world, pl. 90 (1970) – upside down; Bean, Trees and shrubs hardy in the British Isles, edn 8, **1**: 268 (1970); Journal of the Arnold Arboretum **60**: 308 (1979); Grey-Wilson & Matthews, Gardening on walls, pl. 1 (1983) – upside down. Climber to 10 m, evergreen in warmer sites with mild winters, deciduous in cooler regions. Leaflets usually 5, rarely 3 or 4,

3.5–7.5 cm, widest above the middle, margin often wavy. Flowers smelling of vanilla. Male flowers 4–15, sepals 5–9 mm, pale purple, reflexed; stamens 6 or 7. Female flowers usually 2, sepals 3 or 4, 9–15 mm, dark brownish purple; carpels 5–7. Fruit 6.5–10 cm, sausage-shaped, grey-violet or purplish, containing many black or red-brown seeds embedded in white pulp. *China, Japan & Korea (naturalised in part of eastern USA).* H4. Spring.

3. A. × pentaphylla (Makino) Makino. A hybrid between *A. quinata* and *A. trifoliata* which is intermediate between the parents. The leaves usually have 4 or 5 leaflets (rarely 6 or 7) which are widest at or below the middle. The flowers are slightly fragrant and the sepals of the males are 3–4.5 mm. *Japan.* H4. Spring.

LXXVI. MENISPERMACEAE

Climbers, often woody, or erect shrubs or small trees, usually dioecious. Leaves alternate, without stipules, usually with long stalks, simple, entire, toothed or lobed, deciduous or evergreen. Flowers unisexual, in racemes or panicles, each usually with a bract and 2 bracteoles. Sepals 3–many, often in 2 whorls. Petals 0, 3 or 6. Stamens 3, 6 or more numerous, free or united; anthers opening by longitudinal or transverse slits. Staminodes often present in the female flowers. Carpels 1–6, free, each containing a single, marginally attached ovule, and each borne on a short stalk which elongates in fruit. Fruit a collection of stalked drupes (occasionally only 1 from each flower developing). Seeds weakly to conspicuously horseshoe-shaped.

A family of 67 genera and about 450 species, of which only a small number is cultivated, mainly for the sake of their leaves and fruits rather than their small, inconspicuous flowers. Because of this, flowering times in Europe are uncertain, and are not given for many of the species included here. The classification of the family is problematic, and the genera are difficult to identify.
Literature: Diels, L., Menispermaceae, *Das Pflanzenreich* **46** (1910).

Habit. Woody climber with persistent aerial branches: **1–3,6,7**; shrub or small tree: **5**; herbaceous climber without woody, persistent, aerial branches: **4,5,7**.
Leaves. Entire, angled or toothed: **1,2,4–7**; weakly 3-lobed: **5**; with 5 or more

distinct lobes: **3,6,7**;. Hairless when mature: **2,5–7**; hairy, at least beneath when mature with densely matted hairs: **1**; hairy at least beneath when mature, hairs not matted: **3–5,7**.
Petals. Absent: **2,4**; present: **1,3,5–7**.
Stamens. Nine or more, free: **6,7**; nine or more, united: **2**; six, free: **1,3,5**; three or six, united: **1,4**.
Anthers. Opening by longitudinal slits **1,4,7**; opening by transverse slits: **2,3,5,6**.

1a.	Plant an erect shrub or small tree	**5. Cocculus**
b.	Plant a climber	2
2a.	Leaves covered beneath with densely matted hairs	**1. Chondodendron**
b.	Leaves hairless beneath, or, if hairy, the hairs not matted	3
3a.	Plant herbaceous, producing annual, climbing stems from a thickened rootstock	4
b.	Plant woody, with persistent, climbing, aerial branches	6
4a.	Stamens 9 or more, free	**7. Menispermum**
b.	Stamens united	**4. Diosocoreophyllum**
6a.	Stamens 6, free	**3. Jateorhiza**
b.	Stamens 9 or more, free or united	7
7a.	Stamens united	**2. Anamirta**
b.	Stamens free	8
8a.	Anthers opening by longitudinal slits	**7. Menispermum**
b.	Anthers opening by transverse slits	**6. Sinomenium**

1. CHONDODENDRON Ruiz & Pavon
J. Cullen
Woody climbers. Leaves covered with matted hairs beneath, somewhat cordate at the base. Panicles axillary. Sepals 6–18, the outer hairy outside. Petals 6, sometimes very small. Stamens 3 or 6, free or united; anthers opening by longitudinal slits. Carpels 6. Fruit of 1–6 drupes. Seeds strongly horseshoe-shaped.

A genus of about 8 species from C & S America.
Literature: Krukoff, B. A. & Barneby, R. C., Supplementary notes on American Menispermaceae, VI, *Memoirs of the New York Botanical Garden* **20**(2): 34–41 (1970).

1a.	Stamens 3, united for most of their length; leaves absolutely entire	**3. microphyllum**
b.	Stamens 6, free; leaves often slightly scolloped	2
2a.	Petals 1–2 mm	**1. platyphyllum**
b.	Petals 0.2–0.5 mm	**2. tomentosum**

1. C. platyphyllum (St. Hilaire) Miers. Illustration: Das Pflanzenreich **46**: 80 (1910).
Large climber. Leaves to 10 cm or more, often slightly scolloped. Panicles often very compound. Sepals 9. Petals 6, 1–2 mm. Stamens 6, free. Drupes to 1.3 cm. *SE Brazil*. G2.

2. C. tomentosum Ruiz & Pavon. Illustration: Bentley & Trimen, Medicinal plants **5**: pl. 11 (1876).
Large climber. Leaves to 15 cm, often slightly scolloped. Sepals 9. Petals 6, 0.2–0.5 mm. Stamens 6, free. Drupes to 1.2 cm, purple. *Northern S America, from Panama to Bolivia & S Brazil*. G2.

3. C. microphyllum (Eichler) Moldenke.
Large climber. Leaves to 10 cm, margins not at all scolloped. Sepals numerous. Petals 6, to 1 mm. Stamens 3, united. *Brazil (Bahia)*. G2.

2. ANAMIRTA Colebrook
J. Cullen
Tall woody climber. Leaves hairless, somewhat cordate at the base. Panicle large, hanging. Sepals 9–12, petals absent. Stamens 10–many, all fused and borne in a more or less spherical head; filaments not obvious, anthers opening by transverse slits. Carpels 3 (rarely to 5). Fruit of 1–5 shortly stalked drupes. Seeds horseshoe-shaped.

A genus of a single species which is rarely cultivated; it should be treated in the same way as the tropical species of *Cocculus.*

1. A. cocculus (Linnaeus) Wight & Arnott (*A. paniculata* Colebrook). Illustration: Das Pflanzenreich **46**: 109 (1910).
Leaves to 27 × 23 cm. Panicles 20–50 cm, hanging from woody shoots. Drupes spreading, to 1.3 cm. *Tropical E Asia, from India to New Guinea.*

3. JATEORHIZA Miers
J. Cullen
Tall woody climbers, usually conspicuously hairy. Leaves large, palmately lobed with usually 5 lobes. Male flowers in panicles, females in racemes. Sepals 6. Petals 6. Stamens 6, free (in the one cultivated species), anthers opening by transverse slits. Carpels 3, hairy. Drupes ovoid. Seeds only slightly horseshoe-shaped.

A genus of 2 species from tropical Africa; one of the species is occasionally grown and should be treated like the tropical species of *Cocculus*. The generic name is

often incorrectly spelled 'Jatrorhiza' or 'Iatrorrhiza'.

1. J. palmata (Lamarck) Miers (*Cocculus palmatus* (Lamarck) de Candolle; *J. columba* (Roxburgh) Miers). Illustration: Botanical Magazine, 2790, 2791 (1830).
Root swollen and tuber-like. Stems and leaf-stalks with coarse hairs adpressed to the surface. Leaves broadly rounded, 15–35 × 18–40 cm, with a few hairs or hairless. Male flowers stalkless. females shortly stalked. Sepals to 3 mm. Petals to 2 mm. Drupes 2–2.5 cm, greyish black. *SE Africa, from Mozambique to South Africa (Natal)*. G2.

4. DIOSCOREOPHYLLUM Engler
J. Cullen
Herbaceous climber without persistent, aerial, woody stems. Stems often conspicuously hairy. Leaves very variable but not lobed. Flowers in axillary racemes. Sepals 6. Petals absent. Stamens 6, closely united, the anthers opening by longitudinal slits. Carpels 3 or 6. Seeds only slightly horseshoe-shaped.

A genus of about 5 species from tropical Africa. Cultivation as for *Cocculus*

1. D. cumminsii (Stapf) Diels. Illustration: Das Pflanzenreich **46**: 180 (1910).
Stem and leaf-stalks covered with coarse, blackish brown hairs. Leaves to 12 cm, ovate, cordate at the base, toothed, with dark brown hairs on both surfaces. Sepals yellow, to 3.5 mm. *Tropical Africa*. G2.

5. COCCULUS de Candolle
J. Cullen
Climbers, shrubs or small trees. Leaves very variable, often hairy beneath. Panicles axillary. Sepals 6, often hairy outside. Petals 6, small, often notched or bifid at the apex. Stamens 6, free, anthers opening by transverse slits. Carpels 3 or 6. Fruit of 1–6 drupes. Seeds horseshoe-shaped.

A genus of about 7 species, mainly from tropical E Asia and Africa, but 1 from temperate N America. They are easily grown in good soil, and may be propagated by root cuttings.
Literature: Forman, L.L., Menispermaceae of Malaysia: IV Cocculus A.P. de Candolle, *Kew Bulletin* **15**: 479–487 (1961).

1a. Shrub or small tree; leaves evergreen, lanceolate or narrowly elliptic, with 3 main veins, hairless **3. laurifolius**
b. Climber; leaves deciduous, ovate to broadly ovate to almost circular, sometimes weakly 3-lobed, veins not as above, hairy beneath **2**

2a. Petals entire or slightly notched at apex; drupes red **1. carolinus**
b. Petals deeply bifid at apex; drupes purple to black **2. orbiculatus**

1. C. carolinus (Linnaeus) de Candolle. Illustration: Justice & Bell, Wild flowers of North Carolina, 69 (1968).
Slender climber. Leaves deciduous, ovate to broadly ovate to almost circular, sometimes weakly 3-lobed, hairy beneath. Panicles axillary. Petals entire or slightly notched at apex. Carpels 6. Drupes red. *SE USA (Virginia and Illinois to Florida and Texas)*. H5.

2. C. orbiculatus (Linnaeus) de Candolle (*C. trilobus* (Thunberg) de Candolle). Illustration: Das Pflanzenreich **46**: 228 (1910); Botanical Magazine, 8489 (1913); Kitamura & Okamoto, Coloured illustrations of trees and shrubs of Japan, 160 (1977).
Very similar to *C. carolinus* but petals deeply bifid (rarely with 4 teeth) at apex, drupes purple to blackish. *E Asia from the Himalaya to Japan and south to the Philippine Islands*. G2.

The use of the name *C. orbiculatus* is clarified by Forman, Kew Bulletin **22**: 349–374 (1968).

3. C. laurifolius de Candolle. Illustration: Das Pflanzenreich **46**: 240 (1910).
Erect shrub or small tree. Leaves evergreen, hairless beneath, glossy above, lanceolate to narrowly elliptic, with 3 prominent main veins. Petals 6, very small, deeply bifid. Carpels 3. Drupes blackish. *Tropical and subtropical E Asia from India and the Himalaya to China and Japan, south to Indonesia (Java)*. G1.

6. SINOMENIUM Diels
J. Cullen
Woody climber. Leaves variable, often angled or lobed. Panicles axillary. Sepals 6, hairy outside. Petals 6. Stamens 9 (rarely 12), each anther opening by a transverse slit at the top. Carpels 3. Drupes compressed. Seeds horseshoe-shaped.

A genus of a single, very variable species. It can be grown in any good soil, and may be propagated by division or by seed.

1. S. acutum (Thunberg) Rehder & Wilson (*S. diversifolium* (Miquel) Diels). Illustration: Gardeners' Chronicle **52**: 411 (1912).
Leaves variable, ovate, lobed or not, cordate or truncate at the base, acuminate at the apex, 6–12 cm. Sepals 1–2.5 mm. Petals wider than long. Drupes black. *E Asia*. H4. Summer.

7. MENISPERMUM Linnaeus
J. Cullen

Woody climbers or perennials with woody stocks giving rise to annual, climbing stems. Leaves variable, sometimes slightly peltate. Flowers in axillary panicles. Sepals 4–10, rather irregularly arranged. Petals 6–9, also irregularly arranged. Stamens 12–18 or more, anthers opening by longitudinal slits. Carpels 2–4. Drupes compressed. Seeds horseshoe-shaped.

A genus of 2 species, both occasionally cultivated. They are easily grown in good soil, though require sun to produce flowers and fruits. They may be propagated by seed or by division.

1a. Climbing stems woody below,
   persistent **1. canadense**
 b. Climbing stems annual, not woody
   2. dauricum

1. M. canadense Linnaeus. Illustration: Botanical Magazine, 1910 (1917); Das Pflanzenreich **46**: 256 (1910); Menninger, Flowering vines of the world, f. 127 (1970). Tall woody climber. Leaves broadly ovate to almost circular, often 5-angled or -lobed, cordate at base and sometimes slightly peltate, 5–20 cm. Sepals 6–10, 1.5–2.5 mm. Petals *c.* 1 mm. Stamens 10–24. Drupes purple. *Eastern N America from Quebec and Manitoba to Oklahoma and Georgia.* H4. Summer.

2. M. dauricum de Candolle. Illustration: Bailey, Standard cyclopedia of horticulture, 2034 (1916).
Rhizome woody, giving rise to annual, climbing stems. Leaves ovate, cordate at the base and slightly peltate, entire or with 3–9 shallow lobes, 7–10 cm. Sepals 4–6 (rarely to 8), 1.5–3.5 mm. Petals 6–8, 1.5–3 mm. Stamens usually about 12. Drupes black. *E Asia.* ?H2. Summer.

LXXVII. NYMPHAEACEAE

Annual or perennial aquatic herbs with short and erect or long and creeping rhizomes rooting at the nodes; sometimes compact, corm-like rhizomes also present. Mature leaves spirally arranged on rhizome, submerged, floating or above water, long-stalked, peltate or with a deep sinus, simple (submerged leaves rarely finely divided), broadly ovate, kidney-shaped or circular, leathery; veins usually prominent beneath. Juvenile submerged leaves sometimes whorled or opposite, narrow and membranous. Flowers solitary, axillary on long stalks, bisexual, radially symmetric, mostly floating or above water; with usually many, spirally arranged parts. Sepals 4–6, sometimes petal-like, free or slightly joined to ovary at base. Petals 3–5 or numerous, often merging with staminodes and stamens. Stamens often with appendages. Ovary of 3 or more, free to partly or fully fused carpels, superior, to half or fully inferior. Carpels rarely individually sunk in pits in the persistent, corky receptacle. Styles absent; stigmas borne directly on top of the free carpels, or in the form of rays on top of ovary. Ovules 1–many, apical, marginal or scattered. Fruits many-seeded, berry-like, rarely nut-like, often ripening under water.

A cosmopolitan family of *c.* 9 genera and over 100 species, found throughout temperate and tropical regions in still and slow-moving water. Many have very showy flowers and are widely cultivated. Two genera with free carpels and scattered ovules, *Cabomba* and *Brasenia* are sometimes placed in the separate family Cabombaceae. *Nelumbo*, in which the free carpels are sunk in the corky receptacle, is sometimes split off in the *Nelumbonaceae*.

1a. Submerged leaves very finely divided;
   floating leaves simple or absent
   5. Cabomba
 b. All leaves simple 2
2a. Leaves with a deep sinus, reaching or
   almost reaching stalk 3
 b. Leaves peltate, lacking a sinus; stalk
   far from margin 4
3a. Veins of leaf pinnate; ovary superior
   2. Nuphar
 b. Veins of leaf radiating from top of
   stalk; ovary half-inferior
   1. Nymphaea
4a. Leaves spiny above or beneath 5
 b. Leaves not spiny 6
5a. Leaves spiny above and beneath,
   margin flat **4. Euryale**
 b. Leaves spineless above, margin with a
   vertical rim **3. Victoria**
6a. Leaves floating, to 10 cm across
   6. Brasenia
 b. Leaves above water, 30–100 cm
   across **7. Nelumbo**

1. NYMPHAEA Linnaeus
J.C.M. Alexander

Perennial water-plants with submerged, rooted corms or creeping rhizomes. Leaves alternate, floating or rarely above the water, broadly ovate to circular, deeply cleft at base into 2 lobes, one on either side of the long stalk, somewhat leathery, entire, wavy or toothed; juvenile leaves narrower and thinner, submerged. Flowers bisexual, solitary, showy, fragrant, day- or night-opening, long-stalked, floating or held above water. Sepals 4 (rarely 5), green or sometimes petal-like. Petals numerous, often brightly coloured, ovate or obovate to elliptic, narrower towards centre of flower and grading into stamens. Stamens very numerous, sometimes with coloured appendages. Ovary superior, of 3–35 fused or partially fused carpels, with stigmatic rays on top corresponding to number of cells in ovary; ovules numerous. Fruit many-seeded, ripening under water by spiral contraction of the flower-stalk after flowering.

A cosmopolitan genus of *c.* 50 species, found in still and slow-moving water in a wide variety of habitats, in tropical and temperate regions. It has a long history in cultivation and there is a very large number of often brightly coloured cultivars. One of the best known breeders was Joseph Latour-Marliac (1830–1911) who performed numerous crosses and created many cultivars, several of which are still popular today. Unfortunately he was secretive about the parentage of these cultivars and many of those that are commonly grown cannot be assigned to any particular species.

Most species enjoy full sun and thrive in water less than 1 m deep, of an even temperature. They are most easily managed in small submerged tubs or baskets. They are gross feeders and perform well in fibrous loam enriched with cow-dung or other sources of nitrogen. They can be raised from seed (which is planted by rolling it in balls of clay which can then be dropped into the pool), but are more conveniently propagated by division, or from plantlets in the viviparous species. The tropical species will only thrive in water above about 20 °C. Many become dormant in cold or dry conditions and persist as corms or rhizomes.

Literature: Conard, H.S., *The Waterlilies* (1905); Swindells P., *Waterlilies* (1983).

1a. Petals entirely yellow when flowers
   fully open 2
 b. Petals white, blue, pink, red, or
   purple, sometimes yellowish at the
   base 6
2a. Flowers to 5 cm across; leaves to
   6 cm wide **14. xhelvola**
 b. Flowers 6 cm across or more; leaves
   10 cm wide or more 3

3a. Flowers open at night
23. **amazonum**

b. Flowers open in the day 4

4a. Sepals green with brown margins
11. **mexicana**

b. Sepals pale yellow 5

5a. Petals rich canary-yellow; leaves marked with purple and bronze
19. × **marliacea** 'Chromatella'

b. Petals pale yellow; leaves plain green
12. **citrina**

6a. Tender, flowers day-opening; radial walls of ovary separable into 2 distinct layers 7

b. Tender and usually night-opening or hardy and day-opening; radial walls of ovary not 2-layered 16

7a. Leaves entire or very slightly wavy 8

b. Leaves toothed, scolloped or distinctly wavy 12

8a. Plantlets produced at junction of leaf-stalk and blade 9

b. Plantlets not produced 10

9a. Leaves green above; petals bright blue; stamens cream-coloured
7. **micrantha**

b. Leaves brownish green above; petals pale blue; stamens yellow
8. × **daubenyana**

10a. Flowers 5.5 cm across or less; petals 10 or fewer; leaves less than 10 cm across 9. **baumii**

b. Flowers 7 cm across or more; petals 12 or more; leaves more than 15 cm across 11

11a. Leaves 30–40 cm wide, spotted with purple beneath; flowers pale blue; flower-buds conical with straight sides
6. **caerulea**

b. Leaves 20 cm wide or less, purple beneath; flowers pale violet; flower-buds ovate 2. **elegans**

12a. Leaves 15 cm wide or less, petals 16 or fewer 13

b. Leaves 25 cm wide or more, petals 16 or more 14

13a. Petals purple or lilac; stamens purple
5. **colorata**

b. Petals blue, rarely pink or white; stamens pale yellow with blue tips
4. **stellata**

14a. Leaves purple beneath; stamens more than 300 1. **gigantea**

b. Leaves green beneath, sometimes spotted or blotched with purple; stamens much fewer than 300 15

15a. Petals white, 16–20; stamens c. 60
3. **flavovirens**

b. Petals blue, 20–40; stamens 120–225
10. **capensis**

16a. Sepals with prominent veins; tender, usually night-opening 17

b. Sepals with obscure veins; hardy and day-opening 18

17a. Petals white, sometimes pink outside
21. **lotus**

b. Petals red 22. **rubra**

18a. Leaves dark red to brown beneath 19

b. Leaves green beneath, sometimes flushed reddish 20

19a. Flowers 7–15 cm across; leaves 12–25 cm 17. **odorata**
(19. × **marliacea** and 20. × **laydekeri** also key out here)

b. Flowers 2.5–5 cm across; leaves 5–10 cm 13. **tetragona**
(20. × **laydekeri** also keys out here)

20a. Flowers not fragrant, mostly held above water; leaf-stalks with brown stripes, blades always green
18. **tuberosa**
(19. × **marliacea** and 20. × **laydekeri** also key out here)

b. Flowers fragrant, floating; leaf-stalks green, blades reddish at first 21

21a. Leaf-lobes often overlapping, with main veins curved towards each other
16. **candida**

b. Leaf-lobes not overlapping, with more or less straight, diverging main veins
15. **alba**
(19. × **marliacea** and 20. × **laydekeri** also key out here)

1. N. gigantea Hooker. Illustration: Botanical Magazine, 4647 (1852); Parey's Blumengärtnerei, 611 (1958); Graf, Tropica, 704 (1978); Morley & Toelken, Flowering plants in Australia, 48 (1983). Leaves 45–60 cm or more, ovate or elliptic to circular, wavy and sharply toothed, thick and leathery, green above, pinkish brown becoming purple beneath; lobes touching or overlapping, finely toothed; stalks to 2.5 cm wide. Flowers 12.5–30 cm across (to 40 cm in nature), day-opening, scentless, 10–25 cm above water; stalks to 35 × 1–2.5 cm. Sepals 4, ovate to elliptic, rounded, green bordered with blue or purple outside. Petals 18–51, longer than sepals, blue, purple at base, obovate, obtuse, thin and brittle, narrower towards centre. Stamens 350–750, bright yellow; carpels 12–20, easily separable. *Tropical Australia & New Guinea*. G2. Summer.

Various forms are on sale including the white-flowered var. **alba** (Bentham & Muller) Landon, and the pink-flowered forma **rosea** Bentham & Muller.

2. N. elegans Hooker. Illustration: Botanical Magazine, 4604 (1851); Rickett, Wild flowers of the United States 3: 98 (1969); Correll & Correll, Flora of the Bahama archipelago, 525 (1982). Leaves 15–18 cm, broadly ovate to circular, entire, sometimes slightly wavy or weakly toothed at base, thin, dark green above, purple beneath, both surfaces spotted and streaked with black; lobes well separated, with straight or concave margins; stalks pale brown. Flowers 7–13 cm across, day-opening, fragrant, 12–18 cm above water; stalks slender; buds ovate. Sepals 4 (rarely 5), lanceolate, obtuse to acute, dark green with black lines and dots. Petals 12–24, pale violet, white at base, ovate-lanceolate, obtuse. Stamens 80–145, yellow with blue tips; carpels 15–25; fruit 3–3.5 cm across, spherical, flattened, pale green. *N Mexico & Texas to Guatemala*. G1. Spring–summer.

3. N. flavovirens Lehman (*N. gracilis* misapplied). Illustration: Botanical Magazine, 7781 (1901); Conard, The waterlilies, pl. 6 (1905); Bailey, Standard cyclopedia of horticulture, 2312 (1916). Leaves 30–45 cm, ovate to more or less circular, wavy or toothed (rarely entire), mid-green above, paler (rarely red) beneath; sinus closed, lobes acute to rounded; stalk 60–150 × c. 1.2 cm. Flowers 10–20 cm across, day-opening, very fragrant, 20–30 cm above water. Sepals 4, lanceolate, acute, with 2 or 3 short, black lines. Petals 16–20, white (pink, blue or purple in cultivars), greenish outside, narrow, acuminate, firm. Stamens c. 60, deep yellow; carpels 12–15; fruit almost spherical, c. 2.2 cm across. *Mexico; ?S America*. G2. Summer.

Many coloured cultivars and hybrids, especially from crossing with *N. capensis*, are available.

4. N. stellata Willdenow. Illustration: Botanical Magazine, 2058 (1819); de Thabrew, Popular tropical Aquarium plants, 157 (1981); Swindells, Waterlilies, 65 (1983). Leaves 12–15 cm, elliptic to circular, very irregularly wavy (rarely entire), green, occasionally with brown blotches above, pink to purple beneath; sinus usually open. Flowers 5–12 cm across, day-opening, well above water. Sepals 4, dotted, pale green outside. Petals 11–14, pale blue (rarely white), yellowish white at base, lanceolate, acute, shorter than sepals. Stamens 33–54, pale yellow with blue tips; carpels 10–17. *S & SE Asia*. G2. Summer.

Var. **cyanea** (Roxburgh) Hooker & Thomson has bigger blue flowers. Illustration: Ledbetter, Water gardens, 62 (1979). Var. **versicolor** (Roxburgh) Hooker & Thomson has pink flowers with more numerous stamens. Illustration: Botanical Magazine, 1189, (1809).

5. N. colorata Peter. Illustration: Gartenflora **84**: 301 (1935); Graf, Tropica, 706 (1978); Ledbetter, Water gardens, 57 (1979).
Differs from *N. stellata* in the overlapping leaf-lobes, violet anthers, and sepals bright blue in upper half. *E Africa.* G2. Spring–Summer.

6. N. caerulea Savigny (*N. capensis* misapplied; *N. stellata* misapplied). Illustration: Letty, Wild flowers of the Transvaal, pl. 68 (1962); Graf, Tropica, 704 (1978).
Leaves 30–40 cm across, ovate to circular, entire, slightly wavy at base, soft and thin, dark green above, pale green with purple spots and tinged with pink near margin beneath; sinus usually closed; lobes more or less acute; stalks dull brownish green, flattened and *c.* 6 mm wide above. Flowers 7–15 cm across, day-opening, slightly scented, stalks 18–33 cm × 6–8 mm, dull brownish green; buds conical with straight sides. Sepals 4, broadly lanceolate, rounded, dark green with dense purplish-black lines and spots. Petals 14–20, pale blue, white towards base, lanceolate. Stamens 50–73, with pale blue appendages; carpels 14–21. *N & tropical Africa.* G2. Summer.

Var. **albiflora** Caspary has white flowers and sepals without lines and spots.

7. N. micrantha Guillemin & Perrott. Illustration: Botanical Magazine, 4535 (1850); Graf, Tropica, 706 (1978); Heywood (ed.) Flowering plants of the world, 44 (1978).
Similar to *N. caerulea* but mature leaves only *c.* 7 × 5 cm, with spreading lobes, pale green above, bearing plantlets at the junction of leaf-blade and stalk. Flowers 10–23 cm across; petals bright to pale blue or white. *Coastal W Africa & Cape Verde Islands.* G2. Spring–autumn.

8. N. x daubenyana Daubeny. Illustration: Hay & Synge, Dictionary of garden plants in colour, 72 (1969); Graf, Tropica, 706 & 707 (1978).
Differs from its parents (*N. caerulea* and *N. micrantha*) in its brownish green leaves with dark brown blotches. Flowers rarely more than 12 cm across, pale blue, fragrant, infertile. *Garden Origin.* G2.

Produces plantlets in the same manner as *N. micrantha.*

9. N. baumii Rehnelt & Henkel. Illustration: Gardeners' Chronicle **48**: (1916).
Leaves 2–3 cm across, more or less circular, entire, shining green with a reddish sheen, spotted with violet below. Flowers to 2.5 cm across, white to very pale blue, fragrant. *Angola.* G2. Summer.

The smallest member of the genus and possibly no longer in cultivation; perhaps best regarded as a variety of *N. heudelotii* Planchon.

10. N. capensis Thunberg (*N. stellata* misapplied; *N. caerulea* misapplied.)
Illustration: Botanical Magazine, 552 (1802); Perry, Flowers of the world, 197 (1972); Graf, Tropica, 706 (1978); Blundell, The wild flowers of Kenya, pl. 47 (1982).
Leaves 25–40 cm, very broadly ovate to circular, very wavy-edged, thin and soft, green on both sides, spotted and blotched with purple beneath when young; lobes overlapping, finely tapered, acute; stalks 60–150 × 1 cm. Flowers 15–20 cm across, day-opening, fragrant, *c.* 22 cm above water; stalks *c.* 60 cm, dark green. Sepals 4, narrowly triangular, thin and fleshy, dark greenish yellow outside. Petals 20–30, bright blue, thin. Stamens 120–225; carpels 24–31. Fruit spherical, flattened, *c.* 5.5 × 3.5 cm. *S & E Africa, Madagascar.* G2 Summer.

Var. **zanzibariensis** (Caspary) Conard has larger, deeper blue flowers. *Zanzibar.* Illustration: Botanical Magazine, 6843 (1885); Hay & Synge, Dictionary of garden plants in colour, 71 (1969); Graf, Tropica, 706 (1978).
'Azurea' has pale blue flowers; leaves spotted with purple.

11. N. mexicana Zuccarini (*N. flava* Leitner). Illustration: Botanical Magazine, 6917 (1887); Rickett, Wild flowers of the United States **3**: 98 (1969); Swindells, Waterlilies, 30 (1983).
Leaves 10–13 cm across, floating or to 12 cm above water, ovate to circular, entire to wavy or toothed, blotched with brown above, at least when young, deep purple beneath (green in aerial leaves); lobes touching or overlapping, acute or obtuse, deeply indented on outer margin; stalks dark green. Flowers 6–13 cm across, day-opening, faintly scented, floating or to 12 cm above water; stalks 16–150 cm × 3–6 mm. Sepals 4, elliptic to lanceolate-ovate, green with pale brown margins

outside. Petals 12–23, pale to bright yellow. Stamens *c.* 50, yellow; carpels 7–10. *USA (Florida), Texas & Mexico.* G1–H5. Summer.

12. N. citrina Peter. Illustration: Gartenflora **84**: 301 (1935).
Differs from *N. mexicana* in its slightly larger, fragrant flowers to 15 cm across. Leaves to 25 cm across, more or less circular with rounded lobes, green on both sides. Petals *c.* 22; stamens *c.* 125; carpels *c.* 23. *E Africa.* G2. Spring–summer.

Perhaps best regarded as a variety of *N. stuhlmannii* (Schweinfurth) Gilg.

13. N. tetragona Georgi (*N. pygmaea* Aiton). Illustration: Botanical Magazine, 1525 (1813); Bailey, Standard Cyclopedia of horticulture, 2313 (1916); Swindells, Waterlilies, 30 (1983).
Leaves 5–10 × 3.5–7.5 cm, ovate to elliptic, entire, slightly leathery, dark green above (with brown blotches when young), dull red beneath; lobes diverging, often overlapping near base; stalks very slender, 10–30 cm × 4 mm. Flowers 2.5–5 cm, day-opening, scented or scentless, floating; stalks to 4 mm wide. Sepals 4, oblong to ovate or lanceolate, obtuse or acute, green outside. Petals 8–17, pure white (sometmes with faint purple lines), thin, oblong to ovate or lanceolate. Stamens 12–16 (American) or *c.* 40 (Eurasian), golden yellow; Carpels 6–8. *NE Europe, N Asia to China, Japan and Himalaya; N America (Quebec to Minnesota, Michigan, Idaho & Washington).* H2. Summer.

14. N. × helvola Anon. (*N. pygmaea* Aiton var. *helvola* Marliac; *N. tetragona* Georgi 'Helvola'). Illustration: Perry, Flowers of the world, 196 (1972); Ledbetter, Patio ponds, 76 (1982); Swindells, Waterlilies, 53 (1983).
Leaves 6 cm long at most, red blotched with brown on both sides; flowers 2–5 cm across, canary-yellow, sterile. *Garden Origin.* H4. Summer.

The hybrid between *N. tetragona* and *N. mexicana.*

15. N. alba Linnaeus. Illustration: Nicholson et al., Oxford book of wild flowers, 66 (1960); Graf, Tropica, 706 (1978); Keble Martin, New concise British flora, pl. 15 (1982).
Leaves 10–30 cm, ovate to circular, entire, sometimes shallowly notched at apex, leathery, dark green above, yellowish to reddish green beneath; lobes parallel or diverging, with more or less straight, diverging main veins. Flowers 10–20 cm

across, day-opening, scentless or faintly scented on first day, floating. Sepals 4, lanceolate, green flushed with reddish brown. Petals 20–25, white, pink or red, broadly ovate. Stamens 64–100 or more, anthers yellow to orange; carpels usually 12–20, stigma flat; fruit spherical, occasionally flattened. *Europe, NW Africa, Middle East & Caucasus to Kashmir.* H1. Summer.

Var. **rubra** Lonnroth has red flowers. *Sweden.* Illustration: Botanical Magazine, 6736 (1884); Conard, The waterlilies, pl. 15 (1905).

'Candidissima' (possibly *N. alba* × *N. candida*) is a robust variant having large flowers and leaves with overlapping lobes. A pink variant of this is known as 'Candidissima Rosea'.

Many species names have been given to different populations of *N. alba* in the past.

16. N. candida Presl & Presl. Illustration: Conard, The waterlilies, 173 (1905); Swindells, Waterlilies, 24 (1983).
Very similar to *N. alba* but smaller in all its parts. Leaf-lobes touching or overlapping with main veins curved towards each other. Flowers 6.5–7.5 cm across; petals 15–18; stamens 32–70; carpels usually 6–14, stigma strongly concave. *North C & E Europe, N Spain to Caucasus & W Siberia.* H1. Summer.

Nos. **13, 15 & 16** are not easy to distinguish and often hybridise in nature.

17. N. odorata Aiton. Illustration: Botanical Magazine, 819 (1805); Ami des jardins **735**: 44 (1957); Hay & Synge, Dictionary of garden plants in colour, 161 (1969); Swindells, Waterlilies, 30 (1983).
Leaves 12–25 cm across, circular, entire, shallowly notched at tip, thick and leathery, dark green and smooth above, rough and usually purplish red beneath; lobes touching or spreading, obtuse; stalks 30–180 cm, reddish green to dark purplish red. Flowers 7–15 cm across, day-opening, very sweetly and strongly scented, to 15 cm above water; stalks slender. Sepals 4, ovate-lanceolate, green flushed with reddish brown outside. Petals 23–32, pure white, elliptic to ovate or lanceolate. Stamens 55–106; carpels 13–25, with incurved appendages.

Var. **minor** Sims is smaller in all its parts and has purple sepals. *E USA.* Illustration: Botanical Magazine, 1652 (1814). Var. **gigantea** Tricker is larger in all its parts. Leaves upturned at margin near apex, bright red beneath. Flowers almost scentless. *SE USA, Mexico, Cuba & Guyana.*

Var. **rosea** Pursh has pink or red flowers. *E USA.* Illustration: Botanical Magazine, 6708 (1883).

18. N. tuberosa Paine. Illustration: Botanical Magazine, 6536 (1881); Perry, Flowers of the world, 196 (1972); Swindells, Waterlilies, 33 & 34 (1983).
Leaves 12–38 cm across, almost circular, entire, slightly leathery, green on both sides; sinus narrow, lobes shortly acuminate; stalks 30–200 cm × 6–9 mm. Flowers 10–23 cm across, day-opening, scentless or faintly scented, floating or to 15 cm above water. Sepals 4, plain green. Petals *c.* 20, brilliant white, lanceolate to obovate or spathulate. Stamens *c.* 50–100; carpels *c.* 14. *North-east N America.* H1. Summer.

19. N. × marliacea Marliac.
This name covers a complex group of hardy hybrids probably derived from hybridisation between forms of *N. alba, N. odorata, N. tuberosa & N. mexicana.* Many variants have been named; only a few of these are mentioned here:

'Albida': Flowers white flushed with pink, fragrant; sepals pinkish; stamens yellow; leaves red or purple beneath.

'Carnea': Flowers pale flesh-pink, darker at base, white in young plants, fragrant; sepals rose-pink; leaves purplish when young. Illustration: Ledbetter, Patio ponds, 77 (1982).

'Chromatella' (*N. tuberosa* Paine var. *flavescens* invalid) has yellow flowers; flower-stalks and leaf-stalks striped with red. Illustration: Graf, Tropica, 706 (1978); Ledbetter, Water gardens, 61 (1979); Swindells, Waterlilies, 46 & 47 (1983).

'Flammea': Flowers deep red, flecked with white.

'Rosea': Similar to 'Carnea' but young leaves purplish red, later green. Illustration: Ledbetter, Patio ponds, 77 (1982).

20. N. × laydekeri Marliac.
A complex group of hybrids involving *N. alba, N. tetragona* and possibly *N. mexicana.* They are generally similar to *N. × marliacea* but are less robust and have almost circular leaves usually mottled with brown; stamens orange-yellow.

'Fulgens': Flowers crimson with red stamens.

'Liliacea': Flowers pinkish purple with orange stamens.

'Purpurata': Flowers crimson with orange-red stamens; leaves unmottled above. Illustration: Swindells, Water gardening, 25 (1981).

21. N. lotus Linnaeus. Illustration: Conard, The waterlilies, pl. 16 (1905); Flowering plants of Africa 39: pl. 1541 (1969).
Leaves 20–50 cm across, almost circular, wavy and acutely toothed to spiny, stiff, shiny dark green above, greenish to purplish brown beneath; lobes diverging to slightly overlapping. Flowers 15–25 cm across, night- or day-opening, almost scentless, held above water; stalks 6–20 mm wide. Sepals 4, broadly ovate, rounded, green with prominent white veins outside. Petals *c.* 20, white, sometimes tinged with pink outside, rounded, firm. Carpels *c.* 30, styles yellow, incurved. Fruit 6–9 cm across. *Egypt to Tropical & SE Africa, SE Asia; NW Romania.* G2. Summer–autumn.

Var. **dentata** (Schumacher & Thonning) Nichols. Flowers pure white; petals very narrow. *C Africa.* Illustration: Botanical Magazine, 4257 (1846); Graf, Tropica, 705 (1978).

Var. **thermalis** (de Candolle) Tucson. Hardly distinct and probably just a geographical variant. *NW Romania (hot springs).*

'Dentata Superba' (*N. lotus* var. *dentata* misapplied) has flowers 30–35 cm across and hairy leaves. Illustration: Hay & Synge, Dictionary of garden plants in colour, 72 (1969)

22. N. rubra Roxburgh & Salisbury. Illustration: Botanical Magazine, 1280 (1810); Hay & Synge, Dictionary of garden plants in colour, 72 (1969); Perry, Flowers of the world, 197 (1972); Graf, Tropica, 704 & 707 (1978).
Differs from *N. lotus* in its dark purplish red flowers and its obscurely spotted, reddish to greenish brown leaves. *India.* G2. Summer.

Much used in hybridisation. 'Rosea' has paler green leaves with more obvious spots. Illustration: Botanical Magazine, 1364 (1811); Graf, Tropica, 706 (1978).

23. N. amazonum Martius & Zuccarini. Illustration: Botanical Magazine, 4823 (1854); Conard, The waterlilies, pl.19 (1905).
Leaves *c.* 17 × 14 cm, ovate to elliptic, rounded, entire to slightly wavy, thin and soft, green with faint brown spots above, brownish red beneath; lobes usually slightly overlapping; stalks purplish, with a ring of hairs at the top. Flowers 10–12 cm across, night-opening, faintly scented, floating; stalks 40–50 cm × 6–8 mm, brownish green. Sepals 4, rounded, slightly keeled, dark green with purplish lines and dots outside. Petals 16–20, very pale yellow

becoming darker, thick. Stamens 130–200, pale yellow; carpels 25–35, stigmas with club-shaped appendages. Fruit almost spherical, *c.* 4 cm across. *Mexico to Brazil.* G2. Spring–summer.

2. NUPHAR Smith
J.C.M. Alexander
Aquatic perennial herbs with stout, creeping rhizomes. Leaf-blades narrowly to broadly ovate or circular, with a deep basal sinus, not peltate, long-stalked; floating leaves leathery, submerged leaves membranous. Flowers more or less spherical, yellow and green, held above water. Sepals 4–6, broadly ovate to circular, the outer green, the inner yellow tinged with red or green. Petals numerous, yellow, much smaller than sepals, linear to narrowly oblong or spathulate, with nectaries on outer surface. Stamens numerous with broad filaments, borne on receptacle below ovary. Ovary superior, of 5–20 joined carpels, each with many ovules; stigmas in the form of rays on top of the ovary, styles absent. Fruit ripening above water, flask-shaped, berry-like.

About 25 species from northern temperate regions; generally less showy than *Nymphaea* but very much easier to establish and maintain. Cultivation conditions as for the hardy species of *Nymphaea.*

1a. Leaves narrowly ovate, more than 2.5
    times longer than wide   **4. japonica**
 b. Leaves broadly ovate to circular, less
    than 2 times longer than wide   2
2a. Stigmatic rays 10 or fewer; leaves
    4–14 cm   **2. pumila**
 b. Stigmatic rays 12 or more; leaves
    10–40 cm   3
3a. Sepals 6; leaves usually above water,
    lobes at right angles   **3. advena**
 b. Sepals 5; leaves all floating, lobes at
    an acute angle   **1. lutea**

1. N. lutea (Linnaeus) Smith. Illustration: Ross-Craig, Drawings of British Plants 2: pl. 3 (1948); Polunin, Flowers of Europe, pl. 18 (1969); Garrard & Streeter, The wild flowers of the British Isles, pl. 4 (1983). Rhizome 3–8 cm thick. Floating leaves 12–40 × 8–30 cm, ovate-oblong, with a deep sinus; lobes *c.* half leaf-length, at an acute angle; submerged leaves broadly ovate to circular, cordate at base. Flowers 4–6 cm across, with an unpleasant smell; stalks to 2 (rarely 3) m. Sepals 5, 2–3 cm, broadly ovate, bright yellow inside. Petals 7–10 mm, broadly spathulate. Stigmatic disc wider than ovary, usually with 15–20

rays. Fruit 3.5–6 cm. *Europe, N Asia & N Africa.* H2. Summer.

2. N. pumila (Timm) de Candolle. Illustration: Takeda, Alpine Flora of Japan, pl. 54 (1963); Keble Martin, New concise British Flora, pl. 5 (1982); Garrard & Streeter, The wild flowers of the British Isles, pl. 4 (1983).
Smaller in all its parts than *N. lutea.* Rhizome 1–3 cm thick. Floating leaves 4–14 × 3.5–13 cm, broadly ovate. Flowers 1.5–3.5 cm across. Sepals 4 or 5, circular. Stigmatic disc scarcely wider than ovary, usually with 8–10 rays. Fruit 2–4.5 cm. *N, C & E Europe, USSR (W Siberia), Japan.* H2. Summer.

3. N. advena (Aiton) Aiton. Illustration: Botanical Magazine, 684 (1803); Rickett, Wild flowers of the United States 1: 105 (1966); Gartenpraxis for 1978(8): 383. Leaves erect, 15–30 × 12–23 cm, usually well above water, but floating if water very deep, broadly ovate, shiny; lobes spreading at *c.* 90°. Flowers 2.5–4 cm across; sepals usually 6, *c.* 3 cm, ovate to circular. Petals *c.* 20, *c.* 8 mm, yellow tinged with red. Stamens usually more than 200. Stigmatic rays 12–24. *E & C USA, Mexico & W Indies; naturalised in S England.* Spring–summer.

4. N. japonica de Candolle. Illustration: Brüggeman, Tropical plants, pl. 197 (1957); Ohwi, Flora of Japan, 435 (1965). Rhizomes stout. Floating leaves 20–40 × 5–12 cm, narrowly ovate, arrow-shaped at base, hairy beneath when young; submerged leaves narrow, very thin and translucent, with curled edges. Flowers 5–8 cm, held well above water; sepals 5, *c.* 2.5 cm; petals spathulate, *c.* 8 mm. Stigmatic disc with *c.* 11 rays. *Japan.* H2. Summer

3. VICTORIA Lindley
J.C.M. Alexander
Annual or perennial aquatic herbs with short, stout rhizomes. Mature leaves floating, circular, very large, smooth above, with prominent spiny veins beneath; margin upturned, forming a continuous rim; juvenile leaves linear to ovate or circular, submerged or floating. Flowers held above water, 1 m or more across. Sepals 4; petals numerous, white becoming pink or red on second day, oblong, obtuse, usually longer than sepals, merging with staminodes and stamens. Stamens to *c.* 200; ovary inferior, of many fused carpels; ovules numerous. Fruit large, berry-like, prickly; seeds many, dark green.

A genus of 2 very similar species from tropical S America, grown for the spectacular size of their leaves and flowers; in temperate regions usually treated as annuals because of the difficulty in overwintering them. Cultivation conditions are similar to those for the tropical species of *Nymphaea*, though *Victoria* should be allowed as much sunlight as possible. Propagation is generally from seed which should be sown in pots in shallow water during the spring. *V. amazonica* requires a temperature of *c.* 30 °C for germination; *V. cruziana* requires *c.* 20 °C.

1a. Leaves hairless beneath; upturned rim
    to 10 (rarely to 15) cm **1. amazonica**
 b. Leaves softly hairy beneath; upturned
    rim 15–20 cm   **2. cruziana**

1. V. amazonica (Poeppig) Sowerby (*V. regia* Lindley). Illustration: Botanical Magazine, 4275 & 4278 (1847); Morley & Everard, Wild flowers of the world, pl. 169 (1970); The Garden **103**: 122 (1978); Kew Magazine 4(2): 82 (1987).
Leaves 1–2 m across, reddish purple and spiny beneath; rim of larger leaves to 10 (rarely to 15) cm. Sepals prickly, reddish purple outside. Seeds ellipsoid, 7–8 cm × 5.5–6 mm. *Amazon Basin, Guyana, French Guiana & Surinam.* G2. Summer–autumn.

2. V. cruziana Orbigny (*V. regia* misapplied; *V. trickeri* invalid). Illustration: Journal of the Royal Horticultural Society **91**: 160 (1966).
Leaves green, softly hairy and spiny beneath; rim of larger leaves 15–20 cm. Sepals prickly at base only, green outside. Seeds almost spherical, 8–10 mm. *Brazil (Parana), N Argentina & Paraguay.* G2. Summer.

V. 'Longwood Hybrid' (*V. amazonica* × *V. cruziana*) is occasionally grown.

4. EURYALE Salisbury
J.C.M. Alexander
Perennial herbs; rhizome stout; all parts densely covered in short, curved spines. Mature leaves floating, to 1.5 m across, more or less circular, peltate, dark green and wrinkled above, purple and with prominent, spongy ribs beneath; margin flat; stalk long. Young leaves submerged, folded inwards, slipper-like, bursting through an enclosing membrane. Flowers, 4–6 cm across, often submerged, usually failing to open. Sepals 4, green outside, yellow within; petals numerous, shorter than sepals, red to purple or blue. Stamens numerous, all fertile, in groups of 8; ovary

inferior, of 8 fused carpels, styles absent. Fruit berry-like, 8-celled, prickly, with many seeds.

A genus of a single species from still and slow-moving water in Asia. Because of its large size it is often grown as an annual. Cultivation conditions are similar to those for *Victoria*.

1. E. ferox Salisbury. Illustration: Botanical Magazine, 1447 (1812); Cook, Water plants of the world, 336 (1974).
N India, China, Japan & Taiwan. G2. Summer.

5. CABOMBA Aublet
Perennial aquatic herbs with long rooting stems. Leaves stalked, of 2 types: submerged leaves opposite or in whorls, circular in outline, very finely divided into linear or narrowly spathulate lobes; floating leaves usually alternate, peltate, entire (rarely divided), tapered at both ends. Flowers solitary, axillary on long stalks, floating or above surface. Sepals 3, petal-like. Petals 3, indistinctly divided into narrow claw and broad blade, with 2 small side lobes near base (in ours). Stamens 3–6; carpels 2–4 (rarely several), free, narrowly pear-shaped, tapered to a short style; stigma with papillae; ovules usually 3, scattered, pendulous. Fruit indehiscent, leathery.

About 7 species from still and slow-moving water in the warm temperate New World; usually grown as oxygenators in aquaria or small ponds. They prefer lime-free water and can be propagated from seeds or by division; detached pieces will survive in water for several weeks. Literature: Fassett, N.C., A monograph of Cabomba, *Castanea* **18**: 116–128 (1953).

1a. Flowers white, blue or purple; lobes of submerged leaves very narrowly spathulate, broadest towards tip　　2
 b. Flowers yellow; lobes of submerged leaves linear　　3
2a. Petals white or blue, occasionally purple-edged, rounded at tip
　　　　　　　1. caroliniana
 b. Petals purple, shallowly notched at tip
　　　　　　　2. pulcherrima
3a. Young internodes red-hairy; lobes of submerged leaves with distinct midrib
　　　　　　　3. australis
 b. Young internodes yellow-hairy; lobes of submerged leaves without distinct midrib　　**4. aquatica**

1. C. caroliniana Gray (*C. viridifolia* invalid; *C. aquatica* de Candolle not Aublet). Illustration: Parey's Blumengärtnerei, 608

(1958); Rickett, Wild flowers of the United States **2**: 166 (1967); Stodola, Encyclopedia of water plants, 242 (1967).
Stem to 2 m, red-hairy above when young. Submerged leaves 2–5 cm wide; lobes narrowly spathulate, 0.3–0.8 mm wide near tip, narrower towards base, sometimes marked with red; midrib indistinct. Floating leaves 1.5–2.5 mm wide, linear to oblong, entire or sagittate at base. Flowers 7–11 mm long; petals obovate, rounded at tip, white, sometimes bordered with purple; side-lobes yellow. *E, C & S USA*. G1–H5.

2. C. pulcherrima (Harper) Fassett (*C. caroliniana* Gray var. *pulcherrima* Harper). Similar to *C. caroliniana*; sepals and petals shallowly notched at tip; petals purple. *SE USA*. G1.

3. C. australis Spegazzini.
Young internodes red-hairy. Lobes of submerged leaves 0.3–0.6 mm wide, linear, marked with red lines; midrib distinct. Floating leaves 1–1.5 mm wide. Flowers 8–12 mm long; petals very broadly obovate, rounded at tip, white; side-lobes yellow. *S Brazil, Paraguay, Uruguay, E Argentina & Chile*. G1–G2.

4. C. aquatica Aublet. Illustration: Stodola, Encyclopedia of water plants, 239 (1967). Young internodes yellow-hairy. Lobes of submerged leaves 0.1–0.4 mm wide, linear, unmarked; midrib lacking. Floating leaves often absent, broadly elliptic, 1–2 cm wide (rarely narrow as in other species). Flowers 4.5–10 mm long; sepals yellow; petals yellow, triangular, rounded at tip. *Guyana, Surinam, French Guiana, adjacent Brazil*. G2. Summer.

6. BRASENIA Schreber
J.C.M. Alexander
Perennial aquatic herbs, covered in mucilage; stems rhizomatous, branching and rooting, bearing creeping runners. Leaves alternate, to 10 cm, ovate, peltate, entire, all floating (except in seedlings), shiny bright green with red margins above, usually rough and purplish beneath. Flowers solitary, axillary on long stalks, 1–1.5 cm across, floating, lying sideways on the surface at night. Sepals 3 or 4, free; petals 3 or 4, free, linear, purple. Stamens 18–36 with slender filaments; carpels 4–18, free, oblong to ovoid; ovules scattered. Fruit an indehiscent, leathery, beaked pod, usually with 2 seeds.

One sporadically distributed, cosmopolitan species of still and slow-moving, often deep water. The flowers are not of any great merit but the plant is

effective at pool margins, though rather difficult to establish. Propagation is from offsets or seed.

1. B. schreberi Gmelin (*B. peltata* Pursh; *Hydropeltis purpurea* Michaux). Illustration: Botanical Magazine, 1147 (1809); Rickett, Wild flowers of the United States **5**: 105 (1971); Cook, Water plants of the world, 169 (1974).
Tropical Asia, Africa, Australia & America: N America & Japan. G1–2. Summer.

7. NELUMBO Adanson
J.C.M. Alexander
Large, vigorous aquatic herbs, with spreading rhizomes bearing tubers and rooting at the nodes; all parts with white latex. Leaves alternate, hairless, kidney-shaped to circular, floating when young, to 2 m above water and funnel-shaped when mature; veins prominent; stalks to 2 m or more, spiny. Flowers held well above water; sepals 4 or 5, greenish, persistent; petals 10–25, soon falling; stamens to *c.* 200, with appendages; sepals, petals and stamens attached below receptacle. Carpels usually 12–30, free but individually sunk in pits in the stiff, corky receptacle; ovules solitary, apical. Fruits very hard, nut-like.

A genus of 2 species from warm-temperate and tropical Asia and the New World, with a very long history of cultivation for ornamental, religious and culinary purposes. They require water 1–2.5 m deep with underlying loam, enriched with nitrogen, *c.* 40 cm deep. The plants are very strong-growing, to 20 m per year in favourable conditions, and thus do not blend well with other aquatic vegetation; they can be grown in cooler areas if the rhizomes are protected from frost, but perform better in warm conditions. Propagation is either from seed, which may need scarifying, or by division.

1a. Leaves usually wavy; flowers at about same level as leaves; fruits almost spherical　　　　**1. lutea**
 b. Leaves not usually wavy; flowers held above leaves; fruits ellipsoid
　　　　　　　2. nucifera

1. N. lutea (Willdenow) Persoon (*N. pentapetala* (Walter) Fernald). Illustration: Botanical Magazine, 3753 (1840); House, Wild flowers, pl. 55 (1961); Correll & Correll, Aquatic and wetland plants of southwestern United States **2**: f. 447 (1975).
Leaves to 60 cm across, not usually wavy. Flowers yellow, 10–25 cm across, fragrant, at about same level as leaves. Fruits more

or less spherical. *S & E North America, Mexico, West Indies, C America to Colombia; widely introduced.* H5–G1. Summer.

'Flavescens' (*N. flavescens* Bailey) has smaller flowers and leaves with central red spot.

2. N. nucifera Gaertner (*N. speciosa* Willdenow). Illustration: Botanical Magazine, 903 (1806), 3916 & 3917 (1842); Masefield et al., The Oxford book of food plants, pl. 33 (1969); Perry, Flowers of the world, 192, 193 (1972).
Leaves to 1 m across, often wavy at margin. Flowers white tipped with pink, *c.* 30 cm across, very fragrant, held well above leaves. Fruits ellipsoid. *SW Asia to India, Japan & Australia; widely introduced; naturalised in S & SE Europe.* H5–G1. Spring–summer.

Cultivated for its edible fruits and rhizomes; regarded as sacred in India and China for many centuries.

'Alba' and 'Alba Grandiflora' have white flowers; 'Roseum' has deep pink flowers; 'Shiroma' has flowers cream tinged with green.

LXXVIII. CERATOPHYLLACEAE

Aquatic herbs. Leaves whorled, without stipules. Flowers unisexual, usually solitary. Perianth with 1 whorl of united segments. Ovary superior. Fruit a nut.

A family containing 1 cosmopolitan genus only.

1. CERATOPHYLLUM Linnaeus
V.A. Matthews
Rootless, submerged, aquatic herbs, perennating by buds. Stems branched with only 1 branch at a node. Leaves stalkless, in whorls of 6–12 (usually 8–10, thread-like, rather rigid and brittle, each leaf forked 1–4 times, the ultimate segments with 2 rows of tiny teeth and 2 apical bristles. Flowers unisexual, male and female borne on the same plant, usually solitary in a leaf-axil. Perianth with 8–13 linear lobes which are united at the base. Stamens 10–20; filaments absent or short; anther-connectives prolonged at the top into 2 spines. Ovary 1-celled. Fruit a nut with a terminal spine and with or without 2 basal spines.

About 30 species have been named, but recent study suggests that there are only 2 distinct species, both of which have a

number of local variants. This recent concept is followed here.

The plants must be grown under water and will tolerate almost no emergence. They are suitable for aquaria and garden pools, and are good oxygenators. Propagation is by cuttings.
Literature: Wilmot-Dear, M., Ceratophyllum revised – a study in fruit and leaf variation, *Kew Bulletin* 40: 243–271 (1985).

1a. Leaves forked 1 or 2 times, the ultimate segments with many, obvious, marginal teeth
1. demersum
 b. Leaves forked 3 or 4 times, the ultimate segments with few, rather inconspicuous marginal teeth
2. submersum

1. C. demersum Linnaeus. Illustration: Ross-Craig, Drawings of British plants 27: pl. 42 (1970); Correll & Correll, Aquatic and wetland plants of southwestern United States 2: 914 (1975); Sainty & Jacobs, Waterplants of New South Wales, 88–90 (1981); Kew Bulletin 40: 260C,D,H (1985).
Stems 30–150 cm, sometimes to 2 m. Leaves forked 1 or 2 times, dark or bright green, the segments linear, flattened, with many obvious marginal teeth. Nuts 4–5.5 mm, ovoid or ellipsoid, black, with a terminal spine at least as long as the nut, and 2 basal spines of variable length; surface more or less smooth, gland-dotted. *Cosmopolitan.* H1. Summer–autumn.

2. C. submersum Linnaeus. Illustration: Stodola, Encyclopedia of water plants, 71 (1967); Ross-Craig, Drawings of British plants 27: pl. 43 (1970); Cook, Water plants of the world, 178 (1974); Kew Bulletin 40: 260A (1985).
Stems 20–100 cm. Leaves forked 3 or 4 times, light or bright green, the segments linear, flattened, with few, rather inconspicuous marginal teeth. Nuts 3–5 mm, spherical or ellipsoid, with or without a short terminal spine and with or without 2 basal spines, with a distinct marginal rim; surface usually warty-papillose. *Cosmopolitan.* H1. Summer–autumn.

Var. **echinatum** (Gray) Wilmot-Dear (*C. echinatum* Gray; *C. demersum* Linnaeus var. *echinatum (Gray) Gray*). Illustration: Kew Bulletin 40: 260N–R, 263A–C (1985). Nuts always with terminal and basal spines and with a marginal wing to 1 mm wide, formed from the joined bases of lateral teeth or spines.

LXXIX. SAURURACEAE

Perennial, often aromatic herbs usually with short, thick rhizomes and creeping stolons bearing erect or ascending stems with conspicuous nodes. Leaves alternate or all basal, simple, with pinnate or palmate veins; stipules sheathing, partly fused to the leaf-stalks. Flowers radially symmetric, bisexual, in loose or dense terminal or leaf-opposed racemes or spikes, often with a whorl of petal-like bracts at the base (when the whole inflorescence resembles an *Anemone* flower). Smaller bracts may be present below each flower. Sepals and petals lacking. Stamens 3 (rarely 4), 6 or 8, filament-bases often fused to carpels. Carpels 3–5, superior or inferior and sunk in the inflorescence-axis, free to partly or completely fused; styles free. Free carpels each with 2–4 ovules and forming individual, dehiscent, capsule-like fruits. Fused carpels with 6–10 parietal ovules per cell, forming a thick, occasionally slightly fleshy capsule, opening at the apex.

A small family of moisture-loving and aquatic plants from Asia and N America, containing 5 genera and about 7 species.

1a. Flowering heads without petal-like bracts at the base   **3. Saururus**
 b. Flowering heads with petal-like bracts at the base   2
2a. Leaves mostly basal, with pinnate veins   **1. Anemopsis**
 b. Leaves alternate on stem, with palmate veins   **2. Houttuynia**

1. ANEMOPSIS Hooker
J.C.M. Alexander
Leaves mostly basal; stalks 4–18 cm, hairy; blades slightly shorter, elliptic to oblong, obtuse to rounded, truncate to cordate at base, gland-dotted, slightly hairy on margins, with *c.* 6 pairs of pinnate side-veins. Stems to 50 cm, erect, woolly, round in section, with a stalkless, clasping leaf in the upper half and 1–3 smaller leaves in its axil. Flowers in a dense, conical head 1–4 cm long, with *c.* 6 unequal, persistent, whitish, spreading, petal-like bracts at its base; bracts 1–3 cm, some spotted with red, becoming brown and bent downwards in fruit. Each flower, except the lowest, with a small, white, spathulate bract at its base. Stamens 6, free, anthers 2-celled. Styles 3 (rarely 4), ascending, blunt, narrowly conical. Ovary sunk in inflorescence-axis, 1-celled, with 15–25 ovules.

A genus of 1 species from western N America, whose inflorescence with basal bracts resemble a single *Anemone* flower. It thrives in shallow, still water, bog-gardens or very moist soil, and can be grown in containers in ponds. It prefers alkaline conditions and can be propagated by division in spring, from plantlets on the stolons or from seed.

1. A. californica Hooker. Illustration: Botanical Magazine, 5292 (1862); Rickett, Wild flowers of the United States **4**: pl. 79 (1970); Heywood, Flowering plants of the world, 39 (1978); Everett, New York Botanical Gardens illustrated encyclopedia of horticulture, 166 (1980).
SW USA, Mexico. H5. Summer–autumn.

2. HOUTTUYNIA Thunberg
J.C.M. Alexander
To 60 cm, foetid. Leaves alternate, stalks 1–5 cm, often red; stipules 1.5–2.5 cm, blades bluish green, gland-dotted, ovate, acute to slightly acuminate, cordate at base, 3.5–9 × 3–8 cm, with *c.* 5 hairy, palmate veins, margins often red. Flowers in dense cylindric, terminal heads 1–3 cm long, with 4–6 whitish, oblong to obovate, spreading bracts at the base; bracts 1–3 cm × 7–15 mm, withering but persistent. Inflorescence-stalk hairless, 2–3 cm. Stamens 3 (rarely 4), filaments fused to base of ovary. Ovary 1-celled, of 3, partly fused carpels, bearing 3 (rarely 4) hairy, curled styles. Fruit a capsule, splitting above, at the base of the styles. Seeds numerous, small.

A genus of 1 species from the East Himalayas, Taiwan, Japan and Java whose inflorescences with large basal bracts resemble some species of *Cornus*. Seed is sometimes developed without fertilisation taking place. It performs well in shady conditions in shallow water or moist soil, making effective, though rather late-growing ground-cover. It can become rather invasive and should not be planted near small specimen plants. It can also be grown in drier and more open situations when it is generally less vigorous. It is easily propagated by division in the spring or from cuttings taken in the summer, or from seed, which should be planted soon after harvesting.

1. H. cordata Thunberg. Illustration: Botanical Magazine, 2731 (1827); Hay & Beckett, Reader's Digest encyclopedia of garden plants and flowers, 346 (1971); Heywood, Flowering plants of the world, 39 (1978).
E Himalaya, China, Taiwan, Japan, Indonesia (Java). H3. Summer.

3. SAURURUS Linnaeus
J.C.M. Alexander
Leaves alternate; blades with palmate veins converging towards the tip. Stipules indistinct. Flowers in moderately dense, slender racemes, without petal-like bracts at the base. Stamens 6–8; filaments free from carpels. Carpels 3–5, fused at base, with 2–4 ovules per carpel; styles free, curved. Fruit spherical, slightly fleshy, splitting into indehiscent, 1-seeded units.

A genus of 2–4 species from N America and Asia. Cultivation conditions are generally the same as for *Anemopsis*, though they prefer a position in full sun.

1a. Filaments slender; anthers above stigmas   **1. cernuus**
  b. Filaments thick; anthers below stigmas   **2. chinensis**

1. S. cernuus Linnaeus (*S. lucidus* Donn). Illustration: Hortus Third, 1008 (1976); Dutta, Water gardening indoors and out, 97 (1977); Everett, New York Botanical Gardens illustrated encyclopedia of horticulture, 3063 (1982).
Stems to 1.5 m, erect, simple or few-branched, downy naked, below, leafy above. Leaf-blades lanceolate to ovate, acuminate, cordate at base, 7.5–15 × 2–9 cm, dark green, downy when young, with 5–9 veins; stalks shorter than blades. Inflorescence-stalks downy, 3–8 cm; racemes few, to 30 × 1.5 cm, arching at tip. Flowers white, fragrant, each with a small bract borne on or fused to the flower-stalk. Stamens 3–7, with long, white, slender filaments, anthers above the stigmas. Carpels 3 or 4, stigmas curved outwards. Fruit wrinkled and warty. *Eastern N America.* H3. Summer.

2. S. chinensis (Loureiro) Baillon (*S. cernuus* Thunberg not Linnaeus). Illustration: Perry, Water gardening, edn 3, 114 (1961); RHS Dictionary of gardening, 1875 (1965); Li et al., Flora of Taiwan **2**: 555 (1976).
Similar to *S. cernuus.* Upper leaves often pale yellow; racemes *c.* 12 cm; flowers yellowish white; filaments short, anthers below stigmas. *India, China, Korea, Japan, Vietnam, Ryukyu and Philippine Islands.* H3. Summer.

LXXX. PIPERACEAE

Erect or climbing shrubs, small trees or soft, fleshy terrestrial or epiphytic herbs. Leaves simple, stalked or rarely almost stalkless, alternate, opposite or whorled, entire, palmately or pinnately veined, stipules lacking or adhering to the stalk. Inflorescences solitary or multiple, simple or compound spikes or slender racemes; axillary, terminal or leaf-opposed. Flowers bisexual or unisexual, small, in the axils of variously shaped but mostly more or less peltate bracts. Perianth absent; stamens 1–10, free or attached near the base of the ovary, anthers 2-celled, opening longitudinally; ovary superior, stalkless or stalked (occasionally becoming stalked as the fruit develops), 1-celled, with a single basal ovule. Fruit a small drupe.

A family of 8 genera and *c.* 3000 species widely distributed in the tropics but particularly concentrated in central and northern South America. Several species have been cultivated for their attractive foliage over many years while many others have been introduced from the wild for a rather brief period of cultivation before disappearing into obscurity.

Many species have a peppery or pungent odour which is often persistent after drying. This is due to the widespread occurrence of resin- or oil-containing sacs in the tissues. Literature: Trelease, W. & Yuncker, T.G., *The Piperaceae of northern South America* (1950); Yuncker, T.G., The Piperaceae of Brazil I, *Hoehnea* **2**: 19–366 (1972); Yuncker, T.G., The Piperaceae of Brazil II, *Hoehnea* **3**: 29–284 (1973); The Piperaceae of Brazil III, *Hoehnea* **4**: 71–413, (1974); The Piperaceae of Brazil IV, *Hoehnea* **5**: 125–145 (1975); McKendrick, M., The genus Peperomia: taxonomy and cultivation, *The Plantsman* **9**: 163–189 (1987).

1a. Plants herbaceous, fleshy; stigma 1, stamens 2; the stem in cross-section with scattered vascular bundles only   **1. Peperomia**
  b. Plants more or less woody, rarely fleshy; stigmas 2–4, stamens 2–6; the stem in cross-section with a continuous ring of vascular tissue in all but the very young growth   **2. Piper**

1. PEPEROMIA Ruiz & Pavon
G. Argent
Erect to prostrate, tough, fleshy or succulent herbs with scattered vascular bundles. Leaves without stipules. Inflorescence a spike with a thick fleshy axis, sometimes these spikes aggregated into compound panicle-like inflorescences. Flowers subtended by round or peltate

bracts which may be loosely or densely arranged, bisexual, with 2 stamens and a single stigma near the apex of the ovary.

A genus estimated to have up to 1000 species with the distribution of the family. Most Peperomias do particularly well in an open, freely draining compost kept rather dry and a hot, humid atmosphere. Their requirement for heat caused them to lose favour when the price of heating rose sharply; many of the thinner-leaved species, particularly, will not stand draughts and low humidities commonly associated with the household environment. Nevertheless they are attractive foliage plants which do well in poor light and several of the thicker-leaved species will tolerate dry conditions, at least for periods. These thicker-leaved species are very suitable for intermittent staged displays and the toughest will make handsome foliage plants for the house. The selection of species for inclusion has been difficult for it is suspected that various species move in and out of favour rather erratically and new collections can appear from the wild and be propagated very quickly. They are easily propagated by stem-cuttings or more slowly from young leaf-cuttings in the manner used for African Violets. Flowering times have not been given as the flowers are mostly insignificant and erratically produced, though most species flower best towards the end of the summer.

1a. Leaves opposite or whorled, 2 or more, at least at some nodes   2
 b. Leaves alternate, only 1 per node   15
2a. Largest leaves cordate at base or peltate   **23. fraseri**
 b. Largest leaves tapering or rounded at base   3
3a. Leaf-blades more than twice as long as broad   4
 b. Leaf-blades less than twice as long as broad   8
4a. Leaves acutely pointed   **3. puteolata**
 b. Leaves obtuse or rounded   5
5a. Leaves less than 1 cm wide, pinnately or obscurely veined   6
 b. Leaves more than 1 cm wide, palmately 3-veined   **4. pereskiifolia**
6a. Erect plants with pinnately veined leaves   **7. galioides**
 b. Spreading plants with obscurely veined leaves   7
7a. Leaf-blades to 1.2 cm **6. microphylla**
 b. Leaf-blades more than 1.5 cm
                              5. rhombea

8a. Largest leaves less than 1 cm wide   9
 b. Largest leaves more than 1 cm wide
                                       10
9a. Erect plant with red stems **8. rubella**
 b. Prostrate plant with green stems
                            6. microphylla
10a. Prostrate plants with tough, square or markedly angled stems; leaves prominently longitudinally striped with discoloured tissue around the main veins   11
 b. Erect or spreading plants with fleshy, rounded or only weakly angled stems; leaves not longitudinally striped although the veins themselves may be distinct   12
11a. Stem diameter of the ultimate internode *c.* 1 mm; leaves to 3 cm
                         2. quadrangularis
 b. Stem diameter of the ultimate internode more than 2 mm; leaves more than 3 cm   **1. dahlstedtii**
12a. Stems hairless or with minute glands only   **4. pereskiifolia**
 b. Stems hairy   13
13a. Largest leaves less than 1 cm
                              8. rubella
 b. Largest leaves more than 1 cm   14
14a. Largest leaf-stalks to 3 mm, lower leaves succulent   **9. verticillata**
 b. Largest leaf-stalks more than 8 mm, lower leaves not succulent
                               10. blanda
15a. Leaves purse-shaped, folded vertically with the two halves adhering at the edges   **11. dolabriformis**
 b. Leaves not purse-shaped, not or hardly folded, the two halves never adhering   16
16a. Stalks of upper leaves much longer than blades   17
 b. Stalks of upper leaves as long as or shorter than blades   25
17a. Leaf-bases tapering   18
 b. Leaf-bases rounded, cordate or leaf peltate   19
18a. Leaf-blades bent back at right angles or further to stalks; veins at least partly pinnate   **22. maculosa**
 b. Leaf-blades standing erect, nearly in the same plane as stalks; veins entirely palmate   **24. ornata**
19a. Leaves pinnately veined, several main veins arising well above the base
                             22. maculosa
 b. Leaves palmately veined, the main veins all radiating from close to the point where the stalk is attached   20
20a. Leaves sometimes variegated but never in longitudinal stripes   21

 b. Leaves longitudinally striped above, alternately silvery or grey-green and darker green   24
21a. Leaves smooth or depressed to 0.5 mm at the main veins only   22
 b. Leaves deeply wrinkled, depressed to more than 1 mm at both main and secondary veins   23
22a. Leaf-stalks not grooved; inflorescence terminal, regularly branched, spikes white   **23. fraseri**
 b. Leaf-stalks grooved; inflorescence lateral, irregularly branched, spikes green   **24. ornata**
23a. Leaves deeply folded into V-shaped wrinkles whose depth often exceeds their width, the upper surface dark green (except for white areas of variegated variants)   **25. caperata**
 b. Leaves folded into broadly U-shaped wrinkles which are much broader than deep, the upper surface grey-green   **26. griseo-argentea**
24a. Leaves peltate with a smoothly rounded base   **27. argyreia**
 b. Leaves cordate to auriculate, distinctly indented towards the stalk
                            28. marmorata
25a. Leaves circular, to 1 cm   26
 b. Leaves mostly distinctly pointed, more than 1 cm   27
26a. Bushy, more or less erect plants
                               14. orba
 b. Slender, creeping plants
                          21. rotundifolia
27a. Leaves peltate   **22. maculosa**
 b. Leaves not peltate   28
28a. Leaves and stems entirely covered in white woolly matted hairs
                              13. incana
 b. Leaves and stems hairless or with straight hairs   29
29a. Largest leaves less than 1 cm wide
                             12. nivalis
 b. Largest leaves more than 1 cm wide
                                       30
30a. Leaf base tapered   31
 b. Leaf base rounded, cordate or auriculate   36
31a. Leaves finely hairy near the margin
                             16. velutina
 b. Leaves hairless except sometimes on the stalks   32
32a. Leaves with long decurrent margins which almost join the stem; stalks 5 mm or less   **17. clusiifolia**
 b. Leaves without long decurrent margins, stalks more than 1 cm   33
33a. Stalks of the lower leaves more than 5 cm   **22. maculosa**

b. Stalks of the lower leaves less than
3 cm   34

34a. Leaf-blades to 4 cm   35

b. Leaf-blades more than 8 cm
18. obtusifolia

35a. Leaf-blades with broad silvery stripes
above, red beneath   **15. metallica**

b. Leaf-blades green or variegated but
not as above   **20. glabella**

36a. Leaf-blades with conspicuous pale
central stripes the length of the upper
surface   37

b. Leaf-blades without conspicuous pale
central stripes the length of the upper
surface, although sometimes variously
variegated   38

37a. Stems hairy   **16. velutina**

b. Stems hairless   **15. metallica**

38a. Basal leaves forming a rosette almost
without internodes   **23. fraseri**

b. Basal leaves not forming a rosette,
internodes more than 5 mm   39

39a. Stems completely hairy   **14. orba**

b. Stems hairless or with 2 lines of
minute hairs   40

40a. Leaf-bases broadly tapering to
rounded, tips obtuse or rounded
20. glabella

b. Leaf-bases cordate, tips acute with a
long point   **19. scandens**

1. P. dahlstedtii de Candolle (*P. forsteri*
invalid). Illustration: Graf, Exotica, edn 11,
1932 (1982); Hoehnea 4: 292, f. 345 (1974).
Prostrate plant to 40 cm, stems square or
markedly angled, 3–5 mm in diameter,
hairless, rather tough, glossy, reddish.
Leaves in whorls of mostly 3 or 4, blade
elliptic, leathery, hairless, green and
minutely gland-dotted above, 3–5 ×
2–2.5 cm, palmately 3-veined for the
length of the leaf, the veins paler when
viewed against the light, protruding as
prominent ridges beneath; apex broadly
pointed to rounded, base tapering; leaf-stalk
5–8 × *c.* 2 mm, minutely hairy on the
margins, reddish, shallowly grooved.
Inflorescence terminal, without spike-
bracts, a group of up to 4 spikes; spike-stalk
hairless, 2–2.5 cm × 2–3 mm; spike
10–12 cm × 3–5 mm, yellowish green,
Brazil. G2.

Closely related to the following species
and differing in its larger size.

2. P. quadrangularis (Thompson) Dietrich
(*P. angulata* Humboldt, Bonpland & Kunth).
Illustration: Transactions of the Linnaean
Society 9: pl. 21, f.1 (1808); Trelease &
Yuncker, The Piperaceae of northern South
America, f. 450 (1950); Graf, Exotica, edn
8, 1382 (1976).

Prostrate plant to 20 cm, stems square or
angled, minutely hairy, tough and wiry,
reddish. Leaves in whorls of 2–4, blade
broadly elliptic to circular, leathery to
succulent, green above and beneath,
although young leaves may be pinkish
green with pale green veins, 1.2–3 ×
1–2.2 cm, palmately 3-veined for the
length of the leaf, the veins only very
slightly protruding beneath or quite
smooth, apex mostly rounded, occasionally
broadly pointed, base rounded; leaf-stalk
2–4 × *c.* 1 mm, minutely hairy, weakly
grooved. Inflorescence axillary or terminal,
the axis with 2 scale-like bracts half way
up; spike-stalk 1.5–2 × *c.* 1 mm; spikes
1.5–2.5 cm × 1.5–2 mm, green. *West
Indies, Panama and northern S America*. G2.

Smaller and less vigorous than
P. dahlstedtii, it is a more delicate looking
plant.

3. P. puteolata Trelease (*P. cuspidilimba*
misapplied; *P. haughtii* Trelease & Yuncker).
Illustration: Perry, Flowers of the world,
227 (1972); Graf, Exotica, edn 11, 1933,
1941, also 1941 as P. cuspidilimba (1982).
Prostrate plant to 40 cm, stems square or
markedly angled, hairless, rather tough,
glossy, reddish. Leaves in whorls of 3–4,
blade elliptic or narrowly elliptic, stiff and
leathery, hairless, green and minutely
gland-dotted above, darker above than
beneath, but conspicuously patterned
above with pale green areas around the
veins, 5–8 × 1.7–3 cm, palmately
3–5-veined for the length of the leaf, the
veins sunk into grooves above, and raised
as prominent ridges beneath, apex acute or
somewhat acuminate, with a narrowly
rounded point, base tapering; leaf-stalk
4–6 × *c.* 2 mm, minutely hairy on the
margins, reddish, grooved. Inflorescence of
1–4 spikes, both terminal and lateral, each
with or without a minute scale-like spike-
bract, spike-stalk 1–1.5 cm × *c.* 3 mm,
hairless, red; spikes 5–8 × *c.* 4 mm, green.
Colombia. G2.

A very handsome foliage plant.

4. P. pereskiifolia (Jacquin) Humboldt,
Bonpland & Kunth. Illustration: Jacquin,
Icones plantarum rariorum 2: 270 (1786);
Hooker, Exotic Flora 1: pl. 67 (1823);
Trelease & Yuncker, The Piperaceae of
northern South America, f. 467 (1950);
Graf, Exotica, edn 11, 1936, 1939 (1982).
Spreading plant to 30 cm, stems rounded to
weakly angled or grooved, hairless or
minutely gland-spotted, tough, dull reddish
green, to 5 mm in diameter. Leaves in
whorls of 2–4, blade elliptic to broadly

obovate, succulent, hairless, green,
sometimes with a thin red edge, 2.5–6 ×
1–3.5 cm, rather obscurely palmately
3-veined, the veins more or less level with
the leaf surface, slightly paler than the
surrounding leaf when viewed against the
light; apex slightly acuminate with a
rounded point or broadly to narrowly
pointed, base tapered; leaf-stalk 4–6 ×
1.5–2 mm, minutely hairy at the margins,
grooved, flushed red with minute spots.
Inflorescence terminal, spikes solitary or
2–3 together, without distinct spike-bracts,
spike-stalks hairless, 4–6 cm × *c.* 2.5 mm,
green; spike 1–1.5 cm × *c.* 3 mm, green.
Venezuela & Colombia. G2.

5. P. rhombea Ruiz & Pavon. Illustration:
Ruiz & Pavon, Flora Peruvianae et Chilensis
1: pl. 46, f. b (1798); Trelease & Yuncker,
The Piperaceae of northern South America,
f. 458, (1950).
Spreading plant to 30 cm, stems weakly
4-sided, finely hairy, tough, to *c.* 3 mm in
diameter, glossy green. Leaves mostly in
whorls of 4, occasionally 1–7, blade elliptic
or diamond-shaped, becoming succulent,
smooth, minutely hairy when very young,
becoming hairless, green, 1.6–2.3 cm ×
6–10 mm, with a slightly paler but obscure
mid-vein and sometimes two arching
laterals visible, apex obtuse to rounded,
base tapered; leaf-stalk 2–3 × *c.* 1 mm,
minutely hairy, not grooved, green.
Inflorescence a solitary terminal spike with
usually several lateral spikes in the upper
leaf axils, without special spike-bracts;
spike-stalk minutely hairy, 5–15 ×
c. 1 mm, green; spike 1.5–3 cm × *c.* 2 mm,
green. *West Indies, northern S America*. G2.

A good substitute for plastic foliage.

6. P. microphylla Humboldt, Bonpland &
Kunth (*P. hoffmannii* de Candolle).
Illustration: Humboldt, Bonpland & Kunth,
Nova genera et species plantarum 1: pl. 15
(1815); Saunders' Refugium Botanicum 1:
pl. 41 (1869); Graf, Exotica, edn 11, 1940
(1982).
Prostrate plant to 30 cm, stems rounded,
finely hairy, tough, *c.* 2 mm in diameter,
green. Leaves in whorls, mostly 4 together,
blade broadly elliptic to obovate, at first
leathery, becoming succulent with age,
hairless, green, 8–11 × 5–7 mm, veins
obscure except for the mid-vein in the
lower half of the leaf, apex rounded, base
broadly tapered to rounded, minutely hairy
on the margin near the apex when young;
leaf-stalk 2–3 × 0.75 mm, minutely hairy,
not grooved, green. Inflorescence terminal,
solitary, with normal leaves at the base,

spike-stalk minutely hairy, 9–13 ×
c. 1.5 mm; spike 1.8–2.4 cm × c. 2.5 mm,
green. *C and NW South America*. G2.

An attractive bright green trailing plant
of easy cultivation.

7. P. galioides Humboldt, Bonpland &
Kunth. Illustration: Humboldt, Bonpland &
Kunth, Nova genera et species plantarum
1: pl. 17 (1815); Graf, Exotica, edn 11,
1937 (1982).
Erect branching plant to 40 cm, stems
rounded, finely hairy, tough, to c. 5 mm in
diameter, green. Leaves in whorls, usually
of 4–7 per node; blade obovate on young
shoots, otherwise mostly elliptic, finely
hairy above and beneath, green,
1–2.5 cm × 2.5–9 mm, finely and rather
obscurely pinnately veined, apex broadly
pointed to rounded, base broadly to
narrowly tapered; leaf-stalks c. 2 × 1 mm,
hairy, finely grooved. Inflorescence
terminal, solitary from a leafy whorl, spike-
stalk minutely hairy at the base only,
6–12 × c. 1.5 mm; spike 3–5.5 cm ×
c. 2 mm, green. *Widespread in C & S
America and the West Indies*. G2.

8. P. rubella (Haworth) W. J. Hooker.
Illustration: Hooker, Exotic Flora 1: t. 58
(1823); Graf, Exotica, edn 11, 1932 (1982).
Erect branching plant to 15 cm, stems
rounded, rather sparsely long-hairy,
c. 3 mm in diameter, dark red. Leaves in
whorls of mostly 4–5, the blade elliptic,
succulent, convex, sparsely hairy and pink
beneath, flat or slightly concave, hairless
and pale to dark green above, 5–9 ×
2.5–5 mm, mid-vein only sometimes visible
as a silver line above and a paler line
beneath, apex broadly pointed to rounded,
base broadly tapered to rounded; leaf-stalk
1.5–2 × c. 0.5 mm, minutely hairy, red,
not grooved. Inflorescence terminal and
solitary or a cluster with 2–3 laterals from
the upper leaf-axils; spike-stalk 2–5 ×
c. 0.5 mm; spike 1.5–4.5 cm × c. 1 mm,
red. *West Indies*. G2.

9. P. verticillata (Linnaeus) Dietrich
(*P. pulchella* Dietrich). Illustration: Graf,
Exotica, edn 11, 1934 – as P. pulchella,
1942 (1982).
Erect plant, branches to 50 cm, stems
rounded, shaggy-hairy below, finely white-
hairy above, pale green to pink, to 8 mm in
diameter. Leaves in whorls of usually 5 at a
node; blade softly white-hairy, extremely
variable in shape, size and succulence; at
the base of the plant circular, thickly
succulent, 8 × 8 mm, pale green above
reddish or pink beneath, without distinct
veins; the upper leaves larger, 2.5–3 ×

2–2.5 cm, obovate, leathery, entirely pale
green, palmately 3-veined with the lateral
veins disappearing before reaching the top
of the leaf, the veins clearly paler above and
the mid-vein slightly raised beneath. Leaf-
apex rounded in basal, to broadly pointed
in the upper leaves, base rounded in basal
to broadly wedge-shaped in the upper
leaves, margin flat; leaf-stalk varying from
almost absent at the base of the plant to
3 × 2 mm in upper leaves. Inflorescence a
terminal panicle of numerous spikes with
leafy spike-bracts, spike-stalk hairy,
1–1.7 cm × 1mm, spike 2–2.5 cm ×
1.5–2 mm, green. *West Indies*. G2.

10. P. blanda (Jacquin) Humboldt,
Bonpland & Kunth. Illustration: Humboldt,
Bonpland & Kunth, Nova genera et species
plantarum 1: pl. 3 and pl. 13 – as P.
dissimilis (1815); Hooker, Exotic Flora 1: pl.
21 (1823); Graf, Exotica, edn 11, 1934
(1982).
Erect plant to 40 cm, stems rounded, softly
hairy, rather fleshy towards the base, soft
above, pink to red. Leaves in whorls mostly
of 3, blade broadly elliptic to obovate, thin
and soft, finely hairy above and beneath,
green above with the veins standing out as
a paler pattern, uniformly dark green
beneath, 2.5–4 × 2–3.5 cm, palmately
3-veined, the main vein also pinnately
divided above, the veins slightly impressed
above and raised beneath, apex broadly
pointed, sometimes shortly acuminate, base
broadly tapered to rounded; leaf-stalk
5–17 × c. 1.5 mm, grooved, hairy, pink or
green. Inflorescences terminal and from the
upper leaf-axils to form leafy panicles,
spike-stalk hairy, 1–2 cm × 1–2 mm, pink;
spike 3–8 cm × c. 2 mm, green. *Widespread
in S America, C America north to Florida,
West Indies, C & S Africa, Sri Lanka*. G2.

11. P. dolabriformis Humboldt, Bonpland &
Kunth. Illustration: Humboldt, Bonpland &
Kunth, Nova genera et species plantarum
1: pl. 4 (1815); Graf, Exotica, edn 11, 1935
(1982).
Erect, branching plant to 25 cm, stems
rounded, hairless, thick, to 2.5 cm in
diameter, at first green, later brown with a
fine pattern of vertical silvery marks and
conspicuous horizontal leaf-scars. Leaves
alternate, purse-shaped with the two halves
folded together and fused, succulent,
hairless, pale green with a dark green line
around the line of fusion, 2–3.5 cm ×
8–18 mm (folded), palmately veined with
one rather obscure arching lateral on each
face, mid-vein hardly visible, apex
mucronate, base tapered, wedge-shaped;

leaf-stalk not clearly demarcated, c. 4 ×
2 mm, hairless, green. Inflorescence a
terminal twice-branched panicle of many
spikes each arising from the axil of a
slender pointed scale-leaf 2–4 mm long,
inflorescence-stalk, gland-dotted, spike-stalk
2–4 × c. 1 mm, gland-dotted, spike
2.5–6.5 cm × 1.5–2 mm, green. *Peru*. G2.

Attractive if rather bizarre in appearance.

12. P. nivalis Miquel. Illustration: Lamb &
Lamb, The illustrated reference on cacti
and other succulents 2: 548 (1959); Graf,
Exotica, edn 11, 1943 (1982).
Erect or prostrate plant to 20 cm, stems
rounded, hairless. Leaves alternate,
succulent, folded upwards into a wedge-
shape (but not completely so as in
P. dolabriformis), 7–15 × 5–7 mm, with a
short sharp point and tapered to the
rounded base, veins obscure, leaves bright
green above, whitish to pink beneath; leaf-
stalk 6–10 × c. 1 mm, green. Inflorescence
a terminal panicle of many spikes. *Peru*. G2.

Despite being very succulent it prefers
not to dry out and to remain growing
throughout the year.

13. P. incana (Haworth) W. J. Hooker.
Illustration: Hooker, Exotic Flora 1: pl. 66
(1823); Graf, Exotica, edn 11, 1935 (1982).
Semi-erect to rather spreading plant to
30 cm, stems rounded, tough, green,
covered in white-woolly hairs, to 2.5 cm in
diameter. Leaves alternate, broadly ovate to
rounded, often slightly asymmetric,
leathery, densely white-woolly-hairy,
3–5.5 × 2.5–5 cm, mid-vein slightly raised
beneath and visible throughout the blade
length; lateral veins obscurely palmate,
visible in the basal half of the leaf only
when held against the light; apex broadly
pointed, rounded or occasionally
mucronate with the point turned
downwards, base rounded to cordate; leaf-
stalk 6–20 × c. 3 mm, pale green, densely
white-woolly-hairy, shallowly grooved.
Inflorescences axillary and terminal,
subtended by reduced, lanceolate, leaf-like,
less densely hairy, glandular bracts; spikes
mostly solitary but occasionally 2 or more
to a branch tip, spike-stalk white-woolly-
hairy, 1.5-2 cm × c. 4 mm, spike
1.1–2.2 cm × c. 4 mm, green with purple
anthers. *SE Brazil*. G2.

14. P. orba Bunting. Illustration: Baileya
14: 64, 66 & 67 (1966); Graf, Exotica, edn
11, 1935 – 'Princess Astrid'; 1937 - 'Pixie'
(1982).
Erect bushy plant to 15 cm. Stems rounded,
fleshy, to 5 mm in diameter, hairy, dull

green with red flecks. Leaves alternate, blade ovate to elliptic, smooth, leathery, softly hairy, green, mid-vein distinct to the apex, slightly raised beneath, obscurely pinnately veined at least in the larger leaves, apex acute, occasionally obtuse, base rounded; leaf-stalk 6–15 × 1.5–2 mm, hairy, pale green, minutely spotted with red, shallowly grooved. Inflorescence terminal, solitary, usually with 1 reduced, semi-sheathing leaf at the base; spike-stalk hairy, 7–23 × c. 2 mm, pink; spike 3–14 cm × c. 3 mm, dotted with red. *Unknown origin.* G2.

Commonly offered for sale as 'Princess Astrid', which is the adult form of the plant. A juvenile form with a profusion of short stems with much smaller leaves is sold as 'Pixie' or 'Teardrop'. These forms easily revert to adult foliage and have to be carefully pruned to maintain their shape.

15. P. metallica Linden & Rodigas. Illustration: Illustration Horticole 39: pl. 157 (1892); Graf, Exotica, edn 11, 1933 (1982).
Erect plant to 15 cm, stems rounded, fleshy, hairless, to 4 mm in diameter, pink or red. Leaves alternate; blade elliptic, succulent, hairless, brownish green with a broad silvery-green stripe above, reddish or pink beneath, 2–3 × 1–1.5 cm, palmately 3-veined for almost the length of the leaf; apex acutely pointed; base broadly tapered to rounded; leaf-stalk 4–9 × c. 2 mm, red, hairless, shallowly grooved. Inflorescence terminal, solitary, without spike-bracts; spike-stalk 8–15 × c.2 mm, red; spike 3–5 cm × c. 3 mm, pink. *Peru.* G2.

A very decorative species which is unfortunately one of the more delicate, requiring high temperatures and humidity.

16. P. velutina Linden & André (*P. bicolor* Sodiro). Illustration: Illustration Horticole 19: pl. 89 (1872); Trelease & Yuncker, The Piperaceae of northern South America, f. 505, 506 (1950); Graf, Exotica, edn 11, 1931 (1982).
Erect plant to 20 cm, stems round, white-hairy, fleshy, deep red. Leaves alternate, blade broadly elliptic, obovate or almost circular, leathery when young, becoming succulent, hairy above mostly around the margin and on the main veins, hairless beneath and minutely white-dotted, dull green above with shining silvery stripes on the veins, bright pinkish red beneath, 2–4.5 × 1.6–3.5 cm, palmately 5–7-veined, the basal pair often short and weak, often pinnately veined in the upper part of the leaf, the veins showing up green

when viewed against the light from underneath, more or less level with the leaf surface, the main veins very slightly protruding beneath; apex mostly obtuse, rarely broadly acute or rounded; base broadly tapered, rounded or weakly cordate, margin flat, often pink or silver; leaf-stalk 5–8 × c. 2 mm, pink, hairy, grooved above. Inflorescence terminal or lateral in upper axils when the spikes appear paired, spike-bracts leaf-like, spike-stalk red, hairless or sparsely hairy, 5–10 × c. 1 mm; spike 4–10 cm × c. 2 mm, pinkish or brownish green. *Ecuador.* G2.

A most attractive foliage plant which unfortunately is very sensitive to low humidity which causes the leaves to curl. The common variant of the plant in cultivation has small leaves and agrees most closely with *P. bicolor* but there seems to be no good reason to recognise this as separate from *P. velutina* since the dimensions overlap and there seems to be no good structural difference.

17. P. clusiifolia (Jacquin) W. J. Hooker. Illustration: Botanical Magazine, 2943 (1829); Graf, Exotica, edn 11, 1093, 1935 (variegated), 1941 (1982).
Erect plant to 25 cm, stems rounded, green to purple, hairless, tough and fleshy to c. 1 cm in diameter. Leaves alternate, blade obovate or elliptic, succulent, hairless, green or purplish (often with red or purple margin), minutely gland-dotted on the upper surface, 4–11 × 2–3.5 cm, mid-vein distinct to the apex, grooved above, very broad, red and raised at the base of the leaf beneath, otherwise not raised and hardly coloured or conspicuous; obscurely pinnately veined; apex broadly pointed to rounded, sometimes slightly indented, base tapered with the narrowing blade often clasping the stem; leaf-stalk c. 2–3 × 4–6 mm, broadly grooved above, hairless, usually dark red and minutely pitted. Inflorescence terminal or lateral, solitary, sometimes on short lateral shoots with reduced leaves, spike-stalk hairless, red, 2–3.5 cm × c. 2 mm, spike 2.5–3.5 cm × c. 3 mm, pale-green. *West Indies, possibly also Venezuela.* G2.

A tough species, resistant to adverse conditions and occurring as several cultivars with bushier habit or various colourings.

18. P. obtusifolia (Linnaeus) Dietrich (*P. tithymaloides* Dietrich; *P. magnoliaefolia* (Jacquin) Dietrich). Illustration: Jacquin, Icones plantarum rariorum 3: pl. 213 (1793); Trelease & Yuncker, The

Piperaceae of northern South America, f. 594, 595, 596 & 597 (1950); Hoehnea 4: f. 442, 446 & 446a (1974); Graf, Exotica, edn 11, 1933, 1934, 1937 & 1943 (1982).
Plant at first erect, becoming spreading to prostrate, stems round, hairless, tough, green- or red-flecked, to 1 cm in diameter. Leaves alternate, blade elliptic, broadly elliptic or obovate, occasionally almost circular, tough, smooth and leathery, hairless but often gland-dotted, especially beneath, dull green above, paler beneath, 5–15 × 3–8 cm, palmately mostly 5–7-veined near the base, pinnately veined towards the top, the basal veins strong, more than half the length of the leaf, the upper veins straight or curved, all veins level with the leaf surface except the mid-vein which is raised beneath, apex obtuse to rounded or indented, base broadly to narrowly tapered; leaf-stalk 1–3 cm × c. 3 mm, green, pink or purple, grooved above. Inflorescence mostly terminal, sometimes lateral from the upper leaf-axils, solitary or appearing paired, spike-bracts absent or present as lanceolate or boat-shaped scales, spike-stalk 1.5–5 cm × c. 5 mm, usually red, minutely glandular or hairless; spike 6–12 cm × c. 4–5 mm, green, anthers white. *Northern S America, C America north to Mexico, West Indies.* G2.

P. magnoliaefolia has been included within this highly variable and widespread species. It is supposed to differ in having hairless spike-stalks and a thicker and more gently curved hook to the beak of the fruit than *P. obtusifolia* (in the narrow sense), which has minutely hairy spike-stalks and an abruptly hooked beak to the fruit. There are numerous cultivars differing in habit and leaf coloration; this is one of the toughest species for decorative use in dry atmospheres.

19. P. scandens Ruiz & Pavon (*P. serpens* misapplied). Illustration: Graf, Exotica, edn 11, 1933 (1982).
Trailing plant to 50 cm, stems rounded, hairless, fleshy, pale green or with light to heavy red flecking. Leaves alternate; blade lanceolate, leathery, smooth, pale green, hairless, 3–5 × 1.8–3.2 cm, inconspicuously and more or less palmately 5-veined, the inner pair of veins arising well above the base of the leaf, only the midrib impressed above and raised beneath, apex drawn out into a long fine point, base rounded to cordate, minutely hairy at the edge of the reflexed margin; leaf stalk 8–25 × c. 2 mm, pale green heavily flecked with red. Inflorescence terminal, solitary,

usually in the axil of a thread-like spike-bract; spike-stalk hairless, 5–8 × c. 2.5 mm, spike 7–9 cm × c. 3 mm, green with purple anthers. *West Indies, Panama to Brazil and Peru.* G2.

The application of the name *P. scandens* to the plant sometimes referred to as *P. serpens* is unsatisfactory. It is certainly not *P. scandens* or *P. serpens*. It seems best for the time being to use the name *P. scandens* until more detailed work can resolve the problem. This is one of the most commonly cultivated species in its variegated form and can look most attractive either trained up supports or as a hanging basket plant.

20. P. glabella (Swartz) Dietrich. Illustration: Trelease & Yuncker, The Piperaceae of northern South America, pl. 511 (1950); Hoehnea **4**: f. 405 (1974); Graf, Exotica, edn 11, 1933 (1982) – as 'Variegata'; Beckett, RHS encyclopaedia of house plants, 386 (1987).
Erect to sprawling plant to 20 cm. Stems rounded, hairless except for 2 lines of minute hairs running up the stem from each side of the leaf junction, soft and fleshy, glossy, red. Leaves alternate, blade broadly elliptic to slightly obovate, fleshy, hairless, green, commonly with black glandular dots, 2.5–5 × 2.5–3.5 cm, palmately 3–5-veined, the veins slender and disappearing just above mid-leaf, apex rounded to obtusely pointed, base broadly tapered to rounded; leaf stalk 8–10 × c. 2 mm, hairless, red, broadly grooved. Inflorescence terminal and lateral from the axils of upper leaves, solitary except apparently double from a terminal position, spike-stalk hairless, often with scale-like bracts at the base, 4–7 × 2.5–3 mm, green; spike 4.5–12 cm × 2.5–3.5 mm, green. *West Indies, C & S America.* G2.

This widespread and very variable species appears fairly uniform in cultivation but there is still some doubt about the correctness of the name. It is most often seen as a variegated plant but this readily reverts to the totally green state.

21. P. rotundifolia (Linnaeus) Humboldt, Bonpland & Kunth (*P. nummularifolia* Humboldt, Bonpland & Kunth). Illustration: Trelease & Yuncker, The Piperaceae of northern South America, f. 546 (1950); Hoehnea **4**: f. 422 (1974); Graf, Exotica, edn 11, 1932, 1940 (1982).
Creeping plant to 25 cm, stems rounded, hairy or hairless, slender, less than 1 mm in diameter, green. Leaves alternate, blade circular or broadly elliptic, fleshy, hairy, sometimes sparsely so, green, paler

beneath, 2–8 × 2–8 mm, venation obscure, often a trace of a mid-vein in the basal half only, sometimes palmately 3-veined; apex rounded; base rounded, occasionally slightly cordate; leaf-stalk 1–5 × 0.2 mm, hairy, grooved. Inflorescence terminal on short erect shoots, with 1–4 reduced leaves, solitary; spike-stalk usually hairy, occasionally without hairs, 3–7 × c. 0.2 mm, sometimes swollen at the base; spike 5–20 × c. 1 mm, pinkish or green. *Widespread in C & S America, West Indies, South Africa.* G2.

A very delicate species both in form and disposition.

Var. **pilosior** (Miquel) de Candolle (*P. prostrata* Masters & Moore). Illustration: Gardeners' Chronicle 1: 716 (1879); Hoehnea **4**: f. 449 (1974) Graf, Exotica, edn 11, 1932, 1937, 1940 – as *P. nummularifolia* (1982); Plantsman **9**: 186 (1987). More densely hairy than the above and with a distinct pale green net-like pattern on the leaves above. *SE Brazil.* G2.

More vigorous and certainly more attractive than var. *rotundifolia*; plants found in cultivation are probably the direct descendants of the original introduction by Mr B. S. Williams.

22. P. maculosa (Linnaeus) W.J. Hooker (*P. variegata* Ruiz & Pavon; *P. pseudovariegata* de Candolle; *P. sarcophylla* Sodiro). Illustration: Hooker, Exotic Flora **2**: pl. 92 (1825); Trelease & Yuncker, The Piperaceae of northern South America, f. 644 (1950); Graf, Exotica edn 11, 1938 – as *P. elongata*, 1943 (1982).
Sprawling to erect plant to 20 cm, stems rounded, hairless, green heavily flecked with dark red, to 1.5 cm in diameter. Leaves alternate, blade elliptic to ovate, thick and leathery, minutely hairy or hairless, minutely spotted above and beneath, dark glossy green, paler and slightly duller beneath, 10–17 × 5–9 cm, more or less palmately 5–7-veined, the inner pair usually arising c. 2 cm from the base of the leaf, pinnately veined in the upper half with 5–8 pairs of smaller veins mostly crossing to the two inner arching veins; the main vein raised beneath, otherwise all veins level with the leaf surface; apex acute, often slightly acuminate, base rounded to broadly tapered, sometimes peltate with the lower margin up to 1 cm from the leaf-stalk; leaf-stalk 5–12 cm × 4–6 mm, hairless or minutely hairy, conspicuously mottled red and green, weakly grooved in the upper 2–3 cm. Inflorescences terminal, single or

paired, often with 1 small boat-shaped bract at the base; spike-stalk hairless but with some minute glands, 2–5 × c. 5 mm; spike 2–2.6 cm × 5–7 mm, dark purple when young, becoming green with age. *West Indies, Panama and northern S America.* G2.

A very handsome, glossy foliage plant which as understood here is very variable although individual cultivars may be constant. Most obvious is the variation of the leaves from peltate to non-peltate; this can occur on one plant although in those plants called *P. pseudovariegata* the leaves are shorter, broadly ovate and uniformly peltate; this appears to be only an extreme form of *P. maculosa*.

23. P. fraseri de Candolle (*P. resedaeflora* Linden & André). Illustration: Botanical Magazine, 6619 (1882). Graf, Exotica, edn 11, 1935 & 1938 (1982); Beckett, RHS encyclopaedia of house plants, 385 (1987); Plantsman **9**: 179 (1987).
Erect rosette plants to 40 cm at flowering. Stems rounded but grooved lengthwise, minutely hairy, dull red, to 1.5 cm in diameter. Leaves alternate in the basal rosettes, whorled above where the stem elongates; blade broadly ovate to almost circular, leathery, minutely gland-dotted on the upper surface otherwise hairless, green above, the veins becoming purplish on older leaves, pale green beneath with irregular pink spotting and bright red to pink veins, 2.5–4.5 × 2.5–4 cm, palmately 3–5-veined almost to the edge of the leaf, the veins slightly depressed above and raised beneath, the apex broadly pointed, rarely rounded, base cordate, the margin flat or slightly reflexed in the upper half, upper margin shallowly wavy; leaf-stalk 2–7 cm × 1.5–3 mm, dull red or pink, not grooved. Inflorescence a terminal panicle of numerous spikes sometimes with a few leafy bracts near the base, each of the ultimate spikes in the axil of a small thread-like bract, spike-stalk hairless, 6–8 × 1 mm shining white, spike 2–3 × 1.5 mm, white or greenish white. *Ecuador.* G2.

A very handsome and striking plant which has for long attracted attention. It is probably the only species that could be claimed to be grown for its flowers which in addition to looking attractive are pleasantly scented. It is not an easy house-plant but grows freely in a warm greenhouse.

24. P. ornata Yuncker. Illustration: Brittonia **8**: 62 (1954); Hoehnea **4**: f. 414 (1974); Graf, Exotica, edn 11, 1933 (1982).
Erect plant to 20 cm, stems rounded, hairless, dark green, to 2 cm in diameter.

Leaves alternate, crowded towards the ends of the stems, blade ovate, elliptic or sometimes almost circular, stiff and fleshy, without hairs but minutely glandular above, 3.5–6.5 × 2.8–4 cm, palmately 5-veined for about three-quarters of the length of the leaf; veins the same colour or sometimes darker than the upper surface, slightly raised beneath and often tinged with red, apex broadly pointed to rounded, base broadly tapered, rounded or cordate, margin distinctly reflexed; leaf-stalk 3–8 × 2–3 mm, pink to red, clearly grooved above. Inflorescence lateral, mostly compound with a few irregular leafy spike-bracts, some small linear scales, and 1–5 spikes, spike-stalk with minute glands, 3–20 × 1–1.5 mm, flower-spike 2–3.5 cm × 2–3 mm, the flowers rather sparsely arranged, yellowish green. *S Venezuela & N Brazil.* G2.

25. P. caperata Yuncker. Illustration: Botanical Magazine, n.s., 367 (1960); Graf, Exotica, edn 11, 1933 – 'Emerald Ripple', 'Tricolor' & 'Little Fantasy', 1938 – 'Variegata', 1939 (1982); Rochford & Gorer, The Rochford book of house plants, f. 390 (1963).
Erect plant to 20 cm, stems rounded, tough, hairless, to 1 cm in diameter, dark green or purplish. Leaves alternate, blade cordate, often peltate, leathery, dark green above, paler green beneath, minutely hairy above, hairless beneath, 2.5–4 × 2–3 cm; palmately 7–9-veined, the veins deeply impressed in V-shaped folds in the blade above and equally strongly raised beneath, apex broadly pointed to almost rounded, base auriculate or rounded, 4–8 mm from the leaf-stalk insertion when peltate, margin entire or irregularly and finely toothed, flat to minutely reflexed; leaf-stalk 2.5–12 cm × 3–4 mm, hairless, green to dull red, broadly and shallowly grooved with 2 fine vertical lines running down the inner face. Inflorescence solitary, lateral, without spike-bracts, spike-stalk hairless, 5–7 cm × *c.* 2 mm; spike 2.5–5 cm × 3mm, pale green. *Probably Brazil.* G2.

Probably the most commonly cultivated of the species. It is a most attractive foliage plant which is tough and fairly resistant to dry conditions. Several cultivars have arisen including large- and small-leafed variants and variegated plants mottled with white or flushed with pink.

26. P. griseo-argentea Yuncker (*P. hederaefolia* Anon.). Illustration: Graf, Exotica, edn 11, 1934, 1935 – 'Blackie', and P. hederaefolia (1982).

Erect plant to 20 cm, stems rounded, tough, hairless, to 1 cm in diameter, dark green. Leaves alternate, blade cordate, often peltate, leathery, grey-green above, pale green beneath, hairless, 2.5–5 × 2.2–4 cm; palmately usually 7-veined, the veins deeply impressed in the blade and strongly raised beneath, the apex broadly pointed to rounded, base auriculate or rounded, 2–8 mm from the leaf-stalk insertion when peltate, margin entire or very slightly and irregularly wavy, flat to minutely reflexed; leaf-stalk 4–6 cm × *c.* 3 mm, hairless, pale green to pink, rounded or minutely grooved above with 2 fine vertical lines running down the inner face. Inflorescence solitary, lateral, without bracts, spike-stalk hairless, 7–9 cm × *c.* 2.5 mm; spike 5–8 cm × *c.* 3 mm, pale green. *Of unknown origin.* G2.

Another common and very attractive foliage plant which is still most commonly cultivated as *P. hederaefolia.*

27. P. argyreia Morren (*P. peltifolia* de Candolle misapplied; *P. sandersii* de Candolle). Illustration: Botanical Magazine, 5634 (1866); Graf, Exotica, edn 11, 1931 (1982); Beckett, RHS encyclopaedia of house plants, 385 (1987); Plantsman 9: 176 (1987).
Erect plant to 20 cm, stems rounded, very short, the internodes hardly elongated, hairless, fleshy, dark red. Leaves alternate, broadly ovate, peltate, concave, leathery, hairless, silvery grey above with dark green stripes along the main veins, uniformly pale green beneath, 5–9 × 4–7 cm, palmately *c.* 11-veined with about half the veins disappearing towards the top of the leaf, the veins level with the leaf surface above, slightly raised as very fine lines beneath, apex mostly broadly pointed, sometimes slightly acuminate or rounded, base smoothly rounded, 1–2 cm from the leaf-stalk junction, the margin flat or slightly reflexed; leaf-stalk 8–15 cm × 4–6 mm, minutely granular-glandular, dark red with long green flecks, not grooved. Inflorescence terminal although apparently lateral, branched or solitary with mostly 1 small, scale-like spike-bract, spike-stalk minutely glandular, 3.5–4.5 cm × 1.5–2 mm, red; spike 5–8 cm × *c.* 2.5 mm, green. *Northern S America.* G2.

One of the most attractive foliage plants in the genus, moderately tough; it has held its place in horticulture for over 100 years.

28. P. marmorata J. D. Hooker (*P. verschaffeltii* Lemaire). Illustration: Botanical Magazine, 5568 (1866); Illustration Horticole 16: pl. 598 (1869);

Graf, Exotica, edn 11, 1931, 1937 – P. verschaffeltii, 1939 – 'Silver Heart' (1982).
Erect plant to 30 cm with a short, thick, rounded, hairless stem to 1 cm in diameter. Leaves alternate, blade ovate, leathery, hairless, upper surface with minute translucent dots, patterned with silvery grey or white between green areas around the veins, 7–12 × 3.5–7 cm, palmately 5–7-veined for the length of the leaf, the veins slightly impressed above and slightly raised beneath, apex obtuse to acute, base deeply cordate or auriculate, the lobes overlapping, margin slightly reflexed; leaf-stalk 5–6 × *c.* 4 mm, hairless, very slightly grooved near the top. Inflorescence axillary or terminal, solitary or 2–3 together, with peltate bracts at the base; spike-stalk *c.* 5 cm × 5 mm, spike 1–1.8 cm × *c.* 5 mm, green. *S Brazil.* G2.

The plant often known as *P. verschaffeltii* has leaves less deeply cordate at the base, the lobes not overlapping, but is only a minor variant. Much less commonly found in cultivation than formerly, it is an attractive plant but difficult to cultivate well and has probably been superseded by tougher plants more tolerant of lower temperatures.

2. PIPER Linnaeus
G. Argent

Mostly tough woody climbers or erect shrubs, occasionally small trees, rarely herb-like, often with swollen nodes. Leaves alternate, stipules absent or attached to the leaf-stalk, blade often attached to the stalk asymmetrically (1 side lower than the other). Inflorescence a leaf-opposed or axillary cylindric spike, occasionally a compound inflorescence of several spikes. Flowers subtended by a variable but often concave floral bract, bisexual (new world) or unisexual (old world) with 2–10 stamens, stigmas mostly 3, occasionally 2–4.

A genus of up to 2000 species which is treated here in a broad sense.

The species mostly like hot, humid conditions with well-drained compost which should neither dry out nor become too wet. They are tolerant of poor light and the variegated species particularly burn easily in direct sun and should be shaded. None are grown for their flowers, since these may be erratically or continuously (or never) produced. They can be handsome foliage plants when grown well, but, because of their rather exacting requirements, are not especially popular. They can be easily propagated by stem-cuttings or by layering in conditions of high humidity.

1a. Leaves peltate   **8. ornatum**
  b. Leaves not peltate   2
2a. Climbing or prostrate plants with thin, wiry stems   3
  b. Erect shrubs with self-supporting woody stems   8
3a. Leaves very thin and membranous, dark green spotted with red and white   **7. porphyrophyllum**
  b. Leaves leathery, mid-green, without an obvious coloured pattern   4
4a. Leaves auriculate, the basal sinus more than 5 mm deep in at least some leaves, the lobes widely spaced   **3. longum**
  b. Leaves tapered, rounded or cordate at the base, when lobed the sinus less than 5 mm deep and the lobes usually side-by-side or overlapping   5
5a. Fruit-stalk longer than mature fruit   **2. cubeba**
  b. Fruit-stalk shorter than mature fruit or absent   6
6a. Female spikes with the flowers closely massed together, in fruit becoming a solid, cylindric mass   **5. betle**
  b. Female spikes with clearly separated flowers which develop into separate, rounded fruits   7
7a. Female spikes erect, to 4 cm   **4. sylvaticum**
  b. Female spikes hanging, 5–15 cm   **1. nigrum**
8a. Leaves more than twice as long as broad   **9. aduncum**
  b. Leaves to 1.5 times as long as broad   9
9a. Plant to 75 cm, young stems with 6–8 longitudinal, frilled wings   **10. magnificum**
  b. Plant 2–7 m, young stems smooth, without longitudinal wings   **6. methysticum**

1. P. nigrum Linnaeus. Illustration: Botanical Magazine, 3139 (1832); Bentley & Trimen, Medicinal plants 4: f. 245 (1877); Purseglove, Tropical crops 2: 445 (1968); Masefield et al., The Oxford book of food plants, 129 (1969).
Climbing plant to 4 m. Stems rounded, hairless, with swollen nodes. Leaf-blade broadly ovate to heart-shaped, green, hairless, 5–13 × 3–9 cm, more or less palmately or pinnately 5–9-veined, the lateral veins arising in the lower half of the leaf (sometimes almost at the base); apex acute to obtuse, often shortly acuminate, base broadly tapered, rounded or weakly

cordate; stalk 1–3.5 cm × 1.5–2 mm, deeply grooved. Inflorescence leaf-opposed, hanging, solitary; spike-stalk hairless, 1–2 cm × *c.* 1.5 mm; spikes mostly bisexual, 3.5–11 cm × 2.5–10 mm. Fruits loosely arranged, ripening dark red, 4–6 mm in diameter. *India (western Ghats); naturalised in Assam and northern Burma.* G2.

The wild plants probably all have unisexual flowers (plants monoecious or dioecious) but there has been strong selection of bisexual variants for cultivation, thus giving higher yields of the black and white peppers of commerce. Black pepper is produced from green but mature fruit that is dried whole, white pepper from ripened red or yellowish fruits soaked and then washed to remove the outer layers before drying. The plant is not particularly ornamental but is sometimes grown as a curiosity and considered difficult to fruit under greenhouse conditions.

2. P. cubeba Linnaeus filius (*Cubeba officinalis* Miquel). Illustration: Hayne, Getreue Darstellung und Beschreibung der in Arzneykunde gebrauchlichen Gewächse 14: t. 8 (1843); Greshoff, Nuttige Indische Planten, t. 37 (1897); Macmillan, Tropical gardening and planting, 432 (1910).
Climber to 3 m. Stems round, smooth, hairless, nodes somewhat swollen. Leaf-blade elliptic or lanceolate, green above, paler beneath, smooth, 12–16 × 5–7 cm, pinnately veined with 1–2 pairs of basal lateral veins arching to about half-way up the leaf; apex acute, often acuminate; base broadly tapered, rounded or slightly cordate, often asymmetric; stalk 8–14 × *c.* 2 mm, grooved above. Inflorescence leaf-opposed, solitary, at first erect, later hanging. Spike-stalk 2–4 cm × 2–3 mm, hairless; spikes unisexual (plants dioecious); male spikes 7–10 cm × 4–5 mm, female spikes 5–10 cm × 6–30 mm, fruits loosely arranged on stalks which are much longer than the reddish brown berries. *Indonesia.* G2.

Occasionally grown as a curio or for economic interest. The dried, unripe fruits called cubebs are used medicinally.

3. P. longum Linnaeus. Illustration: Plenck, Icones Plantarum Medicinalium 1: t. 26 (1788); Bentley & Trimen, Medicinal plants 4: t. 244 (1877); Die Natürlichen Pflanzenfamilien 3(1): 9 (1887).
Slender climbing plants, 2–4 m. Stems angled or fluted when dry, fairly hairy, nodes not or only slightly swollen.

Leaf-blade broadly lanceolate or lanceolate-elliptic, green above, paler beneath, smooth, minutely hairy beneath, especially on the veins, densely gland-dotted, 5.5–10 × 3–5.5 cm, palmately 5–7-veined with usually only the central 3 veins reaching beyond mid-leaf; apex broadly acute to obtuse, rarely slightly acuminate, base deeply auriculate, the sinus 6–15 mm deep, very slightly asymmetric; stalk 5–20 × *c.* 1 mm, grooved. Inflorescence leaf-opposed, erect, solitary, spike-stalk minutely hairy, in male spikes 2–4 cm, in female 8–15 mm; spikes unisexual (plants dioecious), the male 3–9 cm × 2–3 mm, the female 1.5–2.5 cm × 5–8 mm, fruits very densely arranged. *Eastern Himalaya (subtropical parts).* G2.

The fruits are used locally as a pepper and the roots are used medicinally. It is not particularly ornamental but can form an elegant plant.

4. P. sylvaticum Roxburgh (*P. betle* Linnaeus, in part). Illustration: Basu, Indian Medicinal Plants, t. 823 (1918); Graf, Exotica, edn 11, 819 (1982).
Climber to 4 m. Stem rounded, green, longitudinally lined when dry, becoming woody, nodes somewhat enlarged, up to twice the width of the young stems. Leaf-blade ovate to heart-shaped, dark green above, paler beneath, rough, glandular-hairy on the main veins, gland-dotted on minor veins beneath, 6–12 × 3.5–6.5 cm, more or less palmately veined with usually 5 main veins arising in the lower half of the leaf and arching upwards; apex acute, short- or long-acuminate, base broadly tapered, rounded or slightly cordate, often slightly asymmetric; stalk 6–15 × 1–1.5 mm, grooved. Inflorescence leaf-opposed, erect, solitary, spike-stalk 3–15 × 1–2 mm, minutely glandular-hairy. Spikes unisexual (plants dioecious), male 4–7 cm × 2–4 mm, female 1.3–2.5 cm × 5–7 mm, fruits densely arranged but individually distinguishable. *Eastern Himalaya (subtropical parts).* G2.

Occasionally used as a condiment, this species is also found in variegated variants; leaves with silver markings and a pink sheen are reported. It makes an attractive plant for baskets and indoor trellis-work.

5. P. betle Linnaeus. Illustration: Botanical Magazine, 3132 (1832); Purseglove, Tropical crops 2: 439 (1968); Graf, Exotica, edn 11, 821 (1982); Lötschert & Beese, Collins' guide to tropical plants, f. 208 (1983).
Dioecious climber to 5 m. Stems woody,

rounded, the nodes hardly swollen, commonly and easily producing adventitious roots, hairless. Leaf-blade heart-shaped or broadly ovate, green, hairless, smooth, somewhat leathery, 9–18 × 5–10 cm, palmately 5–9-veined with strong, arching lateral veins all arising in the basal half of the leaf; apex acute or often acuminate, base rounded to cordate; stalk 7–20 × *c*. 2 mm, hairless, with small, pointed stipules extending from a third to half its length. Inflorescence leaf-opposed, solitary, spike-stalk 1.5–2.5 cm × *c*. 1.5 mm; male spike 8–15 cm × *c*. 5 mm, female 4–7 cm × 6–10 mm, pale glaucous green. *India to Malay peninsula.* G2.

Widely cultivated in SE Asia for the leaves and spikes which are chewed with lime and betel nut (fruit of the palm *Areca catechu*). Occasionally grown as a hot-house plant but of no particular ornamental value.

6. P. methysticum Forster. Illustration: Delessert, Icones Selectae Plantarum 3: t. 89 (1837); Gardeners' Chronicle 153: 293 (1963).
Erect shrub 3–4 m. Stem rounded, hairless, with swollen nodes. Leaf-blade heart-shaped to almost circular, deep green above, paler beneath, smooth, minutely granular-hairy above, finely hairy beneath, 10–25 × 10–17 cm, palmately 9–12-veined, most of the lateral veins reaching to mid-leaf; apex rounded, obtuse or broadly acute, occasionally slightly mucronate, base deeply auriculate, the sinus 1–5 cm deep; stalk 1.5–5 cm, broadly winged for over half its length, the wings with membranous edges, more or less hairy. Inflorescence leaf-opposed or terminal, solitary or several together; spike-stalk 1–2 cm × *c*. 2 mm, sparsely hairy. Spikes unisexual (plants probably dioecious), 10–14 cm × 3–5 mm. *South Pacific area.* G2.

Various cultivars are grown, mostly differing in the degree of leaf-coloration. The narcotic drink 'kava' is prepared from the roots by the people of the Pacific, particularly in Fiji. It is a coarse plant, not well suited to temperate horticulture.

7. P. porphyrophyllum (Lindley) N.E. Brown (*Cissus porphyrophylla* Lindley). Illustration: Nicholson, Illustrated dictionary of gardening 3: 147 (1884–88); Graf, Exotica, edn 11, 822 (1982).
Very extensive, weakly climbing or more often creeping shrub, commonly to 8 m in the wild. Stem wiry, reddish, rounded and grooved longitudinally, often with lines of

hairs. Leaf-blade broadly heart-shaped to almost circular, dark green above, spotted with red and white, purplish beneath with the pattern showing through the very thin texture, 10–15 × 9–12 cm, palmately 3–7-veined with the 3 central veins reaching to the upper part; apex obtuse, usually shortly mucronate, rarely almost rounded, base cordate to auriculate, the sinus to 2 cm deep; stalk 2.5–4 cm × *c*. 3 mm, grooved. Inflorescence leaf-opposed, spike-stalk *c*. 2.5 cm, hairy; male spikes *c*. 12 cm × 4 mm, female *c*. 3 cm × 8 mm. *Malay peninsula, Borneo.* G2.

A very handsome plant of difficult cultivation without hot-house conditions. The inflorescence is rarely produced, or at least seen, even in the wild; it probably flowers high in the forest trees.

8. P. ornatum N.E. Brown (*P. crocatum* misapplied). Illustration: Graf, Exotica, edn 11, 818, 819, 821, 822 (1982); Beckett, RHS encyclopaedia of house plants, 395 (1987).
Extensive, weakly climbing or creeping shrub to 5 m. Stems wiry, dark green or reddish, rounded, hairless. Leaf-blade broadly heart-shaped to almost circular, peltate, the leaf-stalk attached 1–2 cm from the lower margin, finely mottled with dark green, silver and pink above, reddish purple beneath, hairless, 6–13 × 5–10 cm, palmately 5–7-veined with the central 3 veins reaching to the upper part of the leaf; apex obtuse to almost rounded or acute and attenuate, base rounded to slightly cordate; stalk 6–9 cm × *c*. 3 mm, grooved above, when young with stipules to 1.2 cm. Inflorescences not recorded, probably rarely produced. *Indonesia (Sulawesi – formerly Celebes).* G2.

A spectacular foliage plant which requires continuous hot-house conditions as the leaves drop readily if the temperature falls for even a short time. It is sometimes known as *P. crocatum* and may be sold under this name, but genuine *P. crocatum* is an erect, densely hairy shrub from Peru which is unlikely to be offered for sale. Other variants of *P. ornatum*, differing in leaf-shape and markings, are often grown.

9. P. aduncum Linnaeus (*P. angustifolium* Ruiz & Pavon var. *cordulatum* de Candolle; *P. elongatum* Vahl var. *cordulatum* de Candolle). Illustration: Jacquin, Icones Plantarum Rariorum 2: t. 210 (1786); Trelease & Yuncker, The Piperaceae of northern South America, f. 215 (1950); Lilloa 27: pl. 16 (1953).
Erect, free-standing shrub or small tree to

6 m. Stems rounded with swollen nodes, longitudinally striped, hairless or hairy when young. Leaf-blade narrowly lanceolate or elliptic, deep even green above, paler green beneath, rough, often finely hairy, corrugated after drying, 10–15 × 2–3.5 cm, pinnately veined with the lateral veins curving upwards; apex narrowly acute or acuminate, sometimes rounded at the extreme point, base cordate to auriculate, often asymmetric; stalk very short, 1–3 × 2–3 mm, finely hairy. Inflorescence leaf-opposed, solitary, spike-stalk 6–12 × 2–3 mm, hairless or hairy. Spike bisexual, 6–15 × 3–5 mm, green or yellowish green, uniformly curved. *Throughout wet tropical America and the West Indies, naturalised in parts of SE Asia and the Pacific area.* G2.

The hairy variants of the species may occasionally be offered for sale as *P. angustifolium*; they are fairly tough plants provided reasonably high temperatures can be maintained.

10. P. magnificum Trelease (*Artananthe magnifica* Linden; *P. bicolor* Yuncker). Illustration: Graf, Exotica, edn 11, 822 (1982).
Erect shrub to 1 m. Stems winged vertically. Leaf-blade ovate, broadly elliptic to almost circular, deep glossy green above, bright purplish red beneath with contrasting white veins and margin, the surface quilted, hairless above, hairy beneath, 8–22 × 7–16 cm, more or less palmately or pinnately 9–15-veined, the arching veins arising in the lower half of the leaf; apex rounded or broadly pointed, base cordate to auriculate, the sinus 1–3 cm deep; stalks 3–4 cm, ridged longitudinally and with broad wings which sheath the stem at the base. Inflorescence leaf-opposed, terminal, solitary, hanging; spike-stalk winged, 1–1.5 mm, reddish purple. Spike bisexual, 3–4 cm × *c*. 5 mm. *Peru.* G2.

Called the laquered pepper on account of its highly glossy, varnished-looking leaves, it can make a very handsome foliage plant.

LXXXI. CHLORANTHACEAE

Herbs, shrubs or trees. Leaves simple, stalked, opposite, with pinnate venation, with stipules. Flowers (floral units) in simple or branched spikes or cymes, bisexual or unisexual, small, rudimentary, with differing interpretations as to structure. Perianth present or absent, when present, sepal-like. Stamens 1–3, united to one another and to the ovary. Ovary superior or inferior, 1-celled, with 1 or a few pendulous ovules. Fruit fleshy.

A family of 5 genera and 60–70 species distributed in 2 separate areas: tropical and subtropical Asia and tropical America. Not a significant horticultural family, grown mostly for botanical interest in the rudimentary structure of the flowers (which, because of differences in interpretation are referred to below as 'floral units') and the curiously 'simple' structure of the stems. The scented flowers are used in China to add fragrance to tea and the roots are used medicinally. Literature: P'ei, C., Chloranthus of China, *Sinensia* 6(6): 665–688 (1935); Swamy, B. & Bailey, I.W., Sarcandra, a vesselless genus of the Chloranthaceae, *Journal of the Arnold Arboretum* 31: 117–129 (1950).

1a. Fruits red or orange when mature; the floral unit with 1 club-shaped, fleshy stamen with 2 pairs of anther-sacs (see Figure 34(1), p. 313)
 1. Sarcandra
 b. Fruits white when mature; the floral unit with a 3-pointed, more or less flattened staminal appendage with 4 pairs of anther-sacs (see Figure 34(2), p. 313)
 2. Chloranthus

1. SARCANDRA Gardner
G. Argent
Herbs or shrubs, without vessels in the wood. Flowers bisexual, each with a single, stalkless, triangular, concave, persistent bract which clasps the base of the flower when young. Perianth absent. Stamen 1 with 2 pairs of anther-sacs, white, fleshy and club-shaped. Ovary ovoid with a single pendulous ovule, surmounted by a single, stalkless, rounded stigma. Fruit a drupe.

1. S. glabra (Thunberg) Nakai (*Chloranthus brachystachys* Blume; *C. glaber* (Thunberg) Makino). Figure 34(1), p. 313. Illustration: *Sinensia* 6(6): 672, f. 3 (1935); *Journal of the Arnold Arboretum* 31: 121, f. 3 (1950); Heywood, *Flowering plants of the world*, 36 (1978).
Erect, more or less shrubby plant to 2 m. Branches rising in succession from a horizontal, woody rhizome, freely branched but ending in inflorescences and of limited duration. Stems green, hairless, rounded, slightly swollen at the nodes. Leaves ovate, lanceolate or elliptic, leathery, hairless, green, 6–18 × 2–7 cm, apex narrowly acute, base narrowly tapering, the margin coarsely toothed with slender, pointed

teeth; stalk 1–1.5 cm × *c*. 2 mm, grooved. Inflorescence a group of spikes, each 1–4 cm long and with up to 10 minute floral units. Fruit a red or orange drupe. *S India, S China and throughout most of the SE Asian archipelago*. H5–G1.
A bright green foliage plant.

2. CHLORANTHUS Swartz
G. Argent
Similar to *Sarcandra*, differing most notably in the possession of vessels in the wood and the blade-like, rather than club-shaped stamen which has 4 pairs of anther-sacs.

1. C. spicatus (Thunberg) Makino (*C. inconspicuus* Swartz). Figure 34(2), p. 313. Illustration: *Sinensia* 6(6): 670, f. 2 (1935). Differing from *Sarcandra glabra* most obviously in the obtusely pointed leaves, the more highly branched inflorescence with larger spikes, each usually with at least 10 floral units and the white fruit and stamen-structure as described in the key. *Japan*. G1.

LXXXII. ARISTOLOCHIACEAE
Perennial herbs and climbers, occasionally with woody stems. Leaves simple, stalked, alternate. Flowers solitary or in axillary clusters, often foetid, bisexual. Perianth radially or bilaterally symmetric, more or less petal-like, 3-lobed at apex or with a single, unilateral lobe. Stamens 6 or 12, in 1 or 2 whorls, filaments free or united with stylar column. Styles 6, free or united to form a column with 6-lobed stigma. Ovary inferior, 6-celled, placentation axile, ovules numerous in each cell. Fruit a capsule.

A family of 7 genera and about 600 species, mostly tropical and subtropical, a few from temperate regions.

1a. Small herbs with rhizomes; flowers terminal; stamens 12 **1. Asarum**
 b. Erect herbs or climbers; flowers axillary; stamens 6 **2. Aristolochia**

1. ASARUM Linnaeus
A. Brady
Herbs with creeping rhizomes rooting at nodes. Flowers solitary, terminal, resin-scented. Perianth radially symmetric, 3-lobed, persistent in fruit. Stamens 12 in 2 whorls, filaments free (rarely fused). Styles 6, free or united into a column. Fruit nearly spherical, opening irregularly. Seeds boat-shaped.

A genus of 75 species native of north temperate regions. They are often planted in a shady part of the rockery. Propagation is by division of the rhizomes in spring or autumn. The genus has been divided into several smaller genera and some American botanists recognise the genus *Hexastylis* Rafinesque, placing within it all those species which possess 6 free styles, reserving *Asarum* for species with styles united into a single column. The flowers are pollinated by flies.

1a. Styles 6, free   2
 b. Styles fused into a column   4
2a. Leaves circular to spear-shaped, to 13 cm   **1. arifolium**
 b. Leaves kidney- or heart-shaped, not more than 8 cm   3
3a. Perianth-lobes erect or spreading, the whole perianth 1–2.5 cm
 3. virginicum
 b. Perianth-lobes incurved, the whole perianth 2.5–5 cm   **2. shuttleworthii**
4a. Perianth-lobes acuminate   5
 b. Perianth-lobes not acuminate   6
5a. Leaves mottled, 10–13 cm across, hairless beneath   **4. hartwegii**
 b. Leaves not mottled, downy beneath
 5. caudatum
6a. Plant deciduous   **7. canadense**
 b. Plant evergreen   **6. europaeum**

1. A. arifolium Michaux (*A. grandiflorum* Klotzsch; *Hexastylis arifolia* (Michaux) Small). Illustration: Hooker, *Exotic Flora* 1: 40 (1844); *Journal of the Royal Horticultural Society* 101: 335 (1976). Evergreen. Leaves circular to spear-shaped, thick, usually mottled, 10–13 cm, stalks to 20 cm. Flower-stalk stout. Perianth 2–5 cm, urn-shaped, constricted at throat; lobes not acuminate. Styles 6. *SE USA*. H2. Late spring.

2. A. shuttleworthii Britton & Baker (*A. grandiflorum* Small; *A. macranthum* (Shuttleworth) Small; *Hexastylis shuttleworthii* (Britton & Baker) Small). Illustration: Britton & Brown, *Illustrated Flora of the northern United States* 1: 539 (1896).
Evergreen. Leaves heart- or kidney-shaped, 2.5–8 cm across, thick, usually mottled, stalks to 20 cm. Mature perianth 2.5–5 cm long, 1.5–2.5 cm in diameter, mottled with violet inside, lobes not acuminate, incurved. Styles 6. *SE USA*. H3. Summer.

3. A. virginicum Linnaeus (*A. shuttleworthii* misapplied; *Hexastylis virginica* (Linnaeus) Small). Illustration: *Journal of the Royal Horticultural Society* 101: 335 (1976).

Similar to *A. shuttleworthii* except that the perianth is 1–2.5 cm long and 8–15 mm in diameter, lobes erect or spreading. *SE USA*. H3. Summer.

Frequently sold as *A. shuttlewothii*.

4. A. hartwegii Watson. Illustration: Munz, California mountain wild flowers, 11 (1969).

Evergreen. Rhizome short. Leaves heart-shaped, mottled with white, 10–13 cm across, stalks hairy, 5–15 cm. Perianth brown-purple, to 1.5 cm in diameter, lobes drawn out into tails, 3–7 cm. Style 1. *W USA (Oregon, California)*. H4. Early summer.

5. A. caudatum Lindley. Illustration: Parsons, Wild flowers of California, 311 (1897); Hitchcock et al., Wild flowers of the Pacific Northwest 2: 105 (1964).

Evergreen. Rhizomes slender, long. Leaves heart- or kidney-shaped, not mottled, 2–10 cm, downy beneath; stalks 3–15 cm. Perianth brown-red, lobes drawn out into tails, 2.5–8 cm. Style 1. *Western N America (British Columbia to California)*. H3. Summer.

6. A. europaeum Linnaeus. Illustration: Smith & Sowerby, English Botany, 1083 (1802); Stodola, Volak & Bunney, The illustrated book of herbs, 79 (1985).

Evergreen. Rhizomes short, hairy, with brown scales at base. Leaves circular or kidney-shaped, margins slightly wavy, 6–8 cm, green, not mottled, glossy; stalk to 12 cm. Perianth dull brown or green-purple, hairy outside, lobes not acuminate. Style 1. *Europe*. H2. Spring.

7. A. canadense Linnaeus. Illustration: Botanical Magazine, 2769 (1827); Gleason, Illustrated Flora of the northeastern United States and adjacent Canada 2: 61 (1952).

Deciduous. Rootstock aromatic, smelling of ginger. Leaves kidney-shaped, not mottled, 5–18 cm, downy; stalk to 30 cm. Flower-stalks very short. Perianth brown-purple, 2–5 cm across, bell-shaped, lobes not acuminate. Style 1. *Eastern N America*. H3. Summer.

2. ARISTOLOCHIA Linnaeus
A. Brady & C. Gorman

Perennial herbs or woody climbers. Leaves deciduous or evergreen, often heart-shaped or hastate at base. Flowers often unpleasantly and strongly scented, axillary, solitary, clustered or in racemes; inflorescence-stalk usually with prominent bracts at base or below ovary. Perianth bilaterally symmetric, tubular, deciduous, with swollen base; tube bent, often constricted at mouth. Stamens usually 6. Fruit a capsule.

About 200 species both tropical and temperate, many with a twining habit. The species prefer a fertile soil high in organic matter that does not dry out. In glasshouses they do better planted out, but are satisfactory in large containers. Propagation is by seed, summer cuttings or by root cuttings from tuberous-rooted species such as *A. clematitis*.

Literature: Pfeifer, H.W., Revision of the north and central American hexandrous species of Aristolochia (Aristolochiaceae), *Annals of the Missouri Botanical Garden* 53: 115–196 (1966).

1a. Herbaceous perennial   2
  b. Woody climber   3
2a. Flowers 2–8 together   **12. clematitis**
  b. Flowers solitary   **13. rotunda**
3a. Young parts covered with felt
                             1. chrysops
  b. Young parts hairless   4
4a. Deciduous   5
  b. Evergreen   8
5a. Flowers appearing before the leaves
                            2. californica
  b. Leaves appearing before the flowers
                                         6
6a. Flowers brown-purple
                           3. macrophylla
  b. Flowers yellow-green   7
7a. Flower-stalk without a bract
                            4. tomentosa
  b. Flower-stalk with an oval bract above
     the middle   **5. moupinensis**
8a. Capsule pear-shaped; perianth-blade
    1-sided   **6. fimbriata**
  b. Capsule not pear-shaped; perianth-
     blade not 1-sided   9
9a. Flowers more than 20 cm   10
  b. Flowers less than 20 cm   11
10a. Leaves downy; perianth tailed, tail to
     3 m; flowering in summer
                           7. grandiflora
  b. Leaves hairless; perianth not tailed;
     flowering in autumn   **8. labiata**
11a. Flowering in early summer
                          9. sempervirens
  b. Flowering in mid–late summer   12
12a. Flowers 10–20 cm, 2-lipped
                           10. cymbifera
  b. Flowers 3–4 cm, not 2-lipped
                            11. littoralis

1. A. chrysops (Stapf) Hemsley (*Isotrema chrysops* Stapf). Illustration: Botanical Magazine, 8957 (1923).

Climber to 5 m. Stems woody, all young parts covered with felt. Leaves very variable in shape and size, broadly ovate, heart-shaped, hastate or lanceolate, 10–15 × 4–7 cm, upper surface greyish-downy. Flowers usually solitary, tube recurved, narrow at about mid-point; all downy and yellow outside, inside yellow and hairless. Perianth-blade 2–3 cm across, 3-lobed, sometimes reflexed, marked off from the tube by projections, the rim golden yellow, with a ring of brown-purple in the centre; stalks slender, flexuous, to 5 cm. Capsule indistinctly ribbed, to 8 cm. *China (Hubei)*. H3. Summer.

2. A. californica Torrey. Illustration: Munz, A California Flora, 964 (1959); Annals of the Missouri Botanical Garden 53: 142 (1966).

Deciduous climber. Stems glaucous and twining, to 4 m. Leaves heart-shaped, circular, bluntly acuminate at apex, 3–10 × 4–12 cm, glaucous. Flower solitary, appearing before the leaves; stalk 1–2 cm with a small ovate bract near the middle. Perianth tubular, inflated, glaucous, 3 cm, purple, the blade 3-lobed. Capsule ribbed, cylindric, to 6 × 2.5 cm. *W USA (California)*. H4. Early spring.

Similar to *A. tomentosa* except the flower is less glaucous, larger and more inflated. It differs from *A. macrophylla* in the absence of the large bract on the flower-stalk.

3. A. macrophylla Lamarck (*A. durior* Hill; *A. sipho* L'Héritier). Illustration: Botanical Magazine, 534 (1801); Gleason, Illustrated Flora of the northeastern United States and adjacent Canada 2: 63 (1952); Annals of the Missouri Botanical Garden 53: 140 (1966).

Deciduous woody climber. Stems to 20 m. Leaves circular to kidney-shaped, 7–45 × 7–50 cm; stalk 2.5–8 cm. Flowers solitary, to 8 cm, tube U-shaped, hairless, blade 3-lobed, yellow-green, the lobes brown-purple, to 2.5 cm across; flower-stalks clasped by an oval bract in the lower third. Capsule cylindric, *c.* 8 cm. *E USA*. H2. Summer.

There has been much discussion about the use of this name; Pfeifer argues that Hill's illustration shows *Bignonia capreolata* Linnaeus, and that the name *A. macrophylla* should be used instead of *A. durior*.

4. A. tomentosa Sims. Illustration: Botanical Magazine, 1369 (1811); Gleason, Illustrated Flora of the northeastern United States and adjacent Canada 2: 63 (1952); Annals of the Missouri Botanical Gardens 53: 143 (1966).

Deciduous woody climber. Stems to 25 m.

Leaves heart-shaped, 8–15 × 8–20 cm, apex rounded, downy beneath. Flowers solitary, 4–5 cm long, green-yellow, tube U-shaped, downy outside, with narrow, purple mouth, blade 3-lobed, reflexed, to 2 cm across, yellow. Flower-stalks without a bract. Fruit woody, finely downy, cylindric, 6–8 × 4–6 cm. *SE USA*. H2. Summer.

5. A. moupinensis Franchet. Illustration: Botanical Magazine, 8325 (1910).
Deciduous climber. Stems downy at first. Leaves heart-shaped, 10–12 × 6–10 cm, slightly pointed, downy above, densely downy beneath; stalks to 2.5 cm. Flowers solitary; stalks slender, to 5 cm, with an oval bract one-third of the way down. Perianth-tube 4–5 cm, pale green, flattened, downy, bent back exposing the yellow mouth and 3 spreading lobes which are yellow with purple spots and green towards the margins. Capsule to 5 cm, narrowly 6-winged. *W China*. H3. Summer.

6. A. fimbriata Chamisso (*A. ciliosa* Bentham). Illustration: Botanical Magazine, 3756 (1839).
Climber to 2 m, or prostrate, hairless, stems slender. Leaves heart-shaped, glaucous beneath, 3–5 × 10 cm. Flowers solitary, tube green, strongly curved, inflated at base, expanding into a 1-sided heart-shaped blade to 2.5 cm across, green-brown outside, purple-brown veined with yellow, ciliate with long glandular hairs. Young capsule pear-shaped. *Brazil*. G2. Early autumn.

7. A. grandiflora Swartz (*A. foetens* Lindley; *A. gigantea* Hooker; *A. gigas* Lindley). Illustration: Edwards's Botanical Register 21: t. 1824 (1836), 28: t. 60 (1842); Botanical Magazine, 4221 (1846); Annals of the Missouri Botanical Garden 53: 163 (1966).
Climber to 3 m. Leaves heart-shaped, sometimes acuminate at tip, 8–15 × 10–20 cm; stalks long. Flowers solitary,

hanging, with offensive odour, tube U-shaped, inflated, yellow-green with a deep purple ring at the mouth, blade ovate, heart-shaped, 20–50 cm across, with a long hanging tail up to 3 m, white veined with purple. Capsule cylindric, to 10 × 4 cm. *C America, West Indies*. G2. Summer.

8. A. labiata Willdenow (*A. brasiliensis* Martius & Zuccarini; *A. ornithocephala* Hooker). Illustration: Edwards's Botanical Register 8: t. 689 (1822); Botanical Magazine, 4120 (1844); Annals of the Missouri Botanical Garden 53: 157 (1966).
Evergreen climber to 10 m. Leaves hairless, light green, heart-to kidney-shaped, 7–15 × 7–12 cm; stalk 6–10 cm, with heart-shaped stipules. Flower solitary, 20–30 cm long, dingy yellow, veined purple, tube inflated, blade 2-lipped, upper lip violet inside, narrow, to 10 cm, beak-like, lower lip kidney-shaped, to 10 × 20 cm, lined with deep purple. Capsule cylindric, to 8 × 3 cm. *Brazil*. G2. Autumn.

9. A. sempervirens Linnaeus (*A. altissima* Desfontaines). Illustration: Botanical Magazine, 1116 (1808).
Evergreen, woody, stems to 5 m, climbing or rarely prostrate. Leaves to 10 × 6 cm, ovate to ovate-lanceolate or heart-shaped, hairless, margins wavy, shiny. Flowers solitary, 2–5 cm long, yellow inside, pale yellow striped with purple outside; tube curved, spherical at base, ovary and flower-stalk downy. Capsule ovoid, 1–4 × 3–5 cm. *S Greece, Italy*. H4. Early summer.

A. baetica Linnaeus is similar but the leaves are glaucous; flowers black-purple or brown-purple outside. *S Spain, Portugal*. H5. Early summer.

10. A. cymbifera Martius & Zuccarini (*A. labiosa* Ker Gawler). Illustration: Botanical Magazine, 2545 (1825).
Evergreen, hairless twiner to 6 m. Leaves heart- to kidney-shaped, to 8 cm. Flowers solitary with offensive odour, creamy white, margin blotched maroon; perianth

sac-like, expanded into a 2-lipped blade; shorter lip lanceolate, acuminate, the larger rounded, kidney-shaped, 10–15 × 10–15 cm, cream with purple veins, throat purple covered with hairs. *Brazil*. G2. Summer.

Pfeifer (reference above) suggests that this is a cross between *A. ringens* Vahl and *A. labiata*.

11. A. littoralis Parodi (*A. elegans* Masters). Illustration: The Garden 29: 576 (1886); Botanical Magazine, 6909 (1886).
Slender, hairless, woody climber. Leaves ovate, heart-shaped, 5–8 cm, hairless above, glaucous beneath. Flowers solitary, tube inflated, 3–4 cm, yellow-green; blade to 5 × 5 cm, white, veined with purple outside, rich purple-brown inside with irregular white marks, throat golden-yellow; stalk 2–6 cm, slender. Capsule 4.5–25 cm. *S America*. G2. Late summer.

12. A. clematitis Linnaeus. Illustration: Gleason, Illustrated Flora of the northeastern United States and adjacent Canada 2: 63 (1952); Stodola, Volak & Bunney, Illustrated book of herbs, 71 (1985).
Deciduous herbaceous perennial. Rhizome creeping, branched. Stems to 1 m, simple, hairless. Leaves 3–5 cm, broadly ovate, hairless, auricles rounded, stalks 3–4 cm. Flowers axillary, 2–8 together, 2–3 cm long, pale yellow outside, yellow with brown veins inside. Ovary hairless; flower-stalk short. Capsule 2.5–3 cm. *Europe*. H2. Summer.

13. A. rotunda Linnaeus. Illustration: Bonnier, Flore complète 9: t. 540 (1927); Polunin, Flowers of Europe, 61 (1969).
Herbaceous perennial. Rootstock tuberous, spherical. Stems erect, 15–20 cm, hairless. Leaves oval to kidney-shaped, 2–7 cm, with a deep sinus at the base, stalkless or stalks not more than 1.5 cm. Flowers solitary, yellow, 3–5 cm, hairy outside, blade dark brown. *S Europe*. H5. Spring.

GLOSSARY

abscission-zone. A predetermined layer at which leaves or other organs break off.

achene. A small, dry, indehiscent, 1-seeded fruit, in which the fruit-wall is of membranous consistency and free from the seed.

acuminate. With a long, slender point.

adpressed. Closely applied to a leaf or stem and lying parallel to its surface but not adherent to it.

adventitious. (1) Of roots: arising from a stem or leaf, not from the primary root derived from the radicle of the seedling. (2) Of buds: arising somewhere other than in the axil of a leaf.

aggregate fruit. A collection of small fruits, each derived from a single carpel, closely associated on a common receptacle, but not united. *Ranunculus* and *Rubus* provide familiar examples.

alternate. Arising singly, 1 at each node; not opposite or whorled (figure 39(2), p. 420).

anastomosing. Describes veins of leaves which rejoin after branching from each other or from the main vein or midrib.

anatropous. Describes an ovule which turns through 180° in the course of development, so that the micropyle is near the base of the funicle (figure 42(2), p. 423).

annual. A plant which completes its life-cycle from seed to seed in less than 1 year.

anther. The uppermost part of a stamen, containing the pollen (figure 41(3 & 4), p. 422).

apetalous. Describes a flower without a corolla (petals).

apical. Describes the attachment of an ovule to the apex of a 1-celled ovary (figure 42(8), p. 423).

apiculate. With a small point.

apomictic. Reproducing by asexual means, though often by the agency of seeds, which are produced without the usual sexual nuclear fusion.

arachnoid. Describes hairs which are soft, long and entangled, suggestive of cobwebs.

areole. See Cactaceae, p. 203.

aril. An outgrowth from the region of the hilum, which partly or wholly envelops the seed; it is usually fleshy.

ascending. Prostrate for a short distance at the base but then curving upwards to that the remainder is more or less erect; sometimes used less precisely to mean pointing obliquely upwards.

attenuate. Drawn out to a fine point.

auricle. A lobe, normally 1 of a pair, at the base of the blade of a leaf, bract, sepal or petal.

awn. A slender but stiff bristle on a sepal or fruit.

axil. The upper angle between a leaf-base or leaf-stalk and the stem that bears it (figure 39(1), p. 420).

axile. A form of placentation in which the cavity of the ovary is divided by septa into 2 or more cells, the placentas being situated on the central axis (figure 42(10), p. 423).

axillary. Situated in or arising from an axil (figure 39(1), p. 420).

back-cross. A cross between a hybrid and a plant similar to one of its parents.

basal. (1) Of leaves: arising from the stem at or very close to its base. (2) Of placentation: describes the attachment of an ovule to the base of a 1-celled ovary (figure 42(6 & 7), p. 423).

basifixed. Attached by its stalk or supporting organ by its base, not by its back (figure 41(3), p. 422).

berry. A fleshy fruit containing 1 or more seeds embedded in pulp, as in the genera *Berberis*, *Ribes* and *Phoenix*. Many fruits (such as those of *Ilex*) which look like berries and are usually so called in popular speech, are, in fact, drupes.

biennial. A plant which completes its life-cycle from seed to seed in a period of more than 1 year but less than 2.

bifid. Forked; divided into 2 lobes or points at the tip.

bilaterally symmetric. Capable of division into 2 similar halves along 1 plane and 1 only (figure 41(9), p. 422).

bipinnate. Of a leaf: with the blade divided pinnately into separate leaflets which are themselves pinnately divided (figure 39(18), p. 420).

blade. A broadened part, furthest from the base, of a petal, corolla or similar organ, which has a relatively narrow basal part – the claw or tube (figure 41(5 & 6), p. 422).

bract. A leaf-like or chaffy organ bearing a flower in its axil or forming part of an inflorescence, differing from a foliage-leaf in size, shape, consistency or colour (figure 40(2), p. 421).

bracteole. A small, bract-like organ which occurs on the flower-stalk, above the bract, in some plants.

bulbil. A small bulb, especially one borne in a leaf-axil or in an inflorescence.

calyx. The sepals; the outer whorl of a perianth (figure 41(1), p. 422).

campylotropous. Describes an ovule which becomes curved during development and lies with its long axis at right angles to the funicle (figure 42(4), p. 423).

capitate. Compact and approximately spherical, head-like.

capitulum. An inflorescence consisting of small flowers (florets), usually numerous, closely grouped together so as to form a 'head', and often provided with an involucre.

capsule. A dry, dehiscent fruit derived from 2 or more united carpels and usually containing numerous seeds.

carpel. One of the units (sometimes interpreted as modified leaves) situated in the centre of a flower and together constituting the gynaecium or female part of the flower (ovary). If more than 1, they may be free or united. They contain ovules and bear a stigma (figure 41(1 & 2), p. 422).

carpophore. See *Silene*, p. 179.

caruncle. A soft, usually oil-rich appendage attached to the seed near the hilum.

catkin. An inflorescence of unisexual flowers, made up of relatively conspicuous, usually overlapping bracts, each of which subtends a small apetalous

419

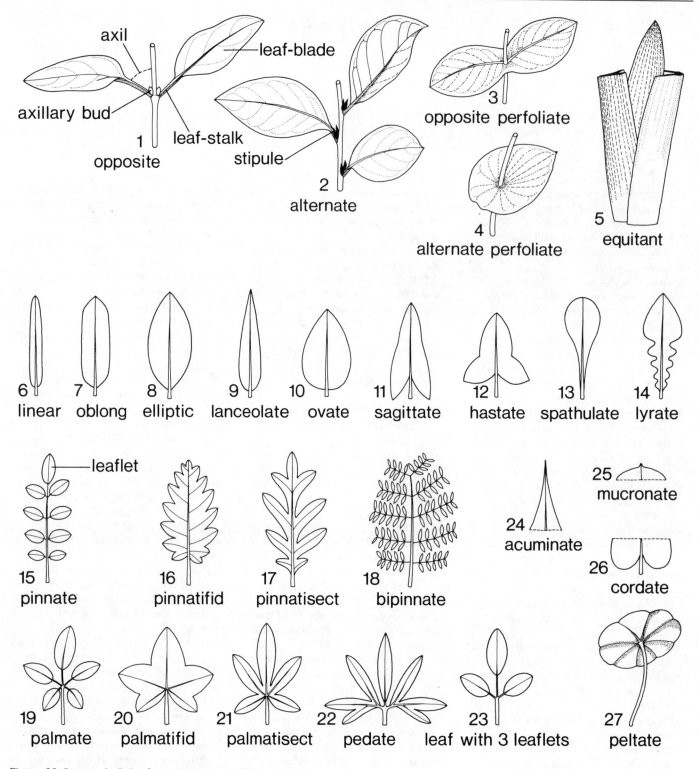

Figure 39. Leaves. 1–5, Leaf insertion types. 6–14, leaf-blade outlines. 15–23, Leaf dissection types. 24–26, Leaf apex and base shapes. 27, Attachment of leaf-stalk to leaf-blade.

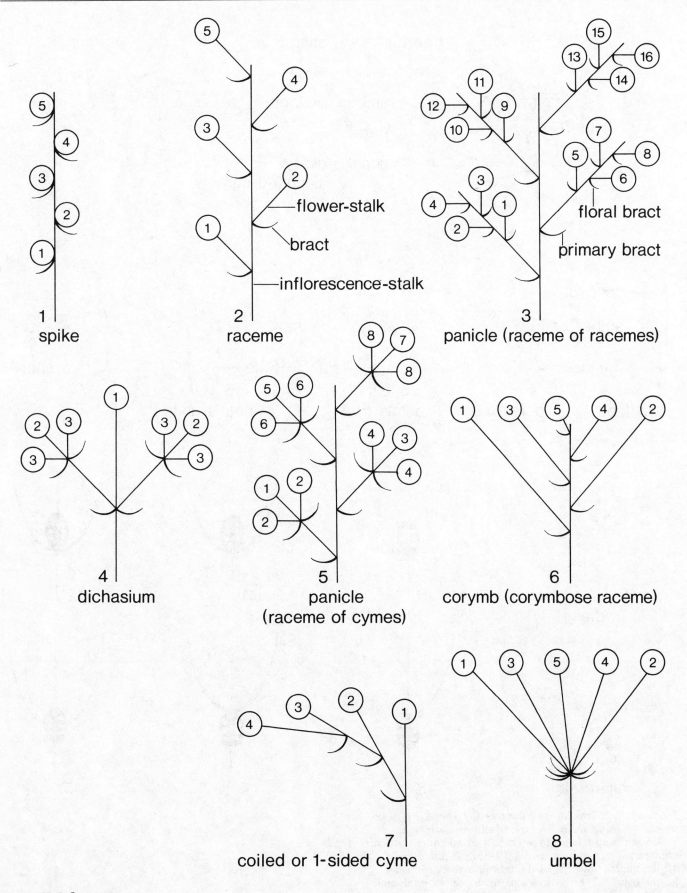

Figure 40. Inflorescences.

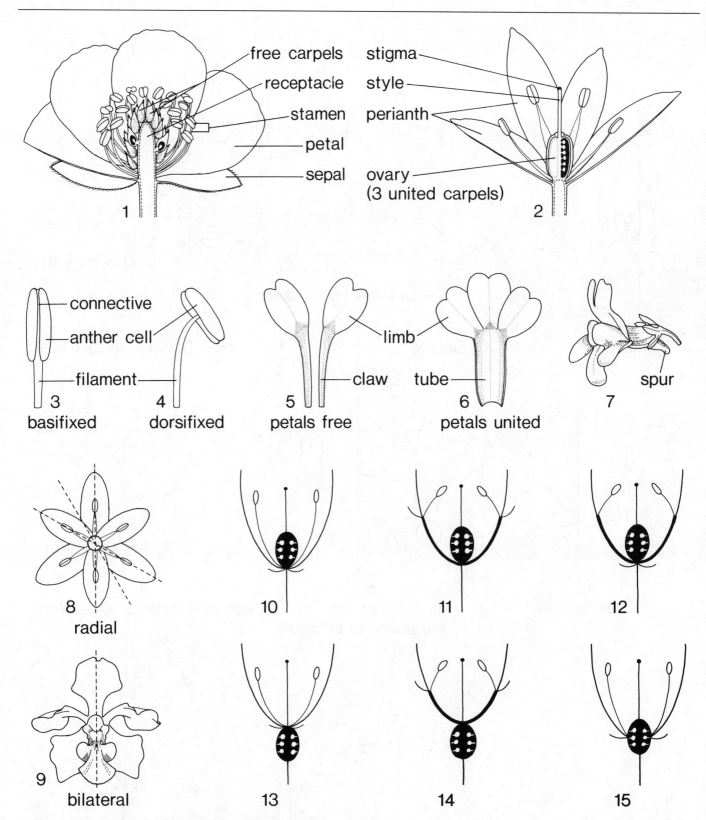

Figure 41. 1, 2, Two flowers illustration floral parts. 3, 4, Two stamens showing alternative types of anther attachment. 5–7, Some terms relating to petals. 8, 9, Floral symmetry, planes of symmetry shown by broken lines. 10–15, Position of ovary. 10–12, Superior ovaries. 13, 14, Inferior ovaries. 15, Half-inferior ovary. 11, Perigynous zone bearing sepals, petals and stamens. 12, Perigynous zone bearing petals and stamens. 14, Epigynous zone bearing sepals, petals and stamens.

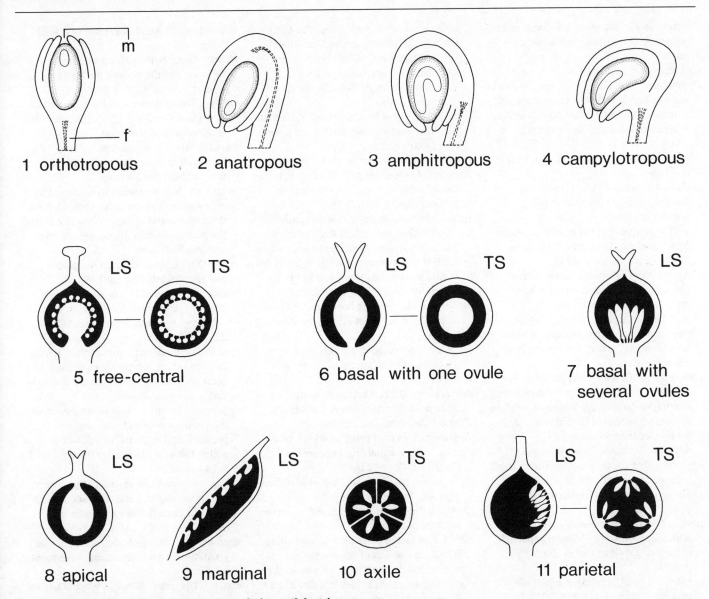

Figure 42. Ovules and placentation. 1–4, Ovule forms (f, funicle; m, micropyle). 5–11, Placentation types (LS, longitudinal section; TS, transverse section).

flower or a group of such flowers; catkins are generally pendent, but some are erect.

cephalium. See Cactaceae, p. 206.

chromosome. One of the small, thread-like or rod-like bodies consisting of nucleic acid and containing the genes, which appear in a cell nucleus shortly before cell division.

ciliate. Fringed on the margin with usually fine hairs.

cladode. A branch which takes on the function of a leaf (the leaves being usually vestigial).

claw. The narrow base of a petal or sepal, which widens above into the limb or blade (figure 41(5), p. 422).

cleistogamous. Describes a flower with a reduced corolla which does not open but sets seed by self-pollination.

clone. The sum-total or the plants derived from the vegetative reproduction of an individual, all having the same genetic constitution.

compound. (1) Of a leaf: divided into separate leaflets. (2) Of an inflorescence: bearing secondary inflorescences in place of single flowers. (3) Of a fruit: derived from more than 1 flower.

compressed. Flattened from side to side.

connective. The tissue which separates the 2 lobes of an anther, and to which the filament is attached (figure 41(3), p. 422).

cordate. Describes the base of a leaf-blade which has a rounded lobe on either side of the central sinus (figure 39(26), p. 420).

corolla. The petals: the inner whorl of a perianth (figure 41(1), p. 422).

coronal scale. See Silene, p. 179.

corymb. A broad, flat-topped inflorescence. In the strict sense the term indicates a raceme in which the lowest flowers have stalks long enough to bring them to the level of the upper ones (figure 40(6), p. 421), but the term corymbose is often used to indicate a flat-topped inflorescence.

cotyledon. One of the leaves preformed in the seed.

crisped. (1) Of hairs: strongly curved so that the tip lies near the point of attachment. (2) Of leaves, leaflets or petals: finely and complexly wavy.

cristate. With elevated, irregular ridges.

cultivar. A variant of horticultural interest or value, maintained in cultivation, and not conveniently equatable with an infraspecific category in botanical classification. A cultivar may arise in cultivation or may be brought in from the wild. Its distinguishing name can be Latin in form, e.g. 'Alba', but is more usually in a modern language, e.g. 'Madame Lemoine', 'Frühlingsgold', 'Beauty of Bath'.

cupule. A group of bracts, united at least at the base, surrounding the base of a fruit or a group of fruits.

cyme. An inflorescence in which the terminal flower opens first, other flowers being borne on branches which arise below it (figure 40(5 & 7), p. 421).

cystolith. A concretion of calcium carbonate found within the cells of the leaf in some plants; they can sometimes be seen when the leaf is viewed against the light, or felt as tiny hard lumps when the leaf is drawn between finger and thumb.

decumbent. More or less horizontal for most of its length but erect or semi-erect near the tip.

decurrent. Continued down the stem below the point of attachment as a ridge or ridges.

dehiscent. Splitting, when ripe, along 1 or more predetermined lines of weakness.

dichasial. Resembling a dichasium.

dichasium. A form of cyme in which each node bears 2 equal lateral branches (figure 40(4), p. 421).

dichotomous. Divided into 2 equal branches; regularly forked.

dioecious. With male and female flowers on separate plants.

diploid. Possessing in its normal vegetative cells 2 similar sets of chromosomes.

disc. A variously contoured, ring-shaped or circular area (sometimes lobed) within a flower, from which nectar is secreted.

dissected. Deeply divided into lobes or segments.

distylic. Having flowers of different plants either with long styles and shorter stamens or with long stamens and shorter styles.

dorsifixed. Attached to its stalk or supporting organ by its back, usually near the middle (figure 41(4), p. 422).

double. Of flowers: with petals much more numerous than in the normal wild state.

drupe. An indehiscent fruit in which the outer part of the wall is soft and usually fleshy but the inner part is stony. A drupe may be 1-seeded as in Prunus or Juglans, or may contain several seeds, as in Ilex. In the latter case, each seed is enclosed in a separate stony endocarp and constitutes a pyrene.

drupelet. A miniature drupe forming part of an aggregate fruit.

ellipsoid. As elliptic, but applied to a solid body.

elliptic. About twice as long as broad, tapering equally both to the tip and the base (figure 39(8), p. 420).

embryo. The part of a seed from which the new plant develops; it is distinct from the endosperm and seed-coat.

endocarp. The inner, often stony layer of a fruit-wall in those fruits in which the wall is distinctly 3-layered.

endosperm. A food-storage tissue found in many seeds, but not in all, distinct from the embryo and serving to nourish it and the young seedling during germination and establishment.

entire. With a smooth, uninterrupted margin; not lobed or toothed.

epicalyx. A group of bracts attached to the flower-stalk immmediately below the calyx and sometimes partly united with it.

epigeal. The mode of germination in which the cotyledons appear above ground and carry on photosynthesis during the early stages of establishment.

epigynous. Describes a flower, or preferably the petals, sepals and stamens (or perianth and stamens) of a flower in which the ovary is inferior (figure 41(13 & 14), p. 422).

epigynous zone. A rim or cup of tissue on which the sepals, petals and stamens are borne in some flowers with inferior ovaries (figure 41(14), p. 422).

epiphyte. A plant which grows on another plant but does not derive any nutriment from it.

exocarp. The outer, skin-like layer of a fruit-wall in those fruits in which the wall is distinctly 3-layered.

farina. The flour-like wax present on the stem and leaves of many species of Primula and of a few other plants.

fastigiate. With all the branches more or less erect, giving the plant a narrow tower-like outline.

filament. The stalk of a stamen, bearing the anther at its tip (figure 41(3 & 4), p. 422).

filius. Used with authority names to distinguish between parent and offspring when both have given names to species, e.g. Linnaeus (C. Linnaeus, 1707–1788), Linnaeus filius (C. Linnaeus, 1741–1783, son of the former).

floret. A small flower, aggregated with others into a compact inflorescence.

follicle. A dry dehiscent fruit derived from a single free carpel and opening along one suture.

free. Not united to any other organ except by its basal attachment.

free-central. A form of placentation in which the ovules are attached to the central axis of a 1-celled ovary (figure 42(5), p. 423).

fruit. The structure into which the gynaecium is transformed during the ripening of the seeds; a *compound fruit* is derived from the gynaecia of more than one flower. The term 'fruit' is often extended to include structures which are derived in part from the receptacle (*Fragaria*), epigynous zone (*Malus*) or inflorescence-stalk (*Ficus*) as well as from the gynaecium.

funicle. The stalk of an ovule (figure 42(1), p. 423).

fusiform. Spindle-shaped; cylindric but tapered gradually at both ends.

gamete. A single sex-cell which fuses with one of the opposite sex during sexual reproduction.

gland-dotted. With minute patches of secretory tissue usually appearing as pits in the surface, as translucent dots when viewed against the light, or both.

glandular. (1) Of a hair: bearing at the tip a usually spherical knob of secretory tissue. (2) Of a tooth: similarly knobbed or swollen at the tip.

glaucous. Green strongly tinged with bluish grey; with a greyish waxy bloom.

glochid. See Cactaceae, p. 203.

graft-hybrid. A plant which, as a consequence of grafting, contains a mixture of tissues from 2 different species. Normally the tissues of 1 species are enclosed in a 'skin' of tissue from the other species.

gynaecium. The female organs (carpels) of a single flower, considered collectively, whether they are free or united.

gynophore. The stalk which is present at the base of some ovaries (and the fruits developed from them).

half-inferior. Of an ovary: with its lower part inferior and its upper part superior (figure 41(15), p. 422).

haploid. Possessing in its normal vegetative cells only a single set of chromosomes.

hastate. With 2 acute, divergent lobes at the base, as in a mediaeval halberd (figure 39(12), p. 420).

haustorium. The organ with which a parasitic plant penetrates its host and draws nutriment from it.

herb. A plant in which the stems do not become woody, or, if somewhat woody at the base, do not persist from year to year.

herbaceous. Of a plant: possessing the qualities of a herb, as defined above.

heterostylic. Having flowers in which the length of the style relative to that of the stamens varies from one plant to another.

hilum. The scar-like mark on a seed indicating the point at which it was attached to the funicle.

hybrid. A plant produced by the crossing of parents belonging to 2 different named groups (e.g. genera, species, subspecies, etc.). An F_1 hybrid is the primary product of such a cross. An F_2 hybrid is a plant arising from a cross between 2 F_1 hybrids (or from the self-pollination of an F_1 hybrid).

hydathode. A water-secreting gland immersed in the tissue of a leaf near its margin.

hypocotyl. That part of the stem of a seedling which lies between the top of the radicle and the attachment of the cotyledon(s).

hypogeal. The mode of germination in which the cotyledons remain in the seed-coat and play no part in photosynthesis.

hypogynous. Describes a flower, or, preferably the petals, sepals and stamens (or perianth and stamens) of a flower in which the ovary is superior and the petals, sepals and stamens (or perianth and stamens) arise as individual whorls on the receptacle (figure 41(10), p. 422).

incised. With deep, narrow spaces between the teeth or lobes.

included. Not projecting beyond the organs which enclose it.

indefinite. More than 12 and possibly variable in number.

indehiscent. Without preformed lines of splitting, opening, if at all, irregularly by decay.

inferior. Of an ovary: borne beneath the sepals, petals and stamens (or perianth and stamens) so that these appear to arise from its top (figure 41(13 & 14), p. 422).

inflorescence. A number of flowers which are sufficiently closely grouped together to form a structured unit (figure 40, p. 421).

infraspecific. Denotes any category below species level, such as subspecies, variety and form. To be distinguished from *subspecific*, which means relating to subspecies only.

integument. The covering of an ovule, later developing into the seed-coat. Some ovules have a single integument, others 2.

internode. The part of a stem between 2 successive nodes.

involucel. A whorl of united bracteoles borne below the flower in some Dipsacaceae.

involucre. A compact cluster or whorl of bracts around the stalk at or near the base of some flowers or inflorescences or around the base of a capitulum; sometimes reduced to a ring of hairs.

keel. A narrow ridge, suggestive of the keel of a boat, developed along the midrib (or rarely other veins) of a leaf, sepal or petal.

laciniate. With the margin deeply and irregularly divided into narrow and unequal teeth.

lanceolate. 3–4 times as long as wide and tapering more or less gradually towards the tip (figure 39(9), p. 420).

layer. To propagate by pegging down on the ground a branch from near the base of a shrub or tree, so as to induce the formation of adventitious roots.

leaflet. One of the leaf-like components of a compound leaf (figure 39(15), p. 420).

legume. The typical fruit of the Leguminosae, formed from a single carpel and opening down both sutures; legumes are, however, variable, some indehiscent, others breaking into 1-seeded segments, etc. *lenticel*. A small, slightly raised interruption of the surface of the bark (or the corky outer layers of a fruit) through which air can penetrate to the inner tissues.

linear. Parallel-sided and many times longer than broad (figure 39(6), p. 420).

lip. A major division of the apical part of a bilaterally symmetric calyx or corolla in which the petals or sepals are united; there is normally an upper and lower lip, but either may be missing.

lyrate. Pinnatifid or pinnatisect, with a large terminal and small lateral lobes.

marginal. Of placentation: describing the placentation found in a free carpel which contains more than 1 ovule.

medifixed. Of a hair: lying parallel to the surface on which it is borne and attached to it by a stalk (usually short) at its mid-point.

mericarp. A carpel, usually 1-seeded, released by the break-up at maturity of a fruit formed from 2 or more joined carpels.

mesocarp. The central, often fleshy layer of a fruit-wall in those fruits in which the wall is distinctly 3-layered.

micropyle. A pore in the integument(s) of an ovule and later in the coat of a seed (figure 42(1), p. 423).

monocarpic. Flowering and fruiting once, then dying.

monoecious. With separate male and female flowers on the same plant; male flowers may contain non-functional carpels (and vice versa).

monopodial. A type of growth-pattern in which the terminal bud continues growth from year to year.

mucronate. Provided with a short, narrow point at the apex (figure 39(25), p. 420).

mycorrhiza. A symbiotic association between the roots of a green plant and a fungus.

nectary. A nectar-secreting gland.

neuter. Without either functional male or female parts.

node. The point at which 1 or more leaves or flower parts are attached to an axis.

nut. A 1-seeded indehiscent fruit with a woody or bony wall.

nutlet. A small nut, usually a component of an aggregate fruit.

obconical. Shaped like a cone, but attached at the narrow end.

oblanceolate. As lanceolate, but attached at the more gradually tapered end.

oblong. With more or less parallel sides and about 2–5 times as long as broad (figure 39(7), p. 420).

obovate. As ovate, but attached at the narrower end.

obovoid. As ovoid, but atttached at the narrower end.

ochrea. A sheath, made up of the stipules of each leaf-base, which surrounds the stem in many species of Polygonaceae; the ochrea is often scarious.

opposite. Describes 2 leaves, branches or flowers attached on opposite sides of the axis at the same node.

orthotropous. Describes an ovule which stands erect and straight (figure 42(1), p. 423).

ovary. The lower part of a carpel, containing the ovules(s) (i.e. excluding style and stigma); the lower, ovule-containing part of a gynaecium in which the carpels are united (figure 41(2), p. 422).

ovate. With approximately the outline of a hen's egg (though not necessarily blunt-tipped) and attached at the broader end (figure 39(10), p. 420).

ovoid. As ovate, but applied to a solid body.

ovule. The small body from which a seed develops after pollination (figure 42, p. 423).

palmate. Describes a compound leaf composed of more than 3 leaflets, all arising from the same point, as in the leaf of *Aesculus*; also used to described similar venation in simple leaves (figure 39(19), p. 420).

palmatifid. Lobed in a palmate manner, with the incisions pointing to the place of attachment, but not reaching much more than halfway to it (figure 39(20), p. 420).

palmatisect. Deeply lobed in a palmate manner, with the incisions almost reaching the base (figure 39(21), p. 420).

panicle. A compound raceme, or any freely branched inflorescence of similar appearance (figure 40(3 & 5), p. 421).

papillose. Covered with small blunt protuberances (papillae).

parietal. A form of placentation in which the placentas are borne on the inner surface of the walls of a 1-celled ovary, or rarely, in a similar manner in a septate ovary (figure 42(11), p. 423).

pectinate. With leaves, leaflets, or hairs in regular, eye-lash-like rows.

pedate. With a terminal lobe or leaflet, and on either side of it an axis curving outwards and backwards, bearing lobes or leaflets on the outer side of the curve (figure 39(27), p. 420).

peltate. Describes a leaf of other structure with the stalk attached other than at the margin (figure 39(27), p. 420).

perennial. Persisting for more than 2 years.

perfoliate. Describes a pair of stalkless opposite leaves of which the bases are united, or a single leaf in which the auricles are united so that the stem appears to pass through the leaf or leaves (figure 39(3 & 4), p. 420).

perianth. The calyx and corolla considered collectively, used especially when there is no clear differentiation between calyx and corolla; also used to denote a calyx or corolla when the other is absent (figure 41(2), p. 422).

pericarpel. See Cactaceae, p. 206.

perigynous. Describing a flower, or, preferably, the petals, sepals and stamens (or perianth and stamens) of a flower in which the ovary is superior and the petals, sepals and stamens (or perianth and stamens) are borne on the margins of a rim or cup which itself is borne on the receptacle below the ovary (it often appears as though the sepals, petals and stamens (or perianth and stamens) are united at their bases) (figure 41(11 & 12, p. 422).

perigynous zone. The rim or cup of tissue on which the sepals, petals and stamens (or perianth and stamens) are borne in a perigynous flower (figure 41(11 & 12), p. 422).

petal. A member of the inner perianth-whorl (corolla) used mainly when this is clearly differentiated from the calyx. The petals usually function in display and often provide an alighting place for pollinators (figure 41(1), p. 422).

phyllode. A leaf-stalk taking on the funtion and, to a variable extent, the form of a leaf-blade.

pinnate. Describes a compound leaf in which distinct leaflets are arranged on either side of the axis (figure 39(15), p. 420). If these leaflets are themselves of a similar compound structure, the leaf is termed *bipinnate* (similarly *tripinnate*, etc.).

pinnatifid. Lobed in a pinnate manner, with the incisions reaching not much more than halfway to the axis (figure 39(16), p. 420).

pinnatisect. Deeply lobed in a pinnate manner, with the incisions almost reaching the axis (figure 39(17), p. 420).

pistillode. A sterile ovary in a male flower.

placenta. A part of the ovary, often in the form of a cushion or ridge, to which the ovules are attached.

placentation. The manner of arrangement of the placentas.

pollen-sac. One of the cavities in an anther in which pollen is formed; each anther normally contains 4 pollen-sacs, 2 on either side of the connective, those of each pair separated by a partition which shrivels at maturity.

polyploid. Possessing in the normal vegetative cells more than 2 sets of chromosomes.

proliferous. Giving rise to plantlets or additional flowers (Cactaceae) on stems, leaves or in the inflorescence.

protandrous. With anthers beginning to shed their pollen before the stigmas of the same flower are receptive.

protogynous. With stigmas becoming receptive before the anthers in the same flower shed their pollen.

pulvinus. A swollen region at the base of a leaflet, leaf-blade or leaf-stalk.

pyrene. A small nut-like body enclosing a seed, 1 or more of which, surrounded by fleshy tissue, make up the fruit of, for example, *Ilex*.

raceme. An inflorescence consisting of stalked flowers arranged on a single axis, the lower opening first (figure 40(2), p. 421).

radially symmetric. Capable of division into 2 similar halves along 2 or more planes of symmetry (figure 41(8), p. 422).

radicle. The root preformed in the seed and normally the first visible root of a seedling.

raphe. A perceptible ridge or stripe, at one end of which is the hilum, on some seeds.

receptacle. The tip of an axis to which the floral parts, or perigynous zone (when present), are attached (figure 41(1), p. 422).

reflexed. Bent sharply backwards from the base.

rhizome. A horizontal stem, situated underground or on the surface, serving the purpose of food-storage or vegetative reproduction or both; roots or stems arise from some or all of its nodes.

rootstock. The compact mass of tissue from which arise the new shoots of a herbaceous perennial. It usually consists mainly of stem tissue, but is more compact than is generally understood by rhizome.

runner. A slender, above-ground stolon with very long internodes.

sagittate. With a backwardly-directed basal lobe on each side, like an arrow-head (figure 39(11), p. 420).

samara. A winged, dry, indehiscent fruit or mericarp.

saprophytic. Dependent for its nutrition on soluble organic compounds in the soil. Saprophytic plants do not photosynthesise and lack chlorophyll; some plants, however, are partially saprophytic and combine the two modes of nutrition.

scale-leaf. A reduced leaf, usually not photosynthetic.

scape. A leafless flower-stalk or inflorescence-stalk arising usually from near ground level.

scarious. Dry and papery, often translucent.

schizocarp. A fruit which, at maturity, splits into its constituent mericarps.

scion. A branch cut from one plant to be grafted on the rooted stock of another.

seed. A reproductive body adapted for dispersal, developed from an ovule and consisting of a protective covering (the seed-coat), an embryo, and, usually, a food-reserve.

semi-parasite. A plant which obtains only part of its nutrition by parasitism.

sepal. A member of the outer perianth whorl (calyx) when 2 whorls are clearly differentiated as calyx and corolla, or when comparison with related plants shows that a corolla is absent. The sepals most often function in protection and support of other floral parts (figure 41(1), p. 422).

septum. An internal partition.

shrub. A woody plant with several stems or branches arising from near the base, and of smaller stature than a tree.

simple. Not divided into separate parts.

sinus. The gap or indentation between 2 lobes, auricles or teeth.

spathulate. With a narrow basal part, which towards the apex is gradually expanded into a broad, blunt blade.

spicate. Similar to a spike.

spike. An inflorescence or subdivision of an inflorescence, consisting of stalkless flowers arranged on a single axis (figure 40(1), p. 421).

spur. An appendage or prolongation, more or less cylindric, often at the base of an organ. The spur of a corolla or single petal or sepal is usually hollow and often contains nectar (figure 41(7), p. 422).

stamen. The male organ, producing pollen, generally consisting of an anther borne on a filament (figure 41(1), p. 422).

staminode. An infertile tamen, often reduced or rudimentary or with a changed function.

stellate. Star-like, particularly of branched hairs.

stigma. The part of a style to which the pollen adheres, normally differing in texture from the rest of the style (figure 41(2), p. 422).

stipule. An appendage, usually 1 of a pair, found beside the base of the leaf-stalk in many flowering plants, sometimes falling early, leaving a scar. In some cases the 2 stipules are united; in others they are partly united to the leaf-stalk.

stock. A rooted plant, often with the upper parts removed, on to which a scion may be grafted.

stolon. A far-creeping, more or less slender, above-ground or underground rhizome giving rise to a new plant at its tip and sometimes at intermediate nodes.

stoma. A microscopic ventilating pore in the surface of a leaf or other herbaceous part.

style. The usually slender, upper part of a carpel or gynaecium, bearing the stigma (figure 41(2), p. 422).

subtend(ed). Used of any structure (e.g. a flower) which occurs in the axil of another organ (e.g. a bract); in this case the bract subtends the flower.

subulate. Narrowly cylindric, and somewhat tapered to the tip.

sucker. An erect shoot originating from a bud on a root or a rhizome, sometimes at some distance from the parent plant.

superior. Of an ovary: borne at the morphological apex of the flower so that the petals, sepals and stamens (or perianth and stamens) arise on the receptacle below the ovary (figure 41(10–12), p. 422).

suture. A line marking an apparent junction of neighbouring parts.

sympodial. A type of growth-pattern in which the terminal bud ceases growth, further growth being carried on by a lateral bud.

tendril. A thread-like structure which by its coiling growth can attach a shoot to something else for support.

terete. Approximately circular in cross-section; not necessarily perfectly cylindric, but without grooves or ridges.

tetraploid. Possessing in its normal vegetative cells 4 similar sets of chromosomes.

throat. The part of a calyx or corolla transitional between the tube and limb or lobes.

trichome. See Cactaceae, p. 203.

triploid. Possessing in its normal vegetative cells 3 similar sets of chromosomes.

tristylic. Having flowers of different plants with long, short or intermediate-length styles; the stamens of each flower are of 2 lengths which are not the same as the style-length of that flower.

truncate. As though with the tip or base cut off at right angles.

tuber. A swollen underground stem or root used for food-storage.

tubercle. A small, blunt, wart-like protüberance.

turion. A specialised perennating bud in some aquatic plants. consisting of a short shoot covered in closely packed leaves, which persists through the winter at the bottom of the water.

umbel. An inflorescence in which the flower-stalks arise together from the top of an inflorescence-stalk: this is a simple umbel (figure 40(8), p. 421). In a compound umbel the several stalks arising from the top of the inflorescence-stalk terminate not in flowers but in secondary umbels.

undivided. Without major divisions or incisions, though not necessarily entire.

urceolate. Shaped like a pitcher or urn, hollow and contracted at or just below the mouth.

vascular bundle. A strand of conducting tissue, usually surrounded by softer tissue.

vein. A vascular strand, usually in leaves or floral parts and visible externally.

venation. The pattern formed by the veins in a leaf.

versatile. Of an anther: flexibly attached to the filament by its approximate mid-point so that a rocking motion is possible.

vessel. A microscopic water-conducting tube formed by a sequence of cells not separated by end-walls.

viviparous. Bearing young plants, bulbils or leafy buds which can take root; they can occur anywhere on the plant and may be interspersed with, or wholly replace, the flowers in an inflorescence.

whorl. A group of more than 2 leaves or floral organs inserted at the same node.

wing. A thin, flat extension of a fruit, seed, sepal or other organ.

xerophytic. Drought-tolerant. Can also describe the environment in which drought-tolerant plants live.

BIBLIOGRAPHY

Index to books cited

This list provides minimal bibliographic details for all books cited in *The European Garden Flora*, volumes I, II & III, as an aid to users who wish to make reference to them. The details given are: author's surname(s) and initials, title in full (unless excessively long), edition used, number of volumes, place of publication and date(s) of publication. References to these works in the main body of the text are generally shorter than those given here. In the case of some recently published, illustrated or popular works, which have been translated or published simultaneously in several languages, only those editions used by authors of accounts are included; the details of such publication are often difficult to ascertain and no attempt has been made at completeness. Some multi-volume Floras are listed under their titles rather than under their editors, as are works with no author indicated. The minimum necessary cross-referencing is also provided.

Abrams, L. & Ferris, R.S., *Illustrated flora of the Pacific States, Washington, Oregon and California*, 4 volumes, Stanford (1923–60).

Aichele, D., *Wild flowers*, London (1975).

Allan, H.H., *Flora of New Zealand*, volume 1, Wellington (1961).

Ames, O., *Orchidaceae*, 5 volumes, Cambridge, Mass. (1905–20).

Ames, O., Hubbard, F.T. & Schweinfurth, C.H., *The genus Epidendrum in the United States and middle America*, Cambridge, Mass. (1936).

Andersohn, G., *Kakteen und andere Sukkulenten*, Niedernhausen (1982); English edn, translated by Haag, M.E., *Cacti and succulents*, London (1983, 1984).

Anderson, E.B., Balfour, A.P., Finnis, V., Fish, M., Nicholson, B.E. & Wallis, N., *The Oxford book of garden flowers*, London (1973).

Anderson, E.F., *Peyote, the divine cactus*, University of Arizona (1980).

Annesley, H., *Beautiful and rare trees and plants*, London (1903).

Arechavaleta, J., *Flora Uruguaya*, Montevideo (1901–11).

Armstrong, M., *Field book of western wild flowers*, New York (1915).

Ary, S. & Gregory, M., *The Oxford book of wild flowers*, Oxford (1962).

Aston, H.I., *Aquatic plants of Australia*, Melbourne (1973).

Aubréville, A. & Leroy, J.-F., later Morat, P. (eds), *Flore du Gabon*, 26 volumes, Paris (1961–).

Backeberg, C., *Die Cactaceae*, 6 volumes, Jena (1958–62).
Das Kakteenlexicon, Jena (1966), edn 2 (1970), edn 3 with appendix by Haage, W. (1976); English edn translated (from German 3rd edn) by Glass, L., *Cactus Lexicon*, Poole (1977).

Baglin, D. & Mullins, B., *Australian wildflowers in colour*, Sydney (1969).

Bailey, L.H., *The standard cyclopedia of horticulture*, 7 volumes, New York (1916–17); popular edition, 3 volumes, New York (1935).

Baillon, H.E., *Histoire des plantes*, 13 volumes, Paris (1868–94).

Baker, J.G., *Handbook of the Amaryllideae including the Alstroemerieae and Agaveae*, London (1888).

Ball, J.S., *Southern African epiphytic orchids*, Johannesburg & London (1978).

Banerjee, M.L. & Thapa, B.B., *Orchids of Nepal*, Calcutta (1978).

Barber, P. & Phillips, C.E.L., *The trees around us*, London (1975).

Barkhuizen, B.P., *Succulents of southern Africa with special reference to the succulent families found in the Republic of South Africa and Southwest Africa*, Cape Town (1978).

Bärtels, A., *[Das grosse Buch der] Gartengehölze*, Stuttgart (1981).

Barthlott, W., *Kakteen*, Stuttgart (1977); English edn, translated by Glass, L., *Cacti*, Cheltenham (1979).

Basu, B.D., *Indian medicinal plants*, Delhi (1918).

Bateman, J., *Second century of orchidaceous plants*, London (1867).

Batten, A., *Flowers of southern Africa*, Sandton (1986).

Batten, A. & Bokelmann, H., *Wild flowers of the eastern Cape Province* (1966).

Baum, B., *Oats: wild and cultivated*, Ottawa (1977).

Bayer, M.B., *The new Haworthia handbook*, Kirstenbosch (1982).

Bean, W.J., *Trees and shrubs hardy in the British Isles*, 4 volumes, edn 8, London (1970–80).

Beaugé, A., *Chenopodium album et espèces affines: étude historique et statistique*, Paris (1974).

Bechtel, H., Cribb, P. & Launert, E., *Orchideenatlas*, Stuttgart (1980).
Manual of cultivated orchid species, Poole (1981).

Beckett, K.A., *The concise encyclopaedia of garden plants*, London (1983).
The RHS Encyclopaedia of house plants including conservatory plants, London (1987).

Benson, L., *The cacti of Arizona*, edn 3, Tucson (1969).
The cacti of the United States and Canada, Stanford (1982).

Benson, L. & Darrow, R.A., *Trees and shrubs of the southwestern deserts*, edn 3, Tucson (1981).

Bentley R. & Trimen, H., *Medicinal plants; being descriptions with original figures of the principal plants employed in medicine and an account of the characters, properties and uses of their parts and products of medicinal value*, 4 volumes, London (1875–80).

Berger, A., *Mesembrianthemen und Portulacaceen*, Stuttgart (1908). *Die Agaven*, Jena (1915).

Berhaut, J., *Flore illustrée du Sénégal*, 6 volumes, Dakar (1971–).

Bichard, J.D. & McClintock, D., *Wild flowers of the Channel Islands*, London (1975).

Bicknell, C., *Flowering plants and ferns of the riviera and neighbouring mountains drawn and described*, London (1885).

Birdsey, M.L., *The cultivated aroids*, Berkeley, California (1951).

Black, J.M. (revised and edited by Jessop, J.P.), *Flora of South Australia*, edn 3, part 1, Adelaide (1978).

Black, P.M., *The complete book of orchid growing*, London (1980).

Blatter, E., *Beautiful flowers of Kashmir*, 2 volumes, London (1927–28).

Bloom, A., *Perennials for your garden*, Nottingham, edn 3 (1975). *Conifers for your garden*, Nottingham (1972).

Blume, C.L., *Flora Javae*, Bruxelles (1828).

Blundell, M., *The wild flowers of Kenya*, London (1982).

Boland, D.J., Brooker, M.I.H., Chippendale, G.M., Hall, N., Hyland, B.P., Johnston, R.D., Kleinig, D.A. & Turner, J.D., *Forest trees of Australia*, edn 4, Melbourne (1984).

Bolus, H., *Orchids of South Africa (Icones Orchidacearum austro-africanarum extra-tropicarum)*, 3 volumes, London (1893–1914).

Bolus, H.M.L., *Notes on Mesembryanthemum and some allied genera*, 3 volumes, Cape Town (1928–58).

Bolus, H.M.L., Barclay, D. & Steer, E.J., *A second book of South African flowers*, Cape Town (1936).

Bonnier, G., *Flore complète illustrée en couleurs de France, Suisse et Belgique*, 13 volumes, Paris (1911–34).

Borg, J., *Cacti*, London (1937), edn 2 (1951).

Bose, T.K. & Bhattarcharjee, S.K., *Orchids of India*, Calcutta (1981).

Boom, B.K., *Nederlandse dendrologie (Flora Cultuurgewassen 1)*, edn 5 (1965).

Boswell, J.T. (alias Syme, J.T.I.) (ed.) *English botany*, edn 3, London (1886–92).

Botschantzeva, Z.P. (translated & edited by H.O. Varekamp), *Tulipa: taxonomy, morphology, cytology, phytogeography and physiology* (1982).

Bowles, E.A., *A Handbook of Narcissus*, London (1934). *A handbook of Crocus and Colchicum for gardeners*, London (1924), edn 2 (1952).

Bramwell, D. & Bramwell, Z., *Wild flowers of the Canary Islands*, London (1974).

Brandis, D., *Indian trees*, London (1906).

Braun, L.E., *The vascular Flora of Ohio*, vol. 1, *The Monocotyledoneae*, Columbus, Ohio (1967).

Bravo, H., *Las Cactáceas de Mexico*, Mexico City (1937); edn 2, with Sanchez-Mejorada, H. (1978).

Brison, F.R., *Pecan culture* (1974).

Britton N.L. & Brown, A., *An illustrated flora of the northern United States and Canada*, 3 volumes, New York (1896–98), edn 2, (1927–).

Britton, N.L. & Rose, J.N., *The Cactaceae*, 4 volumes, Washington (1919–23).

Brown, N.E. and others, *Mesembryanthema*, privately published, undated.

Brüggeman, L., *Tropical plants and their cultivation*, London (1957).

Buchanan, J., *Indigenous grasses of New Zealand*, Wellington (1880).

Buishand, T., Houwing, H.P. & Jansen, K., *The complete book of vegetables*, Leicester & London (1986).

Burbidge, N.T. & Gray, M., *Flora of the Australian Capital Territory*, Canberra (1970).

Burbidge, F.W., *The Narcissus: its history and culture*, London (1875).

Butcher, R.W., *A new illustrated British flora*, 2 volumes, London (1961).

Callen, G., *Les conifères cultivés en Europe*, 2 volumes, Paris (1976–77).

Camus, A., *Les chênes: monographie du genre Quercus*, 3 volumes, Paris (1934–54). *Le chataignier*, Paris (1929).

Camus, E.G., *Les Bambusées*, Paris (1913).

Candolle, A. de, *Monographiae Phanerogamarum*, 9 volumes, Paris (1878–96).

Cansdale, G.S., *The black poplars and their hybrids cultivated in Britain*, Oxford (1938).

Castroviejo, S., Lainz, M., López Gonzalez. G., Montserrat, P., Muñoz Garmendia, F., Pavia, J. & Villar, L. (eds), *Flora Iberica*, 2 volumes, Madrid (1986–).

Cave, N.L., *The iris*, London (1959).

Ceballos A., Casas, J.F. & Garmendia, F.M., *Plantas silvestres de la peninsula iberica*, Barcelona (1980).

Chabaud, B., *Les palmiers de la Côte d'Azur*, Paris (1915).

Chaudun, V., *Ornamental conifers*, London (1956).

Cheeseman, T.F., *Illustrations of the New Zealand Flora*, 2 volumes, Wellington (1914).

Chickering, C.R., *Flowers of Guatemala*, Norman, Oklahoma (1973).

Chippindal, L.K.A. & Cook, A.O., *Grasses of southern Africa*, 3 volumes, Salisbury (1976–78).

Christiansen, S.M., *Grasses, sedges and rushes in colour*, Poole (1979).

Clapham, A.R., Tutin, T.G. & Warburg, E.F., *Flora of the British Isles*, Cambridge, edn 1 (1952), edn 2 (1962).

Clark, L.J., *Wild flowers of British Columbia*, Sidney, British Columbia (1973).
(edited by Trelawny, J.G.), *Wild flowers of the Pacific north west from Alaska to northern California*, Sidney, British Columbia (1976).

Classified list and International Register of Daffodil names, London (1954–).

Clay H.F. & Hubbard, J.C., *The Hawaii garden, tropical exotics*, Honolulu (1977).

Clements, F.E., *Plant physiology and ecology*, New York (1907).

Clements, F.E. & Clements, E.S., *Rocky mountain flowers*, White Plains & New York (1914).

Clinton-Baker H. & Jackson, A.B., *Illustrations of new conifers*, Hertford (1935)

Cochrane G.R., Fuhrer, B.A., Rotherham, E.R. & Willis, J.H., [*Australian flora in colour:*] *Flowers and plants of Victoria*, Syndey & London (1973).

Cockayne, L., *The vegetation of New Zealand*, being vol. 14 of Engler, A. & Drude, O., *Die Vegetation der Erde*, Leipzig (1928).

Cogniaux, C.A. & Goossens, A., *Dictionnaire Iconographique des Orchidées*, many fascicles, Paris (1896–1903).

Cohen, V.A., *A guide to Pacific coast Irises*, London (1967).

Collett, H., *Flora Simlensis*, Calcutta (1902).

Conard, H.S., *The Waterlilies*, Washington (1905).

Condit, I.J., *Ficus, the exotic species*, California (1969).

Cook, C.D.K., *Water plants of the world: a manual for the identification of freshwater macrophytes*, The Hague (1974).

Cooper, D., *A field guide to New Zealand native orchids*, Wellington (1981).

Copeland, E.B., *Genera filicum: the genera of ferns*, Waltham (1947).

Correa, M.N., *Flora Patagonica*, Buenos Aires (1969).

Correll, D.S. & Correll, H.B., *Aquatic and wetland plants of southwestern United States*, Stanford (1972, revised 1975).
Flora of the Bahama Archipelago, Vaduz (1982).

Coste, H.J., *Flore descriptive et illustrée de la France, de la Corse et des contrees limitrophes, avec une introduction sur la flore et la vegetation de la France*, 3 volumes, Paris (1901–06).

Costermans, L., *Native trees and shrubs of south-eastern Australia*, Adelaide (1981).

Costin A.B., Gray, M., Totterdell, C.J. & Wimbush, D.J., *Kosciusko alpine flora*, Melbourne (1979).

Couret, P., [*Joyas de las*] *Orquideas Venezolanas*, Caracas (1977).

Court, D., *Succulent flora of southern Africa*, Rotterdam (1981).

Courtenay B. & Zimmerman, J.H., *Wildflowers and weeds: a guide in full colour*, New York (1972).

Courtenay-Latimer B. & Smith, G.G., *The flowering plants of the Tsitsikama forest and coastal national park*, South Africa (1967).

Coventry, B.O., *Wild flowers of Kashmir*, 3 volumes, London (1923–30).

Craig, R.T., *The Mammillaria handbook*, Pasadena (1945).

Cribb, P.J. & Leedal, G.P., *The mountain flowers of Tanzania*, Rotterdam (1982).

Croat, T.B., *Flora of Barro Colorado Island*, Stanford, California (1978).

Cronquist, A., Holmgren, A.H., Holmgren, N., Reveal, J. & Holmgren, K., *Intermountain flora*, 3 volumes, New York & London (1977).

Crouzet, Y., *Les bambous*, Dargaud (1981).

Csapody, V. & Toth, I., *A colour atlas of flowering trees and shrubs*, Budapest (1982).

Cullmann, W., Götz, E. & Gröner, G., *Kakteen*, edn 5, Stuttgart (1984); English edn, translated by Thomas, K.M., *The Encyclopedia of Cacti*, Sherbourne (1986).

Cunningham, G.M., Mulham, W.E., Milthorpe, P.L. & Leigh, J.H., *Plants of western New South Wales*, Condobolin, NSW (1981).

Curtis, W., *Student's Flora of Tasmania*, 4 volumes, Hobart (1956–79).

Curtis, W. & Stones, M., *Endemic flora of Tasmania*, London (1967–73).

Dahlgren, R.M.T. & Clifford, T.H., *The monocotyledons: a comparative study*, London (1982)

Dallimore W. & Jackson, B.A., *Handbook of Coniferae and Ginkgoaceae*, edn 4, London (1966).

Danesch, E. & Danesch, O., *Orchideen Europas, Südeuropa*, Bern (1969).
Orchideen Europas, Mitteleuropa, edn. 3, Bern (1972).
Orchideen Europas, Ophrys-hybriden, Bern (1972).

Darnell, A.W., *Hardy and half-hardy plants; illustrations and descriptions of beautiful and interesting plants suitable for culture in the British Isles*, 2 volumes, Hampton Wick (1929–32).

Das Pflanzenreich (ed. Engler, A.), 107 volumes, Leipzig (1900–53).

Davidson, W., *Illustrated dictionary of house plants*, London (1983).

Davis, P.H. (ed.), *Flora of Turkey and the east Aegean Islands*, 8 volumes, Edinburgh (1965–84).

Davis, R. S. & Steiner, M.L., *Philippine orchids*, New York (1972).

Dean, B.E., Mason, A., & Thomas, J.L., *Wild flowers of Alabama and adjoining states*, Alabama (1983).

Decaisne, J., *Les Clématites a fleur tubuleuses*, Paris (1881).

Delessert, B., *Icones selectae plantarum*, 5 volumes, Paris (1820–46).

Den Ouden, P. & Boom, B.K., *Manual of cultivated conifers*, The Hague (1965).

Die Natürlichen Pflanzenfamilien, ed. Engler, A. & Prantl, K., many volumes, Leipzig (1887–1915), edn 2 (1924–).

Dietrich, H., *Bibliographia Orchidacearum*, Jena (1980).

Dockrill, A.W., *Australian indigenous orchids*, Sydney (1969).

Dressler, R.L., *The Orchids: natural history and classification*, Cambridge, Mass. (1981).

Dressler, R.L. & Pollard, G.E., *The genus Encyclia in Mexico*, Mexico City (1974).

Dunsterville, G.C.K. & Garay, L.A., *Venezuelan orchids illustrated*, 6 volumes, London (1959–76).

Dutta, R., *Water gardening indoors and out*, London (1977).

Dyer, R.A., Codd, L. E., de Winter, B. & Rycroft, H.B., *Flora of southern Africa*, volume 1, Pretoria (1966).

Dykes, W.R., *The genus Iris*, Cambridge (1915), reprinted 1974.
A handbook of garden irises, London (1924).

Dykes, W.R., edited and illustrated by Dykes, K.E., *Notes on tulip species*, London (1930).

Edlin, H.L., *The tree key. A guide to identification in garden, field and forest*, London (1978).

Edlin, H.L. & Nimmo, H., *The world of trees*, London (1974).

Edwards, C., *Delphiniums*, London (1981).

Elias, T.S., *The complete trees of North America*, New York (1980).

Eliovson, S., *South African wild flowers for the garden*, edn 4, Capetown (1965).
Namaqualand in flower, Johannesburg (1972).
Wild flowers of southern Africa, Braamfontein (1980).

Elliot, R.W. & Jones, D.L., *Encyclopedia of Australian plants suitable for cultivation*, 4 volumes, Melbourne (1980–).

Elliott, R., *The genus Lewisia*, Alpine Garden Society (1977).

Elwes, H.J., *A monograph of the genus Lilium*, London (1877–80).

Elwes, H.J. & Henry, A., *Trees of Great Britain and Ireland*, Edinburgh (1906–13).

Encke, F., Buchheim, G. & Seybold, S., *Zander's Handwörterbuch der Pflanzennamen*, edn 13, Stuttgart (1984).

Engelmann, G., *Cacti of the boundary*, Washington (1858).

Erickson, R., *Australian flora in colour: Flowers and plants of Western Australia*, Sydney & London (1973).

Evans, A., *The peat garden and its plants*, London (1974).

Everard, B. & Morley, B.D., *Wild flowers of the world*, London (1970).

Everett, T.H., *Living trees of the world*, London (1969).
    The New York Botanical Gardens illustrated encyclopedia of horticulture, New York (1980–81).

Fabian A. & Germishuizen, G., *Transvaal wild flowers*, Johannesburg (1982).

Fedtschenko, O., *Kritische Übersicht über die Gattung Eremurus*, Cramer Plant Monograph Reprints **3**, Lehre (1968).

Feinbrun, N. & Zohary, M., *Flora of the land of Israel, Iconography*: drawn by Ruth Koppel, Jerusalem (1940).

Felsko, E., *A book of wild flowers*, Oxford (1956).
    Portraits of wild flowers, Oxford (1959).

Fenaroli, L., *Flora Mediterranea*, 2 volumes, Milano (1962, 1970).

Fernald, M.L., (ed), *Gray's manual of botany*, edn 8, New York, etc. (1950).

Fiori A. & Paoletti, G., *Iconographia Florae Italicae; ossia, Flora Italiano illustrata con 4236 figure d'assieme e 12540 di analisi ...*, Padova (1895–99); edn 3, Firenze (1933).

Fischer, F.E.L. & Meyer, C.A., *Sertum Petropolitanum*, St. Petersburg (1846).

Fisher, F.J.F., *The alpine Ranunculi of New Zealand*, Wellington (1965).

Fitschen, J., Gehölzflora, edn 7, Heidelberg (1977).

Fitter, A.H., *New generation guide to the wild flowers of Britain and northwest Europe*, London (1987).

Fitter, R. & Fitter, A.H., *The wild flowers of Britain and northern Europe*, London (1974).

Flora Iranica (ed. Rechinger, K.H.), Graz (1963–).

Flora Malesiana, many volumes, Djakarta, Dordrecht, Boston, Lancaster, etc. (1950–).

Flora Neotropica, many volumes, New York & London (1968–).

Flora of tropical East Africa (ed. Turrill, W.B. & Milne-Redhead, E.), London (1952–).

Flora republicae popularis sinicae, many volumes, Beijing (1959–).

Flore d'Afrique central, formerly *Flore du Congo Belge et du Ruanda-Urundi* (ed. Robyns, A.), Bruxelles (1973).

Flora SSSR, ed. Komarov, V.L., 32 volumes, Moscow (1934–64).

Fogg, H.G.W. – see Witham Fogg, H.G.

Fogg, J.M. & McDaniel, J.C., *Check-list of the cultivated Magnolias*, Mount Vernon, Virginia (1975).

Foster, F.G., *The Gardener's fern book*, Princeton (1964).

Fournet, J., *Flore illustrée des phanerogames de Guadeloupe et de Martinique*, Paris (1978).

Fowlie, J.A., *The genus Lycaste*, Pomona, California (1970).
    Brazilian bifoliate Cattleyas and their color varieties, Pomona, California (1977).

Franchet, A., *Plantae Davidianae*, Paris (1888).

Francis, W.D., *Australian rain-forest trees including notes on some of the tropical rain forests and descriptions of many tropical species*, edn 3, Canberra (1970).

Freeman-Mitford, A.B., *The bamboo garden*, London & New York (1896).

Galbraith, J., *Collins field guide to the wild flowers of south-east Australia*, Sydney & London (1977).

Garcke, K.H. (ed. von Weihe, K.), *Illustrierte Flora*, edn 23, Berlin (1972).

Gardner, C.A., *Wildflowers of Western Australia*, edn 2, Perth (1973).

Garrard, I. & Streeter, D., *The wild flowers of the British Isles*, London (1983).

Genders, R., *Bulbs. A complete handbook of bulbs, corms and tubers*, London (1973).

Gentry, H.S., *Agaves of continental North America*, Tucson (1982).

George, A.S., *The Banksia book*, New South Wales (1984).
    An introduction to the Proteaceae of Western Australia, Kenthurst, New South Wales (1984).

Gibson, J.M., *Wild flowers of Natal (coastal region)*, Durban (1975).
    Wild flowers of Natal (inland region), Durban (1975).

Gilmour, J. & Walters, S.M., *Wild flowers*, London (1954).

Gleason, H.A. *[The new Britton & Brown] Illustrated flora of the north-eastern United States and adjacent Canada*, 3 volumes, New York (1952–74).

Gledhill, E., *Eastern Cape veld flowers*, Capetown (1969), reprinted (1977).

Godfrey, R. & Wooten, J.W., *Aquatic and wetland plants of the southeastern United States*, 2 volumes, Athens, Georgia (1979–81).

Gombocz, E., *Monographia generis Populi*, Budapest (1908).

Goulandris, N.A. & Goulimis, C.N., *Wild flowers of Greece*, Kifissia (1968).

Gould, F.W., *The grasses of Texas*, College Station, Texas (1975).

Goulding, J. H., *Fanny Osborne's flower paintings*, Auckland (1983).

Graaff, J. de & Hyams, E., *Lilies*, London (1967).

Graf, A.B., *Exotica*, Rutherford, New Jersey, edn 2 (1959), edn 3 (1963), edn 11 (1982); International series, edn 4 (1985).
    Exotic plant manual, edn 4, Rutherford, New Jersey (1974).
    Tropica, Rutherford, New Jersey (1978).

Grandidier, A., *Histoire physique, naturelle et politique de Madagascar*, Atlas, 4 volumes, Paris (1886–97).

Greshoff, M., *Nuttige Indischen Planten*, Buitenzorg (1884–1900).

Grey, C.H., *Hardy bulbs, including half-hardy bulbs and tuberous and fibrous rooted plants*, 3 volumes, London (1938).

Grey-Wilson, C., *The genus Iris subsection Sibiricae*, London (1971).

Grey-Wilson, C. & Blamey, M., *Alpine flowers of Britain and Europe*, London (1979).

Grey-Wilson, C. & Mathew, B., *Bulbs: the bulbous plants of Europe and their allies*, London (1981).

Grey-Wilson, C. & Matthews, V.A., *Gardening on walls*, London (1983).

Grierson, A.J.C. & Long, D.G., *Flora of Bhutan*, several volumes, Edinburgh (1983–).

Griffith, A.N., *Collins' guide to alpines*, London (1964).

Grounds, R., *Ornamental grasses*, London (1979).

Grove, A. & Cotton, A.D., *Supplement to Elwes' Monograph of the genus Lilium*, London (1934-40).

Guinea López, E. & Vidal Box, C., *Parques y jardines de España*, Madrid (1969).

Guittoneau, G.-G. & Huon, A., *Connaitre et reconnaitre la flore et la vegetation méditeranéennes*, Rennes (1983).

Gürke, M. & Vaupel, F. – see Schumann, K.M.

Hackel, E., *Monographia Festucorum europeum*, Kassel & Berlin (1882).

Hall, A.D., *The genus Tulipa*, London (1940).

Hamer, A.H., *Wild flowers of the Cape*, Cape Town (1926).

Hamer, F., *Las Orquideas de El Salvador*, 2 volumes, San Salvador (1974).

Hara, H., *Photo-album of plants of Eastern Himalaya*, Tokyo (1968).

Harrison, C.R., *Ornamental conifers*, Newton Abbot (1975).

Harrison, R.E., *Know your lilies*, Cape Town (1970).
Climbers and trailers, Wellington (1973).

Harrison, S.G. – see Masefield et al.

Hart, C. & Raymond, C., *British trees in colour*, London (1973).

Hartmann, W.L., *Introduction to the cultivation of orchids*, London (1965).

Haskin, L.L., *Wild flowers of the Pacific coast*, Portland, Oregon (1934).

Haustein, E., *Der Kosmos-Kakteenführer*, Stuttgart (1983).

Hawkes, A.D., *Encyclopedia of cultivated orchids*, London (1965).

Hay, R. & Beckett, K.D., *Readers' Digest encyclopaedia of garden plants and flowers*, London (1978).

Hay, R., McQuown, G.F.R. & Beckett, K., *Dictionary of indoor plants in colour*, London (1974), edn 2 (1985).

Hay, R. & Synge, P.M., *Dictionary of garden plants in colour*, London (1969).

Hayne, G.F., *Getreue Darstellung und Beschreibung der in Arzneykunde gebrauchlichen Gewächse*, 14 volumes, Berlin (1805–43).

Healy A.J. & Edgar, E., *Flora of New Zealand*, vol. 3, Wellington (1980).

Hecht, H., *BLV Handbuch der Kakteen*, München (1982).

Hegi, G., *Illustrierte Flora von Mitteleuropa*, edn 1, 7 volumes, München (1906–66), edn 2, Hamburg & Berlin (1964–).

Henrard, J.T., *A monograph of the genus Digitaria*, Leiden (1950).

Herbel, D., *Alles über Kakteen*, München (1978).

Herklots, G., *Flowering tropical climbers*, Folkestone (1976).

Herre, H., *The genera of Mesembryanthemaceae*, Cape Town (1971).

Hess, H.E., Landolt, E. & Hirzel, R., *Flora der Schweiz und angrenzender Gebiete*, 3 volumes, Basel (1967–72).

Hessayon, D.G., *The indoor plant spotter*, Waltham Cross (1985).

Heukels, H. & Van der Meijden, R., *Flora van Nederland*, edn 20, Groningen (1983).

Heywood, V.H. (ed.), *Flowering plants of the world*, Oxford (1978).

Heywood, V.H. & Chant, S.R., *Popular encyclopedia of plants*, Cambridge (1982).

Hill, D.S., *Figs (Ficus spp.) of Hong Kong*, Hong Kong (1967).

Hirao, H., *Colour encyclopaedia of cacti*, Tokyo (1979).

Hitchcock, C.L., Cronquist, A., Ownbey, M. & Thomson, J.W., *Vascular plants of the Pacific northwest*, 5 volumes, Seattle (1955–71).

Hodgson, M. & Paine, R., *A field guide to Australian wild flowers*, 2 volumes, Adelaide (1971–77).

Hoehne, F.C., *Flora Brasilica*, 48 volumes, Sao Paulo (1940–).
Iconographia de orchidaceas do Brasil, Sao Paulo (1949).

Hogg, R. & Johnson, G.W., *Wild flowers of Great Britain*, London (1866).

Holliday, I., & Hill, R., *A field guide to Australian trees*, Adelaide (1969).

Holmboe, J., *[Studies on] The vegetation of Cyprus*, Bergen (1914).

Holttum, R., *A revised Flora of Malaya*, volume 1, *Orchids of Malaya* (1951), volume 2, *Ferns* (1954), Singapore.

Hooker, W.J., *Exotic flora*, 3 volumes, London (1823–27).
Flora Boreali-Americana, London (1829).

Hora, B. (ed), *Oxford encyclopaedia of trees of the world*, Oxford (1981).

Hortus Third (ed. by staff of L.H. Bailey Hortorium, Cornell University), New York & London (1976).

Hoshizaki, B.J., *Fern grower's manual*, New York (1975).

Hosie, R.C., *Native trees of Canada*, Ottawa (1979).

Hough, R.B., *Handbook of the trees of the northern states and Canada*, New York (1947).

House, H.D., *Wild flowers of New York*, New York (1918).
Wild flowers, New York (1961).

Houtzagers, G., *Het geschlacht Populus*, Wageningen (1937).

Howard, R.A., *Flora of the Lesser Antilles*, 3 volumes, Jamaica Plain, Mass. (1974).

Hubbard, C.E., *Grasses*, Harmondsworth, edn 1 (1954), edn 2 (1968).

Hulme, M, *Wild flowers of Natal*, Pietermaritzburg (1955).

Humbert, H., *Flore de Madagascar*, numerous parts, Paris (1939–).

Humboldt, A.F.H., Bonpland, A. & Kunth, C.S., *Nova genera et species plantarum*, 7 volumes, Paris (1815–25).

Humphries, C.J., Press, J.R. & Sutton, D.A., *[The Hamlyn guide to] Trees of Britain and Europe*, London (1981).

Hunt, P.F. & Grierson, M., *Country Life book of orchids*, London (1978).

Hutchinson, J., *The families of flowering plants*, edn 2, 2 volumes, Oxford (1959), edn 3 (1973).
The genera of flowering plants, 2 volumes, Oxford (1964, 1967).

Hutchinson, J. & Dalziel, J.M., *Flora of west tropical Africa*, 2 volumes, London (1927–36); edn 2, ed. by Keay, R.W.J., 3 volumes, London (1954–72).

Hutchison, P.C. & Kimnach, M., *Icones plantarum succulentarum*, Cactus & Succulent Society of America (1956–60).

Huxley, A., *Mountain flowers in colour*, London (1967).

Huxley A. (ed.), *Garden perennials and water plants in colour*, London (1970).
Deciduous garden trees and shrubs in colour, London (1979).
World guide to house plants, London (1983).

Huxley, A. & Gilbert, R. (eds) *Success with house plants*, New York & Montreal (1979).

Huxley, A. & Taylor, W., *Wild Flowers of Greece and the Aegean*, London (1977).

Iconographia Cormophytorum Sinicorum, 5 volumes, Beijing (1972–76).

Innes, C., *The complete handbook of cacti and succulents*, London (1977).

Ito, Y., *The Cactaceae*, Tokyo (1981).

Jackson, W.P.U., *Wild flowers of Table Mountain*, Cape Town (1977).

   Wild flowers of the fairest Cape, Cape Town (1980).

Jacobi, G.A., *Von Versuch zu einer systematischen Ordnung der Agaven*, Hamburg (1864).

Jacobsen, H., translated by Raabe, H., *Handbook of succulent plants*, 3 volumes, London (1960).

   Das Sukkulentenlexicon, Jena (1970); English edn, translated by Glass, L., *Lexicon of succulent plants*, London (1974).

Jacobsen, N., *Akvarieplanter i farver*, Copenhagen (1977).

   Cryptocoryner, Copenhagen (1979).

Jacquin, N., *Icones plantarum rariorum*, 3 volumes, Vienna (1781–93).

Jafri, S.M. & El-Gadi, A., *Flora of Libya*, many fascicles, Tripoli (1976–).

Japan Succulent Society, *Colour encyclopedia of succulents*, Tokyo (1981).

Jávorka S. & Csapody, V., *[A Magyar flora kepekben:] Iconographia Florae hungaricae*, Budapest (1929–34).

Jayaweera, D.M.A., *Orchidaceae* in Dassanayake, M.D., *A revised handbook of the Flora of Ceylon*, volume 2, New Delhi (1981).

Jefferson-Brown, M., *Modern lilies*, London (1965).

Jelitto L. & Schacht, W., *Die Freiland Schmuckstauden*, Stuttgart (1963); edn 3, ed. by Jelitto, L., Schacht, W., & Fessler, A. (1985).

Jeppe, B., *Natal wild flowers*, Cape Town & London (1975).

   South African Aloes, Cape Town (1969).

Jepson, W.L., *A flora of California*, 3 volumes, San Francisco (1909–43).

Jermy, C., & Tutin, T.G., *Sedges of the British Isles*, edn 2, London (1982).

Johnstone, G.H., *Asiatic Magnolias in cultivation*, London (1955).

Jordan A. & Fourreau, J., *Icones ad Floram Europeae novo fundamento instaurandam spectantes*, 3 volumes, Paris (1866–1903).

Jessop, J.P., *Flora of central Australia*, Sydney (1981).

Justice, W.S. & Bell, C.R., *Wild flowers of North Carolina*, N Carolina (1968).

Kaier, E., *Indoor plants in colour*, London (1961).

Kaye, R., *Hardy ferns*, London (1968).

Keble Martin, W., *The concise British flora in colour*, London (1965), edn 2 (1969) – *The new concise British flora in colour*.

Keng, H., *Orders and families of Malayan seed plants*, Kuala Lumpur (1969).

Kerner von Marilaun, A., & Oliver, F.W., *Natural history of plants*, Glasgow, Edinburgh & Dublin (1894–1902).

Kidd, M.M., *Wild flowers of the Cape peninsula*, London & Cape Town (1973).

Kimnach, M. – see Hutchison, P.C.

King H.J. & Burns, T.E., *Wild flowers of Tasmania*, Milton, Tasmania (1969).

Kirk, J.W.C., *A British garden flora*, London (1927).

Kitamura, S. & Murata, G., *Coloured illustrations of the woody plants of Japan*, Osaka (1977).

Kitamura, S., Murata, G. & Koyama, T., *Alpine plants of Japan*, Osaka (1978).

   Coloured illustrations of herbaceous plants of Japan (Monocotyledoneae), Osaka (1970–81).

Kitamura, S. & Okamoto, S., *Coloured illustrations of trees and shrubs of Japan*, Osaka (1959), reprinted (1977).

Klein, L., *Gartenblumen 1: Frühlingsblumen*, Heidelberg (1979).

Kohlhaupt, P., *I fiori della Dolomiti*, edn 3, Bolzano (1984).

Kolakovski, A.A., *Flora Abkhazii*, 4 volumes, Abgiz (1938–49).

Komarov, V. & Klobukava-Alisova, G., *Opredelitel' rastenii dle nevostochnogo kraya*, 2 volumes, Leningrad (1931–32).

Koorders, S.H., *Excursionsflora von Java*, 2 volumes, Jena (1911).

Krainz, H., *Neue und seltene Sukkulenten*, Musingen (1946).

Krainz, H. (ed.), *Die Kakteen*, 63 Leiferungen (loose-leaf parts), Stuttgart (1956–75).

Krempin, J.L., *1000 decorative plants*, London & Canberra (1983).

Krüssmann, G., *Handbuch der Nadelgehölze*, Berlin & Hamburg (1972).

   Handbuch der Laubgehölze, edn 2, Berlin (1976); English edn, translated by Epps, M., *Manual of cultivated broad-leaved trees and shrubs*, London (1986).

Kunkel, G., *Flowering trees in tropical gardens*, The Hague & London (1978).

   Arboles y arbustos de las Islas Gran Canarias, Las Palmas (1981).

Kunkel, G. & Kunkel, M., *Flora de Gran Canaria*, 3 volumes, Las Palmas (1974–).

Kurata, S., *Illustrated important forest trees of Japan*, 5 volumes, Tokyo (1968–73).

Kupper, W. & Linsenmaier, W. (translated by Little, J.), *Orchids*, London (1961).

Lamb E. & Lamb, B.M., *Illustrated reference on cacti and other succulents*, 5 volumes, Poole (1955–78).

   Pocket encyclopaedia of cacti and succulents, London (1969).

Landwehr, J., *Wilde Orchideen van Europa*, 2 volumes, 'S Graveland (1977).

Langlois, A.C., *Supplement to Palms of the world*, Gainesville (1976).

Lasser, T., *Flora of Venezuela*, 15 volumes, Caracas (1964–70).

Laubenfels, D.J. de, *Gymnosperms*; volume 4 of Aubréville, A. & Leroy, J.-F., *Flore de la Nouvelle Calédonie et dependances*, Paris (1972).

Lavallée, A., *Les Clématites a grandes fleurs*, Paris (1884).

Lawson, A.H., *Bamboos*, London (1968).

Leathart, S., *Trees of the world*, London (1977).

Ledbetter, G.T., *Water gardens*, Sherborne (1979).

   [The water gardens book of] Patio ponds, Blagdon (1982).

Ledebour, C.F. von, *Icones plantarum novarum vel imperfecte cognitarum, floram rossicam imprimis altaicam*, 5 volumes, Riga (1829–34).

Leeburn, M.E., *Garden lilies: their selection and cultivation*, London (1963).

Lemaire, C., *Cactaceae aliquot novarum*, Paris (1838).

Le Roux A. & Schelpe, E.A.C.L.E., *Namaqualand and Clanwilliam: South African wild flower guide*, Botanical Society of South Africa (1981).

Leslie, A.C., *The international Lily register*, London (1982).
  The international Dianthus register, edn 2, London (1983).

Letty, C., *Wild flowers of the Transvaal*, Pretoria (1962).

Levyns, M.R., *A guide to the flora of the Cape peninsula*, Cape Town (1966).

Li, H.L., *Woody Flora of Taiwan*, Philadelphia & Narberth (1963).

Li, H.L., Liu, T.S., Huang, T.C., Koyama, T. & De Vos, C.E., *Flora of Taiwan*, 5 volumes, Taipei (1976–79).

Lindley, J., *Collectanea botanica*, London (1821).

Lindman, C.A.M., *Nordens Flora*, Stockholm (1964).

Link, H.F. & Otto, F., *Icones plantarum rariorum*, Berlin (1828).

Linley K. & Baker, B., *Flowers of the Veld*, Salisbury (1972).

Lippert, W., *Fotoatlas der Alpenblumen: Blütenpflanzen der Ost und Westalpen*, München (1981).

Liu, T.S., *A monograph of the genus Abies*, Taipei (1971).

Lodewijk, T., ed. by Buchan, R., *The book of tulips*, London (1979).

Lötschert, W. & Beese, G., *Collins' Guide to tropical plants*, London (1983).

Lousley, J.E. & Kent, D.H., *Docks and knotweeds of the British Isles*, London (1981).

Luer, C., *Native orchids of the United States and Canada excluding Florida*, New York (1975).

Maatsch, R., *Das Buch der Freilandfarne*, Berlin (1980).

MacKelvey, S.D., *Yuccas of the south western United States*, 2 parts, Jamaica Plain (1938, 1947).

Mackenzie, W.F., *Freesias*, London (1957).

Macmillan, A.J.S., *Christmas cacti/Weihnachtskakteen*, Erlenbach (1985).

Macmillan, H.F., *Tropical gardening and planting with special reference to Ceylon*, London, edn 1 (1910), edn 3 (1925), edn 5 (1952).

Macself, A.J., *Ferns for garden and greenhouse*, London (1952).

McClure, F., *The bamboos*, Cambridge Mass. (1966).

McCurrach, J.C., *Palms of the world*, New York (1960).

Mace, T., *Notocactus: a review of the genus incorporating Brasilicactus, Eriocactus and Wigginsia*, edn 3, privately printed (1975).

Maekawa, F., *The wild orchids of Japan in colour*, Tokyo (1971).

Magrini, G., *Le Conifère*, Milan (1967).

Maheshwari, P. & Biswas, C., *Cedrus*, New Delhi (1970).

Maiden, J.H., *Forest Flora of New South Wales*, 5 volumes, Sydney (1902–13).

Maire, R., *Flore d'Afrique du nord*, 15 volumes, Paris (1952–80).

Makino, T., *An illustrated Flora of Nippon, with the cultivated and naturalized plants*, Tokyo (1942).
  An illustrated flora of Japan, Tokyo, edn 1 (1948), edn 2 (1961).
  Makino's new illustrated flora of Japan, Tokyo (1963).

Mansfeld, R., *Verzeichnis landwirtschaftlicher und gärtnerischer Kulturpflanzen*, Berlin, 4 volumes (1986).

Mark A.F. & Adams, N.M., *New Zealand alpine plants*, Wellington, Sydney & London (1973).

Marloth, R., *The flora of South Africa*, 4 volumes, Cape Town (1913–32).

Marshall Cavendish Encyclopedia of gardening (ed. Hunt, P.F.), London (1968–70).

Marshall, W.T. & Bock, T.M., *Cactaceae*, Pasadena (1941).

Martin, K. W. – see Keble Martin, W.

Martin, M.J. & Chapman, P.B., *Succulents and their cultivation*, London (1977).

Martius, C.F. von, *Flora Brasiliensis*, 15 volumes, Leipzig (1840-1906).

Masefield, G.B., Harrison, S.G. & Nicholson, B.E., *The Oxford book of food plants*, Oxford (1969).

Mason, H., *Western Cape sandveld flowers*, Cape Town (1972).

Mathew, B. & Baytop, A., *The bulbous plants of Turkey*, London (1984).

Mathew, B., *Dwarf bulbs*, London (1973).
  The larger bulbs, London (1978).
  The Iris, London (1981).
  P.J. Redouté: Lilies and related flowers, London (1982).
  The crocus, a revision of the genus Crocus (Iridaceae), London (1982).

Matthew, K.M., *Illustrations on the flora of the Tamil Nadu Carnatic*, Tiruchipalli (1983).

Matuda, E. & Lujan, I.P., *Las plantas mexicanas del genero Yucca*, Toluca (1980).

Maw, G.A., *A monograph of the genus Crocus*, London (1886).

Maximowicz, C.J., *Flora Tangutica*, Petersburg (1889).

Megaw, E., *Wild flowers of Cyprus*, London (1973).

Meikle, R.D., *Willows and poplars of Great Britain and Ireland*, London (1984).

Menninger, E.A., *Flowering vines of the world*, New York (1970).

Michotte, F., *Agaves et Fourcroyas, Culture et exploitation*, Paris, edn 3 (1931).

Millar, A., *Orchids of Papua New Guinea*, Canberra (1978).

Miller, H. & Lamb, S., *The oaks of North America*, Naturegraph USA (1985).

Mirov, N.T., *The genus Pinus*, New York (1967).

Mitchell, A., *Conifers in the British Isles*, Forestry Commission, London (1972).
  Field guide to the trees of Britain and northern Europe, London (1974).
  The complete guide to trees of Britain and northern Europe, illustrated by D. More, Limpsfield (1985).

Mitchell, A. & Wilkinson, J., *Collins' hand guide to the trees of Britain and northern Europe*, London (1978).

Miyabe, K. & Kudo, Y., *Icones of the essential forest trees of Hokkaido*, Hokkaido (1920–31), edn 2 (1984).

Moeller, H., *What's blooming where on Tenerife?*, Puerto de la Cruz (1968).

Moggi, G. & Guignolini, L., *Fiori da balcone e da giardino*, Milano (1982).

Mohlenbrock, R.H., *The illustrated flora of Illinois, flowering plants*, Carbondale & Edwardsville (1967–78).

Molon, G., *Le Yucche*, Milan (1914).

Moore, D.M., *Flora of Tierra de Fuego*, Oswestry (1983).

Moore L.B. & Irwin, J.B., *The Oxford book of New Zealand plants*, Wellington (1978).

Morat, P. – see Aubréville, A. & Leroy, J.-F.

Morgenstern, K.D., *Sansevierias in pictures and words*, Kempton (1979).

Moriarty, A., *Wild flowers of Malawi*, Cape Town & London (1975).

Morley, B. D. & Toelken, H.R., *Flowering plants in Australia*, Adelaide (1983).

Moss, C.E., *The Cambridge British Flora*, 2 volumes, Cambridge (1914–20).

Mossberg, B. & Nilsson, S., *Orchids of northern Europe*, Harmondsworth (1979).

Muenscher, W.C., *Aquatic plants of the United States*, Ithaca, New York & London (1944).

Muhlberg, H., *Das grosse Buch der Wasserpflanzen*, Hanau/Main (1980).

Muñoz Pizarro, C., *Sinopsis de la Flora Chilena; claves por la identificacion de familias y generos*, Santiago (1959).
  Flores silvestres de Chile, Santiago (1966).

Munz, P.A., *A California Flora*, Berkeley & Los Angeles (1959).
  California mountain wild flowers, Berkeley & Los Angeles (1963).

Nakai, T., *Flora Sylvatica Koreana*, 22 parts, Keijyo (1915–39).

Nehmeh, M., *Wild flowers of Lebanon*, Beirut (1978).

Nel, G.C., *Lithops*, Stellenbosch (undated, appeared 1946).

Nel, G.C. (ed. Jordaan, P.G. & Shurly, E.W.), *The Gibbaeum handbook*, London (1953).

Nelson, E., *Gestaltwandel und Abbildung erörtet am Beispiel der Orchideen Europas ... der Gattung Ophrys*, Chernex-Montraix (1962).
  Monographie und Ikonographie der Orchidaceen-Gattung Dactylorhiza, Zürich (1976).

Nicholls, W.H., *Orchids of Australia*, Melbourne & London (1969).

Nicholson, B.E. & Clapham, A.R., *The Oxford book of trees*, London (1975).

Nicholson, G., *Illustrated dictionary of gardening*, London (1884-88).

Niehaus T.F. & Ripper, C. L., *Field guide to Pacific States wildflowers* Boston, Mass. (1976).

Niering W.A. & Olmstead, N. C., *Audubon society field guide to North American wildflowers: Eastern region*, New York (1979).

Noailles, Le Vicomte de & Lancaster, R., *Mediterranean plants and gardens*, Nottingham (1977).

Northen, R.T., *Home orchid growing*, London (1962).
  Miniature orchids, New York (1980).

Ochse, J.J. & Bakhuizen van den Brink, R.C., *Vegetables of the Dutch East Indies*, Amsterdam (1931).

Osten, C., *Notas sobre Cactàceas*, Montevideo (1941).

Olmos, J.F.B., *Los cactus y las otras plantas suculentas*, Valencia (1977).

Oudshoorn, W., *126 Cacti and other succulents in colour*, Guildford & London (1977).

Pabst, G.F.J. & Dungs, F., *Orchidaceae Brasilienses*, 2 volumes, Hildesheim (1975–77).

Page, C.N., *The ferns of Britain and Ireland*, Cambridge (1982).

Pal, B.P. & Swarup, V., *Bougainvilleas*, New Delhi (1974).

Palmer, E., *A field guide to the trees of southern Africa*, London (1977).

Palmer, E. & Pitman, N., *Trees of southern Africa*, Cape Town (1972).

Parey's Blumengärtnerei, (ed. Encke, F.), Berlin (1958–61).

Pastor, J. & Valdés, B., *Revision del genero Allium (Liliaceae) en la peninsula Iberica e Islas Baleares*, Sevilla (1983).

Parsons, M.E., *The wild flowers of California*, San Francisco (1897).

Pearse, R.O., *Mountain splendour*, Cape Town (1978).

Perrin, H. & Boulger, G.S., *British flowers*, London (1914).

Perry, F., *Water gardening*, edn 3, London (1961).
  Flowers of the world, London (1972).
  The water garden, London (1981).

Petersen R.T. & McKenny, M., *Field guide to wildflowers of north-eastern and north-central North America*, Boston, Mass. (1968).

Petrova, E., *Flowering bulbs*, London, New York, Sydney & Toronto (1975).

Peyritsch, J., *Aroideae Maximilianae*, Wien (1879).

Pfeiffer, L. & Otto, F., *Abbildung und Beschreibung blühende Kakteen*, 2 volumes, Kassel (1840).

Phillips, R., *Wild flowers of Britain*, London (1977).
  Trees in Britain, Europe and North America, London (1978).

Pignatti, S., *Flora d'Italia*, 3 volumes, Bologna (1982).

Pilbeam, J.W., *Mammillaria: a collector's guide*, London (1981).
  Haworthia and Astroloba, a collector's guide, London (1983).
  Sulcorebutia and Wiengartia: a collector's guide, London (1985).

Pizzetti, M., *The Macdonald encyclopedia of cacti*, London & Sydney (1985); Italian translation: *Piante grasse; le cactaceae*, Milan (1985).

Plenck, J.J. von, *Icones plantarum medicinalium*, 8 volumes, Vienna (1788–1812).

Pocock, M.R., *Ground orchids of Australia*, London (1972).

Polunin, O., *Flowers of Europe: a field guide*, London (1969).
  Flowers of Greece and the Balkans: a field guide, London (1980).

Polunin, O. & Everard, B., *Trees and bushes of Europe*, London (1976).

Polunin, O. & Huxley, A., *Flowers of the Mediterranean*, London (1965).

Polunin, O. & Smythies, B.E., *Flowers of southwest Europe: a field guide* (1973).

Polunin, O. & Stainton, J.D.A., *Flowers of the Himalayas*, Oxford (1985).

Polunin, O. & Walters, M.G., *A guide to the vegetation of Britain and Europe*, Oxford (1985).

Poor, J.M., *Plants that merit attention*, Oregon (1984).

Porsild, A.E., *Rocky mountain wild flowers*, illustrated by Lid, D.T., Ottawa (1974).

Pradhan, U.C., *Indian orchids*, 2 volumes, Kalimpong (1976–79).

Prance, G.T. & Elias, T.S., *Extinction is forever*, New York (1977).

Pratt, A., *Grasses, sedges and ferns of Great Britain*, edn 3, London (1873).

Prime, C.T., *Lords and ladies*, London (1960).

Proctor, G.R., *Flora of the Cayman Islands*, London (1984).

Proctor, M.C.F. & Yeo, P.F., *The pollination of flowers*, London (1970)

Purseglove, J.W., *Tropical crops*, 2 volumes, London (1968, 1972).

Puttock, A.G., *Bulbs and corms*, London (1958).

RHS Dictionary of gardening, London, 4 volumes and supplement, edn 1 (1951–56), edn 2 (1956–69).

Radford A.E., Ahles, H.E. & Bell, C.R., *Manual of the vascular flora of the Carolinas*, edn 2, Chapel Hill, (1968).

Rasetti, F., *I fiori dell Alpi*, Roma (1980).

Rauh, W., *Bromeliads for home garden and greenhouse*, English edn, Poole (1979).

Rauh, W., *Die grossartige Welt der Sukkulenten*, Hamburg & Berlin (1967); English edn, translated by Kendall, H.L., *The wonderful world of succulents*, Washington (1984).
Schöne Kakteen und andere Sukkulenten, Berlin (1978).

Rausch, W., *Lobivia*, Vienna (1975).

Raven, J. & Walters, S.M., *Mountain flowers*, London (1956).

Rayzer, G., *Flowering cacti, a colour guide*, English edn, Newton Abbot & London; published simultaneously in several languages, e.g Italian, *Cactus in fiore* (1984).

Reader's Digest field guide to the wild flowers of Britain, London (1981).

Redouté, P.J., *Les Liliacées*, 8 volumes, Paris (1802–16).

Regel, E.A., *Turkestan Flora* (1876).

Reichenbach, H.G.L., *Icones florae Germanicae et Helveticae*, 25 volumes, Leipzig, (1838–50).
Xenia Orchidacearum, 3 volumes, Leipzig (1858–1900).
Beiträge zu einer Orchideenkunde Central-Amerika's, Hamburg (1866).

Reisigl, H. & Danesch, E. & O., *Flore méditerranéenne*, Lausanne (1979).

Reitz, P.R. (ed.), *Flora ilustrada Catarinense*, many volumes, Itajai, Santa Catarina (1965–).

Reynolds, G.W., *The Aloes of South Africa*, Cape Town, edn 4 (1974).
The Aloes of tropical Africa and Madagascar, Mbabane (1966).

Rice E.G. & Compton, R.H., *Wild flowers of the Cape of Good Hope*, Kirstenbosch (1951).

Richens, R.H., *Elm*, Cambridge (1983).

Rickett, H. W. *Wild flowers of the United States*, 6 volumes, New York (1967–71).

Richter, W., *Orchideen*, Neudamm (1969).

Riha J. & Subik, R., translated by Habora, D., ed. by Beckett, K.A. & Beckett, G., *Illustrated encyclopedia of cacti and other succulents*, London (1981).

Ritter, F., *Kakteen in Südamerika*, 4 volumes, Spangenburg (1979–81).

Rittershausen, B. & Rittershausen, W., *Orchids in colour*, Poole (1979).

Rivière, A. & Rivière, C., *Les bambous*, Paris (1878).

Rix, M., *Growing bulbs*, London (1983).

Rix, M., & Phillips, R., *The bulb book: a photographic guide to over 800 hardy bulbs*, London (1981).

Robert P.A. & Schroeter, C., *Alpine flowers*, London (1938).

Robinson, W., *The English flower garden*, London (1883).

Rochford, T. & Gorer, R., *The Rochford book of house plants*, London (1963).

Rodriguez, R., Matthei, O. & Quezada, M., *Flora arbórea de Chile*, Concepcion (1983).

Roles, S.J., *Illustrations to Clapham, Tutin & Warburg, Flora of the British Isles*, Cambridge, edn 1 (1957–65), edn 2, 4 volumes (1983).

Roscoe, W., *Monandrian plants*, Liverpool (1824–28).

Rose, F., *The wild flower key: a guide to plant identification in the field with and without flowers*, London (1981).

Ross-Craig, S., *Drawings of British plants*, 31 parts, London (1948–74).

Rotherham E.R., Briggs, B.G., Blaxell, D.F. & Carolin, R.C., *Flowers and plants of New South Wales and southern Queensland*, Sydney & London (1975).

Rourke, J.P., *The Proteas of southern Africa*, Capetown (1982).

Rousseau, F., *The Proteaceae of South Africa*, Capetown (1970).

Rowley, G.D., *Illustrated encyclopedia of succulent plants*, London (1978).

Royle, J.F., *Illustrations of the botany of the Himalayan mountains*, 2 volumes, London (1833–40).

Ruiz, H. & Pavon, J.A., *Flora Peruvianae et Chilensis*, Madrid (1798–1802).

Rupp, H.M.R., *Orchids of New South Wales*, edn 2, Sydney (1969).

Rushforth, K.D. *The Mitchell Beazley pocket guide to trees*, London (1980).
Nature library – Trees, London (1983).

Sainty, G.R. & Jacobs, S.W.L., *Water plants of New South Wales*, Sydney (1981).

Salmon, J.T., *Field guide to the alpine plants of New Zealand*, Wellington (1968).
New Zealand flowers and plants in colour edn 2, Wellington (1976).
The native trees of New Zealand, Sydney (1980).

Sanchez, O.S., *Flora del valle de Mexico*, Nueva Santa Marta (1969).

Sander, D., *Orchids and their cultivation*, edn 9, Poole (1979).
Sanders' List of orchid hybrids, St Albans (1905–).

Sargent, C.S., *Silva of North America*, 14 volumes, Boston (1891–1902).
Forest Flora of Japan, Boston & New York (1894).
Trees and shrubs, 2 volumes, Boston (1905–13).
(ed.), *Plantae Wilsonianae*, 3 volumes, Cambridge (1913–17).
Manual of the trees of North America, edn 2, Boston & New York (1922).

Satake, Y., Ohwi, J., Kitamura, S, Watari, S. & Tominari, T., *Wild flowers of Japan*, 3 volumes, Tokyo (1981–82).

Šavulescu, T., *Flora Republicii Populare Romane*, 11 volumes, Bucureşti (1952–66).

Schacht, W., *Rock gardens and their plants*, London (1963).

Schauer, T., *A field guide to the wild flowers of Britain and Europe*, London (1982).

Schauer, T. & Caspari, C., *Pflanzenführer*, München (1978).

Schelpe, E., *An introduction to the South African orchids*, London (1966).

Schimper, A.F.W. (translated by Fisher, W.R.), *Plant geography upon a physiological basis*, London (1903).

Schlechter, R., *Die Orchideen*, Berlin, edn 1 (1914), edn 2, ed. Miethe. E. (1927), edn 3, ed. Brieger, F.G., Maatsch, R. & Senghas, K. (1972–).

Schneider, C.K., *Illustriertes Handbuch der Laubholzkunde*, 3 volumes, Jena (1904–12).

Schumann, K.M., *Gesamtbeschreibung der Kakteen*, Neudamm (1877–79), *Nachtrage, 1898–1902*, Neudamm (1903). continued by Gürke, M. & Vaupel, F., *Blühende Kakteen*, 3 volumes, Neudamm (1904–21).

Schultes, R.E., *Native orchids of Trinidad and Tobago*, Oxford (1960).

Schwantes, G. (ed. by Shurly, J.W.), *The cultivation of the Mesembryanthemaceae*, London (1954).

Scott, C.L., *The genus Haworthia (Liliaceae), a taxonomic revision*, Johannesburg (1985).

Seabrook, P. & Rochford, T.C., *Plants for your home*, Nottingham (1974).

Seidenfaden, G. & Smitinand, T., *Orchids of Thailand*, Bangkok (1960).

Shaw, G.R., *The genus Pinus*, Cambridge, Mass. (1914).

Sheehan, T. & Sheehan, M., *Orchid genera illustrated*, New York (1979).

Shibuya, R., *Intercrossing among pink Calla, white-spotted Calla and yellow Calla*, Tokyo (1956).

Shuttleworth, F.S., Zim, H. & Dillon, G.W., *Orchids*, New York (1970).

Sibthorp, J. & Smith, J.E., *Flora Graeca*, 10 volumes, London (1806–40).

Siebold, P.F. von, *Flora Japonica sive plantae quas in Imperio Japonico collegit, descripsit ex parte in ipsis locis pingendas curavit*, 2 volumes, (1835–70).

Sim, T.R., *Flowering trees and shrubs for South Africa*, Johannesburg (1919).

Simmonds, N.W., *The evolution of the bananas*, London (1962). *Bananas*, edn 2, London (1966).

Simpson, F.W., *[Simpson's] Flora of Suffolk*, Ipswich (1982).

Sjogren, E., *Acores flores*, Horta (1984).

Skelsey, A., *Orchids*, Time-Life Encyclopedia of Gardening, Netherlands (1979).

Skvortsov, A.K., *Ivi SSSR: sistematicheskii i geograficheskii obzor*, Moscow (1968).

Smith, D., *Freesias*, London (1979).

Smith, J.E., *Exotic botany*, 2 volumes, London (1804–05).

Smith, J.E. & Sowerby, J., *English botany*, 36 volumes & 5 supplements, London (1790–1863).

Smith, J.J., *Orchideen von Java*, Leiden, 2 volumes (1905–14).

Spellenberg, R., *Audubon Society field guide to North American wildflowers – western region*, New York (1979).

St Barbe Baker, R., *The redwoods*, edn 2, London (1945).

Stanley, T.D. & Ross, E.M., *Flora of south-eastern Queensland*, Brisbane (1983).

Stary, F., *Poisonous plants*, English translation by Kuthanova, O., London, New York (1983).

Step, E., *Favourite flowers of garden and greenhouse*, 4 volumes, London (1896–97).

Stern, F.C., *Snowdrops and snowflakes: a study of the genera Galanthus and Leucojum*, London (1956).

Stewart, J. & Campbell, B., *Orchids of tropical Africa*, London (1970).

Stewart, J. & Hennessy, E.F., *Orchids of Africa*, London (1981).

Stodola, J., *Encyclopedia of water plants*, New York (1967).

Stodola, J., Volak, J. & Bunney, S., *The illustrated book of herbs*, London (1984), New York (1985).

Stojanoff, N. & Stefanoff, B., *Flora Bulgarica*, Sofiya, edn 1 (1925), edn 3 (1948).

Stoker, F., *A gardener's progress*, London (1938).

Stokes, S.G., *The genus Eriogonum: a preliminary study based on geographical distribution*, San Francisco (1936).

Stork, A.L., *Tulipes sauvages et cultivées*, Série documentaire 13 des Conservatoire et Jardin Botaniques de Genève (1984).

Stout, A.B., *Daylilies; the wild species and garden clones, both old and new, of the genus Hemerocallis*, New York (1934).

Strid, A., *Wild flowers of mount Olympus*, Kifissia (1980). *Mountain Flora of Greece*, Cambridge (1986).

Stubbendieck, J., Hatch, S.L. & Hirsch, K.J., *North American range plants*, edn 3, Lincoln, Nebraska (1986).

Stuntz, M.F., Voth, P.D., Hall, E.A., Flory, W.B., Tuggle, H.I. & Monroe, W.E., *Hemerocallis checklist*, American Hemerocallis Society (1957).

Sudworth, G.B., *Forest trees of the Pacific slope*, Washington (1908).

Sundermann, H., *Europaische und Mediterrane Orchideen*, Hanover (1975).

Suzuki, M., *Alpine plants in eastern Japan*, Tokyo (1982).

Suzuki, S., *Index to Japanese Bambusaceae*, Tokyo (1978).

Sweet, H.R., *The genus Phalaenopsis*, Pomona, California (1980).

Sweet, R., *The British flower garden*, 7 volumes, London (1823–38). *The ornamental flower garden and shrubbery*, 4 volumes, London (1854).

Swindells, P., *Ferns for garden and greenhouse*, London (1971). *Waterlilies*, London (1983).

Syme, J.T.I. – see Boswell, J.T.

Synge, P.M., *Collins' Guide to bulbs*, London (1961). *Lilies: a revision of Elwes' monograph of the genus Lilium and its supplements*, London (1980).

Täckholm, V., *Students' flora of Egypt*, edn 2, Beirut (1974).

Takeda, H., *Alpine Flora of Japan*, Osaka (1960–63).

Taylor, A.W., *Wild flowers of the Pyrenees*, London (1971). *Wild flowers of Spain and Portugal*, London (1972).

Taylor, N.P., *The genus Echinocereus*, Kew (1985).

Terasaki, R., *Nippon shokubutsu zufu*, Tokyo (1933).

Testu, C., *Conifères de nos jardins*, Paris (1970).

Thabrew, W.V. de, *Popular tropical aquarium plants*, Cheltenham (1981).

Thomas, G.S., *Plants for ground cover*, London (1970). *Perennial garden plants: or the modern florilegium*, London (1976), edn 2 (1982). *The art of planting: or the planter's handbook*, London (1984).

Thompson, H.S., *Alpine plants of Europe together with cultural hints*, London (1911). *Sub-alpine plants, or flowers of the Swiss woods and meadows*, London (1912).

Thompson, P.A., *Orchids from seed*, Kew (1979).

Thonner, F., *The flowering plants of Africa*, London (1915).

Thunberg, C.P., *Flora Japonica*, Leipzig (1784).

Todaro, A., *Hortus Botanicus Panormitanus; sive, Plantae novae vel criticae quae in horto botanico Panormitano coluntur descriptae et iconibus illustratae*, Palermo (1876–78).

Tosco, U., *Mountain flowers*, London (1978).

Townsend, C.C. & Guest, E. (eds), *Flora of Iraq*, several volumes, London (1966–).

Traub, H.P. & Moldenke, H.N., *Amaryllidaceae: Tribe Amarylleae*, Stanford, California (1949).

Traub, H.P., *The Amaryllis manual*, New York (1958).

Trauseld, W.R., *Wild flowers of the Natal Drakensberg*, Cape Town (1969).

Tredgold M.H. & Biegel, H.M., *Rhodesian wild flowers*, Salisbury (1979).

Trelease, W., & Yuncker, T.G., *The Piperaceae of northern South America*, 2 volumes, Urbana (1950).

Treseder, N.G., *Magnolias*, London & Boston (1978).

Treseder, N.G. & Blamey, M., *The book of Magnolias*, London (1981).

Turrill, W.B., *Supplement to Elwes' Monograph of the genus Lilium*, London (1960).

Tutin, T. G., Heywood, V.H., Burges, N.A., Valentine, D.H., Walters, S.M. & Webb, D.A. (eds), *Flora Europaea*, Cambridge, 5 volumes (1964–80).

Uberto – see Tosco, U.

Useful plants of Japan, Agricultural Society of Japan, Tokyo, 4 volumes (1891–95).

Valdés, B. – see Pastor J. & Valdés, B.

Valentine E.F.& Cotton, E.M., *Illustrations of the flowering plants and ferns of the Falkland Islands*, London, (1921).

Van Laren, A.J., *Succulents other than cacti*, Los Angeles (1935).

Van Royen, P., *Orchids of the high mountains of New Guinea*, Hirschberg (1980).

Van Steenis, C.G.G.J., *Mountain flora of Java*, Leiden (1972).

Veitch, J. & Sons, *Manual of orchidaceous plants*, London (1887-94).

Vellozo, J.M. da C., *Florae Fluminensis*, Rio de Janeiro (1825).

Vogts, M., *South Africa's Proteaceae*, Cape Town (1982).

Walden B.M. & Hu, S.Y., *Wild flowers of Hong Kong around the year*, Hong Kong (1977).

Waldstein, F.A. & Kitaibel, P., *[Descriptiones et] Icones Plantarum Rariorum Hungariae*, 3 volumes, Wien (1802–12).

Wallich, N., *Plantae Asiaticae Rariores*, 3 volumes, London (1832).

Warburton B. & Hamblen M. (eds), *The world of irises*, Wichita, Kansas (1978).

Weber, W.A., *Rocky mountain flora*, Boulder, Colorado (1972).

Wehrhahn, W., *Die Gartenstauden*, 2 volumes, Berlin (1931).

Welch, H.J., *Manual of dwarf conifers*, New York (1979).

Wendelbo, P., *Tulips and irises of Iran*, Tehran (1977).

Weniger, D., *Cacti of Texas and neighbouring states*, Austin (1984).

Werdermann, E., *Brasilien und seine Saulenkakteen*, Neudamm (1933).

Whistler, W.A., *Coastal plants of the tropical Pacific*, Kauai, Hawaii (1980).

Whitehead, S.B., *Carnations today*, London (1956).

Wickham, C., *The house plant book: a complete guide to creative indoor gardening*, London (1977).

Wiggins, I. L., *Flora of Baja California*, Stanford, California (1980).

Wight, R., *Icones Plantarum Indiae orientalis*, 6 volumes, Madras (1838–53).

Wildeman, E. de, *Icones selectae Horti Thenensis*, Bruxelles (1889–1909).

Études de systématique et de geographie botanique sur la flore du Bas- et du Moyen-Congo, 3 volumes, Bruxelles (1903–12).

Williams, A.E., Williams, J.G. & Arlott, N., *Field guide to the orchids of Britain and Europe*, London (1978).

Williams, B.A. & Kramer, J., *Orchids for everyone*, New York (1980).

Williams, B.S., *The orchid grower's manual*, edn 7, London (1897, reprinted 1982).

Williamson, G., Drummond, R.B. & Grosvenor, R., *Orchids of south central Africa*, London (1977).

Willis, J. H., Fuhrer, B.A. & Rotherham, E.R., *Field guide to the flowers and plants of Victoria*, Terrey Hills, NSW (1975).

Witham Fogg, H.G., *Bulbs*, London (1980).

Wit, H.C.D. de, *Aquarium plants* (English ed.), London (1964). *Aquarienpflanzen*, Stuttgart (1971). *Aquarium Planten*, 4th edn, Baarn (1983). (ed.), Boedjin, K.B., *Plants of the world*, 3 volumes, London (1963-65).

Woodcock, H.B.D. & Stearn, W.T., *Lilies of the world: their cultivation and classification*, London (1950).

Woolward, F.H., *The genus Masdevallia*, London (1890-96).

Wrigley J.W. & Fagg, M., *Australian native plants: a manual for their propagation, cultivation and use in landscaping*, Sydney & London (1979).

Yatabe, R., *Iconographia Florae Japonicae*, Tokyo (1893).

Yung, C., *Illustrated manual of Chinese trees and shrubs*, Nanking (1937).

Zhang, J. (ed.), *The alpine plants of China* (English translation of *Chung-kuo kao shan chih wu*), Beijing & New York (1982).

Zichmanis, Z. & Hodgins, J., *Flowers of the wild: Ontario and the Great Lakes region*, Toronto (1982). Zohary, M., *Flora Palaestina*, 8 volumes, Jerusalem (1966–86).

INDEX

Synonyms and names mentioned only in observations, are printed in *italic* type

macdonaldiae (Hooker) Britton & Rose, 213
megalanthus (Vaupel) Moran, 213
pteranthus (Link & Otto) Britton & Rose, 213
setaceus (de Candolle) Werederman, 213
spinulosus (de Candolle) Britton & Rose, 213
Semiaquilegia Makino, 332
ecalcarata (Maximowicz) Sparague &
    Hutchinson, 332
simulatrix Drummond & Hutchinson, 332
Semnanthe Brown, 151
lacera (Haworth) Brown, 151
Seticereus chlorocarpus (Kunth) Backeberg, 234
icosagonus (Schumann) Backeberg, 234
roezlii (Haage) Backeberg, 234
Setiechinopsis mirabilis (Spegazzini) De Haas, 239
Silene Linnaeus, 179
acaulis (Linnaeus) Jacquin, 181
    'Correvoniana', 181
alba, 181
alpestris Jacquin, 181
armeria Linnaeus, 181
californica Durand, 180
coeli-rosa (Linnaeus) Godron, 181
    'Oculata', 181
compacta Fischer, 181
dioica (Linnaeus) Clairvaux, 181
elisabethae Jan, 180
elongata Bellardi, 181
exscapa Allioni, 181
fimbriata Sims, 180
fruticosa Linnaeus, 180
gallica Linnaeus var. quinquevulnera
    (Linnaeus) Mertens & Koch, 181
hifacensis Willdenow, 180
hookeri Torrey & Gray, 180
keiskii Miquel, 180
    forma minor Takeda, 181
longiscapa Kerner, 181
maritima Withering, 180
mollissima (Linnaeus) Persoon, 180
multifida (Adams) Rohrbach, 180
nutans Linnaeus, 180
pendula Linnaeus, 181
pumilio Wulf, 184
pusilla Wadstein & Kitaibel, 181
quadridentata in the sense of Hayek, 181
quadrifida Linnaeus, 181
schafta Hohenacker, 181
sicula Ucria, 179
uniflora Roth, 180
vallesia Linnaeus subsp. vallesia, 180
virginica Linnaeus, 180
vulgaris Linnaeus subsp. maritima
    (Withering) Löve & Löve, 180
    subsp. vulgaris, 180
zawadskii Herbich, 180
Sinofranchetia (Diels) Henry, 397
chinensis (Franchet) Hemsley, 397
Sinomenium Diels, 399
acutum (Thunberg) Rehder & Wilson, 399
diversifolium (Miquel) Diels, 399
Sinopodophyllum Ying, 396
emodi (Wallich) Ying, 396
Sitodium altile Parkinson, 87
Soehrensia bruchii (Britton & Rose) Backeberg,
    241
formosa (Pfeiffer) Backeberg, 242

grandis (Britton & Rose) Backeberg, 241
Soleirolia Gaudichaud-Beaupré, 105
soleirolii (Requien) Dandy, 105
Solisia pectinata (Stein) Britton & Rose, 292
Spinacia Linnaeus, 197
oleracea Linnaeus, 197
Spraguea Torrey, 175
multiceps Howell, 175
umbellata Torrey, 175
    var. caudicifera Gray, 175
Stauntonia de Candolle, 397
hexaphylla (Thunberg) Decaisne, 397
latifolia (Wallich) Wallich, 397
Stenocactus (Schumann) Hill, 282
coptogonus (Lemaire) Hill, 283
crispatus (de Candolle) Hill, 283
multicostatus (Schumann) Hill, 283
obvallatus (de Candolle) Hill, 283
ochoterenaus Tiegel, 283
phyllacanthus (Dietrich & Otto) Hill, 283
sulphureus (Dietrich) Bravo, 283
vaupelianus (Werdermann) Backeberg, 283
Stenocarpus Brown, 115
cunninghamii invalid, 115
salignus Brown, 115
sinuatus Endlicher, 115
Stenocereus Riccobono, 229
beneckei (Ehrenberg) Buxbaum, 229
dumortieri (Scheidweiler) Buxbaum, 229
eruca (Brandegee) Gibson & Horak, 229
marginatus (de Candolle) Berger & Buxbaum,
    227
pruinosus (Pfeiffer) Buxbaum, 229
queretaroensis (Weber) Buxbaum, 229
stellatus (Pfeiffer) Riccobono, 229
thurberi (Engelmann) Buxbaum, 229
Stetsonia Britton & Rose, 223
coryne (Salm-Dyck) Britton & Rose, 223
Stoeberia Dinter & Schwantes, 170
beetzii (Dinter) Dinter & Schwantes, 170
carpii Friedrich, 170
littlewoodii Bolus, 170
Stomatium Schwantes, 154
geoffreyi Bolus, 154
meyeri Bolus, 154
niveum Bolus, 154
trifarium Bolus, 154
Strombocactus Britton & Rose, 278
disciformis (de Candolle) Britton & Rose, 278
Suaeda Scopoli, 198
fruticosa invalid, 198
vera Gmelin, 198
Sulcorebutia arenacea (Cardenas) Ritter, 245
bicolorispina invalid, 245
candiae (Cardenas) Buining & Donald, 244
canigueralii (Cardenas) Buining & Donald, 245
cylindrica Donald, 244
flavissima Rausch, 245
glomeriseta (Cardenas) Ritter, 244
hoffmanniana (Backeberg) Backeberg, 245
lepida Ritter, 245
mentosa Ritter, 245
rauschii Frank, 245
steinbachii (Werdermann) Backeberg, 245
tarabucoensis Rausch, 245
taratensis (Cardenas) Buining & Donald, 245
vizcarrae (Cardenas) Donald, 244

Talauma salicifolia Siebold & Zuccarini var.
    concolor Miquel, 311
Talinum Adanson, 173
caffrum (Thunberg) Ecklon & Zeyher, 174
calycinum Engelmann, 174
okanoganense English, 174
paniculatum (Jacquin) Gaertner, 173
spinescens Torrey, 174
teretifolium Pursh, 174
Tanquana hilmarii (Bolus) Hartmann & Leide,
    154
prismatica (Schwantes) Hartmann & Leide,
    154
Tasmannia Brown, 314
aromatica Brown, 314
Telopea Brown, 113
'Bradwood Brilliant', 114
mongaensis Cheel, 114
oreades Mueller, 114
speciosissmum (Smith) Brown, 114
truncata (Labillardière) Brown, 114
    forma lutea Gray, 114
Tephrocactus pentlandii (Salm-Dyck) Backeberg,
    298
Tetracentraceae, 324
Tetracentron Oliver, 324
sinense Oliver, 324
Thalictrum Linnaeus, 347
adiantifolium Besser, 348
alpinum Linnaeus, 347
anemonoides Michaux, 349
angustifolium misapplied, 348
aquilegiifolium Linnaeus, 348
    'Album', 348
arenarium Butcher, 348
calabricum Sprengel, 348
chelidonii de Candolle, 349
coreanum Léveillé, 349
delavayi Franchet, 349
    'Album', 349
diffusiflorum Marquand & Airy-Shaw, 349
dipterocarpum Franchet, 349
dipterocarpum misapplied, 349
elatum Boissier, 348
fendleri Engelmann, 349
flavum Linnaeus, 348
    subsp. glaucum (Desfontaines) Battandier,
    348
foeniculaceum Bunge, 348
foetidum Linnaeus, 348
gallicum Rouy & Foucaud, 348
glaucum Desfontaines, 348
'Hewitt's Double', 349
ichangense Oliver, 348
javanicum Blume, 349
kemense Fries, 348
kiusianum Nakai, 348
lucidum Linnaeus, 348
majus Crantz, 348
minus Linnaeus, 348
orientale Boissier, 348
psilotifolium invalid, 348
pubescens Schleicher, 348
purpureum Shang, 348
reniforme Wallich, 349
rochebrunianum Franchet & Savatier, 349
rugosum Aiton, 348